Batteries

Volume 1

Manganese Dioxide

Batteries

Edited by

Karl V. Kordesch

Union Carbide Corporation
Battery Products Division
Research Laboratory
Parma, Ohio

Volume 1

Manganese Dioxide

MARCEL DEKKER, INC.   New York   1974

MARCEL DEKKER, INC.

305 East 45th Street, New York, New York 10017

LIBRARY OF CONGRESS CATALOG CARD NUMBER: 73-82702

ISBN: 0-8247-6084-0

Current printing (last digit):

10 9 8 7 6 5 4 3 2 1

PRINTED IN THE UNITED STATES OF AMERICA

CONTENTS

Chapter 3

ELECTROCHEMISTRY OF MANGANESE DIOXIDE AND
    PRODUCTION AND PROPERTIES OF ELECTROLYTIC
    MANGANESE DIOXIDE (EMD)   385

Akiya Kozawa

Chapter 4

MAGNESIUM BATTERIES   521

Donald B. Wood

ADDITIONAL VOLUMES OF THIS SERIES ARE IN PREPARATION

CONTRIBUTORS TO THIS VOLUME

RICHARD HUBER, Varta A. G., Ellwangen, West Germany.*

KARL V. KORDESCH, Union Carbide Corporation, Battery Products Division, Research Laboratories, Parma, Ohio.

AKIYA KOZAWA, Union Carbide Corporation, Battery Products Division, Research Laboratories, Parma, Ohio.

DONALD B. WOOD, Power Sources Technical Area, U.S. Army Electronics Technology and Devices Laboratory (ECOM), Fort Monmouth, New Jersey.

*Presently retired.

FOREWORD TO THE SERIES

The technology of batteries has been considered a conservative part of electrochemistry, with some scientists studying the electrode processes with sophisticated methods, and some engineers building automobile batteries and flashlight cells following time-honored processes.

This situation has changed in the past ten years. Old battery systems have been improved considerably, and new galvanic cells have been invented, partly due to the space age, but also as a result of progress in technology. New materials and components have been provided by the chemical industry and new requirements have been set by the expanding electronics industry.

Because of these changes, it seemed necessary to collect and publish the available information in a series on batteries. This specialized literature is often difficult to obtain, even for technical men. A critical selection is necessary to avoid getting lost in details. The first volume is entitled "Manganese Dioxide," dedicated to the most important primary battery system. Future plans are to discuss secondary batteries, mainly the lead-acid and nickel-cadmium systems. Rechargeable batteries play a large role in our daily lives, and may become especially important for use in electric vehicles.

As the series continues we will describe some of the most recent innovations: nonaqueous and solid-state batteries. Fuel cells will again become a topic in connection with the energy crisis, environmental problems, and ultimately with the advent of the power supply of the future: nuclear energy. Hydrogen may ultimately become the means of storing and distributing large amounts of energy. New galvanic cells developed by the scientists of the twentieth century will be based on the achievements of the

electrochemists of today. The purpose of this series is to help distribute
this knowledge and to encourage new work and inventions.

KARL V. KORDESCH

PREFACE

The writing of a book on modern batteries was suggested to the editor by several friends and colleagues. However, it needed the initiative of Dr. Maurits Dekker to start the endeavor.

It is not possible to cover such a large field in a monograph; therefore the editor invited renowned specialists in the battery field to contribute their knowledge and write specific chapters. Soon the material increased to such an extent that it was necessary to narrow the subject so as to be able to handle it in one reasonably sized book. Since several books about different battery systems have appeared in the last few years, it was decided to tackle the most difficult subject -- which up to this time had not been sufficiently covered -- and publish first a book about manganese dioxide batteries, one of the oldest and most widely used systems.

The task of unravelling the overwhelming amount of literature in this field was shared by four authors, each a well-known authority in his field. One chapter was written by the editor.

The objective of this book is to inform. This does not mean to report all the literature available, but to select those sources which will give the reader useful and quickly available information. This task involves the critical absorption of the work of thousands of scientists and engineers, spread over dozens of technical journals, and often deposited in the patent literature in an obscure way.

The task seemed to be important: The manganese dioxide battery is (after the lead-automobile battery) the most widely used portable electric power source for today's gadgets, toys, and electronic conveniences. The sales volume of these batteries is estimated at billions per year, with a high economic impact on the consumer.

Richard Huber, Director of the Varta A. G. production facilities in Ellwangen, Germany (recently retired) wrote the first chapter on Leclanché batteries. This battery type energizes practically all the flashlights in the world, and it has come a long way from its conception in the nineteenth century. Dr. Huber spent most of his professional life in the battery business, and certainly is one of the foremost experts in his field.

The second chapter, on alkaline manganese dioxide zinc cells, was written by the editor, who has spent many years of his technical career at Union Carbide Corporation on projects concerned with this type of battery, which is more powerful than the Leclanché type, but also more expensive.

Akiya Kozawa, also of Union Carbide Corporation, has devoted a long, distinguished career as a scientist to the problem of manganese dioxide, and is eminently suited to author the third chapter. Dr. Kozawa was educated in Japan and has first-hand knowledge of the Japanese battery industry. This fact is shown in his extensive coverage of the Japanese literature, which is not well-known in this country or in Europe because of the language difficulties.

Donald B. Wood has contributed a chapter on magnesium batteries. The MnO_2-Mg system has military applications because of its excellent shelf life, but has not yet found commercial use. However, the number of such batteries produced is large, and their qualities are needed in the field. Mr. Wood is a scientist at the U.S. Army Electronics Technology and Devices Laboratory, Fort Monmouth, New Jersey, and since he is involved with all technical aspects of this battery, he is most qualified to write about it.

The editor's thanks go to the contributors and their organizations which recognized the importance of the technical information needed, and provided it freely.

Miss S. J. Cieszewski, of the Parma Technical Center, read and corrected with great patience the entire manuscript and the authors gratefully acknowledge her help.

The editor also wants to thank his wife for her understanding.

It is hoped that the emphasis on practical information will not offend the

specialist with great familiarity with theoretical subjects. Anyone who
wants to "dig deeper" may use the literature references or electrochemical
textbooks for "brushing up."

The limitation in size of this book was intentional: it is hoped that not
only libraries will be able to afford its price, but that it will find a place
on the bookshelf of the technical man.

To the publisher, for his efforts to produce a quality book at the lowest
possible cost, my sincere thanks.

KARL V. KORDESCH

Batteries

Volume 1

Manganese Dioxide

CHAPTER 1

LECLANCHÉ BATTERIES

Richard Huber

Varta A. G.
Ellwangen, West Germany[*]

---

[*]Presently retired.

1

1. CONSTRUCTION OF LECLANCHÉ CELLS

1.1. HISTORICAL REVIEW

Tracing the stages in the development of galvanic cells from the very beginning to the present day demonstrates the impressive history of the powers of human observation and deduction, which were no less characteristic of the craftsmen of early times than of the modern research scientist equipped with the most sophisticated instrumentation typical of our technological age. The knowledge of galvanic processes dates from a much earlier period in history than we would have dreamt of only a few decades ago. The invention of the first effective generators of electric current is generally associated with names such as Caldani, Galvani, and Alessandro Volta (1800). It is only since the archaeologist W. König [1] suggested a highly probable explanation of the mysterious earthenware vessels that have been found at various Mesopotamian sites that we have realized that galvanic cells may have been known as early as about 500 BC. During excavations at Khujut Rabuah near Bagdad, König discovered vessels in which a copper tube and a centrally disposed iron rod were cemented in position with asphalt (Fig. 1). Still unused sets of the copper and iron parts were also unearthed. König interpreted the find as galvanic cells with reserve electrodes. And indeed if such a vessel is filled with an organic acid (such as wine vinegar or the juice of citrus fruits), a galvanic cell is obtained that functions in precisely the same way as the

FIG. 1. Chaldean earthenware vessel.

Volta cell. It is possible that the very thin gold plating found on ornamental
jewelry of the period, which the experts are at a loss to explain, was deposited in an electrolytic process with the aid of such batteries.

It is not unreasonable to assume that such galvanic cells could have found
a medicinal application, since the beneficial effects of electric shocks
(obtained, for example, from the electric ray) in the treatment of disorders of the peripheral nervous system were already known. Certainly
galvanic phenomena could have been used by the Chaldean magicians in
their religious ceremonies.

However, a proper appreciation of the problems involved in galvanic
processes first became possible after their rediscovery toward the end of
the 18th century. The classification of metals according to their potentials
by Volta, the discovery of polarization phenomena by Schönbein and Ritter,
the enunciation of the laws of electrolysis by Faraday, the theoretical
treatment of electrode reactions by Nernst and Helmholtz, and the contributions of Ohm, Arrhenius, and many more are all milestones on the
way to an ever deepening understanding of the electrochemical processes
occuring in galvanic cells.

The galvanic cell developed by the French engineer G. Leclanché (Fig. 2) in the 1860s forms the technical basis of the modern dry-cell industry. Leclanché, who for political reasons lived in exile in Belgium until the collapse of the second Empire, founded a battery factory in Brussels where his newly developed cell was manufactured. The cell consisted of a glass jar containing an aqueous solution of ammonium chloride. A zinc rod partly immersed in the solution served as the negative electrode; a porous earthenware jar into which was pressed a mixture of manganese dioxide and powdered coke and containing a carbon-

Georges-Lionel Leclanché
1838-1882

FIG. 2. Georges-Lionel Leclanché (1838-1882).

rod current collector was the positive electrode (Fig. 3). The Leclanché
cell, although having a lower voltage (about 1.5 V) and able to supply only
significantly lower currents in comparison with competitive systems already in existence (for example, the Bunsen element), soon became predominant. Manufacture was cheap, the cell was simple and safe in
operation, shelf life was excellent, and the electrical energy available
at medium-strength current drain was adequate. A powerful impetus

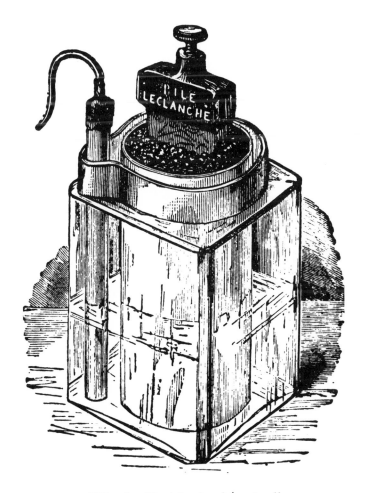

FIG. 3. First Leclanché wet cell.

was given to the manufacture of and demand for the new cell by the rapid
extension of telegraph and telephone networks throughout Europe at this
time. However, a long evolution was still needed to advance from the
wet Leclanché cell of those days to the modern dry cell. The following
stages in development are noteworthy.

a. The zinc rod was replaced by a zinc plate bent into the form of a
cylinder and finally by a zinc can, which, in addition to its electrochemical
function as anode, also served as the cell container. The increased area
of the electrode was equivalent to a reduction of the current density,
which led in turn to an increased efficiency. Furthermore, the use of a
zinc can simplified the cell construction. Originally the zinc cans were
soldered, but today they are formed either by impact extrusion or by
deep drawing.

b. Powdered coke was replaced by graphite, and the porous earthenware cylinder by a paper or gauze wrap. This latter change necessitated
press-forming of prismatic or cylindrical electrodes from the depolarizer
mix. A carbon rod driven into the electrode mix served as the terminal.
The complete electrode is known in the trade as a bobbin. The bobbins
were then wrapped in paper or gauze, which was secured by a few loops
of string.

c. Immobilization of the electrolyte solution was introduced at various
locations more or less simultaneously in about 1880. As has been the
case so often with other inventions, the time was ripe for its appearance.
To quote but a few examples: in 1884, a rechargeable dry cell was reported by the apothecary C. H. Wolf in Blankenese; in 1886, the Danish
dry cell manufacturers Hellesens commenced production [2]; in 1888,
German patents were awarded to Zierfuss [3] in Leipzig and Bender [4]
in Brussels, both relating to processes for the preparation of dry-cell
electrolyte pastes; according to a private communication, the French
company Leclanché began the manufacture of dry cells in the period of
1880-1890. This list could be considerably extended. However, Gassner
is generally credited with the invention of the first dry cell, and this
opinion is repeated over and over again from book to book. In his German
patent 45,250 of 1887, he described an electrolyte paste consisting of an

ammonium chloride solution, thickened to the consistency of paste with
a mixture of zinc oxide and plaster of Paris.

In these early forms of the dry cell, the electrolyte was usually
immobilized by the addition of mixtures of inorganic substances, such
as chalk and calcium chloride, zinc oxide and plaster of Paris, or soluble
silicates to which activating agents were added, which resulted in the
precipitation of gelatinous silicic acid. Silica and other absorbent materials were also used.

The decisive step toward the realization of industrial production was
the recognition of the suitability of naturally occuring carbohydrates, such
as flour and starch, as gelatinizing agents for electrolyte solutions. Even
this improvement cannot be unambiguously attributed to a single inventor.
Jeckel [5] reports that in 1894, Gassner developed a portable lighting
battery in which flour was employed as the gelatinizing agent and which
he demonstrated at the World's Fair in Paris in 1900. According to
Drotschmann [6] , Paul Schmitt, a manufacturer in Berlin, introduced
the "cooking" process with wheat flour as the gelling agent in 1896.

d. About 1870, Sturgeon described the advantages offered by amalgamation of the zinc anode; and in 1877, Fuller [7] introduced the wax
impregnation of carbon electrodes to prevent the creepage of the electrolyte solution and the subsequent corrosion of the metal connectors.

e. When acetylene black became available, it was found beneficial
to replace the graphite in the cathode mix, at first partly and eventually
completely, by this highly absorbent and conductive material. Mixing
and tamping techniques were adjusted to take full advantage of the
properties of the new material with such success that finally the time-
consuming bobbin wrapping and soaking operations could be discarded.

f. The introduction of portable radio receivers and other new applications led to the design of the layer-built (or flat-cell) battery [8,9] with
a higher output-to-volume ratio than was possible with the standard
cylindrical cell design. In 1921, Pörscke obtained significant improvements with cell systems using magnesium chloride instead of ammonium
chloride as electrolyte [10].

g. To conclude, mention must be made of other significant developments, such as the paper-lined cell, the inside-out cell [11] and large

area cells [12], the steel- [13, 14] and plastic-jacketed [15-18] batteries
which show an improved resistance to electrolyte leakage, and finally the
zinc chloride electrolyte cell system [19] which aims at preventing leakage
by chemical means.

The decisive breakthrough to the modern portable power source was
undoubtedly the result of combining the anode and the cell container (zinc
can) and the gelatinization of the electrolyte. Without these design features,
it would hardly have been possible to achieve the remarkable degree of
miniaturization of the dry cell demanded by the wide variety of applications.

The difficulty (or even impossibility) of making accurate forecasts of
future development is shown by a quotation from a discussion in 1890 of
the (then) recently introduced dry cell. "The inventors and manufacturers
doubtless believe that the introduction of the dry cell will give the primary
cell a new lease on life; to date, this has not been sufficiently appreciated
by the general public. Dry cells have the great disadvantage that they
become weak and there is no possibility of regeneration."

Oddly enough, the end of the dry-cell industry has been repeatedly
predicted: for instance, with the invention of the dynamo, with the appearance of the ac-operated radio receiver on the market, with the triumph of
the transistor over the electronic tube, and last but not least with the growing peaceful application of nuclear energy. Regardless of all the pessimis-
tic prophecies, the dry-cell industry has developed to such an extent
that the total annual world production of cells has reached the proud
figure of several billion.

What are the reasons behind this astonishing expansion, and what
are the characteristics of the Leclanché system that enable it to continue
to satisfy the requirements of the trade?

The continual opening up of new fields of application by the electrical
and electronic industries must be recognized as one of the most potent
factors. In the early days, it was the rapid worldwide expansion of telegraph networks that gave the Leclanché cell a solid basis. Then came the
introduction of portable hand lamps and flashlights. After the end of the
First World War, there was the invention of the battery-operated radio
receiver. The industry has received further stimulation from the tendency

toward portable, "cordless" electrical equipment, and the increasing
degree of automation of technical instruments such as cameras.

Why has the Leclanché cell proved to be the most useful power source
for this multiplicity of electrical equipment? The first secret of its
success is that the inventor ingeniously selected an electrochemical
system combining a high energy density and an adequate current density
with excellent shelf-life characteristics in the unused state.

Second, the necessary raw materials are plentiful and are therefore,
in contrast to other electrochemical systems, relatively inexpensive.

A third advantage lies in the ease with which the manufacturing
processes can be mechanized, which is of paramount importance, in times
when economy demands high production efficiency because the cost of
skilled handwork is continually increasing.

Finally, the cell design lends itself readily to a far-reaching miniaturization. All this considered, a tremendous number of chemical and
engineering problems had to be solved in transforming the original Leclanché wet cell into the sophisticated technical product it is today.

1.2. DISCUSSION OF VARIOUS CELL DESIGNS

1.2.1. Pasted Cells

The oldest form of the dry cell is the pasted cell (Fig. 4a).
It consists of a zinc can (3 in the figure) and a cylindrical cathode (1)
pressed from a mixture of manganese dioxide, ammonium chloride, and
acetylene carbon black, with a centrally disposed carbon rod (2). The
electrolyte (5) is immobilized with flour or starch. The core is physically separated from the can at the base by a cardboard or plastic disc (4).
The zinc can serves as anode and also as cell container. The bottom disc,
which may or may not be impregnated with wax, sometimes has the form
of a shallow dish or star. However, during discharge, the zinc at the
edge of the bottom disc is dissolved more rapidly than in other areas,

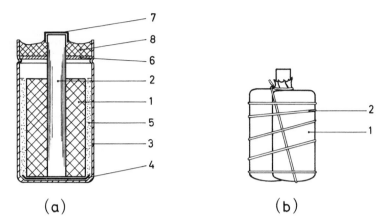

FIG. 4. Pasted cell. (a) Sectional drawing; (b) in tissue-wrapped
bobbin.

leading to perforation of the can. Today it is customary to avoid the
upstanding wall of the cup or star form and to prefer a simple circular
disc or even a square insert; the latter has the additional advantage of
economy. To prevent internal short-circuiting of the cell during assembly,
the lower part of the cylindrical core is protected by a square of thin
Kraft paper. The bobbin with the paper is inserted into the can, which
already contains a measured amount of electrolyte in fluid form. When
the paper is wetted, it adheres to the bobbin, thus preventing the break-
away of mix aggregates. A wax-impregnated, perforated washer (6) is
then applied at the top of the cell, being held in position either by a force
fit or by a bead formed in the wall of the can. A brass cap is then pressed
onto the carbon rod (7). Finally, a hot melt of bitumen or a mixture of
wax and bitumen is poured into the top of the cell to form a seal (8).
Other cell-sealing methods are described in Section 1.3.

The space between the upper surface of the bobbin and the impregnated
sealing washer is known as the expansion chamber and is usually 4-6 cm^3
in volume (D-size cell). Omission of the expansion chamber is not recommended, because such cells tend to swell on storage or to leak during discharge.

The electrolyte used for pasted cells is prepared in the form of a
suspension of cereals in an aqueous solution of zinc and ammonium chlor-

ides. There are two main classes called, respectively, hot- and cold-setting. The hot-setting pastes are so formulated that the gelling of the cereal is achieved by immersing the complete cell to the bead in a hot water bath at 80°-90°C for 1.5-3 min. The cold-setting pastes are so formulated that, although the paste can be stored in the fluid form indefinitely, the gelling action commences as soon as the bobbin is inserted, without any application of heat. By suitable formulation, the rate of the reaction can be controlled so that the gelling time may be varied from about 5 min to several hours (see Section 2 on raw materials). The whole process should be controlled to avoid the appearance of air bubbles in the paste; it should retain a tacky consistency even after prolonged storage. Air bubbles appear in the paste when the electrolyte gels too rapidly, preventing the release of the air displaced from the bobbin by the absorbed fluid.

Since the early cathode-mix formulations contained either very little or no acetylene black, the amount of water they could retain was severely restricted; for good performance, it was necessary to soak the pressed bobbins in electrolyte solution before cell assembly. To prevent disintegration of the bobbin, it was essential to wrap it in paper or gauze (1 in Fig. 4b) held in position by linen thread (2). With the introduction of highly absorbent carbon blacks, this process has become redundant.

1.2.2. Paper-lined Cells

Figures 5a and 5b show diagrammatically the main features of paper-lined cell designs. In principle, they are identical with that of pasted cells, except that the electrolyte paste of the latter is replaced by a paper separator carrying a thin layer of paste.

Paper-lined cells have a greater energy content than pasted cells of the same volume and also a lower internal resistance, since the thickness of the separator layer is only 0.15-0.2 mm compared with 2.5-3.5 mm in pasted cells.

The first production of paper-lined cells used uncoated paper separators, and their shelf-life characteristics were highly unsatisfactory. The

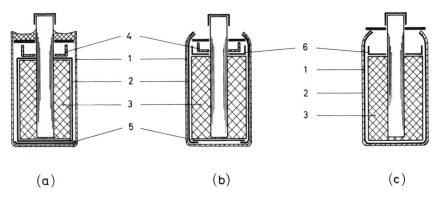

FIG. 5. Paper-lined cells. (a, b) Construction of paper-lined cells;
(c) construction of spin-lined cells.

cathode mix absorbed electrolyte from the paper, and contact corrosion at
the zinc wall proceeded rapidly. Only with the introduction of a lamination
with suitable gelling agents was it possible to obtain adequate wetting of
the zinc electrode and thereby achieve good reliability. Especially important in paper-lined cells is the sealing arrangement, since paper-lined
cells are much more sensitive to moisture loss than pasted cells. The
zinc can must have adequate mechanical strength and be efficiently amalgamated. The compression of the bobbin to obtain contact between the
mix, separator, and anode is also of importance. Badly made cells
generate gas on storage, which results in a fall-off of the flash current
and may even distort the zinc can or disrupt the seal.

There are several patents in existence that protect particular formulations of the laminating paste. Some of these are characterized by the
use of a double laminate. The first layer is an absorbent swelling agent
insoluble in the electrolyte; the second layer has the function of wetting
the zinc surface [20, 21]. Others claim advantages for specific types
of methyl cellulose as a swelling agent.

The manufacturing process is as follows. A cylindrical bobbin (3 in
Fig. 5a) is pressed without the carbon rod and is completely wrapped in a
laminated separator paper (2). The wrapped bobbin is then inserted into
the zinc can and pressed out to the zinc wall (1). Next, the carbon rod
is inserted centrally, and the wrapped bobbin is again lightly compressed.

A plastic spacer (4), a shallow dish with a perforated bottom, supports
the perforated sealing washer and prevents further movement of the
upper portion of the cathode. Sealing compound is applied in the form of
a hot melt.

An alternative method of cell assembly is represented in Fig. 5b.
Here the zinc can is first lined with a laminated separator, and then a
bottom dish (5) of the same material is inserted. The cathode mix is
pressed into the lined can. The compression of the mix after insertion
is accompanied by the inclusion of an oversize paper washer (6) to prevent
contamination of the press tool. After the carbon rod has been inserted,
the laminated liner, which extends to the rim of the can or above, must
be folded over and retained in place by the plastic spacer. The remaining
operations are the same as for the variant described above.

In some designs, the zinc can is swaged after insertion of the carbon
rod and a sealing-compound retaining washer placed externally to the can
(Fig. 5c).

1.2.3. Spin-Lined Cells

A modified form of the paper-lined cell is the spin-lined cell (Fig. 5c),
in which the separator is applied to the inner wall of the rapidly rotating
zinc can as a suspension. The film may dry rapidly enough at room temperature, or the process may be accelerated by a warm air stream, according to the formulation of the suspension. The film obtained is mechanically strong and adheres firmly to the metal.

1.2.4. Inside-Out Cells and Sector Cells

The paper-lined cell lends itself readily to those designs that aim at
making a larger area of the zinc electrode available. There are two
variants, the inside-out cell and the sector design (Fig. 6a, b).

The inside-out cell (Fig. 6a) derives its name from the fact that the
cell container is the positive pole and not, as is usual, the zinc electrode.
This container is formed from graphite containing some binding agent such
as wax and has a low electrical resistance. The cathode mix is pressed

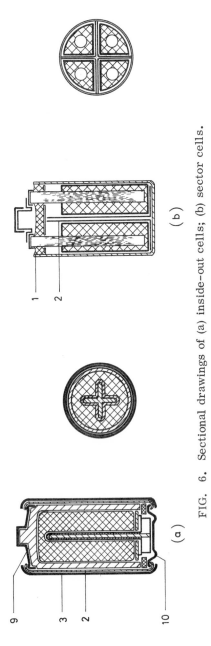

FIG. 6.   Sectional drawings of (a) inside-out cells; (b) sector cells.

directly into the container. The negative electrode (zinc), consisting of two plates at right angles, forming a cross and enclosed in a laminated separator, is then pressed into position. The cell must be "turned around" in polarity for common use. To achieve this in automatic production, the cell is housed in an outer container (3), insulated from the cathode by layer (2) and provided with contacts (9) and (10).

The sector design (Fig. 6b), so called because it consists of four sectors, is built up from four individual paper-lined cells connected in parallel. The design may be realized in two ways. (a) Four complete paper-lined cells are made in the usual cylindrical form, re-formed by pressing into the sector shapes, and inserted into a common cylindrical zinc can (1). The carbon rods are then connected in parallel. (b) The cylindrical bobbins are wrapped in laminated paper, reshaped into sectors, and inserted into a zinc can containing a zinc cross electrode (2). The carbon rods are inserted individually and connected in parallel.

Both the inside-out and the sector type of cell give improved performance on heavy-discharge applications and have less tendency to leak.

1.2.5. Layer-Built Cells

The miniaturization of electronic tubes resulted in a demand for small, compact, high-voltage power sources. This demand was satisfied by the layer-built cell. The prismatic cell form and omission of soldered intercell connections made possible the manufacture of dry cells with a hitherto unattainable energy content per unit volume (8,9).

Fig. 7 illustrates the structure of a layer-built (or flat) cell and of a group of such cells combined into a stack according to four methods. Single-cell construction is best seen in Fig. 7c.

The unit cell is identical in all four variants. The anode is a flat zinc plate (1), coated on one side with a chemically inert but electronically conductive film consisting of carbon and a suitable binding agent. This conductive film may be applied to the zinc as a paint. Several coats are necessary to ensure that the film is not porous, and each coat must be dried before application of the next. The film may also be formed separately as a flexible sheet (2) and rolled on to the zinc under the application of heat and pressure to ensure adhesion.

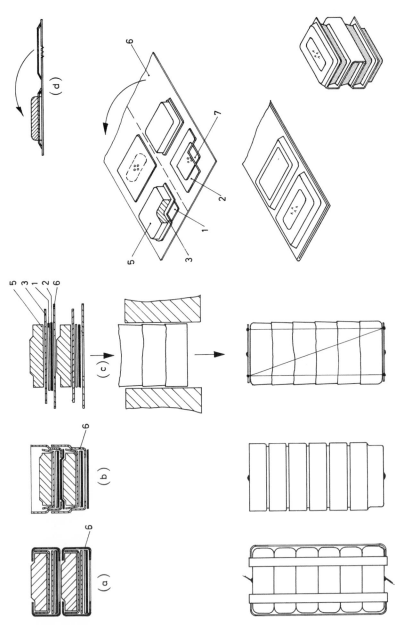

FIG. 7. Various constructions of layer–built cells.

The opposite zinc surface is protected by a laminated separator paper (3) similar to that used in paper-lined cells, the pasted surface being placed in contact with the zinc.

The cathode mix is pressed into the form of a tablet (5), which is partly enclosed in tissue paper (the upper surface of the tablet must be left exposed) to prevent mix aggregates from breaking away, and the whole assembly is then placed on the paper side of the laminated separator.

A complete cell has now been assembled, so far without a cell container, having the surface of the cathode tablet uppermost and the conductive film on the anode at the base. These cells may now be connected in series merely by arranging them one on top of the other.

A little consideration reveals, however, that such a group of cells would be unreliable if electrolyte from one cell should enter another. Then the cathode of the first cell would be directly short-circuited to the anode of the other. It is essential, therefore, that some means of preventing intercell leakage be incorporated in the design. It is in the methods chosen that the four variants differ.

The four variants of Fig. 7 are characterized as follows.

a. The individual cells are enclosed in a shrink-on PVC tube (6).

b. The cell components are placed in a shallow dish (6), extruded from high-impact polystyrene or cellulose acetobutyrate with an opening in the base. The base is then closed by gluing an extra piece of conductive film to the outer surface. The film already attached to the anode is utilized as the contacting member at the inner edge.

c. The cell components are laid on perforated sheets of a suitable plastic material (6).

d. The cell components are laid on a whole sheet of plastic previously treated with adhesive.

With variant (a), the stacked cells are compressed and retained in position by tying. The pressure exerted in the operation seals off the individual cells and maintains electrical contact.

With (b), the open end of the plastic dish must be sealed with wax or some adhesive compound. The stack may be reinforced by tying, but this is not always necessary if a good adhesive is used.

With (c), the group of cells is passed through a heated die. The plastic films are folded upward and heat-sealed. Here again, the group must be tied in order to maintain electrical contact.

With (d), the plastic film (6) is folded over and secured with adhesive. The air is extracted from the individual cell compartments. To obtain electrical contact, a thin metal disc with projecting points (7) is laid between the individual cells and pressure applied so that the points penetrate the plastic film. The envelope containing the cells is then folded together in a zig-zag fashion and the whole assembly retained in position by tying (lower right picture in Fig. 7d).

In layer-built battery manufacture, particular attention must be given to tight sealing to prevent vapor loss from the small cell volumes concerned. Intercell electrical connection can be improved by providing a projecting contact on the cathode tablet.

1.3. EXTERNAL FINISHING OF CELLS

1.3.1. Paper Jacket

Formerly, all cells and batteries were enclosed in paper or cardboard tubes or cartons, which were provided with the manufacturer's label. To improve the general appearance, the containers were printed in several colors and finished with a clear, glossy enamel. The upper finish on individual cells was the bitumen sealing compound, often colored brown or red. At the base of the cell, the exposed part of the zinc can was often polished or sandblasted to improve the electrical contact (Fig. 8a). Today, paper-jacketed cells are manufactured as "low-cost" cells. Figure 8 shows the main components: zinc can (1), electrolyte paste (2), cathode (3), and cardboard cup (5).

In "better" cells, the sealing compound is covered by a plastic or steel lid (Fig. 8b), and the base is protected by a profiled steel disc that is swaged into the cardboard tube (Fig. 8c). The steel parts are generally tin plated. The cardboard jacket is also provided with a

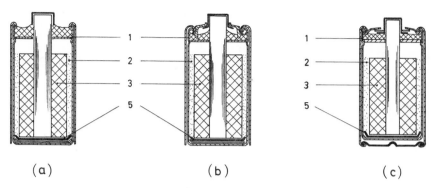

(a)   (b)   (c)

FIG. 8. Various designs of paper-jacketed cells.

printed plastic label, which gives the battery a very attractive, often
metallic, appearance. The cardboard jackets were formerly manufactured in a separate operation and the label was applied to the jacketed
battery. Today, the tubes are made and labeled in a single operation
from paper and plastic sheets helically wound and glued to form endless
tubes, which are then cut to the desired length.

1.3.2. Metal-Clad Cells

The principal advantage of metal-clad cells lies in the efficiency
of the sealing arrangements and the associated excellent shelf-life
characteristics. A further advantage is that such cells do not swell
even after prolonged discharge.

After the expiration of the principal patents [13,14] , the design
was taken up by a number of battery manufacturers throughout the
world, either in the original or in a modified form. Figure 9 shows
metal-clad cells.

The basic cell (exposed zinc can (1), unsealed) is placed in a bitumen
or plastic layer (2) or tube (6) in which a steel disc (10) has already
been positioned to serve as the negative pole. The cell and tube are
sealed off with a bitumen-based sealing compound or a plastic ring (7).
The assembly is then inserted into a lithographed steel jacket (3) and

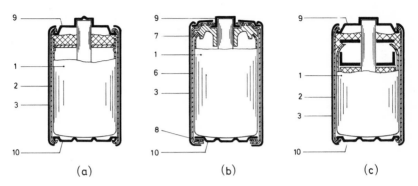

FIG. 9. Various designs of metal-clad cells.

a steel cell lid (9) is pressed onto the carbon rod. The steel jacket is
then swaged, thus retaining the cell lid in position. The jacket is insulated from the lid and bottom disc by the paper tube (Fig. 9a, c).

The maximum volume of a dry cell is fixed by internationally standardized dimensions. In order to obtain the maximum performance
from cells within these dimensions, the thickness of the metal jacket
has been further reduced, the "Mennen" seam dispensed with, and the
jacket trimmed to a butt seam in the case of the smaller sizes and to a
slight overlap, secured by a metal adhesive, in the larger sizes. A
plastic shrink-on tube has replaced the relatively thick layer-board tube.
To save height, the hot-melt bitumen sealing compound has been replaced by a plastic seal (7). Because of the omission of the layer-board
tube, an additional sealing element (8) is necessary at the base of the
cell. These design features have appreciably increased the useful
volume of the zinc can, and consequently the performance of the cell has
improved. The manufacture of such cells is of course somewhat more
complicated and the material costs are higher. Figure 9b illustrates
this type of cell.

1.3.3. Plastic Jacket

This type of finishing is used frequently in western Europe. Figure
10 shows four such jacket designs.

The simple plastic sleeve, which is slipped over the zinc can (1) of the finished cell (Fig. 10a), is applied to the cheapest cells, since these are more likely to leak after prolonged discharge. In the paper-jacketed design, most of the leaking electrolyte is absorbed by the card-board, but with a plastic jacket most of the liquid appears externally. An improvement can be achieved by enclosing the zinc can in a self-sealing adhesive tape, which may also have shrink-on properties. Sometimes the sealing of the positive cap (9) is a problem.

In another design (Fig. 10b, c) a shallow steel dish (10) is secured by a force fit to the bottom of the zinc can. The cell is placed in a plastic cylinder, the bottom edge of which has been turned over by heat treatment. A plastic cap (7) is then applied at the upper end of the cell, replacing the hot-melt sealing compound.

A further variation is obtained by the use of a plastic sleeve into which the cell is placed and the open end clamped tightly to the zinc can by means of an externally applied steel ring (12 in Fig. 10b). A further improvement is shown in Fig. 10c. Here a plastic tube is fitted over the cell. At the top, a plastic sealing element (7) is inserted; at the base, a shallow steel dish is pressed onto the zinc can. The tube is then closed at both ends by externally applied steel rings (12 and 13, with seal ring 8).

In the attempt to obtain the most reliable waterproof sealing between plastic and metal sealing elements, the latter are perforated and the plastic elements formed by injection molding with the metal parts in place. Figure 10d illustrates a dry cell incorporating composite metal-and-plastic sealing elements made in this way.

Finally, it should be mentioned that there is a patent [22] that describes a finished cell completely enclosed in plastic.

2. RAW MATERIALS

2.1. MANGANESE DIOXIDES

Manganese dioxide is one of the most important raw materials in a dry cell, since its properties are fundamental with respect to the ampere-hour

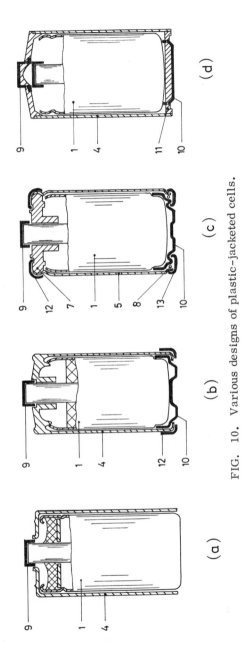

FIG. 10. Various designs of plastic-jacketed cells.

capacity and to the ability to deliver current under load. The name manganese dioxide is applied to a whole series of oxides of four-valent manganese. The individual members of the group are distinguished by varying degrees of hydration, by different crystal structure, and consequently by their characteristic chemical activity.

The types of dioxide in general use in the dry cell industry may be subdivided into four main classes:

a. Natural ores, obtained by open-cast or deep mining operations from natural deposits, usually accompanied by physical concentration processes (for example, sorting, washing, and so on).

b. Chemically activated dioxides, prepared by roasting natural ores and treating the product with acid.

c. Electrolytic manganese dioxides, the products of anodic oxidation of divalent manganese.

d. Synthetic hydrates, which are either obtained as by-products of reactions in which organic radicals are oxidized with $KMnO_4$ (for example, in the preparation of saccharine) or by thermal decomposition of other manganese compounds followed by a wet chemical oxidation process.

2.1.1. Natural Ores

2.1.1.1. Occurrence. Manganese is the eleventh most abundant element in the earth's crust. About 160 different minerals are known to contain manganese, but of these little more than a dozen are important (see Table 1). In various rocks and stones, the element may be present in amounts from 0.01% up to 0.2%. In deep-sea clay, which occurs as a sediment over wide areas of the ocean bed at depths between 1,000 and 2,000 meters, the manganese content is more than 1%. In addition, there are manganese ore deposits in which the manganese content ranges from 10% to more than 50%. These deposits are the result of exceptional geological conditions and are especially important, since it is from such locations that manganese ore can be economically mined.

Iron is about fifty times more abundant in nature than the chemically similar manganese. Both elements usually occur together, and when ex-

TABLE 1

Important Manganese Minerals (Simplified formulas)

		Mn (%)	Hardness	Specific gravity
Psilomelane group	$\alpha\text{-}MnO_2$ (Type: $R_2X_8O_{16}$)	45–60	–	–
Cryptomelane	$R = K$, $X = Mn^{2+}$, Mn^{4+}	–	5–6 (1)	4.3
Hollandite	$R = Ba$, $X = Mn^{4+}$, Fe^{3+}	–	6 (2)	4.5–5
Coronadite	$R = Pb$, $X = Mn^{2+}$, Mn^{4+}	–	5.2–5.6	4.5–5 (6)
Psilomelane	$Ba(Mn^{2+}, Mn^{4+})_9O_{19} \cdot H_2O$	–	5–6 (2)	4.4–4.7
Polianite-pyrolusite	$\beta\text{-}MnO_2$	63	6–7 (2–2.5)	5.0
Ramsdellite	$\gamma\text{-}MnO_2$	63	3	4.7
Manganite	$\gamma\text{-}MnOOH$	62	4	4.3
Rhodochrosite	$MnCO_3$	48	3.5–4.5	3.3–3.6
Braunite	$Mn^{2+}(Mn^{3+}, Fe^{3+})_6SiO_{12}$	50–60	6–6.5	4.7–4.9
Hausmannite	$Mn^{2+}(Mn^{3+})_2O_4$	72	4.8	4.7–5
Bixbyite (Sitaparite)	$(Mn, Fe)_2O_3$	30–40	6	5.0
Jacobsite	$MnFe_2O_4$	24	6	4.8

posed at the earth's surface, there may be some separation due to the greater solubility of manganese compounds, thus leading to enrichment in separate deposits.

The majority of the minerals listed in Table 1 are formed at or at least near the surface. The different manganese dioxide modifications are found principally in weathered deposits, but pyrolusite, manganite, and rhodocrosite also occur in sedimentary formations. In deposits of iron carbonate, manganese may replace iron in amounts between 5 and 8%. Pure deposits of rhodocrosite are rare, but mineral oxides such as braunite, manganite, and hausmannite are commonly of high manganese content. In metamorphic deposits, that is, original formations that have been displaced to great depths and subjected to high temperatures and pressures, bixbyite and jacobsite (a magnetite that contains manganese) are quite common, in addition to braunite, hausmannite, and hollandite.

2.1.1.2. Classification of Manganese Ores. Manganese ore deposits may be economically worked from about 30% Mn content upward, according to the depth of the ore, the available transport facilities, and the requirements of the market in which the product is to be sold. In exceptional cases, ores low in iron can be worked with a manganese content as low as 15%, but the raw extract must be concentrated until it contains 40-50% Mn.

The steel industry is the main consumer of manganese, taking up about 90% of the total production. The remainder is divided equally between the chemical and battery industries. Within the chemical industry, manganese is used in glass, ceramics and pigments. Manganese salts also find application in photography and as disinfectants, bleaching agents, and artificial fertilizers.

All types of ore contain impurities that are detrimental to the quality of the end-product if certain limiting quantities are exceeded. These are silicates, alumina, phosphorus, sulphur, arsenic, and a whole group of heavy metals.

The ores used in the dry-cell industry are of different manganese dioxide modifications and are found at many sites all over the world. However, often they are too low in Mn content or of insufficient purity, so that special treatment of the raw product is usually required.

TABLE 2

World Reserves of Manganese Ores

Country	Reserves (Millions of tons)	Mn content (%)
U.S.S.R.	625	20-45
India	95	~ 40
South Africa	60	~ 45
Brasil	60	~ 45
Morocco	50	~ 45
China	30	20-50
Ghana	10	> 50

2.1.1.3. Extraction and Treatment of Manganese Dioxide for Battery Manufacture. The mining of manganese ores and its subsequent treatment are both dependent on the characteristics of the deposit--extent, type, situation, and so on--and on the nature of the accompanying gangue. Most deposits, especially those in tropical countries, are worked on the open-cast system. In the Caucasian mountains and in the Ukraine, however, deep shaft mining methods are necessary.

Concentration of the product is usually carried out by first crushing the raw ore and following this with a separation process. The crushed ore is suspended in a stream of water, and differences in the rate of sedimentation permit the ore to be separated from rock and other impurities. The concentrate is then dried and finally ground to the required particle size.

Ore containing a high percentage of manganese dioxide is required for the battery industry. Since this material usually occurs in localized, lens-shaped deposits embedded in ore of lower quality, hand-sorting operations are still of importance in the preparation of battery-quality ore. The hand-sorted raw ore is then further processed in a separate unit so that no contamination with metallurgical ore can take place.

The rough-crushed ore is reduced between rollers to particles less than 1/8 in. diam, and these are then further reduced in a grinding unit or a ball mill. During the grinding operation, the product that has passed the mill is treated in a wind sieve and the rougher material returned to the grinder for further processing. By proper adjustment of mill and wind sieve, some control of the particle size spectrum can be achieved.

2.1.2. Activated Manganese Dioxides

Activated manganese dioxide is obtained by chemical treatment of a natural ore. There are several suitable chemical processes known and in use. The basic principle is that only a part of the original dioxide is dissolved from the surface of the particles, and the residual surface is characteristically porous, hydrated, and chemically active.

The manufacturing process consists of heating the ground natural ore to 600° -800° C, with or without the addition of a reducing agent. If natural gas is available, the reduction of the ore can be efficiently carried out as low as 300° C. The lower oxide formed under these conditions is either Mn_2O_3 or Mn_3O_4, depending on the reaction conditions. The product is then treated with hot sulphuric acid. The resulting slurry consists of the activated dioxide as solid phase and an acid solution of manganese sulphate as liquid phase. The subsequent treatment of the slurry offers several possibilities, the most direct approach being a filtering operation. The clear solution can then be used in the preparation of the electrolyte for the electrolytic deposition of manganese dioxide. The filter cake is washed free of acid and dried at carefully controlled temperature. The product is chemical manganese dioxide [23]. Alternatively, the slurry may be fed directly into the electrolyte of an electrolytic process whereby the dioxide formed by anodic oxidation is not collected as a coherent deposit on the electrode but is allowed to settle out and mix with chemically activated solid phase. The filtering and washing operations follow [24]. A third variation of the process is obtained when the slurry is rendered alkaline in the presence of a chemical oxidizing agent (for example, air,

$KMnO_4$, and so on), thereby precipitating manganese oxides of somewhat variable composition on the solid phase of the slurry [25] .

2.1.3. Electrolytic Manganese Dioxide

Electrolytic manganese dioxides (usually abbreviated EMD) are synthetic dioxides prepared by anodic oxidation of $MnSO_4$ in sulphuric acid solution [26]. The starting material for the preparation of the $MnSO_4$ solution is usually rhodocrosite ($MnCO_3$) where this is available locally (Japan, the United States). The natural carbonate is a very favorable material, because it can be dissolved directly in sulphuric acid. Otherwise, a natural dioxide must be used. A reasonably high percentage ore (not less than 75% MnO_2) should be selected, since the increased process costs resulting from the use of low-content ores are greater than the price difference involved in the purchase of high-content ores.

The manufacturing process for EMD may be briefly described as follows. *

A solution of manganese sulfate is prepared by dissolving either rhodocrosite or MnO (obtained from the thermal decomposition of natural dioxide at about 1000° C in the presence of a reducing agent) in sulfuric acid. The clear solution is separated from the insolubles by decantation. Heavy metals are precipitated as sulfides, which are allowed to settle out, and the clear solution is further purified by elevation of the pH and oxidation with air. Under these conditions, iron and other metallic impurities are converted to insoluble hydroxides, which are removed by a filter operation. The purified solution is then electrolyzed, and EMD collects on the anode as a very hard lava-like layer, which has to be broken off, ground to size, washed free of acid, and finally dried.

The composition of the electrolyte must be carefully controlled. The solution is heated to 90°C and pumped through the electrolysis cells at a rate of 3% of the total electrolyte volume per minute, fresh solution

*See Chapter 3 of this book for more details on EMD.

entering the cells at the bottom. The process results in a decreasing manganese and an increasing acid concentration, which must be adjusted. At intervals of 1-2 hr, 10-20% of the working electrolyte volume is removed from the process and the composition corrected by addition of MnO or $MnCO_3$. The solution is returned to the system through a filter unit.

The electrodes are usually made from a lead alloy containing 3% antimony, the anodes being about 3 mm and the cathodes 5 mm thick. The spacing between the working electrode surfaces is about 25 mm. The electrolyzing current density is 9-10 A/ft^2. Higher current densities lead to a gradual dissolution of the anode, with the result that the EMD becomes contaminated with lead. Experience teaches that when EMD (or activated manganese dioxide) contains 0.5% lead the shelf-life of batteries is somewhat restricted because of corrosion at the zinc anode. It is recommended, therefore, that not more than 0.2% lead should be tolerated in EMD.

Instead of the lead-antimony alloy, graphite or titanium is frequently employed for the electrodes.

After the removal of the raw EMD from the electrode, the material is air dried and then ground in a stone mill. Metal cannot be used at this stage because of the sulfuric acid retained in the dioxide. The ground EMD is then subjected to ten washing operations. At each wash, 500 ft^3 of water are used per ton of EMD. Some barium chloride is dissolved in the ninth wash water, and a neutralizing agent is added with the tenth and last wash. The barium chloride precipitates any residual sulfate and the neutralizing agent takes up any free mineral acid and the inherent acidity of the EMD. Sodium carbonate or ammonia is used by some manufacturers, but zinc oxide added in an amount equivalent to 0.5% of the EMD is preferred. Ammonia is particularly disadvantageous in EMD used for dry cells made with magnesium chloride electrolytes.

2.1.4. Synthetic Hydrates

As previously described, chemical activation processes are restricted in effect to the surface layers of a natural ore. Synthetic hydrates, however, are characterized by the fact that the whole oxide is obtained in a highly reactive and usually highly hydrated condition.

There are two distinct types of synthetic hydrate. The first type is obtained from a process in which all the manganese is present in solution in a high state of oxidation (for example, $KMnO_4$). On reduction to the four-valent state, an oxide/hydroxide complex is precipitated in a finely divided and highly hydrated form. The second type is obtained by thermal decomposition of a manganese compound (other than an oxide) so that a finely divided solid oxide is produced; the average valency of the manganese is, however, less than four. The finely divided solid is then suspended in an alkaline oxidizing medium, washed, and dried under precisely controlled conditions in thin layers (for instance, on a drum drier). The final product is a very reactive dioxide somewhat less hydrous than the first type described.

The type-1 synthetic hydrates have characteristics not shared by other modifications, whereas the type-2 synthetics have properties inter - mediate between those of type 1 and EMD. The type-2 hydrates have recently achieved increased importance in the United States and in western Europe.

The basis of the type-1 hydrate manufacture was, until recently, the slurry obtained as a waste product from saccharine production. Here, o-toluol-sulfamide was oxidized in alkaline solution with potassium per - manganate. This process is now of minor importance, and slurries obtained from similar organic radical oxidation processes are also being employed. The slurries must be neutralized and washed free of soluble salts. The product is always very fine and hygroscopic. Its composition is somewhat variable because of the different types of slurry used.

Manganese Chemicals Corporation has developed an interesting pro - duction technique for a type-2 hydrate [27] . Reduced natural manganese dioxide is converted into the soluble manganese amminocarbonate and reprecipitated as carbonate. This is raised to 200°-300° C in an oxygen- rich atmosphere, the result being a product equivalent to a 75% manganese dioxide content. After repeated washing out of residual carbonate and lower oxide, and renewed carbonation and oxidation , a highly active product of about 90% manganese dioxide is obtained.

2.1.5. Chemical and Physical Properties of Manganese Dioxides

Manganese dioxides are by no means chemically pure and well-defined substances. The composition is never quite stoichiometric; there is always a small oxygen deficiency. Natural ores contain the largest proportion of foreign matter, as is to be expected, while EMD and synthetic hydrates are the purest forms.

Impurities contained in manganese dioxides are very largely determined by the origin and the type of ore. Natural ores used in dry-cell production may contain between 10 and 30% foreign matter (largely from stone and rock formations adjacent to the deposit), consisting mainly of oxides and silicates of the alkaline earth metals, aluminum, and iron. In addition, carbonates, phosphates, and sulfates are also found together with heavy metals and elements of the arsenic group. Water must also be included in the impurities, although this may be partly constitutive.

EMD may contain manganese sulfate and sulfuric acid; graphite or lead from the electrodes; ammonium, barium, or zinc salts from the neutralization process; and also trace amounts of heavy metals from the original manganese ore.

Precipitated synthetic hydrates contain a few per cent alkali and relatively large amounts of water.

Apart from the above mentioned impurities typical of the source of the dioxide, other chance inclusions of foreign matter such as wood chips, jute fibre, paper, nails, and traces of oil from the grinding plant are not unknown, although in a well-run production unit this should not happen.

Impurities such as soluble heavy-metal compounds, carbonates, or excess of alkali and ammonium salts in magnesium chloride batteries may have deleterious effects on battery performance. Others, such as insoluble silicates and oxides, may be inert, but undesirable because they reduce the amount of active MnO_2 available in a given volume.

The chemical analysis of manganese dioxides offers no special difficulties. An excellent instrument for determining trace quantities of harmful impurities is the atomic absorption (flame) spectrograph.

In Table 3, several complete analyses of different manganese dioxides are shown. These figures are merely for illustration and are not to be interpreted as typical or mandatory for any particular type of ore.

TABLE 3

Percentage Composition of Various Manganese Dioxides[a]

| Compound | Natural ores | | | EMD | Synthetic hydrate |
	West African	Caucasian	Greek	Mitsui Mining	Permanox
MnO_2	84.1	84.7	74.4	92.7	82.0
MnO	2.5	2.2	3.2	2.4	5.0
SiO_2	3.9	4.6	6.3	-	-
Fe_2O_3	1.4	0.8	1.2	0.06	0.3
Al_2O_3	1.4	0.8	0.9	-	-
CaO	3.4	0.6	5.4	-	0.1
BaO	-	1.4	-	-	-
MgO	0.1	-	0.2	0.2	-
$K_2O + Na_2O$	0.2	0.1	0.3	-	5.9
PbO	-	-	0.17	0.06	-
CuO	0.01	0.02	0.03	-	-
NiO	0.06	0.10	-	-	-
CoO	0.01	-	-	-	-
CO_2	-	0.2	1.5	-	-
P_2O_5	0.2	0.5	0.1	-	-
SO_3	0.03	0.3	0.06	0.9	0.1
H_2O (combined)	2.7	2.0	4.0	3.5	6.0
not determined	0.0	1.7	2.2	0.2	0.6

[a]These results are actual analytical figures obtained on individual samples. They are not to be regarded as average for any specific type.

There are several ways of expressing the active oxygen content of a manganese dioxide sample. The most direct is to determine the amount of active oxygen by the oxalic-acid (or similar) method and then convert this to the equivalent amount of MnO_2. A refinement of this method con-siders also the total analytical amount of manganese (% Mn) present in the ore. The amount of MnO, Mn_2O_3, and MnO_2 can be calculated from the following factors. Subtracting the values shows that not all Mn is available as reducible MnO_2.

$$\Delta \ \% \ MnO \ = (\% \ Mn \cdot 1.291) - (\% \ MnO_2 \cdot 0.816)$$
$$\Delta \ \% \ Mn_2O_3 = (\% \ Mn \cdot 2.874) - (\% \ MnO_2 \cdot 1.816)$$
$$\Delta \ \% \ MnO_2 \ = (\% \ Mn \cdot 1.586) - (\% \ MnO_2)$$

Still another method is to calculate a stoichiometric formula MnO_n in which the index n indicates the average atomic ratio of oxygen to manganese, varying between 1.5 and 2.0. These numbers represent pure Mn_2O_3 and pure MnO_2. Values obtained from actual dioxides lie around 1.95. The index n can be calculated from the following expression.

$$n = 1 + 0.632 \ (\% \ MnO_2/\% \ Mn)$$

in which the $\% \ MnO_2$ is calculated directly from the active oxygen.

Laboratory methods for the preparation of the various modifications of the dioxides, lower oxides, and hydroxides are described in the literature.

2.1.6. Ion-Exchange Properties

Kozawa [28], Gabano et al. [29], and Muller et al. [30] have investigated the ion-exchange properties of γ-MnO_2 for various metal ions. Sasaki [31] and Vosburgh and coworkers [32] have pointed out the importance of the ion-exchange reactions in relation to the measurement of the electrode potential. A detailed discussion of this subject can be found in Chapter 3 of this book.

Summarizing the results of these investigations the following conclusions may be reached:

a. The ratio of the number of zinc ions adsorbed to the number of protons liberated is 1:2. The amount of zinc adsorbed at the saturation level is proportional to the zinc concentration in the solution and the pH, as required by the mass action law. At low pH values, the amount of

zinc adsorbed is also low; at higher pH values, the amount is correspondingly greater. The initial concentration of zinc in the solution operates
in the same sense. Adsorbed zinc ions can be reversibly released by
reduction of the pH of the solution after equilibrium has been established.

    b. The ion-exchange capacity of manganese dioxide is a linear function
of the specific surface (see Section 2.1.9).

    c. The zinc ion adsorption isotherm appears to relate to a typical
chemisorption process.

    d. The metal ions adsorbed by MnO_2 arranged in order of increasing
amount is similar to Hofmeister's classification [33] for ion-exchange
resins:

$$Li^+ < Na^+ < K^+ < Rb^+ < Cs^+ < Mg^{2+} < Ca^{2+}$$
$$< Sr^{2+} < Ba^{2+} < Zn^{2+} < Al^{3+} < Co^{2+} < Ni^{2+} < Cu^{2+}$$

Both classifications are similar, with the exception of the position of Ni^{2+}.

The experimental results summarized above may be explained on the
basis of a model proposed by Kozawa which has found general acceptance
(Fig. 11).

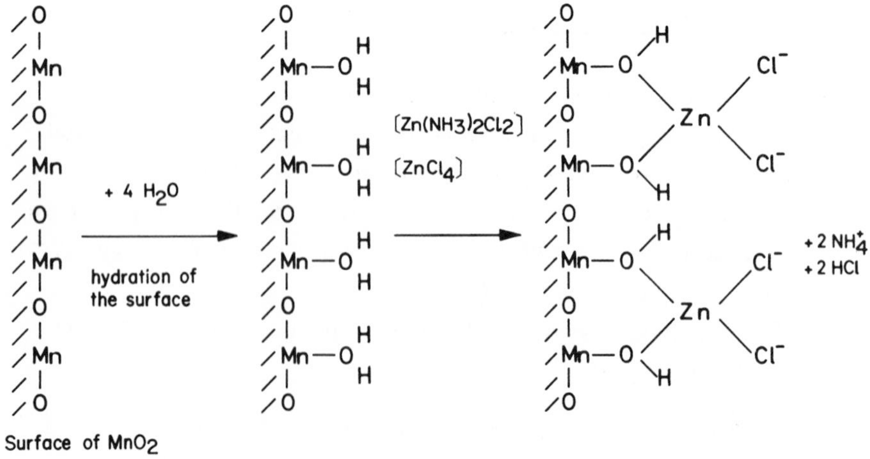

Surface of MnO_2

FIG. 11. Mechanism of the ion-exchange reaction on the surface of
MnO_2 according to Kozawa.

The model explains the weak acid characteristics of highly hydrated manganese dioxides and also the ion-exchange properties. In order that ions, present in the solution as complexes, should be preferentially adsorbed on the MnO_2 surface, the MnO_2-ion complex must be more stable than the ionic complex in the electrolyte. According to the model, adsorbed metal ions form six-membered ring complexes with the MnO_2. From the chemistry of the chelates, it is known that five- or six-membered ring complexes are more stable than complexes with Cl^-, OH^-, NH_3, or H_2O groups as ligands.

2.1.7. Potential Versus pH Characteristics

The different results reported by individual investigators for the dependence of the electrode potential of MnO_2 on pH(Fig. 12) was attributed by Sasaki to the ion-exchange properties of the dioxide. He therefore agitated the dioxide with electrolyte of known pH until equilibrium was attained, measured the electrode potential of the dioxide, and then separated the electrolyte from the solid phase by centrifuging and remeasured the pH. Figure 13 shows the dependence of the electrode potential on the pH of the electrolyte. The broken-line curve shows the potential change before the agitation and the solid line the change after equilibrium had been reached. For a specific electrolyte solution, Sasaki found that there is a particular pH value that does not vary during the agitation with a particular type of MnO_2. This pH value he termed the isoacidic point and found that its value decreased from natural ore to EMD to synthetic hydrates. Thus a low isoacidic pH value is associated with high battery performance of the dioxide.

The electrode potential of the various types of dioxide was found to be, in general, a linear function of the pH in the range pH = 1 to 9, the slope being -0.0587 V/pH unit for natural ores and -0.0570 V/pH unit for EMD. Dioxides that had been treated with acid (HNO_3) show greater slopes, varying between -0.0676 and -0.0698 V/pH unit.

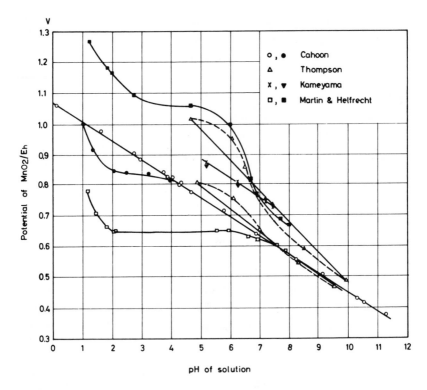

FIG. 12. Relation between MnO_2 potential and pH reported by individual investigators.

2.1.8. Crystal Structure

The various modifications of manganese dioxide used in battery manufacture are characterized by their different crystal structures. There
have been several detailed investigations in recent years of the structure
of many different manganese oxides. Much valuable information has
been collected, but unfortunately the designation of specific types has
become inconsistent. In this chapter, structures similar to cryptomelane
are designated α-MnO_2, pyrolusite β-MnO_2, and ramsdellite γ-MnO_2.

2.1.8.1. The Alpha Group. The α-MnO_2 varieties have the general
formula $R_2Mn_8O_{16} \cdot x\,H_2O$ in which the R may represent Mn^{2+}, Ba^{2+}, K^+,

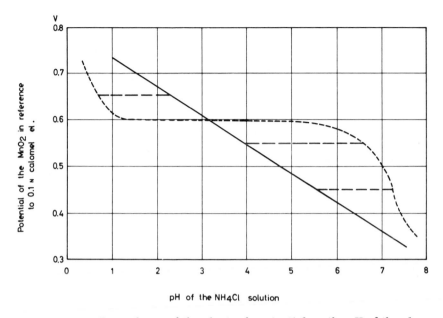

FIG. 13. Dependence of the electrode potential on the pH of the electrolyte.

Li$^+$, and so on, and x represents the amount of combined water, which is generally more than 6%. The oxides also have very pronounced ion-exchange properties. The crystal type is tetragonal, and the unit-cell dimensions are a = 9.82 Å, c = 2.86 Å. The unit cell contains eight molecules of MnO$_2$. Members of this group may be naturally occuring minerals or synthetic hydrates. The structure varies somewhat between individual samples, but may, in general, be regarded as a derivative of the well-defined mineral cryptomelane. Typical are the large lattice spacings of 4.9 and 6.9 Å. Synthetic hydrates included in this group yield very diffuse and weak diffraction patterns.

The value of n in MnO$_n$ may be almost 2, and the oxide may be reduced in homogeneous phase to n = 1.87. Beyond this value, the crystal breaks down to a heterogeneous system with complex components.

2.1.8.2. <u>The Beta Group</u>. The upper limit of oxidation seems to lie at about n = 1.98, and usually there is no combined water present, which accounts for the almost complete absence of ion-exchange properties. The crystal is of the tetragonal (rutile) type with a = 4.42 $\overset{\circ}{A}$ and c = 2.87 $\overset{\circ}{A}$. The unit cell contains two molecules of MnO_2. Members of the group are naturally occuring minerals related to pyrolusite. The x-ray diffraction pattern is very sharp and well defined, with no spacing greater than 3.5 $\overset{\circ}{A}$.

On reduction, β-MnO_2 shows a very narrow phase breadth, the crystal collapsing at n = 1.96. Reduction proceeds in heterogeneous phase, the reduction product being γ-MnOOH, similar in structure to the mineral manganite.

2.1.8.3. The Gamma Group. The gamma group has the general formula $MnO_{1.9-1.96}(x\ H_2O)$, in which x usually represents about 4% combined water. The actual amount of water varies somewhat between samples, but there is some evidence that at least 2% is necessary to stabilize the lattice. This water imparts some ion-exchange properties to the oxide. The crystal is orthorhombic, with a = 4.52 $\overset{\circ}{A}$, b = 9.27 $\overset{\circ}{A}$, and c = 2.86 $\overset{\circ}{A}$. The unit cell contains four molecules of MnO_2. Members of the group are natural ores and EMD. The x-ray diffraction patterns show relatively few lines, some of which are very diffuse and broad. The structure may be regarded as a derivative of the mineral ramsdellite.

Reduction of γ-MnO_2 proceeds in homogeneous phase down to n = 1.5, the phase limit corresponding to α-MnOOH, similar to the mineral groutite. The unit-cell volume increases linearly with the degree of reduction, the final unit-cell dimensions being a = 4.58 $\overset{\circ}{A}$, b = 10.76 $\overset{\circ}{A}$, and c = 2.89 $\overset{\circ}{A}$.

Table 4 gives details of the spacings of the three groups described above. For each group, the parent mineral is shown together with a typical representative of the group.

The phase changes during reduction are easily followed from the electrode potential versus n in MnO_n diagrams. The behavior of the three classes is shown schematically in Fig. 14.

Other designations of structural varieties of manganese dioxide frequently found in the literature are rho (ρ) and delta (δ). The rho variety conforms generally with the gamma group described above, while the delta modification is similar to the alpha types except that the lattice is

TABLE 4

Interplane Spacings in Manganese Dioxide Crystals

d (Å)	Alpha group		Beta group		Gamma group	
	Cryptomelane relative int.	Permanox relative int.	Pyrolusite relative int.	Egyptian ore relative int.	Ramsdellite relative int.	EMD: Mitsui relative int.
6.9	4	–	–	–	–	–
4.9	4	–	–	–	–	–
4.86	–	3	–	–	–	–
4.64	–	–	–	–	5	–
4.30	–	–	–	–	–	–
4.08	–	–	–	–	} 10	} 9
4.07	–	–	–	–	–	
3.76	–	–	–	–	–	–
3.48	1	–	1	–	–	–
3.46	1	–	–	–	–	–
3.14	–	–	10	–	–	–
3.13	–	–	–	10	–	–
3.11	10	–	–	–	–	–
2.60	–	} 10	–	–	–	–
2.52	–		–	–	7	–

TABLE 4 (Continued)

Interplane Spacings in Manganese Dioxide Crystals

d (Å)	Alpha group		Beta group		Gamma group	
	Cryptomelane relative int.	Permanox relative int.	Pyrolusite relative int.	Egyptian ore relative int.	Ramsdellite relative int.	EMD: Mitsui relative int.
2.43	–		–	–		–
2.41	–	10	5	1	5	–
2.40	6		–	–	4	6
2.33	–		–	–	4	–
2.23	–		–	–	–	–
2.21	–	–	1	1	–	–
2.20	4	–	–	–	–	–
2.16	2	–	–	–	–	–
2.13	–	–	3	–	4	–
2.11	–	–	–	1	–	8
1.98	–	–	2	–	–	–
1.97	–	–	–	1	–	–
1.92	–	–	–	–	6	–
1.63	2	–	5	1	6	10

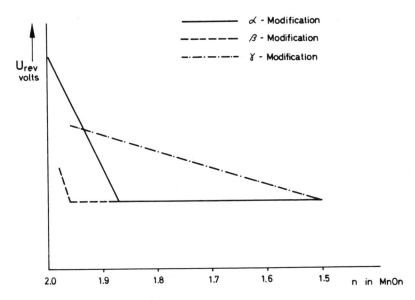

FIG. 14. Potential of manganese dioxide electrode versus n in MnO_n.

very distorted in one direction. All the manganese dioxides mentioned
here show lattice deficiencies to a greater or lesser degree.

2.1.9. Particle Size and Specific Surface

The particle size of ground manganese dioxide for use in dry-cell
manufacture extends over a wide range. The maximum size is rarely
greater than 0.2 mm and the minimum lies at about 0.1 μm. Precipitated
synthetic hydrates , although generally fine, are more uniform, the usual
limits being 1-20 μm.

Well-defined, sharp, and pointed particles are typical of pyrolusite,
and rounded, ill-defined particles are usually found in EMD.

The total surface area of finely divided porous materials is the sum
of the outer geometric area of the individual particles and the inner area
formed by crevices and pores penetrating deep into the grains. Particle-
size determinations yield only a rough estimate of the outer area, and
more precise methods are necessary if the total surface area is to be
assessed.

The BET technique is a gas-absorption method carried out at the temperature of liquid air.[*] The sample of material being tested is first treated to remove any absorbed gas (by heating in a vacuum) and is then allowed to absorb nitrogen at the low temperature. Assuming that a monomolecular layer is formed, the surface area of the substance can be calculated from the amount of gas absorbed, using the known molecular dimensions of nitrogen. An estimate of the pore dimensions can be made by fractional desorption. In order to obtain reproducible results, the technique must be accurately laid down and followed, especially with respect to the initial treatment of the sample.

Comparative values of the specific surface of different classes of manganese dioxide are given in Table 5.

Further individual values are quoted in Table 6.

According to Tvarusko [34] , there is a relationship between the surface area and the amount of adsorbed water; Kozawa and Laurent have shown the relationship between the BET area and the number of adsorbed zinc ions. This latter relationship is dependent on the ion-exchange characteristics of manganese dioxide, which in turn are a function of the crystal structure and the degree of hydration.

2.1.10. General Physical Properties

General experience shows that the natural pyrolusites (for example, Caucasian ore) are the hardest and the cryptomelanes are the softest

TABLE 5

Specific Surface of Various Classes of Manganese Dioxide

Class of MnO_2	Specific surface (m^2/gm)
Natural ores	7–22
EMD	28–43
Synthetic hydrates	30–90

*The abbreviation for Brunauer, Emmett, and Teller is BET.

TABLE 6

Specific Surface of Various Manganese Dioxides

Type of MnO_2	Specific surface (m^2/gm)
Natural ores	
Caucasian	13
Greek (Bailey, fine)	14
Ghana (300-350 mesh)	9.5
Chemically activated	
Ergogen	29
EMD	
Mitsui Mining	29
Perozono	66
Lavinore A	34
Synthetic hydrates	
Permanox	29
Sedema M	81
Sedema WB	57

types of manganese dioxide, EMD being intermediate. The hardness of the ore determines to a large extent the rate at which molding tools wear in production.

The true density of manganese dioxide is difficult to determine directly because of the porous nature of the material. Comparative values are listed in Table 7 with some estimates of the apparent density.

2.2. CONDUCTIVE MATERIALS

Although MnO_2 has slight semiconductive properties, this conductivity is of no practical importance. An electrode containing only the dioxide

TABLE 7

Densities of Manganese Dioxide Types

Type of MnO_2	Density (gm/cm^3)	Apparent density (gm/cm^3)	
		a	b
Natural ores			
Caucasian	4.7	1.4	2.0
West African	4.4	1.8	2.4
Greek	4.2	1.3	2.0
EMD			
Mitsui Mining	4.3	1.8	2.5
Mitsubishi	4.3	1.7	2.4
Synthetic hydrates			
Permanox	3.2	0.8	1.0
Galvanodurit F. 48	2.8	0.8	0.9
Sedema M	-	1.3	1.7

[a]Apparent density, loose packing.

[b]Apparent density, consolidated under own weight.

would not operate efficiently. The depolarizing mix therefore must include an inert electronic conductor, and for this purpose the varieties of carbon known as graphite and carbon black have found universal application.

2.2.1. Graphite

Graphite is a naturally occuring carbon modification with good electrical conductivity, crystallizing in flat hexagonal plates. The crystalites are usually less than 1 mm in diameter. The specific gravity is 2.27 (as determined from x-ray analysis) but this value is rarely found by displacement methods, since the aggregates contain inaccessible pores.

A good battery-quality mineral graphite is found at Kropfmühl, near Passau in West Germany, with about 25% carbon in the raw ore. The mineral occurs in large, lens-shaped inclusions at the boundary zones of gneiss, granite, and limestone and contains sulphides, particularly of zinc and copper, and quartz as the main impurities. Other important deposits have been found in Northern Rhodesia (now Zambia), on the east coast of Madagascar (with 20-30% carbon) and near Skåland in the north of Norway (30% carbon).

Artificial graphites (such as Acheson graphite) are very pure but are usually very fine. The effective conductivity is reduced (contact resistance is increased) and the material is somewhat difficult to process.

The raw natural ore must be concentrated by flotation processes, whereby a product with 90-92% carbon, such as is generally used in battery manufacture, can be readily obtained. Chemical treatment (for example, using hydrofluoric acid) is resorted to only when extremely high purity is required (>99.95% carbon).

Graphite is supplied in a variety of forms, such as large or small flakes, ground and sieved powders, and in micronized form. Ground powders with 15-20% greater than 60 μm are generally acceptable in the battery industry.

2.2.2. Carbon Black

Carbon blacks, made from acetylene gas, are by far the most pre-ferred by the battery industry.

The basic reaction in all manufacturing processes is the following thermal decomposition:

$$C_2H_2 \rightleftarrows 2\,C + H_2 \quad \Delta H = +55\ \text{kcal}$$

At temperatures above $800°$ C, the equilibrium is shifted to the right spontaneously, the evolved heat of reaction being sufficient to maintain the process.

The various methods of industrial production differ in the technique used to control the reaction and in the choice of the physical parameters such as temperature and pressure.

The oldest process is the discontinuous explosion technique developed by Houbon. Acetylene gas, pure or in a rich mixture with other hydrocarbons, is confined in thick walled metal tubes under 3-5 atm pressure and exploded by means of an electric spark. The acetylene decomposed, forming carbon black and hydrogen. The process is difficult to control, is not without danger, and today finds restricted application.

Of greater importance is the continuous thermal decomposition of acetylene under normal pressure, a process first introduced by the Shawinigan Company in Canada. Acetylene is led into the head of a vertical cylindrical retort. To start the reaction, an acetylene/air mixture is burned until the temperature in the reaction zone is sufficient to maintain the thermal decomposition reaction. The acetylene/air mixture is then replaced by pure acetylene or an acetylene-rich hydrocarbon mixture. The carbon black is carried by the gas stream from the retort to a settling chamber, where it is separated from the hydrogen.

Recently furnace blacks have been produced by a special technique. They are finding some acceptance in the battery industry and have the advantage of being in granulated form, which facilitates transport and processing.

Carbon black is an extremely fine, black powder consisting of practically pure carbon. The individual particles are nearly spherical in form, with diameters between a few millimicrons and one micron and form straight and branched chainlike structures. Blacks are classified as high- or low-structure blacks according to the quantity and length of these chains. The elastic and flexible nature of the chains is such that with a relatively low percentage of black in the mix, the manganese dioxide particles are adequately enveloped and a high degree of electronic conductivity is assured.

An important property of carbon black is its absorptive power for electrolyte. Paradoxically, the performance of a dry cell, especially under heavy current drain, is improved by increasing the amount of electrolyte, (wetness) in the bobbin. The amount of electrolyte that can be included in the mix is, however, strictly limited by the mechanical operations that are necessary for production. Mixes containing carbon

black can retain much more electrolyte than mixes without it.   Hence the
importance of the absorptive power for battery technologists.

The absorptive power may be defined in a variety of ways.   One common
method is to determine the volume of a 10% solution of acetone in water that
can be absorbed by 5 gm of black.   The determination is carried out by
adding the solution dropwise to the black; the vessel containing the black
is continuously shaken with a pronounced rotary motion.   As more and
more solution is added, the black tends to form aggregates until at last
there is one coherent sphere of carbon black.   When still more solution
is added, a point is reached at which the sphere breaks down again.   The
absorption number, as it is called, is the volume of solution (in cubic
centimeters) that can be absorbed without breakdown of the sphere.

Carbon black contains about 99.6% carbon, up to 0.3% hydrogen, about
0.1% volatile substances, less than 0.05% ash, about 0.03% sulfur and up
to 0.1% other carbon compounds.   The apparent density lies between 40
and 200 gm/liter.   To facilitate transport, carbon black is usually compressed so that the apparent density is markedly increased.

2.3. CARBON RODS

The carbon rod that serves as the current collector from the positive
electrode is an important constituent of a dry cell.   A poor quality rod
can be the cause of low battery performance and of difficulties in manufacture.

Carbon rods are made from powdered coke and a binding agent.   The
hot plastic mix is extruded on Strang presses to long rods, which are
cooled and then cut to length.   These so-called green rods are then subjected to a slow heating and cooling process in the absence of air whereby
the binding agent is carbonized or vaporized.   The rods are finished by
centerless grinding to the required diameter and to the required degree
of surface roughness before being impregnated.

A good rod should show a certain porosity with respect to gas but must
not permit the electrolyte to reach the metal cap by capillary action,

causing it to corrode. To prevent excessive porosity, carbon rods are
usually impregnated in a vacuum with paraffin wax, mineral oil, or a
resin. A quick method to determine the kind of impregnation is to heat
the rods in a Bunsen flame until it acquires a wet appearance, and then
roll it on a cold glass plate: oil remains fluid, but wax solidifies to a
visible film.

The capillary action in different electrolytes varies considerably, and
a fairly large number of rods must be tested in order to obtain a reliable
assessment of suitability. A more reliable method is to immerse the rods
to about two-thirds of their height in the electrolyte and place the containing vessel for some time in a vacuum. The top surfaces of the rods are
then tested with silver nitrate solution. A white deposit indicates the presence of electrolyte.

The conductivity of the rod is, of course, important. Carbon rods
show specific electrical resistance values between 3 and 5 m$\Omega \cdot$cm, but
this may be reduced to about half by graphitization of the outer surface in
an electric resistance furnace.

Further, a certain degree of mechanical stability is required in order
that the rod may withstand the mechanical operations of tamping, capping,
and swaging. The breaking strength of the rod is tested as follows. The
rod is laid on two supports with a free distance of 5 cm from one to the
other. Then the middle of the rod is loaded with continuously increasing
pressure until the rod breaks. A rod of 8 mm diam (for D-size cells)
should withstand a minimum load of 8.5 kg under these conditions. This
value is normally greatly exceeded.

Finally, the carbon rod must provide good mechanical adhesion between the surface and the depolarizing mix. This is a complex function
of the tamping conditions, the condition of the mix and the porosity and
surface roughness of the rod.

2.4. ELECTROLYTE SALTS

2.4.1. Ammonium Chloride

Technical grade ammonium chloride may contain sodium chloride,
sulphates, and iron impurities, resulting from the manufacturing process.

Sublimed ammonium chloride is extremely pure. When ammonium chloride is dissolved in water, the temperature of the solution falls markedly. To avoid the stickiness of the material caused by humidity during storage in silos, some additives (for example, boron oxide, B_2O_3) are necessary.

2.4.2. Zinc Chloride

This salt is very hygroscopic and must be kept in watertight containers. Basic chlorides, sodium chloride, sulphates, and iron are the usual impurities. Heat is evolved when zinc chloride is dissolved in water. Zinc chloride added to a saturated solution of ammonium chloride prevents the formation of the thick crusts of crystalline ammonium chloride above the liquid surface, typical of the pure solution. The concentration of zinc chloride in the electrolyte influences the gelling rate of electrolyte pastes.

2.4.3. Mercuric Chloride

This salt is also known as sublimate in view of the ease with which it sublimes. It is extremely poisonous and causes inflammation of the nostrils when even minute quantities are inhaled. Many metals may be amalgamated by simple immersion in an aqueous solution of mercuric chloride. An amalgam is a solution of a metal in mercury or vice versa. Mercuric chloride is employed in dry cells to amalgamate the zinc anode and to increase the hydrogen overpotential, thereby reducing the rate of corrosion. Although effective at room temperature, its inhibiting properties are uncertain above $40°C$.

2.4.4. Magnesium Chloride

Magnesium chloride can be obtained in very pure form as the hexahydrate, which is also very hygroscopic. Traces of bromides and iodides are the usual impurities.

2.4.5. Manganese Chloride

This is usually obtained as a pink-colored tetrahydrate containing considerable amounts of impurities, of which nickel and cobalt salts are the most undesirable.

Table 8 lists the more important characteristics of the main electrolyte salts.

2.5. GELLING AGENTS

Gelling agents used to immobilize the electrolyte are in general wheat flour and varieties of starch, such as wheat, potato, and corn starch, used singly or in combination. Besides these natural carbohydrates, modified starches have also been developed in recent years, particularly in the United States, Japan, and France. These modified products are characterized by an increased resistance to oxidation and hydrolysis; pastes incorporating them are more stable and show less gassing and syneresis after prolonged storage. Various cellulose derivatives have also proved advantageous as thickening agents in dry-cell pastes.

2.5.1. Cereals and Related Products

The grain is the fruit of the cereal and contains a single seed. The seed is enclosed in the pericarp, which in turn is surrounded by the husk. The flour-containing part of the grain consists of a very large number of tiny chambers (cells) that contain starch grains. The seed and part of the inner layer of the pericarp constitute the flour. The remainder of the pericarp and the husk are known as the bran.

Different types of cereal have different chemical compositions. Grinding of the grain results in a separation of the starch from the remaining components, the separation being more complete the finer the grinding.

TABLE 8

Properties of Electrolyte Salts

Compound	Formula	Molecular weight	Water of crystallization	Density gm/cm^3	Color of solution	Solubility gm/100 gm solution at 20°C	Gravity of saturated solution Specific gravity	°Baumé
Ammonium chloride	NH$_4$Cl	53.50	·/·	1.54	colorless	27.2	1.079	10.5
Zinc chloride	ZnCl$_2$	136.29	1/1.5/2.5/3/4	2.91	colorless	78.6	1.960	71
Magnesium chloride	MgCl$_2$	95.23	1/2/4/6/8/12	2.33	colorless	35.3	1.295	33
Calcium chloride	CaCl$_2$	110.99	1/2/4/6	1.68	colorless	42.7	1.426	42.5
Manganese chloride	MnCl$_2$	125.84	2/4/6	2.98	rose	42.9	1.472	47
Magnesium bromide	MgBr$_2$	184.15	6/10	3.72	colorless	50.8	1.655	57
Lithium chloride	LiCl	42.40	1/2/3	2.07	colorless	45.3	1.273	31
Mercuric chloride	HgCl$_2$	271.52	·/·	5.42	colorless	6.2	1.052	6.8

After heating to 105°C to remove water, the dry residue is found to contain albumin and other proteins, fats and oils, complex polysaccharides (known as pentosans), raw fibrous materials, and mineral ash in addition to starch. The proteins can be separated from the starch and other components by kneading the dry residue with water into a dough. Further treatment in water permits the starch and other components to be washed away, while the protein remains in the dough as an elastic, ductile mass. Carbohydrates are the chief products in cereals; only small quantities of fats, such as palmitin and olein, are found.

As already stated, the starch is located in the cells of the flour-containing part of the grain. Other carbohydrates, the pentosans, are formed in the pericarp and the husk and are therefore found in increased amounts in bran and in coarse flour.

Enzymes contained in the cereal are located in the innermost layer of the pericarp; when they are retained in the starch or flour, they lead to gas generation in dry cells.

The ash remaining after calcination of the grain contains a relatively high percentage of phosphorus from the phosphoric acid originally present.

The above is summarized in Tables 9-11. Table 9 shows the location of the individual components in the grain. Table 10 indicates the change

TABLE 9

Grain Components

Husk and pericarp	Flour-containing part
Fiber material	Proteins = gliadin and glutenin
Pentosans	Starch
Fat = oil	
Proteins	
Ferments	
Vitamins A, B_1, B_2, E	
Salts = minerals = ash	

TABLE 10

Composition of Milled Grain

No. [a]	Protein	Fat	Starch	Pentosans	Fiber	Ash
00	12.20	1.02	83.60	3.14	0.05	0.45
0	12.53	1.01	82.34	3.19	0.10	0.52
1	12.60	1.32	80.43	3.52	0.18	0.61
2	12.94	1.45	79.12	3.69	0.43	0.92
3	13.45	1.61	76.99	3.87	0.52	0.94
4	14.39	1.99	68.91	3.97	0.93	1.53
5	18.48	2.54	60.83	4.30	1.32	2.87
6	19.97	4.25	51.00	8.97	5.97	4.21

[a]The finest grade of grinding is marked OO, the coarsest with the number 6.

TABLE 11

Rate of Maltose Generation (at 72°C)

Wheat flour, normal	1-2%
Wheat flour, slightly germinated	2.1-2.5%
Wheat flour, germinated	2.6%

in composition as milling proceeds. Table 11 gives the diastatic power (rate of maltose generation) of fresh and aged flour.

The grist is slightly acidic, as are all cereal products. The acidity increases with age, and stored flours can be identified by their pH value. Wheat products should show a pH value between 5.3 and 5.8.

The amount of gas that can be generated from flour is proportional to the amount of fermentable sugars (such as maltose) contained in the product. This quantity may be determined by chemical analysis.

A series of tests have been described in cereal-product literature to

determine whether the flour has been chemically treated to change its
nature. Quantitative determination of the starch content is difficult, but
useful comparative figures are obtainable from viscosity measurements.
On the other hand, the origin of the starch (that is, potato, wheat, corn,
and so on) can be readily determined by microscopic examination in polarized light.

All the characteristics of cereals, flours, and starches described
above are important for the battery technologist, since they influence the
behavior of the electrolyte paste in the finished cell. Experience has
shown that even the weather conditions during the ripening and harvesting
of a cereal with well-known properties can cause abnormal behavior due
to changes in the enzyme activity and formation of the proteins.

2.5.2. Modified Starches

These substances may be prepared by treating natural ungelatinized
starch granules with an ether-forming reagent, such as an aliphatic
dihalide, preferably in alkaline solution. Since oxidation and/or hydrolysis
of the starch proceeds by reaction at the hydroxyl groups, the blocking
of these groups by etherification (or esterification) increases the chemical
stability. The effectiveness of the treatment is dependent on the number
of hydroxyl groups blocked and the nature of the aliphatic radical used
for the reagent.

2.5.3. Cellulose Derivatives

The sodium salts of carboxymethylcelluloses (CMC) are prepared by
the action of sodium monochloracetate on alkali cellulose. The structure
is shown in Fig. 15a. The CMC solutions deposit thin, mechanically
stable, but water-soluble films on drying. Flexibility of the films is
improved by additives such as glycerine, glycol, and polyglycol; insolu-
bility is increased by treatment with aluminum salts or organic acids (for
example, citric acid).

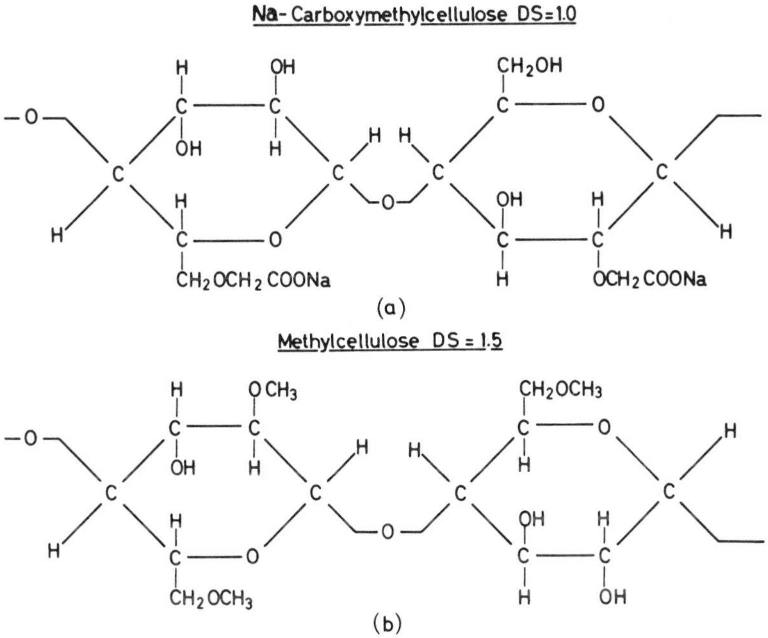

Na-Carboxymethylcellulose DS=1.0

(a)

Methylcellulose DS = 1.5

(b)

FIG. 15. Structural formulas of (a) sodium carboxymethylcellulose;
(b) methylcellulose.

Methylcelluloses may be obtained by the action of methyl chloride on
alkali cellulose. Special properties can be imparted to these products by
subsequent treatment with small quantities of etherifying agents such as
ethyl oxide. These products would be more properly named methylhydrox-
yethylcelluloses. (Fig. 15b)

2.6. ZINC

The anodic material used in the dry cell is an alloy based on zinc.
Fine zinc (zinc content 99.95-99.995%) is prepared industrially either by
a distillation process (for example, New Jersey Zinc Corporation) or by
electrolytic deposition. Only traces of heavy-metal impurities can be

tolerated. The alloy used for the dry-cell anode usually contains some lead and cadmium, both contributing to an improvement in the mechanical and chemical characteristics. Apart from the chemical composition, the physical characteristics of the alloy, such as grain size and surface condition, are of importance in battery technology.

In Table 12 the chemical analyses of the more common alloys used in the dry-cell industry are given.

The resistance of the alloy to corrosion is increased by amalgamation or by a chromate treatment of the surface. In addition, organic inhibitors find some application.

2.6.1. Inhibitors

Morehouse, et al. [35] investigated the inhibiting properties of a whole series of organic and inorganic compounds. Mercury and chromium salts are examples of inorganic inhibitors; the organic inhibitors include compounds with carbonyl groups, heterocyclic nitrogen compounds with a carbon ring attached to the heterocyclic groups, a few commercial products of unknown composition, and some colloidal substances. The protection given by zinc amalgamation at elevated temperature is doubtful. The use of mercury is also limited by the fact that the alloy becomes brittle when a certain level of mercury is exceeded. An undischarged cell with a chromate-protected zinc anode stores well, even at high temperature, but as soon as the cell is partly discharged the protective action of the chromate layer is destroyed. Of the organic compounds examined, the heterocyclic substances gave the best results. Although corrosion was markedly reduced, the authors consider that only colloidal materials would be of practical value, since the other materials either react with the electrolyte or deposit an insoluble film on the zinc surface. In both cases, however, the internal resistance of the cell was increased to such an extent that the electrical performance was also reduced.

The study of the colloids showed that all types of flour delayed the anodic corrosion because of their gluten content. Potato flour was least

TABLE 12

Composition of Zinc Alloys

Type of zinc can	Pb (%)	Cd (%)	Maximum permissible impurities (%)		
			Fe	Cu	Ni and Co etc.
Soldered	0.1–0.6	0–0.3	0.01	0.005	0.001
Drawn	0.06–1.0	0.005–0.007			taken together
Impact extruded	0.1–1.0	0–0.006			0.0005
					individually

effective, since its protein content is very low. The inhibiting action of
gluten is attributed to gliadin and mesonin, both essential constituents of
the protein. However, hydrolysis of flour proteins produces amino acids
that have been found to accelerate rather than inhibit corrosion.

Colloids, such as gelatin, casein, and hemoglobin, and gums, such
as agar-agar, acacia, and catechu, are also claimed to be effective as
inhibiting agents. According to other authors, carboxymethylcellulose,
hydroxymethylcellulose, alginates, and various types of starch have no
protective properties at all. However, very small additions of flour to the
electrolyte paste usually have been found helpful in reducing corrosion.

Aufenast and Muller [36] have confirmed, in principle, the results
quoted above.

2.7. MISCELLANEOUS BATTERY MATERIALS

Of the remaining raw materials for battery manufacture, the sealing
compound and the paper or cardboard containers are of importance. The
sealing compounds must be capable of withstanding temperatures of 60° -
70°C without deformation and must afford satisfactory airtight seal. The
composition varies according to the method of use, but the majority are
based on bitumen, rarely on resin or wax mixtures. Occasionally, mineral
additives are employed to harden and/or color the compound.

For cardboard containers and jackets, water-resistant types are preferred. In recent years plastic containers and jackets have become more
popular.

2.7.1. Bitumen

Only oxidized bitumens are suitable for use in sealing compounds. The
process is usually carried out by blowing air or oxygen through the molten
bitumen (hence the name blown bitumen) whereby a partial dehydrogenation
of the hydrocarbons is effected. The resin content decreases and the

asphaltic components increase. The product, compared with the starting material, has a higher softening point, feels appreciably harder, and the standard penetration depth is reduced (See Table 13.) The amount of change in properties is related to the duration of the blowing operation.

2.7.2. Waxes

The principal components of a macrocrystalline paraffin wax are normal (straight-chain) paraffins, although isoparaffins may be present in amounts up to 30%. Paraffin wax should comply with the general specification found in Table 14.

Further, the wax must be free from chloride and sulfate ions.

For the wax impregnation of flat cell stacks and for the external coating of military batteries, additional special requirements are necessary.

TABLE 13

Physical Requirements for Oxidized Bitumen

Softening point (ring and ball)	$80°$-$90°$C
Dropping point (DIN 51801) Ubbelohde	$92°$-$102°$C
Breaking point according to Fraass	max. $10°$C
Penetration (DIN 51579) at $25°$C	20-30 mm/10
Ductility at $25°$C	3-8 cm

TABLE 14

Specifications for Paraffin Wax

Setting point (DIN 51556)	$52°$-$56°$C
Penetration (DIN 51579)	$\sim$ 20 mm/10
Oil content (DIN 51571)	max. 2%

The wax must adhere tenaciously to the base material. It must be flexible and not prone to cracking. The surface of the wax coating must not be tacky, and the formation of drops (tears) after the dipping process must not be too pronounced. These requirements can be met by microcrystalline waxes.

The microcrystalline paraffin waxes (dropping point $62°$-$92°$C) are obtained from the highly viscous, high-boiling fractions of mineral oils. They consist of branched chain saturated paraffins ($< 30\%$) and branched cyclic paraffins.

For combination with bitumen in sealing compounds, the characteristics shown in Table 15 are generally found suitable. This type of wax is a fossil ester wax found in brown coal.

2.7.3. Resin

Added to a bitumen-based sealing compound, resin imparts brightness to the finish and improves the hardness. The physical properties listed in Table 16 are typical of resins that have been used successfully for many years.

TABLE 15

Characteristics of Montana Wax Type LW-1

Setting point (DIN 51556)	$70°$-$80°$C
Dropping point (DIN 51801) Ubbelohde	$82°$-$86°$C
Penetration (DIN 51579)	max. 3 mm/10

TABLE 16

Characteristics of Resin Additive

Dropping point (Ubbelohde)	$81°$-$88°$C
Softening point (ring and ball)	$71°$-$77°$C
Acid value	165-171
Saponification value	171-177

Sealing compounds based on bitumen, wax, and resin are used when the appearance of the finished seal is important, that is, when the seal is visible from the outside and surface blemishes cannot be tolerated. Such sealing compounds melt sharply and have relatively low viscosities in the molten state, so that they are suitable for machine dispensing without the formation of "strings" and "tears." However, because of the increasing practice of completely enclosing the cell in a steel or plastic jacket, visually perfect sealing is now less important. Since the resin and wax compounds are quite expensive, such compositions are now seldom used.

2.8. SEPARATOR AND INSULATION MATERIALS

The separator for flat or paper-lined cell manufacture is usually a sodium sulfate Kraft paper about 50-100 μm thick and free from metallic inclusions. It is extraordinarily difficult to obtain suitable paper for dry-cell and electrolytic-capacitor manufacture because of this latter requirement. Important here is the number of metallic particles (which originate from the raw pulp or from abrasion of the paper-making equipment) per unit area. The weight percentage may be extremely small, yet the paper may not be usable.

For sealing discs, battery cartons, and other parts, the requirements are not so stringent. It is customary to impregnate with wax any cardboard that may come in contact with the current-carrying elements of the battery. This treatment is not as foolproof as it appears, and consequently it is good practice to avoid contact between the cardboard and any metal part at a higher potential. Failure to observe this simple precaution may result in rapid corrosion of the metal parts involved.

For special applications, the cardboard may be built up into a laminate in which a layer of bitumen, polyethylene, or similar material is sandwiched between the cardboard layers. Such layerboard is practically impervious to electrolyte or moisture and therefore finds widespread application for the manufacture of external and internal tubes for leakproof cells.

Plastic insulation elements, sealing grommets, contact plates, and other parts are now being widely used and are usually made by injection-molding techniques from thermoplastic materials such as polyethylene, polypropylene, and polyamides, or alternatively from thermoplastic elastomers.

Synthetic rubber (urethane caoutchuc) is scarcely economic for small parts because processing costs are high in comparison with the injection-molding method.

Polyvinylchloride, polyethylene (both cheap, elastic materials), and high-impact polystyrene are commonly used for the manufacture of cell jackets. Polystyrene is, however, susceptible to splitting due to internal strain and is not very stable when exposed to ultraviolet light.

Plastic jackets including the top disc may be injection molded in one piece, but they tend to be somewhat conical in form. Alternatively, the cylindrical tube may be extruded and the top disc subsequently welded in. This process is slower but yields precision parts.

Offset printing techniques are used for applying a finish to plastic jackets, wherein polyethylene and polypropylene must be previously flamed. Labeling is also possible, provided that an adhesive corresponding to the basic plastic material is selected. Heat-seal labels are also in use and are warmed to the required temperature during the labeling operation.

Shrink-on tubes or adhesive shrink-films are frequently used for intercell insulation and in some designs of leakproof cells.

3. THEORETICAL CONSIDERATIONS

3.1. GENERAL CELL CHARACTERISTICS AND DEFINITIONS

During the discharge process, reactions in a galvanic cell transform chemical energy (stored in the active electrode material or continuously supplied to the electrode from an external source) into electrical energy.

During the reverse process, that is, the charging of the cell, electrical energy (in the form of the charging current at a given working voltage) is converted into stored chemical energy. Although some heat effects are unavoidably associated with the energy transformations, the electrochem- ical processes proceed directly and not by means of a heat engine.

The cell reactions may be irreversible (as in primary cells) or revers- ible (as in secondary cells or storage batteries). The electrical work performed outside the cell is done at the expense of the energy content of the system (and sometimes with a contribution from the environment). Electrochemical processes are said to occur when (a) the reactants undergo a change in electrical charge, (b) there is a change in concentration at the electrodes of one or more substances without any change in the total amount present, or (c) electrokinetic phenomena are observed. Any of these types of processes may occur in a bulk reaction, the difference being that in the chemical experiment made in a test-tube, the electron transfer proceeds at random wherever the reactants happen to meet each other, whereas in the galvanic cell, the electrons liberated by one reactant proceed along a predetermined path (the external circuit) to the other reactant. If now some electrical apparatus is part of the external circuit, the electron stream (electric current) will perform work on the way from one electrode to the other. In a galvanic cell, the reactants must be arranged as discrete phases, separated physically but able to transfer charges by an ionic conductor. Electronic contact within the cell must be rigorously avoided. The amount of energy that can be obtained from a chemical reaction is dependent only on the initial and final states and can often be calculated from the laws of thermodynamics. However, thermodynamics can yield no information regarding the reaction mechanism or velocity. For these, an investigation of the electrode kinetics is essential.

The electrochemical reactions occurring in most technical primary cells are redox processes. Among these, the combination of oxidation of zinc and reduction of manganese dioxide is the most important. In bio- logical systems, concentration cells and electrokinetic phenomena are the main sources of electric current.

3.1.1. Cell Arrangement

The schematic diagram of a galvanic cell is shown in Fig. 16 for the special case of the Leclanché system.

The active agents, manganese dioxide and zinc (also called electrodes or poles), are spatially separated, but both are in contact with a common aqueous solution of ammonium chloride and zinc chloride. This aqueous solution (called the electrolyte in the following discussion) contains positively charged particles (cations) and negatively charged particles (anions), which are ionic conductors of electricity but possess no electronic conductivity.

3.1.2. Electrochemical Processes

According to Nernst, every substance has a more or less powerful tendency to go into solution in the ionic form. This tendency is called the solution pressure P and is typical of each substance. The ions in solution have a tendency to return to the solid state that is proportional

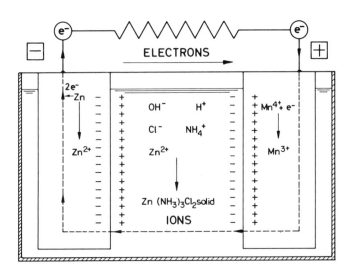

FIG. 16. Schematic layout of a Leclanché cell.

to the osmotic pressure p, which in turn is proportional to the concentration. Thus at the zinc electrode, zinc ions (Zn^{2+}) go into solution as a result of the solution pressure, thereby leaving excess electrons ($2e^-$) in the metal, causing it to become negatively charged. The increasing concentration of Zn^{2+} ions results in an increasing osmotic pressure, which drives zinc ions out of the solution back onto the zinc electrode. Since both processes occur simultaneously, it is readily seen that eventually a dynamic equilibrium will be attained when p has increased to such a value that the rate at which Zn^{2+} ions are leaving the solution is just equal to the rate at which P is driving them into the solution.

Since the zinc electrode is negatively charged (by virtue of the excess electrons), there will be an electrostatic field in close proximity to the electrode surface, causing positive ions to be attracted toward the zinc and negative ions to be repelled. In equilibrium, a condition is reached similar to that of a charged capacitor in which the negative charges on the electrode are balanced by the orientation of positive charges in the electrolyte. The arrangement of the positive and negative charges in this way is called the electric (or Helmholtz) double layer.

A similar situation exists at the manganese dioxide electrode, with the exception that the electrode is positively charged and the ions forming the solution side of the electric double layer are negative.

These dynamic equilibria at the electrodes remain effective until electronic connection is made in the external circuit, thus enabling the excess of electrons on the zinc electrode to flow to the manganese dioxide electrode, where they reduce Mn^{4+} to Mn^{3+} (or MnO_2 to Mn_2O_3). The electrochemical reactions at the electrodes continue as long as the external circuit is closed or until the active agents have been consumed.

3.1.3. Battery Capacity

Faraday's law of electrolysis states that a given amount of electricity deposits equivalent amounts of different substances from their solutions and that the amount of electricity associated with one gram equivalent (or val) of any substance is constant and is 96,494 coulombs or, conventionally, one faraday F. This amount of electricity is the same as that associated with 26.8 ampere-hours (Ah). The weight of active electrode material necessary for the production of one ampere-hour can therefore be easily obtained by dividing the equivalent weight (molecular weight/n) by 26.8 (see Table 17). Here, n is the number of faradays of electrons transferred in the reaction.

3.1.4. Potential, Current Strength, and Internal Resistance

In order to derive the electrode potential theoretically, Nernst considered the problem as follows:

a. The work done by n faradays of electricity passing through a potential change U is given by nFU.

b. The work done by one gram molecule of an ideal gas in changing volume from V_A to V_B against a varying pressure P_G is given by

$$\int_{V_A}^{V_B} P_G \, dV = \int_{V_A}^{V_B} (RT/V) \, dV = RT \ln(V_B/V_A)$$

and changing from volume to pressure, this becomes

$$RT \ln(P_A/P_B)$$

c. Nernst now sets the solution pressure P for P_A and the osmotic pressure p for P_B, and the expression becomes

$$nFU = RT \ln(P/p) \quad \text{or} \quad U = (RT/nF) \ln(P/p)$$

The sign assigned to U is purely a matter of convention, and in cases where $P > p$, it is customary to make U negative. The expression then becomes

$$U = -(RT/nF) \ln(P/p)$$

Since the osmotic pressure is directly proportional to the concentration, or more accurately, to the activity of the ions in dilute solutions ($p = Ka$), the equation can be written in the form

$$U = -(RT/nF) \ln(P/K) + (RT/nF) \ln a$$

TABLE 17

Equivalent Weights of Active Cell Components

Substance	n	Molecular weight (gm)	Equivalent weight (gm)	Weight per ampere-hour (gm/Ah)
MnO_2	1	86.93	86.93	3.24
Zn	2	65.38	32.69	1.22
NH_4Cl	1	53.30	53.50	2.00
$ZnCl_2$	2	136.30	68.15	2.54
$MgCl_2$	2	95.23	47.61	1.78

At constant temperature, the first term is constant. If this is represented by U^0, we obtain

$$U = U^0 + (RT/nF) \ln a$$

When the activity of the ions is unity ($\ln 1 = 0$), then $U = U^0$, the standard electrode potential.

In practice, electrode potentials are determined by measuring the voltage of a cell consisting of the electrode being investigated and some other reference electrode (for example, the calomel electrode, the silver/silver chloride electrode, or the standard hydrogen electrode).

The electrodes of a galvanic cell are often termed the cathode and the anode. These names are derived from the Greek, meaning the way upward and the way downward, respectively. In electrochemistry, the anode is defined as that electrode at which the conventional (positive) current enters the cell and passes from the electrode into the electrolyte. Conversely, positive current passes from the electrolyte into the cathode and then leaves the cell. From the above definition, it will be seen that in an electrolytic cell, to which current is being supplied from an external source, the anode is the positive pole and the cathode the negative, while in the case of a galvanic cell generating current, the anode is the negative and the cathode the positive pole (Fig. 17).

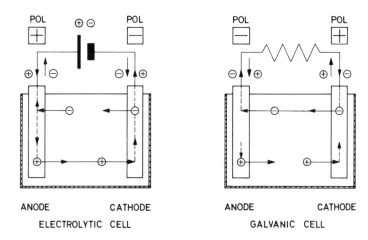

FIG. 17. Schematic diagrams of an electrolytic and a galvanic cell
and the direction of current in them.

The potential of a manganese dioxide electrode is dependent on the
type of dioxide used. Synthetic hydrates and electrolytic and chemically
activated dioxides have a higher potential than naturally occuring materials.
For convenience, a nominal voltage of 1.5 V is assigned to the Leclanché
cell. This value is used in all approximate calculations.

The voltage of a cell that is not delivering any current is called the
open-circuit voltage (o.c.v.) or the off-load voltage U_0. This may be
measured by the Poggendorf compensation method, but for all practical
purposes, the result is sufficiently accurate when a digital voltmeter or
other high-resistance voltmeter is used. When the cell is delivering
current, the voltage is always less than the open-circuit voltage. This
voltage is termed the closed-circuit voltage (c.c.v.), the working voltage,
or the on-load voltage U_ℓ. The greater the current being delivered, the
lower the value of U_ℓ.

Figure 18 shows the meaning of these symbols graphically. The
resistance of the external circuit (for example, connecting wires, flashlight lamp, and switch) is called the external or load resistance R_a. The
resistance of the carbon rod, cathode mix, and electrolyte is called the
internal resistance R_i. The value of U_ℓ and the sum of these two resist-

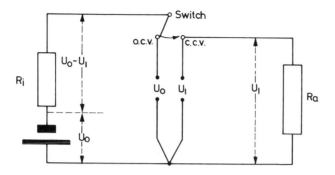

FIG. 18. Circuit diagram of a cell on load.

ances determine the current flowing in the circuit at any moment. If the
cell is shorted, that is, R_a is made virtually zero, then the maximum
current flow that can be generated by the cell, known as the flash current
I_F is produced.

Faraday's law defines the relationship between the quantity of electricity
and the amount of the active materials taking part in the electrochemical
reaction. Another relationship, known as Ohm's law, defines the interdependence of voltage, current strength, and resistance. According to
this law,

$$I \text{ (amperes)} = U \text{ (volts)}/R \text{ (ohms)}$$

and applied to a cell,

$$I = U_0/(R_i+R_a) = U_\ell/R_a$$

A large internal resistance of a cell not only lowers the useful voltage, it
can have important consequences for the performance of electronic equipment. Fortunately, in well-designed equipment, these disadvantages are
usually eliminated by the connection of a capacitor across the power input
terminals.

The electrical output of a dry cell is complex, involving more than the
parameters U_0, U_ℓ, and I_F previously defined. The current/voltage relationship cannot be completely described in terms of the impedance of
the cell, because impedance is not a constant quantity. It is dependent on

the current taken from the battery, the degree of discharge, the ambient temperature, and the age and previous history of the battery. The current flowing in the battery during the experimental measurement of the impedance (during a flash-current test) also affects the values obtained, so that it is not surprising to find that the methods employed, particularly when one measures a varying current, are of extreme significance. The impedance of a cell as determined by ac bridge measurements is not always the same as measured by dc methods.

If a battery with an open-circuit voltage U_0 is loaded with an external resistance R_a, the voltage drop U_ℓ across R_a is always less than U_0 (Fig. 18). This difference $(U_0 - U_\ell)$ is therefore equal to the voltage drop across R_i so that

$$U_\ell /R_a = (U_0 - U_\ell)/R_i \quad \text{and} \quad R_i = R_a [(U_0/U_\ell) - 1]$$

In practical applications of this method, it should be noted that U_0 and U_ℓ must be measured with a high degree of accuracy and also that U_ℓ must be determined within 1 msec in order to avoid undesirable time effects (voltage drop or recovery). For some interrupter methods, such considerations are basic.

The dc resistance R_i is the sum of the (ionic) conductivity of the salts in the electrolyte and in the cathode mix plus the (electronic) conductivity of the mix components and the current collectors. Its value is not only a function of the cell design, formulations, and the choice of raw materials, but is also dependent on the discharge current. R_i increases with the degree of discharge, since the electrochemical processes in the cell change the conductivity of the cathode mix and the electrolyte. A reduction in the ambient temperature also results in an increased value of R_i.

The relationship between the internal resistance and the discharge current is such that the lowest dc resistance values are obtained at the highest drains (Fig. 19, upper curve). The minimum value of the resistance is therefore associated with the highest attainable current drain, that is, the flash current and approximately one finds

$$R_i \approx U_0/I_F$$

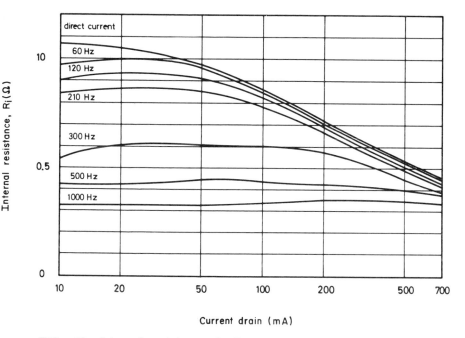

FIG. 19. Internal resistance of a D-size cell at various current drains and frequencies.

The determination of R_i by dc methods has been greatly augmented by the measurement of the impedance at various ac frequencies. In addition to R_i, impedance measurements yield information relating to double-layer capacitance and passivation of the electrodes. Although the data obtained from actual dry cells can scarcely be interpreted quantitatively (since the theoretical treatment assumes plane electrodes) important qualitative deductions have been obtained.

The equivalent circuit shown in Fig. 20 describes the impedance properties of Leclanché cells with sufficient accuracy for most purposes. The dc resistances of the electrolyte, cathode mix, and current collectors are accumulated in R_1. The resistance of the current collectors is very small and remains constant. On the other hand, as has already been explained, the resistance of the electrolyte and mix increases during discharge and to some extent during storage. At the phase boundaries between the electro-

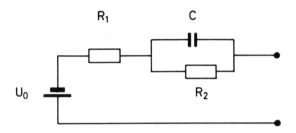

FIG. 20. Circuit diagram of a Leclanché cell.

lyte and the electrodes, electric (or Helmholtz) double layers are present, and passivating films of reduced conductivity also may form with time.

The impedance properties of the whole phase-boundary region are represented in the diagram by a resistance R_2 in parallel with a capacitance C. The passivating films that form on the zinc anode are important contributors to R_2. The formation of such films can be accelerated by means of suitably formulated electrolytes, so that after some storage R_2 can assume very high values. However, such passivating layers are disrupted and degenerate quite rapidly when the cell is placed on discharge. The capacitance C represents the capacitance of the Helmholtz double layers at the electrodes. At high frequencies ($>$20 kHz), the impedance of C is so small that R_2 is virtually short-circuited and the measured impedance is a good estimate of R_1. At frequencies less than 10 Hz, the impedance of C is so high, relative to R_2, that the total impedance approaches the value of R_1 + R_2. At very low frequencies ($<$0.05 Hz), irregularities appear that cannot be explained from Fig. 20, due to diffusion-controlled electrochemical reactions taking place in the cell. By the addition of other electric current elements, or a Warburg resistance to the equivalent circuit, these effects can be better described, but the treatment is too abstract to enter into here.

3.1.5. Conventional Definitions

Cell tension, electromotive force, and electrode tensions have been defined by the Comité International de Thermodynamique et Cinétique

Electrique (CITCE) in association with the International Union of Pure and
Applied Chemistry (IUPAC). *

a. A galvanic cell is defined as a series of electrically conductive
phases, each in contact with the following one. At least one of these
phases is an electrolyte. The two end phases (also called poles or electrodes) are chemically and physically identical but have not necessarily
the same inner potential φ.

The cell is represented according to the convention by listing and
numbering the phases from left to right

$$Zn \ / \ Zn^{2+} \ aq. \ // \ Cu^{2+} \ aq. \ / \ Cu \ / \ Zn$$
$$1 \qquad 2 \qquad\quad 3 \qquad\quad 4 \quad 1'$$

The electric tension of the cell U is equal to the inner potential of the
first electrode φ_1 minus the inner potential of the second electrode $\varphi_{1'}$.

$$U = \varphi_1 - \varphi_{1'}$$

The inner or Galvani potential φ of a phase is related to the outer or
Volta potential ψ and the surface potential χ as follows.

$$\varphi = \psi + \chi$$

Between any two phases, only the Volta potential difference can be
directly measured. The surface and Galvani potential differences are
inaccessible to experimental determination. However, in the case of two
identical phases such as 1 and 1' above, since the surface potentials χ_1
and $\chi_{1'}$ are the same, the difference between the Volta potentials ψ_1 and
$\psi_{1'}$, is equal to the difference between the Galvani potentials φ_1 and $\varphi_{1'}$.

$$U = \varphi_1 - \varphi_{1'} = \psi_1 + \chi_1 - (\psi_{1'} + \chi_{1'}) = \psi_1 - \psi_{1'}$$
$$(\chi_1 = \chi_{1'})$$

*CITCE was recently renamed International Society of Electrochemistry
(ISE).

In a more restricted sense, the term galvanic cell is used to describe a system which, by virtue of spontaneous electrochemical reactions, is able to perform electrical work.

When electrochemical changes are induced in a system by means of electrical energy supplied from an outside source, the system is described as an electrolytic cell and the electrochemical changes as electrolysis.

b. The cell reaction is a heterogeneous chemical reaction between potential-determining particles that takes place if an electric charge is transported through the cell. For example,

$$Zn + Cu^{2+} \rightarrow Zn^{2+} + Cu$$

The direction in which this cell reaction is written determines the sign of its free enthalpy ΔG and of its chemical affinity $A \equiv -\Delta G$.

c. The charge number n of the cell reaction for a given galvanic cell determines the number of positive faradays (F) that are transported within the cell from left to right, if the cell reaction takes place once in the written direction. If a positive charge is transported within the cell from right to left, n is negative.

d. The chemical tension E, the electromotive force, of the galvanic cell is defined by the quotient of the chemical affinity A ($\equiv -\Delta G$) of the written cell reaction and the expression $n \cdot F$:

$$E \equiv A/(n \cdot F) \equiv -\Delta G/(n \cdot F)$$

e. At electrochemical equilibrium in the galvanic cell with no current flowing, the reversible electric cell tension U_{rev} is given by

$$U_{rev} = -A/(n \cdot F) = \Delta G/(n \cdot F) = -E$$

For the inverse cell reaction, the signs of n and of $A \equiv -\Delta G$ are also inverse, but the signs of U_{rev} and of E of the same galvanic cell remain unchanged.

f. The relative electrode tension at zero current or, in brief, the electrode tension of a half-cell such as Zn/Zn^{2+} is actually the electric-cell tension $U \equiv \varphi_A - \varphi_B$ of a reference cell,

$$\text{lead A / Zn / Zn}^{2+} \text{ / H}^+ \text{ / Pt, H}_2 \text{ / lead B}$$

The reference electrode on the right is a standard hydrogen electrode.

g. These electrode tensions at standard conditions represent the "series of tensions".

h. When a galvanic cell is traversed by a current, the electric tension U_I is a function of the direction of the current and of its density:

$$U_I = f(I)$$

The expression

$$\Delta U = U_I - U_{rev} \equiv \eta$$

may be called cell polarization. ΔU is positive if the positive current within the cell goes from left to right, and negative when the current flow is reversed.

The standard electric tension of an electrode, also called the standard electrode potential, is the electric tension of a cell comprising the electrode under study and a standard hydrogen electrode (SHE), the activities of all the substances taking part in the cell reaction being unity. The standard electrode potential of the Zn/Zn^{2+} electrode is, for example, calculated as follows.

Cell scheme:

$$Zn \ / \ Zn^{2+} \ aq. \ // \ H^+ \ aq. \ / \ H_2, \ Pt \ / \ Zn$$
$$a_{(Zn^{2+})} = 1 \quad a_{(H^+)} = 1 \quad p = 1$$
$$1 \qquad 2 \qquad\qquad 3 \qquad\quad 4 \qquad 1'$$

Cell reaction:

$$Zn + 2H^+ \rightarrow Zn^{2+} + H_2$$

$$n = +2 \quad \Delta G = -35.184 \ kcal$$

Standard electrode potential:

$$U^0 = \varphi_1 - \varphi_{1'} = -A/nF = \Delta G/nF = -35.184/2F = -0.763$$

Nonstandard potentials are related to U^0 as shown by the Nernst equation. In general,

$$U = U^0 + (RT/nF) \ln a_{(Zn^{2+})}$$

or, after converting to common logarithms and inserting the values of R, T, n, and F,

$$U = 0.763 + 0.0295 \log a_{(Zn^{2+})} \quad \text{(room temperature)}$$

The equilibrium constant K for the electrode reaction R can be calculated from the free-enthalpy change $\Delta G(R)$ as follows.

$$\Delta G(R) = \Delta G_0(Zn^{2+}) + RT \ln a_{(Zn^{2+})} + RT \ln a_{(H_2)} - 2RT \ln a_{(H^+)}$$

and

$$K = \frac{a_{(Zn^{2+})} a_{(H_2)}}{\left[a_{H^+} \right]^2}$$

and therefore

$$\Delta G(R) = \Delta G_0(Zn^{2+}) + RT \ln K$$

But at equilibrium, $\Delta G(R) = 0$ and

$$\ln K = -\Delta G_0(Zn^{2+})/RT$$

and log K = 25.8 at 25°C.

Inserting this value for K in the above expression and noting that $a_{(H_2)} = 1$ when the hydrogen is developed at atmospheric pressure, equilibrium is attained when

$$\log a_{(Zn^{2+})} = 25.8 - 2 \text{ pH}$$

From the general expression for the zinc electrode potential given above, it may be concluded that this quantity is independent of pH. However, zinc forms complex ions such as $ZnOH^+$, $HZnO_2^-$, ZnO_2^{2-}, and so on in alkaline solution. (Fig. 1 of Chapter 2 of this book.) The actual electrode reaction which proceeds under any particular set of experimental conditions is determined by the relative activities of the various types of zinc containing ions, which in turn is a function of the pH of the electrolyte. In this sense, the zinc electrode potential is not independent of pH although it is constant over a wide pH range.

For the manganese dioxide electrode in slightly acid electrolyte, a similar calculation can be made.

Cell scheme:

$$\text{Pt } MnO_2 \text{ / } Mn_2O_3 \text{ aq. // } H^+ \text{ aq. / } H_2 \text{ Pt / Pt } MnO_2$$
$$\quad\quad 1 \quad\quad\quad\quad 2 \quad\quad\quad\quad 3 \quad\quad 4 \quad\quad 1'$$

Cell reaction:

$$2 \, MnO_2 + 2 \, H^+ + 2 \, e^- \rightarrow Mn_2O_3 + H_2O$$
$$n = -2 \quad \Delta G = -46.78 \text{ kcal}$$

Standard potential:

$$U^0 = \varphi_1 - \varphi_{1'} = \Delta G/nF = -46.78/(-2 \cdot 23.064) = 1.014 \text{ V}$$

Nonstandard potential:

$$U = U^0 + \frac{RT}{nF} \ln \frac{a_{(MnO_2)} \, a^2_{(H^+)}}{a_{(Mn_2O_3)} \, a_{(H_2O)}}$$

Putting the above value of U^0 in this expression, converting to common logarithms, and setting the activities of the solid oxides and water equal to 1, the equation becomes

$$U = 1.014 - 0.0591 \text{ pH}$$

The potential of the manganese dioxide is accordingly a linear function of the pH of the electrolyte, changing by about 60 mV per pH unit.

Further information regarding the potential versus pH relation for the manganese dioxide electrode is to be found in the publications of the following investigators: Holler and Ritcher [37], Thompson [38], and more recently Drotschmann [39], Cahoon [40], McMurdie et al. [41], Sasaki [42], and Johnson and Vosburgh [43].

For the zinc electrode, the work of Brouillet and Jolas [44] is worthy of study.

The chemical reactions in the Leclanché cell may be formulated as shown in Table 18.

3.2. THE CATHODE

The many experimental investigations into the characteristics of the MnO_2 electrode that have been carried out in recent decades have revealed certain criteria that permit a more reliable assessment of the suitability of manganese dioxide ores for use as cathode materials in Leclanché cells. The results obtained from these experiments yield some insight into the

TABLE 18

Chemical Reactions Occurring in the Leclanché Cell

Leclanché Cell with NH_4Cl

Light discharge:

$$2\ MnO_2 + 2\ NH_4Cl + Zn \rightarrow 2\ MnOOH + Zn(NH_3)_2Cl_2$$

Heavy discharge:

$$2\ MnO_2 + NH_4Cl + H_2O + Zn \rightarrow 2\ MnOOH + NH_3 + Zn(OH)Cl$$

Prolonged discharge:

$$6\ MnOOH + Zn \rightarrow 2\ Mn_3O_4 + ZnO + 3\ H_2O$$

Leclanché Cell with $ZnCl_2$

Standard discharge:

$$2\ MnO_2 + 2\ H_2O + ZnCl_2 + Zn \rightarrow 2\ MnOOH + Zn(OH)Cl$$

Prolonged discharge:

$$6\ MnOOH + 2\ Zn(OH)Cl + Zn \rightarrow 2\ Mn_3O_4 + ZnCl_2 \cdot 2\ ZnO \cdot 4\ H_2O$$

reaction mechanisms of the processes occuring on and within the cathode, and however incomplete this may still be, it enables a more direct optimization and adjustment of battery characteristics in the various fields of practical applications than was possible with purely empirical methods.

Feitknecht and his coworkers have, in a series of fundamental researches, significantly increased our knowledge of the structural characteristics of active, gamma type manganese dioxides. Both chemical [45] and electrochemical [46] reduction of the gamma modification proceed in homogeneous phase (that is, the original crystal lattice is converted progressively into another distinct lattice characteristic of the lower-oxide end product). These facts indicate that classical stoichiometric chemistry is too restricted for the interpretation and formulation of the electrochemical electrode reactions.

Potential measurements on beta and gamma manganese dioxide electrodes reported by Wadsley and Walkley [47] point toward a similar conclusion. They found that the electrode potentials decreased continuously as reduction proceeded: for beta, from MnO_2 to $MnO_{1.98}$, and for gamma, from MnO_2 to $MnO_{1.92}$ (Fig. 21). Vosburgh and Pao-Soong-Lou [48] have published similar results for alpha and gamma manganese dioxides (Fig. 22).

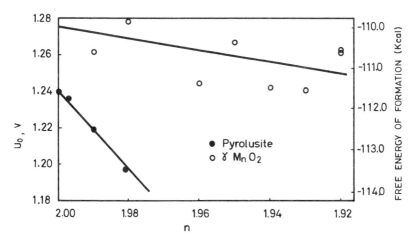

FIG. 21. Relation between the standard potential and the value of n in MnO_n for some manganese dioxides.

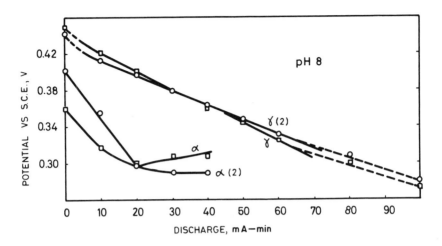

FIG. 22. Relation between equilibrium potential of α- and γ-manganese dioxides and depth of discharge. Electrolyte at pH 8. Circles are averages of two results.

The majority of natural and synthetic chemical compounds show deviations from the theoretical stoichiometric composition. Gattow [49] , for instance, defines manganese dioxide as $MnO_{1.7-2.0}$, disregarding the water or foreign metal content. This fact has led recently to the investigation of the nonstoichiometric aspects of the chemistry of the manganese dioxide minerals.

3.2.1. Phase Analysis of Manganese Dioxide (Bode)

Bode and his coworkers [50] were the first to study the chemical reduction of those varieties of manganese dioxide that are important in the primary cell manufacturing industry from the nonstoichiometric point of view. They found that reduction in homogeneous phase is typical of the alpha modification down to $MnO_{1.88}$, of the beta down to $MnO_{1.96}$, and of the gamma down to $MnO_{1.5}$ (a schematic representation of these results is given in Section 2.1.8, Crystal Structure).

Similar investigations of the electrochemical reduction process were reported by Bell and Huber [51]. The end product of the gamma reduction was found to be α-MnOOH (groutite) and of the beta reduction γ-MnOOH (manganite). Gabano et al. [52], Brouillet et al. [53], and Kozawa et al. [54] have all contributed to this particular field and have arrived at substantially the same conclusions.

3.2.2. The Thermodynamics of Oxide Electrodes (Vetter)

Before proceeding to a discussion of the electrode characteristics of the several types of manganese dioxide, it is helpful to consider Vetter's contribution to the thermodynamic theory of oxide electrodes in general, since here, for the first time, this subject is treated from a nonstoichiometric viewpoint. Further, consideration is given to cases (as in the Leclanché cell) where the oxide is not in contact with its parent metal and is in contact with a solution containing ions of the same metal but of a different valency from those in the solid phase. In order to arrive at an adequate method for the treatment of such systems, Vetter [55] studied the various possible reactions across each relevant phase boundary and derived the necessary equilibrium conditions. In the following discussion, only the case of a metal oxide on an inert electronic conductor is considered, since this is the condition prevailing in the Leclanché cell.

It is, however, worthwhile drawing attention to the fact that in a system consisting of a metal oxide in contact with its parent metal and with an electrolyte, thermodynamic equilibrium can only be attained when the reactions across the phase boundaries are not kinetically inhibited and the electrolyte is saturated with the oxide. This implies that such an oxide must have a definite composition. Vetter refers to such an oxide as the equilibrium oxide and represents its composition by MeO_{n_0} where n_0 is not necessarily a whole number.

3.2.2.1. Metal Oxide, Homogeneous Phase, Inert Conductor (Oxygen-Ion Transfer Between Metal Oxide and Electrolyte). This system may be represented by the phase diagram of Fig. 23.

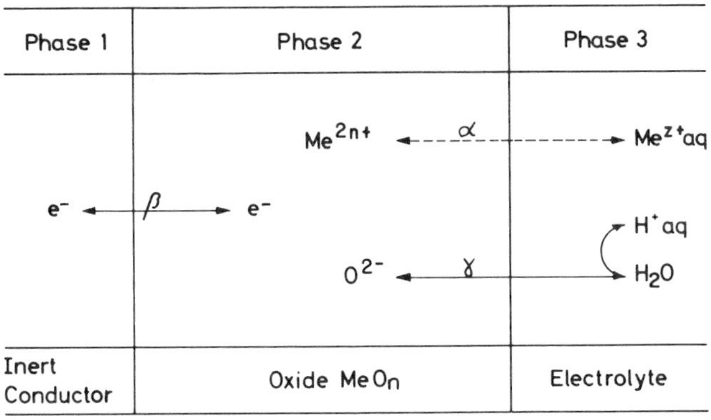

Phase 1	Phase 2	Phase 3
	$Me^{2n+} \xleftarrow{\quad \alpha \quad} Me^{z+}aq$	
$e^- \xleftarrow{\quad} \beta \xrightarrow{\quad} e^-$		H^+aq
	$O^{2-} \xleftarrow{\quad \gamma \quad}$	H_2O
Inert Conductor	Oxide MeO_n	Electrolyte

FIG. 23. Phase diagram. Metal oxide, homogeneous phase, in contact with inert conductor and electrolyte. Oxygen-ion transfer reaction active.

It is assumed that the electronic conductivity throughout phases 1 and 2 is adequate and that the width of the homogeneous phase of the oxide MeO_n is sufficient to allow oxidation or reduction to proceed to equilibrium without the appearance of a new phase. Furthermore, the transfer of oxygen ion across the oxide/electrolyte phase boundary must not be kinetically inhibited, thereby allowing a reversible thermodynamic equilibrium to be established. On the other hand, the passage of metallic ions across the phase boundary 2-3 (reaction α) is assumed to be kinetically retarded to such an extent that equilibrium cannot be reached. Of course there will be a unidirectional flow of metal ions across the phase boundary. If the equivalent current is small in comparison with the exchange current characteristic of reaction γ, this corrosion current will not disturb the oxygen-ion equilibrium significantly and a reversible thermodynamic potential will be set up across the oxide/electrolyte phase boundary.

Coupling the oxide electrode with a hydrogen electrode operating at 1 atm but at the same pH as the electrolyte, the cell

$$Pt \; / \; MeO_n \; / \; H^+ \; aq. \; / \; H_2, \; Pt$$

is obtained with the general electrode reactions

$$\nu \, MeO_n + O^{2-} \rightleftarrows \nu \, MeO_{n+(1/\nu)} + 2e^-$$

$$2 \, H^+ + 2e^- \rightleftarrows H_2$$

and the overall reaction

$$\nu \, MeO_n + H_2O \rightleftarrows \nu \, MeO_{n+(1/\nu)} + H_2$$

where ν denotes any number of moles.

The change in free enthalpy during this reaction $\Delta G(R)$ can be obtained from the difference in the free-enthalpy changes of the two reactions

$$Me + (n + 1/\nu) \, H_2O \rightarrow MeO_{n+(1/\nu)} + (n + 1/\nu) \, H_2, \; \Delta G(n + 1/\nu)$$
$$Me + n \, H_2O \rightarrow MeO_n + n \, H_2, \qquad \Delta G(n)$$

so that

$$\Delta G(R) = \nu \left[\Delta G(n + 1/\nu) - \Delta G(n) \right]$$

Since we are considering a nonstoichiometric oxide in homogeneous phase, $1/\nu$ can be written Δn and made as small as we choose. In the limit, as $\Delta n \rightarrow 0$ (that is, as $\nu \rightarrow \infty$),

$$\lim_{\substack{\Delta n \rightarrow 0 \\ \nu \rightarrow \infty}} \Delta G(R) = \frac{d\Delta G(n)}{dn}$$

Alternatively, $\Delta G(R)$ can be expressed as a function of the free enthalpies of formation from the elements of the oxides ($\Delta G_0(n)$, and so on). Following Bell and Huber [51] and using the same convention regarding Δn,

$$\lim_{\Delta n \rightarrow 0} \Delta G(R) = \frac{d\Delta G_0(n)}{dn} - \Delta G_0(H_2O)$$

The reversible voltage U_{rev} of a cell consisting of the oxide electrode and a hydrogen electrode at 1 atm and at the same pH as the cell electrolyte is then

$$U_{rev} = \frac{\Delta G(R)}{2F}$$

$$= \frac{1}{2F} \frac{d\Delta G_0(n)}{dn} - \frac{1}{2F} \Delta G_0(H_2O)$$

$$= 1.23 \, V + \frac{1}{2F} \frac{d\Delta G_0(n)}{dn}$$

Since the second term in this expression is usually negative, the electrode potential of the oxide electrode may be regarded as that of an oxygen electrode working at reduced partial pressure.

Using the expression for $\Delta G(R)$ derived according to Vetter,

$$U_{rev} = \frac{1}{2F} \frac{d\Delta G(n)}{dn}$$

3.2.2.2. Two Metal Oxides, Heterogeneous System (Two Phases), Inert Conductor.

This system is represented schematically in Fig. 24, graphically in Fig. 25. During the oxidation of an oxide with a limited homogeneous phase width, the upper phase limit n_1 will be reached (point D in Fig. 25). If the process is continued, a new phase with the lower phase limit n_2 will appear (point E in Fig. 25, $n_2 > n_1$), provided that the new phase can crystallize rapidly. Both phases are then in equilibrium. For both oxides, similar considerations apply as for the homogeneous system already described, but with an additional condition that at least one of the three possible reactions across the boundary of the two oxide phases must be active and equilibrium reached and maintained (see Fig. 24).

Actually, two three-sided systems are shown. If active reactions proceed on two sides and equilibrium is maintained, equilibrium is automatically attained on the third side, even though the reaction across the phase boundary may be severely inhibited. The assumptions regarding electronic

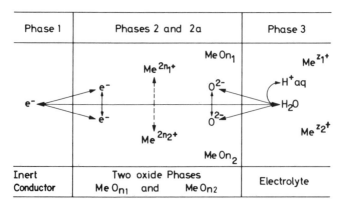

FIG. 24. Phase diagram. Two metal oxides, heterogeneous phases, in contact with inert conductor and electrolyte. Oxygen-ion transfer reactions active.

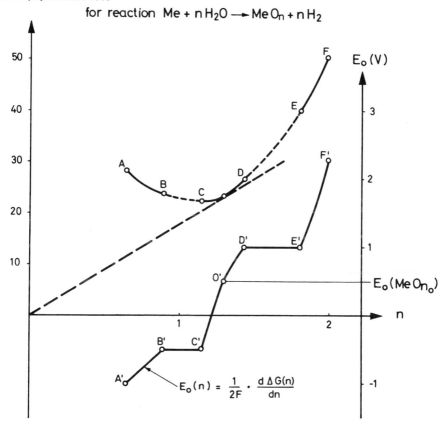

FIG. 25. Free enthalpy of reaction $\Delta G(n)$, according to Vetter, and the associated electrode potential E_0 as functions of n in MeO_n. (Arbitrary illustrative values.)

conductivity and corrosion currents are the same as described previously. Coupled with a hydrogen electrode at the same pH, the overall cell reaction is

$$\frac{1}{n_2 - n_1} \, MeO_{n_1} + H_2O \rightleftarrows \frac{1}{n_2 - n_1} \, MeO_{n_2} + H_2$$

and

$$U_{rev} = 1.23 \text{ V} + \frac{1}{2F} \frac{\Delta G_0(n_2) - \Delta G_0(n_1)}{n_2 - n_1}$$

or alternatively, after Vetter,

$$U_{rev} = \frac{1}{2F} \frac{\Delta G(n_2) - \Delta G(n_1)}{n_2 - n_1}$$

Note that in these expressions, $(n_2 - n_1)$ has a definite value (as distinct from Δn in the homogeneous system), and therefore the differential term is replaced by a difference quotient. Further, since the ΔG and ΔG_0 values are constants, it is implied that the electrode potential remains constant as long as both phases are present and is independent of the relative amounts of the two oxides.

3.2.2.3. <u>Graphical Representation.</u> In the diagram of Fig. 25, $\Delta G(n)$ values have been plotted against n in the range $0.6 < n < 2$. On the curve, the full line indicates homogeneous-phase ranges and the dotted portions are the regions in which heterogeneous systems occur. A companion curve shows the corresponding electrode potentials. It is seen that in the homogeneous phases, the electrode potential decreases continuously with n, whereas it remains constant in the heterogeneous systems.

The composition of the Vetter equilibrium oxide MeO_{n_0} previously referred to can be found from the curve in Fig. 25 by drawing a tangent to the $\Delta G(n)$ curve passing through the origin of the coordinates. The point on the curve touched by the tangent is the point at which $n = n_0$. To prove this, it must first be realized that the electrode potential of the equilibrium oxide is independent of the electronic conductor (parent metal, graphite, platinum, and so on). This potential may be calculated in two ways.

a. We may consider the system to be heterogeneous with the parent metal functioning as the second "oxide" phase. Then

$$U_{rev} = \frac{1}{2F} \frac{\Delta G(n_0)}{n_0}$$

b. We may attribute a small homogeneous phase to MeO_{n_0} and write

$$U_{rev} = \frac{1}{2F} \left[\frac{d\Delta G(n)}{dn} \right]_{at\ n=n_0}$$

It follows, therefore, that

$$\frac{\Delta G(n_0)}{n_0} = \left[\frac{d\Delta G(n)}{dn} \right]_{at\ n=n_0}$$

which is only true at the point on the $\Delta G(n)$ versus n curve where the tangent to the curve passes through the origin of the coordinates.

Further, since

$$\Delta G(n) = \Delta G_0(n) - n\ \Delta G_0(H_2O)$$

it follows that

$$\frac{\Delta G_0(n_0)}{n_0} = \left[\frac{d\Delta G_0(n)}{dn} \right]_{at\ n=n_0}$$

Many interesting conclusions can be drawn from the shape of the $\Delta G(n)$ curve, but for a detailed discussion, reference must be made to the original literature[55].

3.2.2.4. Metal Oxide, Homogeneous Phase, Inert Conductor (Metal-Ion Transfer Between Metal Oxide and Electrolyte). The phase diagram for this system is shown in Fig. 26. It should be noted that the valency of the metal ion is not necessarily the same in the oxide and electrolyte phases. Deviating from the previous discussions, the case is now considered where reactions β and α (metal-ion transfer) are active and are maintained in

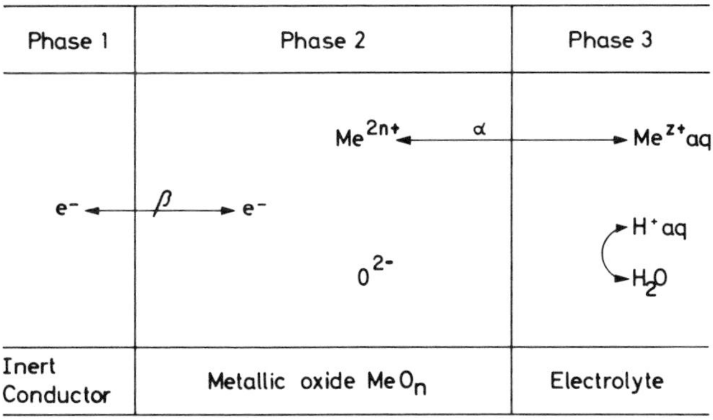

FIG. 26. Phase diagram. Metal oxide, homogeneous phase, in contact with inert conductor and electrolyte. Metal-ion transfer reaction active.

equilibrium. Coupled with a standard hydrogen electrode, a cell is obtained with the electrode reactions

$$\frac{2(n + \Delta n)}{z \, \Delta n} \, MeO_n \;\rightleftarrows\; \frac{2n}{z \, \Delta n} \; MeO_{n+\Delta n} + \frac{2}{z} \, Me^{z+} + 2 \, e^-$$

$$2 \, H^+ + 2 \, e^- \rightleftarrows H_2$$

where z denotes the valency
and the overall reaction

$$\frac{2(n + \Delta n)}{z \, \Delta n} \, MeO_n + 2 \, H^+ \;\rightleftarrows\; \frac{2n}{z \, \Delta n} \; MeO_{n+\Delta n} + \frac{2}{z} \, Me^{z+} + H_2$$

for which the free-enthalpy change is given by

$$\Delta G(R) = \frac{2}{z} \left[\frac{n \Delta G_0 \, (n + \Delta n) - n \Delta G_0(n)}{\Delta n} \right] - \frac{2}{z} \, \Delta G_0(n) + \frac{2}{z} \left[\Delta G_0(Me^{z+}) + \right.$$

$$\left. RT \ln a_{(Me^{z+})} \right]$$

and in the limit, as $\Delta n \to 0$,

$$\lim \Delta G(R) = \frac{2n}{z} \, \frac{d\Delta G_0(n)}{dn} - \frac{2}{z} \, \Delta G_0(n) + \frac{2}{z} \left[\Delta G_0(Me^{z+}) + RT \ln a_{(Me^{z+})} \right]$$

and

$$U_{0, \, H_2} = \frac{\Delta G(R)}{2F} = \frac{n}{zF} \, \frac{d\Delta G_0(n)}{dn} - \frac{\Delta G_0(n)}{zF} + U_{(Me/Me^{z+})}$$

where $U_{(Me/Me^{z+})}$ is the electrode potential of the parent metal in a solution containing Me^{z+} ions at the same activity as in the electrolyte of the cell under consideration.

For the Vetter equilibrium oxide MeO_{n_0} it has been shown in the previous section that

$$n_0 \left[\frac{d\Delta G_0(n)}{dn} \right]_{at \; n=n_0} - \Delta G_0(n_0) = 0$$

so that in this case, the above equation reduces to

$$U_{0, \, H_2} = U_{(Me/Me^{z+})}$$

3.2.2.5. Metal Oxide, Homogeneous Phase, Inert Conductor (Both Oxygen-Ion and Metal-Ion Transfer Reactions). Figure 27 shows the applicable phase diagram. In such a system, when all three reactions (α, β, and γ) are active and in equilibrium, there is complete equilibrium between

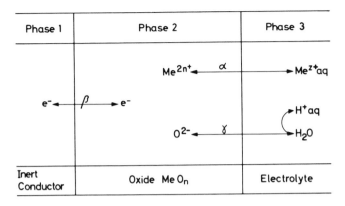

Phase 1	Phase 2	Phase 3

FIG. 27. Phase diagram. Metal oxide, homogeneous phase, in contact with inert conductor and electrolyte. Oxide- and metal-ion transfer reactions active.

the oxide and the electrolyte similar to an ordinary saturation (solubility-product) equilibrium. However, in the case of a solid stoichiometric substance in equilibrium with its saturated solution, the valency of the metal ions in both phases is the same, whereas here equilibrium can be attained even when the valencies are different.

To derive the equilibrium conditions, use is made of the fact that only one value of potential can exist across one phase boundary. The potentials developed by the oxygen-ion transfer reaction and the metal-ion reaction must then be equal. Thus, applying the equation for the oxygen-ion transfer reaction already derived, with reference to a standard hydrogen electrode,

$$U_{0,H_2} \qquad = \qquad U_{0,H_2}$$

Oxygen-ion transfer   Metal-ion transfer

$$1.23V + \frac{1}{2F}\frac{d\Delta G_0(n)}{dn} + \frac{RT}{F}\ln a_{(H^+)} =$$

$$= \frac{n}{zF}\frac{d\Delta G_0(n)}{dn} - \frac{\Delta G_0(n)}{zF} + \frac{\Delta G_0(Me^{z+})}{zF} + \frac{RT}{zF}\ln a_{(Me^{z+})}$$

which can be rearranged to

$$\frac{z - 2n}{2zF} \frac{d\Delta G_0(n)}{dn} + \frac{1}{zF} \Delta G_0(n) = K - 2.303 \frac{RT}{zF} (p \, Me - z \, pH)$$

where

$$K = \frac{1}{zF} \Delta G_0(Me^{z+}) - 1.23V$$

and

$$p \, Me = -\log a_{(Me^{z+})}$$

This equation implies that there is only one specific oxide characterized by the value of n in MeO_n that can exist in equilibrium with an electrolyte in which the values of z, p Me, and pH are predetermined (see Vetter and Jaeger [56].)

Bell has shown [57] that the following two concepts are of paramount importance in Vetter's treatment.

a. In a homogeneous oxide, any local variation of chemical composition is spontaneously equalized over the whole phase.

b. In a heterogeneous system, the oxides involved are in equilibrium; that is, at least one of the three possible equilibria across the solid phase boundary (electron, metal ion, or oxide ion) is active.

In order that the reduction reaction ($n_1 \to n_2$) may proceed spontaneously, $\Delta G(R)$ must be negative and therefore

$$\frac{\Delta G_0(n_1) - \Delta G_0(n_2)}{n_1 - n_2} > \Delta G_0(H_2O) .$$

For a homogeneous phase, at all points on the curve relating $\Delta G_0(n)$ to n the slope must be greater than -56.60 kcal per unit change in n.

Referring now to Fig. 28, let MeO_n and MeO_{n_2} be two oxides placed within the homogeneity range of a MeO_n system. Consider now a mixture of x moles of MeO_{n_1} and y moles of MeO_{n_2}. The average composition of the mixture is

$$\frac{MeO_{xn_1+yn_2}}{x + y}$$

which we can represent by n_*.

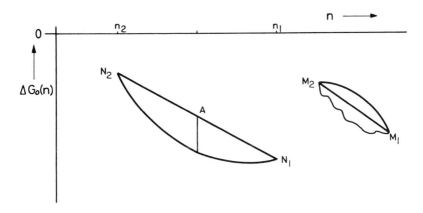

FIG. 28. Free enthalpy of formation versus n for metal oxides MeO$_n$, showing the relationship between a homogeneous phase and a mechanical mixture (arbitrary units).

The average free enthalpy of formation per mole of such a mixture is given by

$$\frac{x\,\Delta G_0(n_1) + y\,\Delta G_0(n_2)}{x + y}$$

and is indicated by the point A on the straight line joining N_1 and N_2 in Fig. 28. It is a fundamental property of a homogeneous phase that such a mixture will spontaneously equilibriate to an oxide MeO$_{n_*}$ of uniform composition. Such a process requires that the free enthalpy of the reaction be negative. The curve joining N_1 and N_2 must therefore pass through a point vertically below A in the figure. It should be noted that N_1 and N_2 may be as close as desired.

Generally, it may be stated that the curve relating $\Delta G_0(n)$ to n joining the two extremes of the homogeneous phase must continually change its slope toward more positive values as n increases. In other words, the second derivative

$$\frac{d^2 \Delta G_0(n)}{dn^2}$$

must be positive. Therefore, such curves as those shown joining M_1 and M_2 in Fig. 28 are excluded.

If a heterogeneous mixture of two oxides (MeO_{n_1} and MeO_{n_2}) is made with the average composition MeO_n , equilibrium implies that there is no other oxide MeO_{n_*} with a more negative free enthalpy (compare homogeneous systems). During reduction, the higher oxide is converted into the lower oxide. This is the only process occuring.

In Fig. 29, RTS is the $\Delta G_0(n)$ versus n curve for a homogeneous oxide phase. A second oxide, independent of the homogeneous system, is represented by the point H. If a mixture of the oxide H and some oxide (say P_1) belonging to the homogeneous phase is made, the free enthalpy per mole of the mixture will be a point on the line HP_1 corresponding to the average value of n. Let this point be Q. Since an oxide exists with the same composition but with a more negative value of the free enthalpy (P_2 in Fig. 29), Q cannot represent an equilibrium state. The two oxides cannot exist side by side, and P_1 will be reduced and H oxidized to P_2. If more of oxide H is added to the system, the same process will be repeated until the moving point P finally coincides with T and the straight line HP coincides with the tangent to the curve HT.

The oxide represented by the point T is the only oxide from the whole homogeneous range RTS that can exist in equilibrium with the oxide H. Any mixture of H and T will be represented by a point on the line HT and is an equilibrium mixture.

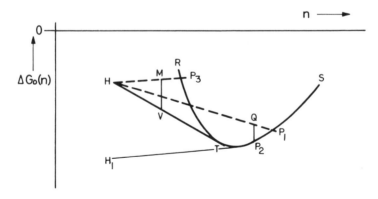

FIG. 29. Free enthalpy of formation versus n for a homogeneous phase and an independent oxide (arbitrary units).

Similar considerations apply to a mixture M of oxides H and P_3 where P_3 is a lower oxide than T. In this case, P_3 will disproportionate, forming H and a higher oxide. Since the average composition must remain constant, the free enthalpy of the mixture moves down the vertical line MV and P_3 down the curve RTS until the final equilibrium position is reached with P_3 coincident with T and HP_3 coincident with the tangent HT.

The important conclusions emerge that the lower oxide H may be in equilibrium with only one oxide T from the whole homogeneous range, and that when such an equilibrium is established, this oxide T becomes the lower limit of the homogeneous phase. In other words, the degree to which an oxide may be reduced in homogeneous phase is determined by the shape of the $\Delta G_0(n)$ versus n curve and the properties of the independent oxide that may be formed.

The above conclusions have been reached by considering the case of an independent oxide of a lower degree than the equilibrium oxide in a homogeneous phase. Obviously, similar conclusions apply to the opposite case, the homogeneous phase being then unstable at the higher degrees of oxidation.

Should the independent oxide be part of a homogeneous phase, the behavior of the system may be deduced by exactly the same reasoning as the foregoing. In Fig. 30, curves FED and CBA represent two homogeneous systems.

Starting from point A, reduction will proceed in homogeneous phase until point B is reached, this being the point of contact with the common

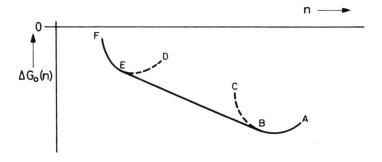

FIG. 30. Free enthalpy of formation versus n for two independent homogeneous phases (arbitrary units).

tangent to the two curves. Further reduction results in the formation of the oxide represented by point E, this being the point of contact of the common tangent to the curve representing the second homogeneous phase FED. Reduction now proceeds in a heterogeneous system until oxide B has been completely converted to E. Finally, a second homogeneous system develops during reduction from E to F. Note that the homogeneous ranges E-D and C-B are both rendered unstable.

In the light of Vetter's theoretical treatment, the following remarks may be made regarding earlier attempts at interpretation of the electrode potential of manganese dioxide.

a. The basic formulation of the "reversible" electrode reaction given in the Pourbaix Atlas

$$MnO_2 + 4\ H^+ + 2\ e^- \rightleftarrows Mn^{2+} + H_2O$$

is really not reversible. Generally it is only possible as an irreversible reaction.

b. Vosburgh and Johnson [58,59] and Neumann and von Roda [60] give as the reversible reaction

$$MnO_2\ (a_4) + H^+ + e^- \rightleftarrows MnOOH\ (a_3)$$

where a_3 and a_4 are activities. From this equation, there follows directly

$$U_{rev} = U_0 + \frac{RT}{F}\ ln\frac{a_4}{a_3} + \frac{RT}{F}\ ln\ a_{(H^+)}$$

This equation can be deduced from Vetter's general theory, but a limiting condition must be assumed, namely that no manganese ions other than Mn^{3+} and Mn^{4+} are relevant to the discussion.

3.2.3. The Thermodynamics of the Manganese Dioxide Electrode (Neumann)

Neumann deposited manganese dioxide electrolytically on platinum electrodes and measured the electrode potential in 5 M NH_4Cl solution in relation to the depth of discharge and the pH of the electrolyte. For the thermodynamic interpretation of the results, the electrodes were treated as single-phase solid solutions of hydrated MnO_2 and $MnO_{1.5}$ in equilibrium with the electrolyte (as referred to under (b) above).

From the equation

$$\log \frac{MnO_2}{MnO_{1.5}} = \frac{U}{0.059} + \frac{\log(1-x)}{xa_{(H^+)}} - \frac{U_{0MnO_2(x)/MnO_{1.5}(1-x)}}{0.059}$$

it is possible, since U_0 is a constant, to evaluate the quantity

$$\varphi\ MnO_2/\varphi\ MnO_{1.5}$$

as a function of x, by simultaneous determination of the equilibrium potential, the pH, and the composition of the electrode. Using the Duhem-Margules equation, the values of $\varphi\ MnO_2$, $\varphi\ MnO_{1.5}$, the respective activities, and the standard potential may be individually calculated.

Table 19 lists the activity coefficients and the related activities found in this manner, and Fig. 31 is a graphic representation of the activity curves in the $MnO_2/MnO_{1.5}$ system.

Figure 32 represents the discharge of a typical manganese dioxide electrode, I being the open-circuit voltage curve and II the on-load voltage. The influence of the activity coefficients of the components on the electrode po-

TABLE 19

Activity Coefficients of $MnO_2/MnO_{1.5}$[a]

a_1	a_2
1.0	0.0
7.1×10^{-1}	1.9×10^{-5}
3.1×10^{-1}	2.1×10^{-3}
1.1×10^{-1}	4.9×10^{-2}
5.6×10^{-2}	1.9×10^{-1}
3.3×10^{-2}	3.6×10^{-1}
2.1×10^{-2}	5.2×10^{-1}
1.3×10^{-2}	6.7×10^{-1}
8.0×10^{-3}	7.9×10^{-1}
3.7×10^{-3}	9.0×10^{-1}
0.0	1.0

[a]See Fig. 31.

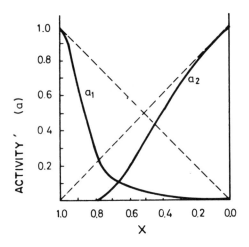

FIG. 31. Activity of each component versus x, the mole fraction of MnO_2 in the system $MnO_2/MnO_{1.5}$ (a_1, MnO_2; a_2, $MnO_{1.5}$).

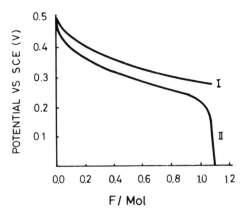

FIG. 32. Electrode potential versus degree of discharge in faradays per mole. I, reversible electrode potential; II, on-load voltage.

tential is shown by the two curves in Fig. 33. The U_{real} curve has been calculated for pH = 0 and includes the proper values of the activity coefficients. The U_{ideal} curve is similar, but with the coefficients set equal to unity. The electrode potential values are, of course, identical with those

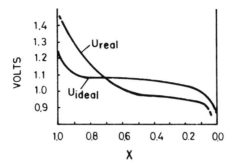

FIG. 33. Calculated equilibrium electrode potential of the $MnO_2/MnO_{1.5}$ system versus the mole fraction of MnO_2. E_{ideal} is based on unit activity coefficients, E_{real} uses activities from Fig. 31. Potentials are referred to a H_2 electrode at the same pH as the electrolyte.

measured against a hydrogen electrode operating at the pH of the cell electrolyte and are plotted against the mole fraction of MnO_2 in the electrode.

3.2.4. The Mechanism of the Cathodic Reaction

The fundamental ideas relating to the cathodic reaction mechanism were first formulated by Keller [61] and more precisely by Coleman. In his paper "Dry Cell Dynamics" [62], Coleman proposed the theory that the potential of a manganese dioxide electrode is determined by the concentration of the reduction products at the electrode/electrolyte phase boundary and presented a model from which the variation in potential during the discharge and recuperation processes could be calculated. According to this theory, the increase in electrode potential during recuperation is explained by the diffusion of the trivalent manganese oxide away from the surface of the individual particles. An equivalent diffusion of the tetravalent oxide occurs from the interior to the surface, whereby the diffusing species in the solid need only be protons and electrons. This theory has been generally accepted.

A quantitative treatment of the diffusion process was presented by Scott [63] based on the Johnson and Vosburgh equation [59]

$$E = E^0 + 0.073 \ln \frac{\text{mole } \% \text{ MnOOH}}{\text{mole } \% \text{ MnO}_2}$$

where E^0 is dependent on pH and has the value 0.4160 at pH 7.5 with reference to the saturated calomel half cell.

Further, it was assumed that MnO_2 is a semiinfinite solid bounded at $x = 0$, the region with $x < 0$ being occupied by electrolyte. The diffusion coefficient D for protons accompanied by electrons was also assumed to be constant. From an analogous problem in the theory of heat conduction, Scott proposed the following equation for the boundary $x = 0$:

$$c = \frac{2F_0}{(D\pi)^{1/2}} \left[t^{1/2} - (t - T)^{1/2} \right]$$

where c is the concentration of MnOOH at a distance x from the surface, F_0 is the number of equivalents of MnOOH being produced per unit area per unit time, D is the diffusion coefficient for protons accompanied by electrons, t is the time elapsed since the start of discharge, and T is the time at which the discharge is stopped and recuperation commences. For $t < T$, terms involving $(t - T)$ are taken as zero.

Using these equations, Scott was able to calculate with reasonable accuracy the discharge and recovery curves reported by Vosburgh and Ferrel [64] by adjustment of the values of F_0 and D to

$$F_0 = 10^{-2} \text{ equivalents cm}^{-2} \text{ sec}^{-1} \quad (i = 9.5 \times 10^{-8} \text{ A cm}^{-2})$$

$$D = 1.2 \times 10^{-18} \text{ cm}^2 \text{ sec}^{-1}$$

Scott also noted that at the time t_s when it would be expected that the surface of the electrode would be saturated with MnOOH ($c = c_s = 5.47 \times 10^{-2}$ equivalents cm^{-3}), there is a slight second inflection in the discharge curve. If a reliable estimate of t_s can be obtained in this way, an independent assessment of the working area of the electrode can be made, since, for a quick discharge, it can be shown that

$$F_0 = c_s / t_s$$

and the area can then be calculated using the known value of the total discharge current. Estimates made in this way show that the working area of the manganese dioxide is only about 12% of the surface area obtained from BET measurements. (see Bell [57] and Kozawa [65]).

Having thus assessed the electrode area, specific values of F_0 and t_s may be found from a medium discharge. Using the second equation above, D may be estimated from

$$D = \left[\frac{-F_0}{c_s}\right]^2 \frac{4t_s}{\pi}$$

Modified forms of the Scott equations were proposed by Kornfeil [66] and Era et al. [67].

The efficiency of a Leclanché cell when discharged to a given cut-off voltage is therefore not only dependent on the amount of MnO_2 present in the electrode but is also significantly influenced by the physical structure of the dioxide and the geometry of the electrode. In other words, mass transfer (or diffusion) processes play an important role in the kinetics of the electrode reaction [62-64, 68-71].

Two types of diffusion process must be taken into consideration. First, there is diffusion in the solid state referred to above, which is dependent on the physical and chemical characteristics of the manganese dioxide. Second, there is diffusion in the liquid state, which is influenced by the composition of the electrolyte and its physical distribution throughout the cathode. The latter is dependent, of course, on the porosity of the electrode.

Brenet and Grund [72] state that the presence of constitutional water favors the movement of protons in the crystal lattice. This point of view is supported by Era and coworkers [73].

The cation exchange properties of hydrated manganese dioxides have already been mentioned. In addition, Muraki et al. [74] propose that anion exchange properties are also present. Accordingly, hydrated manganese dioxide behaves as an amphoteric substance, as shown in the structure diagram.

$$\left[O = Mn\underset{OH}{\overset{O}{\diagdown}}\right]^{-} + H^{+} \quad\longleftarrow\quad O = Mn\underset{OH}{\overset{OH}{\diagup}} \quad\longrightarrow\quad \left[O = Mn\underset{OH}{\overset{}{\diagdown}}\right]^{+} + OH^{-}$$

In acid solutions the dioxide behaves as a weak acid, and in alkaline solutions as a base. Thus MnO_2 exercises a buffering action on the pH of the electrolyte. As the electrolyte pH increases, the OH^- groups become polarized and the bond between the oxygen and hydrogen atoms weakened. Thus protons tend to split off from the hydroxyl groups and diffuse away into the interior of the solid phase. This is a possible explanation for the reduced overpotential of MnO_2 electrodes operating in alkaline media, suggested by Era [73].

Several investigations have been concerned with the amount of water of crystallization (or constitutional water) in manganese dioxide. The total water content falls into two classes, that which is removed at relatively low temperature ($100°$-$110°C$) and which may be reabsorbed by the ore, and that which can only be liberated at high temperature and which is not reabsorbed. It is the second class that is referred to as constitutional water. (See Brenet [75] and Sasaki and Kozawa [76].)

Glemser et al. [77] have shown that hydroxyl groups are present in γ-MnO_2 by means of infrared spectroscopy. In electrolytic dioxides, the content of constitutional water usually lies between 3.8 and 5%, in natural ores between 1 and 4.7%, but intermediate values are also found.

It appears that some relationship between the activity of manganese dioxide and the constitutional water content exists, but the available evidence is not conclusive.

Recapitulating, we may say the following.

a. The characteristics of manganese dioxide must be related to the kinetics of reactions in the solid phase, with special reference to nonstoichiometric chemistry.

b. The electron absorption occurs in the solid phase, whereby Mn^{4+} ions are reduced to Mn^{3+}, inducing an increase in volume.

c. An active oxide must necessarily contain acidic hydroxyl groups. These groups imply the presence of Mn^{3+} ions, which reduce the Fermi level, decrease the electronic conductivity, and act as acceptors. They increase the rate of diffusion of protons in the solid phase.

d. The electrode potential is dependent on the concentration of MnOOH at the oxide/electrolyte phase boundary. This potential may be calculated

from a diffusion equation with reasonable accuracy. The removal of MnOOH
from the electrode surface may be regarded as the rate-determining step in
the overall electrochemical reaction. If the discharge is interrupted, some
of the MnOOH formed at the surface diffuses into the interior, but some
remains at the oxide/electrolyte phase boundary. This surface concentration determines the value of the electrode potential, which, as a consequence
of the theory, must be lower than the value observed in the undischarged condition.

To illustrate what has been said above, Fig. 34 shows discharge curves
of electrolytic MnO_2 electrodes in acid (a), weak acid (b) and alkaline (c)

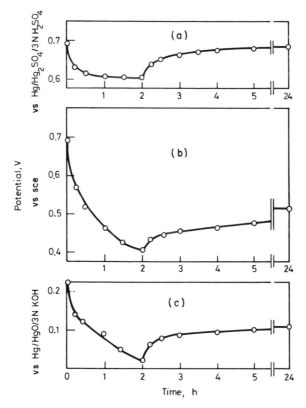

FIG. 34. Polarization and recuperation of the MnO_2 electrode in
various electrolytes. (a) 3 N H_2SO_4; (b) 3 M NH_4Cl, pH 4.7; (c) 3 N KOH.

electrolyte, together with the corresponding open-circuit voltage curves.

In acid solution, MnOOH is removed from the surface by disproportionation (Fig. 34a):

$$2 \text{ MnOOH} + 2 \text{ H}^+ = \text{MnO}_2 + \text{Mn}^{2+} + 2 \text{ H}_2\text{O}$$

The open-circuit voltage will therefore undergo only a rather small change (depending on the concentration of Mn^{2+} ion in the electrolyte) during the discharge process. The overvoltage is defined as the difference between the equilibrium potential after interruption of the discharge and the on-load voltage immediately before the end of discharge. It is less in acid solution (Fig. 34a) than in ammonium chloride (pH = 4.7) solution (Fig. 34b). In strong acid, the on-load voltage remains reasonably constant after a short initial voltage drop. This constant value is reached as soon as the rate of removal of the lower oxide from the electrode surface becomes equal to the rate of formation. In alkaline solution (Fig. 34c), not only the on-load voltage but also the open-circuit voltage decrease as discharge proceeds. The overvoltage is relatively low as in strong-acid solution.

In the weak acid electrolyte, the decrease of the on-load voltage during discharge is large compared with that observed in acid and alkaline electrolytes. The overvoltage is also correspondingly greater (Fig. 34b). In the medium pH range, it can be assumed that MnOOH is removed from the surface by solid-state diffusion assisted by disproportionation. The latter process is of less importance as discharge proceeds, since the pH of the catholyte increases rapidly.

Several investigations have measured the rate of formation of Mn^{2+} ion during discharge [78-82]. Figure 35 illustrates some of these findings.

Electrolytes containing ammonium ion show a lower overvoltage than other solutions [81]. Further, the presence of metal ions in the ammonium chloride solution, such as Zn^{2+}, Ni^{2+}, or Mn^{2+} which form complexes with NH_3 of the form $\text{Me(NH}_3)_n^{2+}$, also decreases the polarization. Era suggests that the ammonium ion functions as a proton donor so the disproportionation of MnOOH can proceed in the absence of hydrogen ions according to the equation

$$2 \text{ MnOOH} + 2 \text{ NH}_4^+ = \text{MnO}_2 + \text{Mn}^{2+} + 2 \text{ NH}_3 + 2 \text{ H}_2\text{O}$$

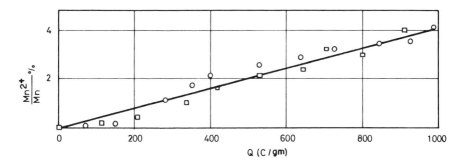

FIG. 35. Ratio of Mn^{2+} to total Mn in solid phase at various stages of discharge (coulombs per gram) of an MnO_2 electrode. Experimental data: squares represent discharge at constant current; circles represent discharge through a constant resistance.

The favorable performance of the complex-ion forming cations can also be explained from the above equation, since ammonia is continually removed in the form of a complex ion. The equilibrium is displaced in favor of the disproportionation reaction. Mg^{2+} ions, for example, which do not form coordination compounds with ammonia, have no influence on the overvoltage. In weak acid electrolytes, therefore, it can be assumed that apart from the solid-state diffusion, a small proportion of the lower oxide (formed during the discharge) is removed by disproportionation.

Hetaerolyte (ZnO · Mn_2O_3) has also been discussed as a possible reaction product [83-85]. From analysis of the cathode and the electrolyte after discharge, Cahoon concluded that both MnOOH and ZnO · Mn_2O_3 are formed. Gabano et al. [52] have shown by chemical analysis as well as by the use of radioactive isotopes that the formation of hetaerolyte is very probable in γ-MnO_2 electrodes at light discharge rates (Fig. 36).

Before closing this treatment of the MnO_2 electrode reaction, reference is again made to Coleman's investigations [62], from which certain fundamental conclusions regarding the distribution of potential and current within the cathode may be drawn. At the beginning of the discharge, it was shown that the potential is uniform throughout the electrode but that the current is unevenly distributed. As discharge proceeds, a condition is approached in which the potential is nonuniform and the current is evenly distributed.

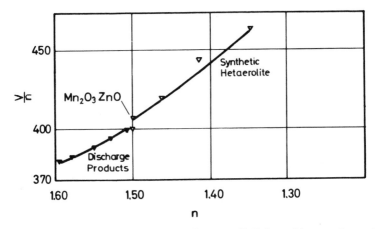

FIG. 36. Hetaerolite. Volume of unit cell V (in cubic angstroms) per unit degree of oxidation n versus n, the degree of oxidation. Solid triangles represent products of discharge; open triangles represent synthetic hetaerolites.

As discharge begins, the current flow is concentrated in the cathode immediately surrounding the carbon rod. As discharge proceeds, the potential in the areas of high current density falls below that of the more distant parts of the cathode, so that these latter regions gradually take over the current production. In this way, an even distribution of the current throughout the electrode is obtained. Cahoon and Heise [79] have published results of investigations in this field. From pH measurements and analysis of various regions of the cathode, similar conclusions can be reached.

Euler [86-88] and Hirai and Fukuda [89] have also made contributions to this subject from studies on equivalent electrical circuits.

More recently, Gabano [52] has reported results obtained from paper-lined D-size dry cells containing γ-MnO_2 as active cathode material. Figures 37-39 illustrate the change in composition of the various layers in the cathode during discharge through a constant resistance of 56 Ω.

All the investigations previously discussed were carried out on relatively large amounts of cathode and cell material. In contrast to this technique, Miyazaki [90] applied electron-probe microanalytical methods to

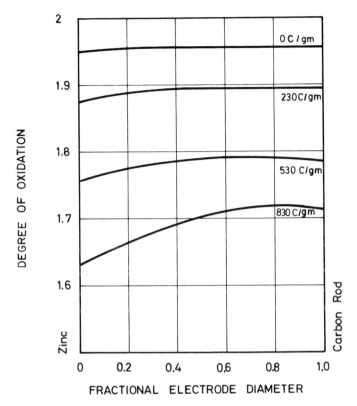

FIG. 37. Degree of oxidation versus depth in a cylindrical cathode (Leclanché type) at various stages of discharge (coulombs per gram).

the examination of the changes occurring during discharge in single particles of ore. Three types of dioxide, a very inactive, a medium quality, and a highly active electrode material, were studied by discharging at light and at heavy currents in D-size cells. The cathode mix was then embedded in epoxy-resin and polished to various depths. Single particle surfaces were then examined with the electron probe. Table 20 lists the end products of the discharge of individual MnO_2, samples according to the Debye-Scherrer analysis. Figures 40-45 (from Ref. [90] picture the results of the electron-probe analyses. An inactive oxide, in contrast to a medium or a highly active material, shows no pores or crevices in the individual particles.

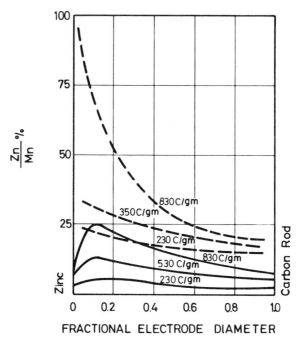

FIG. 38. Ratio of Zn to Mn versus depth in a cylindrical cathode
(Leclanché type) at various stages of discharge (coulombs per gram).
Broken curves, total Zn; full curves, Zn taken up in MnO_2 lattice.

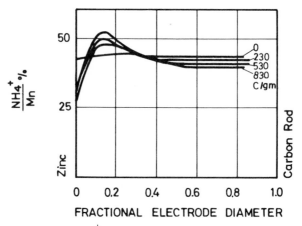

FIG. 39. Ratio of NH_4^+ to Mn versus depth in a cylindrical cathode
(Leclanché type) at various stages of discharge (coulombs per gram).

TABLE 20

Discharge Products of MnO_2 Materials Found by X-Ray Diffraction Technique

Type of MnO_2	Discharge	
	Heavy	Light
Sample I	$MnO_2 + Mn_3O_4$ + very small amount of $ZnCl_2 \cdot 4 Zn(OH)_2$	$MnO_2 + Mn_3O_4$ + very small amount of $ZnCl_2 \cdot 4 Zn(OH)_2$
Sample II	Mn_3O_4 + residual MnO_2 + $ZnCl_2 \cdot 4 Zn(OH)_2$ intense at (003) peak	$ZnCl_2 \cdot 4 Zn(OH)_2$ intense at (003) peak + minor amount of $Mn_2O_3 \cdot ZnO$
Sample III	$Mn_2O_3 \cdot ZnO$	Well-developed $Mn_2O_3 \cdot ZnO$

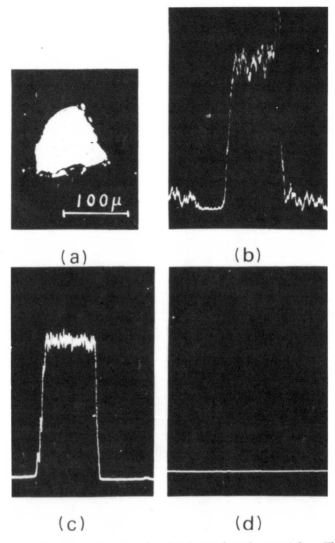

FIG. 40. Cross section (a) of undischarged MnO_2 particle. The relative concentrations of oxygen, manganese, and zinc are shown in (b), (c), and (d).

The manganese and oxygen contents decrease as the activity increases. Low-quality ores do not show noticeable penetration of zinc ions into the crystal lattice. There is, however, penetration of zinc into the particles

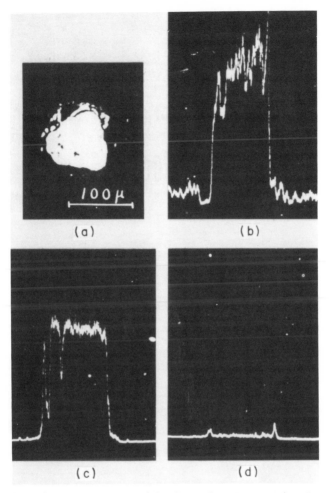

FIG. 41. Cross section (a) of discharged MnO_2 particle (Sample 1). The concentrations of oxygen (b) and manganese (c) are essentially the same as before discharge. Zinc (d) is present only at particle boundaries.

of the medium-quality ore, but without bonding to the manganese. Active ores reveal a uniform distribution of zinc, manganese, and oxygen throughout the particles, suggesting the formation of hetaerolite (ZnO · Mn_2O_3).

These results, although requiring a more quantitative treatment, support in general Gabano's explanation of the electrochemical reduction of

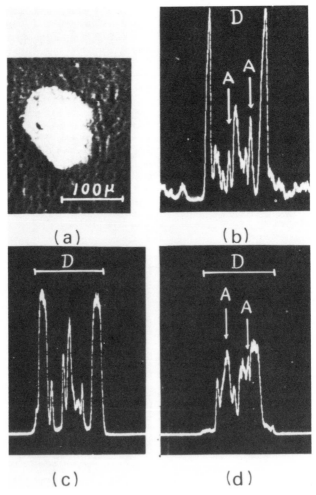

FIG. 42. Cross section (a) of heavily discharged MnO_2 particle
(sample II), showing microcracks. Oxygen and manganese concentrations
(b) and (c) are depleted within particle. Zinc (d) has penetrated to inner
lattice sites.

γ-MnO_2 [52] . According to this theory, the reaction proceeds in two
stages. In the first stage, the crystal lattice expands without disruption
until 600-700 Coulombs have passed for each gram of MnO_2. The crystal-
lographic and chemical characteristics approach more and more closely

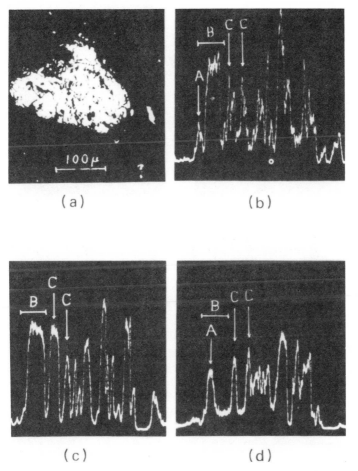

FIG. 43. Cross section (a) of MnO_2 particle (sample II) discharged at low drain, showing microcracks. Oxygen, manganese, and zinc concentrations are shown in (b), (c) and (d). At A, $ZnCl_2 \cdot 4\ Zn(OH)_2$ plus $Mn_2O_3 \cdot ZnO$; at B, zinc replaces manganese; at C, zinc and manganese are present together.

those of Groutite (α-MnOOH). During this stage, the number of hydroxyl groups increases simultaneously with the chemical reduction.

As a first approximation, it can be assumed that the conversion of Mn^{4+} to Mn^{3+} ions is compensated for electrically by the conversion of O^{2-} ions

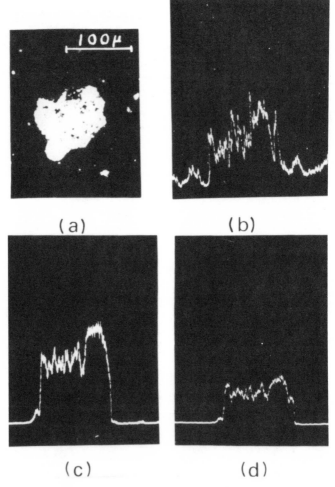

FIG. 44. Cross section (a) of discharged MnO_2 particle (sample III), showing microcracks. The oxygen and manganese concentrations, (b) and (c), are uniformly depleted and have been replaced by zinc (d), indicating the formation of $Mn_2O_3 \cdot ZnO$.

to OH^- groups (the radii of these ions are about the same, 1.45 Å). The increase in volume of the unit cell is a result of the reduction of the Mn^{4+} ion (r = 0.50 Å) to Mn^{3+} ion (r = 0.70 Å). The electrode potential is too high to permit the discharge of protons present in the electrolyte. The

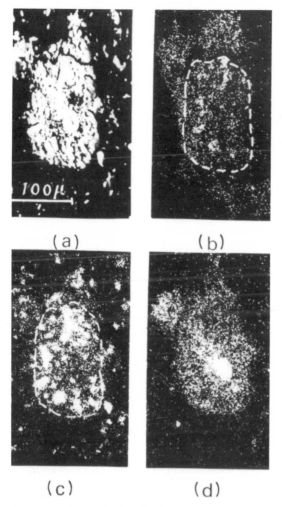

FIG. 45. Cross section (a) of a discharged MnO_2 particle, showing microcracks. The distributions of oxygen and manganese (b) and (c) on the plane surface show isolated accumulations of the elements along the microcracks. The distribution of zinc (d) shows that this element has replaced manganese and oxygen in the central region.

electrons arriving at the electrode from the outer circuit are therefore utilized for the reduction of the tetravalent manganese. To maintain electroneutrality, protons from the electrolyte combine with O^{2-} ions in the solid, thus increasing the number of hydroxyl groups. The diffusion of protons in the crystal lattice will therefore be rate determining. At the same time, zinc appears in the crystal lattice, without, in this first stage, the formation of new compounds. Its presence can be interpreted as the result of ion-exchange processes. The increase in the number of hydroxyl groups results in a corresponding increase in zinc content. Finally, the expanded crystal lattice breaks down and the reaction enters the second stage. The change in crystal structure may be the consequence of the formation of ZnO according to the schematic reaction shown.

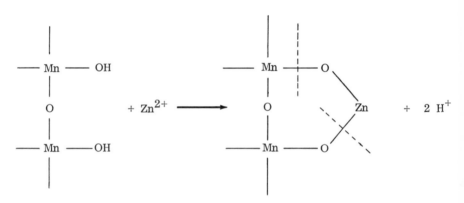

With the progressive reduction of the Mn^{4+} ions, the ZnO content and the number of Mn^{3+} ions increase simultaneously and finally tend to form hetaerolite.

The reduction of γ-MnO_2 can be summarized as follows.

a. Mn^{4+} is reduced to Mn^{3+} in solid homogeneous phase by the arrival of electrons. New hydroxyl groups are formed with protons from the electrolyte.

b. Zn^{2+} ions are introduced into the lattice by ionic exchange.

c. The Mn-O-Zn compound is broken up and zinc oxide is formed.

d. The concentration of Zn^{2+} and Mn^{3+} ions increases in the lattice, leading to the formation of hetaerolite.

In the second stage of the reduction process, the zinc content of the electrode is found to be dependent on the discharge current. When the speed of the electrochemical reaction is low (for example, at low voltage with discharge through a constant resistance) the zinc ion diffuses deep into the lattice. On the other hand, at high speed (for example, discharge at constant high current) there is no time for the zinc to penetrate into the lattice.

At the surface of the electrode facing the separator and electrolyte reservoir, there will be some reduction to divalent manganese as a result of the low pH value in this region.

The electrochemical reduction of β-MnO_2 has not been as intensively investigated as γ-MnO_2. It can be assumed that electrons flow into the electrode and reduce Mn^{4+} ions to Mn^{3+} at the surface of the particles. To maintain electroneutrality, protons will be taken up from the electrolyte and hydroxyl groups formed. This results in the formation of a two-phase heterogeneous system, so that the equilibrium electrode potential remains constant until all of the higher oxide has been reduced.

The solid end products will not necessarily appear in the stoichiometric form. For example, it is possible, as Feitknecht and Marti [91] have shown, that the oxygen/manganese ratio in Manganite can vary from 1.4 to 1.6 (theoretical 1.5). Similar deviations from stoichiometry are known for the isostructural hetaerolite. The equilibration of the oxygen concentration within these nonstoichiometric phases may be achieved by proton diffusion in the lattice or possibly through the liquid phase.

A good summary of the electrochemical behavior of β-MnO_2 is afforded by the potential versus pH diagram in the Pourbaix Atlas [92], which also provides a collection of free-enthalpy data for the relevant reaction products. For similar data relating to γ-MnO_2, reference should be made to Vetter and Jaeger's paper [56].

3.3. THE ANODE

3.3.1. Reactions at the Zinc Anode

There are three requirements for an ideal anode for any galvanic system: a highly negative electrode potential, a small overvoltage even at high current density, and a high degree of stability in the electrolyte (corrosion resistance). These three requirements are found in practice to be mutually exclusive, and the galvanic cell designer is compelled to accept a compromise.

One of the most satisfactory solutions to this problem is to add small quantities of lead and cadmium to the zinc to improve the mechanical handling properties. This is the anode employed in the Leclanché cell. The potential of zinc is well below the thermodynamic stability range of water, and consequently the metal is always subject to chemical corrosion. Amalgamation of the zinc, by increasing the hydrogen overvoltage at the metal surface, reduces the corrosion rate to within acceptable limits. The standard electrode potential is -0.76 V.

The zinc anode is always active, and even at high anodic current densities, there is practically no observable overvoltage. The general behavior of zinc in aqueous solutions can be deduced from the Pourbaix diagram [92]. Brouillet and Jolas [93] have studied the voltage changes at a zinc anode under the special conditions existing in a Leclanché cell; the pH of the electrolyte was varied by addition of KOH solution. The results are shown in Fig. 46. Table 21 lists the pH values, the electrode potentials with respect to a hydrogen electrode, and the experimentally determined slopes of the reaction isotherm. In the pH range 5.1-5.8, the zinc goes into solution as the normal zinc ion. In the range 5.8-7.85, crystals of the composition $ZnCl_2 \cdot 2 NH_3$ form at the zinc surface. At pH values above 7.85, the diammine complex dissolves, forming $Zn(NH_3)_4^{2+}$ ions.

During storage of Leclanché cells, the generation of hydrogen gas can always be observed together with the deposition of insoluble crystals at the zinc surface. These crystals have been identified by x-ray techniques and are in fact $ZnCl_2 \cdot 2 NH_3$, the diammine complex referred to above. At the sites where the diammine formation is proceeding, the pH lies between 5.8 and 7.85; the increase in alkalinity is due to reduction of protons.

3.3.2. Zinc Corrosion

There are two main types of corrosion of zinc anodes in primary cells: general surface corrosion and pitting.

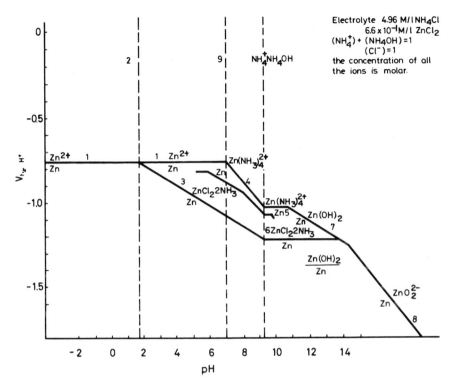

FIG. 46. Potential versus pH for zinc in a $ZnCl_2/NH_4Cl$ electrolyte as used in Leclanché cell manufacture.

TABLE 21

Electrode Potential as a Function of pH Value in a Leclanché Cell

Range of pH	Tension/hydrogen	Experimental slope
5.1-5.8	-0.840	0
5.8-7.85	-0.840 to -0.932	-0.045
7.85-9.3	-0.932 to -1.070	-0.095
9.3-9.65	-1.070	0
9.65-9.85	-1.070 to -1.090	-0.100

General surface corrosion is the usual type of pattern obtained in the absence of any complicating factor and with a more or less uniform anode surface. In ammonium chloride solutions and in the absence of air, the main corrosion reaction is

$$Zn + 2\ NH_4Cl \rightarrow ZnCl_2 \cdot 2\ NH_3 + H_2$$

The speed of this reaction is largely determined by the rate at which hydrogen can be liberated at cathodic centers on the electrode. The presence of metallic inclusions in the zinc surface, at which the hydrogen overvoltage is lower than on the zinc itself, of course accelerates the corrosion rate. On the other hand, the formation of passivating films (with reduced electronic conductivity) inhibits hydrogen discharge.

This reaction also causes the pH at the anode/electrolyte interface to increase, particularly in gelled electrolytes where the diffusion processes are slowed down. We are especially indebted to Feitknecht and coworkers for a study of these passivating films and a description of their chemistry and morphology in relation to their effect on the rate of corrosion of zinc (Feitknecht and Petermann [94], Feitknecht [95]. (See Table 22.)

The formation of pits is by far the most dangerous type of corrosion, since perforation of the zinc may develop very rapidly. This in turn leads to electrolyte leakage and/or drying out of the cell.

The deep pit occurring at the zinc/electrolyte/air boundary in inadequately sealed cells is well known. Further, pitting often occurs along lines of particular stress during can formation. Evans and Davies [96] have shown that pitting occurs even in distilled water if the oxygen supply is deficient. There are many papers on the corrosion rate of zinc in aqueous solutions [97-100], but no satisfactory explanation of pit formation or even more important the means to prevent it has been given. About all we can say is that, in solutions containing ammonium and chloride ions, the passivating films formed on zinc are extraordinarily sensitive and easily ruptured. Any factor producing film rupture may lead to pitting.

A summary of the characteristics of zinc in zinc chloride/ammonium chloride was given by Bell [101]. The corrosion of zinc and zinc alloys has recently been reinvestigated by Krug and Borchers [102]. These authors describe their findings as follows.

TABLE 22

Zinc Corrosion Products[a]

Compound	Formula	Solubility product $K = a_{(Zn^{z+})} a^m_{(OH^-)} a^n_{(Cl^-)}$
Hydroxychloride II	$ZnCl_2 \cdot 4\ Zn(OH)_2$ or $Zn(OH)_{1.6}Cl_{0.4}$	ca 3×10^{-15}
Hydroxychloride III	$ZnCl_2 \cdot 6\ Zn(OH)_2$ or $Zn(OH)_{1.71}Cl_{0.29}$	ca 6×10^{-15}
Amorphous hydroxide	$Zn(OH)_2$	2×10^{-16}
β_1-Hydroxide	$Zn(OH)_2$	2.2×10^{-17}
β_2-Hydroxide	$Zn(OH)_2$	ca 7×10^{-17}
Active zinc oxide	ZnO	8×10^{-17}
Inactive zinc oxide	ZnO	1.3×10^{-17}

[a]After Feitknecht [94, 95].

a. Zinc with a large grain but identical composition to zinc of fine grain corrodes more rapidly. The grain size ranges from $10^5 \, \mu m^2$ to $10^2 \, \mu m^2$.

b. Zinc containing 0.1% cadmium developed hydrogen more rapidly after mechanical deformation than in the recrystallized condition with only isolated, unavoidable twin crystals. The corrosion of the deformed material occurred at innumerable minute pits, whereas the recrystallized zinc showed general surface attack on individual grains.

c. In zinc containing lead, cadmium, and iron simultaneously but in varying amounts, the rate of corrosion decreased as the amount of lead increased. The increase of the cadmium content reduced the corrosion rate only slightly. Corrosion was markedly accelerated by increased iron content, especially at low lead and cadmium contents. With increased lead and cadmium, the influence of the iron was reduced, but the formation of pits was more pronounced.

d. Additives less noble than zinc were preferentially dissolved from the metallic phase. More noble additives with low hydrogen overvoltage accelerate corrosion, whereas those with high hydrogen overvoltage inhibit corrosion. Manganese, magnesium, lithium, gallium, and aluminum additives in 99.99% zinc induced corrosion with increasing concentration. The majority of the zinc samples tested showed no measurable hydrogen generation when preamalgamated.

In a typical alloy (containing 0.065% Cd, 0.28% Pb, and 0.0023% Fe) the addition of cerium only slightly increased the corrosion rate, whereas iron, nickel, and cobalt accelerated corrosion strongly; the effect increased with higher concentration.

e. A fine and uniform distribution of lead in zinc inhibits the corrosion reaction; the effect is more pronounced as the amount of lead is increased. The irregular distribution of large particles of lead increases the rate of corrosion. Such a distribution may arise as a result of a very slow cooling of the melt, or it may be due to lead displacement.

The slow cooling of melts containing iron also favors the formation of large particles of the zinc/iron intermetallic compound, which accelerate corrosion more than the finer particles.

Miyazaki [103] investigated zinc cans from D-size dry cells after
heavy and light intermittent discharge using microanalytic and Debye-
Scherrer techniques. The loss in weight of the anode plotted against the
discharge period is shown in Fig. 47. The products of discharge at the
zinc anode were identified as $ZnCl_2 \cdot 4 \, Zn(OH)_2$ and $ZnCl_2 \cdot 2 \, NH_3$ (see
Figs. 48 and 49). Miyazaki summarized his results as follows.

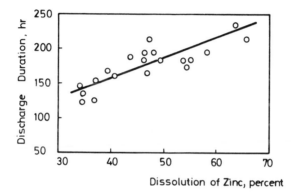

FIG. 47. Rate of dissolution of zinc anode at low current discharge.

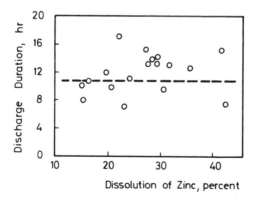

FIG. 48. Rate of dissolution of zinc anode at high current discharge.

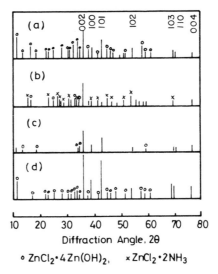

FIG. 49. Debye-Scherrer x-ray analysis of surface of zinc anode in a Leclanché cell after discharge at high current (a and b) and at low current (c and d).

The zinc anode dissolved to a greater extent on light discharge than on heavy load. During the light discharge, the dissolution of the zinc was observed to be approximately proportional to the discharge duration, and the amount of zinc that dissolved was related inversely to the amount of the basic chloride $ZnCl_2 \cdot 4 \, Zn(OH)_2$ formed. Although the anodes before use exhibited an intense (002) orientation, the intensity of this peak was reduced by an amount proportional to the light-discharge hours. During the heavy discharge, the formation of the diammine $ZnCl_2 \cdot 2 \, NH_3$ was characteristic when the initial voltage of the dry cell was low.

Before discharge, mercury was present to a depth of 70-80 μm in the anode surface that had been in contact with the cell electrolyte. The inside wall of the anode with the amalgamated film dissolved at an early stage of discharge. The outer electrode regions having no mercury were left after discharge.

Manganese ions partly migrated from the cathode onto the anode, just as the zinc did from the anode to the cathode. The heavy discharge produced additional pores in the MnO_2 that were larger in diameter than the original micropores. It is implied from a technical point of view that developing means of facilitating the mutual transportation of these ions during discharge will be one of the major concerns for improving the discharge performance of Leclanché dry cells.

3.4. THE COMPLETE CELL

As has been previously mentioned, the Leclanché dry cell represents a good compromise between the partly self-contradictory requirements of maximum electrical performance, excellent storage, and reasonable manufacturing costs. It is well known that the system suffers from polarization. The current-producing processes are unfavorably influenced by various chemical and physical side reactions. The energy yield obtained in heavy drain or continuous discharges is significantly less than that expected theoretically (from the amount of MnO_2 contained in the cell) or that actually obtained by intermittent discharge at low current drains. In the latter case, the polarization effects of the side reactions are counteracted by diffusion processes in the rest periods between discharges.

It is therefore essential to study the nature and the magnitude of the polarization factors in order to contrive an optimal adjustment of the cell to the various conditions of use.

3.4.1. Cell Polarization

When two electrically conducting phases containing a common charged species are brought into contact with each other, an electrical double layer is formed at the phase boundary. Under certain conditions, the phase boundary between metal and electrolyte has the properties of a capacitor. This was explained very early by Helmholtz on the basis of a model known today as the Helmholtz double layer. According to this hypothesis, the electric-

ally charged ions cannot approach the metal surface any closer than a limiting distance, the ionic radius. In this way, a model similar to a two-plate
capacitor was built whereby the value of the dielectric constant of the medium between the charges remained open. The capacitance of such a capacitor should be constant, but experiment showed that in the case of the metal/
electrolyte phase boundary, the capacitance is dependent on the state of
charge and on the electrolyte concentration.

Gouy [104] and Chapman [105] assumed a diffuse space-charge distribution in the electrolyte. Just as in the atmosphere, equilibrium is established between gravitational attraction and thermal dispersion, electrostatic
attraction and thermal displacement are in equilibrium at the metal/electro-
lyte phase boundary. However, the discrepancy between the theoretical and
observed capacitance values was high. Stern [106] combined both models;
he retained the Helmholtz double layer adjacent to the metal surface and
added a diffuse space-charge zone in the electrolyte.

The experimental investigation of the structure of the double layer is
based on sophisticated techniques for the measurement of electrical capacitance and surface potentials. The interpretation of the experimental data
is difficult and requires the application of the Gibbs absorption equation for
surfaces. Graham [107] has contributed much to this field.

3.4.1.1. Kinetics of the Transfer Reaction. In the discussion of the
electrical double layer, the question of how electrically charged species
pass through the phase boundary was not answered.

The effect of the abrupt potential change at the phase boundary on the
rate of transfer of charged species was first calculated by Butler [108].
Subsequently, other authors [109-113] developed analagous methods.
Using an amalgam electrode for illustration, charged-species transfer
means the passage of an atom of the metal from its state of combination in
the amalgam to the state of a solvated ion in the electrolyte whereby n electrons are released and retained in the metallic phase (or the reverse):

$$Me_{(Hg)} \underset{i^-}{\overset{i^+}{\rightleftarrows}} Me^{n+} \cdot aq. + n\, e^-_{(Hg)}$$

An energy barrier must be overcome; consequently, a certain activation
energy is required for the transfer to take place in either direction. The

ion passes through an intermediate condition in which it is no longer held in the metallic lattice but has not yet acquired the full hydration energy. In addition to this purely chemical exchange, there is the effect of the electric field at the phase boundary (Fig. 50). This field accelerates the electrochemical reaction in one direction and inhibits it in the other. The total difference in electrical energy between the initial and final states ΔE_{el} is found from $nF(\Delta\varphi)q$ (q is the factor converting electrical units to calories). The transfer coefficient α is the fraction of the total energy accelerating the reaction in the given direction; the remainder $(1 - \alpha)$ is available to accelerate the reverse reaction. It is obvious that the value of the transfer coefficient must lie between 0 and 1.

To derive the kinetic expressions, it is only necessary to insert the activation energies, suitably modified for the effect of the electric potential, in the usual expressions for reaction speed (Arrhenius equation). From Fig. 50, the activation energies are shown as

$$A^+ = E_A^{\ +} - \alpha nF \, \Delta\varphi \ q$$

and

$$A^- = E_A^{\ -} + (1 - \alpha)nF \, \Delta\varphi \ q$$

Expressing the reaction speed in electrical units as current density i, one finds

$$i^+ = nFK^+ e_{(Me)} \exp - \frac{E_A^{\ +} - \alpha nF \, \Delta\varphi \ q}{RT}$$

$$i^- = nFK^- e_{(Me^{n+})} \exp - \frac{E_A^{\ -} + (1 - \alpha)nF \, \Delta\varphi \ q}{RT}$$

Here K^+ and K^- are reaction rate constants, $e_{(Me)}$ and $e_{(Me^{n+})}$ are concentrations.

Figure 51 shows the magnitude of the anodic and cathodic components of the current in relation to the potential difference $\Delta\varphi$ (assuming that $\alpha = 0.5$). The total external current is the difference

$$i = i^+ - i^-$$

There is one special value of $\Delta\varphi$ at which the component currents are equal in magnitude. This is the equilibrium potential for the reaction.

$$i_0 = i^+ = i^- \quad \text{for} \quad i = 0 \quad \text{and} \quad \Delta\varphi = \Delta\varphi_0$$

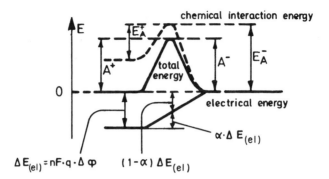

FIG. 50. The effect of a potential difference on the energy associated with the transfer reaction (arbitrary units).

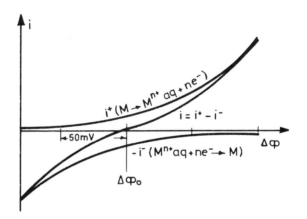

FIG. 51. Current density i versus electrode potential φ when only the transfer reaction overpotential is operative.

Rearranging the equation and inserting $i^+ = i^-$ and $\Delta\varphi_0 = \Delta\varphi$, the Nernst equation for the relation between equilibrium potential and concentration is obtained:

$$\Delta\varphi_0 = \Delta\varphi^0 + \frac{RT}{nF} \ln \frac{e_{(Me^{n+})}}{e_{(Me)}}$$

where

$$\Delta\varphi^0 = \frac{E_A^+ - E_A^-}{nF} + \frac{RT}{nF} \ln \frac{K^-}{K^+}$$

Since Galvani potentials φ are not directly accessible to measurement, it is impossible to calculate absolute reaction-rate constants. It is customary to take the equilibrium condition as the reference point and to consider the effects of potential changes that are directly observable. The difference between the actual and the equilibrium potentials is called overvoltage η and is defined as

$$\eta = \Delta\varphi - \Delta\varphi_0$$

The relation between the transfer current and the overvoltage may be formulated as follows:

$$i = i_0 \left[\exp(\frac{\alpha nF}{RT} \eta) - \exp(-\frac{(1 - \alpha)nF}{RT} \eta) \right]$$

For sufficiently large absolute values of the overvoltage, one of the component currents becomes negligibly small. For instance, if $\eta = 4RT/nF$,

$$i = i_0 e^{4\alpha} \left[1 - e^{-4} \right] \text{ so that } i = i_0 \exp(\frac{\alpha nF}{RT} \eta)$$

and if $\eta = -4RT/nF$,

$$i = i_0 e^{4(\alpha - 1)} \left[e^{-4} - 1 \right] \text{so that } i = -i_0 \exp(-\frac{(1 - \alpha)nF}{RT} \eta)$$

Both of these approximate equations may be expressed in logarithmic form as

$$\ln |i| = a + b\eta \quad \text{if} \quad |\eta| > RT/nF$$

This is an expression known as the Tafel equation.

For very small values of overvoltage, there is a linear relationship between i and η (Ohm's law):

$$i = i_0 nF/RT \, \eta \quad \text{if} \quad |\eta| \ll RT/nF$$

As soon as current flows through the phase boundary, concentration changes arise in both phases. The transfer reaction is therefore always accompanied by mass-transport processes (diffusion, convection, electrical transport).

Optical interference methods have been employed by Ibl et al. [114, 115] to demonstrate the concentration profile. One of their photographs is reproduced in Fig. 52 and shows the concentration profile in the electrolyte immediately next to the electrode. The photograph was made during investigations of boundary layers with natural convection, that is, convection caused only by density differences arising from the passage of current. The result is

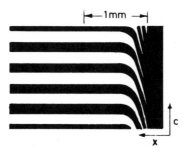

FIG. 52. Interference fringes in front of a Cu electrode in N/10 CuSO$_4$ solution during cathodic current flow and with natural convection (from a photograph taken by N. Ibl with a Jamin interferometer). The fringes show the concentration profile directly.

in good agreement with Nernst's postulate of a diffusion layer adhering to the electrode; the concentration remains constant in the bulk of the electrolyte. The interference bands in the photograph show a rapid decrease in the concentration in the proximity of the electrode.

Frequently, the rate of the total reaction is determined by chemical processes that precede or follow the actual transfer reaction.

A discussion of the experimental techniques follows.

The parameters that can be directly measured during an electrode reaction are the electrode potential and the current density. Every investigation of the kinetics of an electrode reaction is basically an analysis of polarization phenomena.

The classical experimental method consists of the measurement of stationary current density versus potential curves; observations are made only after the establishment of a certain degree of constancy in the electrode system. This condition is determined by the slowest step in the reaction chain. Consequently, it is difficult and sometimes impossible to demonstrate fast reactions. This has led to the development of methods for the investigation of nonstationary processes, whereby it is preferable to keep one of the parameters constant. These methods are especially useful because the concentration polarization (which is difficult to determine accurately) need not be calculated, but may be estimated by an extrapolation method.

An accurate knowledge of the transfer reaction is essential to the understanding of the kinetics of the overall reaction. The determination of the
rate-controlling parameters (that is, exchange current density i_0 and the
transfer coefficient) is of paramount importance. All the methods adopted
for this purpose depend on the division of the total overvoltage into transfer
overvoltage and concentration overvoltage,

$$\eta = \eta_D + \eta_K$$

where η_D is the transfer overvoltage and η_K is the concentration overvoltage. The latter is defined as the difference between the initial equilibrium
potential and the equilibrium potential typical of the actual concentration
conditions obtained at the phase boundary. Thus η_K does not contribute
to the current flow through the system. The driving force for this current
is η_D.

Consider, for example, galvanostatic or potentiostatic experimental
methods. A predetermined definite current is forced through the phase
boundary or a definite overvoltage is imposed on the electrode. The current
flow induces concentration changes; initially, these are negligible, so that
only transfer polarization is operative. Ideally, at $t = 0$, there can be no
concentration polarization. In practice, the changes in potential or current
during the first moments of the experiment are complicated by the process
of charging the double layer. However, as long as the transfer reaction is
not too rapid, it is possible to estimate the initial overvoltage or current
strength by extrapolation of the observed values to $t = 0$.

Similar considerations apply to the passage of alternating current. The
concentration changes induced are small, since the current direction is continually reversed. At infinite frequency, there can be no concentration polarization. This may also be verified by extrapolation methods.

Chemical reactions preceding the transfer reaction are investigated by
study of the limiting current under stationary conditions. The value of the
limiting current is usually a function of the rate mass-transport processes.
When a further reaction precedes the actual transport process (for example,
the breakdown of a complex ion), the limiting current may be less than
would be expected from the rate of mass transport alone. In extreme cases,
a limiting current may be observed without any appreciable decrease in the

amount of the substances that determine the overall reaction. This is termed a reaction limiting current and is characterized by the fact that it is independent of convection (stirring rate, and so on).

A still more powerful method for the investigation of fast preceding reactions is based on the measurement of potential as a function of time at constant current. In principle, this method also is a measurement of reaction limiting currents. Gierst [116, 117] and Delahay [118] have been prominent in the development of this technique.

In the absence of any complicating factors, the potential of an electrode with no external current flowing U_0 is identical to the equilibrium or reversible potential U_{rev}. When current flows through the phase, the electrode potential U_i differs from U_{rev} by an amount η. Therefore, the overvoltage $\eta = U_i - U_{rev}$.

Should several reactions proceed at an electrode (even without the passage of current), the potential U_0 observed is not identical to U_{rev} and is termed a mixed potential U_M. As in the previous case, the passage of current through the system induces a potential change η_P termed the polarization:

$$\eta_P = U_i - U_M$$

Both overvoltage and polarization are functions of the current density and are caused by retardation of the overall electrode reaction. Bonhoeffer, Gerischer, and also Vetter [119] differentiate between the general concept of polarization and the more precisely defined overvoltage. The transfer overvoltage is caused by the transport of charged species through the double layer at the phase boundary. The rate of this reaction is dependent on the potential difference across the double layer. If a chemical reaction is inhibited, current flow induces reaction overvoltage. The chemical reaction may be homogeneous and in the electrolyte phase or heterogeneous at the surface of the solid phase, but in any case its rate constant is independent of the electrode potential. Further, if mass transport to or from the site of the reaction is slow, diffusion overvoltage is induced. Finally, a retardation of the assimilation or release of atoms from a crystal lattice produces crystallization overvoltage.

Other terms frequently found in the literature, such as activation overvoltage and resistance overvoltage, are relegated on theoretical grounds to
the general concept of polarization by the above mentioned investigations.
For instance, Agar and Bowden [120] define activation overvoltage as that
remaining after subtraction of diffusion and resistance overvoltage from the
total. However, this activation overvoltage cannot be interpreted uniformly,
since it consists of elements arising from different causes, each having
different characteristics. A voltage drop across a resistance in the electrode system cannot be regarded as overvoltage, since it does not influence
the electrode reaction in any way. However, it can be a source of error in
U_i from which the true overvoltage is determined ($\eta = U_i - U_{rev}$).

3.4.1.2. Experimental Investigation of Cell Polarization. A detailed
study of the polarization of the complete cell was undertaken by Huber and
Bauer [121]. Pasted D-size cells were discharged through a wide range
of currents; the open-circuit voltage was measured after prolonged recovery
and at various stages of discharge. The cells contained either γ- or β-MnO_2.
The internal resistance, pH, and zinc polarization were determined. Figures 53 and 54 show the voltage after recovery and after an equilibrium
condition had been established. The γ-MnO_2 cells yielded experimental
data in agreement with the theories of Vetter and Neumann, and typical for
a homogeneous phase reaction. Between $MnO_{1.7}$ and $MnO_{1.6}$, the curve
splits into two branches. The earlier voltage step, which has also been
reported by other investigators [52], is presumably due to the formation
of a heterogeneous phase. At very light current drains, the upper branch
of the curve more closely resembles potential measurements on chemically
reduced manganese oxides. The curve for the cells with β-MnO_2 has a
form implying reduction in heterogeneous phase. The first plateau in the
curve can be explained, according to Bode [50], by the presence of water
in the β-MnO_2 crystal lattice.

Figure 55 illustrates the factors contributing to the polarization of the
cell. If the empirically determined discharge curve is corrected for zinc
polarization, ohmic resistance, and pH change, a potential curve is obtained in agreement with the findings of Keller [61] and Coleman [62].
This corrected curve can also be calculated using the diffusion equation
reported by Scott [63]. A comparison of the calculated curve with experi-

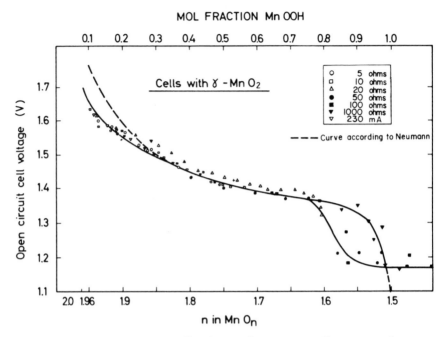

FIG. 53. Open-circuit cell voltage after recuperation versus degree of oxidation (or mole fraction of MnOOH) for γ-MnO$_2$.

mental data shows that the discrepancy between theory and experiment is reduced when the activities estimated by Neumann [60] are used instead of concentrations (see Fig. 56). The remaining discrepancies are attributed to the relatively inaccurate analytical data.

The average total overvoltage is defined as the difference between the areas under the U_{rev} and the actual discharge curves divided by the area under the complete U_{rev} curve (Fig. 57a). Thus the average overvoltage is found from

$$\overline{\eta} = \frac{cm^2 \, U_i - cm^2 \, U_{rev}}{cm^2 \, U_{rev}}$$

where U_{rev} is approximately the open-circuit voltage (o.c.v.) and U_i is the closed-circuit voltage (c.c.v.). If the average total overvoltage is

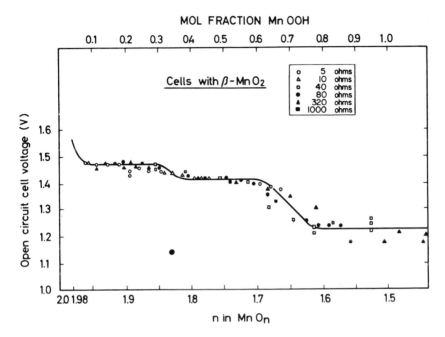

FIG. 54. Open-circuit cell voltage after recuperation versus degree of oxidation (or mole fraction of MnOOH) for β-MnO$_2$.

plotted against the logarithm of the current, a linear relation is obtained similar to the Tafel expression (Fig. 57b).

The results of the investigation indicate that diffusion overvoltage is predominant in the Leclanché cell. Any measures taken to facilitate the diffusion processes result in improvement of the cell characteristics (for example, the use of active manganese dioxides with lattice defects, increased porosity of the cathode mix, increased surface area of the electrodes, well-buffered electrolyte solutions of high ionic conductivity, the use of separators with a minimum retardation of ionic diffusion, and so on).

3.4.2. Influence of Geometry on Cell Performance

The capacity of a dry cell is normally stated in hours of service life or ampere hours and only rarely in watt hours. In view of the variable cell

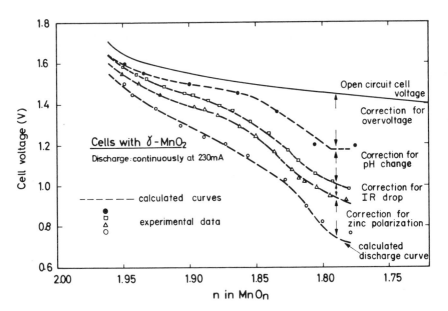

FIG. 55. Derivation of the discharge curve from the open-circuit voltage curve by subtraction of the various forms of polarization (γ-MnO$_2$).

polarization, a stated value for the capacity is useful only when the conditions of discharge and the final cut-off voltage are defined. The battery manufacturer, the general user, and particularly the designer of battery-powered equipment are all interested in an accurate and rapid method of determining the capacity of cells obtained under different discharge conditions. Lawson [122] found that fundamental relationships exist between the obtainable capacity and a series of cell parameters (for example, electrode dimensions) and discharge parameters (current drain, discharge sequence, and so on). However, he did not define the dependence quantitatively.

Huber [121] showed that the capacity in ampere hours obtained from dry cells on continuous discharge through a constant resistance R_a or at constant current I to a cut-off voltage of 0.75 V can be adequately represented by the expression

$$y = k(1-e^{-x})$$

This equation fulfills two conditions: (a) it passes through the origin of the

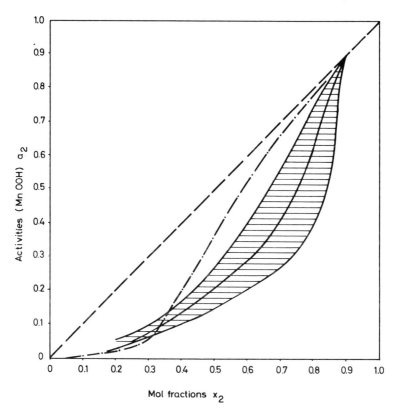

FIG. 56. Activity of MnOOH versus mole fraction. Dot-dash curve
according to Neumann. Shaded area shows the variation of experimental
results (Bauer and Huber).

coordinate system and (b) it tends toward a maximum constant value. These
conditions confirm experimentally determined facts. Under short-circuit
conditions ($R_a = 0$), no useful capacity is obtained. Under the most favorable conditions, the cell cannot deliver more capacity than that predicted
from Faraday's law and the amount of active electrode material. Substituting
the observed capacity for y and the theoretical capacity (grams $MnO_2/3.24$)
for k, the expression becomes

$$Ah = Ah_{th}(1 - e^{-x}) \quad \text{and} \quad \ln \frac{Ah_{th} - Ah}{Ah_{th}} = -x$$

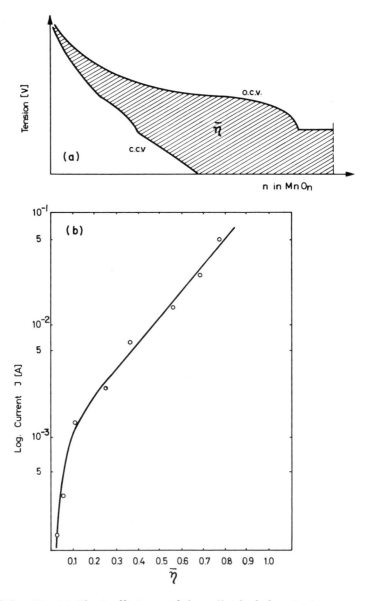

FIG. 57. (a) The inefficiency of the cell (shaded region) is a measure of the polarization $\bar{\eta}$ and may be expressed as a fraction of the total area under the o.c.v. curve. (b) Logarithm of the discharge current I versus fractional inefficiency on polarization $\bar{\eta}$ as defined in (a).

The exponential function is apparently a measure of the inefficiency of the cell. The latter increases as x and R_a decrease, so that a first approximation for x would be KR_a^n. From a study of the experimental data available, Huber showed that n = 1/2 and the expression becomes

$$Ah = Ah_{th}\left[1 - \exp(-KR_a^{1/2}) \right]$$

This is the basic relationship for the capacity calculation shown later in Table 8 (the first equation). The exponent $-KR_a^{1/2}$ can also be written $-K/I^{1/2}$. K is a constant characterizing the physical and chemical properties of the electrodes, R_a or $1/I$ represents the polarizing effect of the current, and the index 1/2 indicates that diffusion processes are major components of the total polarization (see Section 3.4.1.).

Figure 58a shows capacity curves for D-size pasted cells with a very active (curve 1) and a medium (curve 2) cathode mix. Figure 58b illustrates the relationship between the constant K and the discharge resistance R for an active and a medium cathode mix. The capacity values for active mixes are higher and increase more rapidly than the corresponding values for inactive mixes. This is expressed in the formula derived above in greater values for K in active mixes.

From Fig. 58c it will be seen that for extended very light current drain discharges, not only is the maximum limiting output not attained but the capacity curve passes through a maximum and then decreases more or less rapidly according to the quality of the manufacture. Figure 58c shows the capacity curves for cells with excellent, average and poor shelf-life characteristics.

There is a relationship between the K values obtained for cells of different size (but with the same cathode mix and of similar design) reflecting the effect of the cell geometry on performance. In the original paper [123], the geometrical factor was given as $4\pi h$ and interpreted as the ratio of the surface area to the volume of the cathode. This is unsatisfactory, since it leads to false conclusions: for example, that cells with cathodes of the same height but of different diameters would give the same efficiency at the same load (R_a constant). Further study has shown that the factor $4\pi h$ can be replaced with advantage by the expression $2\pi h r_p/(r_p - r_k)$,

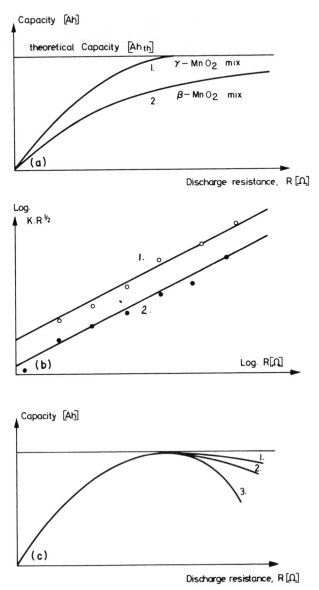

FIG. 58. (a) Capacity versus discharge resistance R. Curve 1 is for
a highly active mix (γ-MnO_2) and 2 is for a less active mix (β-MnO_2).
(b) Log $KR^{1/2}$ versus log R. Active mix 1, inactive mix 2. K is larger
for the active mix. (c) Capacity versus discharge resistance R. Curves 1,
2, and 3 show a progressive fall-off with decreasing shelf stability.

where r_p is the radius of the cathode, r_k the radius of the carbon rod, and h the height of the cathode.

In layer-built cells, $2\pi hr$ is replaced by l x w (length times width) and $r_p - r_k$ by the thickness of the cathode tablet.

The capacity equation then takes the form

$$Ah = Ah_{th} \left\{ 1 - \exp[-K_1 (\frac{hr_p}{r_p - r_k} Ra)^{1/2}] \right\}$$

The factor $(2\pi)^{1/2}$ has been included in the constant K_1. K_1 has the dimensions $ohm^{1/2} cm^{1/2}$ and is the square root of a resistance.

The fraction $hr_p/(r_p - r_k)$ increases when the numerator increases or the denominator decreases. The effect of the geometry is most marked at heavy current drains (R_a small) when an increase in the electrode area (hr_p) or a decrease in the thickness ($r_p - r_k$) improves the cell efficiency. At light current drains, the capacity is primarily determined by the availability of active electrode material.

Figure 59 shows that in double logarithmic coordinates, a plot of K against F (the geometric factor) gives a straight line with the slope $1/2$, in accordance with the relationship $K = K_1 F^{1/2}$. Figure 60 illustrates the capacity of various sizes of dry cells containing natural ore.

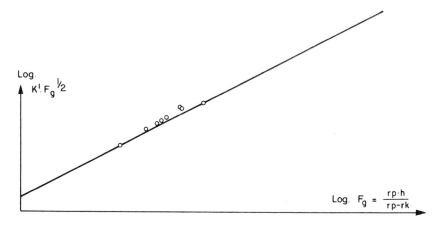

FIG. 59. Log K versus logarithm of geometrical factor F_g [$= hr_p/(r_p - r_k)$]; $K = K^I F_g 1/2$

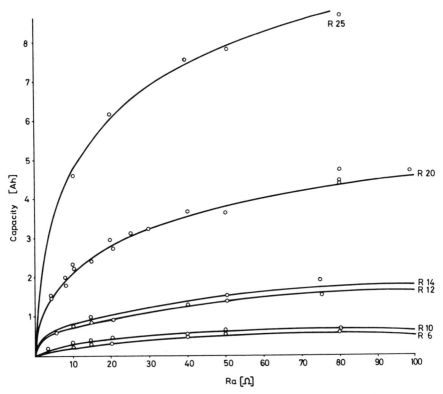

FIG. 60. Capacities of cells of different dimensions but with the same cathode mix versus discharge resistance. The solid curves are calculated using the K value from Fig. 59. Circles represent experimental data.

In summary, the basic principles to be deduced from the capacity equation are as follows.

a. There is a nonlinear relationship between the capacity of a dry cell and the number of equivalents of MnO_2 contained in the cathode. (The anode is usually the cell container; as such, is overdimensioned, and may therefore be omitted from the discussion). Similarly, there is a nonlinear dependence on the value of the load resistance (or the reciprocal of the current strength).

b. The capacity of dry cells with the same chemical properties, containing cathodes with the same diameter but of different height, is directly proportional to the MnO_2 content, provided the discharges are carried out at an equivalent current density. A nonlinear relationship exists for discharges through fixed resistances.

Example (Table 23): The capacity of a dry cell, size R 20 (D-size) discharged through 5 Ω is 1.6 Ah and through 10 Ω 2.13 Ah. The capacity of a similar cell of twice the size, also discharged through 5 Ω, is 4.26 Ah. Thus the output of the larger cell is more than twice that of the R 20 when both are discharged through the same constant resistance, but is exactly double when both cells are discharged at the same current density.

c. Cells having the same geometrical factor F_g but of different sizes have capacities directly proportional to their MnO_2 content.

Example (Table 23): Cells R 15 and R 22, which have approximately the same F_g have proportional ampere-hour capacities.

d. The capacity curves for cells with active cathode mixes have greater slopes than those for inactive mixes. In the capacity equation, K_1 for active mixes is greater than K_1 for inactive ones. (See Fig. 58a, b.)

In the majority of cases, dry cells are used sporadically, and it would be of value to be able to predict the relationship between continuous and intermittent use. A quantitative expression of such a relationship is difficult because secondary reactions occuring in the cell between discharge periods also affect the available capacity. Apart from the generation of hydrogen by the normal corrosion of zinc by the electrolyte, impurities in the cathode mix, inaccuracies in manufacture (defective sealing, off-center cathodes, and so on) may result in loss of capacity on intermittent discharge. In a well-controlled production process, however, there is usually sufficient statistical information available for each type of cell design (cardboard or steel jacket, pasted or paper-lined cans) to enable an empirical relationship between continuous and intermittent discharge capacity to be established.

In accordance with the capacity equation stated before (Fig. 58A), the corresponding curve approaches a constant maximum value assymptotically as the current drain is reduced. Empirical data show, however, that at light-discharge loads, the ampere-hours actually obtained are less than the calculated values and in fact pass through a maximum (see Fig. 58c). The equation, therefore, requires a correcting term such that its value is neg-

TABLE 23

Relationship Between Cell Size and Capacity of Leclanché Cells[a]

	Type							
	R 6	R 10	R 12	R 14	R 15	R 20	R 22	R 25
$F_g = \dfrac{r_p h}{r_p - r_k}$	6.166	3.59	6.53	5.02	7.97	5.82	7.87	10.17
K experimentally determined	0.122	0.092	0.124	0.109	0.139	0.117	0.140	0.156
$K/F_g^{1/2} = K'$	0.0484	0.428	0.0485	0.0486	0.0492	0.0484	0.0499	0.0457
Discharge load	Capacity (Ah)							
(Ω)	R 6	R 10	R 12	R 14	R 15	R 20	R 22	R 25
5	0.180	0.270	0.560	0.550	1.250	1.600	2.420	3.500
10	0.242	0.315	0.750	0.800	1.620	2.123	3.230	4.650
20	0.300	0.490	1.030	1.050	2.110	2.863	4.205	5.915
40	0.428	0.540	1.310	1.360	2.690	3.510	5.340	7.405
80	0.530	0.620	1.600	1.650	3.370	4.410	6.140	8.90
100	6.70	0.720	1.65	1.810	3.500	4.750	6.300	9.280

[a]Pasted cells with β-MnO$_2$ mix.

ligible at heavy currents but increases as the current drain decreases.
Since the reduction in capacity is dependent on the total time on test and on
the design of the cell, both factors must be included in the correction term.
In the expression

$$\Delta Ah = Ah_{th} \exp(-K^{II}R^{-n})$$

where ΔAh is the difference between the calculated and the observed capacity,
n and K^{II} represent the effect of the cell design, and R is the load resistance.
The equation shows that ΔAh is small at heavy loads (R small) and increases
as the current drain decreases (R increasing).

Since the capacity loss is a function of many parameters that cannot be
precisely determined, ΔAh (and therefore K^{II} and n) must be obtained from
statistically significant empirical data.

Dry cells are normally used intermittently, and it would obviously be
useful to have some means of calculating the capacity obtained on intermittent discharge from the continuous performance figures.

Part A of Table 24 contains experimentally determined capacity values
for a specific cell type discharged through different resistances, continuously
and intermittently. From the tabulated results and Fig. 61, the following
can be seen.

a. The curve drawn from the ampere-hour values in column 2 of Table
24A does not approach the theoretical limit (in this case 7.325 Ah) assymptotically, but after reaching a maximum falls off to successively lower values.
The position of the maximum and the degree of fall-off are dependent on
quality parameters of the cell that must be evaluated empirically.

b. On intermittent discharge through the same resistance, the capacity
obtained at first increases as the daily discharge period decreases, passes
through a maximum, and again decreases.

c. As the discharge current decreases, so does the gain in capacity on
intermittent discharge. In fact, a point is reached (about 80 Ω in the table)
when the continuous discharge gives a higher capacity than any of the intermittent discharges.

An explanation of (b) and (c) is to be found in the fact that during the
rest periods of intermittent discharge, the polarization of the cell is partly

TABLE 24

Capacity of D-Size Leclanché Cells

Equation 1: $Ah = Ah_{th} \{ 1 - exp[-K^I R^{1/2}] \}$

Equation 2: $Ah = Ah_{th} \{ 1 - exp[-K^I R^{1/2}] - exp[-K^{II} R^{-n}] \}$

Equation 3: $Ah = Ah_{th} \{ 1 - exp[-K^I (R/C)^{1/2}] - exp[-K^{II}(1/C)^{-1/2}(R/C)^{-n}] \}$

$\Delta Ah = Ah$ (calculated) $-$ Ah (experimental)

$Ah_{th} = 7.325$

$K^I = 0.2$ $R =$ discharge resistance

$K^{II} = 12$ $C = (h/day)/24$

$n = 0.2$

A. Experimental Capacity Data to 0.75 V (Ah)

R (Ω)	24 h/day	8 h/day	4 h/day	2 h/day	1/2 h/day
1.25	1.28	–	–	–	–
2.5	1.87	–	–	–	–
5	2.57	–	3.65	4.25	4.60
10	3.36	4.30	4.90	5.10	4.25
20	4.33	5.65	5.65	5.25	4.30
40	5.75	5.90	6.00	5.00	3.65
80	6.20	6.05	5.65	4.60	3.00
160	7.05	6.30	5.55	4.25	–
320	6.86	6.20	–	–	–
640	7.02	5.50	–	–	–
1000	6.60	–	–	–	–
5000	6.76	–	–	–	–

TABLE 24 (Continued)

B. Calculated Capacity on Continuous Discharge

R (Ω)	Equation 1		Equation 2	
	Ah	ΔAh	Ah	ΔAh
1.25	1.46	+0.18	1.46	+0.18
2.5	1.98	+0.11	1.98	+0.11
5	2.65	+0.08	2.65	+0.08
10	3.44	+0.08	3.43	+0.07
20	4.34	+0.01	4.33	0
40	5.30	-0.45	5.27	-0.48
80	6.11	-0.09	6.07	-0.13
160	6.75	-0.30	6.67	-0.38
320	7.13	+0.27	7.00	+0.14
640	7.27	+0.25	7.08	+0.06
1000	7.30	+0.70	7.05	+0.45
5000	7.325	+0.565	6.74	-0.02

C. Calculated Capacity on Intermittent Discharge (Equation 3)

R (Ω)	8 h/day		4 h/day		2 h/day		1/2 h/day	
	Ah	ΔAh	Ah	ΔAh	Ah	ΔAh	Ah	ΔAh
5	–	–	4.30	+0.65	4.40	+0.15	4.50	-0.10
10	4.70	+0.40	5.10	+0.20	4.85	-0.25	4.30	+0.05
20	5.55	-0.10	5.60	-0.05	5.00	-0.25	4.00	-0.30
40	6.10	+0.20	5.90	-0.10	4.90	-0.10	3.57	-0.08
80	6.45	+0.40	5.82	+0.17	4.68	+0.08	3.10	+0.10
160	6.60	+0.30	5.55	0	4.32	+0.07	–	–
320	6.31	+0.11	–	–	–	–	–	–
640	6.05	+0.55	–	–	–	–	–	–

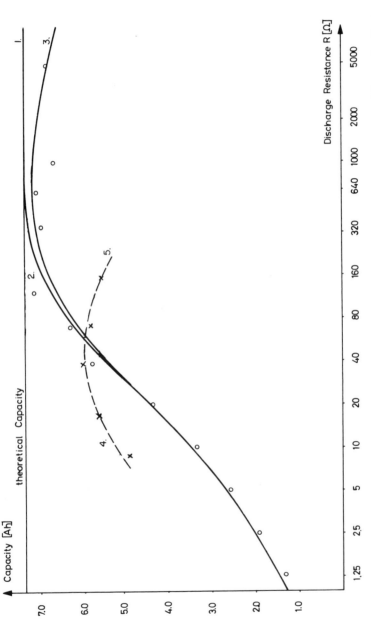

FIG. 61. Capacity of D-size cell vs log of discharge resistance R. Curve 1 is calculated from the first equation; 2 (circles) shows experimental data for continuous discharge; 3 is calculated from the second or third equation with $1/C = 6$ (4 h/day discharges); 5 (crosses) shows the experimental data for 4 h/day discharges at the indicated loads. (See Table 24 for equations and data.)

compensated by diffusion of the reaction products formed at the cathode surface into the interior of the solid dioxide, and by equalization of the concentration gradients in the electrolyte. These processes result in an increased average on-load voltage. In heavy-drain intermittent discharges, the recuperation effect is much more pronounced than on light drains because of the heavier electrode polarization. In fact, at very light drains the polarization is so small that the beneficial effect of recuperation is obscured by the loss of capacity due to storage factors.

In order to correct for shelf losses and the polarization and recuperation factors discussed, the last equation (which is a first approximation valid for continuous discharge only) may be extended as follows (to produce the second equation in Table 24).

$$Ah = Ah_{th} \{1 - \exp[-K^I(F_g R)^{1/2}] - \exp[-K^{II}R^{-n}]\}$$

In this expression, which is still valid only for continuous discharge, the second exponential term defines the fall-off in capacity (ΔAh) on very light prolonged discharges. For heavy discharges (say up to 20 Ω on D cells), ΔAh is negligible and K^I can be found from the expression

$$\ln\left[\frac{Ah_{th} - Ah}{Ah_{th}}\right] = -K^I(F_g R)^{1/2}$$

Note that if the geometrical factor F_g is not accurately known, it may be omitted from the equation and the value of K^I obtained will be automatically adjusted to include its effect.

Having now the value of K^I, Ah may be calculated for large values of R, still neglecting the last term in the equation. As has been said previously, the calculated value will exceed the experimentally observed figure by an amount ΔAh. From the values of ΔAh for various values of R, K^{II} and n may be obtained from the expression

$$\Delta Ah = Ah_{th} \exp[-K^{II}R^{-n}]$$

The final step is now to modify the calculation to take into account the cell performance on intermittent discharge. (Experimental values are given in Fig. 62.) It would appear that the effective load on intermittent discharge could be expressed as the product of the actual resistance R and the reciprocal of the fraction of the day during which the battery is discharg-

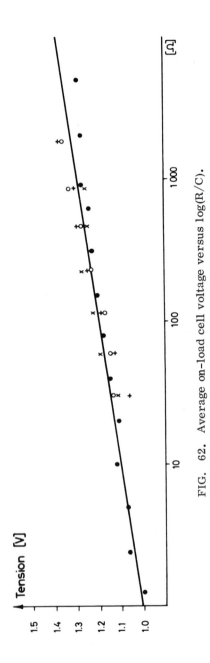

FIG. 62. Average on-load cell voltage versus log(R/C).

ing. Thus, for a 4-h/day discharge through 40 Ω , the effective resistance R_e is found to be

$$R_e = R \frac{1}{C} = 240 \ \Omega$$

where R = 40 Ω and C = 4/24. This relation satisfies the condition that R_e = R for continuous discharges. However, when R_e is substituted for R, the calculated capacities are always greater than the observed values. This means in effect that the "constant" K^{II} assumes a different value for each discharge frequency. The difference between theory and experiment, ΔAh_2, may be expressed in the form

$$\Delta Ah_2 = Ah_{th} \exp[-"K^{II}"R_e^{-z}]$$

or

$$\Delta Ah_2/Ah_{th} = \exp[-p]$$

In Fig. 63, p has been plotted against R_e in double logarithmic coordinates for various discharge frequencies. The straight lines obtained are reasonably parallel, so that the index z is essentially the same as n in the previous capacity. The value of the "constant" K^{II} can be read at R_e = 1.

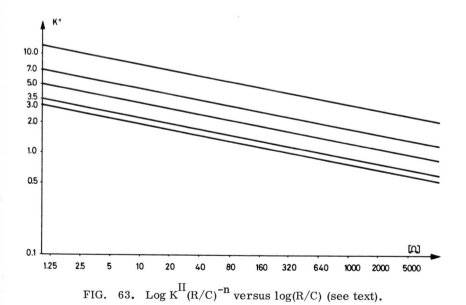

FIG. 63. Log $K^{II}(R/C)^{-n}$ versus log(R/C) (see text).

In Fig. 64, $\log K^{II}$ has been plotted against $\log(1/C)$ and the result shows that the constant at any discharge frequency above 2 h/day, K_f^{II}, may be expressed as

$$K_f^{II} = K^{II}(1/C)^{-1/2}$$

where K^{II} is the value for continuous discharges. For discharge periods less than 2 h/day, the value of K_f^{II} deviates from this relationship, and it is to be expected that its value does not depend only on the discharge frequency. At very low frequencies, the discharge is little more than open-circuit storage of the cell.

In its final form, the capacity equation becomes the third equation in Table 24:

$$Ah = Ah_{th}\{1 - \exp[-K^{I}(R/C)^{1/2}] - \exp[-K^{II}(1/C)^{-1/2}(R/C)^{-n}]\}$$

where Ah is the capacity in ampere hours to 0.75 V/cell, Ah_{th} is the theoretical ampere hour capacity (Faraday's law), K^{I}, K^{II}, and n are to be empirically determined from continuous discharges, R is the actual load resistance, and 1/C is a discharge-frequency factor.

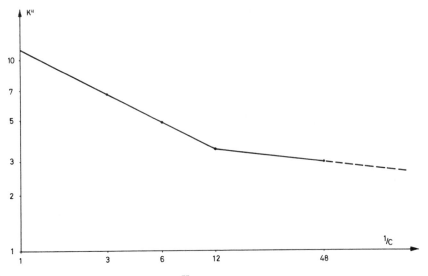

FIG. 64. Log K_f^{II} versus log(1/C) (see text).

In Table 24B, the capacity values calculated from the different equations are compared with experimental data. Figure 61 gives the same information graphically.

Recapitulating, the last equation is an expression that can be used to estimate the ampere hour capacity of Leclanché type dry cells for any discharge frequency greater than 2 h/day and to a cut-off voltage of 0.75 V. Use of the equation permits a rapid estimation of cell quality, since the evaluation of K^I, K^{II}, and n may be achieved with sufficient accuracy from continuous discharges at not more than ten different loads.

Figure 65 presents typical discharge data and shows the agreement with the calculations. Significant departures from the theoretical form of the ampere hour versus resistance curve indicate that the cell is insufficiently

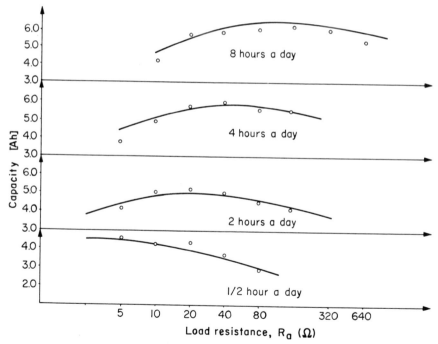

FIG. 65. Capacity of D-size cells versus logarithm of load resistance for various intermittent discharges. The solid curves are calculated from the third equation of Table 24. Circles represent experimental data.

optimized. Deviations are mostly restricted to the heavy-discharge region
and take the form of an elongated S.

The discharge curve for any particular cell may be calculated with the
assistance of the first equation (for the required discharge conditions) whereby the ampere-hour capacity must be approximately determined. From this
curve, the contribution of the four principal polarization factors may be
estimated [119]. Subtraction from the no-load potential curve U_{rev} yields
the on-load voltage curve. A practical example is illustrated later, in
Section 5.

Finally, reference is made to the work of Shephard [124], who has also
published equations for the calculation of potential, ampere-hour capacity,
and watt-hour outputs for all types of galvanic cell.

3.4.3. Influence of Mix Formulation on Cell Performance

The selection of the raw materials and their proper combination into
the cathode mix are factors of primary importance for the electrical output
and quality of dry cells. The shelf life is improved by the use of mix components free from impurities that induce corrosion of the anode; the electrical
performance is dependent not only on the relative amounts of the individual
components but also on the manner in which the mix is prepared. It is a
matter of experience for every dry-cell technologist that the mix composition
must be optimized for the particular application for which the battery is intended. An all-around cell is reminiscent of a decathlon winner, whose average performance in all ten disciplines is high but in the individual branches
only rarely achieves championship level.

In general, at least three varieties of cells are produced: those intended
for light, medium, and heavy current discharge. This implies a marked
differentiation in the mix formulation. For light discharges, a high proportion of MnO_2 is required, and if the end point is low, a high proportion of
NH_4Cl. The water and carbon-black contents are necessarily low; generally,
an inactive natural ore is satisfactory. On the other hand, for high current
discharges, the mix must contain significantly higher proportions of carbon

and water, and the MnO_2 must be highly active to compensate for the reduced amount available. Cells intended for medium current discharge obviously contain mixes that are more or less a compromise between the extremes described above. It is here that mixtures of different types of manganese dioxide ores may be used with advantage.

A study of the relation between mix composition and electrical performance was reported by Bell [123]. The work was based on the assumption that MnO_2 (a), carbon black (b), NH_4Cl (c), $ZnCl_2$ (d), and water (e) are the five main components of a cathode mix.

The 25 mix compositions investigated conformed to the requirements that

$$\text{Components } a + b + c = 100$$
$$d = \quad 0.08a$$
$$e = \quad 1.4b$$

Accordingly, in a triangular coordinate system with components a, b, and c as the variables, each point represents a specific mix composition (Fig. 66).

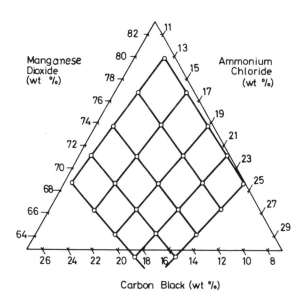

FIG. 66. Composition of experimental mixes.

Both pasted and paper-lined D-size cells were made with each of the mix
formulations and discharged according to the following schedule.

1.25 Ω	10 min/day	(paper-lined cells only)
5.6 Ω	2 h/day	(paper-lined cells only)
100 mA	6 h/day	(pasted cells only)
5 Ω	30 min/day	(pasted cells only)
5 Ω	continuously	(both types)
40 Ω	continuously	(both types)

The electrical performance was expressed as service life (hours) to a
particular end-point. In order that the results should be independent of the
quality of the raw materials actually used, the performance figures were
expressed as percentages of the maximum service life obtained. Finally,
from the results for the individual mixes, lines of equal service life were
calculated and superimposed on the triangular coordinates defining the mix
composition. Four types of diagram were obtained.

Type (a). The performance contours are irregular ellipses. Figures
67 and 68 show these curves, and Table 25 gives the estimated optimum mix
composition for the relevant discharges.

Type (b). The contours are in the form of half-ellipses, open toward
the MnO_2 apex and having vertical major axes (Fig. 69). These curves do
not permit the estimation of the optimum mix composition, but they do demonstrate that for any given MnO_2 content, there are specific NH_4Cl and carbon-black ratios that produce the maximum performance. Further, the
performance increases with increasing MnO_2 content.

Type (c). Here the contours are parallel to the lines of equal carbon-
black concentration. It is not surprising that the flash currents of both types
of cells and the performance on the 5-Ω continuous discharge of paper-lined
cells belong to this class (Fig. 70).

Type (d). The performance contours (Fig. 71) are intermediate between
type (a) and type (c) and were obtained from the 40-Ω continuous test on
paper-lined cells.

The results of this investigation may be summarized as follows.

For each type of discharge and the selected end point, there is a specific

TABLE 25

Estimated Cathode Mix Composition for Maximum Output at Various Discharge Rates

Discharge conditions		Cutoff voltage (V)	Cell construction	Estimated mix composition for maximum output			
(Ω)	(min/day)			Manganese dioxide (wt%)	Carbon black (wt%)	Ammonium chloride (wt%)	
1.25	10	1.0	Paper-lined	66.5	13.5	20	
1.25	10	0.75	Paper-lined	66.5	10.5	23	
5.6	120	1.0	Paper-lined	68.5	11.5	20	
5.6	120	0.75	Paper-lined	68	11.5	20.5	
5	30	1.0	Pasted	68.5	11.5	20	
5	30	0.75	Pasted	74.2	10.5	15.3	
5	Continuous	1.0	Pasted	74.5	14.3	11.2	
5	Continuous	0.75	Pasted	68	13	19	
40	Continuous	1.0	Pasted	76.5	8.5	15	
40	Continuous	0.75	Pasted	69.5	12.7	17.8	

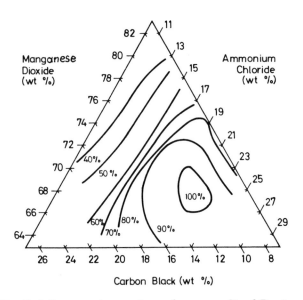

FIG. 67. Relative service contours for paper-lined D-size cells discharged through 5.6 Ω, 2 h/day to 0.75 V as a function of mix composition.

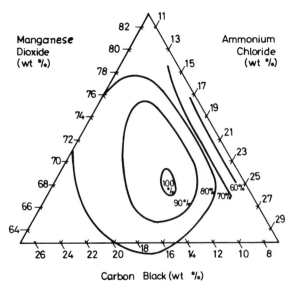

FIG. 68. Relative service contours for pasted D-size cells discharged through 5 Ω continuously to 0.75 V as a function of mix composition.

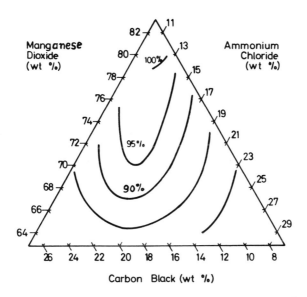

FIG. 69. Relative service contours for pasted D-size cells discharged at 100 mA constant current, 6 h/day to 0.75 V as a function of mix composition.

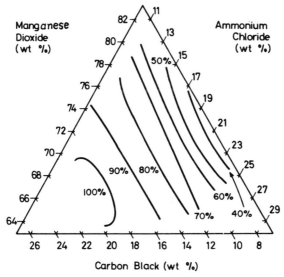

FIG. 70. Relative flash-current contours for paper-lined D-size cells as a function of mix composition.

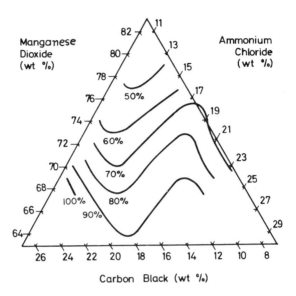

FIG. 71. Relative service contours for paper-lined D-size cells discharged continuously through 40 Ω to 0.75 V as a function of mix composition.

optimum mix composition. Pasted cells appear to be rather more sensitive in this respect than paper-lined cells. In general, mixes for light discharges require a high manganese dioxide content, while mixes for heavy discharge currents demand high water and carbon-black contents and correspondingly less MnO_2.

Since the efficiency of a cathode mix is primarily dependent on the quality of the manganese dioxide, the testing of the latter is of real importance. The former traditional method based on a standard mix formulation is not to be recommended, since it is quite conceivable that an inherently good ore can give a poor performance and be rejected simply because the other mix components are not adjusted to its chemical and physical characteristics.

In order to measure the discharge characteristics of manganese dioxide independently of the mix composition, several test methods have been devised. Cahoon [125], Kornfeil [126], Appelt and Purol [127], Huber and Kändler [71], Schweigart [128], and Bauer and Bell [129] have described particular test procedures. All these methods have the common feature that the dioxide is discharged under conditions that eliminate polarization. If an ore is found to be acceptable according to one test or another, it is still necessary to determine the optimum mix composition, taking into account such characteristics of the ore as particle size, electrolyte absorption power, activity, and so on. One method of determining the optimum composition is illustrated graphically in Fig. 72. The application of mixtures of different types of manganese dioxide ores in cathode mixes is of extreme importance. Such mixtures are justified not only by economic considerations but also technically, since it is often possible by their use to achieve much better adjustment of the properties of the positive electrode to the application in mind than is obtainable with a single variety of ore.

The potential developed by a mixture of ores on reduction is dependent on the relation between the potential and the degree of oxidation of the individual components. In fact, if equilibrium can be established between the phases (which implies that there is no kinetic inhibition), the equilibrium potential of any mixture may be calculated from the properties of the individual oxides concerned.

The mixture of two oxides may be compared to two separate electrodes, each containing only one oxide, joined by an external electronic conductor and immersed in the same electrolyte. Both electrodes will assume the same potential, the equalizing oxidation/reduction current flowing through the electronic conductor. If now the system is partially discharged, the on-load voltage of the composite electrode will fall continuously; both of the subsidiary electrodes will always be at the same potential. Assuming that the current/voltage curves for the two electrodes differ (as will generally be the case), the implication is that the discharge current will be shared unequally and that one oxide will be reduced more than the other. Should the discharge now be interrupted and simultaneously the electronic conductor between the two subsidiary electrodes disconnected, each oxide will assume its own potential, dependent on its degree of oxidation. Again, these potentials will generally be different in value. If the two electrodes are again connected by means of the electronic conductor, an equalizing current will flow until both oxide phases are at the same potential. However, since the

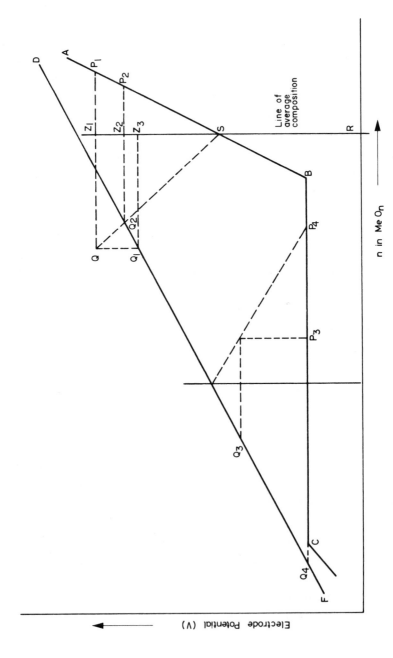

FIG. 72. Electrode potential versus n in MeO$_n$ for two varieties of oxide mixed in the same electrode.

oxygen lost by the one phase is gained by the other, the average composition
of the composite electrode remains unchanged. This latter condition implies
that the two oxides must come to equilibrium at one specific potential.

The method of determining the equilibrium potential is shown in Fig. 72.
ABC and DF represent the potential/composition characteristics of two oxide
systems Q and P. At the end of the discharge period, let the two oxides be
represented by P_1 and Q_1 and let their relative amounts be such that the
average composition is represented by the vertical line RZ_1. As the oxides
change toward equilibrium, Q_1 moves along FD and P_1 along AB. It is easy
to prove that in order to maintain the average composition unchanged, the
points must be such that the ratio of their distances from RZ_1 must be constant and equal to $Q_1Z_3:P_1Z_1$ Then if QQ_1 and P_1Q are parallel to the axes
and QS is drawn cutting DF in Q_2 and RZ_1 in S, Q_2 represents the equilibrium
composition of oxide Q. If Q_2Z_2 is drawn parallel to the axis and meets AB
in P_2, P_2 is the corresponding equilibrium P oxide. The proof is trivial and
is omitted. A second example is shown with the initial oxide Q_3 and P_3 and
the final equilibrium oxides Q_4 and P_4. P_3 and P_4 are representatives of
the heterogeneous phase BC and have equal potentials but varying amounts
of the constituent phases represented by B and C.

The mixing procedure employed in the manufacture of a cathode mix is
of some importance. Obviously, a nonhomogeneous distribution of the individual components will lead to a variation in the battery characteristics.
During the wet mixing process, irreversible changes occur that affect the
consistency of the final mix. Prolonged wet mixing (overmixing) may result
in the mix becoming too wet and unworkable. It is therefore necessary to
adjust the mixing procedure to the mix composition; this is especially true
when a mix with a high proportion of electrolyte is required.

The dry mixing process is usually inefficient, and a mixture of the dry
components is very unstable and tends to separate into the individual constituents if handled in the dry state. It is customary to keep the dry mixing
time short and to restrict the wet mixing time to such a period that the final
mix is workable and overwetting is avoided. It may be advantageous in extreme cases to add the electrolyte (damping solution) in two or more portions.
The design of the mixing equipment is of importance in this respect.

Cathode mixes for cells intended for low current drain discharges are formulated much drier and may be rendered more efficient by prepressing processes that permit a greater weight of mix to be formed into a given volume. Such processes are not applicable to very wet mixes and are in fact somewhat illogical, since a high degree of porosity in the cathode is the ultimate objective.

3.4.4. Application of the Polarization Theory to Leclanché Cells

If the capacity expressed in ampere-hours is plotted in linear Cartesian coordinates against the load resistance, R, curves are obtained that commence in the origin (for R = 0, Ah approximates to 0) and with increasing resistance (decreasing current drain) approach a limiting value. On very low-drain tests, the ampere-hour capacity falls off again because of secondary reactions that reduce to the cell output. As is demonstrated later in this section, such curves can be adequately described by the expression $y = A[1 - \exp(-x)]$. In this expression, the difference between the limiting value A and the value of y for a specific value of x decreases exponentially as x increases. Thus y approaches the limiting value asymptotically.

The introduction of another term $\exp(-z)$ in the above mentioned expression can be used to describe the fall-off in the ampere-hour curve at very light current drains. Finally, the effect of the cell geometry on the degree of polarization is given by the factor $F_g = [hr_p/(r_p - r_k)]^{1/2}$ in the index of the exponential term. The capacity is then expressed by the general equation (from Table 24)

$$Ah = Ah_{th} \{1 - \exp[-K^I(F_g R)^{1/2}] - \exp(-K^{II}R^{-n})\}$$

If only one type of cell is considered, the part of the curve with a positive slope can be expressed by a simplified form of the above equation,

$$Ah = Ah_{th}[1 - \exp(-KR^{1/2})]$$

If $\overline{U}$ represents the average voltage during discharge and $\overline{U}_{rev}$ is the average value of the reversible potential curve, then

$$\overline{U}_{rev} - \overline{U} = \eta \quad \text{and} \quad U_{rev} = \eta_t$$

where η is the polarization of the cell during discharge through the resistance R, and η_t is the total polarization, obtained when R = 0. It follows that

$$\frac{\eta}{\eta_t} = \exp(-KR^{1/2}) = 1 - \frac{Ah}{Ah_{th}} = 1 - \frac{\Delta n}{\Delta n_{max}}$$

and

$$\eta = \eta_t \exp(-KR^{1/2})$$

From this last equation, it is seen that

$$\eta = \eta_t \quad \text{when} \quad R = 0$$

and

$$\eta \to 0 \quad \text{as} \quad R \to \infty$$

The factor $R^{1/2}$ and the form of the final equation (formally an equation for a first-order reaction) give information regarding the nature of the polarization. They indicate that the major component of the polarization of a Leclanché cell is diffusion polarization. In heterogeneous chemical reactions, especially where large areas are involved, the reaction speed is determined by the rate of mass transport from the bulk of the solution through the phase boundary. Applying Fick's first law to this case,

$$\frac{dx}{dt} = \frac{DO}{V\delta} (a - x)$$

is apparently a first-order reaction with $K^I = DO/V\delta$, where D is the coefficient of diffusion, O the area, V the volume of the solution, and δ the thickness of the reaction layer.

Huber and Bauer [121] measured the reversible potential curve and the polarization of commercially available batteries. They showed that diffusion polarization was indeed the major part of the total polarization, and they verified the final equation (see Table 26 and Fig. 55). Considering the relatively large uncertainty in an analytical determination of the individual polarization factors, the agreement of the values of η/η_t and $\exp(-KR^{1/2})$ is satisfactory.

Further, the value of the MnO_2 polarization for various discharge currents and resistances was determined. MnO_2 polarization is considered to be the polarization at the surface of the individual grains of MnO_2 due to the accumulation of lower oxides. The total polarization calculated from the

TABLE 26

Investigation of the Polarization of D-Size Pasted Cells[a]

Discharge resistance (Ω)	$Q = Ah$ to 0.75 V observed	Ah calculated	$x=kR^{1/2}$ observed	x calculated	$1 - \exp(-x)$ observed
2.50	1.96	1.97	0.310	0.315	0.2269
5.00	2.43	2.65	0.400	0.450	0.3300
10.00	3.26	3.40	0.590	0.630	0.4460
20.00	4.70	4.33	1.030	0.900	0.6440
40.00	-	5.27	-	1.280	-
50.00	5.48	5.57	1.390	1.410	0.7510
80.00	-	6.09	-	1.90	-
100.00	6.40	6.31	2.090	2.000	0.8765
240.00	-	6.87	-	3.100	-
500.00	6.90	7.17	2.900	4.500	0.9450
1000.00	7.18	7.28	4.150	6.400	0.9840
ΣU_{rev}	-	-	-	-	-

[a]Experimental data calculated from $Ah = Ah_{th}[1 - \exp(-kR^{1/2})]$.

final equation and the MnO_2 polarization are shown as functions of the discharge resistance in Fig. 73.

The polarization of a Leclanché cell is thus seen to be the sum of the MnO_2 polarization and the factors contributing to polarization in the electrolyte

$$\frac{\eta}{\eta_t} = \exp(-x) = \exp(-x_{MnO_2}) + \exp(-x_{el})$$

Figure 74 shows the relationship between x (or x_{MnO_2}) in $\exp(-x)$ and R, the load resistance. For the exponent of the MnO_2 polarization, one finds $-K^1 R^{1/4}$ [or $-K^1 (R^{1/2})^{1/2}$] so that the polarization equation becomes

$$\eta/\eta_t = \exp(-KR^{1/2})$$
$$= \exp(-K^1 R^{1/4}) + [\exp(-KR^{1/2}) - \exp(-K^1 R^{1/4})]$$

1 - exp(-x) calculated	ΣU (cm^2)	$\Sigma U_r - \Sigma U$ (cm^2)	η/η_t	e^{-x} observed	e^{-x} calculated
0.2702	85.14	277.43	0.765	0.7334	0.7298
0.3592	105.84	256.73	0.709	0.6703	0.6376
0.4674	161.79	200.78	0.554	0.5543	0.5326
0.5934	235.21	127.36	0.352	0.3570	0.4066
0.7220	-	-	-	-	-
0.7559	264.40	98.17	0.2705	0.2490	0.2441
0.8504	-	-	-	-	-
0.8647	317.15	45.42	0.125	0.1237	0.1353
0.9550	-	-	-	-	-
0.9889	341.36	21.21	0.0586	0.0550	0.0111
0.9980	349.80	12.77	0.0352	0.0158	0.003
-	362.57	-	-	-	-

$Ah_{th} = 7.30$ Ah, $k = 0.200$.

that is, the polarization η is the sum of two terms; one describes the polarization in the manganese dioxide, and the other, the polarization in the electrolyte. The polarization in the electrolyte becomes negligibly small at very low current drain (R > 1000 Ω in the R 20 cells investigated). The only remaining source of polarization is then the manganese dioxide. The total polarization approaches zero only slowly at loads greater than 1000 Ω. (See Figs. 73 and 74.)

The polarization of a Leclanché cell can be reduced considerably, therefore, by improving mass transport to or from the site of the reactions. One simple method of achieving this is to increase the electrode area as much as the volume of the cell allows, thus reducing the current density. The sector cell (Fig. 6b) is one form of cell with a large electrode area. In Table 27,

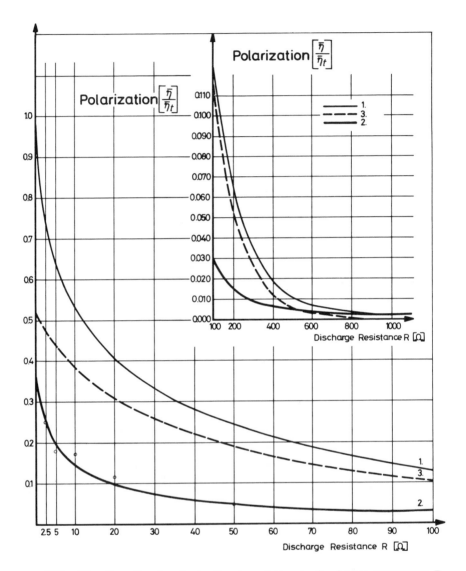

FIG. 73. Fractional polarization in relation to discharge resistance R
(Ω). Curve 1, total polarization; 2, MnO_2 polarization; 3, electrolyte
polarization.

Polarization

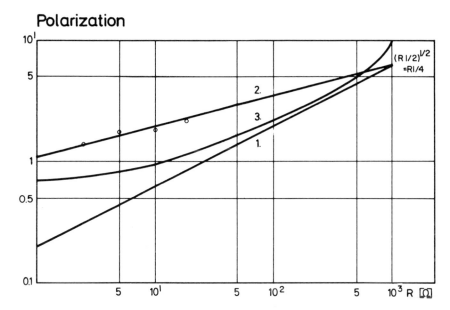

FIG. 74. Log x in exp(-x) versus log R (discharge resistance) for D-size cells. Curve 1, total polarization; 2, MnO_2 polarization; 3, electrolyte polarization.

the capacity of such a cell is compared with that of a standard R 14 cell. Both cells are of equal volume, of paper-lined design, with the same cathode mix formulation. As can be seen from the tabulated data, the capacity of the high-area cell on heavy current discharge is much superior, but at light drain inferior, to that of the standard cell. The latter result is of course a consequence of the reduction of the amount of active materials by reason of the design.

Capacity curves of the type discussed provide a convenient summary for the battery chemist of the performance characteristics of his formulations and cell designs. To compare mixes containing different amounts of manganese dioxide, it is preferable to express the ampere-hours obtained on discharge as Δn in MnO_n. Numerical conversion factors are given in Table 28.

Examination of such capacity curves will show that even on the rising branch, there may be some deviation from the theoretical shape. Such deviations are evidence that the cell is not optimized for the current drains

TABLE 27

Electrical Capacities of a Standard Paper-Lined C-Size Cell
and a Similar Cell with Increased Electrode Area[a]

Discharge resistance (Ω)	Capacity (Ah)		Flash current (A)		Photoflash discharge total (cycles to 0.9 V)	
	Large-area cell	Standard cell	Large-area cell	Standard cell	Large-area cell	Standard cell
1.25	0.580	0.306	20–24	8–10	300	210
2.50	0.707	0.474			$U_{av} = 1.15$ V $U_{av} = 0.98$V	
5.00	0.923	0.719				
10.00	1.230	0.994				
20.00	1.600	1.462				
40.00	1.999	2.052				
80.00	2.172	2.466				

[a]The cathodes of both cell types had the same composition. The discharges were carried out continuously to 0.75 V cut-off.

TABLE 28

Conversion Between Values of Ah and Δn or n in MnO_n

| | | Equivalents | | |
Reaction	Δn	(val)	gm MnO_2/Ah	Ah/gm MnO_2
$MnO_2 \rightarrow MnO$	1.00	2	1.62	0.616
$MnO_2 \rightarrow MnO_{1.33}$	0.67	1.34	2.42	0.413
$MnO_2 \rightarrow MnO_{1.444}$	0.556	1.112	2.92	0.343
$MnO_2 \rightarrow MnO_{1.50}$	0.50	1.00	3.24	0.308

Number of equivalents = number of gram atoms of oxygen used x 2. For example, $MnO_2 \rightarrow MnO_{1.33}$; $\Delta n = 0.67$; number of equivalents = 0.67 x 2 = 1.34.

1 val corresponds to 96,494 Coulombs or 26.8 Ah (= 96,494/3600) for 87 gm MnO_2, 1109 Coulombs or 0.308 Ah/gm MnO_2.

1 val is equivalent to a Δn value of 0.5.

Conversion between Ah and Δn:

$$\Delta n = Ah/(0.616 \text{ x gm } MnO_2) \quad \text{and} \quad Ah = \Delta n \text{ x } (0.616 \text{ x gm } MnO_2)$$

For a given cell type, 0.616 x gm MnO_2 is a constant A, so that

$$\Delta n = Ah/A \quad \text{and} \quad Ah = A \, \Delta n$$

---

involved. In Fig. 75, ampere-hour curves for steel-jacketed D-size cells can be found. In Fig. 75a, the manganese dioxide used was pure electrolytic MnO_2; in Fig. 75b, it was a 2:1 mixture of artificial and natural ores. Curve 1 is the calculated performance, curves 2, 3, and 6 were obtained from paper-lined cells, and curves 4 and 5 apply to pasted cells. A comparison of the individual curves shows that the pasted cells deviate the least from the theoretical curve. The paper-lined cell, represented by curve 6, is based on the electrochemical system $MnO_2/ZnCl_2$ aq./Zn and differs significantly from the Leclanché paper-lined types.

Brouillet and coworkers [93] have studied the S-shaped deviation of the actual curves from theoretical ones. They found that the thinner the paste

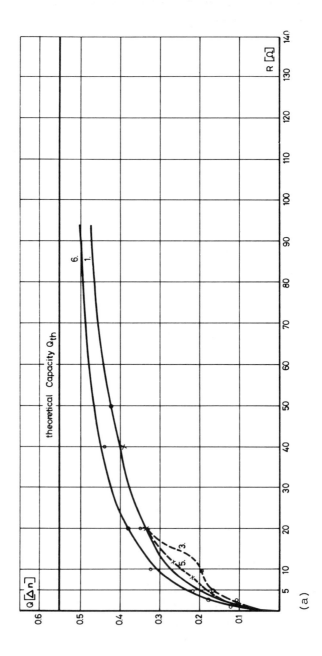

(a)

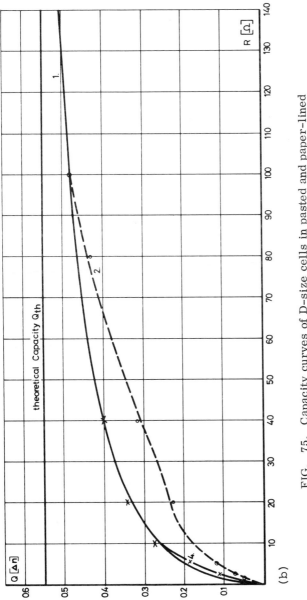

(b)

FIG. 75. Capacity curves of D-size cells in pasted and paper-lined construction. (a) Cathode containing electrolytic manganese dioxide; (b) cathode containing a 2:1 mixture of artificial and natural dioxides. Curve 1, calculated; 2 and 3, paper-lined, Leclanché; 4 and 5, pasted; 6, paper-lined, $ZnCl_2$.

layer, the greater the deviation at high current drains. The explanation, they believe, lies in the decreased conductivity of the paste due to the marked increase in the $ZnCl_2$ concentration. On light-drain discharges, the deviation becomes negligible, and it is assumed that diffusion is rapid enough to prevent the build-ᴜ of extreme $ZnCl_2$ concentrations [private communication]. The fact that 'wet" cathode mixes perform better than "dry" ones on heavy-drain tests corroborates the above point of view. Further, it is a matter of practical experience that the moisture content of mixes, based on finely divided manganese dioxide, is quite critical. In the $ZnCl_2$ paper-lined system, the high moisture content is reflected in the excellent agreement between actual and theoretical performance curves. Finally, it may be mentioned that cells with about 2 mm of paste wall between the zinc and the cathode perform better on a continuous high-drain test (i = 300 mA) than paper-lined cells with a separator thickness of about 0.2 mm.

The study of such performance curves provides suggestions for the optimum application of different types or mixtures of manganese dioxide. For instance, taking curves 2 and 4 in Fig. 75b as examples, it is apparent that the particular cathode mix is much more efficient in the pasted cell (curve 4) than in the paper-lined cell (curve 2). The characteristic of the paper-lined cell can be considerably improved by increasing the moisture content of the mix, but this necessarily implies that the amount of MnO_2 available in the cathode must be reduced. If the cell is to be used only for light drains (for example, for transistor radios), it is significant that the two curves coincide at discharges equivalent to a 60-Ω continuous load or lighter. But since the cathode weight of the paper-lined cell is about 25% higher than that of the pasted cell, the total ampere-hour output of the former is much greater. The conclusion may be reached that a specific mixture of ores can be used advantageously in a paper-lined cell, designed especially for light drain, but cannot be recommended for high-drain or general applications.

In conclusion, the question of the physical significance of K and $R^{1/2}$ in the exponent of the polarization term $\exp(-KR^{1/2})$ may be briefly recapitulated. The factor $R^{1/2}$ is obviously an expression of the dependence of the diffusion polarization on the discharge current. It states quite generally that there is a square-root relationship: The factor K may be regarded as a rate constant for the electrochemical reaction in the cell and, as is shown in Fig. 76, defines the decrease in the polarization as the discharge resistance is increased.

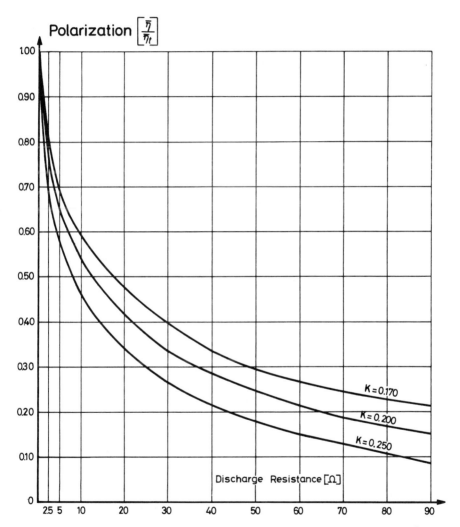

FIG. 76. Polarization curves for various values of K in $\exp(-KR^{1/2})$ (see text).

4. MANUFACTURE OF DRY CELLS

The production of dry cells developed explosively during the first half
of the 20th century. At the turn of the century, the industry was characterized by an enormous number of small, mainly hand-operated production
units with relatively low production figures; today, the scene is dominated
by comparatively few large concerns with sophisticated, fully automatic production units capable of turning out millions of the more popular types.
During the First World War, practically every production operation was
carried out by hand, using primitive tools and jigs. The quality of the final
product was dependent on the experience and skill of the production foreman,
who would develop his own mix formulations empirically and secretively in
some corner of the factory. In the modern plant, with the delegation of
responsibility to a staff of technicians who are expert in a variety of disciplines, there is no place for such a person. Apart from a few standard
machines, the production lines are built around equipment adjusted to the
specifications of the product and are usually designed and developed by the
battery maker. This can only be achieved by close cooperation of machine
designers, electrical and electronic engineers, and battery chemists. The
subsequent adjustment, repair, and maintenance of the production equipment
is an unavoidable task demanding a staff of specialists.

Formerly, raw-material testing could be conveniently carried out by
empirical methods, usually comprising the manufacture and testing of a
limited number of sample batteries. In modern automatic production plants,
rapid and reliable assessment of the suitability of material is essential; the
restrictions imposed by the process are often just as stringent as the electrochemical requirements needed to produce a quality product.

When the daily production of cells can be counted in millions, the loss
of production time implies increased manufacturing costs. The release of
substandard products to the market can result in loss of prestige and increased sales resistance.

The development of efficient raw material, testing of supplied parts, and
the continual improvement of the product are proper functions of the develop-

ment department. In general, this department includes a chemical-physical section, a technological section studying new manufacturing processes, a machine-design development section, and the associated drawing offices and pilot-plant units. Large concerns often maintain central research laboratories for product-orientated fundamental investigations, undisturbed by the day-to-day problems arising in the development departments located at the manufacturing sites. Perhaps one of the most important functions of such research centers is to provide the liaison between industry and pure science, so that progress can be achieved with a minimum of delay.

A significant role in the manufacturing process is played by the inspection and quality-control department. It is expedient to make this department directly responsible to company management and independent of the production and development departments. Its primary function is to ensure that the whole manufacturing process runs in accordance with approved specifications and that deviations from standard manufacturing procedure or from the customary product quality level are revealed as early as possible and corrective action initiated. With millions of cells being manufactured per day, it is essential that the inspection methods be based on mathematical statistics. The careful analysis of inspection data very often yields unsuspected correlation between process parameters and the quality of the product, so that an efficient inspection department contributes markedly to the company image in the trade.

In view of the worldwide commercial relationships and the high production rate of modern battery factories, a well-organized and efficient sales department is indispensable. Since a large and highly automatic factory is rather inflexible, future planning becomes very important not only in the sales area but also regarding administration methods, production techniques, and product development. Last, but not least, a sound financial policy is essential for the marketing success of the concern.

It is obvious that in a large and complex factory, the function and responsibility of each individual must be clearly defined and overlapping avoided. But irrespective of the type of organization plan chosen by a company, it will only be effective to the extent that the company management enforces the implications of the electrochemical system. Also, the human

relations between individual staff members, to each other and to the management, must be such that a real team spirit is encouraged. Each employee
should identify himself with the company and its products. If working conditions are bad, the best of organization schemes will be ineffective. Especially in the dry-cell industry, a staff of skilled and interested employees
is essential, since the mix formulations and manufacturing specifications
must always include a certain tolerance to allow for irregularities in materials and processes. These variations demand adjustment during the
manufacturing process and if they are neglected, a whole production run
may be defective despite "compliance" with the specifications.

4.1. ROUND CELLS

Every manufacturer of importance has developed his own production
process, and it is outside the scope of this book to describe every detail.
In particular, the manufacture of electrolytic manganese dioxide, carbon
rods, contact assemblies, and battery containers will not be described,
although several battery manufacturers have entered one or more of these
fields when it has been considered profitable to do so. Only the traditional
specific battery manufacturing operations will be described.

Figure 77 is a block diagram for a typical manufacture of D-size cells
with steel jackets. In an ideal production layout, the production flow follows
a straight line without direction changes or crossover points. The storage
facilities for raw materials or piece parts are placed as close as possible
to the point of usage to avoid unnecessarily long transport lines. The parent
factories of the large manufacturing companies are seldom good examples
of rational layout. They have grown too rapidly and incorporate technologies
developed at different times. However, many of their smaller, newer sub-
sidiaries are models of rational and functional design.

The arrangement of the individual machines and the interlinkage of
assembly operations are shown directly in Fig. 77 and do not necessitate
further explanation.

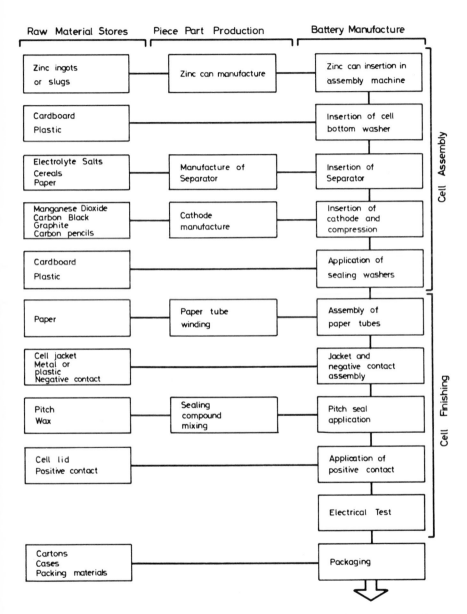

FIG. 77. Block diagram showing stages in the manufacture of steel-jacketed paper-lined cells.

It depends on the personal preference of the machine designer whether the unit manufacturing operations are performed with special tools on single machines independently powered and coupled by conveyor belts, or on complex automatic machines that carry out many operations simultaneously. The latter have the advantage of requiring fewer operators and less floor space, but complex machines frequently break down and production time is lost.

Simple-operation machines are often preferred because of the ease of maintaining the production flow despite individual machine stoppages. Reserve machines can be brought into use on short notice, or buffer stock may be used to bridge the time gap until the breakdown has been rectified. This flexibility may compensate for the greater floor space requirement and increased labor costs. A basic point of machine design is the type of feed to the individual working stations. Years ago, intermittent motion was preferred. An operation would be completed in one place and the part would then be moved to a new station, where it was stopped again until the next operation was performed. Modern design, however, tends toward continuous motion, which implies that the operating units move with the part. Such machines run more smoothly and achieve greater accuracy than the intermittent-feed types, but require higher capital investment. Ultimately, a decision concerning which system will give the best product at lowest cost must be made: several machines working intermittently in parallel or one machine with a multiplicity of operating heads working continuously. (See Fig. 78.)

A modern dry-cell assembly line is usually designed to produce from 200 to 300 cells per minute.

In the next sections, the more important operations typical of dry-cell manufacture will be described.

4.1.1. Cathode Mix

Depending on the production capacity of the factory, the raw materials required for the cathode mix are either stored in the original packaging

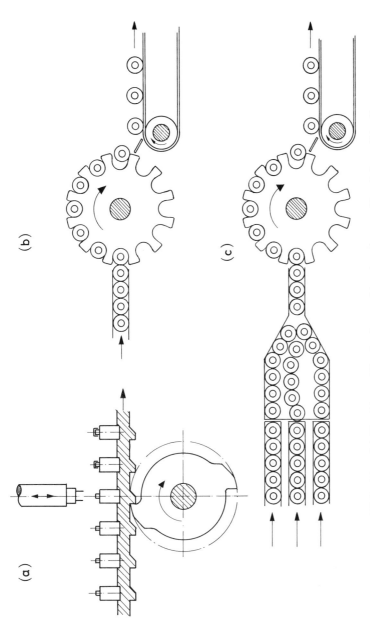

FIG. 78. (a) Machine with single working head operating intermittently. (b) Machine with one working head per cell operating continuously (multiple working heads not shown). (c) Intermittent feed into buffer stock and then into continuously operating unit.

(usually paper or jute sacks) in the warehouse or loose in silos (Fig. 79). This type of silo is divided into an upper and a lower compartment. Material is stored in the upper compartment until the lower one is empty and is then transferred by gravity.

FIG. 79. Storage silos showing filter house at the top and individual pipelines for stock replenishment.

All the material contained in the lower part of the silo can be thoroughly mixed by means of intermittent air blasts. This homogenizing process is particularly useful for natural manganese dioxide ores, which may vary appreciably throughout a sample and render electrical testing (potential) of the product uncertain. The use of silos can only be envisaged when the supply of material can be kept under close control. The advantage to be gained lies in the saving of packaging material and labor costs.

Carbon black is not normally stored in silos because of the large volume required and its tendency to cake during removal. However, given a good bag opening and emptying machine, it may be stored in bunkers and transported by means of an Archimedian screw.

Before the actual mixing operation can begin, weighed amounts of the individual dry components must be transferred to the mixing unit. The weighing operation may be carried out by hand using a standard bascule type weighing machine. Alternatively, a traveling balance may be used which passes under each silo and discharges the correct amount into the mixer. In a more sophisticated plant, the whole operation is automatically controlled by electronic devices using punched cards or tapes. In such units, it is imperative that the electronic equipment, relays, and so on be kept completely protected from corrosive or conductive dust if reliable operation is to be maintained.

Mixing units are of many and varied designs, ranging from the simple horizontally mounted wooden drum with wooden beaters attached to the main shaft, to the rotating drum, to the most intricate equipment including high-speed dispersion units and compression rollers. The mechanical work done on the mix components must be variable over a wide range to adapt to the specific macrostructure of the components. Especially with very wet mixes, the type of mixer employed and the mixing procedure are very significant parameters in determining the quality of the final product. Various mixer designs are shown in Fig. 80.

In some processes, the manganese dioxide ore is ground with graphite in a ball mill before it is used for the preparation of the cathode mix. The object of the operation is to increase the electronic conductivity on the surface of the individual manganese dioxide particles.

(a)

(b)

FIG. 80. (a, b) Double-shaft, single-end-drive mixer, closed (a) and open (b) showing sigma type mixing blades. (c, d) Large capacity vertical mixer (c) and interior view (d) showing mixing and compression rollers. (e, f) Rotary double-cone mixer (e) and interior view showing dispersion blades and electrolyte feed pipe and spray head.

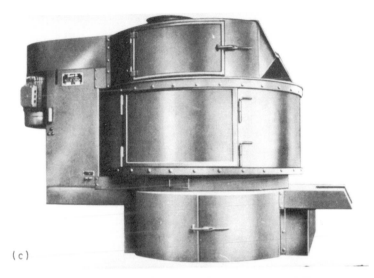

(c)

(d)

FIG. 80. (Continued).

(e)

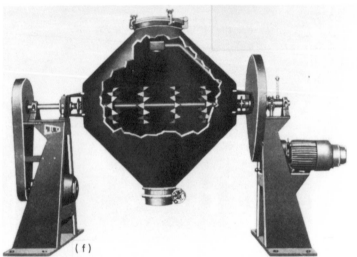

(f)

FIG. 80. (Continued).

Mixing the components in the dry state is a relatively short operation (a few minutes). The specified amount of electrolyte is sprayed into the mix without interruption of the mixing process. Only in the wet condition can the homogeneous distribution of the components be stabilized; therefore the wet mixing operation must not be stopped prematurely. With continued mixing, however, the macrostructure of the mix changes, and the mix becomes very sticky until finally (with excessive wet mixing) a completely unworkable product is obtained. In practice, the mixing procedure must be determined empirically and is dependent on the characteristics of the mix components, the formulation, and the type of mixing equipment available. The efficiency of mixes intended for light-drain intermittent discharge is not sensitive to the amount of electrolyte, the capacity of the cell is largely a function of the weight of the cathode. Prolonged wet mixing times and heavy precompression of the mix can be employed with advantage. For continuous heavy-drain discharge, however, the efficiency is critically dependent on the amount of electrolyte, and both the wet mixing time and the mechanical work exerted on the mix must be reduced to the minimum consistent with reproducibility.

The double cone type of mixer, with its characteristic avoidance of compression of the mix, excels in this field.

In order to achieve final equilibrium, it is customary to store the finished mix in airtight containers one to three days before further processing.

For all the operations concerned with the cathode mix, the equipment designer must bear in mind not only the functional aspects but also the corrosive and abrasive nature of the material being processed.

4.1.2. Electrolyte

The electrolyte of a Leclanché cell is basically an aqueous solution of ammonium and zinc chlorides. To the basic solution, inhibitors (such as mercuric chloride) and/or wetting agents may be added. For the preparation of pastes to be used in pasted cells or for the coating of the separator in paper-lined cells, gelling agents are added to the basic solution. Tra-

ditionally, the gelling agents are natural products such as wheat flour or
corn starch, but in recent decades derivatives such as cross-linked starches,
carboxymethylcellulose, and similar products have found increasing application. For pasted cells, the electrolyte composition is somewhat dependent on the setting process adopted. In hot-setting processes, the concentration of the salts and of the gelling agents must be low enough to avoid
spontaneous setting of the paste when the cathode is inserted. In the cold-
setting systems, this spontaneous reaction is just what is required, but on
the other hand it must not proceed outside the cell; the electrolyte must
not thicken prematurely during storage. Some manufacturers prefer the
use of two electrolytes, which may be stored indefinitely but when mixed
thicken and gel very rapidly. The mixing process may occur immediately
before metering into the cell or actually in the zinc can.

According to composition, the thickening time of the cold-setting group
can be varied from a few minutes to several hours. Slow, cold-setting
pastes have the advantage of a prolonged pot-life and only begin to thicken
in the cell after insertion of the cathode. They produce pastes free from
air bubbles which remain smooth and pliant and show little tendency to swell,
provided that the cathode composition is properly matched. A disadvantage
lies in the impractibility of immediate further processing. A waiting period
of several hours is usually specified to permit the electrolyte paste to set.

As previously described, the electrolyte paste for dry cells consists of
a suspension of gelling agents such as starch or flour in an aqueous salt
solution. The amount of flour or starch added varies from manufacturer
to manufacturer over a wide range (from about 1/2 to 3). About 20-25
parts by weight of cereal are added to 100 parts of electrolyte. The effect
of the individual components of the salt solution and the amount of gelling
agent was studied by Staley and Helfrecht [130]. Figure 81 shows the influence of the weight of starch on the setting time. This period is defined
as the time elapsed (in minutes) between mixing the components and the
point when the reaction vessel could be inverted without loss of paste.

Figure 82 shows the effect of the zinc chloride concentration and the
change in the rate of setting induced by increasing the ammonium chloride
concentration. Curve 4 refers to pure zinc chloride solutions to which a

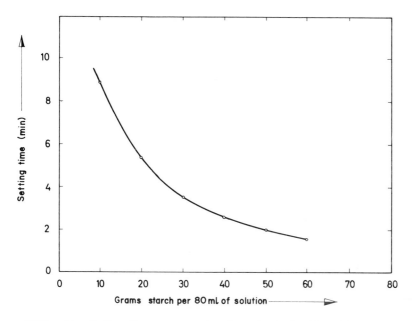

FIG. 81. Setting time versus starch content of cold-setting paste.

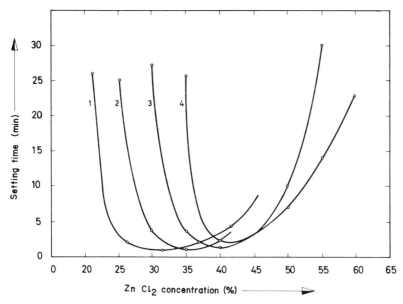

FIG. 82. Setting time versus $ZnCl_2$ concentration of cold-setting paste. See text for explanation.

constant weight of starch was added. The rate of setting increases with
increasing zinc chloride concentration until at about 40%, a minimum in
the setting time is observed. If the concentration of zinc chloride is further
increased, the setting time also increases. With zinc chloride solutions of
concentration less than 35%, a properly set paste was not obtained. The
35% solution formed a white paste in 26 min. Greater concentrations yield
yellowish pastes because of hydrolysis.

Curves 3, 2, and 1 illustrate the behavior of solutions to which 10, 20,
and 25% of ammonium chloride has been added. The general form of the
setting curve is unchanged, but the whole is displaced to the left, showing
that a given setting time can be obtained at lower zinc chloride concentrations if sufficient ammonium chloride is present. However, the acceleration of the gelatinization process by ammonium chloride is less than that
obtained with zinc chloride.

As a result of this study, Staley and Helfrecht proposed a cold setting
paste consisting of two components (A and B) in the tabulation. The composition is given in parts by weight. The solution A contains a high percentage of zinc chloride (about 39%) but no cereal, while the paste B contains all the cereal and only a small amount of zinc chloride (about 6%).
Both components are stable when stored individually. Mixed in the ratio
shown, the paste sets in about 5 min at room temperature.

	Solution A	Paste B
Zinc Chloride	14.3	3.8
Ammonium Chloride	9.3	10.6
Water	13.6	24.1
Gelling agent (starch)	–	24.5
Total	37.0	63.0

Further investigations on electrolyte pastes were undertaken by Hamer
[131]. The characteristics of pastes incorporating different types of starch
and flour were studied. Modified starches and different fractions of natural
flours and starches were included in the study.

A summary of the properties of electrolyte pastes was given more recently (1969) by Huber [132].

Similarly, there are several variations of the hot-setting process. The most common system is to convey the cells through a hot water bath (temperature 65°-90°C). An alternative method is to transport the cells through a tunnel in which hot-air or hot-water jets provide the necessary heat. A more modern variant is the use of medium-frequency induction heating.

Whatever type of paste or heating system is preferred, the electrolyte is prepared at a sufficiently high viscosity to avoid sedimentation during storage. This is easily achieved by pregelatinization of a portion of the gelling agent.

The vessels used for storage of the electrolyte solutions and suspensions are made from earthenware or iron, lined with rubber or plastic. The storage vessels may be sunk into the floor to save space or mounted at a sufficient height to enable the electrolyte to flow by gravity to the point of usage. Where several formulations are in use simultaneously, it may be expedient to store a smaller number of parent solutions from which the actual electrolytes may be made by mixing in various proportions. A problem that arises when electrolyte pastes are transferred by pumping is the inclusion of air in the paste. Unless this air is removed before use, the air bubbles trapped in the gelatinized paste lead to intensified anodic corrosion. Solutions of salts may also be stored in contact with metallic zinc to remove heavy-metal impurities. Chemical methods of checking the composition of the solutions are preferred, since specific-gravity determinations are inherently inaccurate, inapplicable to solutions containing more than one solute, and subject to temperature errors.

For the manufacture of layer-built and paper-lined cells, a coated separator is employed.

The carrier is usually Kraft paper of high purity, such as is used for the manufacture of cables and electrolytic capacitors. The formulation and method of application of the paste coating varies from manufacturer to manufacturer (see Section 2). Figure 83 shows schematically one type of coating and drying equipment. The thickness of the coating is controlled by a doctor plate, an air-stream, or by two rollers, one of which is driven at some chosen speed while the other rotates freely and is placed at an adjust-

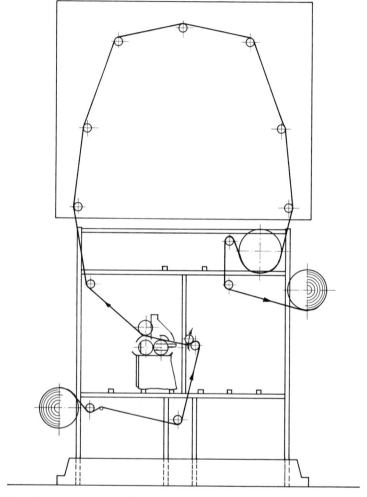

FIG. 83. Schematic outline of equipment for coating paper with cereal paste and drying.

able distance from the driven roller. The coated separator should be flexible and the paste laminate of even density. For the carrier, a paper free from metal inclusions should be selected.

4.1.3. Zinc Cans and Plates

The quality of the zinc cans, and in particular, the shelf characteristics of cells incorporating them, is primarily a function of the zinc alloy used. Not only is the selection of proper components important, but also their homogeneous distribution throughout the metal. Heterogeneous phases such as zinc oxide, dross, and zinc/iron compounds accelerate corrosion.

In the preparation of the specified alloy, a more concentrated alloy is added to the molten zinc and distributed by means of an efficient stirrer. The dross formed at the surface is continually removed. The melt is cast either in ingots or as a continuous strip (Hazelett process), but in either case quick cooling is required to prevent separation of the insoluble components (such as lead) and to obtain a fine grain structure.

The ingots or cast strips are then rolled in several steps to the required thickness. Some rolling mills prefer cross rolling techniques to avoid surface blemishes.

In earlier times, the cans were made from blanks (cut from a zinc strip), which were rolled into shape and secured by soldering, welding, or mechanical locking techniques. The can bottom was punched separately, and inserted into the hollow cylinders before being finally secured in place. In this type of manufacture, it was almost always necessary to apply some kind of internal seal, such as wax or bitumen, to prevent electrolyte leakage later. A further difficulty was the elimination of cans with distorted ("elliptical") cross section.

One modern technique is to draw the cans from sheets on multistage presses. The deep-drawing process can be carried out with high precision and at adequate speed (Fig. 84).

An alternative process is the impact extrusion of zinc slugs on heavy-duty toggle presses. High-quality cans can be made in this way, and a production rate of 150 cans/min is not unusual from a single tool with modern equipment (Fig. 85).

FIG. 84. Multistage press such as is used for deep drawing of zinc cans.

FIG. 85. (a) General view of impact extrusion press. (b) Detail of impact extrusion press showing hexagonal slugs, feeding channel, and impact tool.

FIG. 85 (Continued).

Zinc slugs are either circular or hexagonal. Punches and dies for the manufacture of circular ("round") slugs are easy to make and are therefore cheap. However, 15-25% of the zinc strip must be returned as web scrap to be reprocessed. The hexagonal slugs can be punched from strip with practically no loss of zinc metal. The punching tools are more complex, and inspection of the product more critical.

The zinc slugs delivered from the presses are contaminated with oil, and the edges show some degree of flash formation.

Tumbling with sawdust removes the oil and flash; after passing over a sieve to remove the sawdust, the slugs are lubricated. In addition to the special pastes that have been developed for this purpose, graphite and talc are also employed. The lubricant must be able to withstand the temperatures developed during the impact extrusion process. Extraordinarily small amounts of lubricant are sufficient (about 1 gm per 20 kg of zinc), and excess should be avoided.

It is customary to preheat the slugs in the channel feeding them to the extrusion press. Some manufacturers even prefer heated extrusion tools (the temperatures involved are of the order of 150°C). This may be advantageous when starting up, because the zinc cans tend to stick in cold tools. Pretreatment of the tools with lubricant also helps to avoid this problem.

The zinc slugs must be kept clean and the extrusion tools highly polished for the process to run efficiently. Further, the setting of the machine requires skill and experience.

Figure 86 shows the impact extrusion process diagrammatically and illustrates the results obtained with a badly set machine. Off-center tools or slugs form unsymmetrical cans and may cause tool breakage. Particularly, hexagonal slugs must be kept within small dimensional tolerances if trouble is to be avoided. Figure 87 shows the relationship between various zones in the slug and their ultimate position in the can. If the slug is dirty or if the zinc contains heterogeneous phases such as zinc oxide, very pronounced extrusion parabolas are developed which are prone to rapid corrosion. The grain structure of the zinc in the slug is completely changed during the extrusion process, but the anisotropy of the inhomogeneous phases is retained. Tempering the slugs therefore has no effect on the final structure of the zinc can.

From the extrusion presses, the zinc cans are fed automatically to the trimmers, where they are cut to length. Simultaneously, a bead or rill may be applied if required. (See Fig. 88.)

Finally, it may be remarked that the inner surface of the can must be free from oil or grease, which would induce irregular dissolution of the anode during discharge.

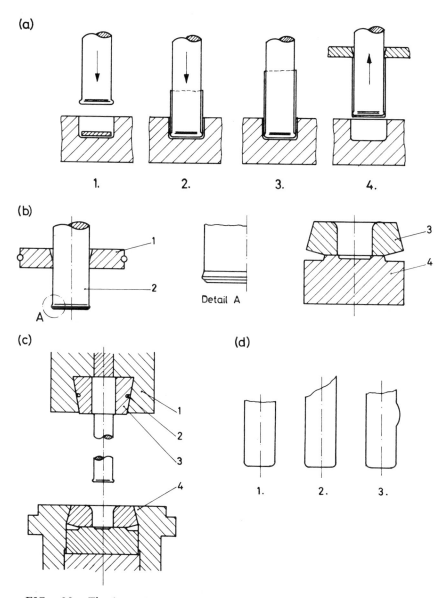

FIG. 86. The impact extrusion process diagrammatically. (a) Four stages in the formation of the can; (b) 1: Can-stripping ring, 2: impact tool, 3: mold, 4: bottom tool; (c) 1: tool holder, 2: adjustment for retaining cone, 3: tool retaining cone, 4: bottom tool holder; (d) 1: well-formed can, 2: asymmetrical can due to badly adjusted tools, 3: distorted can due to off-center slug.

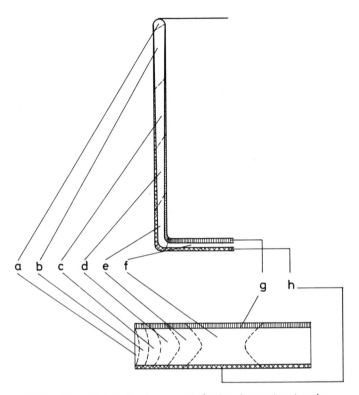

FIG. 87. Metal displacement during impact extrusion.

4.1.4. Cathode Manufacture for Cylindrical Cells

Cathodes (known in the trade also as cores, bobbins, or dollies) may be pressed on single- or double-action presses. The cathode mix, after the specified storage period, is transported to the machine in a closed container and then transferred to the hopper. At each stroke of the machine, about 130% of the required weight of mix is fed into the prepressing mold. The feed may be by hand or by a mechanical device. Overfeeding is deliberate in order to achieve a uniform density and cathodes of equal weight. The various stages in the formation of the cathode are shown in Figs. 89-93. The press tools and ejector may be stationary or rotating, according to the design of the machine. Rotating tools are sometimes used to prevent adhesion of the mix to the tool. A typical cathode press is shown in Fig. 94.

4.1.5. Assembly of Pasted Cylindrical Cells

The design of a pasted cell assembly unit is determined by the type of electrolyte paste employed. The individual operations are

    a. Insertion of a bottom spacer (to separate the electrodes)
    b. Metering the electrolyte paste
    c. Insertion of the bobbin
    d. Paste setting
    e. Application of the seal

The difference in design occurs at stages (b) and (d). With hot-setting pastes, the setting operation involves passing the cells through a water bath (temperature about 90°C, time in bath 2-3 min). Cold-setting pastes may be slow (several hours) or fast (a few minutes). The latter have the advantage that the production process is uninterrupted, but arrangements must be made to prevent the paste from setting while in the metering equipment should the manufacturing process be temporarily stopped. For this reason, two-component electrolytes are often employed; the mixing operation takes place at the moment of insertion into the can. With slow pastes, it is necessary to hold the cells until the paste has set. It is convenient to adjust the formulation so that cells may be further processed on the following day. Whatever type of paste is used, it is necessary to center the bobbin until the paste is sufficiently rigid and to avoid contamination of the zinc in the sealing area.

4.1.6. Assembly of Paper-Lined Cells

This process differs from the assembly of pasted cells considerably because the individual operations are more complicated. Instead of metering a fluid into the cell, a laminated separator (usually paper coated with cereal or other gelling agent) has to be inserted. Furthermore, the cathode employed is mechanically less stable than that used in pasted cells because of the higher moisture content.

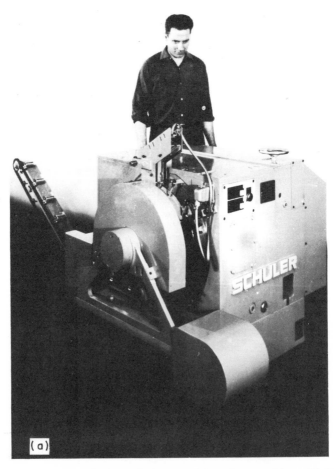

FIG. 88. (a) General view of zinc can trimming machine. (b) Detail
of working heads on can trimmer.

(b)

FIG. 88. (Continued).

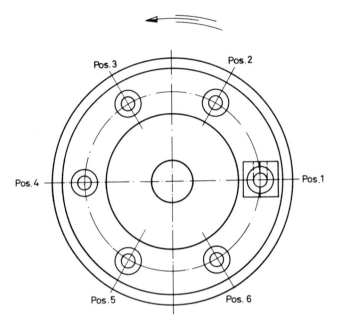

FIG. 89. Cathode manufacture. Front view of mold carrier plate
showing six working stations.

Various processes are known for the manufacture of the separator; some
are patented [133, 134]. By suitable choice of material and method of preparation, separators may be obtained that perform efficiently in the cell.

The most common assembly methods of paper-lined cells are the following.

a. The preformed cylindrical cathode is wrapped completely in the
separator (including the upper and lower surfaces) and inserted into the
zinc can. The insertion of the carbon rod is followed by the final compression of the wrapped cathode in the zinc can.

b. The zinc can is lined with the separator and a bottom disc inserted.
Then follows insertion of a preformed cylindrical cathode, insertion of a
carbon rod, application of a top washer, and final compression.

c. After lining the zinc can with the separator and insertion of a bottom
disc, the cathode mix is extruded directly into the lined can. Insertion of

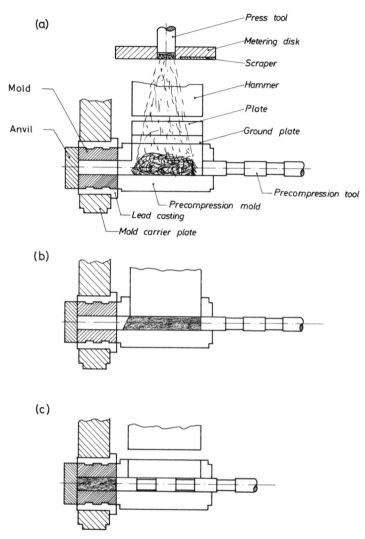

FIG. 90. (a) Position 1. Loose mix is fed into the precompression mold. Hammer is elevated. (b) Hammer falls and closes the mold, lightly compressing the mix. (c) The mix is transferred to the mold and compressed against the anvil.

the carbon rod, application of top washer, and final compression completes the process.

(a)

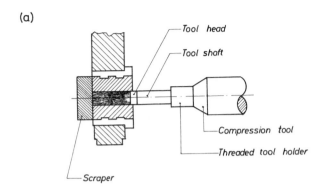

Tool head

Tool shaft

Compression tool

Threaded tool holder

Scraper

(b)

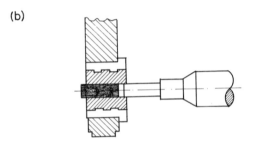

FIG. 91. (a) Position 2. Mix in mold is further compressed. The tool head enters the mold. (b) Position 3. Partial ejection of molded cathode as first stage in trimming to required length.

The process described under (c) has an advantage over the other methods: a uniform density is obtained throughout the cathode, and an excellent contact between separator and cathode mix is assured. On the other hand, the required fluidity of the mix under pressure imposes limits on the choice of raw materials and the mix formulation.

4.2. FLAT CELLS

As in the case of cylindrical cells, the manufacture of flat cells is almost completely automatic. The more common designs have already been described in Section 1. The block diagram reproduced in Fig. 95 is appli-

(a)

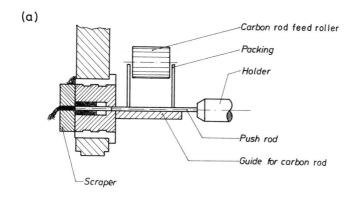

Carbon rod feed roller

Packing

Holder

Push rod

Guide for carbon rod

Scraper

(b)

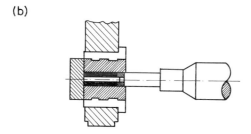

FIG. 92. (a) Position 4. The projecting part of the molded cathode is removed by the scraper during rotation of the mold carrier plate. The carbon rod is inserted. (b) Position 5. The cathode is recompressed to the required length.

cable to all designs and will be described here in relation to flat cells based on rigid cell containers and also to those of flexible construction.

(1) The tamping of tablets from the cathode mix is performed on horizontal or vertical rotary presses.

(2) The raw materials for the conductive foil are mixed and rolled. Or a conductive varnish may be prepared, if it is to be used instead of conductive foil.

(2.1) Final rolling of conductive sheet occurs at this point.

(3) The zinc strip and conductive foil are laminated by compression rolling or alternatively, the zinc is coated with the conductive varnish. Several coats must be applied in order to obtain a nonporous layer.

(a)

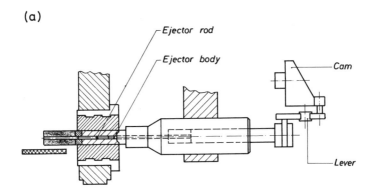

(b)

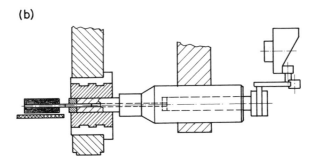

FIG. 93. (a) Position 6. The cathode is ejected from the mold.
(b) Cathode is transferred to conveyor belt.

(3.1) Duplex electrodes (zinc plus coating) are blanked out.

(4) The separator and coating paste are prepared.

(4.1) The separator paper is coated with paste and dried.

(4.2) The coated separator-paper rolls are cut to the required width.

(5) Electrolyte solution is prepared.

(6) The plastic cell containers are inserted in the assembly machine.
Alternatively, in designs based on PVC tubing (flexible construction), the
PVC tube is cut to length and fitted over a rectangular form with a part of
the tube extending out. The tube and carrier then pass through a heating
device so that the free end of the tube shrinks inward. After removal from

FIG. 94. General view of a standard cathode press. Detail of a cathode press, showing molds in the carrier plate, the press tools mounted in the spider, and the carbon-rod container.

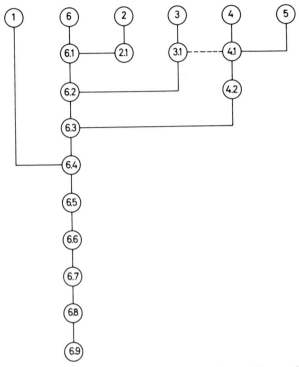

FIG. 95. Block diagram showing stages in flat-cell manufacture.
See text for explanation.

the form, the tube has assumed a shape similar to a rigid cell container.

(6.1) Conductive foil is glued to the base of the rigid cell container.
This operation is omitted with flexible designs.

(6.2) The duplex electrode is inserted into the cell container. If preferred, the two components may be secured with adhesive.

(6.3) The separator paper is inserted. In one variant, the coated paper
is applied directly to the zinc strip already laminated on the opposite surface with conductive foil.

(6.4) The cathode tablet with a protective open paper cup is inserted.

(6.5) Electrolyte is added from a metering pump (not always necessary).

(6.6) The cells are assembled into stacks. In the case of the PVC flex-
ible-tube design, the cells must first pass through a heating device to induce
the tube to shrink around the cell components.

(6.7) The cell stack is immersed in a suitable adhesive so that the individual cells are combined into a unit. Alternatively, the cells may be strapped together.

(6.8) The stack is impregnated with wax.

(6.9) The stacks are assembled into complete batteries.

As in round-cell manufacture, inspection points are included at critical stages of the assembly process.

In flat-cell manufacture, the universal rule of dry cell production loses none of its validity, that is, the cells must be sealed as effectively as possible. Figures 96-101 illustrate the result of neglecting this rule. An anode taken from a poorly sealed flat cell (Fig. 96) clearly shows three zones: (a) the dark edge area, which is dry and corroded; (b) the central area of the plate, which is wet and corroded; and (c) the intermediate area, which is not corroded (bright). The edge corrosion is caused by air diffusing into the cell through the ineffective seal. In Fig. 97, four zinc anodes are shown, taken from unsealed flat cells that had been stored in a vacuum, oxygen, carbon dioxide, and an oxygen/carbon dioxide (1:1) mixture. The corrosive action of oxygen can be readily recognized. The well-known corrosion effect in badly sealed round cells at the zinc/electrolyte/air phase boundary is shown in Fig. 98.

The formation of the zones (b) and (c) in Fig. 96 can be explained by the following experiment. In a single flat cell, two cathode tablets were laid side by side and electronic contact made by means of conductive sheet. One tablet was made from a wet mix and the other from a drier mix. The cell was carefully sealed and after 14 days opened again for examination. The appearance of the zinc anode can be seen in Fig. 99. The potential difference between the two tablets had caused corrosion of the zinc under the moist tablet. In a badly sealed flat cell, it is to be expected that the drying process will proceed from the periphery toward the interior of the tablet. The central portion is thus equivalent to the moist tablet in the experiment described, and the corrosion in zone (c) is due to the potential difference induced in the tablet by the drying process.

Finally, the importance of a uniform coating of the separator paper in both flat cells and paper-lined cells may be illustrated by reference to Figs.

FIG. 96. Anode from a badly sealed flat cell.

100 and 101. Figure 100 shows a zinc anode taken from a cell in which the separator paper was completely uncoated. The corrosion pattern is a reproduction of the fibrous structure of the paper. Figure 101 shows a corroded anode taken from a flat cell in which only half the area of the separator was coated (left). The anode at the right side had a properly coated separator.

4.3. MODIFICATIONS OF THE LECLANCHÉ SYSTEM

Air-depolarized cells and the alkaline manganese dioxide cell may be regarded as modifications of the Leclanché system, but the internal design

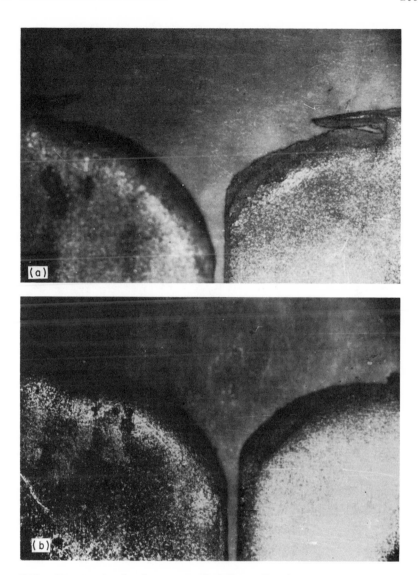

FIG. 97. (a) Anodes from unsealed flat cells after storage in vacuum (left) and in oxygen (right). (b) Anodes from unsealed flat cells after storage in carbon dioxide (left) and a carbon dioxide/oxygen mixture (right).

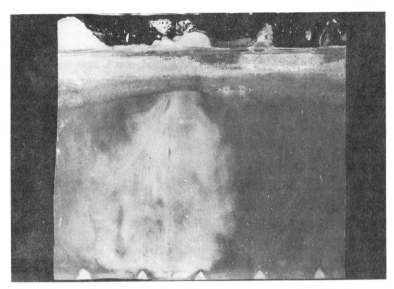

FIG. 98. Corrosion at air line in a pasted cell.

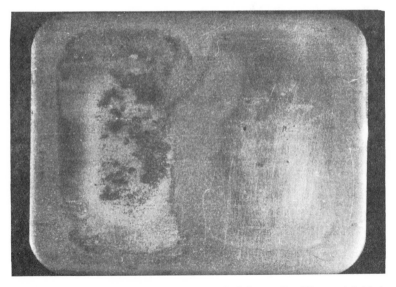

FIG. 99. Zinc anode from a well-sealed flat cell with a wet tablet (left) and a dry tablet (right). See text.

FIG. 100. Zinc anode taken from a flat cell with an uncoated paper separator. The corrosion shows the pattern of the paper fibers.

of these types differs greatly and they should be treated as independent primary cells. However, the magnesium chloride and zinc chloride cells, based on manganese dioxide and zinc electrodes and differing only in the nature of the electrolyte salt employed, are very similar in construction.

4.3.1. The Magnesium Chloride Cell

This system was developed by Rudolf Pörscke in Hamburg in 1916. In the following year, the Pertrix-Chemische Fabriken G.m.b.H. was founded. In 1926, the stock and patent rights were taken over by the Accumulatorenfabrik A.G. (AFA) [10].

The special advantage of the $MgCl_2$ cell lies in its excellent storage properties, not only in the unused condition but also after partial discharge. In the age of round-cell "anode" batteries (that is, high-voltage radio bat-

FIG. 101. Zinc anodes taken from flat cells. Half the area of the separator of the cell on the left was coated. The cell on the right contained a standard, completely coated separator.

teries) and later after the introduction of the layer-built design, this characteristic was of decisive importance.

The reason for this good shelf performance is the formation of a passivating oxide film on the zinc anode as soon as the discharge is interrupted. Zinc passivation is also the cause of the relatively low flash current obtained from $MgCl_2$ cells in comparison with the standard NH_4Cl system. Further, the $MgCl_2$ system requires the use of highly active synthetic chemical manganese dioxide.

The $MgCl_2$ cell may be manufactured in paper-lined, pasted, and flat-cell designs. The formulation and preparation of electrolyte paste are particularly critical, especially the selection of a suitable cereal for the gelling agent. The high current density required by modern electric devices and flashlights has resulted in the decline of this type of cell, but it is expected that it will regain interest in the future since the current requirements for electronic equipment are decreasing steadily.

4.3.2. The Zinc Chloride System

The zinc chloride cell was studied many years ago by the author in an investigation of power sources for high-current-drain applications. The work was eventually put aside. After several years, it was taken up again when it was observed that some experimental cells had given very good electrical results after long storage. In addition, it was established that with zinc chloride electrolyte, the tendency to leak was significantly reduced. Investigations of the leakage characteristics of dry cells [135] at high and extremely low current drains have revealed that two distinct phenomena must be considered. If a cell is short circuited, a liquid is expelled from the interface between the anode and the paste (or separator in paper-lined cells) into the air space after only a few minutes.

This liquid is an approximately 60% solution of $ZnCl_2$ and amounts to 6-8 cm^3 in the case of a D-size cell. If the cell remains under load, the zinc can perforates at the lower level of the air space and the accumulated fluid flows out of the cell before it can be reabsorbed by the cathode. Should the air space be too small, the hydraulic pressure exerted by the extruding liquid may disrupt the cell seal. The whole phenomenon is osmotic in nature and is caused by the paste or separator paper barring the passage of zinc ions. In order that electroneutrality be maintained, hydrated chloride ions are forced to the zinc surface. The increase in volume of the anolyte resulting from this one-way transport of water can only be accommodated by the solution rising into the air space.

At low current discharge, this osmotic effect does not arise, since the concentration profile can be equalized by diffusion processes at a sufficiently high rate. If discharge is continued until the zinc can is almost completely consumed, electrolyte extrusion again occurs because of the increase in volume of the manganese oxide as a consequence of the reduction from the tetravalent to the trivalent state. The volume increase implies a reduction of the pore volume of the cathode, which is further reduced by the deposition of insoluble salts. As a result, toward the end of the discharge, the electrolyte is forced out of the cathode.

The complex salt $Zn(NH_3)_2Cl_2$, formed in regular Leclanché cells during discharge, is insoluble at pH 7 but dissolves as a result of the formation of

$Zn(NH_3)_4Cl_2$ as the pH value rises above this value. The concentrated $ZnCl_2$ solution developed in the anolyte initiates a partial hydrolysis of the cereals present, and the fluidity of the paste increases.

In summary, the processes described lead to leakage of electrolyte through perforations in the zinc can at the end of a fairly heavy discharge.

On light discharge, for example, in an electric clock, the regular Leclanché cell can apparently become quite dry (uniform concentration at pH < 7). However, when the cell is no longer capable of fully operating the winding mechanism of the clock and remains on load, it is surprising how much fluid develops. Should the owner of such a clock neglect to remove the exhausted battery as soon as the clock stops, he may suffer an unpleasant surprise.

An active system such as a galvanic cell develops gas and liquid during the course of the electrochemical reaction and increases in volume because of precipitation of insoluble compounds and the reduction of Mn(IV) to Mn(III); it is quite obvious that such a system cannot be efficiently hermetically sealed. In order to design a completely leakproof cell, it is clear that a waterproof jacket is only the first step. In addition, the cell reaction must be so chosen that water can be combined in some solid compound. There are two reactions known which fulfill this requirement. First, the ammonium chloride electrolyte may be replaced by an alkaline solution (NaOH, KOH). This subject is treated in Chapter 2. Second, ammonium chloride can be omitted from the cell and a concentrated zinc chloride solution used as the electrolyte. Within a certain concentration range, basic zinc salts are formed during the reaction. As is well known, basic zinc compounds crystallize with a relatively high proportion of combined water. Ideally, the reaction may be written as follows.

Anode: $4\ Zn \rightarrow 4\ Zn^{2+} + 8\ e^-$

Cathode: $8\ MnO_2 + 8\ H_2O + 8\ e^- \rightarrow 8\ MnOOH + 8\ OH^-$

Electrolyte: $4\ Zn^{2+} + H_2O + 8\ OH^- + ZnCl_2 \rightarrow ZnCl_2 \cdot 4\ ZnO \cdot 5\ H_2O$

$$\overline{}$$

Overall: $8\ MnO_2 + 4\ Zn + ZnCl_2 + 9\ H_2O \rightarrow 8\ MnOOH + ZnCl_2 \cdot 4\ ZnO \cdot 5\ H_2O$

For the traditional Leclanché cell the following reaction applies.

Anode: $4 \, Zn \to 4 \, Zn^{2+} + 8e^{-}$

Cathode: $8 \, MnO_2 + 8 \, H_2O + 8 \, e^{-} \to 8 \, MnOOH + 8 \, OH^{-}$

Electrolyte: $\underline{4 \, Zn^{2+} + 8 \, NH_4Cl + 8 \, OH^{-} \to 4 \, Zn(NH_3)_2Cl_2 + 8 \, H_2O}$

Overall: $8 \, MnO_2 + 4 \, Zn + 8 \, NH_4Cl \to 8 \, MnOOH + 4 \, Zn(NH_3)_2Cl_2$

Comparing these two overall reactions, it is seen that both result in solid products. In the Leclanché system, water is neither consumed nor produced; in the zinc chloride system, a relatively large amount of water is consumed. It is to be expected, then, that this system will be inherently more leakproof than the Leclanché cell. In order to eliminate the osmotic effect, it is essential that the coated paper separator be replaced by some other material that, on the one hand, wets the zinc anode efficiently and, on the other hand, accelerates or at least does not retard the diffusion of zinc ions away from the anode surface [136].

Zinc chloride cells in paper-lined construction were first marketed in 1968 [137].

5. APPLICATIONS OF LECLANCHÉ DRY CELLS

5.1. IEC RECOMMENDATIONS

In the early days of the dry-cell industry, there were innumerable small-scale manufacturers in operation. The inevitable result was that the market was flooded with a multiplicity of types. Many of these types were electrically almost identical, but the terminals and overall dimensions were sufficiently different to make interchangeability impossible. In order to restrict the further development of this chaotic situation, attempts have been made to achieve some measure of technically and economically desirable standardization. Regional standards were adopted in the United States, Great Britain, Germany, and elsewhere. Still more recently, an international body has been established, the International Electrotechnical Commission (IEC), within which Technical Committee 35 has the task of preparing recommendations for the standardization of dry cells. All member countries have undertaken to embody these recommendations in the national standards, and indeed many have done so. The success of this project would guarantee physical and electrical interchangeability of standard battery types on an international basis, a result that would be invaluable not only for battery users and manufacturers but also for the designers of battery-operated equipment.

Tables 29-31 constitute a summary of the physical and electrical characteristics of most cell and battery types.

5.2. FACTORS RELATED TO BATTERY SELECTION

It is self-evident that the type of battery for use in a particular piece of equipment is selected by the equipment designer. Proper selection of the power source is extremely important for the realization of the potentialities inherent in the design. In addition to electrical capacity, other factors such as internal resistance (impedance), operating voltage, leakage resistance, and shelf life must be considered. The designer is therefore well advised to consult a battery manufacturer at an early stage of the development in order to avoid errors and disappointment. The battery manufacturer will need to know

(a) Operating voltage (maximum and minimum)

(b) Current drain

(c) Frequency of discharge

(d) Capacity or service life desired

(e) Terminal arrangements

(f) Maximum permissable weight

(g) Any unusual features of the application, such as vibration, shock, extreme temperature conditions, and so on.

Wherever possible, an internationally standardized battery should be chosen. This has the advantage that replacement batteries can be obtained throughout the world. Furthermore, these batteries are always more eco-

TABLE 29a

Designation and Approximate Dimensions of Dry Cells and Batteries

IEC[a]	DIN 40 855	British Standard	U.S. Standard	Diameter (mm)	Height (mm)	Volume[b] (cm^3)	Weight[b] (gm)
R 08	–	–	O	11	3	0.3	0.6
R 07	–	–	–	11	5	0.5	1
R 06	R 06	R 06	–	10	22	1.7	3.5
R 03	R 03	R 03	AAA	10	44	3.4	8
R 01	R 01	–	–	11	14	1.3	2.7
R 0	R 0	R 0	(NS)	11	19	2	4
R 1	R 1 (AT)	R 1	N	11	30	3	5.5
R 3	R 3	R 3	–	13.5	25	3.4	7
R 4	R 4	R 4	R	13.5	38	5	10.5
R 6	R 6 (AaT)	R 6 (U 7)	AA	13.5	50	7	15
R 7	R 7	R 7	–	16	17	3.4	7
R 8	R 8 (A)	R 8	A	16	50	10	21
R 9	R 9	–	–	16	6	1.2	2.5
R 10	R 10 (CT)	R 10 (U 6)	(BR)	20	37	11	20
–	–	–	(BF)	20	41	10	20
R 12	R 12 (DT)	R 12 (U 4)	B	20	59	15	35

(continued)

TABLE 29a (continued)

Designation and Approximate Dimensions of Dry Cells and Batteries

IEC[a]	DIN 40 855	British Standard	U.S. Standard	Diameter (mm)	Height (mm)	Volume[b] (cm^3)	Weight[b] (gm)
R 14	R 14 (ET)	R 14 (U 5)	C	24	49	20	45
R 15	R 15	R 15	(CL)	24	70	30	60
R 17	R 17	-	-	25.5	17	9	19
R 18	R 18 (FT)	R 18	(CD)	25.5	83	41	90
R 19	-	-	-	32	17	14	30
R 20	R 20 (JT)	R 20 (U 2)	D	32	61	45	100
R 22	R 22	R 22 (U 1)	E	32	75	58	130
R 25	R 25 (JaT)	R 25	F	32	91	70	160
R 26	R 26 (JbT)	R 26	G	32	105	80	180
R 27	R 27	R 27	J	32	150	120	270
R 40	R 40 (EMT)	R 40	No. 6	64	166	485	1000

[a]International Electrotechnical Commission.

[b]Approximate.

TABLE 29b

Designation and Approximate Dimensions of Dry Cells and Batteries

IEC[a]	DIN 40 855	British Standard	U.S. Standard	Length (mm)	Width (mm)	Thickness (mm)	Approximate volume (cm^3)
F 15	F 15	F 15	F 15	14.5	14.5	3	0.5
F 20	F 20 (BP 1121)	F 20	F 20	24	13.5	2.8	0.9
F 22	F 22	F 22	–	24	13.5	6	2
F 25	F 25	F 25	–	23	23	6	3
F 30	F 30 (BP 1829)	F 30	F 30	32	21	3.3	2.2
F 40	F 40 (BP 1829)	F 40	F 40	32	21	5.3	3.5
F 50	F 50	F 50	F 50	32	32	3.6	3.6
–	F 60	F 60	F 60	32	32	3.8	3.8
F 70	F 70	F 70	F 70	43	43	5.6	10.5
F 80	F 80	F 80	F 80	43	43	6.4	11.7
F 90	F 90	F 90	F 90	43	43	7.9	14.5
F 92	F 92	F 92	F 92	54	37	5.5	11
–	F 92.1	–	–	46	37	5.5	9
–	F 92.2	–	–	52	45	–	–
F 95	F 95	F 95	–	54	37	7.9	16
F 100	F 100 (BP 4558)	F 100	F 100	60	45	10.4	28

[a]International Electrotechnical Commission.

TABLE 30a

Discharge Tests for Portable Lighting Batteries

Test	Discharge resistance (Ω/cell in series)	Intermittent discharge	End point
A	5	5 consecutive days/week A$_1$: 5 min/day according A$_2$: 10 min/day to the type A$_3$: 30 min/day of cell	0.75 V
B	5	5 min/day, 7 days/week	0.9 V
C	4	5 min/day, 7 days/week	0.9 V
D	4	4-min periods at hourly intervals for 8 consecutive hours every day, with 16-h rest periods intervening. (There are eight such discharge periods each day, or a total daily discharge of 32 min.)	0.9 V

Storage before delayed tests: for R 6 and smaller batteries, 3 months; for all other batteries 6 months.

TABLE 30b

Characteristics of Portable Lighting Cells and Batteries

Type	Rated voltage (V)	A_1	A_2	A_3	Test B	Test C	Test D
R 03	1.5	35	-	-	25	-	-
R 6	1.5	75	-	-	-	45	-
R 10	1.5	90	-	-	60	-	-
R 12	1.5	-	210	-	-	150	-
R 14	1.5	-	240	-	-	190	-
R 20	1.5	-	-	690	-	600	480
2 R 10	3.0	90	-	-	60	-	-
2 R 22	3.0	-	-	1080	-	800	660
3 R 12	4.5	-	210	-	-	150	-
3 R 20	4.5	-	-	690	-	600	480
4 R 25	6.0	-	-	1380	-	-	840

The column headers under "Minimum duration (min)" are: Test A (spanning A_1, A_2, A_3), Test B, Test C, Test D.

TABLE 31a

Discharge Tests for Radio Batteries[a]

Test	Discharge resistance (Ω/cell in series)	Intermittent discharge	End point (V/cell in series)
A	40	4 h/day, 7 days/week	0.9
B	75	4 h/day, 7 days/week	0.9
C	150	4 h/day, 7 days/week	0.9

Leakage test: The battery shall show no leakage of electrolyte when the discharge is continued to 0.6 V per cell.

Storage before delayed tests: for R 6 batteries, 3 months; for all other batteries, 6 months.

[a]Transistor sets and similar applications.

TABLE 31b

Characteristics of Radio Cells and Batteries

Type	Rated Voltage (V)	Minimum duration (h)		
		Test A	Test B	Test C
R 6	1.5	–	–	65
R 14	1.5	–	80	–
R 20	1.5	150	–	–
3 R 12	4.5	–	100	–
6 F 22	9	–	–	25
6 F 25	9	–	–	30
6 F 50-2	9	–	40	110
6 F 100	9	–	230	–
6 F 100-3	9	400	–	–

nomical than special designs. The success of newly designed equipment is inextricably related to the reliability, price, and performance of the power source. The designer who presents the battery manufacturer with a completely finished product is taking an unnecessary risk of having to accept a less than optimum choice of battery.

5.2.1. Presentation and Calculation of Battery Capacity

Since the Leclanché system is subject to appreciable polarization, the presentation of capacity data must include details of the discharge conditions.

The total available capacity is directly proportional to the number of equivalents of manganese dioxide, electrolyte salts, and zinc in the cell, but this "theoretical" capacity is modified by such factors as (a) the cut-off voltage, (b) the discharge frequency, and (c) the current drain. All these factors are a direct consequence of polarization, which causes the closed-circuit voltage to decrease continuously during use. In intermittent discharges, the battery is able to recover during the rest periods, so that generally more capacity can be obtained than on continuous use. Further, polarization is related to the current drain by the well-known Tafel equation

$$\eta = a + b \log i$$

so that more effective capacity is available the lighter the current drain. However, for the reasons discussed in Section 3, there is a limit to the increase in capacity obtainable by reducing the current drain and the discharge frequency.

The technical information supplied by battery manufacturers includes the nominal voltage, dimensions, terminal arrangements, weight, and capacity under a variety of discharge conditions. The capacity is usually expressed as service life, sometimes as ampere-hours, and only rarely in watt-hours. Very often, discharge curves are presented for a certain resistive load or current drain.

In order to illustrate discharges that may range from a few hours at high currents to hundreds of hours at light drains in one diagram, a logarithmic abscissa may be used.

Figure 102 shows the same discharge curve in a linear and also in a semilogarithmic coordinate system. Because of the different methods of representing the time scale, the same curves have quite different forms.

Another method of illustrating discharges of different duration is shown in Fig. 103, where the voltage is plotted against n (or Δn) in MnO_n. Such diagrams are doubly useful, since not only do they allow a complete range of discharges at all current drains to be seen at a glance, but also provide information regarding the polarization of the cell at different loads. Since Δn is proportional to the number of ampere-hours delivered by the cell, the abscissa may be directly calibrated in ampere-hours. Such diagrams are simple to construct for discharges at constant current. For discharges through fixed resistances, the calculation is much more complicated (see below). If the amount of manganese dioxide in the cell is known, Table 12 in Section 3 can be used to simplify the conversion of ampere-hours to Δn or vice versa.

A discharge curve is usually prepared by plotting voltage against time in Cartesian coordinates. If the discharge is through a fixed resistance, the ampere-hours delivered may be calculated by one of the following methods.

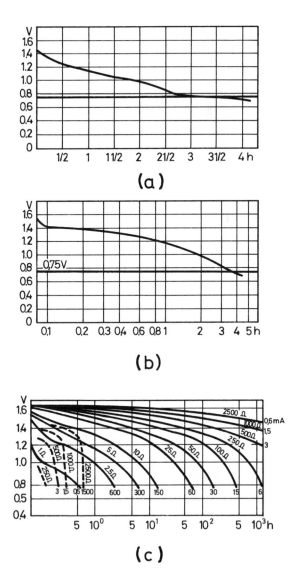

FIG. 102. Discharge curves of D-size cells: (a) in linear Cartesian coordinates (5-Ω curve); (b) in linear/logarithmic coordinates (5-Ω curve); (c) cells discharged through different resistances, linear/logarithmic coordinates. The current (mA) is the load at the beginning of discharging.

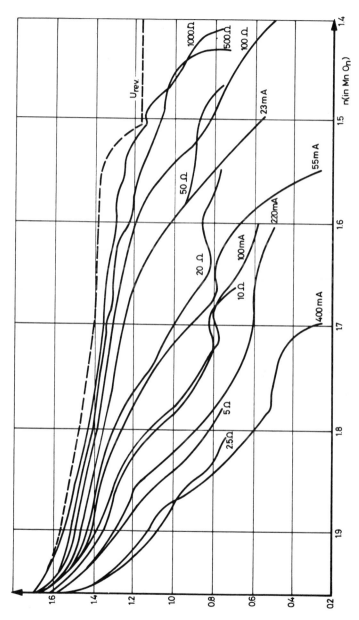

FIG. 103. Discharge curves in voltage versus n in MnO_n diagram.

a. The total service life T is divided into N equal periods (each t) and
the voltage at the end of each period noted $(U_1, U_2, \ldots, U_n)$. U_0 is the
initial on-load voltage and R_e the load resistance. Then the capacity is
given by

$$Ah = \frac{t}{R_e}(U_0 + U_1 + U_2 + \cdots + U_n - \frac{U_0 + U_n}{2})$$

Example: For a given test, let $R_e = 5\ \Omega$ and t = 2 h. Assume that the
service life of the cell ends when the on-load voltage drops to half its initial
value. Suppose the following measurements were obtained.

Duration (h)	0	2	4	6	8
On-load voltage (V)	1.5	1.1	1.0	0.9	0.75

The capacity is

$$Ah = \frac{2}{5}(1.5 + 1.1 + 1.0 + 0.9 + 0.75 - \frac{1.5 + 0.75}{2})$$

$$= 1.65\ Ah$$

The accuracy of the calculation is increased by making t smaller.

b. The area under the voltage versus time curve may also be estimated
(1) by use of a planimeter, (2) by counting the squares under the curve (on
graph paper), or (3) by cutting out the area bounded by the curve and weighing.

It is usual to shorten the voltage axis in drawing discharge curves, for
instance between 1.6 and 0.6 V. The area under the curve, however, must
include the complete ordinate down to zero volts.

Having determined the area under the curve, it is now necessary to
calculate the conversion factor from the scale of the drawing and the load
resistance.

Example: The area under a given curve is found to be 180 cm^2. Conversion factors (from the graph) are

$$1\ V = 10\ cm \quad and \quad 100\ h = 10\ cm$$

Multiplying,

$$100\ Vh = 100\ cm^2$$

If the load resistance R_e is 40 Ω, then

$$\frac{100\ Vh}{40\ \Omega} = 2.5\ Ah = 100\ cm^2$$

$$ampere\text{-}hours\ delivered = \frac{180}{100}(2.5) = 4.5\ Ah$$

c. The foregoing methods of calculating ampere-hours from discharge curves have largely been replaced by the introduction of computer techniques. Most large manufacturers now have computers available that enable service-life, ampere-hours, and watt-hours to be calculated directly from discharge data.

In Figs. 104 and 105, the capacity of a battery is shown graphically in a form to be found in some technical information catalogs. Tabulated values of service life and ampere hours to various cut-off voltages at various drains and under different discharge frequencies are also usually found. In using such information, one must be careful to select the conditions approximating most nearly the anticipated use of the battery.

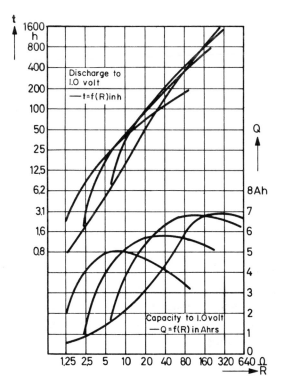

FIG. 104. Electrical capacity of dry cells: service life in hours (upper left) and ampere-hours (lower right).

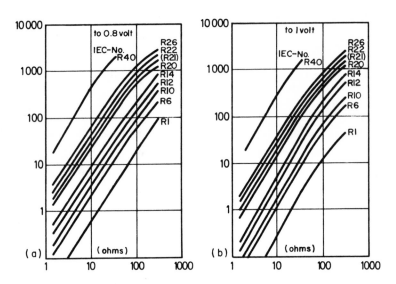

FIG. 105. Service life of various sizes of cells in hours (a) to 0.8 V
and (b) to 1.0 V cut-off voltage.

5.2.2. Causes of Battery Failure and Complaints

As is the case with all mass-produced articles, a certain percentage of
defective products must be accepted in battery manufacture. Even in well-
organized production units equipped with the most modern and sophisticated
machines and employing sensitive quality-control procedures, it is not always possible to eliminate every potentially defective cell. Some causes
of failure that are difficult to detect at an early stage are defective seals,
substandard insulation, porosity in plastic materials, and so on. Such
batteries may eventually give cause for complaint, but the incidence of such
justifiable complaints is negligibly small. Experience shows that in many
cases, batteries are returned as defective and examination fails to reveal
any due cause. Before returning batteries as defective, the user should
therefore consider

a. Why are the batteries considered defective? (Dimensions, leakage,
electrical properties.)

b. How long have they been stored? When and where were they purchased?

c. Have they been improperly handled? (Mechanically damaged, exposed to moisture, stored at high temperature.)

d. Was the equipment in which they were used in good condition?

e. Was the equipment properly used? Switched off after use?

f. Were the batteries inserted properly? (Making good contact, with proper polarity.)

g. Was the complete set of batteries in the equipment renewed, or only part of the set?

h. Have attempts been made to reactivate the batteries? (By heating, by charging.)

It is not the author's intention to claim that defective batteries never find their way to the user, but experience shows that this is a comparatively rare event. If the user would cooperate by considering the points enumerated above, much distress and inconvenience could be avoided.

5.2.3. Recharging Dry Batteries

As previously stated, dry cells cannot be recharged as effectively as storage batteries. It is, however, possible to reactivate a dry cell, provided that certain conditions are fulfilled. One of the most important of these is that discharge and charge should follow each other in regular cycles. It is impossible to recharge a fully discharged dry cell. Furthermore, over-charging must be avoided, since this leads to generation of gas in the cell which may cause swelling. The necessity for regular discharge/recharge cycles makes it virtually impossible for the ordinary battery user to have consistent success with equipment having built-in recharger units.

However, there is theoretical interest in this problem, which justifies the following discussion. Of primary importance is the avoidance of excessive charging voltages, which lead to decomposition of the electrolyte and in some cases to the liberation of chlorine. Several techniques have been proposed; one described by Beer (see Fig. 106) may be taken as

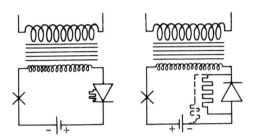

FIG. 106. Circuit diagram of a charging unit for dry cells.

typical. The charging is affected with direct current containing an alterna-
ting-current component. The voltage applied to the cell being charged must
not exceed the voltage of a new and unused cell. The use of an ac component
induces a more regular and compact deposition of zinc on the anode, as has
been demonstrated in the electroplating industry. Similar results were found
with dry cells. Zinc deposited by using pure dc was unevenly distributed
and spongy in texture, and after relatively few cycles the zinc anode was
perforated. With a dc/ac mixture, however, the deposit was much more
evenly distributed and dense. From experience obtained up to the present
time, the following are the essential conditions for successful recharging
of dry cells.

(a) The charging voltage must not exceed 1.7 V/cell.

(b) The charging current should lie between 25 and 35% of the discharge
current.

(c) The charge time should be 4 1/2 to 6 times the discharge period, in
view of the low efficiency of the charging process.

(d) The effect of recharging is greater the shorter the discharge period.
During one discharge period, no more than 10% of the total ampere-hour
capacity of the battery may be withdrawn.

(e) The charging period is most effective if started directly after discharge has ceased.

(f) Too heavily or completely discharged dry cells cannot be recharged.

5.3. ADVICE TO YOUNG BATTERY TECHNOLOGISTS

5.3.1. Sources of Defects in the Manufacture of Dry Batteries

Major errors in production technique seldom present any problem in their detection and elimination. Impure raw materials and defective carbon rods, for instance, are easily recognized. Some faults, however, develop very slowly and become apparent only after storage. When the shelf period is greater than one month, the cause of failure may be difficult to identify. Some of these sources of battery failure are described in the following paragraphs.

In cold-setting pasted cells, it is sometimes observed that after several weeks, the paste contains many gas bubbles and the zinc can is unduly corroded. At this stage, the paste layer is somewhat leathery and separates easily from the core and the metal. This condition arises when the setting time is too short. Assuming that there have been no changes in raw materials or mix specifications, the trouble may be traced to (a) high viscosity of the paste during cell construction (due to age or ambient temperature during storage), (b) high air content of the bobbin (low moisture content or density), or (c) increased rate of setting (temperature during cell construction). The direct cause of the trouble is, of course, that the air present in the bobbin is unable to escape through the paste layer rapidly enough. Examination of cells shortly after the paste has set usually give little or no indication of potential failure.

Similar failures occur in paper-lined cells when the coated separator is not evenly pressed into contact with the zinc can or when folds are formed during the cell assembly process. The paste used to coat the paper separator is usually based on naturally occuring cereals. The degree to which gelation has progressed during the lamination process is important, since the gel must penetrate into the individual paper fibers in order to displace air and prevent pitting at the zinc surface. Examination of faulty cells shows that the paste layer is easily separated from the paper carrier. Steel-jacketed cells having this type of fault develop high internal pressures due to the gas generated. The flash current decreases steadily with age, but it is characteristic that the flash current rises considerably when the cell is perforated and the excess gas is released. The variation in the quality of the paste coating can be caused by unsuitable methods of paste preparation and/or the coating process.

Other causes of failure are found in carelessness in production operations, such as inaccuracy in the wet mixing time during depolarizer mix preparation and improper setting of machines during tamping or during cell assembly. Other examples are poor adjustment of the electrolyte paste volume in pasted cells and defective pressing of the mix cylinder in paper-lined cells. When cells of the same dimensions are made with cathode mixes of different physical properties, it is advisable to allocate the two types to different assembly units. When both types are assembled on the same unit, the necessary changes in the setting of the tools are troublesome and sooner or later may be overlooked, thus leading to a defective product.

The internal layout of a production unit must take into account the possibility of contamination of an intermediate product by airborne impurities arising from some other operation. Thus a paper-coating unit should not be situated in the proximity of a metal grinder. This cause of trouble usually arises sporadically during temporary alterations or repairs.

The efforts of a value engineering team, if not properly directed, may result in a gradual depreciation of quality. In itself, value engineering is a sound and reasonable activity, when directed toward obtaining a given quality at the lowest price. However, when the proposals of such a team are accepted without adequate testing, there is an inherent danger of a fall-off in quality that is particularly difficult to locate and eradicate.

Attempts to increase the electrical capacity of the cell may also result in a more rapid depreciation on the shelf when certain practical rules based on experience are disregarded. The simplest method of increasing the capacity is to increase the volume of the cathode (diameter and/or height). However, should such changes involve the infringement of the safety limits based on manufacturing experience, the final effect can be an unexpected and undesirable reduction in the quality level. Increasing the diameter of the cathode invites mix shorts during cell assembly and internal short-

circuiting during discharge. Increasing the height of the core may lead to
increased leakage when the air space is correspondingly reduced, or alternatively, to inadequate sealing•or cap corrosion.

The following two examples of what can happen even in well-organized
production units are taken from practical experience.

Example 1: In a hot-setting pasted cell process, it was observed that
from a certain point onward, all the bobbins rose 3-4 cm during the setting
process. A check of changes made revealed that a new delivery of ammonium chloride had been made. The raw-material inspection laboratory
accepted the delivery as being of standard quality, on the basis of the usual
sampling procedure. A recheck of the material showed that part of the
consignment consisted of bags of ammonium nitrate.

Example 2: Coinciding with the use of a new shipment of manganese
dioxide, it was found that some cathode mixes showed unusual characteristics.
Examination of the faulty mixes showed that an appreciable amount of oil
had found its way into the product. The source of the oil was eventually
traced to individual bags of ore that had been contaminated by lubricating
oil during the grinding process. Production could not be started again until
a simple test had been devised; each bag of the suspect consignment of ore
was examined and the faulty material identified.

In such cases, there is no alternative but to stop production and ensure
that the suspect raw materials and mixes are available for detailed examination. The object of the foregoing discussion is to impress on the young
battery technologist the importance of accurate observation and of maintaining detailed records of all production changes. In the complexity of the
battery-making process, any change, however small, may have quite unan-
ticipated consequences. Experiments should be limited to the laboratory
and pilot plant and not extended into a production unit.

5.3.2. Storage of Dry Batteries

For delivery of batteries to military departments, it is usually specified
that the individual cells must be stored at least four weeks before battery

assembly. It is recommended that this procedure be followed wherever stringent requirements are necessary, such as export to tropical countries, the introduction of a new battery type, and so on. This preassembly storage is usually sufficient to allow the slowly developing causes of failure to have progressed to such a degree that potentially defective cells can be recognized during electrical testing.

When prolonged storage periods are anticipated, low-temperature storage is advantageous. Normally, temperatures around $0°C$ are satisfactory, but for special purposes $-18°$ to $-20°C$ is preferred. Some purchasing authorities prescribe $-32°C$. An extensive test involving many types of batteries was carried out by the author in which the storage conditions were two years at $-18°C$, followed by one year at $20°C$. Discharge tests proved that immediately after the cold-storage period, the batteries gave capacities equivalent to the initial values and that during the subsequent storage at room temperature the rate of capacity loss was essentially the same as that observed on new batteries (8-12% after 1 yr). In Europe, opinions are divided regarding storage at $-32°C$. Some reports indicate that although the battery remains in good condition during the freezing, the shelf life after thawing out is somewhat restricted. It is plausible that the methods of freezing and thawing, and particularly the nature of the packaging, may be of importance in this type of operation.

ACKNOWLEDGMENTS

The author wishes to express his sincere thanks to the Board of Varta AG for permission to publish this material. He is deeply indebted to Dr. G. Lander for promoting this work and to Mr. G. S. Bell for discussions and for his critical review and translation of the original manuscript.

REFERENCES

1. W. König, Forsch. Fortschr., 14 (1), 8 (1938).

2. Hellesens-Katalog, Paris, 1900; G. Norrie, Hospitals-Tidende (Copenhagen), VI (11), 273 (1888).

3. Zierfuss, German Pat. 49,423 (1888).

4. Bender, German Pat. 48,695 (1888).

5. A. Jeckel, Varta Report, No. 3, 6 (1967).

6. C. Drotschmann, Trockenbatterien, Akadem. Verlagsgesellschaft Becker & Erler, Leipzig, 1945.

7. C. Drucker and A. Finkelstein, Galvanische Elemente und Akkumulatoren, Akademische Verlagsgesellschaft m.b.H., Leipzig, 1932.

8. H. F. French, Proc. I.R.E., 29, 299 (1914).

9. G. J. Nowotny, U.S. Pat. 2,745,894 (1956) (Ray-O-Vac).

10. R. Pörscke, German Pat. 360,660 (1922).

11. J. P. Teas, U.S. Pat. 2,605,299 (1950) (Union Carbide).

12. B. Priebe, German Pat. 1,238,082 (1962) (Varta).

13. H. R. C. Anthony and H. G. Friang, U.S. Pat. 2,392,795 (1946) (Ray-O-Vac).

14. H. R. C. Anthony U.S. Pat. 2,198,423 (1940) (Ray-O-Vac).

15. French Pat. 1,423,094 (1964) (Soc. Piles Wonder).

16. French Pat. 1,507,299 (1966) (Soc. des Accumulateurs Fixes et de Traction).

17. French Pat. 1,557,270 (1968) (Kapsch & Söhne).

18. W. Riedl and W. Wild, German Pat. 1,061,852 (1959) (Varta).

19. W. Krey, British Pat. 1,170,480 (1967) (Varta).

20. N. C. Cahoon, U.S. Pat. 2,534,336 (1950) (Union Carbide).

21. N. C. Cahoon and M. P. Korver, German Pat. 1,046,710 (1956) and U.S. Pat. 3,092,518 (1963) (Union Carbide).

22. French Pat. 1,437,308 (1965) (Ever Ready Co.).

23. German Pat. 709,475 (Leclanché S. A.).

24. German Pat. 711,102 (Leclanché S. A.).

25. German Pat. 630,024 (I. G. Farbenindustrie AG).

26. M. Bolen, B. H. Weil, Literature Search on Dry Cell Technology, Survey of Electrolytic Synthesis of Battery Active Manganese Dioxide, Georgia Inst. Technology, Atlanta, 1949.

27. R. S. Dean and A. L. Fox, German Pat. 922,882 (1955).

28. A. Kozawa, J. Electrochem. Soc., 106, 552 (1959).

29. J. P. Gabano, P. Etienne, and J. F. Laurent, Electrochim. Acta, 10, 947 (1965).

30. J. Muller, F. L. Tye, and L. L. Wood, Proc. 4th Intern. Symp., Brighton, 1964, Pergamon, New York, 1965, p. 201.

31. K. Sasaki, Mem. Faculty Eng. Nagoya Univ. 3 (2) 81, (1951).

32. W. C. Vosburgh, J. Electrochem. Soc., 99, 319 (1952).

33. R. Kunin, Ion Exchange Resins, 2nd ed., Wiley, New York, 1958.

34. A. Tvarusko, Investigation of Manganese Dioxides Surface Area and Water Content, Carl F. Norberg Research Center, Electric Storage Battery Company, Yardley Research Rep. 7499 (12) 2 (September 1963).

35. C. K. Morehouse, W. J. Hamer, and G. W. Vinal, "Effects of inhibitors on corrosion of zinc in dry cells electrolytes," J. Res. Natl. Bur. Std. (U.S.) 40, 151 (1948).

36. F. Aufenast and J. Muller, Proc. 3rd Intern. Battery Symp., Pergamon, New York, 1962, p. 335.

37. H. D. Holler and L. M. Ritcher, Trans. Electrochem. Soc., 37, 607 (1920).

38. P. M. Thompson, Ind. Eng. Chem., 20, 1176 (1928).

39. C. Drotschmann, Z. Elektrochem., 35, 194 (1929).

40. N. C. Cahoon, Trans. Am. Elec. Soc., 58, 177 (1935).

41. H. F. McMurdie, Trans. Electrochem. Soc., 86, 313 (1944).

42. K. Sasaki, Mem. Faculty Eng., Nagoya Univ. 3, (2) (1951).

43. R. S. Johnson and W. C. Vosburgh, J. Electrochem. Soc., 99, 319 (1952).

44. P. Brouillet and F. Jolas, Electrochim. Acta, 6, (1/4), 245 (1962).

45. W. Feitknecht, H. R. Oswald, and U. Feitknecht-Steinmann, Helv. Chim. Acta, 43, VIII, 239 (1960).

46. J. Brenet, in 8th Réunion du CITCE, Madrid 1956 (C.R. CITCE, Ed.), Butterworth, London, 1958 p. 394.

47. A. D. Wadsley and A. Walkley, Electrochem. Soc., 95, 11 (1949).

48. W. C. Vosburgh and Pao-Soong-Lou, Electrochem. Soc., 108 (6), 485, (1967).

49. G. Gattow, Batterien, 15, 201 (1961).

50. H. Bode, A. Schmier, and D. Berndt, Z. Elektrochem., 66, 586 (1962).

51. G. S. Bell and R. Huber, J. Electrochem. Soc., 111, 1 (1964).

52. J. P. Gabano, B. Maurignac, and J. F. Laurent, Electrochim. Acta, 9, 1093 (1964).

53. P. Brouillet, A. Grund and F. Jolas, Compt. Rend. Acad. Sci. Paris, 257, 3166 (1964).

54. A. Kozawa and J. F. Yeager, J. Electrochem. Soc., 112, 959 (1965).

55. K. J. Vetter, J. Electrochem. Soc., 110 (6), 597 (1963).

56. K. J. Vetter and N. Jaeger, Electrochim. Acta, 11, 401 (1966).

57. G. S. Bell, unpublished report, VARTA GmbH, Ellwangen.

58. W. C. Vosburgh and R. S. Johnson, J. Electrochem. Soc., 99, 317 (1952).

59. W. C. Vosburgh and R. S. Johnson, J. Electrochem. Soc., 100, 471 (1953).

60. K. Neumann and E. von Roda, Z. Elektrochem., 96, 347 (1965).

61. A. Keller, Z. Elektrochem., 37, 342 (1931).

62. G. J. Coleman, Trans. Electrochem. Soc., 90, 545 (1947).

63. A. B. Scott, J. Electrochem. Soc., 107, 941 (1960).

64. W. C. Vosburgh and D. T. Ferrel, J. Electrochem. Soc., 98, 334 (1951).

65. A. Kozawa, J. Electrochem. Soc., 106, 79 (1959).

66. K. Kornfeil, J. Electrochem. Soc., 109, 349 (1962).

67. A. Era et al., Electrochim. Acta, 13, 207 (1968).

68. J. P. Brenet, Electrochim. Acta, 1, 231 (1959).

69. J. P. Brenet, Proc. 4th Intern. Symp. Brighton, 1964, Pergamon, New York, 1965, pp. 2, 247.

70. R. Huber, Electrochim. Acta, 2, 258 (1960).

71. R. Huber and J. Kändler, Electrochim. Acta, 8, 265 (1963).

72. J. P. Brenet, A. Grund, Compt. Rend. Hebd. Séance. Acad. Sci. Paris, 240, 1210 (1955).

73. A. Era et al., Electrochim. Acta, 12, 1199 (1967).

74. I. Muraki et al., J. Chem. Ind. Tokyo, 51, 64 (1948).

75. J. P. Brenet, Chimia, 23, 444 (1969).

76. K. Sasaki and A. Kozawa, Denki Kakagu, J. Electrochem. Soc. Japan, 22, 569 (1954).

77. O. Glemser, G. Gattow, and H. Meisiek, Z. Anorg. Allg. Chemie, 309, 121 (1961).

78. S. Gosh and J. P. Brenet, Ber. Bunsenges. Physik. Chem., 67, 723 (1963).

79. N. C. Cahoon and G. W. Heise, Trans. Electrochem. Soc., 94, 214 (1948).

80. N. C. Cahoon, R. S. Johnson, and M. P. Korver, J. Electrochem. Soc., 105, 296 (1958).

81. M. P. Korver, R. S. Johnson and N. C. Cahoon, J. Electrochem. Soc., 107, 587 (1960).

82. W. C. Vosburgh, R. S. Johnson, J. S. Reiser, and D. R. Allenson, J. Electrochem. Soc., 102, 151 (1955) and W. C. Vosburgh, A. M. Chreitzberg, Jr. and D. R. Allenson, J. Electrochem. Soc., 102, 557 (1955).

83. H. F. McMurdie, Trans. Electrochem. Soc., 86, 313 (1944).

84. L. C. Copeland and F. S. Griffith, Trans. Electrochem. Soc., 89, 495 (1946).

85. N. C. Cahoon, J. Electrochem. Soc., 105, 296 (1958).

86. K. J. Euler, Z. Elektrochem. Ber. Bunsenges., 63, 1008 (1959).

87. K. J. Euler, Electrochim. Acta, 7, 205 (1962).

88. K. J. Euler, Z. Angew. Physik, 29, 264 (1970).

89. T. Hirai and M. Fukuda, J. Electrochem. Soc. Japan, 29, 3, E 166 (1961).

90. K. Miyazaki, J. Electrochem. Soc., 116, 1469 (1969).

91. W. Feitknecht and W. Marti, Helv. Chim. Acta, 28, 129 (1945).

92. M. Pourbaix, Atlas D'Equilibres Electrochimiques, Gauthier-Villars, Paris, 1963.

93. P. Brouillet and F. Jolas, Electrochim. Acta, 1, 246 (1959).

94. W. Feitknecht and R. Petermann, Korros. Metallschutz, 19, 181 (1943).

95. W. Feitknecht, Werkstoffe Korros. Mannheim, 16, 15 (1955).

96. U. R. Evans and D. E. Davies, J. Chem. Soc., Part 3, 2607 (1951).

97. H. J. Schmecken, Korrosion, 13, 65 (1960).

98. W. Katz, Metalloberfläche, 11, 125 (1957).

99. C. Drotschmann, Korros. Metallschutz, 19, 188 (1943).

100. G. Masing and G. Mohldenke, Z. Metallk., 4, 406 (1950).

101. G. S. Bell, Electrochim. Acta, 13, 2197 (1968).

102. H. Krug and H. Borchers, Electrochim. Acta, 13, 2203 (1968).

103. K. Miyazaki, J. Electrochem. Soc., 117, 821 (1970).

104. G. Gouy, J. Phys., 4, 9, 357 (1910).

105. D. L. Chapman, Phil. Mag., 25, 775 (1913).

106. O. Stern, Z. Elektrochem., 30, 508 (1924).

107. D. Graham, Chem. Rev. 41, 441 (1947); Z. Elektrochem., 59, 773 (1955).

108. J. A. V. Butler, Trans. Faraday Soc., 19, 729 (1924).

109. T. Erdey-Gruz and M. Volmer, Z. Physik. Chem., Abt. A, 150, 203 (1930).

110. R. W. Gurney, Proc. Roy. Soc. (London), Ser. A, 134, 137 (1931).

111. A. Frumkin, Z. Physik. Chem., Abt. A, 164, 121 (1933).

112. J. Horiuti and M. Polanyi, Acta Physicochim. USSR, 2, 505 (1933).

113. R. Audubert, J. Phys. Radium, 3, 81 (1942); see also K. J. Vetter, Z. Elektrochem., 59, 596 (1955).

114. N. Ibl, W. Rüegy, and G. Trümpler, Helv. Chim. Acta, 63, 1624 (1953); 37, 583 (1954).

115. N. Ibl and R. Müller, Z. Elektrochem., 59, 671 (1955).

116. L. Gierst and A. Juliard, J. Physik, Chem., 57, 701 (1953).

117. L. Gierst, Z. Elektrochem., 59, 784 (1955).

118. P. Delahay and T. Berzins, J. Am. Chem. Soc., 75, 2486 (1953).

119. K. J. Vetter, Elektrochemische Kinetik, Springer, Berlin, 1961.

120. J. N. Agar and J. L. Bowden, Proc. Roy. Soc. (London), 169A, 206 (1939).

121. R. Huber and J. Bauer, Electrochem. Tech., 5, 11, 542 (1967).

122. H. E. Lawson, Trans. Electrochem. Soc., 68, 187 (1936).

123. G. S. Bell, Electrochem. Tech. 5, 513 (1967).

124. C. M. Shephard, J. Electrochem. Soc., 112, 252 (1965).

125. N. C. Cahoon, J. Electrochem. Soc., 99, 343 (1952).

126. F. Kornfeil, J. Electrochem. Soc., 106, 1062 (1959).

127. K. Appelt and H. Purol, Electrochim. Acta, 1, 326 (1959).

128. H. Schweigart, MnO$_2$ Project 6061/4516, Progress Report No. 2, South Africa, 1965.

129. G. S. Bell and J. Bauer, Electrochim. Acta, 14, 453 (1969).

130. W. D. Staley and A. J. Helfrecht, Trans. Electrochem. Soc., 53, 93 (1928).

131. W. J. Hammer, J. Res. Natl. Bur. Stds. (U.S.), 39, 29 (1947) (RP 1810); 40, 251 (1948) (RP 1870).

132. R. Huber, Paper presented at CITCE Symposium, Eindhoven, 1969 (in press).

133. N. C. Cahoon, U.S. Pat. 2,534,336 (1950); German Pat. 1,046,710 (1956).

134. Sindel and R. Huber, German Pat. 1,041,551 (1956).

135. R. Huber, et al., Internal development laboratory reports, Varta G.m.b.H., Ellwangen.

136. W. Krey, German Pats. 1,596,308 (1971) and 1,771,082 (1971).

137. Varta G.m.b.H., "Gold" series, List nos. 280, 281, 282.

CHAPTER 2

ALKALINE MANGANESE DIOXIDE ZINC BATTERIES

Karl V. Kordesch

Union Carbide Corporation
Battery Products Division
Parma, Ohio

1. INTRODUCTION

1.1. HISTORY OF THE ALKALINE MnO_2-Zn CELL (1882-1960)

The first description of an alkaline MnO_2 cell is probably the German patent of G. Leuchs, issued in 1882, claiming a galvanic element consisting of a manganese dioxide-carbon plate, a potassium or sodium hydroxide solution, and a tin plate. The element was "regenerable by an electric current" [1].

S. Yai obtained a U.S. Patent in 1903 on another wet alkaline cell, which employed a cathode consisting of an MnO_2-graphite mix with a carbon rod as collector, potassium hydroxide as electrolyte, a porous separator, and a zinc anode with zinc oxide around it [2].

The first "dry cells" of the alkaline MnO_2 type were patented in 1912 by E. Achenbach. He used a "gelatinous filling of alkaline lye" (starch gel) [3] and KOH or NaOH with 10% MgO [4] as immobilizing agent for the electrolyte. He added HgO to the MnO_2 [5] and realized the advantages of a multi-plycorrugated zinc wire fabric electrode (embedded in jellylike electrolyte) as anode [6]. A. Heil used NaOH in his MnO_2-Zn cells [7]. However, none of these earlier cells achieved commercial importance, and the literature of the decades between the two world wars shows little activity in the field of alkaline MnO_2 cells, while the Leclanché types received much attention [8].

Not until 1936 was another patent obtained for a wet MnO_2-Zn cell [9]. The development of alkaline primary cells continued mainly in the fields of the copper oxide cells, the air-depolarized cells, and the HgO cell [10].

The mercury oxide-zinc cell with caustic electrolyte, known since 1917 (Bronsted), is the nearest relative to the alkaline MnO_2-Zn cell. Many requirements are the same, and the technology of the HgO-Zn cell, which became important during World War II, stimulated renewed interest in the alkaline MnO_2 cell. Ruben obtained numerous patents on HgO cell improvements which directly suggested the construction of similar MnO_2 cells [11-14]. The pioneering work of W. S. Herbert resulted in the first commercial alkaline MnO_2-Zn cell. His patents appeared in the mid 1950s [15-17], and the "crown" cell was first described in 1952 in a short publication [18].

Principal features claimed for the crown cell were higher capacity per unit volume, higher current drain than obtainable from Leclanché cells, no leakage during or after discharge, minimum gas evolution (no venting needed), gradual voltage drop at the end of discharge (in contrast to HgO), and use of a low-cost depolarizer that is insoluble in the electrolyte and has no reaction with the separator (in comparison with Ag_2O cells).

The construction of his cells employed some of the features of present cells. A plain steel can made contact with the depolarizer mass, a mixture of finely divided MnO_2 and graphite pressed into pellet shape. Tin-plated steel was used as current collector to the amalgamated zinc anode, which consisted of a plurality of discs or a cake of pressed zinc powder having sufficient porosity to hold some electrolyte. The main electrolyte carrier was an absorbent separator or a gelled caustic solution. KOH or NaOH with zinc oxide additions were utilized as electrolyte.

However, from today's point of view, Herbert's cells were still low-drain units. They were produced in commercial quantities only as 9-V batteries for transistorized radios. No attempt was made to produce larger cylindrical cells for heavy-drain usage. Indeed, there was a feeling that the system was capable of only low-drain usage.

It took the P. R. Mallory Company several years to achieve some marketing success with these early alkaline MnO_2-Zn cells.

This picture changed considerably after further developments revealed that the alkaline MnO_2-Zn system could deliver far higher currents and still provide more ampere-hour capacity than the Leclanché system. Extensive work was done in the laboratories of the Union Carbide Corporation to achieve this goal.

Essential for the improvements were the changes from the conventional pellet or block (bobbin) type to an (outside) sleeve type cathode [19], the use of additional binders and conductive materials in the MnO_2 mix [20, 21] and the use of improved large-surface, high-porosity powder-zinc anodes with provisions for supplemental electrolyte [22] or suspending zinc particles in gelled KOH electrolyte [23, 24].

The period of 1960-1970, which may be called the modern age of the alkaline MnO_2 cell, brought a very large number of innovations and improvements. These were reflected not only in the new patents but in many serious research publications, probably initiated by the growing importance of this cell type. The details of this work will be the subjects of the main portion of this chapter.

The alkaline MnO_2-Zn cell started as a primary cell, but in recent years the rechargeability of alkaline MnO_2-Zn cells has been developed to such an extent that inexpensive secondary batteries have been marketed.

These have limited cycle life (50-100) but are competitive if one considers the overall cost of the power source to the consumer.

1.2. COMPARISON OF THE ALKALINE MnO_2-Zn CELL WITH OTHER BATTERY SYSTEMS

1.2.1. Voltage Level

A comparison of galvanic cells can be based on the voltage produced. The electrochemist may first look at the theoretical limits given by thermo-dynamic data. The cell reaction potential is related to the Gibbs free-energy change ΔG as given by the equation $\Delta G = -nFE$, where n is the number of equivalents of charge transferred from the anode to the cathode per mole of reaction, and F is the Faraday constant (9.65 x 10^4 Coulomb/equiv or 26.8 Ah per formula weight and 1 e^- change). The standard free-energy change ΔG^o can be obtained from many chemical oxidation and reduction processes outside of a galvanic cell, and the standard potential E^o calculated from ΔG^o is the thermodynamically predicted potential for the reaction as written. In most cases, the equation is simplified because the detailed reaction mech-anism is complicated or not sufficiently known. For the alkaline MnO_2-Zn cell one can write

$$Zn + 2 MnO_2 \rightarrow ZnO + Mn_2O_3$$

for which ΔG^o is -66.2 kcal and E^o is calculated to be 1.44 V. The negative sign of ΔG means that the reaction is power-producing (proceeding sponta-neously in the indicated direction).

In Table 1 (first and second columns), the theoretical potentials [25, 26] and the actual open-circuit voltages of several practical galvanic systems are listed. The noticeable differences are due to the fact that the conditions in actual cells are different from the standard conditions on which the cal-culations are based.

The position of the alkaline Zn and MnO_2 electrodes on the potential scale can be seen from Fig. 1, which is a simplified selection from the Pourbaix diagrams of zinc and MnO_2 [26].

TABLE 1

Comparison Data for Various Galvanic Systems

System	Thermodynamic E° (V)[b]	Terminal OCV	Average CCV	Theoretical energy content (Wh/kg)[c]	Actual energy content of cells[a]			
					(Wh/kg)	(Wh/dm³)	(Wh/lb)	(Wh/in.³)
Alkaline MnO$_2$-Zn	1.44	1.55	1.25	290	77	215	35	3.5
Leclanché cell	1.78	1.58	1.20	280	66	120	30	2
Mercuric Oxide-Zn	1.35	1.35	1.20	229	88-120	300-500	40-50	5-8
MnO$_2$-Mg	2.8	1.9	1.5	392	88-132	120-240	40-60	2-4
O$_2$-(Air) Depol. -Zn	1.6	1.4	1.2	890	176	180-300	80	3-5
Pb-Acid (R)[d]	2.04	2.2	2.0	167	44	60-120	20	1-2
Ni-Cd (R)	1.48	1.35	1.2	220	33-44	80-120	15-20	1.5-2
AgO/Ag$_2$O-Zn (R)	1.82	1.85	1.5	424	120	240-500	55	4-8
NiOx-Zn (R)	1.7	1.8	1.4	320	44	120-150	20	2-2.5
MnO$_2$-Zn (R)	1.44	1.55	1.25	252	1/3 of primary cell due to 30% cycle depth			

[a] Low current drain, ~ c/10 rate. Data reflect variations of cell constructions.
[b] References [25, 26].
[c] Reference [27].
[d] (R) indicates rechargeable cells.

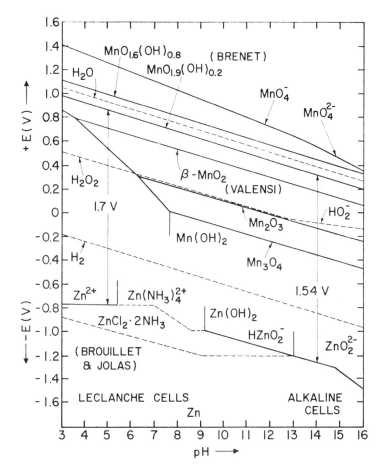

FIG. 1. Simplified Pourbaix diagram of the MnO_2-Zn cell showing electrode potentials as functions of pH [26].

These open-circuit voltages change when a load is applied to the cell because of resistance losses, changes in electrode and electrolyte compositions caused by the current flow, and certain inefficiencies of the reactions. All these effects are usually collected under the name cell polarization.

Components of the polarization losses can be determined and separated by physical and electrochemical methods for which the reader must be referred to corresponding chapters of the electrochemical textbooks listed in Section 7.

However, the shape of the discharge curve under load condition is of
great practical importance. One group of cells (HgO-Zn, Ag_2O-Zn, CuO-Zn,
Air (O_2)-Zn, PbO_2-Pb, Ni-Cd, and so on) has a flat discharge curve with
the voltage dropping slowly over the main portion of the load period, then
rapidly going down as the cell approaches exhaustion. The other group
(including all MnO_2 cells) has a sloping discharge curve in which the slope
varies with construction features and electrolyte choice.

Figure 2 shows examples of HgO and MnO_2 discharge curves. The
degree of the slope depends also on the current-carrying ability of the cell;
for example, the alkaline MnO_2 cells can sustain heavy loads far better
than Leclanché cells.

1.2.2. Output of Cells

The maximum output of different systems is the theoretically calculated
energy content. It can be obtained directly from ΔG^o values by using the
conversion factor 1 kcal = 1.162 kWh or by using the Faraday constant (in

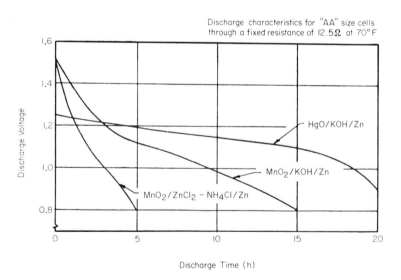

FIG. 2. Typical discharge curves of batteries (HgO, alkaline MnO_2
and Leclanché cells [30].

ampere-hours) and multiplying by the potentials given in Table 1. In each case, the formula weight (239.2 gm for $Zn + 2 MnO_2$, as an example) is the basis of comparison.

The figures obtained are useful only as guidelines, because they assume a package consisting only of the anodic and cathodic materials (100% pure) without electrolyte, cell casing, current collectors, separators, seals, or other items needed to produce a working battery.

The fourth column of Table 1 contains the corresponding maximum energy contents on a weight basis. They have been calculated from the average working voltages of the cells, not from the E^o voltages [27].

Unfortunately, the values of energy densities obtained with actual systems are only one fourth to one tenth of the theoretical figures. And, in addition, they are subject to load conditions and construction features (in somewhat the same manner as the shape of the polarization curve). For comparison, Table 1 contains the actual output data on a weight and volume basis for the listed galvanic cells. These values are derived from commercially sold units and are valid only for low drains. At low currents, the actual energy-content relationship between the cells follows the same sequence as the theoretical figures, indicating a similar "packaging factor" for various systems.

Figure 3 compares different galvanic systems in a diagram that considers the behavior of the cells on load. Originally, this diagram was developed for the comparison of power sources for electric automobiles [28], but it has been extended to cover also systems of the dry-cell field. The added curves represent commercial alkaline MnO_2-Zn cells, heavy-duty and regular Leclanché cells, and HgO cells [29].

It is evident from the curves that the alkaline MnO_2 cell is the best choice of all dry cells under heavy load conditions. The HgO cell has the highest capacity at low current drains. To find the best choice for a given application, the cost picture must also be considered. This will be done in a later chapter.

A comparison of performance between the alkaline MnO_2-Zn cell and other systems for military applications--where cost is not of highest importance--is shown in Fig. 4 [30].

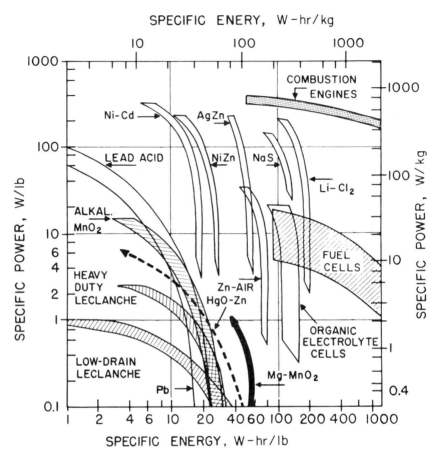

FIG. 3. Chart of specific energy and specific power for battery systems [28].

If alkaline MnO_2 and Leclanché cells are compared at different load levels, it becomes obvious that alkaline cells excel at the higher load levels [31]. This relationship is pictured in Fig. 5.

An even more pronounced difference is shown in a comparison of D-size cell performance on low-temperature duty, as shown in Table 2 [31].

An important consideration for electronic applications is the impedance of cells. Alkaline cells have a far lower resistance, due to the use of KOH

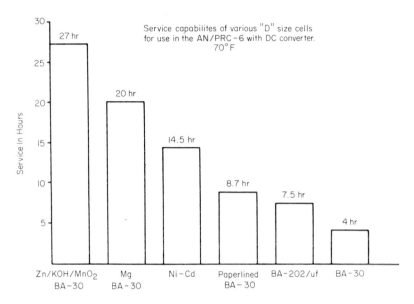

FIG. 4. Alkaline MnO_2-Zn cell compared with other batteries for use in a military application (transceiver) [30].

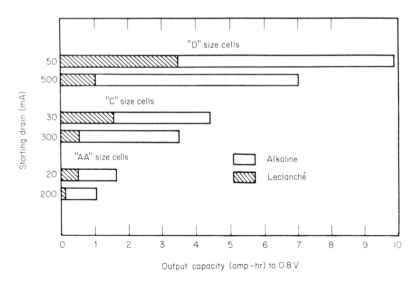

FIG. 5. Alkaline versus Leclanché capacity at room temperature [31].

TABLE 2

Alkaline versus Leclanché Capacity

at Room Temperature and Low Temperatures[a]

Temperature	Service (min)		Capacity (Ah)	
(°F)	Alkaline	Leclanché[b]	Alkaline	Leclanché[b]
70	1200	154	6.93	0.89
32	835	100	4.45	0.58
0	275	50	1.39	0.26
-20	162	9	0.82	0.04

[a]D-size cells, continuous 0.5-A starting drain, service to 0.8 V.
[b]General-purpose flashlight cell.

as electrolyte. From the standpoint of utilization of materials, the efficiency of alkaline cells is also better [32]. These data are collected in Table 3.

The higher efficiency at heavy load levels makes it possible to use smaller size alkaline cells in many applications as well as making possible many new devices requiring high power levels.

1.3. FUTURE POSSIBILITIES

A comparison with other systems would be incomplete if it were not mentioned that the alkaline MnO_2-Zn system is less well developed than other systems, simply because it is not old and less work has been expended on it.

Increases in output per unit weight (or volume) are possible when battery plates are made thinner, mass-transport limitations are removed, and the advantages of the excellent conductivity of KOH electrolyte are fully realized. As will be seen from the progress of battery construction later, battery output also depends greatly on design features. As a rough estimate, it can be said that the actual energy density of a battery is one fourth of the

TABLE 3

Impedance and Efficiency Comparison of Alkaline versus Leclanché Cells

| | D-size cells, continuous discharge at 21°C | |
	Leclanché[a]	Alkaline
Theoretical capacity (Ah)	6.0	10.9
Initial impedance (Ω at 1000 Hz)	0.3	0.04
Final impedance (Ω at 1000 Hz)	2.5	0.045
Hours to 0.65 V at 0.5 Ω	0.5	3.0
1.0 Ω	1.0	6.5
2.25 Ω	3.5	19.0
4.0 Ω	9.0	36.0
10.0 Ω	47.0	85.0
% efficiency to 0.65 V at 0.5 Ω	12	47
1.0 Ω	14	53
2.25 Ω	23	73
4.0 Ω	35	81
10.0 Ω	70	85

[a]General-purpose flashlight cell.

theoretical figure. For the alkaline MnO_2-Zn cell, we should be able to obtain about 70 Wh/kg (35 Wh/lb). This calculation is approximately correct (see Table 1), but only for low current densities. At higher loads, the system becomes far less efficient, as Fig. 3 shows clearly. The active materials in an alkaline MnO_2-Zn cell are as powerful as, for example, in a nickel oxide-Cd cell, but the cell geometry is different. Ni-Cd cells utilize stacks of thin plates, rolled anode-cathode assemblies, small amounts of electrolyte, and good current-collection means; these are a few of their distinguishing features. Recent Ni-Zn cells and Ag_2O-Zn cells show that the Zn anode is not the limiting electrode. Development work will be required on the MnO_2 electrode to lessen the slope of the curve (shown in Fig. 2) and extend it to the 100 W/lb power level and beyond.

Zinc and MnO_2 are far more plentiful than silver, nickel, cadmium, or even lead. With the knowledge that MnO_2 electrodes are rechargeable-- and their performance can undoubtedly be improved--this system should be able to compete effectively for a larger place in the general battery field in the future, not just in the dry-cell class.

It will be impossible to reach capacities like those achieved by zinc-air cells, organic-electrolyte cells, high-temperature molten-salt cells, or fuel cells, but the low cost and practicability aspects greatly favor the alkaline MnO_2-Zn system.

As far as the small-consumer market is concerned, the commercial aspects of alkaline MnO_2-Zn cells are excellent. Recent surveys and predictions mention growth rates higher than those expected in any other battery field. Alkaline-cell production increased four times between 1965 and 1969 [33].

2. PRIMARY ALKALINE MnO_2-Zn CELLS

2.1. BUTTON-TYPE CELLS

2.1.1. The Crown Cell

The first alkaline MnO_2-Zn cell on the market was produced by the Ray-O-Vac Company and was called the "crown cell" because of its crimped cell covers.

Figure 6 shows the cell construction and its components [34]. The cathodic assembly features a small pellet cup into which a disc of MnO_2-graphite mix is molded. A steel cup forms the positive terminal. The anodic part consists of another steel cup (tinned), a zinc pellet (disc of pressed powder), and the electrolyte absorbent fabric or gel. A polymethacrylate spool holds the components together, with its edges serving as a

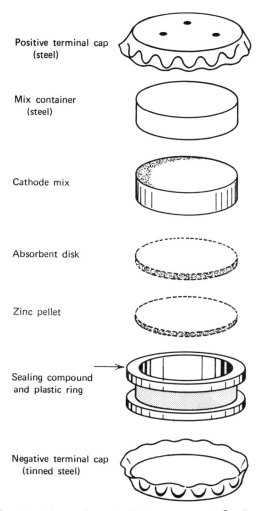

Positive terminal cap
(steel)

Mix container
(steel)

Cathode mix

Absorbent disk

Zinc pellet

Sealing compound
and plastic ring

Negative terminal cap
(tinned steel)

FIG. 6. Construction and parts of the crown cell [34] (courtesy of Electric Storage Battery Corporation).

means of holding the crimped caps and exerting pressure on the elastomeric seals.

Figure 7 shows a view of the assembled crown cell [35]. Crown cells were well suited to series-stack connection by placing several cells into a thermoplastic sleeve and shrinking it to confine the cells.

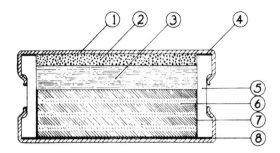

FIG. 7. Cross-sectional view of the assembled crown cell [35]:
(1) tin-plated negative cap, (2) zinc-powder anode, (3) absorbent separator,
(4) sealing material, (5) plastic ring, (6) depolarizer mix, (7) steel can,
(8) positive cap.

An example of the performance of the type 900 C-size crown cell (2.23 cm^3) of 9-V nominal voltage, used for portable radios, is shown in Fig. 8 [34].

2.2. CYLINDRICAL CELLS

2.2.1. Outside-Cathode Cells

2.2.1.1. Principal Construction Features. The bobbin cathode and zinc-can anode as used in Leclanché cells were not satisfactory for alkaline cells, mainly because of mass-transport limitations in the massive cathode and the passivation occurring at relatively low current densities at smooth zinc surfaces.

The crown cell discussed in Section 2.1, which used a zinc disc and a bulky cathode, did not have as high a current output capability as modern alkaline cells have.

A complete change in construction principles had to be made to produce a cell type that could utilize the superior features of the alkaline system.

Discharge cycle: 4 hours/day, 7 days/week
Temperature: 70° F

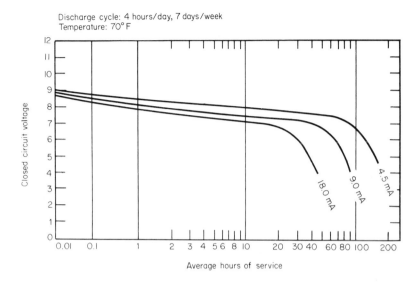

FIG. 8. Service data of the C-900 crown cell (Ray-O-Vac Corporation Engineering Manual) [34].

The outside cathode (arranged as a sleeve inside the cell container) assures a relatively thin MnO_2 electrode with improved diffusion properties, and the zinc-powder anode provides the large area needed to overcome the passivation tendency of the zinc. It can be argued that the same features will improve the Leclanché system; this is true--but only to a limited extent. The lower conductivity of the $NH_4Cl \cdot ZnCl_2$ electrolyte and the irreversible electrode reactions leading to precipitation within the cell make such construction changes less effective in the Leclanché system.

Figure 9 shows the construction of a typical cylindrical alkaline MnO_2-Zn cell as presently produced [36].

The sealed cell consists of a molded cathode, which is formed inside a steel can under high pressure to assure good electrical contact to the outside wall. The can itself may be plain steel or nickel clad (a common practice, depending on the size of the cells and the need for good appearance). For special cells, the can may be plated with a metal that does not easily develop

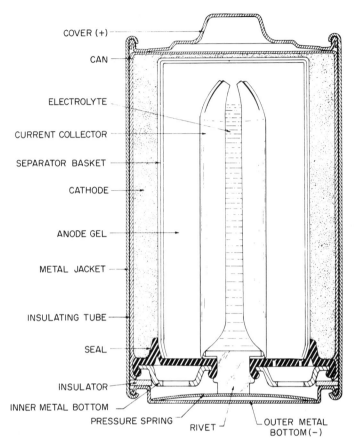

COVER (+)
CAN
ELECTROLYTE
CURRENT COLLECTOR
SEPARATOR BASKET
CATHODE
ANODE GEL
METAL JACKET
INSULATING TUBE
SEAL
INSULATOR
INNER METAL BOTTOM
PRESSURE SPRING
RIVET
OUTER METAL BOTTOM (−)

FIG. 9. Sectional view of an alkaline MnO_2 D-size cell [36] (courtesy of Union Carbide Corporation).

a resistive coating, for example, gold [37]. A carbonized surface has been proposed to assure low resistance over long storage periods [38].

The cathode is composed of MnO_2 mixed with graphite and acetylene black as additional conductors; in some cells, binder materials, for example, Portland cement and fibers, are added to increase strength [39-41].

The cathode is usually lined with a separator sandwich (the "basket" in Fig. 9) consisting of two types of separator materials, one having good absorption of KOH (next to the cathode) and one for preventing zinc penetration (nonwoven fabric).

The gelled zinc-powder anode is inserted into the separator basket and made to fill the space between the separator and the current-collector assembly (brass tongues). A liquid KOH reservoir supplies more electrolyte to the gel-anode during storage and discharge, whenever needed [42].

2.2.1.2. Examples of the Assembly of an Alkaline D-Cell. In Fig. 10, the components needed to produce a D-size cell are shown in a schematic assembly-line layout.

First, the cathode mix is pressed into the steel can and molded. The separator sandwich is assembled and inserted in a parallel line.

The anode cylinder is then placed into the separator basket, followed by insertion of the anode collector assembly. After sealing of the jacket and the cell cover, plates are fitted around the cell in a bottom-up position to conform with the polarity found in Leclanché flashlight cells (positive cap).

2.2.1.3. Cell Sealing. The seal is a very important part of an alkaline cell; it means the difference between a laboratory cell and a marketable, commercially successful battery. A "radial seal" utilizes the resiliency of a plastic material (for example, nylon) to close the cell between the rivet and the can surface [43-45]. Gassing in an alkaline cell is at such a low level that the cell can be sealed hermetically (thereby providing long shelf life). However, cells can be abused (heated, charged by reverse insertion into a series combination, or simply short circuited, which causes somewhere between 10 and 20 A to flow), which requires the provision of a safety vent to prevent high pressure in the cell. A pressure-relief vent valve for a fluid-tight cell, using a resilient deformable ball, is described in Ref. [46].

Most of the cells on the market have these precautions built into the cell, sometimes only in the form of "weak spots" on the can bottom. The cell pictured in Fig. 9 solves the abuse problem with the collector and seal assembly shown in detail in Fig. 11 [47]. It provides a thin section in the plastic cover, which can rupture at fairly low pressure. A gas-permeable membrane may be used [48].

An improved venting mechanism that does not act until a preset pressure is reached is shown in Fig. 12. The inner seal vents when its shape is distorted. In an emergency, the spur tooth pierces the plastic cover imme-

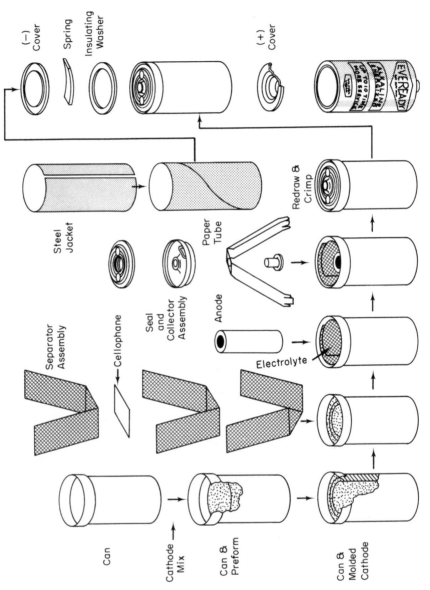

FIG. 10. Assembly of a D-size alkaline MnO$_2$–Zn cell (courtesy of Union Carbide Corporation).

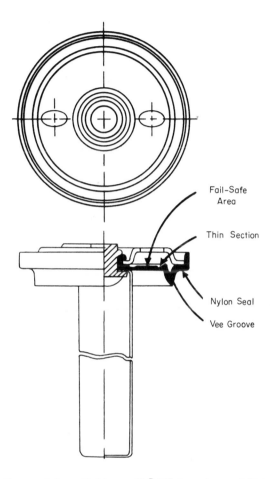

FIG. 11. Top seal for alkaline cell [47] (courtesy of Union Carbide Corporation).

diately. However, the cell is still closed to free air access after venting and can be used for a long time, if the failure was not accompanied by excessive discharge [49].

Of course, such elaborate venting mechanisms can be used only with larger cells (D and C size). The smaller cells (AA and smaller) use simpler devices, for instance, double O-rings (or seal washers) or grooves in the plastic seal member to accommodate safe venting. A typical construction

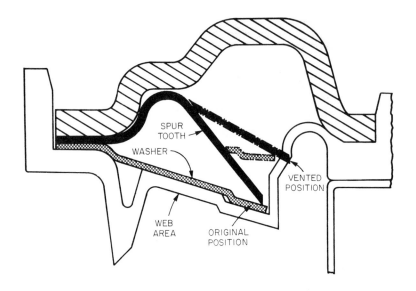

FIG. 12. Improved seal and venting mechanism for D cell [49]
(courtesy of Union Carbide Corporation).

of a leakproof cell is described in Ref. [50].

The closures of alkaline cells have gone through a long line of development, dating back to the early manufacturing of HgO cells. Patents of recent
years featuring a "double top" design [51-54] are concerned with methods
to prevent creepage of caustic to the seal area but at the same time to try
to ensure that enough KOH is available in the separator between cathode
and anode--a difficult and contradictory task.

The use of Na-carboxymethylcellulose (Na-CMC) gelled electrolyte in
combination with powder zinc remedies some difficulties arising from free-
liquid surplus and/or lack of available electrolyte within the anode structure
(see Section 5.3.3).

Other manufacturers provide central reservoirs for the KOH within a
double-pronged current collector, with vents on top of the cells (through
permeable discs) or on the side to vent into the outer jacket (shrink tubing).

2.2.1.4. Cell Constructions. Figure 13 shows four examples of different cell constructions [55-58], which indicate the rather complicated structural designs of alkaline cells.

Efforts to simplify the manufacturing of alkaline cells and to lower the cost are in conflict with the need to produce a safe and good-looking cell. A simple rod as anode collector is used in the cells shown in Fig. 14 [59,60].

The cell of Fig. 14b [60] is remarkable because it does not use a woven or paper separator. A clear starch gel separates the MnO_2 cathode and powder-zinc gel anode structure. There is no safety vent, with the exception of the plastic seal on top of the cell, which may release gas when the pressure becomes high enough to deform it.

2.2.2. Inside-Cathode Cells

The inside-cathode bobbin-type construction of a cylindrical cell is at a disadvantage with respect to mass transport and resistance when larger cells are compared.

In Fig. 15, D-size alkaline MnO_2-Zn cells with both kinds of cathodes (inside and outside) are compared. Cell A has the sleeve type cathode, and cell B contains the bobbin type.

These diagrams are drawn from data obtained with pulse current interrupter equipment and show the polarization of the cell separated from the ohmic resistance of the cell [61]. From these results, it is obvious why the sleeve cathode construction is the better one.

However, when cells are reduced in diameter to, for example, 0.5 in. (AA-size cells and smaller), the differences become less pronounced, and with some increased porosity in the cathode, alkaline MnO_2-Zn cells of the bobbin type show good performance.

Most of the cylindrical cells of the early 1950s (HgO cells) employed inside cathodes with zinc sheet anodes, sometimes perforated or increased in area by some other means [62]. However, the use of zinc powder pressed against the zinc can had already been conceived in 1945 [63]. A modern version of such a cell with MnO_2 instead of HgO is shown in Fig. 16 [64].

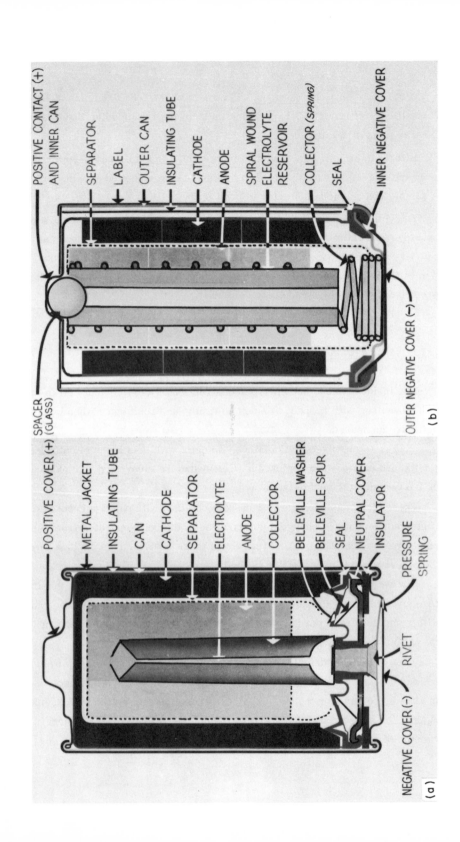

POSITIVE CONTACT (+) AND INNER CAN

SEPARATOR

LABEL

OUTER CAN

INSULATING TUBE

CATHODE

ANODE

SPIRAL WOUND ELECTROLYTE RESERVOIR

COLLECTOR (*SPRING*)

SEAL

INNER NEGATIVE COVER

SPACER (GLASS)

POSITIVE (+)

OUTER NEGATIVE COVER (−)

(b)

POSITIVE COVER (+)

METAL JACKET

INSULATING TUBE

CAN

CATHODE

SEPARATOR

ELECTROLYTE

ANODE

COLLECTOR

BELLEVILLE WASHER

BELLEVILLE SPUR

SEAL

NEUTRAL COVER

INSULATOR

PRESSURE SPRING

RIVET

NEGATIVE COVER (−)

(a)

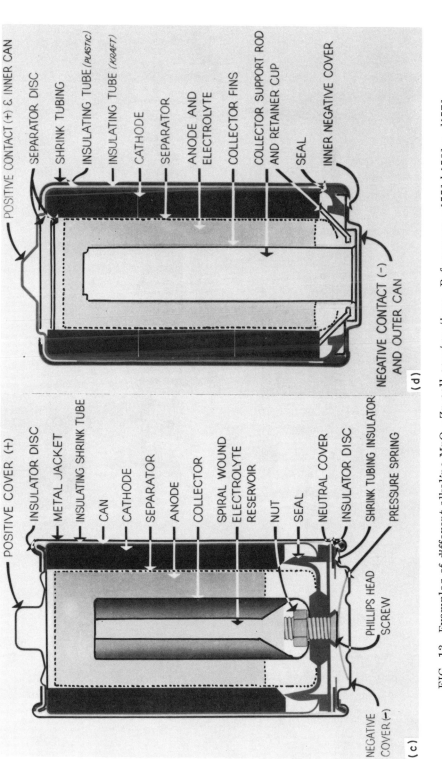

FIG. 13. Examples of different alkaline MnO$_2$–Zn cell constructions. References: a[55];b[56];c,d[57].

POSITIVE CONTACT (+) & INNER CAN
SEPARATOR DISC
SHRINK TUBING
INSULATING TUBE (*PLASTIC*)
INSULATING TUBE (*KRAFT*)
CATHODE
SEPARATOR
ANODE AND ELECTROLYTE
COLLECTOR FINS
COLLECTOR SUPPORT ROD AND RETAINER CUP
SEAL
INNER NEGATIVE COVER
NEGATIVE CONTACT (−) AND OUTER CAN

(d)

POSITIVE COVER (+)
INSULATOR DISC
METAL JACKET
INSULATING SHRINK TUBE
CAN
CATHODE
SEPARATOR
ANODE
COLLECTOR
SPIRAL WOUND ELECTROLYTE RESERVOIR
NUT
SEAL
NEUTRAL COVER
INSULATOR DISC
SHRINK TUBING INSULATOR
PRESSURE SPRING
PHILLIPS HEAD SCREW
NEGATIVE COVER (−)

(c)

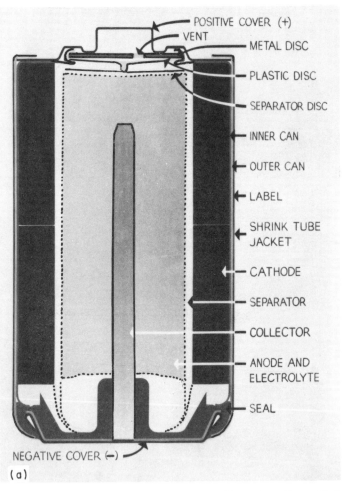

FIG. 14. Examples of alkaline MnO_2-Zn cell construction. References: a, [59]; b, [60].

2.2.3. Miniature Alkaline Cells

The small cell shown in Fig. 17 is typical of the construction of a miniature cell (hearing aid, watch battery).

Because of the small amount of depolarizer material use, miniature cells can economically use expensive metal oxides such as HgO and AgO.

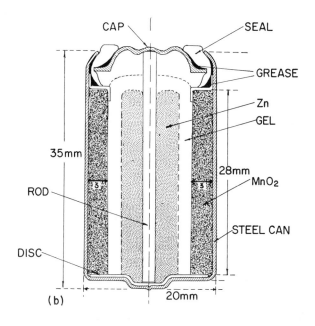

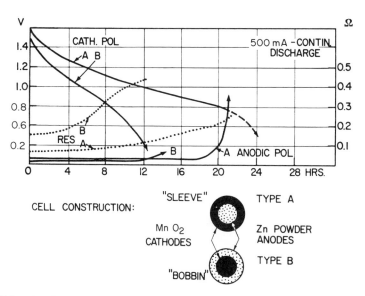

FIG. 15. D-size alkaline MnO_2-Zn cells discharged with 500-mA constant current. Constructions: A (sleeve) and B (bobbin) [61].

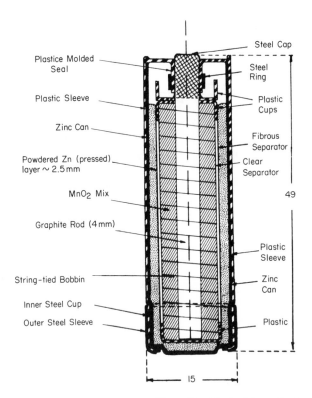

FIG. 16. AA-size alkaline MnO_2-Zn cell with inside cathode [64].

However, the design features and manufacturing steps are very similar to those for larger cylindrical cells.

Usually the cathode is formed in the can to assure good contact. The can may be nickel-plated steel in order to provide a material compatible with the cathode and also to furnish a satisfactory external appearance and electrical contact. The separators commonly used are a membranous material and an electrolyte-absorbent layer of nonwoven fabric. The anode consists of a mixture of zinc particles, mercury, gelling agent, and electrolyte and is in contact with copper-lined or tin-plated anode cup. These metals are used in order to permit amalgamation with the mercury in the anode. A nylon gasket is used between the metal cathode can and the anode

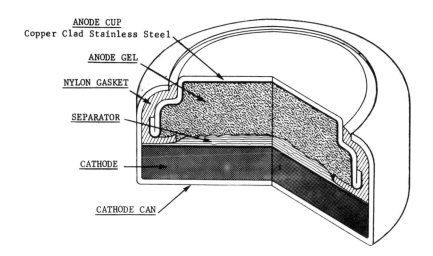

FIG. 17. Construction of a miniature cell [72] (courtesy of Union Carbide Corporation).

cup. This gasket is very important for providing a liquid-tight seal. Note that miniature cells are not provided with a special vent. It is not considered necessary, in view of the small amount of chemicals involved; the gassing rate is extremely low with low levels of impurities and high amalgamation levels (which can be afforded in small cells).

2.3. SPECIAL CELL CONSTRUCTIONS

2.3.1. Reserve Cells

In order to achieve exceptionally long shelf-storage capability, it is sometimes desirable to keep the electrolyte in a separate compartment, from which it is transferred to the cell by some activation step before the cell is put into use. It is important that the activation occur in a short time (less than 1 min, preferably in a few seconds); therefore, easy access of the caustic solution to the cell electrodes must be assured. The breaking of a glass ampule by a mechanical device is one method used in a feasibility study of a reserve D-size cell under government funding [36].

Figure 18 shows the construction of this cell. The zinc powder anode can be a gel-type anode from which all liquid has been removed by evaporation. In order to achieve exceptionally fast penetration of the anodic mass and avoid blocking the deeper layers by swelling of the CMC gel in the surface layer, a coarse anode structure made of agglomerated (dried) particles is advantageous. This anode structure was first suggested for air-depolarized deferred-action cells [65].

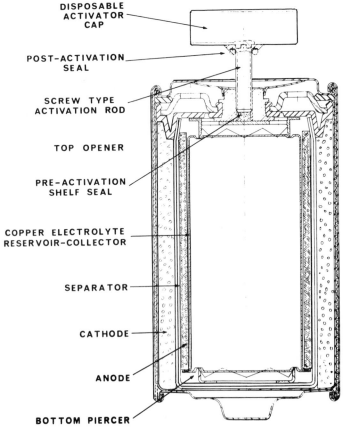

FIG. 18. Alkaline MnO_2 reserve cell [36] (courtesy of Union Carbide Corporation).

In recent years, reserve-type alkaline MnO_2 cells have been made commercially available for small emergency power applications.

In principle, all the methods used for silver oxide-zinc reserve-type military batteries can be applied to the MnO_2-Zn system.

Such methods include automatic action, moving-diaphragm filling, vane-type reservoirs, and balloon-filling [66]. If solid KOH and water are used, the heating effect can help the battery start up at low temperatures.

2.3.2. Increased-Cathode-Area Constructions

Increased-cathode-area constructions improve the utilization of the gelled powder-zinc anode and increase the current-carrying ability of the cathode assembly. Figure 19 shows the combination of an outside cathode with a centrally located inside cathode taking advantage of the available space [67]. The capacity increase obtained is about 30% for a D-size cell (with better depolarizer efficiency at higher loads). A forerunner of this principle is the double-cathode collector construction, which was applied in Leclanché $Zn-MnO_2$ and $Mg-MnO_2$ cells [68].

Figure 20 shows experimental high rate cell designs using thin electrodes [69] allowing a far better utilization of manganese dioxide than possible with present commercial cell constructions.

An interesting attempt to mount flat-plate cells into a cylindrical compartment is described in another patent [70]. This construction results in a better interface-area/volume ratio than in a bobbin cell.

2.4. PERFORMANCE AND APPLICATION

2.4.1. General Considerations

Alkaline MnO_2 cells have been produced to meet a wide range of performance objectives. It is not possible to cover all aspects of performance in this chapter. For details, the reader is advised to consult the technical brochures of the various manufacturers. If differences exist among the data

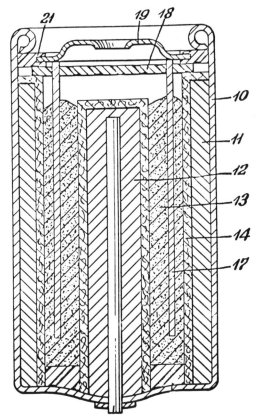

FIG. 19. Cell with double cathode [67]: (10) steel can (+), (11) outside
cathode, (12) central cathode, (13) gelled powder-zinc anode, (14) separator,
(17) anode collector, (18) gas-permeable membrane, (19) terminal plate (-),
(21) insulating washer.

for cells of the same size produced by different companies, it must be
considered that the cell designs might have put emphasis on different properties. Zinc-limited cells usually provide fewer hours service than maxi-
mum-capacity cells, which expose the user to post-discharge gassing and
closure failure; with some cells leakproofness is a high-priority item, at
the expense of efficient volume utilization (space is needed for vents, expansion cavities, multiple casings, and so on). Some manufacturers insist
on high-temperature shelf life (considering transportation and storage in

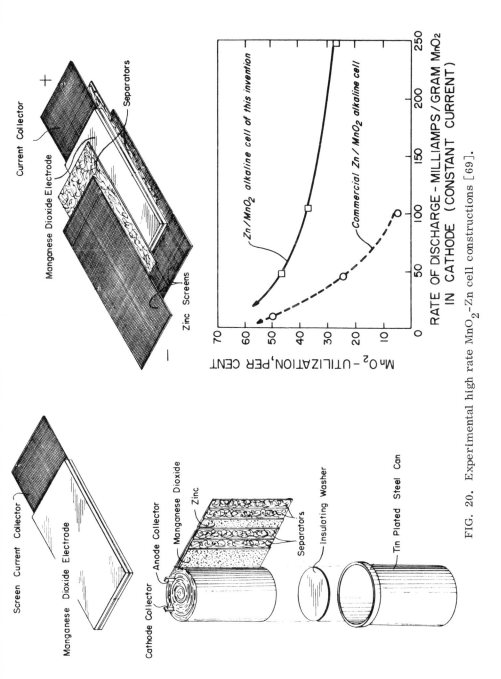

FIG. 20. Experimental high rate MnO_2-Zn cell constructions [69].

summer) and use a higher amalgamation level to decrease anode corrosion, while others believe that a lower-priced product will sell better. These considerations are not noticeable from the outside, and judgment of cell quality is very difficult because it includes many features, depending on the application of the battery. A low-resistance battery designed for a very high current output must be formulated differently from a low-current intermittent-duty power source. The former contains more conductive material (graphite) and thus less active material (MnO_2). Also, the construction features will be aimed at the usage: in general, thin, porous plates excel over heavy plates in performance during load tests, while stronger plates might have a longer life expectancy at low currents. This is especially true for rechargeable batteries.

Low-temperature performance is important for some applications (especially the military); performance under these conditions might be specified with the knowledge that cell operation at high temperatures will be less satisfactory.

Exceptionally good low temperature performance can be expected from thin cathodes used in spirally wound electrode assemblies (see Fig. 20). The large surface reduces the specific current density and therefore the cell shows far less polarization and resistance losses.

The alkaline MnO_2-Zn system comes very close to being a universally applicable battery power source, but it cannot perform best in all the job situations the consumer demands.

2.4.2. Examples of Alkaline MnO_2-Zn Cell Performance

Although different battery manufacturers plot the output data in different ways, it has become general practice to specify a fixed resistance as load and to plot the voltage-time curves with a falling current characteristic.

Figure 21 presents data for D-size cells [71].

Table 4 shows the estimated hours service of a C-size cell at 70°F [72].

Figure 22 shows the room-temperature service of a C-size cell and its performance at lower temperatures. The curve for C-size Leclanché cells at 70°F is included for comparison [36].

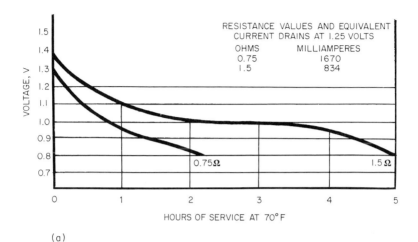

(a)

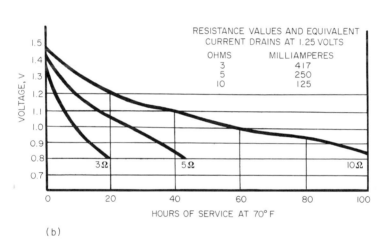

(b)

FIG. 21. Discharge curves of D-size alkaline MnO_2-Zn cells (P. R. Mallory Company Engineering Manual) [71].

Table 5 shows the internal-resistance change of the C-size cell with dropping temperature [72].

Figure 23 compares the AA- and fractional AA-size cells with the Leclanché counterpart of AA-size [36].

KARL V. KORDESCH

TABLE 4

C-Size Cell, Estimated Hours Service at 70°C[a]

Schedule	Starting drains (mA)	Load (Ω)	Cutoff voltage					
			0.7 V	0.8 V	0.9 V	1.0 V	1.1 V	1.2 V
4 h/day	18	83.3	320	300	280	260	230	170
	37.5	40	160	145	130	115	100	70
	60	25	98	88	78	68	58	39
24 h/day	15	100	410	380	335	295	260	205
	30	50	200	185	170	155	125	95
	150	10	34	31	28	25	19	9
	200	7.5	24	22	20	16	12	5
	250	6	18	17	15.5	11.5	8	3.1
	300	5	14.5	13.5	12.3	8.6	5.3	2.1
	375	4	10.8	10.0	9.0	6.0	3.2	1.1
	667	2.25	4.6	4.3	3.2	1.8	0.7	0.2
	1200	1	1.6	1.0	0.6	0.3	0.1	
	2100	0.5	0.4	0.17	0.1			

[a]Reference [72].

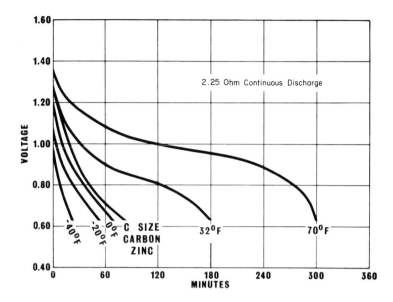

FIG. 22. C-size cell discharge curves for various temperatures, compared with Leclanché cell operated at 70°F [36].

TABLE 5

Internal Resistance of C-Size Cell versus Temperature[a]

Temperature (°F)	Internal resistance (Ω)
70	0.15
32	0.21
20	0.28
0	0.43
-20	0.64
-40	0.96

[a]Reference [72].

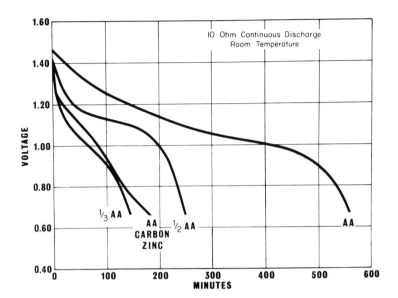

FIG. 23. AA and fractional AA-size alkaline cells versus AA Leclanché cell [36].

2.4.3. Shelf Life

The shelf life of alkaline MnO_2-Zn batteries is excellent. After one-year storage at room temperature or three months at 120°F (approximately 50°C), these batteries can be expected to supply in excess of 90% of their rated capacity.

2.4.4. Abuse Tests

Primary batteries should not be charged and should not be disposed of in a fire, because any rapid excessive pressure increase can cause a closure failure that may lead to personal injury or equipment damage. Other abuse conditions are usually provided for by the cell designer: shorts, continuous heavy discharge, high-temperature storage, high humidity, and temperature cycling (-20°F to +140°F).

2.4.5. Physical Characteristics

The physical characteristics (sizes, weights, and so on) of some representative alkaline MnO_2-Zn primary cells are shown in Table 6. Additional data can be found in the brochures of battery manufacturers [71,72].

2.4.6. Applications and Economics

The alkaline MnO_2-Zn primary battery represents a major advance in portable power sources over the conventional Leclanché battery (often called the carbon-zinc battery because of the carbon-rod collector in the center of the MnO_2 bobbin). The alkaline cell is best suited for heavy and continuous drains and shows its advantages very clearly on the basis of performance per unit of cost.

The cells are hermetically sealed and usually encased in metal, thereby providing a power source not influenced by the environmental conditions. Although the energy density on a weight basis is not much greater than that of the carbon-zinc cell when the latter is discharged at low rates or particularly on an interrupted basis, the alkaline cell has a significant (25%) advantage on a volume basis. The alkaline cell can give up to ten times the service of the common flashlight cell when used in heavy-drain applications such as motion-picture camera cranking, radio-controlled model planes or boats, power tools, toy automobiles, and so on.

In radios and tape recorders, alkaline cells usually last twice as long as carbon-zinc cells; their performance is also better because of the lower internal resistance of the alkaline cells and their flatter discharge curve. The service capacity remains relatively constant when discharge schedules are varied from high rates to low rates. Alkaline cells do not need any recovery periods for optimum service. Where the application requires light service at very low or very high temperatures, the alkaline cell again is superior.

The low resistance and the good storageability of alkaline cells make them preferred for electronic photoflash applications, which require high-rate current pulses for the vibrator or transistor circuits of the voltage

TABLE 6

Physical Characteristics of Alkaline MnO_2-Zn Cells

Cell size	Approx. capacity (Ah)	Suggested current (A)	Diam. (in.)	Diam. (mm)	Height (in.)	Height (mm)	Weight (oz)	Weight (gm)	Vol. (in.³)	Vol. (cm³)
N	0.5	0.08	0.450	11.43	1.120	28.45	0.34	9.64	0.19	3.12
			±0.005	±0.13	±0.010	±0.27				
1/2AA	0.6	0.06	0.563	14.30	1.087	27.61	0.40	11.34	0.27	4.44
AAA	0.6	0.1	0.400	10.16	1.735	44.07	0.40	11.34	0.22	3.61
			±0.010	±0.13	±0.010	±0.26				
AA	1.5	0.15	0.562	14.28	1.969	50.01	0.75	21.26	0.48	7.87
1/2D	3.0	0.3	1.344	34.14	1.203	30.56	2.1	59.54	1.6	26.22
C	4-5	0.5	1.015	25.8	1.930	49.02	2.34	66.34	1.57	25.7
			±0.005	±0.13	±0.010	±0.26				
D	8-10	0.65	1.344	34.14	2.406	61.11	4.5	127.58	3.17	51.96

converter that charges the flash capacitor. The alkaline cell can be compared to the Ni-Cd battery on medium current drains. At extremely high load levels, the Ni-Cd battery excels. The storageability of MnO_2-Zn batteries at high temperatures is very good. Ni-Cd cells lose capacity on extended storage at elevated temperatures.

These same features make the alkaline cells a good choice for small computers, electric shavers, and cassette tape recorders with high-quality stereo amplifiers (which have a large power output to the loudspeakers). Small motors represent a rapidly growing application for alkaline cells. This field includes electric typewriters, toys, hand tools, and portable appliances (such as small TV sets), since the applicability of the alkaline MnO_2 cell has been extended into the secondary-battery field.

3. RECHARGEABLE ALKALINE MnO_2-Zn CELLS

3.1. COMMERCIAL RECHARGEABLE BATTERIES

3.1.1. Performance Characteristics

Several features of the rechargeable alkaline MnO_2-Zn batteries should be recognized in order to make optimum use of this system.

a. The battery is manufactured in a fully charged state and retains its charge for longer periods of time compared to other rechargeable systems.

b. The battery is completely sealed and maintenance free.

c. Alkaline MnO_2-Zn cells should not be discharged below a recommended depth of discharge, depending on the cell design. Usually only one third of the total primary capacity should be removed during cycling to assure best cycle life. Charging must be done with current and voltage controls.

d. Early in cell life, the unused cycling capacity is available as a large reserve capacity, which can be used in an emergency; however, discharge below 0.9 V damages the cell, mainly because the cathode becomes less reversible (because of changes in the MnO_2 structure).

e. Charge retention is far better than experienced with Ni-Cd and lead-acid cells, which need frequent booster charges to stay in charged condition.

f. The available energy per cell decreases with each cycle at a faster rate than that experienced with other secondary batteries. Present commercial MnO_2-Zn rechargeable cells give about 50 cycles until the reserve capacity is used up.

g. Cycle life on a shallow depth-of-discharge regime is often high, for example, up to 200 cycles, depending on the quality of the separator used. Cycle life in this mode is limited by the formation of zinc dendrites, which cause shorting (an anode recharging problem common to all alkaline-zinc systems) and by cathode deterioration (see Section 3.2.1).

Figure 24 shows the discharge characteristics of a 15-V rechargeable TV battery (10 G-cells in series) with a rated cycle capacity of 5 Ah. The discharge curves are shown for each fifth cycle, each curve representing a 4-h discharge across a 9.6 Ω resistor [73]. During its lifetime (approximately 55 cycles), about 3000 Wh have been stored and delivered. Since the battery weighs only 6 lb, the energy density of this system can be stated as 500 Wh/lb, which is at least 20 times higher than that of the primary alkaline MnO_2-Zn cell.

Improvements in rechargeability can be achieved by making the electrodes thinner. Figure 25 shows the cycle life of experimental thin cathodes made in accordance with the teachings of a recent patent [69]. See also Fig. 20.

The charging procedure affects cycle life very much, as will be discussed later. The so-called voltage-limited taper-current charging method with a set voltage limit (1.75 V/cell) is recommended. The charging time should be limited also. Prolonged trickle charge is not required to maintain charge and may tend to form zinc dendrites.

For short periods of time, the discharge currents may be high (up to five times the rated load) without hurting the cell life. An important factor is the energy product (current x time) removed per cycle and the voltage level at which it is withdrawn; between 1.2 and 1.1 V, the battery approaches true reversibility.

The electrical characteristics of some rechargeable alkaline MnO_2-Zn batteries are listed in Table 7.

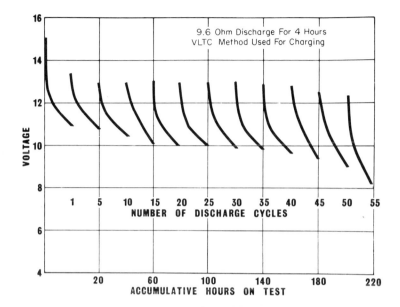

FIG. 24. Discharge characteristics of a rechargeable battery [73]. Discharge is shown for first and every fifth cycle.

3.1.2. Physical Characteristics

Sizes and weights of different commercially used rechargeable alkaline MnO_2-Zn batteries are listed in Table 8.

3.1.3. Differences Between Primary and Secondary Cells

The technological differences between primary and secondary cells at the present time are not great. Use of the same cell design and production equipment for both primary and rechargeable systems is responsible for the relatively low cost of alkaline MnO_2 rechargeable batteries.

The following features distinguish the secondary cells.

a. Cathode structures are improved.

b. Separators are different in order to minimize zinc shorts on recharging. Electrolyte additives may be used for various reasons.

c. Powder-zinc anodes may have larger-area current collectors to assure better deposition of zinc. Means of providing for overcharge are incorporated.

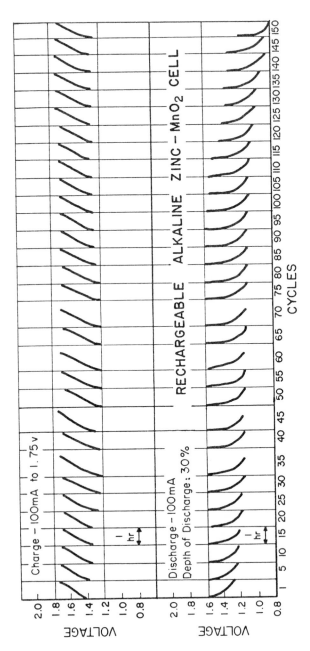

FIG. 25. Cycling characteristics of an experimental thin-electrode alkaline MnO_2-Zn cell [69].

TABLE 7

Electrical Characteristics of Rechargeable Alkaline MnO_2-Zn Batteries[a]

Voltage	No. and size of cells	Rated capac. (Ah)	Max. recommended disch. current (A)
4.5	3 D	2.5	0.625
6	4 G	5	1.25
7.5	5 D	2.5	0.625
13.5	9 G	5	1.25
15	10 G	5	1.25

[a]Courtesy of Union Carbide Corporation [74].

TABLE 8

Physical Characteristics of Rechargeable Alkaline MnO_2-Zn Batteries[a]

Maximum dimensions (in.)				Terminals	Weight	Volume (in.^{3})
Diameter	Length	Width	Height			
1 23/64	–	–	7 5/32	Flat	15 oz	10.1
–	2 25/32	2 25/32	5 11/32	Socket	2 lb 8 oz	39.4
–	2 21/32	1 17/32	7 5/32	Socket	1 lb 9 oz	29
–	8 5/16	2 13/16	5 7/8	Socket	8 lb 8 oz	137
–	8 5/16	2 13/16	5 7/8	Socket	6 lb	137

[a]Courtesy of Union Carbide Corporation [74].

d. Construction modifications include improved vents, better seals, and the new spirally-wound anode-cathode assembly [75].

3.2. SPECIAL FEATURES OF RECHARGEABLE ALKALINE MnO_2-Zn CELLS

3.2.1. Cathode Structures

One of the main problems in rechargeable cells is the slow disintegration of the cathode when the cell is cycled. On discharge, the volume increases, and physical stress can break the cathode structure apart if it is not well supported inside the steel can. For that reason, high molding pressures are normally used.

It is also obvious that outside cathodes withstand the dimensional changes of cycling better than do bobbin cathodes. Addition of binders to the MnO_2-graphite mixes has been used to improve the coherence of the cathodes. Portland cement increases the strength of the structure and also improves the high-temperature stability of cells built primarily for low-temperature use [76]. Addition of steel wool or graphite-coated Dynel fiber produces cathodes with better rechargeability [77]. Mixing with butadienestyrene emulsions (latex) results in stronger, more flexible cathodes [78] with some increase in cell resistance. Polystyrene bonding is described in Ref. [79].

The porosity of cathodes is sometimes too low for high current densities; therefore, it has been proposed to bind small particles of the cathode mix first with a binder, crush the resulting mass, and rebind the agglomerates of larger sizes into a high-void structure [75].

The conductivity of rechargeable cathodes is usually higher because manufacturers use a higher percentage of graphite or metal powder in the mix [80]. A low interfacial resistance between steel can and MnO_2 mix is also important; therefore, the cans are sometimes plated or surface-treated in some manner (see primary-cell constructions, Sections 2.1-2.3). Painting a mixture of $Mn(NO_3)_2$ solution and graphite powder on the collector surface and heating it to $250°$-$300°C$ is said to improve the rechargeability [81].

3.2.2. Separators and Electrolyte Additives

Properties of separators for rechargeable zinc cells are extensively discussed in a later section to indicate the state of the art.

In the presently marketed rechargeable alkaline MnO_2-Zn cells, only separators of the fibrous (nonwoven) types are used. The better and more sophisticated separators used in AgO-Zn batteries have not been used in commercial alkaline MnO_2 cells because of the desire to produce a sealed system. Oxygen produced on charge or overcharge cannot be transported through membrane separators.

Electrolyte additives can be other alkali hydroxides such as NaOH, LiOH, or CsOH, to change the KOH properties, depending on whether zincate precipitation or zinc plating is desired. Additives may also influence cathode rechargeability, as is claimed for the nickel oxide electrode.

Alkaline earth metal hydroxides are generally added for the purpose of changing the solubility of zinc in the caustic solution. Any zincate migrating out of the anodic compartments becomes a source of undesirable zinc deposits on the cathode side or within the separator structure. For this reason, the additives are more effective when they are added to the powder-zinc anode structure so that the zinc stays in the anode in the first place. Also, plating additives are most effective on the larger anodic areas, where they can act as surface-active ingredients and change the shape of the zinc deposits. In gelled electrolytes, the thickening agent also acts as an additive.

3.2.3. Anode Structures

3.2.3.1. Collector Design. Primary cells need only simple contact arrangements for current collection, but for secondary cells, the first requirement for good rechargeability is the existence of a structure onto which the zinc can be plated. Alkaline AgO-Zn rechargeable cells, for example, use silver screens or wires and tin- or silver-plated copper mesh as substrates to assure a large surface. Large, pronged current collectors are used in rechargeable alkaline MnO_2 cells on the market [74].

3.2.3.2. Balancing Problems. The preferred ratio between cathodic and anodic capacity favors the anode (twice the theoretical capacity) so that some zinc is still available as a deposition site after the cell is completely discharged.

However, in a sealed system, a zinc-limited balance must be maintained to prevent the generation of hydrogen on overdischarge. In practice, these requirements are compromised: usually the cells have more total MnO_2 capacity built into the cell, but it is recommended to stop discharge at about one third of this capacity. Since recommendations do not carry much weight with the commercial customer, a cutoff at this point must be enforced. The only method that is effective and also retains a reserve capacity is the use of a voltage-sensitive mechanical relay or Zener diode circuit to limit cell discharge. Unfortunately, this is a relatively expensive solution, which is presently limited to devices that already have electronic control circuitry; for example, TV sets.

Since it is detrimental to the rechargeability of the alkaline MnO_2 cathode for it to be discharged beyond the 1 e$^-$ equivalent, which corresponds to the available capacity above 0.9 V, the voltage-control method is also beneficial from that standpoint.

Still, the simplest method would be to limit the zinc and reduce the cell capacity anodically about 50%. With a large screen collector and a reliably functioning gas vent (see Section 3.2.4.), this proposition may work [80].

Another method is the use of low-normality KOH to force the zinc anode into "premature" passivation, thereby limiting the cell capacity to a predetermined value. Unfortunately, this principle is current-density sensitive [82].

A way to combine the requirements of a large-surface collector and limitation of the zinc capacity more precisely is described in Ref. [83]. In this case, a metal powder is mixed with the zinc powder to provide a large residual surface after anode discharge. Copper and lead act simply as conductors within an immobilized porous anode (bonded with acetylcellulose). Magnesium powder can act as a high-resistance interface as soon as all the zinc has gone into solution and its surface passivates, thereby switching the cell off and even preventing a reversal in series with other cells. However, the poor MnO_2 utilization accompanying these balancing methods is considered to be uneconomical.

3.2.3.3. Overcharge Problems. If the cell is fully charged and still connected to the charging equipment (even on trickle charge), oxygen is produced at the cathode. This oxygen can travel to the large-surface zinc anode and form zinc oxide, which then dissolves and prevents depletion of zincate in the electrolyte, thereby eliminating hydrogen-gas evolution at the anode and producing only heat. This cycle, similar to the overcharge process in hermetically sealed Ni-Cd batteries, can theoretically continue forever.

For this mechanism to function, a surplus of ZnO must be present in the fully charged battery (which is equivalent to a larger anode capacity on charge only) and good oxygen-gas transfer from the cathodic compartment to the powder-zinc anode surface must occur.

Any semipermeable membrane prevents this direct gas communication. A cell with an overeffective separator may reach a dangerously high internal gas pressure on overcharge. Venting helps, as an emergency means, but it disturbs the balance between the active materials.

Several solutions have been suggested. The simplest one, which is still not completely satisfactory, is the use of a fibrous (or nonwoven) separator instead of gas-impermeable membranes; this may lead to early zinc-dendrite short circuits.

The oxygen recombination cycle in the presence of a membrane can be made possible by providing an exposed anode surface in the center of the cell [84]. The flooding of this annular space can be prevented by insertion of a repellent porous plastic cylinder. Figure 26 shows the construction of the cell.

Substituting an ionic cycle for the oxygen-gas phase recombination cycle is feasible if a halogen salt of low decomposition voltage (for example, KBr, KI) is added to the electrolyte. Instead of oxygen gas, molecular iodine or bromine is produced which dissolves and migrates through the separator in the form of the oxyhalogen compound, which acts in the same manner as zinc oxide in preventing H_2 evolution [85]. Unfortunately, most of the halogen finally ends up as iodate or bromate, which is insoluble and is thus removed from the recombination process after 20-50 cycles.

A gas-recombination system that would react with the hydrogen produced on overcharge has been suggested; this system employs a fuel-cell catalyst within the cathodic mass to recombine hydrogen with the MnO_2. In this case, the cathode must have the larger capacity [86]. However, the requirement of a noble-metal catalyst makes this approach uneconomical.

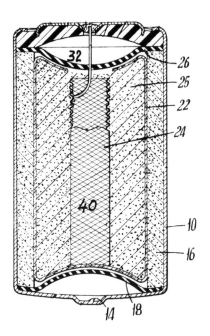

FIG. 26. Rechargeable alkaline MnO_2-Zn cell [84]: (10) steel can, (14) vent, (16) cathode, (18) plastic spacer, (22) cellulosic separator sandwich, (24) copper screen, (25) gelled powder zinc anode, (26) plastic spacer, (32) collector wire, (40) polyethylene sponge.

3.2.4. Construction Features

Improved venting mechanisms, some with H_2-selective membranes, preventing water-vapor losses, have been designed for rechargeable cells. The construction features pictured in Fig. 26 and previously in Fig. 12 are examples. In some of the multicell batteries that are enclosed in a separate metal container, the secondary envelope can take over some of the leakage prevention and safety features.

3.3. METHODS FOR RECHARGING

3.3.1. Voltage-Limited Taper-Current Charging (VLTC System)*

The recommended method for charging the sealed alkaline MnO_2 rechargeable battery is the voltage-limited taper-current system. The batteries are tolerant to variations in charge current characteristics if losses in cycle life are allowed. However, the more common circuits, constant voltage, constant current, and taper current, are not recommended without modifications. In addition, these circuits require some degree of attention by the user in terminating charge or proportioning charge-discharge times to prevent excessive overcharge and loss of battery service. Fundamentally, the battery should be charged proportionate to discharge, the charge current returning about 110-120% of the ampere-hours removed on previous discharge.

Alkaline MnO_2 rechargeable batteries have excellent charge retention characteristics. A new battery is shipped fully charged and has the retention characteristics of a primary battery. The battery should always be discharged before charging. THE BATTERY MUST NEVER BE CHARGED AT HIGH RATES PRIOR TO FIRST DISCHARGE.

Normally, the rated rechargeable ampere-hour capacity is 25-30% of the total discharge capability. Therefore, there is initially a large emergency capacity left; it decreases gradually in present-day commercial cells until, after about 50 cycles, only the rated capacity is left. Beyond that point, the cells are not reliable for critical applications requiring full rated output, but may still be useful for others.

The alkaline MnO_2 rechargeable battery must never be discharged completely. During deep discharge, a secondary electrochemical reaction takes place. This reaction is not reversible and will seriously reduce the battery cycle life. To avoid possible customer complaints of short battery service, it is desirable that the instrument using the battery contain some provision to prevent discharge beyond rated capacity. Conversely, cycle

*This description is partly reprinted from a chapter on the recharging of alkaline MnO_2-Zn batteries in Ref. [74].

life can be increased more than proportionately when less than rated capacity is used during each cycle.

The ampere-hours returned on charge may very with history, number of cycles, and charging method. Extended periods of overcharge may reduce cycle life and should be avoided. The alkaline MnO_2 battery exhibits an appreciable voltage rise during charge. The charge voltage source should be limited to prevent irreversible chemical reactions occurring as a result of high charge potentials.

3.3.1.1. Components of the VLTC Systems. An inexpensive voltage regulator added to the basic constant-current type transformer-rectifier circuit, having a current-limiting resistor between the battery and the regulated output voltage, removes the burden of adjusting charge time to discharge time and provides automatic current control. This charge method is called the voltage-limited taper-current (VLTC) charging method. Though the regulator circuit is somewhat more expensive than other common circuits, the initial cost is more than offset by 50-100% greater cycle life.

The basic characteristics of sealed alkaline MnO_2 rechargeable cells when using VLTC charging are given in Table 9.

TABLE 9

Charging Characteristics of Alkaline MnO_2-Zn Cells[a]

Cell size	Precharge operating voltage at max. current (V)	Maximum initial charging current at 1.3V/cell (A)	Current-limiting resistance (Ω/cell)	Source-voltage limit (V/cell)
D	1.0-1.2	0.6	0.8	1.70-1.75
F	1.0-1.2	0.9	0.5	1.70-1.75
G	1.0-1.2	1.12	0.4	1.70-1.75

[a]Courtesy of Union Carbide Corporation [74].

The charger circuit design is not much different from the usual transformer-rectifier power supply used for, for example, Ni-Cd cells, except
that the voltage setting is different and adjusts to the battery end-of-charge
potential (see Table 9, last column multiplied by the number of cells) with
a regulation span of 2-3%.

The current-limiting resistance (total of resistance, including the control
circuit) should limit the initial current to the 4-6 h rate at the start of charging when the battery voltage is low. Typical starting voltage after withdrawal
of rated capacity is 1.30-1.35 V/cell. The maximum recommended initial
charge current at these voltages are also shown in Table 9.

The advantages of this method of charging are readily apparent.

a. High charge currents flow during the early portion of the charge
period when the battery is best suited for accepting charge and taper to
very low currents near the end of charge, resulting in less critical control
of the time on charge.

b. Currents at end of charge due to active voltage control are small
enough to minimize cell heating and internal gassing, which can be detrimental to battery life.

c. The necessity for the user to calculate the length of charge time
dependent upon the discharge time and rate is avoided. In this method of
charging, the battery tends to accept only the amount of capacity needed to
bring the battery back to the indicated performance level.

d. This method permits the charging of partially discharged batteries,
and prevents serious damage if batteries are inadvertently placed on charge
prior to discharge.

To prevent damage to the cell on charge, the charging potential should
be limited to about 1.7 V/cell with a maximum permissible of 1.75 V/cell.
The voltage rise during charging and the end-of-charge voltage are dependent
on the charge current. The end-of-charge voltage will decrease slightly as
cycle life progresses, as a result of the gradual change in the reversibility
of the system.

The battery voltage after charging will gradually decrease to the nominal
value of 1.5 V/cell when the charger is disconnected and the battery remains

on open circuit. When changing from a charge to a discharge cycle with no time delay between, the battery voltage decay rate will vary with discharge current. With charge currents near 1 A, the battery voltage will fall to the usual closed-circuit values within 5 min.

A charging circuit that fulfills the discussed requirements is shown in Fig. 27.

The permissible deviation of the charge current from the suggested linear taper current varies somewhat with the application and use pattern. Where the application requires repetitive discharge of rated capacity, the charge-current values should fall within the shaded area of Fig. 28, with charge-time adjustment for circuit tolerance if operation should fall in the "minimum" area relative to linear character.

Figure 29 shows the battery voltage and current for a typical 16-h charge period at about mid-cycle life of a 10-cell battery. The area under the charge-current curve is the amount of charge in ampere-hours. In this example, it is approximately 6 Ah.

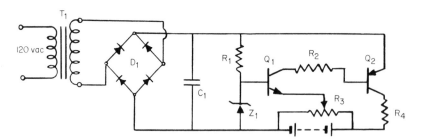

FIG. 27. Charging circuit for 10-cell TV battery [74] (courtesy of Union Carbide Corporation). Parts list: T_1, Stancor RT 201 (input, terminals 1 and 7; connect terminals 3 and 6); D_1, Solitron Devices 2A50; C_1, 500 μF, 50 V dc; Z_1, 1/4 M12Z (Motorola) or 1N963B (Texas Instruments or International Rectifier); Q_1, 2N1304 (RCA or Texas Instruments) or equivalent; Q_2, 2N1557 (Motorola), 2N514 (Texas Instruments), or equivalent; R_1, 4700 Ω, 1/2 W; R_2, 390 Ω, 1/2 W; R_3, 1000 Ω, 1W; R_4, 4Ω, 10W.

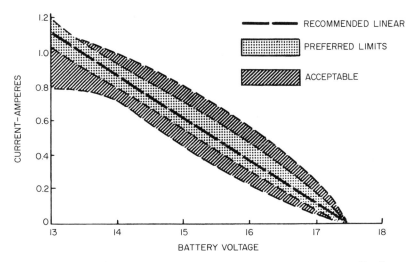

FIG. 28. Ranges of charge currents for 10-cell TV battery [74].

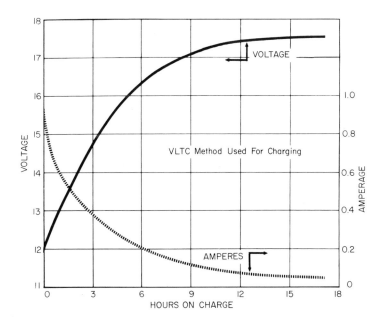

FIG. 29. Charging characteristics of a 10-cell TV battery [74].

In some applications where extended cycle life and improved battery performance are desired, only a portion of the rated battery capacity is used. In these cases, the regulated voltage source can be designed for a somewhat lower value, or the current-limiting resistance can be greater to decrease the rate of charge returned to the battery in a given charge period. A voltage limit between 1.6 and 1.7 V/cell is recommended with current-limiting resistance values between one and two times the recommended design values of Table 9.

3.3.2. Discussion of Other Charging Methods

3.3.2.1. Constant-Current Charging. Constant-current charging and discharging is the method used for investigation of single-electrode behavior and coulombic efficiencies, but it is rarely used in practice unless an automatic time or voltage control circuit is added for protection. An open cell starts to evolve oxygen at about 1.75 V, and when charging is continued, the voltage climbs slowly to 1.9 V and higher, producing hydrogen gas at the anode and oxygen at the cathode, accompanied by manganate and permanganate formation.

Sealed alkaline MnO_2-Zn cells with higher anode capacity can be overcharged at a certain rate (depending on the construction). Because of the oxygen recombination process in the cell, a significant amount of heat is produced and can be sensed easily.

Thermal end-of-charge control is frequently used in Ni-Cd cells for rapid charging, which at the present time is not recommended for commercial alkaline MnO_2-Zn cells.

Thermal sensing and control are simple and reliable. Lock-out thermostats with mechanical reset are practical and inexpensive. Electronic circuit control for charge cut-off at a given temperature of the cell requires only a small sensor (thermistor) in the battery package [74].

3.3.2.2. Voltage-Controlled Charging. Voltage-controlled charging is the commonly used laboratory method. With or without current-limiting devices, it brings the battery to full charge by sensing the end-of-charge voltage and activating a control at the preset value. The current supplied can either be filtered dc or a pulse current. Modern solid-state control circuits can automatically taper, control, or shut off at any desired point of the charge characteristic of the battery.

Circuits for battery charging control based on voltage-level sensing employing solid-state switching are described in the literature [87]. A specific circuit is claimed in Ref. [88].

Special efforts have been made to charge batteries at a high rate in a short time. Principles and methods are described in Ref. [89], and a fast charging circuit with additional heat control is shown in Ref. [90]. Future testing will prove whether these methods are useful for alkaline MnO_2-Zn cells.

Unfortunately, the end voltage is not always known; the reasons are many. Factors such as temperature, electrolyte concentration, general battery history, number of cycles, and so on change the characteristics. One way to mark the end point of the charging process on an otherwise uneventful sloping curve is to include significant steps by incorporation of other oxides into the cathode; for example, silver oxide or nickel oxide.

Figure 30 shows the charge curve of an alkaline MnO_2-Zn cell with 20% nickel oxide added to the MnO_2 [84]. The charging plateau of the battery is raised about 100 mV at an earlier time, as would be expected from the nickel oxide content, and a reproducible nickel oxide end point step is recognizable when O_2 evolution starts. The hydrogen step is delayed by the zinc oxide content of the electrolyte and should not appear in a sealed cell. Addition of silver oxide gives a more pronounced step than nickel oxide, but at least 5-10% must be used to achieve the plateau effect, which poses a cost question.

3.3.2.3. Resistance-Free Charge Potential. A method to determine the end-of-charge that excludes the variation caused by the voltage differential across the resistance components of the cell (varying with temperature, cycle life, and so on) was recently described. The voltage is measured between current pulses [91]. Figure 31 shows the circuit diagram of such a battery charger.

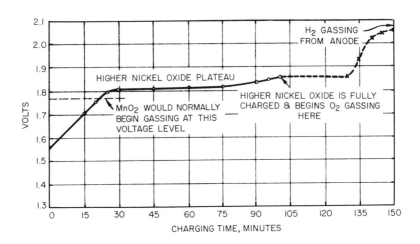

FIG. 30. Charging characteristics of an alkaline MnO_2-nickel-oxide-zinc cell [84].

3.3.2.4. <u>Pulse-Current Charging</u>. Pulse-current charging seems to offer some advantages compared to the conventional filtered-dc battery charging. The charge acceptance seems to be better, even with older batteries [92].

An explanation of the effect of pulse charging may be found in the fact that the shape of zinc deposits is strongly influenced by the current density [93]. Low-current pulse charging delays the appearance of zinc moss and produces more adherent zinc, provided that the on-off times are short (seconds). Higher current densities favor adherent deposits, but increasing temperature also increases the tendency to form mossy zinc deposits. Results of different authors vary, however, and the explanations are not always the same. It is plausible that large-surface zinc dissolves during the off-times faster than more compact zinc, and sizes shift in favor of the larger particles. Also, the distribution of zincate might be improved during the off-times, allowing migration away from the anode. The stirring effects of local heating might also be important. Alternating current superimposed on direct current is also claimed to be beneficial. Asymmetrical ac charg-

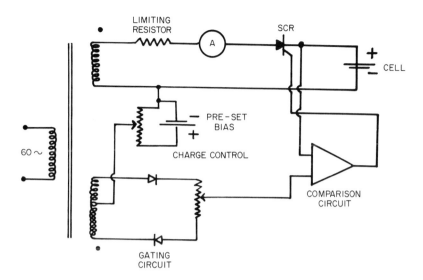

FIG. 31. Circuit diagram of battery-charging equipment eliminating the ohmic resistance components [91]. The battery is charged with 60-Hz pulses to a preset end voltage (MnO_2-Zn cells: 1.65 V). The potential measured between the current pulses is free of the voltage differentials caused by internal and external resistance components. The variability of cell charge and discharge potentials as a function of cycle life--mainly due to resistance changes--can be seen from Fig. 38.

ing of AgO batteries is discussed in Ref. [94]. A report on charge-control devices used for space-designed Ag-Zn cells is given in Ref. [95]. With a two-step voltage control, the cycle life of cells could be increased fivefold.

3.3.3. Morphology of Zinc Deposits

The morphology of zinc deposits as a function of charge methods is extensively documented in Ref. [96]. As far as the application of fundamental data (which are confusing) is possible, the following factors must be considered.

a. Low dc current densities favor mossy deposits; high current densities produce dendrites. Pulse charging may reverse this situation.

b. Temperature increase leads to mossy zinc rather than dendrites.

c. High zincate concentrations in the electrolyte favor formation of mossy zinc.

d. Electrolyte circulation is beneficial for producing more compact deposits.

e. The zinc deposit adheres differently on different substrates. Cu and Ag show poor adherences; Sn and Pb show good adherence [94].

The plate design is also important; cycling of the zinc anode produces shape changes, and preferential dendrite growth was observed at the edges. The depletion of zincate in the center may be the reason for this effect. Material transfer is always observed from top to bottom of a plate, following the gradient of the zincate due to its specific weight [97]. By designing the plate with room for expansion in the center and providing excess at the edges, a convex electrode shape was achieved [98].

3.4. POSSIBLE IMPROVEMENTS OF RECHARGEABILITY

Improvements in rechargeability of the inexpensive MnO_2-Zn system could have far-reaching consequences in the field of commercial batteries, especially for small personal equipment. There are some requirements that must be fulfilled: the cells must be sealed to protect the electronic circuits, the cells must be safe under abuse conditions (reversal, overcharge, slight damage) and should exhibit a good shelf life--initially, between cycles, and discharged. Adding the high-temperature storage requirements of southern areas and the low-temperature performance needs of the north, the tasks seem considerable.

However, primary alkaline MnO_2-Zn cells fulfill many of these requirements and have a capability of about 100 discharge-charge cycles with a reliable cutoff point. D-size cells with a cycle capacity of about 2-3 Ah seem to be a realistic expectation in the near future.

In the following, ways of improvements are indicated--mostly by developments in related fields. Progress in the state of the art of the zinc-silver
oxide battery provides a general outlook for the technology of rechargeable
alkaline batteries with zinc anodes [99].

3.4.1. Cathodes

The poor conductivity of MnO_2 requires the addition of graphite to the
cathode. The internal current distribution (between particles) is dependent
on the type of graphite, particle size, and mixing characteristics [100, 101].
Unknown is the relationship between these parameters and cycle life.

As an example, Fig. 32 shows scanning electron microscope photographs
picturing the appearance of cathode mixtures containing the same electrolytic
MnO_2 but different graphite types.

A lowering of the cell resistance can be achieved by better collector
design. Mass-transport limitations can be remedied by making the cathodes
thinner without losing mechanical strength. The MnO_2 type chosen for the
rechargeable cathode is critical for determining the cycle life and voltage
level. Oxidation of the graphite is certainly a damaging factor, and the
choice of the graphite is again important.

The mode of charging influences the physical integrity of the cathode,
reversibility, and cycle life. Low depths of discharge are beneficial for
increasing the cycle life, but the useful capacity is also reduced. Again,
a thin electrode plate is preferred over a thick, bulky cathode. The MnO_2
utilization automatically improves when the current density is lowered. To
counteract this fact, the geometric surface of a given volume of MnO_2 must
be increased.

Summarizing the points made above, it seems that the most effective
way to arrive at a better cathode is to design the electrodes of the rechargeable alkaline MnO_2-Zn cell like all other secondary-battery electrodes are
constructed, that is, as flat plate, box type units. However, the cost of
doing this may make the system less attractive for some applications.

3.4.2. Anodes

The greatest disadvantage of the zinc anode is the solubility of zinc in
caustic. Any method that lowers the amount of zinc in the electrolyte without

FIG. 32. Comparison of cathode mixes with the same type of MnO_2
but different graphite additions: (a) artificial graphite, (b) natural graphite.
(Scanning electron microscope pictures, 500x.)

substantially reducing the limiting current density is beneficial. Additions of phosphates and borates, for instance, lower the solubility but also induce passivity. $Ca(OH)_2$ and CaO precipitate Ca zincate, which is poorly soluble, and it is possible to trap the zinc in the anode compartment (behind the semipermeable separator) in this manner [102]; $CaSO_4$ added to a powder-zinc polyacrylate KOH gel causes (perhaps) less polarization during the recharging process, but H_2 gas may evolve because of lack of zincate in the electrolyte [103].

The theoretical mechanism of zinc dendrite formation is not clear in spite of many studies [104, 105].

The beneficial effect of lead addition is reported in Ref. [106].

The use of surfactants such as Emulphogene, with and without PVA as a binder, should retard zinc penetration through the cellulosic semipermeable separator [107, 108]. Teflon emulsions added to the zinc decrease migration of the zinc, and the use of oversized negative plates has led to considerable improvements, which are shown in Fig. 33. These capacity and cycle-life improvements are achieved by these measures in AgO-Zn cells. Such achievements give hope that there is still much progress ahead for MnO_2-Zn cells, which have not been studied as extensively as AgO-Zn batteries [99].

Another system that can provide valuable indications of possible improvements is the Ni-Zn battery. A cycle number of 250 has been reported with a deep discharge level [109]. However, these are not yet sealed cells.

Considering the improved charging techniques available, but not practiced in commercial units presently, the anode performance of rechargeable alkaline MnO_2 cells should cease to be the limiting factor.

Last, but not least, the anode collector needs attention.

3.4.3. Separators

The importance of the separator in rechargeable zinc cells will be discussed in detail later. Progress in this area is occasionally promised by new techniques, which then prove disappointing. At the present time, most of the commercial AgO-Zn cells rely on cellulosic membranes.

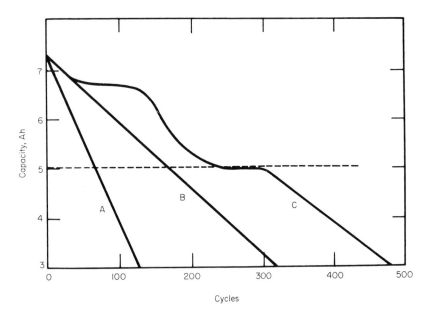

FIG. 33. Capacity as a function of cycle life for dished zinc electrodes: (A) control, (B) extended edges plus Teflon, (C) extended edge plus Teflon plus dished profile [99].

However, space and military cells have proven that, for a price, improvements are feasible. However, such simple measures as, for example, incorporating circulation channels into separator sandwiches have proven to be advantageous [110].

4. THE MANGANESE DIOXIDE ELECTRODE

(by A. Kozawa)

Electrochemistry (basic reactions and behavior in KOH electrolyte) of MnO_2 and the detailed properties of battery cathodes are described in

Chapter 3 of this book [111]. Several points directly related to alkaline MnO_2 cathodes will be summarized in this section, mainly to point out why alkaline MnO_2-Zn cells are capable of higher output than Leclanché cells.

4.1. ELECTROLYTIC MANGANESE DIOXIDE

The manganese dioxide that is currently being used in alkaline MnO_2-Zn cells is almost exclusively electrolytic manganese dioxide (EMD). EMD is produced by depositing MnO_2 on an anode (graphite, Ti, or Pb alloy) from 0.5-1.0 M $MnSO_4$ solution containing some H_2SO_4 (0.1-0.5 M) at 95°-99°C [112, 113]. Today, still a larger amount of natural MnO_2 ore and some chemically produced MnO_2 [114, 115] are consumed for manufacturing Leclanché type cells that use NH_4Cl-$ZnCl_2$ electrolyte. However, a considerable amount of EMD is also used in Leclanché cells, mixed with natural ores in order to upgrade the cell (mainly, to improve the discharge capacity) as needed. EMD is much more expensive than natural ores.

The reason why today's alkaline cells use EMD almost exclusively is its low polarization and high voltage characteristic. Open-circuit voltages (OCV) and typical polarization values measured at 1 mA/100 mg MnO_2 (corresponding to approximately 250 mA per D-size cell) are shown in Fig. 34 for three types of MnO_2 [112]. Each sample (100 mg of MnO_2), mixed with 1 gm of graphite, 2 gm of coke, and abundant 9 M KOH solution, was placed in an experimental cell [113]. The cell was discharged at 1.0 mA continuously for 5-6 h; then the cell was allowed to recover for 17 h and the OCV was measured.

It should be noted that all these MnO_2 samples show the γ-MnO_2 x-ray diffraction pattern. The BET surface areas are also shown. Electrolytic MnO_2 providing a high voltage and low polarization is essential in order to obtain the best cell performance. Another feature of the alkaline cell is its leakproofness, derived from the hermetically sealed construction. In order to be able to seal the cell, metallic and other impurities in MnO_2 must be minimized because they may be dissolved in the KOH electrolyte

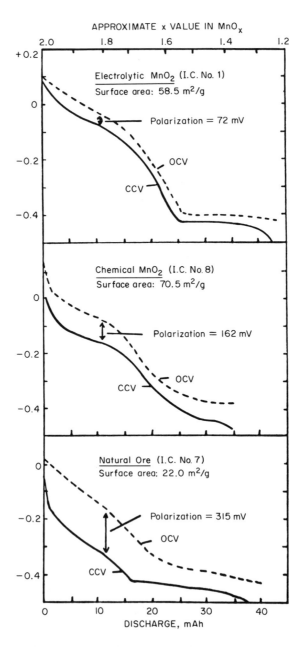

FIG. 34. Open–circuit voltage (OCV) and closed–circuit voltage (CCV) of three types of MnO_2 in 9 M KOH [111].

and could cause hydrogen gas evolution at the zinc anode. Battery-grade EMD is very pure, as shown in Table 10, and the purity level is equivalent to that of reagent grade material.

4.2. CATHODE COMPOSITION

The cathode composition of today's alkaline cells falls in the following range: MnO_2 (EMD), 80-85%; graphite flake and other carbon material, 8-15%; KOH solution (7-9 M), 10-15%; binder and/or other additives (such as cement fibers), 0.3-4%.

The blended mix is usually molded into a steel can under high pressure to form a sleeve. Alkaline MnO_2-Zn cells surpass Leclanché cells in greater ampere-hour capacity and high current capability. The greater capacity is partly attributable to the greater amount of MnO_2 packed in the

TABLE 10

Typical Range of Chemical Analysis of Battery-Grade EMD Samples[a]

MnO_2	$92.0 \pm 0.3\%$	Sb	0.0001%
Total Mn	$60.0 \pm 0.12\%$	As	0.0003%
H_2O (-)[b]	$1.52 \pm 0.20\%$	Mo	0.0002%
Fe	0.008%	Na	0.24%
Pb	0.0007%	NH_3	0.001%
Cu	0.0003%	SiO_2	0.02%
Ni	0.0005%	Insoluble	
Co	0.0008%	to HCl	0.62%
Cr	0.0008%	SO_4^{2-}	0.79%

[a]Reference [116].

[b]H_2O (-) is adsorbed moisture that can be removed by heating at 110°-120°C. Balance is structural water, which is released by heating above 120°C.

same size cell. Table 11 shows the range of MnO_2 contained in today's D-size Leclanché and alkaline cells. As can be seen, alkaline cells contain 30-50% more MnO_2 in the cathode. The main reason why more MnO_2 can be packed in alkaline cells is that graphite is used instead of bulky acetylene black.

Because of the use of graphite in the mix and the high pressure used in the molding process, the alkaline-cell cathode is hard and very compact (less porous) compared to the Leclanché cathode. This also means that the amount of electrolyte retained within the alkaline cathode is much less than in Leclanché cells. Such a small amount of electrolyte in the cathode is sufficient to operate the cell satisfactorily. The reason for this may be ascribed to the following facts.

a. The electrical conductivity of KOH solution (7 M) is roughly twice that of the Leclanché electrolyte. [Conductivity of 7 M KOH is 0.55 $(\Omega \cdot cm)^{-1}$ and 4 M NH_4Cl + 1 M ZnCl is 0.22 $(\Omega \cdot cm)^{-1}$.]

b. OH^- ions are produced by the reduction (or discharge) of the MnO_2:

$MnO_2 + H_2O + e^- \rightarrow OH^- + MnOOH$. Thus the ionic conductivity in the cathode does not decrease as the discharge proceeds.

TABLE 11

Amounts of MnO_2 (EMD) in D-Size Alkaline and Leclanché Cells

	Amount of EMD and Ah	Ore/carbon ratio (wt)	Main carbon material
1. Alkaline cells	37-41 gm 10.5-11.5 Ah[a]	6-9	Graphite
2. Leclanché cells[b]	22-28 gm 5.8-7.7 Ah[a]	5-6	Acetylene black

[a] Based on 0.308 Ah per gram of net MnO_2 according to the one-electron reaction ($MnO_2 + H_2O + e^- \rightarrow MnOOH + OH^-$). Ordinary EMD contains 91% net MnO_2.

[b] Paper-lined cells usually have more MnO_2 than starch paste cells.

In contrast to these situations in alkaline-cell cathodes, Leclanché-cell cathodes need (or should have) as much electrolyte (or water) as possible in order to avoid precipitation of $Zn(NH_3)_2Cl_2$ in the outer layer of the bobbin [117-119]. Therefore, bulky acetylene black is necessary in order to retain as much electrolyte as possible for Leclanché cathodes, even though the packable amount of MnO_2 is thereby considerably reduced.

4.3. CATHODE REACTION AND NATURE OF RECUPERATION

Manganese dioxide is unique and different from other oxides (such as Ag_2O and HgO) in the nature of the electrochemical reduction. From the standpoint of chemical thermodynamics, MnO_2 is reduced in a homogeneous phase, and the potential (OCV) decreases gradually as the discharge proceeds. Ag_2O and HgO are reduced in a heterogeneous phase, and the potential (OCV) is independent of the depth of discharge (or independent of the ratio of the reactant to the product, HgO/Hg or Ag_2O/Ag). This is explained thoroughly in Chapter 3 of this book.

The cathode has a thickness of, for example, 3-4 mm for D-size cells. When the cell is discharged at a high current (500-1000 mA per D-size cell), a portion of the cathode (A in Fig. 35) that is close to the zinc anode will be usually discharged more heavily than the part that is in contact with the steel can (B in the figure), depending on the nature of the conductivity (IR drop) within the cathode. When the discharge is stopped, part A will be oxidized by part B in a local cell reaction, because the potential of part A is lower than that of part B.

In contrast, without recuperation, the active material of an Ag_2O cell or an HgO cell with a sleeve cathode will be discharged gradually from the inner layer to the outer layer.

However, the internal resistance of an MnO_2 cathode steadily increases until it reaches a maximum when x in MnO_x reaches a value of 1.5 to 1.3. This is mainly due to lattice expansion. Figure 36 shows this behavior [120].

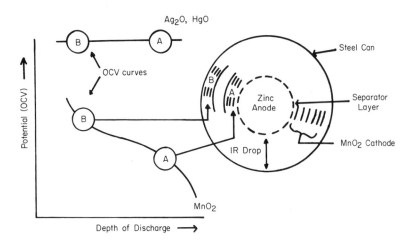

FIG. 35. Illustration of recuperation within the MnO_2 cathode.

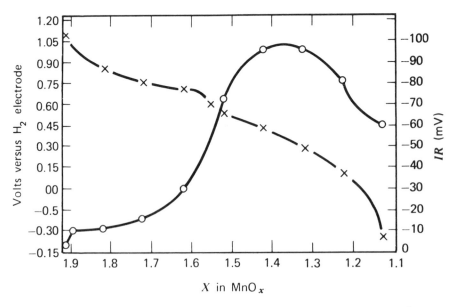

FIG. 36. Discharge of a manganese dioxide electrode at 5 mA/cm^2 in 7 M KOH, showing potential and resistance changes [120].

4.4. RECHARGEABILITY

One unique feature of alkaline MnO_2-Zn cells is the relatively good rechargeability. It is well known that if the cell is not discharged too deeply--for example, if the discharge is stopped at about 25% of the one-electron process $MnO_2 + H_2O + e^- \rightarrow MnOOH + OH^-$ (or roughly 2.5 Ah for a D-size cell)--then 40-50 cycles can be obtained [121]. The maximal recommended discharge current for a D-size cell is 625 mA.

Recently, the rechargeability of MnO_2 itself has been well studied by several investigators [122-125]. Their main results are summarized here.

4.4.1. Effect of KOH Concentration

The effect of KOH concentration was investigated by Kang and Liang [126]. They discharged MnO_2 in 1, 2, 4, 6, and 10 M KOH to -1.0 V versus Hg/HgO in the same KOH solution and then charged the discharged electrode. Rechargeability was found to be better at lower KOH concentrations, as shown in the tabulation.

	x value of the discharged oxide at the end of discharge	x value when the discharged oxide was charged to +0.6 V
1 M KOH	1.51	1.93
2 M KOH	1.45	1.89
4 M KOH	1.35	1.78
6 M KOH	1.24	1.72
10 M KOH	1.14	1.65

These results indicate that MnO_2 cannot be reduced beyond $MnO_{1.51}$ in 1 M KOH, since the solubility of Mn(III) ion is very small in 1 M KOH. Considering the structure, γ-MnO_2 is reduced only to MnOOH, which has the same configuration as γ-MnO_2 and is easily reoxidized electrochemically to the original MnO_2. In 10 M KOH, most of the MnO_2 is reduced to $Mn(OH)_2$,

since Mn(III) ion is produced at a sufficient rate from MnOOH in 10 M KOH
(see Chapter 3 for details).   Once Mn(OH)$_2$ is produced, reoxidation to
MnOOH and further to MnO$_2$ is difficult.

4.4.2. Effect of the Depth of the Discharge

The effect of the depth of discharge was investigated in 7 M KOH by
Ohira and Ogawa [127]. They always charged back 140% of the discharged
ampere-hour capacity. Figure 37 shows the end voltage of the discharge
period versus the cycle number.

The voltage at the end of the discharge cycle decreased only very slowly
when the removed capacity was 25% or less. However, when the discharge
depth was more than 30%, the voltage decreased considerably with increasing
cycle number.

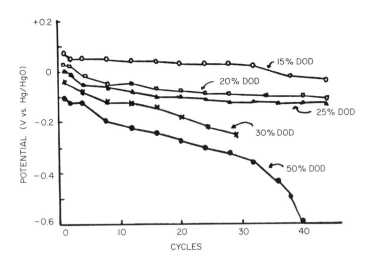

FIG. 37. Depth of discharge (DOD) versus cycle life [127]. Discharge
was carried out to various depths at 30 mA per 1.36 gm MnO$_2$. The
electrode was charged at 15 mA up to 140% of the discharge. The voltage
at the end of each discharge is shown.

In Figure 38, the cycling of an alkaline MnO_2 electrode to a depth of about 11% is shown [128].

The source of failure was determined to be at the interface between collector-wire screen and MnO_2. Film formation (Mn_2O_3) caused a highly increased IR drop (rise in resistance) at the last cycle. Using a second collector, more good cycles could be obtained.

X-ray studies show that some structure other than γ-MnO_2 is found when MnO_2 is discharged completely. Since the γ-MnO_2 lattice expands when it is discharged, the lattice stability seems to be lost gradually with increasing depth, and such a partially discharged γ-MnO_2 lattice tends to transform to another more stable structure. Generally, as long as the γ-MnO_2 structure is maintained, the manganese oxide system seems to be

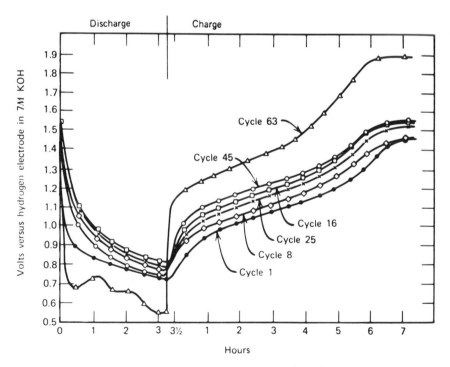

FIG. 38. Alkaline MnO_2 electrode cycle life [126]. Discharge at 5 mA/cm^2 to x = 1.78, charge at 2.5 mA/cm^2 to x = 2 (MnO_2).

rechargeable. In view of the discharge mechanism of γ-MnO_2 (electrons and protons are introduced without changing the basic structure, as described in Chapter 3), this experience can be theoretically confirmed.

It is also known that Leclanché cells are rechargeable to some extent, if the cell is not discharged completely [123, 125, 129].

Actually, the discharge mechanism of MnO_2 in Leclanché electrolyte (NH_4Cl + $ZnCl_2$) is the same as in alkaline cells (to $MnO_{1.75}$). The γ-MnO_2 accepts electrons and protons without changing its basic structure. The rechargeability of Leclanché cells (discharged to only 50% or less) is therefore reasonably explained.

An electron-probe microanalysis of MnO_2 particles in alkaline MnO_2-Zn dry cells during discharge-recharge cycles is described in Ref. [122]. Incorporation of K and Zn into the MnO_2 particles is stated to be the reason for the observed irreversibility on deep discharge.

5. THE ZINC ANODE

5.1. ELECTRODE CHARACTERISTICS

5.1.1. The Electrochemical Reaction Mechanism

The oxidation and dissolution of zinc in alkaline solution is different with pure zinc and amalgamated zinc. In the first case, the reaction mechanism might include zinc "adatoms," strongly adsorbed OH^- ions, $Zn(OH)_2^{2-}$, and $Zn(OH)_2$ species [130]. With mercury-containing zinc, which is nearly exclusively used in galvanic cells, the first reaction [131] is probably

$$Zn_{(Hg)} + 2\,OH^- \rightarrow Zn(OH)_2 \text{ diss. } + 2\,e^-$$

followed by

$$Zn(OH)_2 \text{ diss. } + 2\,OH^- \rightarrow Zn(OH)_4^{2-} \text{ diss. (zincate)}$$

The standard potential of the complete reaction of Zn to $Zn(OH)_4^{2-}$ (versus SHE) is 1.225 V [132]. Variations in mercury content cause small changes in potential (2% Hg: +20 mV). The theoretical capacity of zinc is 49.2 A min/gm (820 Ah/kg).

As the zincate concentration increases, solid products are formed. If the hydroxide-ion concentration is below 6 N, the final precipitate is crystalline $Zn(OH)_2$. Above 8 N hydroxide-ion concentration, ZnO is formed. Exact compositions, especially in intermediate concentration (6-8 N) solutions, are not known.

The following Table 12 summarizes thermodynamic data for zinc and its oxides and hydroxides [133]. The following reactions are the basis of the data:

$$Zn + 2 H_2O \rightarrow Zn(OH)_2 + H_2$$

for zinc hydroxide and

$$Zn + H_2O = ZnO + H_2$$

for the zinc oxide formation.

The solubility product $S = (Zn^{2+})(OH^-)^2$ ranges from 1.43×10^{-16} for α-$Zn(OH)_2$ to 8.41×10^{-18} for ϵ-$Zn(OH)_2$. For inactive zinc oxide, $S' = 9.21 \times 10^{-18}$; for active ZnO, $S' = 6.86 \times 10^{-17}$ [134].

With changing hydroxyl ion concentrations, the potential (tension) of the zinc electrode changes in accordance with

$$E = -0.763 - 0.0295 \text{ pS} + 0.059 \text{ p(OH)} = E_0' + 0.059 \text{ p(OH)},$$

where $E_0 = -0.763$ is the standard potential on the (acidic) hydrogen scale. The E_0' values vary between 1.22 and 1.26 V, depending on the value of S. For that reason, the Pourbaix diagram of zinc in alkaline solutions shows few details [135].

Figure 39 shows the zinc potential as a function of the KOH concentration in aqueous solution saturated with zinc oxide [136, 137]. At the top of the graph, the polarization (in volts) is indicated, starting with 31% KOH concentration, often used for alkaline MnO_2-Zn cells and silver oxide-zinc cells, and extending to a 1% KOH concentration, assuming that the alkalinity could drop that low in the neighborhood of the zinc surface because of diffusion limitations.

TABLE 12

Thermodynamic Data for Zinc and Its Oxides and Hydroxides at 25°C[a]

Formula	State and description	ΔH (kcal/mole)	ΔG (kcal/mole)	ΔS (kcal/mole)
Zn	c	0	0	9.95
Zn^{2+}	aq	-36.43	-35.184	-25.43
ZnO	Inactive BET: 0.5 m^2/gm	-83.17	-76.63	10.5
ZnO	Active BET: 16 m^2/gm	–	-76.40	–
ϵ-$Zn(OH)_2$	c (rhombic)	-153.5	-132.83	(19.9)
α-$Zn(OH)_2$	c (hexagonal)	–	-132.1	–
$Zn(OH)_2$	Amorphous	-150.5	-131.54	–
$Zn(OH)_2^{2-}$	aq	-261.4	-208.6	–

[a]Reference [133].

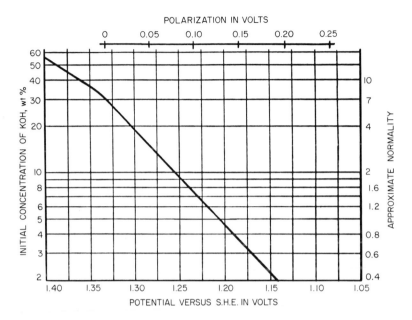

FIG. 39. Zinc potential versus KOH concentration in aqueous solution saturated with zinc oxide [136, 137].

Kinetic measurements show that the overpotential of zinc electrodes on discharge can be represented by Tafel lines (plotting polarization versus current density) and is probably charge-transfer controlled.

Figure 40 shows zinc overpotential versus current-density curves for different caustic concentrations [138].

Figure 41 presents similar data in a linear graph, showing the effect of amalgamation on the zinc polarization [139]. The overvoltage is measured with interrupter equipment [140], and the voltage drop across the resistance of the measuring cell is eliminated. The experiments showed that mercury reduces the overvoltage for the zinc → zincate reaction (charge transfer) and raises the limiting current density (see points a and b in Fig. 41). The fact that mercury increases the overvoltage of zinc against hydrogen evolution (water decomposition) on open circuit (and thereby reduces gassing) is not contradictory.

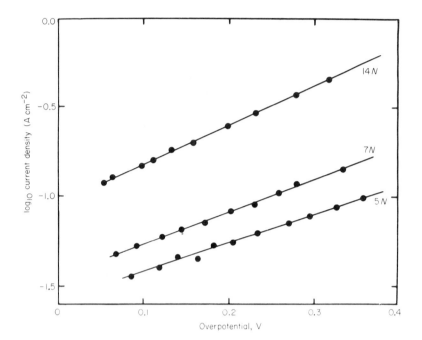

FIG. 40. Tafel lines for smooth zinc electrodes anodically polarized in KOH, 23°C [138].

Hydrogen overvoltage data for the zinc deposition (charge) process are reported in Ref. [141] (Ref. [138] in English) and shown in Table 13 with the coefficients a and b of the Tafel equation $\eta = a - b \log i$.

5.1.2. Zinc Corrosion in Alkaline Electrolytes

Zinc is thermodynamically unstable in contact with caustic solutions. The electrode potential is above the potential of hydrogen; thus zinc reacts with water to evolve H_2 and form a complex ion [142]:

$$Zn + 2\ H_2O + 2\ OH^- = Zn(OH)_4^{2-} + H_2$$

The objective of mercury addition to the zinc is the prevention of wasteful corrosion of the metal. The range of amalgamation used is 0.4% to 8%.

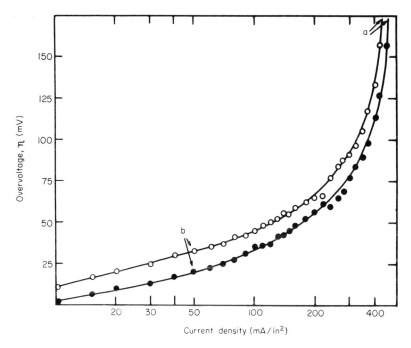

FIG. 41. Overvoltage versus current-density curve for a zinc electrode in 35% KOH, saturated with zinc oxide. Open circles represent clean zinc, solid circles, amalgamated zinc [139].

Gas evolution from zinc in caustic electrolytes is effectively reduced by an increase in the H_2 overvoltage. Mercury can be added by simply mixing it with zinc under caustic, or by displacement or plating reaction in a salt solution.

The effect of amalgamation on the corrosion rate of zinc anode material [143] is shown in Fig. 42. The temperature has a pronounced influence on the corrosion of amalgamated zinc. An increase from 75° to 125°F multiplies the gassing rate 4-7 times. The effect of KOH concentration is quite marked, especially in the low mercury addition range. The gassing rate decreases as KOH concentration increases up to 40%; it is possible that the water activity rather than the OH⁻ ion concentration is responsible. Figure 43 presents such data [144].

TABLE 13

Hydrogen Overvoltage on Zinc in KOH Solutions[c],[d]

KOH	0.1 N	0.2 N	1 N	5 N	7 N	10 N	13 N	14 N
a	-1.394	-1.375	-1.312	-1.230	-1.189	-1.164	-1.230	-1.235
b	0.128	0.130	0.130	0.120	0.118	0.118	0.120	0.125

[c]Reference [141].

[d]The Tafel slope is approximately 120 mV per decade of current density. With increasing alkali concentration, the overvoltage at constant current density decreases.

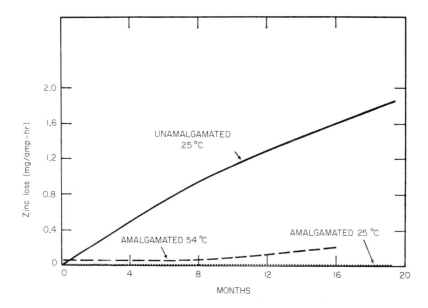

FIG. 42. Corrosion rate of zinc anode material [143].

The addition of zinc oxide to the electrolyte also increases H_2 gassing, as Fig. 44 shows [144]. Other impurities of a nonmetallic nature also affect the hydrogen evolution of zinc in alkaline solution. The effects of salts such as KNO_3 or K_2SO_4 or organic materials such as polyethylene are described in Ref. [145].

Metallic impurities that are harmful are those that cannot easily be amalgamated (such as iron) or metals that lower the overvoltage (such as arsenic, nickel, platinum, and so on).

In some instances, additions of lead can reduce the gassing rate. Indium may act the same way, but the effect of metals other than mercury is questionable. Organic inhibitors have found use as plating (recharging) additives, but as gassing-rate depressing agents, only some gelling agents are claimed to be effective.

5.1.2.1. Intermittent-Discharge Gassing. This type of corrosion is the result of cell operation. Some parts of the zinc are discharged differ-

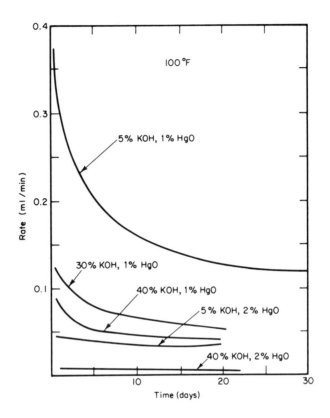

FIG. 43. Gassing rate versus time [144].

ently from others, and potential gradients are built up. These effects are
especially noticeable after long, high-current discharge periods.

    5.1.2.2. Post-Discharge Gassing. If all the depolarizer is used up
and zinc is left in the cell with the electric circuit still closed, the anodic
process will continue and produce gassing. Commercial cells are balanced
to minimize this behavior.

5.1.3. Anode Types

    Zinc anodes can be used in different shapes: plates, rolled sheets,
powders, or fibers. The selection depends on the maximum current drain

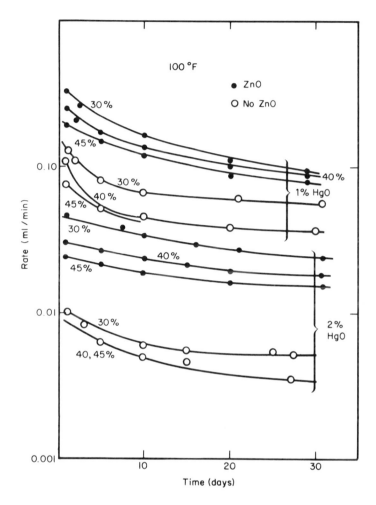

FIG. 44. Effect of ZnO addition on zinc gassing [144].

required from the cell. It is difficult to assess the controlling factors of the diffusion of OH⁻ ions or zincate ions or the concentration gradient in a cell [146]; however, the surface area is generally the governing parameter for efficient utilization of a zinc anode.

The different versions of large-surface zinc anodes and their perform-ance and technology will be discussed later. For general comparison, Fig. 45 shows the performance of a plate anode, a standard powder-zinc

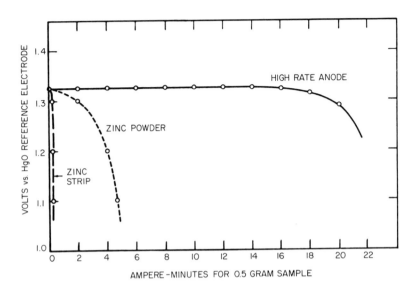

FIG. 45. Performance of zinc anode materials, room temperature test, 430 mA/cm^2 [143].

anode, and a special large-surface anode under extremely severe load conditions [143].

It is interesting to note that amalgamation, which definitely decreases the surface by smoothing out the crystalline zinc structure, actually increases the electrochemical activity by providing more unobstructed (by oxide) metallic surface [139, 146] and by providing conductive amalgam bridges between zinc particles in a powder-zinc anode [147]. Actually, without amalgamation, a compacted powder electrode will not perform at high current densities. The optimum mercury level is about 2-4% Hg [143].

5.1.4. Anode Passivation

The layers of oxides that form on the electrode are usually soft and nonadherent; the electrolyte can diffuse through this type of oxide. Sometimes blue colored oxides are observed, which indicate interstitial zinc in the zinc oxide lattice. At very high current densities, the zinc oxide formed appears black. It is adherent, electronically conductive, nonporous, and essentially insoluble. It can cause rapid electrode failure (passivation).

A detailed study of the reaction products revealed that zinc oxide produced electrochemically is different from chemically produced ZnO [148]; also KOH saturation up to 15% ZnO is possible electrochemically but not chemically (only approximately 8% ZnO at room temperature).

The use of caustic and the production of zinc oxide reach an equilibrium (range) depending on temperature, current drain, and conditions of mass transport in the bulk electrolyte. In the vicinity of the electrode surface, that is, in the pores of a pressed powder-zinc electrode or in the gel structure of a suspended zinc-powder electrode, zinc-complex concentration gradients build up and lower the hydroxide activity very rapidly. Severe diffusion limitations at high current densities lead to a complete loss of available OH^- ions and a breakdown of the zinc potential; this phenomenon is called passivation. It can be reversed by allowing the anode to "rest" on open circuit, unless an insoluble layer of zinc oxide has blocked the surface completely. A schematic plot of the concentration behavior during film formation is shown in Fig. 46 [149]. Average dimensions of the diffusion layer (δ) may be 0.1-0.2 mm; the diffusion coefficient D for zinc ions in zincate solution is about 1 to 2×10^{-5} cm^2 sec^{-1} between 20° and 50°C.

Extensive studies of anodic behavior of zinc in alkaline electrolytes have been performed by Russian authors [150]. Electrolytic zinc deposits and porous active materials or screen collectors were compared (for kinetic behavior and efficiency), and the effects of discharge currents and temperature of the zinc concentrations near the electrodes were determined [151]. Results explain the sharp increase in the tendency toward passivation at higher currents at low temperatures (which is expected) and the fact that at high temperature, passivation occurs more easily at lower current densities. The latter is unexpected, though plausible, if one considers the secondary process of zinc oxide precipitation, that is, formation of a denser structure at lower loads.

Studies of polished vertical zinc electrodes in NaOH revealed two ranges of passivation times: at low current densities, the $\sqrt{t}$ function is dominant, but at higher current densities, linear dependence is observed [152].

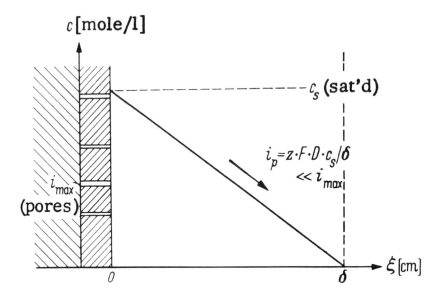

FIG. 46. Schematic plot of the concentration behavior during formation of a porous surface film [149].

With horizontal zinc electrodes, the square-root dependence of the passivation time was established for KOH [153]. This is shown in Fig. 47.

Anodic passivation phenomena have been studied with potentiometric methods [154] and with infrared [155] and nuclear magnetic resonance [156] techniques. The mechanism of film formation has been discussed extensively [157].

It was observed that two types of passivation exist: spontaneous passivation at about 35 mA/cm^2 on smooth surfaces, and long-time passivation occurring after hours of discharge at only 15 mA/cm^2 current density. Polarization and passivation curves of plain sheet zinc, roughened sheet zinc, zinc fabric, and pressed powder tablets (with and without Hg) were compared, and the zinc "area factor," which should be nearly one if the geometric relations are correct, was determined. The agreement was far from ideal [158].

However, convection and diffusion phenomena greatly influence the passivation behavior, and no precise prediction is possible [136].

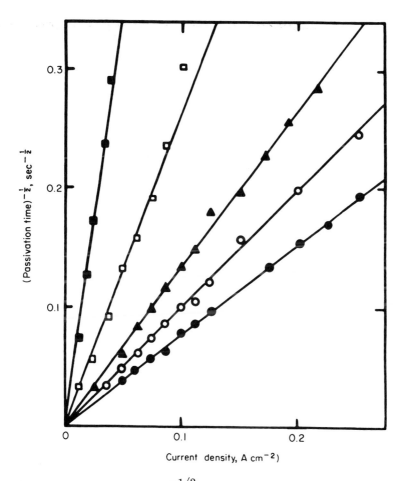

FIG. 47. (Passivation time)$^{-1/2}$ versus current density. 20°C; 1, 2, 3.5, 13.8, and 10 M [153].

Figure 48 shows an interesting effect: the limiting current density reaches a maximum at about 7 N caustic concentration [159].

In alkaline MnO_2-Zn dry cells, the anode is designed to be the service-limiting electrode. The purpose is to avoid postdischarge gassing when the depolarizer is exhausted and the cell remains on load. Passivation of the zinc anode is therefore the characteristic end of discharge in a properly adjusted cell; when the amount of zinc is reduced below a minimum, the

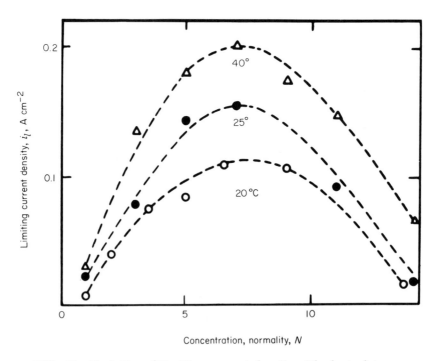

FIG. 48. Variation of limiting current density with electrolyte
concentrations (vertical anodes) [159].

remaining zinc surface cannot sustain the current, and passivation occurs.

Figure 49 shows the voltage-time curves of a D-size cell on intermittent
load. The D-size alkaline MnO_2-Zn cell was discharged on 500 mA (average)
constant current for periods of 7.5 h/day.

In this particular diagram, the measurements were made with pulse-
interrupter equipment, which allows the reading of the closed-circuit voltage
and of the open-circuit voltage between pulses, and the determination of the
polarization of the anode against a reference electrode (resistance free) [140].

The 7.5-h/day discharge regime was chosen to indicate the "recovery"
of the cell overnight. The slope of the cell voltages (resistance-free and
terminal voltage) indicates the MnO_2 characteristic. The relatively flat
curve of the anode (low Zn polarization) extends to the third load period, in
which rapid polarization and finally passivation occurs [139]. The cell

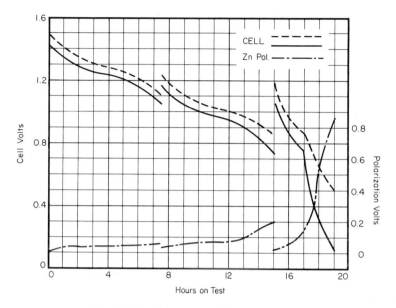

FIG. 49. Intermittent discharge of alkaline cell leading to passivation (Union Carbide Corporation Technical Report).

contains enough MnO_2 to ensure a cathode performance in excess of that of the anode. The internal resistance of the cell can be found by dividing the voltage difference (resistance-free terminal voltage) by the current (0.5 A). Little change occurs until passivation starts. (Initially, the internal resistance is $0.08 \, \Omega$; after 10 h, it is $0.10 \, \Omega$; at the midpoint of the passivation curve, it is $0.70 \, \Omega$).

5.2. SHEET-ZINC ANODES

This group of anodes comprises zinc plates, zinc cans, foils, and other forms, all of such a configuration that the geometric area and the electrolyte-exposed area are approximately identical. "Large-area" anodes can be produced from zinc foil by rolling it up, but this term should not be

confused with the description of an electrode with an "inner" surface such
as porous metals or zinc powder agglomerates have, which is truly large
compared to the physical dimensions of the zinc anode.

In all alkaline cells, the anode material is of high purity, 99.85-99.90%.
It is commercially produced by electroplating (electrorefining) or by dis-
tilling (thermal refining). Sometimes 0.04-0.06% lead is added for greater
corrosion resistance. The amalgamation of zinc foil is a problem because
the embrittlement caused by the mercury prevents handling (rolling, winding) afterward. The difficulty was solved by applying the mercury to the
end of the wound zinc spiral which was the anode in most of the earlier
HgO-Zn cells made following Ruben's designs.

Figure 50 shows a rolled, corrugated zinc-foil anode with a paper spacer
to provide the needed electrolyte [160].

In earlier comparison tests, it was found that corrugated zinc foil anodes
performed better than pressed powder pellets of corresponding geometry.
The reason was (as was later found) the availability of electrolyte from the
adjacent separator (see Fig. 50). In modern cells, this requirement is
taken care of by adding either leachable or heat-decomposable fillers to the
pressed powder pellets.

In general, sheet zinc has been completely replaced by porous zinc
bodies (discs, pellets, cylinders) or suspended zinc powder in gelled elec-

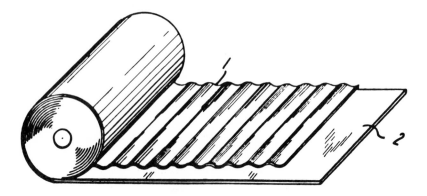

FIG. 50. Wound zinc anode: (1) zinc foil, (2) spacer [160].

trolytes. Today, only large air-zinc cells use amalgamated zinc plates as anodes in caustic solutions (Schumacher [161]).

Zinc cans are used in some alkaline MnO_2-Zn cells, but they serve only as a current collector; powder zinc is pressed against the inside surface (see Section 2.2).

5.3. POWDER-ZINC ANODES

5.3.1. Manufacturing of Zinc Powder

Powder zinc can be produced by a variety of methods. A typical method consists of discharging a thin stream of molten zinc through an opening in the bottom of a melting pot (Fig. 51). The metal falls vertically into the trough of a U- or V-shaped horizontal air jet and is atomized into a collecting system, such as a settling chamber, followed by a cyclone or bag col-

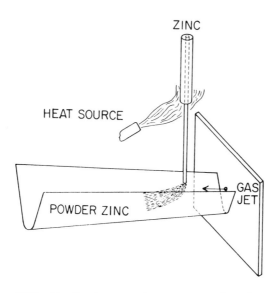

FIG. 51. Production of zinc powder [162].

lector [162]. High-pressure inert-gas atomization [163] is used for the
production of high-temperature metal (alloy) powders; another process uses
water as the dispersing agent [164]. The jet design (orifice) determines
the particle size distribution; the technology is critical, and details are
proprietary. The specifications for a typical zinc powder [165] are shown
in Table 14.

Figure 52 shows graphically how the surface area of this material varies
with the particle diameter (mesh size). The calculated surface areas for
spheres of corresponding diameters are also shown in the same diagram.
Figure 53 presents photomicrographs of zinc powders commonly used in
alkaline MnO_2-Zn cells.

An electrolytic method for producing large-surface zinc powders involves
the precipitation of zinc powders by passing current through solutions or
zinc oxide slurries [166]. Spongy zinc or zinc dendrites, depending on
deposition conditions, can be obtained. This electrolytic method is essentially the deposition process occurring in rechargeable zinc batteries.

5.3.2. Porous Anode Structures

In the course of the development of the HgO-Zn cell, pressed powder-
zinc anodes were brought to a high state of perfection [161]. Herbert used
a pellet pressed from amalgamated zinc powder in the first commercial
alkaline MnO_2-Zn cell, the crown cell [167]. However, the current-carrying
ability of these cells was low at that time. Later, it was recognized that
the powder zinc must not be compressed too much; void volume is essential
to provide for electrolyte storage space.

There are several methods for producing porous zinc anodes from zinc
powder.

a. Metal spraying onto various metal backings produces a deposit that
has a low interconnecting porosity and poor electrode properties.

b. Cold pressing with mercury added to the zinc powder produces "welds"
between the particles even at moderate pressures.

TABLE 14

Product Specifications for Zinc Powder[a]

On 20	0	Pb	0.04-0.055
-20 x 60	25-35	Fe	0.002 max
-60 x 100	25-35	Ni	0.002 max
-100 x 200	25-35	Cd	0.003 max
-200 x 325	7 max	Cu	0.005 max
-325	3 max	Sn	0.001 max
		Met Zn	98.930 max
		ZnO	1.0 max

[a]Reference [165].

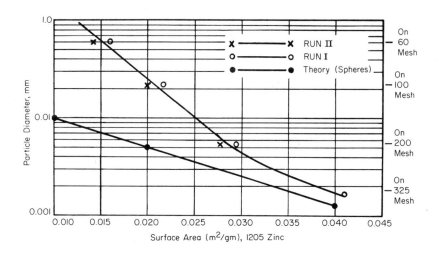

FIG. 52. Surface area versus particle size for zinc powder (Union Carbide Corporation Technical Manual).

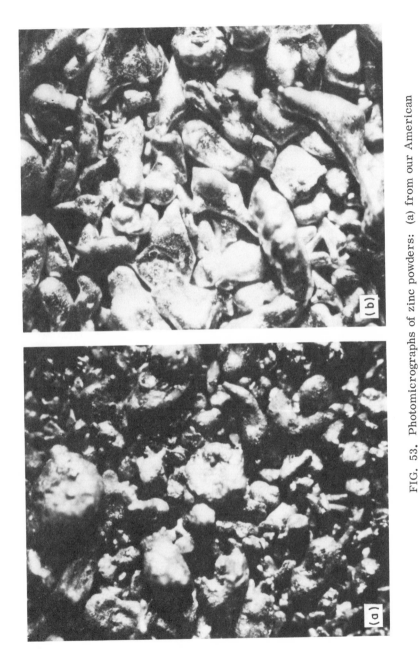

FIG. 53. Photomicrographs of zinc powders: (a) from our American cell, (b) from a Japanese cell (42x).

Figure 54 shows the pressing of a powder-zinc anode into an annular space [168]. The porous sintered bodies produced by this method usually show a low void volume and are not suitable for high current drains unless NH_4Cl (a volatile filler) is used [169].

  c. Cold pressing using a removable filler (for example, NaCl, $NaNO_3$) permits high pressures to be used, and the strength of the compact is good. The filler can be water soluble [170] or volatile (kerosens) [171].

  d. Heat sintering porous zinc electrodes with fillers. Sintering at low pressures to obtain high porosities must be done with zinc particles free of any oxide coating. Hydrogen reduction gives poor results; therefore, a

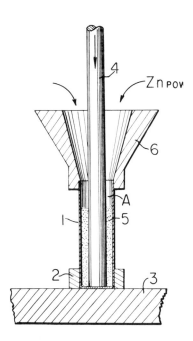

FIG. 54. Method of anode manufacture [168]. Anode cup (1) is seated in ring (2) resting on frame (3). Mandrel (4) is removable, thereby creating a cylindrical space. Amalgamated zinc powder (5) is filled into the cup through funnel (6). Later, the can is drawn to a smaller diameter, thereby compressing the zinc powder.

"fluxing method" was invented, allowing the zinc oxide to be removed at low temperatures [172]. In another process, the porosity is controlled by the amount of filler admixed with the zinc powder; at about 420°C, an ammonium chloride filler can be removed by sublimation [173]. Preliminary bonds are developed between the cleansed zinc particles by sintering in the presence of exothermic compounds such as $N_2H_4 \cdot HCl$ [174].

Especially at low cell temperatures, the effect of porosity is noticeable [169]. Table 15 shows results of capacity determinations at a constant current density of 25 mA/cm^2. The specific area is not as important as the mean pore diameter. The limiting current densities (before passivation) are considerably extended [172]; this feature is especially important for rechargeable zinc anodes.

    e. The bonding method is another manufacturing process for large area zinc electrodes [175]. In this process, amalgamated zinc powder is bonded together by inorganic metal oxides (Zno, CaO) or silicates in alkaline solution. Heating produces the contact between the (amalgamated) particles. These bonded anodes are claimed to be superior to compressed zinc anodes on continuous discharge at low temperatures and to outperform loose zinc powder at room temperature. The bonding agents produce a greater porosity in the anode structure than is possible in compressed anodes without fillers, and the zinc is therefore better utilized. The preferred particle size is -40 to +100 sieve size (U.S. Standard).

    f. Electrolytic reduction of metal compounds. Metal compounds, when subjected to pressure while being electrolyzed, produces very high void-volume plates [176-178].

    Porous zinc anodes in high rate batteries must have high current output, good efficiency, and satisfactory low-temperature behavior. At porosities of 85-90%, all these requirements are fulfilled, but the anode structures are mechanically weak (present use in AgO-Zn cells).

    g. Plastic Bonding. The use of a synthetic binder (polystyrene) in zinc electrodes was described in a 1958 patent [179] aimed at negative plates in silver-zinc batteries. Teflon has since been used as the preferred binder, and very extensive studies have related binder levels and curing temperatures to cycle life and efficiency of operation [180]. These features may be

TABLE 15

Capacity per Volume of Porous Zinc Bodies Discharged at 25 mA/cm^2 at -20°C[a]

Temperature (°C)	i_a = 5 mA/cm^2		i_a = 10 mA/cm^2		i_a = 20 mA/cm^2		i_a = 40 mA/cm^2	
	A (Ah/gm)	B (Ah/cm^3)	A (Ah/gm)	B (Ah/cm^3)	A (Ah/gm)	B (Ah/cm^3)	A (Ah/gm)	B (Ah/cm^3)
20	0.612	2.35	0.538	2.35	0.460	2.20	0.298	2.10
0	0.406	1.96	0.348	1.93	0.230	1.85	0.120	1.73
-20	-	-	0.198	1.19	0.112	0.85	0.034	0.46
-30	0.140	0.83	0.097	0.69	0.048	0.49	0.003	0.25
-40	0.098	0.41	0.042	0.35	0.004	0.25	-	-

[a]Reference [172].

used in future rechargeable batteries with MnO_2 cathodes. Polyacrylate has also been used as a binder [181].

h. Zinc powder with filler materials, pressed into plates and packaged airtight, is a suitable anode structure for reserve-type cells, especially zinc-air cells of the "mechanically rechargeable" [182] or disposable primary type. To obtain high rates, zinc powder (3.5% amalgamated) and a fibrous filler material (12%) were pressed onto copper current collectors. With a zinc particle density of 3.6 gm/cm^3 and a filler density of 0.093 gm/cm^3 (fibrous material 10 μm long, 0.2 μm diam), a calculated zinc porosity of 72% was achieved. Discharge efficiencies of 79.6% at 5 mA/cm^2, 72.9% at 43 mA/cm^2, and 64.3% at 60 mA/cm^2 were obtained [183]. These anodes are activated with KOH. For Al-, Mg-, and Ti-oxides combined with KOH gel structures, see Ref. [181].

Zinc prepared by electrodeposition or by mechanically pulverizing (amalgamated) zinc sheets and impregnated with dry KOH (as filler) can be formed into plates and vacuum sealed. Such anodes can be activated with water [184].

5.3.3. Gel Anodes

The tendency of the zinc anode to passivate at high current drains led first to attempts to increase the surface geometrically (by rolling, corrugating the layers of spiral-wound zinc anodes, and so on). Later, powder zinc was pressed into the required anode shapes and impregnated with electrolyte (see the previous chapters). The main problem, the supply of sufficient electrolyte to the reacting zinc inside the anode, was partly solved by constructing the cell so that liquid reservoirs were utilized. These mechanical approaches were discussed in Section 2.2.

An easy way to provide the anode with sufficient electrolyte and produce a reasonably rigid structure was to use a gelling agent for immobilization of the caustic and mix so many zinc particles into it that electronic conductivity resulted. An anode of that type was first mentioned in 1949 in a patent [185] that describes an electrolyte-anode wafer of a gelatinous mixture comprising

17-20 parts of alkali hydroxide, 50-70 parts of zinc, and 6-10 parts of a gelling agent (Na-carboxymethylcellulose), 2-4 parts of zinc oxide, and the rest water.

The gelled-anode formula of Ref. [186] (1952) is usually considered the first practical use of a powder-zinc/Na-CMC mixture. It was incorporated into some HgO cells. A typical electrolyte is formed of 75 gm KOH (88%), 10 gm ZnO, and 100 gm water, with 2-8 gm Na-CMC added per 100 gm of solution. Good electronic conductivity is achieved when the mixture contains 35-70% Zn. Slugs or slabs can be punched from the gelled material.

A mixture of zinc particles obtained by mechanically crushing (amalgamated) zinc foil with gelling agents (starch, CMC) and fillers (fibers, and so on) and extruded into suitable anode shapes is described in a 1951 patent [187].

The so-called "homogeneous" gel anode is a compromise between good electronic conductivity and maximum pore-volume with optimized zinc particle size (150-200 mesh). Figure 55 relates the percent by volume (or weight) of zinc to the resistivity of an electrolyte/metal-powder paste [147].

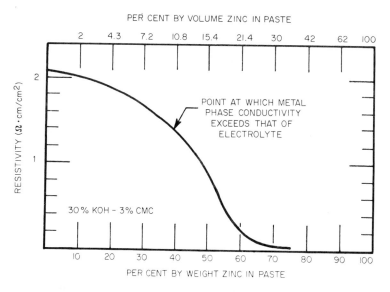

FIG. 55. Resistivity of electrolyte/metal-powder paste [147].

The "homogeneous" mixture lacks high zinc utilization at high current densities. A two-phase anode that consists of a more compact zinc-powder gel and a clear gel reservoir can have up to 90% zinc utilization at high rates.

Twice the service is claimed from nonhomogeneous anodes with viscous gel electrolyte in the center compared to uniform Na-CMC/powder-zinc mixtures. An essential factor is the swelling action of the Na-CMC cylinder, which is inserted into the viscous gel-KOH mixture. This swelling action presses the separator against the cathode [188].

Figure 56 shows a nonuniform, two-phase gel system in a practical cell.

Another patented method of achieving a nonuniform gel with more zinc-powder concentration in the separator direction is to centrifuge a homogeneous gel-anode in the cell by rotation about the length axis [189]. Figure 57 shows this effect of centrifuging the anode gel.

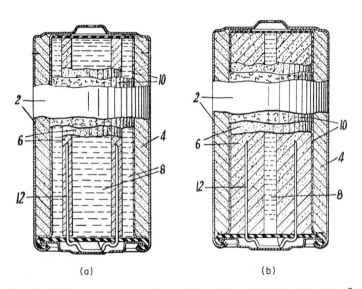

(a)   (b)

FIG. 56. Nonhomogeneous gel-anode system in MnO_2-Zn cells [188]: (2) steel can; (4) cathode; (6) anode cylinder (a) as inserted, (b) after swelling; (8) KOH-Na-CMC gel; (10) separator; (12) anode collector.

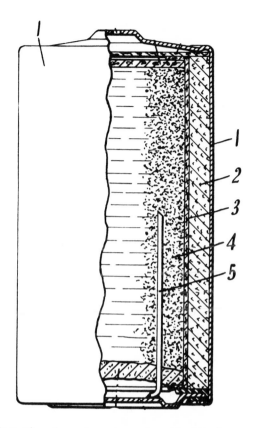

FIG. 57. Centrifuged powder-zinc gel anode [189]: (1) steel can, (2) cathode, (3) separator, (4) centrifuged anode gel, (5) collector.

The same patent also claims a central electrolyte reservoir for KOH (absorbed in liquid-retentive materials) and shows the use of octyl alcohol as a corrosion-inhibiting and creepage-preventing additive.

5.3.4. Rechargeable Zinc Anodes

Rechargeable zinc anodes do not differ in principle from primary-cell anodes; however, the requirements dictate some changes in the anode current collectors (usually screens or expanded metal of copper or silver), in levels of amalgamation (0.25 to 2% only), and in the way the separators are selected and used. Two electrode manufacturing schemes described in the literature [190] will be abstracted here. (It should be noted that these processes are not presently used in alkaline MnO_2-Zn cell technology--they demonstrate the preparation of anodes for silver-zinc cells, but the methods are applicable to any other alkaline-zinc system as well.)

5.3.4.1. Electrodeposited Zinc Negative Plates. The zinc is deposited into frames supplied with grids, mostly of silver wire. The current density used is 120-160 mA/cm^2. The plating solution consists of 45% KOH with 35 $\pm$ 5 gm ZnO per liter. Thorough stirring and temperature control (29°-35°C) are required.

The positive (counter) electrodes are placed on each side of the silver screen to be plated, overlapping its contours. The plating time corresponds to the desired ampere-hours of zinc on the negative plate (1.22 gm Zn = 1 Ah). The plated silver screen is washed free of KOH, excess zinc is removed, and pressure is applied as required for the final plate thickness.

5.3.4.2. Pressed Powder-Zinc Plates. Negative plates for secondary cell usage, requiring 20 cycles in 3-6 months of active life, have 1-2% HgO added to the zinc oxide [191]. Batteries designed for several hundred cycles are made with a Teflon addition [180] and sometimes an Emulphogene solution [192].

a. Pasting procedure. Mix ZnO and HgO, add binders of the dry type, and mix again. Place Viskon paper on silver grids (folds open) in the mold cavity; spread the zinc oxide mix on it. Fold the flaps back and apply pressure until the desired thickness is reached. Add liquid additives; press again. Remove the plate and store it in a polyethylene bag.

b. Conversion of plates. 5% KOH is the electrolyte during the formation procedure. The electrodes are placed in the tank, wrapped in cellophane with the top open, faced with porous separators, and arranged between nickel screen cathodes. Formation is done at a current density of only 1.55 mA/cm^2 for the required conversion time. The plates are washed free of KOH, pressed again to the required thickness, and dried at 50°-55°C. The storage conditions are the same as those for electrodeposited plates, i.e., 15°-30°C under 10% relative humidity.

Details of pasting and conversion quality-control tests for checking the chemicals, plating, and conversion bath are also contained in Ref. [191].

6. ELECTROLYTE SYSTEMS AND SEPARATORS

6.1. ELECTROLYTES

6.1.1. Potassium Hydroxide (KOH)

Potassium hydroxide for battery applications is normally supplied as a 45-50% solution with a specific gravity of 1.46-1.52 at 16°C (60°F) [193,194].

For galvanic cells, the electrical conductivity is the most important parameter. Figure 58 presents conductivity data for different concentrations and temperatures [195].

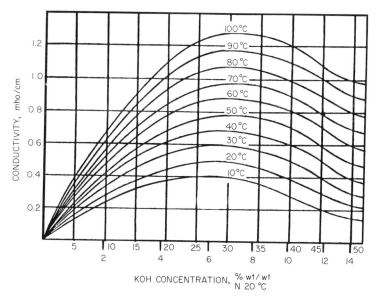

FIG. 58. Electrical conductivity of KOH solutions [195].

Other property data with the sources of information are given in Section 8.

6.1.2. Sodium Hydroxide (NaOH)

Sodium hydroxide was preferred in earlier cells due to its lower price compared with KOH, but the performance of galvanic cells using KOH is in nearly all instances superior to that of cells using NaOH. Therefore, NaOH is seldom used in modern alkaline batteries. NaOH is available in many forms [196]. A 50% solution has a specific gravity of 1.53 at 16°C (60°F). Figure 59 shows conductivity data for various concentrations and temperatures [195]. Mixtures of KOH and NaOH are rarely used. Addition of LiOH has been disclosed [197]. CsOH is claimed to improve low-temperature cells. Other pertinent data are given in the Appendix (with the sources of information).

6.1.3. Zinc Oxide/Hydroxide-Electrolyte Systems

Zinc oxide can be prepared by heating zinc in air or by thermal decomposition of carbonate or other compounds [198]. Its characteristics depend on the method of preparation. At lower temperatures of decomposition, an oxide that is readily soluble is obtained. There are two types of zinc oxides, termed active and inactive, which display differences in surface area (see Table 12). The solubility of ZnO in NaOH and KOH has been determined in many investigations, but with discrepancies in the results due to the different ZnO sources and methods used.

There is a large difference between zinc oxide saturation values of solutions that have been in operating alkaline cells and of chemically prepared solutions. Such solutions often contain twice the amount of zinc in solution, and, on standing, ZnO precipitates slowly. Sometimes it is possible to accelerate the precipitation process by seeding with MgO, silicate, or sulfur compounds [199], but generally such attempts have resulted in reliable means of increasing the electrolyte capacity only if used over very long operating periods. ZnO precipitation results in OH^- ion regeneration and prolonged cell operation (application in "air cells").

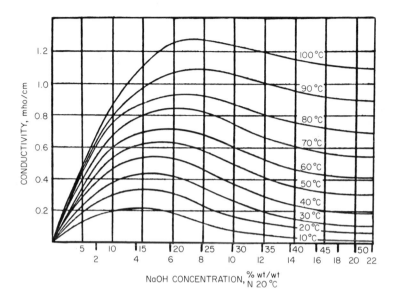

FIG. 59. Electrical conductivity of NaOH solutions [195].

Zinc hydroxide can be prepared chemically by treating zinc salts with the calculated amount of NaOH or KOH or an excess of NH_4OH. Also, dilution of zincate-saturated solutions with water precipitates $Zn(OH)_2$. However, the equilibrium between $Zn(OH)_2$ and ZnO is not certain, especially with temperature variations. Several crystalline forms and one amorphous form have been found. The ϵ-$Zn(OH)_2$ modification is stable [200, 201].

Different colors of anodically prepared oxides or hydroxides have been observed. A blue color indicates a lattice structure containing defects. Semiconductor properties of modified ZnO are probably based on such types of lattice defects [202]. Black zinc oxide or hydroxide may represent a mixture with colloidal zinc (it is highly conductive); this form appears in connection with anodic passivation phenomena.

The $Zn(OH)_2$-ZnO-NaOH system is better known than the KOH system, but still many discrepancies are apparent in the data reported in the literature. Complex formation and related different solubilities are the problem areas.

The solubility diagram shown in Fig. 60 shows the different equilibria believed to exist in the NaOH-Zn(OH)$_2$-ZnO system [203]. The diagram is mainly applicable to commercial wet cells, especially zinc-air cells, which use 5-6 N NaOH, and dry cells containing 8-12 N alkaline solution.

Figure 61 indicates how solubility data change with standing time of the solution. The lower the OH$^-$ ion concentration, the shorter the stabilization time is [204].

A phase diagram for the ZnO-Na$_2$O-H$_2$O system has been drawn from available literature data [205]; however, the existence of solid phases other

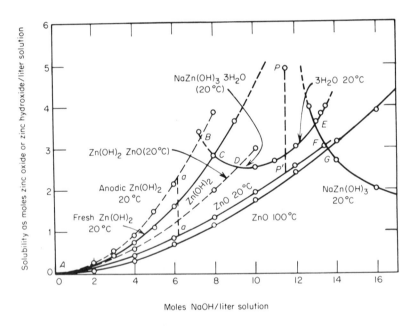

FIG. 60. Solubility of Zn(OH)$_2$ and ZnO in NaOH solutions [203]: (A-B) curve of galvanically saturated solutions; (A-C) solubility of freshly prepared ZnO; (A-D) some of the aged or dehydrated Zn(OH)$_2$; (A-F) ZnO curve at 20° C; (A-G) water-free ZnO curve (100°C); (C-E) A metastable area is indicated by curve of zinc oxide solubility in 8-13 N NaOH. Some solid phases are reported there.

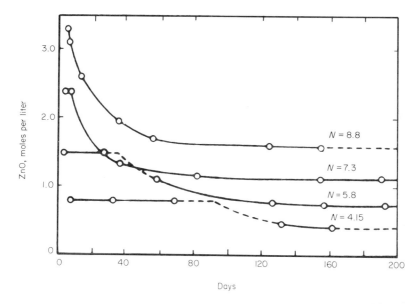

FIG. 61. Change in ZnO concentration with time for saturated solutions of various normalities of NaOH [204].

than ZnO and $Zn(OH)_2$ is in doubt. $NaZn(OH)_3$ and $Na_2Zn(OH)_4$ have been reported as metastable compounds.

The $Zn(OH)_2$-ZnO-KOH system is the one preferably used in alkaline MnO_2-Zn cells. Unfortunately, even less data about equilibrium conditions and complex formation are available.

Figure 62 shows ZnO-KOH solubility data collected from publications of several authors and obtained by different methods [203].

A comparison of the solubility of $Zn(OH)_2$ and ZnO at different temperatures is shown in Fig. 63. The remarkable result is that the ZnO solubility is not temperature sensitive in the range between -3°C and +56°C [205]. Again, solubility data (especially at high temperatures and during preparation of electrolytes for powder-zinc anodes [Section 5.3]) are very time dependent. This behavior is shown in Fig. 64, in which the dissolution of zinc oxide at 145°C is plotted versus time. However, it is confirmed that the temperature dependence is small and the heat of solution close to zero [206].

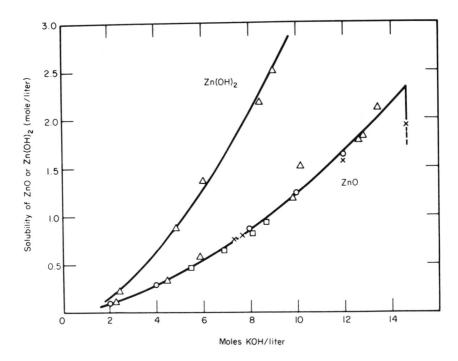

FIG. 62. Solubility of $Zn(OH)_2$ and ZnO in aqueous KOH solutions at 25°C [203].

Specific conductivities of KOH-ZnO solutions as a function of concentration and temperature (from 36° to -66° C) are reported in Ref. [207] and compared with earlier KOH data. The zinc oxide lowers the conductivity, but the main characteristic is maintained. 30% KOH solution shows the best overall low-temperature conductivity. 25% KOH is better between -23° and -36°C, but has a freezing point of -41°C (see the Appendix).

For the KOH-ZnO system, data are very scarce and no solid phase of the type $K_2[Zn(OH)_4]$ has been found [208].

6.1.4. Gelling Agents

The use of gelling agents for the manufacture of gelled powder-zinc anodes was discussed in Section 5.3.3.

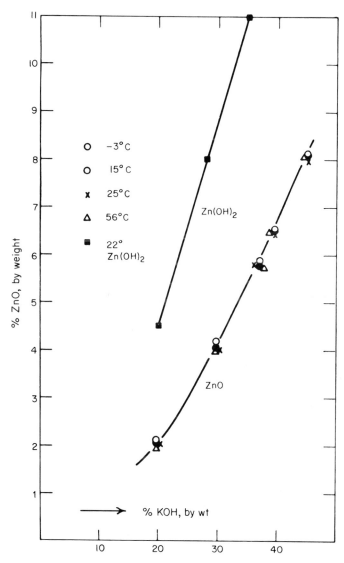

FIG. 63. Solubility of ZnO and $Zn(OH)_2$ in KOH solutions, -3° to 56°C [205].

6.1.4.1. <u>Sodium Carboxymethylcellulose.</u> Sodium carboxymethylcellulose (CMC) is a white, odorless, tasteless, nontoxic solid. Water-soluble
CMC with a degree of substitution of 0.4-1.4 is commercially available;

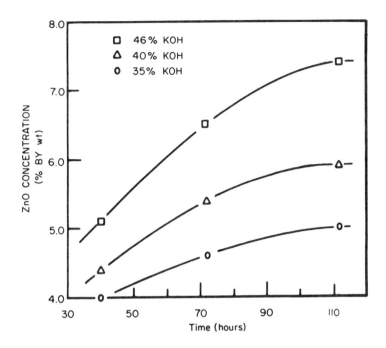

FIG. 64. Dissolution of ZnO in KOH at 145°C [206].

the most common product has a degree of substitution of 0.7–0.8 [209].
Aqueous solutions of CMC are usually thixotropic--a property not desired
for gelled electrolytes--but special production methods can produce materials that are not thixotropic [210, 211].

   CMC is used for stabilizing emulsions, paper sizing, pastes, and is
extensively used in food preparation, pharmaceuticals, and cosmetics
(often called cellulose gum). Its ionic character causes precipitation with
heavy-metal ions, forming salts. This can be prevented by replacing part
of the carboxymethyl group with hydroxyethyl groups, leading to carboxy-
methylhydroxyethylcellulose (CMHEC) [212]. Methyl chloride treatment
results in methylcellulose [213].

   6.1.4.2. *Starches.* Starch is a high-polymeric carbohydrate with an
approximate molecular formula $(C_6H_{10}O_5)n$, where n varies from a few
hundred to over a million. It is made up of the linear polymer amylose and
the branched polymer amylopectin. Granular starches give x-ray diffrac-

tion patterns making it possible to distinguish between cornstarch, potato starch, and sago-tapioca starches [214]. Also, infrared spectra may be used for identification [215].

Starch granules gelatinize when the temperature is raised to $60°-70°C$; at higher temperatures, a sol forms and the shorter linear molecules dissolve. The sol usually gelatinizes on cooling. Caustic alkali additions lower the gelatinizing temperatures and may liquify gels if the starch is not modified. In general, starches containing 15-30% of the linear amylose form a gel on cooling (retrogradation). Starches with only amylopectin form weak gels, linking only at the outer branched parts of the molecules.

The alcohol groups in starch are easily oxidized to form aldehydes, ketones, or carboxyl groups, leading to more stable modifications [216].

The use of starch as electrolyte gelling agent for alkaline cells is not as common as the use of Na-CMC, but starch is utilized in some commercial products either as a clear gel to serve as the separator or in mixtures of starch derivates and CMC for powder-zinc gels.

6.2. SEPARATORS

6.2.1. Requirements for a Separator

The primary function of a separator is to keep the positive and negative cell components apart, preventing electronic shorts but not inhibiting ionic transport in the cell.

Primary batteries have less stringent requirements for separator quality than do secondary cells, which must be able to operate over many discharge-charge cycles. However, in both cases, the strong alkaline electrolyte and the oxidizing action of the manganese dioxide lead to the need for high chemical stability at elevated temperatures (up to $70°C$ for summer storageability) and long life expectancy (several years) combined with mechanical strength and stability (dimensional changes of anode and cathodes during discharge, swelling, and so on).

In secondary alkaline zinc cells, the need to prevent shorts by zinc deposits (dendrites formed during charging) is added. While in primary cells a stable, highly porous separator of low resistance might do very well, in secondary cells, a uniform, microporous separator with selective permeability and oxidation resistance is required.

Many other properties must be considered for production: wettability, wet strength, machine compatibility, and cost.

In general, it can be said that all primary alkaline MnO_2-Zn cells use macroporous separators made from woven, bonded, or felted materials. Secondary cells should use better microporous, polymeric-membrane materials, but these are not compatible with other cell requirements such as recombination or venting of gases produced on charging. Separators used in Ni-Cd cells are well suited for alkaline MnO_2-Zn cells, but generally too high in cost.

The principal types of separators available are the following.

   a. Fibrous separators: fibrous cellulosic and synthetic materials; woven, felted, bonded

   b. Porous-sheet separators: produced by filler removal

   c. Membranous separators: cellulosic--regenerated cellulose, fiber-reinforced cellulose; synthetic--graften membranes, polymeric films

   d. Inorganic separators: inorganic microporous layers with binder materials

6.2.2. Fibrous Separators

Several types of "nonwoven" material suitable for alkaline MnO_2 cells are listed in Table 16 [217].

Besides the role of mechanical spacer, electrolyte absorption is important. Some paper-type separators serve as electrolyte reservoirs between anode and cathode.

A typical nonwoven separator material is Webril, a purified cotton material produced in several thicknesses, length of fibers, different fiber

TABLE 16

Fibrous Separator Materials[a]

Material	Trade name	Structure
Cotton-cellulose	Webril	Nonwoven
95% PVC, 5% PVAC	Vinyon	Nonwoven
Cellulose-Rayon	Viskon	Paper
Cellulose-Rayon	Dexter	Paper
Cellulose-Nylon	Syntosil	Composite
Nylon	Pellon	Nonwoven
Nylon-Dynel	Separators	Fibrous sheets
Dynel	for Ni-Cd	or felts
Polypropylene	cells	

[a]For use of these materials in secondary cells other than MnO_2-Zn, see Ref. [217].

diameter, and various arrangements of fibers (random by air blowing, aligned, layered, and so on) [218].

The bonding processes used to tie the fibers together depend on the materials. Heat and pressure are used, for example, with Dynel and polypropylene fibers, and bonding with adhesives (usually thermoplastics) is applied to cotton and viscose rayon. Solvents may be used with high-melting polymers to aid the fiber-bonding process.

Hydrophobic materials (polypropylene, for example) present problems with liquid adsorption and retention. These properties can be improved by irradiation techniques.

Composite structures are chosen to improve mechanical properties. A typical example is Viskon-Vinyon, which is a mixture of rayon and PVC fibers. This allows the material to be heat sealed and gives good mechanical strength.

Sandwich-layered separators are made from membranous material and woven or felted fibrous synthetics (see Section 6.2.4.).

6.2.3. Porous-Sheet Separators

Porous-sheet separators are listed in Table 17. A typical synthetic separator consisting of a sheet with "holes" is Synpor, a vinyl plastic that has starch grains incorporated into the dough before it is rolled or extruded into a thin film. Also, NaCl has been used as a filler that is later leached out [219]. Unfortunately, these types of separator are quite expensive and are not used in primary alkaline MnO_2-Zn cells. The pore diameter of such materials usually exceeds 1000-10,000 Å, even when polyethylene is used as base material [220].

An ultrafine (UF) porous polymer membrane made by the leaching technique has been recently described: 50% porous, flexible membranes with 40-120 Å pores are obtained by adding sodium benzoate to polyethylene or polypropylene and milling the polymers at elevated temperatures. Leaching is done with water at low temperatures [221].

6.2.4. Semipermeable Membranes

Table 18 lists the membranous separators that are available for use in rechargeable alkaline cells using zinc anodes [217]. These materials find use in silver-zinc batteries and are not presently used in MnO_2 cells.

TABLE 17

Porous Sheet Separators[a]

Material	Name	Pore diameter (Å)
Polyvinylchloride	Synpor	5000-7000
Polyethylene	Porothene S	10000-20000
Dynel	Polypor WA	1000-4000
Dynel	Acropor WA	1000-5000
Polypropylene	UF-Polymer	40-120

[a]For use of these materials in secondary cells other than MnO_2-Zn, see Ref. [217].

TABLE 18

Semipermeable Separators[a]

Material	Trade name	Pore diameter (Å)	Porosity (%)
Regen. cellulose	Cellophane PUD	25-35	>90
Regen. cellulose	Sausage casing	25-35	>90
Modif. cellulose	Permion 600	60-90	50-70
Modif. polyethylene	Permion 300	60-90	50-70
Modif. methylcellulose	30% PVMMA-C-3	60-90	50-70
Polyvinyl alcohol	NAS-5-9107-29	60-90	50-70
Acrylic Porothen	PMA 83/17	20-30	40-60

[a]For use of these materials in secondary cells other than MnO_2-Zn, see Ref. [217].

6.2.4.1. Cellulosic Films. Cotton, which contains only 1-2% noncellulosic impurities, was the standard source for pure cellulose before modern technology could utilize other sources such as wood and produce products of equivalent performance.

The chemistry of cellulose [222, 223]. Cellulose is a disperse polymer of high molecular weight composed of long chains of d-glucose joined together.

The chemical stability of cellulose is less than that desired for a separator; alkaline hydrolysis at elevated temperature splits the structure and exposes more unreacted groups. However, by partial oxidation of the terminal aldehyde groups to carboxyl groups, hydrolysis can be inhibited. Chemical cross-linking can reduce degradation considerably. The ion-exchange properties of cellulose are beneficial for retaining metal impurities.

Cellophane and sausage casings [224]. To produce a film of cellulose, a solution in alkali and carbon disulfide is first prepared (xanthate). A film is generated by extruding the dissolved cellulose into an acidic bath. If the regenerated product is in fiber form, it is called rayon. Films of cellophane are absorbent and retain caustic under swelling, which reduces the wet strength. A film containing cotton lint fibers has a much improved wet strength; it is commercially used for sausage casings and has a high purity.

Cellulose films can also be made from cast nitrocellulose films by denitration from the triacetate by removal of the acetate group (the Fortisan process) and from cuprammonium cellulose.

Modifications of cellulosic materials [217] are used to improve the chemical resistance or the mechanical wet strength. Sodium borohydride treatment reduces the number of reactive aldehyde groups. Formaldehyde acts as cross-linking agent. Grafting with acrylonitrile and methacrylic acid and treatment with antioxidants has also been tried, but this process increases the electrical resistance. Radiation grafting, attaching acrylic acid to the cellulosic structure, and blending with other polymers such as methylcellulose improves the chemical stability [225].

Diffusion data for cellulosic membranes. Measurement of the diffusion rate through membranes, in relation to an effective current density, is needed to estimate the quality of a semipermeable separator. It can then be estimated whether hydroxide diffusion through the membrane is limiting the performance [226].

6.2.4.2. Synthetic Membranes.

Polyethylene-grafted membranes. The procedure used to produce such separators is described in Ref. [227]. The process involves the preparation of a 1-mil-thick film of polyethylene, cross-linking the film by beta radiation to yield a uniform three-dimensional structure, arranging the film with a cheesecloth interlayer, soaking it with the grafting solution, and exposing it to ^{60}Co gamma radiation in situ. Radiation increases the oxidation resistance. The modification with a graft (in this case, methacrylic acid) increases the conductivity in KOH.

Polyvinyl alcohol (PVA) membranes swell in NaOH or KOH electrolytes. PVA is the hydrolysis product of polyvinyl acetate and is quite stable in caustic solutions. Since its mechanical strength is poor, this membrane is used in sandwich (composite) separator systems where the dense gel structure is useful for preventing particle migration (for example, Hg in HgO cells).

Ion-exchange membranes have useful applications as battery separators because of their heavy-metal retention properties and their small pores, which prevent zincate migration. By cross-linking, very tight membranes (still with satisfactory conductivity) are obtained [228].

6.2.5. Inorganic Separators

In order to meet requirements of resistance against oxidation and chemical attack, especially in AgO cells, inorganic separators were developed.

Rigid inorganic discs with pores of 200–500 $\overset{o}{A}$ diam were made from alumina silicate compositions [229]. Pressing at 10,000 lb/in^2 and sintering produced separators that gave improved cycle life after heat sterilization (space-program cells), which could not be done with cellulosic separators.

Flexible inorganic separators were produced by mixing a small amount of polymer material into the inorganic slurry. Casting techniques could be employed with solvent adjustments. The conductivity of special flexible separators could be brought into the cellophane range (3.5 to 10 $\Omega \cdot$ cm). Zincate diffusion was also comparable. Zinc penetration of rigid inorganic separators was slower than with cellophane; flexible inorganic separators were not as good.

6.2.6. Testing of Separator Properties

Screening methods for selection of suitable separators have been developed and compiled in a summary publication by the Air Force Propulsion Laboratory (see Section 7.2).

The tests are concerned with dimensional stability; tensile strength; electrolyte retention; pore-size determination; electrical resistance; degradation by KOH and oxidants, such as AgO; electrolyte diffusion; Ag diffusion; Zn diffusion; and Zn penetration.

7. LITERATURE

7.1. ELECTROCHEMISTRY

a. Modern Aspects of Electrochemistry, (J. O'M. Bockris and B. E.
Conway, eds.), Vols. 1-3, Butterworth, London, 1954-1964; Vols.
4 and 5, Plenum Press, New York, 1966-1969.

b. Modern Electrochemistry, (J. O'M. Bockris and A. K. N. Reddy, eds.),
Vols. 1 and 2, Plenum Press, New York, 1970.

c. New Instrumental Methods in Electrochemistry, (P. Delahay), Wiley-
Interscience, New York, 1954.

d. Advances in Electrochemistry and Electrochemical Engineering, (P.
Delahay and C. W. Tobias, eds.), Vols. 1-9, Wiley-Interscience,
New York, 1961-1971.

e. Electrochemical Kinetics: Theoretical and Experimental Aspects,
(K. J. Vetter), Academic Press, New York, 1967.

f. Electrode Processes, Trans. 1966 Symp. (E. Yeager, H. Hoffman,
and E. Eisenmann, eds.), Electrochemical Society, New York, 1966.

g. The Oxidation States of the Elements and Their Potentials in Aqueous
Solution, (W. Latimer), 2nd ed. Prentice-Hall, Englewood Cliffs, N. J.,
1952.

h. Atlas of Electrochemical Equilibria in Aqueous Solution, (M. Pourbaix),
English ed., Pergamon, New York, 1966.

7.2. BATTERY TECHNOLOGY

a. Alkaline Storage Batteries, (S. U. Falk and A. J. Salkind), Wiley,
New York, 1969.

b. Zinc-Silver Oxide Batteries, (A. Fleischer and J. J. Lander, eds.),
Wiley, New York, 1971.

c. The Primary Battery, (G. W. Heise and N. C. Cahoon, eds.), Vol. 1,
Wiley, New York, 1971.

d. Literature Search on Dry Cell Technology, (M. Bolen and B. H. Weil, eds.), U. S. Signal Corps Engr. Labs., Ft. Monmouth, N. J., Sp. Report No. 27, State Engr. Exp. Sta., Georgia Inst. Technol., 1948.

e. Batteries, Proc. 3rd Intern. Symp., Bournemouth, 1962, (D. H. Collins, ed.), Pergamon, New York, 1963.

   Batteries 2, Proc. 4th Intern. Symp. Brighton, 1964, (D. H. Collins, ed.), Pergamon, New York, 1965.

   Power Sources, Proc. 5th Intern. Symp. Brighton, 1966, (D. H. Collins, ed.), Pergamon, New York, 1967.

   Power Sources 2, Proc. 6th Intern. Symp., Brighton, 1968, (D. H. Collins, ed.), Pergamon, New York, 1970.

   Power Sources 3, Proc. 7th Intern. Symp. Brighton, 1970, (D. H. Collins, ed.), Oriel Press, Newcastle upon Tyne, England, 1971.

f. Characteristics of Separators for Alkaline Silver Oxide-Zinc Secondary Batteries--Screening Methods, (J. E. Cooper and A. Fleischer, eds.), Air Force Aero Propulsion Lab., Dayton, Ohio, 1965.

g. Proceedings, Annual Power Sources Conference (Symposium), 1957-1972, Sponsored by U. S. Army Electronics Command, Ft. Monmouth, N. J.; PSC Publication Committee, Red Bank, N. J. (1966 issue has 10-year index. No conference in 1971.)

h. Advances in Battery Technology Symposium, Vols. 1-3 (1965-1967) (C. Berger, D. Benneon, and H. Recht, eds.), Vol. 4 (1968) (H. Recht, ed.), Vol. 5 (1969) (J. S. Smatko, Chairman), Vol. 6 (1970) (E. L. Littauer and J. E. Oxley, Chairmen); Southern California-Nevada Section. Electrochemical Society.

i. Proceedings of the Symposium on Battery Separators, Columbus Section, Electrochemical Society, 1970.

j. Proceedings on Batteries for Traction and Propulsion, sponsored by Columbus Section of the Electrochemical Society and Battelle Memorial Institute, Columbus, 1972.

k. Electrochemistry of Manganese Dioxide and Manganese Dioxide Batteries in Japan, Vols. 1 and 2, U. S. Branch Office, Electrochemical Society of Japan, 1971.

8. APPENDIX

8.1. POTASSIUM HYDROXIDE

8.1.1. A-1. Composition of KOH*

Chemical Specifications of Standard Grade KOH[a]

	Liquid	Flake/granular
Equivalent KOH	45.0%–50.0%	90.0% min
NaOH	0.04% max	0.10% max
K_2CO_3	0.2% max	1.0% max
KCl as Cl	0.35% max	0.65% max
$KClO_3$	0.0006% max	None
K_2SO_4	0.002% max	0.006% max
Fe	0.0005% max	0.003% max
Si	0.001% max	0.002% max
Ca	0.0005% max	0.001% max
Mg	0.0005% max	0.001% max

[a]Other dry forms available include walnut, broken, powder, crushed, and solid.

Physical Properties of KOH

Molecular weight	56.1
Melting point 45%	$-22°F$
50%	$48°F$
90%	$427°F$
Anhydrous	$716°F$
Weight per gallon 45%	12.18 lb
50%	12.68 lb

*Reference: Hooker Chemical Company.

8.1.2. A-2. Vapor Pressure of KOH Solutions*

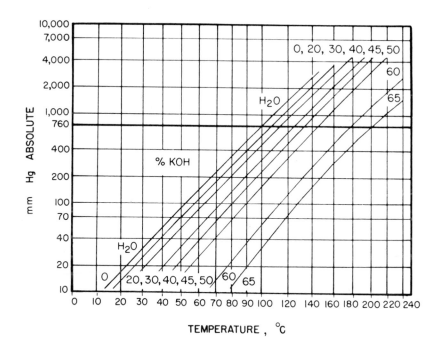

TEMPERATURE , °C

*Reference: International Critical Tables, Vol. III, McGraw-Hill, New York, 1928, p. 373.

8.1.3. A-3. Freezing Point of Aqueous KOH Solutions*

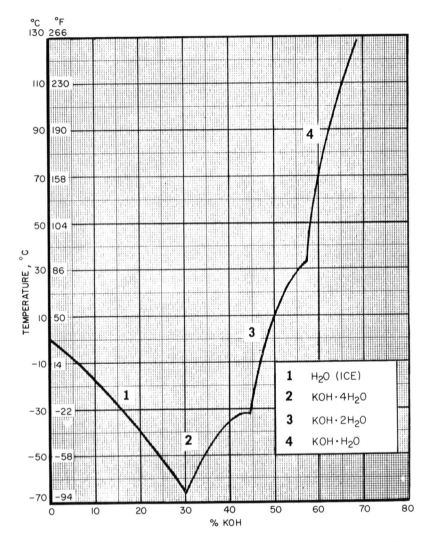

*Reference: International Critical Tables, Vol. IV, McGraw-Hill, New York, 1928, p. 239.

8.1.4. A-4. Concentration Conversions of Aqueous KOH Solutions*

Molality	Density[a]	Wt%	Molarity
0.0000	0.9968	0.0000	0.0000
0.3637	1.0148	1.9999	0.3617
0.7427	1.0328	4.0005	0.7364
1.1377	1.0509	6.0005	1.1239
1.5499	1.0692	8.0007	1.5246
1.9804	1.0877	10.000	1.9387
2.4305	1.1065	12.001	2.3666
2.9016	1.1255	14.001	2.8084
3.3951	1.1445	16.002	3.2645
3.9126	1.1641	18.002	3.7345
4.4560	1.1838	20.002	4.2199
5.0273	1.2036	22.002	4.7199
5.6287	1.2237	24.002	5.2347
6.2625	1.2440	26.002	5.7649
6.9316	1.2655	28.002	6.3155
7.6389	1.2854	30.002	6.8730
8.3878	1.3065	32.002	7.4515
9.1821	1.3278	34.002	8.0466
10.026	1.3494	36.002	8.6586
10.924	1.3714	38.002	9.2879
11.883	1.3935	40.002	9.9348
12.907	1.4160	42.002	10.600
14.005	1.4387	44.003	11.283
15.183	1.4618	46.002	11.985
16.453	1.4852	48.002	12.706
17.824	1.5090	50.003	13.448

[a]Data from G. Akerlof and P. Bender, J. Amer. Chem. Soc. 63, 1085 (1941).

*Table calculated by G. H. Newman, Union Carbide Corporation, Research Laboratory, Parma, Ohio (1964).

8.1.5. A-5. Activity Coefficients of KOH in Water at 25°C*

Molality (mole/1000 gm H_2O)	$\gamma_\pm$[a]	Molarity (mole/liter)	$f_\pm$[b]
0.1	0.798	0.1	0.798
0.2	0.760	0.2	0.760
0.3	0.742	0.3	0.742
0.4	0.734	0.4	0.734
0.5	0.732	0.5	0.732
0.6	0.733	0.6	0.733
0.7	0.736	0.7	0.736
0.8	0.742	0.8	0.742
0.9	0.749	0.9	0.749
1.0	0.756	0.984	0.768
1.2	0.776	1.177	0.791
1.4	0.800	1.366	0.820
1.6	0.827	1.558	0.849
1.8	0.856	1.749	0.881
2.0	0.888	1.937	0.917
2.5	0.974	2.401	1.014
3.0	1.081	2.859	1.134
3.5	1.215	3.311	1.284
4.0	1.352	3.743	1.445
4.5	1.53	4.183	1.65
5.0	1.72	4.610	1.87
5.5	1.95	5.025	2.13
6.0	2.20	5.433	2.43
7.0	2.88	6.233	3.23
8.0	3.77	7.007	4.30
9.0	4.86	7.752	5.64
10.0	6.22	8.454	7.36
11.0	8.10	9.138	9.75
12.0	10.5	9.807	12.9

(continued)

8.1.5. A-5. Activity Coefficients of KOH in Water at 25°C* (Continued)

Molality (mole/1000 gm H_2O)	$\gamma_{\pm}$[a]	Molarity (mole/liter)	$f_{\pm}$[b]
13.0	13.2	10.421	16.5
14.0	15.8	11.031	20.1
15.0	19.6	11.606	25.3
16.0	24.6	12.158	32.4

[a]$\gamma_{\pm}$ is the activity coefficient referred to molality (mole/1000 gm H_2O).

[b]$f_{\pm}$ is the activity coefficient referred to molarity (mole/liter).

*Reference: R. A. Robinson and R. H. Stokes, Trans. Faraday Soc., 45, 612 (1949).

8.1.6. A-6. Relation Between Molality and Molarity and Per Cent
Concentration of Aqueous KOH Solutions at 25°C*

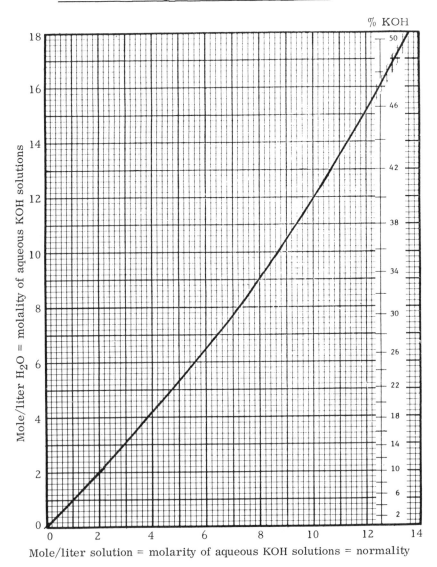

Mole/liter solution = molarity of aqueous KOH solutions = normality

*Handbook data.

8.1.7. <u>A-7. pH versus Normality and Per Cent KOH at 25°C*</u>

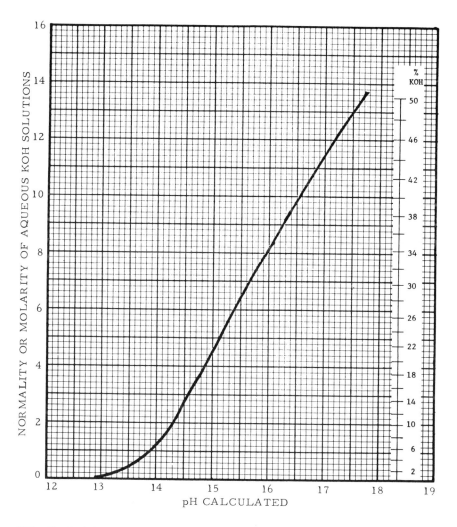

*Handbook data.

8.2. SODIUM HYDROXIDE

8.2.1. A-8. Characteristics of NaOH*

Chemical name	Sodium hydroxide
Chemical formula	NaOH
Molecular weight	40.005
Melting point	318.4°C, 605.0°F
Boiling point	1390°C, 2534°F
Specific gravity	2.130
Refractive index	1.3576
Heat of fusion	72 BTU/lb
	40 cal/gm
Solubility in grams per 100 ml:	
Water 0°C	42
Water 100°C	347

Soluble in alcohol and glycerol; insoluble in acetone and ether

Specifications for NaOH

	Reagent special pellets low in carbonate	Reagent pellets	Reagent 50% solution
Assay (NaOH)	98.0% min	97.0% min	50.0–52.0%
Sodium carbonate (Na_2CO_3)	0.5% max	1.0% max	0.10% max
Chloride (Cl)	0.005% max	0.005% max	0.002% max
Nitrogen compounds (as N)	0.001% max	0.001% max	0.0005% max
Phosphate (PO_4)	0.001% max	0.001% max	0.0005% max
Sulfate (SO_4)	0.003% max	0.003% max	0.001% max
Ammonium hydroxide ppt.	0.020% max	0.020% max	0.010% max
Heavy metals (as Ag)	0.002% max	0.002% max	0.001% max
Iron (Fe)	0.001% max	0.001% max	0.0005% max
Nickel (Ni)	0.001% max	0.001% max	0.0005% max
Potassium (K)	0.020% max	0.020% max	0.010% max

*Reference: J. T. Baker Chemical Company.

8.2.2. A-9. Vapor Pressure of NaOH*

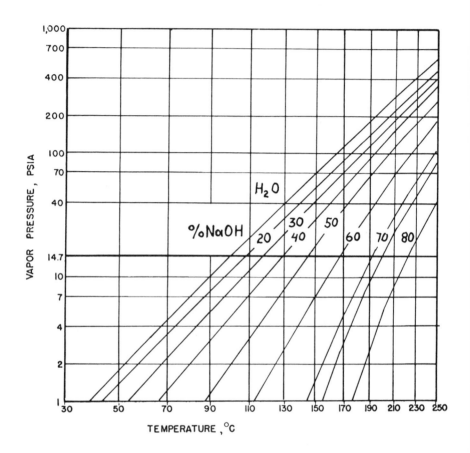

*Reference: International Critical Tables, Vol. III, McGraw-Hill, New York, 1928, p. 370.

8.2.3. A-10. Freezing-Point Curve for Aqueous Solutions of NaOH*

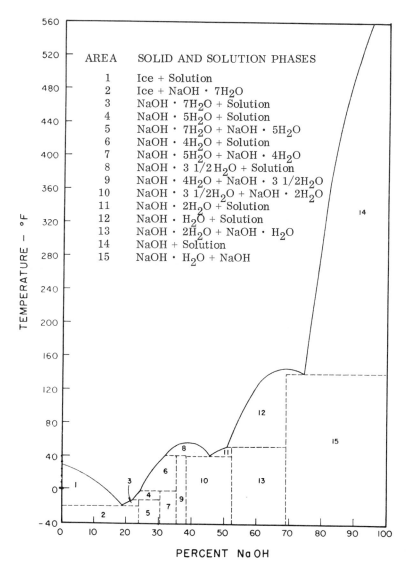

AREA	SOLID AND SOLUTION PHASES
1	Ice + Solution
2	Ice + NaOH · 7H$_2$O
3	NaOH · 7H$_2$O + Solution
4	NaOH · 5H$_2$O + Solution
5	NaOH · 7H$_2$O + NaOH · 5H$_2$O
6	NaOH · 4H$_2$O + Solution
7	NaOH · 5H$_2$O + NaOH · 4H$_2$O
8	NaOH · 3 1/2 H$_2$O + Solution
9	NaOH · 4H$_2$O + NaOH · 3 1/2H$_2$O
10	NaOH · 3 1/2H$_2$O + NaOH · 2H$_2$O
11	NaOH · 2H$_2$O + Solution
12	NaOH · H$_2$O + Solution
13	NaOH · 2H$_2$O + NaOH · H$_2$O
14	NaOH + Solution
15	NaOH · H$_2$O + NaOH

PERCENT Na OH

*Reference: International Critical Tables, Vol. IV, McGraw-Hill, New York, 1928.

8.3. SEPARATORS AND MEMBRANES

8.3.1. <u>A-11. Electrolyte Absorption and Retention of Separator Materials</u>

Separator material	Thickness (mm)	Electrolyte absorbed $\left[\dfrac{\text{wt. electrolyte}}{\text{wt. separator}}\right]$	Electrolyte retained after acceleration (%)
Woven nylon cloth	0.076	0.8	72
	0.051	0.5	89
Autogenetic nylon felt	0.279	1	97
Autogenetic Dynel felt	0.203	8	93
Autogenetic Dacron felt	0.178	4	62
Dynel felt	0.076	0.6	98
Nylon felt	0.254	4	98
Nylon felt on rayon	0.228	6	97
PVA film	0.010	2	95
PVA laminate	0.010	2	97
Nonwoven PVA	0.127	0.6	100
Weak base PVA	0.010	0.7	98
PVA rayon laminate	0.076	0.4	88
PVA laminate	0.178	0.9	98

(continued)

8.3.1. A-11. Electrolyte Absorption and Retention of Separator Materials (Continued)

Separator material	Thickness (mm)	Electrolyte absorbed $\left[\dfrac{\text{wt. electrolyte}}{\text{wt. separator}}\right]$	Electrolyte retained after acceleration (%)
Nonwoven PVA	0.152	0.2	93
Asbestos paper	0.063	6.9	80
Weak acid-irradiated polyethylene film	0.051	0.2	82
Quaternary Amine-irradiated polypropylene film	0.061	1.6	96
Cellophane	0.038	1.9	100
Cellulose felt	0.076	8	93
Cellulose-amide felt	0.279	2	97
Polypropylene felt	0.134	4.4	93
PVC felt	0.134	0.6	91
Polypropylene paper	0.193	0.8	69
	0.180	0.6	95
Nylon paper	0.115	0.2	73
Nonwoven Dynel	0.038	0.6	100

Reference: R. C. Shair and H. N. Seiger in Characteristics of Separators for Alkaline Silver Oxide–Zinc Secondary Batteries (J. E. Cooper and A. Fleischer, eds.), Air Force Aero Propulsion Lab., Wright–Patterson Air Force Base, 1965, p. 25.

8.3.2. A-12. Electrical Resistance of Separator Materials*

Material[a]	Soaking time in 31% KOH at 26°C				
	5 min	10 min	20 min	1 h	24 h
1. Cellulosics	0.090	0.088	0.088	0.088	0.087
2. Unsupported microporous polyethylene	9.3	9.3	3.1	0.34	0.16
3. Unsupported microporous PVC	> 100			12.85	0.68
4. Nylon-supported microporous acrylonitrile-PVC copolymer	0.28	0.27	0.22	0.19	0.15
5. Microporous polyethylene membranes impregnated with cross-linked poly methacrylic acid	> 100		0.49	0.49	0.49

Material	Dry thickness (cm x 10^{-4})	Wet thickness (cm x 10^{-4})	Specific resistance, r (Ω · cm)
1. Cellophane	41	102	6.6
2. Polyethylene	122	124	13
3. PVC	152	147	44
4. Acrylonitrile-PVC copolymer	104	125	11
5. Cross-linked poly methacrylic acid	122	90	54

[a] Surface area, 1 cm², typical values in ohms.

*Reference: A. J. Salkind and J. J. Kelley in Characteristics of Separators for Alkaline Silver Oxide-Zinc Secondary Batteries (J. E. Cooper and A. Fleischer, eds.), Air Force Aero Propulsion Lab.,

ACKNOWLEDGMENTS

The author wishes to thank the management of the Battery Products Division of Union Carbide Corporation for permission to use internal report data and manuals to describe the state of the art of alkaline MnO_2-Zn batteries and provide up-to-date information. Dr. R. A. Powers, director of the Research Laboratories in Parma, read this manuscript critically and provided many valuable suggestions. The many colleagues whose work is reported in this chapter I can thank only by referencing their literature contributions and patents as accurately as possible.

REFERENCES

1. G. Leuchs, German Pat. 24552, (1882).

2. S. Yai, U.S. Pat. 746,227 (1903).

3. E. Achenbach, German Pat. 261,319 (1912).

4. E. Achenbach, German Pat. 279,911 (1913).

5. E. Achenbach, German Pat. 265,590 (1912); U.S. Pat. 1,098,606 (1914).

6. E. Achenbach, U.S. Pat. 1,090,372 (1914).

7. A. Heil, U.S. Pat. 1,195,677 (1916).

8. M. Bolen and B. H. Weil, Literature Search on Dry Cell Technology, U.S. Signal Corps Engr. Labs., Ft. Monmouth, N.J., Sp. Rept. No. 27, State Engr. Exp. Sta., Georgia Inst. Technology, 1948.

9. French Pat. 793,617 (1936).

10. G. W. Heise and N. C. Cahoon, eds., The Primary Battery, Volume 1, Wiley, New York, 1971. Chapters 3, 4, 8 by E. A. Schumacher, Chapter 5 by S. Ruben.

11. S. Ruben, U.S. Pat. 2,422,045 (1947).

12. S. Ruben, U.S. Pat. 2,576,266 (1951).

13. S. Ruben, U.S. Pat. 2,542,574 (1951).

14. M. Friedman and C. E. McCauley, Trans. Electrochem. Soc., 92, 195 (1947).

15. W. S. Herbert, U.S. Pat. 2,650,945 (1953).

16. W. S. Herbert, U.S. Pat. 2,775,534 (1956).

17. W. S. Herbert, U.S. Pat. 2,768,229 (1956).

18. W. S. Herbert, J. Electrochem. Soc., 99, 190 C (1952).

19. P. A. Marsal, K. V. Kordesch, and L. F. Urry, U.S. Pat. 2,960,558 (1960).

20. K. V. Kordesch and R. E. Stark, U.S. Pat. 3,113,050 (1963).

21. K. V. Kordesch, U.S. Pat. 2,962,540 (1960).

22. N. Parkinson, U.S. Pat. 2,822,416 (1958).

23. B. H. King, U.S. Pat. 2,593,893 (1952).

24. E. E. Leger, U.S. Pat. 2,993,947 (1961).

25. W. Latimer, The Oxidation States of the Elements and Their Potential in Aqueous Solution, 2nd ed., Prentice-Hall, Englewood Cliffs, N.J., 1952.

26. M. Pourbaix, Atlas of Electrochemical Equilibria in Aqueous Solution (English ed.), Pergamon, New York, 1966.

27. J. Euler and A. Fleischer, "Primary and Secondary Cells--History and State-of-the-Art Summary," presented at 29th Meeting of AGARD, Propulsion and Energy Panel, Liege, Belgium, June 1967.

28. The Automobile and Air Pollution, Part II, U.S. Dept. of Commerce, Subpanel Report, Dec. 1967, pp. 49-88.

29. Eveready Battery Applications and Engineering Data, Union Carbide Corp., New York, 1971.

30. A. F. Daniel, J. J. Murphy, and J. M. Hovendon in Batteries, Proc. 3rd Intern. Symp., 1962 (Bournemouth), (D. H. Collins, ed.) Pergamon, New York, 1963, pp. 157-169.

31. J. Winger, PSC 18, 80 (1964). (Power Sources Conf. Publ. Committee, Red Bank, N.J.)

32. J. L. S. Daley, PSC 15, 96 (1961).

33. Market Survey of the Battery Industry, Arthur D. Little, Inc., May 1970, p. 35.

34. Ray-O-Vac Company, Madison, Wis. (Electric Storage Battery Corp.)

35. W. S. Herbert, J. Electrochem. Soc., 99, 190 C (1952).

36. H. P. Keating, Advan. Battery Technol. Symp., 1968, Southern Calif.-Nev. Section, Electrochemical Society, 1968, pp. 115-140.

37. S. Ruben, U.S. Pat. 3,066,179 (1962).

38. S. Ruben, U.S. Pat. 3,485,675 (1969).

39. P. A. Marsal, K. V. Kordesch, and L. F. Urry, U.S. Pat. 2,960,558 (1960).

40. K. V. Kordesch, U.S. Pat. 2,962,540 (1960).

41. K. V. Kordesch and R. E. Stark, U.S. Pat. 3,113,050 (1963).

42. E. E. Leger, U.S. Pat. 2,993,947 (1961).

43. R. Carmichael, F. G. Spanur, and G. H. Klun, U.S. Pat. 3,069,489 (1962).

44. J. L. S. Daley, U.S. Pat. 3,042,734 (1962).

45. J. L. S. Daley and E. E. Leger, U.S. Pats. 3,068,312 and 3,068,313 (1962).

46. H. K. Amthor, U.S. Pat. 3,664,878 (1972).

47. J. Winger, U.S. Pat. 3,069,485 (1962).

48. R. Carmichael and A. Vulpio, U.S. Pat. 3,218,197 (1965).

49. F. G. Spanur, U.S. Pat. 3,314,824 (1967).

50. R. E. Ralston and Yung Ling Ko, U.S. Pat. 3,663,301 (1972).

51. F. D. Williams, U.S. Pats. 2,712,565 (1955) and 2,478,798 (1949).

52. R. Colton, U.S. Pat. 2,636,062 (1953).

53. R. R. Clune, U.S. Pat. 3,096,217 (1963).

54. N. Parkinson, U.S. Pat. 2,822,416 (1958).

55. Eveready E93 (U.S.).

56. Mallory MN-1400 (U.S.).

57. National Mallory AM-2 (Japan).

58. Toshiba AM-2 (Japan).

59. Novel (Fuji) AM-1 (Japan).

60. Russian BF-size alkaline MnO_2-Zn cell (USSR-Licenseintorg) (Arrangement for manufacture U.S. Pat. 3,677,824 (1972).

61. K. V. Kordesch and A. J. Marko, J. Electrochem. Soc. 107, 480 (1960).

62. S. Ruben, U.S. Pats. 2,542,574 and 2,542,576 (1951).

63. A. F. Daniel, U.S. Pat. 2,480,839 (1949).

64. Leclanché K-6 (France).

65. K. V. Kordesch, U.S. Pat. 2,935,547 (1960).

66. P. Rhesbeck, in Batteries, Proc. 3rd Intern. Symp., Bournemouth, 1962 (D. H. Collins, ed.) Pergamon, New York, 1963, pp. 419-437.

67. K. V. Kordesch, U.S. Pat. 3,335,031 (1967).

68. R. R. Balaguer, U.S. Pat. 2,903,499 (1959) P.S.C. 20, 90 (1966).

69. K. V. Kordesch and A. Kozawa, Belgian Pat. 793,374 (1973) U.S. Pat. pending.

70. M. D. Kocherginsky et al., U.S. Pat. 3,576,678 (1971), USSR.

71. P. R. Mallory Company, Inc., Tarrytown, N.Y.

72. Eveready Battery Applications and Engineering Data, Union Carbide Corp., New York, 1971.

73. H. P. Keating, Proc. Advan. Battery Technol. Symp., Southern Calif.-Nev. Section, Electrochemical Society, 1968, pp. 116-140.

74. Eveready Battery Applications and Engineering Data, Union Carbide Corp., New York, 1971.

75. L. F. Huf et al, U.S. Pat. 3,734,778 (1973).

76. K. V. Kordesch, U.S. Pat. 2,962,540 (1960).

77. P. A. Marsal, A. Tasch, and L. F. Urry, U.S. Pat. 2,977,401 (1961).

78. K. V. Kordesch and R. E. Stark, U.S. Pat. 3,113,050 (1963).

79. H. Ogawa et al., Japanese Pats. 18,409-1971 and 19,467-1971.

80. Y. Amano et al., U.S. Pat. 3,530,496 (1970).

81. H. Ikeda et al., Japanese Pat. 37,531-1971.

82. N. A. Hampson, M. J. Tarbox, J. T. Lilley, and J. P. G. Farr, Electrochem. Technol. 2, 309 (1964).

83. K. V. Kordesch, U.S. Pat. 3,042,732 (1962).

84. K. V. Kordesch, U.S. Pat. 3,288,642 (1966).

85. K. V. Kordesch, U.S. Pat. 2,991,325 (1961).

86. K. V. Kordesch and L. F. Urry, U.S. Pat. 3,261,714 (1966).

87. SCR-Manual, 4th ed., General Electric Co., Schenectady, N.Y., 1967, Section 8.5.

88. D. R. Grafham, U.S. Pat. 3,310,724 (1967).

89. W. N. Carson, Jr. and R. L. Hadley, Power Sources 2, Proc. 6th Intern. Symp., Brighton, 1968 (D. H. Collins, ed.), Pergamon, 1970, pp. 181-197.

90. A. M. Wilson, U.S. Pat. 3,538,415 (1970) (Texas Instruments, Inc.).

91. K. V. Kordesch, Electrochem. Soc. Meeting, Cleveland, Oct. 1971, Extended Abstract No. 36.

92. H. F. Oswin, U.S. Pat. 3,563,800 (1971).

93. S. Arouete, K. F. Blurton, and H. G. Oswin, J. Electrochem. Soc., 116, 166 (1969).

94. G. A. Dalin, Advan. Battery Technol. Symp., Vol. 3, Southern Calif.-Nev. Section, Electrochemical Society, 1967, p. 31.

95. T. J. Hennigan and K. O. Sizemore, Annual PSC 20, 113 (1966).

96. H. G. Oswin and K. F. Blurton in Zinc-Silver Oxide Batteries (A. Fleischer and J. J. Lander, eds.), Wiley, New York, 1971, Chapter 6.

97. J. E. Oxley, Contract NAS-5-3908, "Improvement of Zinc Electrodes for Electrochemical Cells," 1964-1967, Final Report: N67-30067 (1967).

98. J. Goodkin, 2nd Quarterly Report AD-824719 (December 1967) and 3rd Quarterly Report AD-828224 (January 1968, Yardeney). See also Annual Power Sources Conf. 22, 79 (1968).

99. G. A. Dalin in Zinc-Silver Oxide Batteries (A. Fleischer and J. J. Lander, eds.), Wiley, New York, 1971, Chapter 7.

100. J. Caudle, C. A. Batts, and F. L. Tye in Power Sources 2 (1968) (D. H. Collins, ed.) Pergamon, New York, 1970, p. 319.

101. J. Caudle, D. B. Ring, and F. L. Tye in Power Sources 3 (1970) D. H. Collins, ed.), Oriel Press, Newcastle, 1971, p. 593.

102. W. J. Van der Grinten, U.S. Pat. 3,516,862 (1970).

103. A. Nagamine and Y. Sato, Japanese Pat. 8331-1971.

104. R. W. Powers, Electrochem. Technol. 5, 429 (1967).

105. A. R. Despic, J. Diggle, and J. O. M. Bockris, J. Electrochem. Soc., 115, 507 (1968).

106. K. L. Hapartzumian and R. V. Moshtev in Power Sources 3 (1970) (D. H. Collins, ed.), Oriel Press, Newcastle, 1971, p. 495.

107. J. A. Keralla and J. J. Lander, Electrochem. Technol. 6, 202 (1968).

108. T. P. Dirkse, U.S. Pat. 3,348,973 (1967).

109. F. P. Kober and A. Charkey in Power Sources 3 (1970) (D. H. Collins, ed.), Oriel Press, Newcastle, 1971, p. 309.

110. G. R. Drengler, U.S. Pat. 3,226,260 (1965).

111. A. Kozawa, this book, Chapter 3.

112. K. Takahashi and A. Kozawa, J. Metals, 22, 64 (1970).

113. A. Kozawa in Electrochemistry of Manganese Dioxide and Manganese Dioxide Batteries in Japan, Vol. 1 (S. Yoshizawa, ed.) U.S. Branch of the Electrochemical Society of Japan, Cleveland, 1971, p. 57.

114. T. Honda in Electrochemistry of Manganese Dioxide and Manganese Dioxide Batteries in Japan, Vol. 1 (S. Yoshizawa, ed.) U.S. Branch Office of the Electrochemical Society of Japan, Cleveland, 1971, p. 203.

115. H. Tamura in Electrochemistry of Manganese Dioxide and Manganese Dioxide Batteries in Japan, Vol. 2 (S. Yoshizawa, ed.) U.S. Branch Office of the Electrochemical Society of Japan, Cleveland, 1971, p. 189.

116. EMD imported from Japan.

117. Japanese Patent 1971-7519 (German Pat. Appl. 1970).

118. T. Hirai, Ph.D. Thesis, Kyoto University (1960).

119. Y. Uetani in Electrochemistry of Manganese Dioxide and Manganese Dioxide Batteries in Japan, Vol. 2 (S. Yoshizawa, ed.) U.S. Branch Office of the Electrochemical Society of Japan, Cleveland, 1971, p. 177.

120. D. Boden, C. J. Venuto, D. Wisler, and R. B. Wylie, J. Electrochem. Soc., 114, 415 (1967).

121. Eveready Battery Applications and Engineering Data, Union Carbide Corp., New York, p. 270 (1968), p. 296 (1971).

122. K. Miyazaki in Power Sources 3 (D. H. Collins, ed.), Oriel Press, Newcastle, 1971, p. 607.

123. W. S. Herbert, Electrochem. Technol., 1, 148 (1963).

124. F. Yeaple, Product Engineering, 12, 100 (1965).

125. P. H. Adams, IRE Trans. Component Parts, 5, 76 (1958).

126. H. Y. Kang and C. C. Liang, J. Electrochem. Soc., 115, 6 (1968).

127. T. Ohira and H. Ogawa, Natl. Tech. Rept. 16, 209 (1970).

128. D. Boden, C. J. Venuto, D. Wisler, and R. B. Wylie, J. Electrochem. Soc., 115, 333 (1968).

129. See R. Huber, Chapter 1 of this book, Section 5.2.3.

130. J. P. G. Farr and N. A. Hampson, J. Electroanal. Chem., 13, 433 (1967).

131. Z. Gerischer, Phys. Chem., 202, 302 (1953).

132. W. Latimer, The Oxidation States of the Elements and Their Potential in Aqueous Solution, 2nd ed., Prentice-Hall, Englewood Cliffs, N.J., 1952.

133. H. Bode, V. A. Olipuram, D. Berndt, and P. Ness in Zinc-Silver Oxide Batteries (A. Fleischer and J. J. Lander, eds.), Wiley, New York, 1971, Chapter 2.

134. F. Jolas, Electrochim. Acta, 13, 2207 (1968).

135. P. Brouillet and F. Jolas, Electrochim. Acta, 1, 246 (1959).

136. I. Sanghi and W. F. K. Wynn-Jones, Proc. Ind. Acad. Sci., 47, 49 (1958).

137. I. Sanghi and M. Fleischman, Proc. Ind. Acad. Sci., 49 A, 6 (1959).

138. N. A. Hampson in Zinc-Silver Oxide Batteries (A. Fleischer and J. J. Lander, eds.), Wiley, New York, 1971, Chapter 5.

139. T. P. Dirkse, D. DeWit, and R. Shoemaker, J. Electrochem. Soc., 115, 442 (1968).

140. K. V. Kordesch and A. Marko, J. Electrochem. Soc., 107, 480 (1960).

141. Z. A. Iofa, L. V. Komlev, and V. S. Bagotskii, Zh. Fiz. Khim., 35, 1571 (1961).

142. T. P. Dirkse and F. DeHaan, J. Electrochem. Soc., 105, 311 (1958).

143. R. A. Powers, R. J. Bennett, W. G. Darland, and R. J. Brodd in
Power Sources 2, Proc. 6th Intern. Symp., Brighton, 1968 (D. H.
Collins, ed.), Pergamon, New York, 1970, pp. 461-482.

144. R. N. Snyder and J. J. Lander, Electrochem. Technol., 3, 161 (1965).

145. G. Schneider, Electrochim. Acta, 13, 2223 (1968).

146. T. P. Dirkse in Power Sources 2, Proc. 6th Intern. Symp., Brighton,
1968 (D. H. Collins, ed.), Pergamon, New York, 1970, pp. 411-422.

147. J. L. S. Daley, Annual Power Sources Conf., 15, 96 (1961).

148. P. L. Howard and J. R. Huff in Power Sources 2, Proc. 6th Intern.
Symp., Brighton, 1968 (D. H. Collins, ed.), Pergamon, New York,
1970, pp. 401-410.

149. K. J. Vetter, Electrochemical Kinetics, Academic Press, New York,
1967, p. 776.

150. Collected Papers on Chemical Sources of Currents, Energiya,
Moscow-Leningrad, No. 1 (1966), No. 2 (1967), authors: V. S.
Daniel'-Bek and Z. P. Arkhangelskaya (in Russian). Kinetics of
Electrode Processes, Moscow University, 1952, authors: A. N.
Frumkin, V. A. Bagotskii, Z. A. Iofa, and B. N. Kabanov (in
Russian).

151. Z. P. Arkhangelskaya, G. P. Andreeva, and M. N. Mashevich,
J. Appl. Chem. USSR, 41, 107 (1968); 41, 1640 (1968) in English
(from Zh. Prikl. Khimii, 41). See also: Chem. Abstr., 71, 44914
(1969), from Akkumulyatorn Inst., 1967.

152. R. Landsberg and H. Bartelt, Z. Elektrochem., 61, 1162 (1957).

153. N. A. Hampson and M. J. Tarbox, J. Electrochem. Soc., 110,
95 (1962).

154. T. P. Dirkse, J. Electrochem. Soc., 101, 328 (1954).

155. J. S. Fordyce and R. L. Baum, J. Chem. Phys., 43, 843 (1965).

156. G. H. Newman and G. E. Blomgren, J. Chem. Phys., 43, 2744 (1965).

157. M. W. Breiter, Electrochim. Acta, 16, 1169 (1971).

158. J. Euler, Electrochim. Acta, 11, 701 (1966).

159. N. A. Hampson, M. J. Tarbox, J. T. Lilley, and J. P. G. Farr,
Electrochem. Technol., 2, 309 (1964).

160. R. R. Clune, U.S. Pat. 3,205,097 (1965).

161. The Primary Battery, Vol. 1 (G. W. Heise and N. C. Cahoon, eds.),
Wiley, New York, 1971, Chapters 3, 4, and 8 by E. A. Schumacher,
Chapter 5 by S. Ruben.

162. G. E. Best, U.S. Pat. 2,308,584 (1943).

163. P. L. Probst et al., U.S. Pat. 2,968,062 (1961).

164. P. L. Probst et al., U.S. Pat. 2,460,991 (1949).

165. Manufacturer: New Jersey Zinc Company.

166. J. Goodkin and F. Solomon in Batteries 2, Proc. 4th Symp., Brighton, 1964 (D. H. Collins, ed.), Pergamon, New York, 1965, pp. 475-487.

167. W. S. Herbert, U.S. Pat. 2,650,945 (1953).

168. J. F. Jammet, U.S. Pat. 3,556,861 (1971).

169. R. E. Ralston, U.S. Pat. 3,669,754 (1972).

170. D. H. Morell and D. W. Smith in Power Sources, Proc. 5th Intern. Symp., Brighton, 1966 (D. H. Collins, ed.), Pergamon, New York, 1967, pp. 207-225.

171. British Pat. 1,210,664.

172. F. Przybyla and F. J. Kelly, Power Sources 2, Proc. 6th Intern. Symp., Brighton, 1968 (D. H. Collins, ed.), Pergamon, New York, 1970, pp. 373-387; U.S. Pat. 3,348,976 (1967).

173. L. B. Griffiths and K. H. Krock, U.S. Pat. 3,655,447 (1972).

174. P. R. Mallory & Co., British Pat. 1,211,403.

175. R. W. Fletcher, U.S. Pat. 3,427,203 (1969).

176. C. M. Shepherd and H. C. Langelan, J. Electrochem. Soc., 114, 8 (1967).

177. C. M. Shepherd and H. C. Langelan, J. Electrochem. Soc., 109, 657 (1962).

178. C. M. Shepherd and H. C. Langelan, J. Electrochem. Soc., 109, 661 (1962).

179. P. Garine, U.S. Pat. 2,838,590 (1958).

180. J. Goodkin, Proc. Power Sources Conf., 22, 90 (1968).

181. A. Nagamine, Japanese Pats. 1971-8331, -8332, -8333.

182. R. H. Knapp, Proc. Power Sources Conf., 22, 114 (1968).

183. E. G. Katsoulis and B. Randall, Proc. Power Sources Conf., 22, 120 (1968).

184. A. Charkey and R. DiPasquale, Proc. Power Sources Conf., 22, 117 (1968).

185. L. E. Quinnell, U.S. Pat. 2,483,983 (1949).

186. B. H. King, U.S. Pat. 2,593,893 (1952).

187. A. Marko and K. V. Kordesch, Austrian Pat. 169,782 (1951).

188. E. E. Leger, U.S. Pat. 2,993,947 (1961).

189. P. A. Marsal, K. V. Kordesch, and L. F. Urry, U.S. Pat. 2,960,558 (1960).

190. J. A. Keralla in Zinc-Silver Oxide Batteries (A. Fleischer and J. J. Lander, eds.), Wiley, New York, 1971, Chapter 13.

191. M. N. Yardeney, U.S. Pat. 2,983,777 (1956).

192. J. A. Keralla and J. J. Lander, Electrochem. Technol., 6, 202 (1968).

193. Allied Chemical Company, Caustic Potash Bulletin.

194. Hooker Chemical Company, Standard Grade Caustic Potash.

195. Fuel Cells, Proc. 1st Australian Conf. Electrochem., (T. A. Friend, F. Gutmann, and T. W. Hayes, ed.) Pergamon, New York, 1965, pp. 673, 674.

196. Baker Chemical Company, Caustic Soda Bulletin.

197. M. H. Johnson, U.S. Pat. 3,433,679 (1969).

198. Gmelin, Handbuch d. Anorgan. Chemie, Syst. No. 32, Zink, Verlag Chemie GmbH, Weinheim/Bergstrasse, 1956.

199. E. V. Steffensen, U.S. Pat. 3,649,362 (1972); see also Chem. Abstr. 51, 15220 (1957).

200. W. Feitknecht, Helv. Chim. Acta, 13, 314 (1930).

201. W. Feitknecht and E. Haeberli, Helv. Chim. Acta, 33, 922 (1950).

202. K. Huber, Z. Elektrochem., 48, 26 (1942).

203. E. A. Schumacher in The Primary Battery, Vol. 1, (G. Heise and N. C. Cahoon, eds.), Wiley, New York, 1971, Chapter 3.

204. H. H. Bode, V. A. Olipuram, D. Berndt, and P. Ness in The Zinc-Silver Oxide Battery (A. Fleischer and J. J. Lander, Wiley, New York, 1969, Chapter 2.

205. T. P. Dirkse in The Zinc-Silver Oxide Battery, (A. Fleischer and J. J. Lander, eds.), Wiley, New York, 1969, Chapter 3.

206. W. H. Dyson, L. A. Schreier, W. P. Sholette, and A. J. Salkind, J. Electrochem. Soc., 115, 566 (1968).

207. T. Baker and I. Trachtenberg, J. Electrochem. Soc., 114, 1045 (1967).

208. T. P. Dirkse, J. Electrochem. Soc., 106, 154 (1959).

209. DuPont Sodium CMC, E. I. DuPont de Nemours & Co., Inc., Wilmington, Del., 1970.

210. Hercules Cellulose Gum Properties and Uses, Hercules Powder Co., Wilmington, Del., 1970.

211. ASTM Designation D-1439-617, ASTM Standard, 1961, Part 5.

212. Hercules CMHEC, Hercules Powder Co., Wilmington, Del., 1970.

213. Methocel, Dow Chemical Co., Midland, Mich.

214. B. Zaslow in Starch, Chemistry and Technology, Vol. 1 (R. L. Whistler and E. F. Paschall, eds.), Academic Press, New York, 1965, Chapter 11, pp. 279-287.

215. J. R. Van der Bij and W. F. Vogel, Staerke, 14, 113 (1962).

216. E. T. Hjermstad in Starch, Chemistry and Technology, Vol. 2 (R. L. Whistler and E. F. Paschall, eds.), Academic Press, New York, 1967, Chapter 17, pp. 423-432.

217. S. U. Falk and A. J. Salkind, Alkaline Storage Batteries, Wiley, New York, 1969, Chapter 3.

218. P. N. Dangel, Proc. Symp. Battery Separators, Columbus Section, Electrochem. Soc., 1970, pp. 74-89.

219. E. M. Honey and C. R. Hardy, U.S. Pat. 2,542,527 (1951); British Pat. 576,659 (1946).

220. J. C. Duddy, U.S. Pat. 2,676,929 (1954).

221. J. L. Weininger and F. F. Holub, Proc. Symp. Battery Separators, pp. 122-135, (1970).

222. Cellulose and Cellulose Derivates, Parts 1-3, Vol. 5 of High Polymers, (E. Ott, H. M. Spurlin, M. W. Grafflin, eds.) 2nd ed., Wiley-Interscience, New York, 1955.

223. Encyclopedia of Polymer Science and Technology, Wiley, New York, 1965.

224. G. A. Dalin, Proc. Symp. Battery Separators, Columbus Section, Electrochem. Soc., 1970, pp. 136-151.

225. H. E. Hoyt, Proc. Symp. Battery Separators, Columbus Section, Electrochem. Soc., 1970, pp. 182-186.

226. E. L. Harris in Characteristics of Separators for Alkaline Silver Oxide-Zinc Secondary Batteries--Screening Methods, (J. E. Cooper and A. Fleischer, eds.), Air Force Aero Propulsion Lab., Dayton, 1965, Chapter 9.

227. V. D'Agostino, Proc. Symp. Battery Separators, Columbus Section, Electrochem. Soc., 1970, pp. 224-249.

228. H. P. Gregor in Zinc-Silver Oxide Batteries (A. Fleischer and J. J. Lander, eds.), Wiley, New York, 1971, Chapter 16.

229. C. Berger and F. C. Arrance, U.S. Pats. 3,379,569 and 3,379,570 (1968).

CHAPTER 3

ELECTROCHEMISTRY OF MANGANESE DIOXIDE AND PRODUCTION
AND PROPERTIES OF ELECTROLYTIC MANGANESE DIOXIDE (EMD)

Akiya Kozawa

Union Carbide Corporation
Battery Products Division
Parma, Ohio

Today manganese dioxide is the most widely used cathode material in various practical cells, such as Leclanché cells, alkaline manganese dioxide cells (primary and rechargeable), and magnesium cells. Manganese dioxide is also used in mercury cells and silver cells; it is mixed with HgO or Ag_2O to obtain a sloping discharge curve (to avoid a sudden voltage drop at the end) or to reduce the cost of the cells.

Total world consumption of MnO_2 used for battery production in 1971 is estimated to be as follows: natural MnO_2 ore, 200,000 tons; electrolytic MnO_2 (EMD), 50,000 tons; chemical MnO_2 (CMD), 10,000 tons.

In part 1 of this chapter, the general electrochemical behavior of MnO_2 will be described in relation to the battery reactions.

In part 2, the production process and properties of EMD will be described in detail. The use of EMD for battery manufacture has rapidly increased in recent years, and EMD is expected to be the most useful manganese dioxide in the next ten years. About 50% of the total Leclanché cells manufactured in Japan in 1971 were exclusively EMD cells. Alkaline manganese dioxide cells, which constitute about 20-25% of the U.S. market in value, are made only with EMD.

In part 3, methods for chemical analysis and the determination of the surface area of MnO_2 by zinc-ion adsorption are described. Some useful data on properties of battery-grade manganese dioxide are compiled.

1. ELECTROCHEMISTRY OF MANGANESE DIOXIDE

1.1. INTRODUCTION

The purpose of this section is to provide a concise but up-to-date summary of the electrochemical behavior of manganese dioxide. Electrochemical reactions of MnO_2 in NH_4Cl + $ZnCl_2$ solution, especially in the cathode bobbin, are much more complex than in KOH solution. The behavior of γ-MnO_2 (EMD) and β-MnO_2 in KOH electrolyte will be described first. Then, the discharge mechanisms in Leclanché electrolyte and in magnesium cell electrolyte ($MgClO_4$) will be briefly discussed. The writer hopes that this approach will give the reader a better and clearer understanding of the principal aspects of the electrochemical properties of MnO_2.

No attempt was made to compile all the literature published in this field; this article is based primarily on the author's work. However, essential publications in this field by other workers are listed at the end of the chapter [15-24].

1.2. BASIC DISCHARGE BEHAVIOR OF MnO_2*

A small amount of MnO_2 was deposited anodically from a $MnSO_4$ solution onto a porous spectrographic graphite rod (4.5 mm in diameter and 7.7 cm in length). Discharge curves of this type of electrode in 9 M KOH solution are shown in Fig. 1. At low current densities (curves 1 and 2),

*Reference [1].

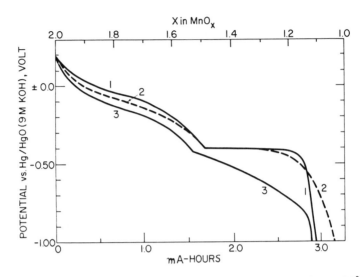

FIG. 1. Discharge curves at constant current in 9 M KOH at 23°C.
The x value in MnO_x was calculated from the amount of electricity drawn.
Curve 1, at 0.11 mA/electrode; curve 2, at 0.33 mA/electrode; curve 3,
at 3.0 mA/electrode. Each electrode had approximately 10.6 mg of MnO_2
[1].

MnO_2 is electrochemically reduced in two steps: at high current densities
(curve 3), the second step becomes less clear. The first step is from
MnO_2 to $MnO_{1.5}$; the potential decreases continuously, producing an
S-shaped curve during this step. The second step is from $MnO_{1.5}$ to
$MnO_{1.0}$; the potential remains almost constant during the major portion
of this step.

   This kind of two-step discharge behavior is not limited to the deposited,
thin-layer type of MnO_2; ordinary battery-grade, powdered manganese
dioxide also exhibits similar discharge curves. Figure 2 shows constant-
current discharge curves of γ-MnO_2 and β-MnO_2 (mixed with carbon
powder) in 9 M KOH [2]. The γ-MnO_2 is an ordinary battery-grade
electrolytic MnO_2; the β-MnO_2 was prepared by heating the γ-MnO_2 in
air at 400°C for 10 days. The cells used in these discharge experiments
are shown in Fig. 3 [2]. As shown in Fig. 2, both γ-MnO_2 and β-MnO_2

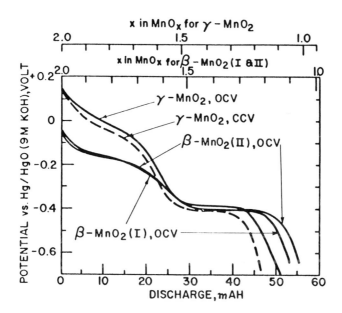

FIG. 2. Discharge curves of γ-MnO_2 and β-MnO_2 [2]. The MnO_2 sample (100 mg) was mixed with 1.0 gm of battery-grade graphite and 2.0 gm of coke and discharged at 0.35 mA for 17 h; the circuit was periodically opened and the final open-circuit voltage (OCV) was obtained after sufficient recovery time. The closed-circuit voltage (CCV) includes the IR drop, which was 10-15 mV. See Ref. [2] for β-MnO_2(I) and β-MnO_2(II).

exhibit two-step discharge curves in 9 M KOH solution, although the voltage level of the first step of β-MnO_2 is considerably lower than that of γ-MnO_2. The useful capacity of ordinary alkaline manganese dioxide cells on the market today lies within the first step of the discharge. In the following sections, the essential mechanisms and magnitude of the polarization of these two steps will be discussed.

1.3. OUTLINE OF THE DISCHARGE MECHANISM

1.3.1. The First-Step Reaction: Reduction from $MnO_{2.0}$ to $MnO_{1.5}$

In the discharge process (electrochemical reduction) that occurs during this step, electrons and protons produced from decomposition of H_2O

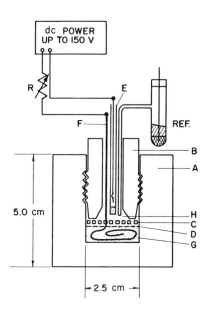

FIG. 3. Discharge cell: A, body of the cell (Lucite); B, threaded
Lucite plug, with hole in center, which is used to keep pressure on G;
C, perforated Teflon disc; D, separator paper (made of synthetic fiber);
E, anode compartment (glass tube with a fine glass frit at the end)
containing a Pt wire electrode; F, gold wire (1 mm diam), with a spiral
at the end which was buried in the mixture (the top straight portion of the
gold wire was covered with a polyethylene tube); G, a mixture of 0.100
gm of MnO_2 + 2.0 gm of coke + 1.0 gm of graphite; H, electrolyte; R,
resistance (25–400 kΩ) [2].

molecules are introduced into the ionic crystal of MnO_2 as shown in Fig. 4
[3]. The concentration of Mn^{3+} ions and OH^- ions in the lattice increases
gradually, and finally the MnO_2 is converted to MnOOH. The valency of
manganese in the original MnO_2 lattice is four, but it is converted to
3 (Mn^{4+} + e^- → Mn^{3+}) as electrons are introduced into the lattice. Since
electrons can move around in the lattice, the positions of Mn^{3+} are not
fixed. The protons (H^+) also jump from one O^{2-} site to another; therefore,
the OH^- positions in the lattice are not fixed either. This reduction or

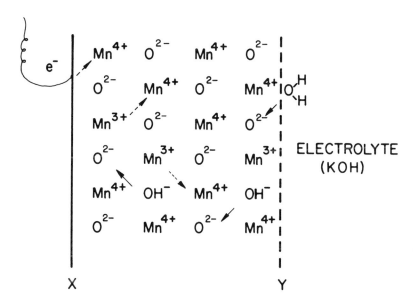

FIG. 4. Schematic presentation of the $(Mn^{4+}-Mn^{3+}-O^{2-}-OH^-)$ system during the discharge. The arrows show the directions of movement of the electrons (dashed arrows) and protons (solid arrows). X represents the MnO_2-graphite interface, Y the MnO_2-KOH electrolyte interface [3].

discharge process may be expressed in Eq. (1), which would imply that the reaction is a conversion of one solid structure (MnO_2) to another ($MnOOH$); but what is really taking place is the introduction of electrons and protons into the lattice without changing its basic (or essential) structure.

$$MnO_2 + H_2O + e^- \rightarrow MnOOH + OH^- \tag{1}$$

In other words, the original MnO_2 is gradually converted to a lower oxide by increasing Mn^{3+} and OH^- concentrations in the lattice while maintaining a homogeneous solid phase during the discharge process. $MnOOH$ is equivalent to $MnO_{1.5}$ except for the extra water ($2MnOOH - H_2O = Mn_2O_3 = MnO_{1.5}$). Because of the nature of the process, this solid $MnO_2 - MnOOH$ system can be best expressed as the $Mn^{4+} - Mn^{3+} - O^{2-} - OH^-$ system, clearly indicating a homogeneous phase. In the $Mn^{4+} - Mn^{3+} - O^{2-} - OH^-$ system, Mn^{3+} and OH^- concentrations are very low prior to discharge and

O^{2-} and Mn^{4+} concentrations are very low at the end of the first step.

1.3.2. The Second-Step Reaction: Reduction of MnOOH to $Mn(OH)_2$*

The second-step reaction consists of three consecutive steps: (1) dissolution of Mn^{3+} ions from MnOOH in the form of complex ions such as $[Mn(OH)_4]^-$ into the electrolyte, (2) electrochemical reduction of $[Mn(OH)_4]^-$ to $[Mn(OH)_4]^{2-}$ (equivalent to $Mn^{3+} + e^- \rightarrow Mn^{2+}$) on the carbon surface, and (3) precipitation of $Mn(OH)_2$ from a saturated solution of $[Mn(OH)_4]^{2-}$. These consecutive steps are shown in the diagram.

MnOOH $\rightarrow$ $[Mn(OH)_4]^- + e^- \rightarrow [Mn(OH)_4]^=$ $\rightarrow$ $Mn(OH)_2$

             dissolution   electrochemical   precipitation

              process   reduction process   process

initial   final solid

solid phase   phase

This process is a heterogeneous phase reaction, because two distinct solid phases are involved, one initial solid phase (MnOOH) being converted to another final solid phase ($Mn(OH)_2$). The electrochemical reaction proceeds through dissolved ionic species.

The discharge mechanism outlined above for the first and second steps is somewhat simplified. It is valid for most of the normal discharge conditions, such as low or moderate current densities at room temperature. At extremely high-current discharges, the γ-MnO_2 structure seems to deteriorate in the later stage of the first step ($MnO_2 \rightarrow MnO_{1.5}$) because of extremely high $[Mn^{3+}]$ concentration on the MnO_2 surface, and a different oxide structure seems to be produced. This means that under such extreme discharge conditions, the reaction does not necessarily proceed in a single or homogeneous phase.

As shown in Fig. 2, the equilibrium potential (OCV) of β-MnO_2 is considerably lower than that of γ-MnO_2. When β-MnO_2 is discharged at

*Reference [4].

higher currents (0.04 mA/mg MnO_2), the second-step reaction begins to take place before the first-step reaction is completed (see Figs. 4 and 5 in Ref. [5]). When the discharge is interrupted at a certain point in the first step and the partially discharged cell is kept at a certain elevated temperature ($45°C$) for a few days to a week, unusual steps (the third and fourth steps) are produced upon subsequent discharge [6]. This will be discussed later, in Section 1.5.3.

1.4. CHEMICAL THERMODYNAMICS OF MnO_2 CELLS

1.4.1. Calculation of the emf (Conventional Treatment)

Calculation of the emf of a cell based on chemical thermodynamics is well known and treated in textbooks on electrochemistry. The calculated cell voltage is often used to estimate or determine the cell reaction by comparing it with the experimentally measured potential. Let us try some calculations for MnO_2-KOH-Zn cells according to the conventional procedure and discuss the results. From the anode reaction [Eq. (2)] and the cathode reaction [Eq. (3)], the overall cell reaction may be written as Eq. (4).

$$Zn + 2\ OH^- = ZnO + H_2O + 2\ e^- \tag{2}$$
$$2\ MnO_2 + H_2O + 2\ e^- = Mn_2O_3 + 2\ OH^- \tag{3}$$

Cell reaction: $\quad Zn + 2\ MnO_2 = ZnO + Mn_2O_3 \tag{4}$
$$\Delta F = -66.5\ \text{kcal},\ \Delta F/2F = 1.44\ V$$

According to Walkley [7], ΔF for reaction (4) is -66.5 kcal; therefore, the cell voltage is found to be 1.44 V. If the final product of the cell reaction is not ZnO and Mn_2O_3 but $ZnMn_2O_4$ (hetaerolite), the cell voltage should be higher. The ΔF value for the formation of $ZnMn_2O_4$ [Eq. (7)] was also obtained by Walkley. The cell voltage is found to be 1.58 V in this case.

$$Zn + 2\ OH^- = ZnO + H_2O + 2\ e^- \tag{5}$$
$$2\ MnO_2 + H_2O + ZnO + 2\ e^- = ZnMn_2O_4 + 2\ OH^- \tag{6}$$

Cell reaction: $Zn + 2\ MnO_2 = ZnMn_2O_4$ (7)

$\Delta F = -72.8$ kcal, $\Delta F/2F = 1.58$ V

The calculation of the ΔF values by Walkley is based on the free energy of $\beta\text{-}MnO_2$. The potential of $\gamma\text{-}MnO_2$ is usually higher than that of $\beta\text{-}MnO_2$ by about 120 mV: compare (a) and (b), taken from Ref. [2].

(a) Potential of $\gamma\text{-}MnO_2$ in 9 M KOH = 0.232 V vs. Hg/HgO (9 M KOH)

(b) Potential of $\beta\text{-}MnO_2$ in 9 M KOH = 0.115 V vs. Hg/HgO (9 M KOH)

(c) ΔE = (a) - (b) = 0.117 V

(d) Cell voltage of alkaline MnO_2 cell using $\gamma\text{-}MnO_2$ = 1.59 V

(e) Cell voltage of alkaline MnO_2 cell using $\beta\text{-}MnO_2$ = 1.47 V [estimated from ΔE (= 0.117 V)]

Thus, the estimated voltage of a cell in which $\beta\text{-}MnO_2$ is used is found to be 1.47 V, which is closer to 1.44 V based on reaction (4) than 1.58 V based on reaction (7). From the cell voltage discussed above, the most likely reaction is reaction (4) rather than (7).

This kind of calculation and discussion is typical of treatment based on chemical thermodynamics. This treatment assumes, without mentioning it, that the cell reaction is a heterogeneous-phase reaction, that is, that the initial state of one solid phase (MnO_2) is converted to another solid phase (Mn_2O_3). As outlined previously, MnO_2 is reduced in one phase, so that this kind of thermodynamic treatment is incorrect in principle, although the agreement is fairly good. The conventional thermodynamic treatment of calculating the cell voltage can be correctly applied to two-phase cell reactions, such as $Ag_2O + e^- \to Ag$, $HgO + e^- \to Hg$, and so on.

1.4.2. Homogeneous and Heterogeneous Phase Reactions

From the standpoint of chemical thermodynamics, anode and cathode reactions in various electrochemical cells can be classified into two categories: homogeneous-phase reactions and heterogeneous-phase reactions. Examples of each category are as follows.

Homogeneous-phase reaction: The electrochemical reduction of γ-MnO$_2$ or β-MnO$_2$ to MnOOH and the cathodic reduction of V$_2$O$_5$ [8] belong in this category. In the case of γ-MnO$_2$, electrons and protons are introduced into the lattice and the reduction proceeds in one solid phase without producing a separate solid phase as the product (as outlined in Section 1.3).

Heterogeneous- (or two-) phase reaction: Reactions in which two (or more) solid phases are involved in the system are called heterogeneous-phase reactions. A typical example is the cathode reaction of silver oxide (Ag$_2$O + H$_2$O + 2 e$^-$ → 2 Ag + 2 OH$^-$) in which the two solid phases involved are Ag$_2$O and Ag. Most of the anode and cathode reactions in practical batteries belong in this category. Examples are

1. Ag$_2$O + H$_2$O + 2 e$^-$ → 2 Ag + 2 OH$^-$
2. HgO + H$_2$O + 2 e$^-$ → Hg + 2 OH$^-$
3. Zn + 2 OH$^-$ → ZnO + H$_2$O + 2 e$^-$
4. MnOOH + H$_2$O + e$^-$ → Mn(OH)$_2$ + OH$^-$
5. Pb + SO$_4$$^{2-}$ → PbSO$_4$ + 2 e$^-$
6. PbO$_2$ + H$_2$SO$_4$ + 2 e$^-$ → PbSO$_4$ + 2 H$_2$O + SO$_4$$^{2-}$

Generally speaking, thermodynamics cannot predict the mechanism of the reaction; it merely indicates the energy change associated with the reaction (or the process). However, thermodynamics can predict whether the reaction is homogeneous or heterogeneous from the shape of the discharge curve (potential decrease on discharge). In the case of a homogeneous-phase reaction, the equilibrium potential or open-circuit voltage (OCV) of the system should decrease continuously with the amount of reduction (or depth of the discharge), while in the case of a heterogeneous-phase reaction, the equilibrium potential of the system should remain unchanged. Vetter [9,10] pointed out that this general conclusion based on chemical thermodynamics can be used as a criterion in the analysis of the nature of battery reactions. In order to determine whether the electrode reaction in question is a homogeneous-phase reaction or a heterogeneous-phase reaction, all we have to do is to measure the OCV at various stages of the discharge. If the equilibrium potential decreases with the depth of the discharge as shown in Fig. 5a, the reaction is taking place in a homogeneous or single solid phase. If the equilibrium potential remains constant regardless of the depth of the discharge (Fig. 5b), the reaction is taking place in a heterogeneous phase; that is, the reactant and the product consist of two phases just like the HgO-Hg system and the Ag_2O-Ag system.

In the case of MnO_2, regardless of the structure (γ- or β- form), the OCV decreases gradually from MnO_2 to $MnO_{1.5}$ (see Fig. 2). Therefore, the reaction in this range must be a homogeneous-phase reaction (the discharge product is not formed as a new separate phase). However, in the essential portion of the second step from $MnO_{1.5}$ to $MnO_{1.0}$, the OCV remains practically constant; therefore, the reaction in this portion must be a heterogeneous-phase reaction. In the discharge of MnO_2, OH^- ions are produced and go into the liquid phase (KOH solution); but since the KOH electrolyte is relatively concentrated, the change in OH^- ion activity is very small. Therefore, no significant change in the liquid phase occurs in this case.

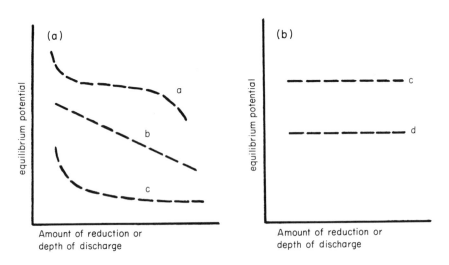

FIG. 5. Two types of electrochemical reduction: (a) reduction in homogeneous-phase (one solid phase) systems (for example, MnO_2, V_2O_5); (b) reduction in heterogeneous-phase systems (for example, Ag_2O, HgO).

The potential change associated with these two types of electrochemical reactions (homogeneous-phase and heterogeneous-phase) is quite analogous to the temperature change when ice is heated. When ice is heated at a constant rate, the temperature increases as shown in Fig. 6. The temperature remains constant whenever the system consists of two phases [ice + water (0°C), water + vapor (100°C)]. However, when the system contains only one phase (ice, water, or vapor alone), the temperature increases continuously. In the electrochemical reduction of MnO_2, the potential decreases continuously when the system consists of one phase and remains constant when the system consists of two phases as shown in Fig. 7. In these two systems, T (temperature) corresponds to E (potential), since both T and E are intensity factors of energy, and Q (heat, or strictly speaking, entropy, Q/T) corresponds to Q' (electricity, amount of discharge), since both Q and Q' are capacitive factors of energy. Since energy can be generally expressed as [intensity factor] x [capacitive factor] (T x Q or E x Q'), we can generalize as follows.

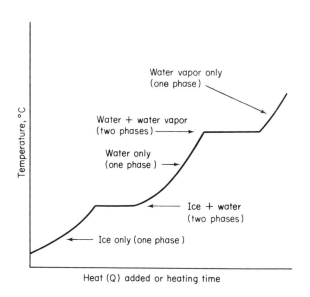

FIG. 6. Temperature of the H_2O system on heating.

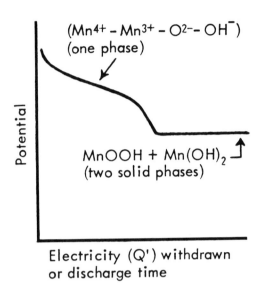

FIG. 7. Electrode potential of the metal oxide system on discharge.

In a single-phase system, the capacitive factor of energy and the intensity factor of energy will change continuously while the reaction is taking place. In a two-phase system, the intensity factor will remain unchanged while the reaction is proceeding from a higher to a lower energy level.

1.4.3. The Homogeneous-Phase Reduction of MnO_2

Although thermodynamics teaches us that the potential of a system decreases continuously when the reaction takes place in one phase, it does not tell us how the potential should decrease. In other words, thermodynamics does not provide any knowledge of the shape of the discharge curve (the equilibrium potential E versus x in MnO_x, abbreviated as E-x). Kozawa and Powers [3] proposed an equation [Eq. (8)] for the E-x relation from the analogy of this (Mn^{4+} - Mn^{3+} - O^{2-} - OH^-) system to a redox system in an aqueous solution, which is a typical homogeneous-phase reaction. It was found that the equation agrees roughly with the experimental results and does describe an essential feature of the MnO_2 system.

Let us discuss a typical homogeneous redox system. When the Fe^{2+}-Fe^{3+} system in an acid solution is titrated with a $KMnO_4$ solution, the potential of a platinum electrode dipped in the solution will change according to the Nernst equation. The titration curve (E versus $[Fe^{2+}] / [Fe^{3+}]$) is a well-known S-shaped curve. In this case, Fe^{2+} and Fe^{3+} ions in the solution (a homogeneous phase) can move freely through the solution, and the electrons contained in the platinum electrode are in equilibrium with the Fe^{2+} and Fe^{3+} ions in the solution.

In the case of the MnO_2 (Mn^{4+} - Mn^{3+} - OH^- - O^{2-}) system, the Mn^{3+} and Mn^{4+} (or O^{2-} and OH^-) ions move freely within the lattice, since protons and electrons can jump from one site to another. The electrons in the graphite that is in contact with the oxide system are in equilibrium with the redox system (Mn^{3+} - Mn^{4+}). It is therefore reasonable to assume that the potential of the oxide system may be expressed as follows.

$$E = E' - \frac{RT}{F} \ln \frac{[Mn^{3+}]solid}{[Mn^{4+}]solid} \tag{8}$$

where $[Mn^{3+}]_{solid}$ and $[Mn^{4+}]_{solid}$ represent the Mn^{3+} and Mn^{4+} concentrations in the oxide system in arbitrary units. In order to compare the E-x curve (Fig. 8) calculated from Eq. (8) to the experimentally obtained E-x curve, the potential at $[Mn^{3+}] = [Mn^{4+}]$ (where half of the oxide is reduced) was taken as the reference point. The two curves (calculated and experimental) were superimposed as shown in Fig. 9. The agreement between the two curves is not perfect, but the two curves agree with each other in their characteristic S-shape [3]. However, the following two points are worth mentioning.

The sharp potential drop at the beginning of the discharge, characteristic of MnO_2, is now understood, on the basis of homogeneous-phase theory. Such a sharp drop in the OCV in the early stages of discharge is not seen in other cell systems such as HgO-Hg, Ag_2O-Ag, PbO_2-$PbSO_4$, and so on, because they are heterogeneous systems.

The initial potential (undischarged cell voltage) has no significance in this theory, since the potential at x = 2.00 (in MnO_x) should be infinity

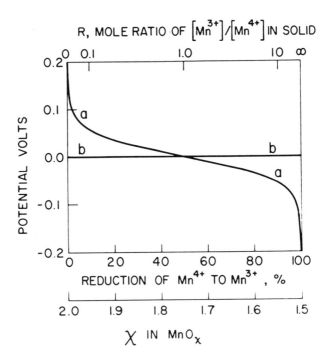

FIG. 8. Calculated equilibrium potential versus x in MnO_x from Eq. (8), which was derived for a one-phase (or homogeneous-phase) reduction [3]. R is related to x values by $x = (2 + 1.5R)/(1 + R)$.

$(E° - \ln[Mn^{3+}] / [Mn^{4+}] = \infty$ because $[Mn^{3+}] \approx 0$ and $[Mn^{4+}] = 1.0)$. Also, the traditional thermodynamic treatment to calculate the potential of the system and to compare it with the measured initial OCV is in principle not correct for a homogeneous-phase system such as MnO_2. The traditional thermodynamic calculation assumes a heterogeneous system having an initial state (one solid phase) and the final state (another solid phase), like HgO and Hg or Hg_2O and Hg, as pointed out previously.

1.4.4. The Mechanism of Heterogeneous-Phase Reactions*

In a heterogeneous-phase reaction (for example, electrochemical reduction of Ag_2O, HgO, or MnOOH), constancy of the equilibrium potential

*Reference [11].

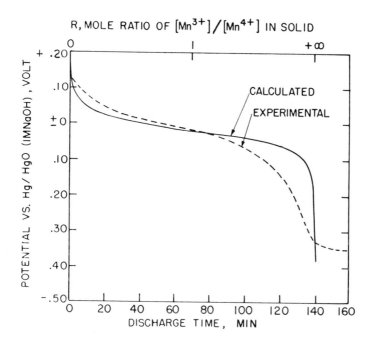

FIG. 9. Comparison of the experimentally measured potential (OCV) with the calculated potential (curve a of Fig. 8). The experimental discharge (equilibrium) potential was measured in 1 M NaOH by discharging a rod electrode at 250 μA (see Ref. [3] for details).

(or OCV) of the system is required by chemical thermodynamics at any stage of the discharge, regardless of the depth of the discharge; this is found to be true in many systems used in practical batteries. As an example, the voltage of a mercury cell (HgO-KOH-Zn system) changes very little from the beginning to the end of its life, in spite of a considerable change in the ratio of the reactant (HgO) to the product (Hg) as the discharge proceeds (see Fig. 10). The cathode reaction in a mercury cell and the Nernst equation corresponding to the reaction are shown below.

$$HgO + H_2O + 2\ e^- \rightarrow Hg + 2\ OH^- \tag{9}$$

$$E = E° - \frac{RT}{2F} \ln \frac{a_{Hg}a^2_{OH}}{a_{HgO}a_{H_2O}} \tag{10}$$

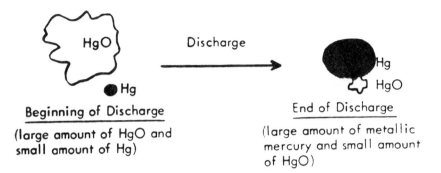

FIG. 10. Schematic representation of the initial state and the final
state of the HgO cathode.

In the mercury cell, the OH^- ion concentration and H_2O concentration should
be kept constant, because OH^- is consumed and H_2O is produced at the zinc
anode in the same amounts as they are produced or consumed at the cathode:

$$Zn + 2\,OH^- \rightarrow ZnO + H_2O + 2\,e^- \tag{11}$$

Therefore, the term $a^2_{OH^-}/a_{H_2O}$ should remain constant through the life
of the cell. Also, since Hg and HgO are solids and the activity values of
solid materials are defined as unity, the values should not change, regardless of the amount of material present. Therefore, the potential is expected
to be constant from the beginning to the end of the discharge according to
Eq. (10) $[a_{Hg} = 1,\ a_{HgO} = 1,\ a^2_{OH^-}/\,a_{H_2O} = const.]$. This conclusion
(constancy of the potential) is a consequence of the Nernst equation, which
was derived from chemical thermodynamics, and agrees with our experience.
This, however, does not give an obvious reason or mechanism for the
constancy of the potential.

The writer would like to propose a simple explanation that provides a
reasonable mechanism for the constancy of the potential in heterogeneous-
phase systems. It is based on the solubility of HgO, Ag_2O [12], or MnOOH
in 7-9 M KOH. These oxides are slightly soluble in concentrated KOH
solutions, and the concentration of the dissolved Hg^{2+}, Ag^+, or Mn^{3+} ion
is in the range of 10^{-3} to 10^{-4} M. In such a system, the observed potential
is probably the potential existing between the metal and the metal ion, for

example, Hg^{2+} - Hg. As long as the KOH solution is saturated with HgO, the Hg^{2+} ion concentration is constant, regardless of the amount of HgO or Hg coexisting; therefore, the potential should be constant from the beginning of discharge to the end, as shown in Fig. 11. This model can explain the constancy of the potential for a number of other systems (Ag_2O-Ag in KOH; $PbSO_4$-Pb in H_2SO_4) equally as well.

In the next sections, it will be shown that this scheme (based on the presence of dissolved ions) is not merely a concept that provides a reasonable understanding of the constancy of the potential, but can be shown to be an actual discharge mechanism in a heterogeneous-phase system. An example, the electrochemical reduction of MnOOH to $Mn(OH)_2$, will be discussed in the following section.

It should be mentioned, however, that in practical alkaline MnO_2 cells, the second-step reaction [MnOOH $\rightarrow$ $Mn(OH)_2$] contributes very little to the useful capacity, even at very low current discharge. Whether the discharge mechanism through dissolved ions is true or not also depends on the discharge condition (current density, temperature, and so on) and should be determined for each system by separate investigation.

1.4.5. Electrochemical Reduction of MnOOH to $Mn(OH)_2$

Since the idea discussed above is based on the existence of some solubility of the metal oxide in concentrated KOH solutions, let us examine the solubility values first. According to the well-known solubility-product concept, the solubility of $Mn(OH)_2$ or $Mn(OH)_3$ should become extremely small as the KOH concentration increases. However, contrary to this, the solubility of Mn(III) and Mn(II) ions increases rapidly with increasing KOH concentration as shown in Figs. 12 and 13 [13]. The reason for this is the formation of complex ions with OH^-, for example, $[Mn(OH)_4]^{2-}$. The predominant form of the complex ion seems to be $[Mn(OH)_4]^{2-}$ in 2-10 M KOH, because the slope of the straight lines (shown in Fig. 13) is 2 ($Mn(OH)_2$ + 2 OH^- $\rightarrow$ $[Mn(OH)_4]^{2-}$.

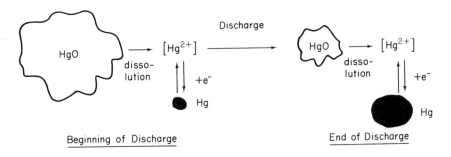

FIG. 11. A possible potential-generating mechanism for the HgO
electrode in KOH electrolyte. The potential is based on a metal-metal ion
$(Hg-Hg^{2+})$ system.

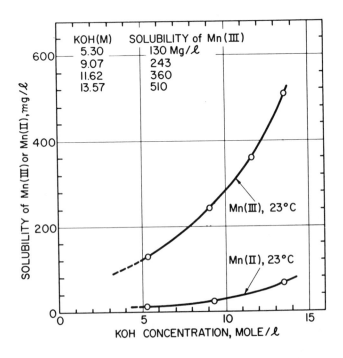

FIG. 12. Solubility of Mn(III) and Mn(II) in KOH [13].

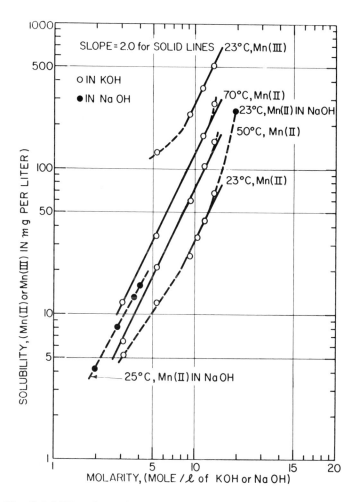

FIG. 13. Solubility of Mn(II) and Mn(III) in KOH and NaOH solutions [13].

Figure 14 shows two possible paths for the electrochemical reduction of MnOOH to $Mn(OH)_2$. On path 1, a solid is reduced directly to another solid, MnOOH to $Mn(OH)_2$; on path 2, discharge takes place through dissolved Mn(III) and Mn(II) ions. The energy change (ΔF) for path 1 should be equal to that for path 2, since the ΔF of the process depends only on the initial state (MnOOH) and the final state $[Mn(OH)_2]$. Steps (b) and (d) in Fig. 14

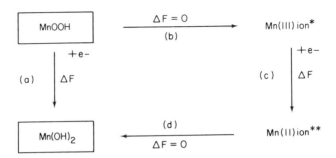

FIG. 14. Two possible paths for reduction of MnOOH to $Mn(OH)_2$:
path 1, direct reduction of MnOOH to $Mn(OH)_2$, (a); path 2, reduction through
dissolved species, (b) → (c) → (d). Mn(III) ion is in equilibrium with MnOOH
(that is, in a solution saturated with MnOOH); Mn(II) ion in equilibrium with
$Mn(OH)_2$ (that is, in a solution saturated with $Mn(OH)_2$).

represent the dissolution of Mn(III) ions and precipitation of Mn(II) ions
respectively. Since Mn(III) and Mn(II) ions are present at saturation, and
the Mn(III) ion or the Mn(II) ion is in equilibrium with MnOOH or $Mn(OH)_2$,
ΔF for the dissolution or the precipitation process is zero. Therefore,
the ΔF value for the reduction of MnOOH to $Mn(OH)_2$ given by

$$MnOOH + H_2O + e^- \rightarrow Mn(OH)_2 + OH^-$$

is equal to that for the reduction of Mn(III) ion to Mn(II) ion given by
$Mn(III) + e^- \rightarrow Mn(II)$ in the saturated solution. The equilibrium potential
of the Mn(III) ion-Mn(II) ion system (which is equal to the $MnOOH-Mn(OH)_2$
system) was found to be 0.395 V (versus Hg/HgO) from the polarographic
half-wave potentials [1, 14] (see Fig. 15).

As can be seen, thermodynamics is not helpful in elucidating the
discharge mechanism, that is, deciding which path is correct for the
reduction of MnOOH. This has to be decided by kinetic measurements,
which will prove that path 2 is the right one.

1.4.6. Kinetic Study of the Second-Step Reaction

In order to study the kinetics of the second-step reaction, cells
containing the $MnOOH-Mn(OH)_2$ system were prepared by reducing $\gamma-MnO_2$

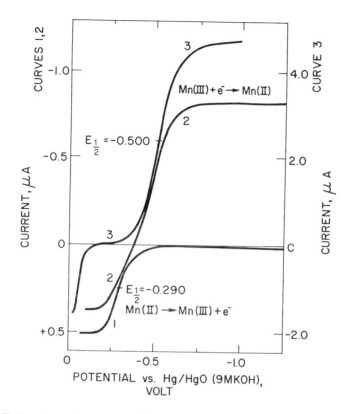

FIG. 15. Polarograms of Mn(II) and Mn(III) in 9 M KOH [1].

electrochemically in 9 M KOH. The reduced system contained $MnO_{1.45}$, which corresponds to a mixture of 90% MnOOH and 10% $Mn(OH)_2$. Polarization curves of such a cell containing $MnO_{1.45}$ were measured (as shown in Fig. 16) for three MnO_2 samples having average particle sizes of 13, 70, and 200 μm diam. These three samples had been prepared from the same deposit of electrolytic manganese dioxide; therefore, the surface area measured by the BET method was approximately the same (39 m^2/gm) for all the samples, regardless of the particle size. Because electrolytic manganese dioxide is very porous in nature, grinding to finer particle sizes does not change the BET surface area appreciably. Since the first-step reaction is a homogeneous-phase (one solid phase) reduction, the electro-

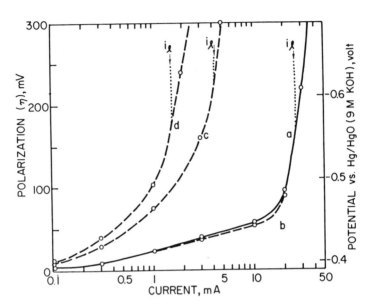

FIG. 16.   Polarization of $MnO_{1.45}$ $[90\%\ MnOOH + 10\%\ Mn(OH)_2]$ in
9 M KOH [4].   The average particle sizes of the oxide (MnOOH) are as
follows: curves a and b, 13 μm; curve c, 70 μm; curve d, 200 μm.

chemical prereduction of the γ-MnO_2 to $MnO_{1.45}$ should not alter appreciably
the pore structure of the original MnO_2 sample.

As shown in Fig. 16, the polarization is relatively low up to a certain
current, but begins to increase rapidly and finally reaches a limiting
current.   The magnitude of the limiting current depends on the particle
size of the original MnO_2 sample; the limiting current value increases with
decreasing particle size [4].

As mentioned in the preceding section, there are two possible paths
(shown in Fig. 14) for the discharge process of MnOOH.   Path 2 consists
of the following three steps: (1) dissolution of Mn(III) ions from MnOOH
and diffusion to the graphite surface, step (b) in Fig. 14; (2) electrochemical
reduction of Mn(III) ion to Mn(II) ion, step (c); and (3) precipitation of
$Mn(OH)_2$ from Mn(II) ions when present in saturation, step (d).   One way
to prove whether this discharge mechanism through the dissolved ion is

correct or not is to compare the observed diffusion-limited currents i_1 to the estimated values based on the superficial (or apparent) surface area. Figure 17 shows a schematic representation of the system composed of porous MnOOH particles and graphite particles in 9 M KOH. The Mn(III) ions that contribute to the current are only those that dissolve from the apparent surface, because the Mn(III) ions dissolving from the inner pore walls need considerable time to diffuse out and reach the graphite interface where the reaction takes place. Since the apparent surface areas of the three samples are considerably different, we can examine with sufficient accuracy whether i_1 is proportional to the apparent surface area or not. The relation between the total apparent surface area S_a and the particle size (radius r) is given by the following equation:

$$S_a = 4\pi r^2 \frac{W/d}{4\pi r^3/3} = \frac{3W}{rd} \approx i_1$$

where W is the weight of the sample, d the density of the oxide, and r the radius of the particle.

The limiting current value i_1 should be proportional to $1/r$, and if S_a should be proportional to i_1, then $i_1 \times r$ should be constant.

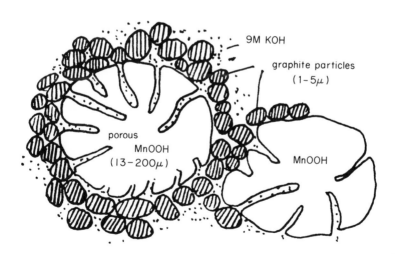

FIG. 17. Schematic representation of a porous MnOOH particle mixed with graphite particles in 9 M KOH electrolyte.

On the other hand, if the solid particle has a fair or good electrical conductivity and the discharge can take place on the entire surface of the pore walls--in other words, if the charge-transfer reaction takes place at the entire solid-solution interface--the limiting current will be proportional to the true surface area. In such a case, the particle size should have no influence, since the total true surface area remains the same, regardless of the particle size.

If the true situation is a mixture of the two mechanisms in a certain proportion, i_1 x r will not be constant. Let us test the experimentally observed i_1 values for the relation. Table 1 shows the results, which indicate reasonably good constancy for the product r x i_1. This means that the observed i_1 values for a wide range of particle sizes can be well explained by the species dissolving from the apparent surface.

The dissolved-species mechanism was previously proposed based on some experimental evidence; namely, the second step in the discharge occurred only at high KOH concentrations or at low KOH concentrations with added triethanolamine (TEA). Both high KOH concentrations and the addition of TEA increase the dissolved Mn(III) ion concentration in KOH electrolyte. The limiting-current data obtained in this study for different particle-size samples and the discussion provided above seems to prove that the dissolved-species mechanism is correct and that the solid-state mechanism (Path 1 in Fig. 14) contributes very little to the discharge current.

TABLE 1

Limiting Current Values i_1 and the Product r x i_1 for γ-MnO_2

r (μm)	i_1 (mA/100 mg MnO_2)	r x i_1
13 $\pm$ 3	26 $\pm$ 3	348
70 $\pm$ 10	4.2 $\pm$ 0.5	294
200 $\pm$ 20	1.5 $\pm$ 0.2	300

1.5. DISCHARGE BEHAVIOR OF MnO_2 UNDER VARIOUS CONDITIONS

In this section, the discharge behavior of MnO_2 under various conditions (varying amounts of carbon powder, particle sizes of MnO_2 powder, concentration of KOH electrolyte, additives to the KOH electrolyte, temperature, effect of interruption of the discharge) will be described. Comments or interpretation will be given briefly in each instance from the standpoint of the discharge mechanisms discussed in previous sections.

Manganese dioxide samples used in these experiments were battery-grade electrolytic manganese dioxide (γ-MnO_2) and also a β-MnO_2 sample which was prepared by heating a γ-MnO_2 sample in air for 10 days. These powdered samples had a particle distribution of 1-50 μm diam and an average size of 20 μm. Physical and chemical properties of these samples are given in Table 2 and Fig. 18.

1.5.1. Effect of the Amount of Carbon Powder and the Particle Size of MnO_2

Manganese dioxide samples mixed with graphite powder (20% by weight of 9 μm or less and 60% of 23 μm or less particle size) and coke was packed with sufficient 9 M KOH solution into an experimental cell shown in Fig. 3. The cell was discharged at a constant current of 1 mA continuously. Discharge curves (shown in Fig. 19) were obtained using different amounts of graphite and coke [6]. It should be noted that the shape of the discharge curve remains unchanged for the first step (from MnO_2 to $MnO_{1.5}$) regardless of the amount of graphite and coke, but the second step (from $MnO_{1.5}$ to $MnO_{1.0}$) in the discharge curve becomes shorter with decreasing amounts of graphite and coke. The reason for this is as follows. Since the first-step reaction is a homogeneous-solid-phase discharge and the MnO_2 sample has fair conductivity (70-100 $\Omega \cdot$ cm resistivity), a small amount of graphite mixed with the MnO_2 sample is sufficient to discharge the entire amount of oxide as long as the current density is not extremely high. The second-step reaction is a heterogeneous-phase process in which the discharge proceeds

TABLE 2

a. Chemical Composition and Electrode Potential of MnO_2 Samples[a]

Sample	O%	Mn%	$MnO_2\%$	x in MnO_x	Potential		
					In 9 M KOH[b] (V)	In 5 M NH$_4$Cl[c] (V)	In 5 M NH$_4$Cl + 2 M ZnCl$_2$[d] (V)
(1) γ-MnO_2	16.66	59.83	90.46	1.956	+0.232 ↑	+0.440[e] →	+0.689 →
(2) β-MnO_2(I)	17.27	62.17	93.78	1.953	-0.015 ↑	+0.196 ↑	+0.460 ↑

[a]Reference [2].

[b]In 9 M KOH vs. Hg/HgO (9 M KOH).

[c]In 5 M NH$_4$Cl (pH 8.8) vs. SCE.

[d]In 5 M NH$_4$Cl + 2 M ZnCl$_s$ + NH$_4$OH (pH 5.0) vs. SCE.

[e]Arrow ↑ indicates the potential tends to increase very slowly with time. Arrow → indicates the potential tends to decrease very slowly with time. No arrow indicates the potential did not change more than a few millivolts in 2 weeks.

b. Physical Properties of MnO_2 Samples

Sample	BET method by N_2 gas (pores less than 600 Å diam)			Hg intrusion method (pores 0.01–100 μm diam)		
	Surface area (m^2/gm)	Total pore volume (cm^3/gm)	Pore volume less than 150 Å $(\%)$	Total pore volume (cm^3/gm)	Average pore diameter (μm)	Density by He gas (gm/cm^3)
γ-MnO_2	39.7	0.05296	59.24	0.2953	2.008	4.410
β-MnO_2(I)	17.4	0.06513	65.98	0.3064	1.561	4.88

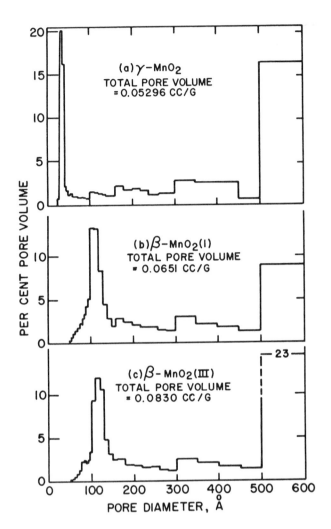

FIG. 18.   Pore-size distribution of γ-MnO$_2$ powder (battery-grade) and β-MnO$_2$, which was made by heating the γ-MnO$_2$ at 400°C for 10 days in air.   See Ref. [2] for details of β-MnO$_2$(I) and β-MnO$_2$(III).   The N$_2$ gas adsorption and desorption method was used in these measurements.   This graph should be read as follows: For example, in (c) β-MnO$_2$(III), the pores between 100 and 110 Å constitute 9.3%, pores between 280 and 300 Å constitute 2.3%, and those between 500 and 600 Å constitute 23%. Therefore, the area under the curve is not proportional to the pore volume.

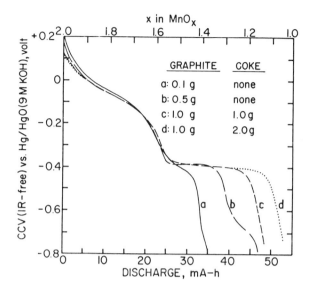

FIG. 19. Influence of the amount of graphite and coke mixed with 100 mg of the MnO_2 sample (γ-MnO_2) on the discharge curve at 1.0 mA constant-current continuous discharge in 9 M KOH [6].

through dissolved Mn(III) ions, which are then reduced electrochemically on the graphite surface. Therefore, when the amount of graphite powder is reduced to a certain value or lower, the manganese dioxide particles will not be separated individually and the discharge current is expected to be reduced. This is obvious from the mechanism discussed in Section 1.4.6.

Figure 20 shows the effect of the particle size of manganese dioxide in the cell in which large amounts of coke (2 gm) and graphite (1 gm) were used with a 100-mg MnO_2 sample. It is clear from Fig. 20 that the particle size has very little influence on the first step, but the capacity of the second step is considerably reduced when the particle size is increased. This is understandable from the discharge mechanisms.

1.5.2. Effect of KOH Concentration

A rod-type electrode (MnO_2 deposited on a graphite rod) was discharged in 0.1 M, 1.0 M, and 9 M KOH at 3.0 mA continuously. The results (Fig.

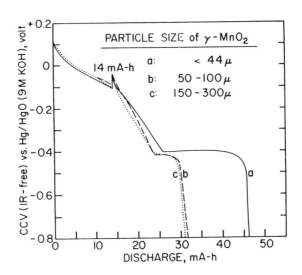

FIG. 20.   Effect of particle size of the MnO$_2$ sample (γ-MnO$_2$).   1 gm
of graphite and 2 gm of coke were mixed with 100 mg of γ-MnO$_2$.   Discharge
was at 1. 0 mA continuously at 23°C, except for a 96-h interruption at the
14-mAh level, during which the electrode was stored at 0°C [6].

21) show that the KOH concentration has very little influence on the first
step in the discharge (MnO$_2$ to MnO$_{1.5}$), but the second step (MnO$_{1.5}$ to
MnO$_{1.0}$) can be obtained only at high KOH concentrations (9 M KOH). Since
the first-step reaction is a solid-state reduction, it is reasonable that there
should be very little effect, but since the second-step reaction proceeds
through dissolved Mn(III) ions and the Mn(III) solubility is very small at low
KOH concentrations (0. 1 M and 1 M) (see Figure 13), such a poor discharge
should be expected. The second step in the discharge at low KOH concentrations can be carried out only at extremely low current densities. The
effect of KOH concentration on the discharge of powdered MnO$_2$ samples
was found to be quite similar to that in Fig. 21 (see Fig. 6 of Ref. [2]).

Figure 22 shows a discharge curve for a rod electrode (having 10. 6 mg
of MnO$_2$) at 250 μA at 23°C in 9 M KOH. When the curve (dashed line in
Fig. 22) obtained in 9 M KOH was compared to a curve (Fig. 9) obtained
in 1 M NaOH under the same discharge conditions, the later portion of the

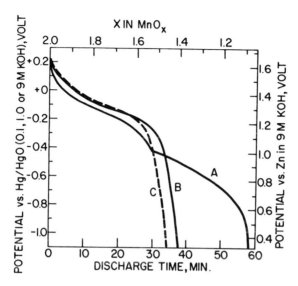

FIG. 21. Effect of KOH concentration on the discharge at a constant current of 3.0 mA. Reference electrodes (Hg/HgO) made with 9, 1, and 0.1 M KOH electrolyte were used for the discharges in 9 (curve A), 1 M (curve B), and 0.1 M (curve C) KOH, respectively [2].

first step (marked X in Fig. 22) exhibited a lower potential than that in 1 M NaOH. This was attributed to faster deterioration of the MnOOH structure in 9 M KOH than in 1 M NaOH, since the Mn(III) ion can dissolve more easily in concentrated electrolyte.

1.5.3. Effect of Additives to KOH

The discharge behavior of MnO_2 was compared in 1 M KOH with and without triethanolamine (TEA) added to the electrolyte. As shown in Fig. 23 [1], the second step, which could not be obtained on discharge in pure 1 M KOH, was obtained on discharge in the presence of 10% TEA in 1 M KOH. TEA is a good complexing agent for Mn(II) and Mn(III) ions, and in the presence of the dissolved Mn(III) ions (as the TEA complex), the second step [MnOOH to $Mn(OH)_2$] takes place on discharge.

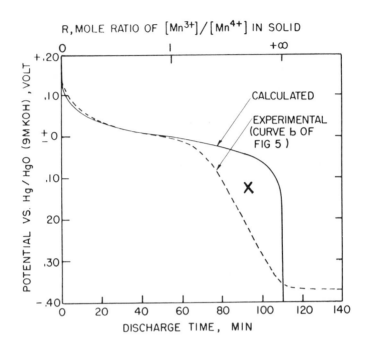

FIG. 22.   OCV of a rod electrode (MnO_2, 10.6 mg) in 9 M KOH at 23°C
and 250 μA intermittent discharge [3].

A discharge curve obtained in 9 M KOH saturated with ZnO shows a
considerably shorter second step compared to that obtained in pure 9 M KOH,
as shown in Fig. 24.   The reason for this is not clear at this time but is
probably due to the fact that the solubility of Mn(III) ion is considerably
reduced in the presence of large amounts of zincate ion (approximately 1 M
in $[Zn(OH)_4]^{2-}$ concentration).

1.5.4.   Effect of Interruption of the Discharge

The powdered γ-MnO_2 samples were discharged in cells (Fig. 3) to
$MnO_{1.71}$ at 1 mA/100 mg MnO_2 continuously at 23°C.   The cells containing
$MnO_{1.71}$ were stored at 0°, 23°, or 45°C for four days and then discharged

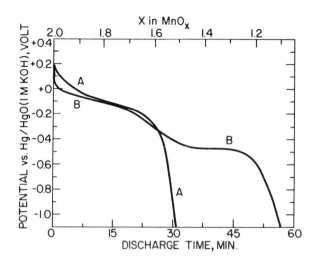

FIG. 23. Effect of triethanolamine (TEA) added to 1 M KOH on the discharge of MnO_2 [1]. A rod electrode (10.6 mg of MnO_2) was discharged at 3.0 mA at 23°C. Curve A, 1 M KOH; curve B, 1 M KOH + 10% TEA.

further. Those discharge curves are shown in Fig. 25. When the partially discharged MnO_2 was stored at 45°C, the subsequent discharge curve showed new steps (II, III, and IV in Fig. 25). However, cells stored at 0° or 23°C showed normal discharge curves. This suggests that at 45°C, the partially reduced (Mn^{4+}-Mn^{3+}-O^{2-}-OH^-) system changes its structure. It was confirmed by discharging Mn_2O_3 and Mn_3O_4 that the new steps (II, III, and IV) are not due to formation of Mn_2O_3 or Mn_3O_4 during the storage period from the partially discharged γ-MnO_2. Similarly, the structural transformation takes place on interruption of the discharge of β-MnO_2 [6]. When MnO_2 was discharged to a lower level (for example, to x = 1.65 in MnO_x) as shown in Fig. 26, such a transformation occurred even at 0°C and 23°C storage. This means that the structure of γ-MnO_2 is not stable in 9 M KOH when it is partially discharged.

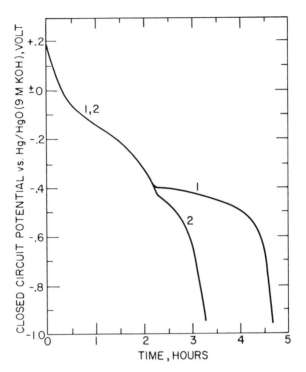

FIG. 24.   Effect of ZnO added to 9 M KOH on the discharge of MnO_2 [3].
A rod electrode (10.6 mg of MnO_2) was discharged in 9 M KOH at 250 μA.
Curve 1, in 9 M KOH; curve 2, in 9 M KOH saturated with ZnO.

1.5.5.   Effect of Temperature*

Discharge curves (CCV and OCV) taken at 23° and 65°C are shown in
Fig. 27.   The CCV curve at 65°C (curve 2 in Fig. 25) is considerably
higher than that at 23°C (curve 1 in Fig. 27) initially, but at a later stage
the voltage levels at the temperatures become approximately the same.
The OCV curve at 65°C (curve 4) is about the same as that at 23°C (curve
3) initially, but later it is much lower than that at 23°C.   This behavior
can be easily interpreted if we assume that the γ-MnO_2 lattice is maintained

*Reference [3].

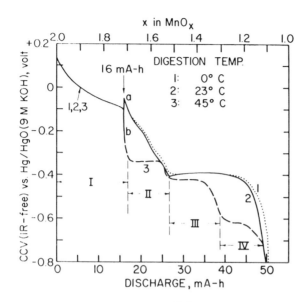

FIG. 25. Effect of storage at 0°, 23°, and 45°C on partially discharged MnO_2 in 9 M KOH. Cells (Fig. 3) containing 100 mg of γ-MnO_2 were discharged at 1.0 mA at 23°C to 16 mAh ($MnO_{1.71}$) and then stored at 0°, 23°, and 45°C for 96 h. Then the cells were further discharged at 1.0 mA at 23°C [6].

in the initial half of the discharge, but in the later half, the structure seems to be transformed into another form (at least on its surface) which exhibits a lower voltage. Such a transformation seems to occur much faster at 65° than at 23°C. The CCV curve in the initial stage is much higher at 65° than at 23°C, because the polarization is less at higher temperatures as long as the same γ-MnO_2 structure is maintained. In the later stage, the OCV is higher at 23° than at 65°C, because at 23°C the γ-MnO_2 structure (a high-voltage oxide structure) is still preserved.

1.6. KINETICS OF THE DISCHARGE PROCESS AND MAGNITUDE OF POLARIZATION

In this section, the results obtained so far concerning the kinetics of the first-step reaction will be summarized and discussed briefly. Also,

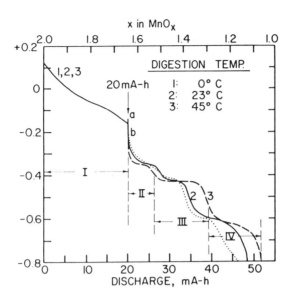

FIG. 26.   Effect of storage at 0°, 23°, and 45°C on partially discharged
MnO_2 [6].   Cells (Fig. 3) containing 100 mg of γ-MnO_2 were discharged at
1.0 mA to 20 mAh ($MnO_{1.65}$) and then stored at 0°, 23°, and 45°C for 96 h.
Then the cells were further discharged at 23°C.

polarization data for battery-grade electrolytic MnO_2 and β-MnO_2 in 9 M
KOH and NH_4Cl electrolytes are summarized.

1.6.1.   Factors that Control the First-Step Reaction

The first step of the reaction taking place at the solid-solution interface
is the decomposition of water ($H_2O \rightarrow H^+ + OH^-$) and injection of protons
into the lattice to form OH^- in the lattice ($O^{2-} + H^+ \rightarrow OH^-$) (see Fig. 4).
Subsequently, H^+ formed from the decomposition of the OH^- jumps from the
OH^- site to an adjacent O^{2-} site.   The essential step in both processes
seems to be the breakage of the O-H bond.   If D_2O is used instead of H_2O
in the solution, the process must be slower, that is, the polarization must
be greater, since D^+ is twice as heavy as H^+.   This was found to be true;

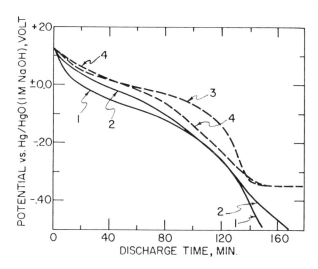

FIG. 27. Effect of temperature on the discharge of MnO_2 (10.6 mg deposited on a graphite rod) in 1 M NaOH [3]. The electrode was discharged at 250 μA, both continuously and intermittently. Curve 1, CCV at a continuous discharge at 23°C; curve 2, CCV at a continuous discharge at 65°C; curve 3, OCV at an intermittent discharge at 23°C; curve 4, OCV at an intermittent discharge at 65°C.

the potential was lower in 9 M KOH + D_2O than in 9 M KOH + H_2O, as shown in Fig. 28. With a powdered γ-MnO_2 sample, the effect of D_2O was measured accurately at much higher current densities than those shown in Fig. 28 in KOD + D_2O. The results are shown in Fig. 29. The polarization is considerably greater in KOD + D_2O than in KOH + H_2O. At high-current discharges (5 or 20 mA per 50 mg of MnO_2 sample), the polarization began to suddenly increase at a certain point and reached a maximum as shown in Fig. 29. This rapid increase in the polarization was interpreted as the formation of a lower oxide structure on the surface of the γ-MnO_2. The reason for the decrease in polarization after the maximum is the onset of the second-step reaction [MnOOH → $Mn(OH)_2$]. When the polarization values at early stages (1-3 mAh for a 50-mg sample, that is, at stages from MnO_2 to $MnO_{1.9}$) are plotted against the current, a straight line is obtained (Fig. 30) for each depth of discharge. The slope is approximately 130 mV/decade.

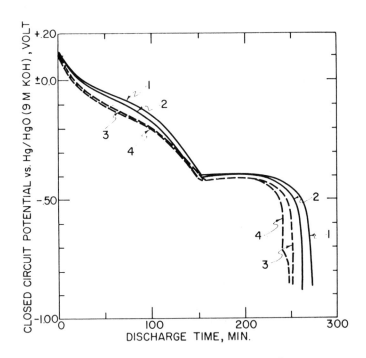

FIG. 28.   Effect of D_2O on the discharge of MnO_2 [3].   A rod electrode
(MnO_2, 10.6 mg) was discharged at 250 μA in 9 M KOH made with H_2O or
D_2O.   Curves 1 and 2, in 9 M KOH + H_2O; curves 3 and 4, in 9 M KOH +
D_2O.

The ratio of the currents (i_{KOH} : i_{KOD}) in KOH and in KOD at the same
polarization value was found to be i_{KOH} : i_{KOD} = 1 : 1.45 $\pm$ 0.1, which is
close to the ratio $1/\sqrt{H}$: $1/\sqrt{D}$ = 1.41.   This suggests that diffusion of D^+
or H^+ (or diffusion-type transport) may control the rate of this process.

Since the slope (130 mV/decade) is close to 120 mV/decade, the rate-
determining step seems to be a one-electron process with an assumption of
α = 1/2 in the following equation.

$$\eta = \frac{RT}{\alpha nF} (\ln \frac{i_0}{i}) = a - b \log i$$

The one-electron process is likely to be the following process:

$$MnO_2 + H_2O + e^- \rightarrow MnOOH + OH^-$$

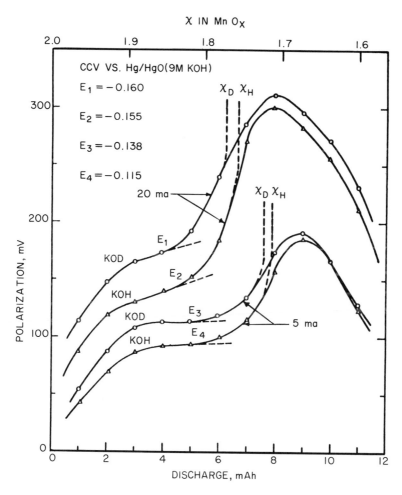

FIG. 29. Polarization of γ-MnO$_2$ in 9 M KOH and 9 M KOD [5]. The measurements were made at 23°C in a cell containing 50 mg of MnO$_2$.

The exchange current i_0 for the process was obtained for γ-MnO$_2$ in 9 M KOH and 9 M KOD and tabulated in Ref. [5].

1.6.2. Polarization Data for Powdered Manganese Dioxide (Battery Grade)

Polarization data for γ-MnO$_2$ and β-MnO$_2$ samples were obtained in 9 M KOH, 9 M KOD, 5 M NH$_4$Cl, and 5 M NH$_4$Cl + 2 M ZnCl$_2$ solution.

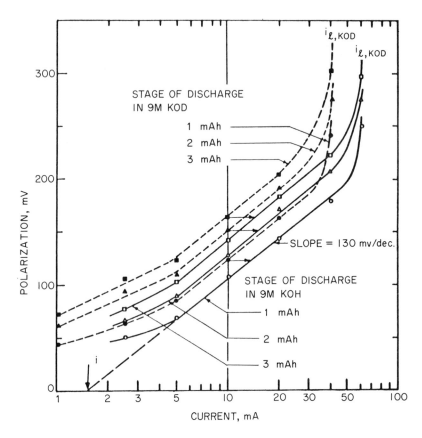

FIG. 30.   Polarization of the γ-MnO_2 sample at early stages of the discharge ($MnO_{1.96} \sim MnO_{1.88}$).   The current is per 50 mg MnO_2 in the cell [5].

Each manganese dioxide sample (100 mg) was packed in a cell (Fig. 3) with a large amount of coke and graphite and abundant electrolyte, and discharged at various constant currents (0.1-4 mA/100 mg of MnO_2) for 10-60 min, sufficient to attain a steady state.   Then the potential (CCV) was measured against a reference electrode (Hg/HgO or SCE) eliminating the IR drop.   The electrode was then maintained on open circuit for sufficient time (15-20 h) before the OCV was measured.   Polarization was obtained as the potential difference between the OCV and the CCV.   Table 3 shows the

TABLE 3

Steady-State Cathodic Polarization[a] (η_S) at 1 mA/100 mg of MnO$_2$ at 4-, 8-, 12-, and 16-mAh Discharge Stages [2]

MnO$_2$	Electrolyte	Stage of the discharge[b]			
		4 mAh $MnO_{1.93}$	8 mAh $MnO_{1.86}$	12 mAh $MnO_{1.79}$	16 mAh $MnO_{1.72}$
(1) γ-MnO$_2$	9M KOH	45	50	50	48
(2) β-MnO$_2$(III)	9M KOH	128	133	130	121
(3) γ-MnO$_2$	9M KOD	57	68	61	63
(4) β-MnO$_2$(III)	9M KOD	137	144	140	135
(5) γ-MnO$_2$	5M NH$_4$Cl (pH 8.8)	53	71	64	73
(6) β-MnO$_2$(III)	5M NH$_4$Cl (pH 8.8)	144	133	129	107
(7) γ-MnO$_2$	5M NH$_4$Cl + 2M ZnCl$_2$ (pH 5.02)	53	63	63	64
(8) β-MnO$_2$(III)	5M NH$_4$Cl + 2M ZnCl$_2$ (pH 5.02)	79	82	76	71

[a] Discharged at 1 mA continuously for 4 h, then the η_S was measured as a potential difference between the closed-circuit voltage corrected for IR drop, and the open-circuit voltage after 10–15 h recovery.

[b] The x values in MnO$_x$ shown are an average of those for γ-MnO$_2$ and β-MnO$_2$.

polarization values for the γ- and β-MnO_2 samples at various depths of discharge. Also, the E-x relation (open-circuit voltage versus x in MnO_x) for the two samples is shown in Figs. 31 and 32. Polarization versus current-density data are shown in Fig. 33, in which scales in mA/gm and mA/m^2 are provided. Details of the experiments and the discussion of these polarization data are provided in Ref. [2].

Two factors may be worth pointing out with respect to the polarization data. First, the polarization values are considerably smaller than those

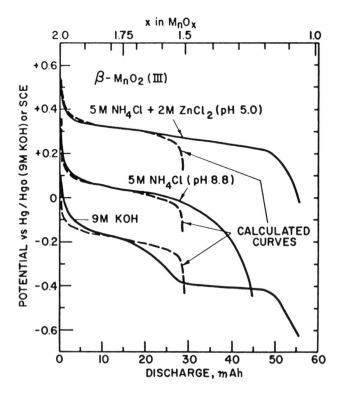

FIG. 31. Comparison of the open-circuit potential of β-MnO_2 with the potential calculated from E = E' - RT/F $\ln[Mn^{3+}]/[Mn^{4+}]$. The Hg/HgO (9 M KOH) reference electrode was used for 9 M KOH electrolyte and SCE was used for the other electrolytes [2].

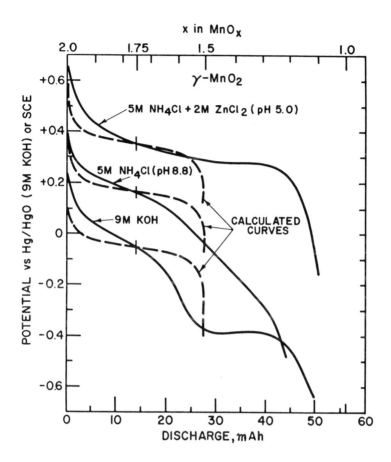

FIG. 32. Comparison of the open-circuit potential of γ-MnO$_2$ with the potential calculated from $E = E' - RT/F \ln[Mn^{3+}]/[Mn^{4+}]$. The Hg/HgO (9 M KOH) reference electrode was used for 9 M KOH electrolyte and SCE was used for the other electrolytes [3].

we might expect from our impressions of the ordinary dry cell; polarization thus measured does not exceed 100 mV at a current equivalent to 400 mA per D-size cell. Second, the polarization of β-MnO$_2$ is not much greater than that of γ-MnO$_2$ in Leclanché electrolytes. Thus the common impression that the Leclanché cell shows great polarization is not necessarily true; the

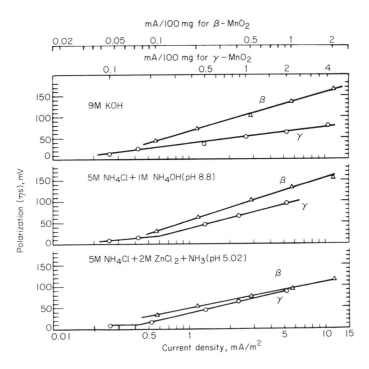

FIG. 33. Polarization characteristics of γ- and β-MnO$_2$ in solutions (2).

voltage drop in an actual dry cell on discharge is largely due to the decrease
in the open-circuit potential and the relatively large IR drop due to the thick
bobbin construction.

1.7. DISCHARGE MECHANISM OF MnO$_2$ IN LECLANCHÉ ELECTROLYTE

It was pointed out in Section 1.4.2. that the shape of the E-x curve
(equilibrium potential versus depth of discharge) can be used as a criterion
for evaluating whether the reaction is taking place in one phase (homogeneous
phase) or in two solid phases (heterogeneous phase). By using this thermodynamic criteria in addition to other factors (pH-potential relation), the

discharge mechanism of the cathode reaction (reduction of MnO_2) in a Leclanché cell can be determined.

In Table 4 are listed four mechanisms so far proposed or considered as possibilities in previous publications. The two evaluation criteria (the slope of the pH-potential relation and potential change or E-x relation) for each mechanism are also shown in Table 4. According to experimental measurements, the slope of the pH-potential relation is 59 mV/pH and the potential (E-x curve) decreases continuously, thereby indicating the reaction to be a one-phase reaction. The only mechanism that is possible from the standpoint of the thermodynamic criteria is the proton-electron mechanism. Since the OCV of γ-MnO_2 decreases continuously to around $MnO_{1.7}$ in $NH_4Cl + ZnCl_2$ electrolyte as shown in Fig. 31, the proton-electron mechanism must be correct for the major portion of the useful discharge capacity of dry cells. Other reactions, such as Mn^{2+} ion dissolution and hetaerolite $(ZnMn_2O_4)$ formation, would take place, depending on the current density and/or electrolyte composition. However, these reactions must be considered as side reactions. It is interesting to note that MnO_2 can be discharged to $MnO_{1.2}$ (far more than a one-electron reduction) in $NH_4Cl + ZnCl_2$ electrolyte as shown in Fig. 31.

NOTE BY THE EDITOR

References 15 to 24 were suggested by A. Kozawa as reading material on the discharge of MnO_2 in Leclanché electrolytes. An extensive treatment of this subject and a detailed bibliography on MnO_2 can be found in Chapter 1 of this book, written by R. Huber.

1.8. DISCHARGE BEHAVIOR OF MnO_2 IN $MgClO_4$ ELECTROLYTE

Magnesium-MnO_2 cells use $MgClO_4$ or $MgBr_2$ solution as electrolyte. The discharge behavior of two manganese dioxide samples (electrolyte MnO_2

TABLE 4

Possible Electrochemical Reactions of the MnO_2 Cathode of Leclanché Cell [2]

Mechanism	Primary electrochemical reaction	Nernst equation at 25°C	Potential change with decreasing x in MnOx	Slope of the pH-potential relation
(1) Proton-electron mechanism	$MnO_2 + H_2O + e^- \rightarrow$ $MnOOH + OH^-$ (one solid phase)	$E = E' - 0.059 \log \dfrac{[Mn^{3+}]solid}{[Mn^{4+}]solid}$ $- 0.059 \, pH$	Decrease (because of one phase)	59 mV/pH
(2) Two-phase mechanism	$2\,MnO_2 + H_2O + 2\,e^- \rightarrow$ $Mn_2O_3 + 2\,OH^-$ (two solid phases)	$E = E' - 0.0295 \log \dfrac{a_{Mn_2O_3}}{a^2_{MnO_2}}$ $- 0.059 \, pH$	Constant (because of two phases)	59 mV/pH
(3) Mn(II) ion mechanism	$MnO_2 + 4\,H^+ + 2\,e^- \rightarrow$ $2\,H_2O + Mn^{2+}$	$E = E' - 0.0295 \log [Mn^{2+}]$ $+ 0.0295 \log a_{MnO_2} - 0.118 \, pH$	Decrease (because of $[Mn^{2+}]$ change)	118 mV/pH
(4) Hetaerolite mechanism	$2\,MnO_2 + Zn^{2+} + 2\,e^- \rightarrow$ $ZnO \cdot Mn_2O_3$ (hetaerolite)	$E = E' - 0.0295 \log \dfrac{a_{ZnO \cdot Mn_2O_3}}{a^2_{MnO_2}}$ $+ 0.0295 \log[Zn^{2+}]$	Constant (if $[Zn^{2+}]$ is constant)	0 mV/pH

and chemical MnO_2) was investigated in 2 M $MgClO_4$ with an experimental cell described previously (Fig. 3) by discharging 5-6 mAh at a time at a constant current (1.0 mA/100 mg MnO_2). The results are shown in Fig. 34.

The general discharge behavior in 2 M $MgClO_4$ solution is quite similar to that in 0.1 M KOH (see Fig. 21); MnO_2 is reduced electrochemically to approximately $MnO_{1.6}$ as a homogeneous phase. After that point, the potential drops to hydrogen-evolution voltage on graphite without showing the second step $[MnOOH \rightarrow Mn(OH)_2]$ that was seen in 9 M KOH (Figs. 1 and 2). As can be seen in Fig. 34, the OCV curve is lower in 2 M $MgClO_4$ solution containing $Mg(OH)_2$ than in 2 M $MgClO_4$ solution to which no $Mg(OH)_2$ has been added. This is partly because of the lower pH value of 2 M $MgClO_4$ (about pH 6.1) than with a $Mg(OH)_2$ suspension (about pH 6.7). The discharge behavior shown in Fig. 34 was also confirmed with practical magnesium cells (C-size) as shown in Fig. 35; at a certain discharge current (75 mA) or less, MnO_2 is reduced to approximately $MnO_{1.63}$ regardless of the discharge mode (intermittent or continuous).

Figure 36 shows polarization of γ-MnO_2 (EMD), β-MnO_2, and chemical MnO_2 samples in 2 M $MgClO_4$ containing $Mg(OH)_2$. Polarization of β-MnO_2 is much higher than that of EMD or CMD, although such a large difference was not seen in 5 M NH_4Cl (pH 8.8) and 5 M NH_4Cl + 2 M $ZnCl_4$ (pH 5.02) solutions (see Fig. 33). The β-MnO_2 was prepared by heating EMD (γ-MnO_2) at 400°C in air for 10 days.

2. PRODUCTION PROCESS AND PROPERTIES OF EMD

2.1. HISTORY OF EMD AND PRESENT PRODUCTION FIGURES

As early as 1918, Van Arsdale et al. [25] reported the preparation of electrolytic manganese dioxide (EMD) by electrolyzing a manganous sulfate solution. They also pointed out that the use of EMD as a dry-cell depolarizer increased the cell discharge capacity. Several authors, including Nichols

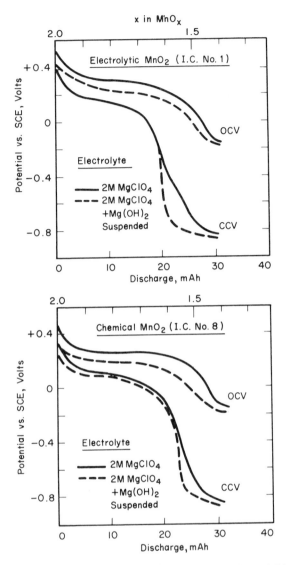

FIG. 34. Discharge behavior of manganese dioxide in 2 M $MgClO_4$ solution with and without $Mg(OH)_2$ suspension. CCV curves are constructed by connecting the IR-free closed-circuit voltage at the end of each 5-6 h continuous discharge at 1.0 mA per 100 mg MnO_2 sample (cell shown previously in Fig. 3). The open-circuit voltage curves were constructed by connecting the open-circuit voltage values that were measured after a 17-h rest period after the discharge of 5-6 mAh. I.C. No. 1 and I.C. No. 8 are International Common Samples (see Section 4.2).

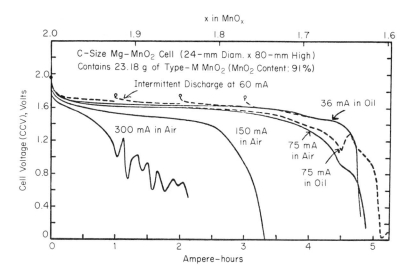

FIG. 35. Discharge capacity of C-size, $Mg-MnO_2$ [2 M $Mg(ClO_4)_2$ electrolyte] cell at constant-current continuous discharge. Some cells were immersed in oil to prevent oxygen ingress.

[26] in 1932, Storey et al. [27] in 1944, and Lee [28] in 1949, described a process by which EMD could be produced on an industrial basis using manganese sulfate solution prepared by leaching rhodochrosite (a mineral form of $MnCO_3$) with sulfuric acid.

In Japan, Kameyama et al. [29] (1934) investigated EMD preparation in which manganous nitrate solution was electrolyzed with alternating current. A detailed investigation of EMD preparation from manganese sulfate with direct current was published by Takahashi [30] (1938). It is interesting to note that, as early as 1929, Inoue et al. [31] obtained a Japanese patent on the use of EMD as a dry-cell depolarizer. Although trials of EMD production for battery use were attempted from time to time since about 1934, production on an industrial scale was not initiated until the end of World War II. At that time, EMD was needed for military batteries. Therefore, the first small plant for EMD production was built in 1944 at the Washizu Plant of Tokyo Shibaura Electric Co. By the time the war ended (1945), it was generally realized that EMD was indispensable for improving the quality of

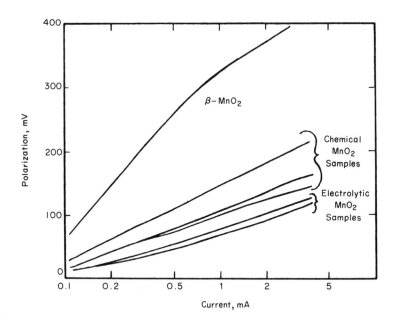

FIG. 36. Polarization of various manganese dioxides in 2 M $Mg(ClO_4)_2$ containing $Mg(OH)_2$ suspension at the 3-mAh stage $(MnO_{1.93})$.

dry cells. A few years after the war (Nov. 1948), the Japanese government (Ministry of Industry and Trade) took the initiative to organize an EMD committee made up of representatives from EMD manufacturers and dry-cell manufacturers. Today (1972), EMD production in Japan is about 45,000 tons per year and about 70% of the world production; 75% of the total production is exported for dry-cell manufacture to various parts of the world. Fig. 37 shows the production and export figures from 1957 to 1970.

2.2. OUTLINE OF THE EMD PRODUCTION PROCESS*

An outline of the production process is shown in Figs. 38 and 39. Most of the Japanese processes are based on rhodochrosite ore. Some details of each step involved in the process are given here.

*Reference [32].

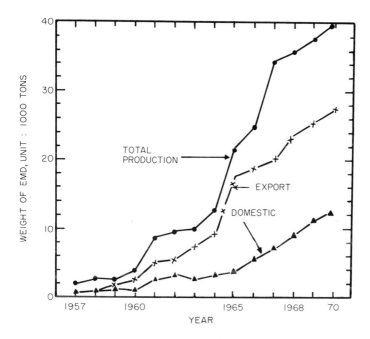

FIG. 37. Production, export and domestic consumption of electrolytic manganese dioxide (EMD) in Japan [94].

The first step is the leaching and purification processes. According to the following equation, 1 mole of H_2SO_4 is needed to leach out 1 mole of $MnSO_4$ from $MnCO_3$:

$$MnCO_3 + H_2SO_4 = MnSO_4 + H_2O + CO_2$$

In practice, the quantity of sulfuric acid used is usually 10% more than the theoretical amount based on the manganese content of the ore, because carbonate and sulfide impurities are present which consume sulfuric acid. The dissolution conditions are as follows: (a) rhodochrosite particle size, < 100 mesh; (b) temperature of solution in tank, $80°-90°C$; (c) H_2SO_4 concentration, 100-150 gm/liter. In order to oxidize Fe^{2+} ion to Fe^{3+} ion in the $MnSO_4$ solution, MnO_2 ore (fine powder) is added and air is bubbled through the solution. Then the pH of the solution is adjusted to 4 to 6 by

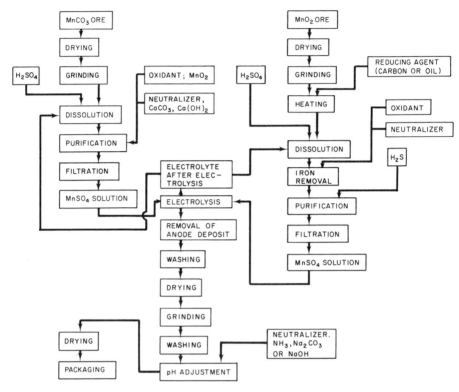

FIG. 38.   Flow diagram of EMD production process from $MnCO_3$ or MnO_2 ore [32].

adding a neutralizing agent [$Ca(OH)_2$ or $CaCO_3$], so that Fe^{3+} ion precipitates out as $Fe(OH)_3$. When the solution is cooled down to room temperature, $CaSO_4$ coprecipitates with $Fe(OH)_3$:

$$Fe_2(SO_4)_3 + 3 \ Ca(OH)_2 = 2 \ Fe(OH)_3 + 3 \ CaSO_4$$

Small amounts of other impurities (Pb^{2+}, Ni^{2+}, Co^{2+}, and so on) are also coprecipitated with $Fe(OH)_3$ or SiO_2. The precipitates are filtered off to obtain a purified $MnSO_4$ solution, which is then fed into the electrolytic cells.

In the electrolytic process, the purified $MnSO_4$ solution is preheated above 80°C and fed into the electrolytic cells. The cells are operated with

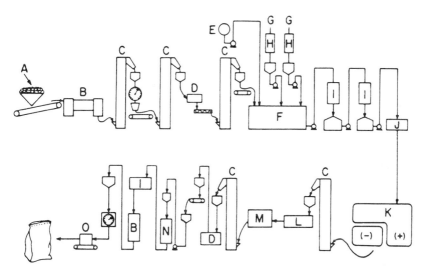

FIG. 39. Flow diagram of EMD process (Tekkosha, Hyuga Plant) [33].
A, rhodochrosite ore; B, dryer; C, basket elevator; D, crusher; E, H_2SO_4
storage; F, dissolution tank; G, ore; H, crusher; I, filter; J, electrolysis
cells; K, anode treatment; L, water washer; M, dryer; N, pH adjustment;
O, packaging.

direct current approximately under the following conditions: (a) electrolyte
concentration--$MnSO_4$, 0.5-1.2 mole/liter; H_2SO_4, 0.5-1.0 mole/liter;
(b) electrolyte temperature, 88°-98°C; (c) anode current density, 0.7-1.2
A/dm^2. As mentioned earlier, titanium, lead alloy, or carbon is used as
anode. The electrolysis conditions depend somewhat on the anode material.
The cell lining can be lead, or an acid-resistant resin (for example, Hypalon
gum, which is chlorosulfonated polyethylene, or fiber-reinforced plastic
[FRP]). Oil or paraffin is floated on top of the electrolyte as a water-vapor
suppressor. Entrainment of the electrolyte in the hydrogen gas produced
at the cathode is also minimized by the oil or paraffin. During electrolysis,
the $MnSO_4$ concentration decreases and the H_2SO_4 concentration increases
in the electrolyte as a result of the following reactions at the anode and the
cathode.

Anode: $Mn^{2+} + 2\ H_2O \rightarrow MnO_2 + 4\ H^+ + 2\ e^-$

Cathode: $2\ H^+ + 2\ e^- \rightarrow H_2$

Overall: $Mn^{2+} + 2\ H_2O \rightarrow H_2 + MnO_2 + 2\ H^+$

The net result is MnO_2 deposition at the anode and H_2SO_4 formation in the electrolyte. The H_2SO_4-rich electrolyte is circulated to the ore-dissolution tank and reused.

The post-electrolysis treatment of the MnO_2 deposit depends on the intended application of the EMD. A typical process for battery use will be described. The thickness of the MnO_2 deposit on the anode depends on the current density and the electrolysis time; however, when the thickness of the deposit reaches 10-30 mm, the anode is usually pulled out of the cell and the deposit is removed by means of mechanical shock. The flake is washed with hot water to remove the oil and electrolyte solution. Then it is dried in a rotary dryer, crushed, and finally washed thoroughly with water. After washing, the powder is neutralized with ammonia, soda ash, or sodium hydroxide solution in order to neutralize free acid and adjust the pH of the final product. After pH adjustment, the EMD powder is filtered and dried. The slurry containing the very fine particles obtained in this filtration process is often used to oxidize Fe^{2+} ion, as mentioned earlier. The final powder is dried with hot air (85°C) and adjusted for particle-size distribution and finally packaged according to particular shipping requirements.

2.3. BRIEF REVIEW OF THE LITERATURE ON EMD

In 1931, Inoue and Haga [34] reported that EMD was an excellent cathode material for dry cells. Around that time, Kameyama and Iida [29] studied the formation of manganese dioxide at platinum anodes by electrolyzing $Mn(NO_3)_2$ or $MnSO_4$ solutions with alternating current. They obtained EMD powder at the bottom of the bath (not deposited on the anode) with 60-70% current efficiency from 0.55 M $Mn(NO_3)_2$ + 0.4 M HNO_3 at 40°C. They

investigated the effect of the ac frequency and bath temperature $(20°\text{-}60°C)$.
The EMD had an MnO_2 content of 93-95%, and they found that under various
discharge conditions, the capacity of the EMD in dry cells was at least
twice as great as that of the natural MnO_2 ore.

   Takahashi [30] made a comprehensive investigation of the factors
involved in the production of EMD by direct-current electrolysis. He
established the best operating conditions, achieving 98% current efficiency
with a cell voltage of 2.0 V, using a Pb-Sb (5%) alloy anode and a graphite
cathode in a 1 M $MnSO_4$ + 0.5 M H_2SO_4 solution at a current density of
1.0 A/dm^2 at 90°C. He also studied in detail the influence of Fe^{2+} ions
present as impurities in the bath, and found that the presence of Fe^{2+} or
Fe^{3+} ions reduced the current efficiency in the electrolysis. The deleterious
effect of iron had also been mentioned by other investigators [25,26], and
the removal of iron from the $MnSO_4$ electrolyte was an important subject
for the EMD manufacturing process, since most rhodochrosite (a $MnCO_3$
mineral) ores contain iron, at least a few per cent as sulfide or carbonate.

   Use of MnO_2 powder for the iron-removal process [by oxidizing Fe^{2+}
to Fe^{3+} and precipitating $Fe(OH)_3$ by raising the pH of the solution] was
described by Yanagihara et al. [35], Matsuno [36], and Creanga [37].
Details of this process were also studied by Muraki [38-40]. Comprehensive
studies on the electrolytic parameters (bath composition, temperature,
current density, additives, and so on) relative to the current efficiency,
cell voltage, and physical and chemical properties of the EMD were made
by Muraki [40-42] with graphite anodes, and more recently by Era et al.
[43-45] with Pb-Sb and graphite anodes, and by Shimizu and Shirahata [46]
using titanium anodes. Similar studies were made by Ogawa et al. [47] with
a lead-alloy anode with emphasis on low-temperature bath, and also by
Tsuruoka et al. [48] with an emphasis on the anode materials (Pb, Pb-Ag,
Pb-Sb, Pb-Ag, Pb-Cd, Pb-Te). Shibasaki [49] studied strains in the EMD
deposits and found that the strain (compression or expansion) depends on the
sulfuric acid concentration in the bath. Deposits that have almost zero
strain were obtained at H_2SO_4 concentrations of 10-20 gm/liter. Codeposition of lead with EMD from a $MnSO_4$ + H_2SO_4 bath containing Pb^{2+} was
investigated by Muraki [50].

Recently, several papers have been published on the deposition mechanism [51-54] and the structure of EMD [55]. Also, a number of papers have been published by Japanese investigators on the electrochemical activity of various manganese dioxides relative to their discharge abilities in Leclanché cells. In these studies, the investigators (Sasaki [56-58], Matsuno [59], Fukuda [60], Hirai [61], Kozawa and Vosburgh [62], Ninagi [63], Ninagi and Miyake [64]) used various EMD samples, some having been heat-treated or chemically modified (autoclaved) prior to testing. Studies of EMD by electron-probe microanalysis, before and after discharge, have been made by Miyazaki [65-72]. Change in electron spin resonance patterns of EMD has been investigated also by Miyazaki [73,74]. Although the contents of these papers [51-74] are interesting, they will not be discussed in this review except in some cases where results are directly related to EMD.

2.4. PURIFICATION OF $MnSO_4$ ELECTROLYTE

The $MnSO_4$ electrolyte prepared by leaching rhodochrosite with hot H_2SO_4 solution is usually purified in the following steps [40]: (1) oxidation of Fe^{2+} ion to Fe^{3+} ion by addition of MnO_2 powder to the solution at $80°\text{-}90°C$ while bubbling air through the solution; (2) neutralization of the solution by $CaCO_3$ in order to precipitate $Fe(OH)_3$ from the solution; (3) filtration of the precipitate, and (4) further purification by the addition of CaS_x or BaS if heavy metal impurities are present. The final solution is adjusted to approximately 136 gm/liter $MnSO_4$ + 20 gm/liter H_2SO_4.

2.4.1. Oxidation of Fe^{2+} Ion

Three kinds of MnO_2 powder (EMD and two natural ores) were tested by Muraki [38-40] for the Fe^{2+} oxidation in acidified $MnSO_4$ solutions ($MnSO_4$: 0-1.1 M) containing 0.2 M $FeSO_4$. The results are shown in Table 5.

TABLE 5

Fe^{2+} Oxidation by MnO_2 [a]

Experiment No.	Kind of MnO_2 [c]	$MnSO_4$ (M)	H_2SO_4 (M)	Fe^{3+} conc. [b] (M)	Fe^{3+} oxidized (%)
1	EMD	1.10	0.424	0.105	52.5
2	EMD	0.55	0.424	0.119	59.5
3	EMD	0.0	0.424	0.117	58.5
4	N_1	1.10	0.424	0.087	43.5
5	N_2	1.10	0.424	0.070	35.0
6	EMD	(1.4– 1.6)	0.053	0.042	20.8
7	EMD		0.106	0.063	31.6
8	EMD		0.159	0.159	41.2
9	EMD		0.265	0.265	51.7
10	EMD		0.424	0.424	53.1

[a] 100 ml of 0.2 M $FeSO_4$ solution containing $MnSO_4$ and H_2SO_4 were mixed with 2.0 gm of MnO_2 powder at 80°C and stirred for 2 h [38,40].
[b] Fe^{3+} concentration after 2 h.
[c] EMD MnO_2 content, 87.2%. N_1 and N_2, natural ores for battery; MnO_2 contents, 75% and 82% respectively.

In these experiments, a 100-ml sample of $MnSO_4$ solution at 80°C was mixed with 2.0 gm of the MnO_2 powder (75% passed through a 200-mesh sieve) and stirred for 2 h at 80°C. The Fe^{2+} concentration was then determined by titration. The following observations were made.

a. The oxidation rate depended upon the kind of MnO_2 being used, with EMD being the best (fastest) oxidizer. Generally speaking, the order of the Fe^{2+} oxidation rates of the MnO_2 samples was quite parallel to the order of their depolarizing activity in dry cells. The oxidation rates increased with decreasing MnO_2 particle size.

b. The $MnSO_4$ concentration had very little effect on the Fe^{2+} oxidation rate (compare experiments 1, 2, and 3 in Table 5). The oxidation rate increased with increasing H_2SO_4 concentration up to around 0.265 M, but then the rate increased very little upon further increases in H_2SO_4 concentration (experiments 6 through 10 in Table 5).

c. The temperature effect on the oxidation rate is seen in Fig. 40. The oxidation rate increased with increasing temperature.

2.4.2. Neutralization and State of Precipitate

Muraki [39,40] studied details of the neutralization process. Solutions containing $MnSO_4$, $Fe_2(SO_4)_3$, and H_2SO_4 were neutralized by NaOH, $MnCO_3$, or $CaCO_3$ at 15° or 60°C. During the course of neutralization, the pH value and light transparency of the solution were measured. The $Fe(OH)_3$ precipitate suspended in the solution at various stages under various neutralization

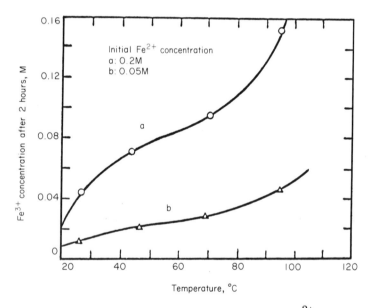

FIG. 40. Effect of temperature on the oxidation of Fe^{2+}. A 100-ml sample of 1.1 M $MnSO_4$ solution containing $FeSO_4$ (0.2 or 0.05 M) and H_2SO_4 (H_2SO_4/Fe^{2+} = 1.6 in molar ratio) was mixed with 2.0 gm of EMD and stirred for 2 h at constant temperature [38,40].

conditions was examined under a microscope. The results were as follows:

a. Regardless of whether $MnSO_4$ was present or not, precipitation of $Fe(OH)_3$ began at pH 2.8 to 3.2 and was completed at pH 3.4 to 3.8. Precipitation of $Mn(OH)_2$ began at pH 5.3 to 7.3 and was completed at pH 9.2 to 9.8.

b. Figure 41 shows the neutralization curves produced by $CaCO_3$ or $MnCO_3$ at 60°C. The beginning and ending points of the precipitation are marked on the curves. The end point was determined not only by the light-transparency measurement but also by Fe^{3+} detection with a KSCN reagent.

c. The state of the $Fe(OH)_3$ precipitate, as observed under a microscope, varies widely, depending on the initial Fe^{3+} ion concentration, temperature of the solution, kind of neutralizing agent (NaOH, $MnCO_3$, or $CaCO_3$), and so on. The $Fe(OH)_3$ precipitate produced by $CaCO_3$ at pH 4.6-5.6 at 60°C was the best from the standpoint of easy and fast filtration.

2.5. CONDUCTIVITY OF $MnSO_4$ ELECTROLYTE

The cell voltage depends on the conductivity of its electrolyte ($MnSO_4$ + H_2SO_4 solution), as well as on the electrode material of the anode and cathode,

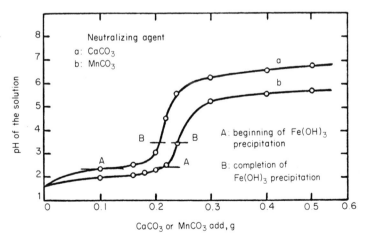

FIG. 41. The pH change and precipitation of $Fe(OH)_3$. A solution containing 0.1 M $MnSO_4$, 0.025 M Fe(III)-sulfate, and H_2SO_4 was neutralized by adding $CaCO_3$ or $MnCO_3$ powder at 60°C [39,40].

and on the distance between the two electrodes. The conductivity values of
the $MnSO_4 + H_2SO_4$ solutions were measured with an ac bridge at 1000 Hz
by Muraki [40,41]. His results are shown in Figs. 42 and 43. It is interesting to note in Fig. 42 that the conductivity decreases with increasing
$MnSO_4$ concentration when the H_2SO_4 concentration is 38 gm/liter or higher.
It should also be noted (in Fig. 43) that the conductivity increases with increasing H_2SO_4 concentration, and that this increase is much greater at
higher temperatures.

Cell voltages under comparable conditions are summarized in Table 6.
It should be noted that cell voltages are significantly higher when lead
anodes are used. Zirconium also seems to be a good cathode material, as
will be mentioned later.

2.6. ELECTRODE MATERIALS

Three anode materials currently in use are graphite, lead alloy (Pb-Sb),
and titanium. Brief comments concerning the advantages and disadvantages

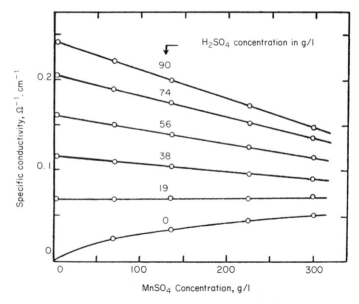

FIG. 42. Specific conductivity of $MnSO_4 + H_2SO_4$ solution at 20°C [41].

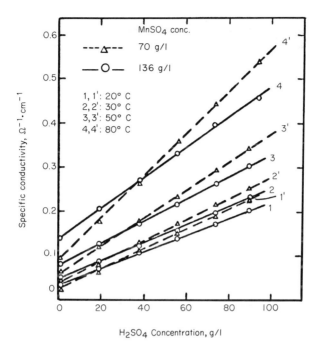

FIG. 43. Specific conductivity of $MnSO_4 + H_2SO_4$ solution at $20°$-$80°C$ [41].

of each anode material are given here.

2.6.1. Graphite Anode

Graphite anodes are used most frequently for research as well as for production. The advantage of this anode is that a relatively high current density (1.5-3.0 A/dm^2) can be applied without passivation or excessive overpotential (Fig. 44) [41]. It is interesting to note in Fig. 44 that the cell voltage increases with increasing electrolyte concentration (see curves 1, 2, and 3). It should be noted that during a long period of electrolysis, the cell voltage increases gradually with a graphite anode, while with a titanium anode, the cell voltage decreases after passing through a maximum (see Fig. 45, which shows pilot-plant data with a 370 x 420 x 3 mm size

TABLE 6

Cell Voltages

Anode material	Cathode material	Cell voltage (V)	Conditions under which cell voltage was measured
1 Graphite	Pb-Sb (7%)	2.15	At 1.0 A/dm^2 in 1 M MnSO$_4$ + 0.5 M H$_2$SO$_4$ at 90° ± 2°C [46]
2 Graphite	Zr	2.15	
3 Titanium	Pb-Sb (7%)	2.20	Cell voltage after a 5-h electrolysis
4 Titanium	Pb	2.20	
5 Titanium	Zr	2.21	
6 Graphite	Pb	2.53	At 1.0 A/dm^2 in 0.9 M MnSO$_4$ + 0.2 M H$_2$SO$_4$ at 82° ± 3°C [40,41]
		2.61	At 1.0 A/dm^2 in 1.49 M MnSO$_4$ + 0.3 M H$_2$SO$_4$ at 82° ± 3°C [40,41]
7 Pb-Sb (8%)	Graphite	2.80	At 0.5-1.25 A/dm^2 in 0.5-0.8 M MnSO$_4$ + 0.15-0.6 M H$_2$SO$_4$ at 80-95°C [45]
Pb-Sb (8%)	Pb-Sb (8%)	2.90	
Graphite	Graphite	2.50-2.63	

titanium electrode [46]). The disadvantages of the graphite electrode are
(a) the electrode is not only slowly corroded electrochemically, but it is
also subject to some wear during the deposit-removal process; (b) the
mechanical shock usually applied to remove the deposit frequently produces
a crack and damages the anode. However, graphite anodes (25-32 mm thick)
recently produced for the EMD process possess almost twice the strength

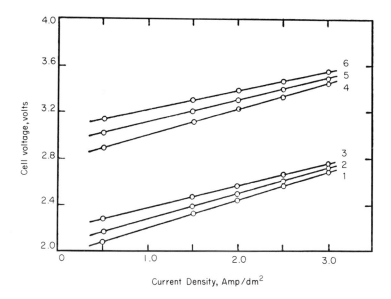

FIG. 44. Cell voltage versus current density [40,41]. Materials: anode, graphite; cathode, pure lead; bath temperature, $83° \pm 3°C$. Bath composition is as follows:

Curve	$MnSO_4$ (gm/liter)	H_2SO_4 (gm/liter)
1	136	20
2	226	30
3	300	45
4	226	0
5	136	0
6	70	0

of the graphite anodes used for brine electrolysis [75]. Thus the new anodes can be used repeatedly (up to 16-18 times on the average) when deposition is carried out for several days on each cycle; also, very thick layers can be deposited on these anodes (for example, a 30-day continuous deposition at about $1.0 \ A/dm^2$ is possible). The life of the graphite anode

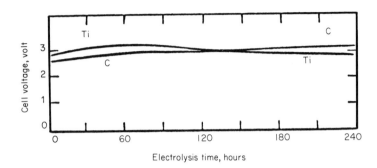

FIG. 45. Cell voltage during 10-day electrolysis [46]. Ti, Ti anode and Pb-Sb (7%) cathode; C, graphite anode and Pb-Sb (7%) cathode. A pilot plant operated at 94°-98°C at 0.9 A/dm^2 with 0.88 M MnSO$_4$ + 0.02M H$_2$SO$_4$ as inlet solution.

is short if excessively high current densities are used or if the electrolysis is carried out at relatively low temperatures [40].

2.6.2. Lead Anode

Pure lead is too soft to be used as anode for the EMD process. Usually, the so-called hard lead, Pb-Sb (5-8%) alloy, is used. An advantage of this electrode is that this anode material can be reused by repeatedly recasting the old electrode. Whether or not the small amount of lead derived from the lead-alloy anode is harmful to dry cells is subject to discussion.

2.6.3. Titanium Anode

Use of titanium anodes in the EMD process has been extensively investigated by Shimizu and Shirahata [46]. The main problems associated with this electrode were poor adhesion of the deposits, and passivation (cell voltage increases to several volts at or above a certain current density, see Fig. 46). Shimizu and Shirahata [46] experimented on this anode with

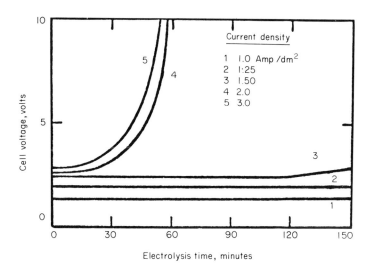

FIG. 46. Cell voltage with time [46]. Titanium anode and lead alloy cathode in 1 M $MnSO_4$ + 0.5 M H_2SO_4 at 90° ± 2°C.

a number of ideas (perforation of the titanium sheet to various degrees, various chemical treatments to remove the oxide film, cylindrical or semicircular shaped electrodes to reduce strain during deposition) and found that a sandblasting treatment gave satisfactory results for the EMD process. When the titanium anode was once properly sandblasted [76], no passivation took place at 1.0-1.2 A/dm^2, the deposit was adherent, and the anode could be used many years without retreatment. A titanium anode is advantageous in that it can be used for years without special care, and it has great resistance to mechanical shock so that the deposit-removal process can be easily automated. The disadvantages are the high initial cost and the careful control needed for the current density and acid concentration.

2.6.4. Cathode Materials

The cathode for the EMD process is usually made of lead, lead alloy (Pb-Sb), or graphite. It was reported [46] that the lead or lead alloy cathodes lose weight during use; this loss was reported to be 0.004-0.1 gm/dm^2 at 1.0 A/dm^2 for 5 h in 1 M $MnSO_4$ containing 0.5 M H_2SO_4. Zirconium was reported to be an excellent cathode material for the EMD process, since there is no corrosion and the cell voltage is the same as that of lead or lead-alloy cathode (see Table 6).

2.6.5. Contact Resistance to Titanium

When a titanium anode is employed, the contact resistance between the titanium and other metals that supply current must be sufficiently small. Figure 47 shows contact resistance values for a few combinations [46]. On the basis of these data, Shimizu suggested that a titanium-coated copper

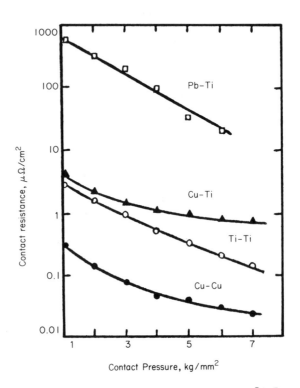

FIG. 47. Contact resistance to titanium [46].

bus bar should be used rather than a lead-coated bus bar, since the Ti-Ti contact resistance is much smaller than that of the Pb-Ti contact.

2.7. ELECTROLYTIC PARAMETERS AND PROPERTIES OF EMD

2.7.1. Current Efficiency versus Current Density

The current-efficiency values obtained by Muraki [41] are shown in Fig. 48, and are based on net MnO_2 determined by chemical analysis. A number of investigators studied current efficiency versus current density: Van Arsdal [25] (101 gm/liter $MnSO_4$ + 150 gm/liter H_2SO_4; 70°-75°C), Nichols [26] (50-150 gm/liter $MnSO_4$ + 0-67 gm/liter H_2SO_4; 70°-75°C),

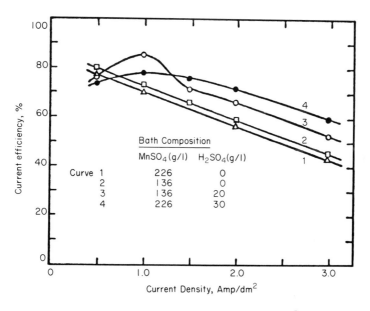

FIG. 48. Current efficiency versus current density [41]. The data were obtained with a graphite anode and a pure lead cathode in the bath composition shown in the figure at 83° ± 3°C.

and Berg [77] (137.5 gm/liter $MnSO_4$ + 150 gm/liter H_2SO_4; 93°C). All
of these investigators have shown that there is a maximum current efficiency
at approximately 1.0 A/dm^2. Takahashi [30] (148-174 gm/liter $MnSO_4$ +
49 gm/liter H_2SO_4; 80°C) and Chakrabarti [78] (200 gm/liter $MnSO_4$ + 50
gm/liter H_2SO_4; 70°-80°C) have shown a gradual decrease in current
efficiency with increasing current density. There is a general agreement
among the workers that the maximum current efficiency is at approximately
1.0 A/dm^2. The effect of $MnSO_4$ concentration is relatively small in the
concentration range of 70-226 gm/liter [41].

2.7.2. Effect of H_2SO_4 Concentration

Figure 49 shows the effect of the molar ratio of $H_2SO_4/MnSO_4$ on
current efficiency [41]. Since the effect of $MnSO_4$ concentration is small,

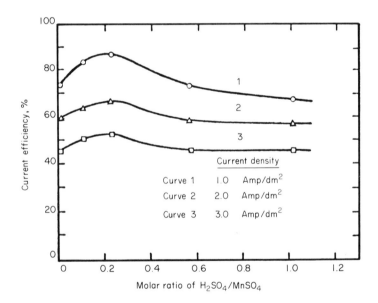

FIG. 49. Effect of H_2SO_4 concentration on current efficiency. Bath
composition: $MnSO_4$, 136 gm/liter; H_2SO_4, 10-90 gm/liter. Electrolytic
conditions are the same as those shown in Fig. 48 [41].

the results are considered to be the effect of H_2SO_4 concentration. The presence of a maximum for each curve is noted in Fig. 49. At an H_2SO_4 concentration of 90 gm/liter or greater, the electrolyte became red colored because of the presence of Mn^{3+} ion during electrolysis; this ion becomes stable, particularly at high H_2SO_4 concentrations and low temperatures.

2.7.3. Effect of Temperature

Current efficiency values obtained at 1.0 A/dm^2 by various investigators are summarized in Fig. 50 [47]. Although these current efficiencies were not measured under the same conditions, we can still see a general trend; generally speaking, the current-efficiency values approached 90-97% at 1.0 A/dm^2 at 90°C or higher, regardless of which anode material or electrolyte concentration was used. At lower temperatures, however, the current efficiency varies considerably, depending on the conditions and the anode material. It is worth noting that the current efficiency is relatively high (72%) even at 30°C under the conditions for curve A (Fig. 50) in which a pure lead anode was used, although a considerable amount of EMD (54% of the total) did not deposit on the electrode but accumulated at the bottom of the cell. The EMD produced at 30°C under the conditions of curve A was γ-MnO_2 and the MnO_2 content was 89%. The depolarizing activity of the γ-MnO_2 measured by the hydrazine method [79] was approximately the same as that of regular EMD produced at 90°C at 1.0 A/dm^2.

2.7.4. Chemical Properties

Table 7 shows MnO_2 and MnO contents, x values in MnO_x, and Pb and SO_4 contents of various EMD samples produced at various current densities and temperatures. As seen in (a) of Table 7, the MnO_2 contents and the x values increase with decreasing current density at 85°C. However, at 90°C [see (c) of Table 7] there is no such clear trend, probably because the MnO_2 content and the x values were already so high that no further variation could take place.

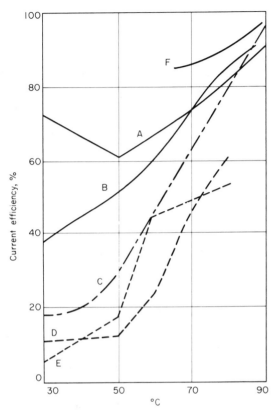

FIG. 50. Current efficiency versus bath temperature at 1.0 A/dm^2 [47]. Conditions for each curve are tabulated below.

Curve	Anode	Cathode	Electrolyte MnSO$_4$	H$_2$SO$_4$	C. eff. based on	Ref.
A	Pb (pure)	Graphite	1.0 M	0	(c)	[47]
B	Graphite	Pb	136 g/l	20 g/l	(b)	[41]
C	Pb-Sb (5%)	Graphite	0.61 M	0.65 M	(a)	[30]
D	Pb (pure)	Stainless steel	0.66 M	0.34 M	(b)	[48]
E	Pb-Sb (6%)	Stainless steel	0.66	0.34	(b)	[48]
F	Ti	Pb-Sb (7%)	1.0 M	0.5 M	(b)	[46]

(a) Based on the deposit being dried at 120°C.

(b) Based on net MnO$_2$ determined by chemical analysis.

(c) Including the MnO$_2$ that fell to the bottom of the cell.

TABLE 7

Chemical Properties of EMD in Relation to Electrolytic Parameters

	Current density (A/dm^2)	Temp. (°C)	MnO_2 (%)	MnO (%)	x in MnO_x	Pb or SO_4 content (%)	Comments
(a)	0.5	85	90.4	2.7	1.96	Pb: 0.005	(a, b) Deposited from a bath
	1.0	85	89.7	2.4	1.97	0.007	containing 136 gm/liter
	1.5	85	88.7	3.4	1.95	0.035	$MnSO_4$ + 20 gm/liter H_2SO_4
	2.0	85	88.4	4.1	1.94	0.098	in a lead-lined steel cell,
	3.0	85	87.3	5.1	1.93	0.128	using a graphite anode and
(b)	1.0	27	64.2	9.5	1.84	–	Pb cathode [41, 42].
	1.0	39	67.8	9.3	1.95	–	
	1.0	50	80.5	5.6	1.92	–	
	1.0	62	86.4	4.5	1.94	–	
	1.0	74	88.7	3.7	1.95	–	
	1.0	83	89.7	2.4	1.97	–	

TABLE 7 (Cont.)

Chemical Properties of EMD in Relation to Electrolytic Parameters

	Current density (A/dm^2)	Temp. $(°C)$	MnO_2 $(\%)$	MnO $(\%)$	x in MnO_x	Pb or SO_4 content $(\%)$	Comments
(c)	0.4	90	91.40	1.67	1.975	–	(c, d) Deposited from a bath
	0.6	90	91.27	1.53	1.979	–	containing 0.8 M $MnSO_4$
	0.8	90	93.03	0.24	1.996	–	+ 0.3 M H_2SO_4 using a Ti
	1.0	90	90.33	3.32	1.956	–	anode and Pb–Sb (7%) cathode
							[46].
(d)	0.90	70	83.0	–	–	–	
	0.90	80	87.1	–	–	–	
	0.90	90	90.3	–	–	–	
(e)	0.5	80–90	–	–	–	SO_4: 1.04	(e) Deposited at a graphite
	0.75	80–90	–	–	–	0.77	anode from a bath containing
	1.0	80–90	–	–	–	0.72	40 gm/liter $MnSO_4$ + 80
	1.5	80–90	–	–	–	0.63	gm/liter H_2SO_4 [44].

As shown in (b) and (d) of Table 7, the MnO_2 content increases with increasing bath temperature. The Pb content increases and the SO_4 content decreases with increasing current density. As will be discussed later, the Pb becomes incorporated in the EMD by codeposition. However, SO_4 in the MnO_2 may be merely trapped in the fine pores of the MnO_2 and/or adsorbed on the surface of the pore wall.

2.7.5. Physical Properties

Table 8 shows some physical properties of EMD prepared under various conditions. The BET surface area increases with increasing current density, whereas the density decreases with increasing current density. Figures 51 and 52 show surface areas of EMD prepared over a wide range of temperatures and current densities [80]. The surface area was determined by the zinc-ion adsorption method [81], which gives results almost parallel to the BET method. As seen in (3) of Table 8, the amount of water driven off at 110°C (mostly physically adsorbed water) [82] increases with increasing current density and decreasing bath temperature; this too is quite parallel to the BET surface area. However, the bound water (weight loss between 110° and 500°C) remains almost unchanged.

Figure 53 shows the shape of the x-ray diffraction peak around 28° (= 2 θ) that is characteristic of γ-MnO_2 (EMD) [83]. We can see a general tendency for the peak to become sharper and higher when the current density is lower and the bath temperature higher. Also, there is a correlation between the BET surface area and the height of the peak, as shown in Fig. 54 [44].

2.7.6. Crystal Structure

Manganese dioxide deposited anodically from $MnSO_4$ solution usually has a so-called γ-MnO_2 structure, but under certain conditions β-MnO_2 is produced together with the γ-MnO_2. Figure 55, given by Era [45], indicates a frequency of formation of γ (γ-MnO_2 only), $\gamma(\beta)$ (γ-MnO_2 probably containing some β-MnO_2), and $\gamma \cdot \beta$ (both γ-MnO_2 and β-MnO_2).

TABLE 8

Physical Properties of EMD in Relation to Electrolytic Parameters

1. Density

Current density (A/dm²)	0.5	1.0	1.5	2.0	3.0
Density of EMD[a] (gm/cm³)	4.30	4.31	4.21	3.58	3.32

2. BET surface area of EMD deposited at 80°–95°C[b] [44]

Current density (A/dm²)	0.5	0.75	1.0	1.5	
BET surface area (m²/gm)	29.0 (90°C)	40.4 (85°C)	36.4 (95°C)	51.5 (80°C)	Graphite anode
	36.8 (85°C)	–	30.9 (80°C)	–	
	28.2 (95°C)	–	44.8 (90°C)	51.9 (85°C)	Lead anode
	–		44.9 (85°C)	–	

3. Relation to water content or weight loss on heating [44]

Anode	Deposition conditions[b] Current (A/dm²)	Temp. (°C)	Wt. loss at 110°C (%)	BET surface area (m²/gm)	Wt. loss between 110° and 500°C (%)
(1) Pb–Sb alloy	0.5	95	1.65	29.2	4.7
(2) Pb–Sb alloy	1.0	90	2.27	44.8	4.3
(3) Pb–Sb alloy	1.5	85	2.44	51.9	4.5
(4) Graphite	0.5	90	1.61	29.0	4.5
(5) Graphite	1.0	85	1.66	40.4	4.5
(6) Graphite	1.5	80	2.54	51.5	4.0

[a]Prepared under the conditions shown in (a) of Table 7.

[b]

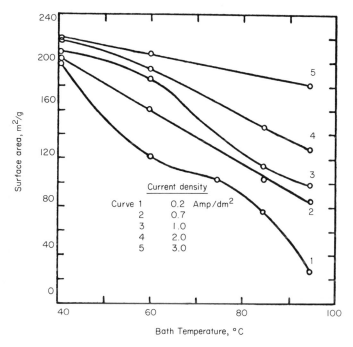

FIG. 51. Surface area of EMD samples prepared under various electrolytic conditions [80]. Bath composition: 1 M $MnSO_4$ + 0.2M H_2SO_4. These EMD samples were prepared by deposition on a platinum anode.

His evaluation criterion for $\gamma\text{-}MnO_2$ is the presence of a diffraction peak around 28° (= 2θ), corresponding to d = 4.0 Å, and that for $\beta\text{-}MnO_2$ is the presence of a peak around 36° (= 2θ), corresponding to d = 3.0 Å. Figure 56 [63] shows the x-ray diffraction diagrams of the structure during the course of the change from $\gamma\text{-}MnO_2$ (EMD) to $\beta\text{-}MnO_2$ upon heating. Extensive x-ray studies of EMD made by Fukuda [84], Nye et al. [85], McMurdie and Golovato [86], Giovanoli et al. [87,88], and Ogawa et al. [47] reported, however, that $\gamma\text{-}MnO_2$ was produced over the entire range of temperature and current density (30° to 90°C; 0.5-3.0 A/dm²) in 1 M $MnSO_4$ without sulfuric acid at a Pb (pure) anode. Storey et al. [27] mentioned formation of $\beta\text{-}MnO_2$ at low temperatures.

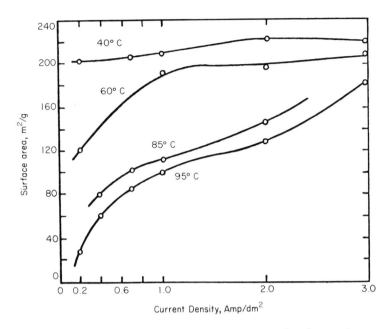

FIG. 52. Surface area of EMD samples prepared under various
electrolytic conditions [80]. Prepared by deposition at a platinum anode
from a bath (1 M MnSO$_4$ + 0. 2 M H$_2$SO$_4$) at the temperatures shown on the
curves.

2.8. DEPOSITION MECHANISM AND RELATED SUBJECTS

Based on recent publications, a brief summary concerning the deposition
mechanism of MnO$_2$ at the anode will be given here to gain some insight
concerning the physical, chemical, and electrochemical properties of EMD
produced under various conditions. Although a few papers [51-53, 89, 90]
have been published on this subject, they are mainly concerned with the
initial state of the deposition at a platinum electrode from dilute solutions
(0. 01-0. 1 M MnSO$_4$) [90] at low temperatures (20°-50°C). The contents
of these papers are not discussed here.

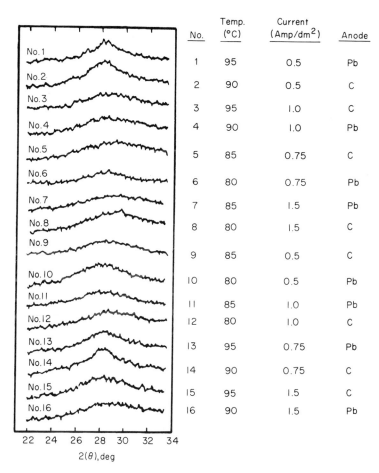

No.	Temp. (°C)	Current (Amp/dm²)	Anode
1	95	0.5	Pb
2	90	0.5	C
3	95	1.0	C
4	90	1.0	Pb
5	85	0.75	C
6	80	0.75	Pb
7	85	1.5	Pb
8	80	1.5	C
9	85	0.5	C
10	80	0.5	Pb
11	85	1.0	Pb
12	80	1.0	C
13	95	0.75	Pb
14	90	0.75	C
15	95	1.5	C
16	90	1.5	Pb

FIG. 53. X-ray diffraction peak corresponding to (110) plane of γ-MnO$_2$ in relation to the electrolytic parameters in 40 gm/liter MnSO$_4$ + 80 gm/liter H$_2$SO$_4$ [44].

2.8.1. Main Anodic Reaction

In the practical industrial process, MnO$_2$ is deposited from 0.5-1.2 M MnSO$_4$ containing 0.2-0.5 M H$_2$SO$_4$ at 90°-98°C with a good current efficiency (85-95%) as a thick MnO$_2$ layer on the anode.

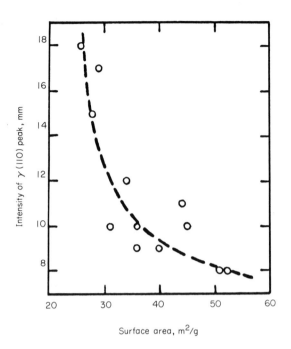

FIG. 54. Surface area versus intensity of the x-ray diffraction peak
[44].

The most likely reaction that takes place under the industrial conditions
is the following one, based on the recent study by Kano et al. [91].

$$Mn^{2+} + 2\ H_2O = MnO_2 + 4\ H^+ + 2\ e^-$$
(12)

The following mechanism, shown in Eqs. (13) and (14), in which Mn^{2+} ion
is first electrochemically oxidized to Mn^{3+} and then the Mn^{3+} ion dispro-
portionates almost instantaneously to Mn^{2+} and MnO_2 was frequently
considered. However, it is not likely that the reaction (13) occurs as a
major reaction, since the actual anode potential during the electrolysis is
far lower than the theoretical value of reaction (13), as will be discussed
below.

$$Mn^{2+} \rightarrow Mn^{3+} + e^-$$
(13)

$$2\ Mn^{3+} \rightarrow Mn^{2+} + MnO_2$$
(14)

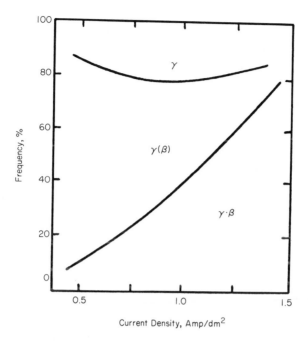

FIG. 55. Crystal structure of EMD produced at various current densities in 40 gm/liter $MnSO_4$, 80 gm/liter H_2SO_4 at a graphite anode at 80°-95°C [45].

The electrochemical reactions that may be involved in the anodic process and their thermodynamic potentials are given in Table 9 [27]. According to these potential values, oxygen evolution can take place along with the reaction (12), but the oxygen evolution should require a considerable over-potential on the MnO_2 surface as well as on the graphite (or PbO_2) surface. This was well demonstrated in Fig. 57 [91]. It should be noted in Fig. 57 that MnO_2 deposition takes place around 0.55 V [very close to the theoret-ical potential shown as reaction (1) of Table 9], whereas, under those conditions, oxygen evolution both on graphite and MnO_2-covered graphite electrodes requires a 350-mV higher potential than the theoretical potential [0.555 V versus Hg/Hg_2SO_4 (1 N H_2SO_4)]. As will be shown in the next section, MnO_2 deposits with good efficiency (95% or so) as long as the

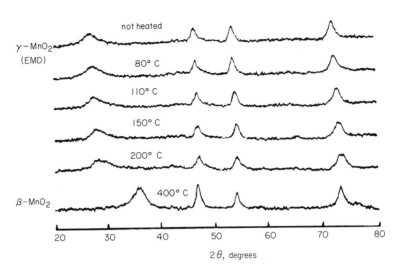

FIG. 56. X-ray diffraction patterns of EMD samples heated in air at various temperatures [63].

electrode potential remains around 0.55 V versus Hg/Hg_2SO_4 (1 N H_2SO_4), according to Kano et al. [91]. At 0.55 V versus Hg/Hg_2SO_4 (1 N H_2SO_4), the oxidation of Mn^{2+} ion to Mn^{3+} is thermodynamically impossible, since that reaction requires at least +0.836 V versus Hg/Hg_2SO_4 (1 N H_2SO_4), as seen in Table 9. Therefore, as long as the current efficiency is high (90% or more) and the anode potential is around 0.55-0.65 V versus Hg/Hg_2SO_4, the major reaction at the anode should be that shown in Eq. (12).

2.8.2. Conditions that Lower the Current Efficiency

Kano et al. [91] investigated the current efficiency of MnO_2 deposition with relation to bath temperature, H_2SO_4 concentration, and $MnSO_4$ concentration, while measuring the anode potential with a Luggin capillary against a Hg/Hg_2SO_4 (1 N H_2SO_4) electrode.

Figures 58-60 show their results. In these experiments, the standard condition was selected as follows:

$$1.0 \text{ M } MnSO_4, \ 75°C$$
$$0.2 \text{ M } H_2SO_4, \ 1.0 \text{ A/dm}^2$$

TABLE 9

Electrochemical Reactions and Their Potentials

Reaction	Potential at 25°C	
	vs. NHE	vs. Hg/Hg_2SO_4 (1 N H_2SO_4)
(1) $Mn^{2+} + 2 H_2O = MnO_2 + 4 H^+ + 2 e^-$	1.23	0.556
(2) $Mn^{2+} = Mn^{3+} + e^-$	1.51	0.836
(3) $Mn^{2+} + 4 H_2O = MnO_4^- + 8 H^+ + 5 e^-$	1.51	0.836
(4) $MnO_2 + 2 H_2O = MnO_4^- + 4 H^+ + 3 e^-$	1.695	1.021
(5) $2 H_2O = O_2 + 4 H^+ + 4 e^-$	1.229	0.555

For each experiment shown in the figures, only one parameter was varied. The current-efficiency values after a 1-h electrolysis were summarized with relation to the anode potential, as shown in Fig. 61. It is clear in this figure that once the anode potential begins to deviate considerably from 0.550-0.650 V versus Hg/Hg_2SO_4 (1 N H_2SO_4), the current efficiency begins to decrease. Such a potential deviation indicates that other electrochemical reactions have begun to take place.

2.8.3. Two Side Reactions

Two side reactions that reduce the current efficiency are Mn^{3+} ion formation and O_2 evolution.

2.8.3.1. $\underline{Mn^{3+} \text{ Ion Formation}}$. When the H_2SO_4 concentration is high and/or the bath temperature is low, the electrolyte becomes reddish in color because of Mn^{3+} ion formation. Mn^{3+} ion formation in $MnSO_4 + H_2SO_4$ solution has been studied by Kano et al. [91] and by Sugimori and Sekine [54] at graphite and platinum anodes, respectively. They believe that MnO_4^- is first produced and that MnO_4^- and Mn^{2+} then react with one another to produce Mn^{3+} ion. In their studies, they measured the absorption spectrum

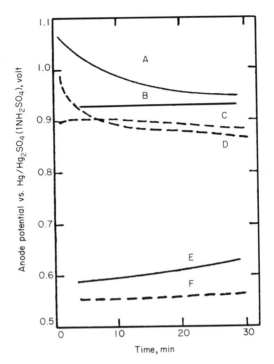

FIG. 57. Anode potential during a 30-min electrolysis at 1.0 A/dm^2 in
0.2 M H_2SO_4 with or without $MnSO_4$ (1.0 M) and 75° and 90°C [91]: A and
D at graphite anode in 0.2M H_2SO_4; B and C at graphite anode on which a
MnO_2 layer was predeposited, in 0.2M H_2SO_4; E and F at graphite anode in
1 M $MnSO_4$ + 0.2M H_2SO_4; solid curves at 75°C, dashed at 90°C.

of the solution. Figure 62 shows the absorption curves of Mn^{2+}, Mn^{3+},
MnO_4^-, and the reddish solution produced by the electrolysis. The Mn^{3+}
concentration cannot be increased beyond a certain value because there is
an equilibrium among the four components

$$Mn^{3+} = Mn^{2+} + MnO_2 + H^+$$

as will be discussed later.

2.8.3.2. Oxygen Evolution. Oxygen evolution is a major side reaction.
Although molecular oxygen can easily oxidize Mn(II) hydroxide to MnO_2 in

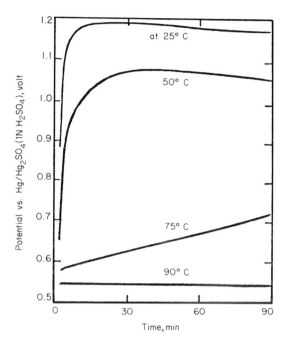

FIG. 58. Effect of temperature on the potential of a graphite anode during a 90-min electrolysis in 1 M $MnSO_4$ + 0.2 M H_2SO_4 at 1.0 A/dm^2 [91].

alkaline media [93], such a reaction does not take place in acid. Percentages of the current consumed for the oxygen evolution are given in Table 10 [91].

2.8.5. <u>Detailed Steps in the Deposition Mechanism</u>

Fleischman et al. [89,90] described more detailed steps possibly involved in the deposition process. Figure 63 shows the proposed scheme [90]. The hydrated Mn(II) ion is electrochemically oxidized to $[Mn(H_2O)n]_{ads}^{3+}$ and/or to $[Mn(H_2O)n]_{ads}^{4+}$, and the absorbed species on the MnO_2 surface are probably different in energy from free Mn^{3+} ion. These species dehydrate successively and finally become a part of the MnO_2 lattice. The dehydration process was considered as the rate-determining step in this process.

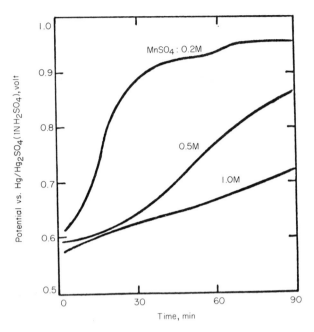

FIG. 59. Effect of MnSO$_4$ concentration on the graphite anode potential during a 90-min electrolysis at 1. 0 A/dm^2 in a 75°C bath containing 0. 2 M H$_2$SO$_4$ [91]. MnSO$_4$ concentration of the bath is shown in the figure.

2.8.5. <u>Equilibrium Among Mn^{3+}, Mn^{2+}, and MnO$_2$</u>

The equilibrium of the following reaction was recently studied by Welsh [92].

$$2\ Mn^{3+} + 2\ H_2O = MnO_2 + Mn^{2+} + 4\ H^+ \qquad (15)$$

According to his results, the reaction seems to attain a true equilibrium from either side; however, the rate from left to right is very slow and very much dependent on the temperature (much faster at higher temperatures), but the rate from right to left is relatively fast. Figures 64 and 65 show the experimentally obtained relation for [Mn^{3+}] versus [Mn^{2+}]$^{1/2}$ and [Mn^{3+}] versus [H$_2$SO$_4$]^2, which are in agreement with the relations (Eqs. 18 and 19) derived from the equilibrium expression (Eq. 16).

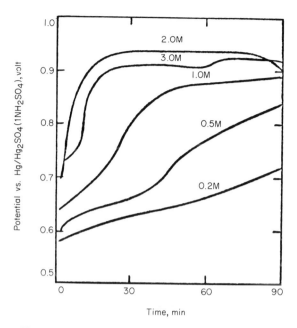

FIG. 60. Change of the graphite anode potential during a 90-min electrolysis in 1 M $MnSO_4$ with varying H_2SO_4 concentration as shown in the figure at 1.0 A/dm^2 at 75°C [91].

$$K = \frac{a_{(MnO_2)} \, a_{(Mn^{2+})} \, a^4_{(H^+)}}{a^2_{(H_2O)} \, a^2_{(Mn^{3+})}} \tag{16}$$

$$K' = \frac{[Mn^{2+}] \, [H^+]^4}{[Mn^{3+}]^2} \tag{17}$$

$$[Mn^{3+}] = K''[Mn^2]^{1/2} \tag{18}$$

$$[Mn^{3+}] = K'''[H^+]^2 \tag{19}$$

Table 11 shows the calculated constant K', which is a reasonably good constant, although activities are not taken into account. Figure 66 shows the change in $[Mn^{3+}]$ concentration with temperature under fixed conditions. Table 11 also gives the change in the K' value with temperature. This

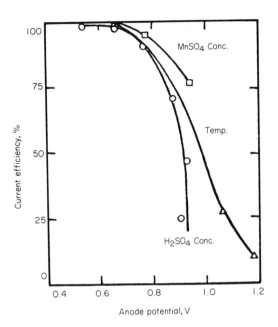

FIG. 61. Current efficiency of MnO_2 deposition in relation to the anode potential with variation in the three bath parameters shown in the figure [91]. This figure was based on the current-efficiency values at various conditions shown in Figs. 58-60.

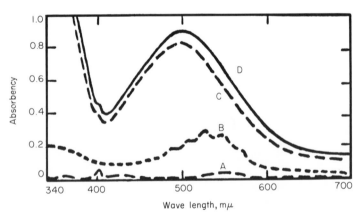

FIG. 62. Absorption spectra of solutions containing various forms of manganese ion [91]: A, 1 M $MnSO_4$ + 3 M H_2SO_4; B, dilute $KMnO_4$ solution; C, solution A after some electrolysis; D, solution A + small amount of $KMnO_4$.

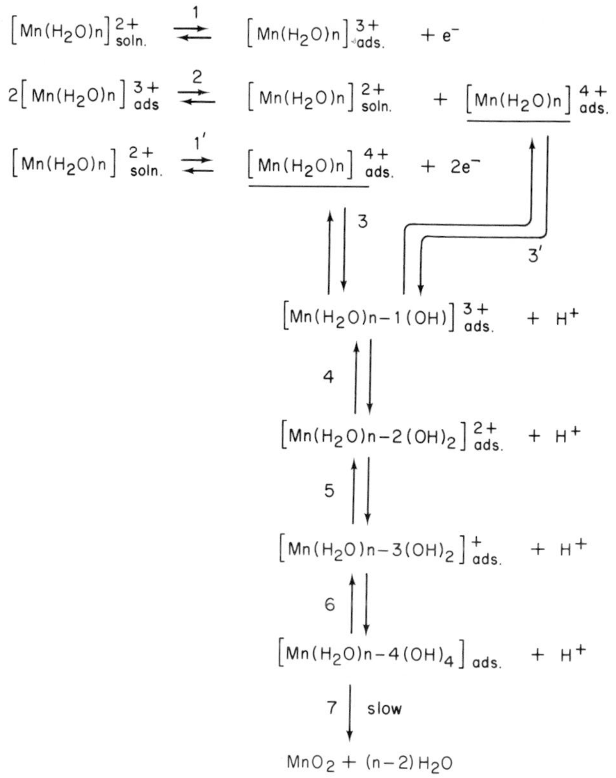

$$[\text{Mn}(\text{H}_2\text{O})n]^{2+}_{soln.} \overset{1}{\rightleftharpoons} [\text{Mn}(\text{H}_2\text{O})n]^{3+}_{ads.} + e^-$$

$$2[\text{Mn}(\text{H}_2\text{O})n]^{3+}_{ads} \overset{2}{\rightleftharpoons} [\text{Mn}(\text{H}_2\text{O})n]^{2+}_{soln.} + \underline{[\text{Mn}(\text{H}_2\text{O})n]^{4+}_{ads.}}$$

$$[\text{Mn}(\text{H}_2\text{O})n]^{2+}_{soln.} \overset{1'}{\rightleftharpoons} \underline{[\text{Mn}(\text{H}_2\text{O})n]^{4+}_{ads.}} + 2e^-$$

3   3'

$$[\text{Mn}(\text{H}_2\text{O})n-1(\text{OH})]^{3+}_{ads.} + \text{H}^+$$

4

$$[\text{Mn}(\text{H}_2\text{O})n-2(\text{OH})_2]^{2+}_{ads.} + \text{H}^+$$

5

$$[\text{Mn}(\text{H}_2\text{O})n-3(\text{OH})_2]^{+}_{ads.} + \text{H}^+$$

6

$$[\text{Mn}(\text{H}_2\text{O})n-4(\text{OH})_4]_{ads.} + \text{H}^+$$

7 | slow

$$\text{MnO}_2 + (n-2)\text{H}_2\text{O}$$

FIG. 63. Hypothetical steps during the MnO_2 deposition [90]: 1 and 1', electrochemical oxidation; 2, disproportionation; 3' and 3-7, dehydration steps.

TABLE 10

Percentage of Current Consumed by Three Reactions

When 1 M MnSO_4 Solution Containing 0.5-3.0 M H_2SO_4 Was Electrolyzed at a Graphite Anode at 1.0 A/dm^2 at 75°C [91]

H_2SO_4 concentration (mole/liter)	Mn^{3+} formation (%)	O_2 evolution[a] (%)	MnO_2 deposited (%)
0.5	0.98	9.33	89.69
1.0	2.03	27.30	70.67
2.0	5.80	47.52	46.68
3.0	15.51	59.82	24.67

[a]There may be some error due to MnO_2 precipitated in the solution.

473

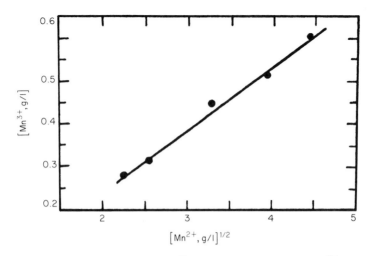

FIG. 64. Relation between $[Mn^{2+}]$ concentration and $[Mn^{3+}]$ concentration at equilibrium. The equilibrium data were obtained from a slurry containing 250 gm/liter of H_2SO_4 and 200 gm/liter of MnO_2 at 15°C [92].

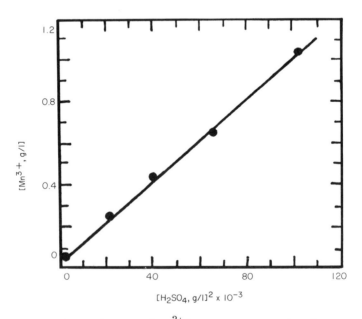

FIG. 65. Relation between $[Mn^{3+}]$ concentration and H_2SO_4 concentration at equilibrium. The equilibrium data were obtained from a slurry containing 20 gm/liter of Mn^{2+} and 200 gm/liter of MnO_2 at 15°C [92].

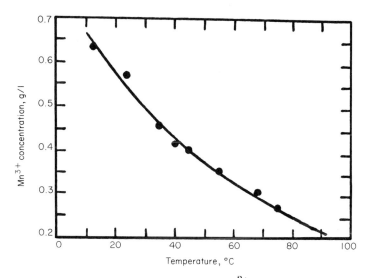

FIG. 66. Equilibrium concentration of Mn^{3+} ion as a function of temperature in a slurry containing 250 gm/liter of H_2SO_4, 20 gm/liter of Mn^{3+} ion (as $MnSO_4$) and 200 gm/liter of MnO_2 [92].

means that the equilibrium shifts to MnO_2 formation (or a decrease in the Mn^{3+} concentration) at higher temperatures. Production of EMD at low temperature, based on the disproportionation of Mn^{3+}, has been described as a continuous process by Welsh [92] and Araki [94].

2.8.6. Physical Properties of EMD

The MnO_2 deposition process can be considered as two competing processes [89]: (a) formation of new nuclei, and (b) growth of the nuclei. Electrolytic manganese dioxides are generally not well crystallized and are highly porous, having a surface area of 40-60 m^2/gm. The surface area of the EMD increases with increasing current density and with decreasing temperature, probably because the rate of the nuclei formation increases at high current densities and low temperatures.

TABLE 11

Values of the Equilibrium Constant[a]

$$K' = [Mn^{2+}][H^+]^4/[Mn^{3+}]^2$$

T (°C)	$[Mn^{2+}]$ (moles/liter)	$[H^+]$ (moles/liter)	$[Mn^{3+}]$ (moles/liter)	K'
15	0.364	5.0	0.0109	1.9×10^6
15	0.263	5.0	0.0091	2.0×10^6
15	0.138	5.0	0.0064	2.1×10^6
15	0.364	7.44	0.0255	1.7×10^6
15	0.364	5.60	0.0145	1.7×10^6
15	0.364	2.40	0.0040	1.5×10^6
				Avg = 1.8×10^6
15	0.364	5.0	0.0111	1.8×10^6
40	0.364	5.0	0.0080	3.6×10^6
60	0.364	5.0	0.0060	6.3×10^6
80	0.364	5.0	0.0045	1.1×10^7
90[b]	0.364	5.0	0.0038	1.6×10^7
100	0.364	5.0	0.0033	2.0×10^7

[a]Reference [92].
[b]Extrapolated Mn^{+3} values.

2.8.7. Codeposition of Pb

Codeposition of Pb in EMD was investigated by Muraki [40, 50]. The solubility of Pb (as $PbSO_4$) was measured in $MnSO_4$ solutions containing 0-90 gm/liter of H_2SO_4 and 0-5 gm/liter of Cl^- ion. The results are shown in Figs. 67 and 68. The solubility of $PbSO_4$ decreases with increasing H_2SO_4 concentrations, but is not much influenced by the $MnSO_4$ concentration. The presence of Cl^- increases the $PbSO_4$ solubility.

Table 12 gives the Pb content of electrolytic MnO_2 deposited under various conditions on graphite electrodes from $MnSO_4 + H_2SO_4$ solution

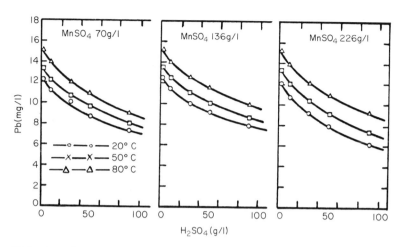

FIG. 67. Solubilities of Pb (as $PbSO_4$) [40,50] in $MnSO_4 + H_2SO_4$ solutions. The $MnSO_4 + H_2SO_4$ solutions were saturated with $PbSO_4$ and the dissolved Pb(II) ion concentration was determined.

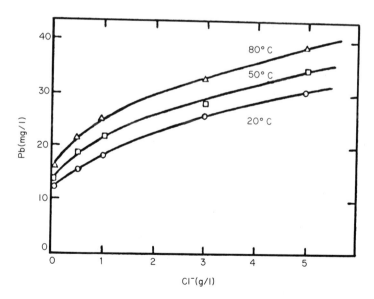

FIG. 68. Solubility of Pb (as $PbSO_4$) in $MnSO_4$ 136 gm/liter solution containing 0-5 gm/liter of Cl^- [40,50].

TABLE 12

Pb Content of EMD Deposited on Carbon Electrodes under Various Conditions
from $PbSO_4$-Saturated $MnSO_4 + H_2SO_4$ Solutions[a]

	$MnSO_4$ (gm/liter)	H_2SO_4 (gm/liter)	Temp. (°C)	Current density (A/dm²)	Average voltage (V)	Current efficiency (%)	Chemical analysis (%)	
							MnO_2	Pb
1	70	10	83	0.5	2.50	68.1	89.7	0.0026
2	136	10	83	0.5	2.45	77.0	90.1	0.0034
3	226	10	83	0.5	2.50	74.8	90.3	0.0031
4	136	0	83	1.0	3.38	73.2	89.5	0.0047
5	136	10	83	1.0	2.73	84.3	88.8	0.0063
6	136	20	83	1.0	2.53	86.7	89.3	0.0074
7	136	50	83	1.0	2.41	75.2	88.5	0.0068
8	136	90	83	1.0	2.38	69.5	87.9	0.0082
9	136	20	27[b]	1.0	3.40	38.2[b]	64.5	0.126
10	136	20	43[b]	1.0	3.08	44.6	66.8	0.043
11	136	20	62	1.0	2.60	62.7	85.7	0.017
12	136	20	83	1.5	2.61	72.5	89.7	0.035
13	136	50	83	1.5	2.63	70.6	88.9	0.033

(continued)

TABLE 12 (Cont.)

Pb Content of EMD Deposited on Carbon Electrodes under Various Conditions from $PbSO_4$-Saturated $MnSO_4 + H_2SO_4$ Solutions[a]

	$MnSO_4$ (gm/liter)	H_2SO_4 (gm/liter)	Temp. (°C)	Current density (A/dm^2)	Average voltage (V)	Current efficiency (%)	Chemical analysis (%)	
							MnO_2	Pb
14	136	20	83	2.0	2.70	67.0	89.1	0.115
15	136	50	83	2.0	2.72	66.5	90.2	0.124
16	136	20	83	3.0	2.86	53.4	88.3	0.138
17	136	50	83	3.0	2.88	52.7	88.7	0.141

[a]References [40, 50].

[b]Not deposited on the electrode, but produced "anode mud," and also the electrolyte become dark red.

saturated with $PbSO_4$ in a lead-lined steel vessel [40,50]. From the
results shown in Table 12, the following factors do not greatly influence
the Pb content: (a) $MnSO_4$ concentration (70-226 gm/liter)--see lines 1 to 3;
(b) H_2SO_4 concentration (0-90 gm/liter)--see lines 4 to 8; (c) temperature
(43°-83°C)--see lines 10 to 12.

The factor that greatly influences the Pb content is current density, as is
shown by the following data from Table 12.

Line	$MnSO_4$ (gm/liter)	H_2SO_4 (gm/liter)	Temp. (°C)	Current density (A/dm^2)	Pb content (%)
6	136	20	83	1.0	0.0074
12	136	20	83	1.5	0.035
14	136	20	83	2.0	0.115
16	136	20	83	3.0	0.138

Other experiments by Muraki [40,50] show that the Pb content changes
within the deposit systematically from the layer next to the lead anode to
the layer in contact with the solution. The Pb content within 9-mm-thick
deposits that were prepared at 1-3 A/dm^2 from the $PbSO_4$-saturated bath
is shown in Fig. 69. The highest Pb content was always in the first layer,
which was in contact with the lead anode.

The reason why the Pb content is high at high current densities and in
the layer next to the lead anode seems to be related to the potential [50].
Theoretical potentials for MnO_2-Mn^{2+} and PbO_2-Pb^{2+} systems are +1.236
V and +1.456 V, respectively. This indicates that the deposition of PbO_2
requires a 220 mV higher voltage than that of MnO_2. Under the normal
electrolysis conditions, which produce MnO_2 at 90-95% current efficiency,
it is not likely that PbO_2 codeposits with MnO_2 at the top layer of the MnO_2
(A in Fig. 69), since the potential is probably below the PbO_2-Pb^{2+} potential.
However, the deposited MnO_2 is porous, and Pb^{2+} ion can slowly diffuse
into the MnO_2 and deposit there because the potential of the interior of the
MnO_2 deposit (B in Fig. 69) should be higher than that of area A. This
explanation is in accordance with the fact that the higher the current density,
the greater is the Pb content in the EMD, since the potential should be

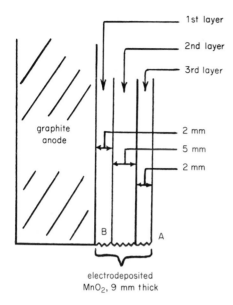

FIG. 69. Lead distribution in the deposited MnO_2 layer [50]. The effect of current density is shown below.

Current density (A/dm^2)	1st layer (%)	2nd layer (%)	3rd layer (%)	average (%)
1	0.021	0.0042	0.0006	0.0074
2	0.363	0.033	0.0068	0.115
3	0.437	0.041	0.0088	0.138

higher at higher current, and this makes the PbO_2 deposition more favorable.

2.9. ELECTROCHEMICAL PROPERTIES OF EMD

Electrochemical properties of EMD samples prepared under various electrolytic conditions have been studied by Muraki and Okajima [42] and

Era et al. [43]. The results obtained by these two investigators are
generally in good agreement; namely, that the most important factors
influencing the electrochemical characteristics (potential, overpotential,
and discharge capacity) of EMD are the current density and the bath
temperature in the electrodeposition process.

2.9.1. The Electrode Potential

The electrode potential of EMD decreases with increasing current
density and also with decreasing bath temperature, as is shown in Table 13.
It should be noted (in Table 7) that the higher the bath temperatures and
the lower the current densities, the higher were the MnO_2 contents and

TABLE 13

Potentials of EMD or EMD-Containing Cells

The EMD Samples Were Prepared at Various Current Densities and

Temperatures as Shown in Footnotes.

Current[a] density (A/dm^2)	Cell[b] voltage	Bath[c] temp. (°C)	Potential[d] vs. SCE
0.5	1.74	80	0.645
1.0	1.74	85	0.658
1.5	1.72	90	0.668
2.0	1.70	95	0.678
3.0	1.68		

[a]EMD was deposited from a solution containing 136 gm/liter $MnSO_4$ +
20 gm/liter H_2SO_4 at 85°C [42].
[b]Tested in C-size Leclanché cells in which EMD samples made at
current densities from 0.5 to 3.0 A/dm^2 were used.
[c]EMD was deposited from a solution containing 40 gm/liter $MnSO_4$ +
80 gm/liter H_2SO_4.
[d]The electrode was prepared by mixing 0.5 gm EMD and 0.2 gm
acetylene black and immersing in 25% NH_4Cl + 10% $ZnCl_2$ (pH 5.0).

x values of MnO_x. The fact that the EMD potential increases with the increasingly higher x value is very reasonable from the standpoint of the potential-generation mechanism (a one-phase system: Mn^{4+}-Mn^{3+}-O^{2-}-OH^-), as explained previously [95]. If the oxide system is a two-phase (or heterogeneous) system (such as Ag_2O-Ag or HgO-Hg), the potential (OCV) should not depend on the purity or MnO_2 content, as is the case for Ag_2O-Ag or HgO-Hg systems.

2.9.2. Overpotential on Discharge

Overpotentials during the discharge of various EMD samples in 25% NH_4Cl + 10% $ZnCl_2$ (pH 5.0) were measured by Era et al. [43] at 20 mA/500 mg of EMD. The overpotential, which is a potential difference between the open-circuit potential and that of the potential during discharge, increases with increasing current density and decreasing bath temperature in the deposition process. The EMD samples made with a Pb anode had a lower overpotential than those made with a graphite anode under comparable conditions [43].

Despite the fact that the EMD samples had lower surface areas (see Table 8 and Figs. 51 and 52) when made at lower current densities, they had lower overpotentials. This fact indicates that surface area is not an important factor in the reaction at the 20 mA/500 mg discharge rate. This can be understood if we are reminded of the nature of the mechanism of the discharge reaction. These overpotentials [43] were measured at an early stage of discharge, and the main electrode reaction at that stage is the following one [96].

$$MnO_2 + H_2O + e^- \rightarrow MnOOH + OH^- \tag{20}$$

The rate-determining step is very likely the proton-diffusion in the MnO_2 lattice, and the electrochemical reaction at the solid-solution interface is not the limiting factor of the reaction rate.

Overpotentials of variously deposited EMD samples were also measured in 0.1 M H_2SO_4 + 0.1 M $MnSO_4$ solutions by Kozawa and Vosburgh [62]. They found that the overpotentials of the manganese dioxides increase with increasing bath temperature and also with decreasing current density during the deposition process, as shown in Fig. 70. The results, as shown in this figure, are quite opposite from those overpotentials measured in 25% NH_4Cl + 10% $ZnCl_2$ (pH 5.0); the main point in this case is that overpotentials are greater whenever EMD samples are prepared under conditions that give lower surface areas. The dependence of overpotential (Fig. 70) on surface area can best be understood from the nature of the electrode reaction in a 0.1 M H_2SO_4 + 1.0 M $MnSO_4$ solution:

$$MnO_2 + 4 H^+ + 2 e^- \rightarrow Mn^{2+} + H_2O \tag{21}$$

In this reaction, Mn^{4+} at the surface of the MnO_2 is reduced to Mn^{2+} and is dissolved into the solution; therefore, the reaction rate should depend on the surface area of the EMD. In other words, the greater the surface area, the smaller should be the overpotential.

In the interpretations given above of the overpotential values of EMD samples produced under various temperatures and current densities,

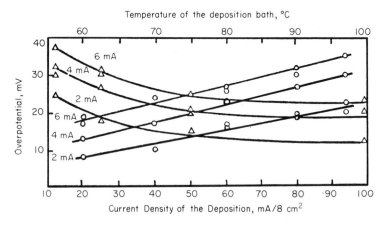

FIG. 70. Overpotentials measured during discharges at 2, 4, and 6 mA/0.2 mmole MnO_2 in 0.1 M H_2SO_4 + 0.1 M $MnSO_4$ at 23°C [62]. The straight lines and the upper scale of abscissa represent MnO_2 deposited at various temperatures and at 25 mA per 8 cm^2. The curves and the lower scale of the abscissa represent MnO_2 made at 90°C and various current densities.

discussion concerning the concentration polarization (pH change in the fine pores) was neglected for simplification. Although it has been pointed out from time to time [61,97] that this type of polarization is undoubtedly important, since the EMD samples contain a large number of fine pores (in the range of 100-200 Å), this factor will not be discussed further here because it is beyond the scope of this writing.

2.9.3. Discharge Capacity

The discharge curves shown in Fig. 71 were obtained by Muraki and Okajima [42] using C-size Leclanché cells made with EMD samples produced at various current densities and cells made with natural MnO_2 ores. We can see that the lower the current density in the deposition, the higher is the discharge capacity. Era et al. [43] have shown that the discharge capacity of EMD samples in Leclanché electrolyte becomes greater with increasing bath temperature.

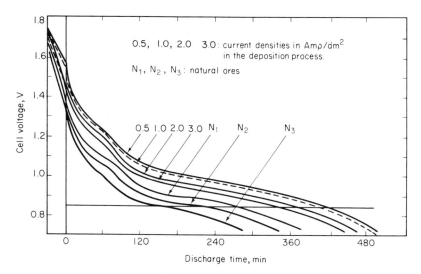

FIG. 71. Discharge curves of C-size cells through 4 Ω resistance at 20°C. The cells contained EMD made at various current densities and natural MnO_2 ores [42].

2.9.4. Effect of Foreign Ions

According to Era et al. [43], up to 0.5 M of Na^+, Mg^{2+}, and Al^{3+} ions added to the electrolyte bath did not modify the electrochemical character-istics of the EMD produced therein. Kozawa and Vosburgh [62] examined the effects of K^+, NH_4^+, Zn^{2+}, Mg^{2+}, and Al^{3+} in the deposition bath. They found that, in the presence of 0.01-1.0 M K^+ or NH_4^+ in the bath (50 gm/liter $MnSO_4$ + 65 gm/liter H_2SO_4), the deposited MnO_2 contained in the range of 4.6-8.3 moles of K^+ or NH_4^+ per 100 gm-atoms of MnO_2; this is very close to a typical cryptomelane in which the K^+ or NH_4^+ content is 8.5%, as shown in Table 14. The MnO_2 containing K^+ or NH_4^+ exhibited poor discharge characteristics when compared with MnO_2 deposited in the absence of NH_4^+ or K^+ (see Fig. 72). The presence of Zn^{2+}, Mg^{2+}, and Al^{3+} in the bath did not modify the discharge characteristics of the EMD samples.

TABLE 14

NH_4^+ or K^+ Content of Electrodeposited MnO_2 from a Bath
(50 gm/liter $MnSO_4$, 65 gm/liter H_2SO_4) Containing $(NH_4)_2SO_4$ or K_2SO_4[a]

$(NH_4)_2SO_4$ or K_2SO_4 (M)	NH_4^+ in MnO_2 (mg/electrode)	K^+ in MnO_2 (mg/electrode)	Moles/100 gm-atoms Mn	
			NH_4^+	K^+
0	0.01	0.06	–	–
0.01	0.10	0.36	3.6	4.6
0.05	0.18	0.47	6.5	6.0
0.10	0.20	0.62[b]	7.3	7.9
0.30	0.22	0.62	7.8	7.9
0.5	–	0.65	–	8.3
1.0	0.24	–	8.6	–

[a]The MnO_2 was deposited at 90°C at 0.3 A/dm^2.
[b]Average of 3 electrodes: 0.62, 0.63, 0.61.

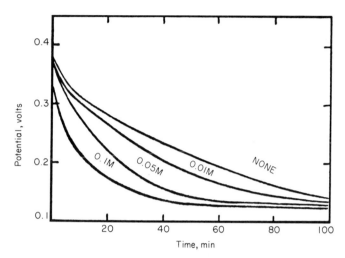

FIG. 72. Discharge curves of MnO_2 deposited from a bath (50 gm/liter $MnSO_4$ + 60 gm/liter H_2SO_4) containing $(NH_4)_2SO_4$ (0-0.1 M as shown in the figure) at 90°C and at 25 mA/8 cm^2 for 30 min [62]. The 0.2 mmole of MnO_2 deposited on the 8-cm^2 graphite rod was discharged at 2 mA in 2 M $NH_4Cl + NH_3$ (pH 7.8) at 23°C.

2.9.5. Heat-Treated EMD Samples

A number of investigators [57,59,62,63,98] studied heat treatment of EMD in relation to discharge characteristics. In general, heat treatment of EMD in air increases the polarization in the initial discharge stage; but under certain conditions during complete discharge, the discharge capacity becomes greater than that of the original unheated EMD. Ninagi's [63,64] results are shown in Fig. 73. The electrode potentials of heated MnO_2 samples were measured in 0.5 M NH_4Cl solution by Sasaki and Kozawa [58], as shown in Fig. 74. Here, it should be noted that the potential did not decrease much until the temperature reached 250°C, but above 250°C it decreased continuously. It is interesting to note that there is no abrupt change in the potential around 450°C where MnO_2 decomposes to Mn_2O_3.

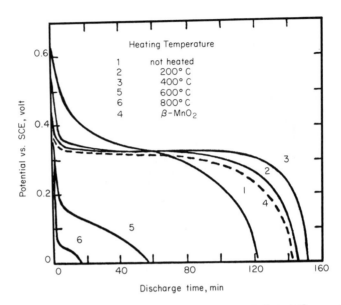

Discharge time, min

FIG. 73. Discharge curves of heat-treated EMD [63, 64] samples.
EMD (γ-MnO$_2$) was heat-treated at the temperatures shown in the figure.
Each MnO$_2$ sample (200 mg) was mixed with 40 mg acetylene black and
discharged at 40 mA continuously in a vessel similar to that used by
Cahoon [97] in NH$_4$Cl + ZnCl$_2$ + H$_2$O (25:10:100 in weight ratio) solution.

Muraki [98] showed that properly heated EMD does have greater capacity
under certain conditions in practical dry batteries than does the unheated
EMD.

3. ANALYTICAL METHODS

3.1. DETERMINATION OF Mn, MnO$_2$, AND x IN MnO$_x$ BY CHEMICAL ANALYSIS

The procedure given below has been thoroughly tested, and reproducible
values have been obtained for various MnO$_2$ samples, including the standard

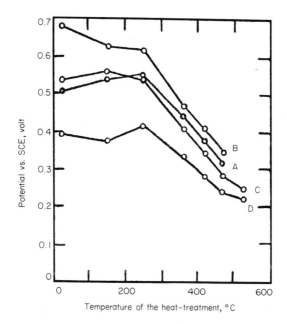

FIG. 74. Change in the electrode potential of MnO_2 as a result of
heat-treatment in air [58]. The potentials were measured in NH_4Cl
(pH 6-7). A, EMD, neutralized; B, EMD, not neutralized; C and D,
natural MnO_2 ores for dry cells.

sample 25b (U.S. National Bureau of Standards) [99]. This method is
based on the potentiometric titration of Mn published by Lingane and Karplus
[100]. The improved procedure eliminates the error due to oxidation of
Mn^{2+} to Mn^{3+} during the neutralization process. After determining the
available oxygen by the ferrous sulfate method, the Mn^{2+} solution is neutralized with $Na_4P_2O_7 \cdot 10 H_2O$ crystals rather than NaOH solution [99].

3.1.1. Procedure

    Transfer a sample (200 $\pm$ 0.5 mg) into a 125-ml Erlenmeyer flask, and
add 25 ml of 0.25 M $FeSO_4$ (containing 3.6 N H_2SO_4) solution. Stir the

mixture with a magnetic stirrer until the MnO_2 sample dissolves completely
at room temperature (it usually takes several minutes). Titrate the solution
with 0.1 N $KMnO_4$ standard solution (buret reading: a ml). Transfer the
whole solution into a 250-ml beaker and add 35-40 gm of reagent-grade
sodium pyrophosphate ($Na_4P_2O_7 \cdot 10\ H_2O$) to neutralize the H_2SO_4 and to
bring the pH value to between 6 and 7. Titrate the solution potentiometrically
with the same 0.1 N $KMnO_4$ standard solution as that used for the available-
oxygen determination (buret reading: a' ml). The potential jump of the
indicator electrode (Pt electrode versus SCE) at the end point takes place
around +0.4 to + 0.5 V.

It is also necessary to determine the number of ml of 0.1 N $KMnO_4$
solution consumed when 25 ml of the unreacted $FeSO_4$ solution are titrated
(buret reading: b ml).

The essential reactions involved in the titrations are:

$$MnO_2\ (Mn^{4+}) + 2\ Fe^{2+} \rightarrow Mn^{2+} + 2\ Fe^{3+}$$

$$5\ Fe^{2+} + Mn^{7+}\ (KMnO_4) \rightarrow 5\ Fe^{3+} + Mn^{2+}$$

$$4\ Mn^{2+} + Mn^{7+}\ (KMnO_4) + 15\ H_2P_2O_7^{2-} \rightarrow 5\ Mn^{3+}\ (\text{pyrophosphate complex})$$

The following calculations can be performed using the titration values
(a, a', b) established before (N is the normality of the $KMnO_4$ solution).

$$\text{mg O in 200-mg sample} = 8N(b - a)$$
$$\text{mg } MnO_2 \text{ in 200-mg sample} = 86.94N(b - a)$$
$$\%\ MnO_2 = 43.47N(b - a)$$
$$\text{mg Mn in 200-mg sample} = 54.94(a' - \frac{a}{4})\frac{N}{5}\ (4)$$
$$\%\ Mn = 27.47(a' - \frac{a}{4})\frac{N}{5}(4)$$
$$\text{x in } MnO_x = 1 + [(b - a)/(a' - \frac{a}{4})\frac{4}{5}]$$

Note: When a sample contains some chloride (for example, a Leclanché
cell cathode), $HgSO_4$ should be added before the first titration in order to
mask the chloride ions. Otherwise, the chloride will interfere with the
$KMnO_4$ titration. The amount of $HgSO_4$ depends on the amount of chloride
in the sample. Addition of 1 gm of $HgSO_4$ should be sufficient in most
cases.

3.2. DETERMINATION OF SURFACE AREA BY THE ZINC-ION ADSORPTION METHOD (ZIA METHOD)

3.2.1. Outline of the Method and Comparison of the Data with the BET Method

Various solid powder materials (SiO_2, Al_2O_3, MnO_2, TiO_2, $BaCO_3$, $NaBiO_3$, $MnPO_4$, and so on) adsorb metal ions when they are shaken with a salt solution. For example, when MnO_2 or SiO_2 powder is shaken with a zinc solution (2 M NH_4Cl containing 0.05-0.1 M ZnO) or a copper salt solution, Zn^{2+} or Cu^{2+} ions are adsorbed and H^+ ions are released. The mole ratio of the released H^+ to the adsorbed Zn^{2+} depends on the material and the surface condition, but the ratio is often very close to a whole integer (1, 2, or 3). In the case of MnO_2, the value is close to 2, and in the case of SiO_2, close to 3. The adsorption reaction is a kind of ion-exchange reaction, as shown below. In the ion-exchange reaction, the hydrated oxide surface acts like a chelating agent (bidentate or tridentate). The details of the adsorption reaction have been investigated by the author [101, 102] and the surface chelation has been proposed as a new adsorption mechanism.

$$
MnO_2 \left\{ \begin{array}{l} -OH \\ \\ -OH \end{array} \right. + Zn^{2+} \ (Cl^- \ \text{complex}) \rightarrow MnO_2 \left\{ \begin{array}{l} -O \\ \\ -O \end{array} \right. \!\!\! Zn \!\!\! \begin{array}{l} Cl \\ \\ Cl \end{array} + 2\,H^+ + 2\,Cl^-
$$

$$
SiO_2 \left\{ \begin{array}{l} -OH \\ -OH \\ -OH \end{array} \right. + Zn^{2+} \ (Cl^- \ \text{complex}) \rightarrow SiO_2 \left\{ \begin{array}{l} -O \\ -O{-}Zn - Cl \\ -O \end{array} \right. + 3\,H^+ + 3\,Cl^-
$$

(Hydrated oxide surface)   (Surface chelates)

It was pointed out previously [101] that under proper conditions, the amount of zinc ion adsorbed ($Zn^{2+}_{ads.}$) is proportional to the surface area of the solid powder (MnO_2, SiO_2, ZrO_2, and so on); therefore, the surface area can be determined by measuring the decrease in the zinc-ion concentration of the

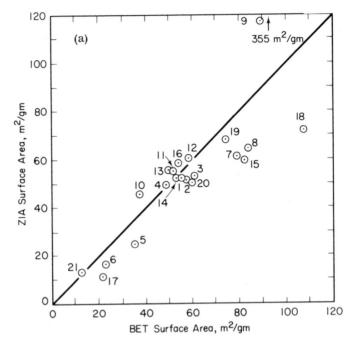

FIG. 75. Surface-area values measured by BET and ZIA methods for
MnO_2 samples No. 1-21 (See Table 15 for description of the samples). In
the BET method, out-gassing was carried out at 80°C under vacuum, overnight. (a) In the ZIA method, each sample (0.40 gm) was shaken with 5.0
cm^3 of 2 M NH_4Cl + 0.11 M ZnO at 75°C for 17 h (overnight). (b) In the
ZIA method, each sample was shaken with the zinc solution at 65°C for 1 h.

solution by simple EDTA titration. This method is suitable for measuring
a large number of samples [103].

The ZIA method has been reinvestigated with various battery-grade
manganese dioxide samples, including ten international common samples
[104]. The results are shown in Fig. 75 and Table 15. The experimental
details will be given in the next section. The surface-area values obtained
by this method depend on the stirring conditions (temperature and time)
because battery-grade manganese dioxides have high porosities and high
surface areas. When samples of this kind were shaken with the prescribed

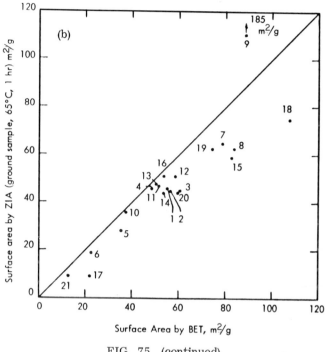

FIG. 75. (continued).

zinc solution overnight at 75°C, the ZIA surface-area values obtained were almost the same as the BET values (see points 1-4, 10-14, and 16 in Fig. 75a). When ground samples were shaken with the zinc solution at 65°C for one hour only, the ZIA surface area values are approximately 20% lower than the BET surface areas, but still the ZIA surface-area values have a good proportionality to the BET values, as shown in Fig. 75b. The dashed line represents the average ZIA values, indicating the deviation from the BET values. When unground samples were shaken at 23°C for 17 h (overnight), the results were practically identical to those obtained with 1 h shaking at 65°C using ground samples (see Fig. 76).

The ZIA method is simple, can process a large number of samples at a time, and eliminates sample heating or vacuum-degassing required by the BET method. Heating may modify the surface area considerably; for

TABLE 15

Surface Areas Determined by ZIA and BET Methods

MnO$_2$ sample	BET method (m^2/gm)	ZIA Method[a]			
		Unground sample		Ground sample	
		75°C, 18 h (m^2/gm)	23°C, 17 h (m^2/gm)	23°C, 1 h (m^2/gm)	65°C, 1 h (m^2/gm)
1. Electrolytic MnO$_2$ for dry cell (<40 μm)	55.1	51.9	40.2	42.0	45.5
2. Electrolytic MnO$_2$ for dry cell (50–100 μm)	56.5	51.6	38.7	39.7	44.9
3. Electrolytic MnO$_2$ for dry cell (150–300 μm)	60.6	53.2	42.8	41.2	44.6
4. Electrolytic MnO$_2$ for dry cell	48.5	49.0	39.5	39.4	45.5
5. No. 4 heated at 200°C in air for 10 days	35.4	24.9	28.0	26.2	28.0
6. No. 4 heated at 400°C in air for 10 days	22.7	16.3	16.3	17.8	18.4
7. Chemically prepared MnO$_2$ (1) for dry cell	78.6	61.3	64.5	64.4	64.4
8. Chemically prepared MnO$_2$ (2) for dry cell	83.7	64.2	67.4	64.0	62.2
9. Activated MnO$_2$[b]	88.8	355.0	175.0	169.0	185.0

10. Electrolytic MnO$_2$ for dry cell	37.3	45.5	31.5	33.2	35.9
11. Electrolytic MnO$_2$ (fine particles)	51.6	54.9	43.8	42.9	46.3
12. I.C.S.[c] No. 1 (EMD made with Ti anode)	58.5	60.4	49.8	46.4	50.8
13. I.C.S. No. 2 (EMD made with Pb anode)	50.1	55.1	44.6	42.8	47.3
14. I.C.S. No. 3 (EMD made with C anode)	53.5	52.0	40.2	37.6	43.8
15. I.C.S. No. 5 (chemical MnO$_2$)	82.5	59.5	64.8	59.5	58.6
16. I.C.S. No. 4 (EMD)	53.9	58.1	49.0	44.6	50.8
17. I.C.S. No. 7 (natural ore, Ghana)	22.0	11.2	8.8	7.9	8.8
18. I.C.S. No. 8 (chemical MnO$_2$)	107.5	72.3	83.3	76.2	74.5
19. I.C.S. No. 9 (EMD, coarse grade)	74.1	67.9	60.4	56.9	62.1
20. I.C.S. No. 10 (EMD)	59.9	50.1	42.8	38.5	43.8
21. I.C.S. No. 11 (chemical MnO$_2$)	12.4	12.8	3.5	6.1	8.8

[a]Recommended procedure uses samples of 400 mg of MnO$_2$ and 5 cm^3 of 2 M NH$_4$Cl + 0.117 M ZnO. (23°C, 17 h) means the sample was shaken in a rotator at 23°C for 17 h.

[b]Special MnO$_2$ for oxidation process of organic compounds (used as a catalyst) obtained from Special Chemical Department of Winthrop Lab., 90 Park Ave., New York, N.Y. 10016.

[c]International Common Sample (I.C.S.) obtained from The International Common Sample Office, P.O. Box 6116, Cleveland, Ohio 44101.

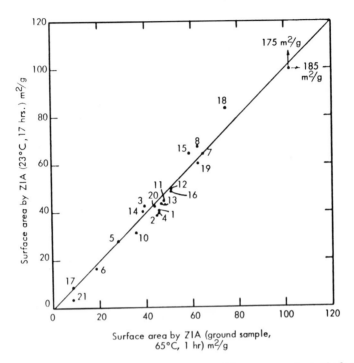

FIG. 76. Correlation of surface areas. Measurements include two variations in shaking conditions of the ZIA method for MnO_2 samples Nos. 1-21 shown in Table 15.

example, the surface-area value for sample No. 9 (Table 15) (which is made for special purposes and has an extremely large surface area) is much lower when obtained by the BET than by the ZIA method. For such a sample, the ZIA method is the only method that can give a reliable result, since the measurement can be done in an aqueous solution at 23°C.

ZIA surface-area values may also be more meaningful for characterizing an extremely porous material in respect to its electrochemical reactivity.

Since 20 to 30 samples can be handled in a day after rotating them over-night in the same rotator, the ZIA method is believed to be best suited for measuring the relative surface area of a large number of oxides manufactured in the same process: for instance, various samples of electrolytic MnO_2 produced for dry-cell use or various TiO_2 materials to be used for paint manufacture. The absolute surface area obtained by the ZIA method may not be accurate, but the relative surface area or the difference in the surface areas should be very reliable and useful.

3.2.2. Recommended Procedure

Battery-grade manganese dioxide samples have relatively large surface areas (20-80 m^2/gm). The content of metallic impurities, which often disturb the EDTA titration, is usually very small. The procedure described here is applicable to various oxides (MnO_2, SiO_2, ZrO_2, TiO_2, Al_2O_3, and others), but particularly suitable for battery-grade MnO_2.

3.2.2.1. General Procedure. Transfer 0.4 gm of a battery-grade MnO_2 sample into a glass tube (Fig. 77) and add 5.00 cm^3 of the zinc solution (2 M NH_4Cl + 0.11 M ZnO). Insert a stopper (with some grease), secure the stopper with a rubber band, and place in a rotator (Fig. 78). Rotate the tubes slowly at 23°C (or 75°C) overnight or shake at 65°C for 1-3 h. Centrifuge the tubes and transfer 2.00 ml of the clear solution to a 50-ml beaker. Add 5.00 ml of the buffer solution and two drops of Eriochrom Black T indicator. Titrate with 0.02 M EDTA solution (buret reading: a ml). Titrate 2.00 ml of the original zinc solution with 0.02 M EDTA solution using the same amount of buffer and indicator as above (buret reading: b ml).

The surface area is calculated using the following equation:

$$\frac{0.02(b - a) \times 5}{0.4 \times 1000 \times 2} \left[\frac{14}{10^{-4}}\right] = 17.5(b - a) \ m^2/gm$$

This equation is based on the specific zinc adsorption value, 0.7×10^{-5} mole/m^2, which was established previously [101]. The above procedure is for samples having the surface area of 30-60 m^2/gm. The amount of sample to be used depends on the surface area of the sample (see later part of this section). The shaking conditions (rotation time and temperature) depend on the purpose. If a surface-area value close to the BET area is desired, the shaking (or rotation) time should be about 18 h at 75°C. If the relative surface-area values among a number of samples are the only interest, any of the following conditions can be used: (a) 17 h (overnight) at 23°C; (b) 1 h at 65°C; (c) 3 h at 75°C. The relative surface-area values obtained by different shaking conditions should depend on the pore size, particle size, and other conditions related to the porous structure.

3.2.2.2. Reagents

a. Zinc solution (2 M NH_4Cl + 0.11 M ZnO). Reagent grade NH_4Cl (53.94 x 2 = 107 gm) was dissolved in about 900 cm^3 of preheated distilled water (45°-50°C) and then finally diluted to a 2.0 M solution. The NH_4Cl solution was heated to 50°-55°C in a beaker with a cover (watch-glass), and 9.0 gm of ZnO powder was added. The solution was kept stirred at 60°-65°C for 15-30 min in order to dissolve the ZnO completely. The solution was then allowed to cool and stored in a stoppered jar or bottle overnight at room temperature. Some white material, probably zinc hydroxide and other metal hydroxides, precipitated out to the bottom. The clear portion was used without disturbing the bottom sludge. The clear portion can be transferred to another bottle by decantating or pipeting. For example, decanting a U-tube through the side arm will prevent disturbing the bottom solution containing the white material. Filtration to remove the bottom sludge is not recommended because the sludge passes through most filter papers. The zinc concentration is checked by EDTA titration and should be 0.11 to 0.12 M.

b. 0.02 M EDTA standard solution. This solution was prepared by dissolving 33.62 gm of disodium salt into 1 liter of distilled water.

c. Eriochrom Black T solution. This is an indicator solution and is prepared by dissolving 0.2 gm of Eriochrom Black T powder into a mixture of 25 ml of methyl alcohol and 75 ml of triethanolamine.

d. Buffer solution. This buffer solution was prepared by dissolving 54 gm of NH_4Cl and 350 ml of concentrated ammonia solution (28-30% NH_3) into 1 liter of distilled water.

e. Cleaning solution. In order to remove residues of manganese dioxide samples from glassware used in the ZIA method, a solution containing 0.5-1.0 M H_2SO_4 and 2-3% H_2O_2 is recommended. MnO_2 powder reacts rapidly with this solution [105].

3. 2. 2. 3. Apparatus

a. A glass tube with a ground glass stopper (Fig. 77). The bottom of
the tube should be round, and the volume should be 6-6.5 cm^3. Application
of small amounts of high-temperature grease to the ground joint is recommended to avoid water evaporation during the overnight rotation at 75°C.
At room temperature, practically no water loss was observed during 17 h.
A strong rubber band is used to secure the stopper. The grease must be
removed before refilling the glass tube with the next sample; suitable
solvents or a KOH solution (for silicones) can be used.

b. Centrifuge. Any laboratory centrifuge can be used, since a high
rotation speed is not required.

c. 5 cm^3 and 2 cm^3 pipets and 10 ml buret.

d. Rotating machine. The glass tubes containing the sample powder
and the zinc solution must be rotated effectively in order to attain adsorption
equilibrium as fast as possible. For this purpose, a rotator (as shown in
Fig. 78) is recommended. Slow rotation (8-10 rpm) is desirable, so that
the rising air effectively stirs the entire mixture.

Addition of a few glass balls (2-4 mm diam) in the glass tube helps to
achieve efficient mixing when the rotational speed is faster than 20 rpm.

3. 2. 2. 4. Comments and Suggestions

a. Amount of sample. The amount of sample to be used depends on the
surface area. In order to obtain an accurate buret reading, the value (b - a)

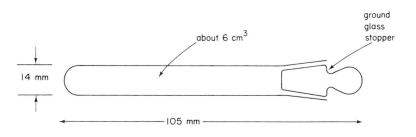

FIG. 77. Glass tube.

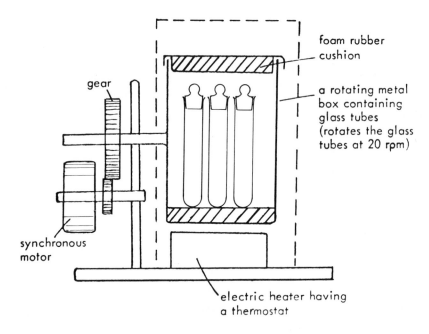

FIG. 78.   Rotator for stirring.

should be at least 2.0 cm^3 to keep the error within a few per cent. Using 0.01 M EDTA instead of 0.02 M EDTA is not recommended because the color change by one drop (0.02-0.03 cm^3) is not sharp. The amount of sample required for the procedure and the equations to be used for the surface-area calculation are given below.

Range of surface area (m^2/gm)	Amount of sample to be used (mg)	Equations to be used (m^2/gm)
60–200	200	35(b – a)
30–60	400	17.5(b – a)
10–30	800	8.75(b – a)

b. <u>Sample grinding.</u> If the average particle size of a MnO_2 sample is 30-50 μm or less, grinding is not necessary. If a sample is 100-200 μm or greater, grinding may be performed before the measurement. Since battery-grade manganese dioxide is porous and brittle, a 500-mg sample having a particle size of 70-200 μm can be ground to 10 μm or less within 5 min by hand grinding in a ceramic mortar with a pestle. It is worth mentioning that some electrolytic MnO_2 samples have a considerable amount of closed pores, and the surface area will increase considerably because the grinding will open the closed pores. For example, a certain EMD sample having an average particle size of 30 μm was ground to 5-8 μm size and the surface area increased by 5-7 m^2/gm. Samples having few or no closed pores and with a surface area greater than 10 m^2/gm may be ground to finer particles without changing their surface areas considerably. The reasoning is made clear by the following geometrical calculation: If a nonporous cube of solid material of 1-cm^3 volume is ground into 10^9 cubes with 10 μm sides (600 μm^2 surface, the surface area increased from 6 cm^2 to 0.6 m^2, representing only a minor addition to any surface area of 10 m^2, as found above. However, the smaller particles will reach the adsorption equilibrium faster.

c. <u>Prewashing for metal impurities.</u> Most battery-grade MnO_2 samples do not require washing. However, small metallic impurities may disturb the titration color change (end point). Even such small amounts of Cu as may come from the water-distillation apparatus will disturb the EDTA titration. MnO_2 samples containing CaO, MgO, and so on, and those having various metal ions adsorbed on the surface by an ion-exchange reaction may be washed with 0.1-0.3 M H_2SO_4 (at 40°-50°C) and subsequently treated with 1-2 M NH_4Cl solution (pH 2-3). Washing fine powder material with distilled water often presents a problem, because it forms a colloidal system and is difficult to filter or centrifuge. In order to avoid such a problem, the use of NH_4Cl solution instead of water for washing samples is helpful, since it seems to act as a coagulant.

d. <u>Reference sample.</u> A reference sample that has a known ZIA surface area should be included in a series of samples. The result of this control test will determine whether the conditions (shaking, and so on) are satisfactory.

4. APPENDIX

4.1. DATA ON BATTERY-GRADE MANGANESE DIOXIDE

Today's battery-grade electrolytic MnO_2 is inexpensive (approximately 35-40¢/kg) and has the following physical properties.

a. Electrical resistivity: 50 to 100 Ω · cm (measured with powder sample in a tablet form under high pressure)
b. BET surface area: 40-50 m^2/gm
c. Pore size: 40-60 Å diam
d. Pore volume (<150 Å pores): 0.032-0.05 cm^3/gm
e. Density: true density: 4.0-4.3 gm/cm^3; apparent (tapping) density: 2.2-2.3 gm/cm^3

Typical EMD products: Table 16 shows a series of typical battery-grade EMDs currently on the market. The particle-size distribution of these samples is shown in Fig. 79 and Table 17.

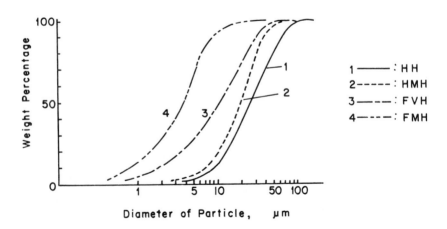

FIG. 79. Particle-size distribution of battery-grade EMD. (Courtesy of Tekkosha Co. Ltd.)

TABLE 16

Examples of Battery-Grade Manganese Dioxide[a]

| | Composition (wt%) | | | |
	Sample 1	Sample 2	Sample 3	Sample 4
MnO_2	92.00	91.69	91.97	91.84
Total Mn	60.12	60.10	59.93	59.67
H_2O	1.49	1.64	1.63	1.29
Fe	0.008	0.008	0.008	0.011
Pb	0.0007	0.0007	0.0007	0.0006
Cu	0.0003	0.0003	0.0003	0.0003
Ni	0.0005	0.0005	0.0005	0.0005
Co	0.0008	0.0008	0.0008	0.0008
Cr	0.0008	0.0008	0.0008	0.0008
Sb	0.0001	0.0001	0.0001	0.0001
As	0.0003	0.0003	0.0003	0.0003
Mo	0.0002	0.0002	0.0002	0.0002
Na	0.24	0.24	0.24	0.24
NH_3	0.001	0.001	0.001	0.001
SiO_2	0.02	0.02	0.02	0.02
SO_4^{-2}	0.79	0.79	0.79	0.80
PH	6.70	6.70	6.70	6.50
BET surface area	$3.2 \text{ m}^2/\text{gm}$	$53.0 \text{ m}^2/\text{gm}$	$56.3 \text{ m}^2/\text{gm}$	$60.2 \text{ m}^2/\text{gm}$
X-ray structure	$\gamma\text{-}MnO_2$	$\gamma\text{-}MnO_2$	$\gamma\text{-}MnO_2$	$\gamma\text{-}MnO_2$
Potential at pH 6	0.800 V	0.786 V	0.776 V	0.760 V

[a]These MnO_2 samples were produced commercially for dry-cell use by an electrolytic method. Samples 1, 2, 3, and 4 correspond to Tekkosha's HH, HMH, FVH, and FMH respectively (see Table 17).

TABLE 17

Particle-Size Distribution of Battery-Grade EMD

Range	HH[a]	HMH	FVH	FMH
(μm)	(wt%)	(wt%)	(wt%)	(wt%)
>74	7.6	–	–	–
74–61	7.4	–	–	–
61–44	11.9	3.6	2.2	–
44–30	16.7	22.2	11.4	0.1
30–20	18.9	23.0	15.0	0.2
20–10	23.6	32.1	24.1	6.7
<10	13.9	19.1	47.3	93.0

[a]HH, HMH, FVH, and FMH correspond to samples 1-4 in Table 16.

4.2. INTERNATIONAL COMMON SAMPLES (ICS)*

4.2.1. Purpose of the Common Sample

Manganese dioxide has been used in the Leclanché cells since 1866. Despite a number of research papers published, the physical and chemical properties of manganese dioxide required or desirable for the battery cathode have not been well established. The electrochemical behavior of manganese dioxide in the battery is influenced by crystal structure, surface area, pore-size distribution, shape and size of the particle, electrical conductivity, surface condition, chemical composition, various impurities (H_2O and foreign elements) in the structure, and various defects in the structure. Reproduction of identical manganese dioxide samples does not appear to be a simple task, at least within the near future, even though the

*Written by Akiya Kozawa and R. A. Powers, Union Carbide Corp., Consumer Products Division, P.O. Box 6116, Cleveland, Ohio 44101, U.S.A. in January 1967, revised January 1970 and September 1973.

conditions of the preparation are thoroughly specified in a written form. At the present status of our knowledge, it is impossible to reach unanimous agreement on standard samples of MnO_2 for electrochemical activity. Under these circumstances, it seems necessary to have international common samples as a control or a reference in order to study, evaluate, and compare various manganese dioxides for battery use.

Technical people in battery manufacturing and in manganese dioxide production need to compare the electrochemical behavior and the physical and chemical properties of various manganese dioxides. Data obtained by one laboratory (or data described in one paper) can only be reasonably compared to those obtained by another laboratory (or to those described in another paper) if the common samples are included in these two sets of data.

Suppose someone measures a certain property of manganese dioxides. If he includes an international common sample in addition to the samples prepared in his laboratory, others who have been working on manganese dioxide in industry and other universities can evaluate the new results more easily and properly.

Once research results (publications) on various manganese dioxides are accumulated for comparison with data on the common samples, this knowledge will be more useful and valuable than information published without reference to such common samples. Therefore, all people who are doing research on manganese dioxide have a need for such common samples, not only for confirmation of their analytical methods for determining the chemical and physical properties, but also as a reference sample for the measurement of unknown properties such as electrochemical activity for battery cathodes.

4.2.2. Preparation and Distribution

The number of common samples is not limited. Hopefully, if an appropriate new type of manganese dioxide is introduced to the market, the supplier will offer it to the International Common Sample Office under the conditions specified below. The new material will be added to the list,

and subscribers to the Bulletin of ICS-MnO$_2$ (Bulletin of the International
Common Samples of Manganese Dioxide) will be notified.

4.2.2.1. Conditions for Adding a Common Sample

a. Common samples are usually donated by manganese dioxide manufacturers. Only under special circumstances will samples be purchased by
the ICS Office. The ICS Office will assign a specific number (ICS No. XX)
for each sample. A sample should be prepared in at least a 200 kg batch
and homogeneously mixed by the best method. From this homogeneous
sample, at least 200 bottles containing 1.0 kg each should be prepared.

b. Each bottle containing approximately 1.0 kg should be sealed with
a plastic tape or other appropriate method, and the bottle labeled with the
number (ICS No. XX) assigned by the ICS Office.

c. Each sample should be prepared from a single lot of manganese
dioxide. In other words, the sample should not be prepared by intentionally
mixing different kinds or different batches of manganese dioxide.

d. The particle-size distribution should be of a normal battery grade,
unless otherwise specified (for example, ICS No. 9).

4.2.2.2. Storage and Distribution

a. The sample will be stored at room temperature ($23° \pm 2°C$) at the
ICS Office. If any accidental changes in environment occur during storage,
they will be noted in the bulletin.

b. The ICS Office will distribute the samples at an appropriate price
upon request. The price of the common sample will include expenses
incurred for publication of the Bulletin of ICS-MnO$_2$.

4.2.3. Bulletin of ICS-MnO$_2$

The Bulletin will be edited and distributed from time to time as deemed
appropriate by the ICS Office. This bulletin will contain a list of samples

available at that date, a list of publications that have information on the
international common sample, and also appropriate technical data contributed by the supplier as well as from those who have purchased the samples.

This bulletin will be distributed to all persons or companies who have
purchased a common sample.

The address of the ICS Office is

> The ICS Office
> P.O. Box 6116
> Cleveland, Ohio 44101
> U.S.A.

The Bulletin will be edited at the ICS Office, with the help of the following advisory group: Dr. G. S. Bell, Dr. J. P. Brenet, Dr. R. Huber,
Dr. J. F. Laurent, Mr. R. L. Orban, Dr. R. A. Powers, Dr. G. Schneider,
Dr. K. Takahashi, Dr. F. L. Tye, the Chairman of the Cleveland Section
of the Electrochemical Society, and the Chairman of the Battery Division
of the Electrochemical Society. The location of the office, the editorship
of the Bulletin, and the membership of the advisory group are subject to
change.

Neither the Electrochemical Society nor the ICS Distribution Center
sponsored by the Cleveland Section of the Electrochemical Society and the
Battery Division of the Electrochemical Society can be held accountable for
the technical information and behavioral characteristics of samples distributed.

4.2.4. Data on Samples

Table 18 is a list of International Common Samples. Table 19 gives
technical data on these samples.

TABLE 18

List of International Common MnO_2 Samples (Sept. 1, 1973)

ICS No.	Nature of sample	Source	Quantity per bottle
1	Electrolytic MnO_2 made by Ti-anode process	Japan	~ 1 kg
2	Electrolytic MnO_2 made by Pb-anode process	Japan	~ 1 kg
3	Electrolytic MnO_2 made by C-anode process	Japan	~ 1 kg
4	Electrolytic MnO_2	LaPile Leclanché (France)	~ 1 kg
5	Chemically prepared MnO_2	Japan Metals & Chemicals Co.	~ 1 kg
6	Natural ore	(Not ready for distribution)	
7	Natural MnO_2 ore	Ghana Ore	~ 1 kg
8	Chemically prepared MnO_2	Sedema, Belgium	~ 1 kg
9	Electrolytic MnO_2 (coarse grade)	Kerr-McGee Corp., U.S.	~ 1 kg
10	Electrolytic MnO_2	Kerr-McGee Corp., U.S.	~ 1 kg
11	Chemical MnO_2 made by chlorate process	Kerr-McGee Corp., U.S.	~ 100 gm

TABLE 19

Some Data on International Common Samples[a]

ICS No.	Mn^b (%)	MnO_2^b (%)	H_2O (%)	Density (g/cc)	Surface Area (1) (m^2/g)	(2) (m^2/g)
1	60.6	91.2	1.2	4.46	58.5	50.8
2	60.1	91.8	2.2	4.47	50.1	47.3
3	60.0	90.8	1.7	4.41	53.5	43.8
4	59.6	90.4	3.1	4.44	53.9	50.8
5	61.1	92.7	1.5	4.76	82.5	58.6
7	53.4	81.2	0.6	4.25	22.0	8.8
8	60.9	90.6	1.0	4.70	70.5	74.5
9	60.3	91.7	3.5	4.50	74.1	62.1
10	59.2	89.7	2.0	4.42	59.9	43.8
11	60.2	92.8	0.5	4.40	12.4	8.8

[a]Methods: Mn determined by potentiometric titration with $KMnO_4$ standard solution; MnO_2 determined by $FeSO_4$ method; H_2O determined by heating 110°-120°C for 2 h in air; density determined by a picnometer with kerosene at 40°C; surface area determined by (a) BET method, (b) zinc-ion adsorption method [101].

[b]Based on dried sample at 110°-120°C for 2 hours.

4.2.5. Scanning Electron Microscope Studies.

The various types of manganese dioxides can be examined with the Scanning Electron Microscope. Figures 80 to 85 are pictures of the surfaces of differently prepared International Common Samples. The ICS Nos. correspond to the numbers in Table 18. The photographs were taken at the Technical Center of Union Carbide Corporation, Parma, Ohio.

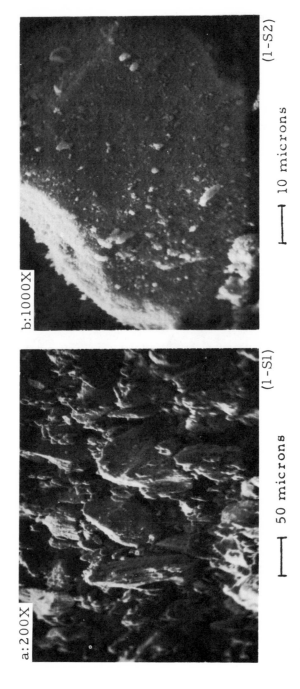

a:200X

b:1000X

(1-S1)

(1-S2)

|—— 50 microns

|—— 10 microns

FIG. 80. I.C.S. No. 1, Electrolytic MnO$_2$ made by Ti-anode process.

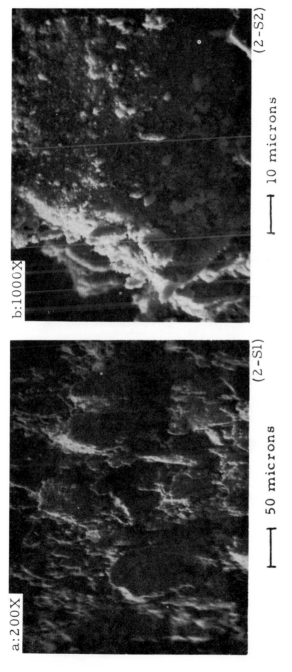

FIG. 81. I.C.S. No. 2, Electrolytic MnO$_2$ made by Pb-anode process.

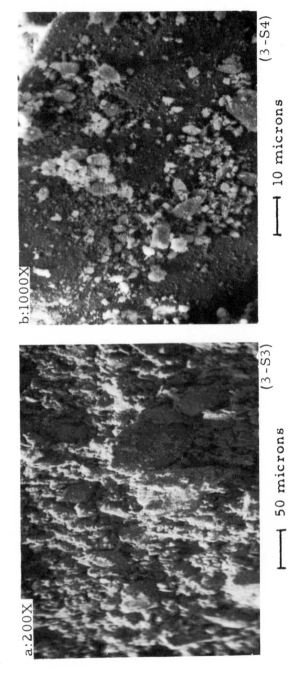

FIG. 82. I.C.S. No. 3, Electrolytic MnO_2 made by C-anode process.

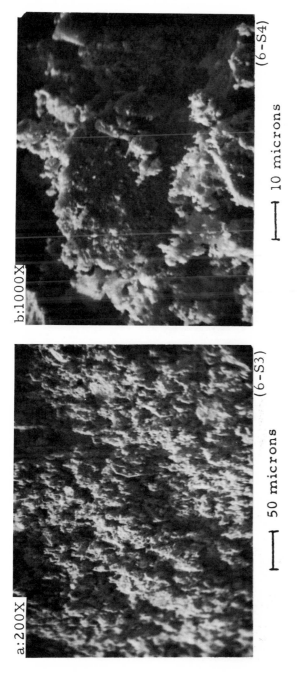

FIG. 83. I.C.S. No.7, Natural MnO_2 ore.

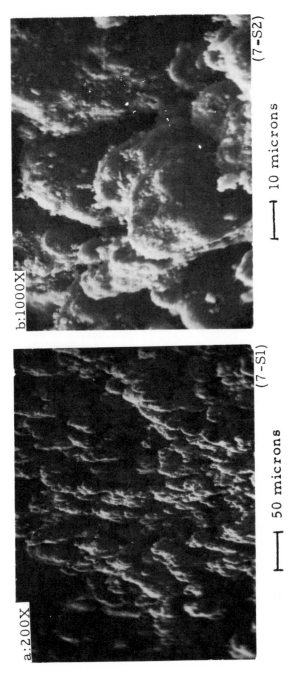

FIG. 84. I.C.S. No. 8, Chemically prepared MnO_2.

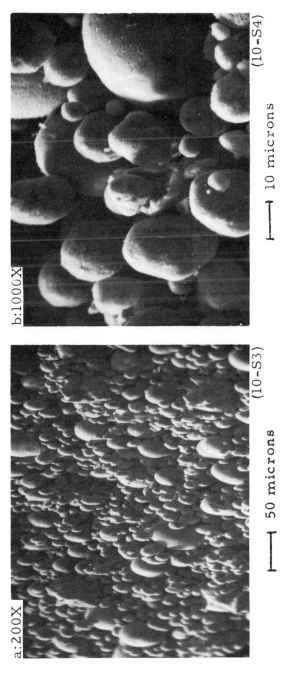

FIG. 85. I.C.S. No. 11, Chemical MnO_2 made by chlorate process.

ACKNOWLEDGMENT

The author wishes to express his appreciation to Mrs. H. M. Joseph
and Mr. W. A. Adams for their help in the preparation of this manuscript.

REFERENCES

1. A. Kozawa and J. F. Yeager, J. Electrochem. Soc. 112, 960 (1965).

2. A. Kozawa and R. A. Powers, Electrochem. Tech. 5, 533 (1967).

3. A. Kozawa and R. A. Powers, J. Electrochem. Soc. 113, 870 (1966).

4. A. Kozawa and J. F. Yeager, J. Electrochem. Soc. 115, 1003 (1968).

5. A. Kozawa and R. A. Powers, J. Electrochem. Soc. 115, 122 (1968).

6. A. Kozawa and R. A. Powers, J. Electrochem. Soc. Japan 37, 31
   (1969).

7. A. Walkley, J. Electrochem. Soc. 97, 209C (1952).

8. K. R. Newby and A. B. Scott, J. Electrochem. Soc. 117, 152 (1970).

9. K. J. Vetter, J. Electrochem. Soc. 110, 597 (1963).

10. K. J. Vetter, Z. Elektrochem. 55, 274 (1951); 66, 577 (1962).

11. A. Kozawa, Dry Batteries, No. 509 (1968).

12. R. F. Amlie and P. Ruetschi, J. Electrochem. Soc. 108, 813 (1961).

13. A. Kozawa, T. Kalnoki-Kis, and J. F. Yeager, J. Electrochem. Soc.
    113, 405 (1966).

14. A. Kozawa, J. Electrochem. Soc. Japan 36, 196 (1969).

15. H. Bode, A. Schmier, and D. Berndt, Z. Elektrochem. 66, 586 (1962).

16. J. P. Brenet, Proc. CITCE 1956 Madrid 8, 394 (1956).

17. A. Era, Z. Takehara, and S. Yoshizawa, Denki Kagaku 34, 483 (1966).

18. W. Feitknecht, H. R. Oswald, and V. Feitknecht-Steinman, Helv. Chim.
    Acta 43, 341 (1961).

19. D. T. Ferrel and W. C. Vosburgh, J. Electrochem. Soc. 98, 334 (1951).

20. J. P. Gabano, C. R. Acad. Sci. Paris 264, 262 (1967).

21. J. P. Gabano, J. Electrochem. Soc. 117, 147 (1970).

22. F. Kornfeil, J. Electrochem. Soc. 109, 349 (1962).

23. A. B. Scott, J. Electrochem. Soc. 107, 941 (1960).

24. W. C. Vosburgh, J. Electrochem. Soc. 106, 839 (1959).

25. G. D. Van Arsdale and C. B. Maier, Trans. Electrochem. Soc. 33, 109 (1918).

26. C. W. Nichols, Trans. Electrochem. Soc. 57, 393 (1932).

27. O. W. Storey, E. Steinhoff and E. R. Hoff, Trans. Electrochem. Soc. 86, 337 (1944).

28. G. A. Lee, J. Electrochem. Soc. 95, 2 (1949).

29. N. Kameyama and H. Iida, J. Chem. Soc. Japan; Ind. Chem. Sect. (Showa 7, August issue), Toyota Kenkyu Iho 1, 87 (1932); 2, 1 (1934).

30. K. Takahashi, Denki Kagaku 6, 227 (1938).

31. H. Inoue and S. Haga, Japanese Pat. 82,344 (Mar. 27, 1929).

32. K. Takahashi and A. Kozawa, J. Metals 22, 64 (1970).

33. T. Nakamura and M. Nakano, Kagaku Kogaku 32, 656 (1968).

34. H. Inoue and S. Haga, Tokyo Kogyo Shikenjo Hokoku 26, 1 (1931).

35. Yanagihara, Teratani, and Ohokubo, J. Japan. Mineral. 59, 463 (1943).

36. S. Matsuno, Japanese Pat. 173,680 (1944).

37. C. Creanga, Rev. Chim. (Bucharest) 5, 53-2-6 (1954) [Chem. Abstr. 49, 15565i (1955)].

38. I. Muraki, J. Chem. Soc. Japan, Ind. Chem. Sect. 61, 659 (1958).

39. I. Muraki and Y. Okajima, J. Chem. Soc. Japan, Ind. Chem. Sect. 62, 163 (1959).

40. I. Muraki, Ph.D. Thesis, Nagoya University, 1961.

41. I. Muraki, J. Chem. Soc. Japan, Ind. Chem. Sect. 63, 2089 (1960).

42. I. Muraki and Y. Okajima, J. Chem. Soc. Japan, Ind. Chem. Sect. 64, 137 (1961).

43. A. Era, Z. Takehara, and S. Yoshizawa, Denki Kagaku 35, 288 (1967).

44. A. Era, Z. Takehara, and S. Yoshizawa, Denki Kagaku 35, 334 (1967).

45. A. Era, Ph.D. Thesis, Kyoto University, 1967.

46. K. Shimizu and I. Shirahata, Furukawa Denko Jiho, No. 43 (May, 1967).

47. H. Ogawa, Y. Amono, and K. Ando, Natl. Tech. Rept. 6, 364 (1960).

48. T. Tsuruoka, K. Shiroki, and R. Asaoka, Denki Kagaku 27, 229 (1959).

49. Y. Shibasaki, J. Chem. Soc. Japan, Ind. Chem. Sect. 57, 181 (1954).

50. I. Muraki, J. Chem. Soc. Japan, Ind. Chem. Sect. 64, 141 (1961).

51. M. Sato et al., J. Chem. Soc. Japan, Ind. Chem. Sect. 71, 104 (1968).

52. M. Sato et al., J. Chem. Soc. Japan, Ind. Chem. Sect. 71, 484 (1968).

53. M. Sugimori and T. Sekine, Denki Kagaku $\underline{37}$, 63 (1969).

54. M. Sugimori and T. Sekine, Denki Kagaku $\underline{37}$, 380 (1969).

55. S. Nishizawa and J. Koshiba, Denki Kagaku $\underline{37}$, 164 (1969).

56. A. Kozawa and K. Sasaki, Denki Kagaku $\underline{22}$, 569, 571 (1954).

57. K. Sasaki and A. Kozawa, Denki Kagaku $\underline{25}$, 273 (1957).

58. K. Sasaki and A. Kozawa, Denki Kagaku $\underline{25}$, 322 (1957).

59. S. Matsuno, J. Chem. Soc. Japan Ind. Chem. Sect. $\underline{44}$, 621,909 (1941);
     $\underline{46}$, 605 (1943).

60. M. Fukuda, Natl. Tech. Rept. $\underline{3}$, 1 (1957); $\underline{4}$, 321 (1958); $\underline{5}$, 1, 127
     (1959). [Denki Kagaku $\underline{27}$, 212 (1959); $\underline{28}$, 67 (1960)].

61. T. Hirai, Ph.D. Thesis, Kyoto University, 1970.

62. A. Kozawa and W. C. Vosburgh, J. Electrochem. Soc. $\underline{105}$, 59 (1958).

63. S. Ninagi, Ph.D. Thesis, Kyoto University, 1968.

64. S. Ninagi and Y. Miyake, Denki Kagaku $\underline{22}$, 574 (1954); $\underline{27}$, 217 (1959).

65. K. Miyazaki, Bull. Chem. Soc. Japan $\underline{41}$, 2785 (1968).

66. K. Miyazaki, J. Electroanal. Chem., $\underline{21}$, 414 (1969).

67. K. Miyazaki, Kogyo Kagaku Zashi, $\underline{72}$, 856 (1969).

68. K. Miyazaki, J. Electrochem. Soc. $\underline{116}$, 1471 (1969), $\underline{117}$, 821 (1970).

69. K. Miyazaki, Kogyo Kagaku Zashi, $\underline{73}$, 291, 900 (1970).

70. K. Miyazaki, Bull. Chem. Soc. Japan, $\underline{41}$, 225, 1730 (1968).

71. K. Miyazaki, Bull. Chem. Soc. Japan, $\underline{41}$, 1730 (1968).

72. K. Miyazaki, Bull. Chem. Soc. Japan, $\underline{42}$, 2046 (1969).

73. K. Miyazaki, 10th Symp. Batteries (Japan), Extended Abstr. No. 1,
     (1969).

74. K. Miyazaki, 23rd Ann. Meeting Chem. Soc. Japan, Abstr. No. 21628,
     (1970).

75. Catalog for NCK M-Series Electrodes, Nippon Carbon Company.

76. K. Shimizu and T. Takehara, U.S. Pat. 3,436,323 (1969).

77. L. Berg and G. E. Lohse, Montana State College Eng. Exp. Sta. Bull.
     $\underline{15}$, 6 (1952).

78. H. K. Chakrabarti and T. Banerjee, J. Sci. Ind. Res. (India) $\underline{12B}$, 211
     (1953).

79. M. Fukuda, Natl. Tech. Rept. $\underline{3}$, 1 (1957).

80. A. Kozawa, Y. Kaneda, and T. Takahashi, paper presented before the
     Primary Battery Committee, Electrochem. Soc. of Japan, Sept. 1961.

81. A. Kozawa, J. Electrochem. Soc. $\underline{106}$, 552 (1959).

82. K. Sasaki and A. Kozawa, Denki Kagaku $\underline{25}$, 115 (1957).

83. Mitsui Mining and Smelting Company, "Mitsui Denman."

84. M. Fukuda, Natl. Tech. Rept. 4, 1 (1958).

85. W. F. Nye, S. B. Levin, and H. H. Kedesdy, Proc. 13th Ann. Power Sources Conf., Atlantic City, N.J., 1959, p. 125.

86. H. F. McMurdie and E. Golovato, J. Res. Natl. Bur. Stand. (U.S.) 41, 589 (1948).

87. R. Giovanoli, R. Maurer, and W. Feitknecht, Helvetica Chim. Acta 50, 1073 (1967).

88. R. Giovanoli, K. Bernhard, and W. Feitknecht, Helvetica Chim. Acta 51, 355 (1968).

89. M. Fleischman and H. R. Thirsk, J. Electrochem. Soc. Japan (Overseas Edition) 28, E-175 (1960).

90. M. Fleischman, H. R. Thirsk, and I. M. Tordesillas, Trans. Faraday Soc. 58, 1865 (1962).

91. G. Kano, M. Masuda, M. Takashima, and O. Nakamura, Denki Kagaku 37, 356 (1969).

92. J. Y. Welsh, Electrochem. Tech. 5, 504 (1967).

93. Y. Kato and T. Matsuhashi, Denki Kagaku 1, 11 (1933).

94. I. Araki, in Electrochemistry of Manganese Dioxide and Manganese Dioxide Batteries in Japan, Vol. 2 (S. Yoshizawa, ed.), U.S. Branch Office of Electrochem. Soc. of Japan, Cleveland, 1971, p. 115.

95. A. Kozawa and R. A. Powers, J. Electrochem. Soc. 113, 870 (1966).

96. A. Kozawa and R. A. Powers, Electrochem. Tech. 5, 535 (1967).

97. N. C. Cahoon and G. W. Heise, J. Electrochem. Soc. 94, 2 (1948).

98. I. Muraki, Japanese Pat. Appl. No. 14512 (published June 20, 1968).

99. A. Kozawa, Mem. Faculty Eng. Nagoya Univ. 11, No. 1-2, 243 (1959).

100. J. J. Lingane and R. Karplus, Ind. Eng. Chem. Anal. Ed. 18, 191 (1946).

101. A. Kozawa, J. Electrochem. Soc. 106, 552 (1959).

102. A. Kozawa, J. Inorg. Nucl. Chem. 21, 315 (1961).

103. A. Kozawa, Y. Kaneda, and T. Takahashi, in Electrochemistry of Manganese Dioxide and Manganese Dioxide Batteries in Japan, Vol. 1 (S. Yoshizawa et al., eds.), U.S. Branch Office of Electrochem. Soc. of Japan, Cleveland, 1971, pp. 78-8 (Figs. 12, 13).

104. International Common Samples of Manganese Dioxide, obtained from the Distribution Center (sponsored by Cleveland Section of the Electrochemical Society), P.O. Box 6116, Cleveland, Ohio 44101, U.S.A.

105. K. Sasaki and A. Kozawa, J. Chem. Soc. Japan, Indust. Chem. Sect. 57, 193 (1954).

CHAPTER 4

MAGNESIUM BATTERIES

Donald B. Wood

Power Sources Technical Area
U.S. Army Electronics Technology
and Devices Laboratory (ECOM)
Fort Monmouth, New Jersey

1. INTRODUCTION

In aqueous solution, magnesium should establish the theoretical voltages according to the following reactions [1]:

$$Mg + 2\ OH \rightarrow Mg(OH)_2 + 2\ e^- \quad E_{base} = -2.67\ V \qquad (1)$$

$$Mg \rightarrow Mg^{2+} + 2\ e^- \qquad E_{acid} = -2.37\ V \qquad (2)$$

In working aqueous electrochemical cells, lower potentials are obtained. The theoretical values are decreased about 1.1 V. Magnesium, as an alloy, is used in two types of primary batteries: the reserve cell, which requires activation by the addition of liquid shortly before use; and the nonreserve or dry cell, which is stored in the activated state for long periods (up to five years) prior to discharge.

The principal chemical compositions of both battery types, which are or could be produced, are presented in Table 1 [2].

1.1. MAGNESIUM ANODE OPERATION

Two reactions take place at the magnesium anode: (a) the reaction that releases the electrons, either Eq. (1) or (2) above, and (b) the corrosion side reaction:

$$Mg + 2\ H_2O \rightarrow Mg(OH)_2 + H_2 \qquad (3)$$

The rate at which these processes interact determines the anode efficiency [3]:

TABLE 1

Chemical Composition of Various Cells with Magnesium Alloy Anodes

Battery type	Cell components		Reaction mechanism
	Aqueous electrolyte	Cathode	
Nonreserve			
Magnesium–manganese dioxide	$Mg(ClO_4)_2$	MnO_2	$Mg + H_2O + 2\,MnO_2 \rightarrow Mn_2O_3 + Mg(OH)_2$
Magnesium–m–dinitrobenzene	$Mg(ClO_4)_2$	m–DNB	$6\,Mg + 8\,H_2O + C_6H_4(NO_2)_2 \rightarrow C_6H_4(NH_2)_2 + 6\,H_2O$
Reserve			
Magnesium–silver chloride	$MgCl_2$	$AgCl$	$Mg + 2\,AgCl \rightarrow MgCl_2 + 2\,Ag$
Magnesium–cuprous chloride	$MgCl_2$	$CuCl$	$Mg + 2\,CuCl \rightarrow MgCl_2 + 2\,Cu$
Magnesium–manganese dioxide	$Mg(ClO_4)_2$	MnO_2	$Mg + H_2O + 2\,MnO_2 \rightarrow Mn_2O_3 + Mg(OH)_2$

$$E = \frac{\text{No. of Faradays obtained in the circuit x 100}}{\text{No. of equivalents of magnesium dissolved}} \tag{4}$$

Anode efficiency influences the cell capacity, since the amount of water available to the cell reactions is usually limited because of the physical constraints of the particular cell.

As shown in Table 2 [4,5], anode efficiency and anode potential are influenced by the composition of the alloy. In Table 3, for 1 N $Mg(ClO_4)_2$, some of the intricacies of the interactions of anode constituents on efficiency are presented. The effect of current density on anode efficiency is shown in Table 4 for AZ21 alloy over various electrolyte concentrations [6]. Cell operating temperature also causes a change in anode efficiency. With AZ21 alloy, it drops from 77% at 70°F to 70% at 0°F in 2.5 N $Mg(ClO_4)_2$ [7].

Because of the inefficiency of the magnesium anode, much heat is evolved during cell operation. The heat originates from two sources: the corrosion reaction that liberates 82.1 kcal/gm mole of magnesium [8], and the IR loss resulting from the difference between the operating and the theoretical voltage [9]. Figure 1 shows a plot of the combined influence of both of these heat sources relative to current drain on an individual cell [10]. The heat generated can be used to advantage at low operating temperatures if it is retained at the battery. As an example, the capacity of a magnesium dry-cell battery at 20°F can be multiplied by a factor of 3 by using insulation.

The hydrogen evolved may create a cell-design problem, since provisions must be made for the gas to escape from the cell and not accumulate.

1.2. DELAYED ACTION

The magnesium anode may not produce a stable voltage as soon as the circuit is closed. In a dry cell, this is particularly important; a typical "delayed action" is shown in Fig. 2, together with the increase in potential when the load is reduced. The voltage buildup may also occur in reserve cells under certain discharge conditions.

The mechanism that causes this delay has not been completely defined. A widely held theory is that the $Mg(OH)_2$ film on the anode breaks down and

TABLE 2

Electrode Potential and Anode Efficiencies of Various Magnesium Alloys in 2 N $MgBr_2$ Saturated with $Mg(OH)_2$ at 2.0 mA/cm^2

Alloys[a]	Composition[b] (%)					Maximum impurities[b] (%)		Anode efficiency (%)	Steady-state anode potential[c] (V)
	Al	Zn	Mn	Ca	Zr	Ni	Fe		
Commercially pure magnesium	–	–	0.15 max.	–	–	0.001	–	63.5	1.327
AZ10A	0.8-1.2	0.3-0.5	0.15 max.	0.1-0.25	–	0.001	0.001	68.3	1.287
AZ21	1.6-2.5	0.8-1.6	0.15 max.	0.1-0.25	–	0.001	0.001	67.0[d]	No data
AZ31	2.5-3.5	0.7-1.3	0.20 min.	0.04 max.	–	0.005	0.005	67.0	1.232
AZ61(5)	6.5	0.7	0.2	–	–	–	–	No data	No data
AZ80A	7.9-9.2	0.2-0.8	0.15 min.	–	–	0.005	0.005	68.9	1.227
ZK60A	–	4.8-6.5	–	–	0.45 min.	–	–	58.3	1.233
M1A	–	–	1.20 min.	0.08-0.14	–	0.01	–	52.6	1.328

[a] Alloys obtained from Dow Chemical Company and White Metal Rolling and Stamping Corporation.
[b] Data supplied by Dow Chemical Company.
[c] Versus saturated calomel electrode.
[d] At 2.85 mA/cm^2 in 2.5 N $MgBr_2$.

TABLE 3

Influence of Anode Composition on Anode Efficiency and Current Density in 1 N Mg(ClO$_4$)$_2$

Alloy composition (%)				Anode efficiency (%) versus current density (mA/in^2)					Maximum efficiency (%)
Al	Ca	Mn	Zn	58	97.5	146	197	294	
<0.03	<0.01	<0.01	<0.001	72%	73%	74%	78%	80%	81
0.51	<0.01	<0.01	<0.001	71	72	71	72	72	73
1.02	<0.01	<0.01	<0.001	64	66	67	68	69	69.5
1.99	<0.01	<0.01	<0.001	77	77	77	76	76	76.5
4.8	<0.01	<0.01	<0.001	67	71	76	78	78	78
0.003	<0.01	<0.01	0.23	71	71	71	75	76	77
0.003	<0.01	<0.01	0.52	74	74	75	77	77	77.5
0.003	<0.01	<0.01	1.0	79	77	78	78	78	78
1.9	<0.01	0.022	1.19	74	75	76	76	77	77.5
1.9	0.036	0.013	1.15	75	75	77	78	78	78
2.0	0.14	0.015	1.25	76	76	78	78	78	78
1.2	0.22	0.037	0.32	30	33	33	38	38	39
2.9	<0.01	0.44	0.90	74	78	77	79	78	79
5.2	<0.01	0.016	1.08	71	72	74	75	76	77

TABLE 4

Effect of Electrolyte Concentration and Current Density

on Anode Efficiency of AZ21 Alloy

Current density (mA/in^2)	Anode efficiency (%) versus $Mg(ClO_4)_2$ concentration					
	1 N	2 N	2.5 N	3 N	4 N	5 N
18	72%	–	75%	–	–	–
10	76	71	79	77	74	69
20	76	–	77	–	–	–
40	78	–	77	–	–	–
57	77	79	77	76	73	72
87	76	78	79	76	73	72
120	75	76	78	73	72	71
195	–	–	78	–	71	–
295	–	–	77	–	–	–
400	–	–	76	–	–	–
495	–	–	75	–	–	–

repairs as the current increases or decreases, respectively. This mechanism controls the anodic voltage behavior [11]. The operating potential, as described by Petrocelli, is a mixed potential composed of the anodic reaction $Mg + 2 (OH)^- \rightarrow Mg(OH)_2 + 2 e^-$ and the cathodic reaction $2 H_2O + 2 e^- \rightarrow H_2 + 2 (OH)^-$ [12]. Perrault [13] argues that if these reactions proceeded directly, the mixed potential would be -1.9 V versus hydrogen (which it is not). It is observed to be -1.32 V in 2 N $Mg(ClO_4)_2$ with pure magnesium. This leads to the hypothesis that some intermediate compounds are formed [14,15] either corresponding to the previously mentioned $Mg(OH)_2$ film repair and breakdown mechanism or perhaps to the formation of a hydride in accordance with the following reactions:

$$Mg \rightarrow Mg^{2+} + 2 e^- \tag{5}$$

$$2 H_2O + 2 e^- \rightarrow 2 OH^- + H_2 \tag{6}$$

$$Mg + 2 H_2O + 2 e^- \rightarrow MgH_2 + 2 OH^- \tag{7}$$

$$MgH_2 + 2 OH^- \rightarrow Mg(OH)_2 + H_2 + 2 e^- \tag{8}$$

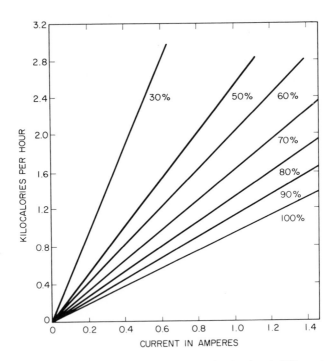

FIG. 1. Heat produced at a magnesium electrode at different anode efficiencies.

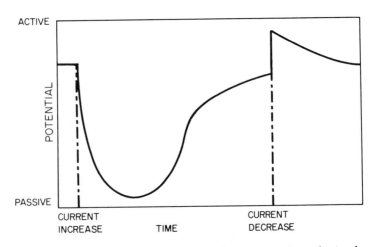

FIG. 2. Generalized voltage transient on magnesium electrode.

Arguments derived from thermodynamic considerations add credence to this reaction path, which also explains phenomena observed relative to the magnesium anode [13].

2. THE MAGNESIUM DRY CELL

Presently, the magnesium dry cell is used as a source of power for military communication equipment. There is no distribution to the civilian market. A formulation that provides a dry cell that meets present military communication requirements is presented in Table 5.

2.1. ANODE

The three anode materials, alloys AZ10A, AZ21, and AZ31, that have received most of the effort during the development of the magnesium dry cell are listed with their various components in Table 2.

TABLE 5

Magnesium Dry-Cell Formulations

Anode:	Magnesium alloy AZ21
Separator:	Uncoated Kraft Nibroc paper
Cathode mix:	86% MnO_2
	10% acetylene carbon black
	3% $BaCrO_4$
	1% $Mg(OH)_2$
Electrolyte:	3.7 N $Mg(ClO_4)_2$ with 0.2 gm/liter Li_2CrO_4

The alloys contain mostly aluminum, zinc, manganese, and calcium as alloying agents. The components interact on various discharge characteristics, and the selection of alloy AZ21 was made after it was found difficult to produce an AZ10 that would not corrode excessively during prolonged storage [16]. The factors that were considered in the development besides corrosion were anode efficiency, delayed action, and the various types of discharges encountered, that is, continuous rates and various interrupted applications. In general, the AZ21 alloy exhibits the low delayed-action characteristics of the AZ10 alloy with the improved anode efficiency of the AZ31. However, the AZ21 is much less sensitive to corrosion under storage conditions than AZ10 [16].

2.2. CATHODE MIX

Manganese dioxide is usually the depolarizer in magnesium dry cells. A chemically processed MnO_2, designated as type M, provides the best capacity in the magnesium dry cell, but an electrolytic type could be employed. A comparison of capacity between an electrolytic MnO_2 and the type M is presented in Table 6, employing ten-day-old cells. Type M is produced by oxidation of manganese oxides by perchlorate under the catalytic influence of the manganous ion [17]. The electrolytic ore can be produced by the electrical deposition of MnO_2 on graphite electrodes in an acidic bath. Certain chemical and physical characteristics of the type M ore and electrolytic ore are presented in Table 7.

Acetylene black is incorporated into the cathode mix to provide conductivity and to aid in binding the required moisture content within the cell. It is the long chainlike structure that allows the acetylene black to retain moisture [18].

Slightly soluble $BaCrO_4$ acts as a chromate reservoir, reducing anode corrosion; it may also have some catalytic influence, resulting in a 7-15% increase in capacity [3]. The $Mg(OH)_2$ is added as a buffering agent and aids storageability of the completed cell [19].

Two of the electrolyte's constituents are essential to support the current-producing reactions: the water required for the anodic reaction (see equa-

TABLE 6

Comparison of Chemical and Electrolytic Manganese Dioxide

in A-Size Cells at 70°F

Ore type and producer	Discharge program	Hours of service to 1.25 V	1.00 V
Manganese dioxide	16.67 Ω	11.7	13.8
Type M	continuous		
(Diamond Shamrock	12.5 Ω, 2 min;	66.8	74.8
Chemical Company)	250 Ω, 18 min		
Electrolytic	16.67 Ω	9.5	10.6
manganese dioxide	continuous		
(American Potash and	12.5 Ω, 2 min;	53.2	55.8
Chemical Corporation)	250 Ω, 18 min		

TABLE 7

Physical and Chemical Characteristics

of Chemical and Electrolytic Manganese Dioxide

Ore type	Electrolytic manganese dioxide[a]	Manganese dioxide Type M[b]
Manganese (wt%)	61.0	60.7
Available oxygen as MnO_2 (wt%)	91.3	90.6
Water at 110°C (wt%)	1.5	1.7
pH	5.1	5.0
Apparent density (gm/in^3)	22.0	24.9

[a]American Potash and Chemical Corporation
[b]Diamond Shamrock Chemical Company

tions in Table 1), and the magnesium perchlorate, which provides the
conductivity and markedly influences the reactions taking place at the anode.
Magnesium bromide solutions may also be used as the electrolyte salt, but
increased delayed action on intermittent types of discharge is usually obtained

The amount of electrolyte present in a cell depends on the acetylene black
content of cathode mix and the amount of mechanical energy applied to the
mix during the dry and wet mixing processes. It is essential that the chain
structure of the carbon black not be broken down, resulting in loss of its
water-absorption quality. The soluble Li_2CrO_4 in the electrolyte acts as a
corrosion inhibitor during storage.

2.3. CATHODE MANUFACTURE

A problem common to each structure employed in making a magnesium
dry cell is the formation of the cathodic mix, which is called a bobbin or a
dolly or, in a flat cell, a mix cake. The dry constituents of the cathodic
mix, the MnO_2, the acetylene black, the $BaCrO_4$, and the $Mg(OH)_2$, are
blended together in a mixer or a ball mill until a homogeneous mix is
obtained, which is then blended with the electrolyte to produce a mixture of
workable consistency. The mixture must be adaptable to the various machines used in bobbin fabrication, that is, it must be capable of being
handled without excessive water and must pass the nozzle used in filling
cells without clogging. Controlling the amount of mixing is important
because variation in the constituents of the cell or in the mixing time must
be adapted to arrive at what is called a good "apparent wetness."

2.4. ROUND-CELL CONSTRUCTION

The round-cell structure is pictured in Fig. 3. The Kraft paper acts as
a physical separator between the cathode mix and the magnesium can; yet it

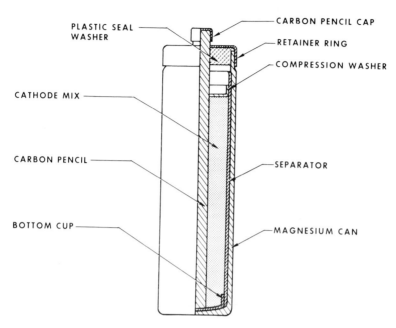

FIG. 3. Magnesium round-cell construction.

absorbs electrolyte from the bobbin and provides the path through which the water migrates to the anode reaction sites. The bottom cup or washer of heavy paper or cardboard is positioned to avoid cathodic reaction in the direction of the cell bottom, resulting in preferential corrosion of the cell side wall [20].

A compression washer keeps the tool used to compact the wet mix in the cell clean. Furthermore, it insures against unfolding of the paper liner, which can cause shorts and corrosion in the air space. The carbon rod, impregnated with paraffin wax, acts as the cathodic current collector and is driven into the cathode mix after the mix is inserted into the lined can. A metal cap, usually brass, is pressed over the carbon rod to provide a point of electrical contact.

The top seal varies in design among the various manufacturers, but it must accomplish two purposes. It must retain the moisture in the cell during storage at temperatures up to 160°F, and it has to provide an escape route for the hydrogen produced during the cell discharge. The hydrogen venting is usually accomplished through a small hole in the plastic top seal washer under the retainer ring. The retainer ring is pressed over the swaged or deformed can, thus requiring some gas pressure before venting occurs. A pitch or wax poured over the plastic top-seal washer may protect against moisture loss, and a pressure blow-out plug is sometimes provided in the plastic insert. The retainer ring is the point used for intercell contacts (by soldering). Development of this type of intercell contact was one of the factors instrumental in making the magnesium dry cell practical. Previously, the so-called magnesium spot weld (not a true weld but an embedment) was used. The spot-weld process was difficult to control because it requires very clean, well-adjusted electrodes.

The magnesium can is formed through a process called impact extrusion. A cylinder of magnesium is cut from rod stock, lubricated with graphite, preheated, inserted in a heated die, and struck with a punch at a temperature of about 630°F [16]. The formed can is stripped from the die and trimmed. In order to achieve a long shelf life of the Mg cell, the can must be cleaned. The can-cleaning process consists of an alkaline soak, then an acid pickle bath of acetic acid-sodium nitrate to remove 0.001 in. of metal, a stabilization in a weak bright pickle (chromic acid and additives), and finally oven drying at 225°F [16]. If the cleaning is not completed satisfactorily, the embedded graphite in the can causes corrosion during storage, resulting in deterioration of cell capacity.

2.5. ROUND-CELL PERFORMANCE

The capacity obtainable from a Mg/MnO_2 dry cell varies with the terminal end voltage per cell, the rate at which the energy is removed, and the temperature of the cell during discharge. The 70°F capacity to 1.25 V for various cell sizes relative to initial current drain is shown in Fig. 4. An end voltage of 1.25 V per cell is considered the optimum for the Mg/MnO_2 dry cell. The dimensions and weights of these cells are presented in Table 8.

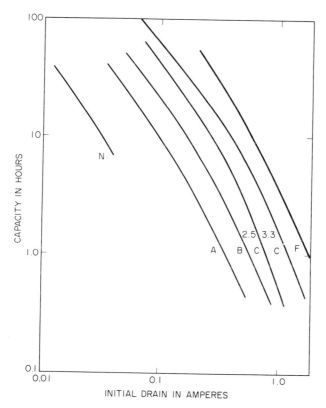

FIG. 4. Capacity of different Mg/MnO_2 cells to 1.25 V end voltage as a function of initial discharge current (70°F).

The interaction of operating temperature and resistive load (to various end voltages) on the capacity of a 3.3 in. C-size cell can be calculated by substituting the respective constants found in Table 9 into the equation

$$\log C = a \log X^2 + b \log X + d \tag{9}$$

where C is the service in hours and X is the value of the resistance in ohms through which the individual cell is discharged, and a, b, and d depend on the end voltage and discharge temperature selected. The cells used to obtain these data were made according to the basic cell formulation presented in Table 5 and the dimensions given in Table 8 for the 3.3 in. C-size

TABLE 8

Magnesium Dry-Cell Sizes and Dimensions

Cell size	Total height (in.)	Cell diameter (in.)	Cell weight (gm)	Mix weight (gm)
N	1.22	0.435	4.9	3.1
Tall N	1.56	0.435	6.9	4.8
R	1.82	0.507	10.1	5.6
A	1.91	0.627	17.0	10.1
B	2.09	0.754	26.5	16.5
2.5-in. C	2.51	0.936	47.0	32.0
3.3-in. C	3.23	0.936	60.0	41.0
D	1.96	1.255	78.0	50.0
F	3.16	1.255	109.0	80.0

TABLE 9

Constants for Capacity Equation

$$\log C = a \log X^2 + b \log X + d$$

for 3.3" C-size Mg/MnO_2 cell[a]

Discharge temperature (°F)	End voltage (V)	a	b	d	Discharge range (h) Low	High
130	1.43	-0.117	1.41	0.156	5	100
130	1.25	0.006	0.958	0.453	7	103
130	1.00	0.107	0.855	0.523	7.6	105
70	1.43	-0.204	1.82	-0.243	2.6	484
70	1.25	-0.224	1.76	-0.040	1.1	490
70	1.00	-0.080	1.33	0.263	2.3	499
20	1.43	-0.150	0.888	-0.662	1	2.1
20	1.25	-0.5	1.8	-0.790	2.6	6.3
20	1.00	-0.54	13.7	7.20	0.5	15.6

[a]C = hours of service, X = resistance in ohms.

cell; 550 ml of electrolyte was used per 1000 gm of dry mix [21]. The cells were produced on equipment capable of fabricating cells in high volume, above 50 cells/min.

In the design of multicell batteries, the end-voltage selected per cell determines the number of cells to be incorporated, as long as the allowable voltage is not exceeded. The optimization of capacity will depend on the current drain and the temperature range over which the battery is to operate. In addition, the heat generated by the cell (due to the inefficient anode) must be considered when the weight of the battery and the rate of current drain warrants.

Example: It was desired to design a 17-Ah battery weighing 10 lb and producing a high current drain for 10% of the discharge period (for example, a drain that would deplete the capacity of the battery in 2 h) followed by a low current drain (for example, 1/30 of the high drain). Result: 130°F is the appropriate temperature at which to discharge a single cell in order to accurately forecast what the battery would provide at 70°F. The calculation leading to this answer can be found in Ref. [21].

The heat generated will influence the battery performance at low temperatures, the magnitude of the influence depending on the rate of current drain and the weight of the battery. Therefore, actual discharge data should be obtained for any set of conditions. A generalized capacity/temperature curve is presented in Fig. 5 as a guide [9], based on a three pound battery having its energy used in 20h at 70°F.

On low-rate discharge, magnesium round cells split open. Usually, this occurs somewhere above the 50-h rate and is dependent upon the thickness of the magnesium can. No water is evident on the outside of the cell when splitting occurs, nor is a normal multicell battery deformed, but splitting influences the cell's further current-producing reaction. Ingested air causes the cell voltage to rise about 0.1 V and also results in an unpredictable increase in capacity.

The rupture is caused by uneven attack on the magnesium can coupled with the pressure from formation of magnesium hydroxide, which occupies about one-and-one-half times the volume of the original magnesium. It expands and presses against the cathode mix, which has hardened appreciably from the loss of water used up in the various cell reactions.

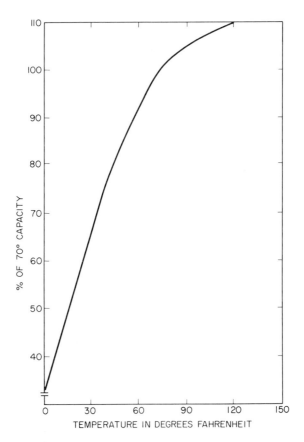

FIG. 5.   Temperature dependence of Mg/MnO_2 dry cell.

2.6. SHELF-LIFE CHARACTERISTICS

An important characteristic of the Mg/MnO_2 dry battery is its ability to withstand high-temperature storage.   Presented in Table 10 are storage characteristics of a communication type battery containing 64 A-size cells made on a production line capable of producing 10,000 batteries per month and employing a seal whose components are inserted mechanically [9]. The data presented are based on 40 random samples from each of six

TABLE 10

Capacity Retention of Production Batteries

| Weeks storage at 160°F | Per cent of initial capacity | | |
	Median	Max.	Min.
5	95	98	88
8	92	95	83
12	83	90	77

monthly runs. Usually, 15 batteries were subjected to each storage period indicated; the median value from each test is shown.

Table 11 presents data after various storage periods at various temperatures from a group of 150 Mg/MnO_2 dry-cell batteries that were fabricated employing hand-operated fabrication equipment; the data indicate what might be accomplished relative to shelf life with the Mg/MnO_2 dry cell [9].

TABLE 11

Capacity Retention of Laboratory-Built Batteries

| Storage conditions | | Per cent of initial capacity |
Temperature (°F)	Period (weeks)	(median value)
113	52	96
130	52	83
145	13	96
145	26	90
160	5	100
160	12	93
160	20	89

2.7. DRY-CELL DELAYED ACTION

The delayed action of the magnesium anode is more pronounced in the "dry" cell as compared to "wet" magnesium batteries. It is defined as the time required for the voltage of the battery to ascend to a useful level. It is influenced by the following factors: (a) operating temperature; (b) the length of time since the last discharge; (c) on intermittent discharge, such as is encountered in communications equipment, the ratio between the transmit drain and the receive drain; (d) the amount of water available for the reactions; (e) the concentration of electrolyte and the type salt employed.

Table 12 presents data obtained during continuous discharge in which the current changes from a 6-h rate for 2 min to a rate 1/20 of that for 18 min; this is encountered in the operation of a field radio, with heavy drain representing the transmit cycle and light drain the receive cycle. Data are also presented for this mode of operation with the drain interrupted for 16h (overnight) every 24h. Here the delay recurs when discharge is resumed. No delay is observed during subsequent cycles unless there is a prolonged interruption of the current.

Since the delayed-action behavior varies with actual operating conditions, characterization is best obtained by actual discharge tests [9].

A method has been devised that should overcome delayed action under any duty cycle. It involves placing a small nickel-cadmium rechargeable battery in parallel with the magnesium battery. The rechargeable battery provides the voltage until the magnesium battery is capable of doing so. In addition, a 25% increase in capacity has been obtained when the magnesium battery is submitted to a current drain at the 10-h rate for 1 min out of every hour with a background current drain at the 1000-h rate [22].

2.8. AC IMPEDANCE

The ac impedance characteristic of a AA-cell employing $Mg/Mg(ClO_4)_2/$ MnO_2 after being discharged to 50% of its rated capacity is presented in

Fig. 6 [10], where it can be seen that impedance is higher at low frequency
and at low drain rates.

2.9. DRY-BATTERY USE

When evaluating a magnesium dry-cell battery for an application, its
operating characteristics must be considered. The delayed-action phenomenon of the voltage is important. The hydrogen evolved should be vented and
sealed out of the electronic package, since hydrogen-air mixtures are flam-
mable above 4.1% and explosive above 18% [23]. The influence of the heat
generated during discharge must be determined, and the ac impedance at
low current drains or when low frequencies are employed must be taken
into account in the design of the using circuitry. In addition, it is reported

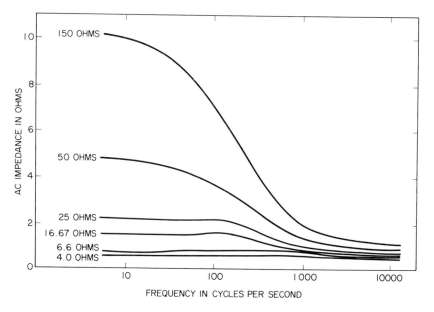

FIG. 6. Frequency dependence of the ac impedance of a $Mg/Mg(ClO_4)_2/$
MnO_2 cell (AA-size) when discharged through different resistances (70°F).

TABLE 12

Starting Delays in Seconds, Continuous versus Discontinuous Discharge[a]

Discharge temperature (°F)	Type	Start	Discharge Cycle[b]							Hours of service
			(6 h)	(14 h)	(22 h)	(30 h)	(38 h)	(46 h)	(62 h)	
10	Cont.	18.9[c]	0.2	-	-	-	-	-	-	17.5
		5.2	0.2	-	-	-	-	-	-	18.5
	Disc.	0.6	2.0	0.75	-	-	-	-	-	16.0
		3.4	4.9	-	-	-	-	-	-	15.5
70	Cont.	0.4	0	0	0	0	0	0	0	72.0
		0.7	0	0	0	0	0	0.2	0.1	71.0
	Disc.	0.60	0.6	0.4	0	0.3	0.3	0.5	0.8	70.5
		0.35	0.8	0.8	0	0.4	0.3	0.5	0.9	71.0

130	Cont.	-	0.2	0.2	-	0.3	0.3	-	-	71.0
		0.1	0.1	0.2	-	0.3	0.4	-	-	80.5
	Disc.	0.0	1.0	0.7	0.2	0.5	0.4	2.0	0.7	63.0
		0.2	0.6	0.3	0.2	0.5	0.3	1.0	0.8	64.0

[a]Continuous discharge: transmit load 2 min, receive load 18 min, 24 h/day; discontinuous discharge: transmit load 2 min, receive load 18 min, 8 h/day (16 h rest). Delays on continuous test are those observed when changing from receive to transmit load. Delays on discontinuous test are those observed at the start of each 8-h test period, beginning with a transmit cycle.

[b]Hours battery has been on discharge are shown in parentheses.

[c]Two values are shown for each condition, indicating the range.

that the chromate inhibition system breaks down after approximately 25% of the capacity is removed, making additional prolonged storage after such usage questionable.

These disadvantages of the magnesium dry-cell battery must be measured against its advantages. When compared with the Leclanché dry-cell system, it provides at least double the capacity while only costing 20% more at the present time.

3. OTHER DRY-CELL STRUCTURES AND SYSTEMS

The inside-out and flat cell structures have been investigated; both may employ sheet magnesium as the anode. The production cost of the inside-out structure and the poorer shelf life of the magnesium flat cell when compared to the round cell structure have limited their use.

3.1. INSIDE-OUT CONSTRUCTION

The inside-out cell structure uses an anode encased in separator material. This combination is driven into the cathode mix. A cup made of carbon forms the electrical contact to the cathode mix. In one structure of this type, called the Balaguer dry cell, a cylinder of magnesium is the anode and a conductive carbon cup with a center rod is used as the cathodic collector [24]. This construction should offer improved high-rate and low-temperature performance characteristics, although this has not been borne out conclusively with operating data. The inside-out construction also offers reproducible low-discharge-rate capability because the cells do not physically rupture as do the more conventional round cells.

3.2. FLAT-CELL CONSTRUCTION

Limited success has been obtained with magnesium flat-cell structures. The shelf life is better than that obtained from a Leclanché dry-cell, which will not usually tolerate storage beyond a week at 160°F [25]. In recently developed flat cells, a thin, flat piece of magnesium covered with a Kraft paper separator is pressed against a cathodic mix cake. A sheet of conductive carbon acts as a cathode collector. The cells are enclosed in a plastic-film envelope. Presented in Table 13 is a comparison of the influence of drain rate on shelf life at various storage temperatures. As the drain rate decreases, the moisture losses that result during storage become less critical and do not have as great an influence on capacity [25].

3.3. m-DINITROBENZENE DRY CELL

Meta-dinitrobenzene (m-DNB) may be used in place of MnO_2 as the depolarizer. The resulting dry cell exhibits characteristics as shown in Fig. 7 [26]. A-size cells of both systems at 70°F are compared. The

TABLE 13

Flat-Cell Shelf Life As a Function of Discharge Rate

Storage conditions		Per cent of initial capacity	
Temperature (°F)	Period (weeks)	High-low current[a]	Constant current[b]
160	4	75	88
160	8	75	82
160	12	32	75
130	13	70	95
130	26	–	95
113	13	69	87
70	13	–	100

[a]High discharge rate for 2 min alternating with low rate for 18 min. Discharge rates: high, 7-9 h; low, 70-90 h.
[b]Discharge rate: 20-70 h.

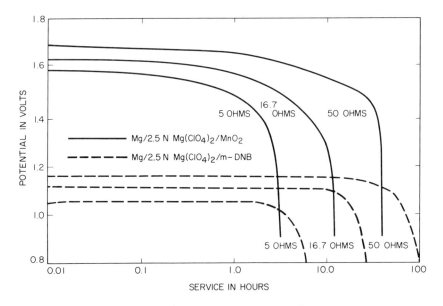

FIG. 7. Comparison of discharge curves at 70°F for A-size cells.

m-DNB system is more sensitive to the influence of lower temperatures
than the MnO_2 system, especially as the current drain increases. Presented
in Table 14 is a comparison of the two systems at 70°F showing their watt-
hours per pound and per cubic inch for various cell sizes [27].

At present, the m-DNB system is suitable for low discharge rates only.
If its operating characteristics can be modified, other applications may be
found. An indication of its promise is that complete reduction of m-dinitrobenzene to m-phenylenediamine will yield 465 Ah/lb of depolarizer-electrolyte
mix, whereas theoretically manganese dioxide can provide only 127 Ah/lb [28].

3.4. MAGNESIUM RESERVE-TYPE BATTERIES

In general, reserve batteries are employed to meet high-rate, low-
temperature power applications where nonreserve batteries cannot be used.
A typical example is a meteorological radio transmitter that operates at

TABLE 14

Energy Density Comparisons Between $Mg/Mg(ClO_4)_2/m$-DNB and
$Mg/Mg(ClO_4)_2/MnO_2$ at the 25- and 100-Hour Rates

Electrochemical system	Cell size	Hourly rate[a]	Energy density	
			(Wh/lb)	(Wh/in^3)
$Mg/Mg(ClO_4)_2/m$-DNB	R(N)	25	44.4	2.2
	A	25	58.2	2.9
	D	25	40.4	1.9
	R(N)	100	58.3	2.9
	A	100	76.0	3.8
	D	100	71.6	3.4
$Mg/Mg(ClO_4)_2/MnO_2$[b]	A	25	45.3	3.9
	D	25	36.2	2.1
	A	100	50.2	3.4
	D	100	52.5	3.1

[a] All cells taken to end voltage of 0.9 V/cell.
[b] Synthetic MnO_2.

altitudes up to 100,000 ft where temperatures may be as low as -60°F. Other uses include lightweight, high-intensity lighting, water detection, and alarm applications [29].

The reserve battery is activated by the addition of water or an aqueous solution of salt such as $Mg(ClO_4)_2$. The length of time it can be stored is unlimited when the cell is dry; once moisture is added, deterioration starts. After activation, the battery must be used within a short time period, since deactivation is impossible. The heat generated during operation (due to the inefficiency of the magnesium anode and heat of solution of the electrolyte salt) does assist the cell in providing energy at low temperature; however, it also causes the loss of cell moisture at high temperature.

Three types of reserve batteries employing magnesium anode will be discussed: magnesium-cuprous chloride, magnesium-silver chloride, and magnesium-manganese dioxide. The cell reactions are presented in Table 1. Unalloyed high-purity (primary) magnesium, alloy AZ61A, and alloy AZ31B are employed as anodes [30].

The magnesium-cuprous chloride battery is the most popular reserve-cell battery type. It has replaced the more expensive magnesium-silver chloride system in applications where weight and volume are not critical.

The positive electrode is cuprous chloride pasted to a copper screen. The electrolyte is formed when the battery is activated with water, causing some of the moderately soluble Cu_2Cl_2 to dissolve [31]. The electrolyte is absorbed into the nonwoven cellulose separator, thus preventing free water. A physical barrier is provided between each pair of cells; in one design, it consists of a strip of copper foil, which also acts as the inter-cell connector.

The magnesium-silver chloride battery provides up to 75 Wh/lb (8 Wh/in.^{3}), while the magnesium-cuprous chloride battery provides up to 30 Wh/lb (1.3 Wh/in.^{3}) [2]. However, the cost of the silver chloride per pound is ten times that of cuprous chloride.

The magnesium-manganese dioxide reserve battery has received much attention as a low-temperature power source for communications and other military applications. The battery tolerates activated storage better than other reserve types, maintaining 75% of rated capacity after 4 days at 75°F. It provides 45 Wh/lb in the temperature range between -40° and 75°F in a 6-lb package, and 33 Wh/lb from a 4-lb package [32].

REFERENCES

1. W. M. Latimer, Oxidation Potentials, Prentice-Hall, Englewood Cliffs, N.J., 1952.

2. R. Glicksman, G. S. Lozier, and C. K. Morehouse, Proc. IRE, 46, 1462 (1958).

3. A. B. Fry, P. F. George, and R. C. Kirk, J. Electrochem. Soc. 99, 323 (1952).

4. R. Glicksman, J. Electrochem. Soc. 106, No. 2, 83 (1959).

5. G. S. Lozier, U.S. Pat. 2,993,946 (1957) (Radio Corporation of America).

6. P. F. King and J. L. Robinson, Dow Chemical Company, Investigation of the Magnesium Anode, 2nd Quarterly Report, DA36-039-SC-88912 (USAECOM); also available as AD 269 524, Jan. 1962.

7. P. F. King and J. L. Robinson, Dow Chemical Company, Investigation of the Magnesium Anode, 1st Quarterly Report, DA36-039-SC-88912 (USAECOM).

8. O. A. Hoagen and K. M. Watson, Chemical Process Principles, Wiley, New York, 1943.

9. D. B. Wood, USAECOM, Magnesium/Manganese Dioxide Dry Cell Batteries; also available as AD 853 863 (USAECOM), Apr. 1969.

10. T. K. Krebs and R. Ryan, Radio Corporation of America, High Capacity Magnesium Batteries, Final Report, DA36-039-SC-85340 (USAECOM), Nov. 1962.

11. P. F. King and J. L. Robinson, J. Electrochem. Soc. 108, 36 (1961).

12. J. V. Petrocelli, J. Electrochem. Soc. 97, 10 (1950).

13. G. G. Perrault, Electroanalyt. Chem. 27, 47 (1970).

14. R. E. Bergeron, E. J. Casey, and G. D. Nagy, Def. Res. Chem. Lab. Dept. No. 158, Project No. D52-54-8008, Ottawa, April 1961.

15. C. F. Kirk and D. A. Vermilyea, J. Electrochem. Soc. 116, 1487 (1969).

16. R. C. Kirk and R. W. Ried, Proc. 14th Ann. Power Sources Conf., P.S.C. Publications Committee, Red Bank, N.J. 1960, p. 125.

17. J. Walsh, U.S. Pat. 2,956,860 (1957) (Manganese Chemical Company).

18. M. Bregazzi, Electrochem. Tech. 5, 507 (1967).

19. D. B. Wood, unpublished work, 1959.

20. J. L. Robinson, private communication, 1970.

21. L. F. Urry, Union Carbide Corporation, Optimization of BA-4270/U

and BA-4386/PRC-25 Magnesium MnO$_2$ Primary Batteries, Final Report, DA28-043-AMC-02596(E); also available as AD 840 285, Sept. 1968.

22. E. Brooks, M. Sulkes, and D. Wood, in Extended Abstracts, Fall 1971 meeting of the Electrochem. Soc., Cleveland, Ohio, p. 18.

23. E. S. Brooks, USAECOM, Study of the Effect of Hydrogen-Air Atmospheres in Radio Sets AN/PRC-25 and AN/PRC-77; also available as AD 877 707 (USAECOM), Nov. 1970, p. 4.

24. R. R. Balaguer, U.S. Pat. 3,405,013 (1968).

25. D. B. Wood, USAECOM, Progress Report on the Magnesium Flat Cell; also available as AD 866 228 (USAECOM), Jan. 1970, p. 13.

26. J. B. Doe, USAECOM, Characteristics of Meta-Dinitrobenzene Dry Cells; also available as AD 684 916 (USAECOM), Jan. 1969, p. 26.

27. J. B. Doe and D. B. Wood, Proc. 22nd Ann. Power Sources Conf. P.S.C. Publications Committee, Red Bank, N.J., 1968, p. 26.

28. J. M. Walbrick, USAECOM, Synthesis and Characterization of Intermediate Compounds Involved in the Electrochemical Reduction of M-Dinitrobenzene; also available as AD 701 879 (USAECOM), Feb. 1970, p. 1.

29. Ray-O-Vac Battery Co., Engineering Brochure, no date.

30. R. H. Williams, Proc. 24th Ann. Power Sources Symp. P.S.C. Publications Committee, Red Bank, N.J., 1970, p. 108.

31. A. L. Almerini, USAECOM, Low Cost High Energy Cathodes for Magnesium Perchlorate Batteries; also available as AD 634 100 (USAECOM), April 1966, p. 6.

32. A. L. Almerini, private communication, 1971.

AUTHOR INDEX

Numbers in parentheses are reference numbers and indicate that an author's work is referred to although his name may not be cited in the text. Underlined numbers give the page on which the complete reference is listed.

A

Achenbach, E. , 243, 375
Adams, P. H. , 311(125), 314(125),
Agar, J. N. 131, 239
Allenson, D. R. , 102 (82), 237
Allied Chemical Company,
  343(193), 383
Almerini, A. L. , 548(31,32), 550
Amand, Y. , 286(80), 288(80), 378
Amlie, R. F. , 402(12), 516
Amono, Y. , 441(47), 455(47),
  456(47), 461(47), 517
Amthor, H. K. , 259(46), 377
Ando, K. , 441(47), 455(47),
  416(47), 517
Andreeva, G. P. , 325(15), 381
Anthony, H. R. C. , 8(13,14),
  19(13,14), 235
Appelt, K. , 158, 239
Araki, I. , 437(94), 475, 519
Arkhangelskaya, Z. P. ,
  325(150,151), 381
Arouete, S. , 298(93), 378
Asaoaka, R. , 441(48), 456(48),
  517
Audubert, R. , 124(113), 238
Aufenast, F. , 58, 236

B

Bagotskii, V. S. , 318(141),
  320(141), 325(150), 380, 381

Baker, T. , 348(207), 383
Balaguer, R. R. , 271(68), 377
  544(24), 550
Banerjee, T. , 454(78), 518
Bartelt, H. , 325(152), 381
Batts, C. A. , 301(100), 379
Bauer, J. , 131, 134(121), 158,
  163, 239
Baum, R. L. , 326(155), 381
Bell, G. S. , 81, 83, 90, 98, 118,
  137(123), 153, 158, 236, 237,
  238, 239
Bender, 6, 235
Bennett, R. J. , 319(143),
  321(143), 324(143), 381
Berg, L. , 454, 518
Berger, C. , 357(229), 384
Bergeron, R. E. , 527(14), 549
Berndt, D. , 80(50), 131(50), 236
  315(133), 316(133), 346(204),
  347(204), 380, 383, (15), 516
Bernhard, K. , 461(88), 519
Berzins, T. , 130(118), 239
Best, G. E. , 331(162), 332(162),
  381
Blomgren, G. E. , 326(156), 381
Blurton, K. F. , 298(93), 299(96),
  378
Bockris, J. O. M. , 303(105), 379
Bode, H. H. , 80, 131, 236,
  315(133), 316(133), 346(204),
  347(204), 380, (15), 516

SUBJECT INDEX

A

Acetylene black, 7, 9, 45-47
Acropor WA, 354
Acrylic porothen, 355
Activated manganese dioxide,
27ff.
Activity coefficient (KOH)
literature, 364-365
Alkaline MnO_2-Zn Cell

abuse tests, 278
applications, 279-281
assembly, 259-260
cell sealing, 259
construction, 256-259,
263-269
gas vents, 259-262
history, 242
increased cathode, 271-273
performance characteristics,
271-278
reserve cells, 269-271
shelf life, 278
sizes, weights, 279-280
Alpha MnO_2, 36, 39

Ammonium chloride, 4, 7, 9,
10, 48
Analytical methods
determination of Mn, MnO_2,
488-491
determination of surface area,
491
zinc-ion adsorption method,
491-500

Anode

collector, 287
definition, 67
for alkaline cells, 287-290

B

Battery selection, 216
failures, 228, 231-233
recharging (Leclanché), 229,
230
storage, 233-234
types, 217-222
Battery technology literature,
358-359
Beta MnO_2, 37, 39-40
Bitumen, 58

C

Calcium chloride, 51
Carbon black, 45-47
Carbon rods (collectors), 47
Carboxymethylcellulose sodium
salt, 54, 55
Cathode
composition, alkaline cells,
307-308
definition, 67
mix production, Leclanché,
178-185
production, Leclanché,
200-202
reaction, alkaline cells,
309-311
Caucasian ore, 32, 42

Synthesis and Release of Adenohypophyseal Hormones

BIOCHEMICAL ENDOCRINOLOGY
Series Editor: Kenneth W. McKerns

STRUCTURE AND FUNCTION OF THE GONADOTROPINS
Edited by Kenneth W. McKerns

SYNTHESIS AND RELEASE OF ADENOHYPOPHYSEAL HORMONES
Edited by Marian Jutisz and Kenneth W. McKerns

Synthesis and Release of Adenohypophyseal Hormones

Edited by

Marian Jutisz

Laboratoire des Hormones Polypeptidiques
Gif-Sur-Yvette, France

and

Kenneth W. McKerns

The International Society for Biochemical Endocrinology
Blue Hill Falls, Maine

PLENUM PRESS • NEW YORK AND LONDON

Library of Congress Cataloging in Publication Data

Main entry under title:

Synthesis and release of adenohypophyseal hormones.

(Biochemical endocrinology)
Proceedings of a conference held in Sept. 1978 at Chateau de Seillac, which was sponsored by the International Society for Biochemical Endocrinology.
Includes index.
1. Pituitary hormones—Congresses. 2. Adenohypophysis—Congresses. I. McKerns, Kenneth W. II. Jutisz, Marian. III. International Society for Biochemical Endocrinology. [DNLM: 1. Pituitary hormones, Anterior—Biosynthesis—Congresses. 2. Pituitary hormones, Anterior—Secretion—Congresses. WK515 S993 1978]
QP572.P52S94   596'.0142   80-96
ISBN 0-306-40247-5

© 1980 Plenum Press, New York
A Division of Plenum Publishing Corporation
227 West 17th Street, New York, N.Y. 10011

Printed in the United States of America

Contributors

Sandor Arancibia, Unite 159 de Neuroendocrinologie, Centre Paul Broca de l'INSERM, 75014 Paris, France

E. C. Augustine, Department of Biochemistry and Biophysics, The Pennsylvania State University, University Park, Pennsylvania 16802

A. J. Baertschi, Department of Animal Biology, University of Geneva, CH-1211 Geneva 4, Switzerland

F. Carter Bancroft, Cellular Gene Expression and Regulation Laboratory, Memorial Sloan-Kettering Cancer Center, New York, New York 10021

Nicholas Barden, Medical Research Council Group in Molecular Endocrinology, Le Centre Hospitalier de l'Université Laval, Quebec G1V 4G2, Canada

Karl Bauer, Max-Volmer-Institut, Abteilung Biochemie, Technische Universität Berlin, 1000 Berlin 10, Germany

Michèle Beaulieu, Medical Research Council Group in Molecular Endocrinology, Le Centre Hospitalier de l'Université Laval, Quebec G1V 4G2, Canada

J. C. Beauvillain, U. 156 I.N.S.E.R.M., Laboratory of Histology, Faculty of Medicine, 59045 Lille, France

Catherine Behrens, Hormone Research Laboratory, University of California, San Francisco, California 94143

Laurence Benoist, Equipe de Recherche Associée au C.N.R.S. No. 070567, Laboratoire de Neurobiologie Moléculaire, Université de Rennes, Campus de Beaulieu, 35042 Rennes, France

Annette Bérault, Laboratoire des Hormones Polypeptidiqes, C.N.R.S., 91190 Gif-sur-Yvette, France

Pierrre Borgeat, Medical Research Council Group in Molecular Endocrinology, Le Centre Hospitalier de l'Université Laval, Quebec G1V 4G2, Canada

Marcia Budarf, Department of Chemistry, University of Oregon, Eugene, Oregon 97403

Eleanor Canova-Davis, Hormone Research Laboratory, University of California, San Francisco, California 94143

François Cesselin, Service de Biochimie Médicale et Laboratoire d'Histologie–Embryologie (C.N.R.S. ERA 484), Faculté de Médecine Pitié-Salpêtrière, 75634 Paris 13, France

M. Chrétien, Clinical Research Institute of Montreal, Montreal H2W 1R7, Canada

M. Ciolkosz, Department of Biochemistry and Biophysics, The Pennsylvania State University, University Park, Pennsylvania 16802

J. A. Coles, Department of Physiology, University of Geneva, CH-1211 Geneva 4, Switzerland

P. Crine Clinical Research Institute of Montreal, Montreal H2W 1R7, Canada

F. Dacheux, Laboratoire de Physiologie Comparée, Faculté des Sciences, 37200 Tours, France; and I.N.R.A.–Station de Physiologie de Reproduction, 37380 Monnaie, France

Jurrien de Koning, Sylvius Laboratories, Department of Pharmacology, Leiden University Medical Center, Leiden, The Netherlands

P. De la Llosa, Laboratoire des Hormones Polypeptidiques, C.N.R.S., 91190 Gif-sur-Yvette, France

Catherine Delarue, Groupe de Recherche en Endocrinologie Moléculaire, Laboratoire d'Endocrinologie, Institut de Biochimie et Physiologie Cellulaire, Faculté des Sciences, Université de Haute-Normandie, 76130 Mont-Saint-Aignan,France

Carl Denef, Laboratory of Cell Pharmacology, Department of Pharmacology, School of Medicine, Campus Gasthuisberg, Katholieke Universiteit Leuven, B-3000 Leuven, Belgium

Paul R. Dobner, Cellular Gene Expression and Regulation Laboratory, Memorial Sloan-Kettering Cancer Center, New York, New York 10021

Maurice P. Dubois, Station "Physiologie de la Reproduction" de l'I.N.R.A., 37380 Nouzilly, France

B. Dufy, Institut National de la Santé et de la Recherche Médicale, U 176, Domaine de Carreire, 33077 Bordeaux, France

P. du Pasquier, Laboratoire de Virologie, Université de Bordeaux II, 33076 Bordeaux, France

Jacques Duval, Equipe de Recherche Associée au C.N.R.S. No. 070567, Laboratoire de Neurobiologie Moléculaire, Université de Rennes, Campus de Beaulieu, 35042 Rennes, France

Alain Enjalbert, Unite 159 de Neuroendocrinologie, Centre Paul Broca de l'INSERM, 75014 Paris, France

Glen A. Evans Division of Endocrinology, Department of Medicine, University of California, San Diego, School of Medicine, La Jolla, California 92093

George Fink, Department of Human Anatomy, University of Oxford, Oxford OX1 3QX, England

H. Fleury, Laboratoire de Virologie, Université de Bordeaux II, 33076 Bordeaux, France

Ernest Follénius, Laboratoire de Cytologie Animale and E.R.A. 492 du C.N.R.S., Université Louis Pasteur, 67000 Strasbourg, France

Jean Fresel, Groupe de Recherche en Endocrinologie Moléculaire, Laboratoire d'Endocrinologie, Institut de Biochimie et Physiologie Cellulaire, Faculté des Sciences, Université de Haute-Normandie, 76130 Mont-Saint-Aignan, France.

M. Friedli, Department of Animal Biology, Universtiy of Geneva, CH-1211 Geneva 4, Switzerland

Jean-Marie Geiger, Institut d'Histologie, Faculté de Médecine, 67085 Strasbourg, France

Giuliana Giannattasio, Department of Pharmacology and CNR Center of Cytopharmacology, University of Milan, 20129 Milan, Italy

C. Gianoulakis, Clinical Research Institute of Montreal, Montreal H2W 1R7, Canada

Brian Gillham, Sherrington School of Physiology and Department of Biochemistry, St. Thomas's Hospital Medical School, London SE1, England

Martin Godbout, Medical Research Council Group in Molecular Endocrinology, Le Centre Hospitalier de l'Université Laval, Quebec G1V 4G2, Canada

E. Goldstein Laboratoire des Hormones Protéiques, Faculté de Médecine, 13385 Marseille 4, France

F. Gossard, Clinical Research Institute of Montreal, Montreal H2W 1R7, Canada

Danielle Gourdji, Groupe de Neuroendocrinologie Cellulaire, Chaire de Physiologie Cellulaire, Collège de France, 75231 Paris 05, France

F. Grisoli, Clinique Neurochirurgicale, Hôpital de la Timone, Marseille, France

M. Guibout, Laboratoire des Hormones Protéiques, Faculté de Médecine, 13385 Marseille 4, France

Edward Herbert, Department of Chemistry, University of Oregon, Eugene, Oregon 97403

Michael Hinman, Department of Chemistry, University of Oregon, Eugene, Oregon 97403

W. C. Hymer, Department of Biochemistry and Biophysics, The Pennsylvania State University, University Park, Pennsylvania 16802

P. Jacquet, Laboratoire des Hormones Protéiques, Faculté de Médecine, 13385 Marseille 4, France

Sylvie Jegou, Groupe de Recherche en Endocrinologie Moléculaire, Laboratoire d'Endocrinologie, Institut de Biochimie et Physiologie Cellulaire, Faculté des Sciences, Université de Haute-Normandie, 76130 Mont-Saint-Aignan, France

Mortyn T. Jones Sherrington School of Physiology and Department of Biochemistry, St. Thomas's Hospital Medical School, London SE1, England

Marian Jutisz, Laboratoire des Hormones Polypeptidiques, C.N.R.S., 91190 Gif-sur-Yvette, France

R. C. Kelsey, East Stroudsburg State College, East Stroudsburg, Pennsylvania 18301

Ashok Khar, Laboratoire des Hormones Polypeptidiques, C.N.R.S., 91190 Gif-sur-Yvette, France

B. Koch, Institut de Physiologie, Université Louis Pasteur, Strasbourg, France

Claude Kordon, Unite 159 de Neuroendocrinologie, Centre Paul Broca de l'INSERM, 75014 Paris, France

Jacob Kraicer, Department of Physiology, Queen's University, Kingston, Ontario, Canada K7L 3N6

D. T. Krieger, Department of Medicine, Division of Endocrinology, Mount Sinai School of Medicine, New York, New York 10029

Fernand Labrie, Medical Research Council Group in Molecular Endocrinology, Le Centre Hospitalier de l'Université Laval, Quebec G1V 4G2, Canada

Lisette Lagacé, Medical Research Council Group in Molecular Endocrinology, Le Centre Hospitalier de l'Université Laval, Quebec G1V 4G2, Canada

François Leboulenger, Groupe de Recherche en Endocrinologie Moléculaire, Laboratoire d'Endocrinologie, Institut de Biochimie et Physiologie Cellulaire, Faculté des Sciences, Université de Haute-Normandie, 76130 Mont-Saint-Aignan, France

Michèle le Dafniet, Equipe de Recherche Associée au C.N.R.S. No. 070567, Laboratoire de Neurobiologie Moléculaire, Université de Rennes, Campus de Beaulieu, 35042 Rennes, France

Philippe Leroux, Groupe de Recherche en Endocrinologie Moléculaire, Laboratoire d'Endocrinologie, Institut de Biochimie et Physiologie Cellulaire, Faculté des Sciences, Université de Haute-Normandie, 76130 Mont-Saint-Aignan, France

M. Lis, Clinical Research Institute of Montreal, Montreal H2W 1R7, Canada

C. Lucas, Laboratoire des Hormones Protéiques, Faculté de Médecine, 13385 Marseille 4, France

B. Lutz-Bucher, Institut de Physiologie, Université Louis Pasteur, Strasbourg, France

Jocelyne Massicotte, Medical Research Council Group in Molecular Endocrinology, Le Centre Hospitalier de l'Université Laval, Quebec G1V 4G2, Canada

Jacopo Meldolesi, Department of Pharmacology and CNR Center of Cytopharmacology, University of Milan, 20129 Milan, Italy

C. Mialhe, Institut de Physiologie, Université Louis Pasteur, Strasbourg, France

John V. Milligan, Department of Physiology, Queen's University, Kingston, Ontario K7L 3N6, Canada

A. Morin, Groupe de Neuroendocrinologie Cellulaire, Laboratoire de Physiologie Cellulaire, Collège de France, 75231 Paris 05, France

G. H. Mulder, Department of Medicine, Division of Endocrinology, Mount Sinai School of Medicine, New York, New York. Present address: Department of Pharmacology, Free University Amsterdam, Amsterdam, The Netherlands

J. Munoz, Department of Ophthalmology, University of Geneva, CH-1211 Geneva 4, Switzerland

L. Olivier, Laboratoire d'Histologie–Embryologie, C.N.R.S. ERA 484, 75634 Paris 13, France

R. Page, Hershey Medical Center, Hershey, Pennsylvania 17033

Vladimir R. Pantić, Institute for Biological Research "Siniša Stanković," Belgrade, Yugoslavia

Thomas L. Paquette, Department of Surgery, Washington University, St. Louis, Missouri 63110

Françoise Peillon, Service de Biochimie Médicale et Laboratoire d'Histologie–Embryologie (C.N.R.S. ERA 484), Faculté de Médecine Pitié-Salpêtrière, 75634 Paris 13, France

Marjorie Phillips, Department of Chemistry, University of Oregon, Eugene, Oregon 97403

Anthony Pickering, Department of Human Anatomy, University of Oxford, Oxford OX1 3QX, England

Mirette Priam, Unite 159 de Neuroendocrinologie, Centre Paul Broca de l'INSERM, 75014 Paris, France

J. Ramachandran, Hormone Research Laboratory, University of California, San Francisco, California 94143

James L. Roberts, Howard Hughes Medical Institute, Endocrine Research Division, University of California School of Medicine, San Francisco, California 94143

Michael G. Rosenfeld, Division of Endocrinology, Department of Medicine, University of California, San Diego, School of Medicine, La Jolla, California 92093

Merle Ruberg, Unite 159 de Neuroendocrinologie, Centre Paul Broca de l'INSERM, 75014 Paris, France

N. G. Seidah Clinical Research Institute of Montreal, Montreal H2W 1R7, Canada

M. Suzanne Sheppard, Department of Physiology, Queen's University, Kingston, Ontario, Canada K7L 3N6

S. S. Spicer, Departments of Anatomy and Pathology, Medical University of South Carolina, Charleston, South Carolina 29403

Madeleine Théoleyre, Laboratoire des Hormones Polypeptidiques, C.N.R.S., 91190 Gif-sur-Yvette, France

Andrée Tixier-Vidal, Groupe de Neuroendocrinologie Cellulaire, Chaire de Physiologie Cellulaire, Collège de France, 75231 Paris 05, France

Marie-Christine Tonon, Groupe de Recherche en Endocrinologie Moléculaire, Laboratoire d'Endocrinologie, Institut de Biochimie et Physiologie Cellulaire, Faculté des Sciences, Université de Haute-Normandie, 76130 Mont-Saint-Aignan, France

Claude Tougard, Groupe de Neuroendocrinologie Cellulaire, Laboratoire de Physiologie Cellulaire, Collège de France, 75231 Paris 05, France

G. Tramu, U. 156 I.N.S.E.R.M., Laboratory of Histology, Faculty of Medicine, 59045 Lille, France

M. Tsacopoulos, Department of Ophthalmology, University of Geneva, CH-1211 Geneva 4, Switzerland

Hannie A. M. J. van Dieten, Sylvius Laboratories, Department of Pharmacology, Leiden University Medical Center, Leiden, The Netherlands

G. Peter van Rees, Sylvius Laboratories, Department of Pharmacology, Leiden University Medical Center, Leiden, The Netherlands

Hubert Vaudry, Groupe de Recherche en Endocrinologie Moléculaire, Laboratoire d'Endocrinologie, Institut de Biochimie et Physiologie Cellulaire, Faculté des Sciences, Université de Haute-Normandie, 76130 Mont-Saint-Aignan, France

Raymonde Veilleux, Medical Research Council Group in Molecular Endocrinology, Le Centre Hospitalier de l'Université Laval, Quebec G1V 4G2, Canada

E. Vila-Porcile, Laboratoire d'Histologie–Embryologie, C.N.R.S. ERA 484, 75634 Paris 13, France

J. D. Vincent, Institut National de la Santé et de la Recherche Médicale, U 176, Domaine de Carreire, 33077 Bordeaux, France

D. L. Wilbur, Departments of Anatomy and Pathology, Medical University of South Carolina, Charleston, South Carolina 29403

W. Wilfinger, University of Cincinnati Medical School, Cincinnati, Ohio 45221

Li-Yuan Yu, Cellular Gene Expression and Regulation Laboratory, Memorial Sloan-Kettering Cancer Center, New York, New York 10021

Antonia Zanini, Department of Pharmacology and CNR Center of Cytopharmacology, University of Milan, 20129 Milan, Italy

Research Council Members of the International Society for Biochemical Endocrinology

Kenneth W. McKerns, M.Sc., Ph.D., President

Asbjorn Aakvaag, Ph.D., Oslo Kommune, Aker Sykehus, Oslo 5, Norway

Hélio Aguinaga, M.D., Director–President, Centro de Pesquisa, Assistência Integrada à Mulher e à Crianca, Rua D. Marina 136, Botofogo, Rio de Janeiro, Brazil

Etienne-Emile Baulieu, M.D., Ph.D., Professeur, Déparement de Chimie Biologique, Faculté de Médecine de Bicêtre, 78, Avenue du Général Leclerc, Université de Paris Sud, 94 Bicêtre, France

John D. Baxter, M.D., Professor, University of California, Endocrine Research Division, 671 HSE, San Francisco, California 94143

George S. Boyd, Ph.D., Professor and Head, Department of Biochemistry, University of Edinburg Medical School, Teviot Place, Edinburgh EH8 9 AG, Scotland

A. Oriol Bosch, M.D., Profesor, Catedra De Endocrinología Experimental, Facultad de Medicina, Universidad Complutense, Madrid 3, Spain

Paul Burgess, L.Th., M.S.P.H., Director, Program in Population and Health, Nova University, 3301 College Avenue, Fort Lauderdale, Florida 33314

A. Kent Christensen, Ph.D., Professor and Chairman, Department of Anatomy, Medical Sciences II Building, University of Michigan, Medical School, Ann Arbor, Michigan 48109

Jean Crabbé, M.D., Professeur, Département de Physiologie, Endocrinologie–Métabolisme, Faculté de Médecine, Université Catholique de Louvain, B 12001 Brussels, Belgium

Derek Denton, M.B.B.S., Director, Howard Florey Institute of Experimental Physiology and Medicine, University of Melbourne, Parkville, Victoria 3052, Australia

Donald Exley, Ph.D., Professor, Biochemistry Department, Queen Elizabeth College, University of London, London, England

Jean Garnier, Ph.D., Laboratoire de Biochimie Physique, Université de Paris—Sud, Bâtiment 433, 91405 Orsay, Cédex, France

Brian P. Setchell, Ph.D., Professor, Agricultural Research Council, Institute of Animal Physiology, Babraham, Cambridge CB2 4 AT, England

Harold Spies, Ph.D., Senior Scientist, Department of Reproductive Physiology, Oregon Regional Primate Center, 505 N. W. 185th Avenue, Beaverton, Oregon 97005

Samuel Spicer, M.D., Professor, Department of Pathology, Medical University of South Carolina, 171 Ashley Avenue, Charleston, South Carolina 29403

Anna Steinberger, Ph.D., Professor, Department of Reproductive Medicine and Biology, 6431 Fannin, Texas Medical Center, Houston, Texas 77025

Emil Steinberger, M.D., Professor and Head, Department of Reproductive Medicine and Biology, 6431 Fannin, Texas Medical Center, Houston, Texas 77025

Clara M. Szego, Ph.D., Professor, Department of Biology, University of California, Los Angeles, California 90024

Paul Talaly, M.D., Professor, The Johns Hopkins University School of Medicine, Department of Pharmacology and Experimental Therapeutics, 725 North Wolfe Street, Baltimore, Maryland 21205

Bun-ichi Tamoaki, Ph.D., Professor, National Institute of Radiological Sciences, 9-1, 4-chome, Anagawa, Chiba-shi, Japan

A. Tixier-Vidal, Ph.D., Neuroendocrinologie Cellulaire, Collège de France, 11 Place Marcelin Berthelot, 75231 Paris, Cédex 5, France

H. J. van der Molen, Professor, Afdeling Biochemie, Erasmus Universiteit Rotterdam, Postbus 1738, Rotterdam, Holland

Claude A. Villee, Ph.D., Andelot Professor of Biological Chemistry, Laboratory of Human Reproduction and Reproductive Biology, Harvard Medical School, 45 Shattuck Street, Boston, Massachusetts 02115

Dorothy B. Villee, M.D., Associate in Endocrinology, Assistant Professor of Pediatrics, Harvard Medical School, The Children's Hospital Medical Center, 300 Longwood Avenue, Boston, Massachusetts 02115

Donald L. Wilbur, Ph.D., Department of Anatomy, Medical University of South Carolina, Charleston, South Carolina 29403

Preface I

In September, 1977, at a conference organized by Dr. Kenneth McKerns in Northeast Harbor, Maine, USA, I was asked by the Editorial Committee of the *Biochemical Endocrinology* series to investigate the possibility of organizing the next meeting in France. I proposed a subject which is in the area of my research interest, and this subject was accepted.

On arriving back in France, I first looked for an appropriate place for the meeting, and the Chateau de Seillac was chosen in accordance with many objective criteria. We know that all who attended the meeting held in Seillac enjoyed this quiet and charming place in the Loire Valley.

The next step was to choose some experts in the field who would contribute to the monograph and present their papers at a conference for the purpose of generating discussions. The action of the local committee, composed of Dr. A. Tixier-Vidal, Dr. Claude Kordon, and me, was crucial in this respect. The local committee proposed the program for the meeting and a list of the majority of contributors to be invited. I wish to thank Dr. Tixier-Vidal and Dr. Kordon for their invaluable assistance.

Originally, we planned to organize this conference only for contributors to the monograph and for council members of the Society. However, several foreign and French colleagues inquired about the possibility of attending the meeting. We are pleased that many of them were able to attend. Some of them presented communications of recent and original data in relevant research areas.

The present volume covers many aspects of cellular and molecular mechanisms related to the synthesis and release of adenohypophyseal hormones. To our knowledge, it is the first time that fundamental biochemical mechanisms of pituitary function and morphological correlates are assembled in the same volume.

I am pleased to say that the conference of Seillac generated many discussions which were recorded and are published here. It also allowed for many private discussions and contacts and even generated collaborative works among those who attended. This was, of course, one of our objectives.

I would like to express my warmest thanks to the chairpersons, contributors, and discussants who contributed to the success of the meeting held in Seillac and to this volume. I also thank my secretary, Madame Hery, whose help was crucial in the organization of this meeting. I am grateful to my laboratory staff, who helped in running the meeting, and also to Madame Jutisz, who, with Madame Hery, organized ancillary local tours and concerts.

Marian Jutisz

Gif-sur-Yvette, France

Preface II

The sixth meeting of the International Society for Biochemical Endocrinology, held in late September, 1978, at the Chateau de Seillac, was considered to be a great success by the participants. The Society is indebted to Dr. Marian Jutisz, Dr. A. Tixier-Vidal, and Dr. Claude Kordon for the social and scientific arrangements of the meeting.

Time was allowed until the end of December for editing of dicussions and revision of manuscripts.

An extensive area of subject matter related to the central theme was covered, as may be seen by the chapters of this monograph. The topics include, for example, structural basis of adenohypophyseal secretory processes; cellular origins of adenohypophyseal hormones; morphological correlates of pituitary function; hormone biosynthesis and transcription mechanisms for FSH, LH, MSH, ACTH, β-lipotropin, and endorphin; methods of labeling hormones; mechanisms involved in hormone release, such as those involving GnRH and TRF; and functional reponses of dispersed cells.

The following served as chairpersons for the meeting: M. Jutisz, S. S. Spicer, A. Tixier-Vidal, J. Meldolesi, F. Peillon, J. D. Baxter, J. Crabbé, C. Villee, P. Jaquet, K. McKerns, C. Kordon, J. Kraicer, C. Labrie, H. Spies, C. Mialhe, and E. Herbert.

The 1979 meeting of the Society was held in September at the Asticou Inn, Northeast Harbor, Maine, on "Reproductive Processes and Contraception." The next meeting, in late September, 1980, will be in Dubrovnic, Yugoslavia, with the topic to be "Brain and Other Hormonally Active Peptides: Their Structure and Function." Professor Vladimir Pantić will be the local host.

Kenneth W. McKerns

Blue Hill Falls, Maine

Acknowledgments

The following Public Organizations and Private Contributors have given financial assistance to conduct this Colloque. The organizing committee is highly thankful to them for their contributions.

- Institut National de la Recherche Agronomique (I.N.R.A.), Paris, France
- Association Naturalia et Biologia, Paris, France
- Centre de Recherche Clin Midy, Montpellier, France
- Institut Choay, Paris, France
- Laboratoire Sandoz, Rueil Malmaison, France
- Produits Roche, Neuilly-sur-Seine, France
- Rhone-Poulenc-Santé, Paris, France

Contents

1

Structural Basis of Adenohypophyseal Secretory Processes 1

Andrée Tixier-Vidal

1. Introduction *1* • 2. Microenvironment of the Adenohypophyseal Cell *1* • 3. Chemical and Cellular Heterogeneity of the Anterior Pituitary Tissue *3* • 4. General Scheme of Structural Organization *3* • 5. Present Concept of the Secretory Cycle *5* • 6. Membrane Traffic in Anterior Pituitary Cells *8* • 7. Summary and Conclusions: Problems for the Future *9* • Discussion *10* • References *11*

2

Immunocytochemical Identification of LH- and FSH-Secreting Cells at the Light- and Electron-Microscope Levels   15

Claude Tougard

1. Introduction *15* • 2. Materials and Methods *17* • 3. Results *18* • 4. Discussion *28* • 5. Summary *32* • Discussion *32* • References *34*

3

Pituitary Polypeptide-Secreting Cells: Immunocytochemical Identification of ACTH, MSH, and Peptides Related to LPH 39

G. Tramu and J. C. Beauvillain

1. Introduction *39* • 2. Methods *41* • 3. Results *42* • 4. Conclusions *61* • Discussion *62* • References *63*

xix

4

Exocytosis and Related Membrane Events   67

E. Vila-Porcile and L. Olivier

1. Introduction *67* • 2. Morphological Data on Granular Exocytosis *68* • 3. Morphological Data on the Exocytosis of Residual Bodies *76* • 4. Morphological Data on the Phenomena of Endocytosis *78* • 5. Problems Raised by Exocytosis in Pituitary Cells *84* • 6. Problems of Endocytosis *89* • 7. Summary *94* • Discussion *94* • References *95*

5

Intracellular Events in Prolactin Secretion   105

Antonia Zanini, Giuliana Giannattasio, and Jacopo Meldolesi

1. Introduction *105* • 2. Biosynthesis and Transport to the Golgi Periphery *106* • 3. Packaging and Maturation *108* • 4. Exocytosis, Crinophagy, and Membrane Recycling *116* • 5. Problems Remaining for the Future *118* • Discussion *120* • References *121*

6

Separated Somatotrophs: Their Use in Vitro and in Vivo   125

W. C. Hymer, R. Page, R. C. Kelsey, E. C. Augustine, W. Wilfinger, and M. Ciolkosz

1. General Introduction *125* • 2. Separation of Somatotrophs from Dispersed Cell Suspensions *129* • 3. Release of Growth Hormone from Isolated Somatotrophs *134* • 4. Functional Heterogeneity of Pituitary Cell Types *144* • 5. Intracranial Implantation of Pituitary Cells *148* • 6. General Conclusions *161* • Discussion *162* • References *164*

7

Pituitary Secretory Activity and Endocrinophagy   167

D. L. Wilbur and S. S. Spicer

1. Introduction *167* • 2. Materials and Methods *168* • 3. Discussion *178* • Discussion *181* • References *184*

8

Ultrastructural Localization of LH and FSH in the Porcine Pituitary 187

F. Dacheux

9

Localization of Anti-ACTH, Anti-MSH, and Anti-α-Endorphin Reactive Sites in the Fish Pituitary 197

Ernest Follénius and Maurice P. Dubois

10

Changes in the Responsiveness to LH-RH in FSH-Treated 4-Day Cycling Female Rats 209

Jean-Marie Geiger

11

Role of Gonadotropin-Releasing Hormone in the Biosynthesis of LH and FSH by Rat Anterior Pituitary Cells in Culture 217

Ashok Khar and Marian Jutisz

12

Processing of the Common Precursor to ACTH and Endorphin in Mouse Pituitary Tumor Cells and Monolayer Cultures from Mouse Anterior Pituitary   237

Edward Herbert, Marjorie Phillips, Michael Hinman, James L. Roberts, Marcia Budarf, and Thomas L. Paquette

13

Biosynthesis of β-Endorphin in the Rat Pars Intermedia   263

P. Crine, F. Gossard, N. G. Seidah, C. Gianoulakis, M. Lis, and M. Chrétien

14

Effect of TRH on the Synthesis and Stability of Cytoplasmic RNAs of GH₃B₆ Cells   285

A. Morin

15

Hormonal Regulation of Prolactin mRNA   295

Glen A. Evans and Michael G. Rosenfeld

19

Enzymatic Degradation of Hypothalamic Hormones at the
Pituitary-Cell Level: Possible Involvement in Regulation
Mechanisms 381

Karl Bauer

20

Methods of Labeling Pituitary Hormones 401

P. De La Llosa

21

Mechanism of Action of Hypothalamic Hormones and
Interactions with Sex Steroids in the Anterior
Pituitary Gland 415

Fernand Labrie, Pierre Borgeat, Martin Godbout, Nicholas Barden, Michèle Beaulieu, Lisette Lagacé, Jocelyne Massicotte, and Raymonde Veilleux

22

23

24

25

Pharmacological Studies of Pituitary Receptors Involved in the Control of Prolactin Secretion   525

Alain Enjalbert, Merle Ruberg, Sandor Arancibia, Mirette Priam, and Claude Kordon

26

The Release of ACTH by Human Pituitary and Tumor Cells in a Superfusion System   543

G. H. Mulder and D. T. Krieger

27

Interaction of Glucocorticoids, CRF, and Vasopressin in the Regulation of ACTH Secretion   561

B. Koch, B. Lutz-Bucher, and C. Mialhe

28

Corticotropin Secretion   587

Mortyn T. Jones and Brian Gillham

29

Modulation of Pituitary Responsiveness to Gonadotropin-Releasing Hormone 617

George Fink and Anthony Pickering

30

The Pattern of LH Release of Rat Pituitary Glands during Long-Term Exposure to LH-RH in Vitro 639

Jurrien de Koning, Hannie A. M. J. van Dieten, and G. Peter van Rees

31

Functional Heterogeneity of Separated Dispersed Gonadotropic Cells 659

Carl Denef

36

Intracellular Recordings from Prolactin-Secreting Pituitary Cells in Culture: Evidence for a Direct Action of Estrogen on the Cell Membrane   765

B. Dufy, H. Fleury, D. Gourdji, A. Tixier-Vidal, P. du Pasquier, and J. D. Vincent

37

Extracellular Potassium Change in the Rat Adenohypophysis: An Indicator of Neurohypophyseal– Adenohypophyseal Communication   775

A. J. Baertschi, M. Friedli, J. Munoz, M. Tsacopoulos, and J. A. Coles

Index   783

Structural Basis of Adenohypophyseal Secretory Processes

Andrée Tixier-Vidal

1. Introduction

The cellular and molecular mechanisms that direct the synthesis and release of adenohypophyseal hormones cannot be discussed without taking account of the structural organization of the corresponding cells. Indeed, structure and function represent two aspects of an integrated system: the living cell. The morphological organization of the anterior pituitary cell reflects the sites of location of the molecular and physiological events involved in adenohypophyseal secretory processes and their regulation. In addition, this structural organization itself suggests many working hypotheses to the cell physiologist and the molecular biologist. Before considering structure at the single-cell level—which can be done only in experimental conditions and *in vitro*—I shall briefly discuss the microenvironment of the adenohypophyseal glandular cells within the anterior pituitary tissue for a complete view of the situation.

2. Microenvironment of the Adenohypophyseal Cell

Two aspects of the microenvironment of the adenohypophyseal cells within the anterior pituitary tissue may be emphasized.

Andrée Tixier-Vidal • Groupe de Neuroendocrinologie Cellulaire, Chaire de Physiologie Cellulaire, Collège de France, 75231 Paris 05, France

2.1. Relationship with Capillaries

The capillaries that penetrate between the epithelial cords of the anterior pituitary tissue are the sites for both the distribution *to the cell* of chemical messengers that regulate their function and the collection *from the cells* of their secretory products. It can be inferred from the structure of the pericapillary spaces (Farquhar, 1961; Vila-Porcile, 1973) that those molecules must cross, successively, several barriers: the capillary endothelial cells, which possess fenestrae covered by thin diaphragms, the basement membrane of the endothelium, the perivascular space, the basement membrane of the epithelial cords, and the plasma membrane of the glandular cells. This transfer can occur in both senses depending on the origin of the molecules. It involves most probably a large variety of mechanisms including pinocytosis and endocytosis, binding at specific sites on plasma membranes, exocytosis, and active or passive transport either through membranes or between cells. Most of these mechanisms are still poorly understood (see Chapter 4).

2.2. Relationship with the Other Cells in the Epithelial Pituitary Cords

The epithelial pituitary cords contain, in addition to glandular cells, agranular or follicular cells (Farquhar, 1957; Vila-Porcile, 1972), the exact role of which is still hypothetical. These cells are associated by complex junctions and display an intense endocytotic activity (Farquhar *et al.,* 1975). They may participate in the transport or uptake of macromolecules within the pituitary tissue as well as in the metabolism of the glandular cells.

The existence of communication, or of transfer of information, between the glandular cells cannot be excluded. In contrast to the case of the follicular cells, the morphological evidence for the existence of specialized junctions between the glandular cells is still very limited. Nevertheless, in some cases, the cell shape and the frequent morphological association of different cell types strongly suggest a possible functional link between them. A good example seems to be that of the prolactin cells in the normal rat pituitary, which display an irregular outline and very often send narrow cytoplasmic extensions around gonadotropic cells. The presence of tight junctions between these two cell types has recently been reported in the rat pituitary (Horvath *et al.,* 1977). The transfer of information between glandular cells of the same type also cannot be excluded. For example, in the quail, the stimulation of thyreotropic cells following thyroidectomy occurs simultaneously in a few independent small clumps of two or three cells before it occurs in a large area of the anterior pituitary (Tixier-Vidal *et al.*, 1972). This finding suggests the transfer of message from cell to cell, the experimental demonstration of which remains difficult.

3. Chemical and Cellular Heterogeneity of the Anterior Pituitary Tissue

The chemical variety of the secretory products of the anterior pituitary corresponds to a cellular heterogeneity. This heterogeneity was long ago postulated by several famous cytologists who discovered that the anterior pituitary cells display various tinctorial affinities that were supposed to be related to the nature of secretory products stored within the cytoplasm. Progress in the chemistry of the anterior pituitary together with that in cytochemistry reinforced this interpretation (see the reviews by Herlant, 1964, 1975). Complete confirmation of the cellular heterogeneity of the anterior pituitary tissue was obtained with the development of immuno-cytochemistry (Nakane, 1970). This led to the proposal of the concept of "one hormone, one cell." This concept is now questioned, however, due to progress in the biochemistry of several anterior pituitary hormones as well as to progress in the specificity of the antibodies that were used for immunocytochemical studies. In some cases, such as the two gonadotropic hormones (LH and FSH) and the polypeptide hormones of the endorphin or β-lipotropin series, there is increasing evidence for several hormones being produced in the same cell (see Chapters 2, 3, and 7).

A practical consequence of the cellular heterogeneity of the anterior pituitary tissue was the search for simplified systems that would enable work to be done with homogeneous populations of cells. This led to considerable development of *in vitro* studies using either dispersed and separated normal anterior pituitary cells (see Chapters 6, 24, and 31) or tumor-derived clonal cell lines (see Chapters 12, 16, and 23).

4. General Scheme of Structural Organization

The general scheme of structural organization of an anterior pituitary cell, as represented in Fig. 1, is the same as for other endocrine cells and even exocrine cells. It is based on electron-microscopic observations that yielded a considerable amount of evidence for the existence of three compartments within the cytoplasm: the rough endoplasmic reticulum (RER), the Golgi apparatus (smooth endoplasmic reticulum, SER), and the membrane-bound secretory granules. In fact, this model is best exemplified by prolactin cells and growth-hormone cells. It may indeed

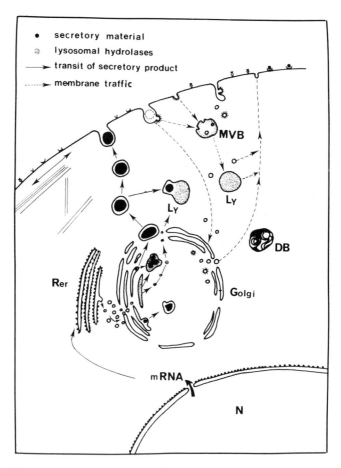

Figure 1. Schematic representation of transit of secretory product and membrane traffic in an ideal anterior pituitary cell in culture. This model is based on partial evidence from prolactin cells and growth-hormone cells. *Transit of secretory product* (→): Cytoplasmic messenger RNA (mRNA) is translated on attached polysomes, and the product of translation is discharged into the rough endoplasmic reticulum (Rer). It is then transported via smooth vesicles, or possibly by direct continuity, to the core of the Golgi zone, where small granules arise from the innermost cisternae. During this step, some biochemical modifications as well as concentration of secretory product occur. In some cases, the secretory products are condensed simultaneously with lysosomal hydrolases or at least with acid phosphatases as revealed by cytochemistry. A separate pathway for lysosomal hydrolases to primary lysosomes may also be postulated. During active secretion, the mature secretory granules move toward the plasma membrane, where they are discharged after fusion of their membrane with the plasma membrane. Alternatively, secretory granules can fuse with lysosomes, where they are degraded (crinophagy). Whether this fusion begins within the Golgi zone or occurs secondarily after complete maturation of the granules is not known, nor are the mechanisms that regulate this phenomenon. *Membrane*

undergo important variations depending on the nature of the secretory product as well as of the functional activity. It was long ago observed that the density of the ribosomes bound to the outer face of the RER is higher in cells that produce proteins than in those that produce glycoproteins (see the review by Herlant, 1964). In the latter, the RER cisternae are often dilated, as in, for example, the gonadotropic cells of the so-called FSH type, the thyreotropic cells after thyroidectomy. In contrast, the RER cisternae are generally flattened in prolactin, as well as in growth-hormone-producing cells. The decrease in number and diameter of the secretory granules that occurs simultaneously with the extension of the RER and of the Golgi zone in highly active pituitary cells is also a classic datum. In this respect, an extreme situation is represented by the cells of the prolactin- and growth-hormone-secreting lines that contain very few small secretory granules (see Chapter 23).

5. Present Concept of the Secretory Cycle

5.1. The Scheme

The first scheme for the secretion of an anterior pituitary hormone was proposed in 1966 for the prolactin cell by Smith and Farquhar (1966). It was based for the most part on the concept already established by Palade and his associates on the exocrine pancreas cell, and it has since been shown to be applicable to most protein-secreting cells (see the review by Palade, 1975). Its main steps, together with their sites of location in the anterior pituitary cell, can be summarized as follows: (1) synthesis on attached polysomes and discharge into the RER; (2) concentration and formation of secretory granules in the center of the Golgi zone; (3) storage in mature secretory granules; (4) extracellular discharge of granules by

---

traffic (---→): The plasma membrane possesses at its outer face various recognition sites (∨,∪,⊔) specific for various ligands (▼, ●, ■). Binding of these ligands on their respective sites in the plasma membrane initiates a series of events that results, finally, in a modification of membrane traffic. Two types of events can be distinguished: (1) After release of secretory granules by exocytosis, their membrane may be either integrated and degraded into lysosomes or recirculated back to the Golgi apparatus by way of small vesicles. (2) The interaction of ligands with their specific binding sites in the plasma membrane induces both a modification of the lateral mobility of membrane components (←→) and stimulation of pinocytosis and endocytosis. This results in the internalization of fragments of plasma membrane that might be either degraded in lysosomes or recirculated by way of small vesicles to the Golgi zone and then to the plasma membrane. A direct shuttle of the vesicles back to the plasma membrane cannot be excluded. Finally, the cytoskeleton [filaments and microtubules (═)] is most probably involved in the various membrane events, although direct evidence for such a connection is still very poor in the anterior pituitary cells.

exocytosis or, alternatively, intracellular fusion of secretory granules in excess with lysosomes (crinophagy). The reader will find more details in the Fig. 1 caption.

The experimental demonstration of this scheme in the anterior pituitary cell is still very incomplete at present, although important progress has been made in recent years (for a discussion, also see Farquhar, 1977).

5.2. Some Biochemical Evidence for the Scheme

Biochemical studies have provided evidence for the synthesis *in vitro* by isolated polysomes, or by cytoplasmic messenger RNA in cell-free systems of the preprolactin (Biswas and Tashjian, 1974; Evans and Rosenfeld, 1976), of pre-growth hormone (Lingappa *et al.*, 1977; Martial *et al.*, 1977) (also see Chapter 16), and of ACTH (see Chapter 12). The processing of precursor product into hormone has been obtained *in vitro* for prolactin and for growth hormone (same authors as above) and by pulse–chase experiments for ACTH and endorphin (see Chapters 12 and 13). The biochemical mechanisms involved in moving and packaging of prolactin granules in the Golgi zone have also been recently studied (see Chapter 5).

5.3. Evidence for the Scheme from High-Resolution Autoradiography

Most of the evidence for the transit of secretory products in anterior pituitary cells comes from quantitative autoradiographic studies of pulse–chase experiments that were mainly performed using [^{3}H]leucine on prolactin or growth hormone cells (Racadot *et al.*, 1965; Tixier-Vidal and Picart, 1967; Howell and Whitfield, 1973; Farquhar *et al.*, 1978). In addition to the first demonstration of a concentration of pulse-labeled protein in the Golgi zone, these studies provided interesting information on the timetable of the transit of secretory proteins in the successive cell compartments. The drainage of the Golgi zone is generally longer in anterior pituitary cells than in an exocrine cell. In addition, it seems to vary with the ultrastructural features and the functional activity: it is less rapid in duck prolactin cells in organ culture, which display dilated RER cisternae and secrete to a relatively low level (Tixier-Vidal and Picart, 1967), than in freshly dispersed prolactin cells taken from estradiol-stimulated rat, which can be presumed to secrete prolactin at a high level (Farquhar *et al.*, 1978). These observations strongly suggest that the speed of transit of secretory protein might be correlated with the turnover time of the hormone. More studies are needed on this aspect of the secretory cycle.

5.4. Evidence for the Scheme from Enzyme Cytochemistry

In addition to high-resolution autoradiography, electron-microscopic enzyme cytochemistry has also provided useful information on the secretory cycle of anterior pituitary cells. The cytochemical localization of acid phosphatases in prolactin cells of the lactating rat after removal of sucklings demonstrated the importance of crinophagy in the regulation of imbalance between synthesis and release of anterior pituitary hormones (Smith and Farquhar, 1966). In addition, the cytochemical visualization of acid phosphatase activities in several anterior pituitary cell types revealed that these activities are concentrated in the innermost Golgi cisternae and sometimes around nascent and even mature secretory granules (Smith and Farquhar, 1966; Farquhar, 1969; Tixier-Vidal and Picart, 1971; Pelletier and Novikoff, 1972). Such findings raise the question of a simultaneous segregation of lysosomal enzymes and secretory products and of the role in this process of the Golgi–endoplasmic reticulum–lysosome (GERL) complex as defended by Novikoff and his associates (see the review by Novikoff, 1976).

5.5. Evidence for the Scheme from Electron-Microscopic Immunocytochemistry

Electron-microscopic immunocytochemistry, by permitting the subcellular localization of hormone immunoreactivity, provided information about the secretory cycle that is complementary to that obtained from high-resolution autoradiography. First of all, it confirmed that the secretory granules are indeed the site of maximum concentration of immunoreactive material, in any cell type (Moriarty, 1973; Nakane, 1975). In addition to secretory granules, other subcellular sites were also found to contain immunoreactive material, at least in some cases.

The content of RER cisternae was found positive in some rat gonadotropic cells, using antisera against ovine or rat LH β (Tougard *et al.*, 1973, 1979) (also see Chapter 2) and in rat thyroidectomy cells with an antiserum against rat TSH β (Moriarty and Tobin, 1976).

A diffuse staining of the cytoplasm was also found, when the "preembedding" method was used, with prolactin cells and gonadotropic cells as well as thyreotropic cells (Tixier-Vidal *et al.*, 1976; Tougard *et al.*, 1973, 1979). Since such a cytoplasmic staining was also observed in degranulated cells, it cannot be considered only as an artifact of fixation. This finding therefore raises the problem of the existence of an extragranular pool of hormones or, at least, of the possibility of the detection of an immunoreactive material at the outer face of attached polysomes during the translation process (see Tixier-Vidal *et al.*, 1976). The role of such an agranular pool of hormones in the secretory cycle remains difficult to explain. It can be noted that pulse–chase experiments performed in the SD1 prolactin cell line, which displays a cytoplasmic staining, provided evidence for prolactin storage in multiple pools, the kinetics of discharge of which differed in basal conditions with respect to stimulation by TRH (Morin *et al.,* 1975).

Curiously enough, the content of the Golgi cisternae was never stained with any cell type despite the presence of positive nascent secretory granules in the core of the Golgi zone. Whether this results from an absence of hormone precursors within the outer cisternae, which might be bypassed, or from an inacessibility of antigenic determinants during the transit into these cisternae remains to be established.

6. Membrane Traffic in Anterior Pituitary Cells

The transport of secretory products within the endoplasmic reticulum of secretory cells obviously involves a primordial role of the membranes. This has been the subject of numerous studies, biochemical as well as morphological, in several secretory cell types. However, for the time being, the available evidence does not justify "a simple, universally applicable schema" (see the review by Meldolesi *et al.,* 1978). In addition, biochemical studies on membranes have been performed mostly in the liver, the exocrine pancreas, the parotid gland, and the adrenal medulla, whereas in the anterior pituitary cell, the available evidence comes almost exclusively from morphological studies.

As summarized in the Fig. 1 caption, two aspects may be distinguished in the analysis of membrane events in anterior pituitary cells: (1) those that are related to the release of secretory granules by exocytosis and (2) those that are related to the interaction of the cell with the various ligands that regulate its secretory activity. The first aspect is the subject of Chapter 4. I will briefly consider the second aspect, which has only recently become the subject of experimental analysis.

The secretion of each anterior pituitary hormone is regulated, positively or negatively, by various hormones originating from both the hypothalamus and the peripheral glands (see Chapter 21). Binding of these ligands on specific binding sites initiates a cascade of events that are not completely understood (see Chapter 22) and that result in the stimulation, or inhibition, of the release and the synthesis of adenohypophyseal hormones. In the case of polypeptide hormones, at least, and perhaps also of some steroid and thyroid hormones, the binding sites are located on the plasma membrane. In view of the present concept of the fluid-mosaic model of the plasma membrane, the effects of such interaction on the mobility of plasma membrane components have been recently investigated by various methods in our laboratory, using as a model system the stimulation by thyroliberin (TRH) of the secretion of prolactin by the SD1 cell line (see Chapter 23).

It has been shown (Brunet and Tixier-Vidal, 1978) that pretreatment of SD1 cells with TRH induces a significant and transient increase of the exposure of surface glycoproteins as measured by the increase in the number of concanavalin A (Con A) binding sites. This effect is specific for TRH and requires that TRH be used under the conditions of equilibrium of binding and maximum occupancy of its high-affinity binding sites. It is not directly related to exocytosis, since it can be observed in the absence of prolactin release and vice versa. Last, since Con A does not interact on TRH binding, this effect might reflect a change in the molecular conformation of TRH receptor after interaction with its ligand.

Evidence for a stimulation of endocytosis by TRH has also been obtained using horseradish peroxidase (PO) as a marker solute of endocytosis. The uptake of PO by SD1 cells involves classic pinocytotic structures and results in the labeling of dense bodies, or lysosomes, that are located in the Golgi zone. Simultaneous exposure to TRH and to PO leads to an increase in the number of labeled dense bodies correlated with an increase in the amount of internalized PO as measured by enzymatic assay. This effect is specific for TRH and seems independent of exocytosis, since it is also observed in the absence of stimulation of prolactin release (Tixier-Vidal *et al.*, 1979). In addition, using Con A as a marker of surface membrane, it was found that SD1 cells internalize 60% of their plasma membrane within 1 hr at 37°C (Tixier-Vidal *et al.*, 1979), which is consistent with other data obtained by other methods in fibroblasts in culture (Steinman *et al.*, 1976) where this strongly suggested the reutilization of part of the internalized membrane fragments (Schneider *et al.*, 1977).

It appears from these studies that the membrane traffic in anterior pituitary cells reflects complex events that are related not only to exocytosis but also to modifications of the lateral and vertical mobility of plasma membrane components as a consequence of interaction of ligands with their specific binding sites. This new aspect of the role of membranes in the anterior pituitary secretory process deserves further work.

7. Summary and Conclusions: Problems for the Future

In this introductory chapter, I have reviewed the structural basis of adenohypophyseal secretory processes. The following points have been briefly considered: the microenvironment of the adenohypophyseal cell, the chemical and cellular heterogeneity of the anterior pituitary tissue, and the general scheme of structural organization. In addition, the present concept of the secretory cycle of the anterior pituitary cell has been outlined and the available evidence for this scheme summarized. Last, membrane traffic related to hormone secretion as well as to cell interaction with specific ligands has been considered.

It is evident from this brief review of the structural basis of adeno-hypophyseal secretory processes that many problems are left for further research, at every step of the secretory cycle. What are the molecular mechanisms involved in regulation of synthesis at both the transcriptional and translational levels? In intracellular transport of secretory material? In hormone release (is there an extragranular pool of hormone? is exocytosis the only means of hormone discharge?)? In regulation of membrane traffic (problem of the recycling of membrane components after exocytosis and endocytosis)? What is the role of the cytoskeleton in adenohypophyseal secretory processes?

Most of these questions have still only partial answers or no answers at all. They are discussed throughout this volume.

ACKNOWLEDGMENTS. I express my thanks to Mrs. Picart, Mr. Pennarun, and Miss A. Bayon for their valued assistance in the preparation of Fig. 1 and of the manuscript.

DISCUSSION

SPICER: You showed, I believe, dilatation of rough ER in cultured cells with sparse secretory granules and found radioautographic evidence that the rate of movement of hormones out of the rough ER was slow. Do these interesting findings prompt the speculation that cells with dilated rough ER are processing the synthetic product through the Golgi slowly and hence contain relatively increased storage of secretion in the ER rather than in the granules? Plasma cells often show such dilated cisternae perhaps when they are not secreting immunoglobulin. One wonders what factors enhance or retard processing of secretion out of the rough endoplasmic reticulum into the Golgi granules.

Concerning the hormone demonstrated in the cytoplasm by ultrastructural immuno-staining, there are problems about cell biologic mechanisms whereby first the unbound ribosomes would function to synthesize and release it in the cytoplasm in these cells and by what means such protein could be released from the cell without defying the general rule against cell plasma membrane having permeability for macromolecules. This is a difficult question to assess, as both the cytochemical and biochemical methods for investigating the problem are subject to diffusion artifact. Routes of secretion directly from rough ER, as possibly occurs in placental syncytiotrophoblast, or in Golgi-derived vesicles bypassing storage in granules, as may occur in pancreas exhausted of granules by cholinergic stimulation or in thyroid follicular cells, are known but do not entail macro-molecules permeating the plasmalemma. Such processes might occur in the GH3 cultured pituitary cells lacking granules.

TIXIER-VIDAL: You raise very important questions which can be answered by speculation only. As to the autoradiographic evidence for a slow rate of movement of molecules in cells with dilated rough ER, I agree with you that such cells display an ability to store their secretory product, or some precursor product, into the rough ER cisternae. Such an interpretation is consistent with the presence of rounded dense droplets in rough ER dilated cisternae of thyreotropic cells as reported by Farquhar in the thyroidectomized rat after breaf thyroxine treatment. The presence in such cisternae of a material immunoreactive with a serum against rat TSHβ has been reported by Moriarty and Tobin. Of course, this does not represent a general situation, but this can occur in some physiological or experimental conditions. The mechanism which regulates the rate of transit of secretory material from the rough ER to the secretory granules is not known in the anterior pituitary cell, at least to my knowledge. It may be related with the turnover of the hormone.

As to the immunostaining of the "cytoplasm," I have to be precise in stating that the reaction product *seems* to be located on polysomes. I must also say that in the case of GH3 cells, most of the ribosomes are in fact bound rather than free. The absence of linear rough cisternae does not mean that there are only free ribosomes. Indeed, attached polysomes extracted from GH3 cells can direct *in vitro* the synthesis of preprolactin according to Biswas and Tashjian. The material which is immunostained in the cytoplasm may correspond to the exposure of antigenic sites of prolactin molecules in the way of being synthesized during the translation process. It may also correspond to the content of small vesicles, the membrane of which is destroyed by the fixation. I agree with you that it is a difficult question to assess because of the lack of satisfactory methods. (See also the answer to Dr. Hymer.)

HYMER: With regard to the GH3 cell, I understand that the granules are much smaller than the "normal" prolactin granules. This "granular pool" must represent a small percentage of total intracellular hormone. My question then is, how is this larger pool released?

TIXIER-VIDAL: In fact, I can answer your question by speculation only, since we have no direct evidence for the subcellular localization of the different pools, or compartments, which have been revealed by the pulse–chase experiments on SD1 cells. If the large pool corresponds to an "agranular," cytoplasmic compartment, there is so far no satisfactory explanation for the mechanism whereby those large molecules can cross the plasma membrane. In order to be in agreement with the "orthodox view" of the Palade concept, one must think that the hormone is transported within the cytoplasm inside small vesicles — which are numerous in the cytoplasm of those cells — and then released from the cells by fusion of vesicles with the plasma membrane. These vesicles might derived from the rough ER. One may also think about a high speed of the formation and transit of secretory granules and vesicles which may make them difficult to follow on electron micrography. It is to be emphasized that these cells do possess granules, even if these granules are very few, and that they have a well-developed Golgi zone. In other words, they display the three compartments of a normal prolactin cell, but the ratio prolactin content/prolactin in the medium is lower than in normal prolactin cells.

REFERENCES

Biswas, D. K., and Tashjian, A. H. Jr., 1974, Intracellular site of prolactin synthesis in rat pituitary cells in culture, *Biochem. Biophys. Res. Commun.* **60**:241.

Brunet, N., and Tixier-Vidal, A., 1978, Increased binding of concanavalin A at the cell surface following exposure to thyroliberin, *Mol. Cell. Endocrinol.* **11**:169.

Evans, G. A., and Rosenfeld, M. G., 1976, Cell-free synthesis of prolactin precursor directed by mRNA from cultured rat pituitary cells, *J. Biol. Chem.* **251**:2842.

Farquhar, M. G., 1957, "Corticotrophs" of the rat adenohypophysis as revealed by electron microscopy, *Anat. Rec.* **127**:291 (abstract).

Farquhar, M. G., 1961, Fine structure and function in capillaries of the anterior pituitary gland, *Angiology* **12**:270.

Farquhar, M. G., 1969, Lysosome function in regulating secretion: Disposal of secretory granules in cells of the anterior pituitary gland, in: *Lysosomes in Biology and Pathology,* Vol. 2 (J. T. Dingle and H. B. Fell, eds.), pp. 462–482, North-Holland, Amsterdam.

Farquhar, M. G., 1977, Secretion and crinophagy in prolactin cells, in: *Comparative Endocrinology of Prolactin* (H. D. Dellman, J. A. Johnson, and D. M. Klachko, eds.), pp. 37–91, Plenum Press, New York.

Farquhar, M. G., Skutelsky, E. H., and Hopkins, C. R., 1975, Structure and function of the anterior pituitary and dispersed pituitary cells: *In vitro* studies, in: *The Anterior Pituitary* (A. Tixier-Vidal and M. G. Farquhar, eds.), pp. 83–135, Academic Press, New York.

Farquhar, M. G., Reid, J. J., and Daniell, L. W., 1978, Intracellular transport and packaging of prolactin: A quantitative electron microscope autoradiographic study of mammotrophs dissociated from rat pituitaries. *Endocrinology* **102**:296.

Herlant, M., 1964, The cells of the adenohypophysis and their functional significance, *Int. Rev. Cytol.* **17**:299.

Herlant, M., 1975, Introduction, in: *The Anterior Pituitary* (A. Tixier-Vidal and M. G. Farquhar, eds.), pp. 3–15, Academic Press, New York.

Horvath, E., Kovacs, K., and Ezrin, C., 1977, Junctional contact between lactotrophs and gonadotrophs in the rat pituitary, *IRCS Medical Science:* Cell and Membrane Biology; Endocrine system; Experimental Animals; Nervous System; Pathology; Physiology **5**:511.

Howell, S. L., and Whitfield, M., 1973, Synthesis and secretion of growth hormone in the rat anterior pituitary. I. The intracellular pathway, its time course and metabolic requirements, *J. Cell Sci.* **12**:1.

Lingappa, V. R., Devillers-Thiery, A., and Blobel, G., 1977, Nascent prehormones are intermediates in the biosynthesis of authentic bovine pituitary growth hormone and prolactin, *Proc. Natl. Acad. Sci. U.S.A.* **74**:2432.

Martial, J. A., Baxter, J. D., Goodman, H. M., and Seeburg, P. H., 1977, Regulation of growth hormone messenger RNA by thyroid and glucocorticoid hormones, *Proc. Natl. Acad. Sci. U.S.A.* **74**:1816.

Meldolesi, J., Borgese, N., De Camilli, P., and Ceccarelli, B., 1978, Cytoplasmic membranes and the secretory process, in: *Membrane Fusion* (G. Poste and G. L. Nicolson, eds.), pp. 510–599, Elsevier/North Holland, Amsterdam.

Moriarty, G. C., 1973, Adenohypophysis: Ultrastructural cytochemistry—A review, *J. Histochem. Cytochem.* **21**:855.

Moriarty, G. C., and Tobin, R. B., 1976, An immunocytochemical study of TSHβ storage in rat thyroidectomy cells with and without D or L thyroxine treatment, *J. Histochem. Cytochem.* **24**:1140.

Morin, A., Tixier-Vidal, A., Gourdji, D., Kerdelhué, B., and Grouselle, D., 1975, Effect of thyrotrope releasing hormone (TRH) on prolactin turnover in culture, *Mol. Cell. Endocrinol.* **3**:351.

Nakane, P. K., 1970, Classification of anterior pituitary cells with immunoenzyme histochemistry, *J. Histochem. Cytochem.* **18**:9.

Nakane, P. K., 1975, Identification of anterior pituitary cells by immunocytochemistry, in:

The Anterior Pituitary (A. Tixier-Vidal and M. G. Farquhar, eds.), pp. 45–64, Academic Press, New York.

Novikoff, A. B., 1976, The endoplasmic reticulum: A cytochemist's view (a review), *Proc. Natl. Acad. Sci. U.S.A.* **73**:2781.

Palade, G. E., 1975, Intracellular aspects of the process of protein secretion, *Science* **189**:347.

Pelletier, G., and Novikoff, A. B., 1972, Localization of phosphatase activities in the rat anterior pituitary gland, *J. Histochem. Cytochem.* **20**:1.

Racadot, J., Olivier, L., Porcile, E., and Droz, B., 1965, Appareil de Golgi et origine des grains de sécrétion dans les cellules adénohypophysaires chez le rat: Étude radioautographique en microscopie électronique après injection de leucine tritiée, *C. R. Acad. Sci.* **261**:2972.

Schneider, Y. J., Tulkens, P., and Trouet, A., 1977, Recycling of fibroblast plasma membrane antigens internalized during endocytosis, *Trans. Biochem. Soc.* **5**:1164.

Smith, R. E., and Farquhar, M. G., 1966, Lysosome function in the regulation of the secretory process in cells of the anterior pituitary gland, *J. Cell. Biol.* **31**:319.

Steinman, R. M., Scott, E. B., and Cohn, Z. A., 1976, Membrane flow during pinocytosis: A stereological analysis, *J. Cell. Biol.* **68**:665.

Tixier-Vidal, A., and Picart, R., 1967, Étude quantitative par radioautographie au microscope électronique de l'utilisation de la DL-leucine-³H par les cellules de l'hypophyse du canard en culture organotypique, *J. Cell. Biol.* **35**:501.

Tixier-Vidal, A., and Picart, R., 1971, Electron microscope localization of glycoproteins in pituitary cells of duck and quail, *J. Histochem. Cytochem.* **19**:775.

Tixier-Vidal, A., Chandola, A., and Franquelin, F., 1972, Cellules de thyroïdectomie et cellules de castration chez la caille japonaise, *Coturnix coturnix japonica:* Étude ultrastructurale et cytoenzymologique, *Z. Zellforsch.* **125**:506.

Tixier-Vidal, A., Tougard, C., and Picart, R., 1976, Subcellular localization of some protein and glycoprotein hormones of the hypothalamo–hypophyseal axis as revealed by the peroxidase-labeled antibody method, in: *Immunoenzymatic Techniques* (G. Feldmann, P. Druet, J. Bignon, and S. Avrameas, eds.), pp. 307–321, North-Holland, Amsterdam.

Tixier-Vidal, A., Brunet, N., and Gourdji, D., 1979, Plasma membrane modifications related to the action of TRH on rat prolactin cell lines, in: *Cold Spring Harbor Conferences on Cell Proliferation,* Vol. 6 (R. Ross and G. Sato, eds.), Chapter 56, Cold Spring Harbor Press, Cold Spring Harbor, N.Y.

Tougard, C., Kerdelhué, B., Tixier-Vidal, A., and Jutisz, M., 1973, Light and electron microscopic localization of the binding sites of antibodies against ovine luteinizing hormone and its two subunits as revealed by the peroxidase labeled antibody technique, *J. Cell Biol.* **58**:503.

Tougard, C., Picart, R., and Tixier-Vidal, A., 1979, Immunocytochemical localization of glycoprotein hormones in the rat anterior pituitary: A light and electron microscope study using antisera against rat β subunits, *J. Histochem. Cytochem.* (in press).

Vila-Porcile, E., 1972, Le réseau des cellules folliculostellaires et les follicules de l'adénohypophyse du rat (pars distalis), *Z. Zellforsch.* **129**:328.

Vila-Porcile, E., 1973, La pars distalis de l'hypophyse chez le rat: Contribution à son étude histologique et cytologique en microscopie électronique, *Ann. Sci. Nat. Zool. Biol. Anim.* **15**:63.

2

Immunocytochemical Identification of LH- and FSH-Secreting Cells at the Light- and Electron-Microscope Levels

Claude Tougard

1. Introduction

The biochemical and functional duality of the two gonadotropic hormones, luteinizing hormone (LH) and follicle-stimulating hormone (FSH), is a fact clearly established. The production sites of these two hormones have been studied for many years by different methods in the rat pituitary and in several other mammalian species. Nevertheless, the problem of the cellular origin of gonadotropic hormones is still controversial.

On the basis of histophysiological correlations, cytologists have described two gonadotropic cell types. With the light microscope, using alcian blue–periodic acid–Schiff (AB-PAS) staining, according to Herlant (1960, 1964), LH cells were red-purple (PAS-positive) and FSH cells were violet (PAS-positive and AB-positive) in the pituitary of several mammalian species. In the rat pituitary, however, this distinction was far less evident (Purves, 1966). With the electron microscope, after the classic studies of Farquhar and Rinehart (1954) on pituitaries of castrated female rats and

Claude Tougard • Groupe de Neuroendocrinologie Cellulaire, Laboratoire de Physiologie Cellulaire, Collège de France, 75231 Paris 05, France

later on of Barnes (1962, 1963) on mice pituitaries, Kurosumi and Oota (1968) distinguished two different cell types in the pituitaries of persistent estrous and diestrous rats: FSH cells characterized by a large size and two types of round secretory granules (200 nm and 300–700 nm in diameter) and LH cells characterized by an ovoid shape and secretory granules of uniform size (200–250 nm in diameter). Thereafter, these morphological criteria were generally used by all cytologists to distinguish gonadotropic cells in rats.

With the advent of immunocytochemistry, this classification has been reconsidered, since the two gonadotropic hormones were found in the same cells using antisera (AS) against whole hormones in the castrated rat (Monröe and Midgley, 1966, 1969; Leleux and Robyn, 1970, 1971; Leleux et al., 1968) and in male rat pituitaries (Nakane, 1970). More recently, knowledge of the quaternary structure of the pituitary glycoprotein hormones, which consists of two subunits (the α subunit common to gonadotropic hormones and thyrotropic hormone and the β subunit specific for each hormone), and the use of AS against β subunits has permitted a more accurate identification of cells synthesizing each of these hormones.

Using for the first time AS against β subunit, we have found in ovariectomized rats (Tougard et al., 1971) and, later, in normal male rats (Tixier-Vidal et al., 1975) that the same cells are immunochemically stained by AS against ovine LHβ and ovine FSH at the light- and electron-microscope levels.

Other studies have confirmed the finding that all gonadotrophs contain both hormones at the light-microscope level in the rat (Herbert, 1975; Bugnon et al., 1977), in man (Pfifer et al., 1973), and in other mammalian species (Herbert, 1976; Beauvillain, 1978). Nevertheless, there is still controversy over this question, since some workers found some gonadotropic cells containing only one hormone at the electron-microscope level in the rat (Nakane, 1970, 1975; Moriarty, 1976) and at the light-microscope level in man (Pelletier et al., 1976) and in the dog (El Etreby and Fath El Bab, 1977).

In this chapter, we review our results obtained with heterologous AS against ovine hormones (Tougard et al., 1973; Tixier-Vidal et al., 1975) and with homologous AS against the β subunits of rat LH and rat FSH (Tougard et al., 1979) in male rat and castrated male and female rat pituitaries. Moreover, to provide complementary information on the cellular origin of LH and FSH, we have followed: (1) the cytogenesis of immunoreactive gonadotropic cells in the rat fetal pituitary (Tougard et al., 1977a) and (2) the quantitative and ultrastructural evolution of gonadotropic immunoreactive cells in long-term primary cultures of dispersed anterior pituitary cells (Tougard et al., 1977b).

2. Materials and Methods

2.1. Preparation of Tissues and Immunochemical Staining

The methods used for preparation of tissues and immunochemical staining have been previously described in detail for adult rat anterior pituitaries (Tougard *et al.*, 1973), fetal rat anterior pituitaries (Tougard *et al.*, 1977a), and monolayers of rat anterior pituitaries (Tougard *et al.*, 1974, 1977b) for both light- and electron-microscopic studies.

Moreover, as concerns normal adult rat anterior pituitaries at the electron-microscope level, two methodological approaches were recently compared (Tougard *et al.*, 1979): the preembedding method on unfrozen, only fixed, 25- to 50-μm thick sections with the peroxidase-labeled antibody technique as previously described (Tougard *et al.*, 1973) and the postembedding method on ultrathin sections according to Moriarty and Halmi (1972), Moriarty and Tobin (1976), and Moriarty and Garner (1977) with the peroxidase–antiperoxidase (PAP) complex (Sternberger *et al.*, 1970).

2.2. Antisera

Antisera against ovine hormones were obtained by B. Kerdelhué (Laboratoire des Hormones Polypeptidiques, CNRS) by immunizing guinea pigs with ovine FSH (oFSH) and ovine LH (oLH) and its two subunits (oLHβ and oLHα) prepared in the same laboratory. These antisera were used in the three systems studied. Antisera against rat LHβ (A-rLHβ) (AFP-2-11-27) and rat FSHβ (A-rFSHβ) (AFP-1-10-25) were used only for adult rat pituitaries.

2.3. Specificity of Immunochemical Staining

To test the specificity of the immunochemical reaction, four types of controls were run at the light- and electron-microscope levels: (C_1) diaminobenzidine (DAB) reaction without any incubation with AS; (C_2) one of the components of the stain was left out; (C_3) specific antiserum was substituted with normal rabbit serum; (C_4) each specific antiserum was preabsorbed for 3 days at 4°C by highly purified pituitary glycoprotein hormones and their subunits (Tougard *et al.*, 1973, 1979; Tixier-Vidal *et al.*, 1975).

3. Results

3.1. Characterization of Antisera

3.1.1. Antisera against Ovine Hormones

Results from radioimmunological studies (Kerdelhué *et al.*, 1971, 1972) and the effects of preabsorption of AS with different antigens on staining (Tougard *et al.*, 1973; Tixier-Vidal *et al.*, 1975) support the following assumptions: (1) A-oLH and A-oLHβ are very specific for LH and LHβ. Only preabsorption by LH and LHβ abolished the staining. Cells that react with both these AS can be considered as LH-secreting cells. (2) As was expected because of the chemical similarity of the α subunits of pituitary glycoproteins, A-oLHα recognizes LH, LHα, FSH, and thyroid-stimulating hormone (TSH), but not LHβ. Cells stained by A-oLHα can be considered as LH, FSH, or TSH cells. (3) A-oFSH is far less specific than A-oLH and A-oLHβ. It shows affinity for FSH, but also for LH and TSH. Cells stained by A-oFSH can be considered as LH, FSH, or TSH cells.

3.1.2. Antisera against Rat Hormones

The effects of preabsorption of A-rLHβ and A-rFSHβ with rat purified β subunits on staining (Tougard *et al.*, 1979) as well as their radioimmunological characterization (kindly provided by Dr. A.F. Parlow) support the assumption that each AS used at the dilution chosen (A-rLHβ, 1:800; A-rFSHβ, 1:20) is very specific for the respective β subunits.

3.2. Control Sections

No staining was observed with any control treatment (see Section 2.3) or when specific antiserum was preabsorbed with homologous antigen.

3.3. Adult Rat Anterior Pituitaries

3.3.1. Light-Microscopic Study

3.3.1a. Staining with A-oLH, A-oLHβ, and A-rLHβ. The cellular binding sites of the three AS were identical. In the normal male rat pituitaries, two cell types were stained: (1) large, rounded cells located in the lateral and anterior regions of the pars distalis and (2) small, oval cells scattered throughout the pars distalis, but more numerous in its posterior part. After DAB destaining and restaining with the AB-PAS, both cell types displayed the same violet color, which indicated that they were AB- and PAS-positive. After restaining with Herlant's tetrachrome, they were stained in blue by aniline blue. These cells, stained by three AS specific for LH and LHβ, may be considered as LH cells.

In castrated male and female rat pituitaries, cells immunochemically stained by the three AS were highly hypertrophied and vacuolized and corresponded to the so-called "castration cells." They were violet after destaining and restaining with AB-PAS.

3.3.1b. Staining with A-rFSHβ. In normal and castrated rat pituitaries, cells immunochemically stained by A-rFSHβ had the same morphological features as those displayed by cells stained by AS against LH and LHβ. The staining affinity of A-rFSHβ was always less than that of A-rLHβ. Nevertheless, in serial sections, it was clear, in different areas of the pars distalis, that the same cells were stained by A-rFSHβ and A-rLHβ (Fig. 1).

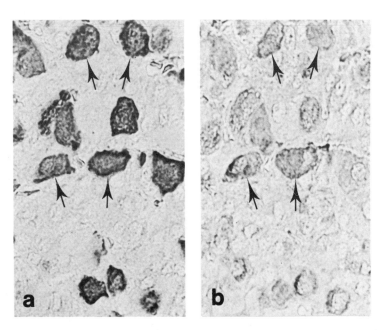

Figure 1. Adjacent 3-μm paraffin sections of a normal male rat pituitary treated with A-rLHβ (1a) and A-rFSHβ (1b). The same cells are stained by the two AS (arrows). The staining intensity is weaker with A-rFSHβ. ×700.

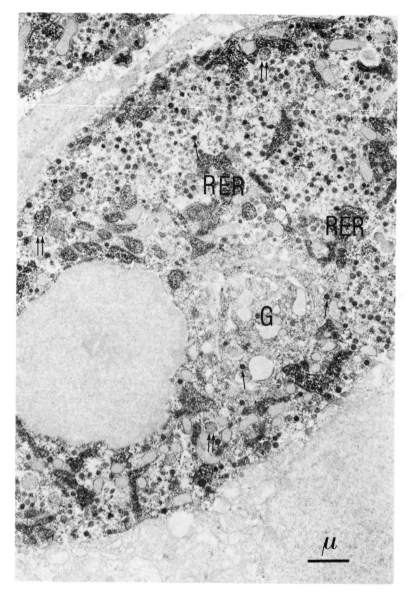

Figure 2. Normal male rat pituitary slice treated with A-rLHβ with the preembedding method. A type A gonadotropic cell with two classes of round secretory granules is immunochemically stained. The reaction product is more abundant over the small secretory granules (arrows) than over the large ones (double arrows). It is also found in dilated RER cisternae (RER). In the Golgi zone (G), the saccules are negative, but a few positive small secretory granules can be seen. ×12,000.

All these cells therefore contained LHβ and FSHβ and may be considered as gonadotropic cells.

3.3.2. Electron-Microscopic Study

3.3.2a. Identification of Positive Cells with A-oLH, A-oLHβ, and A-rLHβ. Cellular and subcellular binding sites were the same with A-oLH, A-oLHβ, and A-rLHβ. The most numerous positive cells (type A cells) were large and rounded and had two classes of secretory granules that differed in diameter (200 nm and 300–700 nm) (Fig. 2). Their rough endoplasmic reticulum (RER) cisternae were dilated and rounded, and their Golgi zone was large and clearly defined. Other positive cells (type B cells), far less numerous than type A cells, were smaller, oval in shape, and contained only small secretory granules (200 nm) (Fig. 3). Their RER cisternae were flattened or not conspicuous. Numerous forms intermediate between these two cell types were seen, mainly large and rounded cells with small secretory granules and dilated RER cisternae (Fig. 4).

All these cell types were stained with the preembedding method as well as the postembedding method (Tougard *et al.*, 1979). Another cell

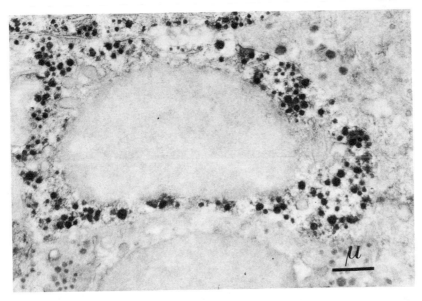

Figure 3. Same material as in Fig. 2 treated with A-oLH. A type B gonadotropic cell with small secretory granules is immunochemically stained. The RER cisternae are not conspicuous. × 12,000.

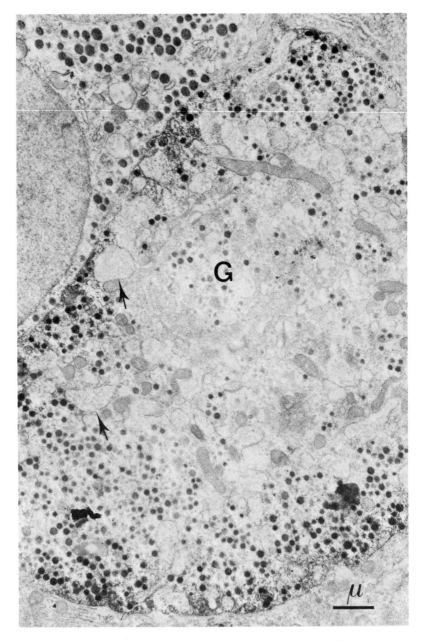

Figure 4. Same material as in Fig. 2 treated with A-oLH. An intermediate form between A and B types displays small secretory granules, dilated ergastoplasmic cisternae (arrows), and a large Golgi zone (G). ×12,000.

type was stained with the postembedding method only. It looked like a corticotropic cell with an angular and stellate shape and peripherally arranged secretory granules (200–230 nm).

3.3.2b. Identification of Positive Cells with A-rFSHβ. The cellular binding sites of A-rFSHβ were identical to those of A-oLH, A-oLHβ, and A-rLHβ. With the postembedding method, on serial ultrathin sections, the same cells were stained by A-rFSHβ and A-rLHβ (Fig. 5).

3.3.2c. Identification of Positive Cells with A-oLHα. Besides staining all gonadotropic cells identified by A-rLHβ and A-rFSHβ, A-oLHα stained another cell type characterized by its polygonal or stellate shape and very small secretory granules (100–150 nm) at the periphery of the cell. These cells have the ultrastructural features of thyrotropic cells detected by immunocytochemistry (Nakane, 1970, 1975; Moriarty and Tobin, 1976; Tougard *et al.*, 1979).

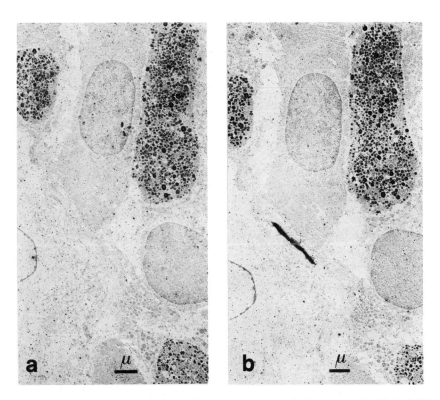

Figure 5. Adjacent ultrathin sections of a normal male rat pituitary treated with A-rLHβ (5a) and A-rFSHβ (5b) with the postembedding method. The same cells are stained by the two antisera. ×6,000.

3.3.2d. Subcellular Binding Sites of Antisera. The subcellular binding sites of AS against LH were the same as those of AS against FSH.

With the preembedding method, the main sites of antigenicity were the secretory granules and the ground cytoplasm. The smallest secretory granules showed the highest staining intensity. The largest secretory granules contained various amounts of reaction product, but always less than did the small secretory granules. In many cells, with A-rLHβ, a fine deposit was also found in dilated RER cisternae (see Fig. 2). In the Golgi area, the saccules were always negative, and only rare, positive small secretory granules could be seen. As concerns the subcellular localization of native LH and its two subunits (LHα and LHβ) in gonadotropic cells, we have not found any difference.

With the postembedding method, only the secretory granules were positive in the gonadotropic cells. The number of PAP molecules was higher in the large granules than in the small granules. On serial ultrathin sections, some of the large secretory granules were stained by A-rFSHβ and A-rLHβ (Fig. 6). A secretory granule may therefore contain the two β subunits.

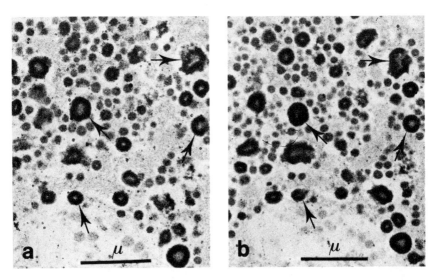

Figure 6. Same material as in Fig. 5 treated with A-rLHβ (6a) and A-rFSHβ (6b) with the postembedding method, showing details of the immunocytochemical staining on the secretory granules. The PAP molecules are localized mainly at the periphery of the large granules. Some of the large secretory granules (arrows) are stained by the two antisera. $\times 19,500$.

3.4. Fetal Rat Pituitaries

3.4.1. Light-Microscopic Study

The first immunoreactive cells were detected after staining with A-oLHβ between 17 and 19 days of gestation (day 1 dpc: the first day postcopulation). Cells immunoreactive with A-oLH appeared 24 hr later. Cells stained with A-oFSH were observed only between 20 and 21 days of gestation. Nevertheless, on 18 dpc, both LH and FSH were detected by radioimmunoassay (RIA) in fetal pituitaries (unpublished observations with Dr. B. Kerdelhué). Between 18 dpc and 19 dpc, the increased number of immunoreactive cells in the fetal pituitary corresponds to a large increase of LH and FSH detected by RIA.

3.4.2. Electron-Microscopic Study

On 16 dpc, immunoreactive cells were detected only with A-oLHβ and displayed very primitive features: they were small with a high nucleus/cytoplasm ratio, but they always contained a few small secretory granules (diameter: 80–120 nm). At this stage, immunoreactive cells did not differ from other negative cells containing some secretory granules. On 17 dpc, positive cells with the same features were found with A-oLHβ as well as with A-oFSH (Fig. 7). Between 18 and 19 dpc, positive cells became more numerous and their ultrastructural organization evolved. At the end of gestation, on 21 dpc, cells immunoreactive with both A-oLHβ and A-oFSHβ displayed a rounded or ovoid shape, an extension of the cytoplasmic area corresponding to a development of ergastoplasmic cisternae that were flattened or slightly dilated, and an enlargement of the Golgi zone. The size of the secretory granules varied from 80 to 150 nm, and in some cells a few larger ones (300 nm) could be found.

3.5. Primary Cultures of Rat Anterior Pituitary Cells

In such primary cultures, it has been shown (Vale *et al.*, 1972; Tixier-Vidal *et al.*, 1973; Steinberger *et al.*, 1973) that LH and FSH medium content fell during the first week, then reached a very low or undetectable level.

3.5.1. Quantitative Evolution of Immunoreactive Gonadotropic Cells

The number of immunoreactive cells with A-oLHβ, A-oLH, and A-oFSH in five experimental series varied from 1 to 15% as a function of the culture series and of duration of culture and with the nature of the AS

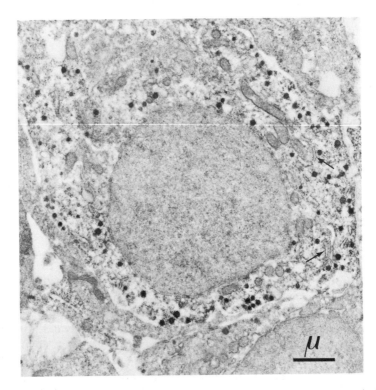

Figure 7. Rat fetal pituitary (stage 18 dpc) treated with A-oLHβ with the preembedding
method. This immunoreactive gonadotropic cell displays a large nucleus and a small
cytoplasmic area, small positive secretory granules (80–120 nm), and scarce and linear
ergastoplasmic cisternae (arrows). ×12,000.

(Tougard *et al.*, 1977b). Nevertheless, it decreased only slightly between
5 and 35 days, and in one experimental series it increased.

In long-term primary cultures (35 or 56 days), positive gonadotropic
cells were still present in the petri dishes, whereas the medium content of
gonadotropic hormones detected by RIA was very low or undetectable:
the FSH level was low, but always higher than the LH level. The hormonal
content of the cells decreased with time in culture, but at 42 days, the
levels of LH and FSH were still detectable and approximately the same
within the cell.

3.5.2. Ultrastructural Evolution of Immunoreactive Gonadotropic Cells

At any time of culture, the ultrastructural organization of immunoreactive cells with A-oLH, A-oLHβ, and A-oFSH was the same. At 5–7
days of culture, positive cells displayed the same ultrastructural features

previously described *in vivo* (see above): type A and B cells and their numerous intermediate forms. Progressive modifications of the ultrastructural organization of immunoreactive cells occurred as time in tissue increased. Finally, a single gonadotropic cell type remained, characterized by small secretory granules (125–150 nm), large dense bodies, a few ergastoplasmic cisternae, and a small Golgi zone (Fig. 8).

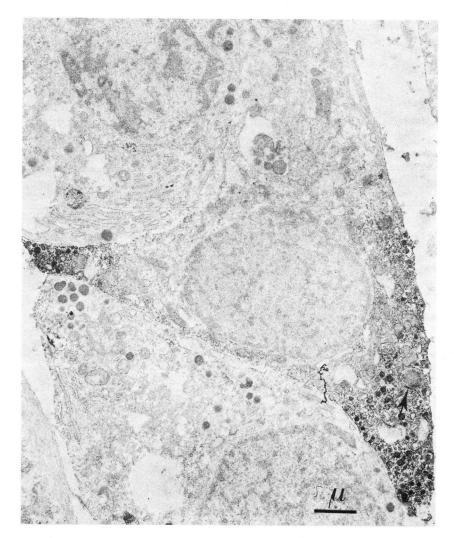

Figure 8. A 26-day monolayer of rat anterior pituitary cells treated *in situ* with A-oLH. This immunoreactive gonadotropic cell displays an angular shape and small secretory granules. The reaction product is concentrated in two areas of the cell: the secretory granules and the cytoplasm. The dense bodies are negative (arrow). ×12,000.

4. Discussion

The results reported herein, obtained by immunocytochemistry with several AS specific for rLH and one AS specific for rFSH, provide information about the cellular origin of the two gonadotropic hormones.

At the light-microscope level, in normal male and castrated rats, all cells violet after AB-PAS-Orange G staining, whether large or small, were immunoreactive with A-oLH, A-oLHβ, and A-rLHβ—antisera specific for rLH and rLHβ—and with A-rFSHβ antiserum specific for rFSHβ. The presence of LH and FSH in the same gonadotropic cells was also reported in other immunocytochemical studies in the rat (Nakane, 1970, 1975; Herbert, 1975; Bugnon *et al.*, 1977), in man (Pfifer *et al.*, 1973; Pelletier *et al.*, 1976), and in other mammalian species (Herbert, 1976; El Etreby and Fath El Bab, 1977; Beauvillain, 1978; Baker and Gross, 1978). However, some of these workers found a small proportion of gonadotropic cells that contained only one hormone, LH or FSH, in the center of the gland in the rat (Nakane, 1970), FSH only in man (Pelletier *et al.*, 1976), and LH only in the dog (El Etreby and Fath El Bab, 1977). The possibility still remains that these cells may contain undetectable levels of the other hormone, depending on the cell content or on the titer of AS.

At the electron-microscope level, AS specific for rLH and rLHβ and AS specific for rFSHβ recognized the type A cells and the type B cells as well as their intermediate forms. These cells displayed the ultrastructural features of gonadotropic cells described by conventional electron microscopy: type A cells looked like the Kurosumi–Oota "FSH cell" and type B cells looked like the Kurosumi–Oota "LH cell" (Kurosumi and Oota, 1968). On serial ultrathin sections, with the postembedding method, the same cells were stained by both A-rLHβ and A-rFSHβ. Therefore, the ultrastructural duality of gonadotropic cells does not correspond to an immunological duality. Similar results were also obtained by Nakane (1970, 1975) and by Moriarty (1976). Our type A cell looked like the type A gonadotropic cell of Nakane (1970, 1975) immunochemically stained by both A-oLH and A-oFSH and the type I of Moriarty (1976) immunochemically stained by A-bovine (b) LHβ and A-rFSHβ. Our type B cell looked like the type II of Moriarty (1976) stained by A-bLHβ and A-rFSHβ. Moriarty has also described another cell type called "type III" (Moriarty, 1976; Moriarty and Garner, 1977) stained by A-ACTH$_{17-39}$, A-rFSHβ, and A-bLHβ. Such a cell was also observed in our study, with the postembedding method only (Tougard *et al.*, 1979) with A-oLHβ, A-rLHβ, and A-rFSHβ. However, since this cell type was never detected in our preembedding method, until this immunoreactive component has been further identified, we will not consider this cell type as a specific gonadotropic cell. In addition, Moriarty (1976), in a quantitative study of immunoreactive cells in the cycling female rat pituitary, found more FSH-positive cells than LH-positive cells at all stages of the sexual cycle, but the ultrastructural features of cells containing only FSH were the same as those of cells containing the two hormones. One may imagine that depending on the physiological conditions, some cells contain either one hormone only or a low level of the other hormone undetectable by the immunocytochemical procedure.

Therefore, for the time being, as well as with respect to both the light and the electron microscope, all workers agree on having found LH and FSH in most of the gonadotropic cells. In no work was there described any cell with distinct histochemical and ultrastructural features that was immunochemically stained by AS directed specifically against LH or FSH.

Arguments in favor of the unicity of gonadotropic cells are also provided by the fetal pituitary and by primary cultures. In the fetal pituitary, at the end of gestation, immunoreactive gonadotropic cells display a single type of ultrastructural organization, whereas LH and FSH were detected by RIA in the pituitary. The morphological heterogeneity of adult gonadotropic cells appeared after birth during the prepubertal period in relation probably to the level of circulating sexual steroids and the level of luteinizing hormone–releasing hormone (LH-RH). In primary cultures, as time increased, the ultrastructural organization of gonadotropic cells evolved, and in long-term cultures a single gonadotropic cell type remained, characterized by small secretory granules, whereas FSH and LH were detectable by RIA in the cells and FSH was preferentially released into the medium.

All these immunocytochemical results support the assumption that both gonadotropic hormones are present in the same cells, and therefore the one cell, one hormone theory may not apply to gonadotropic hormones.

If both LH and FSH can be found in the majority of gonadotropic cells, what is the significance of the morphological heterogeneity of these cells? The existence of numerous forms intermediate between type A and type B cells suggests, as already proposed by Yoshimura and Harumiya (1965), that they might represent different stages of the secretory cycle of one single cell type (Fig. 9). Original arguments for such a cycle were provided by the fetal pituitary, by primary cultures, and by the pituitaries of castrated animals. In the fetus, the first immunoreactive gonadotropic cells contained only small secretory granules. In long-term primary cultures, the single gonadotropic cell type that remained looked like the fetal gonadotropic cell. In some adult pituitaries, some very scarce cells called type C (Tixier-Vidal *et al.*, 1975) and characterized by the ultrastructural organization with very small secretory granules were found. Without immunocytochemistry, these cell types could not be identified as gonadotropic cells. In the pituitaries of castrated animals, the large secretory

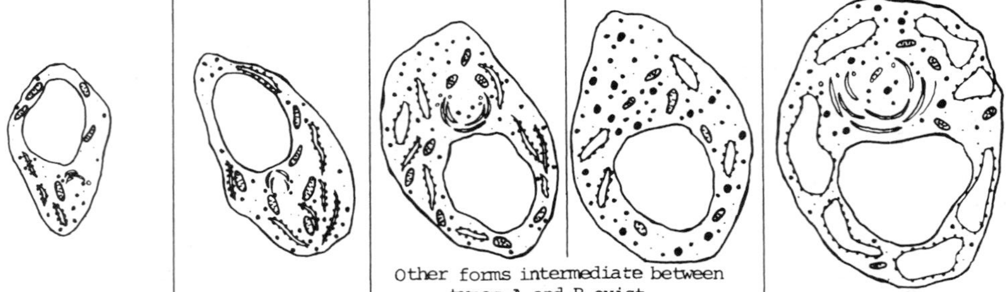

Other forms intermediate between types A and B exist

ULTRASTRUCTURAL ORGANIZATION: SIZE	very small	small	large	large	very large
SHAPE	oval	oval	rounded	rounded	rounded
SECRETORY GRANULES	rounded and very small (50-150 nm)	rounded and small (200 nm)	rounded and small (200 nm)	rounded two sizes: 200 and 300-700 nm very numerous	rounded two sizes: 200 and 300-700 nm
RER CISTERNAE	flattened and not abundant	flattened	flattened or slightly dilated	slightly dilated and infrequent	dilated and numerous
GOLGI ZONE	small or inconspicuous	small	large	large when it is conspicuous	very large
SYSTEM	<u>TYPE C</u> ADULT FETUSES (STAGE 16dpc) PRIMARY CULTURES (Δ 35 DAYS)	<u>TYPE B</u> ADULT FETUSES (STAGE 21dpc) PRIMARY CULTURES (Δ 5 DAYS)	<u>INTERMEDIARY FORM B/a</u> ADULT FETUSES (STAGE 21dpc) PRIMARY CULTURES (Δ 5 DAYS)	<u>INTERMEDIARY FORM b/A</u> ADULT PRIMARY CULTURES (Δ 5 DAYS)	<u>TYPE A</u> ADULT PRIMARY CULTURES (Δ 5 DAYS)

Figure 9. Diagrammatic representation of the different ultrastructural forms of the gonadotropic cell.

granules of type A cells disappeared in cells strongly modified by the castration, hypertrophied with highly dilated ergastoplasmic cisternae (Tougard *et al.,* 1973). In these cells, secretory granules displayed a uniform size (200–250 nm). As suggested in our previous study (Tougard *et al.,* 1973), large secretory granules could represent condensing vacuoles and a presecretory form of the small granules.

Recent results of Denef *et al.* (1978a) are consistent with such a secretory cycle of gonadotropic cells: after separation by unit gravity sedimentation of different fractions of gonadotropic cells from 14-day-old female, 14-day-old male, and adult male rats, they have found LH and FSH by RIA in all these fractions, the majority of gonadotrophs containing both FSH and LH. Moreover, these authors have observed in monolayer cultures of these fractions (Denef *et al.,* 1978b) that "the gonadotroph cell population is heterogeneous not only in terms of the magnitude of the response to LH-RH, but also in terms of the relative proportion of FSH and LH secreted."

Indeed, the simultaneous localization of the two hormones in the same cell does not exclude the possibility of a dissociation in the release of these two hormones. In primary cultures, we have seen that whereas LH and FSH were present in the cells, FSH was preferentially released into the medium. The existence of one hypothalamic factor, LH-RH, that modulates the release of both LH and FSH also favors the presence of the two hormones in the same cell.

One may imagine that the level of each hormone inside the cell as well as the number of LH-RH receptors vary in relation to the cell cycle and are modulated by the circulating steroids. The subcellular localization of antigenicity discussed in detail elsewhere (Tixier-Vidal *et al.,* 1975; Tougard *et al.,* 1979) did not provide information to explain the dissociation of LH and FSH release in response to a physiological stimulation. Some observations must be pointed out: (1) We did not find any difference between the binding sites of AS against the two subunits of ovine LH. (2) In some of the large secretory granules, LHβ and FSHβ were found simultaneously, which implies that the two hormones can be stored in the same granule. Nevertheless, the possibility still remains that LHβ and FSHβ share enough common antigenic determinants so that even AS very specific for each of them are not able to discriminate between LH and FSH molecules when they are localized into the intracellular compartments. Techniques other than immunocytochemistry have to be employed: only knowledge of the mechanisms involved in the biosynthesis of the subunits of the two gonadotropic hormones, their association, and their pathway of secretion will permit verification of the presence of these two hormones in the same cell and understanding of the dissociation of their release in some physiological conditions.

5. Summary

The cellular origin of LH and FSH in the rat pituitary has been investigated by immunocytochemistry at the light- and electron-microscope levels using heterologous antisera against ovine LH, ovine LHβ, ovine LHα, and ovine FSH, and homologous antisera against rat LHβ and rat FSHβ. In normal male rats and castrated male and female rats, the two gonadotropic hormones were detected in the same cells by antisera specific for rat LH and rat FSH. Arguments in favor of the unicity of gonadotropic cells were also provided by the study of immunoreactive cells in the fetal pituitary and in primary cultures of dispersed anterior pituitary cells.

To explain the morphological heterogeneity of these cells, which did not correspond to an immunological duality, a secretory cycle of the gonadotropic cell is tentatively proposed.

ACKNOWLEDGMENTS. The author is very grateful to Dr. B. Kerdelhué and Dr. M. Jutisz for the generous gift of antisera against ovine hormones as well as for their helpful advice. She acknowledges the generosity of Dr. L.A. Sternberger for the peroxidase–antiperoxidase complex and Dr. A.F. Parlow, National Institute of Arthritis, Metabolism and Digestive Diseases, for antisera against rat β subunits. She greatly appreciates the excellent technical assistance of R. Picart, M.F. Moreau, and C. Pennarun and the help of A. Bayon in typing the manuscript. This work was supported by grants from the C.N.R.S. (France, E.R. 89) and from the D.G.R.S.T. (contract 72 7 0 100).

DISCUSSION

SIMON: Does the extragranular intracellular localization of the hormone vary after hormonal stimulation?

TOUGARD: I did not observe clear variations of extragranular localization at the electron-microscope level under LH-RH stimulation in primary cultures. This was difficult to detect because the cytoplasmic reaction products could vary in relation to the observed cell and the antiserum used. Nevertheless, at the light-microscope level, after 15 minutes of LH-RH action, I have observed a transitory decrease in cytoplasmic immunoreactivity related certainly to secretory granules, but perhaps also to extragranular hormone.

In stimulated gonadotropic cells in ovariectomized rats, the reaction product was preferentially localized on both cytoplasm and granules at the cell periphery.

PANTIC: Five types of gonadotropic cells have been identified by Yoshimura *et al.* (*Endocrinol. Jpn.* **24**:185, 1977). You identified three cell types. Do you think it is better to use the term "stages of cell activities" instead of "cell types"?

TOUGARD: Certainly, I think all types of gonadotropic cells identified immunocytochemically correspond to different stages of the secretory cycle of a single gonadotropic cell. If I spoke about cell types, it was to establish correlations between the immunocytochemical affinities of these cells and the previous morphological distinction of LH and FSH cells, but I have found all intermediate forms between these cell types.

MELDOLESI: One might have the impression that a free pool of hormones soluble in the cytoplasm would be present in some secretory systems, such as endocrine cells, and absent in exocrine cells, such as the exocrine pancreas. Actually, this is not the case—all the evidence which has been published in favor of soluble pools in endocrine systems also exists for the exocrine pancreas. For example, the amount of secretory proteins recovered in the final supernatants isolated from pancreas homogenates is considerably larger than that of PRL. Free pools may not exist at all. They might arise from artifactual leakage from intracellular particulate pools. My question concerns the primary culture experiments in which you showed a dissociation between tissue content and release of LH and FSH. I wonder whether you measured hormone biosynthesis?

TOUGARD: I do not exclude the possibility that the immunocytochemical staining in the cytoplasm could correspond to an artifact, but several facts are in favor of a specific phenomenon. Nevertheless, only biosynthesis and biochemical studies, particularly fractionation, could give an answer. I did not measure the hormone biosynthesis in primary culture.

JUTISZ: In your cell culture system, do you have any evidence for exocytosis, either in the presence or in the absence of GnRH?

TOUGARD: In 7-day primary cultures, the LH-RH induced a rapid migration of numerous secretory granules along the plasma membrane, but we have not found classical exocytosis as described by Farquhar in prolactin cells. The density of some secretory granules decreased, as well as their antigenicity. At the same time, the plasma membrane showed small invaginations which could be interpreted as remnants of granule exocytosis. It seems that during granule exocytosis, the electron density of the secretory granule disappears more rapidly in gonadotropic cells then in prolactin cells.

BEAUVILLAIN: By using the preembedding technique, a heavy staining is observed in the small granules of the gonadotropic cells of the rat anterior pituitary, but with the postembedding technique, the staining that you have obtained is more intense on the big granules. How can you explain these differences?

TOUGARD: The decrease of the reaction product in small granules which is observed with the postembedding method might be due to a slight extraction of the hormonal content during dehydration and embedding. For the large granules, the increased reaction product with the postembedding method is difficult to explain. It is more marked at the periphery of the granule, and perhaps it could be due to a diffusion and accumulation of secretory product. The results obtained with the preembedding methods are probably closer to the normal situation than those obtained with the postembedding method.

FINK: Brown-Grant and Kreig (*J. Endocrinol.* **65**:389–395, 1975) administered sodium pentobarbitone to female rats at various times during the afternoon of proestrus. If given before the beginning of the LH and FSH surges, both were blocked. If the neural blocker was given after the beginning of the surges, plasma LH concentration fell according to its half-life, but FSH secretion continued. A similar situation is obtained in the human. Do you have any evidence for a lack of dependence of FSH secretion on LH-RH in your cultures?

TOUGARD: Yes, in long-term primary cultures (3–5 weeks) in the absence as well as in the continuous presence of LH-RH, FSH was preferentially released into the medium and LH was undetectable. The LH and FSH cell contents were approximately the same.

VAUDRY: I would like to raise again the question of the specificity of the antibodies used for immunocytochemical staining. You state very emphatically that LH and FSH originate from the same cells. Therefore, it is very important to make sure of the specificity of the antisera since all of the results were based upon the presumed absence of cross-reactivities between anti-β subunit of LH with FSH and vice versa.

TOUGARD: The specificity has been tested by radioimmunological study carried out by Dr. B. Kerdelhué for antisera against ovine hormone and by Dr. Parlow for antisera against rat β-subunits. The specificity has also been studied by the effects of preabsorption by different doses of antigens on immunochemical staining. It appears from all these results that anti-ovine LH, anti-ovine LHβ, and anti-rat LHβ are very specific, respectively, for oLH, oLHβ, and rat LHβ. Anti-rat FSHβ is also very specific for rat FSHβ. The cross-reactivity between the two hormones in the two systems is lower than 1 in 1000.

FINK: Dr. Moriarty reported (*Nature* **265**:356–358, 1977) that she found ACTH in some FSH-containing cells by immunocytochemistry. This raises a problem of the technique of absorption to check the validity of immunocytospecificity. Antibodies are raised against an antigen, and then the antigen is used to check the specificity of the antibody by absorption, thus defying Newton's third law. The antibodies probably react with only a handful of amino acids or perhaps just a certain conformation. Two questions arise from this: has anyone else seen peptides like ACTH in gonadotrophs and are there any methods other than absorption for checking the validity of immunocytochemistry?

TOUGARD: In answer to your first question, I have not seen peptides like ACTH in gonadotrophs, but I have found, like Moriarty, an immunoreactivity with antisera against ovine LHβ, rat LHβ, and rat FSHβ in some cells which look like ACTH cells when using the postembedding method. I have never found such a cell with the preembedding method. Perhaps dehydration and embedding of the material made accessible some antigenic determinants common to gonadotropic hormones and an ACTH prohormone or to other components of the secretory granules.

Concerning your second question, until now only preabsorption tests and radioimmunological characterization of the antisera are used to test the validity of immunocytochemistry. But these tests are to be performed not only with the homologous antigen, but also with different antigens at different doses. The efficiency of the absorption is improved when it is performed in a solid-phase system.

REFERENCES

Baker, B. L., and Gross, D. S., 1978, Cytology and distribution of secretory cell types in the mouse hypophysis as demonstrated with immunocytochemistry, *Am. J. Anat.* **153**:193.

Barnes, B. G., 1962, Electron microscope studies on the secretory cytology of the mouse anterior pituitary. *Endocrinology* **71**:618.

Barnes, B. G., 1963, The fine structure of the mouse adenohypophysis in various physiological states, in: *Cytologie de l'Adénohypophyse* (J. Benoit and C. Da Lage, eds.), pp. 91–110, Editions du CNRS, Paris.

Beauvillain, J. C., 1978, Caractérisation à l'échelle ultrastructurale des cellules adénohypophysaires individualisées par différentes techniques immunocytochimiques, Thèse d'etat, Université de Lille.

Bugnon, C., Fellmann, D., Lenys, D., and Bloch, B., 1977, Etude cytoimmunologique des cellules gonadotropes et des cellules thyréotropes de l'adénohypophyse du Rat, *C. R. Soc. Biol.* **4**:907.

Denef, C., Hautekeete, E., De Wolf, A., and Vanderschueren, B., 1978a, Pituitary basophils from immature male and female rats: Distribution of gonadotrophs and thyrotrophs as studied by unit gravity sedimentation, *Endocrinology* **103**:724.

Denef, C., Hautekeete, E., and Dewals, R., 1978b, Monolayer cultures of gonadotrophs separated by velocity sedimentation: Heterogeneity in response to luteinizing hormone–releasing hormone, *Endocrinology* **103**:736.

El Etreby, M. F., and Fath El Bab, M. R., 1977, Localization of gonadotrophic hormones in the dog pituitary gland: A study using immunoenzyme histochemistry and chemical staining, *Cell Tissue Res.* **183**:167.

Farquhar, M. G., and Rinehart, J. F., 1954, Cytologic alterations in the anterior pituitary gland following thyroidectomy: An electron microscopic study, *Endocrinology* **55**:857.

Herbert, D. C., 1975, Localization of antisera to $LH\beta$ and $FSH\beta$ in the rat pituitary gland, *Am. J. Anat.* **144**:379.

Herbert, D. C., 1976, Immunocytochemical evidence that luteinizing hormone (LH) and follicle stimulating hormone (FSH) are present in the same cell type in the rhesus monkey pituitary gland, *Endocrinology* **98**:1554.

Herlant, M., 1960, Étude de deux techniques nouvelles destinées à mettre en évidence les differentes catégories cellulaires présentes dans la glande pituitaire, *Bull. Microsc. Appl.* **10**:37.

Herlant, M., 1964, The cells of the adenohypophysis and their functional significance, *Int. Rev. Cytol.* **17**:299.

Kerdelhué, B., Pitoulis, S., and Jutisz, M., 1971, Étude par radioimmunologie de la spécificité des immunsérums de l'hormone lutéinisante (LH) ovine et de ses sous-unités, *C. R. Acad. Sci.* **273**:511.

Kerdelhué, B., Pitoulis, S., and Jutisz, M., 1972, Immunological properties of antisera against ovine luteinizing hormone and its subunits obtained in guinea pigs, in: *Structure–Activity Relationships of Protein and Polypeptide Hormones*, Part II (M. Margoulies and F. C. Greenwood, eds.), p. 396, Excerpta Medica, Amsterdam.

Kurosumi, K., and Oota, Y., 1968, Electron microscopy of two types of gonadotrophs in the anterior pituitary glands of persistent estrous and diestrous rat, *Z. Zellforsh.* **85**:34.

Leleux, P., and Robyn, C., 1970, Etude en immunofluorescence d'une réaction immunologique croisée entre les gonadotropines humaines et les gonadotropines du rat. *Ann. Endocrinol.* **31**:181.

Leleux, P., and Robyn, C., 1971, Immunohistochemistry of individual adenohypophysial cells, in: *In Vitro Methods in Reproductive Cell Biology, Karolinska Symp. Res. Methods Reprod. Endocrinol.* **3**:168–205.

Leleux, P., Robyn, C., and Herlant, M., 1968, Mise en évidence des cellules gonadotropes de l'hypophyse du rat par une méthode d'immunofluorescence, *C. R. Acad. Sci.* **267**:438.

Monröe, S. E., and Midgley, A. R., 1966, Localization of luteinizing hormone in the rat pituitary gland by a cross-reaction with antibodies to human chorionic gonadotropin, *Fed. Proc. Fed. Am. Soc. Exp. Biol.* **25**:315.

Monröe, S. E., and Midgley, A. R., 1969, Immunofluorescent localization of rat luteinizing hormone, *Proc. Soc. Exp. Biol. Med.* **130**:151.

Moriarty, G. C., 1976, Immunocytochemistry of the pituitary glycoprotein hormones, *J. Histochem. Cytochem.* **24**:846.

Moriarty, G. C., and Garner, L. L., 1977, Immunocytochemical studies of cells in the rat adenohypophysis containing both ACTH and FSH, *Nature (London)* **265**:356.

Moriarty, G. C., and Halmi, N. S., 1972, Electron microscopic study of the adrenocorticotropin producing cell with the use of unlabeled antibody and the soluble peroxidase–antiperoxidase complex, *J. Histochem. Cytochem.* **20**:590.

Moriarty, G. C., and Tobin, R. B., 1976, Ultrastructural immunocytochemical characterization of the thyrotroph in rat and human pituitaries, *J. Histochem. Cytochem.* **24**:1131.

Nakane, P. K., 1970, Classifications of anterior pituitary cell types with immunoenzyme histochemistry, *J. Histochem. Cytochem.* **18**:9.

Nakane, P. K., 1975, Identification of anterior pituitary cells by immunoelectron microscopy, in: *The Anterior Pituitary* (A. Tixier-Vidal and M. G. Farquhar, eds.), pp. 45–61, Academic Press, New York.

Pelletier, G., Leclerc, R., and Labrie, F., 1976, Identification of gonadotropic cells in the human pituitary by immunoperoxidase technique, *Mol. Cell. Endocrinol.* **6**:123.

Pfifer, R. F., Midgley, A. R., and Spicer, S. S., 1973, Immunohistologic evidence that follicle-stimulating and luteinizing hormones are present in the same cell types in the human pars distalis, *J. Clin. Endocrinol. Metab.* **36**:125.

Purves, H. D., 1966, The cytology of the adenohypophysis, in: *The Pituitary Gland* (G. W. Harris and B. T. Donovan, eds.), pp. 147–232, Butterworths, London.

Steinberger, A., Chowdhury, M., and Steinberger, E., 1973, Effect of repeated replenishment of hypothalamic extract on LH and FSH secretion in monolayer cultures of rat anterior pituitary cells, *Endocrinology* **92**:12.

Sternberger, L. A., Hardy, P. H. Jr., Cuculis, J. J., and Meyer, H. G., 1970, The unlabeled antibody enzyme method of immunohistochemistry: Preparation and properties of soluble antigen–antibody complex (horseradish peroxidase–antiperoxidase) and its use in identification of spirochetes, *J. Histochem. Cytochem.* **18**:315.

Tixier-Vidal, A., Kerdelhué, B., and Jutisz, M., 1973, Kinetics of release of luteinizing hormone (LH) and follicle stimulating hormone (FSH) by primary cultures of dispersed rat anterior pituitary cells: Chronic effect of synthetic LH and FSH releasing hormone, *Life Sci.* **12**:499.

Tixier-Vidal, A., Tougard, C., Kerdelhué, B. and Jutisz, M., 1975, Light and electron microscopic studies on immunocytochemical localization of gonadotropic hormones in rat pituitary using antisera against ovine FSH, ovine LH and its two subunits, *Ann. N. Y., Acad. Sci.* **254**:433.

Tougard, C., Kerdelhué, B., Tixier-Vidal, A., and Jutisz, M., 1971, Localisation par cytoimmunoenzymologie de la LH, de ses sous-unités α et β, et de la FSH dans l'adénohypophyse de la ratte castrée, *C. R. Acad. Sci.* **273**:897.

Tougard, C., Kerdelhué, B., Tixier-Vidal, A., and Jutisz, M., 1973, Light and electron microscope localization of binding sites of antibodies against ovine luteinizing hormone and its two subunits in rat adenohypophysis using peroxidase-labeled antibody technique, *J. Cell Biol.* **58**:503.

Tougard, C., Picart, R., Tixier-Vidal, A., Kerdelhué, B., and Jutisz, M., 1974, *In situ* immunochemical staining of gonadotropic cells in primary cultures of rat anterior pituitary cells with the peroxidase labeled antibody technique: A light and electron microscope study, in: *Electron Microscopy and Cytochemistry* (E. Wisse, W. T. Daems, I. Molenaar, and P. Van Duijn, eds.), p. 163, North-Holland, Amsterdam.

Tougard, C., Picart, R., and Tixier-Vidal, A., 1977a, Cytogenesis of immunoreactive gonadotropic cells in the fetal rat pituitary at the light and electron microscope levels, *Dev. Biol.* **58**:148.

Tougard, C., Tixier-Vidal, A., Kerdelhué, B., and Jutisz, M., 1977b, Étude immunocytochimique de l'évolution des cellules gonadotropes dans des cultures primaires de cellules antéhypophysaires de rat: Aspects quantitatifs et ultrastructuraux, *Biol. Cell.* **28**:251.

Tougard, C., Picart, R., and Tixier-Vidal, A., 1979, *J. Histochem. Cytochem.* (in press).

Vale, W., Grant, G., Amoss, M., Blackwell, R., and Guillemin, R., 1972, Culture of enzymatically dispersed anterior pituitary cells: Functional validation of a method, *Endocrinology* **91**:562.

Yoshimura, F., and Harumiya, K., 1965, Electron microscopy of the anterior lobe of pituitary in normal and castrated rats, *Endocrinol. Jpn.* **12**:119.

3

Pituitary Polypeptide-Secreting Cells: Immunocytochemical Identification of ACTH, MSH, and Peptides Related to LPH

G. Tramu and J. C. Beauvillain

1. Introduction

Until recently, the anterior lobe of the hypophysis was thought to be responsible for corticotropin (ACTH) elaboration, while melanotropin (MSH) secretion was attributed to the intermediate lobe. Other results, however, have indicated a relationship between the intermediate lobe and corticotropic function. The possibility of corticotropic activity of the neurointermediate lobe had been underlined by Mialhe-Voloss (1958), and Rochefort *et al.* (1959) also came to similar conclusions. Gosbee *et al.* (1970) showed that the intermediate lobe could be changed morphologically by adrenalectomy and concluded that it is implicated in functional ACTH production. Other results indicated the presence of cells in the rostral portion of the intermediate lobe that were morphologically identical to anterior-lobe ACTH cells (Porte *et al.*, 1971; Stoeckel *et al.*, 1971). Furthermore, Kraicer *et al.* (1971) have demonstrated the presence of ACTH in isolated intermediate lobes, in even greater quantities than in the anterior lobe.

G. Tramu and *J. C. Beauvillain* ● U. 156 I.N.S.E.R.M., Laboratory of Histology, Faculty of Medicine, 59045 Lille, France

The use of immunocytochemical techniques has enabled this problem to be studied from a different angle. These techniques, confirming certain results, have also revealed new information. Immunohistochemical ACTH localization was attempted by using antisera against 1–39 ACTH (Marshall, 1951; McGarry and Beck, 1968; Nakane, 1970) or synthetic 1–24 ACTH (Hachmeister and Kracht, 1965; Breustedt, 1968; Hess et al., 1968). These authors had shown that some of the anterior-lobe cells and all the intermediate-lobe cells were stained. However, a cross-reaction with α-MSH, which constitutes the first 13 amino acids of the ACTH sequence, could have explained the intermediate-lobe staining. This doubt has been removed by using specific antisera against 17–39 ACTH, which has no sequence in common with α-MSH. In fact, using these antibodies, the staining was analogous to that obtained with anti-1–24 ACTH antisera (Phifer and Spicer, 1970; Baker and Drummond, 1972; Moriarty, G. C., and Halmi, 1972; Dubois et al., 1973; Naik, 1973). Immunohistochemical techniques have also been widely used to localize α-MSH and β-MSH, peptides known for their high melanotropic activity. α-MSH was found to be localized only in the intermediate lobe (Dubois, 1972a; Stefan and Dubois, 1972), consistent with the high melanotropic activity of this lobe. β-MSH, however, was localized both in some anterior-lobe cells and in the whole of the intermediate lobe. The β-MSH-immunoreactive anterior-lobe cells were shown to be corticotrophs in ovines, bovines, and porcines (Dubois, 1972a) as well as in man (Phifer et al., 1974).

The immunohistochemical techniques have also been used to locate the presence of β-lipotropin (β-LPH), which is a pituitary polypeptide composed of 91 amino acids (Birk and Li, 1964). The cells already shown to secrete ACTH and β-MSH were also shown to secrete this substance (Dubois, 1972b; Dubois and Graf, 1973; Moon et al., 1973; Dessy et al., 1973), the function of which has not yet been well elucidated.

The recent discovery of a morphinelike pituitary peptide, β-endorphin (Cox et al., 1975), which is the C-terminal portion of the β-LPH molecule, has been followed by studies on its cellular localization. β-Endorphin is also present in the corticotrophs of the anterior lobe and in all the intermediate-lobe cells (Bloom et al., 1977; Begeot et al., 1978).

Last, Hughes (1975) isolated two morphinelike pentapeptides in the central nervous system. One of them, methionine-enkephalin, is 61–65 β-LPH. They have been assayed in the hypophysis (Rossier et al., 1977), but not yet morphologically localized.

In short, a cell that was at first thought to be an ACTH-producing cell has been shown to be responsible for producing several hormones or peptide substances.

This chapter is concerned with the cellular localizations of ACTH, MSH, and morphinelike peptides related to LPH (endorphins, enkephalins) in the rodent adenohypophysis. The results discussed herein are concerned mostly with the guinea pig hypophysis, the intermediate lobe of which is well developed, but also with the hypophysis of the lerot (*Eliomys quercinus*), a hibernating mammal with an extremely small intermediate lobe.

2. Methods

2.1. Light Microscopy

In most cases for the immunocytochemical reaction, the pituitaries were immersed in Bouin–Hollande fluid or in formol–sublimate mixture.

For the detection of enkephalins, perfusion with picric acid–p-formaldehyde (PAF) was performed, and tissues were frozen and sectioned on a cryostat (Tramu and Léonardelli, 1979).

Immunohistochemical staining was carried out according to the indirect method previously described (Tramu and Dubois, 1977). The method of antibody elution for the successive localization of two different antigens on the same section (Tramu *et al.*, 1978) was also used.

2.2. Electron Microscopy

The conventional description at the electron-microscope level was performed after individualization of the corticotrophs on adjacent semithin sections according to a method previously described (Beauvillain and Tramu, 1973; Beauvillain *et al.*, 1975). As a fixative, glutaraldehyde–p-formaldehyde or PAF, each followed by osmic acid, was used (Beauvillain and Tramu, 1973; Beauvillain *et al.*, 1975; Beauvillain, 1978). The immunocytochemical staining on ultrathin sections was carried out according to G. C. Moriarty and Halmi (1972).

2.3. Antibodies

The antisera were prepared in rabbits only from synthetic antigens, bound to albumin (Dubois, 1971, 1972a; Tramu and Léonardelli, 1979). Purification and serological study of antibodies have been described earlier (Dubois, 1971; Stefan and Dubois, 1972). The specificity of immune sera in this study was tested by previous absorption with homologous and heterologous antigens. The results agree with a high affinity of the antisera for homologous antigens only. Moreover, antienkephalin antisera have been tested in a radioimmunoassay system (Croix, 1978, personal communication), and the preliminary results ruled out the existence of a cross-reaction with β-endorphin and indicated a high specificity of these antibodies.

3. Results

The results are summarized in Table I.

3.1. Corticotropin Localization

3.1.1. Light Microscopy

Immunohistochemical ACTH detection was obtained using specific antisera against certain synthetic fractions of the ACTH molecule: 1–24 ACTH, 17–39 ACTH, and 25–39 ACTH.

3.1.1a. Guinea Pig. In the anterior lobe, the same cell type was stained by all the antisera. The stainings are all perfectly superimposable: i.e., only the cells stained by one antiserum are all stained by the two other antisera (Fig. 1).

In the intermediate lobe, the anti-1–24 ACTH, anti-17–39 ACTH and anti-25–39 ACTH antisera stained all the cells (Fig. 2). The anti-1–24 ACTH staining does not seem to be due to a cross-reaction with α-MSH, since a preliminary absorption of the antiserum by α-MSH did not prevent the staining. Contrary to what was seen in the anterior lobe, different staining intensities were observed, depending on which antiserum was used: in certain cells, stainings for 1–24 ACTH, 17–39 ACTH, and 25–39 ACTH were analogous, but it should be noted that in other cells, staining for 1–24 ACTH was very poor, whereas anti-17–39 ACTH and anti-25–39

Table I. Summary of Results of Immunohistochemical Stainings for Polypeptide Hormones in Lerot and Guinea Pig Hypophysis[a]

Species and site	ACTH	α-MSH	β-MSH	Endor-phins	4–10 ACTH
Lerot					
Anterior lobe	+	−	+	+	+ or −
Intermediate lobe	−	+	+	+	?
Guinea pig					
Anterior lobe	+	−	−	+	−
Intermediate lobe	+	+	−	+	+

[a] (+) Positive reaction; (−) negative reaction.

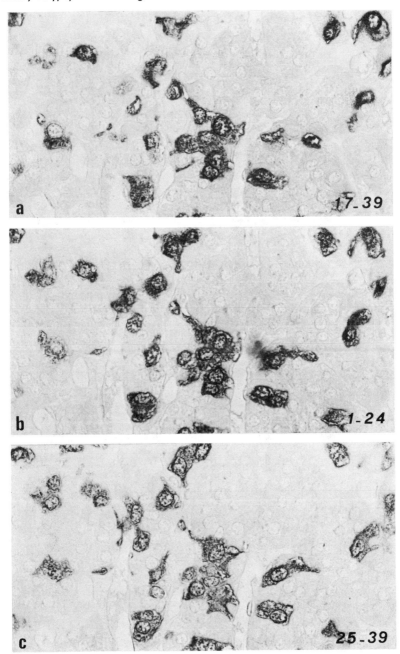

Figure 1. Guinea pig anterior lobe. ×525. Adjacent sections stained for 17–39 ACTH (a), 1–24 ACTH (b), and 25–39 ACTH (c). A precise superimposition is observed with the three antisera.

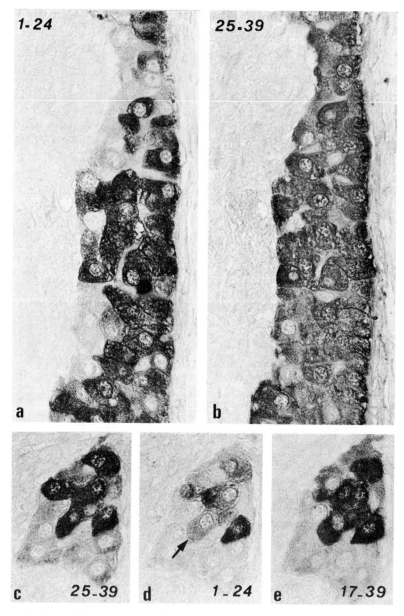

Figure 2. Guinea pig intermediate lobe. ×525. (a, b) Adjacent sections stained for 1–24 ACTH (a) and 25–39 ACTH (b). In (a), all the cells do not react with the same intensity, but in (b), the staining is more homogeneous. (c–e) Intermediate-lobe-cell islet in the posterior lobe. Adjacent sections stained for 25–39 ACTH (c), 1–24 ACTH (d), and 17–39 ACTH (e). Some weakly stained cells (arrow) in (d) react strongly in (c) and (e).

ACTH antisera produced intense stainings (Fig. 2). These last cells may enclose more antigenic determinants related to the C-terminal portion of the ACTH molecule than to its N-terminal portion. These observations tend to show that certain cells elaborate ACTH, but also suggest the presence of a greater quantity of an isolated C-terminal sequence different from the ACTH molecule. This accords with previous observations of Dubois (1972b) showing a dissociation between stainings for 1–24 ACTH and for 17–39 ACTH. The results are in complete agreement with those of Scott *et al.* (1974) concerning the presence of corticotropinlike intermediate peptide (CLIP) (which is the 18–39 amino acid sequence of ACTH) uniquely in the intermediate lobe and agree with those of C. M. Moriarty and Moriarty (1975), who demonstrated the presence of bioactive ACTH in much smaller quantities than CLIP in the intermediate lobe.

3.1.1b. Lerot. As far as ACTH localization is concerned, the anterior lobe of the lerot was quite analogous to that of the guinea pig. Specific antisera against the N-terminal or the C-terminal portion revealed the same cell population. Differences appear in the intermediate lobe, which is very small in the lerot. Stainings for 1–24 ACTH and for 17–39 ACTH or 25–39 ACTH using specific antisera always revealed the same set of cells, which constituted only 20–30% of the lobe-cell population. Furthermore, these intermediate-lobe corticotrophs were different from those producing α-MSH, as will be seen in Section 3.2.

3.1.2. Electron Microscopy

The cell type stained with all the anti-ACTH antisera at the light-microscope level was characterized at the ultrastructural level in the anterior lobe employing a technique of superimposing light and electron images. Depending on the species, important morphological differences were observed. In the guinea pig, the corticotrophs were irregular and often stellate-shaped, and contained few, small granules (1500–2000 Å in diameter) (Fig. 3). In the lerot, however, these cells were always oval and contained many large granules, 2500–3500 Å in the majority of cases (Beauvillain and Tramu, 1973) (Fig. 4). In the corticotrophs of the guinea pig, there were also vesicles, the membranes of which were sometimes discontinuous. These vesicles were often bigger than the granules, and their content was slightly electron-opaque. They were more numerous in PAF–osmium-tetroxide-fixed tissue, suggesting that they may be due to the fixation step (see Fig. 3).

Immunocytochemical techniques were also used directly on ultrathin sections to localize at a subcellular level the antigenic sites reacting with the anti-ACTH antisera. In the lerot and the guinea pig, all the granules were stained. However, mainly in the guinea pig, it has been shown, first,

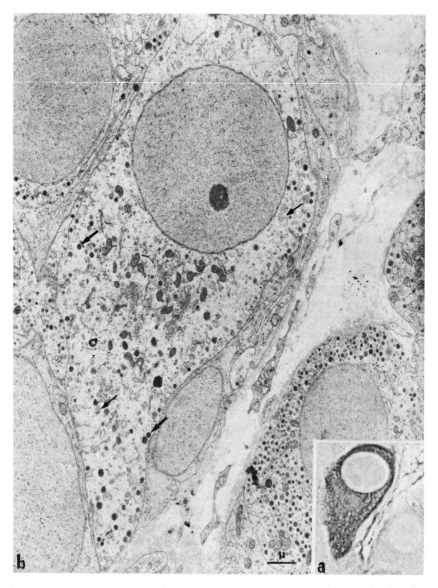

Figure 3. Guinea pig anterior lobe. Superimposition between semithin and ultrathin sections. The semithin section (a) shows a cell stained by anti-17–39 ACTH antiserum. This same cell is seen in the adjacent ultrathin section (b). The granules are about 1500–2000 Å in diameter and show a marked electron density (large arrows). Furthermore, some vesicles with a lower density can be observed (small arrows).

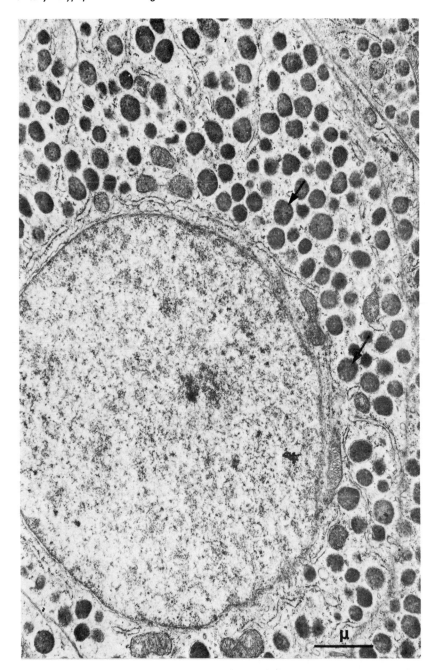

Figure 4. Lerot anterior lobe. Characteristic appearance of a corticotroph. The granules (2500–3500 Å) are very numerous and electron-opaque (arrows). Some of them have a polygonal shape.

that the staining intensity is highly variable from one granule to another and, second, that peroxidase–antiperoxidase (PAP) molecules can be seen outside the granules (Fig. 5a, b). The extragranular staining could reveal an antigen diffusion out of the granules, due to the fixation procedure. In many cases, indeed, clusters of extragranular PAP molecules may be the result of the staining of the vesicles previously described with the conventional method. But there were also other molecules scattered in the rest of the cytoplasm, which were not located, at least in the plane of the section, near the granules. Part of the staining may therefore correspond with normal localizations. In this case, the staining would reveal the hormone itself or its prehormone at a stage preceding the formation of granules, or a hormone form that remains free in the cytoplasm, implying a peculiar secretion method that has never been proved. G. C. Moriarty and Halmi (1972) had also noted this extragranular staining in corticotrophs, but unlike them, we never observed staining of the granules forming within the Golgi saccules, not even with the preembedding technique (Beauvillain, 1978). This absence of Golgi staining has also been reported by Weber *et al.* (1978) and cannot be explained at present, although various hypotheses may be proposed, for example: (1) denaturation or transformation of the hormone at the Golgi level; (2) incomplete maturation (Weber *et al.*, 1978).

3.2. Melanotropin Localization

3.2.1. α-Melanotropin

3.2.1a. Guinea Pig. All the intermediate-lobe cells were clearly stained by anti-α-MSH antiserum. No anterior-lobe cells were stained (Fig. 6c). This corresponds with the observations made in other species (Dubois, 1972a; Stefan and Dubois, 1972; Girod and Dubois, 1977).

The intermediate-lobe cells were also studied by comparing successive stainings with anti-α-MSH and anti-17–39 ACTH antisera. For the two polypeptides, the results were quite superimposable. The same variations in staining intensity were observed with anti-α-MSH and anti-17–39 ACTH antisera (Fig. 7). It seems, therefore, that comparable quantities of α-MSH and of a molecule related to the C-terminal portion of ACTH are present in the same cell; this reinforces the hypothesis that α-MSH and CLIP are the result of a splitting of the ACTH molecule in the intermediate lobe (Scott *et al.*, 1974).

3.2.1b. Lerot. The intermediate lobe in this species is reduced to one or two cell layers (Figs. 8 and 9). Moreover, staining for α-MSH revealed only some of the intermediate-lobe cells (Fig. 8a, c), while in both the guinea pig and the rat, α-MSH is contained in all the intermediate-lobe cells. Although not very numerous, the lerot α-MSH cells were found

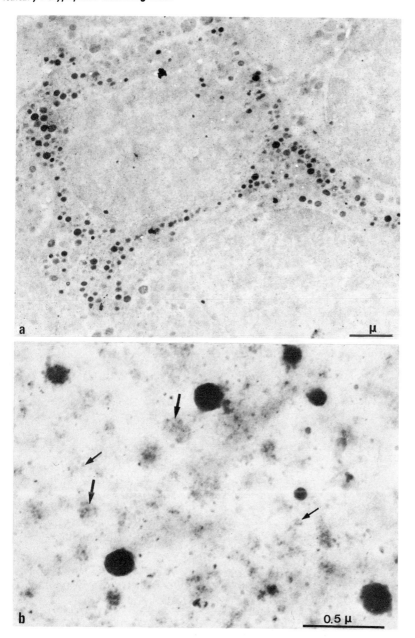

Figure 5. Guinea pig anterior lobe. The immunocytochemical staining for 17–39 ACTH (PAP technique) shows that only one cell type was stained. The staining intensity may vary from one granule to another (a). At a higher magnification (b), a heavy staining is visible on some granules, but PAP molecules are also observed either grouped (large arrows) or scattered in the cytoplasm (small arrows).

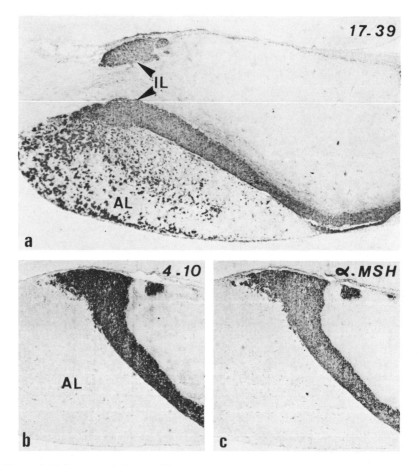

Figure 6. Guinea pig pituitary. ×30. At low magnification, localization of 17–39 ACTH staining (a) on the anterior lobe (AL) and on the whole intermediate lobe (IL). (b, c) Two adjacent sections stained first for 4–10 ACTH (b) and for α-MSH (c). Both antisera stain only the intermediate lobe.

only in the intermediate lobe, and therefore the small volume of the intermediate lobe does not seem to be due, as could be supposed, to a redistribution of its cells to the anterior lobe.

By successive staining for different antigens, the localization of α-MSH was compared with that of ACTH, β-MSH, and β-endorphin. The cells stained by anti-α-MSH antiserum did not react with anti-1–24 ACTH antiserum or with anti-17–39 ACTH antiserum, whereas the latter two antisera stained many cells clearly, both in the anterior lobe and in the rest of the intermediate lobe (Fig. 8b) (Tramu and Dubois, 1977). In the lerot,

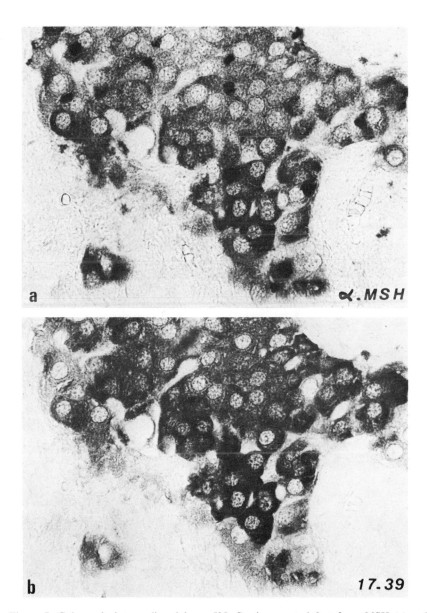

Figure 7. Guinea pig intermediate lobe. ×525. Section reacted first for α-MSH (a) and then, after decolorizing and elution, for 17–39 ACTH (b). The two stainings seem to be comparable.

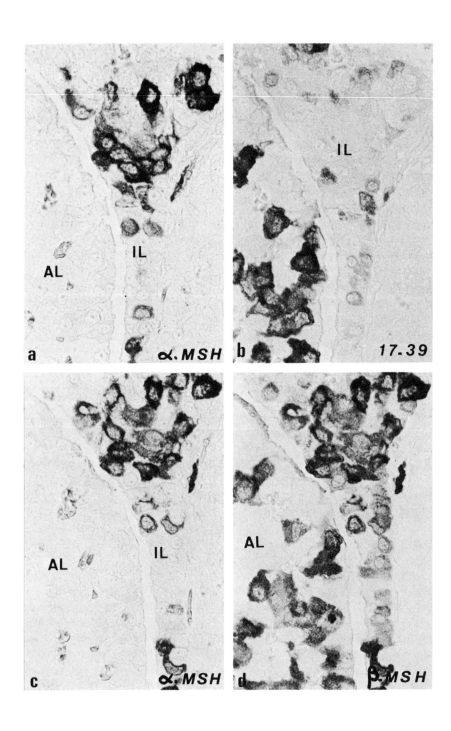

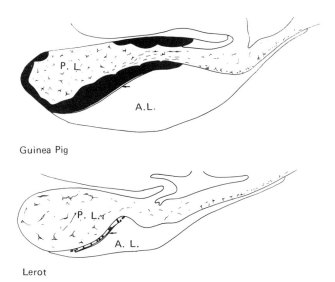

Guinea Pig

Lerot

Figure 9. Diagrams showing sagittal sections of guinea pig and lerot hypophysis reacted for α-MSH. The staining (in black) is localized only in the intermediate lobe. In contrast to the guinea pig, the lerot intermediate lobe is very reduced and is not formed exclusively of α-MSH-producing cells. Vessels are symbolized in the posterior lobe (P.L.); (AL) anterior lobe; (arrows) pituitary cleft.

therefore, α-MSH seems to be elaborated in cells that produce neither ACTH nor a peptide analogous to CLIP, contrary to what is apparently observed for the guinea pig. Possibly CLIP is absent in this species, but there remains the fact that α-MSH, which is present, was never found in cells containing ACTH. Therefore, if α-MSH results from a splitting of the ACTH molecule, the remaining C-terminal part has disappeared. This being quite unlikely, lerot α-MSH may not be the product of ACTH splitting.

Though ACTH has never been found in α-MSH cells, β-MSH and β-endorphin were both present simultaneously with α-MSH. These cells would therefore be responsible for secreting a substance related to LPH and not including the ACTH sequence.

Figure 8. Lerot pituitary. ×525. (a, b) Section reacted first for α-MSH (a) and then, after decolorizing and elution, for 17–39 ACTH (b). α-MSH-containing cells do not react for 17–39 ACTH, but other cells in the intermediate lobe (IL) and in the anterior lobe (AL) are 17–39-ACTH-immunoreactive. 1–24 ACTH staining was shown to be comparable with the 17–39 ACTH staining. (c, d) Another section reacted first for α-MSH (c) and then, after decolorizing and elution, for β-MSH (d). α-MSH-containing cells also react for β-MSH. Moreover, other cells in the intermediate lobe (IL) and in the anterior lobe (AL) are also stainable with anti-β-MSH antiserum. β-Endorphin staining was observed to be comparable with the β-MSH staining.

3.2.2. β-Melanotropin

3.2.2a. Lerot. Anti-β-MSH antiserum stained not only the α-MSH cells, but also all the cells stained by anti-ACTH antisera. Some of the β-MSH-immunoreactive cells of the intermediate lobe were stained much more strongly than others. They correspond to α-MSH cells (Fig. 8c, d). This unexplained observation underlines the uniqueness of the lerot α-MSH cells.

3.2.2b. Guinea Pig. In this species, no really significant staining was observed with either of the two β-MSH antisera used. Only a few intermediate-lobe cells contained large and scarce immunoreactive granules dispersed in the cytoplasm. The antisera do not seem to be at fault, since they give very good results in the lerot, as well as in the rat and in man (unpublished observations). This almost negative result may be due to a very slight cross-reaction with guinea pig β-MSH, or to the fact that only infinitely small quantities of the hormone are present in this species. But it is also possible that β-MSH does not exist as a hormone entity, as suggested by Bachelot *et al.* (1977). In this case, β-MSH, which forms part of the β-LPH sequence, would not be accessible to the antibodies. Although no definitive conclusions can be drawn on this subject, it should be noted that as far as we know, the guinea pig is the only mammalian species the ACTH cells of which do not react for β-MSH.

3.3. Anti-4–10 ACTH Staining

A specific antiserum against the heptapeptide present in ACTH, α-MSH, β-MSH, and β-LPH was used.

3.3.1. Guinea Pig

In this species, staining for 4–10 ACTH was localized only in the intermediate lobe and appeared quite comparable to α-MSH staining (See Fig. 6b, c). To examine the possibility of a cross-reaction with α-MSH, anti-4–10 ACTH staining was compared with that against α-MSH by using the successive localization technique. In some animals, the results could be superimposed (Fig. 10a, b), but in others, they were quite different (Fig. 10c, d), which excludes the possibility that α-MSH was revealed by the anti-4–10 ACTH antiserum. The total lack of staining in the anterior lobe would indicate that the antibodies were not capable of reaching the 4–10 site, which was nonetheless certainly present in the cell. The intermediate-lobe staining for 4–10 ACTH shows, therefore, that a substance different from α-MSH and related to the 4–10 ACTH sequence may be present in this lobe. 4–10 ACTH may even exist as a hormonal entity,

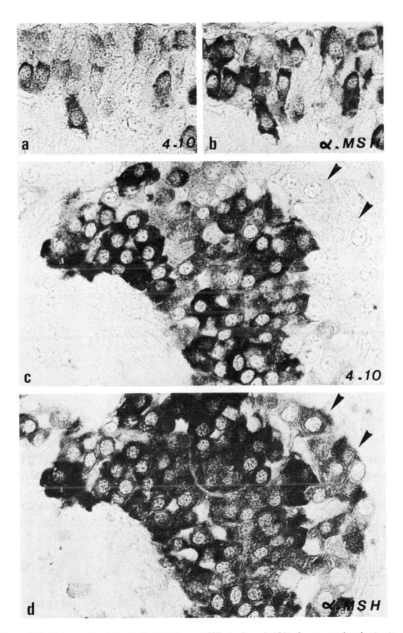

Figure 10. Guinea pig intermediate lobe. ×525 (reduced 10% for reproduction). (a, b) Section reacted first for 4–10 ACTH (a) and then, after decolorizing and elution, for α-MSH (b). Here, the 4–10 ACTH staining is comparable with the α-MSH staining. (c, d) In another animal, the same successive staining procedure shows that 4–10-ACTH-reactive cells (c) can differ greatly from α-MSH-containing cells (d). Certain cells (arrows) do not stain for 4–10 ACTH, but react for α-MSH.

as suggested by its behavioral (DeWied, 1976) and physiological (Drouhault *et al.*, 1975) activities.

3.3.2. Lerot

Specific antibodies against the 4–10 heptapeptide bound to some corticotrophs in the anterior and intermediate lobes. However, not all the ACTH cells are stained, as shown by comparing adjacent sections (Tramu and Dubois, 1977). Immunoreactive cells are more frequent in the caudal than in the anterior portions of the hypophysis. This staining was always quite different from that obtained with anti-α-MSH antiserum. It appears, then, that in the lerot as in the guinea pig, a substance separate from ACTH and from α-MSH was revealed by the anti-4–10 ACTH antiserum.

3.4. Endorphin Localization

3.4.1. Guinea Pig

Stainings for α-endorphin and for β-endorphin revealed all the corticotrophs of the anterior lobe (Fig. 11) and all the intermediate-lobe cells (Fig. 12). These results are consistent with those of Bloom *et al.* (1977) and of Begeot *et al.* (1978).

3.4.2. Lerot

The two anti-endorphin antisera stained both the cells revealed by anti-ACTH antisera and those revealed by anti-α-MSH antiserum. The reaction was therefore identical to that obtained with anti-β-MSH antiserum, but α-MSH cells were not more immunoreactive than others, as observed for β-MSH staining. Since β-MSH and endorphin sequences are included in the β-LPH molecule, the possibility of a cross-reaction cannot be excluded, and anti-β-MSH and anti-β-endorphin antisera might stain β-LPH.

3.5. Enkephalin Localization

Two morphinelike pentapeptides called methionine-enkephalin and leucine-enkephalin have been isolated from the brain (Hughes, 1975). The morphological localization of these peptides was attempted in rat and guinea pig, on tissues fixed by perfusion and cut on a cryostat, since the method used to localize adenohypophyseal polypeptides did not yield very satisfactory results. At the start, our intention was to demonstrate the pentapeptides in the hypothalamus. Hence, nerve endings containing

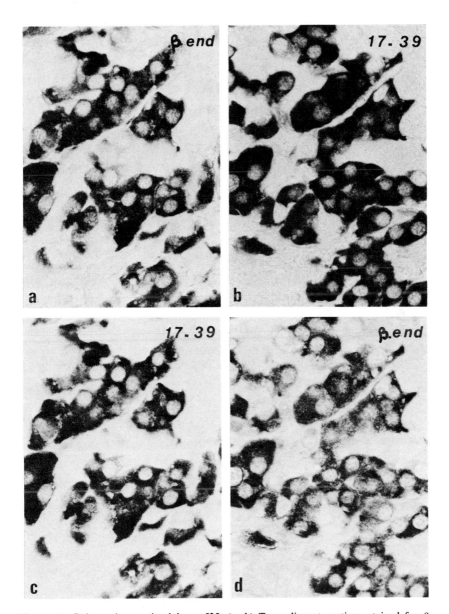

Figure 11. Guinea pig anterior lobe. ×525. (a, b) Two adjacent sections stained for β-endorphin (a) and for 17–39 ACTH (b). (c) The section stained for β-endorphin (a) was decolorized, eluted, and stained again for 17–39 ACTH. The two stainings are identical. (d) The section stained for 17–39 ACTH (b) was decolorized, eluted, and stained again for β-endorphin. The stained cells are identical.

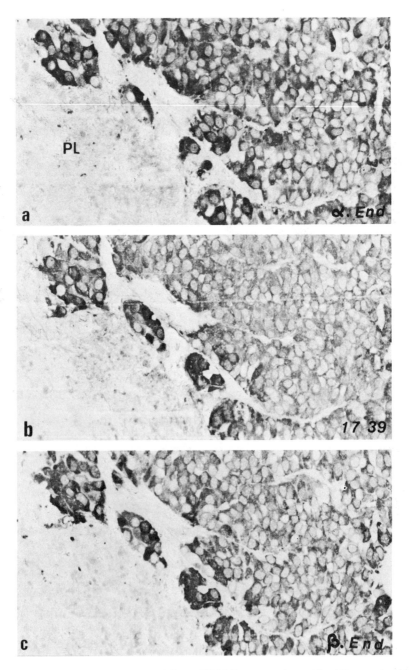

Figure 12. Guinea pig intermediate lobe. ×325. Adjacent sections stained for α-endorphin
(a), 17–39 ACTH (b), and β-endorphin (c). Analogous stainings are observed. (PL) Posterior
lobe.

enkephalins were localized in the external layer of the median eminence (Tramu and Léonardelli, 1979). Furthermore, the antisera used here, raised against either leucine-enkephalin (one antiserum) or methionine-enkephalin (three antisera), gave interesting results concerning the adenohypophysis.

3.5.1. Rat

All the antisera, whether specific to leucine-enkephalin or to methionine-enkephalin, stained the intermediate-lobe cells clearly, and to a lesser degree the corticotrophs of the anterior lobe (Fig. 13a, b). Since methionine-enkephalin represents the 61–65 β-LPH sequence, a possible cross-reaction between anti-methionine-enkephalin antiserum and β-LPH cannot be refuted, but is unlikely, since our antisera do not cross-react with β-endorphin. Furthermore, the anti-leucine-enkephalin antiserum used does not bind significantly to methionine-enkephalin in a radioimmunoassay system. These results suggest that the corticotrophs of the anterior lobe and even more so those of the intermediate lobe produce enkephalins independently of β-endorphin production, but further investigations are necessary to establish whether leucine-enkephalin is really synthesized in the hypophysis.

As well as staining the corticotrophs of the anterior lobe slightly, all the antisera clearly revealed large cells shown to be thyrotrophs by the successive localization method. This surprising result is discussed further below, but it can be mentioned here that the synthesis of the 61–65 β-LPH sequence in thyrotrophs is unlikely and thus the immunohistochemical staining possibly reveals the binding of enkephalins in the cytoplasm of these cells.

3.5.2. Guinea Pig

In this species, the anti-leucine-enkephalin antiserum and two of the anti-methionine-enkephalin antisera revealed the corticotrophs of the anterior lobe and the intermediate-lobe cells. But on the whole, the staining was much less significant than in the rat. In the anterior lobe, the thyrotrophs were stained by all the antisera, confirming the results obtained in the rat. Furthermore, another cell type was also revealed, especially when anti-leucine-enkephalin antiserum was used. These cells, smaller in size and greater in number than the thyrotrophs, were proved to be gonadotrophs (Fig. 13c, d). It should be noted that gonadotrophs located in the anterior part of the hypophysis, close to the median eminence, were always much more immunoreactive than the others. According to exact comparisons that have been carried out, the intracytoplasmic antienkephalin staining did not appear to be directly superimposable with the anti-gonadotropin hormone staining (Tramu and Léonardelli, 1979). The results

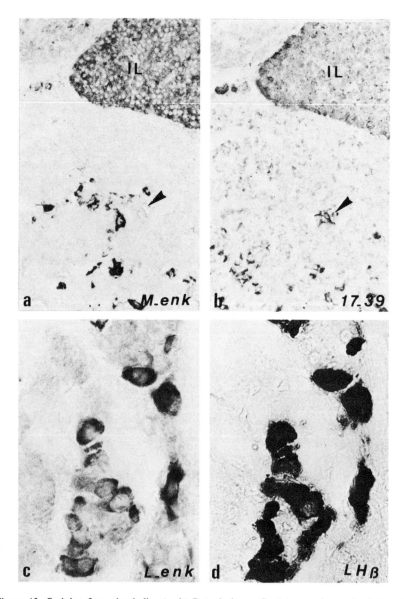

Figure 13. Staining for enkephalin. (a, b) Rat pituitary. Cryostat section stained first for methionine-enkephalin (a) and then, after decolorizing and antibody elution, for 17–39 ACTH (b). Corticotrophs recognized in (b) are weakly stained for methionine-enkephalin (arrows); the intermediate lobe (IL) is immunoreactive with both antisera. Large cells stained in (a) were shown to be thyrotrophs. ×130. (c, d) Guinea pig pars tuberalis. Cryostat section stained first for leucine-enkephalin (c) and then, after decolorizing and antibody elution, for LHβ (d). It can be observed that gonadotrophs are clearly leucine-enkephalin-immunoreactive. ×525.

observed in the guinea pig therefore confirm the thyrotroph staining, but also suggest that gonadotrophs contain an enkephalinlike substance in the cytoplasm. If the reactions are specific, as controls seem to prove, what is the meaning of an enkephalinlike substance in the cytoplasm of cells producing glycoprotein hormones? Since it is most unlikely that enkephalins are produced in these cells, the pentapeptides may enter the cell where they would modulate the synthesis or the release or both, of gonadotropins and thyrotropin. A similar action has been suggested for gonadoliberin (LH-RH) in gonadotrophs by Sternberger and Petrali (1975), who demonstrated the presence of the decapeptide in target-cell cytoplasm. The enkephalins also seem to play an important role in pituitary secretions (Frederickson, 1977), and in particular, enkephalins, like morphine, have an inhibiting effect on LH, FSH, and TSH release (Bruni *et al.*, 1977), but the locus of this action is not yet clear. Authors studying the effects of enkephalins on pituitary secretion believe that their stimulating action on somatotropin and prolactin release is mediated by hypothalamic mechanisms (Cocchi *et al.*, 1977; Rivier *et al.*, 1977; Shaar *et al.*, 1977). Nonetheless, in the case of LH, FSH, and TSH, our immunohistochemical results suggest that enkephalins could bind in the cells, perhaps to modulate the action of the hypothalamic factor on the synthesis or the release or both of hormone according to a mechanism that remains to be established.

4. Conclusions

Taking into consideration the many observations already published (Girod, 1976, 1977) as well as our findings, one can conclude that the polypeptide hormones and associated substances are elaborated by a cellular system comprised both of cells dispersed throughout the anterior lobe and of all the cells in the intermediate lobe, the pituitary localizations of enkephalin constituting a special case. Depending on their localization, the cells are not identical, since the intermediate lobe is specific for producing α-MSH, which has never been observed in the anterior lobe. For the guinea pig, and even more generally, α-MSH and the C-terminal portion of ACTH are present in the same cells of the intermediate lobe. This fact has been the basis for the hypothesis of a splitting of ACTH into α-MSH and CLIP. This phenomenon does not seem to take place in all mammals, since α-MSH-producing cells in the lerot do not appear to elaborate ACTH, but do elaborate β-MSH and β-endorphin, the sequences of which are part of the β-LPH molecule. The genesis of pituitary polypeptides may therefore be different from one species to another.

Considering the number of functionally different peptides elaborated by these cells, the existence of precursor is quite probable. This has been shown to be even more likely by the results of Roberts and Herbert (1977) and Mains *et al.* (1977), who have shown the presence in a pituitary producing ACTH tumor of a substance that includes the ACTH and LPH sequences and has in addition an important unknown sequence. Although such a precursor has not been demonstrated in normal hypophysis, its presence is probable. However, one cannot be sure that this molecule exists as is in all mammalian species.

Another problem is to determine whether the different immunohisto-chemical results observed, depending on species and on anterior-lobe or intermediate-lobe localizations, are due only to variations in the enzymatic splitting of a single precursor.

ACKNOWLEDGMENTS. The authors are grateful to Mr. A. Pillez (CNRS) for skillful technical assistance. They also wish to thank Dr. M. P. Dubois (INRA, Nouzilly, France) for supplying the majority of the antisera used, and Dr. P. A. Desaulles and Dr. W. Rittel (Ciba, Geigy, Basel, Switzerland) for the generous gift of synthetic fractions of ACTH and MSH.

DISCUSSION

RAMACHANDRAN: In what species did you find that 17–39 ACTH is not present in the same cells as α-MSH?

TRAMU: In the dormouse, which is a hibernating mammal. It is interesting to note this difference from other species such as the guinea pig and rat, where they are both found in the same cell.

LABRIE: Concerning your findings of differential labeling of cells for α-MSH and ACTH 17–39, were the antibodies used developed against antigens of the same species? Otherwise, an absence of reaction could be due to a species difference in the antigen. I should also mention that β-endorphin and met-enkephalin have not been found to affect TSH release in anterior pituitary cells in culture in our laboratory or by Vale's group. Endorphins act on pituitary hormone secretion by an action at the suprapituitary level.

TRAMU: Your *in vitro* results concerning the action of β-endorphin and methionine-enkephalin on TSH secretion at the hypothalamic level are very interesting. To answer your first question: the dormouse and guinea pig corticotropins are not available, and so we used antisera against human ACTH. However, it has been shown in our presentation that the anti-17–39 ACTH antiserum will stain cells in the dormouse pituitary which are the same cells stained for 1–24 ACTH. The absence of reaction concerned only the cells which are stained with α-MSH.

VAUDRY: What evidence do you have that your antisera were specific for the hormones you expected to detect? For example, no β-endorphin antiserum has been described which does not cross-react with β-LPH. Are you sure that all of the cells don't contain β-LPH which would be stained by most of the antisera you have used in your study?

TRAMU: Of course it is probable that the antisera against β-endorphin, α-endorphin, and β-MSH, which are contained in the β-LPH molecule, may also cross-react with this entire molecule. In fact, in these observations we wanted to show the presence of β-LPH-related peptides in the ACTH cells.

MELDOLESI: My question concerns the interpretation of immunochemical results in quantitative terms. Don't you think that differences in staining intensity could be due, not only to difference in quantity of the antigen, but also to a variety of other factors such as reactivity of the antigen and antibody, availability of the antigen, and penetration of the antibody which might be variable depending on the state of the cells?

TRAMU: Yes, with the immunohistochemical technique at the light-microscope level, a lot of factors might be evoked to explain why some cells are more stained than others. However, we have already shown by comparison of light and electron micrographs, that the intensity of the staining parallels the quantity of granules.

REFERENCES

Bachelot, I., Wolfsen, A. R., and Odell, W. D., 1977, Pituitary and plasma lipotropins: Demonstration of the artefactual nature of β-MSH, *J. Clin. Endrocrinol. Metab.* **44**:939.

Baker, B. L., and Drummond, T., 1972, The cellular origin of corticotropin and melanotropin as revealed by immunochemical staining, *Am. J. Anat.* **134**:395.

Beauvillain, J. C., 1978, Caractérisation à l'échelle ultrastructurale des cellules adénohypophysaires individualisées par différentes techniques immunocytochimiques (étude chez le cobaye et le lérot), Thèse Sciences, Lille, France.

Beauvillain, J. C., and Tramu, G., 1973, Cellules à activité corticotrope de l'hypophyse du lérot *Eliomys quercinus*: Superpositions des résultats de microscopic optique (immunofluorescence et colorations) et de microscopie électronique, *C. R. Acad. Sci.* **277**:1025.

Beauvillain, J. C., Tramu, G., and Dubois, M. P., 1975, Characterization by different techniques of adrenocorticotropin and gonadotropin producing cells in lerot pituitary (*Eliomys quercinus*), *Cell Tissue Res.* **158**:301.

Begeot, M., Dubois, M. P., and Dubois, P. M., 1978, Immunofluorescent evidence for β-LPH and β-endorphin in normal and anencephalic human fetal pituitary gland, *C. R. Acad. Sci.* **286**:213.

Birk, Y., and Li, C. H., 1964, Isolation and properties of a new, biologically active peptide from sheep pituitary glands, *J. Biol. Chem.* **239**:1048.

Bloom, F., Bettenberg, E., Rossier, J., Ling, N., Leppaluoto, J., Vargo, T. M., and Guillemin, R., 1977, Endorphins are located in the intermediate and anterior lobes of the pituitary gland, not in the neurohypophysis, *Life Sci.* **20**:43.

Breustedt, H. J., 1968, Zur immunohistologischen ACTH: Localisation in der Rattenhypophyse, *Endokrinologie* **53**:1.

Bruni, J. F., Van Vugt, D., Marshall, S., and Meites, J., 1977, Effects of naloxone, morphine and methionine enkephalin on serum prolactin, luteinizing hormone, follicle stimulating hormone, thyroid stimulating hormone and growth hormone, *Life Sci.* **21**:461.

Cocchi, D., Santagostino., A., Gil-Ad, I., Ferri, S., and Muller, E. E., 1977, Leu-enkephalin stimulated growth hormone and prolactin release in the rat: Comparison with the effect of morphine, *Life Sci.* **20**:2041.

Cox, B. M., Opheim, K. E., Teschemacher, H., and Goldstein, A., 1975, A peptide like substance from pituitary that acts like morphine, *Life Sci.* **16**:1777.

Dessy, C., Herlant, M., Chretien, M., 1973, Détection par immunofluorescence des cellules synthétisant la lipotropine, *C. R. Acad. Sci.* **276**:335.

De Wied, D., 1976, Hormonal influences on motivation, learning and memory processes, *Hosp. Pract.* **11**:123.

Drouhault, R., Bethous, C., and Blanquet, P., 1975, Étude *in vitro* de l'activité hypocalcémiante, hypophysphorémiante, hyperlipémiante et immunogène des tronçons peptidiques communs aux hormones ACTH, α et β MSH et hormone β lipotropique, *C. R. Acad. Sci.* **280**:2141.

Dubois, M. P., 1971, Les cellules corticotropes de l'hypophyse des bovins, ovins et porcins: Mise en évidence par immunofluorescence et caractères cytologiques, *Ann. Biol. Anim. Biochem. Biophys.* **4**:589.

Dubois, M. P., 1972a, Localisation cytologique par immunofluorescence des sécrétions corticotropes, α et β mélanotropes au niveau de l'antehypophyse des bovins, ovins et porcins, *Z. Zellforsch.* **125**:200.

Dubois, M. P., 1972b, Nouvelles données sur la localisation au niveau de l'adénohypophyse des hormones popypeptidiques, ACTH, MSH, LPH, *Lille Med.* **17**:1391.

Dubois, M. P., and Graf, L., 1973, Demonstration by immunofluorescence of the lipotropic hormone/LPH in bovine, ovine and porcine adenohypophysis, *Horm. Metab. Res.* **5**:229 (abstract).

Dubois, P., Vargues-Regairaz, M., and Dubois, M. P., 1973, Human foetal hypophysis: Immunofluorescent evidence for corticotropin and melanotropin activities, *Z. Zellforsch.* **145**:131.

Frederickson, R. C. A., 1977, Enkephalin pentapeptides: A review of current evidence for a physiological role in vertebrate neurotransmission, *Life Sci.* **21**:23.

Girod, C., 1976, Histochemistry of the adenohypophysis, in: *Handbuch der Histochemie* (W. Graumann and K. Neumann, eds.), Vol. VIII, Part 4, pp. 205–224, Gustav Fischer Verlag, Stuttgart.

Girod, C., 1977, Apport de l'immunohistochimie à l'étude cytologique de l'adénohypophyse, *Bull. Assoc. Anat.* **62**:21.

Girod, C., and Dubois, M. P., 1977, Mise en évidence par immunofluorescence des cellules corticotropes et des cellules mélanotropes de l'adénohypophyse chez les singes *Erythrocebus patas, Cercopithecus aethiops* et *Papio hamadryas, C. R. Soc. Biol. (Paris)* **171**:367.

Gosbee, J. L., Kraicer, J., Kastin, A. J., and Schally, A. V., 1970, Functional relationship between the pars intermedia and ACTH secretion in the rat, *Endrocrinology* **86**:560.

Hachmeister, U., and Kracht, J., 1965, Antigene Eigenshaften von β 1–24 corticotropin, *Virchows Arch. Pathol. Anat.* **339**:245.

Hess, R., Barratt, D., and Gelzer, J., 1968, Immunofluorescent localization of the corticotrophin in the rat pituitary, *Experientia* **24**:584.

Hughes, J., 1975, Isolation of an endogenous compound from the brain with pharmacological properties similar to morphine, *Brain Res.* **88**:295.

Kraicer, J., Bencosme, S. A., and Gosbee, J. L., 1971, The *pars intermedia* and ACTH secretion in the rat, *Fed. Proc. Fed. Am. Soc. Exp. Biol.* **30**:533.

Mains, R. E., Eipper, B. A., and Ling, N., 1977, Common precursor to corticotropins and endorphins, *Proc. Natl. Acad. Sci. U.S.A.* **74**:3014.

Marshall, J. M., Jr., 1951, Localization of adrenocorticotropic hormone by histochemical and immunochemical methods, *J. Exp. Med.* **94**:21.

McGarry, E. E., and Beck, J. C., 1968, Dosage immunologique de l'ACTH, *Ann. Endrocrinol. (Paris)* **29**:17.

Mialhe-Voloss, C., 1958, Posthypophyse et activité corticotrope, *Acta Endocrinol. (Copenhagen) Suppl.* **35**:1.

Moon, H. D., Li, C. H., and Jennings, B. M., 1973, Immunohistochemical and histochemical studies of pituitary β lipotrophs, *Anat. Rec.* **175**:529.

Moriarty, C. M., and Moriarty, G. C., 1975, Bioactive and immunoactive ACTH in the rat pituitary: Influence of stress and adrenalectomy, *Endocrinology* **96**:1419.

Moriarty, G. C., and Halmi, N. S., 1972, Electron microscopic study of the adrenocorticotropin producing cell with the use of unlabeled antibody and the soluble peroxidase–antiperoxidase complex, *J. Histochem. Cytochem.* **20**:590.

Naik, D. V., 1973, Electron microscopic immunocytochemical localization of adrenocorticotropin and melanocyte stimulating hormone in the pars intermedia cells of rats and mice, *Z. Zellforsch.* **142**:305.

Nakane, P. K., 1970, Classifications of anterior pituitary cell types with immunoenzyme histochemistry, *J. Histochem. Cytochem.* **18**:9.

Phifer, R. F., and Spicer, S. S., 1970, Immunologic and immunopathologic demonstration of adrenocorticotropic hormone in the *pars intermedia* of the adenohypophysis, *Lab. Invest.* **23**:543.

Phifer, R. F., Orth, D. N., and Spicer, S. S., 1974, Specific demonstration of the human hypophyseal adeno-melanotropic (ACTH/MSH) cell, *J. Clin. Endrocrinol. Metab.* **39**:684.

Porte, A., Klein, M. J., Stoeckel, M. E., and Stutinsky, F., 1971, Sur l'existence de cellules de type "corticotrope" dans la pars intermedia de l'hypophyse du rat: Étude au microscope électronique, *Z. Zellforsch.* **115**:60.

Rivier, C., Vale, W., Ling, N., Brown, M., and Guillemin, R., 1977, Stimulation *in vivo* of the secretion of prolactin and growth hormone by β-endorphin, *Endocrinology* **100**: 238.

Roberts, J. L., and Herbert, E., 1977, Characterization of a common precursor to corticotropin and β lipotropin: Cell free synthesis of the precursor and identification of corticotropin peptides in the molecule, *Proc. Natl. Acad. Sci. U.S.A.* **74**:4826.

Rochefort, G. J., Rosenberger, J., and Saffran, M., 1959, Depletion of pituitary corticotrophin by various stresses and by neurohypophysial preparations, *J. Physiol. (London)* **146**:105.

Rossier, J., Vargo, T. M., Minick, S., Ling, N., Bloom, F. E., and Guillemin, R., 1977, Regional dissociations of β endorphin and enkephalin contents in rat brain and pituitary, *Proc. Natl. Acad. Sci. U.S.A.* **74**:5162.

Scott, A. P., Lowry, P. J., Ratcliffe, J. G., Rees, L. H., and London, J., 1974, Corticotrophin-like peptides in the rat pituitary, *J. Endocrinol.* **61**:355.

Shaar, C. J., Frederickson, R. C. A., Dininger, N. B., and Jackson, L., 1977, Enkephalin analogues and naloxone modulate the release of growth hormone and prolactin: Evidence for regulation by an endogenous opioid peptide in brain, *Life Sci.* **21**:853.

Stefan, Y., and Dubois, M. P., 1972, Localisation par immunofluorescence des hormones corticotropes et mélanotropes dans l'hypophyse du rongeur, *Ellobius lutescens, Z. Zellforsch.* **133**:353.

Sternberger, L. A., and Petrali, J. P., 1975, Quantitative immunochemistry of pituitary receptors for luteinizing hormone–releasing hormone, *Cell Tissue Res.* **162**:141.

Stoeckel, M. E., Dellmann, H. D., Porte, A., and Gertner, C., 1971, The rostral zone of the intermediate lobe of the mouse hypophysis, a zone of particular concentration of corticotrophic cells: A light and electron microscopic study, *Z. Zellforsch.* **122**:310.

Tramu, G., and Dubois, M. P., 1977, Comparative cellular localization of corticotropin and melanotropin in lerot adenohypophysis (*Eliomys quercinus*): An immunohistochemical study, *Cell Tissue Res.* **183**:457.

Tramu, G., and Léonardelli, J., 1979, Immunohistochemical localization of enkephalins in median eminence and adenohypophysis, *Brain Res.* **168:**457.

Tramu, G., Pillez, A., and Léonardelli, J., 1978, An efficient method of antibody elution for the successive or simultaneous localization of two antigens by immunocytochemistry, *J. Histochem. Cytochem.* **26:**322.

Weber, E., Voigt, K. H., and Martin, R., 1978, Granules and Golgi vesicles with differential reactivity to ACTH antiserum in the corticotroph of the rat anterior pituitary, *Endocrinology* **102:**1466.

4

Exocytosis and Related Membrane Events

E. Vila-Porcile and L. Olivier

1. Introduction

Exocytosis (sometimes called emiocytosis) is a phenomenon through which the membrane-bounded content of a cellular constituent is transferred from the intracellular compartment to the extracellular compartment, by means of complex membrane processes, without any discontinuity of the cell-limiting surface. In the pituitary, exocytosis is considered as mainly involving the secretory granules and possibly secretory vesicles. However, it can also involve residual bodies, resulting from the lysosomal autophagy of secretory granules, after secretion blockade. Granular exocytosis triggers endocytotic phenomena during which patches of the peripheral membrane as well as a part of the extracellular fluid become internalized.

In this chapter, we will describe in turn the morphological data on exocytosis of secretory granules and residual bodies, and on endocytosis. Following this presentation, we will discuss these phenomena. We will often refer (principally as far as iconography is concerned) to our previous work on the prolactin cell, studied *in vivo*, in lactating rats, during the hormone release triggered by suckling, according to the procedure of Pasteels (1963).

E. Vila-Porcile and *L. Olivier* ● Laboratoire d'Histologie–Embryologie, C.N.R.S. ERA 484, 75634 Paris 13, France

2. Morphological Data on Granular Exocytosis

Table I is a summary of papers on exocytosis in the different pituitary cell types. It shows that exocytosis has been observed more often in somatotrophs and prolactin cells than in other cell types (see Section 5.4.2).

2.1. General Outlook on the Secretory Granule

In normal conditions, pituitary cells hold in their cytoplasm free granules that are the final stage of a synthesizing and packaging process lasting for 55–185 min (Farquhar et al., 1978). This process involves the rough endoplasmic reticulum, the Golgi complex (Racadot et al., 1965; Tixier-Vidal et al., 1965; Porcile et al., 1966; Tixier-Vidal and Picart, 1967; Farquhar et al., 1975, 1978), and in the view of some authors, the Golgi endoplasmic reticulum–lysosome (GERL) complex (Pelletier and Novikoff, 1972; Hand and Oliver, 1977). The secretory granules leave the Golgi apparatus, where they have acquired their definitive structure. At that time, their diameter and shape vary greatly from one cell type to the other and also within a given cell (i.e., gonadotroph). However, they share some characteristics: they constitute a particular compartment within the cell, since their membranes enclose and separate the stored content from the hyaloplasm.

Morphologically, these membranes belong to the general category of biological membranes (unit membrane, with a trilaminar aspect). It is difficult to define their precise relationships to the granule content. These relationships depend on the species, on the cell type, and on the fixation and staining method. Whenever standard methods are employed (aldehydic fixation, osmium postfixation, uranyl–lead staining), a clear and thin lining often separates the core from the granule membrane.

2.2. Successive Phenomena during Exocytosis

It is possible to enumerate three stages during exocytosis: (1) the first one leads to the opening of the granule membrane toward the outside; (2) the second one corresponds to a confrontation between the exteriorized granule content and extracellular medium; (3) finally, in the third stage, the granule and the exocytotic pocket disappear.

2.2.1. First Stage

The first stage involves a contact between the granule membrane and the plasmalemma, followed by the fusion of the membranes at the contact

Table I. Exocytoses in Different Pituitary Cell Types

Authors	Animal	Experimental conditions	Stimulation
		Prolactin cells	
Sano (1962)	Mouse	*In vivo*	0
Herlant (1963)	Mole	*In vivo*	Suckling
Pasteels (1963)	Rat ♀	*In vivo*	Suckling
	♂	Organ culture	0
Girod and Dubois (1965)	Hamster	*In vivo*	Suckling
Smith and Farquhar (1966)	Rat ♀	*In vivo*	Suckling
			Weaning
Tixier-Vidal and Picart (1967)	Duck	Organ culture	0
Smith and Farquhar (1970)	Rat ♀	*In vivo*	Suckling
			Estrogens
Farquhar (1971)	Rat ♀	*In vivo*	Suckling
Shiino *et al.* (1971)	Rat ♀	*In vivo*	Cyclic
			Suckling
Olivier and Vila-Porcile (1972)	Rat ♀	*In vivo*	Suckling
Pelletier *et al.* (1972)	Rat	Incubation	cAMP
Shiino *et al.* (1972a)	Rat	*In vivo*	Suckling
Pasteels (1972a)	Rat	*In vivo*	Stress (ether)
Vila-Porcile *et al.* (1973b)	Rat ♀	*In vivo*	Suckling
			Weaning
Vila-Porcile *et al.* (1973a)	Rat ♀	*In vivo*	Weaning
Farquhar *et al.* (1975)	Rat ♀	*In vivo*	Active cells
Tixier-Vidal (1975)	Rat	Monolayer culture	0
Vila-Porcile (1975)	Rat ♀	*In vivo*	Suckling
Batten *et al.* (1976)	Teleost.	*In vivo*	0
Leatherland and Percy (1976)	Cyclost. Actinopt.	*In vivo*	0
Farquhar (1977)	Rat ♀	*In vivo*	Suckling
Fridberg and Ekengren (1977)	Salmon	*In vivo*	0
Farquhar *et al.* (1978)	Rat ♀	Dispersed cells	Estrogen *in vivo*
Haüsler *et al.* (1978)	Rat ♀	*In vivo*	Suckling
		Somatotrophs	
Farquhar (1961a)	Rat	*In vivo*	0
Farquhar (1961b)	Rat	*In vivo*	0
Kurosumi (1961)	Rat, other mammals	*In vivo*	0
Rennels (1964)	Rat	*In vivo*	Scalding
De Virgiliis *et al.* (1968)	Rat	*In vivo*	Hypothalamic extract
Couch *et al.* (1969)	Rat	*In vivo*	"Purified growth hormone releasing factor"
Girod and Dubois (1969)	Hamster	*In vivo*	Starvation
Coates *et al.* (1970)	Rat	*In vivo*	Hypothalamic extract

(continued)

Table I *(continued)*

Authors	Animal	Experimental conditions	Stimulation
		Somatotrophs (continued)	
Coates *et al.* (1971)	Rat	*In vivo*	Growth hormone releasing factor
Pelletier *et al.* (1971)	Rat	*In vivo*	Lysine vasopressin
		Incubation	Hypothalamic extract
Pelletier *et al.* (1972)	Rat	Incubation	cAMP
Daikoku *et al.* (1973)	Rat	*In vivo*	Hypothalamic extract
Echave Llanos and Gomez Dumm (1973)	Mouse	*In vivo*	Castration Hepatectomy
Wilbur *et al.* (1974)	Rat	*In vivo*	cAMP (intraportal)
Farquhar *et al.* (1975)	Rat	*In vivo*	0
Gas (1975)	Carp	*In vivo*	Starvation
Schofield and Orci (1975)	Ox	Incubation	Pronase
Shiino and Rennels (1975a)	Rat	*In vivo*	Hepatectomy
		Thyrotrophs	
Farquhar (1969)	Rat	*In vivo*	0
		In vivo	Thyroidectomy
Farquhar (1971)	Rat	*In vivo*	0
		In vivo	Thyroidectomy
Pelletier *et al.* (1971)	Rat	*In vivo*	Hypothalamic extracts
		Hemipituitaries	High-K^+ medium
Moguilevsky *et al.* (1973)	Rat	Hemipituitaries	TRF *in vivo*
Shiino *et al.* (1973)		*In vivo*	Thyroidectomy + TRH
Cuerdo-Rocha and Zambrano (1974)	Rat	Organ culture	Radiothyroidectomy in neonates + TRF + puromycin, cyclo- heximide, or actino- mycin
Wilbur *et al.* (1978)	Rat	*In vivo*	TRF (intraportal)
Tixier-Vidal (1978: personal communication)	Rat	Monolayer culture	Tritiated TRH
		Corticotrophs (pars distalis)	
Kurosumi and Kobayashi (1966)	Rat	*In vivo*	0
		In vivo	Adrenalectomy
Rennels and Shiino (1968)	Rat	*In vivo*	Removing left adrenal + cyanoketone
Bunt (1969)	Newt	*In vivo*	Adrenalectomy + metopirone
Kurosumi and Kobayashi (1969)	Rat	*In vivo*	Adrenalectomy
Pelletier *et al.* (1971)	Rat	*In vivo*	Hypothalamic releasing factors

(continued)

Table I *(continued)*

Authors	Animal	Experimental conditions	Stimulation
	Corticotrophs (continued)		
Vila-Porcile (1972)	Rat	*In vivo*	0
Bácsy *et al.* (1976)	Rat	Monolayer culture	0
	Gonadotrophs		
Tixier-Vidal (1965)	Duck	*In vivo*	0
Kurosumi and Oota (1966)	Rat	*In vivo*	Persistent diestrous rats
Farquhar (1971)	Rat	*In vivo*	Castrated rats
Tixier-Vidal *et al.* (1971)	Ewe–lamb	Incubation	LRF (LH-RH)
Shiino *et al.* (1972b)	Rat	*In vivo*	LH-RH
Mendoza *et al.* (1973)	Rat	*In vivo*	LH-RH
Shiino and Rennels (1973)	Rat	*In vivo*	Testosterone in neonates
Zambrano *et al.* (1975)	Rat	Organ culture	LH-RH
Farquhar *et al.* (1975)	Rat	Dispersed cells	Castration
Luborsky-Moore *et al.* (1975)	Rat	*In vivo*	LH-RH (intraportal)
Tixier-Vidal *et al.* (1975a)	Rat	Monolayer culture	LH-RH
Wada (1975)	Quail	*In vivo*	LH-RH
Römmler *et al.* (1978)	Rat	*In vivo*	LH-RH

site and then by the formation of a diaphragm. Either the breaking up or the dissolution of this diaphragm implies the opening of the granule compartment toward the extracellular compartment. The different steps of these events have been detailed in other cellular models with current electron-microscopic methods as well as with freeze-etching: parotid gland (Amsterdam *et al.*, 1969, 1971; Kalina and Robinovitch, 1975; De Camilli *et al.*, 1976), exocrine pancreas, Geuze and Poort, 1973; Kramer and Geuze, 1974; Meldolesi, 1974; Oliver and Hand, 1978), endocrine pancreas (Orci *et al.*, 1973), adrenal medulla (Diner, 1967; Benedeczky and Smith, 1972; Abrahams and Holtzman, 1973; Winkler *et al.*, 1974), neurohypophysis (Douglas *et al.*, 1971; Theodosis *et al.*, 1978), neuromuscular junction (Heuser and Reese, 1973; Ceccarelli *et al.*, 1976), and mast cells (Tandler and Poulsen, 1976; Lawson *et al.*, 1977).

 This step has not been investigated in the pituitary, since freeze-etching has not been frequently used up to now (Daikoku *et al.*, 1973).

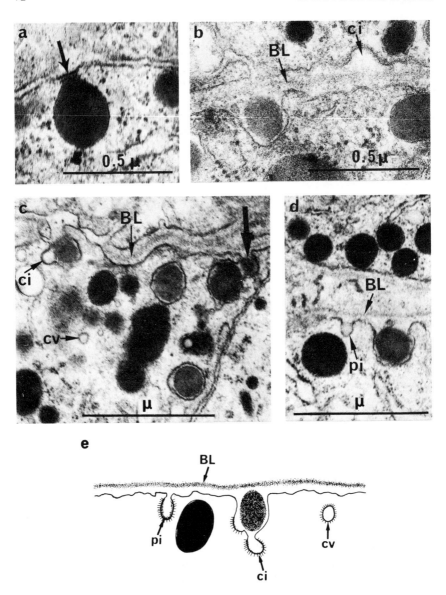

Figure 1. Exocytosis. (a) Contact step: modifications of both granule membrane and plasmalemma (arrow). (b) Omega phase: granule core in its pocket. (c) Multiple exocytoses: two granules in the same pocket (arrow); remodeling of the pocket wall. (d) Remodeling of the plasmalemma. (e) Summarizing schema: the extruded granule core is lighter than the intracellular granule. (BL) Basal lamina; (ci) coated invagination; (cv) coated vesicle; (pi) plasmalemmal invagination.

Using current electron microscopy in our study on several hundred exocytoses, we observed but one contact picture (Fig. 1a). At the contact site, both membranes appear to be "connected by cross-bridges" as described by Benedeczky and Smith (1972) in the adrenal medulla. This low probability of observing contact images on ultrathin sections should be noted. It suggests that a very short-duration phenomenon might be involved on a very limited surface, as Normann (1976) mentioned in the case of the neurohypophysis.

2.2.2. Opened-Granule Phase ("Omega Phase")

It is relatively easy to observe the images of granules opened toward the outside medium, as compared to the first step, which is only postulated. At this stage, the granule membrane enclosing the product is connected to the plasmalemma around the entire periphery of the opening, giving an "omega" image on thin sections (Normann, 1976) (Fig. 1b).

The granule content is now transferred into the extracellular compartment. The granule core is clearly visible inside the exocytotic pocket. Its shape is unaltered. However, two modifications occur: its electron contrast is reduced and its relationships with its former membrane—now the pocket membrane—are different. A clear space of approximately 100 Å or more between the lighter core and the pocket membrane is present. On the other hand, the pocket wall is remodeled through small invaginations (see below). This is probably the stage of longest duration during exocytosis, since it is the most frequently observed.

This omega phase represents the preliminary step in the destruction of the granule assembly; we are aware of its high resistance, and this especially applies to the prolactin granule (Zanini and Giannattasio, 1974; Giannattasio *et al.*, 1975). Obviously, during the omega phase, the extracellular fluid interacts with the granule content. Little is known about this phenomenon. As far as prolactin granules are concerned, we observed after *in vivo* marking of the extracellular medium through the horseradish peroxidase (HRP) technique (cf. Section 4.1) the progressive penetration of this diffusion tracer from the superficial pole of the granule to the achievement of a complete impregnation (Fig. 2).

2.2.3. Granule Disintegration and Disappearance of the Exocytotic Pocket

Although light-contrast granules are seen within the pocket, morphological images of the disappearance of the granules never occur, nor do

a   b

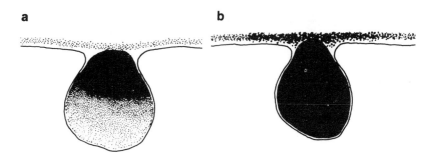

Figure 2. Penetration of HRP into the extruded granule core. (a) Progressive penetration;
(b) complete impregnation.

empty exocytotic pockets. We may conclude that after the preliminary
step, the granule is suddenly solubilized while the pocket membrane is
spreading and becoming completely inserted into the surrounding plasmalemma, except the invaginations mentioned above, which may already
be internalized (cf. Section 4.2).

2.2.4. Localization of Plasmalemmal Sites of Granular Exocytosis

The extrusion sites in the systems of cell cultures *in vitro* are randomly
located on the whole cell periphery (Tixier-Vidal, 1975). This is different
in vivo. In fact, each pituitary cord is a mosaic of different cell types. The
plasmalemma of a given cell is related, through a space of 200 Å or more,
either to the plasmalemma of another endocrine cell of an identical or
different type or to the plasmalemma of a follicular cell. On the other
hand, part of the plasmalemma is adjacent to a basal lamina (Fig. 3). *In
vivo*, exocytoses actually take place along the basal lamina, as shown by
the majority of published images (Ueberberg and Hohbach, 1972) (also cf.
Table I). This has been related to the privileged location of the basal
lamina, between the perivascular space and the epithelial cord (De Virgiliis
et al., 1968). However, this is not the main factor, since in rats there are
basal laminae that are inside the cord between the epithelial cells and are
therefore called "intracordonal basal laminae" (Olivier and Vila-Porcile,
1972; Vila-Porcile, 1973, 1975). In these sites, exocytoses are as numerous
as those against the pericordonal basal laminae (Fig. 4).

In a previous study performed on the prolactin cells in lactating rats,
we showed that 70% of the exocytoses occurred toward the basal laminae,
and therefore the latter determine privileged locations at the plasmalemma
level (Olivier and Vila-Porcile, 1972; Vila-Porcile, 1973). At that level, the

plasmalemma is more contrasted than on the other cellular faces (Olivier and Vila-Porcile, 1972). Also, during the omega phase, the pocket membrane shows the same contrast.

2.2.5. Multiple Exocytoses

Whenever the cell is deeply stimulated, multiple exocytotic pockets are observed (De Virgiliis *et al.*, 1968; Couch *et al.*, 1969; Farquhar, 1971; Pasteels, 1972a). One observes pockets with a polycyclic periphery containing strings of granules. These pictures indicate that in the first pocket a second granule is opening and so on (see Figs. 1c and 4).

2.2.6. Quantitative Data

Up to now, the phenomena of exocytosis have not been quantified. We ignore the time between the start and the end of exocytosis, and the actual number of exocytoses taking place at a given time and for a given

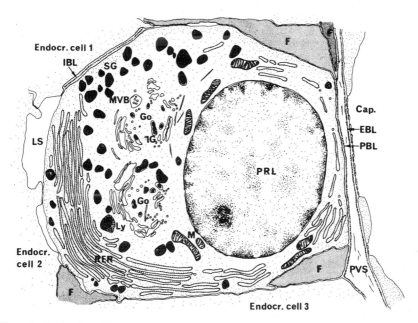

Figure 3. Prolactin cell environment. (Cap.) capillary; (EBL) endothelial basal lamina; (PBL) pericordonal basal lamina; (IBL) intracordonal basal lamina; (PVS) perivascular space; (LS) lacunary space; (F) follicular cell; (SG) mature secretory granule, (MVB) multivesicular body; (Go) Golgi complex; (IG) immature granule; (Ly) lysosome; (M) mitochondrion; (RER) rough endoplasmic reticulum. See the text for a discussion.

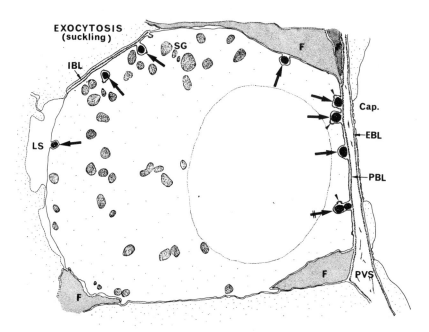

Figure 4. Schema of a prolactin cell stimulated during suckling. Exocytosis sites are shown by heavy arrows. Some pockets bear coated invaginations (arrowheads). One pocket holds two extruding granule cores (crossed arrow). See the Fig. 3 caption for abbreviations.

cell is still vague. This figure probably remains low under nonstimulation; Table I shows that exocytoses are observed mostly in stimulated systems. However, even in these cases, accurate counting of extruding granules is difficult to achieve, since we can rely only on omega images. In any case, a section beside the plane of the opening of the pocket could lead to the erroneous conclusion that the granule is still intracellular since the modifications of both the core and the granule membrane are too light to ascertain the extrusion of the granule.

In experiments performed with HRP, the core of the granule is impregnated with HRP, and its contrast is greatly amplified as compared to that of a nonextruded granule. It is therefore possible to obtain an exact count. On tangential sections (approximately 100 nm thick), we could observe up to 40 exocytoses on a 10.5 μm^2 surface (Fig. 5).

3. Morphological Data on the Exocytosis of Residual Bodies

Blocking of both hormone release and exocytosis by discontinuing the stimulation causes a destruction *in situ* of the granules by lysosomal autophagy (crinophagy) followed by the formation of residual bodies (Smith and Farquhar, 1966). In lactating rats in which prolactin secretion is blocked through weaning, these residual bodies formed by membrane stacks extrude as lamellar bodies (Fig. 6) (Vila-Porcile, 1973; Vila-Porcile *et al.*, 1973a–c). This membrane extrusion never occurs along basal lamina, but always toward lacunary spaces, where the membrane stacks are dissociated and disappear. These phenomena occur in the neighborhood of follicular cells. Such phenomena are also observed, after blocking, in other pituitary cell types. They also are frequently mentioned in organotypic culture in which secretion is decreased, as well as in pituitary tumors, in which hormone-regulating processes are defective (personal observations). This residual membrane exocytosis indicates that under these conditions, the pituitary cell is not able to achieve the degradation of excess membranes.

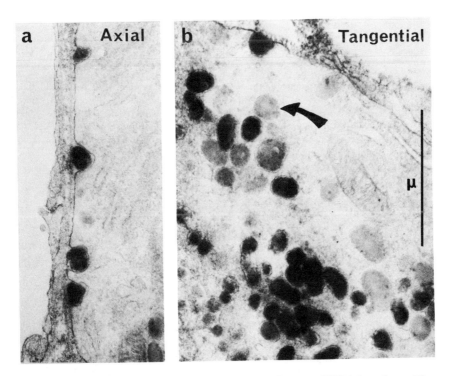

Figure 5. Exocytoses in a stimulated prolactin cell from an HRP-injected rat. The extracellular HRP impregnates the core of the extruding granules, thus enhancing their contrast, as compared to that of intracellular granules (arrow). (a) Axial section; (b) tangential section (note the numerous exocytoses).

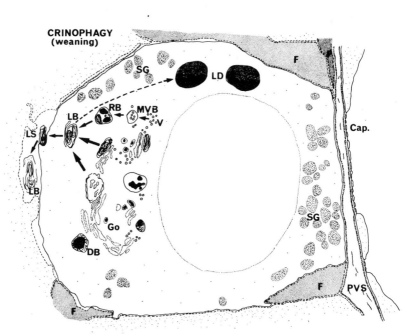

Figure 6. Crinophagy. Summary of the events possibly occurring in a prolactin cell blocked by weaning (see the text). (Cap.) Capillary; (PVS) perivascular space; (F) follicular cell; (V) vesicles; (MBV) multivesicular body; (RB) residual body; (LB) lamellar body; (LS) lacunary space; (Go) Golgi complex; (DB) dense body; (SG) secretory granules. The dotted arrow shows a route to a possible transitory storage of some membrane constituents in lipid droplets (LD).

4. Morphological Data on the Phenomena of Endocytosis

Endocytosis corresponds to the uptake of extracellular medium through plasmalemma motions leading to an internalization of membrane and fluid.

4.1. Methods and Procedures

The morphological study of exocytosis does not require sophisticated experimentation. In contrast, exploration of endocytosis with current techniques of electron microscopy requires the introduction into the extracellular medium of macromolecules detectable at the electron-microscope level. The tracers used are either enzymatic proteins (HRP) secondarily demonstrated through cytochemistry technique or directly visible particulate markers (ferritins) Freeze-etching could be efficient, as observed in other cellular models; however, it has rarely been used for the pituitary until now (Daikoku *et al.*, 1973).

4.2. Observations

Studies of endocytosis are less numerous than for exocytosis. Table II is a recapitulation of presently published observations on the pituitary. In it are summarized the cellular models and the tracers used as well as the main results and conclusions. It is evident from Table II that the conclusions are sometimes discrepant. We studied the prolactin cells of lactating rats (cf. Section 2.2.1). At 10 min after suckling is started, and while the cells are extruding their granules, the animals are injected once intravenously with HRP (15 mg/100 g body weight, i.e., approximately 2 mg/ml blood). The animals are then killed at timed intervals up to 90 min after the HRP injection. The tracer passes through the endothelium, impregnates the basal laminae, and then diffuses within the extracellular space system of the epithelial cords, where it remains detectable for 1 hr or more while its concentration is decreasing (Vila-Porcile and Olivier, 1970) (Fig. 7).

Two modes of endocytosis are observed during these kinetic studies: membrane invaginations leading to the formation of either (1) vesicles or (2) saccules or tubules.

Rounded vesicular invaginations (diameter 100–200 nm) mostly appear on the exocytotic pocket during the omega phase. Several invaginations can be visible on the same pocket. Their walls are strongly contrasted and their cytoplasmic faces bristle-coated. Similar invaginations external to the pockets can be observed on the plasmalemma (see Fig. 1). They might correspond to pockets just spreading in the plasmalemma plane after the dissolution of the granule, immediately following the omega phase. Coated and smooth-surfaced vesicles of equal diameter are observed in the cytoplasm quite near the plasmalemma. In HRP-injected animals, the surface invaginations and the vesicles are filled with exogenous HRP, which evidences their endocytotic nature (Fig. 8). Afterward, these vesicles, filled with HRP, go deeply down into the cytoplasm. Their fate is probably not uniform. Some fuse with multivesicular bodies, in which tracer accumulates; others are found among the Golgi vesicle population.

Saccular invaginations involve large plasmalemmal areas. They appear as large depressions of the surface, closing by fusion of their edges and leading to large vacuoles within the cytoplasm (pinocytosis). These formations move through the cell and take another shape. They look like

Table II. Endocytosis in Pituitary Cells

Ref. No.[a]	Experimental conditions	Tracers	Cells	Intracellular compartments[b]					Conclusions
				EV	GS	NG	MVB	LF	
1	In vitro:Pituitary fragments; stimulation, mbcAMP, 5 mM	HRP, 30 mg/ml	STH	+	+ (inner-most)	+	+	+	". . . transfer of HRP to saccules by the intermediary of smooth vesicles . . . granule membranes are added directly to the Golgi saccules to replace the membranes lost during the formation of secretory granules."
			PRL	+	0	+	+	+(?)	". . . transfer of HRP to the granules during their formation by the intermediary of Golgi vesicles . . . probably involved in the aggregation of small granules into the large irregular granules."
2	In vivo: Sprague-Dawley rats, 200–250 g	HRP, 22–24 mg/ml[c]	STH	+	NR	NR	+	+	"High turnover of growth hormone and prolactin."
			PRL	+	NR	NR	+	+	
			ACTH	+	NR	NR	NR	+	"Turnover slower than that of STH or PRL."
			TSH	+	NR	NR	NR	+	
			Gonadotr.	0	0	0	0	0	"Release by a mechanism other than exocytosis or very low rate of exocytosis."
			Thyroidect. cells	±	NR	NR	NR	±	". . . suggested that TSH molecules are released without having been previously stored in granules."

#	System	Dose	Treatment	EV	GS	NG	MVB	LF	Comments
3	In vitro: Dissociated cells + dbcAMP, 5 mM	HRP, 1 mg/ml	Adrenalect. cells	+	NR	NR	NR	+	Correlation of endocytosis with increased exocytosis.
			Castration cells	0	0	0	0	+	Neither exocytosis nor endocytosis.
			STH	+	± "penultimate cisterna"	0	NR	+	"Membrane relocated to the cell surface during exocytosis is recaptured intact and recirculated back to the Golgi. "Recirculation is restricted . . .to the same cisternae from whence the pieces of membrane originally came."
4	In vivo: Lactating rats, 200 g (Sprague–Dawley)	HRP, 2 mg/ml[c]	PRL	NR	NR	NR	NR	NR	"Whereas the exocytosis is polarized, the endocytosis of HRP occurs on all the cell sides."
5	In vitro: Monolayer cultures	HRP, 2 mg/ml	PRL–SD1	+	0	0	++	++	"Suggests that the fragments of membrane involved in HRP uptake are degraded by lysosomes rather than recirculated in newly formed secretory granules."
			SD1+TRH	+	0	0	++	++	
			SD1+cAMP	+	0	0	+	+	
6	In vivo	HRP, 2 mg/ml[c]	Gonadotr.	±	0	0	±	±	
			+LH–RH	+	0	0	+	+	
			PRL	+	+	+	+	+	Recycling of the granule membranes

(continued)

[a] References: (1) Pelletier (1973); (2) Pelletier and Puviani (1974); (3) Farquhar et al. (1975); (4) Vila-Porcile (1975); (5) Tixier-Vidal et al. (1976); (6) Vila-Porcile and Olivier (1978); (7) Farquhar (1978).

[b] Abbreviations and symbols: (EV) endocytotic vesicles; (GS) Golgi saccules; (NG) neoformed granules; (MVB) multivesicular bodies; (LF) lysosomal formations; (+) marking; (0) no marking; (NR) not reported.

[c] Dose related to blood volume, evaluated according to Creskoff et al. (1963).

Table II. (continued)

Ref. No[a]	Experimental conditions	Tracers	Cells	Intracellular compartments[b]					Conclusions
				EV	GS	NG	MVB	LF	
7	In vitro: Dissociated cells	Native ferritin, 1–5 mg/ml	PRL	±	0 GERL +	0	NR	±	"Native ferritin . . . a content rather than a membrane marker. Cationic ferritin begins as a membrane marker, and in some cases retains its original relation to membrane, but in others it apparently detaches . . . ."
		Cationic ferritin (CF) 0.05–0.5 mg/ml	PRL	+	+ GERL +	+	NR	+	"The route of the incoming vesicle varies for the same tracer depending on its surface charge . . . charge interaction influences the fate of the incoming membrane, since incoming vesicles loaded with negatively charged ferritin fuse only with the vacuolar system or lysosomal system, whereas incoming vesicles loaded with the same protein positively charged fuse with compartments of the secretory pathway."
			STH TSH ACTH Gon.	+	+ GERL +	+	NR	+	

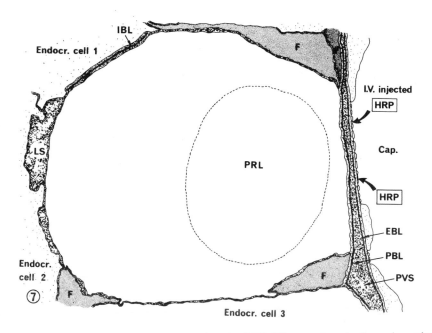

Figure 7. Following an intravenous injection, the HRP diffuses and marks the perivascular and extracellular space system of the pituitary parenchyma. (Cap.) Capillary; (EBL) endothelial basal lamina; (IBL) intracordonal basal lamina; (LS) lacunary space; (PBL) pericordonal basal lamina; (PVS) perivascular space; (F) follicular cell; (PRL) prolactin cell.

flattened saccules or tubules, in which the contrast of the product is increased, suggesting a concentration phenomenon (Fig. 9a). These saccules or tubules are apparently attracted by the Golgi area, and move into the periphery of the stacked cisternae.

In these cells, dense bodies, completely filled with HRP, are always present. They evidently result from endocytosed structures. Probably they originate from multivesicular bodies following concentration phenomena.

Some of these phenomena are very rapid: marked coated vesicles are visible in only 1 min after an intraaortic HRP injection. At 5 min after an intravenous injection, multivesicular bodies are marked and vesicles and saccules filled with the tracer have already reached the periphery of the Golgi area (Fig. 9b). After 1 hr, the multivesicular bodies and dense bodies are still marked.

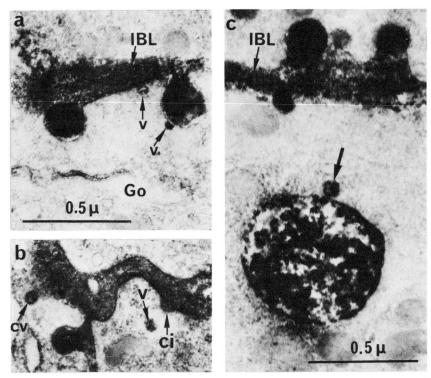

Figure 8. Endocytosis. (a) Intracordonal basal lamina (IBL) and granule cores impregnated by HRP. Two HRP-filled vesicles (v) are visible, one on the pocket wall, the other on the plasmalemma. (b) A coated invagination (ci) and a coated vesicle (cv) at the plasmalemma level. One smooth vesicle (v) is within the cytoplasm. (c) A coated vesicle (arrow) is reaching a multivesicular body, already filled with numerous marked vesicles.

5. Problems Raised by Exocytosis in Pituitary Cells

5.1. Nonmembranous Structures Involved in Exocytosis

To continue a strictly morphological approach, we will leave aside the mechanisms of molecular triggering in exocytosis, as well as the ionic exchanges occurring during granular extrusion. On the other hand, some structures are probably responsible for the granule transfer toward plasmalemma and also for the fusion phenomena and membrane movements during exocytosis. In fact, the polarization of exocytosis implies that granules have to find either a nonrandom way to the privileged membrane or a site they can identify or both. As far as granule transfer is concerned, microtubules as a guidance system and microfilaments as a contractile

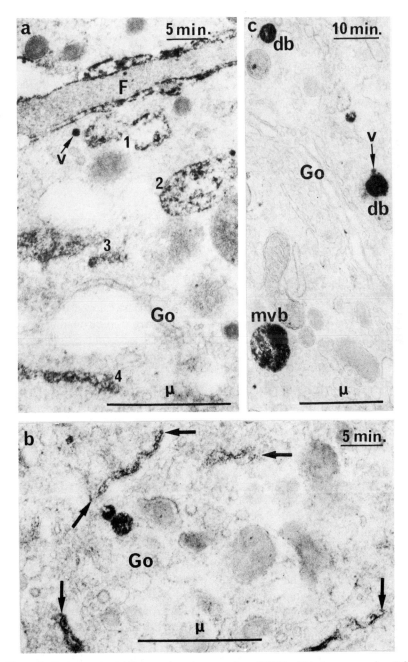

Figure 9. Endocytosis. (a) Progressive concentration of HRP within large vacuoles (1,2,3) leading to the formation of tubules (4) near the Golgi area (Go). (b) Note the labeling of tubules (or saccules) at the periphery of the Golgi area. (c) Fusion of a vesicle (v) with a dense body (db) in the Golgi area. See the text for a discussion.

system have been considered. These structures abound in fact in the pituitary cells (Warchol *et al.*, 1974), and they do hypertrophy in some cases of pathological hypersecretion (in some adenomas) (Olivier *et al.*, 1965, 1975; Horvath *et al.*, 1975). Experimental data suggest the effective role of these structures (Labrie *et al.*, 1973a; Shiino and Rennels, 1975b); however, the arguments advanced are mostly biochemical (Ostlund *et al.*, 1977; Sherline *et al.*, 1977). Therefore, for every type of granular cell, this matter is still widely discussed, as far as the role of the microtubules (Nicoll, 1972; Allison and Davies, 1974; Farquhar *et al.*, 1975) and that of the contractile proteins are concerned (Lawson *et al.*, 1977; also cf. Meldolesi *et al.*, 1978).

5.2. Membrane Events Involved in Exocytosis

Freeze-fracture has not yet been used for the study of fusion phenomena in the pituitary. These techniques, in other cell systems, helped to determine morphological differences between the granule membrane and the plasmalemma (Orci *et al.*, 1973; Orci and Perrelet, 1975; Theodosis *et al.*, 1978). It was even possible to detect in very particular systems some recognition sites (Satir *et al.*, 1973). As yet, in current electron microscopy of the pituitary, we do not have pictures of the fusion of membrane sheets comparable to the pictures demonstrated in other very attractive models (Tandler and Poulsen, 1976; Lawson *et al.*, 1977; Pinto da Silva and Nogueira, 1977). However, when the pocket is opened, the granule membrane is modified and acquires the plasmalemma properties since it then admits a fusion with the membrane of another granule, while mature granule membranes never fuse within the cell.

5.3. Secretion Problems Raised by Exocytosis

The direct consequence of pocket opening is the liberation of the granule core constituants. The hormone release during core disaggregation is unanimously and implicitly conceded. The granules are known to be a hormone storage site, as evidenced by biochemical analyses of isolated secretory granules (Costoff and McShan, 1969; Hymer, 1975) and by the bulk of immunocytochemical results obtained since the work of Nakane (1970). Moreover, exocytosis is known to be concomitant with the measurable hormone release either in the blood or in the culture medium (Pasteels, 1972b; Tixier-Vidal *et al.*, 1976). But the released hormones could not be demonstrated in the extracellular spaces (Parsons, 1977). Most probably, the hormone structure does not allow for the application of present detection methods.

But the destruction of the granule releases more than the hormone.

The granules are in fact heterogeneous, and differ, as well, from one granule type to another. Thus, in the prolactin granule, through cytochemical methods, glycoproteins have been demonstrated at the periphery of granule cores (Rambourg and Racadot, 1968; Porcile *et al.*, 1968; Tixier-Vidal and Picart, 1971; Vila-Porcile, 1972). In membraneless granules (Giannattasio *et al.*, 1975) of these same cells, biochemical analyses have shown other elements together with prolactin: calcium, minor protein components, and sulfated proteoglycans (Giannattasio and Zanini, 1976). Similar constituents have been demonstrated in somatotroph granules (Slaby and Farquhar, 1976). The release of components other than hormone during exocytosis has also been demonstrated for other endocrine glands (Winkler *et al.*, 1974).

The opening of the pocket integrates the granule membrane into the plasmalemma, exposing its endogranular face to the exterior medium, which could support special molecules such as enzymes, e.g., NDPase, observed in prolactin granules (Smith and Farquhar, 1970). In some cell fractionation experiments, the "small granule fraction" could hold, besides hormones, an alkaline protease (Tesar *et al.*, 1969).

The question can be asked: are all the granules undergoing exocytosis hormone secretory granules? The answer cannot be given by current electron microscopy. The hypothesis has been advanced that under some conditions such as nonfunctional pituitary adenomas, secretion may be restricted to the elaboration of granules bearing a hormonally inactive secretory product (Olivier *et al.*, 1975; Tramu *et al.*, 1976).

5.4. Hormone Release without Visualized Exocytosis

Under certain conditions, exocytoses have been observed in such limited numbers that their very existence has been questioned.

5.4.1. Agranular or Scarcely Granulated Cells: Exocytosis from Vesicles?

In some cases, pituitary cells contain a very small number of granules. These granules are of reduced diameter as compared to normal; i.e., they represent but a small quantity of hormone while the cells are still actively secreting. These cases concern long-term-stimulated cells [i.e., thyroidectomy cells (Kurosumi and Oota, 1966; Farquhar, 1971)] or tumor cells [i.e., spontaneous human tumors, experimental tumors either transplanted or continuous cell lines (cf. the review by Olivier *et al.*, 1975)].

Through what mechanisms do these cells manage hormone release? A disturbance of the membrane packaging process is evident (Farquhar, 1971). Numerous vesicles with a clear and apparently nonconcentrated content are present in these cells. The vesicles multiply under the influence of releasing factors [GH3 cells under TRF (Gourdji *et al.*, 1972)]. Possibly the condensation step is bypassed and vesicles find their way directly to the plasmalemma (cf. Meldolesi *et al.*, 1978). Under such conditions, the probability of demonstrating exocytoses with current electron microscopy is very low. We have already mentioned that the step allowing for identification of exocytosis is the omega phase related to the destruction of the granule-dense assembly. The dilution of the vesicle content suggests that the release conditions are similar to those in the presynaptic nervous ending, where it has been particularly difficult to demonstrate exocytoses (Couteaux and Pécot-Dechavassine, 1970; Douglas, 1974; Heuser *et al.*, 1975; Ceccarelli *et al.*, 1976; Dreifuss *et al.*, 1976), probably because these phenomena are very rapid, some of them lasting for only a few milliseconds (Heuser *et al.*, 1975).

We could suppose that all the intermediary steps could be found between the granule exocytosis and the vesicle exocytosis. Several morphological observations have demonstrated variations in the granule diameters, for a given cell and within given limits (Pooley, 1971). In hyperactive cells, either *in vitro* or *in vivo,* the diameter of intracellular granules tends to decrease (Farquhar, 1971; Pasteels, 1972a; Herlant, 1975). On the other hand, authors studying synthesis and hormone release phenomena *in vitro* conclude, from biochemical data, that the most recently synthesized hormone is the first to be released (Swearingen, 1971; Labrie *et al.*, 1973b). Since it is frequently observed that exocytosing granules have a smaller diameter than stored granules, there might be a preferential extrusion of neosynthesized granules that did not undergo all the steps of intragolgian maturation.

5.4.2. Gonadotrophs

The second case in which granular exocytosis is questioned concerns gonadotrophs. This problem is opposite to the former one, for these cells always have a significant content of different types of granules (Tougard *et al.*, 1971); i.e. all the conditions allowing for the observation of exocytoses are present. However, this observation is rather infrequent and requires strong stimulation by castration or LH-RH or both (Shiino *et al.*, 1972b). Even under such experimental conditions, exocytoses cannot be observed in all the cells. Pelletier and Puviani (1974) have gone so far as to question the validity of these observations.

Mechanisms other than exocytosis have been proposed, such as the release of granule content into the hyaloplasm through a rupture of the granule membrane (Herlant, 1963); such a release scheme has been suggested by Costoff (1973) for the corticotrophs. However, the pictures supporting such a hypothesis have been considered as artifacts by Herlant (1975) himself.

It could also be noted that positive immunocytochemical reaction to gonadotroph hormones has been observed in the hyaloplasm of the cells; however, the authors are very cautious as far as interpretation is concerned (Tougard *et al.*, 1971, 1973; Tixier-Vidal *et al.*, 1975b).

Finally, as far as pituitary cells are concerned, Farquhar *et al.* (1975, p. 99) states that " . . . exocytosis is the *only* release mechanism so far demonstrated. *No* alternatives have been found. . . ."

6. Problems of Endocytosis

The works mentioned in Table II demonstrate actual endocytosis in the pituitary. However, its actual modalities and its consequences, as well as its importance according to the cell type, raise numerous problems.

6.1. Relationships between Exocytosis and Endocytosis; Plasmalemmal Sites Involved in Endocytosis

In the pituitary, the data on these relationships are rather scarce, perhaps because most of the studies of endocytosis were made in *in vitro* models, in which exocytoses are apparently not as frequently observed as in *in vivo* models (cf. Table I).

Authors give a general description of endocytosis as invaginations leading to a microvesiculation from the limiting membrane. According to our experiments, this microvesiculation is found, on one hand, at the level of exocytotic pockets during granule extrusion, and, on the other hand, externally to exocytotic pockets. It can be stated that, in the first site, the former granule membrane is involved. As to the second site, pits might also form from granule membranes already entirely inserted into the plasmalemma, which cannot be identified by current methods. In other cell systems, freeze-fracture showed this possibility to be plausible (De Camilli *et al.*, 1976; Orci and Perrelet, 1975, 1978; Theodosis *et al.*, 1978).

Also in our experiments, there was a third site, not yet described as far as we know. It consists in actual pinocytosis engulfments, topographically independent from exocytotic pockets, and internalizing large membrane surfaces devoid of bristle coat. The position of these areas independent from sites of exocytosis does not preclude the participation either of granule membranes previously inserted or of some of their constituents, due to possible migrations related to membrane fluidity.

Theoretically, it would be important to identify the actual location of sites of endocytosis, since they are implicated in the future behavior of internalized membranes. The origin of these membranes could determine their eventual site of (complete or partial) reutilization at the level of the cellular membrane compartments (cf. Meldolesi *et al.*, 1978).

Unfortunately, the present methods of investigation play an important part in the assessment of these phenomena. The characteristics of the marker and the dose used (varying from 1 to 30 mg/ml according to authors) (cf. Table II) certainly influence the events, as do the preparation types. *In vivo*, the HRP injection has a parallel effect on the behavior of the animal (Cotran *et al.*, 1968; Straus, 1977). *In vitro*, where experimental conditions are quite variable (hemipituitaries, monolayer cultures, dispersed cells), Tixier-Vidal *et al.* (1976) mentioned a transitory stimulating effect of HRP, even at low doses. Finally, Farquhar (1978) has demonstrated obvious differences in the behavior of the tracers according to their charge (native or cationic ferritin) (cf. Table II). It is therefore possible that in experiments (including ours), specific and nonspecific phenomena are intermingled.

6.2. Endocytosed Compartment within the Cell

All the endocytosed membranes as well as the fluids taken up constitute a new membrane-bounded compartment added to the former compartments. The assessment of this latter compartment is different depending on the experiment (Table II).

Authors agree on the presence of coated vesicles underlying the plasmalemma, as a result of microvesiculation. The coating of such vesicles has been observed in several other cases: endothelial cells and exocrine, endocrine, and neuroendocrine cells, as well as neurons (Palade and Bruns, 1968; Friend and Farquhar, 1967; Abrahams and Holtzman, 1973; Douglas, 1974). This coat is formed by a latticelike array of short, thin filaments, the chemical composition of which is just beginning to be explored (Woodward and Roth, 1978). Smooth-surfaced vesicles are also observed, and they are generally believed to originate from coated ones, having lost their coat.

Authors also agree on the fusing of a certain number of these vesicles with multivesicular bodies either preexisting or resulting from the vesicular traffic related to endocytosis. In multivesicular bodies, the tracer is either discharged into the matrix or fixed on the external wall of the accumulated vesicles. These multivesicular bodies are considered as secondary lysosomes (De Duve, 1969; Geuze and Kramer, 1974).

Authors also agree on the presence of large, dense bodies filled with the tracer, whatever the tracer. These bodies are believed to be lysosomes. The kinetics of formation of these dense bodies in the pituitary have not yet been investigated.

The incorporation of endocytosed structures into the Golgi apparatus is somewhat in question. It has not been found in every case (cf. Table II). By means of HRP, Farquhar (1971) and Pelletier (1973) have observed Golgi vesicles and saccules containing some tracer. We also observed saccular or tubular structures heavily marked at the periphery of the Golgi area.

The pathway followed by the tracer between the cell surface and the Golgi apparatus is not well known. Most probably, the tracer reaches the Golgi apparatus through a fusion of vesicles derived from the cell surface (Farquhar, 1978). However, in our experiments, tubules directly resulting from pinocytosis and in which HRP was progressively concentrating seemed to originate directly from the cell periphery to become aggregated with the Golgi apparatus (Figs. 9 and 10).

Finally, the most questionable compartment is the granule compartment. Pelletier (1973) has claimed that mature granules could hold HRP. However, he is the only one to have made this observation. On the other hand, with cationic ferritin, which behaves as a membrane tracer, Farquhar has shown convincing pictures in which the ferritin is accumulating at the core periphery of neoformed granules (Farquhar, 1978).

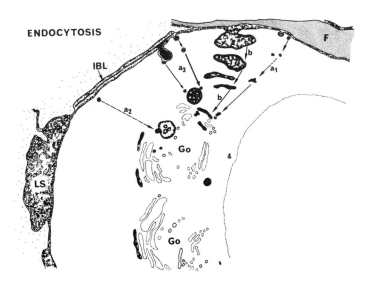

Figure 10. Summary of the possible pathways for endocytosed structures. (a) Microvesiculation (micropinocytosis) and pathways to: (a₁) Golgi (Go) area; (a₂) multivesicular body; (a₃) dense body. (b) Pinocytosis: progressive concentration of large engulfments, reaching the periphery of the Golgi complex, as tubular structures. (IBL) Intracordonal basal lamina; (LS) lacunary space; (F) follicular cell.

6.3 Consequences of Endocytosis

It is therefore obvious that in addition to the traffic of membrane-bounded elements from the cell to the external medium, there is a reverse traffic. What does this reverse traffic mean?

6.3.1. Fate of Endocytosed Membranes

This is a critical and widely discussed question, since three hypotheses can be considered a priori: recirculation, dismantling, or complete degradation of the membranes, as reported by Farquhar et al. (1975) and discussed by Meldolesi et al. (1978). That membranes are recirculated or reused at the Golgi or the GERL level for the formation of new granules is at present generally agreed in many cell models (cf. Douglas, 1974). However, demonstration of this phenomenon by tracers such as HRP is difficult, inasmuch as this molecule behaves like a content marker rather than a membrane marker (Steinman and Cohn, 1972; Oliver and Hand, 1978; Farquhar, 1978). The fate of the membrane can be followed provided that it remains associated to the histochemically detectable tracer. A membrane marker such as cationic ferritin is more favorable, and Farquhar could thus demonstrate membrane reutilization up to the granule level.

Indirect and strictly morphological arguments support this recycling notion. Whenever a prolactin cell is blocked and its granules are degraded in situ instead of being extruded (crinophagy), lamellar bodies are observed in the lysosomes, obviously expressing the slowness and perhaps the inability of these lysosomes to digest the membrane constituents. That these pictures are not visible under normal cell activity suggests that endocytosed membranes escape the process even when entering the lysosomal compartment. We saw in cases of crinophagy that the membrane residues are expelled from the cell (cf. Section 3). This observation suggests that the cell no longer requires membranes on the secretory route and therefore expels the now useless membranes.

Membrane internalization could also provide another effect: internalization of cell-surface receptors and lysosomal destruction of both the ligand and the receptor in lysosomes. Such internalization has been demonstrated in fibroblasts or carcinoma cells for insulin and epidermal growth factor (Carpenter and Cohen, 1976; Gordon et al., 1978; Haigler et al., 1978; Schlessinger et al., 1978). As far as pituitary is concerned, Gourdji et al. (1973) demonstrated the entry of TRH into GH_3 cells. This could play a part in the secretion-regulating process.

6.3.2. Fate of Endocytosed Fluid

Generally speaking, even in cells in which endocytosis has been more particularly studied, such as macrophages (cf. Steinman *et al.*, 1976), the fate of endocytosed fluid is not completely known.

As far as pituitary is concerned, not even the nature of the endocytosed fluid is known. It should be pointed out that microvesiculation takes place at the very level of extruding granules; therefore, the endocytosed fluid should have a rather high concentration of hormones and other constituents just released. On the other hand, due to the mosaic organization of the pituitary parenchyma, the vesicles are also able to internalize hormones released by cell of another type. This phenomenon could play a regulating role. In fact, a feedback control release of prolactin by prolactin has been demonstrated *in vitro* (Herbert, 1978).

6.3.3. Endocytosis Rate; Timing of Endocytotic Phenomena

The membrane surface internalized during exocytosis is certainly equivalent to the membrane surface inserted into the plasmalemma, since the average cell volume is constant. This does not imply that the internalized volume is equivalent to the volume of extruded material, since the surface/volume ratio is lower for a microvesicle than for a granule (cf. discussion in Douglas *et al.*, 1971). We know (Pelletier, 1973; Tixier-Vidal *et al.*, 1976; Herzog and Farquhar, 1977) that endocytosis is increased whenever the cell is stimulated. However, the duration of endocytosis from the beginning of stimulation remains unknown. According to our data, it obviously starts with exocytosis, but we do not know whether it is still carried on after exocytosis is over. Neither do we know the average intracellular life span of endocytosed vesicles before they fuse with the different compartments, even though we know that in some cases less than 5 min is required between the beginning of endocytosis and the observation of marked structures at the Golgi apparatus level. Such times have also been observed *in vivo* in other cells such as the parotid cell (Herzog and Farquhar, 1977). It is also known that in *in vitro* experiments with cationic ferritin, the first marked granules appear in the prolactin cell after 15 min (Farquhar, 1978).

6.4. Importance of Endocytosis in the Various Cell Types

The first question is whether endocytosis exists in various cell types. It has been widely discussed in gonadotrophs. If the secretion of these cells does not imply exocytosis, it logically means that they do not endocytose, according to Pelletier's conclusions (Pelletier and Puviani, 1974). However, in monolayer cultures of dispersed rat anterior pituitary cells, Tixer-Vidal *et al.* (1976) have observed in gonadotrophs an actual endocytosis, which increased after LH-RH stimulation.

The second question concerns endocytotic modalities and the fate of endocytosed structures according to cell type. Taking into consideration differences in granule size and the organization of the Golgi apparatus, endocytosis will probably be different. However, experimental data are lacking despite Pelletier's indications (Pelletier and Puviani, 1974).

7. Summary

It has been clearly demonstrated, at least for somatotrophs and prolactin cells, that secretion implies successive rehandlings at the level of the cell-limiting membrane, through additions and removals. Addition occurs at each opening of a secretory granule to the surface (exocytosis) through a connection of the two membranes, together with the release of the secretory content ("quantal" secretion). Membrane removal, together with the trapping of extracellular fluid, follows exocytosis. It originates endocytotic vesicles and tubules that travel down into the cell and aggregate with multivesicular bodies, lysosomes, and the Golgi apparatus.

Secretion metabolism therefore implies, among other mechanisms, a two-way traffic of two discontinuous flows of membrane-bounded material. The first flow, from the inside to the ouside (exocytotic), brings out concentrated secretion products. The second (endocytotic) flow brings into the cell a fluid, the fate and the physiological significance of which are still unknown, as well as a membrane bulk balancing what has been exocytosed. At least a part of these membranes goes back to the Golgi apparatus, multivesicular bodies, and lysosomes, and could be used in the packaging of new granules (membrane recycling). An internalization of surface receptors followed by lysosomal destruction could also be suggested.

ACKNOWLEDGMENTS. We gratefully acknowledge the technical assistance of Mrs. Y. Drouet and Miss D. De Nève, and the valuable help of Mrs. M. de Venoge, Mrs. A. Combrier, and Dr. J. Grémain in the preparation of the manuscript.

DISCUSSION

DUVAL: Do you think that endocytosis may have something to do with the release of hormones in the gonadotrophs by bringing some of the releasing hormone inside the cell?

VILA-PORCILE: Since the experiments with tracers indicate the possibility of quantities of extracellular fluid and membranes to be internalized, we suggest that the releasing factors follow the same pathways. Thus, they could act on intracellular receptors. Endocytosis may also be involved in the regulation of secretion following metabolism of the internalized elements, possibly explaining long-term reactions.

HYMER: What is the nature of the material in the extracellular fluid which may act directly on the secretion granule?It is possible that proteases may act on the granule in the secretory process?

VILA-PORCILE: The composition of the extracellular fluid has not yet been studied. Its composition may vary as a function of time and activity of the different adenophypophyseal cells. The extracellular fluid contains substances from plasma and also molecules from the metabolism of the pituitary cells themselves, such as hormones, enzymes, and other metabolites. One should keep in mind that *in vivo*, the pituitary endocrine cells are engulfed in a network of follicular cells. The latter have a large membrane surface and exhibit a high endocytotic potency. Our experiments with diffusion tracers have shown that they concentrate molecules coming from the extracellular fluid. It may be suggested that these cells locally modulate the composition of the extracellular fluid. It is thus possible to suppose interactions between the secretory processes and the composition of the extracellular fluid.

REFERENCES

Abrahams, S. J., and Holtzman, E., 1973, Secretion and endocytosis in insulin-stimulated rat adrenal medulla cells, *J. Cell Biol.* **56**:540.

Allison, A.C., and Davies, P., 1974, Interactions of membranes, microfilaments and microtubules in endocytosis and exocytosis, in: *Cytopharmacology of Secretion* (B. Ceccarelli, F. Clementi, and J. Meldolesi, eds.), pp. 237–248, Raven Press, New York.

Amsterdam, A., Ohad, I., and Schramm, M., 1969, Dynamic changes in the ultrastructure of the acinar cell of the rat parotid gland during the secretory cycle, *J. Cell Biol.* **41**:753.

Amsterdam, A., Schramm, M., Ohad, I., Salomon, Y., and Selinger, Z., 1971, Concomitant synthesis of membrane protein and exportable protein of the secretory granule in rat parotid gland, *J. Cell Biol.* **50**:187.

Bácsy, E., Tougard, C., Tixier-Vidal, A., Marton, J., and Stak, E., 1976,Corticotroph cells in primary cultures of rat adenohypophysis: A light and electron microscopic immunocytochemical study, *Histochemistry* **50**:161.

Batten, T.F.C., Ball, J.N., and Grier, H.J., 1976, Circadian changes in prolactin cell activity in the pituitary of the teleost *Poecilia latipinna* in freshwater, *Cell Tissue. Res.* **165**:267.

Benedeczky, I., and Smith, A.D., 1972, Ultrastructural studies of the adrenal medulla of golden hamster: Origin and fate of secretory granules, *Z. Zellforsch.* **124**:367.

Bunt, A.H., 1969, Fine structure of pars distalis and interrenals of *Taricha torosa* after administration of metopirone (SU 4885), *Gen. Comp. Endocrinal.* **12**:134.

Carpenter, D., and Cohen, S., 1976, ^{125}I-labeled human epidermal growth factor: Binding, internalization and degradation in human fibroblasts, *J. Cell Biol.* **71**:159.

Ceccarelli, B., Peluchetti, D., Grohovaz, F., and Jezzi, N., 1976, Freeze fracture studies on changes at neuromuscular junctions induced by black widow spider venom, in: *Electron Microscopy* (Y. Ben Shaul, ed.), pp. 376–378, Tal International, Jerusalem.

Coates, P. W., Ashby, E. A., Krulich, L., Dhariwal, A. P. S., and McCann, S. M., 1970, Morphologic alterations in somatotrophs of the rat adenohypophysis following administration of hypothalamic extracts, *Am. J. Anat.* **128**:389.

Coates, P. W., Ashby, E. A., Krulich, L., Dhariwal, A. P. S., and McCann, S. M., 1971, Fine structure of somatotrophs in the adenohypophysis incubated with growth hormone releasing factor, *Anat. Rec.* **169**:299 (abstract).

Costoff, A., 1973, *Ultrastructure of Rat Adenohypophysis: Correlation with Function,* Academic Press, New York.

Costoff, A., and McShan, W. H., 1969, Isolation and biological properties of secretory granules from rat anterior pituitary glands, *J. Cell Biol.* **43**:564.

Cotran, R. S., Karnovsky, M. J., and Goth, A., 1968, Resistance of Wistar–Furth rats to the mast cell damaging effect of horseradish peroxidase, *J. Histochem. Cytochem.* **16**:382.

Couch, E. F., Arimura, A., Schally, A. V., Saito, M., and Sawano, S., 1969, Electron microscopic studies of somatotrophs of rat pituitary after injection of purified growth hormone releasing factor (G.R.F.), *Endocrinology* **85**:1084.

Couteaux, R., and Pécot-Dechavassine, M., 1970, Vesicules synaptiques et poches au niveau des zones actives de la jonction neuromusculaire, *C. R. Acad. Sci.* **271**:2346.

Creskoff, A. J., Fitz-Hugh, T., and Farris, E. J., 1963, Hematology of the rat: Methods and standards, in: *The Rat in Laboratory Investigation* (E. J. Farris and J. Q. Griffith, eds.), pp. 406–420, Hafner, New York.

Cuerdo-Rocha, S., and Zambrano, D., 1974, The action of protein synthesis inhibitors and thryrotropin releasing factor on the ultrastructure of rat thyrotrophs, *J. Ultrastruct. Res.* **48**:1.

Daikoku, S., Takahashi, T., Kojimoto, H., and Watanabe, Y. G., 1973, Secretory surface phenomena in freeze–etched preparations of the adenohypophysial cells and neurosecretory fibers, *Z. Zellforsch.* **136**:207.

De Camilli, P., Peluchetti, D., and Meldolesi, J., 1976, Dynamic changes of the luminal plasmalemma in stimulated parotid acinar cells: A freeze fracture study, *J. Cell Biol.* **70**:59.

De Duve, C., 1969, The lysosome in retrospect, in: *Lysosomes in Biology and Pathology,* Vol. I (J. T. Dingle and H. B. Fell, eds.), pp. 3–40, North-Holland/Elsevier, Amsterdam and New York.

De Virgiliis, G., Meldolesi, J., and Clementi, F., 1968, Ultrastructure of growth hormone–producing cells of rat pituitary after injection of hypothalamic extract, *Endocrinology* **83**:1278.

Diner, O., 1967, L'expulsion des granules de la médullo-surrénale chez le hamster, *C. R. Acad. Sci. Ser. D* **265**:616.

Douglas, W. W., 1974, Mechanisms of release of neurohypophysial hormones: Stimulus–secretion coupling, in: *Handbook of Physiology,* Sect. 7, Vol. 4, Part 1 (E. Knobil and W. H. Sawyer, eds.), pp. 191–224, American Physiology Society, Washington, D. C.

Douglas, W. W., Nagasawa, J., and Schultz, R., 1971, Electron microscopic studies on the mechanism of secretion of posterior pituitary hormones and significance of microvesicles (synaptic vesicles): Evidence of secretion by exocytosis and formation of microvesicles as a by-product of this process, *Mem. Soc. Endocrinol.* **19**:353.

Dreifuss, J. J., Akert, K., Sandri, C., and Moor, H., 1976, Specific arrangements of sites of exo–endocytosis in the freeze–etched neurohypophysis, *Cell Tissue Res.* **165**:317.

Echave Llanos, J. M., and Gomez Dumm, C. L., 1973, Ultrastructure of STH cells of the pars distalis of castrated mice after hepatectomy, *Z. Zellforsch.* **114**:31.

Farquhar, M. G., 1961a, Origin and fate of secretory granules in cells of the anterior pituitary gland, *Trans. N. Y. Acad. Sci.* **23**:346.

Farquhar, M. G., 1961b, Fine structure and function in capillaries of anterior pituitary gland, *Angiology* **12**:270.

Farquhar, M. G., 1969, Lysosome function in regulating secretion: Disposal of secretory granules in cells of the anterior pituitary gland, in: *Lysosomes in Biology and Pathology* (J. T. Dingle and H. B. Fell, eds.), pp. 462–482, North-Holland/Elsevier, Amsterdam and London.

Farquhar, M. G., 1971, Processing of secretory products by cells of the anterior pituitary gland, in: Subcellular organization and function in endocrine tissues, *Mem. Soc. Endocrinol.* (H. Heller and K. Lederis, eds.), pp. 79–124, Cambridge University Press.

Farquhar, M. G., 1977, Secretion and crinophagy in prolactin cells, *Adv. Exp. Med. Biol.* **80**:37.

Farquhar, M. G., 1978, Recovery of surface membrane in anterior pituitary cells, *J. Cell Biol.* **77**:R35.

Farquhar, M. G., Stutelsky, E. H., and Hopkins, C. R., 1975, Structure and function of the anterior pituitary and dispersed pituitary cells: *In vitro* studies, in: *The Anterior Pituitary Gland* (A. Tixier-Vidal and M. G. Farquhar, eds.), pp. 82–135, Academic Press, New York.

Farquhar, M. G., Reid, J. J., and Daniell, L. W., 1978, Intracellular transport and packaging of prolactin: A quantitative electron microscope autoradiographic study of mammotrophs dissociated from rat pituitaries, *Endocrinology* **102**:296.

Fridberg, G., and Ekengren, B., 1977, The vascularization and the neuroendocrine pathways of the pituitary gland in the Atlantic salmon, *Salmo salar, Can. J. Zool.* **55**:1284.

Friend, D. S., and Farquhar, M. G., 1967, Functions of coated vesicles during protein absorption in the rat vas deferens, *J. Cell Biol.* **35**:357.

Gas, N., 1975, Influence du jeûne prolongé et de la réalimentation sur les cellules somatotropes de l'hypophyse de carpe: Étude ultrastructurale, *J. Microsc. Biol. Cell.* **23**:289.

Geuze, J. J., and Kramer, M. F., 1974, Function of coated membranes and multivesicular bodies during membrane regulation in stimulated exocrine pancreas cells, *Cell Tissue Res.* **156**:1.

Geuze, J. J., and Poort, C., 1973, Cell membrane resorption in the rat exocrine pancreas cell after *in vivo* stimulation of the secretion, as studied by *in vitro* incubation with extracellular space markers, *J. Cell. Biol.* **57**:159.

Giannattasio, G., and Zanini, A., 1976, Presence of sulfated proteoglycans in prolactin secretory granules isolated from the rat pituitary gland, *Biochim. Biophys. Acta* **439**:349.

Giannattasio, G., Zanini, A., and Meldolesi, J., 1975, Molecular organization of rat prolactin granules. I. *In vitro* stability of intact and "membraneless" granules, *J. Cell Biol.* **64**:246.

Girod, C., and Dubois, P., 1965, Étude ultrastructurale des cellules gonadotropes antéhypophysaires, chez le hamster doré (*Mesocricetus auratus* Waterh.), *J. Ultrastruct. Res.* **13**:212.

Girod, C., and Dubois, P., 1969, Recherches en microscopie optique et en microscopie électronique sur les cellules somatotropes antéhypophysaires du hamster doré (*Mesocricetus auratus* Waterh.), *C. R. Acad. Sci.* **268**:2361.

Gorden, P., Carpentier, J. L., Cohen, S., and Orci, L., 1978, Epidermal growth factor: Morphological demonstration of binding, internalization, and lysosomal association in human fibroblasts, *Proc. Nal. Acad. Sci. U.S.A.* **75**:5025.

Gourdji, D., Kerdelhué, B., and Tixier-Vidal, A., 1972, Ultrastructure d'un clone de cellules hypophysaires sécrétant de la prolactine (clone GH3): Modifications induites par l'hormone hypothalamique de libération de l'hormone thyréotrope (TRH), *C. R. Acad. Sci.* **274**:437.

Gourdji, D., Tixier-Vidal, A., Morin, A., Pradelles, P., Morgat, J. L., Fromageot, P., and Kerdelhué, B., 1973, Binding of a tritiated thyrotropin-releasing factor to a prolactin secreting clonal cell line (GH3), *Exp. Cell Res.* **82**:39.

Haigler, H., Ash, J. F., Singer, S. J., and Cohen, S., 1978, Visualization by fluorescence of the binding and internalization of epidermal growth factor in human carcinoma cells A-431, *Proc. Natl. Acad. Sci. U.S.A.* **75**:3317.

Hand, A. R., and Oliver, C., 1977, Relationship between the Golgi apparatus, GERL, and secretory granules in acinar cells of the rat exorbital lacrimal gland, *J. Cell Biol.* **73**:399.

Häusler, A., Rohr, H. P., Marbach, P., and Flückiger, E., 1978, Changes in prolactin secretion in lactating rats assessed by correlative morphometric and biochemical methods, *J. Ultastruct. Res.* **64**:74.

Herbert, D. C., 1978, Feedback control of *in vitro* prolactin release by prolactin, *Anat. Rec.* **190**:419 (abstract).

Herlant, M., 1963, Apport de la microscopie électronique à l'étude du lobe antérieur de l'hypophyse, in: *Cytologie de l'Adénohypophyse* (J. Benoit and C. Da Lage, eds.), pp. 73–90, CNRS, Paris.

Herlant, M., 1975, Introduction, in: *The Anterior Pituitary* (A. Tixier-Vidal and M. G. Farquhar, eds.), pp. 1–19, Academic Press, New York.

Herzog, V., and Farquhar, M. G., 1977, Luminal membrane retrieved after exocytosis reaches most Golgi cisternae in secretory cells, *Proc. Natl. Acad. Sci. U.S.A.* **74**:5073.

Heuser, J. E., and Reese, T. S., 1973, Evidence for recycling of synaptic vesicle membrane during transmitter release at the frog neuromuscular junction, *J. Cell Biol.* **57**:315.

Heuser, J. E., Reese, T. S., and Landis, D. M. D., 1975, Preservation of synaptic structure by rapid freezing, *Cold Spring Harbor Symp. Quant. Biol.* **40**:17.

Horvath, E., Kovacs, K., Stratmann, I. E., and Ezrin, C., 1975, Two distinct types of microfilaments in the cytoplasm of human adenohypophyseal cells, in: *Electron Microscopic Concepts of Secretion: Ultrastructure of Endocrine and Reproductive Organs* (M. Hess, ed.), pp. 287–298, Wiley, New York.

Hymer, W. C., 1975, Separation of organelles and cells from the mammalian adenohypophysis, in: *The Anterior Pituitary* (A. Tixier-Vidal and M. G. Farquhar, eds.), pp. 137–180, Academic Press, New York.

Kalina, M., and Robinovtich, R., 1975, Exocytosis couples to endocytosis of ferritin in parotid acinar cells from isoprenalin stimulated rats, *Cell Tissue Res.* **163**:373.

Kramer, M. F., and Geuze, J. J., 1974, Redundant cell membrane regulation in the exocrine pancreas cells after pilocarpine stimulation of the secretion, in: *Cytopharmacology of Secretion* (B. Ceccarelli, F. Clementi, and J. Meldolesi, eds.), pp. 87–97, Raven Press, New York.

Kurosumi, K., 1961, Electron microscopic analysis of the secretion mechanism, *Int. Rev. Cytol.* **71**:1.

Kurosumi, K., and Kobayashi, Y., 1966, Corticotrophs in the anterior pituitary glands of normal and adrenalectomized rats as revealed by electron microscopy, *Endocrinology* **78**:745.

Kurosumi, K., and Kobayashi, Y., 1969, Corticotrophs in the rats' anterior pituitaries under certain experimental conditions, *Gunma Symp. Endocrinol.* **6**:213.

Kurosumi, K., and Oota, Y., 1966, Corticotrophs in the anterior pituitary glands of gonadectomized and thryroidectomized rats as revealed by electron microscopy, *Endocrinology* **79**:808.

Labrie, F., Gauthier, M., Pelletier, G., Borgeat, P., Lemay, A., and Gouge, J. J., 1973a, Role of microtubules in basal and stimulated release of growth hormone and prolactin in rat adenohypophysis *in vitro, Endocrinology* **93**:903.

Labrie, F., Pelletier, G., Lemay, A., Borgeat, P., Barden, N., Dupont, A., Savary, M., Côté, J., and Boucher, R., 1973b, Control of protein synthesis in anterior pituitary gland, *Karolinska Symp. Res. Methods Reprod. Endocrinol.* **6**:1.

Lawson, D., Raff, M. C., Gomperts, B., Fewtrell, C., and Gilula, N. B., 1977, Molecular events during membrane fusion: A study of exocytosis in rat peritoneal mast cells, *J. Cell Biol.* **72**:242.

Leatherland, J. F., and Percy, R., 1976, Structure of the nongranulated cells in the hypophyseal rostral pars distalis of cyclostomes and actinopterygians, *Cell Tissue Res.* **166**:185.

Luborsky-Moore, J. L., Poliakoff, S. J., and Worthington, W. C. Jr., 1975, Ultrastructural observations of anterior pituitary gonadotrophs following hypophysial portal vessel infusion of luteinizing hormone releasing hormone, *Am. J. Anat.* **144**:549.

Meldolesi, J., 1974, Secretory mechanism in pancreatic acinar cells: Role of cytoplasmic membranes, in: *Cytopharmacology of Secretion* (B. Ceccarelli, F. Clementi, and J. Meldolesi, eds.), pp. 71–97, Raven Press, New York.

Meldolesi, J., Borgese, N., De Camilli, P., and Ceccarelli, B., 1978, Cytoplasmic membranes and the secretory process, in: *Membrane Fusion. Cell Surface Reviews,* Vol. 5 (G. Poste and G. L. Nicolson, eds.), Elsevier/North-Holland, Amsterdam and New York.

Mendoza, D., Arimura, A., and Schally, A. V., 1973, Ultrastructural and light microscopic observations of rat pituitary LH-containing gonadotrophs following injection of synthetic LH-RH, *Endocrinology* **92**:1153.

Moguilevsky, J. A., Cuerdo-Rocha, S., Christot, J., and Zambrano, D., 1973, The effect of thyrotrophic releasing factor on different hypothalamic areas and the anterior pituitary gland: A biochemical and ultrastructural study, *J. Endocrinol.* **56**:99.

Nakane, P. K., 1970, Classifications of anterior pituitary cell types with immunoenzyme histochemistry, *J. Histochem. Cytochem.* **18**:9.

Nicoll, C. S., Discussion in Pasteels (1972b), p. 279.

Normann, T. C., 1976, Neurosecretion by exocytosis, *Int. Rev. Cytol.* **46**:1.

Oliver, C., and Hand, A. R., 1978, Uptake and fate of luminally administered horseradish peroxidase in resting and isoproterenol stimulated rat parotid acinar cells, *J. Cell Biol.* **76**:207.

Olivier, L., and Vila-Porcile, E., 1972, Les lames basales des cordons épithéliaux du lobe antérieur de l'hypophyse chez le rat, *C. R. Soc. Biol.* **166**:1413.

Olivier, L., Porcile, E., de Brye, C., and Racadot, J., 1965, Étude de quelques adénomes hypophysaires chez l'homme en microscopie électronique, *Bull. Assoc. Anat. (Nancy)* **127**:1258.

Olivier, L., Vila-Porcile, E., Racadot, O., Peillon, F., and Racadot, J., 1975, Ultrastructure of pituitary tumor cells: A critical study, in: *The Anterior Pituitary* (A. Tixier-Vidal and M. G. Farquhar, eds.), pp. 231–276, Academic Press, New York.

Orci, L., and Perrelet, A., 1975, *Freeze–Etch Histology: A Comparison between Thin Sections and Freeze–Etch Replicas,* Springer-Verlag, Berlin.

Orci, L., and Perrelet, A., 1978, Ultrastructural aspects of exocytic membrane fusion, in: *Membrane Fusion: Cell Surface Reviews,* Vol. 5 (G. Poste and G. L. Nicolson, eds.), pp. 630–651, Elsevier/North-Holland, Amsterdam and New York.

Orci, L., Malaisse-Lagae, F., Ravazzola, M., and Amherdt, M., 1973, Exocytosis–endocytosis coupling in the pancreatic beta-cells, *Science* **181**:561.

Ostlund, R. E., Leung, J. T., and Kipnis, D. M., 1977, Muscle actin filaments bind pituitary secretory granules *in vitro, J. Cell Biol.* **73**:78.

Palade, G. E., and Bruns, R. R., 1968, Structural modulations of plasmalemmal vesicles, *J. Cell Biol.* **37**:633.

Parsons, J. A., 1977, Discussion in Farquhar (1977), p. 92.

Pasteels, J. L., 1963, Recherches morphologiques et expérimentales sur la sécrétion de prolactine, *Arch. Biol. (Liège)* **74**:439.

Pasteels, J. L., 1972a, Morphology of prolactin secretion, in: *Lactogenic Hormones: Ciba Found. Symp.* (G. E. W. Wostenholme and J. Knight, eds.), pp. 241–255, Churchill Livingstone, Edinburgh and London.

Pasteels, J. L., 1972b, Tissue culture of human hypophyses: Evidence of a specific prolactin in man, in: *Lactogenic Hormones: Ciba Found. Symp.* (G. E. W. Wostenholme and J. Knight, eds.), pp. 269–286, Churchill Livingstone, Edinburgh and London.

Pelletier, G., 1973, Secretion and uptake of peroxidase by rat adenohypophyseal cells, *J. Ultrastruct. Res.* **43**:445.

Pelletier, G., and Novikoff, A. B., 1972, Localization of phosphatase activities in the rat anterior pituitary gland, *J. Histochem. Cytochem.* **20**:1.

Pelletier, G., and Puviani, R., 1974, Permeability of capillaries to different tracers and uptake of horseradish peroxidase by the secretory cells in rat anterior pituitary gland, *Z. Zellforsch.* **147**:361.

Pelletier, G., Peillon, F., and Vila-Porcile, E., 1971, An ultrastructural study of granule extrusion in the anterior pituitary of the rat, *Z. Zellforsch.* **115**:501.

Pelletier, G., Lemay, A., Béraud, G., and Labrie, F., 1972, Ultrastructural changes accompanying the stimulatory effect of N^6-monobutyryl adenosine $3',5'$ monophosphate on the release of growth hormone (GH), prolactin (PRL) and adrenocorticotropin hormone (ACTH) in rat anterior pituitary gland *in vitro, Endocrinology* **91**:1355.

Pinto da Silva, P., and Nogueira, M. L., 1977, Membrane fusion during secretion: A hypothesis based on electron microscope observation of *Phytophtora palmivora* zoospores during encystment, *J. Cell Biol.* **73**:161.

Pooley, A. S., 1971, Ultrastructure and size of rat anterior pituitary secretory granules, *Endocrinology* **58**:400.

Porcile, E., Racadot, J., and Olivier, L., 1966, Étude radioautographique du transit intracellulaire des protéines constitutives des grains de sécrétion dans les cellules de l'adénohypophyse, *J. Microsc.* **5**:73a.

Porcile, E., Olivier, L., and Racadot, J., 1968, Identification des types de grains de sécrétion de l'adénohypophyse au microscope électronique par la méthode acide chromique–acide phosphotungstique, *J. Microsc.* **7**:51a.

Racadot, J., Olivier, L., Porcile, E., and Droz, B., 1965, Appareil de Golgi et origine des grains de sécrétion dans les cellules adénohypophysaires chez le rat: Étude radioautographique en microscopie électronique après injection de leucine tritiée, *C. R. Acad. Sci.* **261**:2972.

Rambourg, A., and Racadot, J., 1968, Identification en microscopie électronique de six types cellulaires dans l'antéhypophyse du rat à l'aide d'une technique de coloration par le mélange acide chromique–phosphotungstique, *C. R. Acad. Sci.* **266**:153.

Rennels, E. G., 1964, Electron microscopic alterations in the rat hypophysis after scalding, *Am. J. Anat.* **114**:71.

Rennels, E. G., and Shiino, M., 1968, Ultrastructural manifestations of pituitary release of ACTH in the rat, *Arch. Anat. Histol. Embryol.* **51**:575.

Römmler, A., Seinsch, W., Hasan, A. S., and Haase, F., 1978, Ultrastructure of rat pituitary LH gonadotrophs in relation to serum and pituitary LH levels following repeated LH-RH stimulation, *Cell Tissue Res.* **190**:135.

Sano, M., 1962, Further studies on the theta cell of the mouse anterior pituitary as revealed by electron microscopy, with special reference to the mode of secretion, *J. Cell Biol.* **15**:85.

Satir, B., Schooley, C., and Satir, P., 1973, Membrane fusion in a model system: Mucocyst secretion in *Tetrahymena, J. Cell Biol.* **56**:153.

Schlessinger, J., Schechter, Y., Willingham, M. C., and Pastan, I., 1978, Direct visualization of binding, aggregation and internalization of insulin and epidermal growth factor on living fibroblastic cells, *Proc. Natl. Acad. Sci. U.S.A.* **75**:2659.

Schofield, J. G., and Orci, L., 1975, Release of growth hormone from ox pituitary slices after pronase treatment. *J. Cell Biol.* **65**:223.

Sherline, P., Lee, Y. C., and Jacobs, L. S., 1977, Binding of microtubules to pituitary secretory granules and secretory granule membranes, *J. Cell Biol.* **72**:380.

Shiino, M., and Rennels, E. G., 1973, Ultrastructural observations of gonadotrophin release in rats treated neonatally with testosterone, *Tex. Rep. Biol. Med.* **31**:215.

Shiino, M., and Rennels, E. G., 1975a, Ultrastructural observations of growth hormone (STH) cells of anterior pituitary glands of partially hepatectomized rats, *Cell Tissue Res.* **163**:343.

Shiino, M., and Rennels, E. G., 1975b, Vinblastine-induced microtubular paracrystals in prolactin cells of anterior pituitary gland of lactating rats, *Am. J. Anat.* **144**:399.

Shiino, M., Rennels, E. G., and Williams, M. G., 1971, Ultrastructural observations of pituitary release of prolactin in the rat by suckling stimulation, *Anat. Rec.* **169**:427 (abstract).

Shiino, M., Williams, M. G., and Rennels, E. G., 1972a, Ultrastructural observations of pituitary release of prolactin in the rat by suckling stimulus, *Endocrinology* **90**:1.

Shiino, M., Arimura, A., Schally, A. V., and Rennels, E. G., 1972b, Ultrastructural observations of granule extrusion from rat anterior pituitary cells after injection of LH-releasing hormone, *Z. Zellforsch.* **128**:152.

Shiino, M.. Williams, M. G., and Rennels, E. G., 1973, Thyroidectomy cells and their response to thyrotrophin releasing hormone (TRH) in the rat, *Z. Zellforsch.* **138**:327.

Slaby, F., and Farquhar, M. G., 1976, The major polypeptides of somatotrophic and mammotrophic granules from the rat anterior pituitary, *J. Cell Biol.* **70**:92a.

Smith, R. E., and Farquhar, M. G., 1966, Lysosome function in the regulation of the secretory process in cells of the anterior pituitary gland, *J. Cell Biol.* **31**:319.

Smith, R. E., and Farquhar, M. G., 1970, Modulation in nucleoside diphosphatase activity of mammotrophic cells of the rat adenohypophysis during secretion, *J. Histochem. Cytochem.* **18**:237.

Steinman, R. M., and Cohn, Z. A., 1972, The interaction of soluble horseradish peroxidase with mouse peritoneal macrophages *in vitro, J. Cell Biol.* **55**:186.

Steinman, R. M., Brodie, S. E., and Cohn, Z. A., 1976, Membrane flow during pinocytosis: A stereologic analysis, *J. Cell Biol.* **68**:665.

Straus, W., 1977, Altered renal cortical reabsorption of protein and urinary excretion of sodium in relation to vascular leakage induced by horseradish peroxidase, *J. Histochem. Cytochem.* **25**:215.

Swearingen, K. C., 1971, Heterogeneous turnover of adenohypophysial prolactin, *Endocrinology* **89**:1380.

Tandler, B., and Poulsen, J. H., 1976, Fusion of the envelope of mucous droplets with the luminal plasma membrane in acinar cells of the cat submandibular gland, *J. Cell Biol.* **68**:775.

Tesar, J. T., Koenig, H., and Hugues, C., 1969, Hormone storage granules in the beef anterior pituitary. I. Isolation, ultrastructure and some biochemical properties, *J. Cell Biol.* **40**:225.

Theodosis, D. T., Dreifuss, J. J., and Orci, L., 1978, A freeze fracture study of membrane events during neurohypophysial secretion, *J. Cell Biol.* **78**:542.

Tixier-Vidal, A., 1965, Caractères ultrastructuraux des types cellulaires de l'adéno hypophyse du canard mâle, *Arch. Anat. Micr. Morph. Exp.* **54**:719.

Tixier-Vidal, A., 1975, Ultrastructure of anterior pituitary cells in culture, in: *The Anterior Pituitary* (A. Tixier-Vidal and M. G. Farquhar, eds.), pp. 181–229, Academic Press, New York.

Tixier-Vidal, A., and Picart, R., 1967, Étude quantitative par radioautographie au microscope électronique de l'utilisation de la D-L leucine ^{3}H par les cellules de l'hypophyse du canard en culture organotypique, *J. Cell Biol.* **35**:501.

Tixier-Vidal, A., and Picart, R., 1971, Electron microscopic localization of glycoproteins in pituitary cells of duck and quail, *J. Histochem. Cytochem.* **19**:775.

Tixier-Vidal, A., Fiske, S., Picart, R., and Haguenau, F., 1965, Radioautographie au microscope électronique de l'incorporation de leucine tritiée par l'adénohypophyse du canard en culture organotypique, *C. R. Acad. Sci. Ser. D* **261**:1133.

Tixier-Vidal, A., Kerdelhué, B., Bérault, A., Picart, R., and Jutisz, M., 1971, Action *in vitro* du facteur hypothalamique de libération de l'hormone lutéinisante (LRF) sur l'antéhypophyse d'agnelle. II. Étude ultrastructurale des tissus incubés, *Gen. Comp. Endocrinol.* **17**:33.

Tixer-Vidal, A., Gourdji, D., and Tougard, C., 1975a, A cell culture approach to the study of anterior pituitary cells, *Int. Rev. Cytol.* **41**:173.

Tixier-Vidal, A., Tougard, C., Kerdelhué, B., and Jutisz, M., 1975b, Light and electron microscopic studies on immunocytochemical localization of gonadotropic hormones in the rat pituitary gland with antisera against ovine FSH, LH, LHα and LHβ, *Ann. N. Y. Acad. Sci.* **254**:433.

Tixier-Vidal, A., Picart, R., and Moreau, M. F., 1976, Endocytose et sécrétion dans les cellules anté-hypophysaires en culture: Action des hormones hypothalamiques, *J. Microsc. Biol. Cell.* **25**:159.

Tougard, C., Kerdelhué, B., Tixier-Vidal, A., and Jutisz, M., 1971, Localisation par cytoimmunoenzymologie de la LH, de ses sous-unités α et β, et de la FSH dans l'adénohypophyse de la ratte castrée, *C. R. Acad. Sci.* **273**:897.

Tougard, C., Kerdelhué, B., Tixier-Vidal, A., and Jutisz, M., 1973, Light and electron microscope localization of binding sites of antibodies against ovine luteinizing hormone and its two subunits in rat adenohypophysis using peroxidase labeled antibody technique, *J. Cell Biol.* **58**:503.

Tramu, G., Beauvillain, J. C., Mazucca, M., Fossati, P., Martin-Linquette, A., and Christiaens, J. L., 1976, Dissociation des résultats obtenus en immunofluorescence avec des antisérums anti-ACTH dans trois cas d'adénomes "chromophobes" sans hypercorticisme, *Ann. Endocrinol. (Paris)* **37**:55.

Ueberberg, H., and Hohbach, C., 1972, Zur Frage der Exocytose im Vorderlappen der Hypophyse, *Z. Zellforsch.* **132**:287.

Vila-Porcile, E., 1972, La pars distalis de l'hypophyse chez le rat: Contribution à son étude histologique et cytologique en microscopie électronique. Thesis, Université Pierre et Marie Curie, Paris, 114 pp.

Vila-Porcile, E., 1973, La pars distalis de l'hypophyse chez le rat: Contribution à son étude histologique et cytologique en microscopie électronique, *Ann. Sci. Nat. Zool. (Paris)* **15**:61.

Vila-Porcile, E., 1975, Morphological and functional relationships between the different compartments of the rat anterior pituitary, in: *Proceedings of the Xth International Congress of Anatomy*, Tokyo (E. Yamada ed.), p. 24, Science Council of Japan, Tokyo.

Vila-Porcile, E., and Olivier, L., 1970, Utilisation de la peroxydase comme traceur de diffusion pour l'étude des espaces périvasculaires et intercellulaires de l'adénohypophyse, in: *VIIe Congres International de Microscopie Electronique*, Vol.

3, Grenoble, (P. Favard, ed.), pp. 43–44, Société Française de Microscopie Electronique, Paris.

Vila-Porcile, E., and Olivier, L., 1978, Exocytose et endocytose dans la cellule à prolactine au cours de la sécrétion: Étude cinétique *in vivo* au moyen de la peroxydase, *J. Microsc. Biol. Cell.* **32:**26a.

Vila-Porcile, E., Olivier, L., and Racadot, O., 1973a, Sites membranaires d'expulsion des corps résiduels lysosomiaux des cellules à prolactine au cours de la post-lactation chez la ratte, *J. Microsc.* **17:**107a.

Vila-Porcile, E., Olivier, L., and Racadot, O., 1973b, Exocytose polarisée des corps résiduels lysosomiaux des cellules à prolactine dans l'adénohypophyse de la ratte en post-lactation, *C. R. Acad. Sci. Ser. D* **276:**355.

Vila-Porcile, E., Olivier, L., and Racadot, O., 1973c, Functional polarization of the prolactin cell: An electron microscopic study in the female rat during lactation and after weaning, in: *Proceedings of the International Symposium on Human Prolactin, Int. Congr. Ser.* **308:**56.

Wada, M., 1975, Cell types in the adenohypophysis of the Japanese quail and effects of injection of luteinizing hormone releasing hormone, *Cell Tissue Res.* **159:**167.

Warchol, J. B., Herbert, D. C., and Rennels, E. G., 1974, Localization of microfilaments in prolactin cells of the rat anterior pituitary gland, *Cell Tissue Res.* **155:**193.

Wilbur, D. L., Worthington, W. C., and Markwald, R. G., 1974, Ultrastructural observations of anterior pituitary somatotrophs following pituitary portal vessel infusion of dibutyryl-cAMP, *Am. J. Anat.* **141:**139.

Wilbur, D. L., Yee, J. A., and Raigue, S. E., 1978, Hypophysial portal vascular infusion of TRH in the rat: An ultrastructural and radioimmunoassay study, *Am. J. Anat.* **151:**277.

Winkler, H., Schneider, F. H., Rufener, C., Nakane, P. K., and Hörtnagl, H., 1974, Membranes of adrenal medulla: their role in exocytosis, in: *Cytopharmacology of Secretion* (B. Ceccarelli, F. Clementi, and J. Meldolesi, eds.), pp. 127–139, Raven Press, New York.

Woodward, M. P., and Roth, T. F., 1978, Coated vesicles: Characterization, selective dissociation and reassembly, *Proc. Natl. Acad. Sci. U.S.A.* **75:**4394.

Zambrano, D., Cuerdo-Rocha, S., and Bergmann, I., 1975, Ultrastructure of rat pituitary gonadotrophs following incubations of the gland with synthetic LH-RH, *Cell Tissue Res.* **150:** 179.

Zanini, A., and Giannattasio, G., 1974, Molecular organization of rat prolactin secretory granules, in: *Cytopharmacology of Secretion* (B. Ceccarelli, F. Clementi, and J. Meldolesi, eds.), pp. 329–339, Raven Press, New York.

5

Intracellular Events in Prolactin Secretion

Antonia Zanini, Giuliana Giannattasio, and Jacopo Meldolesi

1. Introduction

Considerable experimental evidence, obtained during the last several years, indicates that secretion of prolactin (PRL) is carried out in pituitary mammotrophs by means of a series of complex interconnected events (analogous to those occurring in other protein-secreting cells) that ultimately result in hormone discharge by exocytosis (Fig. 1). This chapter will focus primarily on one such event, namely, the packaging of PRL within secretion granules. The others, namely, the biosynthesis of the hormone and its transport to the Golgi periphery, exocytosis, and crinophagy, as well as the recycling of membranes from the cell surface, will be dealt with only briefly, because they have been covered in detailed reviews that have been published recently or will appear soon (Farquhar *et al.*, 1975; Palade, 1975; Farquhar, 1977; Meldolesi *et al.*, 1978; De Camilli *et al.*, 1979; Herzog and Miller, 1979). In addition, specific problems related to these events are discussed in other chapters of this volume.

Antonia Zanini, Giuliana Giannattasio, and *Jacopo Meldolesi* • Department of Pharmacology and CNR Center of Cytopharmacology, University of Milan, 20129 Milan, Italy

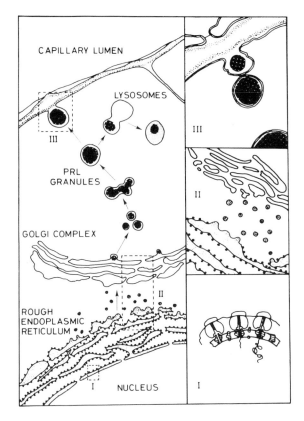

Figure 1. Schematic representation of PRL synthesis, transport, storage, and release. Three moments of the secretory process are shown in insets I–III: (I) Synthesis of PRL by bound polyribosomes (attached by their large subunits to specific receptors localized in RER membranes) and transmembrane growth of PRL polypeptide chains; (II) translocation of PRL molecules from the transitional elements to the Golgi complex by means of small vesicles; (III) two moments of exocytosis: fusion of granule membrane with the plasmalemma and expulsion of granule content into the extracellular space. Left-hand panel redrawn after Farquhar *et al.* (1975).

2. Biosynthesis and Transport to the Golgi Periphery

In the animal species so far investigated, the rate of PRL biosynthesis by far exceeds that of the proteins destined to turn over within mammotroph cells. Like all other secretory proteins, PRL is synthesized only by bound polyribosomes, which are attached to the outer surface of the rough endoplasmic reticulum (RER) (see Palade, 1975). The link occurs between large ribosome subunits on one side and specific receptors, localized exclusively in RER membranes, on the other (Kreibich *et al.*, 1978a, b) (Fig. 1, I). The hollow structure of these receptors makes possible the transmembrane growth of PRL polypeptide chains, which results eventually in the segregation of the finished hormone molecules within the RER cisternae. Recently, it has been demonstrated that up to a defined stage of their elongation, growing chains of PRL (as well as those of many other secretory proteins) bear an extra polypeptide sequence covalently attached at the hormone N terminus. This sequence, which is now designated the signal peptide, is cleaved either before or immediately after chain termination by an endopeptidase localized in ER membranes (Evans *et al.*, 1977; Lingappa *et al.*, 1977; Maurer and McKean, 1978). The discovery of signal peptides for many secretory proteins has fostered the hypothesis that the interaction of these peptides with the ribosome receptors of RER membranes would be the initial event of polyribosome attachment, which therefore would be specified by the translation products and not by the mRNAs (Blobel and Dobberstein, 1975; Campbell and Blobel, 1976).

Following their synthesis and segregation within RER cisternae, PRL molecules are transported vectorially to the Golgi complex. The rate of this transport is relatively rapid (Farquhar *et al.*, 1978). Thus, in isolated rat mammotrophs pulse-labeled for 5 min with [³H]-L-leucine, radioactive PRL molecules were first detected by Farquhar *et al.* (1978) in Golgi stacks after 5 min of chase incubation. After an additional 10 min in chase medium, over two thirds of the label had moved from the RER to the Golgi area, where it was detected both within the stacks and in immature granules. No information exists about the mechanisms of this drainage, except that it is energy-dependent (Labrie *et al.*, 1973). However, Jamieson and Palade (1968), working on pancreatic acinar cells in which ATP synthesis had been impaired by various means, demonstrated that secretory proteins were still transported from the sites of synthesis to transitional elements (partially rough and partially smooth cisternae located at the Golgi periphery), but were unable to cross the RER–Golgi boundary. Thus, the energy demand may be not for driving proteins within RER cisternae, but for maintaining the continuity (whether physical or only functional is still debated) between RER and Golgi complex. In conclusion, no specific driving forces that might be responsible for the movement of newly synthesized PRL molecules to the Golgi periphery have ever been identified. On the other hand, since hormone molecules are most probably soluble while in the RER cisternae, they would be expected to move at random by simple diffusion. Random diffusion could be converted into vectorial movement provided that an adequate trapping mechanism restricted movement of hormone molecules at a geometrically specified site of the system. These considerations suggest that in pituitary mammotrophs, binding sites for PRL could exist within the transitional elements.

The nature of these sites is still undefined. Recently, Jamieson (1979) has proposed that they might be accounted for by true receptors, localized at the lumenal surface of the membrane. Binding of secretory proteins to these receptors would result in their accumulation, envisaged as a prerequisite for their subsequent translocation to the Golgi complex. An alternative possibility is the following. As discussed in the next section, ample evidence indicates that within granules, secretion products are not soluble but assembled in a solid-state structure. The assembly process, which will be referred to herein as "packaging," is known to take place in the Golgi complex, but it might start already in transitional elements. Were this the case, a unique process, i.e., packaging, would be responsible for a number of apparently different events in secretory cells: (1) trapping of secretion products in transitional elements; (2) appearance of immature granules in Golgi cisternae and vacuoles; and (3) maturation of secretion granules by progressive concentration of their segregated content.

3. Packaging and Maturation

As already anticipated in the previous section, intracellular transport of PRL entails a change in the physical state of the hormone, from soluble (in the RER) to structured in a solid supramolecular organization (in secretion granules). This conclusion was reached some years ago when we demonstrated that isolated rat PRL granule cores remain aggregated even after solubilization of their limiting membrane by treatment with mild detergents (Giannattasio *et al.*, 1975). Among the detergents we used, the best results were obtained with Lubrol PX. Using rat PRL granules stripped of their limiting membrane by treatment with the latter detergent (membraneless granules), we were able to demonstrate that the unusual stability of intact PRL granules depends not on the limiting membrane, but on the organization of the segregated content. In fact, intact and membraneless granules are equally unaffected by *in vitro* incubation in a variety of different conditions: exposure to media of different tonicity and ionic strength, to mono- and divalent cations, and to chelating agents (Giannattasio *et al.*, 1975). Both intact and membraneless granules are stable over a wide pH range: from 3.5 to 6.5 (Fig. 2). However, at pH between 7 and 8, an interesting difference emerged: intact granules are still stable, while membraneless granules are partially solubilized by incubation in isotonic sucrose buffered at pH 7.4 with 10 mM phosphate buffer (~40% solubilization in 1 hr at 24°C). In addition, solubilization of the granule matrix seems to depend also on the ionic composition of the incubation fluid, since physiological salt solutions, such as Krebs–Ringer bicarbonate (KRB) buffer, are much more effective (~80% in 1 hr at 37°C) than phosphate-buffered isotonic sucrose and monovalent ion solutions

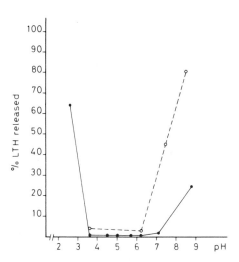

Figure 2. Effect of pH on the stability of intact (●——●) and membraneless (○———○) PRL granules isolated from rat pituitary glands. Preparations of isolated granules were resuspended in 0.32 M sucrose and the pH was adjusted to the specified values by the addition of phosphate buffer (0.01 M, final concentration) and incubated for 1 hr at 24°C. The PRL solubilization was estimated by polyacrylamide gel electrophoresis of both the pellets (obtained by centrifuging the incubated granules) and the supernatants. Experimental details in Giannattasio *et al.* (1975). Reproduced with permission from Giannattasio *et al.* (1975).

(Fig. 3). When pelleted preparations of partially dissolved membraneless granules were studied by conventional electron microscopy, they were found to be accounted for primarily by networks of long, twisted filaments, approximately 4 nm in diameter, that were often in continuity with recognizable, partially disarranged granule matrixes (Fig. 4).

The degree of stability of PRL granules is not the same over the entire population present in mammotroph cells. Rather, stability seems to increase progressively during the maturation of the organelles. The first indication of this phenomenon came from experiments in which membraneless PRL granules were isolated from pituitary tissue slices labeled *in vitro* with [³H]-L-leucine for 90 min. We found (Table I) that the specific radioactivity of PRL solubilized during the first minutes of incubation at pH 7.4 was much higher than that of the hormone solubilized later on, suggesting that young granules, which contain the hormone synthesized in

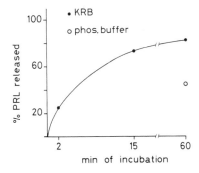

Figure 3. *In vitro* release of PRL from rat membraneless granules incubated at 37°C, in either Krebs–Ringer bicarbonate (KRB) or 0.32 M sucrose buffered with 0.01 M phosphate. In both cases, the pH of the incubation fluid was 7.4. The PRL solubilization was estimated as described by Giannattasio *et al.* (1975).

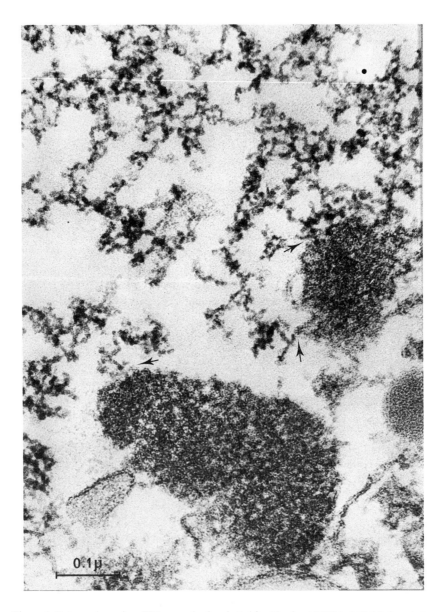

Figure 4. Rat membraneless PRL granules incubated for 60 min at 24°C in 0.01 M phosphate buffer, pH 8.5. Most of the material sedimented by centrifugation from these preparations is accounted for by a network of thick, twisted filaments that are often continuous with partially disarranged PRL granules (arrows). Reproduced with permission from Giannattasio *et al.* (1975).

Table I. Release of Prolactin and ^{3}H Radioactivity from Rat
Membraneless Granules Incubated at 37°C for Different Time
Intervals in Krebs–Ringer Bicarbonate Buffer[a]

Time (min)	PRL release (%)	^{3}H radioactivity release (%)	Specific activity of solubilized PRL (dpm/mg PRL)
2	25	56	866.700
15	73	86	532.800
60	82	92	470.600

[a] The granules were isolated from pituitary slices labeled *in vitro* with [^{3}H]leucine for 90 min. For further experimental etails see Giannattasio *et al.* (1975).

the course of the experiment, are more fragile than those preexisting within mammotroph cells. However, this interpretation was questionable because the results could have been substantially affected by experimental artifacts, such as, for instance, the uneven penetration of the tracer amino acid, oxygen, and substrates into the incubated tissue. Therefore, to check the interpretation, we carried out double-label experiments. Rat pituitary tissue slices were labeled for two 10-min periods, first with [^{3}H]-L-leucine and then with [^{14}C]-L-leucine. Our protocol also included two incubations in chase medium: a period of 75 min interposed between the two pulse incubations and a final chase period of 15 min after the ^{14}C labeling, before homogenization of the slices. From the available information on the timetable of the intracellular transport of PRL in rat mammotrophs, we expected the ^{14}C- and ^{3}H-labeled PRL molecules to be located at the end of the incubations primarily in immature and mature granules [types I and III+IV of Farquhar *et al.* (1978)], respectively. Double-labeled PRL membraneless granules were isolated and tested *in vitro* at pH 7.4, and the ^{14}C/^{3}H ratio of the solubilized hormone was determined. The ratio in the hormone solubilized during the first 2 min of incubation was much higher than that in the input granule preparation (1.26 vs. 0.69). In contrast, after 60 min of incubation, no difference was evident between the input and solubilized fractions. These data confirm that new granules are indeed solubilized at a much faster rate than old granules.

An important question that might be asked at this point concerns the mechanisms by which PRL is packaged within granules in such a stable organization. In recent years, the increased knowledge of the composition of secretion granules as well as the more detailed understanding of the physiological role of the Golgi complex (the cell structure in which the change of the physical state of secretion products is known to take place) have led to the proposal of different hypotheses to account for the packaging process. Essentially, two functions of the Golgi complex are now suspected to be implicated: ion transport and glycosaminoglycan and glycoprotein synthesis.

In relation to ion transport, it should be mentioned that at least in some cell systems, the RER membrane is known to be very permeable (see Meldolesi *et al.*, 1978). Thus, the ionic environment within the RER cisternae is most probably the same as in the surrounding cytoplasm. The situation in the Golgi complex appears different. Although detailed permeability studies have not been carried out yet, the association of specific ion pumps with Golgi membranes in a variety of secretory systems (Selinger *et al.*, 1970; Argent *et al.*, 1975; Baumrucker and Keenan, 1975) and the cytochemical demonstration in many cell types, including pituitary mammotrophs, of high cation concentrations within Golgi cisternae (Clemente and Meldolesi, 1975; Stoeckel *et al.*, 1975; Reith, 1976) suggest that the internal ionic environment is peculiar and possibly specifically regulated. During the last few years, divalent cations, especially Ca^{2+}, have been considered as good candidates for playing a major role in the packaging of secretion products. This hypothesis is based, on one hand, on the well-known ability of Ca^{2+} ions to bring molecules together by acting as bridges between adjacent negative charges, and on the other, on the fact that all secretion granules so far investigated (especially those of the adrenal and parotid glands) contain fairly large amounts of Ca^{2+} (from 20 to 80 nmol/mg protein).

Recent developments in the field, however, cast some doubts on the possibility that interaction with Ca^{2+} might be the sole mechanism of packaging of secretion products. In fact, Flashner and Schramm (1977) have recently succeeded in releasing the bulk of the segregated Ca^{2+} by incubating isolated parotid secretion granules in an isotonic fluid containing ethyleneglycoltetraacetic acid (EGTA) together with the Ca^{2+} ionophore A23187, with no apparent derangement of the structure of the organelles. Moreover, in rat membraneless granules, the molar Ca^{2+}/PRL ratio is less than 0.5 (Zanini and Giannattasio, 1974), and therefore the direct binding of all hormone molecules to Ca^{2+} appears impossible. These observations, however, do not rule out the possibility that Ca^{2+} is but one of the agents responsible for the packaging process, as will be further elaborated below.

On the other hand, the involvement of glycosaminoglycans and glycoproteins has been postulated on the basis of the following considerations: (1) Macromolecular carbohydrates (in most cases not yet characterized in biochemical terms) are minority components of many and perhaps all secretory granules (for a review, see Giannattasio *et al.*, 1979a). (2) Completion of sugar side chains and attachment of sulfate residues to glycosaminoglycans and glycoproteins take place in the Golgi complex, at the same time as the packaging process. This concomitance might not be fortuitous. In contrast, the introduction of negative charges could be instrumental in the interaction of sugar macromolecules with the other secretion products and finally result in the insolubilization of the granule matrix.

To test experimentally the hypothesized role of sugar macromolecules in PRL granules, we carried out a detailed biochemical characterization of the membraneless organelles isolated from rat and bovine pituitaries (Giannattasio and Zanini, 1976; Zanini *et al.*, 1979). We found that the hormone is not the only component of the granule matrix; in fact, a number of other proteins, present in very low amount, were detected by sodium dodecyl sulfate (SDS) polyacrylamide gel electrophoresis (Fig. 5A). Many of these components were identified as glycoproteins on the

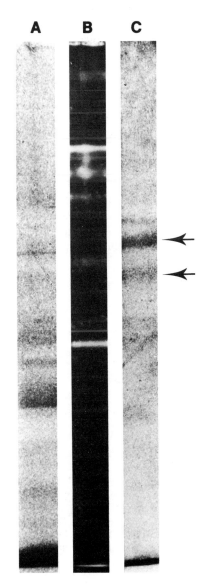

Figure 5. SDS 7.5% polyacrylamide slab gel of bovine membraneless PRL granules. (A) Coomassie Blue staining; (B) [^{125}I]concanavalin A fluorography; (C) "stains all" staining. Arrows indicate bands stained blue by the "stains all" procedure, which might be accounted for by sialoglycoproteins. Gels of this porosity separate proteins of molecular weight greater than 30,000 daltons. Thus, in this system, PRL migrates with the front. Experimental details will be specified elsewhere (Zanini *et al.*, 1979).

basis of their [125I]concanavalin A binding property (Fig. 5B) and two of them as acidic (possibly sialoglycoproteins) because they are stained blue by the "stains all" procedure (Fig. 5C). Labeling experiments with [35S]sulfate demonstrated that at least some of these glycoproteins are sulfated. Very low amounts of glycosaminoglycans (major component: heparan sulfate) were also detected within the matrix of PRL granules. Additional evidence that sulfated glycosaminoglycans and glycoproteins are copacked with the hormone within granules comes from kinetic experiments in which cow pituitary slices were incubated together with [3H]-L-leucine (to label PRL) and with [35S] inorganic sulfate (to label sulfated macromolecules) (Giannattasio *et al.*, 1979b). We found that the rate of transport of the hormone and that of sulfated glycosaminoglycans and glycoproteins throughout mammotroph cells are approximately the same.

To obtain information on the nature of the possible interaction between PRL and complex carbohydrates, membraneless granules, isolated from rat pituitary slices double-labeled with [3H]-L-leucine and [35S]sulfate, were suspended in media of varying pH (between 6 and 8.5) and monovalent ion concentration (between 0.08 and 0.77 M NaCl), and the release of the 3H-labeled hormone and 35S-labeled macromolecules during incubation was determined. As in the previous experiments, the release of PRL proved to be independent of the monovalent ion concentration of the medium and strictly dependent on the pH (Fig. 6). In contrast, solubilization of sulfated macromolecules was much less dependent on the pH (Fig. 6) but was greatly increased by raising the Na+ concentration of the incubation fluid (Table II).

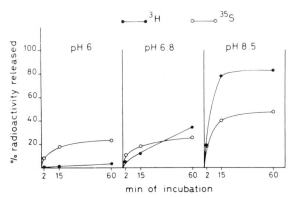

Figure 6. *In vitro* release of 3H- and 35S-labeled macromolecules from double-labeled rat membraneless PRL granules incubated at 24°C for various time intervals, in 0.32 M sucrose buffered at different pHs. The granules were isolated from pituitary slices double-labeled *in vitro* with [3H]leucine and [35S]inorganic sulfate for 50 min. For further experimental details see Giannattasio *et al.* (1975).

Table II. Release of ³H- and ³⁵S-Labeled
Macromolecules from Rat PRL Membraneless
Granules Incubated for 1 hr at 37°C in Krebs–Ringer
Bicarbonate Buffer Containing NaCl at Various
Concentrations[a]

Incubation medium	³H-labeled macromolecule release (%)	³⁵S-labeled macromolecule release (%)
KRB 0.08 M NaCl	78	17
KRB 0.15 M NaCl	92	37
KRB 0.77 M NaCl	86	70

[a] The granules were isolated from pituitary slices labeled *in vitro* with [³H]leucine for 90 min. For further experimental details see Giannattasio *et al.* (1975).

These results are consistent with the idea that the integration of PRL and sulfated macromolecules in the granule matrix occurs by ionic bonding. The hormone would bind through positive charges, which are neutralized when the pH is raised from 6.5 to 7.4, while the binding of polyanions through their negative sulfate residues would be interfered with by Na^+ ions. However, it should be acknowledged that the results we have reported on the dependence of the granule matrix aggregation on environmental conditions are still preliminary and have to be strengthened and expanded before solid conclusions are achieved. In addition, there is no proof that PRL and sulfated macromolecules are bound to each other in the granule matrix. Reconstitution experiments suggesting the existence of a direct link of some secretory proteins to glycosaminoglycans have recently been reported for pancreatic zymogen granules by Reggio and Dagorn (1978); no experiments of this type have ever been carried out on pituitary mammotroph granules.

In conclusion, it is only fair to recognize that our knowledge of PRL packaging is still fragmentary and contradictory. On one hand, the low level of macromolecular polyanions and their molecular heterogeneity, as well as the nonparallel release of PRL and sulfated macromolecules in the course of *in vitro* dissolution of membraneless granules, argue against the simple direct binding of all segregated hormone molecules to polyanions. On the other, the lack of effect of salt solutions and chelating agents in releasing PRL from membraneless granules suggests that divalent cations cannot be solely responsible for the granule structure. Thus, packaging might be a complex, probably multifactorial process. In other words, the environmental conditions existing within the Golgi cisternae, taken as a whole, rather than individual events, would be instrumental in the change in physical state of the hormone. It could be that in this process, macromolecular polyanions are not the direct organizers of the granule matrix, but rather play a role in the formation of condensation cores, around which the interaction of PRL molecules could eventually develop. This hypothetical interpretation appears consistent with cytochemical observations suggesting that glycosylated molecules are concentrated at the surface of the granule core (Rambourg and Racadot, 1968).

4. Exocytosis, Crinophagy, and Membrane Recycling

Release of PRL to the extracellular space occurs by exocytosis. Images of discharged granule cores in various stages of dissolution, lying in pocket indentations of the plasmalemma, are quite common, especially at the vascular pole of mammotroph cells (see Farquhar *et al.*, 1975, 1978) (also see Chapter 4). The high frequency of these images is due, on one hand, to the high turnover of the hormone, and on the other, to the stability of the granule matrix, which is solubilized slowly after discharge.

Granule discharge can occur not only to the extracellular space but also to the lysosomal compartment by a process designated as crinophagy (see Farquhar *et al.*, 1975; Farquhar, 1977). In this case, the granule membrane fuses, not with the plasmalemma, but with the membrane of a lysosome. Crinophagy occurs primarily when the secretory activity of mammotroph cells is suddenly reduced, e.g., when pups are separated from the mother. In other conditions, it constitutes a relatively rare event. Following their engulfment in lysosomes, granule cores remain recognizable for several hours, until the lysosomes involved are transformed into residual bodies.

In the past, the existence of discharge mechanisms other than exocytosis was repeatedly suggested. However, claims of direct, molecular release of the hormone from a soluble cytoplasmic pool, which were advanced to account for *in vitro* kinetic experiments (Swearingen, 1971), have not been substantiated by more recent studies. Analogously, it now seems unlikely that newly assembled granules are discharged preferentially with respect to older granules, as previously suggested (Swearingen, 1971; Labrie *et al.*, 1973). In fact, when the release of PRL was investigated adequately (Meldolesi *et al.*, 1972) it was found to occur according to first-order kinetics, as was to be expected if all secretion granules had equal opportunity to undergo the discharge process. This conclusion is also supported by the observation that the granules caught during exocytosis are morphologically quite heterogeneous: the relative proportion between young, polymorphic (type II) granules and older ovoidal or spherical (type III and IV) granules is approximately the same within the cell and in the indentations of the plasmalemma. The previous results in apparent contradiction with this conclusion can be explained on the basis of the heterogeneity, not of the intracellular hormone pool, but of the population of mammotroph cells. In fact, it is now clear that especially when tested *in vitro*, some mammotrophs have low activity of hormone synthesis and release, as a consequence either of damage during tissue dissociation (isolated cells) (see Chapter 6) or of insufficient penetration of oxygen and substrates (tissue slices) (Farquhar *et al.*, 1975).

During exocytosis, the granule membrane is incorporated into the plasmalemma. However, the enlargement of the cell surface is only transient, because the excess membrane is rapidly removed. During the last decade, the removal process and the ultimate fate of the membrane of discharged granules were investigated extensively, in a variety of secretory cell systems, by following with time the uptake and distribution within cells of nonpermeant tracer molecules (such as ferritin, horseradish peroxidase, and dextran) added to the extracellular fluid. Until recently, the results obtained had fostered divergent interpretations, especially as to the fate of the granule membrane. In fact, in all systems investigated, the tracers were picked up at the cell surface by vesicles or small vacuoles or both. However, in some systems, tracer uptake was followed exclusively by accumulation within lysosomes, suggesting that all vesicles were destroyed after recycling; in other systems, including rat mammotroph cells (Pelletier, 1973; Farquhar *et al.*, 1975), it was followed by the appearance of tracer in Golgi stacks and immature granules as well, suggesting a partial reutilization in the formation of new granules (for a detailed review see Meldolesi *et al.*, 1978).

In the course of the last year, thanks to the work of Farquhar and Herzog (Herzog and Farquhar, 1977; Farquhar, 1978; Herzog and Miller, 1979), it has become clear that most previous tracer results need to be reinterpreted. In fact, a basic assumption of these studies, namely, that the intracellular distribution of externally applied tracers corresponds to the distribution of both the content and the membrane of the recycled vesicles, proved to be incorrect. All the traditional tracers have no affinity for membranes, and remain in the content. Their distribution after uptake therefore corresponds to that of the soluble molecules interiorized within the vesicles, which does not necessarily coincide with that of the membranes. A new tracer, cationic ferritin, which, being positively charged, has affinity for membranes (which are negatively charged), has a completely different distribution. As studied by Farquhar (1978) in dissociated mammotrophs from estrogen-treated female rats, after vesicular uptake at the cell surface, the label appears rapidly in the Golgi zone, in vesicles, in rigid cisternae, and in cisternal stacks (Fig. 7). In the latter, the tracer particles cluster preferentially at the dilated rims of the innermost cisterna and adhere to the surface of forming PRL granules. Within 30 min, the tracer appears in polymorphic type II granules and within 60 min in mature granules. In granules, the tracer molecules always appear attached to the surface of the dense content, rather than to the membrane. Lysosomes are also labeled by cationic ferritin. However, their labeling is relatively low, much lower than that observed with anionic ferritin, which has no affinity for membranes (Farquhar, 1978). These results strongly suggest that the membrane of discharged PRL granules is recycled within the cell and reused to bind new granule cores. The finding that in granules the positively charged tracer adheres preferentially to the content is a further indication that the negatively charged molecules of the latter are superficially located. Finally, the differential recovery of cationic and anionic ferritins in lysosomes and in Golgi and Golgi-derived structures is open to various interpretations. In fact, it is possible that the final destiny of vesicles recycled from the plasmalemma might be determined by the nature of the segregated tracer (Farquhar, 1978). Alternatively, all recycled vesicles might first fuse with lysosomes, then detach from lysosomes and move to the Golgi area. As a consequence of this multistep movement, tracers of the content type would remain mostly trapped at the first station, while tracers with high affinity for membranes would be transported to Golgi and then to the granules. The fact that fusion of vesicles with lysosomes is not necessarily a dead end, i.e., that vesicles can be recycled and reused after fusion with lysosomes, has recently been demonstrated in other cell systems (Steinman *et al.*, 1976; Schneider *et al.*, 1977).

5. Problems Remaining for the Future

A number of questions concerning the cellular events in PRL secretion remain to be answered by future work. Some of these questions, which concern specifically the packaging of the hormone in secretion granules, are discussed below.

First of all, it would be important to know in more detail where packaging occurs in mammotroph cells. Everybody agrees that the Golgi complex is involved; however, the Golgi is quite heterogeneous, and the study of its individual components has not been carried out in pituitary mammotrophs. In addition, it is possible that packaging is initiated already in transitional elements.

Other interesting questions concern the permeability properties of Golgi membranes and the microenvironment existing within Golgi cisternae. Although the technology is available (see, for instance, Johnson and

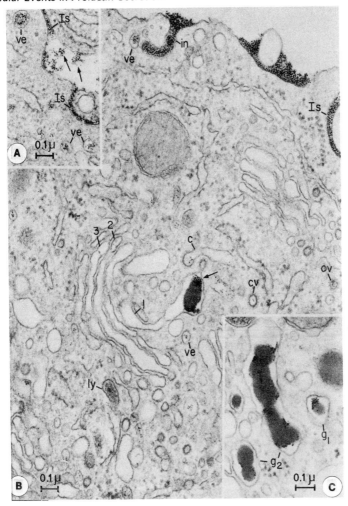

Figure 7. Mammotrophs from estrogen-treated females incubated with cationized ferritin (0.1 mg/ml). (A) Initially (after 15 min of incubation), cationized ferritin is seen binding to the cell membrane or within numerous vesicles (ve) located in the cytoplasm near the cell membrane. Cationized ferritin molecules are aggregated on the free cell surface (arrows), but form a regular layer (one or two molecules deep) in the intercellular spaces (Is). Inside the vesicles, the cationized ferritin is also aggregated. (B) Later (after 60 min), cationized ferritin aggregates are still seen along the free cell surface (upper right), within membrane invaginations (in), and within smooth (ve) or coated (cv) vesicles. Some molecules are also found within the stacked Golgi cisternae and around a forming secretion granule, as well as in a lysosome (ly). Here, cationized ferritin molecules are present in three (1–3) of the stacked Golgi cisternae. It is concentrated at the periphery of a granule (arrow) condensing within the innermost cisterna. Several molecules are also present within the coated tip of another smooth cisterna (c). (C) At 60 min, cationized ferritin is seen within immature type I (g_1) and type II (g_2) prolactin granules, sticking to the periphery of the dense content. Reproduced with permission from Farquhar (1978).

Scarpa, 1976), studies aimed at answering specifically these questions have never been carried out on secretory cell systems.

A long series of experiments is needed to explore in detail the role of the minority sugar polyanions of the granule content (glycosaminoglycans and glycoproteins) in the packaging process. These include: reconstitution experiments aimed to identify *in vitro* the optimal conditions for PRL–sugar polyanion interaction; the study of the effects of specific inhibitors, known to affect the synthesis of either glycoproteins or glycosaminoglycans, on the intracellular transport and packaging of PRL; a deeper investigation on the localization of individual sugar polyanions within the granule matrix. Finally, the fate of the sugar polyanions of the granule matrix should also be clarified. We know that in other systems, at least part of the ^{35}S-labeled macromolecules are released to the extracellular space along with the regular secretion products (Berg and Austin, 1976; Berg, 1978). However, the nature of these discharged macromolecules is unknown. In a recent study on bovine mammotrophs, we have obtained evidence that at least part of the sulfated macromolecules of the PRL granule core are not discharged during exocytosis, but are retained in the cells (Giannattasio *et al.*, 1979b). The further route taken by these molecules remains to be clarified. Several possibilities are open. For instance, they might remain caught at the cell surface and possibly be returned to the cell interior by membrane recycling.

ACKNOWLEDGMENTS. We thank Dr. M. G. Farquhar for granting us permission to reproduce Fig. 7. The photographic assistance of Mr. P. Tinelli and F. Crippa is gratefully acknowledged.

DISCUSSION

TIXIER-VIDAL: Have you excluded the possibility that your membraneless-granule preparation may retain carbohydrates from the inner face of the membrane granule?

MELDOLESI: Our evidence is not adequate to determine whether the carbohydrates are peripheral components of the granule membrane or whether they are authentic secretion products.

RAMACHANDRAN: I wonder if it is possible to invoke a simpler explanation for the stability of the membraneless granule at pH 6 and the decrease of stability at higher pH. Prolactin has an isoelectric point around pH 6 and is very insoluble at this pH. Are the membraneless granules stable at pH values below 6?

MELDOLESI: The membraneless granules are stable over a wide pH range (3.5–6.5). Thus, the stability of the granules seems to be a specific phenomenon and not due just to isoelectric precipitation.

SPICER: Histochemistry provides an additional approach to this question. The dialyzed iron reagent shows staining in the periphery of granules of many cell types at the electron-microscope level. This staining probably indicates the sulfated complex carbohydrate shown biochemically by the authors. At the light-microscope level, histochemical methods show sulfated mucosaccharides in thyrotropin, but no acid complex carbohydrate in acidophiles. Can the cohesiveness of isolated granules be correlated with these differences, or do the acid glycoconjugates possibly have some other biological significance than the proposed regulation of diffusion of secretory products into Golgi cisternae and their organization in secretory granules?

MELDOLESI: No biochemical data are available for secretory granules of cells found in small numbers in the pituitary gland. I do not think that it is true that the higher the sulfated complex carbohydrate the stronger the internal interaction of granule components. One cannot disregard the possible importance of the majority components of the granules — the hormones, which, of course, are different in difference granules. The cytochemistry technique may be limited in that the reagents may not penetrate to the core of the secretory granules.

REFERENCES

Argent. B. E., Smith, R. K., and Case, M. R., 1975, The distribution of bivalent cation–stimulated adenosine triphosphate hydrolysis and calcium accumulation by subcellular particles, *Biochem. Soc. Trans.* **3**:713.

Baumrucker, C. R., and Keenan, T. W., 1975, Membranes of mammary gland. X. Adenosine triphosphate–dependent calcium accumulation by Golgi apparatus–rich fractions from bovine mammary gland, *Exp. Cell Res.* **90**:253.

Berg, N. B., 1978, Sulfate metabolism in the exocrine pancreas. II. The production of sulfated macromolecules by the mouse exocrine pancreas, *J. Cell Sci.* **31**:199.

Berg, N. B., and Austin, B. P., 1976, Intracellular transport of sulfated macromolecules in parotid acinar cells, *Cell Tissue Res.* **165**:215.

Blobel, G., and Dobberstein, B., 1975, Transfer of proteins across membranes. I. Presence of proteolytically processed and unprocessed nascent immunoglobulin light chains on membrane-bound ribosomes of murine myeloma, *J. Cell Biol.* **67**:835.

Campbell, P. N., and Blobel, G., 1976, The role of organelles in the chemical modification of the primary translation products of secretory proteins, *FEBS Lett.* **72**:215.

Clemente, F., and Meldolesi, J., 1975, Calcium and pancreatic secretion. I. Distribution of calcium and magnesium in the acinar cells of the guinea pig pancreas, *J. Cell Biol.* **65**:88.

De Camilli, P., Zanini, A., Giannattasio, G., and Meldolesi, J., 1979, Synthesis, intracellular transport, packaging and release of growth hormone and prolactin in normal and tumoral pituitary cells, in: *Pituitary Microadenomas* (G. Faglia, M. A. Giovanelli, and R. M. MacLeod, eds.), Academic Press, New York (in press).

Evans, G. A., Hucko, J., and Rosenfeld, M. G., 1977, Preprolactin represents the initial product of prolactin mRNA translation, *Endocrinology* **101**:1807.

Farquhar, M. G., 1977, Secretion and crinophagy in prolactin cells, in: *Comparative Endocrinology of Prolactin* (H. D. Dellmann, J. A. Johnson, and D. M. Klachko, eds.), pp. 37–94, Plenum Press, New York.

Farquhar, M. G., 1978, Recovery of surface membrane in anterior pituitary cells: Variations in traffic detected with anionic and cationic ferritin, *J. Cell Biol.* **77**:R35.

Farquhar, M. G., Skutelski, E. H., and Hopkins, C. R., 1975, Structure and function of the anterior pituitary and dispersed pituitary cells: *In vitro* studies, in: *The Anterior Pituitary* (A. Tixier-Vidal and M. G. Farquhar, eds.), pp. 84–128, Academic Press, New York, San Francisco, and London.

Farquhar, M. G., Reid, J. J., and Daniell, L. W., 1978, Intracellular transport and packaging of prolactin: A quantitative electron microscope autoradiographic study of mammotrophs dissociated from rat pituitaries, *Endocrinology* **102:**296.

Flashner, Y., and Schramm, M., 1977, Retention of amylase in the secretory granules of parotid gland after extensive release of Ca^{++} by ionophore A-23187, *J. Cell Biol.* **74:**789.

Giannattasio, G., and Zanini, A., 1976, Presence of sulfated proteoglycans in prolactin secretory granules isolated from the rat pituitary gland, *Biochim. Biophys. Acta* **439:**349.

Giannattasio, G., Zanini, A., and Meldolesi, J., 1975, Molecular organization of rat prolactin granules. I. *In vitro* stability of intact and "membraneless" granules, *J. Cell Biol.* **64:**246.

Giannattasio, G., Zanini, A., and Meldolesi, J., 1979a, Complex carbohydrates of secretory organelles, in: *Complex Carbohydrates of Nervous Tissue* (R. U. Margolis and R. K. Margolis, eds.), pp. 327–345, Plenum Press, New York.

Giannattasio, G., Zanini A., Rosa, P., Meldolesi, J., Margolis, R. U., and Margolis, R. K., 1979b, Molecular organization of prolactin granules. III. Intracellular transport of glycosaminoglycans and glycoproteins of bovine prolactin granule matrix *J. Cell Biol.* (submitted).

Herzog, V., and Farquhar, M. G., 1977, Uptake of horseradish peroxidase in the Golgi complex of stimulated acinar cells of the parotid gland, *Proc. Natl. Acad. Sci. U.S.A.* **74:**5073.

Herzog, V., and Miller, F., 1979, Membrane retrieval in secretory cell systems, in: *Secretory Mechanisms* (C. C. Duncan and C. R. Hopkins, eds.), Cambridge University Press (in press).

Jamieson, J. D., 1979, Intracellular transport and discharge of secretory proteins: Present status and future perspectives, in: *Transport of Macromolecules in Cellular Systems* (S. Silverstein, ed.), pp. 273–288, Dahlem Konferenzen, Berlin.

Jamieson, J. D., and Palade, G. E., 1968, Intracellular transport of secretory proteins in the pancreatic exocrine cell. IV. Metabolic requirements, *J. Cell Biol.* **39:**589.

Johnson, R. G., and Scarpa, A., 1976, Ion permeability of isolated chromaffin granules, *J. Gen. Physiol.* **68:**601.

Kreibich, G., Ulrich, B. L., and Sabatini, D. D., 1978a, Proteins of rough microsomal membranes related to ribosome binding. I. Identification of ribophorins I and II, membrane proteins characteristic of rough microsomes, *J. Cell Biol.* **77:**464.

Kreibich, G., Freienstein, C. M., Pereyra, B. N., Ulrich, B. L., and Sabatini, D. D., 1978b, Proteins of rough microsomal membranes related to ribosome binding. II. Cross-linking of bound ribosomes to specific membrane proteins exposed at the binding sites, *J. Cell Biol.* **77:**488.

Labrie, F., Pelletier, G., Lemay, A., Borgeat, P., Barden, N., Dupont, A., Savary, M., Côté, J., and Boucher, R., 1973, Control of protein synthesis in anterior pituitary gland, *Acta Endocrinol. Suppl.* **180:**301.

Lingappa, V. R., Devillers-Thiery, A., and Blobel, G., 1977, Nascent prehormones are intermediates in the biosynthesis of authentic bovine pituitary growth hormone and prolactin, *Proc. Natl. Acad. Sci. U.S.A.* **74:**2432.

Maurer, R. A., and McKean, D. J., 1978, Synthesis of preprolactin and conversion to prolactin in intact cells and a cell-free system, *J. Biol. Chem.* **253:**6315.

Meldolesi, J., Marini, D., and Demonte-Marini, M. L., 1972, Studies on *in vitro* synthesis and secretion of growth hormone and prolactin. I. Hormone pulse labeling with radioactive leucine, *Endocrinology* **91:**802.

Meldolesi, J., Borgese, N., De Camilli, P., and Ceccarelli, B., 1978, Cytoplasmic membranes and the secretory process, in: *Membrane Fusion* (G. Poste and G. L. Nicolson, eds.), pp. 509–627, Elsevier/North-Holland, Amsterdam.

Palade, G. E., 1975, Intracellular aspects of the process of protein secretion, *Science* **189**:347.

Pelletier, G., 1973, Secretion and uptake of peroxidase by rat adenohypophyseal cells, *J. Ultrastruct. Res.* **43**:445.

Rambourg, A., and Racadot, J., 1968, Identification en microscopie électronique de six types cellulaires dan l'antéhypophyse du rat a l'aide d'une technique de coloration par le mélange acide chromique–phosphotungstique, *C. R. Acad. Sci.* **266**:153.

Reggio, H., and Dagorn, J. C., 1978, Ionic interactions between bovine chymotrypsinogen A and chondroitin sulfate A.B.C., *J. Cell Biol.* **78**:951.

Reith, E. J., 1976, The binding of calcium within the Golgi saccules of the rat odontoblasts, *Am. J. Anat.* **147**:267.

Schneider, Y. J., Tulkens, P., and Trouet, A., 1977, Recycling of fibroblast plasma membrane antigens internalized during endocytosis, *Trans. Biochem. Soc.* **5**:1164.

Selinger, Z., Naim, E., and Lasser, M., 1970, ATP-dependent calcium uptake by microsomal preparations from rat parotid and submaxillary glands, *Biochim. Biophys. Acta* **137**:326.

Steinman, R. M., Scott, E. B., and Cohn, Z. A., 1976, Membrane flow during pinocytosis: A stereological analysis, *J. Cell Biol.* **68**:665.

Stoeckel, M. E., Hinderlag-Gertner, C., Dellmann, H. D., Poste, A., and Stutinsky, F., 1975, Subcellular distribution of calcium in mouse hypophysis. I. Calcium distribution in the adeno- and neurohypophysis under normal conditions, *Cell Tissue Res.* **157**:307.

Swearingen, K. C., 1971, Heterogeneous turnover of adenohypophyseal prolactin, *Endocrinology* **89**:1380.

Zanini, A., and Giannattasio, G., 1974, Molecular organization of rat prolactin secretory granules, in: *Advances in Cytopharmacology*, Vol. 2 (B. Ceccarelli, F. Clementi, and J. Meldolesi, eds.), pp. 329–339, Raven Press, New York.

Zanini, A., Giannattasio, G., Nussdorfer, G., Margolis, R. K., Margolis, R. U., and Meldolesi, J., 1979, Molecular organization of prolactin granules. II. Characterization of glycosaminoglycans and glycoproteins of bovine prolactin granule matrix *J. Cell Biol.* (submitted).

6

Separated Somatotrophs: Their Use in Vitro and in Vivo

W. C. Hymer, R. Page, R. C. Kelsey, E. C. Augustine, W. Wilfinger, and M. Ciolkosz

1. General Introduction

Since the original discovery of growth hormone (GH) by Drs. Herbert Evans and Joseph Long in 1921, this molecule has been a subject of keen interest to biochemical endocrinologists. There are a number of reasons for this interest. To state but a few: (1) the molecule does not have a single, well-defined target organ, and its biological and metabolic effects are quite numerous (e.g., regulation of growth as well as metabolism of protein, carbohydrate, and fat); (2) given the availability of several assay systems, the molecule is well suited to structure–function studies; (3) the molecule has an established clinical usefulness for the treatment of hypopituitary children in addition to its suggested use in the treatment of ulcers, muscular dystrophy, and coronary-prone hypercholesterolemic patients (Li, 1975); and (4) the cell that produces GH, viz., the pituitary somatotroph, offers an interesting model for investigation of mechanisms of intracellular processing and secretion of peptide hormones in addition

W. C. Hymer, E. C. Augustine, and *M. Ciolkosz* • Department of Biochemistry and Biophysics, The Pennsylvania State University, University Park, Pennsylvania 16802 *R. Page* • Hershey Medical Center, Hershey, Pennsylvania 17033 *R. C. Kelsey* • East Stroudsburg State College, East Stroudsburg, Pennsylvania 18301 *W. Wilfinger* • University of Cincinnati Medical School, Cincinnati, Ohio 45221

to the study of the regulation of somatotroph function. This chapter deals with the last issue.

There are obviously many ways to approach the study of somatotroph function. The results from these approaches, taken together, are beginning to give us a fairly good picture of how this cell type works. However, as is usually the case, one must constantly be aware of the limitations of the experimental methods used. For example, the elegant studies by Tannenbaum and Martin (1976) have revealed a remarkable periodicity in circulating GH in the rat (Fig. 1). *In vivo* administration of secretory agents (e.g., arginine, TRH, prostaglandin) will, in certain species, considerably modify such secretory patterns. Usually, such studies employ an immunological procedure for the determination of circulating GH levels, and as pointed out by Li (1977), ". . . in order for such methods to have absolute validity, it is necessary that the antigenic site and the active center in the molecule be identical." This may not always be so. A case in point arises from the work of Ellis and Grindeland (1974), in which it was demonstrated that the concentration of rat or human plasma GH measured by the tibia test was quite different from the value obtained by radioimmunoassay (RIA). In a very real sense, progress toward the identification and elucidation of the structure of GH-releasing factor (GRF) has been hampered by this very problem. Furthermore, some have argued that concurrent quantitation of GH-receptor activity in GH-responsive tissues is required for meaningful interpretation of plasma GH levels. Nevertheless, few would disagree that measurement of plasma GH levels is the key to understanding somatotroph function in normal and disease states.

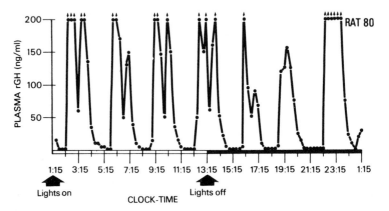

Figure 1. Twenty-four hour secretory pattern of growth hormone (GH) release in the rat. From Tannenbaum and Martin (1976).

To cite another example of different approaches used to study mechanisms of GH synthesis and secretion, the hemipituitary incubation procedure originated by Saffran and Schally (1955) was popular for many years. Indeed, the recent studies by Stachura *et al.* (1978) continue to provide interesting information on intracellular GH pools and their control by hypothalamic factors. Nevertheless, the hemipituitary procedure suffers from well-documented limitations such as variation among individual glands and poor diffusion of essential nutrients, which can rapidly lead to central tissue necrosis. A particularly good example of this latter effect, taken from Farquhar *et al.* (1975), is shown in the radioautogram in Fig. 2. This hemipituitary was incubated for 4 hr in a mixture of tritiated amino acids (100 μCi/ml). A gradient of decreasing grain concentration is obvious, and clearly documents incomplete penetration of substrate. Yet, under these same conditions, many investigators have shown that incorporation rates of radiolabeled amino acids into protein, and nucleotides into nucleic acids, are linear up to 4 hr.

In the early 1970s, a few different groups began to utilize suspensions of single cells, obtained by enzymatic dispersion of pituitary tissue, as models to study hormone synthesis and secretion. By 1977, studies utilizing such preparations were not hard to find. A critical examination of the literature, however, reveals disturbingly few common threads in the methodologies used and results obtained. In terms of the dissociation procedures themselves, a careful evaluation of the importance of the following variables, as they relate to the quality of the product obtained, has not yet been done: (1) type, concentration, and purity of enzyme(s) used (e.g., trypsin, collagenase, neuraminidase); (2) ionic composition of dissociation medium [e.g., Krebs–Ringer bicarbonate glucose (KRBG), minimum essential medium (MEM), others with and without bovine serum albumin (BSA) and with and without divalent cations] used; and (3) severity of the necessary mechanical procedures involved (e.g., tissue mincing, tissue dispersion with a Pasteur pipette) and physical nature of the dissociation used (e.g., vessel or stirring bars or both). The quality and yield of the cells obtained will obviously be dependent on such variables. In general, most investigators report cell yields of 1–2.5 $\times$ 10^6 cells/rat pituitary ($\approx$ 40–70% on a DNA basis) and greater than 90% viability (usually determined by the relatively insensitive, but simple, trypsin blue exclusion test). Although not usually described in the literature, problems relating to cell lysis, blebbing, and reaggregation sometimes occur. Another difficulty relates to the responsiveness of acutely dispersed cells to hormone-releasing agents. This problem will be discussed later in this chaper as well as at various points throughout this volume. Suffice it to say at this point that most investigators believe that acutely dispersed cells are relatively insensitive to secretagogues because of the damage done to

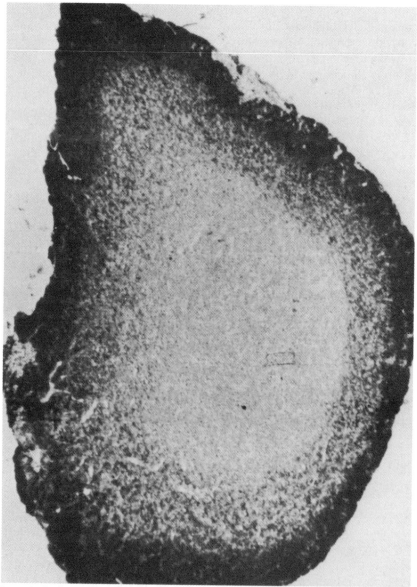

Figure 2. Radioautogram of a rat hemipituitary incubated 4 hr in a mixture of tritiated amino acids (100 μCi/ml). Note gradient of decreasing grain concentration, indicating incomplete penetration of amino acids. From Farquhar *et al.* (1975).

cell-surface receptors during the dissociation process. Accordingly, various "receptor recovery periods" in serum-supplemented media are usually called for prior to challenge with the stimulatory agent in serum-free media. It is easy to understand how different results can be obtained from quite similar experiments. Nevertheless, dispersed cell systems obviously circumvent many of the problems encountered with the use of intact tissue *in vitro*. Given the qualifications listed above, a cursory review of the literature will show that cells are extremely useful for studying synthesis and release of pituitary hormones. In addition, they have the added advantage of serving as the starting material for cell-separation studies.

This list of approaches to the study of somatotroph function is obviously incomplete. Yet it should give the reader an idea of what has been done, as well as some of the problems associated with such approaches. In this chapter, our own approaches will be considered not only in terms of the kind of data they generate, but also for their potential usefulness in future studies on the pituitary.

2. Separation of Somatotrophs from Dispersed Cell Suspensions

Somatotrophs make up approximately 30–35% of the cell types found in freshly prepared pituitary cell suspensions. Accordingly, we initially directed our efforts toward separation of this cell type. Of the several cell-separation techniques available, the most fully developed were those that utilized differences in cell size or in cell density as separation parameters.

2.1. Velocity Sedimentation at Unit Gravity (1g)

2.1.1. Theoretical Principle

In this technique, sometimes referred to as STAPUT, the cells settle through a shallow stabilizing gradient of some nonviscous material, usually BSA, sucrose, or serum. The theoretical basis for this separation method is considered in the original paper of Peterson and Evans (1967) and in the later study of Miller and Phillips (1969). As pointed out by these latter authors, the sedimentation of a spherical particle (which acutely dispersed cells are) in a shallow gradient depends primarily on (1) the density difference between the particle and its surrounding medium and (2) the square of the radius of the particle. Since the latter may often represent a sizable quantity, most investigators using this technique suggest that separations are based primarily on differences in cell size. The pituitary cell, however, may offer a special case, since its cytoplasmic secretion granules probably convey appreciable density (Hymer, 1975). In fact, we have shown that the degree of granulation in the prolactin cell can profoundly affect its sedimentation rate in the 1G system (Hymer *et al.*, 1974).

2.1.2. Operation of the Sedimentation Chamber and Representative Results Obtained

The chamber and gradient generating devices are shown in Fig. 3. The lucite sedimentation chamber has a 500-ml capacity, is 11 cm in o.d. and 11 cm high, and was designed for the separation of relatively small numbers of cells. Experiments were carried out at room temperature under conditions where mechanical vibrations and thermal variations were kept to a minimum to maintain stability of the shallow gradient. The gradient was generated with two beakers (B), one containing 300 ml of 1% and the other 300 ml of 3% BSA prepared in Medium 199 at pH 7.3. Vessel (C) contained 60 ml of 0.3% BSA in Medium 199. The gradient material entered the chamber (A) through a three-way valve (E). The flow rate was regulated at (F), and gradient solutions were mixed by magnetic stirrers (D). Freshly dispersed pituitary cells were applied on top of the threaded baffle (G) through the opening in the top of the chamber; the cells were then quickly (3–5 min) lifted into the chamber by the incoming gradient

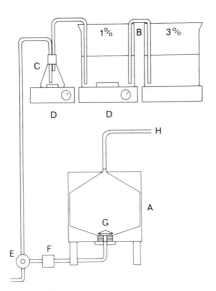

Figure 3. Apparatus for sedimentation of adenohypophyseal cells at unit gravity. See the text for details of operation. From Hymer (1975).

solution, after which time the flow rate was decreased such that it took 1 hr to fill the chamber. During this time, cells settled into the gradient due to gravity. After the chamber had been filled, an additional settling time of 15–85 min was used, depending on the experiment. During this time, four to six cell bands were consistently seen at different levels in the gradient. Cell fractions were collected from the top of the chamber (H) by displacement of the gradient with a 7% sucrose solution introduced from a vessel (not shown in Fig. 3) via the three-way valve (E).

Under the conditions described above, the BSA under the initial cell layer rose rapidly from 0.3 to 1% in the first 60 ml of the gradient. It then became linear with a slope of 0.0037% BSA/ml for the remaining distance, and total gradient of 0.3–2.4% BSA was generated.

Initially, the sample layer was smoothly distributed over the gradient as a thin layer because of the design of the baffle and conical bottom of the chamber. As originally indicated by Peterson and Evans (1967) and later stressed by Miller and Phillips (1969), an important factor in determining the degree of cell enrichment is the concentration of cells in the original suspension. Above a certain critical concentration, rapidly settling cells will cause vertical streams at the bottom of the initial cell band, presumably due to an increased density at that point. This cell-streaming phenomenon is undesirable because it increases the width of the initial cell band, thereby diminishing resolution. Under the conditions described above, streaming occurred above a concentration of 1.4×10^6 cells/ml. In the usual experiment, 10×10^6 pituitary cells were applied, but effective enrichments have been achieved with as many as 20×10^6 cells. Gradients of sucrose, sucrose plus BSA, and fetal calf serum have also been tried, but BSA in Medium 199 has given the best results.

After 1.25 hr sedimentation, cells in the original suspension were distributed as shown in Fig. 4 (top panel). A major cell peak at fraction 12 (i.e., 120 ml into the gradient), with a progressive decrease in cell numbers to fraction 32, was observed. The positions of these cell peaks were virtually identical from run to run; however, the number of cells in each peak was variable. Routinely, 99% of the RBCs were recovered in fraction 4. A majority of the somatotrophs sedimented to an area encompassed by fractions 18–32 (Fig. 4, bottom panel). From 60 to 70% of the cells in this region of the gradient were somatotrophs. Chromophobes, basophils, and unknowns contaminated this fraction to about an equal extent ($\approx 10\%$). In our experience, it has not been possible to further purify somatotrophs by this technique. Nevertheless, a 2 to 3-fold enrichment is useful for certain studies. We have, for example, obtained similar results with the separation of somatotrophs from the human pituitary gland by this simple procedure (Hymer *et al.*, 1976).

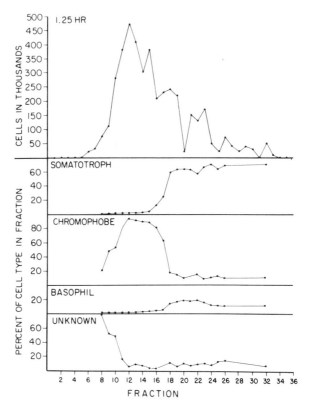

Figure 4. *Top panel:* Distribution profile of separated pituitary cells after 1.25 hr sedimentation at unit gravity. *Bottom panels:* Localization of specific cell types expressed as percentages of cell types in consecutive 10-ml fractions. From Hymer *et al.* (1972).

2.2. Continuous Density-Gradient Centrifugation

We next used procedures described by Shortman (1968), originally developed for separation of antibody-forming precursor cells, to further purify the somatotrophs in fractions 18–32 from the 1G step (see Fig. 4). The pooled cells were dispersed in an isotonic, linear gradient of BSA (14–28%) and centrifuged at 2000g for 45 min. Separations achieved by this procedure were based entirely on differences in buoyant cell density. After the cells were centrifuged to isodensity, three cell peaks were consistently recovered (Fig. 5). The first peak was at density 1.0601 ± 0.0007 g/cm^3; the second, at 1.0695 ± 0.0007 g/cm^3; the third, at 1.0761 ± 0.0007 g/cm^3 (S.E.M., 6 experiments). Differential counts of the cells banding between 1.0705 and 1.0850 g/cm^3 revealed that 80–92% were somatotrophs. Approximately 65% of the somatotrophs applied to the gradient had a density greater than 1.070 g/cm^3 and could therefore be separated in good purity. Those somatotrophs banding at lower density appeared partially degranulated by both light and electron microscopy (see later). Interestingly, a majority of the basophils that cosedimented with the somatotrophs in the 1G system banded at densities 1.0575–1.0680 g/cm^3. In some experiments, basophils made up 50% of the cells in this density region. We speculate that the apparent density differences between these cell types may be partially explained on the basis of granule densities in their cytoplasm. Chromophobes appear to be the least dense cell type, since they were found at the top of the gradient.

The results of morphological studies have shown that the isolated somatotrophs retain normal morphology at both the light- and electron-microscope levels (Hymer *et al.*, 1972; Snyder, G., *et al.*, 1977). These results, coupled with the findings that the somatotrophs (1) incorporated

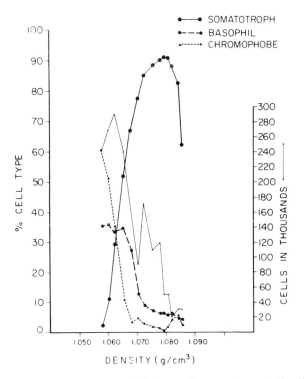

Figure 5. Distribution and percentages of pituitary cell types after centrifugation in a linear gradient of 14–28% BSA. These cells were obtained by pooling fractions 18–32 (see Fig. 4) from the 1G system and concentrating them prior to density-gradient centrifugation. From Hymer *et al.* (1972).

radiolabeled amino acids into protein in linear fashion and (2) released immunoassayable GH when challenged with dibutyryl cAMP and partially purified preparations of GRF (Kraicer and Hymer, 1974), have documented that these preparations provide a valuable model to study intracellular mechanisms of GH release. On the negative side, however, a major disadvantage of the methods used was simply that they took too long (9 hr after the rats were killed) to obtain the purified cells.

2.3. Discontinuous Density-Gradient Centrifugation

In an effort to shorten the separation procedures, we took advantage of our finding that somatotrophs had a density in the range of 1.070–1.076 g/cm^3. The procedure that was eventually developed involved centrifugation of freshly dispersed cells through two consecutive discontinuous density gradients of BSA. The first consisted of two BSA solutions of differing densities: lower, 1.087 g/cm^3; upper, 1.068 g/cm^3. Freshly dispersed cells were layered on the upper solution (1.068 g/cm^3) and centrifuged. Those cells that collected at the 1.068–1.087 g/cm^3 BSA interface were harvested and recentrifuged through a second discontinuous gradient consisting of densities 1.071 and 1.087 g/cm^3. Cells that sedimented to the 1.071–1.087 g/cm^3 interface were found to consist of 80–92% somatotrophs (for further details, see Snyder, G., et al., 1977). The total time required to obtain these cells was approximately 4 hr after the rats were killed.

Recovery of immunoassayable GH from such gradients showed that approximately 40% of the hormone originally applied to the gradient was associated with the somatotrophs of density greater than 1.071 g/cm^3. Approximately another 40% was associated with lightly granulated somatotrophs that were not sufficiently dense to break through the original layering interface, i.e., that had a density of less than 1.068 g/cm^3. The remaining GH was associated with damaged somatotrophs in the cell pellet. To differentiate between the two cell bands containing GH, we have referred to the less dense cell as the type I somatotroph, and the more dense as the type II somatotroph.

Obviously, this shorter procedure has proved more convenient in terms of testing the responsiveness of the cells to various GH secretagogues.

3. Release of Growth Hormone from Isolated Somatotrophs

We originally used methodologies associated with short-term incubation systems to study GH release from the isolated somatotrophs. Basically, this involved incubation in defined media (Medium 199 or Hanks balanced salts, both containing 0.1% BSA) in the presence or the absence of secretagogues in 95% : 5% air : CO_2 mixture at 37°C. After incubation, the tubes were centrifuged, and GH in the cells, as well as the supernatant fraction (representing released GH), were determined by highly specific RIA. Data, expressed as percentage release [i.e., (GH in medium/GH remaining in cells after incubation + GH in medium) × 100], have been taken as a measure of the responsiveness of the cells to each agent tested.

We were first interested to compare levels of *basal* GH release. We found that GH release from incubated glands, dispersed unpurified cells, and isolated somatotrophs was, after 1 hr, 2.9 ± 0.2, 4.3 ± 0.5, and 9.2 ± 2.6%, respectively (Snyder, G., and Hymer, 1975). This result suggested that dispersed somatotrophs were "more leaky" than their tissue counterparts. It also suggested that the longer times required for separation might also elevate basal release.

3.1. Short-Term Incubation in the Presence and in the Absence of Secretagogues

Data briefly described in this section come from several published papers, to which the interested reader is referred for additional details.

3.1.1. Dibutyryl Cyclic Adenosine Monophosphate (dbcAMP)

We have used doses of 0.03–6 mM dbcAMP in different published studies. In one, the 6-mM dose caused a 2.8-fold increase in GH release from hemipituitaries and a 2.4-fold increase from type II somatotrophs in the same study (Snyder, G., and Hymer, 1975). In another, the 6-mM dose evoked a 2.1- and 2.3-fold increase from somatotrophs isolated after 1G and continuous density-gradient centrifugation, respectively (Kraicer and Hymer, 1974). These responses were all highly statistically significant, and show that this cyclic nucleotide has GH-releasing activity from both intact glands and isolated cells. In another study (Snyder, G., *et al.*, 1977), we compared the responsiveness of type I and II somatotrophs to doses of cyclic nucleotide ranging from .03 to 3.0 mM for periods up to 2 hr. As shown in Fig. 6, release was usually dependent on the quantity of nucleotide added. Responses were often obtained with the 0.3-mM dose. When evaluated on a percentage release basis, the type I cells often appeared to release more of their intracellular stores than their denser counterparts.

3.1.2. High K+

Several investigators have shown that a 5-fold elevation in K^+ in the medium will promote GH release from hemipituitary tissue, presumably through an effect on membrane permeability. In several experiments,

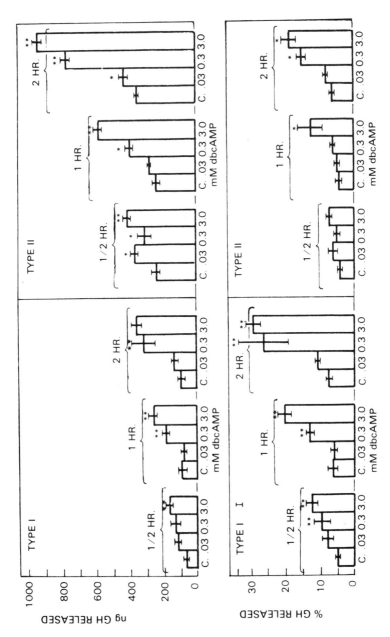

Figure 6. Effect of dbcAMP from type I and type II somatotrophs in short-term incubation. Data are presented as the average nanograms GH released/50,000 cells (*top*) and as the average percentage of the total recovered GH released to the medium (*bottom*). Vertical brackets represent standard error of the mean. Data were analyzed by analysis of variance and Dunnett's Multiple Comparison Test. *$p < 0.05$; **$p < 0.01$. From G. Snyder *et al.* (1977).

however, we were unable to show that high K^+ had any effect whatsoever on GH release. We must consider the possibility that the permeability characteristics of the somatotrophs might be altered during the dissociation procedure. More work is required on this aspect of the problem.

3.1.3. Somatostatin (SIF)

Type I and II somatotrophs incubated for 30, 50, and 120 min in media containing 10^{-11}, 10^{-9}, or 10^{-7} M SIF had no effect on GH release. However, SIF was capable of virtually completely blocking the dbcAMP-stimulated GH release from the isolated somatotrophs (Snyder, G., *et al.*, 1977). Since SIF presumably acts via receptors on the plasma membrane of the somatotroph, its failure to suppress basal GH release from acutely dispersed cells might be attributed to trypsin-induced damage of SIF receptors. Stachura (1976) has reported that SIF can block dbcAMP-stimulated GH release from hemipituitary tissue *in vitro*. Our results with cells are obviously similar. Although we have no critical data to explain these different results, it is possible that SIF has more than one site of action.

3.1.4. Thyroxine (T₄) and Hydrocortisone (HC)

At doses of 10^{-6} to 10^{-8} M, T_4 and HC had no effect on GH release from isolated somatotrophs in short-term incubation.

3.1.5. GH-Releasing Factor (GRF)

In a few experiments, a partially purified extract of ovine hypothalamus, rich in GRF, was added to media containing isolated somatotrophs. The characteristics of this preparation, obtained by Fawcett, have been described (Milligan *et al.*, 1972). At doses that evoked significant hormone release from hemipituitaries, release from isolated somatotrophs was also significantly elevated (25–55%) (Kraicer and Hymer, 1974). Such data, while preliminary, support the notion that the isolated cells have GRF receptors. When the structure of GRF is finally elucidated, progress in this area will obviously be more rapid.

3.2. Cell Culture in the Presence and in the Absence of Secretagogues

The interested reader is referred to previously published material for additional data covering the topics in this section.

3.2.1. "Basal" GH Release

When type I or II somatotrophs were cultured for a period of 1 month, the content of GH in both the medium and the cells fell in linear fashion with time. The data in Table I, which represent the average of three experiments, give an accounting of GH recoveries from the lightly and heavily granulated cells during the 30-day culture period. It is obvious that the dense somatotrophs simply release their intracellular GH stores in culture. Thus, recovery of hormone was 83% (relative to that originally plated in the dish), and 99.8% of the recovered hormone was found in the medium. On the other hand, the lightly granulated somatotrophs produced hormone in culture. Recovery of GH was 375%, and 99.5% of this GH was also found in the medium. All the GH produced by the type I cells was produced in the first week in culture. Such results demonstrate a fundamental difference in the GH synthetic–secretory capabilities of these two preparations. In other experiments, we studied the histology and ultrastructure of the cultured somatotrophs. The lightly granulated cells, which stained a very pale yellow (Herlant's stain) at the start of culture, stained an intense yellow after 1 week. At the electron-microscope level, these cells became more heavily granulated during the early culture period. The heavily granulated cells, on the other hand, tended to be somewhat smaller and contained fewer granules than at the onset of culture.

3.2.2. Dibutyryl Cyclic Adenosine Monophosphate (dbcAMP)

As with acutely dispersed cells (see Section 3.1.1), addition of dbcAMP (final concentration 3 mM) to the culture medium provoked GH release from both lightly and heavily granulated cells. This response was especially noticeable during the first 3 days of culture (2.3- and 1.2-fold increases for type I and II, respectively). At the end of culture, intracellular

Table I. Summary of Growth Hormone Kinetics through 30 Days in Culture[a]

Parameter	Type I somato- trophs	Type II somato- trophs
Initial GH seeded (ng)	1815	13,450
Total GH released after 30 days (ng)	6795	11,185
Total GH remaining in cells after 30 days (ng)	30	22
Total GH recovered in medium and cells (ng)	6825	11,207
Recovery (%)	375	83

[a] From G. Snyder et al. (1977).

Table II. Effect of Somatostatin on Growth Hormone Release and Production in Cultured Type I and Type Somatotrophs[a]

Culture	0- to 3-Day medium (ng GH)	4- to 6-Day medium (ng GH)	Total GH released (ng)	Cell GH content (ng)	Total GH (ng)
Type I					
Control	1154 ± 100	1596 ± 122	2750 ± 108	2128 ± 154	5180 ± 44
10^{-7} M SIF	453 ± 26[b]	846 ± 71[b]	11299 ± 96[b]	4686 ± 375[b]	5985 ± 470
Type II					
Control	2157 ± 217	682 ± 186	3140 ± 346	2612 ± 93	5752 ± 428
10^{-7} M SIF	1187 ± 217[b]	565 ± 39	1752 ± 88[b]	3451 ± 47[b]	5203 ± 48

[a] From Snyder *et al.* (1977).
Data are presented as ng GH 50,000 cells and are the means of three replicate samples ± S.E.M..
Statistical differences between the means of control and SIF conditions were determined by Student's
t test.

stores were virtually depleted (90%) in the lightly granulated cells, but not nearly to the same extent ($\approx$50%) in the denser somatotrophs. These responses were roughly similar to those seen with acutely dispersed cells, in which, the reader will recall (see Section 3.1.1), the type I cells tended to release more of their intracellular stores than their denser counterparts (see Fig. 6). Our experimental design did not permit analysis of precisely when the cyclic nucleotide stimulated GH release. However, it is likely that it happened in the very early stages of the first 3-day period, since this agent was ineffective in the second 3-day period.

3.2.3. Somatostatin (SIF)

While SIF was ineffective in blocking GH release from somatotrophs in short-term incubation (see Section 3.1.3), the data in Table II show that it was effective on somatotrophs in culture. Thus, release from the type I and II cells was significantly depressed (53 and 45%, respectively) over the 6-day culture period with 10^{-7} M SIF. Suppression tended to be greater during the early culture period. The type I cells seemed more responsive to SIF. As shown in Table II, intracellular GH was significantly elevated. However, the quantity of total GH recovered was not different from the vehicle-treated controls, indicating the SIF acted on GH secretion and not on GH synthesis.

Since SIF presumably works via a receptor-mediated event, it would seem logical to invoke the notion of receptor regeneration during culture. Brazeau *et al.* (1973) originally utilized the pituitary cell culture technique in the assay of SIF. They reported that the extent of 10^{-7} to 10^{-10} M SIF-induced inhibition of GH release was on the order of 60–90%. Our data

(Table II) are consistent with this earlier report. They also begin to provide an idea as to the responsiveness of the different somatotrophs to this interesting molecule.

3.2.4. Hydrocortisone (HC)

Some time ago, Tashjian's group reported that the addition of 10^{-6} M HC to cultures of rat GH_3 pituitary tumor cells increased GH production 4 to 5-fold (Bancroft *et al.*, 1969). About that same time, Kohler *et al.* (1968) also reported similar results in primary pituitary cell cultures. We were therefore interested to see the effect of this agent on isolated somatotrophs, and soon found that HC stimulated GH synthesis, but not release, when added to the medium of cultured rat pituitary cells over a 6-day period. This response occurred over a wide dose range of the steroid (10^{-6} to 10^{-10} M). As shown in Fig. 7, a similar response was found in the lightly granulated somatotrophs. This result was highly repeatable. Interestingly, HC had no effect on GH production in the type II cells.

3.3. Effects of Medium Formulation and Serum on Somatotroph Function in Culture

The experiments described in Section 3.2 involved culture of cells in Medium 199 supplemented with 20% fetal calf serum. In a recently completed study (Wilfinger *et al.*, 1979), we tested the effects on release of prolactin (PRL) and GH from rat pituitary cells of the following: (1) lower serum content (5%); (2) horse vs. calf serum; and (3) 20 different types of commercially available media. In these experiments, 25,000 cells were cultured in dishes for 9 days, and an *in vitro* production index of hormone was calculated by determining the ratio of the hormone recovered from both medium and cells during culture divided by the quantity of hormone originally seeded into the dish. In this study, we found that medium formulation had a marked effect on PRL synthesis–secretion. However, as shown in Table III, neither medium formulation nor serum type had such effects on somatotroph function. Calf serum seemed slightly superior to horse serum. The effects of different concentrations of fetal calf serum in the 20 media on somatotroph function would be of interest. Such studies have not yet been done.

In summary, our data, as well as those of many others, show that it is extremely difficult to maintain actively secreting somatotrophs in conventional monolayer for any significant period of time. A good part of this difficulty can be attributed to our lack of knowledge about the chemical nature of native GRF.

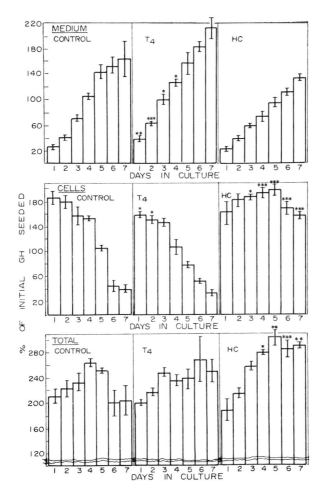

Figure 7. Effects of 10^{-6} M T_4 and 10^{-6} M HC on release (*top*), cell content (*middle*), and total recovered GH (*bottom*) from type I somatotrophs in culture. Data represent the means of three cultures ± S.E.M. (vertical brackets) for each two experiments and are expressed as the percentage of GH seeded (1184 ng GH, mean of three experiments) at the beginning of the culture. Comparisons are made by Student's *t* test on a daily basis between treated and control cultures. *$p < 0.05$; **$p < 0.01$; ***$p < 0.001$.

3.4. Release of Growth Hormone from Perfused Somatotrophs

In 1974, Lowry (1974) described a simple, yet powerful technique to study the dynamics of hormone release from pituitary cells immobilized in columns of Biogel. The reader is referred to Chapter 26 of this volume for

Table III. Effect of Medium Formulation and Serum
Supplements on Growth Hormone Production during a 9-Day
Culture Period[a]

	Medium	GH recovered in medium and cells	
		Initial GH cell content	
		5% Horse serum	5% Calf serum
1.	MEM, Alpha	1.17 ± 0.11	$1.55^c \pm 0.14$
2.	MEM, Dulbecco's	1.12 ± 0.10	$1.62^c \pm 0.17$
3.	RPMI 1640	0.85 ± 0.11	1.14 ± 0.07
4.	MEMe	0.93 ± 0.12	1.32 ± 0.14
5.	Williams's E	0.98 ± 0.08	$1.36^c \pm 0.09$
6.	Ham's F12	0.89 ± 0.08	1.34 ± 0.17
7.	BMEe	0.79 ± 0.12	1.18 ± 0.14
8.	Neuman, Tytell	0.97 ± 0.12	$1.42^c \pm 0.19$
9.	McCoy's 5A	0.96 ± 0.11	1.34 ± 0.11
10.	BMEh	0.78 ± 0.12	$1.45^c \pm 0.20$
11.	Leibovitz L15	0.96 ± 0.10	$1.62^c \pm 0.18$
12.	Ham's F10	0.76 ± 0.13	1.21 ± 0.16
13.	MEMh	0.88 ± 0.09	$1.68^c \pm 0.34$
14.	CGM	0.89 ± 0.12	$1.38^c \pm 0.16$
15.	CMRL 1066	1.07 ± 0.10	$1.63^c \pm 0.13$
16.	NCTC 135	0.87 ± 0.07	
17.	Medium 199h	0.97 ± 0.13	1.17 ± 0.18
18.	Swim's S-77	0.72 ± 0.09	$1.36^c \pm 0.17$
19.	Waymouth 87/3	0.96 ± 0.10	$1.53^c \pm 0.17$
20.	Waymouth 752/1	0.98 ± 0.11	$1.74^c \pm 0.16$

[a] From Wilfinger *et al.* (1979). Data are means $\pm$ S.E.M. ($N = 6$).
[c] Indicates that this value is significantly different from other values with the same superscript.

operational details of this method. The methods that we use in this procedure are very similar to those reported by Lowry and Mulder.

We were interested to compare, in a general way, the kind of data obtained in Section 3.1. with the Biogel system. As seen in Fig. 8 (top), "basal" GH release from approximately 5×10^6 cells packed in a Biogel column was relatively constant 30–60 min after the experiment was started. The response of cells to 2-min pulses (Fig 8, bottom, width of arrow) of graded doses of dbcAMP showed the magnitude of GH release to be dose-dependent. The response to dbcAMP in this system has been highly repeatable. Perfusion with theophylline, a phosphodiesterase inhibitor, also promoted GH release. Moreover, hormone release induced by dbcAMP was potentiated by concurrent perfusion with theophylline. Finally, sodium butyrate alone had little effect on GH release. This

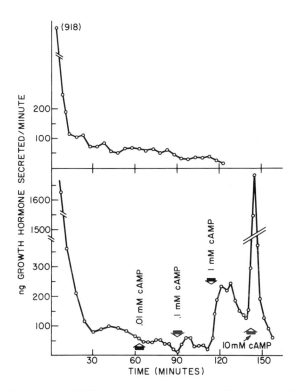

Figure 8. *Top:* Basal release of GH from approximately 5×10^6 pituitary cells immobilized in a Biogel P-2 column. *Bottom:* Response of cells in a Biogel column to 2-min pulses of graded doses of dbcAMP.

important experimental control indicated that the activity in dbcAMP could not be attributed to butyric acid (see Fig. 9).

The versatility of the Biogel column technique is also shown by the data in Fig. 10. In this experiment, three columns, each containing 5×10^6 cells, were run concurrently. The response to repeated 2-min pulses of 1 mM dbcAMP is evident in Fig. 10A. Inclusion of 10^{-7} SIF in the column buffer dampened basal release and prevented dbcAMP-stimulated GH release (Fig. 10B). Finally, perfusion with SIF immediately after dbcAMP dampened the response to the cyclic nucleotide. Eventually, however, large "rebound" peaks of GH were often encountered (Fig. 10C). Stachura (1977) showed, using a rat hemipituitary perfusion system in combination with SIF and dbcAMP, that the gland contained "an immediately release-able pool" of GH that was preferentially blocked by SIF. Despite this

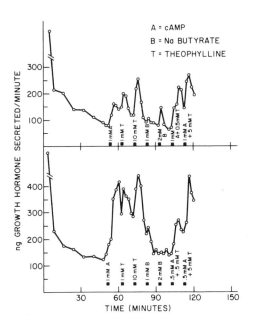

Figure 9. GH secretory profiles from 5×10^6 pituitary cells immobilized in Biogel after 1-min pulses of various agents (see the legend).

block, however, GH still continued to accumulate in the releaseable pool. GH release after SIF withdrawal then far exceeded the rate expected on the basis of dbcAMP stimulation alone. Obviously, the results in Fig. 10C are consistent with Stachura's results.

In summary, the Biogel column method has proved to be an extremely useful way to study the dynamics of GH release. Comparison of the data with those in short-term incubation (see Section 3.1) show, in general, quite similar results. The technique has a number of obvious advantages over static incubation systems. It can be expected that the method will gain acceptance and be much more widely used in future studies on the pituitary.

4. Functional Heterogeneity of Pituitary Cell Types

The evidence discussed in Section 3 indicated that lightly granulated somatotrophs responded to some GH-active agents rather differently from the heavily granulated somatotrophs. Farquhar *et al.* (1975) were probably the first to suggest that physiological heterogeneity might exist within the somatotroph pool. Their suggestion was based on differential labeling of somatotrophs (uptake of [^{14}C]leucine) as demonstrated by radioautography.

A few other studies have now been reported that also indicate that functional activity within a population of pituitary cells producing a given hormone may be heterogenous. For example, a few years ago we offered evidence that the magnitude of PRL-secreting activity by isolated mammotrophs during a 14-day culture period depended on the size and degree of granulation in this cell type (Hymer *et al.*, 1974). Injection of estradiol to female rats will markedly increase intracellular PRL content, whereas ovariectomy has the opposite effect. However, not all mammotrophs were affected to the same degree by these treatments. This is reflected in the cell-sedimentation data in Fig. 11. The *in vitro* PRL secretory capacity of the mammotrophs in these different fractions revealed significant heterogeneity. Thus, large, heavily granulated mammotrophs isolated from estrogen-primed animals (Fig. 12, fraction VI) released approximately 6 times as much hormone as the smaller cells in fraction II. Note that release is normalized to the same number of mammotrophs seeded. Since DNA synthesis is extremely low (or nonexistent) in cultured mammotrophs (Snyder, J., *et al.*, 1976, 1978), the differential responsiveness probably cannot be attributed to differences in mammotroph numbers in the dish.

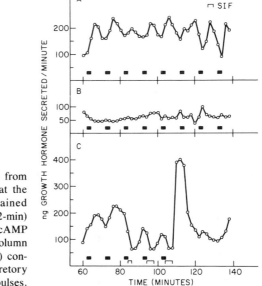

Figure 10. GH secretory profiles from three pituitary cell columns run at the same time. Each column contained 5 × 10^6 cells. (A) Periodic pulses (2-min) with 1 mM dbcAMP; (B) same dbcAMP pulses as in (A), except that the column buffer (medium 199 + 0.1% BSA) contained 10^{-7} M SIF; (C) GH secretory responses after dbcAMP and SIF pulses.

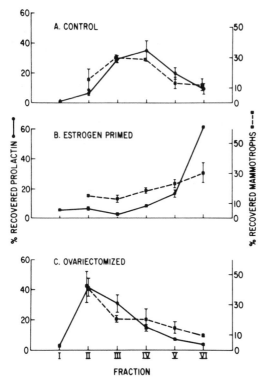

Figure 11. Distribution patterns of mammotrophs and PRL in six cell-separation experiments. Tissue sources: (A) Glands from rats in proestrus in one experiment and diestrus in the other; (B) glands from rats receiving 20 μg estradiol benzoate/day for 5 days prior to kill (two experiments); (C) glands from rats ovariectomized for 18 days prior to kill (two experiments). The points represent the means of the two experiments within each treatment, and the vertical brackets represent the ranges in values for the two experiments. The data express the percentage of recovered immunoassayable PRL in gradient regions I through VI, or the percentage of recovered mammotrophs in these same regions as determined by immunoperoxidase staining. The mean numbers of mammotrophs recovered from the cell separation gradients were: normal, 3.1×10^6; estrogen-primed, 3.5×10^6; ovariectomized, 4.3×10^6. The mean nanograms of PRL recovered were: normal 80,000; estrogen-primed, 100,000; ovariectomized, 95,000. From Hymer et al. (1974).

Thus, one is led to conclude that not all mammotrophs work equally well within the PRL-producing cell population. We have obtained similar evidence for mammotrophs isolated from the human pituitary gland (Hymer et al., 1976).

Recently, a report from Moriarty's laboratory (Leuschen et al., 1978) provided evidence that functional heterogeneity might also exist within the thyrotroph population. After sedimentation at unit gravity, cells in the top two fractions consisted of 60–80% thyrotrophs. In culture, however, only the medium from the smaller, lighter thyrotrophs contained significant levels of TSH. Thus, there appears to be a good degree of similarity between the activity of the less granulated thyrotroph and somatotroph in cell culture as well as a lack of activity of their better-granulated counterparts. Note, however, that such may not be true for mammotrophs.

Finally, the recent report of Denef et al. (1978) offers evidence for heterogeneity within the gonadotroph population. Since this material is

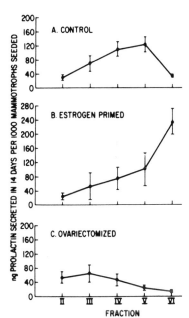

Figure 12. Amount of PRL secreted by 1000 mammotrophs from different gradient regions after a 14-day culture period. The data represent the means of two experiments within each treatment, and the vertical brackets represent the ranges in values in the two experiments. For other details, see the Fig. 11 caption. Fraction I consisted primarily of debris and nonviable cells and therefore was not cultured. From Hymer *et al.* (1974).

detailed by Denef in Chapter 31 of this volume, suffice it to say that convincing evidence is presented to show that not all gonadotrophs have the same secretory capacity in response to a maximally effective dose of LH-RH. Functional differences in gonadotrophs, as with other isolated pituitary cell types, are probably related to differences in cell size or density or both. Denef's suggestion of "cooperative phenomena between cells of the same type" to account for heterogeneity is an intriguing one and certainly merits further work.

In an *in vivo* sense, what is the meaning of physiological heterogeneity in pituitary cells? Of course, one can only speculate about this question at this time. However, it may be well to recall the elegant cytological studies of the early workers, for clues to the answer may be found therein. For example, Severinghaus (1937) was among the first to postulate the existence of a secretory cycle within the pituitary cell itself. A degranulated cell might be representative of one in the hormone synthetic phase; a well-granulated cell, of one in a storage phase, and so on. It seems logical enough to propose that cells in different stages of this cycle would show differential responsiveness to hormone secretory agents. On the other hand, the level of circulating hormone usually represents only a small percentage of the total glandular content. One could thus alternatively invoke the notion of a pool of cells within the gland that are active under "basal" conditions. The remainder would represent reserve stores to be called upon in times of need. New, imaginative approaches to the study of pituitary cell function are required. One such approach that in our opinion has this potential is described in the next section.

5. Intracranial Implantation of Pituitary Cells

Anyone who is interested in growth hormone will find that the account by Leslie Bennett (1975) of Herbert Evans's research career and the account by James Leathem (1977) of Phillip Smith's studies make for delightful reading indeed. One can easily imagine the excitement in the laboratories of these scientists when the importance of the pituitary to animal growth was first being discovered. As Bennett writes:

> . . . in his (Evans) Harvey Lecture he illustrated a modest degree of gigantism of a female rat chronically treated with crude extract. [See Fig. 13.] More famous and found in many physiology texts is the picture of littermate dachshund dogs . . . one of these dachshunds had been treated from the sixth to the thirty-second week of life and showed not only an increase in body size but an even more remarkable overgrowth of skin around the neck and extremities. For many years in Dr. Evans' laboratory there were maintained one after another, littermate pairs of hypophysectomized rats. One untreated weighing 50 gm or less, and the other chronically treated with growth hormone weighing perhaps 400 to 600 gm.

Li (1977) has recently pointed out that some of these early observations form the basis for GH bioassays that still enjoy use to this day. Thus, weight increases induced by injection of GH into (1) the mature plateaued intact rat, (2) the hypophysectomized rat, and (3) the dwarf mouse all give some index of GH potency of the preparation.

An alternative to replacement GH therapy in the hypophysectomized rat is, of course, pituitary implants. The well-recognized need for vascular connections between the implant and the hypothalamus arose, in part, out of the double-transplantation experiments of Nikitovitch-Winer and Everett (1958), in which it was shown that rats in pituitary failure resulting from transplantation to the kidney capsule could be corrected by retransplantation of the same pituitary to the region beneath the basal hypothalamus. The experiments of Smith (1963) also showed that growth of hypophysectomized rats was reinitiated after transplantation of the pituitary back into its original site beneath the hypothalamus. In 1963, Halasz et al. (1963) demonstrated that reasonable growth of hypophysectomized rats occurred when glands were implanted in the hypothalamic hypophysiotropic area. However, implantation in other CNS or peripheral sites (including the cerebral ventricles) did not result in good growth. Finally, relatively meager growth responses were obtained after implantation into

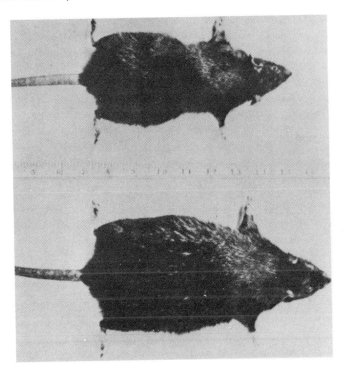

Figure 13. A Harvey Society Lecture illustration from 1924 showing a littermate pair of female rats, one treated and one not treated with bovine anterior lobe substance. The animal on the bottom illustrates the degree of gigantism produced in this initial series of experiments. From Bennett (1975).

the anterior chamber of the eye, intramuscularly or subcutaneously. For example, Gittes and Kastin (1966) reported a log dose relationship between growth and number of glands implanted intramuscularly, and by extrapolation concluded that 750 glands were required for normal growth.

The report by Shiino *et al.* (1977) provides a modern "twist" to these earlier transplantation studies. Using a cell line of pituitary anlage originally derived from epithelial cells of Rathke's pouch, several clonal strains have been developed. One of these (2A8) releases ACTH, GH, and PRL (but not glycoprotein hormones) in culture. These cells are essentially agranular. When they were implanted into the preoptic area or median eminence of the hypothalamus, cellular redifferentiation apparently occurred, since several anterior pituitary cell types were observed in the grafts (Fig. 14). As shown in Fig. 15, the body weights of hypophysectomized rats receiving 1×10^6 2A8 cells in the hypothalamus reflected functional somatotrophs in the implant. Although the possibility is not stated by Shiino and co-

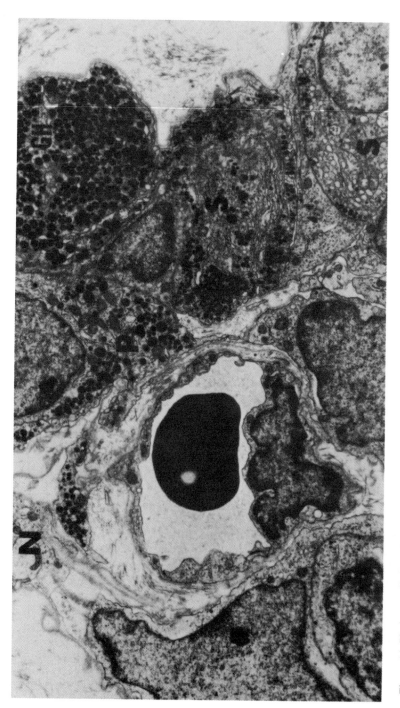

Figure 14. Pituitary cells from 2A8 clone after transplantation to the median eminence. Three typical granular cells and a capillary can be seen. (P) PRL cell; (GH) somatotroph; (S) small granule-containing cell (gonadotroph); (N) nerve fiber. ×3600 (reduced 25% for reproduction). From Shiino et al. (1977).

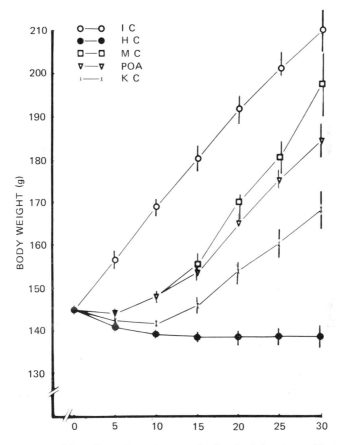

Figure 15. Body weights of host (hypophysectomized) animals bearing multipotential clone (2A8) under kidney capsule or in hypothalamus. (IC) Intact control (7 rats); (HC) hypox control (7 rats); (MC) median eminence (5 rats); (POA) preoptic area (5 rats); (KC) kidney capsule (5 rats). From Shiino *et al.* (1977).

workers, the minimal growth rate of animals receiving implants in the kidney capsule might be attributed to growth-promoting activity inherent in the PRL molecule.

There is increasing evidence that cerebrospinal fluid (CSF) contains various hypothalamic hormones that may participate in the regulation of pituitary function. A few years ago, we began to study body growth of hypophysectomized rats implanted with pituitary cells in the cerebral ventricles. Results from this study, recently published (Weiss *et al.*, 1978), are briefly described below.

5.1. Intraventricular Implantation

In our usual experimental protocol, 1–3×10^6 acutely dispersed pituitary cells were implanted into the left lateral ventricle of rats hypophysectomized at approximately 30 days of age. They were often implanted via an indwelling cannula positioned stereotaxically in the ventricle, but were sometimes implanted using a Hamilton syringe with a fixed needle of precalibrated length (4.0 mm) directly into the lateral ventricle.

As shown in Fig. 16, growth (% of weight gain) of animals bearing 1×10^6 cells was similar, during the first 3-week postimplantation period, to sham-hypophysectomized littermates. As shown in the insets in Fig. 16, growth was proportional to the number of cells implanted. Animals that received 3×10^6 cells doubled their body weight over a 90-day period! A number of tests were used to show that the increased weight reflected

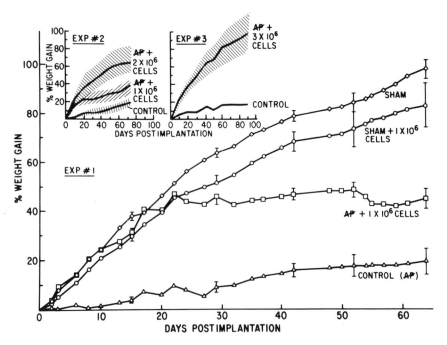

Figure 16. Expt. 1: Percentage increase in body weight of approximately 30-day male hypox rats receiving a 10 μl intraventricular injection of either "mock" CSF (control) or 1×10^6 pituitary cells from 70-day-old donors (bottom two curves) or sham-hypox littermates $\pm 1 \times {}^*10^6$ cells (top two curves). Each line corresponds to the weight gain of 4 animals; vertical brackets represent S.E.M. Effect of implantation of $1 \times$ and $2 \times$ (Expt. 2) or 3×10^6 cells (Expt. 3) on weight gain is shown in inserts. From Weiss et al. (1978).

real growth. For example, tibial, femoral, and pelvic bone lengths were 31.5 ± 0.29, 26.0 ± 0.33, and 29.6 ± 0.24 mm, respectively, for controls and 33.8 ± 0.88, 28.9 ± 0.24, and 33.3 ± 0.44 mm for experimentals (1×10^6 cells, 30 days postimplantation; $p < 0.05$). Increased long-bone lengths were also easily seen in radiographs. Moreover, analysis of total body composition showed that weight gains of control animals could be attributed entirely to fat, whereas the experimental rats showed significant increased deposition of both protein and fat.

Results of other studies using this basic approach demonstrated that:

1. Younger recipients grew better than older ones.
2. Cells from pituitaries of older donors gave better responses than equivalent cell numbers from glands of younger donors.
3. Purified somatotrophs (type II cells) gave a positive response.
4. Castration did not affect the response, a result that suggested that anabolic steroids were probably not involved.
5. Positive responses were not obtained when cells were placed either in the anterior chamber of the eye or intraperitoneally. Live, intact cells were required, since heat-killed cells or a 100,000g pituitary cell particle fraction (containing GH granules) gave negative responses.

Finally, in a few pilot experiments, we found that implantation of cells from one strain of rats into hypophysectomized donors of a second strain resulted in an abbreviated response that we attributed to immune cells in the CSF.

5.2. Encapsulated Cells: Artificial Pituitary Units

Several years ago, Amicon Corporation (Lexington, Massachusetts) marketed a product, called "hollow fibers," that consists of a very thin anisotropic Diaflo membrane and a thicker spongy layer of the same polymer with increasingly larger openings. These hollow fibers are made of a polyvinyl chloride–acrylic copolymer (XM-50) with an internal diameter of 100 μm and controlled pore sizes with a nominal molecular-weight cutoff at 50,000.

Use of such fibers for the production of artificial pancreas units is becoming more widespread in the research laboratory. In one approach (Tze *et al.*, 1976), pancreatic islets were attached to the surface of a polyethylene-encapsulated XM-50 hollow fiber. The unit was then implanted into a streptozotocin-induced diabetic rat in such a way that blood flowed through the lumen of the hollow fiber. Insulin, by virtue of its molecular size, permeated the pores of the fiber and entered the bloodstream. The usefulness of the implant in restoring normoglycemia over a

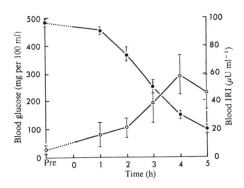

Figure 17. Decline in blood glucose levels in diabetic rats after the implantation of an artificial endocrine pancreas. (Pre) Sample taken before ether anesthesia for the implantation procedure; (0) represents the connection of the artificial endocrine pancreas as a shunt in the circulatory system. The values expressed are the means + S.E. (vertical brackets) blood glucose (•) and insulin (○) levels of 8 diabetic rats receiving such implants. From Tze et al. (1976).

4-hr period is demonstrated by the data in Fig. 17. As pointed out by Tze et al. (1976):

> . . . the potential importance of this approach is to provide functional pancreatic islets in the diabetic recipient which can supply insulin, glucagon and possibly other islet factors, in response to physiological needs and ultimately ameliorate insulin-requiring diabetes and perhaps prevent the development of complications. Islets within the unit may be protected from immune destruction by circulating host's lymphocytes and antibodies, by virtue of the selective permeability of the molecular size of the synthetic capillaries. Thus, the artificial endocrine pancreas unit would mimic an immunologically privileged site. The system, if successful in the allogeneic rat, would have clinical potential.

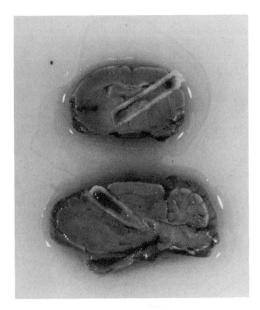

Figure 18a. Sections of brains of two animals receiving capsules coronally (top) or parasagitally (bottom).

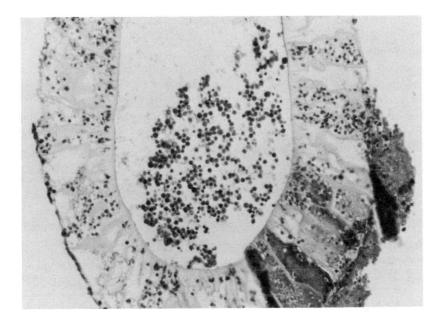

Figure 18b. Section of capsules 1 day postimplantation. Pituitary cells (linear) are single.

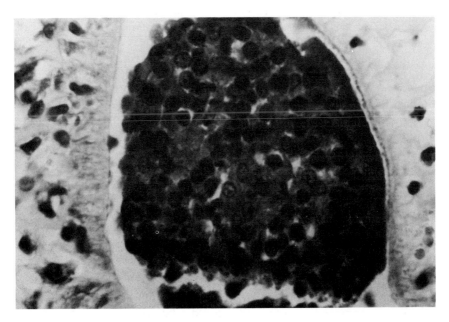

Figure 18c. Section of capsule 7 days postimplantation. Cells are clumped into a structure resembling intact pituitary tissue.

In light of the experimental data discussed in Section 5.1, we reasoned that a generally similar approach, applied to the pituitary cell intracranial implantation system might prove interesting. We further reasoned that insertion of the cells into the *lumen* of the hollow fiber might be more beneficial simply because in this configuration the polyethylene shell used by Tze *et al.* (1976) could be eliminated. Accordingly, 10-mm fibers were sealed at both ends by heat after the cells had been injected into the lumen (8 μl). A single capsule was then implanted either coronally or parasagittally by way of a small hole drilled in the skull. Even though the capsule occupied a large area of the brain, it had no obvious visible effect on either the behavior or the physical appearance of the recipient. Shown in Fig. 18 are thick sections of the fibers implanted coronally (top) or parasagittally (bottom). This preparation is typical of many that have been examined. In our experience, none of the fibers get into the hypothalamus. However, they often approach, or are in contact with, a ventricular surface. Our experience to date with the histology of the encapsulated cells is also shown in Fig. 18. At 1 day after implantation, the cells were single, but after 7 days, they had aggregated and assumed some of the characteristics of intact pituitary tissue. Detailed morphological analyses by immunocytochemistry and transmission and scanning electron microscopy are currently being done.

The growth curves of hypophysectomized male rats implanted with pituitary capsules (Fig. 19) showed that pieces of intact tissue, as well as acutely dispersed cells, were capable of stimulating significant growth. Controls received empty capsules. In this experiment, the growth rates plateaued after about 1 month postimplantation. We were somewhat surprised to find that the intact tissue gave a positive response in light of the limitations associated with the use of pituitary tissue (see Fig. 2). Even though the unit cannot be "hooked up" to the circulating system in the brain, it would seem that sufficient oxygenation takes place to support functional cells. The growth curve in Fig. 20 showed that implantation of pituitary cell units into hypophysectomized female rats also gave a good response. Finally, implantation of pituitary cells from (1) a different strain of rats (Fisher 344), (2) sheep, or (3) pieces of human pituitary tissue (postmortem) all elicited positive responses (Figs. 21 and 22). These latter results suggest that the capsules may provide an immunologically privileged site for the tissue.

5.3. Suppression of the in Situ Pituitary after Intraventricular Pituitary Cell Implantation

The reader may have noticed a curious result in Fig. 16: implantation of freshly dispersed pituitary cells into the lateral ventricle of intact rats caused a slight but significant *suppression* of growth. In another series of

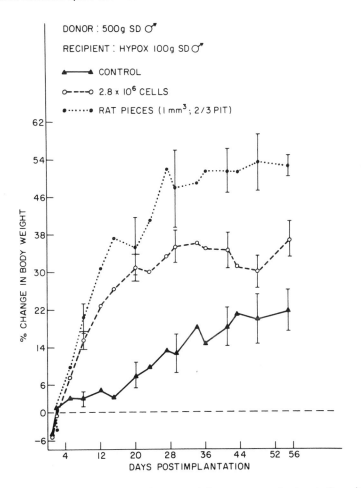

Figure 19. Growth curves of hypox male rats receiving empty capsules (control) or pituitary capsules implanted coronally.

experiments designed for a very different purpose, we obtained data that were reminiscent of these earlier results. In this newer series, 1×10^6 pituitary cells were implanted intraventricularly into either sexually immature or sexually mature female rats of two different strains (Sprague–Dawley and Fisher 344). At 10 days postimplantation, the animals were weighed and killed, and the pituitaries were analyzed for GH content by RIA. As shown in Table IV, the implantation of cells consistently suppressed growth of the immature rats of both strains by 30–40%. Not unexpectedly, the cells had no effect on body weights of the sexually mature plateaued animals. The pituitary glands of the recipient animals showed remarkable decreases in both total weight and total protein

Table IV. Effect of Intraventricular Implantation of 1×10^6 Cells on the *in Situ* Pituitary

Treatment	Number of cells implanted	Change in body weight (%)	Pituitary		
			Weight (mg)	Protein (µg)	GH (mg)
Immature SD Expt. 1					
Control	—	56.1 ± 4.6	7.5 ± 0.6	205.4 ± 15.2	617.7 ± 123.8
Experimental	1.2×10^6	40.5 ± 1.4[a]	4.1 ± 0.1[a]	130.4 ± 7.8[a]	285.9 ± 25.0[b]
Immature SD Expt. 2					
Control	—	46.3 ± 4.2	7.6 ± 0.3	214.2 ± 18.5	650.4 ± 49.4
Experimental	2.0×10^6	31.5 ± 4.7[a]	5.2 ± 0.4[a]	151.6 ± 12.7[b]	320.7 ± 64.1[b]
Mature SD Expt. 1					
Control	—	21.1 ± 2.4	11.0 ± 0.7	344.6 ± 15.4	935.4 ± 49.4
Experimental	0.9×10^6	26.0 ±2.0	7.9 ± 0.3[a]	246.5 ± 7.3[a]	545.2 ± 72.6[a]
Mature SD Expt. 2					
Control	—	15.0 ± 2.4	12.1 ± 1.0	385.8 ± 25.1	1498.3 ± 216.1
Experimental	1.0×10^6	12.5 ± 3.3	8.7 ± 0.7[a]	298.8 ± 21.4[b]	761.8 ± 103.6[b]
Immature F-344 Expt. 1					
Control	—	46.8 ± 2.6	5.4 ± 0.2	157.0 ± 8.5	564.3 ± 12.5
Experimental	0.9×10^6	27.3 ± 1.8[a]	3.7 ± 0.1[a]	94.9 ± 5.1[a]	192.4 ± 16.2[a]
Immature F-344 Expt. 2					
Control	—	52.3 ± 4.2	5.7 ± 0.4	174.5 ± 12.0	606.3 ± 33.9
Experimental	1.3×10^6	29.8 ± 2.1[a]	3.6 ± 0.1[a]	101.8 ± 1.8[a]	211.9 ± 11.3[a]
Mature F-344 Expt. 1					
Control	—	-1.7 ± 0.5	12.4 ± 0.8	381.6 ± 28.0	775.5 ± 64.1
Experimental	1.3×10^6	1.0 ± 0.6	9.4 ± 0.2[a]	256.7 ± 14.0[a]	426.8 ± 33.8[a]

[a] $P < 0.01$.
[b] $P < 0.05$.

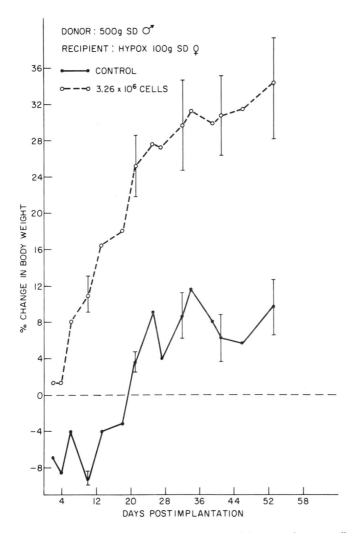

Figure 20. Growth curve of hypox female rats receiving capsules coronally.

content. The glands of the immature animals showed reductions in weight and protein content of approximately 36%. Interestingly enough, glands from the older rats were suppressed by approximately 28%, even though body weights were unaffected. GH contents of the glands were also drastically affected. Thus, glands from the immature rats had 50–60% less GH than control rats that had received vehicle only. In the older animals, GH contents were suppressed 40–50%.

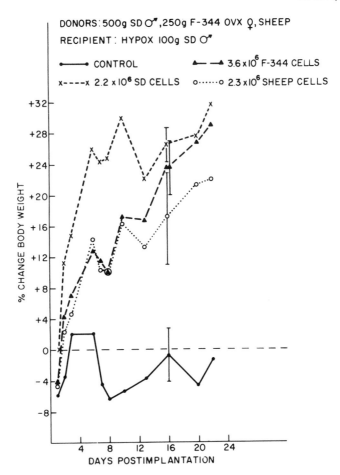

Figure 21. Growth curves of hypox male rats after coronal implantation of capsules containing different types of pituitary cells.

We are aware of a few isolated reports in the literature that suggest that GH released from cells or tissues at other than its normal site (e.g., from animals bearing transplantable pituitary tumors) have smaller *in situ* glands with lower intracellular GH stores (Daughaday *et al.*, 1975). A particularly dramatic example can also be found in the report from Daughaday's laboratory (Garland and Daughaday, 1972). In this study, subcutaneous implantation of tapeworms (which release a "GH-like" molecule) into rats produced effects on the *in situ* pituitary that were of the same order of magnitude as found in our study (Table IV). A feedback suppression of GH on the *in situ* pituitary most readily accounts for such

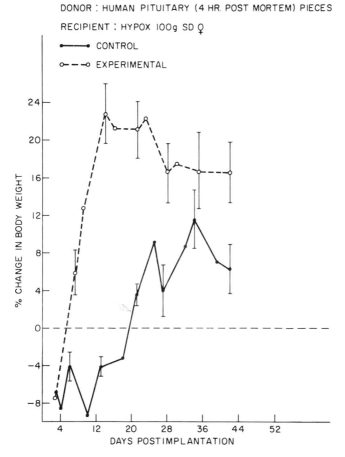

DONOR : HUMAN PITUITARY (4 HR. POST MORTEM) PIECES

RECIPIENT : HYPOX 100g SD ♀

●——● CONTROL

○– –○ EXPERIMENTAL

Figure 22. Growth curve of hypox female rats receiving capsules containing human pituitary tissue.

findings. The specificity of this response, evaluated by RIA of the other pituitary hormones, is currently being studied.

6. General Conclusions

In this chapter, we have attempted to show that the somatotrophs that are present in freshly prepared suspensions of single pituitary cells serve as useful models for the study of the regulation of GH cell function. Studies have shown that dissociated somatotrophs:

1. Retain staining properties characteristic of those in intact tissue. Furthermore, they also retain ultrastructural integrity at the electron-microscope level.
2. Retain GH during dissociation.
3. Actively exclude dye.
4. Synthesize protein (and protein hormone—see the text).
5. Respond to a variety of secretagogues by releasing GH *in vitro* in dose-related fashion.
6. Promote growth of hypophysectomized rats after implantation *in vivo*.

Several investigators have also used freshly dispersed pituitary cells as the starting material for their separation into various hormone-producing cell types. Success has been reasonably good, with reports of 70–90% purity not uncommon. Careful accounting of the cells during separation has revealed a rather surprising finding: the heterogeneity that exists within a class of cells (e.g., somatotrophs, thyrotrophs) in terms of size or density or both may very well carry over to their functionality. The true meaning of such functional heterogeneity will obviously require careful study in the future.

Finally, the implantation of live pituitary cells back into properly chosen recipient animals provides an added dimension for their study. The ease with which these implanted cells can be recovered for further *in vitro* study (i.e., via the capsule technique) appears to make this approach all the more attractive.

It seems clear that the pituitary becomes less and less of a "black box" as time goes along. With ever-increasing new ways to approach the study of pituitary gland function, the future looks exciting indeed.

ACKNOWLEDGMENT. Some of the work described in this chapter was supported in part by Grant Number 23248 from the National Cancer Institute and by Contract NAS9-15566 from the National Aeronautics and Space Administration.

DISCUSSION

LABRIE: Did you study the morphology of the pituitary cell implanted intraventricularly to see if mammotrophs become predominant? Also, did you measure prolactin levels in brain?

HYMER: Such morphological studies are currently under way. Prolactin levels in the brain or serum of the implanted animals have not been measured.

TIXIER-VIDAL: Do the implanted cells divide? Also, what was the CO_2 pressure? Why did you do a 9-day study rather than a short period, say 0 to 24 hours?

HYMER: Our impression is that the implanted cells do not divide, since we have not observed mitotic activity. We have not done the appropriate studies with [³H]thymidine. The CO_2 pressure was 8% for those media with high bicarbonate and 5% for the others. In regard to the long-term study, since a large majority of the studies aimed at defining the molecular effects of secretagogues on pituitary cells are done on cultured cells, we felt it important to do the experiments over a reasonably long culture period. It seems important to us that use of culture medium which permits optimal (i.e., maximal) hormone release and synthesis should be used in such experiments. Use of media with suboptimal characteristics with or without secretagogues may generate data which are difficult to interpret and repeat. Finally, I should add that we have done a few experiments comparing different media in a short-term system. The results were entirely comparable to those obtained over long-term culture.

FINK: In your intact rats bearing a capsule that contains cells, could the effects on the recipient's pituitary be due to an immune phenomenon—e.g., autoimmune? I always try to avoid an ultra-short-loop feedback explanation if possible.

HYMER: We don't think that the suppression of the *in situ* pituitary can be explained by an autoimmune phenomenon. In all experiments where suppression was obtained, animals of the same strain and often littermates were used as pituitary donors. The most reasonable explanation for the suppression rests in an interaction between the implant and the *in situ* gland, i.e., short-loop feedback.

KRAICER: How much of the weight effects might be due to changes in food intake or food utilization?

HYMER: Hypophysectomized rats exhibit a postoperative weight gain which can be attributed entirely to lipid. Furthermore, forced feeding of hypoxed rats will not lead to an increase in protein deposition. Thus, our findings that the implanted animals show a gain in protein strongly suggests that the increase in body weight can be attributed to the somatotrophs in the implant. Finally, we have not seen any evidence of increased food intake in the experimental groups.

SPICER: In respect to Dr. Fink's question, the late Dr. R.F. Phifer, in our laboratory, showed the development of an autoimmune response in the pars intermedia of the rabbit injected with 17–39 ACTH over prolonged intervals. This was done to prepare rabbit antibody to 17–39 ACTH. The autoimmune resonse was visualized by simple histologic examination of the pituitaries of the immunized rabbit. The question also comes up as to whether the growth hormone cells in the implanted tubes are eventually rejected or survive the host's immune mechanisms.

HYMER: The cells survive more than 40 days.

SPICER: If, as you indicate, they are not affected by the host's antibodies, how is it that growth hormone can diffuse out of the tube implant, but antibodies do not diffuse into the tube? Is it a matter of molecular sieving by an appropriate pore size in the tube wall?

HYMER: Yes. The question about diffusion of antibodies and hormones in and out of the capsule is made clear in the text of my manuscript. Briefly, growth hormone has a molecular weight of 22,500 and its antibody 150,000. The fiber of molecular weight 50,000 excludes the antibody, but not growth hormone.

MELDOLESI: Did you identify the proteins synthesized by your type II somatotrophs? Are isolated cells less efficient than tissue fragments when implanted in the brain within the hollow fibers?

Hymer: Although the type II somatotrophs are among the most active in incorporating radioactive amino acids into TCA-precipitable protein, we have no information as to the types of molecules synthesized. In the experiments comparing hypoxed rats receiving capsules with pituitary pieces or intact cell, we made an effort to keep the number of cells comparable (2.8×10^6). It is not possible to be really quantitative. Nevertheless, the data indicate that pituitary pieces are capable of promoting growth comparable to that obtained with dispersed cells. We need to determine the extent of cellular necrosis in the pieces.

REFERENCES

Bancroft, F. C., Levine, L., and Tashjian, A., 1969, Control of growth hormone production by a clonal strain of rat pituitary cells, *J. Cell Biol.* **43**:432.

Bennett, L., 1975, Endocrinology and Herbert M. Evans, in: *Hormonal Proteins and Peptides*, Vol. III (C. H. Li, ed.), pp. 247–272, Academic Press, New York.

Brazeau, P., Vale, W., Bengus, R., Ling, N., Butcher, M., Rivier, J., and Guillemin, R., 1973, Hypothalamic peptide that inhibits the secretion of immunoreactive pituitary growth hormone, *Science* **179**:77.

Daughaday, W., 1975, Regulation of growth by endocrines, *Annu. Rev. Physiol.* **37**:211.

Denef, C., Hautekeete, E., and Dewals, R., 1978, Monolayer cultures of gonadotrophs separated by velocity sedimentation: Heterogeneity in response to luteinizing hormone-releasing hormone, *Endocrinology* **103**:736.

Ellis, S., and Grindeland, R. E., 1974, in: *Advances in Human Growth Hormone Research* (S. Raiti, ed.), DHEW Publ. No. 74-612, pp. 409–424.

Farquhar, M. G., Skutelsky, E. H., and Hopkins, C. R., 1975, Structure and function of the anterior pituitary and dispersed pituitary cells: *In vitro* studies, in: *The Anterior Pituitary* (A. Tixier-Vidal and M. G. Farquhar, eds.), pp. 84–135, Academic Press, New York.

Garland, J., and Daughaday, W., 1972, Feedback inhibition of pituitary GH in rats infected with *Spirometia*, *Proc. Soc. Exp. Biol. Med.* **139**:497.

Gittes, R. F., and Kastin, A. J., 1966, Effects of increasing numbers of pituitary transplants in hypophysectomized rats, *Endocrinology* **78**:1023.

Halasz, B., Pupp, L., Vhalrik, S., and Tima, L., 1963, Growth of hypophysectomized rats bearing pituitary transplant in the hypothalamus, *Acta Physiol. Acad. Sci. Hung.* **23**:287.

Hymer, W. C., 1975, Separation of organelles and cells from the mammalian adenohypophysis, in: *The Anterior Pituitary* (A. Tixier-Vidal and M. G. Farquhar, eds.), pp. 137–180, Academic Press, New York.

Hymer, W. C., Kraicer, J., Bencosme, S., and Haskill, J., 1972, Purification of somatotrophs from the rat adenohypophysis by velocity and density gradient centrifugation, *Proc. Soc. Exp. Biol. Med.* **141**:966.

Hymer, W. C., Snyder, J., Wilfinger, W., Swanson, N., and Davis, J., 1974, Separation of pituitary mammotrophs from the female rat by velocity sedimentation at unit gravity, *Endocrinology* **95**:107.

Hymer, W. C., Snyder, J., Wilfinger, W., Bergland, R., Fisher, B., and Pearson, O., 1976, Characterization of mammotrophs separated from the human pituitary gland, *J. Natl. Cancer Inst.* **57**:995.

Kohler, P. O., Bridson, W., and Rayford, P. L., 1968, Cortisol stimulation of growth hormone production by monkey adenohypophysis in tissue culture, *Biochem. Biophys. Res. Commun.* **33**:834.

Kraicer, J., and Hymer, W. C., 1974, Purified somatotrophs from rat adenohypophysis: Response to secretagogues, *Endocrinology* **94:**1525.

Leathem, J., 1977, Hypophysectomy and Phillip E. Smith, in: *Hormonal Proteins and Peptides*, Vol. IV (C. H. Li, ed.), pp. 175–192, Academic Press, New York.

Leuschen, M. P., Tobin, R., and Moriarty, M., 1978, Enriched populations of rat pituitary thyrotrophs in monolayer culture, *Endocrinology* **102:**509.

Li, C. H., 1975, The chemistry of human pituitary growth hormone: 1967–1973, in: *Hormonal Proteins and Peptides*, Vol. III (C. H. Li, ed.), pp. 1–40, Academic Press, New York.

Li, C. H., 1977, Bioassay of pituitary growth hormone, in: *Hormonal Proteins and Peptides*, Vol. IV (C. H. Li, ed.), pp. 1–41, Academic Press, New York.

Lowry, P. J., 1974, A sensitive method for the detection of corticotropin releasing factor using a perfused pituitary cell column, *J. Endocrinol.* **62:**163.

Miller, R., and Phillips, R., 1969, Separation of cells by velocity sedimentation, *J. Cell. Physiol.* **73:**191.

Milligan, J., Kraicer, J., Fawcett, C. P., and Illner, P., 1972, Purified growth hormone releasing factor increases ^{45}Ca uptake into pituitary cells, *Can. J. Physiol. Pharmacol.* **50:**613.

Nikitovitch-Winer, M., and Everett, J. W., 1958, Functional restitution of pituitary grafts retransplanted from kidney to median eminence, *Endocrinology* **63:**916.

Peterson, E., and Evans, W., 1967, Separation of bone marrow cells by sedimentation at unit gravity, *Nature (London)* **214:**824.

Saffran, M., and Schally, A. V., 1955, The release of corticotropin by anterior pituitary tissue *in vitro*, *Can. J. Biochem.* **33:**408.

Severinghaus, A. E., 1937, Cellular changes in the anterior hypophysis with special reference to its secretory activities, *Physiol. Rev.* **17:**556.

Shiino, M., Ishikawa, H., and Rennels, E., 1977, *In vitro* and *in vivo* studies on cytodifferentiation of pituitary clonal cells derived from epithelium of Rathke's pouch, *Cell Tissue Res.* **181:**473.

Shortman, K., 1968, The separation of different cell classes from lymphoid organs, *Aust. J. Exp. Biol. Med. Sci.* **46:**375.

Smith, P. E., 1963, Postponed pituitary homotransplants into the region of the hypophysial portal circulation in hypophysectomized female rats, *Endocrinology* **73:**793.

Snyder, G., and Hymer, W. C., 1975, A short method for the isolation of somatotrophs from the rat pituitary gland, *Endocrinology* **96:**792.

Snyder, G., Hymer, W. C., and Snyder, J., 1977, Functional heterogeneity in somatotrophs isolated from the rat anterior pituitary, *Endocrinology* **101:**788.

Snyder, J., Wilfinger, W., and Hymer, W. C., 1976, Maintenance of separated rat pituitary mammotrophs in cell culture, *Endocrinology* **98:**25.

Snyder, J., Hymer, W. C., and Wilfinger, W., 1978, Culture of human pituitary prolactin and growth hormone cells, *Cell Tissue Res.* **191:**379.

Stachura, M., 1976, Mechanism of somatostatin action: Influence of synthetic SRIF and dibutyryl cyclic AMP on GH release by rat pituitaries *in vitro*, *Proc. Annu. Meet. Endocrinol. Soc.* **58:**A-133.

Stachura, M., 1977, Interaction of somatostatin inhibition and dibutyryl cyclic AMP or potassium stimulation of growth hormone release from perfused rat pituitaries, *Endocrinology* **101:**1044.

Stachura, M., Szabo, M., and Frohman, L. A., 1978, Multiphasic effect of porcine stalk median eminence extract on growth hormone release from perfused rat pituitaries, *Endocrinology* **102:**1520.

Tannenbaum, G. S., and Martin, J. B., 1976, Evidence for an endogenous ultradian rhythm governing growth hormone secretion in the rat, *Endocrinology* **98**:562.

Tze, W., Wong, F., Chen, L., and O'Young, S., 1976, Implantable artificial endocrine pancreas unit used to restore normoglycemia in the diabetic rat, *Nature (London)* **264**:466.

Weiss, S., Bergland, R., Page, R., Turpen, C., and Hymer, W. C., 1978, Pituitary cell transplants in the cerebral ventricles promote growth of hypophysectomized rats, *Proc. Soc. Exp. Biol. Med.* **159**:409.

Wilfinger, W., David, J., Augustine, E. C., and Hymer, W. C., 1979, Effects of culture conditions on prolactin and growth hormone production by rat anterior pituitary cells, *Endocrinology* **105**:530.

7

Pituitary Secretory Activity and Endocrinophagy

D. L. Wilbur and S. S. Spicer

1. Introduction

Recent advances in quantitative electron-microscopic autoradiography, cytochemistry, and immunoassay techniques have provided more precise information on the location and kinetics of the intracellular processing of adenohypophyseal hormones than has been possible previously using other systems. These studies have elucidated many of the biochemical and morphological pathways in the synthesis, intracellular transport, release, and endocrinophagy of adenohypophyseal hormones. The growth-hormone (GH) cell or somatotroph has served as the model for our laboratory investigations on the mechanism(s) by which secretagogues stimulate hormone synthesis and release from the adenohypophysis.

Our present knowledge of somatotropin synthesis and release has been obtained mostly from *in vitro* studies. Even when *in vivo* techniques were used, they were severely handicapped by technical limitations. The major problem is the difficulty in exposing the hypothalamus and hypophysis cerebrii (pituitary gland) sufficiently to enable one to perform critical experiments on the intact hypothalamo–hypophyseal unit. These structures lie deep within the cranium and are protected by the brain from above and guarded from below by bones of the skull and oral cavity as well as overlying muscles and other tissues. The *in vivo* techniques used

D. L. Wilbur and *S. S. Spicer* • Departments of Anatomy and Pathology, Medical University of South Carolina, Charleston, South Carolina 29403

were further handicapped by other inherent variables such as systemic dilutions of the stimuli, hypothalamic influences, and the period of time involved before an effect could be observed.

The *in vivo* technique used in this study, the infusion of secretagogues directly into a single portal vessel, has the distinct advantage of minimizing the inherent variables listed above. This technique of hypophyseal portal vessel infusion allows one to follow the route of the infused substance directly into the adenohypophysis. Additional benefits of this technique are such that it allows one to observe the particular area of the anterior pituitary that is being infused directly, while the minimization of systemic dilutions allows the infusion of lower, more physiological doses.

The purpose of this chapter is to provide a review of the very recent literature on the secretory process of somatotroph cells and to project some of the unresolved questions that remain for future experimentation.

2. Materials and Methods

Adult male Wistar rats weighing 200–240 g were housed in an environmentally controlled room with two rats in each cage and maintained on Purina laboratory food and water *ad libitum*. The animals were anesthetized with 10% urethane (1200 mg/kg), and the anterior pituitary gland and stalk portal vessels were exposed surgically by a modification of the method described by Worthington (1955) (Fig. 1).

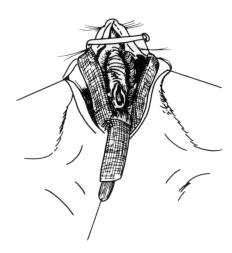

Figure 1. Surgical approach to hypothalamo–hypophyseal portal vessels.

A Leitz micromanipulator was used to place the microcannula into a portal vessel. A Harvard infusion pump was used to regulate flow of infusate through the microcannula at a rate of 1–2 μl/min (Porter *et al.*, 1970). The medium for the secretagogues was physiological saline buffered to pH 7.4, containing a small amount of lissamine green added for ease in determining the distribution of the infused portal vessel. Synthetic thyrotropin (TRH) obtained from Beckman Bioproducts was dissolved in the medium to give a final concentration of 5 ng/μl. A new bovine hypothalamic peptide possessing immunoreactive somatotropin-releasing activity currently begin characterized by Nair *et al.* (1978b) was dissolved in the medium to give a final concentration of 5 ng/μl. N^6-$O^{2'}$-dibutyryl adenosine-3'-5'-cyclic phosphate monosodium-5-H_2O (dbcAMP, mol.wt. 582.6, Calbiochem) was dissolved in the medium to give a final concentration of 4.6×10^{-7} M dbcAMP. Synthetic luteinizing hormone–releasing hormone (LH-RH) (Glu-His-Trp-Ser-Tyr-GLY-Leu-Arg-Pro-Gly-NH_2) was prepared by a solid-phase method and purified as described previously (Matsuo *et al.*, 1971) with minor modification to improve the yield. The superactive analogues, His^2-Me-D-Ala^6- and D-Ser^6-desGly^{10}-LH-RH-ethylamide (superactive analogues I and II), were synthesized and characterized by methods described elsewhere (Coy *et al.*, 1974; Nair *et al.*, 1978a; Nair, unpublished). These analogues were 25–30 times more potent than LH-RH. LH-RH or its superactive analogues were dissolved in saline to a final concentration of 15 ng/μl. Tissue was collected after sacrificing animals by guillotine at 1, 5, 15, 30, or 60 min after infusion. There were four experimental animals for each time period.

Three animals in each time period were infused with saline and lissamine green only. Additional controls consisted of sham-operated animals without infusion, unoperated animals, and tissue from infused glands in an area outside the distribution of the infused vessel (Fig. 2).

Before and after cannulation of a portal vessel and administration of infusates, a 1-ml sample of blood was collected from the femoral vein into a heparinized syringe. The blood was emptied into a heparinized collection tube in an ice bath and centrifuged. The serum was separated and stored at -40°C for radioimmunoassay (RIA).

Anterior pituitary glands were separated into infused portions and noninfused portions, then cut into pieces of 1 mm^3. Tissue was fixed in 6% glutaraldehyde with 1% hydrogen peroxide (Peracchia and Miller, 1972) in 0.1 M cacodylate buffer, pH 7.3, for 1 hr at room temperature, followed for 1 hr at 4°C by 6% glutaraldehyde without hydrogen peroxide added. The tissue was subsequently rinsed, postfixed in cacodylate-buffered 1% osmium tetroxide, dehydrated in graded alcohols, embedded in Epon 812, sectioned, doubly stained with uranyl acetate and lead citrate (Venable and Coggeshall, 1965), and viewed with a Hitachi 11E-2 microscope.

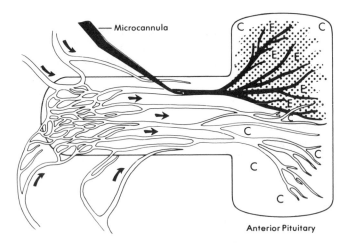

Figure 2. A typical connulated hypophyseal portal vessel indicating the area of distribution of the green infusate medium among the capilliary beds of the anterior pituitary gland. (E) Secretagogue-infused anterior pituitary tissue; (C) noninfused adjacent tissue.

Semithin sections stained with toluidine blue were used to identify corresponding thin sections.

Serum GH and luteinizing hormone (LH) levels were determined by RIA using the NIH rat-GH and LH-RIA kits.

2.1. Light-Microscopic Identification

The adenohypophysis is thought to contain at least five morphologically distinct hormone-secreting cell types: corticotrophs, gonadotrophs, mammotrophs, somatotrophs, and thyrotrophs. These secretory cell types can be distinguished using several approaches, either singly or in combination. Classic acid–base staining procedures such as the alcian blue–periodic acid–silver–Orange G method (Phifer *et al.,* 1973) and the aldehyde thionine–PAS-Orange G technique, as modified by Ezrin and Murray (1963), are used for the differentiation of basophil subclasses. The acidophil subclasses, which consist of mammotrophs and somatotrophs, can be distinguished by a variety of classic stains such as the azocarmine–Orange G technique of Brookes (1968). Immunoenzyme and immunocytochemical techniques (Phifer *et al.,* 1970; Nakane, 1970; Sternberger *et al.,* 1970; Moriarty, 1973; Baker and Gross, 1978; Bowie *et al.,* 1978) have clearly distinguished and helped to identify the somatotroph cell in a variety of species and in both sexes. The immunoenzyme and immunocytochemical techniques allowed an assessment of cell function in relation to the cell types. These techniques differentiated biochemical differences among the cells and focused new attention on many new biochemical similarities. In addition, these techniques demonstrated that the cells varied in appearance with the physiological state of the animal. Possibly the most important contribution of the new techniques was the demonstration by Phifer *et al.* (1974) that the one cell, one hormone theory was a fallacy. The number of hormones/secretory products secreted by the ACTH/MSH cell alone is staggering. Additional evidence has been presented to show that LH and FSH not only exist in the same cell but also are found within the same secretory granule.

Somatotroph cells have been found to lie in groups in certain areas of the anterior lobe (Daniel, 1966; Purves, 1966; Halmi *et al.*, 1975; Baker, 1974) or to be dispersed evenly throughout the gland (Nakane, 1970, 1975). The somatotroph cells are usually situated along sinusoids and vary in shape from oval to pyramidal with an approximate diameter of 10–13 μm (Nakane, 1970).

2.2. Electron-Microscopic Studies

2.2.1. Control Animals

The most common secretory cell type observed in the adenohypophysis was the somatotroph. Somatotroph cells were easily identified by their large numbers of dense, rounded secretory granules measuring approximately 350 nm in maximal diameter. A typical ovoid somatotroph is shown in Fig. 3, and corresponds to the description of others (Rinehart and Farquhar, 1953; Coates *et al.*, 1971; Costoff, 1973; Wilbur *et al.*, 1974, 1975, 1978). Rough endoplasmic reticulum (RER) was extensive, generally nondistended, and polarized often to one surface. An exhaustive examination of numerous sections revealed that only vacuoles and vesicles were associated with the Golgi complex. The Golgi lamellae were extensive but nondilated. Secretory granules varied in number from cell to cell, but were always distributed heterogeneously throughout the cell. Granules varied in diameter from 200 to 400 nm. Granule extrusion was rarely encountered in glands from untreated and sham-operated animals. Controls that had been infused with saline exhibited more exocytotic activity, but the number of exocytotic vesicles was small and varied from cell to cell. Acid phosphatase activity was restricted primarily to vacuoles and primary lysosomes.

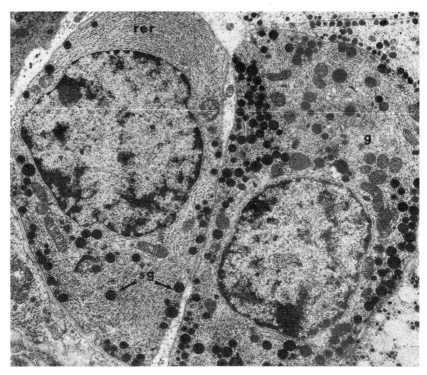

Figure 3. Somatotrophs from sham operated control animal. Rough endoplasmic reticulum (rer), Golgi complexes (g), and secretory granule (sg) size and distribution are characteristic of somatotrophs from control animals. Note the presence of coated vesicles along the plasma membrane. ×13,066 (reduced 48% for reproduction). From Wilbur et al. (1975).

2.2.2. Experimental Animals

Group I. Animals infused with dbcAMP for 30 min

1. Exocytotic activity did not occur in greater frequency than observed in controls.
2. RER involvement was greatest at 1 min following infusion, and gradually decreased to control levels 60 min following infusion.
3. Golgi complex involvement was greatest in the 30-min-postinfusion group, but an increase in Golgi-complex-associated structures such as microtubules and small smooth-surfaced and coated vesicles was seen as early as 15 min postinfusion.
4. Lysosomal activity was observed to be greatest at 60 min after the infusion period.

5. Serum GH was measured by RIA, and the results are shown in Table I.

Group II. Animals infused with dbcAMP for 1 min

1. Exocytotic activity was observed in the 1-, 5-, 15-, and 30-min groups in greater frequency than in infusion controls. The maximal amount of activity was observed at the earliest time period studied (Fig. 4).
2. RER involvement was greatest in the 15-min-postinfusion group. The RER was still distended 60 min postinfusion.
3. Golgi-complex involvement was maximal at 30 min postinfusion. An increase in the number of smooth-surfaced and coated vesicles was

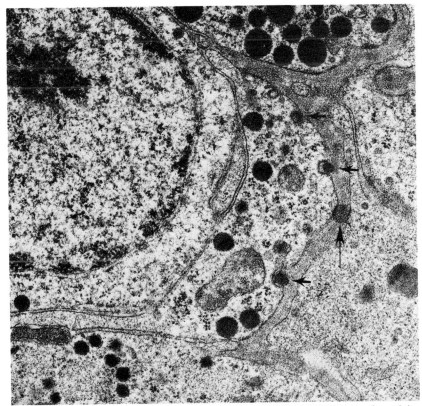

Figure 4. Somatotroph 1 min following 1-min infusion of dbcAMP. Exoctyotic activity (arrows) was most intense during this time period. Often the granule core (long arrow) would be seen intact within the intercellular space. ×38,920 (reduced 36% for reproduction). From Wilbur *et al.* (1974).

Table I. Serum GH (ng/ml) Before and After Hypophyseal Portal Vessel Infusion of Secretagogues[a]

Secretagogue	Preinfusion	Postinfusion				
		1 min	5 min	15 min	30 min	60 min
dbcAMP 30-min infusion	19.2 ± 2.0	34.3 ± 10.5[b]	22.3 ± 3.7	26.0 ± 4.8	20.8 ± 3.1	17.1 ± 2.5
1-min infusion	14.5 ± 2.1	32.7 ± 6.1[b]	79.0 ± 2.1[b]	38.3 ± 10.7[b]	31.7 ± 4.6[b]	13.3 ± 4.1
TRH	2.4 ± 0.4	9.0 ± 1.7[b]	16.3 ± 3.2[b]	7.7 ± 0.8[b]	6.0 ± 3.5	2.5 ± 0.5
New GH-RH	21.7 ± 4.7	69.9 ± 9.8[b]	62.0 ± 6.8[b]	55.7 ± 4.7[b]	32.6 ± 4.8[b]	19.2 ± 2.9

[a] Values are ng/ml NIAMDD-rat-GH-RP-1 and are expressed as the means ± S.E.M., 4 animals/group.
[b] $p < 0.01$, paired Student's t test and two-way ANOVA.

observed near the Golgi complex and along the plasma membrane as early as 5 min.

4. Serum GH levels were determined by RIA, and the results are shown in Table I.

Group III. Animals infused with hypophysiotropic hormones

1. Animals were infused with TRH, LH-RH, or a new growth hormone–releasing hormone (GH-RH).
2. Exocytotic activity was observed at all time periods studied (1, 5, 15, 30, and 60 min postinfusion). Extensive and multiple release of granules occurred at 1, 5, and 15 min postinfusion (Figs. 5 and 6) and decreased in frequency at 30 and 60 min postinfusion.
3. Golgi complex (Fig. 7), RER (Fig. 8), and lysosomal activities were similar to those observed at the same time intervals following the infusion of dbcAMP for 1 min.

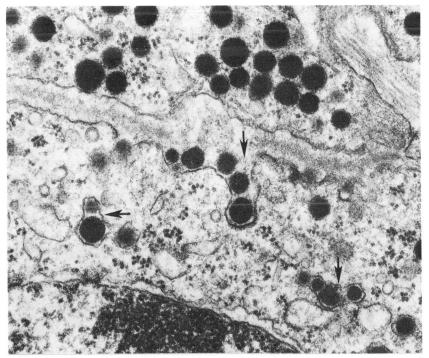

Figure 5. Somatotrophs 1 min following infusion of GH-RH displayed massive exocytotic activity, often with multiple granules coalescing and emerging through a common secretory packet (arrows). ×46,148 (reduced 44% for reproduction).

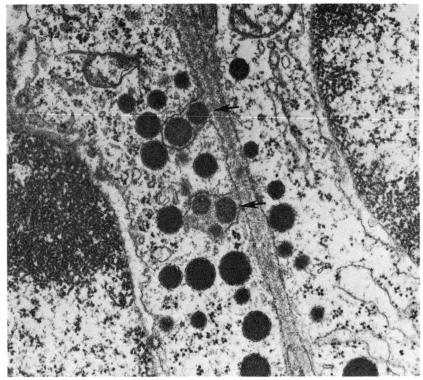

Figure 6. Somatotrophs 1 min following infusion of TRH displayed massive exocytotic activity, often with many granules coalescing and emerging through a common channel (arrows). ×43,368 (reduced 44% for reproduction). From Wilbur *et al.* (1978).

4. Serum GH and LH levels were determined by RIA, and the results are shown in Tables I and II.

Group IV. Animals infused with superactive analogues of the hypophysiotropic hormones

1. Animals were infused with His^2-Me-D-Ala^6- and D-Ser^6-desGly^{10}-LH-RH-ethylamide.
2. Several differences were apparent at 30 min between gonadotrophs of LH-RH- and analogue-infused pituitaries. The RER of analogue-stimulated cells was often greatly dilated and frequently coalesced to form large vacuoles filled with a dense, somewhat granular material. The Golgi area of analogue-stimulated gonadotrophs, which was often greatly enlarged compared to controls and LH-RH-stimulated gonadotrophs, contained many dense granules usually of the smaller sizes,

Table II.
Serum LH (ng/ml) before and after Hypophyseal Portal Vessel Infusion of LH-RH or Its Superactive Analogues[a]

Time postinfusion	LH-RH (30 ng/ml)			Analogue (3 ng/ml)		
	Before	After	p value[b]	Before	After	p value[b]
1 min	28.3 ± 10.9	173.8 ± 11.1	< 0.001	−	−	−
30 min	36.8 ± 11.3	156.7 ± 22.7	< 0.001	19.0 ± 2.7	421.7 ± 120.2	0.001
180 min	18.5 ± 2.1	30.5 ± 2.1	NS	16.5 ± 1.2	530.0 ± 105.1	0.001

[a] Values are ng/ml NIAMDD-rat-LH-RP-1 A and are expressed as the means ± S.E.M., 3 animals/group.
[b] Paired Student's *t* test.

condensing vacuoles, small coated vesicles, and a larger number of dilated cisternae. Exocytotic activity was much greater in analogue-stimulated cells (Luborsky-Moore *et al.*, 1978).

3. At 3 hr after LH-RH infusion, gonadotrophs resembled controls. Analogue-stimulated gonadotrophs were highly activated at 3 hr. The

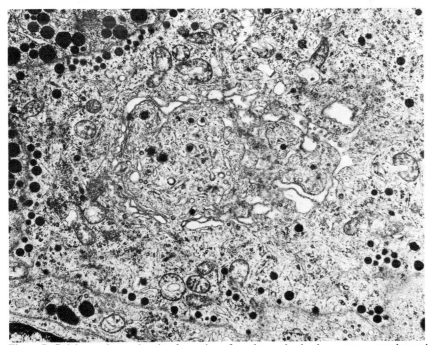

Figure 7. Golgi complexes showing formation of newly synthesized secretory granules and numerous microtubules associated with many smooth and coated vesicles 30 min following TRH infusion. ×35,028 (reduced 50% for reproduction).

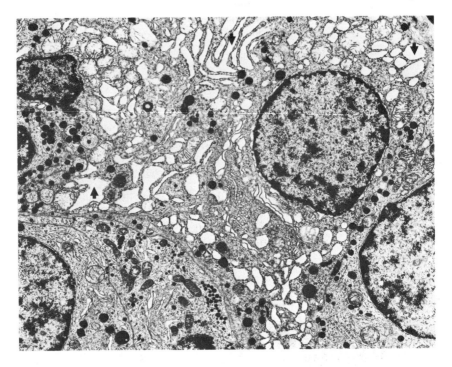

Figure 8. Somatotrophs 15 min following infusion of GH-RH displayed greatly dilated cisternae of the RER (arrows). ×26,688 (reduced 50% for reproduction).

RER and Golgi complex were similar to those observed at 30 min postinfusion. Exocytosis was still frequent and often involved a larger area of the cell surface than observed at earlier time periods. The release of larger granules, which previously were infrequently seen undergoing release, appeared to be slightly increased at 3 hr. The granule content of stimulated gonadotrophs was never observed to be completely depleted.
4. Serum LH was measured by RIA, and the results are shown in Table II.

3. Discussion

These studies indicate that the initial response following the infusion of secretagogues into a hypophyseal portal vessel is one of granule release. Exocytotic activity was observed consistently in most anterior pituitary cells at 1 and 5 min following infusion. Profiles of individual granules were often observed in various stages of the exocytotic process. In many other instances, several secretory granules were observed to coalesce into either large secretory packets or long channels containing many exocytotic images. This latter observation was due to the alignment of the secretory granules. A nonspecific exocytotic response was often observed following the infusion of the various secretagogues. This response was also observed in saline-infused controls, but the number of exocytotic vesicles was small and varied from cell to cell. Immunoassay serum levels indicated that the saline-infused response was not significantly different from sham-operated and noninfused control animals. A detailed morphometric analysis of this cellular event should provide information as to the frequency and specificity of this response.

Extensive dilations of the RER were observed in all cell types of the anterior pituitary, except for corticotrophs. The corticotrophs' lack of response may reflect a different level of sensitivity within the exocytotic and protein-synthesis compartments of these cells, since exocytosis was frequently observed. The extensive development of the RER at the early time period is consistent with studies implicating this organelle as the site of protein synthesis (Jamieson and Palade, 1971; Howell and Whitfield, 1973; Farquhar et al., 1978). Within 60 min after infusion, the RER would resemble the RER of control cells, thus indicating a quick emptying or transfer of newly synthesized protein from this compartment to another. The presence of fibrillar material in the dilated cisternae of the RER may represent newly synthesized protein, since this material is seen only after treatments that are known to stimulate the cells and disappears when the cells are incubated with protein-synthesis inhibitors (Farquhar and Rinehart, 1954; Moguilevsky et al., 1973).

Golgi complexes were maximally stimulated in these studies within 30 min following the infusion of the secretagogues. An increase in the number of smooth-surfaced and coated vesicles was observed near the Golgi complex and along the plasma membrane as early as 5 min postinfusion. Images were very often recorded showing coated vesicles in and near the plasma membrane of actively secreting cells, suggesting that the cell may utilize excessive plasma membranes, acquired from frequent granule release, in the formation of these vesicles. Recent findings by Farquhar (1978) demonstrate that the surface membrane is removed from the cell surface in vesicles and that the intracellular pathway taken by these vesicles is influenced by the net charge of the tracer. Thus, not only is the physiological state of the cell a factor in the regulation of vesicle traffic, but also the work of Farquhar (1978) clearly demonstrates that the charge of the tracer may also determine the vesicles' eventual fate. The ultrastructural changes observed in these studies, in the Golgi complex within 15 min postinfusion, together with additional observations showing the presence of secretory granules within Golgi complexes presumably in the formative stage, and increase TPPase localization in the Golgi lamellae (Wilbur *et al.*, 1978), provide additional evidence that this is a period of intense synthetic activity. At 60 min following the infusion of the secretagogues, Golgi complexes often contained vesicular lamellae, indicating that the Golgi complex may be slower than the RER in its compartmentalization of the newly synthesized protein and reutilization of its surface membrane.

The increased number of multilamellated and multivesicular bodies along with an increase in the number of acid-phosphatase-reactive bodies appears to be correlated with both granule depletion and continuing synthesis. These bodies may, as proposed by Smith and Farquhar (1966), provide a mechanism for degradation of excess secretory products and indicate an increase in membrane turnover rates associated with the Golgi complex or plasma membrane or both. Under the experimental conditions of these studies, many of the normal cellular processes may be greatly accelerated. Therefore, the increase in multilamellated and multivesicular bodies may represent a process of removing worn-out cellular organelles such as endoplasmic reticulum, plasma membrane, or mitochondria, as was reported by Dougherty (1973) in frog parathyroid cells.

Concerning the specificity of action of the various secretagogues used in these studies, dbcAMP was observed to be very nonspecific in stimulating exocytic activity; TRH was shown to stimulate exocytotic activity in thyrotrophs, gonadotrophs, and somatotrophs; and the new GH-RH consistently stimulated exocytotic activity in somatotrophs. These findings suggest that some common features may be shared by these cells, since amino acid sequence analysis of the releasing hormones for these cells indicates that one such common feature may be the pyroGlu-His sequence found on the N terminus. Exocytotic activity was seen in some corticotrophs, but this may be due to the stress of the surgical procedure, because there were fewer ACTH cells displaying exocytosis than observed after the infusion of dbcAMP. The new GH-RH appeared to be more specific for somatotrophs. The specificity of the LH-RH superactive analogues appears to be specific for gonadotrophs (Luborsky-Moore *et al.*, 1978), since only the FSH-type gonadotroph (Kurosumi and Oota, 1968; Moriarty, 1973) from analogue-infused pituitaries displayed exocytotic activity 3 hr following infusion. The possible mechanism for the prolonged activity of the superactive analogues of LH-RH may be due to changes in the structure and binding properties of the molecule rather than differences in metabolism, since the half-life of the D-Ala ethylamide analogues and native LH-RH is approximately the same (10–15 min) *in vivo* (Nair *et al.*, 1978a).

A probable explanation for our findings that dbcAMP stimulated a nonspecific exocytotic response may be found in the length of the infusion period and in the amount of time before a blood sample is collected. We found that following a 30-min infusion of dbcAMP, levels of GH at 30 and 60 min postinfusion are not significantly different from control levels. Furthermore, electron-microscopic examination of anterior pituitary tissue from rats at these time periods was similar to controls. Because the 30-min infusion may have masked exocytotic activity, our experiments were designed to collect tissue and plasma at the earliest possible time periods following a 1-min infusion.

The preceding studies outline our present understanding of the cellular events in somatropin secretion and its control. Several areas of research in need of future research have been exposed:

- How are coated vesicles involved in secretion and synthesis of hormones?
- Are microtubules and/or microfilaments involved with secretion and/or synthesis of hormones?
- What permits the secretory granule membrane to fuse with the plasma membrane? With other secretory granule membranes? With lysosomes?
- How is endocrinophagy controlled?
- How can different hormones be stored within the same secretory granule? What triggers their specific release?

These and many other questions await scientific investigation. With the experimentation and testing of better ideas along with advances in technology, these and other unsolved questions will be resolved.

ACKNOWLEDGMENTS. The authors wish to express their sincere thanks to Mr. Steve Raigue and Mrs. Kit Hargrove for their excellent technical assistance in this work, to Dr. R.M.O. Nair for his generous gifts of the various secreatagogues used in this work, and to Mrs. Janie Nelson for her secretarial assistance in the preparation of this manuscript. Figure 3 is reproduced with the permission of S. Karger Publishers, Basel, Switzerland. Figures 4 and 6 are reproduced with the permission of The Wistar Institute Press, Philadelphia, Pennsylvania.

DISCUSSION

TIXIER-VIDAL: I am surprised that following a stimulation of exocytosis, you found a modification of the Golgi zone after seeing a dilation of the cisternae of the RER. Could you comment on that in terms of regulation of the secretory process?

WILBUR: The route of secretion fits that proposed by Palade (1966, 1975) on the exocrine pancreas and by Farquhar (1971, 1977, 1978) on the intracellular transport and packaging of prolactin in the anterior pituitary. Perhaps the confusion stems from some authors distinguishing between the peripheral elements and the cisternae of the Golgi complex. The Golgi complex consists of a number of compartments, and if we consider coated and smooth-surfaced vesicles as solely belonging to the Golgi complex, then I could say we see changes in the Golgi complex at even earlier times. We see many new-coated and smooth-surfaced vesicles in the proximity of the Golgi complex at 5 and 10 minutes. I place the majority of Golgi changes—dilation of cisternae, condensation and formation of secretory granules, as observed in our studies—as occurring between 15 and 30 minutes following the administration of secretagogues. In other words, there is a short "lag" period of about 10 minutes between activation of RER elements as indicated by dilation of RER cisternae and the overall activation of the Golgi complex as indicated by increased numbers of smooth and coated vesicles, dilation of cisternae, condensation of granules, and formation of granules along with increased thiamine pyrophosphatase activity.

KRAICER: What were the concentrations of secretagogues used and how do they compare with those found or suggested as being in the hypophyseal–portal circulation?

WILBUR: The concentrations of the various secretagogues range from 3 ng/ml for the LH-RH analogue, 30 ng/ml for LH-RH, 5 ng/μl for TRH, and 5 to 10 ng/μl per min for 1 min for the new GH-RH. I believe the concentration of LH-RH agrees with those levels reported by Fink (see Chapter 29 in this volume). I am not aware of reported levels for the superanalogue LH-RH, TRH, or our new GH-RH.

DUVAL: I am surprised that you have found a secretory response after infusing such a low concentration of dbcAMP. In the slide, 15 minutes after injection of dbcAMP, I could not see any exocytosis pattern.

WILBUR: The concentration of dbcAMP used was 1.4 to 4.6 $\times$ 10^{-7} M, and I don't think of this amount as being so low. In our pilot studies using dbcAMP, we infused the compound for 30 minutes as described by Porter et al. (1970). Our findings indicated that following a 30-minute infusion of dbcAMP, levels of radioimmunoassayable growth hormone were not significantly different from controls at 5, 15, 30, and 60 minutes following infusion. Furthermore, electron-microscope examination of anterior pituitary tissue of rats at the same time periods was similar to controls. Because the 30-minute infusion may have masked exocytic activity, we designed our experiments to collect tissue and plasma at the earliest possible time periods following a 1-minute infusion. Our ultrastructural and immunoassay results showed GH levels significantly elevated ($p < 0.05$) over preinfusion levels at 1, 5, 15, and 30 minutes, but not at 60 minutes. The slide you are referring to represents a low-magnification transmission-electron micrograph of dbcAMP-infused tissue 30 minutes following infusion. We have not examined all the functional cell types following dbcAMP; therefore, we cannot say dbcAMP stimulates exocytosis in all cell types. What we can say is that it stimulates the release of GH.

VAUDRY: Have you infused, or are you planning to infuse, inhibiting factors such as somatostatin to study the fine-structural modifications of the pituitary cells?

WILBUR: We have not published data on somatostatin infusion or other inhibiting factors. We are presently studying SIF and MSH-IF in our laboratory.

MELDOLESI: One should realize that when cells are stimulated, a number of phenomena occur such as K$^+$ efflux and Ca^{2++} influx. These phenomena, especially in heavily stimulated cells, might themselves be responsible for ultrastructural changes or fixation problems. In addition, in talking of changes of structure such as the ER and Golgi, which undergo drastic changes in geometry, one should use with caution words such as hypertrophy which have direct quantitative meaning. In fact, in other cases, claims of Golgi and ER hypertrophy, which were based on subjective screening of the pictures, were not confirmed by morphometric analysis of the data.

WILBUR: We are well aware of the influence of K^+ efflux, Ca^{2++} influx, and redistribution. For more detailed discussion of these phenomena, see the appropriate chapters in this monograph. In regrads to fixation, we have tried a number of fixatives such as (a) 2.5% glutaraldehyde and 0.1 M cacodylate, pH 7.4; (b) Karnovsky's high osmolarity fixative, 5% glutaraldehyde–4% paraformaldehyde; (c) 3% glutaraldehyde in 0.067 M cacodylate, pH 7.4; or (d) 6% glutaraldehyde in 0.1 M cacodylate. In each case where one compared the morphologic appearance of pituitaries of control animals (controls were sham-operated), surgical controls, saline-infused controls, and tissue collected from secretagogue-infused animals—but outside the area of infusion—vs. secretagogue-infused animals, we could easily observe the difference. Furthermore, you should be aware of the double-blind aspects of our study. The pituitaries are cannulated by Wilbur and given to an EM technician. The EM technician codes the tissue after it has been sectioned and stained. I don't know what the tissue represents at the time I examine it for ultrastructural observations. In every case after an exhaustive examination of tissue, blocks, and grids, the ultrastructural changes between controls vs. secretagogue-infused are so striking one can easily identify the secretagogue-infused tissue. We are well aware of the need for morphometric analysis of the data and are presently undertaking this enormous task.

SPICER: A cell packaging secretory product and at the same time transporting acid phosphatase and other acid hydrolases to secondary lysosomes appears to require specialized Golgi elements for each of these two different activities. Possibly conventional Golgi cisternae could mediate the former and GERL the latter. The function of GERL in packaging secretory product or recycling cell membranes to lysosomes or some other identified cell activity seems unclear at present.

VAUDRY: Has Dr. Hymer or Dr. Kraicer tried during the purification of somatotrophs to characterize other cell fractions, especially thyrotrophs or corticotrophs? Since corticotrophs are found only in low numbers in anterior pituitary and considering the difficulties inherent in ACTH measurement, this approach would facilitate the study of the mechanisms controlling ACTH secretion.

KRAICER: We initially reported on the purification of thyrotrophs from glands of PSV-treated rats in 1973. This treatment was used to increase the number and size of the TSH-producing cells, thereby permitting effective separations (up to 60% purity) by the IG technique. A year ago, Dr. Moriarty (Childs) also reported studies on the functional behavior of thyrotrophs isolated by the IG technique. Corticotrophs, on the other hand, have received less attention. We are currently attempting their isolation from the bovine gland.

MIALHE: Is the endocytosis mechanism compatible with the concept of receptor sites, adenyl cyclase activation, and cyclic AMP activity?

LABRIE: This endocytosis process is apparently a general phenomenon in most cell types. It is likely that the components that become internalized include not only receptors and associated hormones, but also adenyl cyclase components. They may well continue to be active for some time inside the cell.

References

Baker, B. L., 1974, Functional cytology of the hypophysial pars distalis and pars intermedia, in: *The Pituitary Gland and Its Control—Adenohypophysis, Handbook of Physiology,* Endocrinology IV, Part 1 (R.O. Greep and E.B. Astwood, eds.), pp. 45–80, American Physiology Society, Bethesda, Maryland.

Baker, B. L., And Gross, D. S., 1978, Cytology and distribution of secretory cell types in the mouse hypophysis as demonstrated with immunocytochemistry, *Am. J. Anat.* **153:**193–216.

Bowie, E. P., Ishikawa, H., Shiino, M., and Rennels, E. G., 1978, An immunocytochemical study of a rat pituitary multipotential clone, *J. Histochem. Cytochem.* **26:**94–97.

Brookes, L. D., 1968, A stain for differentiating two types of acidophil cells in the rat pituitary, *Stain Technol.* **43:**41.

Coates, P. W., Ashby, E. A., Krulich, L., Dhariwal, A. P. S., and McCann, S. M., 1971, Morphologic alterations in somatotrophs of the rat adenohypophysis following administration of hypothalamic extracts, *Am. J. Anat.* **128:**389–412.

Costoff, A., 1973, *Ultrastructure of Rat Adenohypophysis: Correlation with Function,* Academic Press, New York.

Coy, D. H., Coy, E. J., Schally, A. V., Vilchez-Martinez, J., Hirotsu, Y., and Arimura, A., 1974, Synthesis and biological properties of (D-Ala^6-des-Gly-NH$^{10}_2$)-LH-RH ethylamide, a peptide with greatly enhanced LH and FSH releasing activity, *Biochem. Biophys. Res. Commun.* **57:**335–337.

Daniel, P. M., 1966, The anatomy of the hypothalamus and pituitary gland, in: *Neuroendocrinology* (L. Martini and W.F. Ganong, eds.), Vol. I, pp. 15–80, Academic Press, New York.

Dougherty, W. J., 1973, Ultrastructural changes in the secretory cycle of parathyroid cells of winter frogs (*Rana pipiens*) after pituitary homoimplantation, *Z. Zellforsch.* **146:**167–175.

Ezrin, C., and Murray, S., 1963, The cells of the human adenohypophysis in pregnancy, thyroid disease, and adrenal cortical disorder, in: *Cytologie de l'Adenohypophyse* (J. Benoit and C. DaLage, eds.), pp. 183–200, Editions du C.N.R.S., Paris.

Farquhar, M. G., 1971, Processing of secretory products by cells of the anterior pituitary gland, *Mem. Soc. Endocrinol.* **19:**79–122.

Farquhar, M. G., 1977, Secretion and crinophagy in prolactin cells, in: *Comparative Endocrinology of Prolactin,* Vol. 80 (H.D. Dellman, J.A. Johnson, and D.M. Klachko, eds.), *Advances in Experimental Medicine and Biology,* pp. 37–94, Plenum Press, New York.

Farquhar, M. G., 1978, Recovery of surface membrane in anterior pituitary cells: Variations in traffic detected with anionic and cationic ferritin, *J. Cell. Biol.* **77**(3):R35–R42.

Farquhar, M. G., and Rinehart, J. F., 1954, Cytologic alterations in the anterior pituitary gland following thyroidectomy: An electron microscopic study, *Endocrinology* **55:**857–876.

Farquhar, M. G., Reid, J. J., and Daniell, L. W., 1978, Intracellular transport and packaging of prolactin: A quantitative electron-microscopic autoradiographic study of mammotrophs dissociated from rat pituitaries, *Endocrinology* **102:**296–311.

Halmi, N. S., Parsons, J. A., Erlandsen, S. L., and Duello, T., 1975, Prolactin and growth hormone cells in the human hypophysis: A study with immunoenzyme histochemistry and differential staining, *Cell Tissue Res.* **158:**497.

Howell, S. L., and Whitfield, M., 1973, Synthesis and secretion of growth hormone in the rat anterior pituitary. I. The intracellular pathway, its time course and energy requirements, *J. Cell. Sci.* **12:**1–21.

Jamieson, J. D., and Palade, G. E., 1971, Synthesis, intracellular transport and discharge of secretory proteins in stimulated pancreatic exocrine cells, *J. Cell Biol.* **50**:135–158.

Kurosumi, K., and Oota, Y., 1968, Electron microscopy of two types of gonadotrophs in the anterior pituitary glands of persistent estrous and diestrous rats, *Z. Zellforsch. Mikrosk. Anat.* **85**:34–46.

Luborsky-Moore, J. L., Nair, R. M. G., Poliakoff, S. J., and Worthington, W. C., Jr., 1978, Stimulation of gonadotrophs by pituitary portal vessel infusion of superactive LH-RH analogues: An ultrastructural study, *Neuroendocrinology* **26**:93–107.

Matsuo, H., Arimura, A., Nair, R. M. G., and Schally, A. V., 1971, Synthesis of the LH and FSH-releasing hormone by the solid phase method, *Biochem. Biophys. Res. Commun.* **45**:822–827.

Moguilevsky, J. A., Cuerdo-Rocha, S., Christot, J., and Zambrano, D., 1973, The effect of thyrotrophic releasing factor on gland: A biochemical and ultrastructural study, *J. Endocrinol.* **56**:99–109.

Moriarty, G. C., 1973, Adenohypophysis: Ultrastructural cytochemistry—A review, *J. Histochem. Cytochem.* **21**:855–894.

Nair, R. M. G., Sagel, J., Colwell, J. A., Mathur, R. S., Powers, J. M., Luborsky-Moore, J. L., and Worthington, W. C., 1978a, Modern concepts on the structure–activity relationships and the mechanism of action of gonadotropin-releasing hormone and its superactive analogues, in: *Hypothalamic Hormones—Chemistry, Physiology, and Clinical Applications* (D. Gupta and W. Voelter, eds.), pp. 21–45, Verlag Chemie, Weinheim, New York.

Nair, R. M. G., DeVillier, C., Barnes, M., Antalis, J., and Wilbur, D. L., 1978b, A bovine hypothalamic peptide possessing immunoreactive growth hormone-releasing activity, *Endocrinology* **103**:112–120.

Nakane, P. K., 1970, Classification of anterior pituitary cell types with immunoenzyme histochemistry, *J. Histochem. Cytochem.* **18**:9–20.

Nakane, P. K., 1975, Identification of anterior pituitary cells by immunocytochemistry, in: *The Anterior Pituitary* (A. Tixier-Vidal and M.G. Farquhar, eds.), pp. 45–61, Academic Press, New York.

Palade, G. E., 1966, Structure and function at the cellular level, *Am. Med. Assoc.* **198**:815–825.

Palade, G. E., 1975, Intracellular aspects of the process of protein secretion, *Science* **189**:347–358.

Peracchia, C., and Miller, B. S., 1972, Fixation by means of glutaraldehyde–hydrogen peroxide reaction products, *J. Cell Biol.* **53**:234–238.

Phifer, R. F., Spicer, S. S., and Orth, D. N., 1970, Specific demonstration of the human hypophyseal cells which produce adrenocorticotropic hormone, *J. Clin. Endocrinol. Metab.* **31**:347.

Phifer, R. F., Midgley, A. R., and Spicer, S. S., 1973, Immunohistologic and histologic evidence that FSH and LH are present in the same cell type in the human pars distalis, *J. Clin. Endocrinol. Metab.* **36**:125–142.

Phifer, R. F., Orth, D. N., and Spicer, S. S., 1974, Specific demonstration of the human hypophyseal adrenocortico–melanotropic (ACTH/MSH) cell, *J. Clin. Endocrinol. Metab.* **39**:684.

Porter, J. C., Mical, R. S., Kamberi, J. A., and Grazia, Y. R., 1970, A procedure for the cannulation of a pituitary stalk portal vessel and perfusion of the pars distalis in the rat, *Endocrinology* **87**:197–201.

Purves, H. D., 1966, Cytology of the adenohypophysis. in: *The Pituitary Gland,* Vol. I, *Anterior Pituitary* (G. W. Harris and B. T. Donovan, eds.), pp. 147–232, Butterworths, London.

Rinehart, J. F., and Farquhar, M. G., 1953, Electron microscopic studies of the anterior pituitary gland, *J. Histochem. Cytochem.* **1**:93–113.

Smith, R. E., and Farquhar, M. G., 1966, Lysosomal function in the regulation of the secretory process in cells of the anterior pituitary gland, *J. Cell Biol.* **31**:319–347.

Sternberger, L. A., Hardy, P. H., Jr., Cuculis, J. J., and Meyer, H. G., 1970, The unlabelled antibody enzyme method of immunohistochemistry: Preparation and properties of soluble antigen–antibody complex (horseradish peroxidase–anti-horseradish peroxidase) and its use in identification of spirochetes, *J. Histochem. Cytochem.* **18**:315.

Venable, J. H., and Coggeshall, R., 1965, A simplified lead citrate stain for use in electron microscopy, *J. Cell Biol.* **25**:407–408.

Wilbur, D. L., Worthington, W. C., Jr., and Markwald, R. R., 1974, Ultrastructural observations of anterior pituitary somatotrophs following pituitary portal vessel infusion of dibutyryl cAMP, *Am. J. Anat.* **141**:147–154.

Wilbur, D. L., Worthington, W. C., Jr., and Markwald, R. R., 1975, An ultrastructural and radioimmunoassay study of anterior pituitary somatotrophs following pituitary portal vessel infusion of growth hormone releasing factor, *Neuroendocrinology* **19**:12–27.

Wilbur, D. L., Yee, J. A., and Raigue, S. E., 1978, Hypophysial portal vessel infusion of TRH in the rat: An ultrastructural and radioimmunoassay study, *Am. J. Anat.* **151**:277–294.

Worthington, W. C., Jr., 1955, Some observations on the hypophysial portal system in the living mouse, *Bull. Johns Hopkins Hosp.* **97**:343–357.

8

Ultrastructural Localization of LH and FSH in the Porcine Pituitary

F. Dacheux

1. Introduction

In the porcine pituitary, Mirecka and Pearse (1971), using the immunofluorescence technique, concluded that the majority of gonadotrophs contained both gonadotropins. However, Herlant and Ectors (1969) and Herlant (1972) described, at the electron-microscope level, two cells types as LH- and FSH-secreting cells. Recently, using specific antibodies against β subunits of porcine LH and FSH, we have clearly shown in the porcine pituitary that LH and FSH are contained within the same cell (FSH/LH cell), but that the concentration of LH and FSH can vary from cell to cell (Dacheux, 1978). In the FSH/LH cells, the finding of granules positive for both LHβ and FSHβ demonstrated that two hormones could be stored in the same granules (Dacheux, 1978). Using the same technique on dissociated cells from porcine pituitary, Batten and Hopkins (1978) obtained similar results. The aim of the study presented in this chapter was to determine the precise localization of FSHβ, LHβ, and LHα in the gonadotropic cells and to determine whether all the granules contain the same proportion of LH and FSH.

F. Dacheux • Laboratoire de Physiologie Comparée, Faculté des Sciences, 37200 Tours, France; and I.N.R.A.–Station de Physiologie de Reproduction, 37380 Monnaie, France

2. Materials and Methods

Male porcine pituitaries were fixed in glutaraldehyde or in picric acid–formaldehyde without OsO_4 postfixation, since osmium tetroxide destroys the immunoreactivity of $FSH\beta$ (Dacheux, 1978). Antisera against porcine $LH\alpha$, $LH\beta$ (A-$pLH\alpha$ and A-$pLH\beta$), and $FSH\beta$ subunits (A-$pFSH\beta$) as well as antisera against rat $LH\alpha$, $LH\beta$, and $FSH\beta$ were used. The antisera were kindly provided as follows: A-$pLH\alpha$ and A-$pLH\beta$ by Dr. M.P. Dubois (I.N.R.A., Tours, France); A-$pFSH\beta$ by Dr. J. Closset (Université de Liège, Belgique); and A-$rLH\alpha$, A-$rLH\beta$, and A-$rFSH\beta$ by Dr. A.F. Parlow (National Institute of Arthritis, Metabolism and Digestive Diseases, Bethesda, Maryland). The immunocytochemical reaction was performed by the peroxidase–antiperoxidase (Sternberger et al., 1970) unlabeled antibody method (Petrali et al., 1974) as previously described (Dacheux and Dubois, 1976). The specificity of the antisera was tested with antisera absorbed with different doses of glycoprotein hormones (Dacheux, 1978). Numerous consecutive ultrathin serial sections were collected to observe the same granules on adjacent sections stained for (1) $FSH\beta$, (2) $LH\beta$, and (3) $LH\alpha$.

3. Results

3.1. FSH/LH Cells

The observation of corresponding areas of the same cell stained for $LH\beta$ and $FSH\beta$ showed that very often the same granules were strongly positive for both $FSH\beta$ and $LH\beta$ (Fig. 1A and B). However, in the cells observed, there were always some rare granules weakly stained for one hormone ($LH\beta$ or $FSH\beta$) but strongly reactive for the other ($FSH\beta$ or $LH\beta$) (Fig. 1A and B). We have verified that this variation of intensity was not the result of a tangential section of the granule before its disappearance from the plane of the section from one section to the next. Thus, the proportion of FSH and LH within the secretory granules could vary, but only in a very small number of granules. In the FSH/LH cells, the granules positive for both $LH\beta$ and $FSH\beta$ were also reactive for $LH\alpha$ (Fig. 2A–C).

3.2. "Intermediate" Cells, "LH-Only Cells," and "FSH-Only Cells"

Some cells strongly reactive for $FSH\beta$ were weakly stained for $LH\beta$ (intermediate cells) (Fig. 3A and B, cell 2) or were negative for $LH\beta$ (FSH-only cells) (Fig. 3A and B, cell 3). Other cells strongly reactive for

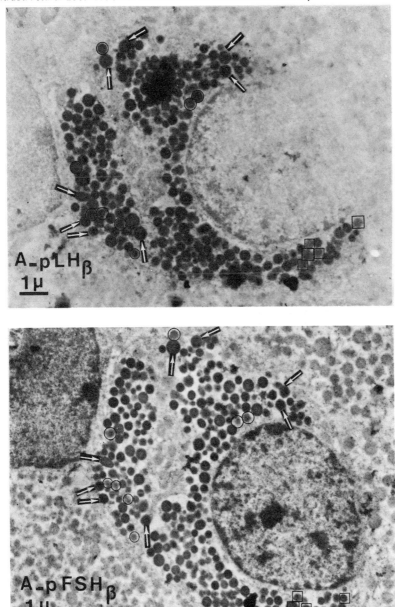

Figure 1. Serial sections of the gonadotropic cell treated with A-pLHβ and A-pFSHβ show that numerous granules are strongly reactive for both FSHβ and LHβ (granules indicated by arrows). However, some granules weakly stained for LHβ are strongly reactive for FSHβ (granules enclosed in squares), and other granules weakly stained for FSHβ are strongly reactive for LHβ (granules enclosed in circles).

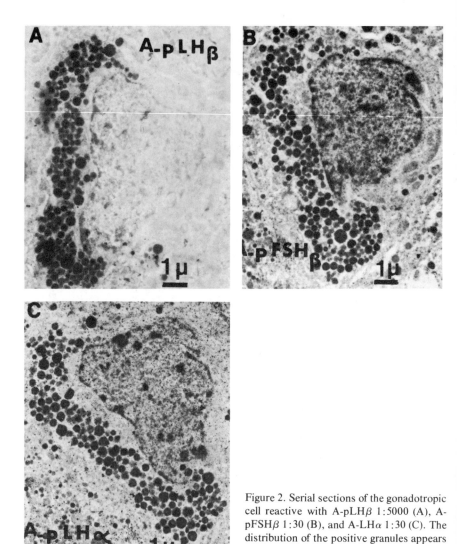

Figure 2. Serial sections of the gonadotropic cell reactive with A-pLHβ 1:5000 (A), A-pFSHβ 1:30 (B), and A-LHα 1:30 (C). The distribution of the positive granules appears identical.

LHβ were slightly stained or negative for FSHβ (LH-only cells). Thus, the proportion of LH and FSH may vary from cell to cell. In these cells, for one hormone, all the granules presented the same staining intensity (Fig. 3A and B, cell 2). These results indicated that the proportion of LH and FSH was similar (or varied in the same manner) for all granules of the same cell. The LH-only cells and FSH-only cells also contained LHα in the same granules (Fig. 4).

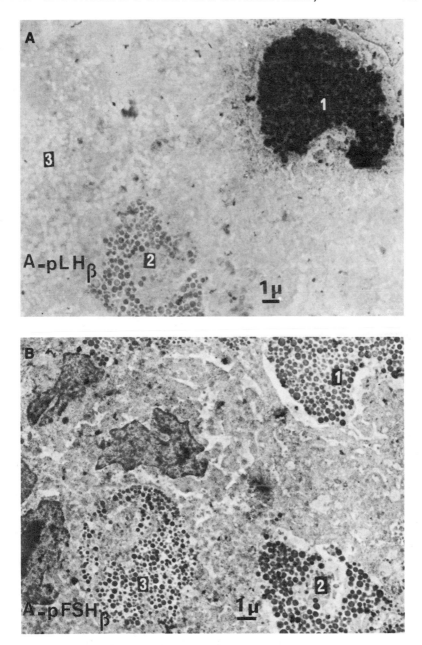

Figure 3. Serial sections treated with A-pLHβ 1:5000 (A) and with A-pFSHβ 1:30. (B). Cell 1 is strongly stained for both LHβ (A) and FSHβ (B); cell 2 is positive for FSHβ (B) but weakly stained for LHβ (A) cell 3 is positive only for FSHβ (B) and is negative for LHβ (A).

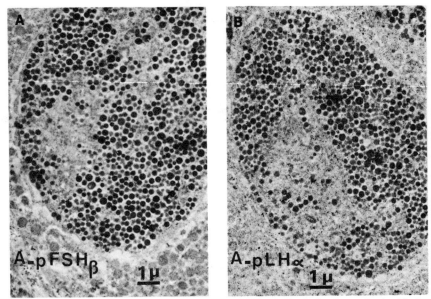

Figure 4. An "FSH-only cell" reactive with A-pFSHβ 1:30 (A) is also stained with A-pLHα 1:30 (B).

4. Discussion

In the present study, the specificity of the immunoperoxidase staining procedure can be regarded as established. We have previously shown that both A-pLHβ and A-pFSHβ were specific for their respective β subunits. Staining was inhibited only by the specific antigen and was not abolished after absorption with other glycoprotein hormones (Dacheux, 1978). Our investigation indicates that in the porcine pituitary, there are several populations of gonadotropic cells: (1) very numerous FSH/LH cells; (2) some intermediate cells; (3) rare LH-only cells; (4) rare FSH-only cells. In the porcine pituitary, most of the gonadotropic cells are responsible for the production of both LH and FSH. Our results are in agreement with studies in other species, including rat (Nakane, 1970; Tixier-Vidal et al., 1975; Herbert, 1975; Denef et al., 1976), fox (Bugnon et al., 1974), monkey (Herbert, 1976), man (Phifer et al., 1973; Robyn et al., 1973; Pelletier et al., 1976), and dog (El Etreby and Fath El Bab, 1977). However, some rare cells containing either LH or FSH confirm the results of Nakane (1970), Moriarty (1976), Pelletier et al. (1976), and El Etreby and Fath El Bab (1977), who explained this in terms of subpopulations of LH- or FSH-specific cells. In the FSH/LH cells, most of the secretory granules strongly reactive for both hormones show that the proportion of LH and FSH is similar for many granules, although some rare granules seem to contain less LH or less FSH. In some cells (intermediate cells, LH-only cells, FSH-only cells), the proportion of LH and FSH can vary from cell to cell, but the equal staining intensity of the granules indicates that the concentration of LH and FSH is similar (or varies in the same manner) for all the granules. In conclusion, our observations indicate that in the porcine pituitary, most of the gonadotropic cells contain both LH and FSH, but the proportion of LH and FSH in the secretory granules may vary from cell to cell but not among the granules of a single cell.

5. Summary

Using the peroxidase–antiperoxidase unlabeled antibody method and antisera against β subunits of porcine and rat LH and FSH, it was clearly shown in the porcine pituitary that most of the gonadotropic cells are responsible for the synthesis of both LH and FSH. In the FSH/LH cells, the same secretory granules strongly reactive for both hormones show that the proportion of LH and FSH is similar for many granules. However, in some cells (intermediate cells, LH-only cells, FSH-only cells), the proportion of LH and FSH within the secretory granules can be different but is identical for all the granules of the same cell.

DISCUSSION

TOUGARD:　What is the respective titer of each antiserum used: anti-pLHβ and anti-pFSHβ?

DACHEUX:　For immunocytochemical staining purposes, the titers were 1:5000 for anti-pLHβ and 1:30 for anti-pFSHβ.

C. VILLEE:　Is it possible that there is really only one type of gonadotroph? Perhaps the cells staining with only anti-FSH have discharged their LH just before the tissue was fixed and the cells staining with only anti-LH have discharged their FSH.

DACHEUX:　Various cell populations have been observed: numerous "LH and FSH cells" containing both the two hormones in the same granules; some "intermediate cells," in which the amount of one or the two hormones is very low; rare "FSH-only cells"; and rare "LH-only cells."
　　If all the granules containing both LH and FSH have been released from the "LH and FSH cells," there would be only the rare "FSH-only granules" or rare "LH-only granules" left. In fact, when we observe "FSH-only" cells or "LH-only" cells, the granules are quite abundant. These observations would not fit with your concept.

SPICER:　What proportion of gonadotrophs do the FSH- and LH-secreting cells comprise? Also, what percentage of cells are secreting only FSH or LH?

DACHEUX:　Most of the gonadotrophic cells contain both LH and FSH; only some cells contain FSH alone, some cells contain LH alone.

SPICER:　It seems unlikely, in reply to Dr. Villee's question, that LH cells are those that recently secreted their FSH. Since there is no granule heterogeneity in gonadotrophs, but only a single uniform granule population, no mechanism can be imagined by which the gonadotrophs containing both hormones in all their granules can secrete FSH and retain LH.

REFERENCES

Batten, T. F. C., and Hopkins, C. R., 1978, Discrimination of LH, FSH, TSH and ACTH in dissociated porcine anterior pituitary cells by light and electron microscope immunocytochemistry, *Cell Tissue Res.* **192:**107.

Bugnon, C., Lenys, D., Kerdelhué, B., Dessy, C., and Fellmann, D., 1974, Étude cytoimmunologique des cellules gonadotropes du renard roux, *C. R. Soc. Biol.* **168:**814.

Dacheux, F., 1978, Ultrastructural localization of gonadotrophic hormones in the porcine pituitary using the immunoperoxidase technique, *Cell Tissue Res.* **191:**219.

Dacheux, F., and Dubois, M. P., 1976, Ultrastructural localization of prolactin, growth hormone and luteinizing hormone by immunocytochemical techniques in the bovine pituitary, *Cell Tissue Res.* **174:**245.

Denef, C., Hautekeete, E., and Rubin, L., 1976, A specific population of gonadotrophs purified from immature female rat pituitary, *Science* **194:**848.

El Etreby, M. F., and Fath El Bab, M. R., 1977, Localization of gonadotrophic hormones in the dog pituitary gland, *Cell Tissue Res.* **183:**167.

Herbert, D. C., 1975, Localization of antisera to LH and FSH in the rat pituitary gland, *Am. J. Anat.* **144:**379.

Herbert, D. C., 1976, Immunocytochemical evidence that luteinizing hormone (LH) and follicle stimulating hormone (FSH) are present in the same cell type in the rhesus monkey pituitary gland, *Endocrinology* **98:**1554.

Herlant, M., 1972, Ultrastructure des cellules gonadotropes de l'hypophyse chez les mammifères, in: *Hormones Glycoprotéiques Hypophysaires* (M. Jutisz, ed.), pp. 5–29, Inserm, Paris

Herlant, M., and Ectors, F., 1969, Les cellules gonadotropes de l'hypophyse chez le porc, *Z. Zellforsch.* **101:**221.

Mirecka, J., and Pearse, A. G. E., 1971, Localization of FSH and LH-producing cells in the pig adenohypophysis by an immunohistochemical technique, *Folia Histochem. Cytochem.* **9:**365.

Moriarty, G. C., 1976, Immunocytochemistry of the pituitary glycoprotein hormone, *J. Histochem. Cytochem.* **24:**846.

Nakane, P. K., 1970, Classifications of anterior pituitary cell types with immunoenzyme histochemistry, *J. Histochem. Cytochem.* **18:**9.

Pelletier, G., Leclerc, R., and Labrie, F., 1976, Identification of gonadotropic cells in the human pituitary by immunoperoxidase technique, *Mol. Cell. Endocrinol.* **6:**123.

Petrali, J. P., Hinton, D. M., Moriarty, G. C., and Sternberger, L. A., 1974, The unlabeled antibody enzyme method of immunocytochemistry: Quantitative comparison of sensitivities with and without peroxidase complex, *J. Histochem. Cytochem.* **22:**782.

Phifer, R. F., Midgley, A. R., and Spicer, S. S., 1973, Immunohistologic and histologic evidence that follicle-stimulating hormone and luteinizing hormone are present in the same cell type in the human pars distalis, *J. Clin. Endocrinol. Metab.* **36:**125.

Robyn, C., Leleux, P., Vanhaelst, L., Golstein, J., Herlant, M., and Pasteels, J. L., 1973, Immunohistochemical study of the human pituitary with anti-luteinizing hormone, anti-follicle stimulating hormone and anti-thyrotrophin sera, *Acta Endocrinol. (Copenhagen)* **72:**625.

Sternberger, L. A., Hardy, P. H., Jr, Cuculis, J. J., and Meyer, H. G., 1970, The unlabeled antibody enzyme method of immuno-histochemistry: Preparation and properties of soluble antigen–antibody complex (horseradish peroxidase–antiperoxidase) and its use in identification of spirochetes, *J. Histochem. Cytochem.* **18:**315.

Tixier-Vidal, A., Tougard, C., Kerdelhué, B., and Jutisz, M., 1975, Light and electron microscopic studies on immunocytochemical localization of gonadotropic hormones in the rat pituitary gland with antisera against ovine FSH, LH, LHα and LHβ, *Ann. N. Y. Acad. Sci.* **254:**433.

9

Localization of Anti-ACTH, Anti-MSH, and Anti-α-Endorphin Reactive Sites in the Fish Pituitary

Ernest Follénius and Maurice P. Dubois

1. Introduction

Studies of the relationships among ACTH, MSH, and lipotropin under way for several years have recently yielded important conclusions. More and more arguments are accumulating in favor of common pathways for the synthesis of a whole family of hormonal peptides including ACTH, MSH, β-lipotropin (β-LPH) and the endorphins.

In mammals, Li et al. (1965) have shown that the β-LPH molecule contains the sequence corresponding to β-MSH, and according to Guillemin et al. (1976), sequences of its COOH-terminal part are identical to α-endorphin (β-LPH 61–76) or β-endorphin (β-LPH 61–91). As proposed by Mains et al. (1977) and Allen et al. (1978), pituitary tumor cells in culture release several peptides similar in molecular weight to "big ACTH"— ACTH, β-LPH and β-endorphin. Even nonpituitary tumor tissue, causing the ectopic ACTH syndrome, contains immunoreactive ACTH, lipotropins, and α and β-endorphins (Orth et al., 1978). But the cellular origin of these peptides could not be traced very far with these methods. Only with cell isolation from the rat intermediate lobe has it been possible to localize

Ernest Follénius • Laboratoire de Cytologie Animale and E.R.A. 492 du C.N.R.S., Université Louis Pasteur, 67000 Strasbourg, France   *Maurice P. Dubois* • Station "Physiologie de la Reproduction" de l'I.N.R.A., 37380 Nouzilly, France

biochemically the synthesis of the high-molecular-weight precursor, of β-LPH and of β-endorphin in this cell type (Crine et al., 1978).

For cell localization, immunocytology has long been making significant contributions. It indicates a possible relationship between the synthesis of ACTH and that of MSH, which had also been postulated as a result of biochemical studies (cf. Lowry and Scott, 1977). From their immunoreaction, both peptides were localized in pars intermedia cells under the light and electron microscopes. Their coexistence in the same cell type has been demonstrated in mammals (Dubois, 1972a,b; Moriarty and Halmi, 1972; Naik, 1973; Roux and Dubois, 1976) and in amphibians (Doerr-Schott and Dubois, 1972). Moreover, they occur together in the same cell type of the human fetal pituitary (Bégeot et al., 1978). In some species of mammals, β-LPH has been revealed in both corticotropic and melanotropic cells (Dubois, 1972b; Dessy et al., 1973; Dessy and Herlant, 1974; Moon et al., 1973; Pelletier et al., 1977; Bégeot et al., 1978) and also in plasma (Bachelot et al., 1977). The presence of ACTH and β-LPH in the same secretory granules (Moriarty, 1973; Pelletier et al., 1977) suggested the possibility of a common metabolic origin.

It was therefore of particular interest to localize the endorphins, which, as mentioned earlier, represent parts of the β-LPH molecule. Soon after their characterization (Guillemin et al., 1976), they were localized in intermediate-lobe cells and in some anterior-lobe cells of the rat (Bloom et al., 1977; Dubois, 1977).

We sought to take advantage of the immunocytological approach to localize the various fish peptides cross-reacting with anti-ACTH, anti-MSH, and anti-α-endorphin prepared from antigens of mammalian or synthetic origin. This approach has obvious limits due to problems of immunospecificity in heterologous systems, especially with fixed cells. The fixation technique may modify the native peptides, especially by fragmentation of high-molecular-weight precursors, thus making immuno-cytological detection easier. Immunocytological findings, duly checked, may, however, prove very useful in preliminary discussions on the phylogenetic extent of the synthetic pathways leading to ACTH, MSH, and endorphin-like substances in our oldest and most primitive vertebrates. The immunological reactions between fish pituitary hormones and anti-mammalian hormone antibodies indicate common sequences, the extent and biological significance of which need checking.

2. Immunocytochemical Techniques

In our studies, ACTH, α-MSH, and endorphin were detected immu-nocytologically with Coon's indirect method (cf. Dubois, 1972a). Details of the preparation of the β-1–24 ACTH antiserum (AS), the anti 17–39

ACTH AS and the α-MSH and β-MSH AS are given in the same study. The endorphin AS were produced in rabbits by multiple intradermal injections of a purified preparation of porcine α-endorphin (Dr. Guillemin, San Diego, California), coupled with human serum albumin and emulsified with complete Freund's adjuvant (vol./vol.). The cross-reaction of the α-endorphin AS with β-LPH is far from negligible (cf. Follénius and Dubois, 1978a). For the immunocytological detection of endorphin, it was important to ensure that the α-endorphin AS could not react with either ACTH or MSH present in the same cells. To exclude this possibility, a series of slides with three groups of sections were incubated as follows: group 1: incubation with α-endorphin AS pretreated with 12, 22, or 45 μg of β-1–24 ACTH (Ciba, Basel) per milliliter of diluted AS; group 2: incubation with dilute (1:40–1:400) α-endorphin AS; group 3: incubation with α-endorphin AS pretreated with 12, 22, or 45 μg of α-MSH (Bachem). The absence of inhibition of the immunoreaction in these assays excludes any possible cross-reactions and also the presence of antibodies against ACTH or α-MSH contaminants, which could be present in the purified immunogen used for the preparation of the antiserum.

3. Immunocytological Observations

3.1. Immunocytological Detection of ACTH-like Peptides in the Fish Pituitary

In the fish pituitary, cytophysiological studies were first applied in localizing the supposed corticotropic cells in a layer several cells thick, bordering the dorsal and caudal aspects of the rostral pars distalis (cf. the review by Ball and Baker, 1969). In most species, this layer contains only corticotropic cells, which is very conducive to immunocytochemical detection of immunoreactive ACTH. In fact, the detection of the immunoreactive ACTH-like peptides in these cells, which stain with alizarin blue of Herlant's tetrachrome (Herlant, 1960), or with lead hematoxylin (Olivereau, 1964, 1970), both considered specific for corticotropic cells, was the final step in their characterization. A very strong reaction occurred in these cells with an anti β-1–24 ACTH (Synacthen Ciba) in the following species: *Onchorhynchus nerka* (McKeown and van Overbeeke, 1969); *Tinca tinca* (Romain *et al.*, 1974); *Gasterosteus aculeatus, Carassius auratus, Lebistes reticulatus, Salmo irideus,* and *Perca fluviatilis* (Follénius and Dubois, 1976a; Follénius *et al.*, 1978); *Boops salpa* (Malo-Michèle *et al.*, 1976); and *Anguilla anguilla* (Olivereau *et al.*, 1976a).

The cytoplasm of the corticotropic cells becomes strongly fluorescent (Figs. 1 and 2), and the nucleus remains negative. The immunoreaction is due to the content of the fine secretory granules, as shown by electron

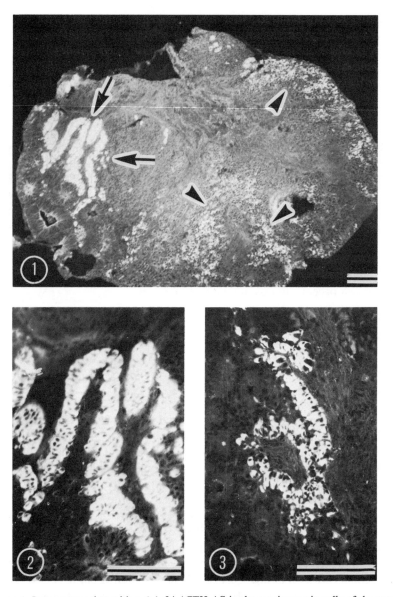

Figure 1. Immunoreaction with a β-1–24 ACTH AS in the corticotropic cells of the rostral pars distalis (thin arrows) and in the MSH cells of the pars intermedia (broad arrows) of the trout pituitary. Calibration bar: 100 μm.

Figure 2. Corticotropic zone of the rostral pars distalis of the trout pituitary. Immunoreaction with β-1–24 ACTH AS. Calibration bar: 100 μm.

Figure 3. Corticotropic zone of the rostral pars distalis of the trout pituitary. Immunoreaction with an α-endorphin AS. The corticotropic cells display a strong fluorescence. Calibration bar: 100 μm.

microscopy (Follénius *et al.*, 1978). The specificity of this reaction was assessed by two means: (1) no immunoreaction occurred in these cells when an AS treated with 1–24 ACTH was used, nor was there any reaction in the absence of the specific AS from the first incubation; (2) in all cases, the reaction was restricted to the cells displaying affinity for alcian blue or lead hematoxylin. This cytological localization constitutes a further important criterion for the specificity of the reactions. We obtained a positive reaction only with the AS against the 1–24 sequence of ACTH, but Olivereau *et al.* (1976a) and Malo-Michèle *et al.* (1976) also observed reactions with a 17–39 ACTH AS.

From these immunocytological observations, the existence of common sequences between mammalian and fish ACTH seems highly probable, as with selacian ACTH, the biochemical composition of which is closely related to that of human ACTH in its 1–19 part (Lowry and Scott, 1975, 1977). The extent and localization of these common sequences have yet to be established in teleost fishes by precise immunological and biochemical studies. Immunological reactions between trout ACTH and anti-ACTH also occur in radioimmunoassay (RIA) systems, which has enabled the study of the evolution of synthetic activity in culture systems (Scott and Baker, 1974, 1975).

3.2. Immunological Detection of α-MSH and β-MSH in Fish Pituitary

The intermediate lobe of the fish pituitary contains two cell types: one stains with lead hematoxylin (type 1), the other with periodic acid–Schiff (PAS) (type 2). Biological assays enabled Baker (1965) to localize the secretion of the melanophore-stimulating hormone in the intermediate lobe. Treatment with metopirone and reserpine (Olivereau, 1965, 1972) produces a concomitant darkening of the skin and degranulation of the lead-hematoxylin-positive cells. The same cell type changes in correlation with the adaptation to black and white backgrounds and was therefore considered as probably melanotropic (Baker, 1972). Contradictory statements on this point (cf. Follénius and Dubois, 1974) rendered necessary a more direct characterization of the MSH-producing cells. In *Gasterosteus aculeatus* and in *Perca fluviatilis*, a clear immunocytological response was observed in the type 1 cells (Follénius and Dubois, 1974, 1976b). In the perch intermediate lobe, treated with an anti α-MSH AS to reveal the MSH cells (Fig.4) and then stained with the alcian blue PAS technique, the PAS-positive cells are obviously not immunoreactive (Fig. 5). Similar comparisons made in *Tinca* (Romain *et al.*, 1974) and *Anguilla* (Olivereau *et al.*, 1976a) have shown that the anti-α-MSH-reactive cells correspond to the cells staining with lead hematoxylin. In the trout (Fig. 6), the immunoreaction with anti-α-MSH is not as important as in the

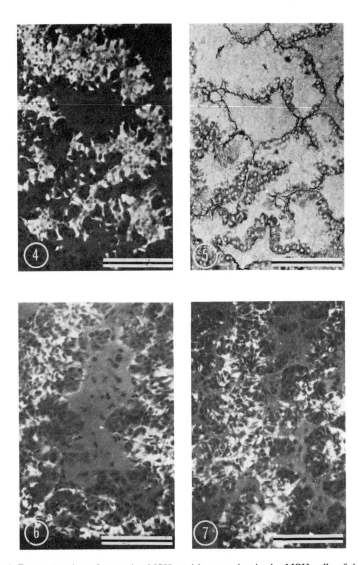

Figure 4. Demonstration of an anti-α-MSH-positive reaction in the MSH cells of the perch intermediate lobe. Calibration bar: 100 μm.

Figure 5. Staining of the same section as in Fig. 4 with alcian blue PAS allows establishment of the correspondence between immunocytological reaction and staining. The dark PAS-positive cells are not immunoreactive, whereas the alcian-blue-staining cells display a bright immunoreaction. Calibration bar: 100 μm.

Figure 6. Pars intermedia of the trout pituitary. Immunoreaction with an α-MSH AS in cells of type 1. Calibration bar: 100 μm.

Figure 7. Pars intermedia of the trout pituitary. Immunoreaction with an α-endorphin AS in cells of type 1. Calibration bar: 100 μm.

perch. Cells displaying a strong immunoreaction are scattered among others with weak reactions. Some cell clusters devoid of any reaction correspond to the type 2 intermediate-lobe cells.

Attempts to localize β-MSH in fish intermediate-lobe cells by immunocytology have not succeeded. No reaction was observed in *Tinca* (Romain *et al.*, 1974) or *Gasterosteus* or *Perca* (Follénius and Dubois, 1974, 1976b); in *Boops salpa* (Malo-Michèle *et al.*, 1976); in *Anguilla anguilla* (Olivereau *et al.*, 1976a); or in various Salmonidae (Olivereau *et al.* 1976b). Immunoreaction for β-MSH was also lacking in two different RIA systems (Shapiro *et al.*, 1972; Scott and Baker, 1975). But Pezalla *et al.* (1978) were able to detect peptides reacting with β-MSH AS in pituitary extracts from two species of teleosts. They used gel filtration of the pituitary extracts from the coho salmon and the Pacific hake to isolate peptides that migrate and behave like β-MSH, but no β-LPH was found. β-LPH was also absent from the intermediate lobe of the tench, where no immunocytological reaction was observed with porcine β-LPH AS (Romain *et al.*, 1974).

These differences between fish and other vertebrates as to the occurrence of β-MSH and β-LPH have still to be explained. Too few species have so far been examined to enable any broad generalization involving a particular specificity of the teleost fishes in this regard. Fish β-LPH may be converted more rapidly to active fragments than in mammals (Pezalla *et al.*, 1978), or its breakdown may occur during the fixation procedure, or during the extraction for biochemical analysis (Liotta *et al.*, 1978). Thus, the immunological or immunocytological detection of fragments corresponding to its COOH-terminal part is of special interest.

3.3. Coexistence of Anti-ACTH and Anti-MSH Immunoreactivity in Intermediate-Lobe Cells of the Fish

As in other vertebrates, ACTH immunoreactivity in intermediate-lobe cells has been observed in many species. The immunoreaction in the MSH cells is, however, clearly weaker than in the ACTH cells. Reactions with a β1–24 ACTH AS were observed in the MSH cells of *Tinca tinca* (Romain *et al.*, 1974) and *Carassius auratus* and *Salmo irideus* (Fig. 1) (Follénius and Dubois, 1976a). Immunoreaction with a 17–39 ACTH AS also occurs in *Anguilla* (Olivereau *et al.*, 1976b) and in *Boops salpa* (Malo-Michèle *et al.*, 1976). The specificity of this reaction was checked by pretreating the 17–39 ACTH AS with 25–39 ACTH, which prevented the immunocytological reaction. ACTH or CLIP or both may be present in the MSH cells. These *in situ* reactions corroborate previous studies on the production of ACTH-like peptides by trout intermediate lobes in culture (Scott and Baker, 1975).

3.4. Immunological Detection of Endorphins in the Fish Pituitary

Both immunocytological and immunoassay methods have been applied to localize and characterize peptides similar to the endorphins in the pituitary of some species of fishes. With α-endorphin AS two distinct zones of the trout pituitary (Follénius and Dubois, 1978a) display positive reactions. The first clearly corresponds to the corticotropic zone (see Fig. 1) of the rostral pars distalis, where all the cells are strongly fluorescent (see Fig. 3). Comparison with the picture obtained using β1–24 ACTH AS (see Fig. 2) reveals that the same cell type displays both immunocytological reactions. The cells adjoining the anterior neurohypophysis are particularly fluorescent, whereas the deeper cells are often only weakly fluorescent. These differences may be related to the importance of the intracellular accumulation of secretory granules that was observed with the electron microscope. The second positive zone belongs to the intermediate lobe, which is well developed in this species. Cell clusters of various shapes are intermingled with the numerous nerve processes of the posterior neurohypophysis (see Fig. 7). The immunoreaction with an α-endorphin AS is in general less intense than in the corticotropic cells. There are also great differences in the intensity of the reaction from one cell to another and from individual to individual, suggesting the possibility of important physiological variations. The cytoplasm of the positive cells is not particularly abundant. The immunoreactive cells form the centers of the parenchymal tissue cords, the boundaries of which are lined with nonreactive type 2 cells. These same cell clusters display a positive reaction with α-MSH AS and α-endorphin AS. This immunoreaction to α-endorphin is immunocytologically specific, as shown by inhibition assays (Table I). No reaction occurs when an α-endorphin AS treated with α-endorphin is used, or when incubation with the specific antiserum is omitted. Furthermore, only a very slight inhibition is noted when the α-endorphin AS is treated with either β1–24 ACTH or α-MSH.

No reaction with α-endorphin AS was observed in the intermediate-lobe cells of *Carassius auratus* and *Cyprinus carpio*. This lack of immunocytological reaction of the MSH cell content might result from initial differences in composition (absence of cross-reaction) or from differences in preservation after fixation.

Our immunocytological results on the presence of peptides reacting with α-endorphin AS can be compared with those obtained by biochemical analysis (Pezalla *et al.*, 1978). In pituitary extracts from the Pacific hake and from the coho salmon, immunoreactive β-endorphin has been isolated, but its cellular origin has not been established. In addition to the two potential sources of immunoreactive endorphin, i.e., the corticotropic and melanotropic cells, a third one is highly probable, as we have recently

Table I. Immunocytological Reactions in Corticotropic and Melanotropic Cells of the Trout (*Salmo irideus* Gibb)

Antiserum	Dilution	Treatment of AS with:	Corticotropic cells (rostral pars distalis)[a]	Melanotropic cells (pars intermedia)[a]
Anti-α-1–24 ACTH	1:40–1:100	—	+++	++
Anti-α-1–24 ACTH	1:40–1:100	β-1–24 ACTH	−	−
Anti-α-MSH	1:100	—	±	++
Anti-α-MSH	1:100	α-MSH	−	−
Anti-α-endorphin	1:40–1:100	—	+++	++
Anti-α-endorphin	1:40–1:100	α-Endorphin	−	−
Anti-α-endorphin	1:40	β-1–24 ACTH	+++	++
Anti-α-endorphin	1:40	α-MSH	+++	++

[a] Intensity of the immunocytochemical reaction: (+++) strong; (++) normal; (+) weak; (±) faint; (−) absence of reaction.

shown in *Carassius auratus* and *Cyprinus carpio* (Follénius and Dubois, 1978b). In these species, anti-α-endorphin immunoreactivity was localized on nerve fibers stemming from the lateral part of the nucleus lateralis tuberis and terminating in the distal posterior neurohypophysis. Furthermore, anti-met-enkephalin immunoreactivity was detected on another important nerve fiber tract of the posterior neurohypophysis (Follénius and Dubois, 1979). The origin of the peptides reacting with the endorphin AS requires much finer analysis than was previously believed.

4. Concluding Remarks

Immunocytological methods have provided the first information on the occurrence of peptides reacting with α-endorphin AS in the fish pituitary. Their localization in the corticotropic and melanotropic cells is particularly interesting. Despite the difficulties in identifying which peptides are involved in the immunocytological reactions, the data suggest that in fishes, as in mammals, relationships among the syntheses of ACTH, MSH, and endorphins may exist. It has now to be established whether the different peptides derive from a higher-molecular-weight precursor, as is now suggested for some mammals. The fact that only anti-α-MSH and anti-α-endorphin reactions were observed is of rather limited significance, but deserves further study. The immunocytological reaction reveals only a given sequence, which may be a part of a bigger molecule. For this reason, immunocytological studies, so useful in localizing immunoreactive peptides, can afford only preliminary information for more basic biochemical studies. As shown in the studies on fishes, the exact localization of the peptide sources is one of the prime requisites for work on the nervous regulation involved in the control of their synthesis and release.

DISCUSSION

KRAICER: In regard to the specificity of the endorphin localization, how did you rule out that the immunological localization was not to LPH rather than endorphin?

FOLLÉNIUS: This was not ruled out.

REFERENCES

Allen, R. G., Herbert, E., Hinman, M., Shibuya, H., and Pert, C. B., 1978, Coordinate control of corticotropin, β-lipotropin and β-endorphin release in mouse pituitary cell cultures, *Proc. Natl. Acad. Sci. U.S.A.* **75**:4972.

Bachelot, I., Wolfsen, A. R., and Odell, W. D., 1977, Pituitary and plasma lipotropins: Demonstration of the artifactual nature of β-MSH, *J. Clin. Endocrinol.* **44**:939.

Baker, B. I., 1965, The site of synthesis of the melanophore stimulating hormone in the trout pituitary, *J. Endocrinol* **32**:397.

Baker, B. I., 1972, The cellular source of melanocyte stimulating hormone in *Anguilla* pituitary, *Gen. Comp. Endocrinol* **19**:515.

Ball, J. N., and Baker, B. I., 1969, The pituitary gland: Anatomy and histophysiology, in: *Fish Physiology* (W. S. Hoar and D. J. Randall, eds.), Vol. 2, pp. 1–110, Academic Press, New York.

Bégeot, M., Dubois, M. P., and Dubois P. M., 1978, Localisation par immunofluorescence de l'hormone β-lipotrope (β-LPH) et de la β-endorphine dans l'antéhypophyse de foetus humains normaux et anencéphales, *C. R. Acad. Sci. Ser. D* **286**:213.

Bloom, F., Battenberg, E., Rossier, J., Ling, N., Leppaluoto, J., Vargo, T., and Guillemin, R., 1977, Endorphins are located in the intermediate and anterior lobes of the pituitary gland, not in the neurohypophysis, *Life Sci.* **20**:43.

Crine, P., Gianoulakis, C., Seidah, N. G., Gossard, F., Pezalla, P. D., Lis, M., and Chrétien, M., 1978, Biosynthesis of β-endorphin from lipotropin and a larger molecular weight precursor in rat pars intermedia, *Proc. Natl. Acad. Sci. U.S.A.* **75**:4719.

Dessy, C., and Herlant, M., 1974, Localisation comparée de l'immunserum antilipotropine, anticorticotropine et antimélanotropine au niveau de l'hypophyse de porc, *C. R. Acad. Sci. Ser. D* **278**:1923.

Dessy, C., Herlant, M., and Chrétien, M., 1973, Detection par immunofluorescence des cellules synthetisant la lipotropine, *C. R. Acad. Sci. Ser. D* **276**:335.

Doerr-Schott, J., and Dubois, M. P., 1972, Identification par immunofluorescence des cellules corticotropes et mélanotropes de l'hypophyse des amphibiens, *Z. Zellforsch.* **132**:323.

Dubois, M. P., 1972a, Localisation cytologique par immunofluorescence des secrétions corticotropes, α-et β-melanotropes au niveau de l'adénohypophyse de bovins, ovins et porcins, *Z. Zellforsch.* **125**:200.

Dubois, M. P., 1972b, Nouvelles données sur la localisation au niveau de l'adénohypophyse des hormones popypeptidiques: ACTH, MSH, LPH, *Lille Med.* **17**:1391 (Engl. abstr., p. 1418).

Dubois, M. P., 1977, Observation cited in Bloom *et al.* (1977).

Follénius, E., and Dubois, M. P., 1974, Immunocytological localization and identification of the MSH producing cells in the pituitary of the stickleback (*Gasterosteus aculeatus* L.), *Gen. Comp. Endocrinol.* **24**:203.

Follénius, E., and Dubois, M. P., 1976a, Etude immunocytologique des cellules cortico-tropes de plusieurs espèces de poissons téléostéens: *Gasterosteus aculeatus* L., *Carassius auratus* L., *Lebistes reticulatus* P., *Salmo irideus* Gibb et *Perca fluviatilis* L., *Gen. Comp. Endocrinol.* **28**:339.

Follénius, E., and Dubois, M. P., 1976b, Étude cytologique et immunocytologique de l'organisation du lobe intermédiaire de l'hypophyse de la perche, *Perca fluviatilis* L., *Gen. Comp. Endocrinol.* **30**:462.

Follénius, E., and Dubois, M. P., 1978a, Immunocytological detection and localization of a peptide reacting with an α-endorphin antiserum in the corticotropic and melanotropic cells of the trout pituitary (*Salmo irideus* Gibb), *Cell Tissue Res.* **188**:273.

Follénius, E., and Dubois, M. P., 1978b, Distribution of fibres reacting with an α-endorphin antiserum in the neurohypophysis of *Carassius auratus* and *Cyprinus carpio, Cell. Tissue Res.* **189**:251.

Follénius, E., and Dubois, M. P., 1979, Différenciation immunocytologique de l'innervation hypophysaire de la carpe a l'aide de serums anti metenképhaline et anti-α-endorphine *C. R. Acad. Sci. Ser. D.* **288**:639.

Follénius, E., Doerr-Schott, J., and Dubois, M. P., 1978, Immunocytology of pituitary cells from teleost fishes, *Int. Rev. Cytol.* **54**:193.

Guillemin, R., Ling, N., and Burgus, R., 1976, Endorphines, peptides d'origine hypothal-amique et neurohypophysaire à activité morphinomimétique: Isolement et structure moléculaire de l'α-endorphine, *C. R. Acad. Sci. Ser. D* **282**:783.

Herlant, M., 1960, Étude critique de deux techniques nouvelles destinées a mettre en évidence les différentes catégories cellulaires présentes dans la glande pituitaire, *Bull. Microsc. Appl.* **10**:37.

Li, C. H., Barnafi, L., Chrétien, M., and Chung, D., 1965, Isolation and amino-acid sequence of β-LPH from sheep pituitary glands, *Nature (London)* **208**:1093.

Liotta, A. S., Suda, T., and Krieger, D. T., 1978, β-lipotropin is the major opioid-like peptide of human pituitary and rat pars distalis: Lack of significant β-endorphin, *Proc. Natl. Acad. Sci. U.S.A.* **75**:2950.

Lowry, P. J., and Scott, A. P., 1975, The evolution of vertebrate corticotropin and melanocyte hormone, *Gen. Comp. Endocrinol.* **26**:16.

Lowry, P. J., and Scott, A. P., 1977, Structural relationship and biosynthesis or cortico-tropin, lipotropin and melanotropin, in: *Melanocyte Stimulating Hormone: Control, Chemistry and Effects* (F. J. H. Tilders, D. F. Swaab, and T. B. van Wiersma-Greidanus, eds.), *Frontiers in Hormone Research,* Vol. 4, pp. 11–17, S. Karger, Basel and New York.

Mains, R. E., Eipper, B. A., and Ling, N., 1977, Common precursor to corticotropin and endorphins, *Proc. Natl. Acad. Sci. U.S.A.* **74**:3014.

Malo-Michèle, M., Bugnon, C., and Fellmann, D., 1976, Étude cyto-immunologique des cellules corticotropes et cortico-mélanotropes de l'adénohypophyse de la saupe *Boops Salpa* L. (téléosteen marin) dans différentes conditions (variations de salinité et couleur de fond), *C. R. Acad. Sci. Ser. D* **283**:643.

McKeown, B. A., and van Overbeeke, A. P., 1969, Immunohistochemical localization of ACTH and prolactin in the pituitary gland of adult migratory sockeye salmon (*Onchorhynchus nerka*), *J. Fish. Res. Board Can.* **26**:1837.

Moon, H. D., Li, C. H., and Jennings, B. M., 1973, Immunohistochemical and histochem-ical studies of pituitary β-lipotrophs, *Anat. Rec.* **175**:529.

Moriarty, G. C., 1973, Adenohypophysis: Ultrastructural cytochemistry—A review, *J. Histochem. Cytochem.* **21**:855.

Moriarty, G. C., and Halmi, N. S., 1972, Adrenocorticotropin production by the interme-
diate lobe of the rat pituitary: An electron microscopic immunohistochemical study,
Z. Zellforsch. **152**:1.
Naik, D. U., 1973, Immunohistochemical localization of adrenocorticotropin and melanocyte stimulating hormone in pars intermedia of rat hypophysis, *Z. Zellforsch.* **142**:289.
Olivereau, M., 1964, L'hématoxyline au plomb permet-elle l'identification des cellules
corticotropes de l'hypophyse des téléostéens, *Z. Zellforsch.* **63**:496.
Olivereau, M., 1965, Action de la métopirone chez l'*Anguilla* normale et hypophysectom-
isée en particulier sur le système hypophyse-cortico-surrénalien, *Gen. Comp. Endocrinol.* **5**:109.
Olivereau, M., 1970, La coloration de l'hypophyse avec l'hématoxyline au plomb; Données
nouvelles chez les téléostéens et comparaison avec les résultats obtenus chez les
autres vertébrés, *Acta Zool. (Stockholm)* **51**:229.
Olivereau, M., 1972, Action de la réserpine chez l'*Anguilla*. II. Effet sur la pigmentation
et le lobe intermédiaire: Comparaison avec l'effet de l'adaptation sur fond noir, *Z.
Anat. Entwicklungsgesch.* **137**:30.
Olivereau, M., Bugnon, C., and Fellmann, D., 1976a, Localisation cytoimmunologique de
α-MSH et d'ATCH dans les cellules hypophysaires colorables avec l'hématoxyline au
plomb chez l'*Anguille*, *C. R. Acad. Sci. Ser. D* **283**:1321.
Olivereau, M., Bugnon, C., and Fellmann, D., 1976b, Identification cytoimmunologique de
deux catégories cellulaires dans le lobe intermédiaire de l'hypophyse des Salmonidés:
Présence d'ATCH et d'α-MSH, *C. R. Acad. Sci. Ser. D* **283**:1441.
Orth, D. N., Guillemin, R., Ling, N., and Nicholson, W. E., 1978, Immunoreactive
endorphins, lipotropins and corticotropins in a human nonpituitary tumor: Evidence
for a common precursor, *J. Clin. Endocrinol. Metab.* **46**:849.
Pelletier, G., Leclerc, R., Labrie, F., Cote, J., Chrétien, M., and Lis, M., 1977, Immunohistochemical localization of β-lipotropic hormone in the pituitary gland, *Endocrinology* **100**:770.
Pezalla, P. D., Craig Clarke, W., Lis, M., Seidah, N. G., and Chrétien, M., 1978,
Immunological characterization of β-lipotropin fragments (endorphin, β-MSH and N
fragment) from fish pituitaries, *Gen. Comp. Endocrinol.* **34**:163.
Romain, R., Bugnon, C., Dessy, C., and Fellmann, D., 1974, Étude cytoimmunologique
des cellules hématoxyline au plomb positives de l'adénohypophyse de la tanche (*Tinca
tinca* L.), *C. R. Soc. Biol.* **168**:1245.
Roux, M., and Dubois, M. P., 1976, Immunohistochemical study of the pars intermedia of
the mouse pituitary in different experimental conditions, *Experientia* **32**:657.
Scott, A. P., and Baker, B. I., 1974, *In vitro* release of corticotrophin by the pars intermedia
of the rainbow trout pituitary, *J. Endocrinol.* **61**:24.
Scott, A. P., and Baker, B. I., 1975, ACTH production by the pars intermedia of the
rainbow trout pituitary, *Gen. Comp. Endocrinol.* **27**:193.
Shapiro, M., Nicholson, W. E., Orth, D. N., Mitchell, W. M., Island, D. P., and Liddle,
G. W., 1972, Preliminary characterization of the pituitary melanocyte stimulating
hormones of several vertebrate species, *Endocrinology* **90**:249.

10

Changes in the Responsiveness to LH-RH in FSH-Treated 4-Day Cycling Female Rats

Jean-Marie Geiger

1. Introduction

Previous work (Chateau, 1971) showed that FSH treatment at the onset of 4-day cycles caused an increased number of follicles to develop until the stage of proestrus with respect to untreated control females. FSH treatment also induced the rupture of an excessive number of follicles (superovulation) and the formation of a great number of corpora lutea with included oocytes (superluteinization). No changes in the timing of the "critical period" of the cycle were observed during the afternoon of proestrus (Geiger and Chateau, 1976). However, LH and FSH surges appeared to be decreased compared to control females (Geiger and Plas-Roser, 1976). Our aim was, then, to determine whether this decline in gonadotropin release was related to changes in the pituitary responsiveness to LH-RH. The study presented in this chapter was also designed to compare plasma estradiol-17β concentration in FSH-treated and nontreated control females.

Jean-Marie Geiger ● Institut d'Histologie, Faculté de Médecine, 67085 Strasbourg, France

2. Materials and Methods

For the study, 3- to 4- month-old virgin female Wistar rats weighing 180–220 g from our colony (strain WII) were used. They were exposed to a normal rhythm of natural lighting and fed with a commercial laboratory food and water *ad libitum*. Only rats that experienced two or three successive 4-day cycles were used. Vaginal sequence consisted of diestrus I, diestrus II, proestrus, and estrus.

A first group of animals consisted of 18 females, which were divided into two lots of 9 animals each. The first lot was subcutaneously injected with two successive dosages of 150 μg/100 g body weight, FSH P101 I in equivalent of NIH-FSH-S1 (FSH P 101 I = 9.47 × NIH-FSH-S1) on diestrus I at 5 P.M. and on diestrus II at 9 A.M. The second lot remained uninjected. In both lots, the females were perfused on the afternoon of the next proestrus with either saline or an LH-RH–saline solution.

Immediately after 35 mg/kg pentobarbital was injected at 1:30 P.M., two catheters were inserted, one into the left saphenous vein and the other into the right carotid artery. Synthetic LH-RH (Stimu-LH Roussel) dissolved in saline was administered through the saphenous-vein catheter over a 1-hr period between 3 and 4 P.M. The infusion rate was 0.62 ml/hr. The solution infused contained either 322 ng/ml LH-RH (3 females) or 80.5 ng/ml LH-RH (3 females). The remaining 3 females were perfused with saline only.

All blood samples were collected from the intracarotid catheter. Before blood collection was started, a very small amount of heparin (500IU) was injected through the right saphenous vein. Four blood samples of about 0.8–1 ml were collected successively at 3, 3:30, 4, and 5 P.M. All blood samples were centrifuged at 4°C and stored at −20°C until they were assayed for LH and FSH. Plasma LH and FSH concentrations were measured by double antibody radioimmunoassay using the method described by Kerdelhué *et al.* (1972, 1973). FSH assay was performed using NIAMDD anti-rat-FSH-S6. The NIAMDD rat-FSH-RP1 (2.1 × NIH-FSH-S1) was used as standard, and a purified laboratory rat FSH preparation was used as tracer. LH assay was performed using an antiserum against ovine LH and a laboratory rat preparation (1.2 × NIH-LH-S1) as standard and tracer. The sensitivity of LH and FSH was as follows: LH = 0.05 ng in terms of NIH-LH-S1; FSH = 10 ng in terms of NIH-FSH-S1. The determinations were made in duplicate for LH, and the results were averaged.

A second group of 72 females was used to measure plasma estradiol-17β concentration in 6 lots of either FSH-treated or uninjected control females (6 animals in each lot) during a period extending from the morning of diestrus II to the afternoon of proestrus. Blood was collected with a heparinized syringe from the inferior vena cava after light ether anesthesia. Following centrifugation at 4°C, plasma was stored at −20°C until it was assayed. Estradiol-17β was estimated after ether extraction and further purification on a Sephadex LH 20 column by radioimmunoassay using [6, 7-³H]estradiol (1 μCi/ml) and an antibody against estradiol ([³H]estradiol RIA-KIT, Biomérieux). Unbound hormone was removed from dextran-coated charcoal. The radioactivity of the bound fraction was measured by an Intertechnique SL 30 liquid scintillation counter. An amount of 2.5–3.5 ml plasma was used for each dosage.

3. Statistics

Variance analysis was used for comparing estradiol plasma concentrations following logarithmic transformation of the data.

4. Results

4.1. Action of LH-RH in FSH-Treated Female Rats

As shown in Fig. 1 (top), FSH-treated females completely failed to respond to a dose of 25 ng/100 g body weight LH-RH, which appeared capable of increasing plasma LH within 30 min. in non-FSH-treated females. It may also be noted that the dose of 100 ng/100 g body weight caused a slight increase in plasma LH in only 1 of 3 FSH-treated females while exerting strong effects within 60 min. in non-FSH-treated females.

The effects of LH-RH on plasma FSH in FSH-treated females are presented in Fig. 1 (bottom). It may be seen that doses of 25 and 100 ng/100 g body weight, capable of causing an increase in plasma FSH concentration within 30 and 60 min, respectively, in non-FSH-treated females, remained without effect in FSH-treated females.

4.2. Changes in Blood Estradiol-17β Concentrations in FSH-Treated Females

Table I indicates that plasma estradiol-17β concentration was significantly lower at midnight during the night following diestrus II in FSH-treated than in non-FSH-treated females ($F_{10}^1 = 8.04$, $p < 2.5\%$). Conversely, a higher plasma estradiol-17β concentration was noted at 8 A.M. on proestrus in FSH-treated than in non-FSH-treated females ($F_{10}^1 = 8.60$, $p < \pm 2.5\%$).

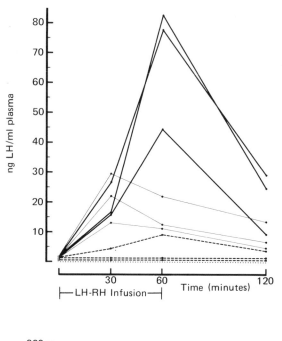

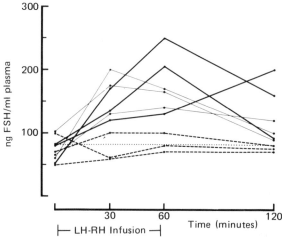

Figure 1. Individual plasma LH *(top)* and FSH *(bottom)* surges in rats treated with FSH at the onset of 4-day cycles and infused with LH-RH after blockade with pentobarbital at 13:00 on proestrus. Controls: 100 ng LH-RH/100 g b.w. (————); 25 ng LHRH/100 g b.w. (————). FSH treatment: 100 ng LH-RH/100 g b.w. (– – – –). The dotted line corresponds to females treated either with saline or with 25 ng LHRH following FSH treatment.

Table. I. Changes in Blood Estradiol-17β Concentration following FSH Treatment
at the Onset of 4-Day Cycles in the Rat

| | Estradiol-17β (pg/ml ± S.E.M.) | | | | | |
| | Diestrus II | | | Proestrus | | |
Treatment[a]	8:00	17:00	24:00	8:00	14:00	18:00
Uninjected females	10.7 ± 1.8	11.9 ± 4.1	33.0 ± 9.2	18.2 ± 5.3	39.7 ± 16.0	46.0 ± 20.3
FSH-treated females	8.6 ± 1.2	13.2 ± 5.4	21.7 ± 5.8	38.3 ± 18.6	47.1 ± 25.7	33.8 ± 12.3

[a] There were 6 animals in each group.

5. Discussion

The results reported here strongly suggest that the decrease in the responsiveness to LH-RH of the pituitary gland in FSH-treated female rats accounts for the reduced LH and FSH surge that was shown to occur during the critical period of the cycle in these FSH-injected females. The changes in plasma estradiol-17β concentration that have been observed in FSH-treated females are probably involved in the mechanisms leading to a decreased pituitary sensitivity to LH-RH. Kalra (1975) showed that ovarian estradiol-17β secretion from 23:00 to 3:00 during the night following diestrus II strictly controlled the triggering of LH release in the afternoon of proestrus. Moreover, much evidence has been provided in favor of the sensitization of the pituitary gland to LH-RH by estrogens in the rat (Arimura and Schally, 1971; Weick *et al.*, 1971; Aiyer and Finck, 1974; Vilchez-Martinez *et al.*, 1974; Greeley *et al.*, 1975). Therefore, it may be that the decline in estradiol-17β concentration that occurred at midnight during the night following diestrus II in FSH-treated females impaired the pituitary response to LH-RH on the afternoon of proestrus. Another point deserves attention. Recently, Turgeon and Barraclough (1977) claimed, in contradiction to the aforementioned authors, that a decrease in plasma circulating estrogens on the afternoon of proestrus would sensitize the pituitary gland to the action of LH-RH. The question then arises whether or not in experimental conditions, the high level of plasma estradiol-17β observed at 8:00 on proestrus in FSH-treated females as compared to the controls was responsible for serum changes in pituitary responsiveness to LH-RH. Finally, the role of progesterone cannot be disregarded, although the response of the pituitary gland to LH-RH in estrogen-treated female rats was not shown to be modified following progesterone injection (Libertun *et al.*, 1974). Conversely, Aiyer and Finck (1974) demonstrated that an increased pituitary sensitivity to LH-RH depended on progesterone stimulation of the estrogen-primed pituitary gland. Experiments are in progress to reinvestigate this problem using our experimental model.

6. Summary

LH-RH-induced release of both FSH and LH was shown to be decreased on the afternoon of proestrus in female rats treated with FSH at the onset of 4-day cycles. A decline in the blood estradiol-17β concentration was observed at midnight of diestrus II in FSH-injected females as compared to control animals. Conversely, the blood estradiol-17β concentration appeared higher at 8 A.M. on proestrus in FSH-treated than in control animals. Changes in the positive-feedback action of estradiol on the hypothalamic–pituitary system were considered to be responsible for the decrease in the pituitary responsiveness to LH-RH in FSH-treated females.

Discussion

Fɪɴᴋ: Did you check whether FSH injection produced progesterone release? Also, did you look for luteinization? The point is that even if estrogen does rise, it will not trigger an LH surge or increase pituitary responsiveness if it is acting on a hypothalamus–hypophyseal axis that has been exposed to high progesterone.

Gᴇɪɢᴇʀ: We did not measure progesterone, but we did find luteinization of follicles.

References

Aiyer, M. S., and Fink, G., 1974, The role of sex steroid hormones in modulating the responsiveness of the anterior pituitary gland to luteinizing hormone releasing factor in the female rat, *J. Endocrinol.* **62**:553.

Arimura, A., and Schally, A. V., 1971, Augmentation of pituitary responsiveness to LH releasing hormone (LH-RH) by estrogen, *Proc. Soc. Exp. Biol. Med.* **136**:290.

Chateau, D., 1971, Données nouvelles sur les modalités d'action de FSH sur l'ovaire de la ratte au cours de cycles de 4 jours, *C. R. Soc. Biol.* **165**:688.

Geiger, J. M., and Chateau, D., 1976, Modalités chronologiques et quantitatives de la lutéinisation et de la ponte ovulaire chez des rattes exposées à l'action de FSH au début de cycles de 4 jours, *C. R. Soc. Biol.* **170**:464.

Geiger, J. M., and Plas-Roser, S., 1976, Modification proestrale du taux plasmatique de LH et de FSH chez des rattes exposées a l'action de FSH au début de cycles de 4 jours, *C. R. Acad. Sci.* **283**:1771.

Greeley, G. J., Allen, M. B., and Mahesh, V. B., 1975, Potentiation of luteinizing hormone release by estradiol at the level of the pituitary, *Neuroendocrinology* **18**:233.

Kalra, S. P., 1975, Observations on facilitation of the preovulatory rise of LH by estrogen, *Endocrinology* **96**:23.

Kerdelhué, B., Pitoulis, S., and Berault, A., 1972, Dosage radioimmunologigue des gonadotropines de rat, *Symposium sur les Techniques Radioimmunologiques*, p. 257, INSERM, Paris.

Kerdelhué, B., Catin, S., and Jutisz, M., 1973, New data concerning the plasma levels of prolactin and gonadotropins throughout the estrous cycle in the rat, in: *Human Prolactin* (J. L. Pasteels and C. Robyn, eds.), p. 149, Exerpta Medica, New York.

Libertun, C., Orias, R., and McCann, S. M., 1974, Biphasic effect of estrogen on the sensitivity of the pituitary to luteinizing hormone–releasing factor (LRF), *Endocrinology* **94**:1094.

Turgeon, J. L., and Barraclough, C. A., 1977, Regulatory role of estradiol in pituitary responsiveness to luteinizing hormone releasing hormone on proestrus in the rat, *Endocrinology* **101**:548.

Vilchez-Martinez, J. A., Arimura, A., Debeljuk, L., and Schally, A. V., 1974, Biphasic effect of estradiol benzoate on the pituitary responsiveness to LHRH, *Endocrinology* **94**:1300.

Weick, R. F., Smith, E. R., Dominguez, R., Dhariwal, A. P. S., and Davidson, J. M., 1971, Mechanism of stimulatory feedback effect of estradiol benzoate on the pituitary, *Endocrinology* **88**:293.

11

Role of Gonadotropin-Releasing Hormone in the Biosynthesis of LH and FSH by Rat Anterior Pituitary Cells in Culture

Ashok Khar and Marian Jutisz

1. Introduction

Due to the glycoprotein nature and subunit structure of gonadotropins, studies on the mechanism of biosynthesis of these hormones have been quite complicated. Different laboratories have used various types of approaches, which have or have not involved the incorporation of labeled amino acids or glucosamine by intact rat anterior pituitaries, pituitary halves, or tissue and cell cultures of anterior pituitaries. The effect of gonadotropin-releasing hormone (GnRH) on the biosynthesis and release of luteinizing hormone (LH) and follicle-stimulating hormone (FSH) has also been examined by many authors.

Kobayashi *et al.* (1965) were the first to study the problem of gonadotropin synthesis. Using the technique of autoradiography, they measured the incorporation of [³H]leucine into pituitary gonadotrophs. Samli and Geschwind (1967) did not observe any effect of hypothalamic extracts on the incorporation of either [¹⁴C]leucine or [¹⁴C]glucosamine into LH during a 4-hr incubation of rat pituitaries. Similarly, Jutisz *et al.*

Ashok Khar and *Marian Jutisz* • Laboratoire des Hormones Polypeptidiques, C.N.R.S., 91190 Gif-sur-Yvette, France

(1972) reported that a crude preparation of sheep GnRH had no effect on the incorporation of [³H]leucine into LH after 3-hr incubation of rat pituitary halves.

Some assumptions concerning the effect of GnRH on the synthesis of gonadotropins were also made from experiments carried out in the absence of labeled precursors. It was reported that in a 2-hr incubation of anterior pituitary halves from ovariectomized, estradiol progesterone-treated rats, purified preparations of ovine GnRH as well as a high-potassium medium (51.2 mEq K⁺) and 0.26 mM cyclic AMP had a stimulatory effect on both the release of FSH and the resynthesis of bioassayable FSH (Jutisz and De la Llosa, 1967, 1968). As an interpretation of these results, the existence of two compartments in a gonadotropic cell has been suggested, the first containing FSH directly available for release and FSH being synthesized in the second. The release of FSH from the first compartment would induce synthesis of the hormone in the second compartment (Jutisz *et al.*, 1970).

On the basis of kinetic studies made with rat anterior pituitary halves incubated in a Krebs–Ringer medium, Jutisz and Kerdelhué (1973) were able to postulate that the synthesis rate of FSH is higher than that of LH. This was confirmed further by Labrie *et al.* (1973), who used a monolayer culture system of pituitary cells, and by Chowdhury and Steinberger (1975), who incubated anterior pituitary glands.

An increase in total radioimmunoassayable LH and FSH and bioassayable FSH was also observed in tissue cultures of rat anterior pituitaries in the presence of a purified preparation of porcine GnRH (Mittler *et al.*, 1970). These findings were further confirmed and extended using pure and synthetic preparations of GnRH (Redding *et al.*, 1972). Similarly, using dispersed anterior pituitary cells in monolayer cultures, several authors have reported an increase in the total amount of radioimmunoassayable gonadotropins in the presence of GnRH (Vale *et al.*, 1972; Labrie *et al.*, 1973).

More recently, Liu *et al.* (1976) observed a stimulatory effect of GnRH on the *in vitro* incorporation of [³H]glucosamine into LH in anterior pituitaries from immature male, intact female, and ovariectomized rats incubated for 4 hr at 37°C. GnRH was also found by these authors to increase the release of either [³H]leucine- or [³H]alanine-labeled LH into medium, but this neurohormone had no significant effect on the incorporation of labeled amino acids into total LH in the system. Liu and Jackson (1978) further confirmed that GnRH stimulates incorporation of [³H]glucosamine, but not of [¹⁴C]alanine, into LH by quartered rat anterior pituitaries. They also reported that cycloheximide blocked synthesis of [¹⁴C]alanine-LH and greatly reduced the GnRH-induced synthesis and release of [³H]glucosamine-LH, but that it reduced release of immunoreactive LH by only 25%. Actinomycin D had no effect on GnRH-induced synthesis and release of LH at 2-hr incubation, but significantly reduced both at 4-hr incubation. In the authors' opinion, these data suggested that (1) the LH newly synthesized in response to GnRH has more sugar residues than that released under basal conditions; (2) the GnRH-induced LH release can occur under conditions in which LH synthesis has been blocked; and (3) synthesis of messenger RNA (mRNA) is not required for GnRH-induced LH release or short-term LH synthesis, but seems rather to be necessary for further synthesis and subsequent release of LH.

In contrast to the data cited above, Menon *et al.* (1977) reported that GnRH enhanced the incorporation of both [^{3}H]glucosamine and [^{3}H]amino acid mixture into immunoprecipitable LH when rat hemipituitaries were incubated for 2 hr. Furthermore, the same laboratory (Azhar *et al.,* 1978), using rat anterior pituitary cells in culture, confirmed the results of Liu *et al.* (1976) showing that the incorporation of [^{3}H]proline into LH was unaffected by GnRH, but that the peptide stimulated a 3- to 4-fold increase in [^{3}H]glucosamine incorporation into LH. The agonistic analogue, [desGly-NH$_2$^{10}]-GnRH ethylamide, mimicked the GnRH effects. These authors postulated that GnRH might preferentially stimulate the turnover or incorporation of glucosamine into the carbohydrate portion of LH. GnRH stimulation of LH glycosylation might result either from a direct enzymatic activation or through increased synthesis or decreased degradation of glycosyltransferases.

Recently, we have reported the incorporation of [^{3}H]proline and [^{3}H]glucosamine into LH and FSH by dispersed rat anterior pituitary cells in culture and by rat pituitary homogenates, and the effect of GnRH on this incorporation (Khar *et al.,* 1978). Our results and their interpretation differ somewhat from those published by Liu and Jackson (1978) and by Azhar *et al.* (1978).

The aim of this chapter is to discuss, in view of all available data, the problem of the involvement of and a possible role for GnRH in the biosynthesis of LH and FSH. We have also extended our discussion to include the role of GnRH receptor sites in the effect of the neurohormone on this synthesis.

2. Experimental Methods and Results

2.1. Immunoprecipitation Studies

Anti-rat LHβ and anti-rat FSHβ used in our work for immunoprecipitations were kindly provided by Dr. A. F. Parlow, NIAMDD. According to the specificity data reported for these antisera, they are highly specific for their respective subunits. Binding studies for anti-rLHβ serum with

[^{131}I]-rLHβ at half-saturation in a radioimmunoassay (RIA) system showed that the displacement of 1 ng rLHβ will require, respectively, 100 ng rLH, 160 ng rat thyroid-stimulating hormone (rTSH), 240 ng rTSHβ, 360 ng rLHα, or 1000 ng rFSHβ. The same studies for anti-rFSHβ with [^{131}I]-rFSHβ showed that the displacement of 1 ng rFSHβ will require, respectively, 250 ng rTSHβ or 600 ng rFSH. No cross-reactivity with rLHβ, rLHα, rLH, rTSH, rGH, or rat prolactin was observed in the latter system. Therefore anti-rFSHβ serum seems to be more specific than anti-rLHβ serum.

To determine the exact amounts of antisera to be used for immuno-precipitation studies, different dilutions of antisera were tried. The final optimal dilutions for precipitating the respective subunits were found to be 1:30,000 for anti-LHβ and 1:6000 for anti-FSHβ.

For immunoprecipitation (see Khar et al., 1978), the cells were disrupted in distilled water and transferred into test tubes. Both media and cells were then treated with trichloracetic acid (TCA) for precipitation of proteins, as described previously. The TCA precipitate was dissolved and used for immunoprecipitation. Nonspecifically bound radioactivity was eliminated by treating the solutions first with normal rabbit serum and then with anti-rabbit γ-globulin (Burek and Frohman, 1970).

The immunoprecipitation of LHβ and FSHβ was carried out in phosphate-buffered saline–bovine serum albumin (PBS–BSA) (0.05% BSA) using the specific antisera at the final dilutions indicated above. After 4 days at 4°C, the complex antigen–antibody was precipitated with anti-rabbit γ-globulin for 2 days at 4°C. The washed precipitate was dissolved in Soluene-350 (Packard) and counted in toluene containing 2,5-diphenyloxazole (PPO) and 1,4-di[2-(4-methyl-5-phenyloxazolyl)]benzene (dimethyl-POPOP). The significance of the differences among groups was tested by means of Duncan's new multiple range test.

2.2. Incorporation Studies with Pituitary Homogenates

To investigate the possibility of biosynthesis of gonadotropins in anterior pituitary homogenates, anterior pituitaries from adult rats (irrespective of sex) were homogenized, and the homogenates were incubated at 37°C in minimum essential medium (MEM) at pH 7.35 in the presence of either [^{3}H]proline or [^{3}H]glucosamine and in the presence or absence of GnRH and other test substances.

Figure 1 shows the incorporation of [^{3}H]proline into LH and FSH synthesized by pituitary homogenates. An optimal incorporation into both LH and FSH was observed 30 min after the start of incubation. GnRH at a relatively high concentration of 1 μg/ml had no significant effect on this incorporation. Puromycin and cycloheximide at different concentrations (1–5 mM) completely inhibited the incorporation of labeled proline.

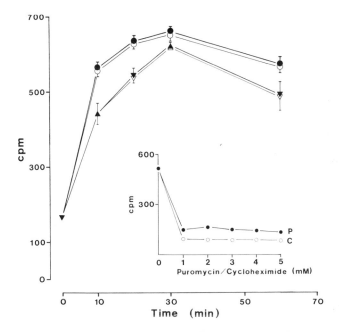

Figure 1. Time course of the incorporation of [³H]proline into LH and FSH by rat pituitary homogenates in the presence (1 μg/ml) or in the absence of GnRH and the effect of different concentrations of puromycin and cycloheximide on this incorporation after 30 min of incubation. (△) LH (controls); (▲) LH (GnRH); (○) FSH (controls); (●) FSH (GnRH); (P) puromycin; (C) cycloheximide.

Pituitary homogenates also incorporated [³H]glucosamine into LH and FSH (Fig. 2). As in the case of proline, the optimal incorporation of labeled glucosamine took place after 30-min incubation, and GnRH (1 μg/ml) had no effect on this incorporation.

These results suggest that the integrity of gonadotropic cells is not necessary for the achievement of LH and FSH biosynthesis, but that integrity of the cells may be required for the effect of GnRH on this biosynthesis. They also show that in the conditions used, GnRH had no effect on the glycosylation of gonadotropins. The decline in the rate of biosynthesis after 30 min may be due to the presence of peptidases in pituitary homogenates (Kochman *et al.*, 1975).

2.3. Incorporation Studies with Dispersed Pituitary Cells

In our laboratory, cells from the anterior pituitaries of the adult rats, irrespective of sex (Wistar laboratory strain), were dispersed first with collagenase and then with viokase (Debeljuk *et al.*, 1978). The dispersed cells were washed with MEM-BSA, pH 7.4.

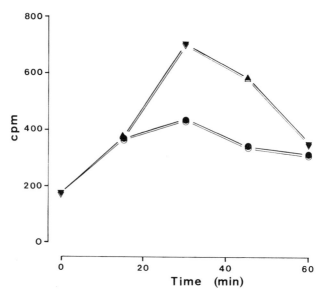

Figure 2. Time course of the incorporation of [³H]glucosamine into LH and FSH by rat pituitary homogenates and the effect of GnRH (1 μg/ml) on this incorporation. (△) LH (controls); (▲) LH (GnRH); (○) FSH (controls); (●) FSH (GnRH).

2.3.1. Freshly Dispersed Cells and "Overnight" Cultures

In the experiments with freshly dispersed cells, washed cells were suspended in medium F-10 and used immediately after the enzymatic treatment.

Table I shows that freshly dispersed cells incubated at 37°C incorporated labeled proline into LH and FSH, but that GnRH had no effect on either release of LH or FSH (results not shown) or incorporation of labeled proline into gonadotropins.

Table I. Incorporation of [³H]Proline into LH and FSH by the Freshly Dispersed Pituitary Cells and the Effect of GnRH[a]

Hormone	GnRH (ng/ml)	Incubation time		
		2 hr	4 hr	6 hr
LH	None	1247 ± 130	3662 ± 547	3160 ± 216
LH	100	1183 ± 154	3569 ± 466	3217 ± 328
FSH	None	1117 ± 53	3021 ± 241	3354 ± 291
FSH	100	1083 ± 22	2986 ± 311	3425 ± 177

[a] Results expressed as counts per minute ± S.E.M. incorporated into cells plus medium.

Since Nakano *et al.* (1976) have shown that when the pituitary cells were dispersed with collagenase only and preincubated "overnight" (22 hr) they responded to stimulation with GnRH, we tried the same procedure. After dispersion with collagenase, cells were washed and suspended in medium F-10 containing fetal calf serum (2.5%), rat serum (15%), *N*,2-hydroxyethylpiperazine-*N*,2-ethanesulfonic acid (HEPES) (25 mM), and a mixture of antibiotics (50 IU penicillin, 50 μg streptomycin, and 2.5 μg fungizone/ml) and incubated for 22 hr at 37°C. Afterward, they were suspended in medium F-10 and used for the incorporation studies.

Table II shows that cells from "overnight" cultures incorporated [³H]leucine into LH and FSH, but that GnRH, as in the previous case, had no effect on this incorporation. Similar observations were made when these cells were incubated with labeled glucosamine.

The lack of effect of GnRH on the incorporation of labeled amino acids and glucosamine into LH and FSH by freshly dispersed cells and "overnight" cultures suggests that the integrity of the receptor sites for GnRH, present in the plasma membrane of the gonadotropic cells, may be necessary for this action. Indeed, binding of the neurohormone to its receptor sites probably represents the first step of its action (Théoleyre *et al.*, 1976). The receptor sites are certainly damaged or disintegrated during the enzymatic treatment, and a 4- to 7-day culture is necessary for the receptor sites to be restored and the cells to be able to respond to GnRH (Vale *et al.*, 1972; Tixier-Vidal *et al.*, 1973). Despite the use of collagenase only for dispersion of pituitary cells and an "overnight" preincubation as proposed by Nakano *et al.* (1976), we were unable to obtain cells that would respond to stimulation with GnRH.

2.3.2. Monolayer Cultures

Pituitary cells, dispersed and washed as described above, were suspended in medium F-10 containing the same sera and other constituents used for "overnight" cultures. The cells were distributed in plastic petri

Table II. Incorporation of [³H]Leucine into LH and FSH by "Overnight" Pituitary Cultures and the Effect of GnRH[a]

Hormone	GnRH (ng/ml)	Incubation time		
		2 hr	4 hr	6 hr
LH	None	726 ± 32	2075 ± 228	3477 ± 336
LH	100	683 ± 13	2134 ± 172	3319 ± 425
FSH	None	570 ± 45	1228 ± 361	1643 ± 326
FSH	100	623 ± 93	1335 ± 213	1793 ± 478

[a] Results expressed as counts per minute ± S.E.M. incorporated into cells plus medium.

dishes (1–2 × 10⁶ cells/2 ml medium) and cultured up to 7 days with a change of medium on day 4. Finally, the cells were incubated with labeled proline or glucosamine (5 μCi/dish) in medium F-10 on day 7.

2.3.2a. Incorporation of [³H]Proline. As shown in Table III, [³H]proline is incorporated into total proteins, LH, and FSH after 4-hr incubation of pituitary cells in culture. The ratios of LH and FSH synthesized vs. total protein radioactivity are 2.0–2.3% and 2.8–3.5%, respectively. Incubation in the presence of GnRH (3 ng/ml) significantly increased these ratios to 5.0–5.3% and 6.3–8.1%, respectively. The amount of [³H]proline incorporated into FSH was higher than that incorporated into LH, and this despite a higher content of proline in the LH molecule than in the FSH molecule. This confirms the earlier reports concerning a higher rate of synthesis for FSH than for LH (Jutisz and Kerdelhué, 1973; Labrie *et al.*, 1973; Chowdhury and Steinberger, 1975).

Figure 3 shows the time course of the incorporation of [³H]proline into LH and FSH from medium and cells and the effect of GnRH on the incorporation of [³H]proline and the release of labeled LH. The incorporation of [³H]proline into both LH and FSH from the cells increased progressively up to 6 hr of incubation. GnRH at an optimal concentration (3 ng/ml) significantly enhanced this incorporation up to 4 hr of incubation; afterward, controls and treated groups reached the same values (in the case of LH, the values for GnRH-treated groups dropped after 4 hr). At the same time, GnRH stimulated the release of labeled LH, but this release was significant only at 4 and 6 hr of incubation. At present, we have no

Table III. Incorporation of [³H]Proline into Total Proteins, LH, and FSH after 4-Hour Incubation of Cultured Pituitary Cells (Medium Plus Cells)[a]

	Treatment			
	Control		GnRH (3 ng/ml)	
Parameter	cpm/dish ± S.E.	%[b]	cpm/dish ± S.E.	%[b]
Experiment I				
[³H]protein (total)	83,284 ± 1,743	—	82,530 ± 1,134	—
[³H]-LH	1,670 ± 24	2.0	4,396 ± 93[c]	5.3
[³H]-FSH	2,320 ± 142	2.8	5,219 ± 236[c]	6.3
Experiment II				
[³H]protein (total)	79,836 ± 1,493	—	80,345 ± 1,686	—
[³H]-LH	1,868 ± 73	2.3	3,986 ± 53[c]	5.0
[³H]-FSH	2,776 ± 23	3.5	6,483 ± 18[c]	8.1

[a] from Khar *et al.* (1978). Courtesy of North-Holland Publishing Company.
[b] With respect to total radioactive proteins.
[c] $p < 0.01$ with respect to the corresponding control group.

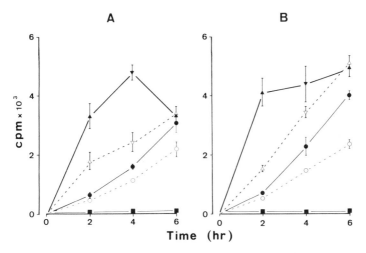

Figure 3. Time course of the incorporation of [³H]proline into LH (A) and FSH (B) by cultured rat pituitary cells in the presence (3 ng/ml) or in the absence of GnRH and the effect of cycloheximide (1 mM) on this incorporation. In the presence of cycloheximide, radioactivity was measured in the cells plus media. (O) Medium (controls); (●) medium (GnRH); (△) cells (controls); (▲) cells (GnRH); (■) medium plus cells (cycloheximide) in the presence or in the absence of GnRH. From Khar *et al.* (1978). Courtesy of North-Holland Publishing Company.

explanation for the cessation of the stimulatory action of GnRH on the incorporation of [³H]proline after 4 hr of incubation. This may be due to the limited amount of [³H]proline introduced into the medium at the beginning of the experiment.

Cycloheximide (1 mM) and puromycin (1 mM) completely inhibited the incorporation of labeled proline into LH and FSH (medium plus cells), both in the presence and in the absence of GnRH.

2.3.2b. Incorporation of [³H]Glucosamine.   Figure 4 shows the time course of [³H]glucosamine incorporation into LH and FSH and the release of labeled gonadotropins in the presence and in the absence of GnRH. The effect of actinomycin D and cycloheximide on the incorporation of [³H]glucosamine and the release of labeled hormones in the presence of GnRH is also shown. In Table IV, the values for the cells and corresponding media were added to evaluate the effect of GnRH and of antibiotics on the total incorporation of [³H]glucosamine. It is clear from these data that [³H]glucosamine was incorporated into LH and FSH during 6 hr of incubation, and that GnRH had a slight but significant effect on this incorporation [into LH after 4-hr and 6-hr incubation, into FSH only after 4-hr incubation (Table IV)]. Starting from 2 hr of incubation, GnRH

Table IV. Incorporation of [³H]Glucosamine into LH and FSH by Anterior Pituitary Cells in Culture and the Effect of GnRH in the Presence or in the Absence of Cycloheximide and Actinomycin D[a]

Hormone	GnRH (ng/ml)	Cycloheximide (mM)	Actinomycin D (nM)	Incubation time		
				2 hr	4 hr	6 hr
LH	None	None	None	1262 ± 77	1846 ± 54	2236 ± 47
LH	9.0	None	None	1391 ± 41	2387 ± 108^b	2781 ± 63^b
LH	9.0	1.00	None	895 ± 31^c	974 ± 23^c	1098 ± 19^c
LH	9.0	None	0.15	1348 ± 92	1630 ± 58^c	2288 ± 57^c
FSH	None	None	None	913 ± 21	1297 ± 35	1684 ± 57
FSH	9.0	None	None	1156 ± 27	1660 ± 72^b	1863 ± 123
FSH	9.0	1.00	None	792 ± 32^c	760 ± 31^c	686 ± 21^c
FSH	9.0	None	0.15	978 ± 53	1220 ± 35^c	1574 ± 54

[a] Results expressed as total counts per minute ± S.E.M. incorporated into cells plus medium.
[b] $p < 0.05$ with respect to the control group.
[c] $p < 0.05$ with respect to the GnRH-treated group.

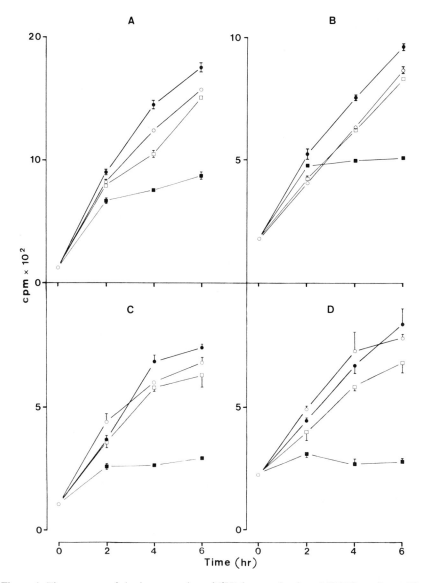

Figure 4. Time course of the incorporation of [³H]glucosamine into LH [(A) medium; (C) cells] and FSH [(B) medium; (D) cells] by cultured rat pituitary cells in the presence (3 ng/ml) or in the absence of GnRH and the effect of cycloheximide (1 mM) and actinomycin D (0.15 nM) on this incorporation. (○) Controls; (●) GnRH; (□) GnRH plus actinomycin D; (■) GnRH plus cycloheximide.

induced release of a significant amount of labeled LH and FSH as compared with controls (Fig. 4A and B).

Cycloheximide completely inhibited the effect of GnRH on the incorporation of [³H]glucosamine into LH and FSH for all the incubation periods studied; inhibition was only partial when compared with the values for the controls after 2 hr of incubation (Table IV and Fig. 4). In fact, cycloheximide did not inhibit, or inhibited only slightly, the release of labeled LH and FSH after 2 hr of incubation (Fig. 4A and B); inhibition was much more pronounced for the incorporation of [³H]glucosamine into LH and FSH from the cells.

Actinomycin D had no effect on the basal incorporation of [³H]glucosamine into LH and FSH, but it significantly inhibited the effect of GnRH on the incorporation of [³H]glucosamine into both LH and FSH (Table IV) as well as the release of labeled gonadotropins from the cells (Fig. 4A and B).

The results with labeled glucosamine are at variance with those observed with labeled proline. Whereas in the latter case GnRH stimulated incorporation of [³H]proline into LH and FSH very significantly (2-fold) up to 4 hr of incubation, incorporation of [³H]glucosamine was more discrete and at the limit of significance only after 4 and 6 hr of incubation (except for FSH after 6 hr of incubation). Thus, it seems doubtful that GnRH directly stimulated incorporation of [³H]glucosamine into LH and FSH. Since GnRH stimulated synthesis of the polypeptide chains of LH and FSH, the small increase in incorporation of [³H]glucosamine in the presence of GnRH may be due to the increased number of polypeptide chains of these hormones available for glycosylation.

On the other hand, cycloheximide only partially inhibited the incorporation of labeled carbohydrate into LH and FSH after 2 hr of incubation. Since cycloheximide very rapidly inhibited the synthesis of polypeptide chains of LH and FSH (see Fig. 3), the only available polypeptide moieties of these hormones for glycosylation were either completed chains or nascent chains, still bound to the ribosome–mRNA complex. Glucosamine was incorporated during the first 2 hr of incubation, either into completed and only partially glycosylated subunits or into nascent subunits. It was recently reported that the nascent chains of a membrane glycoprotein can be glycosylated (Rothman and Lodish, 1977).

3. Discussion

Taking into account information published by others as well as our own results, we will try to answer the three following questions: (1) Are the polypeptide chains of LH and FSH synthesized during the incubation of rat anterior pituitary tissue or cells? (2) Are these chains glycosylated during the incubation period? (3) Is GnRH involved in the synthesis or glycosylation, or both, of LH and FSH polypeptide chains?

The answer to the first two questions is "yes." All the data from the literature concerning the incorporation of labeled amino acids and labeled glucosamine into LH from rat anterior pituitary tissue and cells and our own results related to the incorporation of labeled proline and glucosamine into both LH and FSH suggest that the polypeptide chains of gonadotropins are synthesized and glycosylated *in vitro*.

The main discrepancy between our results and the recent work of Liu *et al.* (1976), Liu and Jackson (1978), and Azhar *et al.* (1978) concerns the third question, i.e., the role of GnRH in the synthesis or glycosylation, or both, of LH polypeptide chains. Whereas the latter authors reported that GnRH stimulates incorporation of labeled glucosamine but not that of labeled amino acids into LH, our results are in favor of the reverse situation. In the conditions used in our laboratory, GnRH significantly stimulated the incorporation of [^{3}H]proline into LH and FSH during the first 4 hr of incubation (see Table III), while under similar conditions, the glycosylation of LH and FSH polypeptide chains with [^{3}H]glucosamine was unaffected or affected only very slightly by GnRH (see Table IV). Since GnRH did not stimulate the incorporation of labeled proline into total proteins but did stimulate this incorporation into both LH and FSH (Table III), another argument in favor of our present findings is provided.

The evaluation of the synthesis of a specific protein using radioactive precursors requires its isolation in a highly purified form and is consequently dependent in the present case on the specificity of antisera and immunoprecipitation methods used for isolation of labeled molecules. It is evident that each of the three groups of workers under consideration employed different antisera and somewhat different immunoprecipitation methods. The NIAMDD antisera used in our work were species-specific and had a high specificity for their respective subunits. According to the data on the specificity of these antisera, the cross-reactivity of different pituitary hormones and derivatives with anti-LHβ was 10:1000 for rLH, 6:1000 for rTSH, 4:1000 for rTSHβ, 3:1000 for rLHα, and 1:1000 for rFSHβ. In the case of anti-rFSHβ, the cross-reactivity was 4:1000 for rTSHβ and 2:1000 for rFSH, but no cross-reactivity with other pituitary hormones was observed. Despite high specificity in RIA systems of the antisera used in our work, we cannot disregard the possibility that an immunoprecipitate might contain some of its parent hormone, in addition to a β subunit of a gonadotropin. Indeed, we found that under our experimental conditions, anti-rLHβ will precipitate approximately 90% of ^{3}H-labeled ovine LH (De la Llosa *et al.*, 1974). Thus, it is possible that in the case of cell extract, an immunoprecipitate is constituted in our work mainly from a corresponding β subunit, and in the case of medium, the parent hormone is mainly immunoprecipitated. In fact, we do not know whether GnRH is able to release only LH and FSH, or their subunits as well. The dose of GnRH (3 ng/ml) usually employed throughout our work was capable of releasing LH and FSH from the cells in amounts about 7–11 times higher, as compared with the control groups after 6-hr incubation (Khar *et al.*, 1978).

Azhar *et al.* (1978) have given only a few details in their paper regarding the antisera used; therefore, it is rather difficult to be certain about the specificity of their immunoprecipitation method. In contrast, antiserum anti-ovine LHβ prepared by Liu *et al.* (1976) seems to have been well characterized by the authors and fairly specific for LH. The main difference between the methodology used by the latter authors and ours concerns the starting material: they incubated rat anterior pituitary quarters and we used rat anterior pituitary cells in monolayer culture. It is possible that pituitary tissue incubated *in vitro* undergoes a progressive necrosis, which starts as early as 30 min after the beginning of incubation, as was demonstrated by Tixier-Vidal *et al.* (1971). Necrosis may be important after a 6-hr incubation of pituitary quarters and could interfere with the final results reported by Liu *et al.* (1976) and Liu and Jackson (1978). Another possibility is that the rate of uptake of labeled precursors may be different in the dispersed cells, where the contact between the cell membrane and the medium is better than with tissue fragments, particularly for cells situated in the inner part of the fragment. The rate of infusion of a labeled precursor may influence the time course of the synthesis.

Another hypothesis proposed by Liu and Jackson (1978) is difficult to accept, i.e., "that GnRH can exert a direct effect on the anterior pituitary to modify the molecular species of LH being synthesized and released." In fact, in the opinion of these authors, "the carbohydrate content of radiolabeled LH released in response to exogenous GnRH differs from that released in the absence of this GnRH." Except for the idea that GnRH stimulates the glycosylation of a precursor molecule of LH, we did not find any report in the literature regarding any major structural heterogeneity in the carbohydrate moiety of LH molecule.

In addition to the results discussed above, our data provide some information on the way GnRH acts on the synthesis of polypeptide chains of LH and FSH. They suggest that the integrity of the gonadotropic cells is not necessary for the synthesis and glycosylation of polypeptide chains, but that the integrity of both GnRH receptors and that of cells may be required for the action of GnRH on the biosynthesis. On the other hand, taking into account the fact that LH contains about twice as much proline as FSH, our data also suggest (see Table III and Fig. 3) that the rate of synthesis of FSH must be at least twice as high as that of LH, thus confirming the earlier reports (Jutisz and Kerdelhué, 1973).

The last point we would like to discuss is the effect of antibiotics on the glycosylation and release of [³H]glucosamine-labeled LH and FSH. The fact that cycloheximide did not inhibit or only slightly inhibited the GnRH-induced release of [³H]glucosamine-labeled LH and FSH up to 2 hr of incubation, but that it inhibited very rapidly and completely the synthesis of polypeptide chains of these glycoproteins (Fig. 3), suggests that gonadotropic cells contain some nonglycosylated or partially glycosylated completed subunits or nascent subunits of LH and FSH that can be glycosylated in the presence of this antibiotic.

On the other hand, actinomycin D had no effect either on the glycosylation of the control groups or on the basal release of [³H]glucosamine-labeled LH and FSH (data not shown), but it did completely inhibit the GnRH-induced release of labeled gonadotropins (see Fig. 4A and B). Our results support the findings of many authors (Aiyer *et al.*, 1974; Pickering and Fink, 1976; Vilchez-Martinez *et al.*, 1976; Edwardson and Gilbert, 1976; De Koning *et al.*, 1976) who postulate that GnRH-induced release of LH and FSH proceeds through two complementary mechanisms, an acute effect that is not affected by inhibitors of protein and RNA synthesis and a priming effect that is markedly inhibited by these inhibitors. Our data support these findings. They suggest that actinomycin D inhibits the priming effect of GnRH, which is probably dependent on protein or RNA synthesis or both.

In conclusion, our results, which are contradictory to those reported by others, suggest that GnRH is involved in the biosynthesis of polypeptide moieties of LH and FSH and that it probably does not stimulate their glycosylation or stimulates it only indirectly. Nevertheless, it is still not clear whether GnRH stimulates synthesis of polypeptide chains of LH and FSH directly or indirectly through an intracellular feedback mechanism coupled to release. More information is necessary before the mechanisms controlling release, synthesis, and glycosylation of LH and FSH can be well understood. In our opinion, this will be possible only by investigating cell-free biosynthesis of gonadotropins using purified mRNA.

4. Summary

In this chapter, we have reviewed the present state of the *in vitro* biosynthesis of LH and FSH and the effects of GnRH on this synthesis. Other investigators, using either incubation of rat pituitary quarters or pituitary cells in culture, have reported that GnRH had no effect on the incorporation of labeled amino acids into LH, but that it stimulated significantly the incorporation of [³H]glucosamine into this hormone.

We investigated the incorporation of [³H]proline or [³H]leucine and [³H]glucosamine into LH and FSH by anterior pituitary cells in monolayer culture. Like others, we found that in our system the polypeptide chains of gonadotropins were synthesized and glycosylated. However, in contrast to the observations of others, our results showed that GnRH stimulates significantly the incorporation of [³H]proline into LH and FSH during the first 4 hr of incubation, whereas the incorporation of [³H]glucosamine under similar conditions was unaffected or affected only slightly by GnRH. The discrepancy between our results and those reported by other authors may be due to the use of different antisera and immunoprecipitation methods and also to the fact that one of the other groups incubated pituitary quarters, whereas we employed dispersed pituitary cells in culture.

In addition, our data show that the structural integrity of gonadotropic cells is not necessary for the synthesis and glycosylation of the polypeptide chains of LH and FSH, but that the integrity of both GnRH receptor sites on the plasma membranes and of cells is necessary for the action of GnRH on the biosynthesis. On the other hand, our results suggest that the rate of synthesis of FSH must be at least twice that of LH.

Cycloheximide completely blocked the synthesis of the polypeptide chains of gonadotropins, but only reduced their glycosylation. In our opinion, this latter observation suggests that gonadotropic cells contain some nonglycosylated or partially glycosylated, completed, or nascent subunits of LH and FSH available for glycosylation. The fact that actinomycin D had no effect on either glycosylation or release of [³H]glucosamine-labeled LH and FSH in the controls may signify that gonadotropic cells contain enough mRNA to ensure synthesis of polypeptide chains of gonadotropins during 6 hr of incubation. Our observation that actinomycin D completely inhibited the GnRH-induced release of [³H]glucosamine-LH and -FSH confirms the findings of many authors, suggesting that the GnRH-induced release of gonadotropins proceeds through two complementary mechanisms, an acute effect, followed by a priming effect. Thus, the priming effect of GnRH on the release is probably dependent on protein or RNA synthesis or both.

ACKNOWLEDGMENTS. We would like to thank Dr. A. F. Parlow and NIAMDD for the antisera against rat LHβ and rat FSHβ. Part of this work has been done in collaboration with Dr. L. Debeljuk. The authors greatly appreciate the excellent technical assistance of Mrs. T. Taverny-Bennardo. One of us (A.K.) is a postdoctoral fellow of the Ministère des

Affaires Etrangères, Paris, and thanks the Foundation Simone and Cino del Duca for partial financial support. This work was supported in part by a grant from DGRST, Paris (Contract Nos. 75-7-1657 and 76-7-0715).

DISCUSSION

DUVAL: You said that the synthesis of protein and RNA was a necessary step for the synthesis and release of LH and FSH. In superfusion experiments, using continuous infusion of LH-RH and cycloheximide pulses, we were able to show that the antibiotic does not inhibit the acute release of the gonadotropin, but only the high secondary release. This sharp inhibition of LH and FSH releases by a short pulse of cycloheximide could be interpreted in terms of the inhibition of the synthesis of a specific protein, but other interpretations are possible. The antibiotic may well have other actions in animal cells; thus, one should be extremely cautious when interpreting release data based on the use of cycloheximide.

JUTISZ: We agree with your comment; however, our data with antibiotics indicate the possible involvement of the synthesis of a protein and/or RNA in the mechanism of action of GnRH.

DENEF: There may be changes in the regulation of synthesis of LH and FSH in response to LH-RH at various times in culture. Have you looked at LH and FSH synthesis on day 3 or 4 in culture? In our experiments with immature rat pituitary cells in culture, on day 3 LH-RH stimulates only the synthesis of FSH and not LH. At later times in culture, LH-RH stimulates the synthesis of both FSH and LH. Have you tested the influence of androgen pretreatment, as androgens induce a selective increase in FSH synthesis?

JUTISZ: We have always used cultures on day 7, since the response to GnRH was optimal then. We have not looked at the synthesis of LH and FSH on day 3 or 4 of culture, nor have we tested the influence of pretreatment with androgens.

LABRIE: You have made your studies after a 6-hour incubation with LH-RH. We have previously found that the optimal time of the stimulatory effect of LH-RH on LH and FSH synthesis was between 12 and 24 hours after neurohormone. LH-RH stimulates an immediate effect on the release of gonadotropins, whereas the effect on synthesis is delayed. Have you investigated the rates of LH and FSH synthesis after longer periods of incubation with the neurohormone?

JUTISZ: Since we use the synthetic medium F-10 without the addition of sera, we had to be cautious about the functional integrity of cells incubated for a longer period of time. Therefore, we did not try our release or incorporation studies beyond 6 hours. We will attempt to check the integrity of cells after 6 hours of incubation.

MELDOLESI: How was the purity of the immunoprecipitate checked to exclude unspecific coprecipitation of other components together with the hormones?

JUTISZ: First, we eliminated the binding of nonspecific radioactivity by treating the samples with normal rabbit serum and anti-rabbit gamma-globulin (Burek and Frohman, 1970). Secondly, we passed our samples through an affinity column (anti-LHβ-Sepharose), and then we did the SDS–acrylamide gel electrophoresis where we observed single bands corresponding to LH or FSH.

BAXTER: The experiments were performed with the continuous presence of radioactive precursor, rather than with pulses of precursor. It is possible that the decrease in the total cellular radioactivity in cells incubated with the releasing factor, as compared with the control, was due to the releasing factor-stimulated depletion of intracellular hormone.

JUTISZ: We have no explanation for this phenomenon; however, the possibility raised by you cannot be excluded.

REFERENCES

Aiyer, M. S., Chiappa, S. A., and Fink, G., 1974, A priming effect of luteinizing hormone releasing factor on the anterior pituitary gland in the female rat, *J. Endocrinol.* **62**:573.

Azhar, S., Reel, J. R., Pastushok, C. A., and Menon, K. M. J., 1978, LH biosynthesis and secretion in rat anterior pituitary cell cultures: Stimulation of LH glycosylation and secretion by GnRH and an agonistic analogue and blockade by an antagonistic analogue, *Biochem. Biophys. Res. Commun.* **80**:659.

Burek, C. L., and Frohman, L. A., 1970, Growth hormone synthesis by rat pituitaries *in vitro:* Effect of age and sex, *Endocrinology* **86**:1361.

Chowdhury, M., and Steinberger, E., 1975, Biosynthesis of gonadotropins by rat pituitaries *in vitro, J. Endocrinol.* **66**:369.

Debeljuk, L., Khar, A., and Jutisz, M., 1978, Effects of gonadal steroids and cycloheximide on the release of gonadotropins by rat pituitary cells in cultures, *J. Endocrinol.* **77**:409.

De Koning, J., Van Dieten, J. A. M. J., and Van Rees, G. P., 1976, LHRH-dependent synthesis of protein necessary for LH release from rat pituitary glands *in vitro, Mol. Cell. Endocrinol.* **5**:151.

De La Llosa, P., Marche, P., Morgat, J. L., and De La Llosa-Hermier, M. P., 1974, A new procedure for labeling luteinizing hormone with tritium, *FEBS Lett.* **45**:162.

Edwardson, J. A., and Gilbert, D., 1976, Application of an *in vitro* perfusion technique to studies of luteinizing hormone release by rat anterior hemi-pituitaries: Self-potentiation by luteinizing hormone releasing hormone, *J. Endocrinol.* **68**:197.

Jutisz, M., and De La Llosa, M. P., 1967, Studies on the release of FSH *in vitro* from rat pituitaries stimulated by hypothalamic follicle-stimulating hormone-releasing factor, *Endocrinology* **81**:1193.

Jutisz, M., and De La Llosa, M. P., 1968, Recherches sur le contrôle de la sécrétion de l'hormone folliculo-stimulante hypophysaire, *Bull. Soc. Chim. Biol.* **50**:2521.

Jutisz, M., and Kerdelhué, B., 1973, *In vitro* studies on synthetic LHRH and its assay in plasma using a radioimmunological method, in: *Hypothalamic Hypophysiotropic Hormones* (C. Gual and E. Rosemberg, eds.), pp. 98–104, Excerpta Medica (I.C.S. Series No. 263), Amsterdam.

Jutisz, M., De La Llosa, M. P., Bérault, A., and Kerdelhué, B., 1970, Le rôle des hormones hypothalamiques dans la régulation de l'excrétion et de la synthèse des hormones adénohypophysaires, in: *Neuroendocrinologie* (J. Benoit and C. Kordon, eds.), pp. 287–301, Colloque National du CNRS (No. 927), Paris.

Jutisz, M., Kerdelhué, B., Bérault, A., and De La Llosa, M. P., 1972, On the mechanism of action of the gonadotropin releasing factors, in: *Gonadotropins* (B. B. Saxena, C. G. Beling, and H. M. Gandy, eds.), pp. 64–71, Wiley, New York.

Khar, A., Debeljuk, L., and Jutisz, M., 1978, Biosynthesis of gonadotropins by rat pituitary cells in culture and in pituitary homogenates: Effect of gonadotropin releasing hormone, *Mol. Cell. Endocrinol.* **12**:53.

Kobayashi, T., Kobayashi, T., Kigawa, M., Mizuno, M., Amenonori, Y., and Watanabe, T., 1965, Autoradiographic studies on [^{3}H]leucine uptake by adenohypophysial cells *in vitro, Endocrinol. Jpn.* **12**:47.

Kochman, K., Kerdelhué, B., Zor, U., and Jutisz, M., 1975, Studies of enzymatic degradation of luteinizing hormone–releasing hormone by different tissues, *FEBS Lett.* **50**:190.

Labrie, F., Pelletier, G., Lemay, A., Borgeat, P., Barden, N., Dupont, A., Savary, M., Côté, J., and Boucher, R., 1973, Control of protein synthesis in anterior pituitary gland, *Acta Endocrinol. (Copenhagen) Suppl.* **180**:301.

Liu, T. C., and Jackson, G. L., 1978, Modifications of luteinizing hormone biosynthesis and release by GnRH, cycloheximide, and actinomycin D, *Endocrinology* **103**:1253.

Liu, T. C., Jackson, G. L., and Gorski, J., 1976, Effects of synthetic GnRH on incorporation of radioactive glucosamine and amino acids into LH and total protein by rat pituitaries *in vitro, Endocrinology* **98**:151.

Menon, K. M. J., Gunaga, K. P., and Azhar, S., 1977, GnRH action in rat anterior pituitary gland: Regulation of protein, glycoprotein and LH synthesis, *Acta Endocrinol. (Copenhagen)* **86**:473.

Mittler, J. C., Arimura, A., and Schally, A. V., 1970, Release and synthesis of luteinizing hormone and follicle-stimulating hormone in pituitary cultures in response to hypothalamic preparations, *Proc. Soc. Exp. Biol. Med.* **133**:1321.

Nakano, H., Fawcett, C. P., and McCann, S. M., 1976, Enzymatic dissociation and short-term culture of isolated rat anterior pituitary cells for studies on the control of hormone secretion, *Endocrinology* **98**:278.

Pickering, A. J. M. C., and Fink, G., 1976, Priming effect of luteinizing hormone releasing factor: *In vitro* and *in vivo* evidence consistent with its dependence upon protein and RNA synthesis, *J. Endocrinol.* **69**:373.

Redding, T. W., Schally, A. V., Arimura, A., and Matsuo, H., 1972, Stimulation of release and synthesis of LH and FSH in tissue cultures of rat pituitaries in response to natural and synthetic LH and FSH releasing hormone, *Endocrinology* **90**:764.

Rothman, J. E., and Lodish, H. F., 1977, Synchronized transmembrane insertion and glycosylation of nascent membrane protein, *Nature (London)* **269**:775.

Samli, M. H., and Geschwind, I. I., 1967, Some effects of the hypothalamic luteinizing hormone releasing factor on the biosynthesis and release of luteinizing hormone, *Endocrinology* **81**:835.

Théoleyre, M., Bérault, A., Garnier, J., and Jutisz, M., 1976, Binding of LHRH to the pituitary plasma membranes and the problem of adenylate cyclase stimulation, *Mol. Cell. Endocrinol.* **5**:365.

Tixier-Vidal, A., Kerdelhué, B., Bérault, A., Picart, R., and Jutisz, M., 1971, Action *in vitro* du facteur hypothalamique FRF sur l'antéhypophyse d'agnelle: Étude ultrastructurale des tissus incubés, *Gen. Comp. Endocrinol.* **17**:33.

Tixier-Vidal, A., Kerdelhué, B., and Jutisz, M., 1973, Kinetics of release of LH and FSH by primary cultures of dispersed rat anterior pituitary cells: Chronic effect of synthetic LH and FSH releasing hormone, *Life Sci.* **12**:499.

Vale, W., Grant, G., Amoss, M., Blackwell, R., and Guillemin, R., 1972, Culture of enzymatically dispersed anterior pituitary cells: Functional validation of a method, *Endocrinology* **91**:562.

Vilchez-Martinez, J. A., Arimura, A., and Schally, A. V., 1976, Effect of actinomycin D on the pituitary response to LHRH, *Acta Endocrinol. (Copenhagen)* **81**:73.

12

Processing of the Common Precursor to ACTH and Endorphin in Mouse Pituitary Tumor Cells and Monolayer Cultures from Mouse Anterior Pituitary

Edward Herbert, Marjorie Phillips, Michael Hinman, James L. Roberts, Marcia Budarf, and Thomas L. Paquette

1. Introduction

Adrenocorticotropin (ACTH) and β-lipotropin (β-LPH) have been shown to be derived from the same high-molecular-weight precursor protein (common precursor) in mouse pituitary tumor cells (AtT-20/D$_{16v}$ cells) (Mains *et al.*, 1977; Nakanishi *et al.*, 1977; Roberts and Herbert, 1977a, b). When messenger RNA from AtT-20 cells is translated in a reticulocyte cell-free system, a single form of the precursor is synthesized with a

Edward Herbert, Marjorie Phillips, Michael Hinman, and *Marcia Budarf* • Department of Chemistry, University of Oregon, Eugene, Oregon 97403   *James L. Roberts* • Howard Hughes Medical Institute, Endocrine Research Division, University of California School of Medicine, San Francisco, California 94143   *Thomas L. Paquette* • Department of Surgery, Washington University, St. Louis, Missouri 63110

molecular weight of 28,500 (28.5K pro-ACTH-endorphin) (Roberts and Herbert, 1977a,b). The use of the "polysome runoff technique" (Dintzis, 1961) has made it possible to arrange the tryptic peptides of ACTH and β-LPH relative to one another (Roberts and Herbert, 1977b) in the 28.5K precursor as shown in the model of structure in Fig. 1. In this model, β-LPH is located at the C terminus of the precursor and ACTH is adjacent to β-LPH near the middle of the molecule, leaving an unidentified sequence of approximately 100 amino acids at the N terminus. A 31,000-molecular-weight form of the precursor has been isolated from AtT-20 cells and shown to have a structure very similar to that depicted in Fig. 1 (Eipper and Mains, 1978). Thus, a single protein contains the sequence of α(1–39)ACTH, α-melanocyte-stimulating hormone (α-MSH), and the component hormones of β-LPH: β-endorphin, β-MSH, and Met-enkephalin.

Although these hormones are synthesized together as a single protein, they differ greatly in their physiological actions. β-Endorphin and Met-enkephalin mimic the action of morphinelike substances in the animal (see the review by Goldstein, 1976). ACTH stimulates the production of glucocorticoids in the adrenal cortex, and α and β-MSH stimulate pigment formation in melanocytes (Ganong, 1977). Under some conditions, release of these hormones is coupled in the animal (Abe *et al.*, 1967, 1969; Guillemin *et al.*, 1977) and in pituitary cells in culture (Allen *et al.*, 1978; Vale *et al.*, 1978).

2. Biosynthesis of the Common Precursor

2.1. Presence of the Common Precursor and Component Hormones of the Precursor in the Pituitary

The anterior and intermediate–posterior lobes of mouse and rat pituitary both contain the common precursor to ACTH and endorphin (Roberts *et al.*, 1978; Eipper and Mains, 1979). However, the distribution of the biologically active peptides derived from the precursor is quite different in the two lobes (Roberts *et al.*, 1978). The predominant peptides

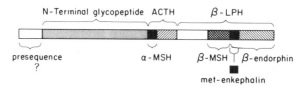

Figure 1. Model of the structure of the common precursor to ACTH and β-endorphin.

in the anterior lobe are α(1–39)ACTH (4.5K ACTH), the glycosylated form of α(1–39)ACTH (13K ACTH), and β-LPH. β-Endorphin is a minor species of peptide in the anterior lobe. The predominant peptides in the intermediate lobe are α-MSH, β-LPH, β-endorphin (Roberts *et al.*, 1978) and α(18–39)ACTH (Scott *et al.*, 1975). Very little enkephalin has been found in either lobe of the pituitary. These studies show that the precursor is processed differently in the two lobes of the pituitary. We would like to know precisely how processing reactions differ in the two lobes of the pituitary and how these reactions are regulated.

As a first step toward answering these questions, we have undertaken a detailed study of processing of pro-ACTH-endorphin in AtT-20/D_{16v} cells. These cells are very favorable material for biochemical studies because of the large amounts of ACTH and endorphins they produce and because of the ease with which they can be managed in culture (Herbert *et al.*, 1978). They are also a good model system for studying processing of the precursor in the anterior pituitary, since they are similar in their responses to regulators and they have almost the same distribution of the molecular-weight forms of ACTH and endorphins as anterior pituitary cells (Allen *et al.*, 1978; Roberts *et al.*, 1978). Nevertheless, since AtT-20 cells are tumorgenic, it is essential to compare processing in these cells with that in normal anterior pituitary cells in culture. These studies are reported in this chapter.

2.2. Processing Events

Processing of the common precursor consists basically of two kinds of events: glycosylation and proteolysis. We have studied glycosylation events by labeling cells with both [^{35}S]-Met and [^{3}H]sugars. In this way, we can identify the forms of ACTH that contain various sugars (glucosamine, mannose, galactose, and fucose), which allows the order of the sugar addition and proteolytic cleavage steps to be determined. An outline of the approach is presented below:

1. AtT-20/D_{16v} cells are labeled with [^{35}S]-Met and a [^{3}H]sugar and then extracted.
2. Cell extracts are immunoprecipitated with specific antisera to the ACTH, β-endorphin, or N-terminal region of the precursor.
3. Proteins in the immunoprecipitates are separated by sodium dodecyl sulfate–polyacrylamide gel electrophoresis (SDS-PAGE) (either tube gel or slab gel electrophoresis).
4. Proteins are eluted from the gels, digested with trypsin and tryptic peptides, and glycopeptides are analyzed by paper electrophoresis or chromatography or both.

Mains and Eipper (1976) have isolated four molecular-weight classes of ACTH by this approach: 31K, 23K, 13K, and 4.5K ACTH. All these size classes of ACTH are glycosylated except 4.5K ACTH (Eipper *et al.*, 1976; Roberts *et al.*, 1978).

When AtT-20 cells are labeled with [³H]glucosamine and [³⁵S]-Met for 2 hr and fractionated with ACTH antiserum (as above) employing SDS-PAGE systems with higher resolving power than those used previously, the 31K, 23K, and 13K classes of molecules are separated into a number of subclasses (Fig. 2). Note that three glycoproteins are resolved in the high-molecular-weight region of the gel (29K, 32K, and 34K). All these glycoproteins contain the tryptic peptides of ACTH and β-LPH, and can be immunoprecipitated with either the β-endorphin or the ACTH antiserum (Roberts *et al.*, 1978). They can also be resolved by slab gel electrophoresis (see Figs. 5 and 11), and they migrate with the same mobility when rerun in the same gel-electrophoresis system (Roberts and Herbert, 1977a), indicating that they are not artifacts of the gel system used.

Results of several kinds of studies suggest that the differences in mobilities of the 29K, 32K, and 34K forms of the precursor are due to differences in carbohydrate content, rather than to differences in peptide

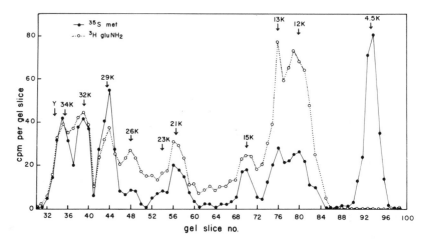

Figure 2. Analysis by SDS-PAGE of ACTH immunoprecipitates labeled with [³⁵S]-Met and [³H]glucosamine. AtT-20/D₁₆ᵥ cells were grown to about half confluency in Falcon Microtest wells in Dulbecco-Vogt minimal essential medium with 10% horse serum (Herbert *et al.*, 1978). The cells were then incubated for 2 hr with 25 μCi of L-[³⁵S]-Met (1000 Ci/mmol) and 125 μCi of [³H]glucosamine (19 Ci/mmol) in 50 μl of modified low-glucose Dulbecco-Vogt minimal essential medium containing 10% horse serum. Cells were extracted, and immunoprecipitated with antiserum Bertha. The specific ACTH-containing proteins were separated by electrophoresis with 12% Biophore tube gels. Dansyl-YADH was included as an internal molecular-weight marker. For further details, consult Roberts *et al.* (1978).

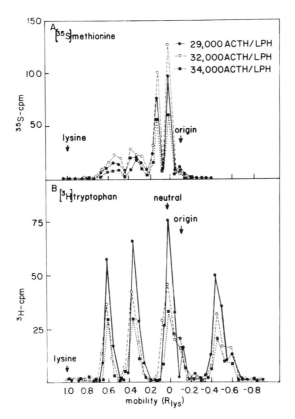

Figure 3. Tryptic peptides of [^{35}S]-Met- and [^{3}H]-Trp-labeled forms of pro-ACTH-endorphin. Labeled pro-ACTH-endorphin was prepared as described in the Fig. 2 caption. After elution of the 29K, 32K, and 34K proteins from SDS gel, they were digested with trypsin, and the digests were analyzed by paper electrophoresis at pH 6.5. Mobility is defined relative to Lys ($R_{Lys} = 1.0$; ϵ-DNP-Lys = 0).

backbones. First, the 32K and 34K forms have more mannose and glucosamine than the 29K form relative to Met residues (Roberts *et al.*, 1978); second, tryptic peptide maps (paper electrophoresis at pH 6.5) of the three forms labeled with a variety of different radioactive amino acids show that the tryptic peptides of the three forms are very similar. The results of such an analysis with [^{35}S]-Met and [^{3}H]-Trp are shown in Fig. 3 (from Roberts *et al.*, 1978). However, when glucosamine-labeled tryptic peptides of the same three forms are examined by this mapping procedure, marked differences are noted (Fig. 4). There are at least three glycopeptides present. All the precursor forms appear to contain a neutral and basic glycopeptide in somewhat variable amounts, whereas only the 32K forms

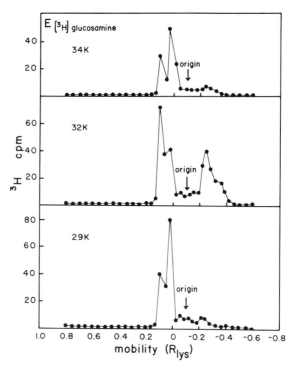

Figure 4. Tryptic peptides of [³H]glucosamine–labeled forms of pro-ACTH-endorphin. AtT-20 cells were incubated with 300 μCi of D-[6-³H]-glucosamine (20 Ci/mmole) for 2 hr. Tryptic peptide analysis was performed as described by Roberts *et al.* (1978).

contain an acidic glycopeptide. Essentially the same results are obtained when [³H]mannose is used as a precursor instead of [³H]glucosamine (results not shown). Further characterization of these glycopeptides has been done by amino acid composition studies (Roberts *et al.*, 1978).

2.3. Biosynthetic Relationships of the 29K, 32K, and 34K Precursor Forms

When AtT-20 cells are labeled for 5 min with a radioactive amino acid, over 90% of the radioactivity that is immunoprecipitable with either ACTH or endorphin antiserum is present in the 29K form of the precursor (Fig. 1 in Roberts *et al.*, 1978). Pulse–chase studies show that label in the 29K form chases much more rapidly than label in the 32K and 34K forms of the precursor (Figs. 2 and 3 in Roberts *et al.*, 1978). Stoichiometric analysis of the flow of label through the precursor forms during the pulse–chase suggests that 29K pro-ACTH-endorphin is the precursor of the 32K and 34K forms.

2.4. Biosynthesis and Structure of Carbohydrate Side Chains

Studies of the composition of the CHO side chains present in various forms of ACTH have shown that the side chains are of the complex type consisting of glucosamine, mannose, galactose, and, in some cases, fucose and possibly sialic acid (Eipper and Mains, 1977; Roberts *et al.*, 1978). Biosynthesis of side chains of this type has been studied in detail in other systems and has been found to occur in two steps. First, a core of *N*-acetylglucosamine and mannose residues is synthesized as a lipid-linked intermediate and then transferred to the protein as a unit (Waechter and Lennarz, 1976). This event takes place in the rough endoplasmic reticulum (Czichi and Lennarz, 1977). Addition of a core group of sugars can be expected to generate a discrete higher-molecular-weight species of the protein. The second step is the trimming back of mannose residues and the addition of peripheral sugars (galactose, fucose, some glucosamine and sialic acid) one residue at a time in the smooth endoplasmic reticulum and Golgi and can be expected to generate a heterogeneous collection of glycoproteins.

Our studies of the biosynthesis of CHO side chains suggest that most of the peripheral sugars are added during or shortly after the precursor forms are converted to lower-molecular-weight forms of ACTH. Thus, the 29K, 32K, and 34K forms of the precursor contain mainly core sugars.

2.5. Proteolytic Processing of Precursors to Lower-Molecular-Weight Forms of ACTH, Endorphin, and N-Terminal Glycopeptides

To study thoroughly the processing of the precursor forms, the intermediates and end products of processing must be identified and characterized. The availability of antisera to three different regions of the precursor (ACTH, endorphin, and N-terminal regions) has made it possible to separate processed fragments derived from these regions of the precursor by immunoprecipitation and SDS–slab gel electrophoresis as shown in Fig. 5. In this experiment, cells were incubated for 2 hr with [^{35}S]-Met, extracted, and the extracts were immunoprecipitated with antisera to β-endorphin or ACTH. Proteins in the immunoprecipitates were fractionated by SDS–slab gel electrophoresis using a modification (O'Farrell, 1975) of the method developed by Laemmli (1970). Lanes 1 and 2 in Fig. 5 show the molecular-weight forms immunoprecipitated with the β-endorphin and the ACTH antiserum, respectively. One can see fairly sharp bands of the 29K and 32K forms of pro-ACTH-endorphin in lanes 1 and 2. Less-well-defined bands corresponding to intermediate forms of ACTH (26K and 20–23K ACTH) and 12–15K ACTH can be seen in lane 2. The latter bands become more diffuse with increasing time of labeling (not shown). The

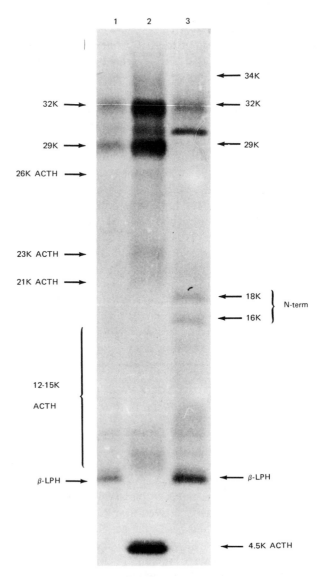

Figure 5. SDS–slab gel analysis of [^{35}S]-Met-labeled ACTH, endorphin, and N-terminal peptides. AtT-20 cells were labeled for 2 hr with [^{35}S]-Met as described in the Fig. 2 caption. Cell extracts were immunoprecipitated with antiserum to ACTH or β-endorphin. An aliquot of the supernatant from the ACTH immunoprecipitation was immunoprecipitated with antiserum to the N-terminal region of the precursor. The immunoprecipitates were analyzed by SDS–slab gel electrophoresis using the O'Farrell (1975) modification of the Laemmli (1970) method. The gels were dried, and autoradiograms were prepared. Lanes 1, 2, and 3 correspond to β-endorphin, ACTH, and N-terminal immunoprecipitates, respectively.

width of the bands appears to be related to the degree of glycosylation of the peptide. The unglycosylated peptides (4.5K ACTH and β-LPH) and the peptides with predominantly core CHO side chains (29K and 32K forms) migrate as sharper bands than peptides with peripheral sugars attached to CHO chains (20–26K and 12–15K ACTH) (Roberts *et al.*, 1978).

The fragments in these bands can be further characterized by tryptic peptide mapping. Comparison of tryptic peptides of a fragment with those of the precursor enables one to identify that part of the precursor from which the fragment is derived. For example, the three radioactive peaks designated 26K, 23K, and 21K in Fig. 2 all appear to be ACTH intermediates, since they contain ACTH and N-terminal peptides found in the precursor but are missing β-LPH peptides (Phillips, unpublished results). Pulse–chase experiments with radioactive amino acids support the idea that the three fragments are intermediates in the biosynthesis of 4.5K and 13K ACTH (Roberts *et al.*, 1978).

2.6. Identification of N-Terminal Glycopeptides

Fragments are present in extracts of AtT-20 cells and in culture medium from these cells that can be immunoprecipitated with a partially purified antiserum to the N-terminal part of the precursor. When the immunoprecipitate is fractionated by SDS-PAGE, two major radioactive peaks with apparent molecular weights of 18K and 16K are detected (Fig. 6). [A β-LPH-like component (12K peak) and a higher-molecular-weight material that is not pro-ACTH-endorphin are also present since the antiserum is not pure.] These fragments can also be resolved by SDS–slab gel electrophoresis as shown in lane 3 in Fig. 5. The tryptic peptide analysis shows that the 18K and 16K fragments (Fig. 6) contain only the tryptic peptides derived from the N-terminal region of the precursor and that the two fragments have a very similar peptide backbone (Phillips *et al.*, unpublished results). However, maps of the glucosamine-labeled tryptic glycopeptides of the two fragments are different (Fig. 7). The 16K fragment has only a neutral glycopeptide, whereas the 18K fragment has both a neutral and a basic glycopeptide. As would be expected, the acidic ACTH glycopeptide (seen in 32K pro-ACTH-endorphin) is not present in either of the N-terminal fragments.

The simplest interpretation of this tryptic glycopeptide analysis is that there are two N-terminal CHO side chains. The 16K form contains only one of these, but the 18K form contains both. The existence of these two forms of the N-terminal fragment suggests that there should be corresponding forms of the pro-ACTH-endorphin precursor. One model is that the 34K form contains two N-terminal side chains and is a precursor to the 18K N-terminal fragment, whereas the 29K and 32K forms contain only

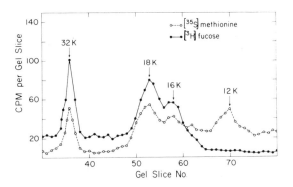

Figure 6. Purification of N-terminal glycopeptides by immunoprecipitation and SDS-PAGE with tube gels. Tumor cells were labeled as described previously with [^{35}S]-Met and [^{3}H]fucose. The N-terminal fragments were immunoprecipitated as described in the Fig. 5 caption, and SDS-PAGE was performed as described by Roberts *et al.* (1978) with 12% Biophore tube gels.

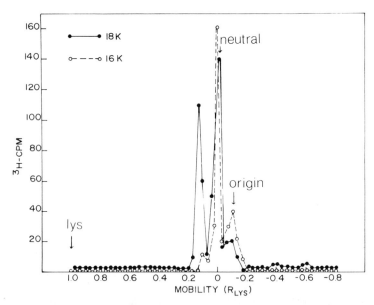

Figure 7. [^{3}H]Glucosamine-labeled tryptic peptides of 18K and 16K N-terminal glycoproteins. The N-terminal glycopeptides were purified by immunoprecipitation as described in the Fig. 5 caption. The immunoprecipitate was fractionated by SDS-PAGE using 12% Biophore tube gels as shown in Figure 6. Analysis of the tryptic peptides was carried out by paper electrophoresis at pH 6.5 as described by Roberts *et al.* (1978).

one N-terminal CHO side chain and are precursors to the 16K N-terminal fragment and its counterpart, a fragment that contains only the basic tryptic glycopeptide. (The latter form has not yet been identified in cell extracts. Perhaps it comigrates with the 18K form during SDS-PAGE.)

2.7. Summary of Processing Pathways

A glycopeptide analysis similar to that described above has been performed with all the intermediates and end products of processing in the AtT-20/D$_{16v}$ cells. These data are summarized in Table I. One can arrange the forms of the precursor, intermediates, and end products listed in Table I into groups depending on the types of tryptic glycopeptides they contain. When this information is combined with the results of the pulse–chase experiments (Roberts *et al.*, 1978) and interpreted in the light of the model of structure in Fig. 1, a complete processing scheme can be constructed for all the components listed in the table. This scheme is shown in Fig. 8.

According to the scheme in Fig. 8, the first glycosylation event is addition of a CHO side chain to the N-terminal region of the precursor to form 29K pro-ACTH-endorphin. The 29K form is at a branch point in the pathway. It can be processed proteolytically to 21K ACTH and β-LPH, or it can be further glycosylated. A CHO side chain can be added to the ACTH region of the precursor to generate the 32K form or to the N-terminal region to generate the 34K form. Since the precursor forms do not appear to contain fucose or sialic acid, their side chains are depicted as core CHO side chains in the diagram.

Table I. Tryptic Glycopeptide Summary

	Basic tryptic glycopeptide ($R_{Lys} = 0.1$)	Neutral tryptic glycopeptide ($R_{Lys} = 0$)	Acidic $\alpha(22-39)$ tryptic glycopeptide
34K ACTH-endorphin	+	+	−
32K ACTH-endorphin	+	±	+
29K ACTH-endorphin	±	+	−
23K ACTH	±	+	+
21K ACTH	+	+	−
12–15K ACTH	−	−	+
18K N-terminal glycoprotein	+	+	−
16K N-terminal glycoprotein	−	+	−
11.7K β-endorphin	−	−	−
3.5K β-endorphin	−	−	−

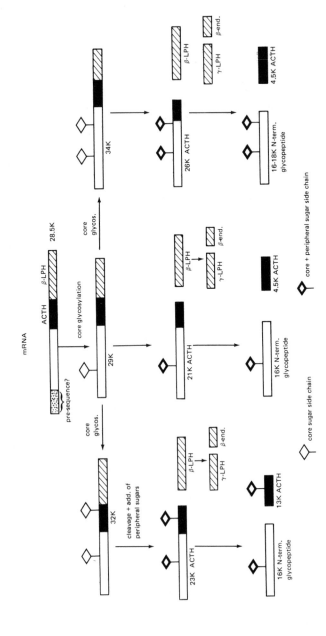

Figure 8. Summary of pro-ACTH-endorphin processing pathways in AtT-20/D$_{16v}$ cells.

The first proteolytic cleavage step generates a β-LPH-like molecule and an ACTH intermediate. Each form of the precursor gives rise to a different ACTH intermediate (with different CHO side chains). Subsequent cleavage of the ACTH intermediates results in formation of 4.5K ACTH, 13K ACTH, and two or three different forms of the N-terminal glycopeptide. Since the ACTH intermediates and the 12–15K forms of ACTH contain peripheral sugars (fucose and galactose), these components are depicted as having complete or nearly complete CHO side chains (containing both core and peripheral sugars).

β-LPH is processed further to β-endorphin with a rather slow time course (Roberts *et al.*, 1978). Neither of these peptides contains sugars at any stage of processing.

2.8. Biosynthesis and Processing of the Common Precursor in Monolayer Cultures of Mouse Anterior Pituitary

It is important to know how closely processing in pituitary tumor cells resembles that in the pituitary. It has already been shown that the molecular-weight distribution (Roberts *et al.*, 1978) of ACTH and the regulation of release of ACTH (Allen *et al.*, 1978) in AtT-20 cells is much more like that in anterior pituitary than in intermediate pituitary from mouse. Thus, a detailed study of processing of the precursor was undertaken in monolayer cultures of mouse anterior pituitary.

Cultures were prepared as described by Allen *et al.* (1978). A number of experiments were first performed to determine whether culturing pituitary cells *per se* alters any of the properties we were interested in studying. We found that there was no significant change in the cultures with regard to the following properties for at least 6 days: (1) ability to release ACTH; (2) level of ACTH in the culture; (3) regulation of ACTH release by hypothalamic extract, vasopressin, or glucocorticoids; and (4) cellular distribution of the molecular-weight forms of ACTH (which was not altered from that found in the anterior pituitary). On the basis of these studies, 4-day cultures were used routinely in the work described below.

Cultures were incubated in serum-free culture medium containing [^{35}S]-Met for varying periods of time. The rate of protein synthesis was measured by determining the amount of [^{35}S]-Met incorporated into acid-insoluble material and was found to be linear for 3–4 hr. The rate of incorporation of [^{35}S]-Met into immunoprecipitable ACTH was linear for at least 6 hr. The cells were extracted, and the forms of ACTH were purified by the fractionation scheme outlined earlier. SDS-PAGE profiles (Fig. 9) of the immunoprecipitated material show the presence of radioactive peaks with mobilities very similar to those of the forms of ACTH seen in tumor cells (see Fig. 2). When a large excess of α_p(1–39)ACTH is

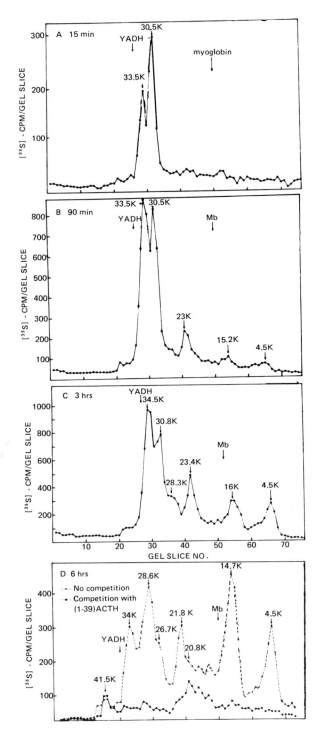

added to the immunoprecipitation mixture, almost all the labeled peaks in the profile disappear, suggesting that these radioactive components contain an ACTH determinant (Fig. 9D).

Figure 9 shows that after 15 min of labeling, about 80% of the radioactivity is in a 30.5K peak and the remainder is in a 33–34K peak. With increasing time of labeling, there is a shift in radioactivity from the 30.5K peak to the 33K peak and the appearance of a small amount of label in lower-molecular-weight forms of ACTH. The precursor–product relationships can be seen more easily by plotting the proportion of total immunoprecipitable label present in each form of ACTH against time as is done in Fig. 10. One can see that label enters the 30.5K form first and then the 33K form. As the proportion of label decreases in these forms, it increases in 23K and 27K forms of ACTH (only 23K is shown in Fig. 10). Finally, label enters 12–15K and 4.5K forms of ACTH. It should also be noted that there is little change in the distribution of label in the forms of ACTH after 12 hr, indicating that steady-state labeling has occurred by that time. Thus, the flow of label in anterior-lobe cells is consistent with the general scheme for ACTH processing in tumor cells. Pulse–chase experiments (results not shown) confirm the sequence of events depicted in Fig. 10.

It is also of interest that only two high-molecular-weight forms of ACTH are seen in anterior-pituitary cells as opposed to three forms in the tumor cells. SDS–slab gel electrophoresis shows that the two high-molecular-weight forms in anterior-pituitary cells (Fig. 11, lane 2,) migrate with the same mobility as the 29K and 32K forms of pro-ACTH-endorphin in tumor cells (Fig. 11, lanes 1 and 3). We cannot detect a significant amount of the tumor-cell 34K precursor form in anterior-pituitary cultures.

2.9. Tryptic Peptide Mapping of the Forms of ACTH in Anterior-Pituitary Cells

Radioactive material was eluted from the SDS gels in Fig. 9 and digested with trypsin (Roberts *et al.*, 1978). The results in Fig. 12A show that the 30.5K and 33K forms from anterior-pituitary cultures have the same [^{35}S]-Met-labeled tryptic peptides as the 29K form of pro-ACTH-

---

Figure 9. Time course of [^{35}S]-Met labeling of the forms of ACTH in monolayer cultures of mouse anterior pituitary. The cultures were prepared as described by Allen *et al.* (1978). After 4 days of culturing, the cells were labeled with [^{35}S]-Met in serum-free and Met-free medium for the times indicated. Cell extracts were prepared and immunoprecipitated with ACTH antiserum. The immunoprecipitates were fractionated by SDS-PAGE with 12% Biophore tube gels.

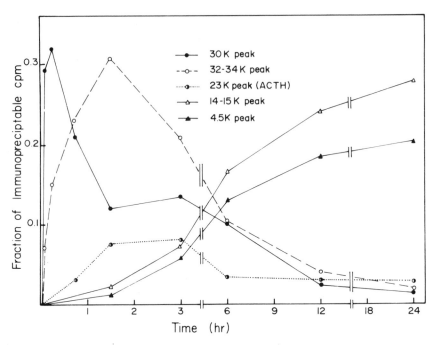

Figure 10. Summary of time course of labeling of forms of ACTH in monolayer cultures of mouse anterior-pituitary cells. The fraction of total ACTH immunoprecipitable counts per minute present in each peak in Fig. 9 plus data from 12-hr and 24-hr incubations (not shown in Fig. 9) are plotted against time.

endorphin isolated from the tumor cells including the $\alpha(1-8)$ACTH and $\beta(61-69)$LPH peptides, providing additional evidence that the 30.5K and 33K components from anterior-pituitary cultures are forms of pro-ACTH-endorphin. Also note that the $\beta(61-69)$LPH tryptic peptide is missing from the profile of the 27K form (Fig. 12), showing that this component does not contain the β-endorphin region of the precursor and is very likely an intermediate form of ACTH. The profile of the ^{35}S-labeled tryptic peptides of the 23K form is almost identical to that of the 27K form, whereas the profile of the 4.5K is missing both the $\beta(61-69)$LPH peptide and the N-terminal peptide (results not shown).

3. Discussion

Three forms of pro-ACTH-endorphin have been identified in AtT-20/ D_{16v} cells by SDS-PAGE and by tryptic peptide mapping studies (29K,

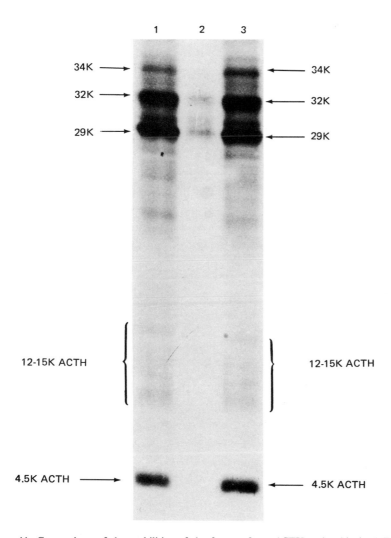

Figure 11. Comparison of the mobilities of the forms of pro-ACTH-endorphin in AtT-20 cells and in monolayer cultures of anterior pituitary by SDS–slab gel electrophoresis. The cell cultures were labeled with [³⁵S]-Met for 1 hr (anterior lobe) or 2 hr (tumor cells) as described in the Fig. 2 caption, except that no glucosamine was used. Cell extracts were prepared and precipitated with ACTH antiserum. The immunoprecipitates were analyzed by SDS–slab gel electrophoresis as described in the Fig. 5 caption. Lanes 1 and 3 are tumor-cell immunoprecipitates; lane 2 is an immunoprecipitate of anterior pituitary cells.

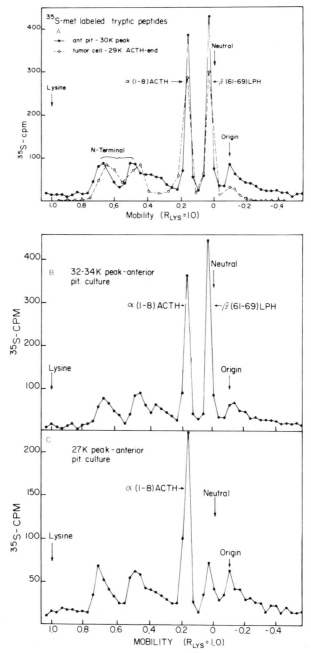

Figure 12. Tryptic peptide maps of [³⁵S]-Met-labeled forms of ACTH in anterior-pituitary cultures and AtT-20/D₁₆ᵥ cultures. Radioactive material in the 30K, 32K, and 27K peaks was eluted from the SDS gel (3-hr incubation in Fig. 10) and digested with trypsin as described by Roberts et al. (1978).

32K, and 34K ACTH-endorphin). The same methods have revealed the presence of three classes of ACTH intermediates (26K, 23K, and 21K ACTH), two forms of endorphin [11.5K endorphin (β-LPH-like) and 3.5K endorphin], and two species of ACTH (the 12–15K molecules and 4.5K ACTH) and two forms of the N-terminal glycopeptides (18K and 16K). Analysis of the CHO content of these forms by double-label experiments and mapping of tryptic glycopeptides (see Figs. 4 and 7) has shown that the 32K precursor, 23K ACTH, and 12–15K ACTH are structurally related, since only these forms contain the α(22–39)ACTH tryptic glycopeptide. It is also proposed in the scheme in Fig. 8 that the 34K precursor, 26K ACTH, and 18K N-terminal form are related, since only these forms are postulated to contain two N-terminal CHO side chains. Pulse–label and pulse–chase experiments with radioactive amino acids suggest that the 32K and 34K forms of the precursor are derived from the 29K form.

These results, together with the knowledge that 11.5K endorphin is located at the carboxy terminus of the precursor molecule and that ACTH is in the middle of the molecule, suggest the processing scheme shown in Fig. 8.

One feature of the model in Fig. 8 that should be emphasized is that alternate pathways are available for processing of the 29K form of the precursor. The alternate pathways lead to the formation of different ACTH end products. Conversion of the 29K form to the 32K form by addition of a CHO side chain and subsequent proteolytic processing of the 32K form results in the formation of 13K ACTH, whereas direct proteolytic processing of the 29K form leads to formation of 4.5K ACTH. At this point, we do not know the reason for the production of two forms of α(1–39)ACTH or what determines the choice of pathway a given 29K molecule takes in the cell.

It should also be pointed out that the proteolytic cleavages occur in a specific order in that β-LPH is cleaved out before the N-terminal glycoprotein and 13K and 4.5K ACTH. Very little of a species is found that is missing only the N-terminal fragment (Eipper and Mains, 1978; Phillips *et al.*, unpublished observations). Therefore, the order of processing in Fig. 8 appears to be mandatory for the major portion of precursor molecules. Perhaps the processing enzymes exist as a membrane complex with a particular spatial arrangement of each enzyme such that the product of the first enzyme reaction in the complex is an obligatory substrate for the action of the second enzyme.

The array of biologically active peptides generated by processing of pro-ACTH-endorphin is staggering. With the exception of α(1–39)ACTH, we know very little about the function of these peptides. Thus, it may seem premature to speculate about the significance of alternate pathways of processing of the precursor. Still, a few observations have been made that are of interest in this connection. First, it has been shown that the release of immunoreactive ACTH and β-endorphin is coupled in mouse and rat pituitary-cell culture systems in the basal and stimulated states (stimulated by hypothalamic extracts) (Allen *et al.*, 1978; Vale *et al.*, 1978). Second, the processing and release of the glycosylated and unglycosylated forms of α(1–39)ACTH appear to be coupled in these states as well. Even though uncoupling of processing and release of these peptides has not been observed, it is not hard to imagine ways in which this could happen at the level of posttranslational processing of the precursor. For example, a block of conversion of ACTH intermediates to end products could reduce the production of both forms of α(1–39)ACTH and the N-terminal glycoproteins without affecting the production of β-LPH and its derivatives. If the block were more specific and affected conversion of only one of the ACTH intermediates (e.g., 23K and not 21K ACTH), then production of the unglycosylated form of α(1–39)ACTH would be reduced without affecting production of the glycosylated form of the molecule. One could also imagine blocks occurring at the branch point that produces either the 32K or the 34K form of the precursor. A search for regulators that uncouple release of ACTH and β-endorphin is under way in our laboratory with these ideas in mind.

Intracellular processing of secretory proteins is a highly compartmentalized process, and the subcellular locations of glycosylating enzymes and proteolytic enzymes are known for many types of cells. Also, the time course of processing of secreted proteins is known in many tissues. Since we know the time course of glycosylation and proteolytic processing of pro-ACTH-endorphin in AtT-20/D$_{16v}$ cells, we can predict where these events take place. Studies with the electron microscope show that AtT-20/D$_{16v}$ cells contain rough and smooth endoplasmic reticulum and secretory granules that resemble those seen in corticotrophs in the anterior pituitary (M. Budarf, unpublished observations; S. Sabol, National Institutes of Health, personal communication). Therefore, processing of protein in the subcellular organelles may be similar in both tumor cells and normal cells.

The appearance of radioactive amino acids in the 29K and 32K forms of the precursor within 5 min of labeling suggests that the addition of core sugars to the precursor occurs in the rough endoplasmic reticulum perhaps while the peptide backbone is being synthesized on the ribosome, as in the case of ovalbumin (Kiely *et al.*, 1976) and immunoglobulin (Bergman and Kuehl, 1977). The appearance of radioactive amino acids in ACTH intermediates and in β-LPH occurs at 10–20 min of labeling, suggesting that the first proteolytic cleavage event occurs in the smooth endoplasmic reticulum or the Golgi or both. Since the CHO side chains of the ACTH intermediates contain fucose, whereas the CHO side chains of the precursors do not, it is likely that this peripheral sugar is added at about the time that proteolytic cleavage of the precursor occurs. Radioactivity does not appear in the forms of $\alpha(1-39)$ACTH and β-endorphin until after 20–40 min of labeling. Hence, the final processing steps probably occur in the secretory granules during their maturation. Processing studies with isolated subcellular fractions would help to test the accuracy of these predictions. For further discussion, consult Roberts *et al.* (1978).

The processing studies that have been done with the monolayer cultures of mouse anterior pituitary indicate that the major pathways depicted in Fig. 8 are not tumor-cell artifacts. It appears that the 30.5K and 33K forms in these cultures are actually equivalent to the 29K and 32K forms of the precursor (respectively) seen in the tumor cells (see Fig. 11). Further studies are required to document the details of processing in the anterior-pituitary cultures.

Finally, it is worth noting that many of the component hormones in the precursor (ACTH, β-LPH, endorphins, and Met-enkephalin) and Leu-enkephalin are found in a variety of places in the brain as well as the pituitary (Pacold *et al.*, 1978; Watson *et al.*, 1978; Zimmerman *et al.*, 1978). Although the source of the brain peptides is not known, it appears unlikely that they come from the pituitary because the levels of these peptides in the brain are not reduced by hypophysectomy (Krieger *et al.*, 1977; Watson *et al.*, 1978). Furthermore, no Leu-enkephalin can be detected in the sequence of the precursor from AtT-20 cells [more than 96% of the precursor chains have the Met-enkephalin sequence (Roberts *et al.*, unpublished results)] and from rat intermediate lobe cells (Seidah *et al.*, 1978).

DISCUSSION

VAUDRY: Evidence has been given for the presence of calcitonin in human carcinoma cells (Abe *et al.*, 1977) and in human pituitary (Deftos *et al.*, 1978a,b). From these results, it would appear that calcitonin would be part of the precursor of part of a bigger molecule.

HERBERT: Several other groups have tested the culture medium and cell extract of AtT-20-D_{160} cells for cross-reactivity with calcitonin antibodies and found no reactivity. This does not exclude the possibility that calcitonin is in the precursor. However, we have been unable to incorporate [^{35}S]cysteine into the precursor in mRNA-directed cell-free synthesizing system and in tumor cells. This argues against calcitonin being part of the precursor molecule, because calcitonin from all species investigated contains at least two cysteine residues. Of course, amino acid sequencing will provide a definitive answer.

CRINE: I have three questions: (1) Since you are relying on immunoprecipitation to isolate your labeled proteins, do you think that you could have missed a larger molecular form of the precursor where the antigenic determinent is masked by secondary structure? (2) Do you think that the different forms of the precursor that you have isolated could be characterized by slightly different sequences? I am asking this question with the problem of the origin of Leu-enkephalin in mind. (3) Do you think that β-endorphin and the various forms of ACTH are always cleaved or released on a 1:1 molar ratio basis?

HERBERT: (1) We have used several different antibodies to ACTH, endorphin, and the N-terminal portion of the precursor for immunoprecipitation, and have not seen any molecular species larger than the 30,000-molecular-weight class of precursors described. However, it is possible that a higher molecular form might have been missed. The size of the ACTH-endorphin mRNA is such as to suggest that if a larger precursor existed, it could not be much bigger than the one described. Further studies of the mRNA size and structure will clear up that point. (2) Yes, it is possible that different sequences of amino acids are present in different precursor molecules and that these sequences dictate which processing pathway a given precursor molecule goes through. One possibility is that the carbohydrate attachment sites differ in different precursor molecules. With regard to the presence of leucine in the Leu-enkephalin part of the precursor [β(65)-LPH] in place of methionine, we can say that less than 3% of the amino acid in this position is leucine. Hence, the precursor does not seem to be a major source of Leu-enkephalin in the pituitary. (3) Our data does not enable us to tell at this point whether the ratio of ACTH released (13K and 4.5K ACTH taken together) to β-endorphin released (as β-LPH and β-endorphin) is the same under all conditions in the tumor-cell cultures. However, in the cells, the ratio of these hormones appears to be close to one under all conditions of incubation studied, including treatment with glucocorticoids at concentrations that inhibit synthesis of the precursor 2- to 3-fold.

LABRIE: At times up to 12 hours of incubation in the rat anterior-pituitary cells in primary culture, dexamethasone leads to a parallel inhibition of β-endorphin plus β-LPH and ACTH release. Parallelism is even found during acute stimulation by CRF or cAMP derivatives. It is possible, however, that upon long-term stimulation or inhibition, processing could be affected and thus lead to different ratios of peptides released. Thus far, we have detected very little effect of dexamethasone on processing of the ACTH-endorphin precursor even after 48 hours of treatment (at which time the release of both endorphin and ACTH is reduced more than 2-fold.

BAXTER: Dr. James Roberts in collaboration with Dr. Herbert and myself found that in AtT-20 cells, the dominant mechanism by which glucocorticoids decreased ACTH production is by decreasing ACTH mRNA. There is little effect of the hormone on the processing. However, there are some minor changes in the higher-molecular-weight forms. The significance of this is currently under investigation.

HERBERT: This is correct. There is a shift in proportion of label in higher-molecular-weight forms of ACTH-endorphin. The shift is being examined in more detail now.

RAMACHANDRAN: The presence of carbohydrate in ACTH is puzzling. You indicated that the carbohydrate moiety is attached through an asparagine residue. The only asparagine residue in mammalian ACTH sequences is not suitable for glycosylation, since the sequence is not present as Thr-X-Asn or Ser-X-Asn. Since there is a great deal of the glycosylated form in the mouse anterior-pituitary cells and since this form is more stable, it is amazing that the glycosylated form has been missed during the isolation of ACTH from pituitary glands.

HERBERT: We do not yet know the sequence of the mouse ACTH in the region where the carbohydrate side chain is attached. The carbohydrate is attached in the C-terminal region where there is an aspartic acid residue, but we do not know whether this could be an asparagine within an attachment site (Ser-X-Asn). In regard to your second comment, it is very surprising that the glycosylated forms of ACTH were not detected earlier. Perhaps there is less glycosylated form of ACTH in the pituitaries of some species (particularly in the pituitaries of those species that were used for the earlier structural work on ACTH, that is, bovine, ovine, sheep pituitaries, etc.).

KRAICER: You mentioned that the electrophoretic patterns were different in the rat PI and the other two models (mouse tumor and PD). Do you have any information as to whether the differences are due only to differences in the carbohydrate moiety, or are there differences in the peptide components as well?

HERBERT: No, we do not have any direct chemical evidence that the differences you cite are due to differences in carbohydrate content only. However, we do know that the separations we obtain on the gels (SDS gels) are very sensitive to carbohydrate content. The 34K form has more carbohydrate than the 29K form, for example. However, it is certainly possible that the differences you refer to are due to differences in protein structure.

HYMER: Could you clarify a point in regard to the released forms of the hormones in the two systems, i.e., normal and tumor pituitary. Are the percentages of the different molecular forms similar? What percentage is released as high-molecular-weight form in both systems? Also, please comment about the granule content of the tumor cells vs. the normal cells.

HERBERT: The distribution of the molecular-weight forms of the hormones in the tumor cells is very similar to that in anterior-pituitary-cell culture and anterior pituitary of the mouse, but very different from that in the intermediate-lobe cultures and intermediate lobe. The percentage of high-molecular-weight forms in the release material is very low in both cultures—of the order of 5–15% of the total hormone released. This is the same percentage that one finds inside the cells in both cultures. Tumor cells contain secretory granules in smaller numbers (per cell) than anterior-lobe corticotrophs.

VAUDRY: Have you any explanation for the presence of a common peptide (ACTH 4–10) in the sequences of ACTH and LPH?

HERBERT: No, I do not have any explanation for a repeat of this sequence in the precursor.

TIXIER-VIDAL: What is the kinetics of appearance of the various molecular forms in the medium as compared to the cell during the chase experiments? Also, do you have any idea of the cell structures that are involved in the successive cleavages?

HERBERT: It takes approximately 45 minutes to 1 hour for labeled hormone to appear in the medium (amino-acid-labeled forms of ACTH or endorphin). Steady-state labeling of all of the forms of ACTH and β-endorphin takes about 8 hours in the tumor cells and 2–3 times as long in primary cultures from anterior pituitary of the mouse. In regard to your second question, EM work by several groups has shown the existence of secretory granules in AtT-20-D_{160} cells that closely resemble, with regard to size and staining intensity, granules in anterior-pituitary cells, except they are organized differently. The tumor cells also have rough and smooth endoplasmic reticulum.

REFERENCES

Abe, K., Nicholson, W. E., Liddle, G. W., Island, D. P., and Orth, D. N., 1967, Radioimmunoassay of beta-MSH in human plasma and tissues, *J. Clin. Invest.* **46:**1609.

Abe, K., Nicholson, W. E., Liddle G. W., Orth, D. N., and Island, D. P., 1969, Normal and abnormal regulation of beta-MSH in man, *J. Clin. Invest.* **48:**1580.

Abe, K., Adachi, J., Miyakawa, S., Tanaka, M., Yamaguchi, K., Tanaka, N., Kameya, T., and Shimosato, Y., 1977, Production of calcitonin, adrenocrticotropic hormone and

-melanocyte-stimulating hormone in tumors derived from amine precursor uptake and decarboxylation cells, *Cancer Res.* **37**:4190–4194.

Allen, R. G., Herbert, E., Hinman, M., Shibuya, H., and Pert, C., 1978, Coordinate control of corticotropin, β-lipotropin, β-endorphin release in mouse pituitary cell cultures, *Proc. Natl. Acad. Sci. U.S.A.* **75**:4972.

Bergman, L. W., and Kuehl, W. M., 1977, Addition of glucosamine and mannose to nascent immunoglobulin heavy chains, *Biochemistry* **16**:4490.

Czichi, U., and Lennarz, W. J., 1977, Localization of the enzyme system for glycosylation of proteins via the lipid-linked pathway in rough endoplasmic reticulum, *J. Biol. Chem.* **252**:7901.

Deftos, L. J., Burton, D., Bone, H. G., Catherwood, B. D., Parthemore, J. G., Moore, R. V., Minnick, S., and Guillemin, R., 1978a, Immunoreactive calcitonin in intermeditate lobe of pituitary gland, *Life Sci.* **23**:743–748.

Deftos, L. J., Burton, D., Catherwood, B. D., Bone, H. G., Parthemore, J. G., Guillemin, R., Watkins, W. B., and Moore, R. Y., 1978b, Demonstration by immunoperoxidase histochemistry of calcitonin in the anterior lobe of the rat pituitary, *J. Clin. Endocrinol.* **47**:457–460.

Dintzis, H. M., 1961, Assembly of the peptide chains of hemoglobin, *Proc. Natl. Acad. Sci. U.S.A.* **47**:247.

Eipper, B. A., and Mains, R. E., 1977, Peptide analysis of a glycoprotein form of adrenocorticotropic hormone *J. Biol. Chem.* **252**:8821.

Eipper, B. A., and Mains, R. E., 1978, Analysis of the common precursor to corticotropin and endorphin, *J. Biol. Chem.* **253**:5732.

Eipper, B. A., and Mains, R. E., 1979, Biosynthesis of ACTH and endorphins in pituitary cell culture systems, *J. Supra-mol. Struct.* **8**:247.

Eipper, B. A., Mains, R. E., and Guenzi, D., 1976, High molecular weight forms of adrenocorticotropic hormone are glycoproteins, *J. Biol. Chem.* **251**:4121.

Ganong, W. F., 1977, Adrenal medulla and adrenal cortex, Chapt. 20, p. 267, and The pituitary gland, Chapt. 22, p. 298, in: *Review of Medical Physiology,* Lange Medical Publications, Los Altos, California.

Goldstein, A., 1976, Opioid peptides (endorphins) in pituitary and brain, *Science* **193**:1081.

Guillemin, R., Vargo, T., Rossier, J., Scott, M., Ling, N., Rivier, C., Vale, W., and Bloom, F., 1977, β-Endorphin and adrenocorticotropin are secreted concomitantly by the pituitary gland, *Science* **197**:1367.

Herbert, E., Allen, R. G., and Paquette, T. P., 1978, Reversal of dexamethasone inhibition of adrenocorticotropin release in a mouse pituitary tumor cell line either by growing cells in the absence of dexamethasone or by addition of hypothalamic extract, *Endocrinology* **102**:218.

Kiely, M. L., McKnight, S., and Schimke, R. T., 1976, Studies on the attachment of carbohydrate to ovalbumin nascent chains in hen oviduct, *J. Biol. Chem.* **251**:5490.

Krieger, D. T., Liotta, A., and Brownstein, M. J., 1977, Presence of corticotropin in brain of normal and hypophysectomized rats, *Proc. Natl. Acad. Sci. U.S.A.* **74**:648.

Laemmli, U. K., 1970, Cleavage of structural proteins during the assembly of the head of bacteriophage T₄, *Nature (London)* **227**:680.

Mains, R. E., and Eipper, B. A., 1976, Biosynthesis of adrenocorticotropic hormone in mouse pituitary tumor cells, *J. Biol. Chem.* **251**:4115.

Mains R. E., Eipper, B. A., and Ling, N., 1977, Common precursor to corticotropins and endorphins, *Proc. Natl. Acad. Sci. U.S.A.* **74**:3014.

Nakanishi, S., Inoue, A., Taii, S., and Shosaku, N., 1977, Cell-free translation product containing corticotropin and β-endorphin encoded by messenger RNA from anterior lobe and intermediate lobe of bovine pituitary, *FEBS Lett.* **84**:105.

O'Farrell, P. H., 1975, High resolution two-dimensional electrophoresis of proteins, *J. Biol. Chem.* **250:**4007.

Pacold, S. T., Kirsteins, L., Hojvat, S., and Lawrence, A. M., 1978, Biologically active pituitary hormones in the rat brain amygdaloid nucleus, *Science* **199:**804.

Roberts, J. L., and Herbert, E., 1977a, Characterization of a common precursor to corticotropin and β-lipotropin: Cell-free synthesis of the precursor and identification of corticotropin peptides in the molecule, *Proc. Natl. Acad. Sci. U.S.A.* **74:**4826.

Roberts, J. L., and Herbert, E., 1977b, Characterization of a common precursor to corticotropin and β-lipotropin: Identification of β-lipotropin peptides and their arrangement relative to corticotropin in the precursor synthesized in a cell-free system, *Proc. Natl. Acad. Sci. U.S.A.* **74:**5300.

Roberts, J.L., Phillips, M., Rosa, P.A., and Herbert, E., 1978, Steps involved in the processing of common precursor forms of adrenocorticotropin and endorphin in cultures of mouse pituitary cells, *Biochemistry* **17:**3609.

Scott, A.P., Lowry, P.J., Ratcliffe, J.G., Rees, L.H., and Landon, J., 1974, Corticotrophin-like pepties in the rat pituitary, *J. Endocrinol.* **61:**355.

Seidah, N.G., Gianoulakis, C., Crine, P., Lis, M., Benjannet, S., Routhier, R., and Chretien, M., 1978,*In Vitro* biosynthesis and chemical characterization of β-lipotropin, γ-lipotropin, and β-endorphin in rat pars intermedia, *Proc. Natl. Acad. Sci. U.S.A.* **75:**3153.

Vale, W., Rivier, C., Yang, L., Minick, S., and Guillemin, R., 1978, Effects of purified hypothalamic corticotropin-releasing factor and other substances on the secretion of adrenocorticotropin and β-endorphin-like immunoactivities *in vitro, Endocrinology* **103:**1910.

Waechter, C. J., and Lennarz, W. J., 1976, The role of polyprenol-linked sugars in glycoprotein synthesis, *Annu. Rev. Biochem.* **45:**95.

Watson, S. J., Richard, C. W., III, and Barchas, J. D., 1978, Adrenocorticotropin in rat brain: Immunocytochemical localization in cells and axons, *Science* **200:**ll80.

Zimmerman, F. A., Liotta, A., and Krieger, D. T., 1978, β-Lipotropin in the brain: Localization in hypothalamic neurons by immunoperoxidase technique, *Cell Tissue Res.* **186:**393.

13

Biosynthesis of β-Endorphin in the Rat Pars Intermedia

P. Crine, F. Gossard, N. G. Seidah,
C. Gianoulakis, M. Lis, and M. Chrétien

1. Introduction

1.1. The β-Lipotropin Precursor Hypothesis

β-Lipotropin (β-LPH) is a 91-amino-acid pituitary peptide the biological importance of which has remained obscure for a long time except for a role as a prohormone. It was discovered and isolated in 1964 from sheep pituitary glands by Li (1964) and Birk and Li (1964). As a fat-mobilizing agent, β-LPH is most active in rabbit adipose tissue, only weakly active in rat, and essentially inactive in mouse (Lohmar and Li, 1967). This lipolytic activity, however, has not been a specific property of lipotropin, since α-melanocyte-stimulating hormone (α-MSH) and β-MSH are 100 times more active. In fact, almost all the hormones from the anterior pituitary show some lipolytic action besides their main biological function.

Figure 1 shows the amino acid sequence of the ovine hormone as established by Li *et al.* (1965).

The isolation of γ-LPH by Chrétien and Li (1967) provided a molecule possessing the structure of an intermediate product between two other known substances, β-LPH and β-MSH (Fig. 2).

P. Crine, F. Gossard, N. G. Seidah, C. Gianoulakis, M. Lis, and *M. Chrétien* • Clinical Research Institute of Montreal, Montreal H2W 1R7, Canada

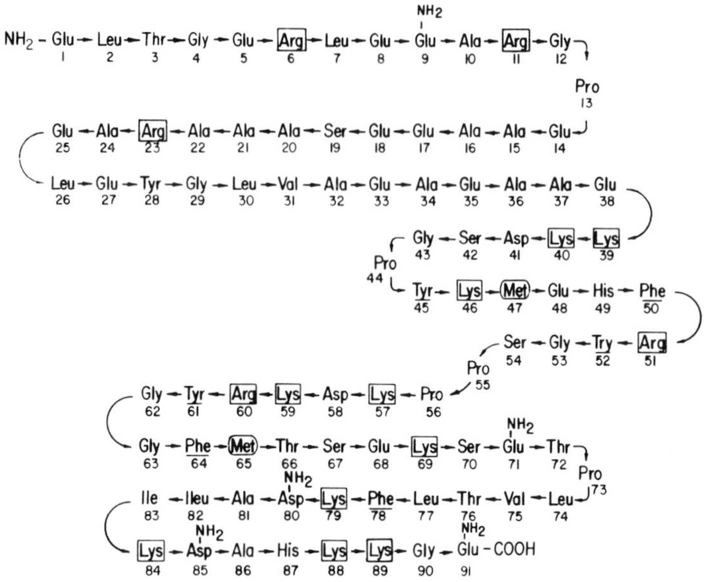

NH₂

NH₂ – Glu ← Leu ← Thr ← Gly ← Glu ← [Arg] ← Leu ← Glu ← Glu ← Ala ← [Arg] ← Gly
  1   2   3   4   5   6   7   8   9   10   11   12

Pro
13

Glu → Ala → [Arg] → Ala → Ala → Ala → Ser → Glu → Glu → Ala → Ala → Glu
 25   24   23   22   21   20   19   18   17   16   15   14

Leu → Glu → Tyr → Gly → Leu → Val → Ala → Glu → Ala → Glu → Ala → Ala → Glu
 26   27   28   29   30   31   32   33   34   35   36   37   38

Gly → Ser → Asp → [Lys] → [Lys]
Pro   43   42   41   40   39
44

Tyr → [Lys] → (Met) → Glu → His → Phe
 45   46   47   48   49   50

Ser → Gly → Try → [Arg]
Pro   54   53   52   51
55

Gly → Tyr → [Arg] → [Lys] → Asp → [Lys] → Pro
 62   61   60   59   58   57   56

NH₂

Gly → Phe → (Met) → Thr → Ser → Glu → [Lys] → Ser → Glu → Thr
 63   64   65   66   67   68   69   70   71   72   Pro
NH₂   73

Ile → Ileu → Ala → Asp → [Lys] → Phe → Leu → Thr → Val → Leu
 83   82   81   80   79   78   77   76   75   74
NH₂   NH₂

[Lys] → Asp → Ala → His → [Lys] → [Lys] → Gly → Glu – COOH
 84   85   86   87   88   89   90   91

Figure 1. Amino acid sequence of ovine β-LPH.

At that time, it was proposed by Chrétien and Li (1967) that β-LPH could be a rather inactive precursor for a more potent peptide such as β-MSH, which is 100 times more active than β- or γ-LPH as far as lipolysis and melanophore stimulation are concerned (Chrétien and Li, 1967). At the same time, Steiner et al. (1967) demonstrated that insulin was synthesized from a larger and relatively inactive precursor called "proinsulin." Chance et al. (1968) published the complete structure of porcine proinsulin and pointed to the presence of two diamino acids at the cleavage sites. The resemblance of these cleavage sites to those proposed for the LPH peptides (Fig. 3) was striking and gave strong support to the β-LPH precursor hypothesis. Since that time, many hormones have been proved

β-LPH	Glu	....Asp	-Ser-	Gly-	Pro-	Try-	Lys-	Met-	Glu-	His-	Phe-	Arg-	Try-	Gly-	Ser-	Pro-	Pro-	Lys-	Asp	Glu(NH₂)
	1	41	42	43	44	45	46	47	48	49	50	51	52	53	54	55	56	57	58	90
γ-LPH	Glu....	Asp-	Ser-	Gly-	Pro-	Try-	Lys-	Met-	Glu-	His-	Phe-	Arg-	Try-	Gly-	Ser-	Pro-	Pro-	Lys-	Asp (NH₂)	
	1	41	42	43	44	45	46	47	48	49	50	51	52	53	54	55	56	57	58	
β-MSH	Asp-	Ser-	Gly-	Pro-	Try-	Lys-	Met-	Glu-	His-	Phe-	Arg-	Try-	Gly-	Ser-	Pro-	Pro-	Lys-	Asp(NH₂)		
	1	2	3	4	5	6	7	8	9	10	11	12	13	14	15	16	17	18		

Figure 2. Relationship between the structure of lipotropic hormones and β-MSH. β-MSH is made of amino acid residues 41–48 of both β- and γ-LPH.

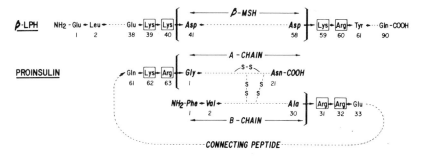

Figure 3. Amino acid sequences of β-LPH and proinsulin at their site of enzymatic cleavage. The sequence of β-LPH and proinsulin reveals the presence of two diamino acids at the site of cleavage by a trypsinlike enzyme.

to be synthesized first as part of a larger prohormone. One of the best-documented cases is pro-PTH, in which the cleavage of the precursor form of the molecule into the final matured product also occurs at the carboxyl side of a pair of dibasic amino acids (Habener *et al.*, 1977).

Until 1975, the C-terminal portion of β-LPH had not attracted too much attention until the astonishing discovery of the enkephalins by Hughes *et al.* (1975). These authors described the isolation from hog brains of two peptides, called "enkephalins," that were agonists of morphine. The first, Tyr-Gly-Gly-Phe-Met, had a methionine residue in the fifth position and was therefore named met-enkephalin. In the second, a leucine residue replaced the methionine in the fifth position. The sequence of met-enkephalin is identical to the fragment containing residues 61–65 of β-LPH. This discovery prompted several groups to look in pituitary glands for endogenous morphinelike substances structurally related to β-LPH.

Several investigators (Chrétien *et al.*, 1976; Bradbury *et al.*, 1976; Li *et al.*, 1976; Li and Chung, 1976) were able to show the presence of such morphinelike substances in human as well as in camel, sheep, and hog pituitaries. Most of these substances have sequences corresponding to the segment 61–91 of their homologous β-LPH and were named "β-endorphin." The segments 61–76 and 61–77 of β-LPH were also extracted from fragments of pig hypothalamus–neurohypophysis (Lazarus *et al.*, 1976; Guillemin *et al.*, 1976). It was observed, moreover, that while β-LPH had no morphinelike activity, even at high doses (Seidah *et al.*, 1977), the portion 61–91 had considerable activity in the morphine bioassay and opiate receptor binding assay. These findings clearly indicate that the relationship between β-LPH and the opioid peptides is more than accidental. The original hypothesis of β-LPH being the prohormone for β-MSH was thus extended, and it was proposed, on the basis of structural considerations, that β-LPH also constitutes the biosynthetic precursor for β-endorphin (Lazarus et al., 1976; Chrétien et al., 1977).

However, before a true precursor role for β-LPH in the biosynthetic pathway of β-endorphin can be established with certainty, careful pulse–labeling and pulse–chase experiments are required.

It is the purpose of this chapter to demonstrate by kinetic labeling studies that in the normal pituitary gland, β-endorphin is indeed synthesized via β-LPH and a higher-molecular-weight protein.

1.2. Choice of an Experimental Model for Studying the Biosynthesis of β-Endorphin

In the vertebrates, the pituitary gland is divided into three anatomical zones called "pars distalis," "pars intermedia," and "pars nervosa." β-LPH immunoreactivity has been found both in the corticotropic cells of the pars distalis and in all cells of the pars intermedia (Pelletier et al., 1977). However, corticotropic and pars-intermedia cells are not predominant in the whole gland, and β-LPH and its related peptides are not major synthesis products of the pituitary (Bertagna et al., 1974). To perform biosynthetic labeling experiments, one has to resort to the use of a particular cell type that would synthesize these hormones in higher quantities.

Immunocytochemical studies have shown that β-LPH immunoreactive peptides occur in all the parenchymal cells of the pars intermedia (Pelletier et al., 1977). Moreover, when an acutely dispersed preparation of bovine pars-intermedia cells was used, it was found that after a 3-hr incubation with radioactive amino acids, large amounts of labeled γ-LPH and β-endorphin could be extracted from the cells (Crine et al., 1977a). The simplicity of the synthesis pattern in the pars intermedia compared to the whole pituitary (Crine et al., 1977b) led us to use this cell preparation for further pulse–chase experiments.

However, the viability of bovine cells was very low, probably due to the long delay between the death of the animal in the slaughterhouse and the beginning of the incubation in the laboratory.

When rat tissue was used, incubation of the tissue could start within a minute after the death of the animal, and a cell viability better than 90% was routinely obtained. Neither rat β-endorphin nor β-LPH had been isolated or characterized at that time, and we were faced with the difficult task of characterizing the biosynthetic rat hormones with only the porcine or ovine peptides as standards.

Two types of experimental approach can be used to solve this problem: (1) the immunological approach, employing antisera raised against synthetic or native hormones to precipitate the biosynthetic radio-active products extracted from the cells; or (2) the sequencing approach, requiring the unambiguous chemical characterization of the neosynthesized labeled peptides by radiosequencing. We chose the latter approach, and as a first step we had to purify and identify the different peptides of interest synthesized by the pars intermedia. This task was facilitated by the fact that in this tissue, 20–30% of the biosynthetic activity of the cells is devoted to the production of β-LPH and its related peptides (Crine *et al.*, 1977a).

2. Isolation and Chemical Characterization of Neosynthesized Rat Intermediate-Lobe Peptides

2.1. β-Endorphin

In our search for the β-LPH-related peptides synthesized by rat intermediate pituitaries, we decided to look first for β-endorphin. In all the species studied so far, this peptide has a very conserved sequence, in contrast to the variable region corresponding to γ-LPH (Chrétien *et al.*, 1977), and it was thought that this peptide would be easier to isolate and characterize.

Figure 4 represents the carboxymethyl (CM)–cellulose chromatography of the [³H]leucine-labeled proteins synthesized after a 3-hr incubation by isolated cells of rat pituitary pars intermedia. Similar patterns were obtained with [³H]lysine-, [³⁵S]methionine-, and [³H]phenylalanine-labeled proteins.

The radioactive peptides recovered in the F3 region where carrier ovine β-endorphin is found was rechromatographed under identical conditions on a shorter (1 × 10 cm) CM–cellulose column for further purification. The purified peptides were then analyzed by polyacrylamide disc gel electrophoresis at pH 4.5 (Fig. 5) and by electrophoresis on sodium dodecyl sulfate (SDS)–urea–polyacrylamide gels (Fig. 6). The electrophoretic pattern obtained on the two types of gels shows that this fraction contains a radioactive peptide with a charge and a molecular weight similar to those of ovine β-endorphin.

Automatic Edman degradations of the purified radioactive peptides were done on a Beckman 890B sequenator, using 150 nmol sperm whale apomyoglobin as a carrier. The thiazolinones collected in butyl chloride were assayed for radioactivity directly in a toluene-base scintillation cocktail. Figure 7 shows that the partial sequence of this peptide could be established as follow: Met 5, Lys 9, Leu 14, 17. Such a sequence is identical to that expected from the known sequences of sheep, camel (Li,

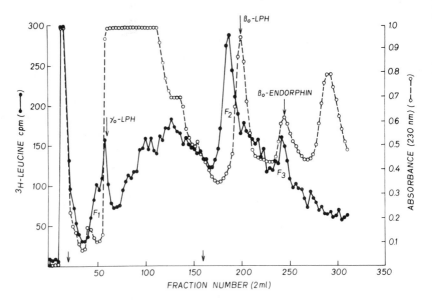

Figure 4. CM–cellulose chromatography of material obtained after extraction of rat pars-intermedia cells incubated *in vitro* for 3 hr with 2 mCi of [³H]leucine. After the incubation, the cells were extracted with 5 M acetic acid containing 0.3 mg/ml phenylmethylsulfonyl fluoride, 0.3 mg/ml iodoacetate, 0.3 mg/ml bovine serum albumin, and 5 mmol unlabeled leucine. The extract was then desalted on a Sephadex G-25 column, and the labeled peptides were chromatographed together with 100 mg sheep pituitary fraction D. The CM–cellulose column (1 × 40 cm) was eluted with 0.01 M NH₄OAc (pH 4.6) for 20 tubes and then with a concave gradient made by introducing 0.1 M NH₄OAc (pH 6.7) through a 250-ml mixing flask (first arrow on the abscissa). This buffer was replaced by 0.2 M NH₄OAc (pH 6.7) at tube 160 (second arrow). The elution positions of ovine γ-LPH, β-LPH, and β-endorphin on this chromatogram are also shown.

1977), and bovine β-endorphin. It can be calculated that there is only one chance out of 10^8 for such a match between any two segments of protein to be merely fortuitous (Dayhoff *et al.*, 1978). It can thus be deduced that the biosynthesized material isolated from fraction F3 on CM–cellulose chromatography (see Fig. 4) is identical in molecular weight, partial sequence, and charge properties to the homologous ovine, camel, and bovine β-endorphins.

2.2. γ- and β-Lipotropins

[³H]-Leu-labeled peptides eluting slightly before the known position of ovine γ-LPH were analyzed next.

Figure 5 shows that a purification scheme similar to that outlined above for β-endorphin yielded a pure radioactive peptide migrating slightly slower than ovine γ-LPH on acidic gels with a molecular weight identical to that of β-LPH on SDS–urea gels (see Fig. 6). Similarly, an F2 peptide resembling β-LPH in its electrophoretic properties could be purified from the CM–cellulose.

The electrophoretic behavior of those peptides, however, provides insufficient evidence to unequivocally identify the material obtained in fractions F1 and F2 as rat γ-LPH and β-LPH, respectively. We decided, therefore, to obtain additional experimental evidence from two sources to ascertain the identify of these peptides:

(1) Comparison of the leucine sequence of the material obtained from [³H]-Leu F1 (Fig. 7A) to that obtained from [³H]-Leu F2 (Fig. 7B) revealed that we were dealing with two peptides with identical NH₂-terminal sequences, namely, Leu 2, 10, 14 in both cases. This result was to be expected, since γ-LPH represents the first 58 amino acids of β-LPH.

(2) F2 peptide should also contain the β-endorphin sequence as its COOH-terminal segment. In an effort to demonstrate this point, the ε-amino groups of lysine in these peptides were blocked via citraconylation,

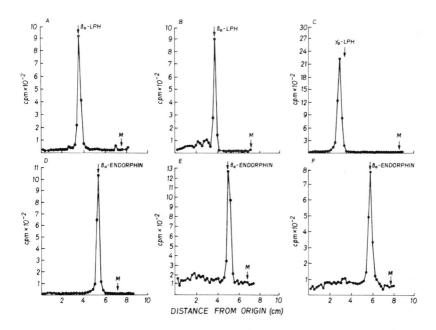

Figure 5. Polyacrylamide disc gel electrophoresis at pH 4.5 of selected repurified fractions, obtained from CM–cellulose columns: (A) [³H]-Leu F2; (B) [³H]-Lys F2; (C) [³H]-Leu F1; (D) [³H]-Leu F3; (E) [³H]-Lys F3; (F) [³⁵S]-Met F3. The migration positions of ovine β-LPH, γ-LPH, and β-endorphin are also shown. The arrow at the end of each gel (M) shows the position of the pyronine tracking dye.

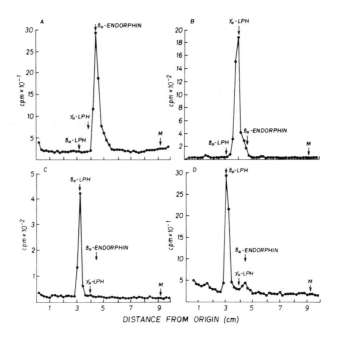

Figure 6. SDS–urea–disc polyacrylamide gel electrophoresis: (A) [^{3}H]-Leu F3; (C) [^{3}H]-Leu F2; (D) [^{3}H]-Lys F2. The positions of ovine β-endorphin, γ-LPH, and β-LPH are shown together with that of the bromophenol blue tracking dye (M).

and the peptides were then selectively cleaved with trypsin at their arginine residues. Following decitraconylation, the resulting peptides were analyzed by acidic gel electrophoresis (Fig. 8). Fractions [^{3}H]-Leu F2 and [^{3}H]-Lys F2 (Fig. 8B and C) released a peptide with $R_f = 0.75$, identical to that of ovine and rat β-endorphin, indicating the presence of this sequence at the COOH terminus of the peptide in fraction F2. Moreover, no rat β-endorphin band ($R_f = 0.75$) was observed when the citraconylated [^{3}H]-Leu F1 peptide was selectively cleaved with trypsin, indicating that this sequence is lacking in this fraction (Fig. 8A). Comparison of Fig. 8A and B shows that the band obtained for [^{3}H]-Leu F1 with $R_f = 0.27$ is also found in the digest of [^{3}H]-Leu F2, indicating a common sequence. Combining these results with the identity of the NH$_2$-terminal sequences of [^{3}H]-Leu F1 and [^{3}H]-Leu F2 and the molecular weights obtained for F1 and F2 led to the conclusion that the material in F2 is rat β-LPH and the one recovered in F1 is γ-LPH.

Incubation of the rat pars-intermedia cells with [^{3}H]leucine also allowed us to assess the possible presence of Leu 5 β-endorphin. In Fig. 7D, no trace of Leu 5 within the sequence of [^{3}H]-Leu β-endorphin could be detected. From the 93% repetitive yield obtained, based on the yields at residue Nos. 14 and 17, it can be calculated that if Leu 5 β-endorphin were present and if it coeluted with Met 5 β-endorphin on CM–cellulose, it could not account for more than 1% of the total β-endorphin content. In double-labeling experiments with [^{35}S]methionine and [^{3}H]phenylalanine, rat γ-LPH was shown to contain phenylalanine at position 6, but could not be labeled with [^{35}S]methionine (data not shown). This result, confirmed by two incubations with [^{35}S]methionine, must mean that rat γ-LPH lacks methionine within its sequence. We will take advantage of this fact later on during pulse–chase studies and structural analysis of the precursor for β-LPH and adrenocorticotropin (ACTH).

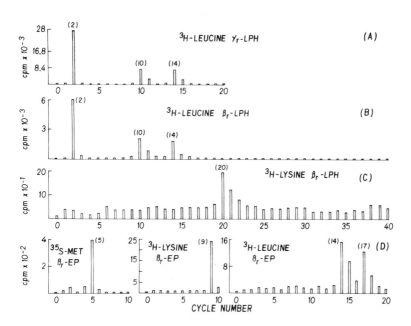

Figure 7. Partial NH$_2$-terminal sequence analysis of labeled peptides. Automatic Edman degradations of the purified peptides were done on an updated Beckman 890B sequenator, using 0.1 M Quadrol as the coupling buffer and 150 nmol sperm whale apomyoglobin as carrier. To stabilize the film, in each sequence a complete preliminary cycle was done without adding the coupling reagent phenylisothiocyanate to the cup (cycle number 0). This was followed by a double coupling for the first cycle only. The thiazolinones collected in butyl chloride were assayed for radioactivity directly in a toluene-base scintillation cocktail (4 g Omnifluor/liter toluene). (A) [^{3}H]-Leu Fl; (B) [^{3}H]-Leu F2; (C) [^{3}H]-Lys F2; (D) [^{35}S]-Met F3 *(left)*, [^{3}H]-Lys F3 *(middle)*, and [^{3}H]-Leu F3 *(right)*. The total inputs of radioactivity on the sequenator cup were: (A) 6 × 10^4 cpm of [^{3}H]-Leu; (B) 4 × 10^4 cpm of [^{3}H]-Met and [^{3}H]-Lys and 4 × 10^4 cpm of [^{3}H]-Leu.

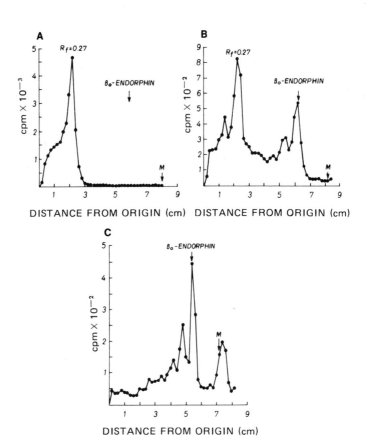

DISTANCE FROM ORIGIN (cm)

Figure 8. Polyacrylamide disc gel electrophoresis at pH 4.5 of total tryptic digest of citraconylated biosynthetic peptides from pars-intermedia cells. The radiolabeled peptide, to which 1 mg sheep β-LPH was added as carrier, was dissolved in 0.5 ml 0.5 M Na₂HPO₄ (pH 8.5), and citraconic anhydride (Eastman) was gradually added with stirring; the pH was kept at 8.5 manually with 5 M NaOH. The total ratio between free amino groups and citraconic anhydride was 1:300 (mol/mol) (as calculated from the 1 mg of β-LPH carrier). The mixture was left to react for 1 hr at room temperature, and then was applied directly to a Sephadex G-25 column (1.5 × 50 cm). The column was eluted with 0.1 M NH₄HCO₃ (pH 8.5) for removal of excess salts and citraconic acid, and the void volume fraction was lyophilized. For trypsin digestion of the citraconylated peptide, a trypsin/peptide ration of 1:100 (wt/wt) was used, and digestion was for 4 hr at 37°C in 0.1 M NH₄HCO₃ (pH 8.5). At the end of the digestion period, ovomucoid trypsin inhibitor (Sigma) was added in an amount equal to 4 times the weight of trypsin. The reaction was left to proceed for 5 min at 37°C, and then the pH was decreased to 3.0 with glacial acetic acid and decitraconylation proceeded for 4 hr at 37°C. The solution was then lyophilized, and the material was all placed on the polyacrylamide gel for electrophoresis and assayed for radioactivity. (A) [³H]-Leu Fl; (B) [³H]-Leu F2; (C) [³H]-Lys F2. The migration positions of β-endorphin (R_f = 0.75) and the pyronine tracking dye (M) are shown.

3. Kinetic Labeling Experiments in Isolated Rat Pars-Intermedia Cells

3.1. Pulse Incubations

Intermediate-lobe cells were incubated with [³H]phenylalanine for 10 min, and the proteins extracted from the cells were desalted on Sephadex G-25 and directly analyzed by SDS–gel electrophoresis (Swank and Munkres, 1971). One major peptide with an apparent molecular weight of 30,000 ± 1500 was found (Fig. 9A) (Crine *et al.*, 1978). This protein represented 25–30% of the total radioactivity recovered on the gel after electrophoresis. It could be immunoprecipitated with either anti-β-MSH serum or anti-ACTH serum as shown by SDS–gel electrophoresis (Fig. 9B and C). In each case, immunoprecipitation of labeled material was shown to be specific, since the radioactive peaks on the gels were abolished by the addition of an excess of cold β-MSH or ACTH to the sample before addition of the corresponding immunoserum. Moreover, a nonimmune rabbit serum failed to precipitate any radioactivity (data not shown).

These immunoprecipitation experiments show that the major labeled synthesis product extracted from the cells after a short 10-min pulse with radioactive amino acids contains antigenic determinants for both ACTH and β-LPH. Since crude antiserum preparations have been used for these experiments, and since identical segments can be found in both the ACTH and the β-LPH molecule, we thought that the nature of the precursor protein should be better documented by a precise and independent chemical method to avoid the problems resulting from nonspecific immunological cross-reactivity.

Figure 10 shows that the 30,000-dalton peptide obtained after a 10-min pulse with radioactive [³⁵S]methionine and purified by SDS–gel electrophoresis contained at least two major radioactive tryptic fragments. One comigrated with the oxidized form of the synthetic peptide corresponding to residues 61–69 of β-LPH. The other comigrated with one oxidized fragment resulting from the tryptic digestion of ACTH 1–10 and was therefore identified as the tryptic fragment 1–8 of ACTH.

3.2. Pulse–Chase Incubations

In another experiment, intermediate-lobe cells were incubated with [³⁵S]methionine for 10 min, after which excess unlabeled methionine (final concentration 2 mM) was added and the incubation continued. Aliquots (150 μl) of the cell suspension were withdrawn at the end of the 10-min labeling period and then 20, 40, 60, 90, and 120 min later. Figure 11 shows that during the chase period with unlabeled methionine, the 30,000-dalton

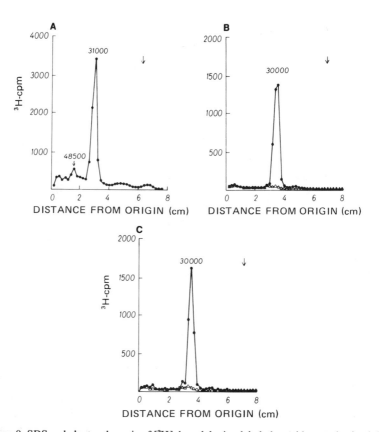

Figure 9. SDS–gel electrophoresis of [³H]phenylalanine-labeled peptides synthesized during a 10-min incubation by rat intermediate-lobe cells. Gels were prepared and used according to Weber and Osborn (1969). The position of the bromophenol blue tracking dye is marked (right arrow). The molecular weights of the different species have been calculated by comparing their relative mobilities with those of standards protein run on a separate gel. (A) Total extracted peptides as obtained directly after desalting on Sephadex G-25; (B, C) immunoprecipitate of the extracted peptide. A 5000-cpm aliquot of the ³H-labeled cell extract that had been previously desalted on Sephadex G-25 was mixed with an excess of either β-MSH antiserum (B) or ACTH antiserum (C). The amount of antiserum necessary for quantitative immunoprecipitation of the labeled material was previously determined by measuring the radioactivity precipitated by increasing amounts of antiserum. The mixture was incubated for 16 hr at 4°C in a phosphate–saline buffer (0.01 M sodium phosphate, 0.15 M NaCl, 0.025 M EDTA) (PBS, EDTA) and 2% Triton X-100, after which an immunoprecipitate was formed by addition of goat antiserum to rabbit immunoglobulins (Calbiochem). The immunoprecipitate was centrifuged at 10,000g for 4 min through a 1-ml sucrose cushion (1 M sucrose in PBS, EDTA, 2% Triton X-100), washed several times with PBS, EDTA, and dissolved by boiling in 50 μl of a solution containing 12 mg tris, 15 mg DTT, and 10 mg SDS/ml. (●) Direct immunoprecipitates; (△) control with excess purified β-MSH (B) or ACTH (C) (10 μg each) added to the sample before immunoprecipitation.

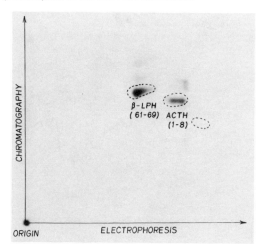

Figure 10. Tryptic mapping of 30,000-dalton precursor labeled with methionine. Radioactive peptides synthesized during a 20-min incubation of rat pars-intermedia cells were separated by SDS–gel electrophoresis. After electrophoresis, the gel was fixed in 25% isopropyl alcohol, 10% acetic acid, and washed extensively with 10% methanol. The gel was then sliced in 1-mm sections, which were then lyophilized and rehydrated with 500 μl of a solution containing 50 μg bovine serum albumin and 0.5 μg trypsin. After an overnight incubation at 37°C, the radioactive material corresponding to the 30,000-dalton molecule was lyophilized and mapped together with 10 μg β-LPH (61–69) (generous gift from Dr R. Guillemin) and with 75 μg of a tryptic digest of ACTH 1–10 (Ciba-Geigy). The standards and the radioactive material were oxidized with performic acid just prior to the mapping on the thin-layer cellulose plate. Electrophoresis was for 1.5 hr at 1000 V in formic acid–acetic acid–water (20:80:900), and chromatography was developed in the organic phase of a mixture containing butanol–acetic acid–water (200:50:250). Standard peptides (circled spots) were developed with cadmium nihydrin, and radioactive peptides were visualized by a 2-week radioautography.

peptide progressively disappeared, giving rise to smaller fragments (Fig. 11A). After 20 min of chase, two fragments were observed: an 18,000-dalton fragment and another smaller peptide comigrating with β-LPH (Fig. 11B). After 40 min of incubation in the presence of unlabeled methionine, a new peptide comigrating with standard β-endorphin appeared (Fig. 11C). As the incubation with unlabeled methionine proceeded, two fragments increased in importance: the 18,000-dalton peptide and the β-endorphin-sized fragment. Meanwhile, the 30,000-dalton protein slowly disappeared as its apparent molecular weight slightly increased. The fragment corresponding to β-LPH seemed to have a transient existence, since it was no longer detected after 90 min. The 40,000- to 50,000-dalton protein remained

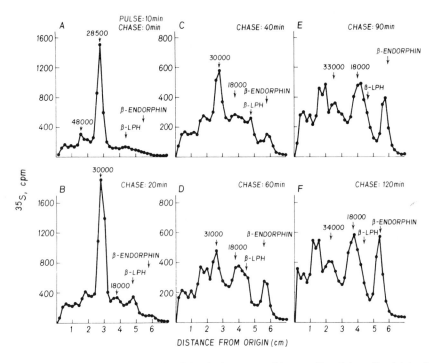

Figure 11. SDS–gel electrophoresis of desalted extracts of intermediate-lobe cells obtained during a pulse–chase labeling experiment. The cells were incubated with 2 mCi/ml [³⁵S]methionine for the first 10 min, after which a 300-fold excess of unlabeled methionine was added to the culture medium and the incubation continued for up to 2 hr. Samples were taken from the incubation medium, at the time indicated in the figures, extracted in 5 N acetic acid, and desalted on Sephadex G-25 before electrophoresis. The positions of standard labeled rat β-LPH and β-endorphin are marked on each curve, as well as the molecular weights corresponding to the positions of the peaks.

constant and represented about 5% of the total counts per minute recovered from the gel throughout the chase incubation.

When cell extracts from the same pulse–chase experiment were also analyzed by acidic gel electrophoresis (Fig. 12), the major molecular species present after 10 min of [³⁵S]methionine incorporation had an R_f of 0.31 (Fig. 12A). During the chase, it diminished and gave rise to three peptides: a first one comigrating with standard labeled rat β-LPH (R_f = 0.50), a second one comigrating with β-endorphin (R_f = 0.75), and a third one that had an R_f of 0.26. At 20 and 40 min (Fig. 12B and C) of chase, the peptide comigrating with standard rat β-LPH was clearly apparent as a well-characterized peak on the acidic gels; however, this species disappeared as the chase proceeded and was barely detectable after 2 hr (Fig. 12F).

A more definite identification of the molecular species characterized by R_f values of 0.75 and 0.50 on the acidic gels was obtained by rerunning those peptides extracted from the acidic gels on SDS–urea gels, where they also comigrated with standard rat β-endorphin and β-LPH, respectively (Fig. 13). It is worth noting here that rat γ-LPH does not contain methionine and therefore could not be recovered among the chase fragments of the [³⁵S]methionine-labeled precursor.

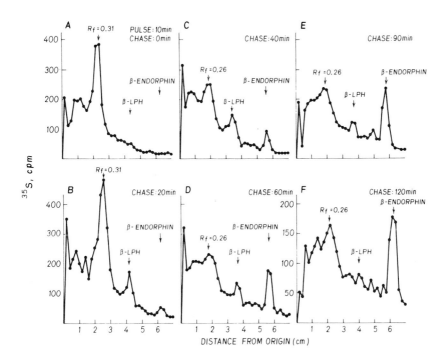

Figure 12. Acid gel electrophoresis of desalted extracts of intermediate-lobe cells obtained during a pulse–chase labeling experiment. Labeling and chase conditions were identical to those described in the Fig. 11 caption. The positions of standard labeled rat β-LPH and β-endorphin are marked on each curve.

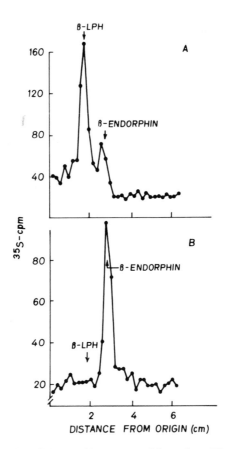

Figure 13. Characterization of the peptides recovered form the acidic gels by SDS–urea electrophoresis. Chase fragments of the 30,000 precursor obtained after a 1-hr chase incubation were analyzed first on acid gels. The material comigrating with β-LPH on the acidic gel was extracted from the gel and was shown to comigrate also with the ovine standard hormone on SDS–urea gels (A). A similar conclusion was also drawn for the β-endorphin-like material recovered on the acidic gel (B).

4. Discussion

Previous studies involving labeled amino acid incorporation in whole pituitary slices had proved that β-endorphin was biosynthesized in the pituitary gland. However, the elucidation of the detailed biosynthetic pathway required the use of a more specific system in which β-endorphin biosynthesis would represent a high percentage of the total protein synthesis. Immunohistochemical studies have shown that β-LPH and β-endorphin are most concentrated in cells of the pars intermedia of the pituitary. These observations led us to reinvestigate the biosynthesis of β-endorphin in isolated pars-intermedia cells. We have shown here that this tissue can indeed synthesize β-LPH together with its two fragments, γ-LPH and β-endorphin. In 10-min pulse experiments, one major labeled protein comprising 25–30% of the total radioactivity recovered from the SDS gels was extracted from the cells. This molecule had a molecular weight of 30,000 ± 1500 and contained antigenic determinants for both ACTH and β-MSH. By peptide mapping of the tryptic fragments, it was shown that this protein contained the segment 61–69 of β-LPH and 1–8 of ACTH. Another minor protein with an apparent molecular weight of 45,000 ± 5000 was also synthesized. Its nature is at present unknown.

When the radioactivity incorporated in the cells within the first 10 min of the incubation was chased for 2 hr by an excess of unlabeled amino acid, several fragments were formed from the initial 30,000-dalton peptide, and these fragments could be separated by SDS–gel electrophoresis as well as by acidic-gel electrophoresis. In both systems, one of the fragments that was produced in high yield comigrated with standard rat β-endorphin (Crine *et al.*, 1978).

When the chase incubation was performed for shorter periods (e.g., 20 min), only two major fragments were formed from the initial 30,000-dalton protein and could be separated by electrophoresis on SDS–gels. One fragment comigrated with standard rat β-LPH, and the other had an apparent molecular weight of 18,000. This peptide has not been fully characterized yet. However, considering the fact that the initial peptide has been shown to contain at least part of the ACTH sequence within its structure, we tentatively propose that this peptide could be a high-molecular-weight form of ACTH.

Following its cleavage from the precursor, the maturation of β-LPH into β-endorphin proceeded rapidly. Even after a 20-min chase, when β-LPH was the predominant form, a small peak of β-endorphin was already present. During this process, γ-LPH is probably formed by the same enzymatic cleavage step. However, rat γ-LPH does not contain any methionine residues and was not detected when intermediate-lobe cells were incubated with this labeled amino acid (Seidah *et al.*, 1978).

Even after a 2-hr chase, a fraction of the initial precursor still remained in the cells, but its apparent molecular weight had increased from 30,000 to 36,000. This result has also been observed in AtT-20 tumor cells, and could be explained by an increased glycosylation of the peptide backbone (Roberts *et al.*, 1978). The extensively glycosylated precursor could represent a stable form that could have its own biological role.

Immunocytochemical studies have shown that ACTH- and β-LPH-immunoreactive peptides occur in all the parenchymal cells of the pars intermedia (Pelletier et al., 1977; Moon et al., 1973; Bloom et al., 1977). Furthermore, when these cells were observed under the electron microscope, staining for ACTH and β-LPH was seen in all the granules and in the structures of the rough endoplasmic reticulum. These results are consistent with the existence of a common precursor for ACTH and β-LPH, a hypothesis that has received strong support from two recent studies using the ACTH-secreting mouse pituitary cell line AtT-20. Mains et al. (1977) used a double-immunoprecipitation technique to isolate labeled proteins from cells incubated with radioactive amino acids. Roberts and Herbert (1977a,b) prepared a cell-free translation product from a poly(A)-containing mRNA fraction obtained from AtT-20 cells. In both cases, the ACTH–β-LPH precursor appeared to be a 28,000- to 31,000-dalton protein that was cleaved into β-LPH, β-endorphin, and several high-molecular-weight forms of ACTH (Mains et al., 1977; Mains and Eipper, 1976, 1978).

The results presented herein indicate that the mechanism of β-endorphin biosynthesis in the rat intermediate lobe is very similar to that observed for ACTH and β-LPH biosynthesis in AtT-20 cells. The experiments described in this chapter conclusively show that the pars intermedia of the pituitary is a highly specialized tissue for the synthesis of the high-molecular-weight precursor that is transformed into β-endorphin with β-LPH as an intermediate.

During the course of this work, we developed a new approach for characterizing radioactive neosynthesized peptides that did not require the availability of homologous nonlabeled purified material. The first step involved the purification of the peptides synthesized in high yield by the pars intermedia. The purified radioactively labeled proteins were then analyzed by sequential automatic Edman degradation. The partial sequences obtained using several preparations from incubations done with different radioactive amino acids were then used to provide an unambiguous characterization of the labeled hormones of interest.

After having purified and characterized the various peptides related to β-LPH, we used them as markers for the identification of the peptides generated from a high-molecular-weight precursor during pulse–chase incubation. Characterization of the chase fragments both by SDS–polyacrylamide gel electrophoresis and by acidic-gel electrophoresis allowed the unambiguous identification of β-LPH and β-endorphin in the chase fragments.

Such a methodology can be used further to study other fragments generated by the splitting of the initial precursor. In particular, it would be very interesting to determine whether rat ACTH is present in the radioactive cell extracts. This characterization would enable us to identify the other intermediates involved in the maturation of the common precursor proposed for ACTH and β-LPH.

ACKNOWLEDGMENTS. This work was supported by Medical Research Council of Canada grants (MA-6612 and PG2).

DISCUSSION

RAMACHANDRAN: Did I understand you correctly when you said that you don't see any α-LPH because there is no methionine in this segment?

CRINE: Yes, there is no methionine.

RAMACHANDRAN: This means there is no methionine in rat β-MSH. I think this is the first mammalian β-MSH lacking methionine.

HERBERT: We find no methionine in β-MSH in the mouse in agreement with what Dr. Crine has said about β-MSH in the rat. Do you find any cysteine incorporated into the precursor form in the intermediate-lobe cultures of the rat?

CRINE: We have not done that experiment.

HERBERT: Do you have any leucine in place of methionine in the enkephalin portion of β-LPH or β-endorphin in the intermediate lobe of rat pituitary?

CRINE: We find less than 1% amino acid in this position that could be leucine. It is almost all methionine.

VAUDRY: Have you seen during your pulse–chase experiments the appearance of other opiate compounds such as γ-endorphin, α-endorphin, or methionine enkephalin?

CRINE: The smallest form of opiatelike peptide that we have observed is β-endorphin.

VAUDRY: So your results would indicate that cleavage of the β-endorphin molecule does not occur in pituitary cells, but takes place outside of the pituitary?

CRINE: I would not say that. It is possible that smaller forms of pituitary peptides could be generated by artifactual proteolytic cleavage during extraction. Unless you take certain precautions to avoid such proteolytic degradation, you cannot draw any conclusions about the significance of the presence of a peptide in a given tissue.

JONES: You are doubtlessly aware of the controversy concerning the source of the β-endorphin found in the brain. Several people have suggested that it comes from the pituitary gland. Have you looked for the biosynthesis of β-endorphin in brain tissue?

CRINE: Yes, we have. So far the results have been negative. It is possible that the technique that we have used and which is essentially the one which has been developed for the pars-intermedia cell is not sensitive enough for studying this problem in the brain.

KERDELHUÉ: Could you detect the presence of calcitonin in the big molecular form?

CRINE: We have tried immunoprecipitation of the precursor with anticalcitonin serum, but find nothing.

TIXIER-VIDAL: Did you check the cell viability at the end of the chase experiment?

CRINE: No, what we checked is the linearity of the incorporation of radioactive amino acid in the TCA-precipitable counts recovered inside the cells. It is linear for at least 3 hours. Moreover, the kinetics of the maturation of the precursor seems to fit well with other models such as proinsulin and pro-PTH.

REFERENCES

Bertagna, X., Lis, M., Gilardeau, C., and Chrétien, M., 1974, Biosynthèse *in vitro* de la béta-LPH de boeuf, *Can. J. Biochem.* **52**:349.

Birk, Y., and Li, C. H., 1964, Isolation and properties of a new, biologically active peptide from sheep pituitary glands, *J. Biol. Chem.* **239**:1048.

Bloom, F., Battenberg, E. O., Rossier, J., Ling, N., Leppeluoto, J., Vargo, T. M., and Guillemin, R., 1977, Endorphins are located in the intermediate and anterior lobes of the pituitary gland, not in the neurohypophysis, *Life Sci.* **20**:43.

Bradbury, A. F., Smyth, D. G., Snell, G. R., Birdsell, N. J. M., and Hulme, E. C., 1976, C-fragment of lipotropin has a high affinity for brain opiate receptors, *Nature (London)* **260**:793.

Chance, E. R., Ellis, R. M., and Bromer, W. W., 1968, Porcine proinsulin: Characterization and amino-acid sequence, *Science* **160**:165.

Chrétien, M., and Li, C. H., 1967, Isolation, purification and characterization of gamma-lipotropic hormone from sheep pituitary glands, *Can. J. Biochem.* **45**:1163.

Chrétien, M., Benjannet, S., Dragon, N., Seidah, N. G., and Lis, M., 1976, Isolation of peptides with opiate activity from sheep and human pituitaries: Relationship to beta-lipotropin, *Biochem. Biophys. Res. Commun.* **72**:472.

Chrétien, M., Seidah, N. G., Benjannet, S., Dragon, N., Routhier, R., Motomatsu, T., Crine, P., and Lis, M., 1977, A beta-LPH precursor model: Recent developments concerning morphine-like substances, *Ann. N. Y. Acad. Sci.* **297**:84.

Crine, P., Benjannet, S., Seidah, N. G., Lis, M., and Chrétien, M., 1977a, *In vitro* biosynthesis of beta-endorphin, gamma-lipotropin, and beta-lipotropin by the pars intermedia of beef pituitary glands, *Proc. Natl. Acad. Sci. U.S.A.* **74**:4276.

Crine, P., Benjannet, S., Seidah, N. G., Lis, M., and Chrétien, M., 1977b, *In vitro* biosynthesis of beta-endorphin in beef pituitary glands, *Proc. Natl. Acad. Sci. U.S.A.* **74**:1403.

Crine, P., Gianoulakis, C., Seidah, N. G., Gossard, F., Pezalla, P. D., Lis, M., and Chrétien, M., 1978, Biosynthesis of beta-endorphin from beta-lipotropin and a larger molecular weight precursor in rat pars intermedia, *Proc. Natl. Acad. Sci. U.S.A.* **75**:4719.

Dayhoff, M. D., Hunt, L. T., Barker, W. C., and Schartz, R. M., 1978, Protein sequence data file, National Biomedical Research Foundation, Washington, D.C.

Guillemin, R., Ling, N., and Burgus, R., 1976, Endorphines, peptides d'origine hypothalamique et neurohypophysaire à activité morphinomimétique: Isolement et structure moléculaire de l'alpha-endorphine, *C. R. Acad. Sci. Ser. D.* **282**:783.

Habener, J. F., Kemper, B. W., Rich, A., and Potts, J. T., Jr., 1977, Biosynthesis of parathyroid hormone, *Recent Prog. Horm. Res.* **33**:249.

Hughes, J., Smith, T. W., Kosterlitz, H. W., Fothergill, L. A., Morgan, B. A., and Morris, H. R., 1975, Identification of two related pentapeptides from the brain with potent opiate agonist activity, *Nature (London)* **258**:577.

Lazarus, L. H., Ling, N., and Guillemin, R., 1976, Beta-lipotropin as a prohormone for the morphinomimetic peptides, endorphins, and enkephalins, *Proc. Natl. Acad. Sci. U.S.A.* **73**:2156.

Li, C. H., 1964, Lipotropin, a new active peptide from pituitary glands, *Nature (London)* **201**:924.

Li, C. H., 1977, Beta-endorphin: A pituitary peptide with potent morphine-like activity, *Arch. Biochem. Biophys.* **183**:592.

Li, C. H., and Chung, D., 1976, Isolation and structure of an untriakontapeptide with opiate activity from camel pituitary glands, *Proc. Natl. Acad. Sci. U.S.A.* **73**:1145.

Li, C. H., Barnafi, L., Chrétien, M., and Chung, D., 1965, Isolation and amino acid sequence of beta-LPH from sheep pituitary gland, *Nature (London)* **208**:1093.

Li, C. H., Chung, D., and Doneen, B. A., 1976, Isolation, characterization and opiate activity of beta-endorphin from human pituitary glands, *Biochem. Biophys. Res. Commun.* **72**:1542.

Lohmar, P., and Li, C. H., 1967, Isolation of bovine beta-lipotropic hormone, *Biochim. Biophys. Acta* **147**:381.

Mains, R. E., and Eipper, B. A., 1976, Biosynthesis of adrenocorticotropic hormone in mouse pituitary tumor cells, *J. Biol. Chem.* **251**:4115.

Mains, R. E., and Eipper, B. A., 1978, Coordinate synthesis of corticotropins and endorphins by mouse pituitary tumor cells, *J. Biol. Chem.* **253**:651.

Mains, R. E., Eipper, B. A., and Ling, N., 1977, Common precursor to corticotropins and endorphins, *Proc. Natl. Acad. Sci, U.S.A.* **74**:3014.

Moon, H. D., Li, C. H., and Jennings, B. M., 1973, Immunohistochemical and histochemical studies of pituitary beta-lipotropins, *Anat. Rec.* **175**:529.

Pelletier, G., Leclerc, R., Labrie, F., Côté, J., Chrétien, M., and Lis, M., 1977, Immunohistochemical localization of beta-lipotropic hormone in the pituitary gland, *Endocrinology* **100**:770.

Roberts, J. L., and Herbert, E., 1977a, Characterization of a common precursor to corticotropin and beta-lipotropin: Cell-free synthesis of the precursor and identification of corticotropin peptides in the molecule, *Proc. Natl. Acad. Sci. U.S.A.* **74**:4826.

Roberts, J. L., and Herbert, E., 1977b, Characterization of a common precursor to corticotropin and beta-lipotropin: Identification of beta-lipotropin peptides and their arrangement relative to corticotropin in the precursor synthesized in a cell-free system, *Proc. Natl. Acad. Sci. U.S.A.* **74**:5300.

Roberts, J. L., Phillips, M., Rosa, P. A., and Herbert, E., 1978, Steps involved in the processing of common precursor forms of adrenocorticotropin and endorphin in cultures of mouse pituitary cells, *Biochemistry* **17**:3609.

Seidah, N. G., Lis, M., Gianoulakis, C., Routhier, R., Benjannet, S., Schiller, P., and Chrétien, M., 1977, Morphine-like activity of sheep beta-LPH and of its tryptic fragments, *Can. J. Biochem.* **55**:35.

Seidah, N. G., Gianoulakis, C., Crine, P., Lis, M., Benjannet, S., Routhier, R., and Chrétien, M., 1978, *In vitro* biosynthesis and chemical characterization of beta-lipotropin, gamma-lipotropin, and beta-endorphin in rat pars intermedia, *Proc. Natl. Acad. Sci, U.S.A.* **75**:3153.

Steiner, D. F., Cunningham, D., Spiegelman, L. S., and Aten, B., 1967, Insulin biosynthesis: Evidence for a precursor, *Science* **157**:697.

Swank, R. T., and Munkres, K. D., 1971, Molecular weight analysis of oligopeptides by electrophoresis in polyacrylamide gel with sodium dodecyl sulfate, *Anal. Biochem.* **39**:462.

Weber, K., and Osborn, M., 1969, The reliability of molecular weight determinations by dodecyl-sulfate–polyacrylamide gel electrophoresis, *J. Biol. Chem.* **244**:4406.

14

Effect of TRH on the Synthesis and Stability of Cytoplasmic RNAs of GH_3B_6 Cells

A. Morin

1. Introduction

Rat prolactin continuous cell lines offer invaluable models for studying the mechanism of action of thyroliberin (TRH) on prolactin (PRL) secretion in a homogeneous population of target cells (Tashjian and Hoyt, 1972; Tixier-Vidal *et al.*, 1975). One of these cell lines, GH_3B_6, is a subclone of the GH_3 strain derived from a transplantable pituitary tumor induced by chronic estrogen treatment (Tashjian *et al.*, 1968). As reported by Tashjian *et al.* (1970, 1971) and Gourdji *et al.* (1972) for GH_3 cells, GH_3B_6 cells secrete both PRL and, at a low level, growth hormone (GH), and they respond to TRH by an increase of PRL secretion.

Previous studies (Morin *et al.*, 1975) provided evidence for TRH having a biphasic effect on PRL secretion: an early effect, increasing PRL release from a stored PRL pool, and a delayed stimulating effect on de novo PRL synthesis. Simultaneously with this increase of PRL secretion, TRH induces a decrease of GH secretion (Tashjian *et al.*, 1971; Tashjian and Hoyt, 1972).

The intracellular mechanisms of the stimulating effect of TRH on PRL synthesis have been further studied. Biswas and Tashjian (1974) reported

A. Morin • Groupe de Neuroendocrinologie Cellulaire, Laboratoire de Physiologie Cellulaire, Collège de France, 75231 Paris 05, France

that PRL was synthesized predominantly on membrane-bound polysomes in GH_3 cells, and that TRH treatment for 3 days induces a large enhancement of their synthetic capacity. Using a cell-free protein-synthesizing system derived from wheat embryos, Evans and Rosenfeld (1976) showed that polyadenylic-acid-rich messenger RNAs [poly(A)-mRNAs] extracted from GH_3 cells directed synthesis of a PRL precursor that was processed into PRL by addition of a "GH_3 cell extract." This messenger activity coding for pre-PRL was stimulated 3- to 6-fold in TRH-pretreated cells, suggesting that regulation of PRL synthesis by TRH was, at least in part, at a transcriptional level. This was recently confirmed using complementary DNA probes to quantify specific mRNA sequences coding for PRL (Evans *et al.*, 1978).

Our purpose in the experiments reported in this chapter was to study the effect of TRH on the synthesis and stability of cytoplasmic poly(A)-mRNAs in GH_3B_6 cells by measuring the incorporation of [^{3}H]uridine.

2. Materials and Methods

2.1. Cell Culture

GH_3B_6 cells were grown, as a monolayer at 37°C, without antibiotic, in Ham F-10 solution supplemented with non-heat-inactivated sera: 15% of horse serum and 2.5% of fetal calf serum. These culture conditions permitted obtaining the maximum rate of PRL synthesis.

2.2. TRH Treatment

The greatest TRH increase of PRL synthesis was obtained when GH_3B_6 cells were precultivated during 4 days and then incubated with TRH (27 nM) during 48 hr without renewal of the medium.

2.3. RNA Labeling

Our investigation procedure was essentially that described by Buckingham *et al.* (1974).

Cytoplasmic RNA synthesis was estimated by labeling GH_3B_6 cells with 100 μCi [^{3}H]uridine (30 Ci/mmol) for the last 4 hr of the control or TRH 48-hr incubation.

Cytoplasmic RNA stability was estimated by pulse–chase experiments. GH_3B_6 cells were labeled with 400 μCi [^{3}H]uridine (30 Ci/mmol) for the last 2 hr of the control or TRH 48-hr incubation. Then, they were incubated for 45 min in fresh medium containing glucosamine chloride (10 mM) and uridine (100 μg/ml), with or without TRH (27 nM). The glucosamine treatment permitted reduction of the large uridine precursor pools and carrying out of proper RNA chase experiments. After this treatment, a chase incubation with fresh medium supplemented or not with TRH (27 nM) was performed for various periods up to 30 hr. At the end of the pulse–labeling or at the appropriate time of the chase incubation, RNAs were extracted and analyzed.

2.4. RNA Extraction

The cells were washed once with Ham F-10 solution and twice with phosphate-buffered saline, and then harvested in a solution containing 0.14 M NaCl, 10 mM tris-HCl (pH 7.5), and 5 mM MgCl₂; at 4°C, 0.1% NP 40 detergent was added and the cells were allowed to lyse for 10 min with gentle stirring.

After centrifugation to remove the nuclei, EDTA was added to the supernatant to a final concentration of 5 mM and SDS up to 0.2%. Cytoplasmic RNAs were extracted and deproteinized, at least three times, with an equal volume of mixing "chloroform–phenol (vol./vol.), 1% isoamyl alcohol." The aqueous phase was precipitated with 2.5 volumes of ethanol, and stored at −30°C.

2.5. RNA Analysis

Cytoplasmic RNA precipitate was pelleted, dissolved in a solution containing 0.1 M NaCl, 10 mM HEPES (pH 7.5), 1 mM EDTA, and 0.1% SDS, then layered onto a 15–30% sucrose gradient and centrifuged in an SW 41 Ti rotor at 27,000 rpm for 17 hr at 20°C. The optical-density profile of the gradient was analyzed at 254 nm, and 0.8- to 1-ml fractions were collected.

Next, 100 μl of each fraction was counted immediately for all cytoplasmic RNA radioactivity.

Then, most cytoplasmic mRNAs containing an adenosine-rich sequence in association with the informational RNA were selected. For this purpose, poly(A)-RNAs contained in 500 μl of each fraction were separated by hybridization to poly(U)–glass fiber filter according to Sheldon *et al.* (1972). The radioactivities of total cytoplasmic RNAs as well as poly(A)-RNAs measured in each fraction were summed. We also calculated the rate of radioactivity incorporated into the poly(A)-mRNAs with respect to that incorporated into total cytoplasmic RNAs in the same fraction [(poly(A)-RNA/total cyt-RNA)%].

3. Results

3.1. Pulse–Labeling Experiments

Whatever the duration of TRH treatment (24, 48, or 72 hr), the treatment induced a 3- to 6-fold decrease of radioactivity incorporated into cytoplasmic RNAs (Table I).

In contrast, TRH treatment had no effect or a slight stimulating effect on radioactivity incorporated into poly(A)-mRNAs (Table I).

The rate of incorporated radioactivity [(poly(A)-RNA/total cyt-RNA)%] was increased 2- to 5-fold in TRH-treated cells (Table I).

These data suggest that TRH decreases the rate of synthesis of most cytoplasmic RNAs while increasing the rate of snythesis of cytoplasmic poly(A)-mRNAs. The fact that TRH decreases the radioactivity in total cytoplasmic RNA is not in contradiction to this conclusion, since it is well know that cytoplasmic poly(A)-RNAs represents only about 1% of total cytoplasmic RNAs.

Optical-density profiles showing the distribution of cytoplasmic RNA species along 15–30% sucrose gradients displayed two major peaks corresponding to 18 S and 28 S RNA of ribosomal subunits, respectively (Fig. 1). These profiles were not modified by TRH treatment of GH_3B_6 cells. The profiles showing the distribution of radioactivity associated with cytoplasmic RNAs along the same sucrose gradients were similar to optical-density profiles with two peaks corresponding to 18 S and 28 S,

Table I. Effect of TRH on [^{3}H]Uridine Incorporation in Cytoplasmic RNAs and Poly(A)-mRNAs of GH_3B_6 Cells[a]

Treatment	Cytoplasmic RNA (dpm)	Poly(A)-mRNA (dpm)	$\dfrac{\text{Poly(A)-mRNA}}{\text{Cyt-RNA}}\%$
Control	4,221,000	14,100	0.3
TRH (24 hr)	1,255,000	21,800	1.7
Control	871,000	14,800	1.7
TRH (48 hr)	232,000	13,400	5.8
Control	3,076,000	15,800	0.5
TRH (48 hr)	1,437,000	23,600	1.6
Control	1,052,000	14,800	1.4
TRH (72 hr)	266,000	12,400	4.7

[a] GH_3B_6 cells were plated at a density of 2×10^6 cells per 10-cm petri dish. Three dishes were used for each experimental series. TRH (27 nM) was added to the culture medium for the indicated intervals of time. Cytoplasmic RNAs were labeled and extracted as described in Section 2. Poly(A)-containing RNAs were separated by hybridization onto poly(U)–glass fiber filters.

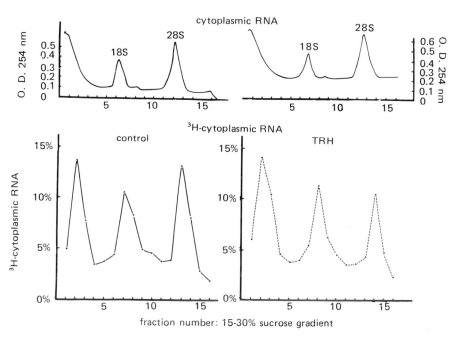

Figure 1. Effect of TRH on synthesis of the different species of cytoplasmic RNAs. GH₃B₆ cells were plated as described in the Table I footnote and treated with TRH (27 nM) for 48 hr. Cytoplasmic RNAs were labeled, extracted, and analyzed as described in Section 2. The distribution of labeled cytoplasmic RNAs along sucrose gradients was referred to the optical-density profiles, which permitted the location of 18 S and 28 S ribosomal RNA.

respectively (Fig. 1). They were also not modified by TRH treatment of the cells.

In contrast, the profiles showing the distribution of radioactivity associated with poly(A)-mRNAs were different and modified more or less depending on the duration of TRH treatment. After a 48-hr incubation of GH₃B₆ cells with TRH, a 5-fold-increased incorporation was observed at the level of a 12–15 S peak, whereas the other peaks (22–26 S region) decreased simultaneously (Fig. 2).

For all fractions along sucrose gradients, TRH increased the ratio [(poly(A)-RNA/total cyt-RNA)%] (Fig. 3). Moreover, this increase was 10- to 15-fold for the 12–15 S peak instead of 2- to 3-fold as in the 22–26 S region.

These results point out that TRH induces a preferential increase of the rate of synthesis of the 12–15 S poly(A)-mRNA among cytoplasmic RNAs.

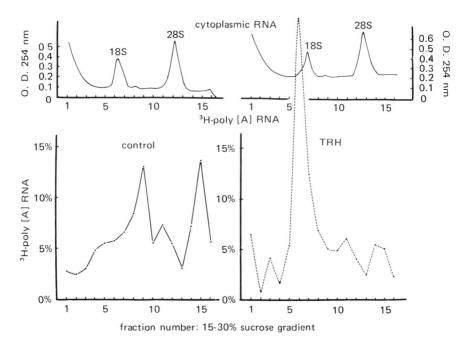

Figure 2. Effect of TRH on synthesis of the different species of cytoplasmic poly(A)-mRNAs. The poly(A)-containing mRNAs were isolated from the same GH₃B₆ cells as in Fig. 1 (see section 2). The identification of 18 S and 28 S ribosomal RNAs from the optical-density profiles provided an internal control to the attribution of "S" values to poly(A)-mRNAs.

3.2. Pulse–Chase Experiments

In chase experiments, the half-life of the 12–15 S poly(A)-mRNA was estimated by determining the radioactivity associated with poly(A)-RNA in the 12–15 S peak, at the end of the pulse–labeling and after increasing time intervals of chase–incubation in the presence or absence of TRH.

The radioactivity resulting from the incorporation of [³H]uridine into 12–15 S poly(A)-RNA as expressed per microgram of cellular proteins, DNA, or RNA was plotted as a function of time, using a logarithmic scale (Fig. 4). The values obtained for control cells can be approximately fitted with a line having a slope that corresponds to a half-life of 10 hr. For TRH-treated cells, the half-life was increased more than 2-fold to 23–25 hr at least for up to 22 hr, and then the difference decreased. The same half-life values were found whether the radioactivity was expressed as a function of RNA or DNA or cellular protein content. These data suggest

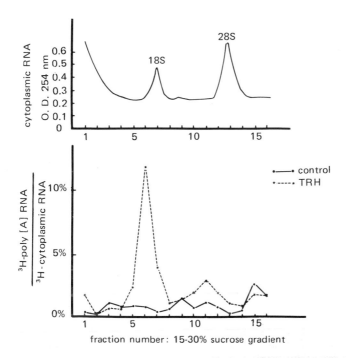

Figure 3. Effect of TRH on rate of synthesis of poly(A)-mRNAs with respect to total cytoplasmic RNA [(poly(A)-RNA)/(total Cyt-RNA)]%. These data were obtained from those of Figs. 1 and 2.

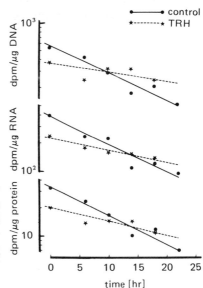

Figure 4. Effect of TRH on the half-life ($t_{1/2}$) of the 12–15 S poly(A)-RNA. GH_3B_6 cells were plated as described in the Table I footnote, and pulse–chase experiments were carried out as described in Section 2. Plotting estimated and calculated ($t_{1/2}$: log 2/slope) determinations of half-lifes gave equivalent values: control, 10 hr; TRH, 23–25 hr.

that TRH induces, within 48 hr, a stabilization of the 12–15 S poly(A)-mRNA.

4. Discussion

Evans *et al.* (1977), who isolated the mRNA coding for pre-PRL from GH$_3$ cells, showed that it migrates at approximately "12–13 S" using rate zonal centrifugation. Taking these data into account, the increased synthesis and the longer half-life of the "12–15 S" poly(A)-mRNA that we observed in this study after a 48-hr treatment of GH$_3$B$_6$ cells by TRH can be interpreted as a stimulation of synthesis and a stabilization of the mRNA coding for PRL. Both effects participate in the TRH-induced increase of PRL synthesis by these target cells.

The increased synthesis of the "12–15 S" mRNA is in agreement with the recent findings of Evans *et al.* (1978), who showed that PRL mRNA sequences increased from 1.1 to 4.5% of cytoplasmic poly(A)-RNAs within 48 hr after addition of TRH. To complete our experiments, we intend now to demonstrate with specific complementary DNA the presence of sequences coding for PRL in the "12–15 S" poly(A)-mRNA.

The decreased incorporation of [^{3}H]uridine into cytoplasmic RNAs in TRH-treated cells could be attributed, at least in part, to a chronic reduction of the uridine nucleotide intracellular pool as shown by Martin and Tashjian (1978). Consequently, the enhancement of poly(A)-mRNA synthesis might be underestimated.

Our findings of an increased incorporation of [^{3}H]uridine into the putative mRNA for PRL, together with the evidence obtained by Evans and co-workers (Evans and Rosenfeld, 1976; Evans *et al.*, 1977, 1978) using another approach, favor a direct action of TRH at a transcriptional nuclear level. In addition, the evidence that we obtained for a stabilization of the putative mRNA for PRL strongly suggests another effect of TRH, at a posttranscriptional cytoplasmic level.

5. Summary

The mechanism of stimulation of PRL synthesis in cultured rat pituitary cells (GH$_3$B$_6$) by TRH was investigated by [^{3}H]uridine incorporation into cytoplasmic poly(A)-containing RNAs. Simultaneously with the increase of PRL synthesis within 48 hr, TRH modified the distribution of radioactivity associated with different species of poly(A)-mRNAs and induced a preferential increase of snythesis of "12–15 S" poly(A)-mRNA, a putative messenger for pre-PRL.

The stability of this "12–15 S" poly(A)-mRNA was investigated by chase experiments: after a 48-hr treatment of GH_3B_6 cells by TRH, the half-life of the "12–15 S" poly(A)-mRNA is increased more than 2-fold, showing a stabilization of the messenger for pre-PRL.

These data strongly suggest both a transcriptional and a posttranscriptional effect of TRH.

ACKNOWLEDGMENTS. We are grateful to Dr. B. Croizat (Laboratoire de Biochimie Cellulaire, Collège de France, Paris) for teaching the technical procedures used in this work and for his helpful advice, and to M.F. Moreau for her excellent technical assistance. This work was supported by grants from the DGRST (Contract No. 77 7 0492).

DISCUSSION

DUVAL: Did you check the incorporation of radioactive uridine into the nuclear RNAs to see whether some modification in the processing of pre-rRNA and pre-mRNA occurs?

MORIN: No, we just studied cytoplasmic RNAs, but the analysis of nuclear heterogeneous RNAs would provide evidence for thyroliberin stimulating the synthesis of prolactin mRNA directly at the transcriptional level.

DUVAL: Did you check the specificity of the TRH effect by adding any other peptide to your controls?

MORIN: No, we did not.

REFERENCES

Biswas, D, K., and Tashjian, A. H., Jr., 1974, Intracellular site of prolactin synthesis in rat pituitary cells in culture, *Biochem. Biophys. Res. Commun.* **60**:241.

Buckingham, M. E., Caput, D., Cohen, A., Whalen, R. G., and Gros, F., 1974, The synthesis and stability of cytoplasmic messenger RNA during myoblast differentiation in culture, *Proc. Natl. Acad. Sci. U.S.A.* **71**:1466.

Evans, G. A., and Rosenfeld, M. G., 1976, Cell-free synthesis of a prolactin precursor directed by mRNA from cultured rat pituitary cells, *J. Biol. Chem.* **251**:2842.

Evans, G. A., Hucko, J., and Rosenfeld, M. G., 1977, Preprolactin represents the initial product of prolactin mRNA translation, *Endocrinology* **101**:1807.

Evans, G. A., David, D. N., and Rosenfeld, M. G., 1978, Regulation of prolactin and somatotropin mRNAs by thyroliberin, *Proc. Natl. Acad. Sci. U.S.A.* **75**:1294.

Gourdji, D., Kerdelhué, B., and Tixier-Vidal, A., 1972, Ultrastructure d'un clone de cellules hypophysaires secrétant de la prolactine (GH3): Modifications induites par l'hormone hypothalamique de libération de l'hormone thyréotrope (TRF), *C. R. Acad. Sci. Ser. D* **274**:437.

Martin, T. F. J., and Tashjian A. H., Jr., 1978, Thyrotropin-releasing hormone modulation of uridine uptake in rat pituitary cells: Evidence that uridine phosphorylation is regulated, *J. Biol. Chem.* **253**:106.

Morin, A., Tixier-Vidal, A., Gourdji, D., Kerdelhué, B., and Grouselle, D., 1975, Effect of thyrotropin-releasing hormone (TRH) on prolactin turnover in culture, *Mol. Cell. Endocrinol.* **3**:351.

Sheldon, R., Jurale, C., and Kates, J., 1972, Detection of polyadenylic acid sequences in viral and eukaryotic RNA, *Proc. Natl. Acad. Sci. U.S.A.* **69**:417.

Tashjian, A. H., Jr., and Hoyt, R. F., Jr., 1972, Transient controls of organ specific functions in pituitary cells in culture, in: *Molecular Genetics and Developmental Biology* (M. Sussman, ed.), pp. 353–387, Prentice-Hall, Englewoods Cliffs, New Jersey.

Tashjian A. H., Jr., Yasumura, Y., Levine, L., Sato, G. H., and Parker, M. L., 1968, Establishment of clonal strains of rat pituitary tumor cells that secrete growth hormone, *Endocrinology* **82**:342.

Tashjian A. H., Jr., Bancroft, F. C., and Levine, L., 1970, Production of both prolactin and growth hormone by clonal strains of rat pituitary tumor cells: Differential effects of hydrocortisone and tissue extracts, *J. Cell Biol.* **47**:61.

Tashjian A. H., Jr., Barowsky, N. J., and Jensen, D. K., 1971, Thyrotropin releasing hormone: Direct evidence for stimulation of prolactin production by pituitary cells in culture, *Biochem. Biophys. Res. Commun.* **43**:516.

Tixier-Vidal, A., Gourdji, D., and Tougard, C., 1975, A cell culture approach to the study of anterior pituitary cells, *Int. Rev. Cytol.* **41**:173.

15

Hormonal Regulation of Prolactin mRNA

Glen A. Evans and Michael G. Rosenfeld

1. Introduction

The neuroendocrine regulation of pituitary hormone synthesis and secretion is a complex process in which polypeptide hormones, bioamines, and steroid hormones play a part. Various hormones have been documented to exert stimulatory or inhibitory effects on the biosynthesis and secretion of specific pituitary hormones. The molecular processes underlying these hormonally induced alterations are less well understood, but current data suggest that significant differences may exist in the actions of steroid and polypeptide hormones at the molecular level.

Steroid hormones are postulated to regulate synthesis of specific proteins via translocation of the steroid–receptor complex into the nucleus and subsequent alterations in the pattern of gene transcription. The mechanisms by which polypeptide hormones induce specific protein synthesis in their target tissues is less well defined, but alterations in the concentrations of specific cytoplasmic messenger RNAs (mRNAs) appear to play a significant role. The first step in the interaction of a polypeptide

Glen A. Evans and *Michael G. Rosenfeld* • Division of Endocrinology, Department of Medicine, University of California, San Diego, School of Medicine, La Jolla, California 92093

hormone with the target tissue takes place at the cell surface when the hormone binds to a cell-surface receptor. For some hormones, this results in increased intracellular concentrations of cyclic AMP (cAMP), activation of protein kinase, and subsequent phosphorylation of specific cellular protein substrates. However, for other polypeptide hormones, such a sequence of events has not been demonstrated. It has recently been demonstrated that polypeptide hormones can alter the levels of specific cytoplasmic mRNAs in parallel with altering the biosynthesis of specific proteins. Lines of rat pituitary tumor cells in permanent culture (GH) have provided excellent model systems for the study of hormone action on the pituitary gland, in which both steroid and polypeptide hormonal effects are found. This family of closely related cell strains possesses unique properties for the study of the regulation of prolactin synthesis at the genetic level. The use of GH cells in the study of hormone action has recently been reviewed (Martin and Tashjian, 1977). GH cells synthesize and secrete into the media prolactin and growth hormone, which are apparently identical to the hormones isolated from rat pituitary glands. The rate of prolactin biosynthesis may achieve 50 μg/10^6 cells per 24 hr, which represents approximately 1–2% of the total protein synthesized by the cells under basal conditions. GH cells do not produce detectable quantities of other pituitary hormones such as ACTH, TSH, FSH, or LH. In contrast to the intact pituitary, GH cells store very little prolactin or growth hormone, and cellular production (micrograms of hormone appearing in the media per unit time) closely approximates the rate of hormone synthesis. Prolactin synthesis is modulated by the presence of several hormones when these hormones are added at physiological concentrations. Thyroliberin (TRH, thyrotropin-releasing hormone) at 10^{-10} to 10^{-8} M induces prolactin synthesis up to 10 times the basal level. Estradiol or estrogen analogues at 10^{-11} to 10^{-8} M were reported to increase prolactin production up to 5-fold over background. Addition of these hormones simultaneously decreases slightly the rate of growth hormone biosynthesis. Several related lines of GH cells were developed that varied in the basal and stimulated rate of prolactin production. GH_3 cells produce both prolactin and growth hormone at approximately equimolar basal rates of 10 μg/10^6 cells per day. GH_4 cells produce relatively more prolactin than growth hormone. The GC cell line produces growth hormone at a basal level of 50 μg/10^6 cells per day, but produces little or no prolactin. GH cells can be maintained in tissue culture in completely defined medium in the absence of serum, which precludes any possible effects of unknown serum factors (Hayashi and Sato, 1976). The induction of prolactin synthesis by TRH in GH cells provides a model system in which to study the action of polypeptide hormones on gene expression.

2. In Vitro Synthesis of Prolactin

Rat prolactin is a single polypeptide chain of 198 amino acids that has a molecular weight of 26,000 as calculated from the complete amino acid sequence (Shome and Parlow, 1977). The *in vitro* synthesis of rat prolactin by translation of isolated cellular mRNA has been reported by several groups. Messenger RNA from GH_3 cells (Evans and Rosenfeld, 1976) and from rat pituitaries (Maurer *et al.*, 1976) produces immunoreactive translation products in the wheat germ cell-free protein-synthesizing system. On one-dimensional sodium dodecyl sulfate–polyacrylamide gel electrophoresis (SDS-PAGE), the immunoprecipitable translation product migrates with an apparent molecular weight of 28,000. Since authentic rat prolactin migrates with an apparent molecular weight of 24,000, it is apparent that the intracellular precursor of prolactin, preprolactin, contains an obligatory amino-terminal leader sequence of molecular weight 2000–4000. Maurer *et al.* (1977) have determined the position of several amino acids in the leader sequence and determined the length of the leader sequence as 29 amino acids, with an estimated molecular weight of 4000. Bovine pituitary prolactin has also been shown to contain an amino-terminal leader sequence of 30 amino acids (Lingappa *et al.*, 1977).

In the wheat germ translation system, authentic prolactin is generated as a translation product only when microsomal membranes are included in the reaction mixture during translation (Lingappa *et al.*, 1977; Evans and Rosenfeld, 1976). This finding is compatible with the signal hypothesis proposed by Blobel and Dobberstein (1975), in which the leader polypeptide containing hydrophobic amino acids is necessary for membrane binding and directional transport of the nascent peptide chain into the lumen of the endoplasmic reticulum.

Recently, it has been shown that some pituitary hormones may be synthesized as much larger precursors containing two or more hormones that are generated by posttranslational processing. A 31,000-molecular-weight protein synthesized by mRNA from ACTH-producing cells has been shown to contain both ACTH and β-lipotropin (Mains and Eipper, 1976; Roberts and Herbert, 1977). High-molecular-weight forms of immunoreactive prolactin have been reported, but may reflect protein–protein interactions (Rogol and Chrambach, 1975). Analysis of the *in vitro* translation product, however, has documented that prolactin is processed from preprolactin and not from a precursor of yet higher molecular weight. Cell-free protein synthesis in the presence of $[^{35}S]$methionyl-tRNA$_i$ generates labeled preprolactin with label incorporated into the amino-terminal methionine, but not into internal methionine residues (Evans *et al.*, 1977). Since methionyl initiator tRNA donates the initiating amino acid in all nascent peptide chains, preprolactin is suggested to represent the initial translation product and is not the product of processing of a larger precursor molecule.

A more precise analysis of GH-cell translation products can be made by two-dimensional polyacrylamide–isoelectric focusing gels. Using this technique, up to 2000 individual proteins can be separated according to molecular weight and isoelectric point (O'Farrell, 1975). On two-dimensional gels, authentic rat prolactin shows microheterogeneity in that it is resolved to two spots of identical molecular weight but differing in isoelectric point (Fig. 1A). Immunoreactive rat preprolactin also exhibits microheterogeneity (Fig. 1B), resolving into two clearly separate spots differing by one charge unit, and each migrating at a pI more basic than that of prolactin. The basis of this microheterogeneity is unknown, but could result either from posttranslational modification or from two molecular forms of prolactin mRNA differing slightly in sequence.

The cytoplasmic content of prolactin mRNA can be indirectly determined by analysis of mRNA-directed protein synthesis in a cell-free reaction. From 1 to 2% of the radioactivity is incorporated into preprolactin when cell-free protein synthesis is directed by mRNA from unstimulated GH_3 cells. When mRNA from TRH-stimulated cells is used and translation products are analyzed by immunoprecipitation or two-dimensional gel analysis, the amount of preprolactin synthesized increases 2- to 6-fold, equivalent in magnitude to the TRH-induced increase in prolactin production (Evans and Rosenfeld, 1976; Dannies and Tashjian, 1976). Messenger RNAs from several GH cell lines show mRNA levels by cell-free translation that are proportional to the amount of prolactin being produced by the cells. GH_4 cellular mRNA produces relatively greater amounts of preprolactin than GH_3 mRNA, and GH_4 cells produced 4 to 5 times the amount of prolactin produced by GH_3 cells. GC cells, which produce no detectable levels of prolactin, contain undetectable amounts of preprolactin in mRNA-directed translation products (Evans et al., 1978). Cell-free translation of prolactin mRNA is inhibited by addition of 7-methyl guanosine monophosphate, indicating that prolactin mRNA probably contains the [7mGppp-Pu] cap structure that has been documented for many mammalian and viral mRNAs.

3. Preparation of Prolactin Complementary DNA Probes for Molecular Hybridization

Complementary DNA (cDNA) probes for specific nucleic acid sequences are extremely powerful tools for the analysis of gene expression in that they allow the accurate quantitation of specific mRNA sequences

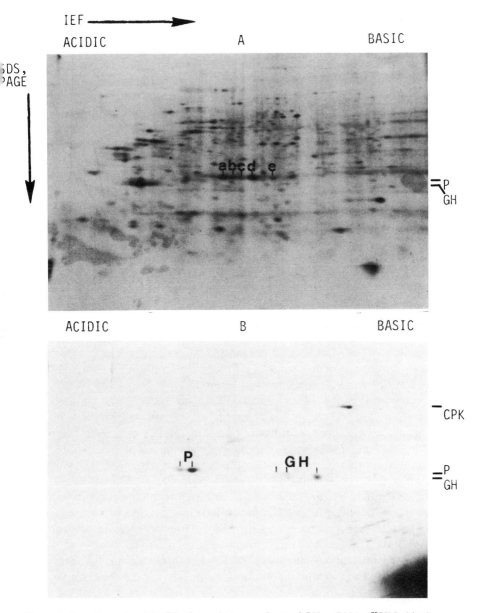

Figure 1. Two-dimensional PAGE of translation products of GH_4 mRNA. [^{35}S]Methionine-labeled proteins from cell-free reactions using wheat germ embryo lysate were analyzed by isoelectric focusing in the horizontal dimension and by SDS-PAGE in the vertical dimension. Spots *a* and *b* are selectively precipitated with prolactin-specific antiserum and represent preprolactin. Spots *c*, *d*, and *e* are selectively precipitated with growth hormone-specific antiserum and represent presomatotropin. The positions of authentic rat prolactin (P) and growth hormone (GH) are shown in parallel gel stained for protein.

at levels as low as several molecules per cell. They have played an important part in defining the mechanisms of steroid hormone action (Monahan *et al.*, 1977). As a general technique for the preparation of cDNA probes, a specific mRNA must be obtained in relatively pure form, generally attained by repetitive sizing techniques. From the pure mRNA, a cDNA is synthesized using avian myoblastosis virus reverse transcriptase primed with oligo(dT$_{12-18}$). This produces molecules of varying length from the 3' termini of the mRNA that have high specific radioactivity when labeled with [^{3}H]- or [^{32}P]deoxyribonucleotides. After separation of the cDNA from the mRNA template material, the cDNA can be used to determine specific nucleic acid sequences by molecular hybridization.

From cell-free translation experiments, several properties of the prolactin mRNA can be deduced. The coding capacity required to synthesize a 198-amino-acid protein with a 29-amino-acid leader sequence is 681 nucleotides. On rate zonal sedimentation through denaturing formamide–urea–sucrose density gradients, prolactin mRNA migrates at 12 S, corresponding to a molecular length of about 1000 nucleotides (Evans *et al.*, 1977). Since prolactin mRNA contains a poly(A) tail at the 3' end, as evidenced by its binding to oligo(dT)–cellulose, approximately 70–100 nucleotides correspond to polyadenylic acid. Moreover, prolactin mRNA would be expected to contain an additional 100–200 noncoding nucleotides. A prolactin cDNA probe that contained the entire sequence of prolactin mRNA, exclusive of the poly(A) tail, would therefore be expected to be between 800 and 900 nucleotides in length.

The purification of prolactin mRNA to serve as a template for generation of specific cDNA was impractical because all available prolactin-producing cell lines also produce large quantities of growth hormone, and the molecular weights would be quite similar. Moreover, since both hormones contain heterogeneous poly(A) tails, their mRNAs would not exhibit a discrete absolute length. In addition, it is difficult to prevent significant mRNA degradation during the requisite chromatographic and sizing steps required for mRNA purification.

An alternate approach was devised for the preparation of prolactin cDNA that did not require the purification of translationally active prolactin mRNA (Evans and Rosenfeld, 1979). Total cytoplasmic mRNA from GH$_4$ cells was used as a template for cDNA synthesis. Under the conditions of synthesis, high-molecular-weight cDNA with an average length of 1000 nucleotides was produced. On denaturing polyacrylamide gels, a cDNA band of 840 nucleotides in length was isolated and found to be specific cDNA for prolactin mRNA. The cDNA was further purified by two cycles of preparative hybridization. The 840-nucleotide band was hybridized to mRNA from GH$_4$ cells, and the rapidly annealing, and therefore highly abundant, material was isolated by binding to hydroxyapatite. The mRNA is removed by alkaline hydrolysis and the cDNA then hybridized to mRNA from GC cells. GC mRNA contains low or no prolactin mRNA, but otherwise produces identical translation products on two-dimensional gels. The nonannealing material, or single-stranded cDNA, is recovered by hydroxyapatite chromatography, the mRNA hydrolyzed with base, and the resulting material characterized. This final step serves to eliminate low-abundancy contaminants and to remove contaminating growth hormone cDNA sequences.

The resulting cDNA probe has the following characteristics: (1) a length of 840 ± 20 nucleotides migrating as a single band on denaturing polyacrylamide gels (Fig. 2); (2) is hybridized to greater than 98% completion to mRNA from prolactin-producing cells over a range of 2 log eR_0t; (3) exhibits a sharp melting profile with a T_m of 88°C; (4) does not hybridize to mRNA from GC cells, indicating an absence of cross-reactivity with growth hormone mRNA; and (5) hybridization of the cDNA probe to GH_4 mRNA in cDNA excess specifically abolishes translation of prolactin mRNA, confirming the identity of the cDNA probe.

4. Regulation of Prolactin Messenger RNA by Thyroliberin

The use of prolactin-specific cDNA confirms that addition of TRH to cultured GH_4 cells results in an increase in the cytoplasmic concentration of prolactin mRNA. Hybridization of prolactin cDNA to mRNA from cells stimulated for increasing periods of time results in an increase in the rate of hybridization consistent with an increase in prolactin mRNA concentration from 1 to 5% of the total cellular mRNA (Fig. 3). In the same experiment, hybridization to a cDNA probe specific for rat growth hormone indicates a small decrease in growth hormone mRNA concentration. This effect is consistent with the 10–50% decrease in growth hormone production following addition of TRH to cultured GH_4 cells.

GH_4 cells in a basal unstimulated state produce and discharge into the medium about 15,000 molecules of prolactin per cell per minute. The cells contain about 100,000 TRH binding sites on the cell surface that are saturated within 30 min of the addition of 10 nM TRH to the medium (Dannies and Tashjian, 1976). An increase in prolactin production is detectable in the medium at 4 hr after TRH addition. At 4–6 hr after TRH addition, an increase in the concentration of prolactin mRNA is detectable by hybridization with prolactin cDNA; prolactin mRNA levels increase linearly to a maximum of 10,000–20,000 prolactin mRNA molecules per cell at 48–72 hr (Fig. 4). The rate of prolactin production, as determined

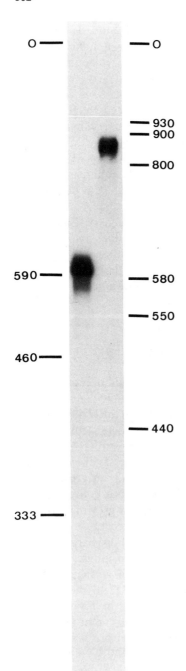

Figure 2. Analysis of purified prolactin cDNA on a denaturing polyacrylamide gel after purification by gel elution and two cycles of preparative hybridization. Markers represent single-stranded DNA fragments of known length from phage lambda DNA or rabbit β-globin cDNA. A rabbit β-globin cDNA of 590 nucleotides in length is shown on the left for comparison.

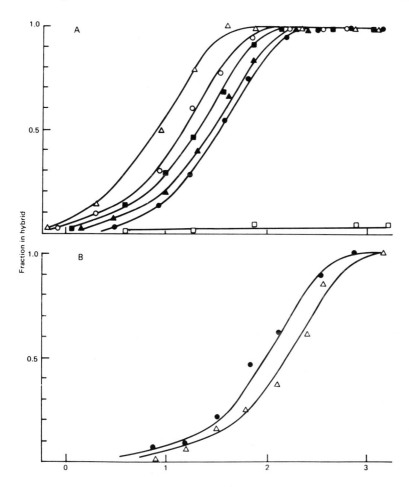

Figure 3. Hybridization analysis of GH$_4$ mRNA. (A) Prolactin cDNA was hybridized to cytoplasmic RNA from GH$_4$ cells in the basal state (●) or cells treated for 6 hr (▲), 12 hr (■), 24 hr (○), or 48 hr (△) with TRH (20 ng/ml). (B) In the same experiment, a growth hormone cDNA probe was hybridized to RNA from control (●) cells and cells treated for 48 hr (△) with TRH. R_0t is the product of the RNA concentration in moles of nucleotide per liter and time in seconds.

by radioimmunoassay for prolactin in the medium, increases linearly to a maximum rate of production at 48–72 hr. The rate of production of growth hormone and the growth hormone mRNA concentration concurrently decrease approximately 50% at 72 hr after addition of TRH.

In a large series of determinations of prolactin mRNA concentration by molecular hybridization, the rate of prolactin production is linearly

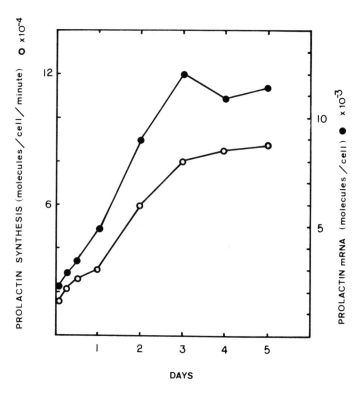

Figure 4. Kinetics of TRH action on GH₄ cells. TRH was added to cultures of GH₄ cells and prolactin synthesis determined by radioimmunoassay of prolactin accumulated in the medium. Prolactin mRNA concentration was determined by hybridization of total cytoplasmic mRNA to a [³H]prolactin cDNA probe under conditions of RNA excess.

related to the cytoplasmic concentration of prolactin mRNA (see Fig. 5). This relationship appears to hold over the range of prolactin production of $0.2-3$ μg prolactin 10^6 cells per 24 hr and prolactin mRNA concentrations of $0.085-6\%$ of the total poly(A)-containing mRNA using GH₃, GH₄, and GC cell strains. Therefore, it appears that the rate of prolactin synthesis is directly related to the cytoplasmic concentration of prolactin mRNA. The primary effect of TRH on prolactin biosynthesis would therefore appear to be mediated through an increase in prolactin mRNA concentration; such an increase could be mediated at a transcriptional or a posttranscriptional level. Translational regulation of prolactin synthesis in the absence of altered mRNA levels does not appear to play a major role in TRH action under most *in vitro* circumstances.

5. The Prolactin Gene

It is possible to assay directly for specific gene number in DNA of mammalian cells when specific cDNA probes of sufficient purity are available. The prolactin-specific cDNA allows the determination of the number of prolactin genes present in GH$_4$ cells. This can be most accurately determined by measuring the rate and final extent of hybridization of the prolactin cDNA probe to its DNA complement in probe excess. Under these conditions, a doubling of gene number results in a doubling of the saturating amount of cDNA (Old *et al.*, 1976). Saturation hybridization experiments show that the amount of DNA complementary to prolactin cDNA is equivalent to the presence of 2–6 prolactin DNA copies per genome, or 1–3 copies per haploid genome (Fig. 6). This value is consistent with the 1–2 copies per haploid genome found with other unique genes such as globin. If the number of genes does slightly exceed 1–2 per haploid genome, it could result from the random aneuploidy found in GH cells (Sonnenschein *et al.*, 1970). However, it is clear from this analysis that the prolactin gene is of low copy number in GH cells.

TRH action on GH cells results in an increase in cytoplasmic prolactin mRNA. Though the mechanism by which this occurs is not known, it

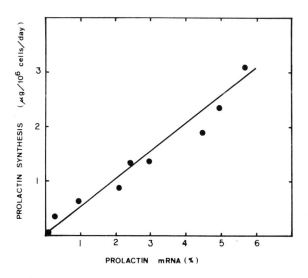

Figure 5. Comparisons of prolactin mRNA concentration and prolactin production in GH$_3$, GH$_4$, and GC cell lines under various conditions of hormonal stimulation.

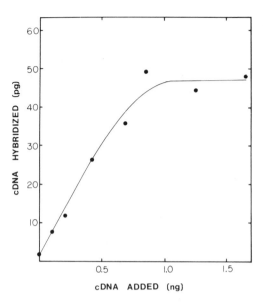

Figure 6. Hybridization of prolactin cDNA to GH₄ cellular DNA in cDNA excess. Increasing amounts of prolactin cDNA were hybridized to 100 μg sheared denatured DNA from unstimulated GH₄ cells. The amount of cDNA hybridized in the reaction was determined by binding to hydroxyapatite in 0.12 M phosphate buffer at 60°C.

could result from increased rate of gene transcription, decreased degradation of cytoplasmic prolactin mRNA, increased rate of mRNA transport or processing, or an increased gene copy number. This latter possibility has recently been found to occur in mammalian cells under certain conditions. Methotrexate induces an increase in dihydrofolate reductase and in cytoplasmic dihydrofolate reductase mRNA through a proportional increase in the number of dihydrofolate reductase genes (Alt *et al.*, 1978), presumably by favoring selection of cells in which gene duplication has randomly occurred. Gene amplification also occurs with ribosomal genes in amphibian and other oocytes (Brown and Dawid, 1968). To investigate this possibility for TRH action, gene copy number in GH₄ cells was determined after TRH induction of prolactin mRNA. Gene copy number with 72 hr of TRH stimulation is essentially identical to basal gene copy number. Amplification of the prolactin gene is therefore not an explanation of TRH effect on prolactin mRNA levels.

6. Effects of Thyroliberin on Other Messenger RNAs

Several lines of evidence suggest that TRH action on the mRNA of GH cells is not a generalized phenomenon, but is highly specific. First, there is no detectable increase in total mRNA content or total mRNA synthesis during TRH stimulation. The recovery of poly(A)-mRNA for GH cells is identical regardless of the presence or absence of TRH. Second, two specific mRNA probes are available for mRNAs known to be under hormonal control. TRH specifically increases prolactin mRNA while growth hormone mRNA simultaneously decreases slightly. Third, analysis of proteins synthesized by hormone-treated and control cells in the wheat germ cell-free translation system by two-dimensional gel electrophoresis showed a specific increase in preprolactin synthesized, without major changes in other proteins. By such analysis of mRNA-directed cell-free protein synthesis, the protein products of no more than five additional mRNAs were quantitatively altered by addition on TRH; more than 500 proteins were analyzed by this technique. Therefore, as has been seen with several other hormones, TRH alters the intracellular levels of mRNAs coding for only a relatively small number (<1%) of specific proteins.

Though TRH appears to exert a regulatory influence on prolactin mRNA, it is clear that other agents can also influence the rate of prolactin production by GH cells. Hydrocortisone, estradiol-17β, dibutyryl cAMP, thyroid hormone, ergocryptine, prostaglandins, and various estrogen analogues all affect prolactin production to some extent (Martin and Tasjian, 1977; Dannies *et al.*, 1977; Gautvik and Kriz, 1976). Since cell cultures usually require the addition of serum to support growth of cells, factors other than added hormones could be responsible for supporting prolactin production or as permissive factors for TRH action. This effect could be tested by performing TRH stimulation experiments in a completely defined serum-free medium. The rate of production of prolactin by GH_4 cells under these conditions and the cytoplasmic mRNA level declined to below the serum-supported level when maintained under serum-free conditions (Evans *et al.*, 1978). Nevertheless, TRH addition results in a 2- to 3-fold increase in prolactin production and cytoplasmic mRNA addition above the basal level. TRH alone can increase prolactin synthesis via increasing intracellular prolactin mRNA levels. However, additional factors present in serum may be required for maintaining the high basal rate of prolactin production by GH_3 cells in various experiments (Martin and Tashjian, 1977).

7. Conclusion

In the absence of exogenous hormone, GH cells produce prolactin through a distinct series of steps. A finite number of discrete prolactin genes are transcribed, producing a pre-mRNA molecule. The primary transcript is presumably processed, removing from the pre-mRNA any noncoding internal sequences and resulting in a colinear coding sequence. The mRNA is then polyadenylated at the 3′ end, "capped" at the 5′ end, and transported across the nuclear membrane. The mRNA is bound by ribosome subunits with the action of specific protein initiation factors, and translation is initiated. When a sufficient length of peptide chain is synthesized to expose the leader sequence, the polyribosome complex binds to the cytoplasmic surface of the endoplasmic reticulum. Following directional transport of the nascent chain across the membrane and subsequent cleavage of the leader sequence inside the rough ER, the final protein product is vesiculated and secreted.

TRH action is mediated by an increase in the concentration of cytoplasmic prolactin mRNA. Although activation of adenylate cyclase, and an increase in cAMP concentration and protein-kinase-mediated phosphorylation of cytoplasmic and nuclear protein, has been suggested, there is no clear evidence for cAMP mediation of TRH effects in this system (Martin and Tashjian, 1977). TRH has relatively discrete effects on gene expression in GH cells. While intracellular levels of prolactin mRNA are increased, and levels of growth hormone mRNA appear to decrease, both the total content of poly(A)-mRNA and its kinetic complexity remain stable. TRH results in quantitative changes in less than 1% of total mRNA species, and has no effect on prolactin gene dosage. The exact sites and biochemical mechanisms by which TRH acts to increase prolactin mRNA levels have yet to be elucidated.

ACKNOWLEDGMENTS. This investigation was supported by the National Institutes of Health Research Grant USPHS AM18477 from the National Institute of Arthritis, Metabolism and Digestive Diseases. G.A.E. is a Predoctoral Trainee, supported by grants and funds from Training Grant PSGM07198 from the National Institute of General Medical Sciences. M.G.R. is a recipient of Research Career Development Award AM00078 from the National Institute of Arthritis, Metabolism and Digestive Diseases.

REFERENCES

Alt, F. W., Kellems, R. E., Bertino, J. R., and Schimke, R. T., 1978, Selective multiplication of dihydrofolate reductase genes in methotrexate-resistant variants of cultured murine cells, _J. Biol. Chem._ **253:**1357.

Blobel, G., and Dobberstein, B., 1975, Transfer of proteins across membranes: Presence of proteolytically processed and unprocessed nascent immunoglobulin light chains on membrane-bound ribosomes of murine myeloma, _J. Cell Biol._ **67:**835.

Brown, D. D., and Dawid, J. B., 1968, Specific gene amplification in oocytes, _Science_ **160:**272.

Dannies, P. S., and Tashjian, A. H., 1976, TRH increases prolactin mRNA activity in the cytoplasm of GH cells as measured by translation in a wheat germ cell-free system, _Biochem. Biophys. Res. Commun._ **70:**1180.

Dannies, P. S., Yen, P. M., and Tashjian, A. H., Jr., 1977, Anti-estrogenic compounds increase prolactin and growth hormone synthesis in clonal strains of rat pituitary cells, *Endocrinology* **101**:1151.

Evans, G. A., and Rosenfeld, M. G., 1976, Cell-free synthesis of a prolactin precursor directed by mRNA from cultured rat pituitary cells, *J. Biol. Chem.* **251**:2842.

Evans, G. A., and Rosenfeld, M. G., 1979, Regulation of prolactin mRNA analyzed using a specific cDNA probe, *J. Biol. Chem.* **254**:8023.

Evans, G. A., Hucko, J., and Rosenfeld, M. G., 1977, Preprolactin represents the initial product of prolactin mRNA translation, *Endocrinology* **101**:1807.

Evans, G. A., David, D. N., and Rosenfeld, M. G., 1978, Regulation of prolactin and somatotropin mRNAs by thyroliberin, *Proc. Natl. Acad. Sci. U.S.A.* **75**:1294.

Gautvik, K. M., and Kriz, M., 1976, Effects of prostaglandins on prolactin and growth hormone synthesis and secretion in cultured rat pituitary cells, *Endocrinology* **98**:352.

Hayashi, I., and Sato, G. H., 1976, Replacement of serum by hormones permits growth of cells in a defined medium, *Nature (London)* **259**:132.

Lingappa, V. R., Devillers-Thiery, A., and Blobel, G., 1977, Nascent prehormones are intermediates in the biosynthesis of authentic bovine pituitary growth hormone and prolactin, *Proc. Natl. Acad. Sci. U.S.A.* **74**:2432.

Mains, R. E., and Eipper, B. A., 1976, Biosynthesis of adrenocorticotropic hormone in mouse pituitary tumor cells, *J. Biol. Chem.* **251**:4115.

Martin, T. F. J., and Tashjian, A. H., Jr., 1977, Cell culture studies of thyrotropin-releasing hormone action, in: *Biochemical Actions of Hormones,* Vol. IV (G. Litwack, ed.), pp. 270–312, Academic Press, New York.

Maurer, R. A., Stone, R., and Gorski, J., 1976, Cell-free synthesis of a large translation product of prolactin mRNA, *J. Biol. Chem.* **251**:2801.

Maurer, R. A., Gorski, J., and McKean, D. J., 1977, Partial amino acid sequence of rat pre-prolactin, *Biochem. J.* **161**:189.

Monahan, J. J., Harris, S. E., and O'Malley, B. W., 1977, Analysis of cellular messenger RNA using complementary DNA probes, in: *Receptors and Hormone Action,* Vol. I (B. W. O'Malley and L. Birnbaumer, eds.), pp. 297–329, Academic Press, New York.

Old, J., Clegg, J. B., Weatherall, D. J., Ottolenghi, S., Comi, P., Giglioni, B., Mitchell, J., Tolstoshev, P., and Williamson, R., 1976, A direct estimate of the number of human γ-globin genes, *Cell* **8**:13.

O'Farrell, P. H., 1975, High resolution two-dimensional electrophoresis of proteins, *J. Biol. Chem.* **250**:4007.

Roberts, J. C., and Herbert, E., 1977, Characterization of a common precursor to corticotropin and β-lipotropin: Identification of β-lipotropin peptides and their arrangement relative to corticotropin in the precursor synthesized in a cell-free system, *Proc. Natl. Acad. Sci. U.S.A.* **74**:5300.

Rogol, A. D., and Chrambach, A., 1975, Radioiodinated human pituitary and amniotic fluid prolactins with preserved molecular integrity, *Endocrinology* **97**:406.

Shome, B., and Parlow, A. F., 1977, Human pituitary prolactin: The entire linear amino acid sequence, *J. Clin. Endocrinol. Metab.* **45**:1112.

Sonnenschein, C., Richardson, U. I., and Tashjian, A. H., Jr., 1970, Chromosomal analysis, organ-specific function and appearance of six clonal strains of rat pituitary tumor cells, *Exp. Cell Res.* **61**:121.

16

Pregrowth Hormone Messenger RNA: Glucocorticoid Induction and Purification from Rat Pituitary Cells

F. Carter Bancroft, Paul R. Dobner, and Li-Yuan Yu

1. Introduction

For several years, our laboratory has been engaged in studies of the biosynthesis of growth hormone by the rat pituitary. These studies have had two goals, which are clearly interdependent. The first goal has been the elucidation of the steps involved in the synthesis of growth hormone. Thus, we hope ultimately to be able to provide a complete description at the molecular level of the structure and transcription of the growth hormone gene, the structure of the resultant RNA transcript, the processing of this transcript into mature cytoplasmic growth hormone messenger RNA (mRNA), the structure of the primary protein product of the translation of this mRNA, and the processing of this primary translation product into mature growth hormone. The second goal has involved an understanding of the hormonal regulation of the expression of the growth hormone gene. Studies with a number of model systems have suggested that steroid hormones regulate specific gene expression at a pretransla-

F. Carter Bancroft, Paul R. Dobner, and *Li-Yuan Yu* • Cellular Gene Expression and Regulation Laboratory, Memorial Sloan-Kettering Cancer Center, New York, New York 10021

tional level, via a general mechanism that involves localization in the nucleus of steroid bound to a specific cytoplasmic receptor (reviewed in Yamamoto and Alberts, 1976). We showed some years ago that glucocorticoid hormones such as hydrocortisone stimulate the production of growth hormone by cultured rat pituitary cells (Bancroft *et al.*, 1969). Our second goal has thus been to determine both the step(s) in the production of growth hormone at which glucocorticoid hormones act and the molecular mechanisms whereby the stimulation by glucocorticoids is exerted.

Recently, we have made progress in our elucidation of the process by which growth hormone mRNA is translated to yield mature growth hormone. Thus, in experiments involving translation of rat pituitary cell mRNA in the wheat germ cell-free translation system, we have detected the synthesis of a probable precursor of growth hormone, designated pregrowth hormone (Sussman *et al.*, 1976). Experiments involving either treatment of pituitary cells with the protease inhibitor L-1-tosylamide-2-phenyl-ethylchloromethyl ketone (TPCK) (Sussman *et al.*, 1976) or the cell-free incubation of pituitary cell polysomes (Spielman and Bancroft, 1977) have yielded additional evidence that pregrowth hormone is the physiological precursor of growth hormone and that cleavage of pregrowth hormone to growth hormone occurs during synthesis. More recently, in experiments carried out in collaboration with the laboratory of Günter Blobel of Rockefeller University, we have determined the amino-terminal sequence of rat pregrowth hormone and have shown that the precise conversion of rat pregrowth hormone to growth hormone can be achieved *in vitro* (Sussman-Berger *et al.*, in prep.).

We have also made progress in our understanding of the action of glucocorticoids on growth hormone production by pituitary cells. We have shown that the increased production observed earlier (Bancroft *et al.*, 1969) arises from an increase in the rate of growth hormone synthesis, rather than from other possible mechanisms such as decreased degradation or increased secretion of growth hormone by pituitary cells (Yu *et al.*, 1977). This observation then led naturally to the question of whether the increase in growth hormone synthesis under glucocorticoid stimulation arose from an increase in the efficiency of translation of preexisting pregrowth hormone mRNA or from an increase in the amount of pregrowth hormone mRNA in the cytoplasm of pituitary cells. In this chapter, we describe how we have gone about answering this question. In addition, we describe our use of pituitary cells grown in tissue culture as a source from which to isolate pregrowth hormone mRNA, which has in turn permitted us to use the techniques of recombinant DNA technology to create a tool that promises to be extremely useful in further studies of the nuclear events involved in the production of pregrowth hormone mRNA and the regulation of this process by glucocorticoid hormones.

2. Experimental System and Procedures

2.1. Experimental System

The investigations described herein were carried out with either of two clonal strains of rat pituitary tumor cells, designated GH₃ and GC. The GH₃ cells produce both growth hormone and prolactin (Tashjian *et al.*, 1970; Martin and Tashjian, 1977). In our experience, the GH₃ cells exhibit a greater stimulation by exogenous hormones of growth hormone synthesis than do the GC cells; hence, the GH₃ cells have been employed for all our studies in this area. The GC cells are a subclone of the GH₃ cells, obtained as follows: During mass culture of GH₃ cells in suspension culture in 1970, a culture was repeatedly observed to produce growth hormone but no detectable prolactin, as assayed by microcomplement fixation (F. C. Bancroft and A. H. Tashjian, Jr., unpublished observations). These cells were cloned, and a clone designated GC was selected. The GC cells synthesize more growth hormone than the GH₃ cells, and the rate of prolactin synthesis by the GC cells is at most 1% of the rate of growth hormone synthesis (Sussman-Berger, 1978). Furthermore, the GC cells have been observed to contain either undetectable amounts of preprolactin mRNA (Evans *et al.*, 1978) or at most 0.1% as much preprolactin mRNA as pregrowth hormone mRNA (Sussman-Berger, 1978). Hence, in our work on pregrowth hormone, and on the purification of pregrowth hormone mRNA, we have employed the GC cells. The virtual absence of preprolactin mRNA from the GC cells has been a particularly useful property during the purification of the pregrowth hormone mRNA from these cells, since the mRNAs for rat growth hormone and rat prolactin are very similar in size (unpublished observations). Either GH₃ or GC cells were grown in suspension culture as described previously (Bancroft, 1973; Yu *et al.*, 1977). Where indicated, dexamethasone phosphate (5×10^{-7} M) was added to the growth medium of the GC cells 3–5 days prior to cell fractionation.

2.2. Experimental Procedures

Procedures that have not been described previously or have been modified are briefly described in this section.

2.2.1. Messenger RNA Preparation

RNA isolated as described by Tushinski *et al.* (1977) was, unless specified otherwise, subjected to chromatography on oligo(dT)–cellulose (Collaborative Research, type T-3) as described by Aviv and Leder (1972), except that the RNA was heated (85°C, 5 min) and rapidly cooled prior to salt addition and application to the column, and was subjected to two cycles of binding and elution.

2.2.2. Cell-Free Translation Systems

Incubation of RNA in the presence of ascites extracts was essentially as described by Bancroft et al. (1973), except that the $Mg(CH_3COO)_2$ concentration was 2.5 mM.

Incubation of RNA in the presence of wheat germ extracts (See Fig. 6) was as described previously (Sussman et al., 1976), with the following modifications: reaction volumes were 50 μl, the concentration of dithiothreitol was 2.4 mM [the value of 24 mM reported in Sussman et al. (1976) was a typographical error], 20 mM KCl plus 36 mM $K(CH_3COO)$ was used in place of 56 mM KCl, spermine was added to 16 μM, and the radioactive amino acid was [^{3}H]leucine (1 mCi/ml, 60 Ci/mmol).

Preparation of rabbit reticulocyte lysates was essentially as described by Schimke et al. (1974). Treatment of the lysates with micrococcal nuclease was as described in a modification (Kamine and Buchanan, 1977) of the original procedure (Pelham and Jackson, 1976). Incubation of RNA in the presence of the nuclease-treated reticulocyte lysates was as follows (Schimke et al., 1974): The final assay system contained 40% (vol./vol.) mRNA-depleted reticulocyte lysate, 75 mM KCl, 2mM $Mg(CH_3COO)_2$, 820 μM GTP, 12 mM creatine phosphate, 43 units/ml creatine kinase, 19 unlabeled amino acids at a concentration of 43 μM, 320 μCi/ml [^{35}S]methionine (580 Ci/mmol) or, where indicated, 872 μCi/ml [^{35}S]methionine (695 Ci/mmol), and various amounts of mRNA as indicated. Reaction volumes of 25 or 50 μl were incubated at 30°C for 60 min.

Small duplicate aliquots of any of the reaction mixtures described above were removed at the beginning and end of the incubation, treated with KOH (to degrade aminoacylated transfer RNA) and Cl_3CCOOH (Bancroft et al., 1973), and total acid-insoluble radioactivity was determined (Yu et al., 1977). After incubation, reaction mixtures were either analyzed by sodium dodecyl sulfate–polyacrylamide gel electrophoresis (SDS-PAGE) (Tushinski et al., 1977) or subjected to indirect immunoprecipitation as described below.

2.2.3. Immunoprecipitation of Growth-Hormone-Related Proteins

2.2.3a. Direct Immunoprecipitation. Products of RNA translation in the presence of Krebs II ascites extracts were immunoprecipitated directly with baboon antiserum to rat growth hormone plus rat growth hormone carrier as described previously (Yu *et al.*, 1977). All other immunoprecipitations were performed according to the indirect technique described below.

2.2.3b. Indirect Immunoprecipitation. Immunoprecipitation of the products of cell-free protein synthesis in the presence of wheat germ extracts or nuclease-treated reticulocyte lysates was performed by the indirect immunoprecipitation procedure of Kessler, employing fixed *Staphylococcus aureus* (Cowan I Strain) bacteria for the precipitation of soluble antibody-antigen complexes (Kessler, 1975). All procedures were at 4°C. Duplicate aliquots (5–10 μl) of reaction mixture were diluted with 200 μl phosphate-buffered salts. A 1.0-μl volume of either baboon antiserum to rat growth hormone or a preimmune serum from the same animal was then added. Following incubation overnight, the mixture received 50 μl of a 10% (wt./vol.) suspension (in phosphate-buffered salts containing 1% sodium deoxycholate, 1% Brij-58, 0.5% Nonidet P-40, 1 mg/ml bovine serum albumin, 2 mM leucine, and 5 mM methionine) of *Staphylococcus aureus* (Cowan I Strain) bacteria (obtained from the New England Enzyme Center), which had been formalin-fixed and heat-killed as described by Kessler (1975). Following incubation for 20 min, the mixture was layered over 0.5 ml of a sucrose–detergent solution, prepared as described previously (Yu *et al.*, 1977), except that NaCl was added to a final concentration of 0.5 M, and the bacteria–antibody–antigen complexes were pelleted by centrifugation (15,000g, 2 min). The pellet was resuspended, washed once with phosphate-buffered salts, then suspended in 100 μl SDS sample buffer (Sussman *et al.*, 1976) and heated at 95°C for 2 min to dissociate the bacteria–antibody–antigen complexes. Following removal of the bacteria by centrifugation (15,000g, 2 min), the supernatant was either analyzed by gel electrophoresis or treated with KOH and Cl_3COOH (Bancroft *et al.*, 1973), followed by determination of acid-insoluble radioactivity.

3. Results

3.1. Do Glucocorticoid Hormones Cause an Accumulation of Pregrowth Hormone Messenger RNA?

To answer this question, the wheat germ translation system was employed to measure the relative concentrations of pregrowth hormone mRNA in the cytoplasm of cells grown in the absence or presence of the synthetic glucocorticoid dexamethasone. We had found earlier that about

3 days is required for the stimulation by dexamethasone of growth hormone synthesis to reach a maximum (Yu *et al.*, 1977). Hence, GH$_3$ cells were grown for 3 days in the presence or absence of dexamethasone (10^{-5} M), and a portion of the cells in either culture were employed for measurement of specific growth hormone synthesis (growth hormone synthesis ÷ total protein synthesis) as described by Yu *et al.* (1977). Total cytoplasmic RNA was isolated from the remaining cells in either culture and incubated in the wheat germ cell-free system, as described by Tushinski *et al.* (1977). The relative concentration of pregrowth hormone mRNA in either RNA preparation was then estimated by measurement of the amount of pregrowth hormone that had been synthesized, as follows: Following direct immunoprecipitation of the reaction mixtures with antiserum to rat growth hormone, the immunoprecipitates were dissociated, and analyzed by SDS-PAGE. It should be noted that immunoprecipitation of pregrowth hormone by this technique is only about 30% complete (Tushinski *et al.*, 1977). However, since this value was quite reproducible (S.D. = 15%), and since only relative pregrowth hormone synthesis (control compared with induced cells) is significant for the question being asked here, no correction needed to be made for this incomplete immunoprecipitation. It should be noted that we have recently developed an *indirect* immunoprecipitation technique in which the same antiserum will quantitatively precipitate pregrowth hormone (see Section 3.3.2).

The result of such an experiment is shown in Fig. 1. Calculations, using the results shown in Fig. 1A plus measurements of total protein synthesis, of the values of specific growth hormone synthesis in control and induced cells yielded values of 0.75 and 8.6%, respectively. Hence, in this experiment, dexamethasone induced an 11.5-fold stimulation of specific growth hormone synthesis. Calculations using the results shown in Fig. 1B showed that in the same experiment, dexamethasone induced a 12.4-fold stimulation of pregrowth hormone mRNA, as assayed by the ability of this mRNA to direct the synthesis of pregrowth hormone in wheat germ extracts.

This result thus shows that dexamethasone does induce an increase in the cytoplasmic concentration of pregrowth hormone mRNA in the GH$_3$ cells. The excellent agreement observed between the induction of growth hormone synthesis and of pregrowth hormone mRNA further suggests that the induction in the GH$_3$ cells occurs *entirely* at some step preceding pregrowth hormone mRNA translation. Observation of an agreement between these values in another induction experiment in which partially purified pregrowth hormone mRNA was translated in the wheat germ system (Tushinski *et al.*, 1977), as well as in similar experiments with the GC cells (unpublished observations), further supports this conclusion.

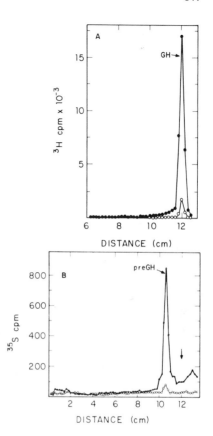

Figure 1. Dexamethasone induction of growth hormone synthesis and translatable pregrowth hormone mRNA. GH$_3$ cells were grown for 3 days in the absence (○) or presence (●) of 10^{-5} M dexamethasone. (A) Induction of growth hormone synthesis. After incubation of a portion of the cells from either culture with [³H]leucine, growth hormone synthesis was measured by SDS-PAGE of immunoprecipitates of cytoplasmic lysates, as described by Yu *et al.* (1977). (B) Induction of translatable pregrowth hormone mRNA. Total cytoplasmic mRNA was isolated from the remaining cells in both cultures and incubated in wheat germ extracts containing [³⁵S]methionine. Immunoprecipitates of 25-μl aliquots were then analyzed by SDS-PAGE. The two gels were aligned on the basis of the positions of internal ³H-labeled growth hormone markers (unlabeled arrow). Reproduced from Tushinski *et al.* (1977), by courtesy of the Editors of the *Proceedings of the National Academy of Sciences of the U.S.A.*

3.2. Use of Dexamethasone Induction to Identify Pregrowth Hormone Messenger RNA

The results presented above suggested that it might be possible to identify pregrowth hormone mRNA by comparison of the bands obtained following PAGE of RNA isolated from control and induced cells. GH$_3$ cells were incubated in the presence or absence of 10^{-5} M dexamethasone for 3 days. Pregrowth hormone mRNA was then partially purified from the cells in either culture as follows: We had shown previously that at least 95% of the pregrowth hormone mRNA in the GC cells is contained in the "membrane fraction" (Bancroft *et al.*, 1973), which is prepared from total cytoplasm by a crude differential centrifugation technique (Bancroft, 1973). Hence, RNA was isolated from the membrane fraction of control or induced cells. Following partial purification of mRNA from these RNA preparations by fractionation on oligo(dT)–cellulose, equal quantities of the bound RNA from control or induced cells were analyzed by formamide–PAGE, as described by Tushinski et al. (1977). It can be seen in Fig. 2 that, aside from the contaminating ribosomal RNA bands obtained in these early experiments involving oligo(dT)–cellulose fractionation, two major RNA bands labeled A and B were observed. Since the induction of band B (5- to 8-fold) was comparable to the induction of translatable pregrowth hormone mRNA in this experiment (5.6-fold) (see Tushinski et al., 1977), it seemed probable that band B corresponded to pregrowth hormone mRNA. To prove this directly, bands A and B were eluted from a gel of the type shown in Fig. 2 and translated in the wheat germ system. Analysis of the resulting products showed that band B codes for pregrowth hormone, and thus contains pregrowth hormone mRNA (Tushinski et al., 1977). Band A did not code for pregrowth hormone, and its identity is unknown. The molecular weight of pregrowth hormone mRNA was then estimated by employing formamide–PAGE to compare the mobility of band B with the mobilities of RNA markers, yielding a value of 3.6×10^5 (Tushinski et al., 1977).

3.3. Purification of Pregrowth Hormone Messenger RNA

In the investigations described in the previous section, we showed that dexamethasone caused an increase in the cytoplasmic concentration of functional (i.e., translatable) pregrowth hormone mRNA. It was not, however, possible to determine with certainty whether dexamethasone acts to increase the physical concentration of this mRNA (i.e., the concentration of growth-hormone-specific mRNA sequences) in the cytoplasm of the GH_3 cells, although the results shown in Fig. 2 suggest that this is the case. Furthermore, it was not possible to determine whether the accumulation of this mRNA in the cytoplasm arises primarily from a stimulation by dexamethasone of the rate of transcription of the growth hormone gene or from an effect on some step in the processing by the pituitary cell of the initial transcript of this gene into mature cytoplasmic mRNA.

To investigate these questions effectively, it is necessary to produce a highly specific complementary DNA (cDNA) probe for growth-hormone-specific RNA. The availability of purified pregrowth hormone mRNA would make the production of such a cDNA probe possible. For example, a cDNA copy of the purified pregrowth hormone mRNA could be incorporated into a suitable bacterial plasmid, yielding a source of virtually unlimited amounts of this cDNA in a pure form.

As described in the previous section, we were able to employ the induction by dexamethasone of pregrowth hormone mRNA in the GH_3

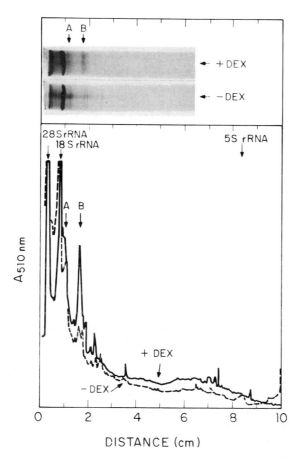

Figure 2. Analysis by formamide-PAGE of partially purified mRNA from untreated and dexamethasone (DEX)-treated GH₃ cells. Membrane fraction (Bancroft, 1973) RNA was isolated from untreated or dexamethasone-treated cells. Equal aliquots (18 μg) of the RNA from either preparation that bound to oligo(dT)–cellulose was analyzed by formamide–PAGE, and the gels were stained, destained, and scanned at 510 nm. Reproduced from Tushinski *et al.* (1977), by courtesy of the Editors of the *Proceedings of the National Academy of Sciences of the U.S.A.*

cells to identify by formamide–PAGE a band (band B in Fig. 2) that contained pregrowth hormone mRNA. However, since the recovery of RNA from this band was so low, we were unable to determine the purity of the pregrowth hormone mRNA in this band (for further discussion of this point, see Tushinski *et al.,* 1977). Furthermore, because of the similarities in the sizes of the mRNAs for growth hormone and prolactin, the GH₃ cells are not a favorable starting material for the purification of pregrowth hormone mRNA. Finally, it was clear that the incomplete precipitation of pregrowth hormone by direct immunoprecipitation with antiserum to growth hormone referred to above would create difficulties in assaying the purity of pregrowth hormone mRNA following various fractionation procedures. We therefore instituted the series of investigations described below, in which the GC cells were employed as starting material for the purification of pregrowth hormone mRNA. Furthermore, an indirect immunoprecipitation procedure was developed that yielded quantitative precipitation of pregrowth hormone, and thus permitted direct estimates by cell-free translation of the purity of pregrowth hormone mRNA at each stage of purification.

3.3.1. Development and Application of a Purification Scheme

It should be noted that the purification of pregrowth hormone mRNA has been facilitated by the absence of problems arising from endogenous RNAase activity in the GC and GH_3 cells. Thus, we have found that either intact functional pregrowth hormone mRNA (Sussman et al., 1976) or intact polysomes (Spielman and Bancroft, 1977) could be readily isolated following cellular fractionation procedures, without addition of RNAase inhibitors or recourse to other special procedures to prevent RNA degradation. To prevent problems from exogenous RNAase activity, we have performed all steps beginning with lysis of cells under sterile conditions. Thus, all containers were autoclaved and all solutions sterilized with diethylpyrocarbonate (Palmiter, 1974).

Based on our previous and preliminary observations, the following scheme for purification of pregrowth hormone mRNA was developed: For the reasons stated in Section 2.1, the GC cells were employed as starting material for the purification. Since preliminary experiments showed that dexamethasone yielded a 2- to 4-fold stimulation of the rate of growth hormone synthesis (and hence presumably also of pregrowth hormone mRNA levels) in the GC cells, the cells were grown in dexamethasone prior to cell fractionation. We had observed previously (Bancroft et al., 1973) that virtually all the detectable pregrowth hormone mRNA in the GC cells is in the membrane fraction. Hence, RNA was prepared from this cellular fraction, with a yield of about 2 mg RNA/10^9 cells. Since we had shown previously that in common with most mammalian mRNAs, pregrowth hormone mRNA contains a 3' homopoly(A) sequence (Tushinski et al., 1977), oligo(dT)–cellulose chromatography was employed to separate pregrowth hormone mRNA from cellular RNA such as ribosomal RNA, which lacks a poly(A) sequence. Further purification of pregrowth hormone mRNA then involved the application of techniques that fractionate RNA molecules on the basis of their size, as described below.

RNA prepared as described above was analyzed by sucrose-gradient centrifugation, and the RNA in individual fractions was assayed for pregrowth hormone mRNA activity by translation in the Krebs ascites cell-free system. The Krebs ascites system was employed at this point because the product of translation of pregrowth hormone mRNA in this system is growth hormone (Bancroft *et al.*, 1973), which is completely precipitable by direct immunoprecipitation with antiserum to growth hormone (Yu *et al.*, 1977). Pregrowth hormone mRNA was observed to sediment as a homogeneous species with a sedimentation coefficient of approximately 12 S (Fig. 3), in good agreement with our earlier estimate of its size (Tushinski *et al.*, 1977). [It should be noted that the contaminating ribosomal RNA observed in this early experiment was largely abolished in later experiments (see Fig. 8) in which improved procedures for oligo(dT)–cellulose chromatography (see Section 2.2) were employed.] When fractions 18–20 from this experiment were analyzed by formamide–PAGE, a major RNA band was observed (data not shown), the migration of which relative to RNA markers was similar to that observed previously for pregrowth hormone mRNA (Tushinski *et al.*, 1977).

3.3.2. Development of a Translational Assay for Pregrowth Hormone Messenger RNA Purity

The results described in the preceding section indicated that pregrowth hormone mRNA had been extensively purified by the procedures employed. Suitable techniques then had to be developed to assay its purity by translation *in vitro*. The Krebs II ascites system was not suitable, due to its high level of endogenous protein synthesis (Bancroft *et al.*, 1973). The nuclease-treated reticulocyte lysate system, which exhibits little or no endogenous protein synthesis (Pelham and Jackson, 1976), (also see Figs. 5, 6, and 9), was therefore employed for this purpose. However, the product of pregrowth hormone mRNA translation in this system is pregrowth hormone (see Fig. 9), which, as has been mentioned earlier, is only incompletely precipitated by direct immunoprecipitation with antiserum to rat growth hormone. Since antiserum raised against pregrowth hormone is not available, development of a quantitative translational assay for pregrowth hormone mRNA purity required an immunoprecipitation technique in which antiserum against growth hormone could be employed for the complete and specific precipitation of pregrowth hormone.

Preliminary experiments indicated that the indirect immunoprecipitation technique employing fixed *Staphylococcus aureus* bacteria (see Section 2.2) led to complete precipitation of pregrowth hormone. Hence, the specificity and completeness of this technique were examined further. The specificity of this technique was demonstrated by the observation that

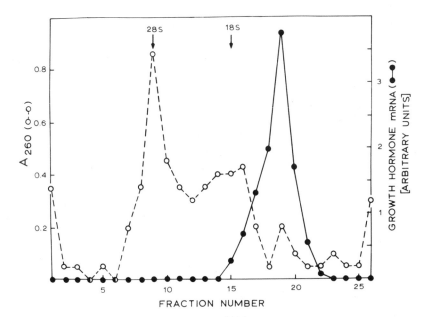

Figure 3. Sucrose-gradient ultracentrifugation of RNA isolated from the membrane fraction of GC cells grown in the presence of dexamethasone, and enriched for poly(A+) mRNA. RNA (190 μg) was enriched for poly(A+) mRNA by chromatography on type T2 oligo(dT)–cellulose as described by Aviv and Leder (1972). Linear gradients of 15–30% sucrose in 10 mM Tris (pH 7.4), 10mM EDTA, 0.1 M NaCl, and 0.2% SDS were prepared. RNA to be analyzed was dissolved in the same buffer (lacking sucrose), disaggregated by heating at 85°C for 2 min, cooled quickly, and loaded onto the gradient. Centrifugation was at 22,000 rpm for 20 hr at 22°C in an SW40 rotor. Following collection of fractions and measurement of absorbance at 260 nm in a spectrophotometer, each fraction received 10 μg/ml E. coli tRNA, and the RNA was precipitated by addition of 2 volumes of ethanol and storage overnight at −20°C. The precipitated RNA was prepared for translation by washing twice with70% ethanol and twice with 95% ethanol, evaporating residual ethanol, and dissolving in 25 μl H_2O. Aliquots (5 μl) were then translated in the Krebs ascites cell-free system. Pregrowth hormone mRNA activity was then determined by direct immuno-precipitation, followed by slab-type SDS-PAGE of the dissociated immunoprecipitates and fluorography of the dried gels as described by Tushinski et al. (1977), and determination of the area under the resulting growth hormone peak as described by Harpold et al. (1978). (o) A_{260}; (•) growth hormone mRNA activity.

radioactivity can be quantitatively displaced from immunoprecipitated pregrowth hormone by competition with unlabeled growth hormone (Fig. 4). We have also shown previously by gel analysis that only pregrowth hormone is detected in an indirect immunoprecipitate of a mixture of cell-free translation products (Harpold et al., 1978). The completeness of this technique for the precipitation of pregrowth hormone is demonstrated by the observation that when the most highly purified pregrowth hormone mRNA preparation was incubated in the mRNA-depleted reticulocyte lysate system, 90–100% of the pregrowth hormone synthesized was immunoprecipitated (see Fig. 9 and Table I).

3.3.3. Assay of Relative Pregrowth Hormone Messenger RNA Activity at Various Stages of Purification

The techniques described in the previous section were employed to assay the purity of pregrowth hormone mRNA following dexamethasone stimulation of GC cells and the application of procedures for the fractionation of cells and of RNA. All incubations were performed employing mRNA concentrations in a range (0–30 μg/ml) that was observed (Fig. 5) to yield a linear relationship between mRNA concentration and protein synthesis. The results are shown in Table I. Pregrowth hormone mRNA activity in the cytoplasm of the GC cells was 15% of the total poly (A+) mRNA activity. Incubation of the cells in the presence of dexamethasone increased this value to 32%. Isolation of membrane fraction poly(A+) mRNA from dexamethasone-treated cells further increased this value to 61%. Analysis of this mRNA on a sucrose gradient yielded a single peak of pregrowth hormone mRNA activity, the peak fraction of which yielded a ratio of pregrowth hormone mRNA activity to total mRNA activity of 73%.

The purity of pregrowth hormone mRNA recovered following sucrose-gradient sedimentation as described above was further investigated by SDS-PAGE of the total products of the translation of this RNA in the

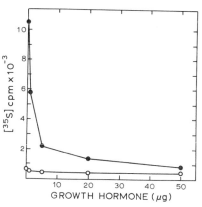

Figure 4. Competition by unlabeled growth hormone with the indirect immunoprecipitation of radioactive pregrowth hormone. Poly(A+) mRNA was prepared from the membrane fraction of GC cells grown in the presence of dexamethasone and incubated at a final RNA concentration of 30 μg/ml in the mRNA-depleted reticulocyte lysate system containing [^{35}S]methionine. Following incubation, aliquots (2.5 μl) received various amounts of cold rat growth hormone. The aliquots were then subjected to indirect immunoprecipitation with fixed *Staphylococcus aureus* plus either baboon antiserum to growth hormone (●) or a preimmune serum from the same animal (○). The immunoprecipitated acid-insoluble radioactivity was then determined as described in Section 2.2.

Table I. Purification of Pregrowth Hormone Messenger RNA[a]

Source of poly (A+) mRNA	Amount mRNA incubated (μg)	Total mRNA activity/5 μl (cpm $\times$ 10^{-3})	Pregrowth hormone mRNA activity/5 μl (cpm $\times$ 10^{-3})	Pregrowth hormone mRNA/total mRNA (%)
None	—	0.28	—	—
Cytoplasm (− dexamethasone)[b]	1.5	69.7	10.7	15
Cytoplasm (+ dexamethasone)[b]	1.5	85.3	27.7	32
Membrane fraction (+ dexamethasone)	3.0	21.7	13.2	61
After sucrose-gradient centrifugation	—[c]	26.8	19.6	73
After gel electrophoresis	—[d]	3.78	3.37	89

[a] Poly(A+) mRNA preparations were incubated in a reaction mixture (50 μl) containing mRNA-depleted reticulocyte lysates plus [³⁵S]methionine (320 μCi/ml, 580 Ci/mmol). Total acid-insoluble radioactivity following KOH treatment (see Section 2.2) in a 5-μl aliquot taken at 0 min was subtracted from the value of this quantity in a 5-μl aliquot taken at 60 min. Total mRNA activity was calculated by subtracting from this value obtained in the same fashion in the absence of exogenous RNA. Aliquots (5 μl), taken at 60 min, were subjected to indirect immunoprecipitation as described in Section 2.2. After addition of 2 μl of cold reticulocyte lysate to the proteins recovered following removal of the bacteria, acid-insoluble radioactivity following KOH treatment was determined. This value was corrected for nonspecific precipitation by subtraction of the value obtained when another 5-μl aliquot was immunoprecipitated using preimmune serum in place of immune serum, to yield a value for pregrowth hormone mRNA activity. Nonspecific precipitation was found to represent 3–10% of the total mRNA activity. Each value represents the average of two separate incubations (except for the RNA recovered following gel electrophoresis) of the RNA preparation, and of duplicate aliquots taken during each incubation.
[b] For these incubations, 872 μCi/ml of [³⁵S]methionine (695 Ci/mmol) was employed.
[c] Following sucrose-gradient centrifugation of 50 μg RNA and location of the fraction containing the maximum pregrowth hormone mRNA activity as described in the Fig. 6 caption, 20% of the RNA recovered from the fraction was incubated.
[d] 10% of the RNA recovered from the gel depicted in Fig. 8 was incubated.

mRNA-depleted reticulocyte lysate system (Fig. 6A). A single major protein product (6.1 cm) plus a small amount of a lower-molecular-weight protein (7.8 cm) were observed. Since both these products were immunoprecipitated by antiserum to growth hormone (Fig. 6B), the material at 7.8 cm in Fig. 6A probably arises from a small amount of premature termination in this translation system. The very small peak at 11.3 cm in Fig. 6A is an artifact at the position of globin, arising from the very high globin concentration in the reticulocyte lysates. No synthesis of globin, or of any other proteins, was detected in this translation system in the absence of exogenous mRNA (Fig. 6C).

The possibility was then investigated that the apparent high degree of

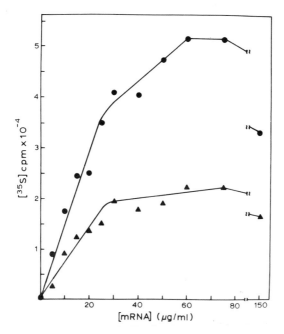

Figure 5. Translation of poly(A+) mRNA prepared from GC cells or reticulocytes in the mRNA-depleted reticulocyte cell-free system. RNA was prepared either from the membrane fraction of GC cells grown in the presence of dexamethasone or from Mg^{2+}-precipitated reticulocyte polysomes (Palmiter, 1974), as described in Section 2.2. Following isolation of poly(A+) mRNA by chromatography on oligo(dT)–cellulose, the mRNA preparations were incubated at the indicated concentrations in the mRNA-depleted reticulocyte lysate system containing [^{35}S]methionine. Aliquots (5 μl) were removed at 0 and 60 min, and the acid-insoluble radioactivity following KOH treatment was determined. The value in the 0-min aliquot was subtracted from the value in the 60-min aliquot to yield each quantity shown. The quantities shown have *not* been corrected by subtraction of endogenous incorporation of radioactivity into acid-insoluble material. (▲) GC cell mRNA; (●) reticulocyte (i.e., globin) mRNA.

purity of sucrose-gradient-purified pregrowth hormone mRNA observed in this experiment might be due to preferential translation by reticulocyte lysates of pregrowth hormone mRNA. Pregrowth hormone mRNA purified in this fashion was incubated in a cell-free system containing wheat germ extract, and the total products were examined by SDS-PAGE (Fig. 7A). Again, a single major protein product was detected. This product was precipitated by antiserum to growth hormone (Fig. 7B), and hence is by definition pregrowth hormone (Sussman *et al.*, 1976). The lower-molecular-weight proteins observed in small amounts in the total products (Fig. 7A) were also detected in the immunoprecipitated products (Fig. 7B). Hence, these proteins are probably the products of premature termination of the

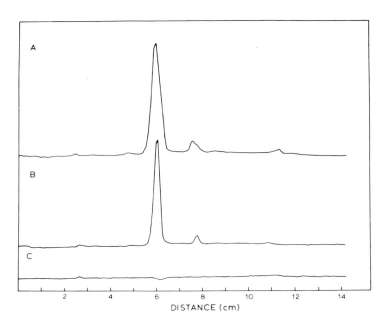

Figure 6. Analysis by SDS-PAGE of the products of translation in mRNA-depleted reticulocyte lysate system of pregrowth hormone mRNA purified by sucrose-gradient ultracentrifugation. Poly(A+) mRNA was prepared as described in Section 2.2 from the membrane fraction of GC cells grown in the presence of dexamethasone, and an aliquot (50 μg) was analyzed by sucrose-gradient ultracentrifugation as described in the Fig. 3 caption. Following fractionation, the fraction containing the maximum pregrowth hormone mRNA activity was located by measurement of the total acid-insoluble radioactivity coded for in the reticulocyte lysate translation system by a small aliquot of the RNA in each fraction. An aliquot (25%) of the RNA in the peak fraction was then incubated in a 50-μl reaction mixture containing mRNA-depleted reticulolcyte lysate plus [^{35}S]methionine. An aliquot (2 μl) of the total reaction mixture was boiled in SDS sample buffer and applied to a slab gel. Another aliquot (2 μl) was subjected to indirect immunoprecipitation, and the total immunoprecipitated material was dissociated by boiling in SDS sample buffer and applied to the gel. In addition, 2 μl of a reaction mixture incubated in the absence of exogenous RNA was boiled in SDS sample buffer and applied to the gel. Following electrophoresis until the marker dye had migrated 14 cm, the dried gel was fluorographed and the individual lanes scanned as described by Harpold et al. (1978). (A) Total products synthesized in the presence of peak fraction mRNA; (B) immunoprecipitate of products synthesized in the presence of peak fraction mRNA; (C) total products synthesized in the absence of exogenous RNA.

translation of pregrowth hormone mRNA in the wheat germ system, similar to what we have reported previously (Sussman et al., 1976). Thus, the apparent high degree of purity of sucrose-gradient-purified pregrowth hormone mRNA in the previous experiment (see Fig. 6) does not appear to be due to a specific preferential translation of pregrowth hormone mRNA by reticulocyte lysates.

The purity of pregrowth hormone mRNA recovered following form-amide–PAGE was assayed as follows: Membrane fraction poly(A+) mRNA from dexamethasone-treated GC cells was analyzed on a sucrose gradient. The three fractions containing the most pregrowth hormone mRNA activity were pooled and subjected to formamide–PAGE. After the gel was scanned at 258 nm, the center portion of the prominent peak observed (Fig. 8) was excised. The RNA was then recovered and incubated in the mRNA-depleted reticulocyte lysate system. SDS-PAGE analysis of

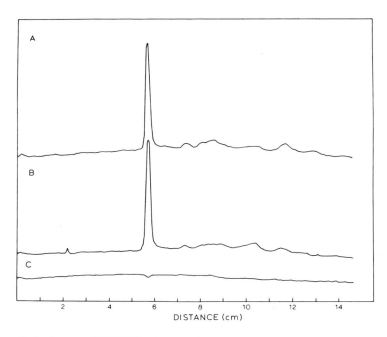

Figure 7. Analysis by SDS-PAGE of the products of translation in the wheat germ system of pregrowth hormone mRNA purified by sucrose-gradient ultracentrifugation. Poly(A+) mRNA (130 μg) was prepared by oligo(dT)–cellulose chromatography of membrane fraction RNA from GC cells grown in the presence of dexamethasone and analyzed on a sucrose gradient as in Fig. 6. The fractions containing the maximum growth hormone mRNA activity were identified by translation of a small aliquot of the RNA in each fraction in the wheat germ translation system, followed by gel analysis of the protein products. RNA from the three peak fractions was pooled, and an aliquot (7%) of this RNA was incubated in a reaction mixture (25 μl) containing wheat germ extracts plus [³H]leucine. An aliquot (2 μl) of the total reaction mixture, the immunoprecipitate of 2 μl of the reaction mixture, and 2 μl of a reaction mixture incubated in the absence of exogenous RNA were analyzed on a slab gel as described in the Fig. 6 caption. (A) Total products synthesized in the presence of the RNA from the pooled peak fractions; (B) immunoprecipitate of products synthesized in the presence of RNA from the pooled peak fractions; (C) total products synthesized in the absence of exogenous RNA.

Figure 8. Further purification of pre-growth hormone mRNA by forma-mide–PAGE. Preparation of poly(A+) mRNA from the membrane fraction of GC cells grown in the presence of dex-amethasone, analysis of 120 μg of the mRNA on a sucrose gradient, and loca-tion of the fractions containing the max-imum pregrowth hormone mRNA activ-ity were performed as described in the Fig. 6 caption. The three peak fractions were pooled, and an aliquot (22%) was analyzed by formamide–PAGE (Tushin-ski *et al.*, 1977). The gel was then scanned at 258 nm, and the segment between the dashed lines was located and excised as described by Tushinski *et al.* (1977). The RNA in this segment was then recovered and translated (see Fig. 9). The apparent broad peak at 7–8 cm is due to an aberration in the gel. The labeled arrows indicate the positions of *E. coli* ribosomal RNA markers analyzed in a parallel gel.

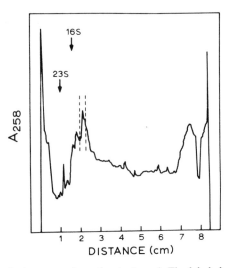

the total translation products of the eluted RNA yielded a single major peak (Fig. 9) that was completely immunoprecipitated by antiserum to growth hormone (Fig. 9C; see the Fig. 9 caption). The small peak at 2.4 cm in Fig. 9A was also seen in Fig. 9B; hence, this material represents the product of a small amount of residual mRNA in the reticulocyte lysates. Since the small peak at 7.1 cm in Fig. 9A was also detected in the immunoprecipitate (Fig. 9C), this material probably represents the product either of nicked pregrowth hormone mRNA or of premature termination of the translation of this mRNA. The small broad peak at 10.2 cm in Fig. 9A is, as described above, an artifact at the position of globin.

The data in Table I show that in the RNA eluted from a formamide–polyacrylamide gel, pregrowth hormone mRNA represented 89% of the total mRNA activity.

4. Conclusions

In analyzing the molecular events involved in the stimulation by a hormone of the expression of a particular gene, a crucial early question to be answered is whether the hormone acts to increase the cellular concen-tration of the mRNA product of this gene. We have described herein our evidence that the synthetic glucocorticoid dexamethasone acts on cultured rat pituitary cells to increase the cytoplasmic concentration of functional pregrowth hormone mRNA. Similar results and conclusions have also been reported recently by other laboratories (Martial *et al.*, 1977; Shapiro *et al.*, 1978). Thus, it seems clear that the glucocorticoid stimulation of growth hormone synthesis in this system occurs at a pretranslational level.

Further analysis of the action of dexamethasone on the expression of the growth hormone gene in this system will involve a determination of the step(s) in the cellular synthesis and metabolism of pregrowth hormone mRNA at which this hormone acts. As a first step in the construction of

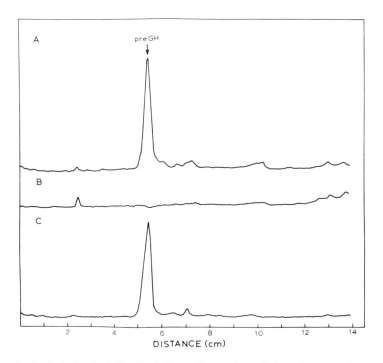

Figure 9. Analysis by SDS-PAGE of the products of translation of pregrowth hormone mRNA recovered from a formamide–polyacrylamide gel. RNA was recovered from the gel segment between the dashed lines in Fig. 8 as described by Longacre and Rutter (1977). An aliquot (10% of the recovered RNA) was then translated in a reaction mixture (25 μl) containing mRNA-depleted reticulocyte lysates plus [^{35}S]methionine. An aliquot (2 μl) of the total reaction mixture, the immunoprecipitate of 2 μl of the reaction mixture, and 2 μl of a reaction mixture incubated in the absence of exogenous RNA were then analyzed on a slab gel, and the gel was fluorographed and scanned, as described in the Fig. 6 caption. The ratio of the area under the major peak in C to the area under the major peak in A was found to be 0.99. (A) Total products synthesized in the presence of the recovered RNA. The arrow shows the position of pregrowth hormone synthesized in the wheat germ translation system, and analyzed in an adjacent lane. (B) Total products synthesized in the absence of exogenous RNA. (C) Immunoprecipitate of products synthesized in the presence of recovered RNA.

a specific cDNA probe to be employed in this analysis, we have purified pregrowth hormone mRNA from the cytoplasm of a clonal strain of rat pituitary tumor (GC) cells.

The major purpose for the purification of pregrowth hormone mRNA was to obtain a purified RNA preparation suitable for use as a template for the synthesis and cloning by recombinant DNA technology of a DNA copy. Since under the conditions employed for cDNA synthesis only mRNA [i.e., poly(A+) RNA] molecules serve as templates, an assay was required that would measure the purity of pregrowth hormone mRNA relative to other GC cell mRNAs at various stages of the purification procedure. The development of such an assay required first a cell-free mRNA translation system that is efficient, reliable, and dependent on exogenous mRNA for protein synthesis. The wheat germ cell-free translation system did not appear suitable, since in our hands this system exhibits variable amounts of both endogenous protein synthesis and premature termination of translation. However, as described in this chapter, we have found that the mRNA-depleted reticulocyte translation system, prepared as described in Section 2.2, is very well suited to this purpose.

The product of the translation of pregrowth hormone mRNA in the presence of an mRNA-depleted reticulocyte lysate is indistinguishable in size from pregrowth hormone synthesized in the wheat germ cell-free translation system (see Fig. 9), and hence is probably the same protein product as we have detected previously in experiments employing the latter translation system (Sussman et al., 1976).

Since the product of translation of pregrowth hormone mRNA in the reticulocyte lysate system is pregrowth hormone, development of an assay for the purity of pregrowth hormone mRNA relative to other mRNAs required the use of an immunoprecipitation technique in which antiserum raised against rat growth hormone could be employed for the quantitative precipitation of rat pregrowth hormone. The observation that about 90% of the radioactive protein coded for by the most highly purified pregrowth hormone mRNA preparation obtained was specifically precipitated when the indirect immunoprecipitation technique described here was employed (Table I) suggests that this technique is nearly quantitative for the immunoprecipitation of pregrowth hormone. When the total cell-free products of the most highly purified pregrowth hormone mRNA preparation and the immunoprecipitate of an equal aliquot of the total products were analyzed by SDS-PAGE, the areas under the resulting pregrowth hormone peaks were equal (Fig. 9). This result shows that the indirect immunoprecipitation technique does in fact quantitatively precipitate pregrowth hormone. These observations suggest that the incomplete precipitation of pregrowth hormone by the direct immunoprecipitation technique employed previously (Tushinski *et al.*, 1977) arose from the rapid formation of the immunoprecipitate, leading to partial exclusion of pregrowth hormone from the precipitate.

Using the assay described above, the purity of pregrowth hormone mRNA at various stages of the purification has been measured (see Table I). The procedure began with GC cells in which pregrowth hormone mRNA activity represents about 15% of total cytoplasmic poly(A+) mRNA activity. The greatest enrichment was achieved by growing the cells in the presence of dexamethasone, and then isolating poly(A+) mRNA from the membrane fraction, each step resulting in about a two-fold enrichment of pregrowth hormone mRNA relative to other mRNAs. The successive application of sizing techniques (sucrose-gradient centrifugation plus formamide–PAGE) produced a further enrichment of pregrowth hormone mRNA, yielding a final preparation in which pregrowth hormone mRNA represented 89% of the total mRNA activity (Table I).

The nature of the assay employed in the studies presented herein is such that it is not possible to estimate directly the achieved enrichment of pregrowth hormone mRNA relative to total cytoplasmic RNA. However, this quantity may be estimated indirectly as follows: No RNA peaks migrating at the expected positions of 18 S or 28 S ribosomal RNA were detected in the A_{258} scan of pregrowth hormone mRNA purified to the formamide–PAGE stage (see Fig. 8), suggesting that the purification procedures preceding this step were successful in eliminating most of the contaminating ribosomal RNA. Furthermore, chromatography on oligo(dT)–cellulose, performed as described in Section 2.2, of total cytoplasmic RNA yields about 3% of the input RNA in the bound fraction (data not shown). Thus, pregrowth hormone mRNA probably represents about 0.45% (15% × 3%) of the total RNA in the cytoplasm of the GC cells. Assuming, as described above, that the final pregrowth hormone mRNA preparation is not contaminated to a significant extent by non-mRNA RNA, the calculated overall enrichment of pregrowth hormone mRNA achieved was about 200-fold (89%/0.45%).

Recently, we have employed pregrowth hormone mRNA purified as described in this chapter to construct a growth-hormone-specific cDNA probe. Pregrowth hormone mRNA purified through the sucrose-gradient step in Table I was used as a template for the synthesis of double-stranded cDNA. The recently developed techniques of recombinant DNA technology were then employed to insert this cDNA into a bacterial plasmid, and to amplify the resulting recombinant plasmid by growth in a suitable bacterial host (Harpold *et al.*, 1978). As a result, we now have available virtually unlimited amounts of growth-hormone-specific cDNA. Furthermore, although the pregrowth hormone mRNA preparation employed as starting material for this procedure was not totally pure, the nature of the cloning process is such that the cloned growth-hormone-specific cDNA is in fact absolutely pure.

We are at present employing this cloned growth hormone-specific cDNA as a probe in investigations of a number of aspects of the hormonal regulation of the growth hormone gene. We are attempting to determine whether the stimulation by dexamethasone of the amount of pregrowth hormone mRNA in the cytoplasm of the GH_3 cells arises from a stimulation of the rate of transcription of the growth hormone gene or from an effect on some step in the processing or degradation by the cells of the initial transcript of this gene. We are also asking similar questions concerning the interesting synergistic stimulatory effects of thyroid and glucocorticoid hormones on pregrowth hormone mRNA concentrations in cultured rat pituitary cells, which have been observed recently (Martial *et al.*, 1977; Shapiro *et al.*, 1978). Finally, since we have found previously that there appears to be a lag of about 6 hr before a significant stimulation by dexamethasone of growth hormone synthesis by the GH_3 cells can be detected (Yu *et al.*, 1977), we are using this probe to investigate further the early events that occur during this apparent lag period.

In conclusion, there are now available clonal strains of rat pituitary cells in which expression of the growth hormone gene is under hormonal control, and virtually unlimited amounts of absolutely pure growth-hormone-specific cDNA. Thus, it is now possible to obtain detailed information of fundamental interest about both the expression in pituitary cells of the growth hormone gene and the mode of action of hormones that regulate this expression.

ACKNOWLEDGMENTS. We are grateful to Dr. Phyllis Sussman-Berger for advice on the use of the wheat germ translation system and of fixed *S. aureus* in the indirect immunoprecipitation of pregrowth hormone. We thank Ms. Lorna Bauerle for excellent technical assistance, and Miss Julia Davis for typing the manuscript. P.R.D. and L.-Y.Y. are Faculty Predoctoral Fellows of Columbia University. This work was supported by Grants GM-21000 and GM-24442 from the National Institutes of Health, and Core Grant CA-08748 from the National Cancer Institute.

REFERENCES

Aviv, H., and Leder, P., 1972, Purification of biologically active globin messenger RNA by chromatography on oligothymidylic acid–cellulose, *Proc. Natl. Acad. Sci. U.S.A.* **69**:1408.

Bancroft, F. C., 1973, Intracellular location of newly synthesized growth hormone, *Exp. Cell Res.* **79**:275.

Bancroft, F. C., Levine, L., and Tashjian, A. H., Jr., 1969, Control of growth hormone production by a clonal strain of rat pituitary cells: Stimulation by hydrocortisone, *J. Cell Biol.* **43**:432.

Bancroft, F. C., Wu, G. -J., and Zubay, G., 1973, Cell-free synthesis of rat growth hormone, *Proc. Natl. Acad. Sci. U.S.A.* **70**:3646.

Evans, G. A., David, D. N., and Rosenfeld, M. G., 1978, Regulation of prolactin and somatotropin mRNAs by thyroliberin, *Proc. Natl. Acad. Sci. U.S.A.* **75**:1294.

Harpold, M. M., Dobner, P. R., Evans, R. M., and Bancroft, F. C., 1978, Construction and identification by positive hybridization–translation of a bacterial plasmid containing a rat growth hormone structural sequence, *Nucleic Acids Res.* **5**:2039.

Kamine, J., and Buchanan, J. M., 1977, Cell-free synthesis of two proteins unique to RNA of transforming virions of Rous sarcoma virus, *Proc. Natl. Acad. Sci. U.S.A.* **74**:2011.

Kessler, S. W., 1975, Rapid isolation of antigens from cells with a Staphylococcal protein A–antibody absorbent: Parameters of the interaction of antibody– antigen complexes with protein A, *J. Immunol.* **115**:1617.

Longacre, S. S., and Rutter, W. J., 1977, Isolation of chicken hemoglobin mRNA and synthesis of complementary DNA, *J. Biol. Chem.* **252**:2742.

Martial, J. A., Seeburg, P. H., Guenzi, D., Goodman, H. M., and Baxter, J. D., 1977, Regulation of growth hormone gene expression: Synergistic effects of thyroid and glucocorticoid hormones, *Proc. Natl. Acad. Sci. U.S.A.* **74**:4293.

Martin, T. F. J., and Tashjian, A. H., Jr., 1977, Cell culture studies of thyrotropin-releasing hormone action, in: *Biochemical Actions of Hormones*, Vol. IV (G. Litwack, ed.), pp. 269–312, Academic Press, New York.

Palmiter, R. D., 1974, Magnesium precipitation of ribonucleoprotein complexes. Expedient techniques for the isolation of undegraded polysomes and messenger ribonucleic acid, *Biochemistry* **13**:3606.

Pelham, H. R. B., and Jackson, R. J., 1976, An efficient mRNA-dependent translation system from reticulocyte lysates, *Eur. J. Biochem.* **67**:247.

Schimke, R. T., Rhoads, R. E., and McKnight, G. S., 1974, Assay of ovalbumin mRNA in reticulocyte lysates, in: *Methods in Enzymology*, Vol. 30F (L. Grossman and K. Moldave, eds.), pp. 694–701, Academic Press, New York.

Shapiro, L. E., Samuels, H. H. and Yaffe, B. M., 1978, Thyroid and glucocorticoid hormones synergistically control growth hormone mRNA in cultured GH_1 cells, *Proc. Natl. Acad. Sci. U.S.A.* **75**:45.

Spielman, L. L., and Bancroft, F. C., 1977, Pregrowth hormone: Evidence for conversion to growth hormone during synthesis on membrane bound polysomes, *Endocrinology* **101**:651.

Sussman, P. M., Tushinski, R. J., and Bancroft, F. C., 1976, Pregrowth hormone: Product of the translation *in vitro* of messenger RNA coding for growth hormone, *Proc. Natl. Acad. Sci. U.S.A.* **73**:29.

Sussman-Berger, P. M., 1978, Discovery and characterization of the precursor of rat growth hormone, Ph.D. thesis, Department of Biological Sciences, Columbia University, New York.

Tashjian, A. H., Jr., Bancroft, F. C., and Levine, L., 1970, Production of both prolactin and growth hormone by clonal strains of rat pituitary tumor cells: Differential effects of hydrocortisone and tissue extracts, *J. Cell Biol.* **47**:61.

Tushinski, R. J., Sussman, P. M., Yu, L.-Y., and Bancroft, F. C., 1977, Pregrowth hormone messenger RNA: Glucocorticoid induction and identification in rat pituitary cells, *Proc. Natl. Acad. Sci. U.S.A.* **74**:2357.

Yamamoto, K. R., and Alberts, B. M., 1976, Steroid receptors: Elements for modulation of eucaryotic transcription, *Annu. Rev. Biochem.* **45**:721.

Yu, L.-Y., Tushinski, R. J., and Bancroft, F. C., 1977, Glucocorticoid induction of growth hormone synthesis in a strain of rat pituitary cells, *J. Biol. Chem.* **252**:3870.

17

Adenohypophyseal Cell Specificities and Gonadal Steroids

Vladimir R. Pantić

The relationship of the adenohypophysis to the central nervous system is characterized by Harris (1948) as that of "a gland under nervous control, but lacking a nerve supply." However, nerve terminals are identified only throughout the pars intermedia (PI). The adult innervation pattern of type A (neurosecretory) and type B (adrenergic) is achieved by the end of the first postnatal week of C_3H mice. Type B terminals formed a synapselike contact with the glandular cells, indicating that the primary innervation is supplied by adrenergic neurons (Jarskar, 1977). Meurling and Björklung (1970) observed that the PI cell body (synthesis pole) and the cell apex (release pole) received innervation from the neurosecretory and catecholamine fibers.

Many papers, monographs, and reviews deal with the cytology and cytochemistry of specific pituitary-cell properties during the development of various species of animals and man (Herlant, 1964; Herlant and Pasteels, 1967; Fawcett *et al.*, 1969; Nakane, 1970; Pantić, 1975), but only some of the results will be mentioned here.

Immunoreactive adrenocorticotropin (ACTH) appeared in the primordial cell outgrowth of Rathke's pouch on the 16th day of the rat fetal life.

Vladimir R. Pantić • Institute for Biological Research "Siniša Stanković," Belgrade, Yugoslavia

This hormone was detected on the polysomes before the appearance of secretory granules by Sétáló and Nakane (1976). They identified prolactin cells in some fetuses at the same stage of development and on subsequent days. However, the onset of thyroid-stimulating hormone (TSH), luteinizing hormone (LH), and growth hormone (GH) synthesis was observed in cells of Rathke's pouch on the 14.5-day stage of embryonic life by Negm (1970). He suggested that the PI cells of the rat pituitary in the perinatal period of development may have the ability to secrete TSH and somatotropin (STH).

The differentiation of the α-melanocyte-stimulating hormone (MSH)-, β-MSH-, ACTH-, α- and β-endorphin-, and γ- and β-lipotropin (LPH)-containing cells in pars distalis (PD) and PI of the fetal rat hypophysis has been reported by Dupouy and Chatelain (1979). They revealed that the differentiation of ACTH-, MSH-, and LPH-producing cells does not need the presence of the fetal hypothalamus or of the upper nervous structure.

Considering the pituitary-cell categorization from a functional standpoint, Herlant (1962, 1964) proposed the histochemical criterion as the more appropriate one. He recognized two cell types in the PD: (1) acidophils or serous cells producing simple proteins [prolactin (PRL), GH, and ACTH]; (2) mucoprotidic cells as basophils producing glycoprotein hormones [TSH, follicle-stimulating hormone (FSH), and LH]. Fawcett *et al.* (1969) pointed out the size and shape of the granules as the most dependable criterion for the recognition of cell types. The development, organization, and nature of the granular endoplasmic reticulum (GER), mitochondria, Golgi complex, and other structures are useful criteria (Tixier-Vidal and Gourdji, 1970; Pantić, 1975).

Having in mind the value of the tinctorial and morphological cell characteristics, Pantić (1974a) considered the pituitary cells of various vertebrates from the point of view of their common cellular features and specific ability of certain cell types to synthesize hormones from the same ancestral molecule. This consideration was based on the fact, established by Li (1972), that the amino acid sequences of pituitary hormones have certain common features overlapping biological activity. The intention was to combine these cell properties with the tinctorial and morphological criteria, expecting them to be useful in the further elucidation of the regulation of specific biochemical events during pituitary cell differentiation, their specificities enabling the production of two or more specific hormones during ontogenesis or after cell retrodifferentiation. Thus, Pantić (1975) proposed that the adenohypophyseal cells, as genetically programmed cells, differentiate from chromophobes into three main groups of specific cells: (1) ACTH and MSH cells; (2) luteotropic hormone (LTH) and STH cells; (3) glycoprotein-hormone-producing cells. To facilitate better understanding of the pituitary-cell-specific reaction to gonadal steroids, some of the cell properties will be mentioned below.

Chromophobes without granules, known as follicular, stellate, or marginal cells, have been described in the vertebrate pituitary (Vila-Porcile, 1972). Most of the chromophobes, so classified by the use of light microscopy, have some specific granules, and it is rather difficult to estimate whether they are the results of accumulation of secretory products or the results of phases of degranulation. Phagocytic activity by stellate cells was observed by Farquhar *et al.* (1975).

The six hormones of the anterior pituitary of male rats were localized using the peroxidase-labeled antibody method: GH, ACTH, LTH, and TSH cells were found in separate cells. FSH and LH were frequently found in the same cell. TSH cells were scarce at the periphery of the gland. The anterior ventral portion of the gland contained few or no GH, ACTH, PRL, or TSH cells, but it was abundant in gonadotropic hormone (GTH) cells. In the area near the intermediate lobe, GH, ACTH, and TSH cells were not found (Nakane, 1970).

Yoshimura *et al.* (1977) identified five types of basophils differentiated as a result of a sequential, ultrastructural transformation of I-type cells that represent immature cells to V-type basophil cells. The working hypothesis of Soji *et al.* (1977) is that various types of basophils may not become independent of one another by modifying their shape and property according to the different phases of the secretory cycle, i.e., the synthesizing, storing, secreting, and resting phases. They suggested possible synergystic or countervailing influences of the releasing hormones on the secretion of the other trophic hormone, which is contradictory to the concept that "one releasing hormone is responsible for one target cell."

During the last two years, pituitary hormones have also been detected outside the adenohypophysis, mainly in the brain neurons. β-LPH localization in the hypothalamic nuclei supports the possibility that the brain β-LPH may be a precursor for the opiatelike molecule and other peptides that may be involved in neuromodulation or neurohormonal activities (Zimmerman *et al.*, 1978). The β-LPH contains a sequence corresponding to MSH with its NH-terminal part identical to the α- and β-endorphins isolated and characterized by Guillemin *et al.* (1976, 1977). These peptides act like opiates (Guillemin, 1977), and their presence is established in fish pituitary (Follénius and Dubois, 1978), in pig neurohypophysis and intermediate lobe (Guillemin *et al.*, 1976), and in other animals as well.

By use of radioimmunoassay or bioassay, ACTH (Krieger *et al.*, 1977), GH (Tan Pacold *et al.*, 1977), LTH (Fuxe *et al.*, 1977), and α-MSH (Dubé and Pelletier, 1977; Pelletier and Dubé, 1977) have been found in brain extracts.

The PI, as a lobe connected with nerve terminals, is dependent on peptides produced by brain neurons, and on hormones of "target" organs as well. During the stages of hypothalamic hyperactivity, the involution of the PI was observed after water deprivation or continuous illumination, or during the course of lactation. The cells lose their vacuoles, they are stained intensely with periodic acid–Schiff (PAS), and the number of basophils increases. Hypoplasia of the PI, in the case of hypothalamic hypoactivity or section of the hypophyseal stalk, was observed as well (Legait, 1963).

Considering all these recent data, there is no doubt that brain neurons are involved not only in the biosynthesis and release of releasing hormones (or factors) and inhibiting hormones (or factors), but also in the production of adenohypophyseal hormones. New and improved methods allow better understanding of the role of brain neurons in the biosynthesis of releasing and inhibiting peptides and pituitary hormones through the developmental and adult stages of animals and man. It has been shown that during early fetal life, brain neurons are able to synthesize hormones such as luteinizing hormone–releasing hormone (LH-RH), e.g., on the 45th day of guinea pig fetal life (Barry and Dubois, 1975), in the hypothalamic nuclei of human fetuses from the 13th gestational week until birth, and so on. Nevertheless, many questions closely connected with the mechanisms involved in the regulation of biosynthesis and release of these hormones have yet to be answered.

2. Aim of Investigation

The main purpose of our investigations was to follow the sensitivity of the adenohypophyseal cells of some teleost fishes and of rats and pigs in responding to signals originating from gonadal steroids administered during embryogenesis, during the so-called "critical" period, gonadal maturation, or adulthood and old age. Attention was paid to the glandular cells, their proliferation, differentiation, ultrastructure, pathways of synthetic and secretory activities, and other characteristics (Fig. 1).

At a time when many scientific programs are directed to obtaining more information concerning cell specificity, we expect that these investigations can contribute to further discussions related to the question whether the term "one cell, one hormone" is acceptable during cell differentiation, activity, or retrodifferentiation. For that reason, and to contribute to further discussion related to pituitary-cell specificities and reaction, an attempt has been made to answer such questions as the following:

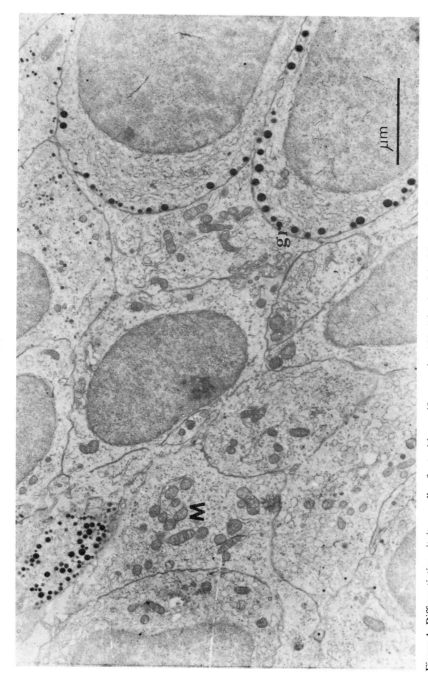

Figure 1. Differentiating pituitary cells of rat with specific granules (gr), mitochondria (M), and other cell organelles appearing in their cytoplasm.

- Among many environmental factors, might gonadal steroids play an important role in influencing the realization of genetically programmed pathways?
- Which cell types or cell groups are the most sensitive to gonadal steroids, and what is the character of their changes at the cellular and biochemical levels?
- What difference is there in the reaction of pituitary cells after treatment of various animal species with a single dose of gonadal steroids, or after gonadectomizing of the newborn animals that lose these hormones?
- Is the reaction of the examined species specific, and what are the similarities and dissimilarities in the response to the same steroid?
- Is there any sex difference in response to gonadal steroids, and if so, at which stage of development and how is it expressed?
- Do PI cells react to the sufficiency and deficiency of gonadal steroids, and if so, how is the reaction expressed?
- Could pretreatment with estrogen or testosterone propionate modulate the sensitivity of pituitary cells to gonadotropin-releasing hormone (GRH)?
- Could gonadal steroids be used as protectors from X-irradiation?
- Is the specific cell reaction dose-dependent?
- Finally, is the effect of these hormones on the pituitary cells expressed directly or via the hypothalamus and other nervous centers?

For these investigations, various doses of estrogen, estrogen and progesterone, or testosterone were administered to males and females of the animal species examined.

3. Effect of Estrogen on Fish Pituitary Cells

Gonadotropic cells of the pituitaries of 15.5-month-old carp treated as mature 12-month-olds with a single dose of estrogen, or estrogen and progesterone, appeared more vacuolized than in the controls. The degree of vacuolation seemed more clearly pronounced in animals treated with estrogen and progesterone. The cytoplasm of gonadotropic cells of the peripheral part of the mesadenohypophysis was more granulated in treated animals than in the controls. The results obtained up to now show that gonadotropic cells of carp reacted to a single dose of gonadal steroids, under our experimental conditions, to a lesser extent than the chicken and the mammals examined (Pantić and Lovren, 1977b).

The hyperplasia of the chromophobes, hypertrophy of LTH cells, and diminution of STH cells occur in adult carp, as well as in *Carassius carassius* kept in aquaria and treated with repeated doses of estrogen and progesterone. The response of LTH cells of male carp to female gonadal steroids was similar to the reaction of this cell type in mammals: GER was circularly oriented, secretion and synthesis of LTH were stimulated (Fig. 2), and light and dark STH cells were present (Pantić and Sekulić, 1978).

4. Long-Term Effect of a Single Dose of Estrogen on Chicken Pituitary Cells

4.1. Administration of Estrogen during Embryogenesis

Examination of the pituitary cells of roosters and hens up to the age of 8.5 months showed that when the animals were treated with a single dose of 500 gammas of estrogen during embryogenesis, a reciprocal relationship was observed between the number of LTH and GTH cells, so that when the number of chromophobes, intermediate (transition stage), and LTH cells increased, the number of GTH cells decreased. In parallel, STH cell differentiation was slowed (Pantić and Škaro, 1974). These and other examinations proved the pituitary cells to be most sensitive to the exogenous estrogen during early embryogenesis.

4.2. Administration of Estrogen after Hatching

In the pituitaries of chickens at the age of 14, 30, or 90 days treated with estrogen on day 4 or 8 after hatching, the reciprocal interrelationship between the gonadotrophs and LTH cells was expressed in the animals treated up to day 8 after hatching (Škaro-Milić and Pantić, 1976, 1977). LTH cells of the pituitaries of 1- and 3-month-old chickens treated in that period with estrogen were increased in number, GER cavities in their cytoplasm were cysternoid, and in some of them a single vacuole appeared, giving an impression of "signet" LTH cells. In the pituitaries of the same animals, small-sized "signet" gonadotrophs were far less numerous than in the corresponding controls.

In the pituitaries of 3-month-old roosters treated with estrogen on day 13 after hatching, no clear signs of a reciprocal interrelationship between LTH and GTH cells were observed, showing that at that age, pituitary cells are less sensitive than in the so-called "critical" period.

A large number of dark cells, like gonadotrophs, observed in chickens, treated with gonadal steroids after hatching, especially on day 4, seems to represent a regressive form of these cells.

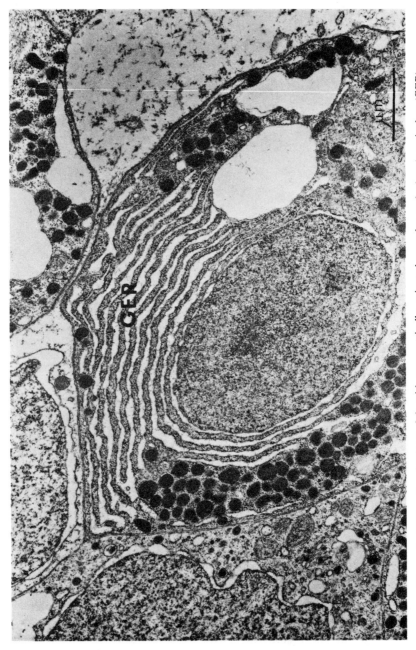

Figure 2. STH cells of teleost fish with concentrically oriented granular endoplasmic reticulum (GER).

5. Pituitary Pars Anterior Cells of Neonatally Estrogen-Treated Animals

5.1. Rats

A prolonged stimulative effect of estrogen administered to newborn rats on LTH synthesis and release was expressed in cell hypertrophy, proliferation of endoplasmic reticulum (ER), increase of corresponding band density, and higher rate of [^{14}C]leucine incorporation. The degranulation of STH cells, the significant slowing of the corresponding hormone band density, and the reduction of [^{14}C]leucine incorporation was considered to be a result of the estrogen effect of decreasing the synthesizing and secretory rate of these cells at the time of sacrifice of the animals. The reaction of both LTH and STH cells was also more pronounced than 4 months after treatment, and was more clearly evident in male than in female rats (Pantić and Genbačev, 1972).

Pantić (1971) considered the proliferation of GER in the LTH and STH cells, clearly visible 15 days and later in neonatally treated male and female rats, as a result of the direct stimulative effect of estrogen (Fig. 3).

The LTH band of pituitary homogenate became more pronounced simultaneously with an increase in the number of hypertrophied LTH cells, but even so it could not be identified in the corresponding control (Pantić and Genbačev, 1969, 1970).

Examinations of the synthesizing and secretory capacity of LTH and STH pituitary cells of rats treated with estrogen on day 10 and sacrificed 15 and 30 days later showed that pituitary LTH and STH cells of 10-day-old rats were still sensitive to estrogen. Clear signs of an increased synthesizing activity were observed as evidenced by the ultrastructural properties, LTH band density, and incorporation of [^{14}C]leucine (Genbačev *et al.*, 1977). However, if the animals were treated on day 15 with estrogen and sacrificed 15 or 30 days later, besides the incorporation of concetrated [^{3}H]estrogen in the pituitary cells of 15-day-old female and male rats, no differences in the synthesis and accumulation of LTH in the pituitaries of rats treated at that time and of corresponding controls were observed.

From our previous results, it was clear that as a consequence of estrogen treatment (Fig. 4), while the number of LTH cells is increased and their activity is stimulated, a deficiency of GTH cells is clearly expressed (Pantić and Genbačev, 1970, 1972). Vigh *et al.* (1978) established that the deficiency of both FSH- and LH-containing cells was severe and specific, being pronounced in males after a single and in females after two repeated, but equal, doses of estrogen. The immunocytochemically reactive FSH- and LH-containing cells were smaller in number and size, appearing as cells accumulating, but not releasing, their hormone content.

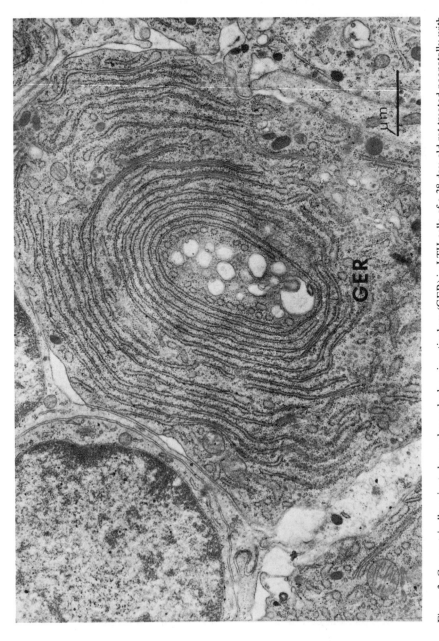

Figure 3. Concentrically oriented granular endoplasmic reticulum (GER) in LTH cells of a 38-day-old rat treated neonatally with estrogen.

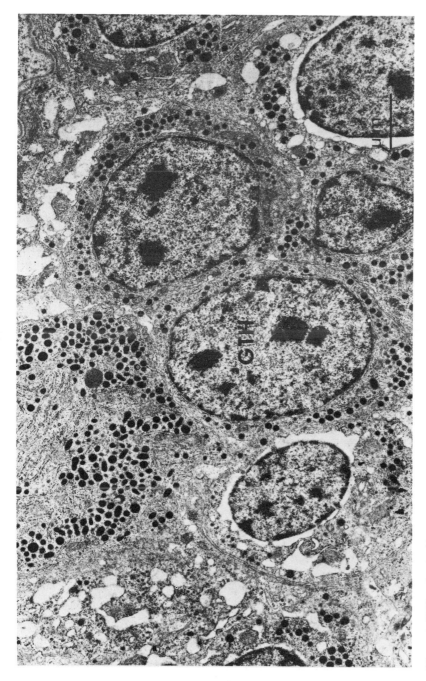

Figure 4. GTH cells of a 38-day-old male rat treated neonatally with estrogen and progesterone, with the specific granules localized in the vicinity of the plasma membrane.

Prolactin cells were increased in number, and some of them were hyper-trophic and filled with immunoreactive granules. This confirmed our previous results that pituitary cells of males are more sensitive to estrogen than those of females (Pantić and Genbačev, 1972; Genbačev and Pantić, 1975; Pantić, 1974b, 1975).

By use of the double-staining method, it was rather difficult to establish differences between the distribution and the number of ACTH and STH cells per unit area in the pituitaries of estrogen-treated and control rats (Fig. 5).

5.2. Pigs

Pituitary GTH cells in 1- to 6-month-old pigs, intact or castrated on the first day of life, or treated on the same day with a single dose of female gonadal steroids, have been examined by Pantić and Gledić (1977b). Piglets were chosen for examination of the pituitary-cell reaction to the androgen deficiency after castration and the sufficiency of gonadal steroids, since Meussy-Dessolle (1975) established a high concentration of testos-terone propionate in the plasma of 15- to 17-day-old piglets and progressive increased testosteronemia in pigs from 4 to 6 months old, appearing analogously to its appearance in man.

As is known, hyperplasia and hypertrophy of GTH cells occur in the pituitaries of all animals gonadectomized on the first day of life and is expressed up to the age of sexual maturity. However, a small number of "signet" GTH cells with mainly slightly hypertrophic Golgi zones were observable in animals aged from 4 to 6 months. We tried to explain these findings as a result of the effect of extragonadal steroids produced in the adrenals (Pantić and Kilibarda, 1978).

6. Pars Intermedia Cells of Estrogen-Treated Animals

Focusing their attention on the interdependence of the cytological characteristics of PI cells and gonadal steroids, Pantić and Šimić (1977a,b) investigated the effects of these hormones administered as a single dose to fishes or repeated doses to neonatal rats, and as a single dose to piglets on the first day of life.

6.1. Carp and Carassius carassius

Studying the differences in PI cell development, organization, and distribution and their sensitivity to the deficiency or sufficiency of gonadal steroids, Pantić and Šimić (1977a) observed that as a result of the effect of

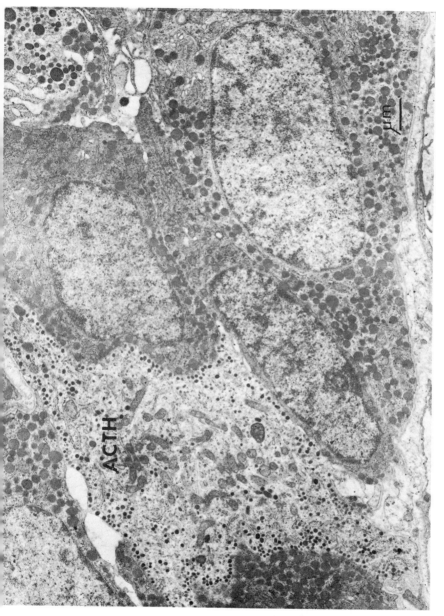

Figure 5. ACTH and dark STH cells in the pituitary of a 38-day-old male rat treated neonatally with estrogen and progesterone.

female gonadal steroids administered before the spawning period to *C. carassius*, the lead hematoxylin cells were enlarged, the volume of their nuclei increased, their cytoplasmic organelles were less developed, and a number of cells appeared swollen. It also seems that PAS+ cells of carp, as the most differentiated cells, are more sensitive than chromophobes to gonadal steroids.

6.2. Rats

Studying the long-term effect of estrogen administered as a single dose or as three repeated doses during the neonatal period of rats, Pantić and Šimić (1977a) established that the size of PI cells of 70-day-old rats was greatly diminished; the number of cells was decreased to such an extent that in some parts only a single cell layer or two cell layers were present. The changes in PI cells observed in fish treated with gonadal steroids in January, and in rats neonatally treated with gonadal steroids, are catabolic in nature and are expressed more strongly in the animals treated with repeated doses than in those treated with single doses, but are clearly pronounced if the rats are treated during the so-called "critical" period. However, the strong reaction of PI cells is as well expressed in the rats treated after day 15 with gonadal steroids. PI cell sensitivity to hypothalamic hormones is suggested, in various papers, as ACTH release from the PI being stimulated by the substance(s) originating from the hypothalamus and the "secretagogue" being neither vasopressin nor the same ACTH-releasing hormone that will stimulate ACTH release from the PD (Kraicer and Morris, 1976).

6.3. Male Pigs

After having established a close interrelationship between the PI development, mating, and hunting of deer (Pantić and Šimić, 1975), as well as the above-described regressive reaction of the pituitary cells of teleosts and rats to gonadal steroids (Pantić and Šimić, 1977a), we attempted to study the specificities of pig PI cells, their sensitivity to the deficiency or sufficiency of gonadal steroids, their specificities in the nature of the response, and the pathway of differentiation when the males were castrated on the first day of life or when they were treated with a single dose of female gonadal steroids. We observed that the PI cells of newborn piglets are sensitive and react specifically to the lack of these steroids that occurs after castration, and that they are also sensitive to the exogenous steroids.

As a result of steroid deficiency during the neonatal period in castrated piglets, an increased number of vacuolated hypertrophic GTH cells appeared, not only inside the gonadotropic region, but also outside this region and in the PI as well (Pantić and Šimić, 1977b) (Fig. 6).

The regression of PI is observable in rats and pigs. However, apart from the fact that both animal species were treated neonatally with gonadal steroids, and pigs were castrated on the first day of life, one must bear in mind that the hypothalamo-pituitary system in rats is sexually undifferentiated, and in piglets differentiation occurs during intrauterine fetal development. In examining the pituitary cells of 1-day-old intact piglets, we established the presence of chromophobes as the most numerous cells and of scarce PAS+ cells. Chromophobes predominated in 1-month-old piglets, PAS+ cells were less numerous, and GTH cells could be observed.

The PI of pigs neonatally treated with gonadal steroids is less developed than in intact and castrated animals.

In the other parts of the PI, chromophobes and PAS+ cells predominate. Numerous LTH cells in this lobe are clear signs of these cells' specific reaction to gonadal steroids. The degree of the regressive changes of PI cells in the pituitaries of 4- to 6-month-old animals treated with female gonadal steroids was most intensely pronounced.

7. Sexually Maturing, Adult, and Old Rats Treated with Estrogen

Adult male intact or gonadectomized rats treated with 50, 100, or 500 μg estradiol dipropionate were sacrificed 15, 30, and 60 days after the onset of the experiment. The results obtained so far have shown that chromophobes are the cells most sensitive to estrogen. Hyperplasia was most pronounced during the first 2 weeks of treatment, and later it was no longer proportional to the dose and duration of the treatment. Describing the synthetic and secretory processes in the large chromophobes and LTH cells, Pantić and Genbačev (1969) pointed out that synthetic processes predominate in most of the cells during the first 2 weeks. Suppressive processes become more evident in the chromophobes and LTH cells with prolongation of the treatment, and they are expressed in the regression of the nucleolus, in the dissolution of the GER and other cytoplasmic components, and in the heterogeneity of the lysosomal bodies. In both intact and gonadectomized rats, estrogen stimulates differentiation of LTH cells and their synthetic capacities.

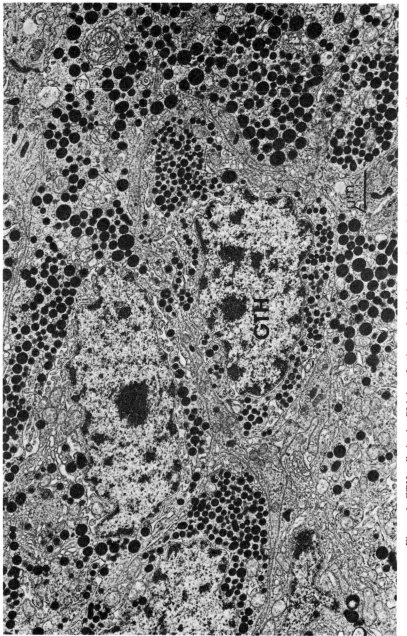

Figure 6. GTH cells in the PI lobe of a 6-month-old pig gonadectomized on the first day of life.

8. Effects of Testosterone Propionate on Pituitary Cells of Newborn, Maturing, Adult, and Old Rats

Pituitary GTH cells of rats treated with a single dose of testosterone propionate in the neonatal period, or in repeatedly treated rats at the time of meiotic division in the first wave of spermatogenesis, showed that GTH-cell differentiation was slowed so that these cells reached the corresponding value for the age of sexual maturity in intact rats 20 days later, i.e., at 90 days of age. The GTH cell volume corresponded to the highest values obtained for these cells in intact 40- and 50-day-old rats. The glands of treated animals were extremely hyperrhemic, and their capillaries were dilated, reaching a diameter as great as 40 μm (Pantić and Gledić, 1977a).

Examining GTH pituitary cells of rats repeatedly treated from day 19 to day 29, Pantić and Gledić (1977a) established that 24 hr after the last injection, the number of these cells was extremely small in comparison to the controls, and they were completely degranulated.

9. Sex-Dependent Reaction of Gonadectomized Rats to Estrogen

To determine whether the sex differences in the reaction of LTH and STH cells to estrogen remain, we examined rats gonadectomized at the age of 30 days, i.e., at the time when, as we know it, the maturation of the adult receptor system for Oe takes place.

The results obtained by Genačev *et al.* (1977) indicate the existence of sex differences in the reaction of LTH cells to gonadectomy, showing that female pituitaries were more sensitive than male ones to the lack of gonadal steroids. LTH synthesis was stimulated much more in the pituitaries of gonadectomized females than in those of male rats, while the inhibition of STH synthesis was more pronounced in estrogen-treated, castrated, 30-day-old rats.

10. Role of Gonadal-Steroid Pretreatment in Modulation of Pituitary-Cell Responses

The secretion of an androgen by infantile testes is responsible for the differentiation of the male type of hypothalamic mechanisms related to gonadotropic secretion, and treatment of newborn castrated male or of normal prepubertal female rats with testosterone propionate produces the tonic male type of gonadotropin secretion in the adult (Barraclough, 1967).

Barraclough and Gorski (1961) suggested that the particular malfunction in androgen-sterilized rats was the failure of the hypothalamus to initiate the ovulatory surge of gonadotropin.

LH cells in the so-called "sex zone" tended to be hypersensitive, their number in males being about twice that in intact animals, but these alterations did not occur after ovariectomy. Testosterone propionate injected into intact newborn animals suppressed numerical development of LH cells, especially in females. These findings showed the involvement of sex steroids in sexual differences in morphological development of LH cells in newborn animals (Matsumura and Daikoku, 1978).

For better understanding of the mechanisms of GRH action, we pretreated rats with gonadal steroids. GTH cells of animals pretreated with testosterone propionate and treated with GRH appeared as degranulated cells. A lot of them lost their staining affinity, so that it was rather difficult to identify them as specific GTH cells. This can partly be explained by the results of Spona (1975a,b) showing that sex steroids affect gonadotropin release induced by LH-RH at the pituitary plasma membrane level and that the mechanisms controlling LTH-induced gonadotropin release are different in the male and female.

11. Properties of Transplantable Pituitary Mammotropic Tumor Cells

Two types of specific cells could not be distinguished as specific cell types in transplantable pituitary tumor cells producing LTH and ACTH. In the cytoplasm of these tumor cells, polysomes predominated, GER and Golgi zones were undeveloped, and secretory granules were greatly reduced in number. An increased synthesis and release of ACTH and LTH, and no ability to accumulate their product in the form of specific granules, is characteristic for these tumor cells (Pantić et al., 1971b) (Fig. 7).

Examination of the ultrastructural properties of pituitary cells and their ability to synthesize, accumulate, and secrete LTH, STH, and ACTH in mammotropic-tumor-bearing rats at different times after transplantation (Pantić et al., 1971a) established that in rats bearing mammotropic tumors for 20 days, chromophobic cells were as a single cell or in groups, and in the vicinity of the PI, they were polarized, forming small follicles with microvilli on the apical cell surface. Signs of the inhibition of synthetic activity were visible in LTH cells during the first 3 weeks of tumor growth, and the hormone content was significantly decreased and nearly undetectable 2 months after tumor transplantation. Besides the variability in the number of STH cells and in the degree of granularity of their cytoplasm,

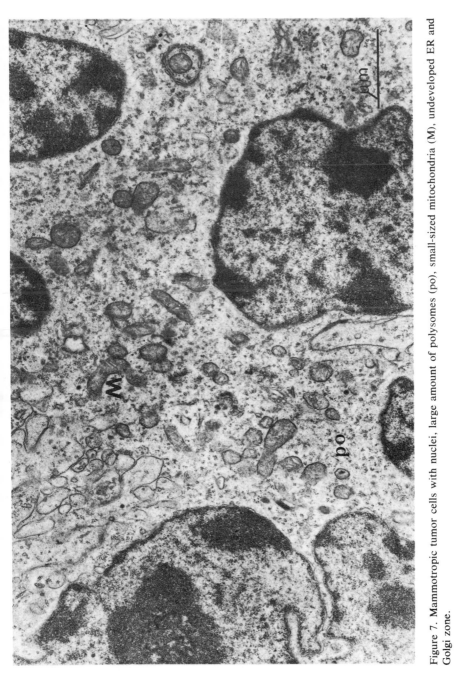

Figure 7. Mammotropic tumor cells with nuclei, large amount of polysomes (po), small-sized mitochondria (M), undeveloped ER and Golgi zone.

the tendency for degranulation and decrease of band density was expressed, so that in rats with older tumors, the STH band was diffused. Some ACTH cells and a reduced number of gonadectomy cells were observed in animals bearing tumors for 10 and 20 days. However, from 38 days on, GTH cells appeared as gonadectomy cells.

12. How Do Cells Evolving as Producers of the Same Ancestral Molecule React to Gonadal Steroids?

The results obtained up to now undoubtedly show that a single dose or repeated doses of estrogen, estrogen and progesterone, or testosterone stimulate the mitotic rate of chromophobes and their differentiation into LTH cells (Fig. 8). An increase in the number of LTH cells is followed by a higher value of their synthesizing and secretory capacity observed in all the species examined and under our experimental conditions.

Similar or the same structural properties of hormones were established in species that are phylogenetically apart for LTH (Emart and Wilhelmi, 1968) and for STH (Farmer et al., 1976). Very few differences between LTH and STH cells were observed by Follénius (1968). By cytological criteria, both cells have similar characteristics, such as eosinophilia, highly developed GER, and accumulation of lipoproteins and hormone. In the largest specific granules, their intracellular events are involved in the regulation of the biosynthesis of two hormones as a single polypeptide chain (Fig. 9); these cells could clearly be identified as being of two types. However, both these types of cells produce hormones evolving from the same common ancestral, prolactinlike molecule (Li, 1972). These and other data encourage us (Pantić, 1974b, 1975) to study the similarities and differences in the reaction of both cell types to gonadal steroids.

Considering LTH and STH cells as one cell group, we could add some more data closely related to the STH cell reaction. In carp and C. carassius treated with repeated doses of estrogen, LTH cells were stimulated, STH cells were smaller in size than in the corresponding controls, and the effect was more clearly expressed if the animals were treated with estrogen and progesterone in combination (Pantić and Sekulić, 1978). It seems to us that transformation of STH into LTH cells occurs, and that both hormones might be synthesized in the same cell type. The changes observed in these cells in the pituitaries of roosters treated with a single dose during embryogenesis or after hatching (Pantić and Škaro, 1974) and in rats and piglets (Pantić and Gledić, 1977b), such as smaller-sized cells, the development of GER, diminished granularity, decreased amount of STH in both cells and medium, and other changes, are undoubtedly clear signs of an interdependence of these two cell types in response to gonadal steroids, showing that LTH cells are stimulated at the expense of STH-cell activities.

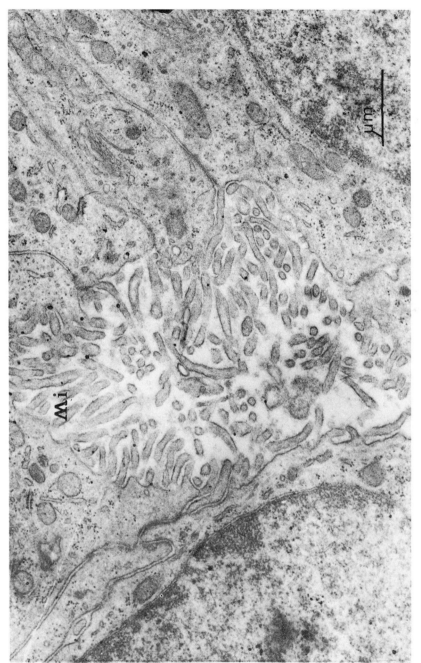

Figure 8. Pituitary cells of a 38-day-old male rat with microvilli (Mi) on the apical part, forming a follicle-like structure.

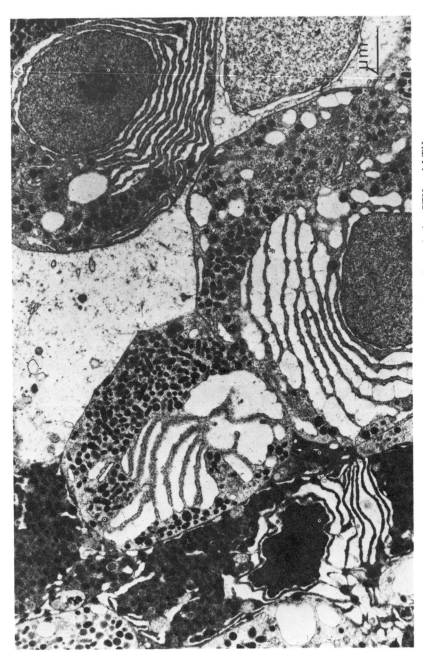

Figure 9. Subcellular organization of fish granular cells producing STH and LTH.

There exists interdependence in the reaction to gonadal steroids in the second group of cells producing glycoprotein hormones, but it is far less pronounced: while a deficiency of GTH cells is expressed clearly in males treated with a single dose and females treated with two doses of estrogen, TSH cells could still be clearly identified (Pantić and Gledić, 1979; Vigh *et al.*, 1978). A decreased number of both GTH and, to a lesser extent, TSH cells was observed in adult and old rats chronically treated with testosterone propionate. However, to answer these issues, further studies are in progress.

Since the changes in PI cells of deer during the hunting and mating periods have been observed from the point of view of the stressogenic reaction of these animals, Pantić and Šimić (1975) placed more emphasis on PI-cell reaction to gonadal steroids. MSH and ACTH cells predominate in this lobe in most animals (Pavlović and Pantić, 1975; Pantić *et al.*, 1978). Simultaneously, an inhibition of spermatogenesis, metaplasia, and other changes were clearly expressed in fish (Pantić and Lovren, 1977a), roosters (Pantić and Kosanović, 1973), and rats and pigs (Pantić and Gledić, 1977b, 1978). Nevertheless, the differentiation of these genetically programmed cells seems to be independent of the gonadal steroids. There are data in favor of this assumption; for example, with the lack of androgen in neonatally castrated piglets, GTH cells differentiate, and the size of this lobe is diminished in rats treated with estrogen (Pantić and Šimić, 1977a) (Fig. 10).

13. General Conclusions

The administration of a large dose of female gonadal steroids during chick embryogenesis, during the critical period of the rat's development, or to the neonatal pig has the following mainly long-term effects on the anterior pituitary cells: (1) *stimulative,* increasing glandular cell proliferation expressed as hyperplasia of chromophobes and LTH with a prolonged rate of synthesizing and secretory capacity for PRL. Stimulation of LTH production in the pituitary cells of all the examined animals treated with gonadal steroids occurs at the expense of STH-cell activities; (2) *inhibitory,* on the differentiation of GTH cells, expressed in the paucity of both FSH- and LH-containing cells and permanent alteration of testicular development. The effect of gonadal steroids is obviously expressed in all the animal species examined, and in the adults chronically treated, but the reaction with long-term consequences was most clearly pronounced when the animals were treated during embryogenesis or the "critical" period.

Differentiating PI cells are sensitive to the lack or sufficiency of gonadal steroids altering the programmed pathway. As a result of the

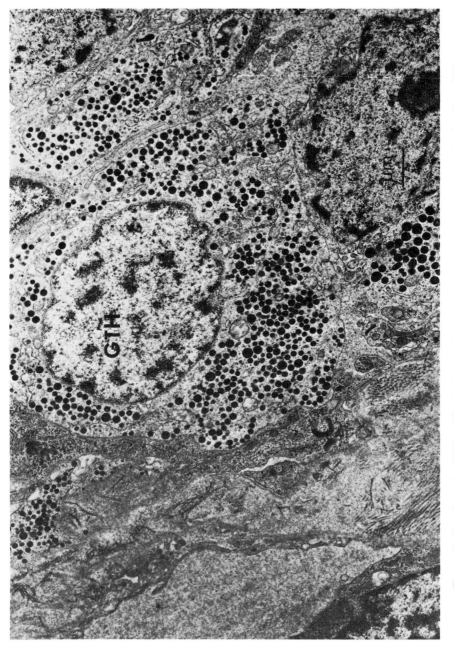

Figure 10. Gonadotropic (GTH) cell in the PI lobe of a 6-month-old pig gonadectomized on the first day of life.

androgen deficiency that arises in growing pigs gonadectomized as newborns, GTH cells differentiated, being numerous in the cranial part of this lobe. The appearance of these cells as hypertrophic and their structural properties are the same as in the pars anterior, and a small number of "signet" cells in 6-month-old pigs is undoubtedly a result of extragonadal production of gonadal steroids, e.g., in the adrenal cortex, and elsewhere.

The character of the reaction of pituitary cells to sufficiency of gonadal steroids in the neonatal period of life is similar in all the species of animals examined, but the degree of cell response is less expressed in fish than in chickens, rats, and pigs.

The intracellular mechanisms controlling the glandular cells' capacities for synthesis, accumulation, and release are affected by gonadal steroids, altering subcellular organization of cell organelles, especially ER, Golgi, and specific granules, to such an extent that the cells lose their specificities for production of one hormone. Thus, the concept "one cell, one hormone" has to be verified.

The administration of gonadal steroids as a single dose, during embryogenesis, or into neonatal animals provides a very convenient model for further elucidation of the mechanisms of action of hormones on the adenohypophyseal cells, brain neurons, and "target" cells.

Discussion

TIXIER-VIDAL: What is your evidence for a transformation of somatotrophic cells into prolactin cells in normal pituitaries?

PANTIĆ: This is based on the somatotropic cell staining affinity and subcellular organization of cell organelles, especially on the variability and properties of the granules in the cells of animals treated with estrogen.

DENEF: Neonatal androgen treatment of female rats has been shown not to change plasma levels of FSH in adult life. In contrast, your studies showed a decrease in the number of immunochemically identifiable gonadotrophs. Could you comment on this? Have you measured plasma gonadotropin levels in the neonatally treated animals?

PANTIĆ: Our presentation is based mainly on the effect of androgen on the gonadotropic cells in the pituitary of male animals. Their number was decreased in all animals examined, i.e., in males neonatally treated with single doses of testosterone, in rats repeatedly treated from day 19 to day 29 of life, and in the same species of animals chronically treated as adults. We have not measured gonadotropins in the plasma in the neonatally treated animals.

SPICER: You mentioned a change in chromophobes in response to estradiol treatment. Chromophobes have often been found to contain immunocytochemically demonstrable hormone, and so it is a question whether the chromophobes that increase after estradiol represent a functionally definable cell type.

PANTIĆ: Both chromophobes and prolactin cells are increased in number and size. However, so-called "chromophobes," after their response to estrogen treatment, represent various stages of prolactin-cell differentiation rather than a specific cell type.

REFERENCES

Barraclough, C. A., 1967, Modifications in reproductive function after exposure to hormones during the prenatal and early postnatal period, *Neuroendocrinology* 2:61–99.

Barraclough, C. A., and Gorski, R. A., 1961, Evidence that the hypothalamus is responsible for androgen induced sterility in the female rat, *Endocrinology* 68:68–79.

Barry, J., and Dubois, M. P., 1975, Immunofluorescence study of LRF neurons in primate, *Cell Tissue Res.* 164:163–178.

Dubé, J., and Pelletier, G., 1977, Immunohistochemical localization of alpha-melanocyte stimulating hormone (α-MSH) in the rat brain, *Endocrinology* 100:A319.

Dupouy, J. P. and Chatelain, A., 1979, Differentiation of βMSH, αMSH, ACTH, α and β endorphin, γ, βLPH containing cells in the pars distalis and pars intermedia of the fetal rat hypophysis immunocytological study, in: *Proceedings of the International Symposium on Neuroendocrine Regulatory Mechanisms*, pp. 219–229, Serbian Academy of Science and Arts, Belgrade.

Emart, E. W., and Wilhelmi, E. A., 1968, Immunochemical studies with prolactin-like fractions of fish pituitaries, *Gen. Comp. Endocrinol.* 11:515–527.

Farmer, S. W., Parkoff, H., Hayashidat, Bewley, T. A., and Li, C. H., 1976, Purification and properties of Teleost growth hormone, *Gen. Comp. Endocrinol.* 30:91–100.

Farquhar, M. G., Reid, J. J., and Daniell, L. W., 1975, Intracellular transport and packaging of prolactin: Quantitative electron microscope autoradiographic study of mammotrophs dissociated from rat pituitary, *Endocrinology* 102:296.

Fawcett, D. W., Long, J. A., and Jones, A. L., 1969, The ultrastructure of endocrine glands, *Recent Prog. Horm. Res.* 25:315–380.

Follénius, E., 1968, Analyse de la structure fine chez differents types de cellules hypophysaires des poissons Teleosteens, *Pathol. Biol.* 16:619–632.

Follénius, E., and Dubois, M., 1978, Immunocytological detection and localization of a peptide reacting with an α-endorphin antiserum in the corticotropic and melanotropic cells of the trout pituitary (*Salmo irideus* Gibb), *Cell Tissue Res.* 188:273–283.

Fuxe, F., Hökfelt, T., Eneroth, P., Gustafsson, J. A., and Skett, P., 1977, Prolactin-like immunoreactivity: Localization in nerve terminals of rat hypothalamus, *Science* 196:899–900.

Genbačev, O., and Pantić, V., 1975, Pituitary cell activities in gonadectomized rats treated with oestrogen, *Cell Tissue Res.* 157:273–282.

Genbačev, O., Pantić, V., and Ratković, M., 1977, Activities of luteotropic (LTH) and somatotropic (STH) cells in the pituitaries of rats treated with oestrogen on day 10 or 15 after birth, in: *Problems in Comparative and Experimental Morphology and Embryology*, pp. 149–155, Bulgarian Academy of Science, Sofia.

Guillemin, R., 1977, Endorphins, brain peptides that act like opiates, *N. Engl. J. Med.* 296:226–228.

Guillemin, R., Ling, N., and Burgus, R., 1976, Endorphines, peptides d'origine hypothalamique et neurohypophysaire a activité morphinomimétique: Isolement et structure moléculaire de l'α-endorphine, *C. R. Acad. Sci. Ser. D* 282:783–785.

Guillemin, R., Vargo, T., Rossier, J., Monick, S., Ling, N., Rivier, C., Vale, W., and Bloom, F., 1977, β-Endorphins and adrenocorticotropin are secreted concomitantly by the pituitary gland, *Science* 197:1367–1369.

Harris, G. W., 1948, Electrical stimulation of hypothalamus and the mechanism of neural control of the adenohypophysis, *J. Physiol. (London)* 107:418–429.

Herlant, M., 1962, Quelques notions récentes sul l'histophysiologie de l'hypophyse, *Biol. Med. (Paris)* 51:205–222.

Herlant, M., 1964, The cells of the adenohypophysis and their functional significance, *Int. Rev. Cytol.* 17:299–382.

Herlant, M., and Pasteels, J. L., 1967, Histophysiology of human anterior pituitary, *Methods Achiev. Exp. Pathol.* **3**:250–305.

Jarskar, R., 1977, Electron microscopical study on the development of the nerve supply of the pituitary pars intermedia of the mouse, *Cell Tissue Res.* **184**:121–132.

Kraicer, J., and Morris, A. R., 1976, In vitro release of ACTH from dispersed rat pars intermedia cells. I. Effect of secretagogues, *Neuroendocrinology* **20**:79–96.

Krieger, D. T., Liotta, A., and Brownstein, M.J., 1977, Presence of corticotropin in brain of normal and hypophysectomized rats, *Proc. Natl. Acad. Sci. U.S.A.* **74**:648–652.

Legait, E., 1963, *Cytophysiologie du Lobe Intermediaire de l'Hypophyse de Mammiferes*, pp. 91–105, Editions du CNRS, Paris.

Li, H. C., 1972, Hormones of the adenohypophysis, *Proc. Am. Philos. Soc.* **116**:365–382.

Matsumura, H., and Daikoku, S., 1978, Quantitative observations of the effect of sex-steroids on the postnatal development of LH cells, *Cell Tissue Res.* **188**:491–496.

Meurling, P., and Björklund, A., 1970, The arrangement of neurosecretory and catechol-amine fibers in relation to the pituitary intermedia cells of the skate, *Raja radiata, Z. Zellforsch.* **108**:81–92.

Meussy-Dessole, N., 1975, Variation quantitative da la testosterone plasmatique chez le porc male, de la naissance a l'age adult, *C. R. Acad. Sci.* **281**:1875–1878.

Nakane, P. K., 1970, Classifications of anterior pituitary cell types with immunoenzyme histochemistry, *J. Histochem. Cytochem.* **18**:9–20.

Negm, I. M., 1970, Development of the intermediate lobe of the rat pituitary and differentiation of its cells, *Acta Anat.* **77**:422–437.

Pantić, V., 1971, Ultrastructure of the endocrine glandular cells and protein synthesis, *Bull. T. CCXXXI Acad. Serbe Sci. Arts* **24**:273–283.

Pantić, V., 1974a, The cytophysiology of thyroid cells, *Int. Rev. Cytol.* **38**:153–243.

Pantić, V. R., 1974b, Gonadal steroids and hypothalamo–pituitary–gonadal axis, *INSERM* **32**:97–118.

Pantić, V. R., 1975, The specificity of pituitary cells and regulation of their activities, *Int. Rev. Cytol.* **40**:153–195.

Pantić, V., and Genbačev, O., 1969, Ultrastructure of pituitary lactotropic cells of oestrogen treated male rats, *Z. Zellforsch.* **95**:280.

Pantić, V., and Genačev, O., 1970, Ultrastructure of pituitary luteotropic (LTH) and somatotropic (STH) cells of rats neonatally treated with oestrogen, *Septieme Congress International Microscopie Electronique* (Grenoble), pp. 569–570.

Pantić, V., and Genbačev, O., 1972, Pituitaries of rats neonatally treated with oestrogen. I. Luteotropic and somatotropic cells and hormone content, *Z. Zellforsch.* **126**:41–52.

Pantić, V., and Gledić, D., 1977a, Reaction of pituitary gonadotropic cells and testes to testosterone propionate (TP), *Bull. T. XVI Acad. Serbe Sci. Arts* **15**:132–146.

Pantić, V., and Gledić, D., 1977b, Long term effect of gonadal steroids on pig pituitary gonadotropic cells and testes, *Bull. T. LX Acad. Serbe Sci. Arts* **16**:67–80.

Pantić, V., and Gledić, D., 1978, Gonadotropic and germ cells reaction to oestrogen, in: *Proceedings of the International Symposium on Neuroendocrine Regulatory Mechanisms*, pp. 87–94, Serbian Academy of Science and Arts, Belgrade.

Pantić, V., and Kilibarda, M., 1978, Adrenal cortex of rats treated with oestradiol neonatally and during sexual maturity, *Acta Vet. (Belgrade)* **28**:1–16.

Pantić, V., and Kosanović, M., 1973, Testes of roosters treated with a single dose of oestradiol dipropionate, *Gen. Comp. Endocrinol.* **21**:108–117.

Pantić, V., and Lovren, M., 1977a, The effects of female gonadal steroids on carp pituitary gonadotropic cells and oogenesis, *Folia Anat. Iugoslav.* **7**:25–34.

Pantić, V., and Lovren, M., 1977b, Examination of testes of *Carassius carassius* treated with choriogonadotropin or female sexual steroids, *Folia Anat. Iugoslav.* **6**:73.

Pantić, V., and Sekulić, M., 1978, Pituitary prolactin and somatotropic cells of teleostea treated with gonadal steroids or choriogonadotropin, *Acta Vet.* **28**(2):71–80.

Pantić, V., and Šimić, M., 1975, Pars intermedia of the deer pituitary, *Arch. Sci. Biol.* **26**(1–2):15–18.

Pantić, V., and Šimić, M., 1977a, Effect of gonadal steroids on pituitary pars intermedia cells of some teleostea and rat, *Bull. T. LX Acad. Serbe Sci. Arts* **16**:17–40.

Pantić, V., and Šimić, M., 1977b, Sensitivity of the pituitary pars intermedia to castration or gonadal steroids, *Bull. T. LXI Acad. Serbe Sci. Arts* **16**:67–80.

Pantić, V., and Škaro, A., 1974, Pituitary cells of roosters and hens treated with a single dose of oestrogen during embryogenesis or after hatching, *Cytobiologie* **9**(1):72–83.

Pantić, V., Genbačev, O., Milković, S., and Ožegović, B., 1971a, Pituitaries of rats bearing transplantable MtT mammotropic tumor, *J. Microsc.* **3**:405–415.

Pantić, V., Ožegović, B., Genbačev, O., and Milković, S., 1971b, Ultrastructure of transplantable pituitary tumor cells producing luteotropic and adrenocorticotropic hormones, *J. Microsc.* **12**:225–232.

Pantić, V., Sekulić, M., Lovren, M., and Šimić, M., 1978, Neurosecretory and pars intermedia cells of fish and mammals, in: *Neurosecretion and Neuroendocrine Activity Evolution, Structure and Function* (W. Bargmann, A. Oksche, A. Polenov, and B. Scharrer, eds.), pp. 257–259, Springer-Verlag, Berlin.

Pavlović, M., and Pantić, V., 1975, The adenohypophysis in the Teleostea *Alburnus albidus* and *Alosa fallax* in different phase of sexual cycle, *Acta Vet.* **25**(4):163–178.

Pelletier, G., and Dubé, D., 1977, Electron microscopic immunohistochemical localization of α-MSH in the rat brain, *Am. J. Anat.* **150**:201–206.

Sétáló, G., and Nakane, P. K., 1976, Functional differentiation of the fetal anterior pituitary cells in the rat, *Endocrinol. Exp.* **10**:155–166.

Škaro-Milić, A., and Pantić, V., 1976, Gonadotropic and luteotropic cells in chickens treated with oestrogen after hatching, *Gen. Comp. Endocrinol.* **28**:283–291.

Škaro-Milić, A., and Pantić, V., 1977, Pituitary cells of roosters treated with a single dose of testosterone propionate (TP) after hatching, *Folia Anat. Iugoslav.* **6**:33–44.

Soji, T., Sato, S., Shishiba, Y., Igarashi, M., Shioda, T., and Yoshimura, F., 1977, Chronic effect of TRH and LRH upon a series of basophils along with the serum and pituitary TSH, LH and FSH concentration, *Endocrinol. Jpn.* **24**(1):19–39.

Spona, J., 1975a, Some structural requirements for LH-RH actions, *Endocrinol. Exp.* **9**:159–165.

Spona, J., 1975b, Sex steroids influence LH-RH–receptor interaction, *Endocrinol. Exp.* **9**:167–176.

Tan Pacold, S., Hojvat, S., Kirsteins, L., Jarzagary, L., Kisla, J., and Lawrence, A. M., 1977, Brain growth hormone: Evidence for the presence of production of biologically active GH-like immunoreactivity from the amygdaloid nucleus, *Clin. Res.* **25**:299A.

Tixier-Vidal, A., and Gourdji, D., 1970, Synthesis and renewal of proteins in duck anterior hypophysis in organ culture, *J. Cell Biol.* **46**:130–136.

Vigh, S., Sétáló, G., Török, A., Pantić, V., Flerkö, B., and Gledić, D., 1978, Deficiency of FSH and LH cells in rats treated with oestradiol in the early postnatal life, *Bull. T. LXI Acad. Serbe Sci. Arts* **17**:1–8.

Vila-Porcile, E., 1972, La reseau des cellule folliculo-stellaires et les follicules de l'adenopypophyse du rat (pars distalis), *Z. Zellforsch.* **129**:328–369.

Yoshimura, F., Soji, T., Kumagai, T., and Yokoyama, M., 1977, Secretory cycle of the pituitary basophils and its morphological evidence, *Endocrinol. Jpn.* **24**:185.

Zimmerman, E. A., Liotta, A., and Krieger, D.T., 1978, β-Lipotropin in brain: Localization in the hypothalamic neurone by immunoperoxidase technique, *Cell Tissue Res.* **186**:393–398.

18

Development of a Photoaffinity Probe for Adrenocorticotropin Receptors

J. Ramachandran, Eleanor Canova-Davis, and Catherine Behrens

1. Introduction

Despite the enormous interest in the study of polypeptide hormone receptors, progress has been slow owing to the extremely small concentrations of receptors in target tissues. While some significant progress has been made in the case of peptide hormones that are acidic in character, such as human chorionic gonadotropin (Dufau *et al.*, 1975) and prolactin (Shiu and Friesen, 1974), the identification of the specific receptors of the basic polypeptide hormones has proved to be much more difficult. Peptide hormones, especially those with isoelectric points above pH 8 [e.g., adrenocorticotropin (ACTH), melanocyte-stimulating hormone (MSH), glucagon], display a strong tendency to bind to a variety of inert materials as well as nonreceptor components of the target tissue. Since such binding appears specific by the criteria generally employed to define receptors and since the number of such nonreceptor sites greatly exceeds that of specific receptors (Cuatrecasas *et al.*, 1975), the task of detecting and characterizing the physiologically relevant receptor is a formidable one. Fractionation of the plasma-membrane components of the target cell on the basis of binding of the radioactive hormone may result in the isolation of nonre-

J. Ramachandran, Eleanor Canova-Davis, and *Catherine Behrens* • Hormone Research Laboratory, University of California, San Francisco, California 94143

ceptor components that may display high affinity for the hormone. Other approaches to the identification of the specific receptor are therefore necessary.

The method of affinity labeling of the receptor under physiological conditions holds considerable promise. A reactive group is introduced into the hormone in such a manner that the affinity of the hormone for the receptor is not diminished significantly. The modified hormone can then be attached covalently to the receptor provided a functionality of appropriate chemical reactivity is present at the hormone-binding region of the receptor. There are, however, two major drawbacks to conventional chemical-affinity labeling, namely, the requirement for the reactive group not to be destroyed by water or by reaction with other nucleophils on the polypeptide hormone prior to reaching the receptor, and second, the requirement of a nucleophilic group at the hormone-binding region of the receptor. Both these limitations can be overcome by the use of a photogenerated species for labeling the binding site. In this approach, a chemically inert but photochemically labile group is introduced into the hormone or other ligand. Irradiation of the hormone–receptor complex at the appropriate time results in the formation of a very reactive species that reacts indiscriminately with whatever chemical groups are present at the binding region of the receptor. The principles of photoaffinity labeling and its application to a number of biological systems are discussed in an excellent review by Bayley and Knowles (1977).

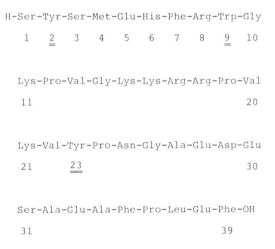

Figure 1. Structure of human adrenocorticotropin. There are two tyrosine residues, at positions 2 and 23, and a single tryptophan residue at position 9.

For the identification of the receptor molecule by the method of photoaffinity labeling, it is necessary to perform two modifications on the polypeptide hormone. A suitable radioactive label as well as a photoreactive group must be introduced into the hormone molecule without significantly altering the affinity of the hormone for its physiological receptor. In this chapter, we describe our work on the modification of the adenohypophyseal hormone ACTH (Fig. 1) to produce a photoaffinity probe for its receptors. In the first part, the preparation of specifically tritiated ACTH of high specific radioactivity and full biological activity is described. The introduction of the photoreactive group and characterization of the modified hormone are discussed in the second section. Binding of tritiated ACTH to isolated rat adipocytes and photoaffinity labeling of ACTH receptors on rat adipocyte plasma membranes are presented in the final section.

2. Preparation of Tritiated Adrenocorticotropin

2.1. Choice of Radioactive Label

In most studies of hormone–receptor interactions, ^{125}I-labeled ligands are used because of the high specific activity of this isotope. ^{125}I-labeled ACTH peptides were employed to investigate the binding to extracts of a mouse adrenal tumor (Lefkowitz *et al.*, 1970) and to isolated rat adrenal cells (McIlhinney and Schulster, 1975) as well as rat adrenal homogenate fractions (Ontjes *et al.*, 1977). Our early studies of structure–activity relationships of ACTH indicated that iodination of ACTH with I_3^- resulted in a marked reduction of biological activity (Ramachandran and Cervantes, unpublished observations). Complete loss of biological activity after iodination with chloramine-T has also been reported (Greenwood *et al.*, 1963; Landon *et al.*, 1967; Rees *et al.*, 1971). More recently, preparation of [^{125}I]-ACTH with 50% of the steroidogenic activity has been reported by McIlhinney and Schulster (1974). The loss of activity following iodination was attributed to oxidation of the methionine residue in the molecule, and regeneration of the methionine residue by reduction with cysteine has been reported (Rae and Schimmer, 1974). Recent studies of synthetic human ACTH analogues containing 3,5-diiodotyrosine residues in place of tyrosine have shown that introduction of the iodine atoms into the tyrosine residue at position 2 causes a 98% loss of biological activity (Lemaire *et al.*, 1977). Similar results were reported earlier by Lowry *et al.* (1973), who studied iodo derivatives of ACTH[1-24]. McIlhinney and Schulster (1974) also found that iodination of ACTH peptides by the milder lactoperoxidase method resulted in the predominant modification of Tyr^2. In view of all this, we chose to employ tritium as the radioactive label for ACTH. We felt that this would provide a labeled hormone of sufficient radioactivity with full retention of biological activity.

2.2. Introduction of Tritium into Adrenocorticotropin

Specifically tritiated ACTH was prepared by catalytic dehalogenation of 3,5-diiodo-Tyr^{23} ACTH$^{1-39}$ or 3,5-diiodo-Tyr^{2,23} ACTH$^{1-39}$ (Lemaire et al., 1977) in the presence of tritium (Ramachandran and Behrens, 1977). Introduction of tritium into the 3 and 5 positions of the phenolic ring of tyrosine residues of peptides by catalytic dehalogenation has been employed to prepare tritiated angiotensin (Morgat et al., 1970a), oxytocin (Morgat et al., 1970b), and vasopressin (Pradelles et al., 1972). Brundish and Wade (1973) reported the preparation of tritiated ACTH$^{1-24}$ by this procedure.

Catalytic dehalogenation of 3,5-diiodo-Tyr^{23} ACTH$^{1-39}$ in the presence of palladium oxide and tritium gas in 0.1 N acetic acid resulted in quantitative removal of iodine. The product was fully active and identical to synthetic human ACTH, but the specific radioactivity was only 5–8 Ci/mmol, or approximately 10% of the theoretical radioactivity attainable. The low specific radioactivity was traced to the rapid exchange of iodine on the tyrosine residue with hydrogen atoms from water in the presence of the catalyst. The solvent for the tritium–halogen exchange was modified to minimize the concentration of exchangeable hydrogen atoms. This was accomplished by dissolving the iodotyrosyl peptide in the minimum amount of 0.1 N acetic acid and diluting with aprotic, polar solvents such as hexamethyl phosphoramide and dimethyl-formamide. Catalytic dehalogenation in the minimally protic medium (1% H_2O) yielded tritiated ACTH of high specific radioactivity (80% of the theoretical value) (Ramachandran and Behrens, 1977).

2.3. Characterization of Tritiated Adrenocorticotropin

[3,5-^{3}H]-Tyr^{23} ACTH$^{1-39}$ and [3,5-^{3}H]-Tyr^{2,23} ACTH$^{1-39}$ were prepared by catalytic dehalogenation of 3,5-diiodo-Tyr^{23} ACTH$^{1-39}$ and 3,5-diiodo-Tyr^{2,23} ACTH$^{1-39}$, respectively. The tritiated hormone was purified by gel filtration on Sephadex G-25 and then by ion-exchange chromatography on carboxymethyl-cellulose. The major product that emerged in the position of synthetic human ACTH was obtained in the overall yield of 30%. The product was found to be homogeneous by paper electrophoresis (Fig. 2). The radioactive peptide comigrated with synthetic human ACTH. The location of tritium in the radioactive hormone was ascertained by electrophoresis of the tryptic digest of the labeled hormone. In the case of

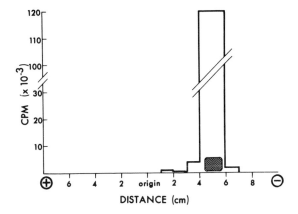

Figure 2. Electrophoresis of [3,5-³H]-Tyr²³ ACTH¹⁻³⁹. Synthetic human ACTH¹⁻³⁹ (0.1 mg) was mixed with tritiated ACTH isolated after carboxymethylcellulose chromatography and subjected to electrophoresis on Whatman No. 3 paper in pyridine–acetate buffer, pH 3.7, for 4 hr at 400 V. The peptide was visualized by staining with ninhydrin. The paper was cut into 1-cm strips, mixed with scintillant, and counted.

[3,5-³H]-Tyr²³ ACTH¹⁻³⁹, more than 80% of the radioactivity in the tryptic digest migrated in the position of ACTH²²⁻³⁹. The remainder of the radioactivity was associated with undigested and partially digested hormone (Ramachandran and Behrens, 1977). In the case of [3,5-³H]-Tyr²,²³ ACTH ¹⁻³⁹, the radioactivity was equally distributed in the tryptic peptides migrating in the positions of ACTH¹⁻⁸ and ACTH²²⁻³⁹. The ultraviolet absorption spectrum of the radioactive hormone was identical to that of synthetic human ACTH, indicating that the iodine atoms on the tyrosine residues were quantitatively removed. The amino acid composition of an acid hydrolysate of [3,5-³H]-Tyr²,²³ ACTH¹⁻³⁹ agreed well with that expected for human ACTH.

The biological activity of the tritiated hormone preparations were assessed in two isolated cell systems. The ability of [3,5-³H]-Tyr²³ ACTH¹⁻³⁹ to stimulate glycerol release (lipolysis) in isolated rat adipocytes (Fig. 3A) or corticosterone synthesis in isolated rat adrenocortical cells (Fig. 3B) was indistinguishable from that of synthetic human ACTH. The potency of [3,5-³H]-Tyr²,²³ ACTH¹⁻³⁹ was similarly found to be identical to that of synthetic human ACTH in both assays.

The specific radioactivity of [3,5-³H]-Tyr²³ ACTH¹⁻³⁹ was found to be 46.6 Ci/mmol, and that of [3,5-³H]-Tyr²,²³ ACTH¹⁻³⁹ was 90 Ci/mmol , or approximately 80% of the theoretical radioactivity attainable. [3,5-³H]-Tyr²,²³ ACTH¹⁻³⁹ was employed in all the binding studies and photoaffinity

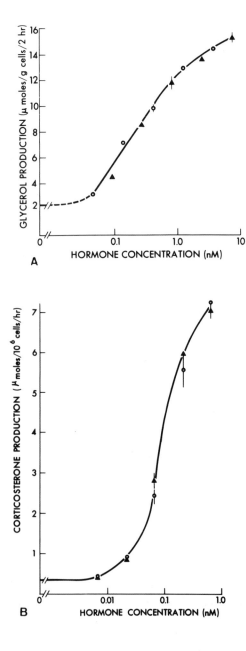

Figure 3. Stimulation of (A) lipolysis (glycerol release) in isolated rat fat cells and (B) corticosterone production in isolated rat adrenocorticol cells by synthetic human ACTH (▲) and ³H-ACTH (○).

labeling experiments to be described and will be designated as [³H]-ACTH. The specific radioactivity of [³H]-ACTH amounted to 200 dpm/fmol.

3. Conversion of Adrencorticotropin into a Photoaffinity Probe

3.1. Choice of Site in the Hormone

Having introduced the radioactive label into the ACTH molecule with full retention of biological activity, and therefore affinity for the receptor, we turned to the problem of introducing a suitable photoreactive group. The ε-amino group of lysyl residues is the site that is generally modified in polypeptides for the introduction of a photoreactive moiety (Levy, 1973; Ji, 1977; Das *et al.*, 1977). In the case of ACTH, however, both the ε-amino group of lysine residues and the α-amino group at the N-terminal are unsuitable for modification, since alteration of these sites lowers the biological activity considerably (Ramachandran, 1973). In fact, there are few sites in the ACTH molecule that can be modified without loss in biological activity. Analysis of concentration–response curves of a large number of analogues of ACTH has shown that in most cases the decrease in activity is due to a decrease in affinity, since the analogues produce the same maximal response as ACTH at higher concentrations. Furthermore, isolation of the modified hormone containing a single photoreactive group is a difficult task when there are a number of similar or identical functionalities present. It is therefore desirable to modify that amino acid residue that is unique. We have previously found that selective modification of the single tryptophan residue at position 9 in ACTH by reaction with *o*-nitrophenyl sulfenyl chloride resulted in an analogue that acted as an inhibitor of ACTH. This analogue, *o*-nitrophenyl sulfenyl ACTH (NPS-ACTH), was found to be a potent inhibitor of ACTH-induced lipolysis in isolated rat adipocytes (Ramachandran and Lee, 1970a) as well as adenylate cyclase activity in rat fat cell ghosts (Ramachandran and Lee, 1970b). Subsequently, NPS-ACTH was shown to inhibit ACTH-induced adenylate cyclase stimulation in rat adrenocortical cells (Moyle *et al.*, 1973) and in rat adrenal homogenate fractions (Ramachandran *et al.*, 1976). These results suggested that the tryptophan residue in ACTH is intimately involved in the interaction of the hormone with the receptor. Since ACTH is a long, flexible coil, it is important to introduce the photoreactive group in the segment of the hormone directly interacting with the receptor. The single tryptophan residue in ACTH is therefore a suitable site for introducing the photosensitive function.

3.2. Preparation of Photoreactive Aryl Sulfenyl Chlorides

The photogenerated species that are most useful in photoaffinity-labeling studies are carbenes and nitrenes (Bayley and Knowles, 1977). Since tryptophan was chosen as the site of modification, we decided to synthesize an aryl sulfenyl chloride containing an azido group as the nitrene precursor, since this could then be introduced into ACTH under mild conditions. To ascertain that introduction of a substituent such as the azido group into the aryl sulfenyl group would not affect the affinity of the hormone, we prepared three more analogues modified at the tryptophan residue and tested their biological properties. 2,4-Dinitrophenyl sulfenyl ACTH (DNPS-ACTH), 2 nitro-4-carboxyphenyl sulfenyl ACTH (NCPS-ACTH), and 2-nitro-4-carbamidophenyl sulfenyl ACTH (NCMPS-ACTH) were prepared and characterized (Canova-Davis and Ramachandran, 1976). All three were found to be potent inhibitors of the activation of rat adrenal adenylate cyclase by ACTH (Glossmann and Ramachandran, unpublished observations). Two new aryl sulfenyl chlorides, namely, 2-nitro-4-azidophenyl sulfenyl chloride (NAPS-Cl) and 2,4-dinitro-5-azido-phenyl sulfenyl chloride (DNAPS-Cl), were synthesized (Ramachandran and Canova-Davis, 1977; Canova-Davis and Ramachandran, in prep.). The synthetic scheme for DNAPS-Cl is shown in Fig. 4. NAPS-Cl was found to be much less reactive than NPS-Cl or DNPS-Cl. DNAPS-Cl, on the other hand, was highly reactive. Both were used for modification of ACTH.

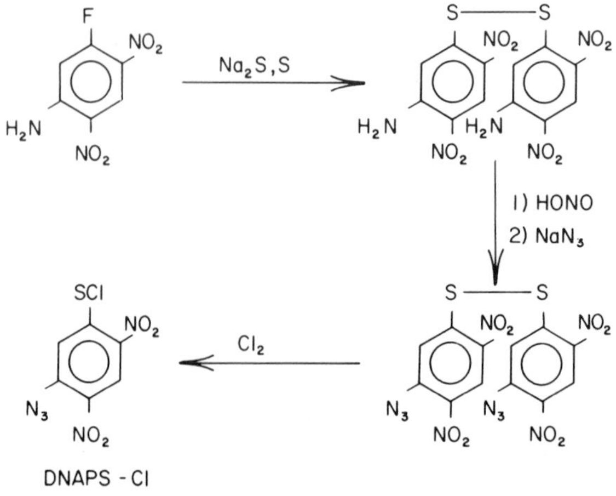

Figure 4. Scheme for the synthesis of 2,4-dinitro-5-azidophenyl sulfenyl chloride (DNAPS-Cl).

3.3. Preparation and Characterization of [2,4-Dinitro-5-azidophenyl sulfenyl]-Trp⁹ ACTH¹⁻³⁹ (DNAPS-ACTH)

Introduction of the photoreactive DNAPS group into ACTH was accomplished by allowing the hormone to react with DNAPS-Cl in 90% acetic acid. An excess of methionine was included in the reaction mixture to act as a scavenger and protect the methionine residue in the hormone from oxidation (Canova-Davis and Ramachandran, 1976). DNAPS-ACTH was isolated by gel filtration and purified by ion-exchange chromatography on carboxymethylcellulose.

Amino acid analysis of an enzyme digest of DNAPS-ACTH showed that tryptophan was quantitatively modified and that the methionine residue was intact (Canova-Davis and Ramachandran, 1976). Peptide maps of tryptic digests of DNAPS-ACTH and ACTH were identical except for the absence of the peptide Trp-Gly-Lys-Pro-Val-Gly-Lys in the map of the modified hormone. Instead, a single, new, yellow, ninhydrin-positive spot was observed. The peptide eluted from this spot had the amino acid composition of the missing tryptophan peptide. Further evidence of modification at a single site in ACTH was obtained from the ultraviolet absorption spectra. The extinction coefficient of DNAPS-ACTH at 360 nm was found to be comparable to that of the model compound acetyl-DNAPS-tryptophan amide.

The presence of a photosensitive group in DNAPS-ACTH was revealed by the alteration in the ultraviolet absorption spectrum after photolysis. When DNAPS-ACTH was mixed with a crude ACTH-binding protein, FI, isolated from acid–acetone extracts of ovine pituitary gland (Canova-Davis and Ramachandran, 1976) and photolyzed with radiation of wave-length greater than 300 nm, covalent attachment of the hormone was observed. After photolysis, the binding protein–DNAPS-ACTH complex was subjected to gel filtration on Sephadex G-50 in 1 M formic acid to facilitate separation of noncovalently bound peptide. Spectral analysis of the binding protein, FI, showed that the ratio of absorbance at 360 : 280 nm had increased from a value of 0.064 to 0.168 after photolysis (Table I), indicating the covalent attachment of DNAPS-ACTH. Electrophoretic analysis of the binding protein after photolysis also revealed that DNAPS-ACTH was covalently attached (Ramachandran and Canova-Davis, 1977).

It is known that the dinitrophenyl sulfenyl group can be cleaved from the indole ring of the modified tryptophan residue by reaction with mercaptoethanol (Wilchek and Miron, 1972). This procedure was employed to reveal the presence of ACTH covalently bound to the pituitary binding protein. When the photolyzed complex was treated with mercaptoethanol prior to electrophoresis, a peptide with the mobility of ACTH reappeared (Ramachandran and Canova-Davis, 1977). The reductive cleavage releases

Table I. Photolabeling of the Pituitary Binding Protein FI[a]

Treatment	Ratio of absorbance at 360 nm/280 nm	
	FI + NAPS-ACTH	FI + DNAPS-ACTH
Unphotolyzed	0.066	0.064
Photolyzed	0.104	0.168
Photolyzed and treated with mercaptoenthanol	0.129	—

[a] Spectral analysis was performed after gel filtration of the binding protein–DNAPS-ACTH complexes on Sephadex G-50 in 1 M formic acid to facilitate separation of noncovalently bound peptides.

ACTH by the formation of 2-thioltryptophan, and the nitrophenyl thio group is left on the binding protein. This was confirmed by spectral analysis of the binding protein after treatment with mercaptoethanol (Table I). The ratio of absorbance 360:280 nm increased further as expected, since the binding protein retained the chromophore that absorbs at 360 nm but lost the 280 nm absorbing capacity of the ACTH molecule.

DNAPS-[^{3}H]-ACTH was similarly prepared by reaction of [^{3}H]-ACTH with DNAPS-Cl. The photosensitivity of DNAPS-[^{3}H]-ACTH was also checked by photolysis in the presence of the ACTH-binding protein, FI. Radioactivity eluted with the binding protein on Sephadex G-50 only after photolysis, indicating covalent attachment.

4. Photoaffinity Labeling of Adrenocorticotropin Receptors on Rat Adipocytes

4.1. Interaction of [^{3}H]Adrenocorticotropin with Rat Adipocytes

There has been only one report so far on the interaction of labeled ACTH with isolated rat adipocytes. Lang et al. (1974) investigated the binding of the analogue Phe^2,4,5-dehydro [4,5-^{3}H]norvaline ^{4}ACTH$^{1-24}$, which had a specific radioactivity of 7.42 Ci/mmol and a biological potency 10% that of ACTH$^{1-24}$. Their results indicated the presence of multiple orders of binding sites with association constants ranging from 1×10^8 to 3×10^3 M^{-1}.

In the present study, [^{3}H]-ACTH was incubated with isolated rat adipocytes at room temperature, and the unbound hormone was separated by a careful procedure using filtration through unipore polycarbonate membranes that exhibited the minimum amount of binding of the tritiated hormone (Behrens and Ramachandran, in prep.). Specific binding of [^{3}H]-

ACTH was determined by subtracting the amount of radioactive hormone bound in the presence of a large excess of nonradioactive ACTH.

Binding of [³H]-ACTH was found to be rapid, reaching a maximum within 5 min. Binding was found to be highly specific, since insulin, glucagon, and the very basic peptide β-endorphin did not displace [³H]-ACTH at concentrations equivalent to that of ACTH causing a 62% displacement. The binding of [³H]-ACTH to rat adipocytes and the stimulation of lipolysis as a function of hormone concentration were measured in the same preparations. Binding of [³H]-ACTH was found to be saturable and was almost superposed on the lipolytic response curve. From the concentration required for half-maximal binding, and apparent dissociation constant of 2×10^{-9} M was estimated. Using the estimate of 7800 adipocytes/mg dry weight of cells (Lang *et al.*, 1974), it can be calculated that there are approximately 9000 specific binding sites for ACTH per adipocyte. This is vastly lower than the estimate of 2.12×10^6 sites/cell reported by Lang *et al.* (1974). However, in view of the high specific radioactivity of [³H]-ACTH, the specificity of the binding, the saturability, and the close parallel between the measured binding and the lipolytic response, there is good reason to believe that the binding observed is directly related to the physiological function of the hormone.

4.2. Interaction of [³H]Adrenocorticotropin with Rat Adipocyte Ghosts and Plasma Membrane

Since adipocytes float and are too fragile to subject to vigorous mixing procedures, we decided to perform the photoaffinity-labeling experiments on ghosts or plasma-membrane preparations derived from the adipocytes. The effects of ACTH and DNAPS-ACTH on rat adipocyte ghost adenylate cyclase activity are shown in Table II. Whereas ACTH stimulated adenylate cyclase in ghosts, DNAPS-ACTH had no effect even at concentrations 4 times that of ACTH. On the other hand, DNAPS-ACTH caused a 58% inhibition of ACTH-induced adenylate cyclase stimulation, showing that the photosensitive probe could bind to the ACTH receptors on rat adipocyte ghosts. The inhibition was competitive, since a higher concentration of ACTH reduced the degree of inhibition by DNAPS-ACTH.

It is apparent from the results in Table II and earlier studies (Ramachandran and Lee, 1976) that the concentration of ACTH required for stimulation of adenylate cyclase activity in adipocyte ghosts is much higher than the concentration of hormone required for eliciting lipolysis or cyclic AMP accumulation in intact adipocytes. Since nonreceptor interactions increase with increasing concentration of hormone, the use of ghosts or plasma-membrane preparations for photoaffinity labeling posed a problem. Therefore, the binding of [³H]-ACTH to rat adipocyte ghosts and purified

Table II. Effects of DNAPS-ACTH on Rat Adipocyte Ghost Adenylate
Cyclase

	Hormone	Concentration	cAMP production (% increase over basal)	Inhibition (%)
I.	ACTH	3×10^{-7} M	80	—
		1×10^{-6} M	131	—
		3×10^{-6} M	110	—
II.	DNAPS-ACTH	4.4×10^{-6} M	0	—
III.	ACTH	1×10^{-6} M		
	+		55	58
	DNAPS-ACTH	4.4×10^{-6} M		
IV.	ACTH	3×10^{-6} M		
	+		79	28
	DNAPS-ACTH	4.4×10^{-6} M		

plasma-membrane preparations was examined. Binding of [³H]-ACTH to rat adipocyte ghosts is shown in Fig. 5A. In this case, bound and free hormone were separated by centrifugation in a microfuge. Although no stimulation of adenylate cyclase activity can be observed in such preparations at ACTH concentrations below 10^{-8} M, binding of [³H]-ACTH appeared to saturate around 7×10^{-9} M. A similar pattern of binding was obtained when [³H]-ACTH was incubated with purified adipocyte plasma membranes (Fig. 5B). These results indicate that loss of sensitivity to ACTH observed in ghosts and membrane preparations may not be due to a loss or change in the receptor, but rather may be the consequence of other factors such as altered coupling of receptor and adenylate cyclase. These results also suggest that near-physiological concentrations of the photoreactive derivative of ACTH could be used in the photoaffinity-labeling experiments, thereby minimizing interactions of the hormone derivative with nonreceptor components of the plasma membrane.

4.3. Photolysis of DNAPS-[³H]Adrenocorticotropin: Rat Adipocyte Plasma-Membrane Complex

Rat epididymal fat cell plasma membranes were isolated according to the procedure of Harwood *et al.* (1973) and incubated in the dark with DNAPS-[³H]-ACTH (1×10^{-8} M) at 0°C for 18 hr in the presence and absence of unlabeled ACTH (10^{-4} M). The reaction mixtures were irradiated for 10 min at 0°C in a Rayonet photochemical reactor with RUL-3000 Å ultraviolet lamps. Bovine serum albumin (BSA) (0.5%) was included in the medium to act as a scavenger for photoreactive ACTH molecules not bound to the receptor. Following photolysis, the membranes were thoroughly washed with phosphate-buffered saline (PBS) containing 0.5% BSA

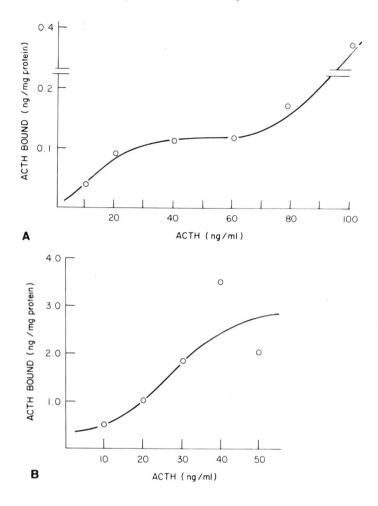

Figure 5. (A) Binding of [³H]-ACTH to rat adipocyte ghosts. Hormone and ghosts were incubated overnight at 4°C. Bound and free hormone were seperated by centrifugation in a microfuge. Radioactivity not displaced by an excess of unlabeled ACTH (5 × 10⁻⁵ M) has been subtracted to obtain the specific binding. Each incubation contained 200 μg ghost protein. (B) Binding of [³H]-ACTH to purified rat adipocyte plasma membranes. Specific binding was measured as described in (A). Each incubation contained 60 μg membrane protein.

to remove as much of the noncovalently bound hormone analogue as possible. The membranes were then solubilized in 2% sodium dodecyl sulfate (SDS) as described by Laemmli (1970) and analyzed by electrophoresis on polyacrylamide slab gels in the presence of SDS according to Ames (1974). The results are shown in Fig. 6. Despite the careful washing,

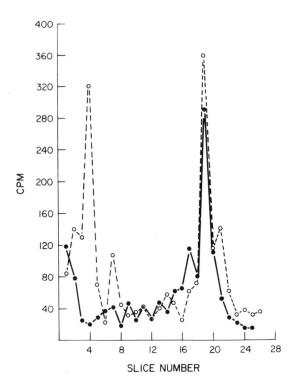

Figure 6. SDS-polyacrylamide gel electrophoresis of rat adipocyte plasma membranes photolyzed in the presence of DNAPS-³H-ACTH. Plasma membranes (50 μg) were incubated with DNAPS-³H-ACTH (50 ng) in 0.3 ml PBS containing 0.5% BSA for 18 hr at 0°C. The samples were photolyzed for 10 min at 0°C, washed with PBS–0.5% BSA (5×), dissolved in 2% SDS, and subjected to electrophoresis according to Ames (1974). (○) Photolyzed in the absence of unlabeled ACTH, (●) photolyzed in the presence of 10^{-4} M ACTH.

some radioactivity attributable to DNAPS-[³H]-ACTH or its photoproduct was present, as shown by the radioactive peak near the front. Radioactive peaks were also found near the origin. The presence of a major peak of radioactivity near the origin clearly showed that the DNAPS-[³H]-ACTH was covalently linked to a macromolecular component of the rat fat cell plasma membrane. When the photolysis was conducted in the presence of an excess of ACTH, there was no radioactivity in the region corresponding to this component (Fig. 6). No incorporation of radioactivity was observed in other controls that were not irradiated. The region of the gel that incorporated radioactivity also stained for carbohydrate, suggesting that the macromolecular component interacting with DNAPS-[³H]-ACTH may be a glycoprotein. It was further found that the radioactivity associated with this component could be removed by treatment with mercaptoethanol, as shown also for the product of photolysis of DNAPS-ACTH complex with the pituitary binding protein (see Section 3.3).

From the binding of [^{3}H]-ACTH to rat fat cell plasma membranes (see Fig. 5B), it can be estimated that 500 fmol of ACTH are bound per milligram protein. It can be inferred from the results shown in Fig. 6 that approximately 100 fmol of DNAPS-[^{3}H]-ACTH are covalently attached per milligram protein of plasma membranes. Thus, 20% of the sites are labeled by this procedure. It is highly encouraging that the photoaffinity probe selectively labeled a significant proportion of one plasma-membrane component that is likely to be the receptor for ACTH.

5. Conclusions and Projections

It has been shown that by judicious choice of the radioactive label and the site of chemical modification, it is possible to prepare a photoaffinity probe for ACTH receptors that retains the high affinity of the hormone for the receptor. The photosensitive agent, DNAPS-[^{3}H]-ACTH, was found to attach covalently to unique sites on rat adipocyte plasma membrane photolysis. This covalent labeling of specific sites on the adipocyte membrane was prevented when an excess of unlabeled ACTH was present during photolysis. These results indicate, but do not prove, that the membrane component labeled by the photoaffinity probe is the specific receptor for ACTH. To prove this point, it is necessary to demonstrate a correlation between the degree of labeling and inactivation of the physiological response. Owing to the decreased sensitivity of rat adipocyte membrane preparations to ACTH, it will be quite difficult to obtain such a correlation in this system. The isolated rat adrenocortical cell as well as plasma-membrane preparations are highly sensitive to physiological concentrations of the hormone, and photoaffinity labeling of ACTH receptors in this system should prove highly useful for identifying and characterizing these receptors.

The covalent attachment of the radioactive hormone to the receptor would provide a convenient marker for the subsequent isolation of the receptor. Solubilized plasma-membrane preparations obtained by treatment with nonionic detergents can be fractionated and the radioactive hormone–receptor complex isolated. The radioactive hormone may be removed by treatment with mercaptoethanol, leaving only the dinitrophenyl thio group on the receptor. Alternatively, plasma-membrane preparations photolabeled with nonradioactive DNAPS-ACTH can be treated first with mercaptoethanol to remove the hormone. The receptor that retains the dinitrophenyl thio group can be isolated by affinity chromatography using antidinitrophenyl antibodies. Since the receptor is the only component of the plasma membrane tagged with the dinitrophenyl thio group by the photolabeling and reduction procedures, this approach should lead to the isolation of the receptor in a high degree of purity.

It is also evident that the photoaffinity probe DNAPS-[³H]-ACTH should be useful for investigating ACTH receptors on other target cells. In this connection, it may be noted that secretion of ACTH by the clonal strain of mouse pituitary tumor cells, AtT 20/D 16, may be controlled by an ultrashort-loop negative-feedback mechanism (Richardson, 1978). DNAPS-[³H]-ACTH may be of use in identifying putative ACTH receptors in these cells as well as normal pituitary corticotrophs.

ACKNOWLEDGMENTS. The authors wish to thank Professor C. H. Li for his interest and Dr. Martin Shetlar for the loan of the photoreactor. This work was supported by grants from the National Institutes of Health (GM 2907 and CA 16417).

DISCUSSION

SPICER: Are the tritium atoms on the hormone firmly bound so that the labeled hormone could be studied by radioautography and its uptake by the target cell followed?

RAMACHANDRAN: The tritium atoms are introduced into the 3 and 5 positions of the phenyl ring of tyrosine residues and are very stable. This labeled hormone should be very useful for autoradiographic studies.

DE LA LLOSA: What happens to the adenylate cyclase when the ACTH derivative is linked covalently to the receptor? Is it maintained continuously stimulated?

RAMACHANDRAN: DNAPS-ACTH is an antagonist. It inhibits the action of ACTH on adenylate cyclase of rat adipocyte ghosts. Therefore, after photoaffinity labeling, the derivative will permanently inhibit the action of ACTH.

KORDON: Do all the photolabeled analogues that you have tested have antagonist properties in receptor studies?

RAMACHANDRAN: Yes, all the aryl sulfenyl derivatives of ACTH that we have tested are antagonists of the action of ACTH on adenylate cyclase of rat adipocytes and adrenocortical cells. The o-nitrophenyl sulfenyl of Gln⁵-α-MSH, however, is an agonist which is three times more potent then Gln⁵-α-MSH in the stimulating melanophores and rabbit adipocytes. We expect that the photoreactive derivative of α-MSH (that is, DNAPS-α-MSH) would be an agonist.

CRINE: Did you look for ACTH-binding sites in the brain? Also, could you see any difference between binding sites for α-MSH and ACTH?

RAMACHANDRAN: We have not done any studies with brain; however, we have provided Drs. Akil and Watson at Stanford University with tritiated ACTH, and they are looking for possible binding sites in rat brain. We have not evaluated the possible difference in binding between α-MSH and ACTH.

GOURDJI: You have provided convincing data that [³H]-ACTH was fully biologically active on intact cells and that DNAPS-ACTH was an antagonist of ACTH on adipocyte ghosts. Why did you choose only ghosts and/or purified membrane preparations for subsequent steps in identifying ACTH-binding sites with the DNAPS-ACTH? In view of the possibility of internalization of peptide hormones, aren't intact cells and even living cells an even better tool?

RAMACHANDRAN: We used adipocyte plasma membranes for our initial photolabeling studies because this is a simpler system to analyze by SDS-electrophoresis. We also felt that the efficiency of photolysis may be lower if intact fat cells are used, since they float. I agree that the use of intact cells is important for studying internalization of the hormone, and we plan to do such studies with rat adrenocortical cells in suspension as well as in primary culture.

REFERENCES

Ames, G. F.-L., 1974, Resolution of bacterial proteins by polyacrylamide gel electrophoresis on slabs, *J. Biol. Chem.* **249**:634.

Bayley, H., and Knowles, J. R., 1977, Photoaffinity labeling, *Methods Enzymol.* **46**:69.

Brundish, D. E., and Wade, R., 1973, Synthesis of [3,5-^{3}H$_2$-Tyr^{23}]-β-corticotropin-(1-24)-tetracosapeptide, *J. Chem. Soc. Perkin Trans. I* **1973**:2875.

Canova-Davis, E., and Ramachandran, J., 1976, Chemical modification of the tryptophan residue in adrenocorticotropin, *Biochemistry* **15**:921.

Cuatrecasas, P., Hollenberg, M. D., Chang, K.-J., and Bennett, V., 1975, Hormone receptor complexes and their modulation of membrane function, *Recent. Prog. Horm. Res.* **31**:37.

Das, M., Miyakawa, T., Fox, D. F., Pruss, R. M., Aharonov, A., and Herschman, H. R., 1977, Specific radiolabeling of cell surface receptor for epidermal growth factor, *Proc. Natl. Acad. Sci. U.S.A.* **74**:2790.

Dufau, M. L., Ryan, D. W., Baukal, A. J., Catt, K. J., 1975, Gonadotropin receptors, *J. Biol. Chem.* **250**:4822.

Greenwood, F. C., Hunter, W. M., and Glover, J. S., 1963, the preparation of ^{131}I-labelled human growth hormone of high specific radioactivity, *Biochem. J.* **89**:114.

Harwood, J. P., Löw, H., and Rodbell, M., 1973, Stimulatory and inhibitory effects of guanyl nucleotides on fat cell adenylate cyclase, *J. Biol. Chem.* **248**:6239.

Ji, T. H., 1977, A novel approach to the identification of surface receptors, *J. Biol. Chem.* **252**:1566.

Laemmli, U. K., 1970, Cleavage of structural proteins during the assembly of the head of bacteriophage T4, *Nature (London)* **227**:680.

Landon, J., Livanou, T., and Greenwood, F. C., 1967, the preparation and immunological properties of ^{131}I-labelled adrenocorticotropin, *Biochem. J.* **105**:1075.

Lang, U., Karlaganis, G., Vogel, R., and Schwyzer, R., 1974, Hormone–receptor interactions: Adrenocorticotropic hormone binding site increase in isolated fat cells by phenoxazones, *Biochemistry* **13**:2626.

Lefkowitz, R., Roth, J., Pricer, W., and Pastan, I., 1970, ACTH receptors in the adrenal: Specific binding of ACTH-^{125}I and its relation to adenyl cyclase, *Proc. Natl. Acad. Sci. U.S.A.* **65**:745.

Lemaire, S., Yamashiro, D., Behrens, C., and Li, C. H., 1977, Adrenocorticotropin. 51. Synthesis and properties of analogues of the human hormone with tyrosine residues replaced by 3,5-diiodotyrosine, *J. Am. Chem. Soc.* **99**:1577.

Levy, D., 1973, Preparation of photoaffinity probes for the insulin receptor site in adipose and liver cell membranes, *Biochim. Biophys. Acta* **322**:329.

Lowry, P. J., McMartin, C., and Peters, J., 1973, Properties of a simplified bioassay for adrenocorticotropic activity using the steriodogenic response of isolated adrenal cells, *J. Endocrinol.* **59**:43.

McIlhinney, R. A. J., and Schulster, D., 1974, Preparation of biologically active ^{125}I-labelled ACTH by a simple enzymic radioiodination utilizing lactoperoxidase, *Endocrinology* **94**:1259.

McIlhinney, R. A. J., and Schulster, D., 1975, Studies on the binding of ^{125}I-labeled corticotropin to isolated rat adrenocortical cells, *J. Endocrinol.* **64**:175.

Morgat, J. L., Hung, L. T., Candinaud, R., Fromageot, P., Bockaert, J., Imbert, M., and Morel, F., 1970a, Peptidic hormone interactions at the molecular level: Preparation of highly labelled ³H oxytocin, *J. Labelled Compd.* **6**:276.

Morgat, J. L., Hung, L. T., and Fromageot, P., 1970b, Preparation of highly labelled [³H] angiotensin II, *Biochim. Biophys. Acta* **207**:374.

Moyle, W. R., Kong, Y.-C., and Ramachandran, J., 1973, Steriodogenesis and cyclic adenosine 3',5'-monophosphate accumulation in isolated rat adrenal cells, *J. Biol. Chem.* **248**:2409.

Ontjes, D., Kirkways, D., Mahaffee, D. D., Zimmerman, C. F., and Gwynne, J. T., 1977, ACTH receptors and the effect of ACTH on adrenal organelles, *Ann. N. Y. Acad. Sci.* **297**:295.

Pardelles, P., Morgat, J. L., Fromageot, P., Camier, M., Bonne, D., Cohen, P., Bockaert, J., and Jard, S., 1972, Tritium labelling of 8-lysine vasopressin and its purification by affinity chromatography on sepharose bound neurophysins, *FEBS Lett.* **26**:189.

Rae, P., and Schimmer, B. P., 1974, Iodinated derivatives of adrenocorticotropic hormone, *J. Biol. Chem.* **249**:5649.

Ramachandran, J., 1973, Structure and function of adrenocorticotropin, in: *Hormonal Proteins and Peptides* (C. H. Li, ed.), Vol. II, pp. 1–28, Academic Press, New York.

Ramachandran, J., and Behrens, C., 1977, Preparation and characterization of specifically tritiated ACTH, *Biochim. Biophys. Acta* **496**:321.

Ramachandran, J., and Canova-Davis, E., 1977, Synthesis and use of photoreactive arylsulfenyl chlorides, in: *Peptides—Proceedings of the Fifth American Peptide Symposium* (M. Goodman and J. Meienhofer, eds.), pp. 553–555, Wiley, New York.

Ramachandran, J., and Lee, V., 1970a, Preparation and properties of the o-nitrophenyl-sulfenyl derivative of ACTH: An inhibitor of the libpolytic action of the hormone, *Biochem. Biophys. Res. Commun.* **38**:507.

Ramachandran, J., and Lee, V., 1970b, Divergent effects of o-nitrophenyl sulfenyl ACTH on rat and rabbit fat cell adenyl cyclases, *Biochem. Biophys. Res. Commun.* **41**:358.

Ramachandran, J., and Lee, V., 1976, Divergent effects of adrenocorticotropin and melanotropin on isolated rat and rabbit adipocytes, *Biochim. Biophys. Acta* **428**:339.

Ramachandran, J., Kong, Y. C., and Liles, S., 1976, Effects of ACTH and its o-nitrophenyl sulfenyl derivative on adrenocortical function in vivo, *Acta Endocrinol.* **82**:587.

Rees, L. H., Cooke, D. M., Kendall, J. W., Allen, C. F., Kramer, R. M., Ratcliffe, J. G., and Knight, R. A., 1971, A radioimmunoassay for rat plasma ACTH, *Endocrinology* **89**:254.

Richardson, U. I., 1978, Self-regulation of adrenocorticotropic secretion by mouse pituitary tumor cells in culture, *Endocrinology* **102**:910.

Shiu, R. P. C., and Friesen, H. G., 1974, Solubilization and purification of a prolactin receptor from the rabbit mammary gland, *J. Biol. Chem.* **249**:7902.

Wilchek, M., and Miron, T., 1972, The conversion of tryptophan to 2-thioltryptophan in peptides and proteins, *Biochem. Biophys. Res. Commun.* **47**:1015.

19

Enzymatic Degradation of Hypothalamic Hormones at the Pituitary-Cell Level: Possible Involvement in Regulation Mechanisms

Karl Bauer

1. Introduction

Within the concert of regulatory mechanisms balancing the functions of an organism according to the needs of the body, there are principally two control levels. On one hand, the biological effect exerted by a given concentration of an active substance is dependent on the various physiological parameters that collectively determine the "responsiveness" of the target. On the other hand, the mechanisms regulating the hormone concentrations at physiologically appropriate levels are evidently of fundamental importance, since under given physiological conditions the physiological response is, within certain limits, directly correlated with the concentration of the biologically active substance that becomes effective at the target site. Among these mechanisms, the controlled inactivation of a biologically active substance is an eminently important event. This becomes evident in the case of certain pathological disorders in which loss

Karl Bauer • Max-Volmer-Institut, Abteilung Biochemie, Technische Universität Berlin, 1000 Berlin 10, Germany of control over destruction might be the primary cause of certain endocrine diseases (Knight *et al.*, 1973).

At the site of the pituitary, there are two hormone levels that have to be controlled, namely, the concentration of the hypothalamic hypophysiotrophic hormones and the concentration of the hypophyseal hormones that are available for release.

2. Degradation of Hypophyseal Hormones

After the isolation and identification of the hypophyseal hormones, several studies were undertaken to investigate whether peptidases or proteases might be involved in the processes regulating the intracellular hormone concentrations, which are generally determined by the rate of synthesis, the rate of secretion, and the rate of degradation. For this system, degradation might act as a security device following the principle of an unsaturated system: the more substrate available, the more degraded. For the hypophyseal hormones, such a system operating as a secondary-level control mechanism was first described by Dr. Farquhar and her associates (Smith and Farquhar, 1966; Farquhar, 1969, 1971). Under conditions when hormonal discharge is suppressed, excess secretory granules are taken up and disposed of in lysosomes by a process known as "crinophagy" (also see Chapter 7). Pituitary lysosomes are known to contain proteases and peptidases capable of degrading pituitary hormones down to oligopeptides and dipeptides (McDonald *et al.*, 1971). After diffusion through the lysosomal membrane into the cell sap, the oligopeptides will be further degraded into the constituent amino acids, which could then be restored to the metabolic pool for reutilization (McDonald *et al.*, 1971; Farquhar, 1971).

Several of the pituitary peptidases and proteases have been characterized in more detail (McDonald *et al.*, 1971), and further studies have been undertaken to determine enzymatic activities under various physiological conditions. First, Meyer and Clifton (1956) observed that pituitary alkaline protease was increased after long-term treatment with estrogen. Since this enzymatic activity was also observed within the small-granule fraction, it has been suggested that this enzyme might be involved in the release mechanisms of the corresponding hormones (Perdue and McShan, 1962; Tesar *et al.*, 1969). When female rats were injected with estradiol for 20 days, Vanha-Pertulla (1969) observed significant changes for acid protease(s) and dipeptidyl arylamidases I and III, as well as for several aminopeptidases. Testosterone decreased and castration increased these enzymatic activities in male rats.

Since endocrine conditions are likely to cause changes in the functions of various cell types, it is extremely difficult to interpret these biochemical data in terms of physiological functions, especially for a tissue as heterogeneous as the pituitary. This is true the more so as these enzyme activities when determined in total tissue extracts cannot be related to the specialized processes in cellular events. It seems more likely that the fluctuations of these total enzymatic activities do reflect the altered metabolic activities in homeostatic adaptation processes.

In contrast to this, however, it is generally assumed that proteolytic or glycolytic enzymes, or both, within the secretory granules or at extragranular pool sites might represent important control elements. During the processes leading to maturation of the granules, such enzymes (which might be closely related to the lysosomal enzymes from the enzymological point of view) most likely generate the active principle from the zymogen (the prohormone) by a process of limited proteolysis. Furthermore, it is conceivable that by specific cleavages of secretory materials, additional peptides could be formed that might act as signals for triggering further events such as the selective release of mature granules or the selection of granules for crinophagic disposal. So far, however, the identification and characterization of these enzymes still warrant careful investigations, which it is to be hoped might provide a basis for an understanding of intracellular hormone regulatory mechanisms.

3. Degradation of Hypothalamic Hypophysiotropic Hormones by Hypophyseal Tissue Enzymes

The isolation and characterization of the hypothalamic hypophysiotropic hormones as oligopeptides aroused new interest in investigating whether hypophyseal tissue enzymes may play an important function by inactivating the neuropeptides at their target site. One might conceive that the signal for switching off the stimulation by the relefact is not only determined by dissociation constant of the hormone–receptor complex, but might also be influenced by the enzymatic inactivation of the relefact. In this case, the degree of stimulation and duration of action could be modulated by neuropeptide-degrading enzymes.

First, however, one might ask whether such enzymes are present in pituitary tissue. For thyroliberin (TRF), it was known that this tripeptide is not degraded by general proteolytic enzymes. This fact is quite understandable from the structural features of the tripeptide. The cyclized amino terminus protects the peptide against the action of aminopeptidases, while the amidated carboxy terminus hinders the action of carboxypeptidases. In addition, due to the proline residue, internal cleavages by endopeptidases are rather restricted.

3.1. Degradation of Thyroliberin

3.1.1. Enzymatic Fragmentation of TRF

To study the degradation of TRF, radiolabeled TRF was used as tracer. After incubation with the hypophyseal tissue homogenate or extracts thereof, the reaction mixture could be resolved by thin-layer chromatography. The radiolabeled fragments were localized by radioscanning and identified by cochromatography with authentic peptides (Fig. 1). When TRF, radiolabeled in the histidine or proline moiety, was used as tracer, deamido-TRF and His-Pro-NH$_2$ could be identified as primary cleavage products. The dipeptideamide could be detected only as a minor metabolite, since this compound cyclizes readily with formation of histidyl-proline diketopiperazine ($_\Gamma$His-Pro$_\lnot$). This conversion is a well-known reaction, characteristic for proline-containing dipeptide derivatives (Rydon and Smith, 1956). As a minor fragment, His-Pro could be detected, which apparently represents the common secondary cleavage product. Also, only a small amount of the radiolabeled end product proline or histidine could be detected.

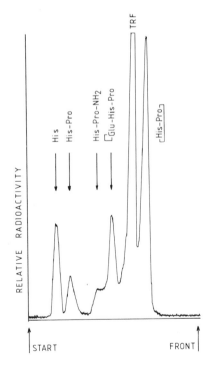

Figure 1. Fragmentation of TRF by adenohypophyseal tissue enzymes. The extract was prepared by homogenizing 50 mg adenohypophyseal tissue in 2 ml buffer (100 mM potassium phosphate buffer, pH 7.4, containing 2mM DTT and 1 mM EDTA). After centrifugation at 27,000g for 30 min, 20 μl of the supernatant was incubated for 60 min at 37°C with 4 μCi TRF-[^{3}H-His] in 40 μl buffer. The incubate was resolved by thin-layer chromatography on silica gel using the solvent system butanone-2–propanol–pyridine–acetic acid–water (4:4:4:2). The radiolabeled fragments were localized by radioscanning and the cochromatographed synthetic materials by staining with Pauly's reagent.

Among the theoretically possible radiolabeled fragments, only [^{3}H] prolineamide or pyroGlu-His could not be detected. Since [^{3}H]-Pro-NH$_2$ was not observed even in appropriate trapping experiments for which unlabeled prolineamide was added to the incubation mixture, this result indicates that under these incubation conditions, the histidyl-proline bond of TRF or His-Pro-NH$_2$ is apparently not subject to hydrolytic cleavage by pituitary enzymes. On thermodynamic grounds, one cannot expect to find glutamyl-containing TRF fragments under the incubation conditions used, since an eventually occurring hydrolysis of the lactam bond should be an energy-dependent reaction as has been shown for the conversion of pyroglutamate to glutamic acid. For this reaction, a concomitant cleavage of ATP to ADP and inorganic phosphate could be observed (Van der Werf *et al.*, 1971).

3.1.2. Pathway of TRF Catabolism

As indicated by the fragmentation pattern, the degradation of TRF follows a complex scheme. To elucidate the pathway of these enzymatic events, the tissue extract was fractionated by gel filtration on Sephadex G-150. When the column effluents were tested for the various enzymatic activities using highly specific and sensitive radiochemical tests, the elution profile of four enzymes could be monitored. These results and further enzyme chemical and kinetic investigations indicate that the catabolism of TRF by adenohypophyseal enzymes follows the proposed pathway shown in Fig. 2.

The fragmentation is initiated by two enzymes. E$_1$ catalyzes the hydrolysis of TRF at the pyroglutamyl-histidine bond and therefore likely represents a pyroglutamate aminopeptidase. By cochromatography with standard proteins, a molecular weight of 28,000 has been estimated for this enzyme, while for E$_2$, the TRF-deamidating enzyme, an apparent molecular weight of 76,000 has been determined by the same method. While E$_1$ also catalyzed the hydrolysis of deamido-TRF with formation of His-Pro, the deamidation of His-Pro-NH$_2$ is catalyzed by E$_3$, but not by the TRF-deamidating enzyme. E$_3$ exhibits the characteristics of glycyl-proline-β-naphthylamidase (Hopsu-Havu and Sarimo, 1967), an enzyme that has been characterized recently as a postproline dipeptidyl aminopeptidase (Yoshimoto *et al.*, 1978). The hydrolysis of His-Pro to the free amino acids is catalyzed by E$_4$, an enzyme that presumably represents a proline dipeptidase. An enzyme that catalyzes the formation of histidyl-proline diketopiperazine could not be detected when the fractionated tissue extract was monitored for such an enzymatic activity. By enzyme kinetic analysis it could be demonstrated that the cyclization of His-Pro-NH$_2$ proceeds at a constant rate independent of enzyme or substrate concen-

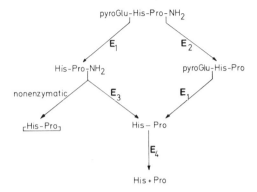

Figure 2. Pathway of TRF catabolism. (E_1) Pyroglutamate aminopeptidase; (E_2) postproline cleaving enzyme; (E_3) postproline dipeptidyl aminopeptidase; (E_4) proline dipeptidase.

tration. From these findings, it is concluded that the formation of His-Pro is exclusively due to the nonenzymatic cyclization reaction.

3.1.3. Characterization of the TRF-Inactivating Enzymes

Within the degradation pathway, pyroglutamate aminopeptidase and the "TRF-deamidating enzyme" represent the key enzymes that catalyze the initial reactions and thus the biological inactivation of TRF. It was therefore of considerable enzymological as well as physiological interest to study these enzymes in more detail. After further purification by conventional chromatographic procedures (ammonium sulfate fractionation, ion-exchange chromatography on DEAE cellulose, gel filtration, and adsorption chromatography on hydroxylapatite), these enzyme preparations were tested for their substrate specificity.

3.1.3a. Pyroglutamate Aminopeptidase. It could be shown that the partially purified TRF-metabolizing pyroglutamate aminopeptidase also catalyzed the hydrolysis of deamido-TRF, the hydrolysis of the dipeptides pyroGlu-His and pyroGlu-Ala, as well as the liberation of β-naphthylamine from the chromogenic substrate pyroglutamyl-β-naphthyl-amide. When luliberin (LH-RF) ([^{3}H]pyroglutamyl) was incubated with this enzyme preparation, a single exponential relationship with time could be observed for the liberation of pyroglutamic acid. A TRF- or LH-RF-specific pyroglutamate aminopeptidase could not be observed, however (Nowak and Bauer, unpublished). This demonstrates that both TRF and LH-RF are hydrolyzed at the pyroglutamylhistidine bond by a general hypophyseal pyroglutamate aminopeptidase that was first described by Mudge and Fellows (1973).

3.1.3b. Characterization of the "TRF-Deamidating Enzyme." The enzyme catalyzing the deamidation of TRF has been descriptively termed "TRF-deamidating enzyme." This designation could denote, however, neither that this enzyme might be specific for TRF nor that it might represent a deamidating enzyme. For the characterization of this enzyme with regard to substrate specificity, competitive inhibitor studies were first performed. It could be shown that the formation of radiolabeled deamido-TRF is not inhibited by amino acid amides, dipeptideamides, or even tripeptideamides such as pyroGlu-His-Leu-NH$_2$. This clearly demonstrated that this enzyme is apparently not a general deamidating enzyme. On the other hand, the deamidation of TRF was competitively inhibited not only by TRF and analogues such as pyroGlu-His-Pro-NHCH$_3$, pyroGlu-Phe-Pro-NH$_2$, and pyroGlu-His-Pro-Gly-NH$_2$, but also orders of magnitude more effectively by biologically active peptides such as LH-RF, angiotensin II, oxytocin, and vasopressin. This indicates that this enzyme obviously is not TRF-specific.

When the structural features of these peptides were compared, it became evident that the proline moiety represents the only structural feature these peptides share in common. This indicated that the proline residue is specifically recognized by this enzyme. By the identification of the enzymatically formed cleavage products, the specific cleavage of the Pro-X bond could indeed be demonstrated for various peptides such as pyroGlu-His-Pro-Gly-NH$_2$, angiotensin II, the collogenase substrate Z-Gly-Pro-Leu-Gly-Pro, and others. For LH-RF, the specific cleavage at the Pro-Gly peptide bond could be demonstrated by the identification of glycineamide and the LH-RF 1–9 fragment as the only enzymatic cleavage products (Knisatschek and Bauer, 1978; Knisatschek *et al.*, 1979).

On the basis of the information gathered from these experiments, the "TRF-deamidating enzyme" could be characterized as a postproline cleaving endopeptidase. A postproline cleaving enzyme from kidneys that exhibits characteristics similiar although not identical to those of the hypophyseal tissue enzyme has recently been described by Koida and Walter (1976). Originally, this enzyme was discovered when the degradation of the neurohypophyseal hormones by hormone-responsive tissues was tested (Walter *et al.*, 1971).

3.1.4. Tissue Specificity of the TRF-Inactivating Enzymes

3.1.4a. Pyroglutamate Aminopeptidase. For various tissues of the rat (brain, liver, kidneys, pancreas, lung, and muscle), the activity of pyroglutamate aminopeptidase was determined by following the liberation of β-naphthylamine from the chromogenic substrate pyroglutamyl-β-naphthylamide and by determining the degradation of TRF to His-Pro-NH$_2$. In agreement with the literature (Szewczuk and Kwiatowska, 1970), the presence of this enzyme in all these tissues could be observed. There was no indication that in some tissues there might be an additional enzyme that exhibits specificity toward the neuropeptide. At this point, it might be interesting to add that serum contains a potent enzymatic activity that catalyzes the enzymatic cleavage of the pyroglutamyl-histidine bond of TRF (Bauer *et al.*, 1978). In contrast to the general tissue enzyme, however, serum pyroglutamate aminopeptidase does not catalyze the hydrolysis of pyroglutamyl-β-naphthylamide or other pyroglutamyl-containing dipeptides nor even of the structurally closely related LH-RF (Nowak and Bauer, unpublished). In addition, the serum enzyme exhibits a molecular weight of 280,000, while a molecular weight of 28,000 has been estimated for tissue pyroglutamate aminopeptidase.

 3.1.4b. Postproline Cleaving Enzyme. This enzymatic activity was also observed in all the tissues tested (brain, liver, kidneys, pancreas, lung, and muscle). Within various brain regions, the highest activity was found in the pineal gland. It might be interesting to add that in these tissue extracts, the activity of this enzyme accounted for only one third of the specific activity that could be determined for the adenohypophyseal tissue extracts.

3.1.5. Cellular and Subcellular Distribution of the TRF-Inactivating Enzymes

 So far, the distribution of pyroglutamate aminopeptidase and postproline cleaving enzyme among the adenohypophyseal cell population has not been determined. As far as can be seen from our studies (in collaboration with Drs. A. Faivre-Baumann, D. Gourdji, F. de Vitry, and A. Tixier-Vidal, Collège de France, Paris), however, qualitative differences cannot be expected. When TRF was incubated with extracts of TRF-responsive GH_3 tumor cells, we observed the same fragmentation pattern as seen for the degradation of TRF by whole pituitary tissue extract. In addition, these enzymes exhibited the same enzyme chemical characteristics as the tissue enzymes (e.g. pyroglutamate aminopeptidase is highly sensitive towards the -SH blocking agent 2-iodoacetamide, but not toward diisopropyl fluorophosphate, while postproline cleaving enzyme exhibits opposite characteristics).

 Both enzymatic activities could also be detected when extracts of various hypothalamic cell lines were tested. It might be noteworthy, however, that among these extracts pronounced differences with regard to specific activities as well as with respect to the relative activities between these two enzymes could be observed.

Since we found both enzymatic activities even in extracts of frog eggs, these findings strongly suggest that this pyroglutamate aminopeptidase as well as postproline cleaving enzyme are apparently basic cell constituent enzymes.

The subcellular distribution of these enzymes still awaits careful investigation. Both enzymes are found mainly in the 100,000g supernatant, while only a small proportion ($\approx$ 2%) remains associated with the particulate fraction. Whether the low levels of these enzymatic activities present in the particulates may be of special physiological importance remains to be investigated.

3.2. Fragmentation of Luliberin

Rapid degradation of LH-RF by various tissues has been reported (Kochman *et al.*, 1975; Marks and Stern, 1974, 1974; Koch *et al.*, 1974; for reviews, see Jeffcoate, 1977; Marks, 1978). It has been suggested (Koch *et al.*, 1974) that the enzymatic inactivation of LH-RF is due mainly to the specific cleavage of the glycyl-leucine bond. Hudson *et al.* (1976), however, assume on the basis of their studies that the cleavage of the internal peptide bond takes the place of the tyrosyl-glycine bond. Our studies indicate that there is not just one cleavage site, but rather that the fragmentation of LH-RF follows a more complex pattern. For this, LH-RF, radiolabeled in the pyroglutamyl residue, was used as tracer. After incubation with hypothalamic or hypophyseal tissue extracts, the reaction mixture was resolved by ion-exchange chromatography. The radiolabeled fragments could then be localized by radioscanning. As shown in Fig. 3, several radiolabeled fragments could be observed.

This pattern and further enzyme kinetic analysis indicated that there are at least three radiolabeled primary fragments. Besides the expected fragments pyroGlu and desglycineamide-LH-RF (the products that are formed by the characterized pyroglutamate aminopeptidase and postproline cleaving enzyme), a third fragment was observed. The chromatographic properties of this radiolabeled fragment indicated that this split product is formed by internal peptide-bond cleavage. This fragment could not be identified directly since, in contrast to LH-RF itself, the primary split products are subject to secondary reactions catalyzed by the carboxypeptidases and aminopeptidases present in the tissue extracts. With a purified enzyme preparation, however, the selective cleavage of the tyrosyl-glycine bond could be demonstrated (Bauer *et al.*, 1979).

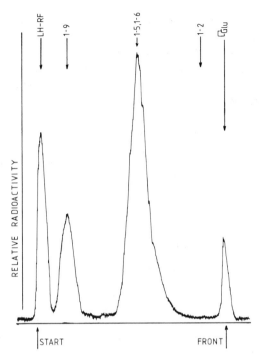

Figure 3. Fragmentation of LH-RF by adenohypophyseal enzymes. The enzyme extract was prepared as described in the Fig. 1 caption. A 10-μl sample of the supernatant was incubated for 30 min with 5 μCi LH-RF ([³H]pyroglutamyl) in 30 μl buffer. The incubate was resolved by ion-exchange chromatography on cellulose Whatman phosphate paper P-81 (ascending, 16 cm). The radiolabeled fragments were localized by radioscanning and the cochromatographed synthetic materials by staining with Pauly's reagent. This fragmentation pattern indicates the formation of three primary fragments. With highly purified enzyme preparations, the selective cleavage of the pyroglutamyl-histidine bond by the pyrogluta-mate aminopeptidase, the hydrolysis of the prolyl-glycine bond by the postproline cleavage enzyme, and the cleavage of the tyrosyl-glycine bond by a so far uncharacterized enzyme (Bauer *et al.*, 1977) could be demonstrated.

3.3. Degradation of Somatostatin

The degradation of somatostatin by adenohypophyseal enzymes has not yet been demonstrated in detail. Studies on the inactivation of somatostatin by brain enzymes *in vitro* (Marks and Stern, 1975) indicate that the Phe-Phe bond at positions 6 and 7, the Trp-Lys bond at positions 8 and 9, and the Thr-Phe bond at positions 10 and 11 are susceptible to enzymatic hydrolysis. This indicates that for the degradation of this tetradecapeptide also, more than one enzyme system is available.

4. Reflections on the Physiological Role of Neuropeptide-Degrading Enzymes

It is the ultimate goal of these studies to elucidate whether at given sites these neuropeptide-degrading enzymes do serve an important physiological function. While the rapid inactivation of these peptides by liver and kidney enzymes likely serves the rapid clearance of these peptides from the peripheral circulation, so far the physiological role of these enzymes in neuropeptide-responsive tissues remains obscure. Reflections on their physiological role therefore remain only speculative and hypothetical. Theoretically, one might conceive that these enzymes could be of physiological importance as (1) inactivators and eventually also as regulators of preexisting activities or (2) generators of new activities.

4.1. Peptidases as Inactivators and Regulators of Preexisting Activities

In several laboratories, studies have been undertaken to investigate whether peptidases could serve an important physiological role by regulating the concentration of the active substance at various levels. For the well-known rapid inactivation of thyroliberin by peripheral and portal blood, it has been suggested that the enzymatic degradation of the tripeptideamide may be involved in regulatory mechanisms determining the number of molecules that become available to the target sites at the pituitary. Following the fragmentation of radiolabeled TRF by serum enzymes, it could be shown that the inactivation of TRF is catalyzed by an enzyme that degrades TRF in pyroglutamic acid and His-Pro-NH$_2$. This enzyme exhibits enzymatic–chemical characteristics distinctly different from those of the known pyroglutamate aminopeptidases catalyzing the same reaction. Furthermore, and most importantly, it could be shown that this enzyme exhibits an extraordinarily high degree of substrate specificity. The TRF-degrading serum enzyme does not catalyze the degradation of luliberin, neurotensin, or the substrates of the known pyroglutamate aminopeptidases pyroglutamyl alanine, pyroglutamyl histidine, etc. The strict substrate specificity of this enzyme (which may adequately be termed "thyroliberinase") suggests that this enzyme might serve specific physiological functions. This notion is supported by the observation that the activity of this enzyme is controlled by thyroid hormones. Following the degradation of TRF by rat serum, significant fluctuations of the TRF-degrading enzyme activity could be observed after physiological manipulation (Redding and Schally, 1969; White *et al.*, 1976; Bauer, 1976). The decreased enzymatic activity in serum of hypothyroid rats could be restored after treatment with thyroid hormones. It might be mentioned parenthetically that in man an association of enzyme activity with thyroid function could not be observed (Bassiri and Utiger, 1972; Jeffcoate and White, 1974). Such a relationship might be obscured, however, by the wide interpatient variations of enzyme activities (Jeffcoate and White, 1974). Since by these manipulations only the activity of the "thyroliberinase" specifically altered while the activities of other serum enzymes remained unchanged, these results might indicate that the degradation of TRF could even form part of the mechanisms involved in hormonal feedback within the hypothalamic–pituitary–thyroid axis.

In contrast to the TRF-degrading serum enzyme, the function of tissue peptidases as potential regulators of neuropeptide concentrations still remains obscure. This may be due to fundamentally different preconditions. It is easily conceivable that enzymes need not necessarily be substrate- or cell-specific to fulfill specific functions at given sites. The selectivity and specificity are most likely inherent in the mechanisms controlling the availability of the neuropeptide. Receptors or carriers, for example, likely do play the most decisive role and likely also represent the limiting factors within regulatory mechanisms while enzymatic activity represents only a secondary control element. Within this context, some parellelism between the physiological role of neuropeptide-degrading enzymes and the monoamine-metabolizing enzymes can be seen. In this regard, it becomes obvious that the physiological function of these peptidases might become apparent under certain physiological disorders, but if the enzyme activity does not represent the limiting factor within the regulatory system, the physiological role of these enzymes cannot be easily evaluated by determining the activity of these enzymes under various physiological conditions.

There are additional difficulties which have to be accounted for when enzyme activities are determined in total tissue extracts. Since neuropeptides can be degraded by various tissue enzymes, significant changes in the rate of neuropeptide inactivation, which might be observed after physiological manipulation, do not neccesarily indicate that the tissue peptidases have regulatory functions. The observed overall effect might reflect homeostatic alterations of several peptidases rather than specific changes in the activity of a specific neuropeptide-degrading enzyme. On the other hand, in case the activity of a tissue enzyme indeed represents the limiting factor within a regulatory unit, the experimental evaluation may also be extremely difficult. Significant changes in enzyme activities at a given site might be obscured by the complexities one has to account for when enzyme activities are determined in total tissue extracts (cell heterogeneity, compartmentalization of enzyme and substrate, and so forth).

The same problem is also implicit for studies designed to evaluate the function of neuropeptide-degrading enzymes in the mechanisms controlling the intracellular concentration of the neuropeptides at their place of synthesis. Apparently, more refined analytical tools have to be developed for further studies. So far, not even the fundamental question whether the neuropeptides do become available to these enzymes has been answered. The determination of the subcellular distribution of these enzymes, at present under investigation in several laboratories, might not answer this question finally as long as we do not know the final sites where mediation takes place. We do not know, for example, how TRF might reach the nucleus or get out of the nucleus (Bournaud *et al.,* 1977). We also do not know how LH-RF could reach the secretory granules (Sternberger and Petrali, 1975), and we do not know the fate of an eventually internalized hormone–receptor complex. It is to be hoped that progress in various fields will provide the basis to determine whether these enzymes might serve distinct physiological functions at given sites. However, it should also be mentioned that there are some aspects indicating that neuropeptide-degrading enzymes might indeed participate in the mechanisms transducing the information of the relefact at the trophic cells of the pituitary. It is striking, for example, that the superactive analogues of LH-RF do exhibit structural alterations exactly at the enzymatic cleavage sites, namely, at the pyroGlu-His bond at positions 1 and 2, the Tyr-Gly bond at positions 5 and 6, and the Pro-Gly bond at positions 9 and 10. It seems unlikely that this effect reflects only changes favorable for receptor interactions. In fact, increased binding to plasma membranes could not be observed for the highly active LH-RF analogue D-Phe[6]desGly[10]LH-RF-ethylamide (Heber and Odell, 1978). Further characterization of the neuropeptide-degrading enzymes in combination with receptor binding studies might provide a basis not only for the rational design and synthesis of potent analogues useful for physiological and clinical studies or for therapeutical application, but also for further studies on the molecular mechanisms of action of the relefact at the trophic cells of the pituitary. These analogues in turn, could provide a most useful tool for evaluating whether the neuropeptide-degrading enzymes do play a physiological role in these mechanisms.

4.2. Proteolytic Enzymes as Generators of New Activities

It has been clearly recognized in recent years that the fundamental importance of proteolytic enzymes in biological regulatory systems is not restricted to the control of metabolism. Diverse functions of these enzymes in the regulation of other physiological phenomena could be explored. In the process of zymogen activation, new physiological functions can be generated in response to physiological stimuli by the selective cleavage of peptide bonds (for an excellent review, see Neurath and Walsh, 1976). The specificity of the activation process is dependent on the conformation of the zymogen as well as that of the enzyme. Provided a biological test is available for assaying the activity of the generated principle, a direct correlation of both factors can easily be demonstrated in some cases, but not in others where additional control elements are involved. Intercellular communication, for example, adds another element of control, as is known for the generation of angiotensin or for the activation of pancreatic trypsinogen by enterokinase that is secreted from the brush border of the small intestine (Maroux et al., 1971). The conversion of corticotropin, lipotropin, and opioid peptides presumably follows the same principle. It might even be conceivable that multiple physiological products may be formed from one common precursor to express multiple messages in a concerted manner. Here again, intercellular communication between substrate and the protease conceivably could add another control element. In contrast to reversible regulatory systems suited for maintaining a steady state of intermediary metabolism, zymogen activation by limited proteolysis is an essentially irreversible reaction because it is an exerogenic reaction that is likely to induce a prompt response to a given signal.

In this regard, fragments might not only represent intermediates for final disposal but also contain important physiological information for the balance of physiological functions. Among the TRF fragments observed under the *in vitro* incubation conditions, ⌐His-Pro¬ could be detected as a major fragment. It should be noted that this product is also observed as a TRF fragment after incubation of TRF with hypothalamic, brain, or other tissue extracts, and also in small amounts after incubation with serum (Bauer et al., 1978). We became especially interested in searching for possible physiological effects of this fragment, since this substance, as a lactam-ring-containing cyclic compound, exhibits some important structural features of pharmacologically active drugs (e.g., ergotamines) (Fig. 4). Therefore, the effect of this fragment on prolactin release was tested. When the TRF-responsive cells of the GH$_3$ tumor cell line were incubated with ⌐His-Pro¬, we observed that this substance inhibited the basal secretion of prolactin in a dose-related manner to approximately 50% of

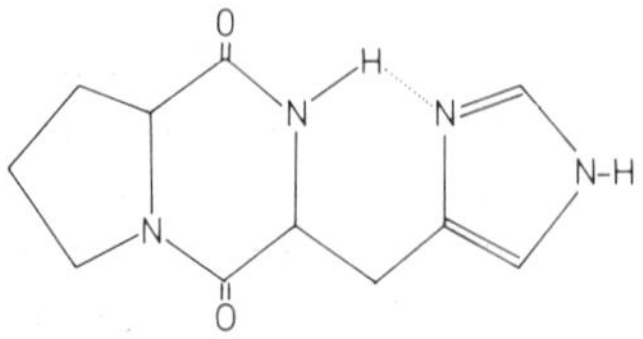

Figure 4. Structure of histidyl-proline diketopiperazine (⌐His-Pro¬) (Bauer et al., 1978).

the control levels. When this compound was injected into proestrous rats, a significant inhibition of the TRF-induced release of prolactin could be observed *in vivo* (Bauer *et al.*, 1978). Another example of the biological effectiveness of the cyclized product was recently reported by Prasad *et al.* (1977). These workers observed that ⌐His-Pro⌐ antagonizes ethanol-induced sleeping time 8 times more effectively than TRF itself.

It may be conceivable that the generation of a counterbalancing principle might be especially important for the balanced secretion of those hormones not being feedback-regulated by their target hormones. It is furthermore conceivable that limited proteolysis might even represent a more general principle for the generation or integration or both, of multiple physiological effects evoked by one stimuli. In this case, the special anatomical features within the structural network of communication as well as the biochemical features of the target cells might represent the important control elements. So far, however, it remains to be investigated whether the effects observed for histidyl-proline diketopiperazine do reflect a physiological rather than a pharmacological effect. It is unknown whether by the enzymatic degradation of TRF an effective concentration of this substance could be formed at given sites or whether an effective concentration of this substance could become accumulated at specific sites. So far, we do not even know whether this product represents a naturally occurring TRF metabolite.

5. Conclusions and Perspectives

So far, the mechanisms of action of the hypothalamic hypophysiotropic hormones at their target sites, the trophic cells of the pituitary, have not been fully elucidated. The mechanisms regulating the transduction of information subsequent to the primary interaction of the relefact at the plasma membranes still remain obscure. Proteolytic enzymes have been clearly recognized in recent years to represent important control elements that can turn other reactions on and off by generating or destroying their catalysts. To elucidate whether the hypophyseal neuropeptide-degrading enzymes might represent such a control element, it is necessary to delineate first the enzymatic degradation of these peptides and to characterize these enzymes.

By following the fragmentation of TRF and LH-RF, it could be shown that there are apparently several enzyme systems catalyzing the inactivation of the neuropeptides. So far, two enzymes could be identified, namely, a pyroglutamate aminopeptidase that hydrolyzes the pyroglutamyl-histidine bond of TRF and LH-RF and a postproline cleaving enzyme that inactivates TRF by deamidation of the tripeptideamide and also catalyzes the degradation of LH-RF by hydrolyzing the proline-glycine bond at positions 9 and 10. Both enzymes apparently are not specific for the neuropeptides, but rather are general peptidases that also hydrolyze other synthetic substrates or biologically active peptides exhibiting either an amino-terminal pyroglutamyl residue or a proline-X bond. In addition, a rather ubiquitous distribution of these enzymes could be observed. These findings might provide new aspects for further studies. The elucidation of the actual molecular sites of enzyme action should be of special interest for the design of synthetic analogues that in turn might provide a most useful tool for further investigations. For example, the determination of receptor affinity constants together with pharmacokinetic investigations and the determination of the enzymatic inactivation rate might provide new parameters that might be helpful in understanding the physiological role of these enzymes. With the identification of these enzymes and the development of specific biochemical tests, additional tools are available for further studies. Besides the determination of enzymatic activities under various physiological conditions or in various pathological disorders, further characterization of these enzymes should enable us to search for specific enzyme inhibitors suitable for physiological and pharmacological studies. In addition, the characterization of these enzymes should enable us to synthesize enzyme-specific substrates useful not only for the determination of enzymatic activities but also for further histochemical and cytochemical studies. Studies complementary to these should become feasible by using antibodies against these enzymes that could be prepared once these enzymes have been isolated. Furthermore, the identification of the enzymatically formed fragmentation products should enable us to test these fragments for biological activity to evaluate the possible function of proteolytic enzymes as generators of new activities.

It is hoped that the enzyme chemical studies might provide a basis for further investigations to evaluate the still-unknown physiological role of neuropeptide-degrading enzymes at various sites.

ACKNOWLEDGMENTS. For support and helpful discussions, I thank B. Horsthemke, H. Knisatschek, P. Nowak, J. Sievers, and Dr. J. Salnikow. I thank Dr. H. Kleinkauf for his continuing interest and support and gratefully acknowledge the help and stimulation I received from Drs. A. Tixier-Vidal, A. Faivre-Bauman, D. Gourdji, and F. de Vitry, Collège de France, Paris. I thank Drs. H. Lévine, J. L. Morgat, and P. Fromageot, CEN Saclay, for the generous gifts of radiolabeled peptides. I thank the Deutsche Forschungsgemeinschaft for financial support.

DISCUSSION

JONES: Does T_3 influence the enzymatic degrading activity in the pituitary? I also want to say that we have found that histidyl-proline-diketopiperazine releases ACTH from the pituitary of basal hypothalamic-lesioned rats. I don't see how this observation could fit into your scheme of things.

BAUER: Daily subcutaneous injections of 10 μg T_3 100 g body weight did not result in a change in the specific enzyme activity of either the pyroglutamate or the postproline cleaving enzyme. It would be interesting to see whether the enzyme activities in the thyrotrophs, for example, are altered.

KRAICER: You indicate clearly that control of the rate of removal of a hormone may be just as important as the rate of secretion in regard to the levels of circulating hormones. Are there any data that the rate of removal of hypothalamic–hypophysiotropic hormones is altered under defined physiological or pharmacological conditions?

BAUER: I am not aware of any data.

KORDON: One would expect a degrading enzyme playing a role in regulatory mechanisms to have a higher affinity then that of the total peptidase population, which ranges between 10^{-5} and 10^{-4} M. Did you measure the apparent affinity of some of the TRH-degrading enzymes that you have purified?

BAUER: For TRF as substrate, the hypophyseal enzymes exhibit a K_m: (1) pyroglutamate aminopeptidase, $K_m = 8 \times 10^{-5}$M; (2) postproline cleaving enzyme, $K_m = 4.1 \times 10^{-4}$ M. In agreement with the characterization of postproline cleaving enzyme as an endopeptidase, we observe a higher affinity for LH-RF. LH-RF inhibits the deamidation of TRF about 100 times more effectively than TRF itself (competition assay; deamidation of [^{3}H]-TRF). The TRF-specific pyroglutamate aminopeptidase present in serum (distinctly different from the tissue pyroglutamate aminopeptidase) exhibits a K_m of 5.6×10^{-5} M for TRF as substrate.

KORDON: We have shown that enzymes degrading TRH and LH-RH are not found in the same subcellular fractions as the peptides themselves. The enzymes are mostly recovered from the supernatant fractions, whereas the peptides are almost exclusively synaptosomal.

BAUER: Two to five percent of both enzyme activities is found associated with the pellet fraction, while 95–98% is contained in supernatant (100,000g). Fractionation studies are in progress to determine whether enzyme activities are accociated with plasma membrane, etc.

JAQUET: Have you checked the effects of pro-his-diketopiperazine on TSH basal release, and after TRH administration on TSH release?

BAUER: No effects on the TSH level could be observed, neither on the basal nor on the TRF-induced TSH level.

DENEF: Is his-pro-diketopiperazine a dopamine-receptor agonist?

BAUER: That is an interesting question which is presently under investigation.

ENJALBERT: We observed that diketopiperazine elicited a dose-dependent inhibition of prolactin release from incubated hemipituitaries of male rats. The maximal inhibition was about 20%, and the concentration of diketopiperazine producing half-maximal inhibition was in the range of 10^{-9} M.

REFERENCES

Bassiri, R., and Utiger, R. D., 1972, Serum inactivation of the immunological and biological activity of TRH, *Endocrinology* 91:657–667.

Bauer, K., 1976, Regulation of degradation of TRH by thyroid hormones, *Nature (London)* 259:591–593.

Bauer, K., Gräf, K. J., Faivre-Baumann, A., Beier, S., Tixier-Vidal, A., and Kleinkauf, H., 1978, Inhibition of prolactin secretion by histidyl-proline diketopiperazine, *Nature* 274:174–175.

Bauer, K., Horsthemke, B., Knisatschek, H., Nowak, P., and Kleinkauf, H., 1979, Degradation of luliberin (LH-RF) by brain and pituitary enzymes, *Hoppe Seyler's Z. Physiol. Chem.* 360:229.

Bournaud, F., Gourdji, D., Mongongu, S., and Tixier-Vidal, A., 1977, [³H]-Thyroliberin (TRH) binding to nuclei isolated from a pituitary clonal cell line (GH₃), *Neuroendocrinology* 24:183–194.

Farquhar, M. G., 1969, Lysosome function in regulating secretion: Disposal of secretory granules in cells of the anterior pituitary gland, in: *Lysosomes in Biology and Pathology* (J. T. Dingle and H. B. Fell, eds.), Vol. 2, pp. 462–482, North-Holland, Amsterdam.

Farquhar, M.G., 1971, Processing of secretory products by cells of the anterior pituitary gland, *Mem. Soc. Endocrinol.* 19:79–122.

Heber, D., and Odell, W. D., 1978, Pituitary receptor binding activity of active, inactive, superactive and inhibitory analogs of gonadotropin-releasing hormone, *Biochem. Biophys. Res. Comm.* 82:67–73.

Hopsu-Havu, V. K., and Sarimo, S. R., 1967, Purification and characterization of an aminopeptidase hydrolyzing glycly-proline-β-naphthyl-amide, *Hoppe-Seyler's Z. Physiol. Chem.* 348:1540–1550.

Hudson, D., Pickering, A., McLoughlin, J. L., Matthews, E., Sharpe, R., Fink, G., McIntyre, I., and Szelke, M., 1976, The synthesis and metabolic fate of 3-[³H]Trp, 9-[¹⁴C]Pro-LHRH, VI International Congress of Endocrinology, Hamburg, Abstract No. 43.

Jeffcoate, S. L., 1977, The hormonal peptides of the hypothalamus, in: *Topics in Hormone Chemistry* (W. R. Butt, ed.), Vol. 1, pp. 13–47, Ellis Horwood, Chichester, New York, Brisbane, and Toronto.

Jeffcoate, S. L., and White, N., 1974, The inactivation of thyrotropin releasing hormone by plasma in thyroid disease, *Clin. Endocrinol.* 4:231.

Knight, E. B., Baylin, S. B., and Foster, G. V., 1973, Control of polypeptide hormones by enzymatic degradation, *Lancet* 2:719–723.

Knisatschek, H., and Bauer, K., 1978, Deamidation of thyroliberin (TRF) by a post-proline cleaving endopeptidase, *12ᵗʰ FEBS Meeting, Dresden*, Abstract No. 2819.

Knisatschek, H., Bauer, K., and Kleinkauf, H., 1979, Characterization of thyroliberin deamidating enzyme as a post proline cleaving enzyme, *Hoppe Seyler's Z. Physiol. Chem.* 360:303–304.

Koch, Y., Baram, T., Chobsieng, P., and Fridkin, M., 1974, Enzymic degradation of luteinizing hormone–releasing hormone (LH-RH) by hypothalamic tissue. *Biochem. Biophys. Res. Commun.* 61:95–103.

Kochmann, K., Kerdelhué, B., Zor, U., and Jutisz, M., 1975, Studies of enzymatic degradation of luteinizing hormone–releasing hormone by different tissues, *FEBS Lett.* 50:190–194.

Koida, M., and Walter, R., 1976, Post-proline cleaving enzyme, *J. Biol. Chem.* 251:7593–7599.

Marks, N., 1978, Biotransformation and degradation of corticotropins, lipotropins and hypothalamic peptides, in: *Frontiers in Neuroendocrinology* (W. F. Ganong and L. Martini, eds.), Vol. 5, pp. 329–377, Raven Press, New York.

Marks, N., and Stern, F., 1974, Enzymatic mechanisms for the inactivation of luteinizing hormone-releasing hormone (LH-RH), *Biochem. Biophys. Res. Commun.* **61:**1458–1463.

Marks, N., and Stern, F., 1975, Inactivation of somatostatin (GH-RIH) and its analogs by crude and partially purified rat brain extracts, *FEBS Lett.* **55:**220–224.

Maroux, S., Baratti, J., and Desnuelle, P., 1971, Purification and specificity of porcine entereokinase, *J. Biol. Chem.* **246:**5031–5039.

McDonald, J. K., Callahan, P. X., Ellis, S., and Smith, R. E., 1971, Polypeptide degradation by dipeptidyl aminopeptidase I (cathepsin C) and related peptidases, in: *Tissue Proteinases* (A. J. Barrett and J. T. Dingle, eds.), pp. 69–107, North-Holland, Amsterdam.

Meyer, R., and Clifton, K., 1956, Effect of diethyl stilbestrol on the quantity and intracellular distribution of pituitary proteinase activity, *Arch. Biochem. Biophys.* **62:**198–209.

Mudge, A. W., and Fellows, R. E., 1973, Bovine pituitary pyrrolidone-carboxyl peptidase, *Endocrinology* **93:**1428–1434.

Neurath, H., and Walsh, K. A., 1976, Role of proteolytic enzymes in biological regulation (a review), *Proc. Natl. Acad. Sci. U.S.A.* **73:**3825–3832.

Perdue, J. F., and McShan, W. H., 1962, Isolation and biochemical study of secretory granules from rat pituitary glands, *J. Cell Biol.* **15:**159–172.

Prasad, C., Matsui, T., and Peterkofsky, A., 1977, Antagonism of ethanol narcosis by histidyl-proline diketopiperazine, *Nature (London)* **268:**142–144.

Redding, T. W., and Schally, A. V. 1969, Studies on the thyrotropin-releasing hormone (TRH): Activity in peripheral blood, *Proc. Soc. Exp. Biol. Med.* **131:**420–425.

Rydon, N. H., and Smith, P. W. G., 1956, Polypeptides: The self condensation of the esters of some peptides of glycine and proline, *J. Chem. Soc.* **1956:**3642–3650.

Smith, R. E., and Farquhar, M. G., 1966, Lysosome function in the regulation of the secretory process in cells of the anterior pituitary gland, *J. Cell Biol.* **31:**319–347.

Sternberger, L. A., and Petrali, J. P., 1975, Quantitative immunocytochemistry of pituitary receptors for luteinizing hormone–releasing hormone, *Cell Tissue Res.* **162:**141–176.

Szewczuk, A., and Kwiatkowska, J., 1970, Pyrrolidonyl peptidase in animal, plant and human tissues, *Eur. J. Biochem.* **15:**92–96.

Tesar, J. T., Koenig, H., and Hughes, C., 1969, Hormone storage granules in the beef anterior pituitary, *J. Cell. Biol.* **40:**225–235.

Van der Werf, P., Orlowski, M., and Meister, A., 1971, Enzymic conversion of 5-oxo-L-proline (L-pyrrolidone carboxylate) to L-glutamate coupled with cleavage of adenosine-triphosphate to adenosine-diphosphate, a reaction in gamma-glutamyl cycle, *Proc. Natl. Acad. Sci. U.S.A.* **68:**2982–2985.

Vanha-Pertulla, T., 1969, Aminoacyl and dipeptidyl arylamidases (aminopeptidases) of the pituitary gland related to function, *Endocrinology* **85:**1062–1069.

Walter, R., Shlank, H., Glass, J. D., Schwartz, I. L., and Kerenyl, T. D., 1971, Leucylglycineamide released from oxytocin by human uterine enzyme, *Science* **173:**827–829.

White, N., Jeffcoate, S. L., Griffiths, E. C., and Hooper, K. C., 1976, Effect of thyroid hormone–degrading activity of rat serum, *J. Endocrinol.* **71:**1–7.

Yoshimoto, T., Fischl, M., Orlowski, R. C., and Walter, R., 1978, Post-proline cleaving enzyme and post-proline dipeptidyl aminopeptidase, *J. Biol. Chem.* **253:**3708–3716.

20

Methods of Labeling Pituitary Hormones

P. De La Llosa

1. Introduction

Labeling of pituitary hormones has become an essential technique in endocrinological research, extremely valuable in the study of the mechanism of action of hormones and of their biosynthesis and metabolism. To label these proteins, radioactive atoms are introduced into the purified hormones by means of various chemical reactions. The most widely employed radioactive atoms are radioiodine and tritium. The specific activity of carbon-14 is low for this kind of study in most cases. Attempts have also been made to label hormones with carbon-11, a short-lived radionuclide, which can be detected externally with a gamma camera, a great advantage for *in vivo* studies.

The high specific radioactivity of iodine and its moderate half life (60 days for iodine-125) have drawn the interest of research workers toward this radioelement. Radioiodination is the most widely used method of labeling at present. The reaction is generally obtained by producing "active iodine" in the medium by means of a chemical oxidizing agent or by enzymatic action. Iodination can also be performed with iodine chloride, although this method is less efficient if very high specific activity is required. The iodination occurs chiefly on the phenolic nucleus of tyrosine and rarely on histidine imidazole. Other sites in the protein that can react

P. De La Llosa • Laboratoire des Hormones Polypeptidiques, C.N.R.S., 91190 Gif-sur-Yvette, France

with iodine are the sulfhydryl groups and tryptophan (Cohen, 1968). The oldest and most widely used procedure is that which uses chloramine T as the oxidizing agent. However, its disadvantages are becoming more and more evident. Another procedure, based on the action of lactoperoxidase on iodide in the presence of small amounts of hydrogen peroxide, is being employed increasingly more for this reason.

2. Radioiodination Methods

2.1. Radioiodination with Chloramine T

The pioneer work in this area was done by Greenwood et al. (1963). Those procedures that have been reported and recommended more recently are the modifications that were introduced in the method of Greenwood and co-workers to avoid the deleterious action of chloramine T. Indeed, chloramine T has been proposed for the specific oxidations of methionine residues to methionine sulfoxide in proteins by Shechter et al. (1975). In the case of bovine luteinizing hormone (LH), it has been shown (De La Llosa et al., 1977) that this modification can be achieved almost quantitatively without significant oxidation of the cystine residues and with only moderate destruction of the terminal fucose (30%). The oxidation of the methionine residues of LH results in complete suppression of biological activity.

Hovever, by reducing the exposure of the protein to chloramine T in the radioiodination procedures, most of the biological activity can be preserved, at least in the case of human chorionic gonadotropin (hCG) and LH (Midgley, 1966; Lunenfeld and Eshkol, 1967; Dufau et al., 1972; Kammerman and Canfield, 1972; Leidenberger and Reichert, 1972a,b). These last authors have emphasized the importance of time and temperature for obtaining a biologically active iodinated protein (20 sec and 2–3°C are the optimal conditions with a chloramine T/hormone ratio of 2:1, according to these authors) (see Table I).

However, if the incorporation of iodine into the protein is less than 1 atom per molecule, the interpretation of the results may be complicated.

Unfortunately, follicle-stimulating hormone (FSH) and thyrotropin appear to be more fragile (Reichert and Bhalla, 1974). Iodination of FSH by the chloramine T technique has been reported by several authors. The study of Bell et al. (1974) has shown, however, that a considerable fraction of this hormone is transformed into subunits or an aggregated form during the iodination procedure, even under the conditions that lead to incorporation of less than 1 iodine atom per molecule. A disconcerting fact is that it is just this aggregated material that exhibits the maximum of activity in

Table I. Modification of Conditions for Iodination of Proteins[a]

Modifi-cation	Chloramine T(μg)/ hormone (μg) ratio	Temper-ature (°C)	Time of exposure before addition of NA-metabisulfite (sec)	Specific activity (mCi/μg protein)
A	12 : 1	21–23	120	120–160
B	2 : 1	2–3	20	50–100
C	4 : 1	21–23	60	105[b]
D	1 : 2	2–3	15	10–20

[a] From Leidenberger and Reichert (1972a).
[b] Based on a single iodination (see Fig. 1).

the radioligand receptor test. For these reasons, most researchers now prefer to resort to the peroxidase procedure (also see Miyachi and Chrambach, 1972).

In Chapter 18, Dr. Ramachandran and co-workers have exposed the difficulties encountered in the iodination of ACTH. Other hormones, however, seem to support radioiodination by chloramine T without special complications. As is known, the method of Greenwood *et al.* (1963) was first applied to somatotropin. With slight modifications, application of this method to this hormone or to prolactin still gives satisfactory results (see, for example, Ranke *et al.*, 1976).

2.2. Radioiodination by the Enzymatic Method

Because of several disadvantages of the chloramine T oxidation technique, the enzymatic method of iodination was proposed as an alternative procedure by Marchalonis (1969) and by Thorell and Johansson (1971). In this procedure, lactoperoxidase, iodide, and the hormone are dissolved in a buffer, and the reaction is initiated by the addition of a small amount of hydrogen peroxide. This method has been applied to various pituitary hormones (see, for example, Miyachi *et al.*, 1972; Shiu and Friesen, 1974; Rolland *et al.*, 1976). In the case of iodinated thyrotropin, 80% of the biological activity seems to be preserved (Jaquet *et al.*, 1974). Better specific binding of iodinated FSH to rat testis receptors is observed when iodination is performed by this method (Kettelslegers and Catt, 1974). A detailed study of the iodination of porcine LH by lactoperoxidase was undertaken by Combarnous and Maghuin-Rogister (1974). They were able to show that the tyrosine residue located in position 21 in the α subunit is the most reactive residue. Even after incorporation of two iodine atoms into the molecule, the iodinated derivative exhibits 60% of its biological activity. If the iodination is continued, tyrosines α92 and α93 also react. But the heptaiodolutropin (in which the tyrosine 41 is also labeled) retains only 12% of its bioactivity (Fig. 1). The tyrosines of the β subunit are iodinated in the dissociated subunit, but do not react when the β subunit is associated with the α subunit, a feature previously reported by Yang and Ward (1972), who employed the chloramine T method.

Since the α subunit is common to lutropin, thyrotropin, and follitropin, it can be presumed that tyrosine $\alpha21$ is also iodinated immediately in these hormones (excepting, of course, in the case of human hormones, in which the corresponding position is occupied by a phenylalanine residue). Whether the tyrosine residues located in the β subunit can also be easily labeled in these hormones at the same time is a point that has not yet been examined.

3. Tritiation Methods

Even if iodination is the most widespread method of labeling and gives satisfactory results in a number of cases, the problems encountered in the radioiodination procedures gave impetus to search for other labeling procedures. Furthermore, in some cases the iodinated derivatives are not

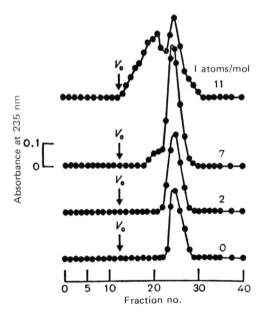

Figure 1. Gel-filtration patterns of porcine LH substituted with increasing amounts of iodine on Sephadox G-100. From Combarnous and Maghuin-Rogister (1974).

suitable for the study of some particular problems. For example, iodination of the β-LPH tyrosines prevents enzymatic cleavage in position 60–61. In any case, the corroboration of the results obtained with iodinated derivatives with those obtained using other kinds of derivatives is of great interest. Even when the iodinated derivatives are active, some problems may arise concerning the equivalence (quantitative and qualitative) of the native hormone and the derivative. This point has incited polemics and controversies in the case of glucagon, for example (Rodbell *et al.,* 1971; Bromer *et al.,* 1973; Lin *et al.,* 1975; Desbuqois, 1975; Von Schenk and Jeppson, 1977).

Thus, several attempts have been made to prepare radioactive derivatives by other methods, especially by introducing tritium. Dr. Ramachandran has discussed the catalytic hydrogenation of iodinated derivatives. I shall describe three other techniques that are easy to perform and can be employed in any conventional laboratory. The first example is based on the reversible oxidation of galactose. Second, the best results with the glycoprotein hormones are obtained by the oxidation–reduction of sialic acid. Finally, modification of the lysine residues by reductive alkylation appears to be compatible with the preservation of the biological activity of some hormones, and this reaction affords a new and promising way of labeling them with both tritium and carbon-11.

3.1. Tritiation by Oxidation of Sugar Residues Followed by Reduction with Borohydride

The galactose oxidase method of tritiating protein, which was first applied to ceruloplasmin (Morell *et al.,* 1966), is based on the reaction scheme shown in Fig. 2. It involves oxidation of the terminal galactose by the highly specific enzyme galactose oxidase, followed by reduction with tritiated borohydride. This technique has been applied successfully to thyrotropin (Winand and Kohn, 1970) (confirming at the same time the existence of terminal galactose in this hormone). Unfortunately, the number of terminal galactose residues in the glycoprotein hormones is generally very small (maximum of 1 residue in bovine thyrotropin, for instance), and consequently the number of tritium atoms that can be introduced is not very large. Nevertheless, labeling of thyrotropin by this method has been very useful in investigations on the binding properties of this hormone (Amir *et al.,* 1973; Moore and Wolf, 1974).

The other procedure for the introduction of tritium into the sugar moiety of glycoprotein hormones is illustrated in Fig. 3. Sialic acid is first oxidized by periodate, and the derivative obtained is then reduced by tritiated borohydride (Van Lenten and Ashwell, 1971). Several glycoprotein hormones contain terminal sialic acid and can therefore be tritiated by this

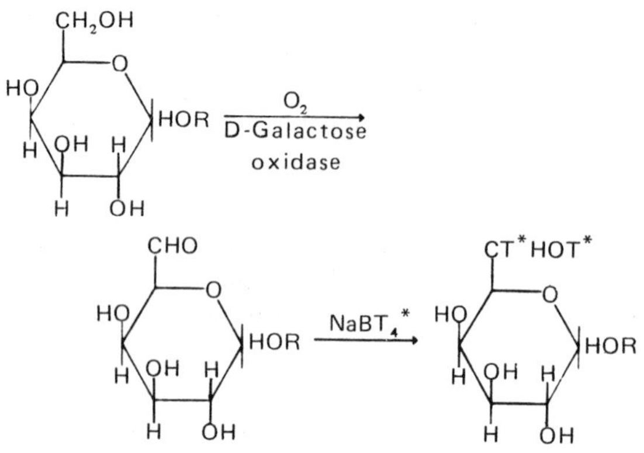

Figure 2. Reaction scheme of the oxidation of terminal galactose by D-galactose oxidase followed by reduction with tritiated borohybrid. From Winand and Kohn (1970).

method. In the case of hCG, 80% of the biological activity was preserved (Vaitukaitis *et al.*, 1971a), but only 50% for human FSH (see Table II) (Vaitukaitis *et al.*, 1971b). Human CG oxidized by periodate was less active than hCG that was oxidized and subsequently reduced by borohydride. Periodate oxidizes the methionine residues to methionine sulfoxide, which can be reduced in turn to methionine. In the case of bovine LH, periodate oxidation in conditions very similar to those employed in the hCG tritiation procedure (20°C, 30 min, ratio periodate/LH = 35) is accompanied by the loss of biological activity (De La Llosa *et al.*, 1977). In these conditions, disulfide bonds are not oxidized, but the terminal fucose is destroyed, as is galactose, during the labeling of hCG.

The low content of sialic acid in sialoprotein hormones limits the amount of tritium introduced, but these derivatives have been useful in studies on the specific binding of FSH to testis (Means, A. R., and Vaitukatis, 1972) or on the distribution of hCG in the ovary (Ashitaka *et al.*, 1973).

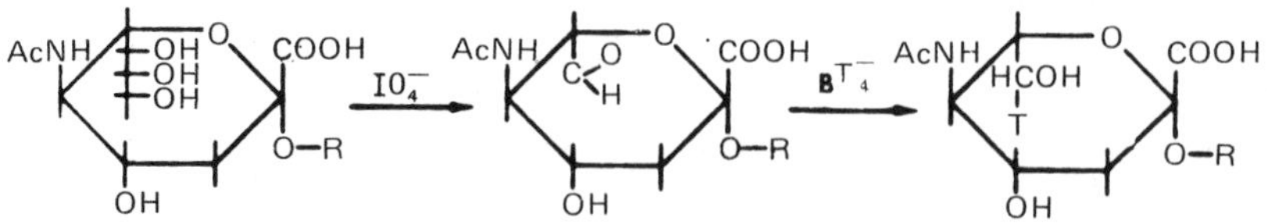

Figure 3. Chemical conversion of glycosidically bound sialic acid to the tritiated 5-acetamido-3,5-dideoxy-L-arabino-2-heptulosonic acid derivative. From Van Lenten and Ashwell (1971).

Table II. Biological Activity of Tritiated FSH and Its
Stability[a]

	Time after labeling (months)			
	0	1	2	4
Native FSH	—	—	2975	2750
			(1838–6265)	(2150–3225)
[³H]-FSH	1252	—	1444	884
	(822–1768)		(998–1982)	(633–1144)

[a] From Vaitukaitis *et al.* (1971b). Values are expressed in IU/mg; figures in parentheses represent 95% confidence limit of potency estimate.

3.2. Tritiation by Reductive Alkylation

A method recently proposed for the tritiation of pituitary hormones is the reductive alkylation of the lysine residues, a method that can be used in proteins and peptides without a sugar moiety and that allows the introduction of a large number of tritium atoms. Reductive alkylation is achieved by hydrogenation of Schiff's bases. Using borohydride, G. E. Means and Feeney (1968) made this reaction much easier and more practicable in protein chemistry. When reductive alkylation is performed with formaldehyde, the dimethyl derivative is usually obtained. When acetaldehyde, acetone, and the like are used, only the monoalkylated derivative is formed. The reactions leading to the dimethylated lysine are summarized in Fig. 4.

When the degree of methylation is over 50%, the proportion of monoalkylated derivative is very low (<10%). Since the properties of the methylated lysine are not very different from those of the free amino acid, they must be identified by means of special buffers containing isopropanol. The α amino group of the peptide chain is also alkylated; however, since the corresponding derivatives do not react with ninhydrin, they cannot be directly identified during amino acid analysis. No other group is methylated. The only secondary reaction that may occur during reductive alkylation is the hydrogenation and scission of some very labile disulfide bonds. If necessary, these bonds can be restored by dialysis against the oxidized form of glutathione. The pK value of dimethylated amino groups is about 0.5 lower than the pK of lysine (Table III). Methylation of lysine is thus a slight chemical modification that can be introduced in this amino acid residue. But even if very slight, this modification may have important repercussions on the biological activity in some cases. For this reason, application of this method is limited, and it is necessary to control the biological activity of the derivative obtained.

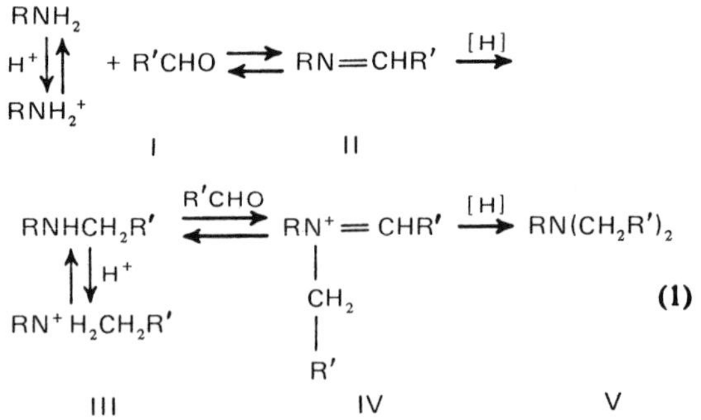

$$RNH_2$$

$$H^+ \updownarrow \quad + R'CHO \rightleftharpoons RN{=}CHR' \xrightarrow{[H]}$$

$$RNH_2^+$$

I   II

$$RNHCH_2R' \xrightleftharpoons{R'CHO} RN^+{=}CHR' \xrightarrow{[H]} RN(CH_2R')_2$$

$$\updownarrow H^+$$

$$RN^+H_2CH_2R' \qquad\qquad CH_2 \qquad\qquad\qquad (1)$$

$$R'$$

III   IV   V

Figure 4. Reaction of amino groups with carbonyl compounds in the presence of sodium borohydrid. From G. E. Means and Feeney (1968).

LH is a good example of this tritiation method. Its α subunit contains 10 lysine residues. In the ovine and bovine species, the β subunit contains two residues, but neither of them seems to be essential for the biological activity (De La Llosa et al., 1974a). Figure 5 shows the diagram of gel filtration of tritiated methylated LH on Biogel P-100 (De La Llosa et al., 1974b). As can be seen, the reaction results in quantitative yields; no dissociation of the hormone into subunits or aggregation of the protein molecules is observed. The degree of alkylation and specific radioactivity of two different preparations are shown in Table IV.

Comparison of the specific radioactivity of preparation 73 with that of the tritiated $NaBH_4$ employed shows that methylation was achieved only on the ϵ and α NH_2 groups of the peptide chain, which confirms that the amino groups of the osamines are blocked (acetylated), as expected.

Methylated LH exhibits in vivo (OAAD test) a biological activity somewhat higher than that of native LH. But its stereidogenic potency in vitro, in the case of biosynthesis of progesterone or testosterone, is not significantly different from that of native LH. Its stability is shown in Table V. About half its biological activity is lost after 8 months of storage

Table III. Values for pK of Methylated
Amino Groups[a]

Compound	pK
α-N-Acetylysine	10.8
α-N-Acetyl-ϵ-N,N-dimethyllysine	10.4
Alanine	9.8
α-N,N-Dimethylalanine	9.2

[a] From G. E. Means and Feeney (1968).

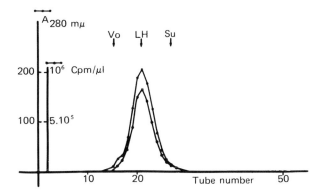

Figure 5. Gel filtration of tritiated methylated LH on a Biogel P-100 column. Arrows indicate the elution volumes corresponding to the void volume (Vo), native luteinizing hormone (LH), and LH subunits (Su). From De La Llosa *et al.* (1974).

at $-196°C$. At this time, some evolution of the K_D value of association of the hormone to the receptors of corpora lutea is observed. Tritiated methylated LH (freshly prepared) has been used in our laboratory to investigate the binding properties of LH to its receptors (De La Llosa-Hermier *et al.*, 1976a,b). The results obtained were similar to those reported by other authors using iodinated hormone. It was therefore possible to put together a radioligand receptor assay to study the relationship between chemical structure and biological activity (De La Llosa-Hermier *et al.*, 1977). This derivative was also employed to study the association of LH with gangliosides present in the corpora lutea (as well as in the plasma-membrane fraction and in the cytosol fraction). This association, which occurs not only between gangliosides and native LH but also between gangliosides and the LH subunits or inactive derivatives, appears to be very different from the very specific association observed in the case of the membrane receptors (De La Llosa and Durosay, 1977).

Tritiation by reductive methylation can also be applied to hCG. The

Table IV. Degree of Alkylation and Specific Radioactivity of Tritiated Methylated LH[a]

Methylated derivative	Degree of alkylation (%)			Radioactivity (Ci/mmol)
	Dimethyl Lys	Monomethyl Lys	Free Lys	
Preparation 72	60	10	30	98
Preparation 73[b]	84	6	10	35

[a] From De La Llosa *et al.* (1974b).
[b] The specific radioactivity of the tritiated NaBH₄ used in this experiment was 6.51 Ci/mmol.

Table V. Biological Activity of Tritiated Methylated LH[a]

Period of storage at $-196°C$	Relative potencies in terms of native LH	Relative potencies in terms of unlabeled methylated LH[b]
1st month	1.69 (0.83–3.4)	1.12 (0.54–2.3)
	2.07 (0.88–4.8)	
8th month	1.00 (0.55–1.82)	0.62 (0.34–1.10)

[a] From De La Llosa et al. (1974b). Relative potencies and 95% confidence limits are expressed in terms of either native LH or methylated LH.
[b] The relative potency of this preparation with respect to native LH was measured in three separate bioassays: March 1972, 1.43 (1.01–2.02); May 1973, 1.76 (0.65–4.8); March 1974, 1.52 (0.68–3.46).

methylated hCG exhibits about 80% of the bioactivity of the native hormone (De La Llosa and Hermier, unpublished results). On the other hand, this procedure has also been used for introducing carbon-11 atoms in hormones (Marche et al., 1975). Unfortunately, this method of tritiation can not be extended to other related hormones. Thus, methylated thyrotropin with 50% of alkylated lysine residues loses most of its biological activity (De La Llosa and Hennen, unpublished results). The β subunit of this hormone is much richer in lysine residues, some of which may play an important role in the biological activity. Similarly, 98% of the biological activity of purified bovine FSH (as measured by its ability to stimulate ovarian adenylate cyclase) is suppressed by methylation (C. Tertrin-Clary and P. de La Llosa, unpublished data). The presence —or the absence— of lysine and its role in the interaction with the receptors is perhaps one of the elements involved in evolution that permitted the narrow specialization and the emergence of divergent physiological activities in this group of closely related molecules.

In conclusion, the preparation of radioactive derivatives of pituitary hormones can be undertaken today by several procedures. Radioiodination is the most widely used, and it often gives satisfactory results. The use of lactoperoxidase for iodination avoids secondary reactions and reduces the formation of damaged products. This method is preferable in any case to the chloramine T procedure.

Tritiated derivatives of several pituitary hormones are also available. They constitute complementary tools and, in some cases, show advantages over the iodinated derivatives.

4. Summary

The preparation of radioactive derivatives of pituitary hormones can be undertaken today by several procedures. Radioiodination is the most widely used, and it often gives satisfactory results. The use of lactoperoxidase for iodination avoids secondary reactions and reduces the formation of damaged products. This method is preferable in any case to the chloramine T procedure.

Tritiation of glycoproteins can be achieved by oxidation of terminal sugars (galactose or sialic acid), followed by reduction with tritiated borohydride. Another procedure of tritiation is reductive methylation of lysine residues. The advantages and disadvantages of these methods are discussed in this chapter.

Discussion

RAMACHANDRAN: Do you get complete methylation of the LH?

DE LA LLOSA: No, we can obtain 90% methylation, but some lysines in the β subunit, for instance, are resistant and are not completely methylated even after repeating the alkylation.

RAMACHANDRAN: Then do you separate the unmodified hormone from the methylated species?

DE LA LLOSA: No, we don't separate. We probably have a heterogeneous population of methylated LH molecules, a fraction of them methylated and some not methylated on lysine 20, for instance.

RAMACHANDRAN: With ACTH, we find there is peptide-bond cleavage at the sites where tritium is introduced. Because of low specific activity, you have to introduce many methyl groups. I think the chemical damage increases with increase in the number of sites of tritiation.

DE LA LLOSA: No differences were observed between methylated LH and methylated tritiated LH as determined by Biogel P-100 filtration or by amino acid composition. Maybe the loss of biological activity in storage is due to slight changes in conformation.

KERDELHUE: Is [³H] methylated LH more potent than native LH in terms of biological acitvity? Did you find similar results in regard to cyclic AMP production?

DE LA LLOSA: In OAAD test *in vivo*, methylated LH is more potent than native LH; as it is also in terms of cyclic AMP production using plasma membranes prepared from immature rat ovaries. In the plasma membranes, the difference is abolished if EDTA is added to the incubation medium. This means, perhaps, that calcium is involved in the stimulation process of this adenylate cyclase by LH. In regard to *in vitro* steroid production (either progesterone or testosterone), no difference is observed between the potency of native LH and methylated LH.

JUTISZ: Did you try to methylate LH subunits and to reassociate them?

DE LA LLOSA: The dissociated LH subunits can be methylated separately. The methylated β subunit, which contains only 2 lysine residues, reassociates with the intact α subunit with good yield, but not the methylated α subunit with intact β subunit. The α subunit contains ten lysine residues, and the association of the methylated derivative (85% of the lysines are modified) with the intact β subunit occurs only with 50% yield. In any case, the dimers obtained by the association of a methylated subunit and an intact subunit exhibit lower biological activity than the products obtained by direct methylation of the dimer.

JUTISZ: Do you think that the bad yield of reassociation or a lesser activity of reassociated product may be due to the fact that ϵNH_2 of lysine are necessary to the binding of the two subunits?

DE LA LLOSA: It seems more probable that the methylation of the lysine residues in dissociated subunits is accompanied by some distortion of the conformation of the molecule. Reassociation is not absolutely prevented, but becomes less easy to be achieved, and the recombination products probably do not recover entirely the native conformation and, consequently, the biological activity. The conformation of the dissociated subunits is certainly more fragile than that of the dimer, as evidenced by the incomplete recovery of the biological activity when the dissociation of the subunits is achieved using acidic pH.

REFERENCES

Amir, S. M., Carraway, T. F., Jr, Kohn, L. D., and Winand, R. J., 1973, The binding of thyrotropin to isolated bovine thyroid plasma membranes, *J. Biol. Chem.* **248**:4092.

Ashitaka, Y., Tsong, Y. Y., and Koide, S. S., 1973, Distribution of tritiated human chorionic gonadotropin in superovulated rat ovary, *Proc. Soc. Exp. Biol. Med.* **142**:395.

Bell, J., Benveniste, R., Schwartz, S., and Rabinowitz, D., 1974, Iodinated FSH:Components and their properties, *Endocrinology* **94**:952.

Bromer, W. W., Boucher, M. E., and Patterson, J. M., 1973, Glucagon structure and function. II. Increased activity of iodoglucagon, *Biochem. Biophys. Res. Commun.* **53**:134.

Cohen, L. A., 1968, Group-specific reagents in protein chemistry, *Annu. Rev. Biochem.* **37**:695.

Combarnous, Y., and Maghuin-Rogister, G., 1974, Luteinizing hormone: Relative reactivities of tyrosil residues of the porcine hormone towards iodination, *Eur. J. Biochem.* **42**:13.

De La Llosa, P., and Durosay M., 1977, Interaction of lutropin with gangliosides from bovine corpora lutea, *Acta Endocrinol. Suppl.* **212**:124.

De La Llosa, P., Durosay, M., Tertrin-Clary, C., and Jutisz, M., 1974a, Chemical modification of lysine residues in ovine luteinizing hormone: Effect on biological activity, *Biochim. Biophys. Acta* **342**:97.

De La Llosa, P., Marche, P., Morgat, J. L., and De La Llosa-Hermier, M. P., 1974b, A new procedure for labelling luteinizing hormone with tritium, *FEBS Lett.* **45**:162.

De La Llosa, P., El Abed, A., and Roy, M., 1977, Oxidation of methionine residues in lutropin, *Z. Physiol. Chem.* **358**:1241.

De La Llosa-Hermier, M. P., Hermier, C., and De La Llosa, P., 1976a, Binding of tritiated methylated luteinizing hormone to bovine *corpus luteum* receptors, *Acta Endocrinol.* **83**:393.

De La Llosa-Hermier, M. P., Hermier, C., and De La Llosa, P., 1976b, Mise au point d'un dosage de l'hormone luteinisante (LH) basé sur l'inhibition competitive de l'interaction entre l'hormone tritié et ses récepteurs de gonades: Application à l'étude de relations structure hormonale–activité de liaison, *J. Physiol. (Paris)* **72**:24B.

De La Llosa-Hermier, M. P., De La Llosa, P., and Hermier, C., 1977, Studies of the binding activity to different gonadal receptors of ovine luteinizing hormone (LH) after chemical modification of lysine residues, *Gen. Comp. Endocrinol.* **31**:302.

Desbuquois, B., 1975, Iodoglucagon: Preparation and characterization, *Eur. J. Biochem.* **53**:569.

Dufau, M. L., Catt, K. J., and Tsuruhara, T., 1972, Biological activity of human chorionic gonadotropin released from testis binding sites, *Proc. Natl. Sci. U.S.A.* **69**:2414.

Greenwood, F. C., Hunter, W. M., and Glover, J. S., 1963, The preparation of ¹³¹I-labelled human growth hormone of high specific radioactivity, *Biochem. J.* **89**:1963.

Jaquet, P., Hennen, G., and Lissitzky, S., 1974, Enzymatic radioiodination of porcine thyroid-stimulating hormone, *Biochimie* **56**:769.

Kammerman, S., and Canfield, R. E., 1972, The inhibition of binding of iodinated human chorionic gonadotropin to mouse ovary *in vivo*, *Endocrinology* **90**:384.

Kettelslegers, J. M., and Catt, K. J., 1974, Receptor binding properties of ¹²⁵I-hFSH prepared by enzymatic iodination, *J. Clin. Endocrinol. Metab.* **39**:1159.

Leidenberg, F., and Reichert, L. E., Jr, 1972a, Studies on the uptake of human chorionic gonadotropin and its subunits by rat testicular homogenates and interstitial tissue, *Endocrinology* **91**:135.

Leidenberg, F., and Reichert, L. E., Jr., 1972b, Evaluation of a rat testis homogenate radioligand receptor assay for human LH, *Endocrinology* **91**:901.

Lin, M. C., Wright, D. E., Hruby, V. I., and Rodbell, M., 1975, Structure–function relationships in glucagon: Properties of highly purified des-His¹-, monoiodo, and [des Asn²⁸,Thr²⁹] (homoserine²⁷)-glucagon, *Biochemistry* **14**:1559.

Lunenfeld, B., and Eshkol, A., 1967, Immunology of human chorionic gonadotropin (HCG), *Vit. Horm.* **25**:137.

Marchalonis, J. J., 1969, An enzymic method for the trace iodination of immunoglobulins and other proteins, *Biochem. J.* **113**:299.

Marche, P., Marazano, C., Maziere, M., Morgat, J. L., De La Llosa, P., Comar, D., and Fromageot, P., 1975, ¹¹C-labelling of ovine luteinizing hormone by reductive methylation, *Radiochem. Radioanal. Lett.* **21**:53.

Means, A. R., and Vaitukaitis, J., 1972, Peptide hormone "receptors": Specific binding of ³H-FSH to testis, *Endocrinologie* **90**:39.

Means, G. E., and Feeney, R. E., 1968, Reductive alkylation of amino groups in proteins, *Biochemistry* **7**:2192.

Midgley, A. R., 1966, Radioimmunoassay: A method for human chorionic gonadotropin and human luteinizing hormone, *Endocrinology* **79**:10.

Miyachi, Y., and Chrambach, A., 1972, Structural integrity of gonadotropins after enzymatic iodination, *Biochem. Biophys. Res. Commun.* **46**:1213.

Miyachi, Y., Vaitukaitis, J., Nieschlag, E., and Lipsett, M. B., 1972, Enzymatic radioiodination of gonadotropin, *J. Clin. Endocrinol. Metab.* **34**:23.

Moore, W. V., and Wolff, J., 1974, Thyroid-stimulating hormone binding to beef thyroid membranes: Relation to adenylate cyclase activity, *J. Biol. Chem.* **249**:6255.

Morell, A. G., Van Den Hamer, C. J. A., Scheinberg, I. H., and Aswell G., 1966, Physical and chemical studies on ceruloplasmin: Preparation of radioactive, sialic acid free ceruloplasmin labeled with tritium on terminal D-galactose residues, *J. Biol. Chem.* **241**:3745.

Ranke, M. B., Stanley, C. A., Rodbard, D., Baker, L., Bongiovanni, A., and Parks, J. S., 1976, Sex differences in binding of human growth hormone to isolated rat hepatocytes, *Proc. Natl. Acad. Sci. U.S.A.* **73**:847.

Reichert, L. E., Jr., and Bhalla, V. K., 1974, Development of a radioligand tissue receptor assay for human follicle-stimulating hormone, *Endocrinology* **94**:483.

Rodbell, M., Kraus, H. M. J., Pohl, S. L., and Birnbaumer, L., 1971, The glucagon-sensitive adenyl cyclase system in plasma membranes of rat liver, *J. Biol. Chem.* **246**:1861.

Rolland, R., Gunsalus, G. L., and Hammond, J. M., 1976, Demonstration of specific binding of prolactin by porcine corpora lutea, *Endocrinology* **98**:1083.

Shechter, Y., Burstein, Y., and Patchornik, A., 1975, Selective oxidation of methionine residues in proteins, *Biochemistry* **14**:4497.

Shiu, R. P. C., and Friesen, H. G., 1974, Properties of a prolactin receptor from the rabbit mammary gland, *Biochem. J.* **140:**301.

Thorell, J. I., and Johansson, B. G., 1971, Enzymatic iodination of polypeptides with ^{125}I to high specific activity, *Biochim. Biophys. Acta* **251:**363.

Vaitukaitis, J., Hammond, J., Ross, G. T., Hickman, J., and Ashwell, G., 1971a, A new method of labeling human chorionic gonadotropin for physiological studies, *J. Clin. Endocrinol. Metab.* **32:**290.

Vaitukaitis, J. L., Sherins, R., Ross, G. T., Hickman, J., and Ashwell, G., 1971b, A method for the preparation of radioactive FSH with preservation of biological activity. *Endocrinology* **89:**1356.

Van Lenten, L., and Ashwell, G., 1971, Studies on the chemical and enzymatic modification of glycoproteins: A general method for the tritiation of sialic acid-containing proteins, *J. Biol. Chem.* **246:**1889.

Von Schenk, H., and Jeppson, J. O., 1977, Preparation of monoiodotyrosine-13-glucagon, *Biochim. Biophys. Acta* **491:**503.

Winand, R. J., and Kohn, L., 1970, Relationship of thyrotropin to exophthalmic-producing substance: Purification of homogenous glycoproteins containing both activities from [3H]-labeled pituitary extracts, *J. Biol. Chem.* **245:**967.

Yang, K. P., and Ward, D. N., 1972, Iodination of ovine luteinizing hormone and its subunits, *Endocrinology* **91:**317.

21

Mechanism of Action of Hypothalamic Hormones and Interactions with Sex Steroids in the Anterior Pituitary Gland

Fernand Labrie, Pierre Borgeat, Martin Godbout,
Nicholas Barden, Michèle Beaulieu,
Lisette Lagacé, Jocelyne Massicotte, and
Raymonde Veilleux

1. Introduction

Although peripheral hormones have been shown for many years to play a major role in the control of adenohypophyseal activity in man and experimental animals, *in vivo* approaches could not distinguish between hypothalamic and pituitary sites of action. This area of research has been much facilitated by the development of the pituitary cell culture system (Vale *et al.*, 1972; Labrie *et al.*, 1973). In fact, adenohypophyseal cells in primary culture have been extremely useful not only for assessment of the biological activity of analogues of thyrotropin-releasing hormone (TRH),

Fernand Labrie, Pierre Borgeat, Martin Godbout, Nicholas Barden, Michèle Beaulieu, Lisette Lagacé, Jocelyne Massicotte, and *Raymonde Veilleux* • Medical Research Council Group in Molecular Endocrinology, Le Centre Hospitalier de l'Université Laval, Quebec G1V 4G2, Canada

luteinizing hormone–releasing hormone (LH-RH), and somatostatin (Labrie *et al.*, 1973, 1976a, b; Belanger *et al.*, 1974), but also for determination of the characteristics of interaction between hypothalamic and peripheral hormones at the adenohypophyseal level (Drouin *et al.*, 1976a,b; Drouin and Labrie, 1976a,b).

This chapter will attempt to summarize the evidence obtained so far on the effect of three synthetic hypothalamic hormones, namely, TRH, LH-RH, and somatostatin, as well as one catecholamine, dopamine (DA), on cyclic AMP accumulation in anterior pituitary gland. Since the characteristics of binding of TRH and properties of cyclic-AMP-dependent adenohypophyseal protein kinase and some of its substrates have been described in recent reviews (Labrie *et al.*, 1975a,b), these aspects will not be included in this discussion.

Since it is known that LH-RH stimulates the secretion of both luteinizing hormone (LH) and follicle-stimulating hormone (FSH) (Borgeat *et al.*, 1972; Labrie *et al.*, 1973), the divergence frequently observed *in vivo* between the rate of secretion of the two gonadotropins can be best explained by differential effects of gonadal steroids at the pituitary level on the secretion of these two hormones. Emphasis will thus be given to the specific effects of androgens, estrogens, and progesterone on basal and LH-RH-induced secretion of LH and FSH in anterior pituitary cells in culture. Data describing the effects of "inhibin" of testicular and ovarian origin at the pituitary level on gonadotropin secretion will also be presented. These *in vitro* studies will be complemented by the *in vivo* experiments aimed at dissociating the steroid feedback effects at the hypothalamic and pituitary levels.

Since estrogens are known to be potent stimulators of prolactin secretion, our studies of the interaction of estrogens with dopaminergic action at the pituitary level both *in vitro* and *in vivo* will then be presented. Estrogens were found to act directly at the pituitary level and, more surprisingly, to have potent antidopaminergic activity on prolactin secretion.

2. Stimulatory Effect of LH-RH and TRH on Pituitary Cyclic AMP Accumulation

The observations that theophylline and cyclic AMP derivatives (Labrie *et al.*, 1973; Ratner, 1970) have a stimulatory effect on LH release and that theophylline potentiates the effect of a crude preparation of FSH-releasing hormone on FSH release (Jutisz and Paloma De la Llosa, 1970) already suggested that cyclic AMP plays a role in the control of gonadotropin secretion. Definite proof that the adenylate cyclase system is a mediator of the action of LH-RH had to be obtained, however, by measurement of adenohypophyseal adenylate cyclase activity or cyclic AMP concentration under the influence of the pure neurohormone.

It is now well know that addition of LH-RH leads to stimulation of cyclic AMP accumulation in rat anterior pituitary gland *in vitro* (Borgeat *et al.*, 1972, 1974a; Kaneko *et al.*, 1973; Makino, 1973; Naor *et al.*, 1975a). The concentration of LH-RH required for half-maximal stimulation of cyclic AMP accumulation is 0.1–1.0 ng/ml, or 0.1–1 nM LH-RH. When LH-RH analogues having a spectrum of biological activity ranging between 0.001 and 500–1000% of the activity of LH-RH itself were used, the same close parallelism between stimulation of cyclic AMP accumulation and both LH and FSH release was found under all experimental conditions (Borgeat *et al.*, 1974a). That LH-RH exerts its action by activating adenylate cyclase and not by inhibiting cyclic nucleotide phosphodiesterase is indicated by the observation that a similar effect of the neurohormone is observed in the presence or absence of theophylline (Borgeat *et al.*, 1972).

The possibility of developing a contraceptive method based on inhibitory LH-RH analogues has led to the synthesis of many such substances, some of which are potent inhibitors of LH-RH action both *in vivo* (Ferland *et al.*, 1975b) and *in vitro* (Labrie *et al.*, 1976c). The availability of such LH-RH antagonists offered the possibility of investigating the correlation between their inhibitory effect on LH-RH-induced cyclic AMP accumulation and LH and FSH release.

As an example, Fig. 1 shows the inhibitory effect of increasing concentrations of [D-Phe^2, D-Leu^6] LH-RH on cyclic AMP accumulation and LH and FSH release in rat anterior pituitary gland *in vitro*. The close correlation observed between inhibition of LH-RH-induced cyclic AMP accumulation and LH and FSH release adds strong support to the concept of an obligatory role of the adenylate cyclase system as mediator of LH-RH action in the anterior pituitary gland.

As additional direct evidence for a stimulatory effect of LH-RH on pituitary adenylate cyclase activity, the neurohormone has been found to stimulate cyclic AMP formation in rat anterior pituitary homogenate (Deery and Howell, 1973) and membrane fractions (Makino, 1973). A stimulatory effect of LH-RH on adenylate cyclase activity has also been reported in homogenate from the ventral lobe of the pituitary of the dogfish (Deery and Jones, 1975).

Since gonadotrophs represent about 5% of the total cell population in the anterior pituitary gland, it is not surprising that addition of LH-RH leads to only a 100-300% stimulation (over control) of anterior pituitary cyclic AMP concentration (Borgeat *et al.*, 1972, 1974a; Labrie *et al.*, 1973; Naor *et al.*, 1975a,b). To induce a significant increase of total cyclic AMP

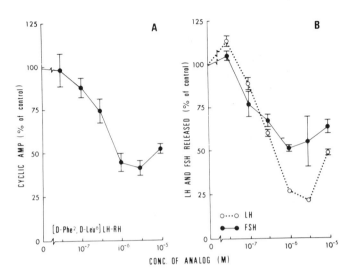

Figure 1. Effect of increasing concentrations of [D-Phe²,D-Leu⁶]-LH-RH on 3 nM LH-RH-induced cyclic AMP accumulation (A) and LH and FSH release (B) in male rat hemipituitaries *in vitro*.

accumulation, LH-RH must then stimulate specific cyclic AMP formation at least 20- to 60-fold in gonadotrophs.

We have recently found (Drouin *et al.*, 1976b) that estrogens increase the sensitivity of the LH responsiveness to LH-RH by a direct action at the pituitary level, while androgens have the opposite effect. Such a gonadal-hormone-induced change of pituitary responsiveness to LH-RH may explain why pituitaries obtained from male rats show a consistent increase of pituitary cyclic AMP levels under the influence of LH-RH (Borgeat *et al.*, 1972, 1974a; Labrie *et al.*, 1973, 1975a,b; Naor *et al.*, 1975a,b), while no significant effect could be observed using female rat pituitaries (Borgeat, Beaulieu, and Labrie, unpublished observations). The higher sensitivity to LH-RH in female animals is expected to require lower changes of cyclic AMP levels to induce LH release, while higher changes of the intracellular cyclic AMP concentration are likely to be needed in male pituitaries.

Suggestions against an obligatory role of cyclic AMP as mediator of LH-RH action pertain to the findings that nonspecific agents leading to changes of total pituitary cyclic AMP accumulation (such as theophylline, prostaglandins or inhibitors of their synthesis, cyclic AMP derivatives, and cholera toxin) did not always lead to parallel changes of LH release and cyclic AMP concentration (Naor *et al.*, 1975a; Sundberg *et al.*, 1976). Since it is now clear that prostaglandins are not involved in LH-RH action at the level of the anterior pituitary gland (Drouin and Labrie, 1976a; Drouin *et al.*, 1976b; Labrie *et al.*, 1976a; McCann *et al.*, 1976) and, as mentioned earlier, gonadotrophs represent only 5% of the total cell population in the anterior pituitary gland, it is not surprising that the changes of cyclic AMP levels observed with the aforementioned compounds take place in cell types other than gonadotrophs. In fact, somatotrophs represent approximately 50% of the total adenohypophyseal cell population and are highly sensitive to all the substances tested in the aforementioned studies (Barden *et al.*, 1976; Drouin and Labrie, 1976a; Labrie *et al.*, 1975b; 1976a,b). All these negative attempts to correlate changes of cyclic AMP levels with alterations of LH release can be explained by the lack of specificity of the substances used, which did not take into account the heterogeneity of the pituitary cell population. In fact, while the aforementioned nonspecific compounds could also be acting in other cell types, their action in somatotrophs could, by itself, explain all the reported changes of cyclic AMP levels that were not accompanied by specific effects on LH secretion (Naor *et al.*, 1975a,b; Sundberg *et al.*, 1976).

Although the changes of cyclic AMP levels were of relatively small magnitude, a significant increase (30% over control) was measured after 15 min of incubation with TRH, while a maximal effect at 50% over control was found after 2 hr of incubation (Labrie *et al.*, 1975a,b). As found previously with LH-RH for LH and FSH release, the changes of cyclic AMP levels induced by TRH were accompanied by parallel changes of thyroid-stimulating hormone (TSH) release. Since the experiments were performed in the presence of 5 mM theophylline, it is likely that the observed changes of cyclic AMP concentrations are secondary to parallel modifications of adenylate cyclase activity, rather than to inhibition of cyclic nucleotide phosphodiesterase.

3. Inhibitory Effect of Somatostatin and Dopamine on Pituitary Cyclic AMP Accumulation

Since we had found that a purified fraction of growth hormone–releasing hormone (GH-RH) led to a marked stimulation of pituitary cyclic AMP accumulation and growth hormone (GH) release (Borgeat *et al.*, 1973), it was of interest to study the effect of somatostatin on pituitary cyclic AMP accumulation. It was then found that somatostatin led to a rapid inhibition of cyclic AMP accumulation in anterior pituitary gland *in vitro* (Borgeat *et al.*, 1974b; Kaneko *et al.*, 1973), this inhibitory effect being accompanied by a marked inhibition of both GH and TSH release (Borgeat *et al.*, 1974b).

Since GH- and TSH-secreting cells account for 50–70% of the total adenohypophyseal cell population in adult male rats, the 50% inhibition of cyclic AMP accumulation in total pituitary tissue suggests an almost complete inhibition of cyclic AMP accumulation in the GH- and TSH-secreting cells. The inhibitory effect of somatostatin is observed under both basal and prostaglandin E_2- (PGE$_2$) or theophylline-induced conditions, thus suggesting an inhibitory action of somatostatin on adenylate cyclase activity.

The correlation observed between inhibition of cyclic AMP levels and GH release is further illustrated by an experiment performed in pituitary cells in primary culture (Fig. 2). It can be seen that the approximate 10-fold increase of cyclic AMP levels induced by 10^{-5}M PGE$_2$ is 60% inhibited by somatostatin at a half-maximal (ED$_{50}$) value of 0.3 nM. Under the same experimental conditions, GH release is 90–95% inhibited at maximal concentrations of somatostatin. It can also be observed that the inhibitory effect of somatostatin on cyclic AMP levels and GH release is measured at the same ED$_{50}$ value (0.3 nM). The absence of a significant effect of

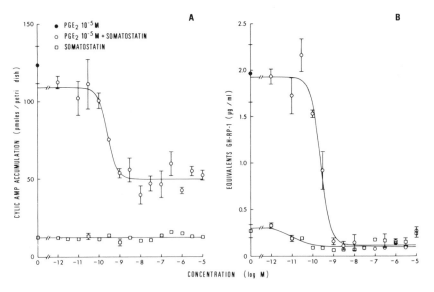

Figure 2. Effect of increasing concentrations of somatostatin alone (□) or in the presence of 10^{-5} M PGE$_2$ (○) on cyclic AMP accumulation (A) and GH release (B) in rat anterior pituitary cells in primary culture. Cells were incubated for 30 min in the presence of the indicated substances as described previously (Labrie *et al.*, 1973), and cyclic AMP was measured by radioimmunoassay.

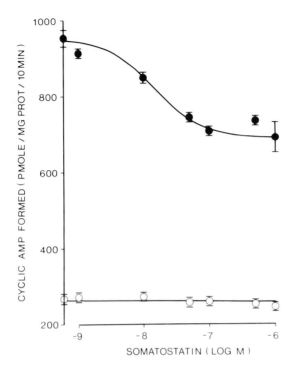

Figure 3. Inhibition of pituitary adenylate cyclase activity by somatostatin. Rat pituitary homogenate (100 μg protein) was incubated in 80 mM tris-maleate, pH 7.4, containing 0.2 mM EGTA, 10 mM theophylline, 4 mM MgSO$_4$, 5 × 10^{-6} M GTP, 0.15 mM cyclic AMP, and 1 mM [^{32}P]-ATP (5 × 10^6 dpm) for 10 min at 30°C, and the cyclic AMP formed was isolated on successive columns of Dowex and alanine. Addition of somatostatin caused no inhibition of basal adenylate cyclase activity (○), but at concentrations of 10^{-8} M or higher it caused a significant decrease ($p < 0.01$) of the PGE$_1$-stimulated adenylate cyclase activity (●) with an ED$_{50}$ value of 1.5 × 10^{-8} M.

somatostatin on basal cyclic AMP levels in cells in culture in the presence of an approximate 50% inhibitory effect on basal GH release can probably be explained by the presence of fibroblasts in the culture. As illustrated in Fig. 3, a similar inhibitory effect of somatostatin on PGE$_1$-induced adenylate cyclase activity was observed in rat anterior pituitary gland homogenate, a half-maximal inhibition ($p < 0.01$) being observed at 15 nM somatostatin.

Besides its intrinsic interest, this system could be advantageous as a model for studies of the mechanisms of action of peptides having opposite effects on cyclic AMP accumulation in the same cell type.

4. Inhibitory Effect of Dopamine on Pituitary Cyclic AMP Accumulation

Much evidence obtained in the rat indicates that dopamine (DA) secreted by the tuberoinfundibular system is the main factor involved in the control of prolactin secretion (MacLeod and Lehmeyer, 1974; Bishop *et al.*, 1972; Takahara *et al.*, 1974; Shaar and Clemens, 1974). According to these data, DA released from nerve endings in the median eminence is transported to the pituitary prolactin-secreting cells by the hypothalamo–adenohypophyseal portal blood system. In support of such a physiological role of DA at the pituitary level on prolactin secretion, DA has recently been measured in portal blood (Ben-Jonathan *et al.*, 1977), and a typical dopaminergic receptor has been characterized in anterior pituitary gland (Caron *et al.*, 1978; Labrie *et al.*, 1978; Creese *et al.*, 1977; Calabro and MacLeod, 1978; Brown *et al.*, 1976).

The first suggestive evidence for a role of cyclic AMP in the control of prolactin secretion originated from the observations that a cyclic AMP derivative (Lemay and Labrie, 1972) or theophylline, an inhibitor of cyclic nucleotide phosphodiesterase (Lemay and Labrie, 1972; Parsons and Nicoll, 1970; Wakabayashi *et al.*, 1973), stimulated prolactin release. These data obtained with theophylline and cyclic AMP derivatives already suggested that the cyclic nucleotide has a stimulatory role in the control of prolactin secretion.

More convincing evidence supporting a role of cyclic AMP in the action of DA on prolactin secretion had to be obtained, however, by measurement of adenohypophyseal adenylate cyclase activity or cyclic AMP concentration under the influence of the catecholamine. As illustrated in Fig. 4, addition of 100 nM DA to male rat hemipituitaries led to a rapid inhibition of cyclic AMP accumulation, a maximal effect (30% inhibition) being already obtained 5 min after addition of the catecholamine. Thus, while DA is well known to stimulate adenylate cyclase activity at the level of the striatum (Kebabian, 1977; Kebabian *et al.*, 1973), its effect at the adenohypophyseal level in intact cells is inhibitory.

The data presented so far clearly show that two stimulatory hypothalamic hormones, TRH and LH-RH, lead to parallel stimulation of cyclic AMP accumulation and specific hormone release, while one inhibitory peptide, somatostatin, and a catecholamine (DA) lead to parallel inhibition of cyclic AMP accumulation and hormone release. Such findings strongly suggest that changes of adenylate cyclase activity are involved in the mechanism of action of these three peptides and DA in the anterior pituitary gland.

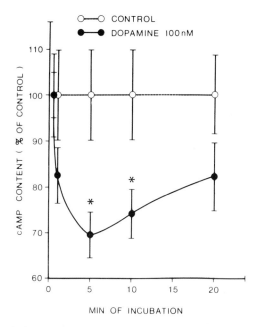

Figure 4. Effect of DA (100 nM) on cyclic AMP accumulation in male rat anterior pituitaries. The experiment was performed as described previously (Borgeat *et al.*, 1972).

5. Interactions among LH-RH, Sex Steroids, and Inhibin in the Control of LH and FSH Secretion

Although the influence of the hypothalamus on the secretion of both gonadotropins is probably exerted exclusively through LH-RH, it is well recognized that gonadal steroids can have a marked influence on LH and FSH secretion.

The recent observation that LH-RH can potentiate the LH response to subsequent injection of the neurohormone (Aiyer *et al.*, 1973; Castro-Vasquez and McCann, 1975; Ferland *et al.*, 1976) illustrates that it is almost impossible to distinguish between hypothalamic and pituitary sites of steroid action under *in vivo* conditions. In fact, a stimulatory effect of gonadal steroids on LH-RH secretion should lead to an increased LH responsiveness to the neurohormone (in the absence of any direct effect of the steroid at the pituitary level), while the opposite situation should follow the inhibitory effect of a steroid on LH-RH secretion.

As shown in Fig. 5, preincubation of male rat anterior pituitary cells

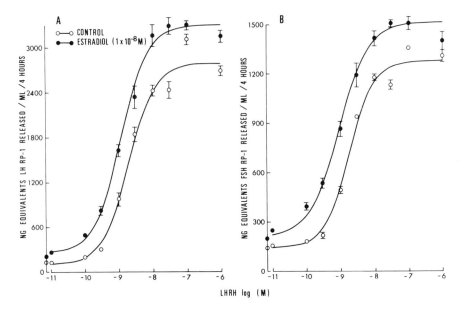

Figure 5. Effect of increasing concentrations of LH-RH on LH (A) and FSH (B) release by anterior pituitary cells in primary culture preincubated for 40 hr in the presence of 1×10^{-8} M 17β-estradiol (•) or control medium (○). Anterior pituitary cells were obtained from adult male rats. The response to LH-RH was performed during a 4-hr period after preincubation in the presence or absence of the estrogen. The results are presented as means ± S.E.M. of data obtained from triplicate dishes.

for 40 hr in medium containing 1×10^{-8} M 17β-estradiol (E_2) increased the LH responsiveness to LH-RH. The LH-RH concentration required to produce a half-maximal stimulation (ED_{50}) of LH release is decreased by E_2 pretreatment from 2.30 ± 0.03 to 1.20 ± 0.01 nM ($p < 0.01$). It can also be seen in Fig. 5A that preincubation with E_2 increased the basal LH release from 120 ± 8 to 205 ± 10 ng LH-RP-1/ml per 4 hr ($p < 0.01$). Moreover, in this and similar experiments performed with adenohypophyseal cells obtained from male and female animals (Drouin *et al.*, 1976b), the maximal LH response to LH-RH is slightly, but not significantly, increased. E_2 pretreatment increased both the basal FSH release and the maximal response of the hormone to LH-RH (Fig. 5B). Similar effects have been previously obtained in anterior pituitary cells obtained from female rats (Drouin *et al.*, 1976b).

This stimulatory effect of E_2 at the adenohypophyseal level may well be responsible, at least partly, for the increased LH and FSH sensititity to LH-RH observed at proestrus in the rat (Ferland *et al.*, 1975a; Gordon and Reichlin, 1974) and during the preovulatory period in the human (Nillius and Wide, 1971).

As illustrated in Fig. 6, pretreatment with 10^{-8} M testosterone led to a marked inhibition of the LH responsiveness to LH-RH, the LH-RH ED_{50} being increased from 1 to 3 nM in the presence of the androgen (p < 0.01). It can be seen that the androgen did not affect basal LH release, but slightly decreased the maximal response to the neurohormone. In contrast to the LH data, it can be seen that in the same experiment, testosterone did not significantly affect the LH-RH ED_{50} (1 nM) for FSH release. Both the spontaneous and maximal release of FSH were, however, slightly (30–40%) but consistently increased after androgen pretreatment (p < 0.01).

These data clearly show that androgens have not only specific but also opposite effects at the pituitary level on LH and FSH secretion. In fact, pretreatment of pituitary cells with androgens can markedly inhibit the LH response to LH-RH, while the effect on FSH secretion is stimulatory. Qualitatively similar results have been obtained when anterior

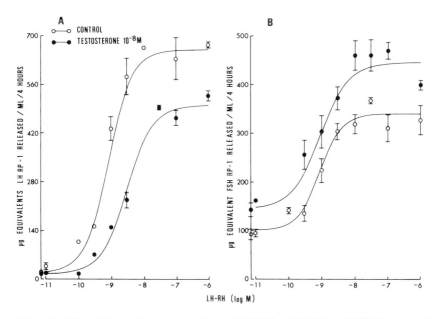

Figure 6. Effect of increasing concentrations of LH-RH on LH (A) and FSH (B) release by anterior pituitary cells in primary culture preincubated for about 40 hr in the presence (●) or absence (○) of 1×10^{-8} M testosterone. Results are expressed as means ± S.E.M. of triplicate determinations.

pituitary cells obtained from male or female rats were used. These findings can offer an explanation for the observations in the rat (Swerdloff *et al.,* 1972) and in man (Swerdloff and Odell, 1968) of a greater sensitivity of LH than FSH release to the inhibitory action of androgen administration *in vivo.*

The data summarized above show differential and specific effects of sex steroids on LH and FSH secretion: while estrogens stimulate both basal and LH-induced secretion of both LH and FSH, androgens and progesterone (in the presence of estrogens) inhibit LH but stimulate FSH secretion. It thus appears that the action of the three classes of sex steroids on FSH secretion at the adenohypophyseal level is exclusively stimulatory. These data provide some support for the concept first proposed by McCullagh (1932) of an inhibitory substance of testicular origin that could be involved in the specific inhibition of FSH secretion.

As illustrated in Fig. 7B, incubation of female rat anterior pituitary cells for 72 hr in the presence of increasing concentrations of Sertoli cell

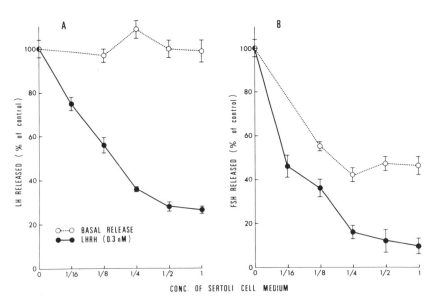

Figure 7. Effect of increasing concentrations of Sertoli cell culture medium (days 5–8 in culture, 35-day-old animals, 1.5 mg protein/7.5 ml culture medium) on basal (○) and 0.3 nM LH-RH-induced (●) LH (A) and FSH (B) release. The response to LH-RH was performed during a 4-hr period after a 72-hr preincubation in the presence of the indicated concentrations of Sertoli cell culture medium. The results are presented as percentage of control (mean ± S.E.M. of triplicate determinations). Hormone release in control cells under basal and LH-RH-induced conditions was: 25 ± 2 and 710 ± 21 (LH) and 20 ± 2 and 110 ± 5 (FSH) ng/ml, respectively.

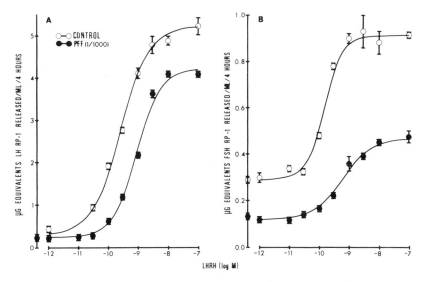

Figure 8. Effect of porcine follicular fluid on LH (A) and FSH (B) dose–response curves to LH-RH in anterior pituitary cells in culture. Cells were incubated for 40 hr in the presence (●) or absence (○) of porcine follicular fluid (PFF) at a 1:1000 dilution, while the LH-RH response was performed during a 4-hr period. The results are presented as mean ± S.E.M. of triplicate determinations.

culture medium (days 5–8 in culture) led to a maximal 45% inhibition of spontaneous FSH release, while no effect was observed on basal LH release. It can be seen in Fig. 7, however, that the LH-RH-induced release of both gonadotropins was inhibited by Sertoli cell culture medium.

That porcine follicular fluid treated with dextran-coated charcoal to remove endogenous steroids exerts specific effects similar to those of Sertoli cell culture medium on spontaneous gonadotropin secretion is illustrated in Fig. 8. It can be seen in Fig. 8 that porcine follicular fluid has no effect on spontaneous LH release, while it leads to an approximate 70% inhibition of spontaneous FSH secretion. It can also be seen that the inhibitory effect of porcine follicular fluid is much less specific when LH-RH-stimulated gonadotropin secretion is studied instead of spontaneous release. In fact, although LH-RH-induced FSH secretion is inhibited to a higher degree than that of LH at maximal concentrations of LH-RH, a significant inhibition of LH-RH -induced LH release can also be observed in the presence of follicular fluid. "Inhibin" leads to an increase of the LH-RH ED_{50} value for LH release from 0.3 to 1 nM, while the maximal response to the neurohormone is 25% reduced ($p < 0.01$). A similar decrease of the sensitivity of the FSH response to LH-RH is also observed in the presence of porcine follicular fluid, while the maximal FSH response to LH-RH is 50% reduced. It should be mentioned that identical results were obtained when anterior pituitary cells were prepared from animals ovariectomized 3 weeks previously. Although the concentration of porcine follicular fluid used in this experiment (1:1000) was maximal, a significant inhibition of LH-RH-induced LH and FSH release was observed at a 1:50,000 dilution.

The interaction among estrogens, dihydrotestosterone (DHT), and porcine follicular fluid is illustrated in Fig. 9. It can be seen that a 48-hr incubation with E_2 (10 nM) led to a stimulation of the LH and FSH responses to LH-RH, while similar preincubation with DHT (10 nM) led to a marked inhibition of the LH response to LH-RH. Preincubation with porcine follicular fluid led to a marked inhibition of the LH-RH-induced release of both LH and FSH in the absence of steroids and completely abolished the stimulatory effect of E_2 on the secretion of the two gonadotropins. Moreover, addition of both porcine follicular fluid and DHT led to a greater inhibition of LH-RH-induced LH release than that observed in the presence of porcine follicular fluid or DHT alone.

The present data clearly show specific and differential effects of estradiol, androgens, progesterone, and (a) substance(s) of rat Sertoli cell

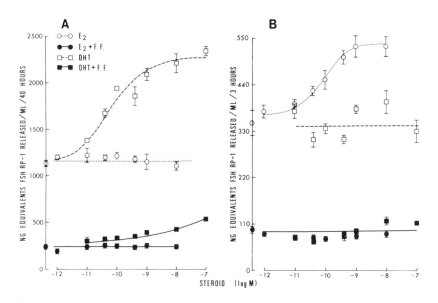

Figure 9. Effect of increasing concentrations of estradiol (E_2) (○), (●) and DHT (□), (■) in the presence (●, ■) or absence (○, □) of porcine follicular fluid (F.F.) on the LH (A) and (FSH) (B) responses to 0.1 nM LH-RH. LH-RH was present during a 3-hr incubation period after a 40-hr preincubation with the indicated steroids or porcine follicular fluid.

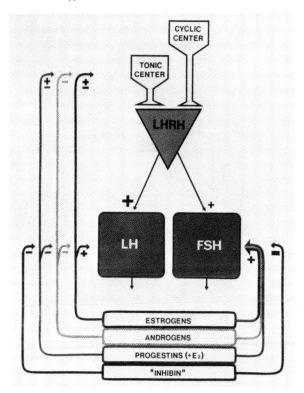

Figure 10. Schematic representation of the interactions among LH-RH, estrogens, androgens, progestins, and "inhibin" in the control of LH and FSH secretion in the rat.

and porcine follicular fluid origin on LH and FSH secretion by a direct action at the anterior pituitary level (Fig. 10). These findings suggest that testicular and ovarian "inhibin" could interact with sex steroids and LH-RH in the differential control of LH and FSH secretion and explain the changes of ratio of LH and FSH secretion frequently observed in man (Franchimont *et al.*, 1975; Grumbach *et al.*, 1974) and experimental animals (Ferland *et al.*, 1976).

6. Interactions between Sex Steroids and Dopamine in the Control of Prolactin Secretion at the Pituitary Level

In vivo treatment with estrogens is well known to lead to a stimulation of prolactin secretion both in man (Frantz *et al.*, 1972; Yen *et al.*, 1972) and in the rat (Ajika *et al.*, 1972; Chen and Meites, 1970; De Léan *et al.*,

1977; Fuxe *et al.*, 1969). At least part of this stimulatory effect of estrogens is likely to be due to a direct action at the pituitary level on prolactin secretion (Lu *et al.*, 1971; Nicoll and Meites, 1964; Raymond *et al.*, 1978). Recently, we have found that E_2 not only stimulates basal and TRH-induced prolactin secretion in rat anterior pituitary cells in primary culture, but also, somewhat surprisingly, could reverse almost completely the inhibitory effect of DA agonists on prolactin release (Caron *et al.*, 1978; Raymond *et al.*, 1978). This chapter extends these previous findings and investigates in more detail the antidopaminergic action of estrogens and their interaction with progestins and androgens on the pituitary DA receptor controlling prolactin secretion.

The interest in such studies is strengthened by the recent observation that the antidopaminergic action of estrogens, first observed at the anterior pituitary level (Labrie *et al.*, 1978; Raymond *et al.*, 1978), also takes place in the central nervous system. In fact, estrogen treatment decreases the circling behavior induced by apomorphine administration in rats having a unilateral lesion of the entopeduncular nucleus (Bédard *et al.*, 1978) and inhibits the apomorphine-induced accumulation of acetylcholine in rat striatum (Euvrard *et al.*, 1978). Moreover, clinical studies have recently

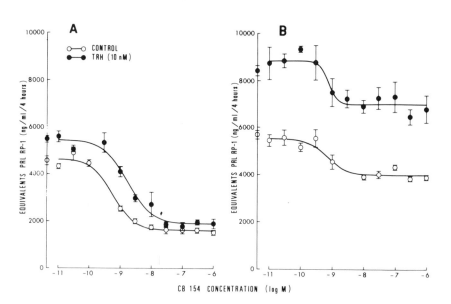

Figure 11. Effect of 17β-estradiol on the prolactin response to increasing concentrations of the DA agonist CB-154 in rat adenohypophyseal cells in culture. The cells were preincubated for 5 days in the presence (B) or absence (A) of 1 nM 17β-estradiol before a 4-hr incubation in the presence (●) or absence (○) of 10 nM TRH and the indicated concentrations of CB-154.

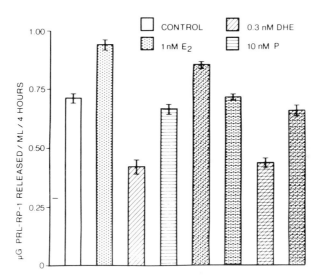

Figure 12. Effect of preincubation for 10 days with 17β-estradiol (E$_2$, 1 nM) or progesterone (P, 10 nM) alone or in combination on spontaneous prolactin release in the presence or absence of dihydroergocornine (DHE, 0.3 nM). Prolactin release was measured during a 4-hr incubation.

shown that estrogens have a beneficial effect on L-dopa- and neuroleptic-induced dyskinesias (Bédard *et al.*, 1977). It is thus hoped that the pituitary cell culture system can be used as a model for other less accessible brain dopaminergic systems.

A detailed analysis of the inhibitory effect of estrogens on the activity of the DA receptor is presented in Fig. 11. It can be seen that CB-154 leads to a maximal 70% inhibition of prolactin release at an ED$_{50}$ value of approximately 3 nM (Fig. 11A) in both the presence and the absence of 10 nM TRH in control cells. However, preincubation for 5 days with E$_2$ (Fig. 11B) led to a small stimulation (approximately 20%) of spontaneous prolactin release, while the maximal response to TRH was increased 70%. The most dramatic effect of estrogen treatment was observed in the presence of CB-154: the 70% inhibition of prolactin release induced by the DA agonist in control cells was reduced to 20% in E$_2$-treated cells.

Since progestins and androgens are well known to exert antiestrogenic activity at the uterine level, we next studied the possibility of a similar effect on prolactin secretion in rat anterior pituitary cells in culture. As illustrated in Fig. 12, while preincubation for 10 days with 10 nM proges-terone alone had no effect on prolactin release, the stimulatory effect of E$_2$ was 40–50% reversed by the progestin in both the presence and the absence of the dopamine agonist dihydroergocornine (DHE). It can also

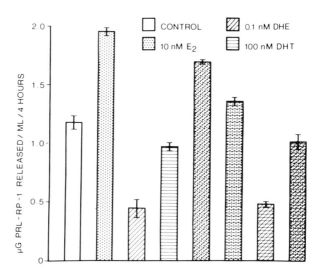

Figure 13. Effect of preincubation for 10 days with 17β-estradiol (E$_2$, 10 nM) or 5α-dihydrotestosterone (DHT, 100 nM) on prolactin release alone or in combination in the presence or absence of dihydroergocornine (DHE, 0.1 nM). Prolactin release was measured during a 4-hr incubation.

be seen in Fig. 13 that DHT exerted antiestrogenic effects almost super-imposable on those observed with progesterone.

Following our *in vitro* data showing a potent antidopaminergic effect of estrogens on prolactin secretion, it then became of interest to investigate whether such a potent activity of estrogens occurs under *in vivo* conditions. The present study was facilitated by our recent findings that the endogenous inhibitory dopaminergic influence on prolactin secretion can be eliminated by administration of opiates, thus making possible study of the effect of exogenous dopaminergic agents without interference by endogenous DA. As illustrated in Fig. 14, the subcutaneous administration of 100 or 400 μg DA completely prevented the increase of plasma prolactin levels following morphine injection in rats ovariectomized 2 weeks previously. Treatment with estradiol benzoate (20 μg/day) for 7 days (Fig. 14B) led to a stimulation of basal plasma prolactin levels from 14 $\pm$ 1 to 56 $\pm$ 8 ng/ml and to a marked increase of the maximal plasma prolactin response to morphine from 215 $\pm$ 60 to 2175 $\pm$ 390 ng/ml. The most interesting finding was, however, that the 100 and 400 μg doses of DA that could maintain plasma prolactin levels at undetectable levels after morphine injection in control rats led to only 40 and 85% inhibition of prolactin levels, respectively, in animals treated with estrogens.

These studies clearly demonstrate that estrogens have potent antidopaminergic activity on prolactin secretion, not only in anterior pituitary cells in culture, but also *in vivo,* the effect being qualitatively similar in both female and male animals. As reflected by an increase of the ED_{50} value of DA agonists, the *in vitro* effect of estrogens was due to a decreased sensitivity of prolactin release to DA action at the anterior pituitary level. Such findings indicate that higher concentrations of DA in the hypothalamo–hypophyseal portal blood system are likely to be required to inhibit prolactin secretion under conditions of high estrogenic influence. The almost complete reversal of the inhibitory effect of low doses of DA by estrogen treatment clearly indicates an important interaction between estrogens and DA at the adenohypophyseal level.

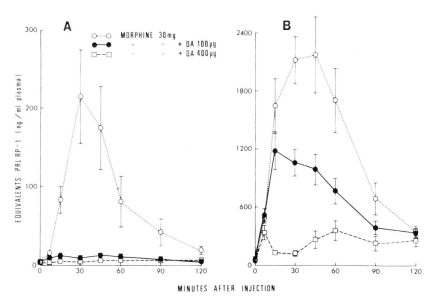

MINUTES AFTER INJECTION

Figure 14. Effect of estrogen treatment on the inhibitory effect of depamine (DA) on prolactin release in the female rat. Adult Sprague–Dawley rats ovariectomized 2 weeks previously were injected subcutaneously (s.c.) with estradiol benzoate (10 μg, twice a day) for 7 days or with the vehicle alone (0.2 ml 0.1% gelatin in 0.9% NaCl) before insertion of a catheter into the right superior vena cava under anesthesia (Surital, 50 mg/kg body weight, intraperitoneally). Two days later, undisturbed freely moving animals were injected s.c. with morphine sulfate (30 mg) alone or in combination with DA (100 or 400 μg). Blood samples (0.7 ml) were then withdrawn at the indicated time intervals for measurement of plasma prolactin concentration. Data are expressed as mean ± S.E.M. of duplicate determinations of samples obtained from 8–10 animals per group.

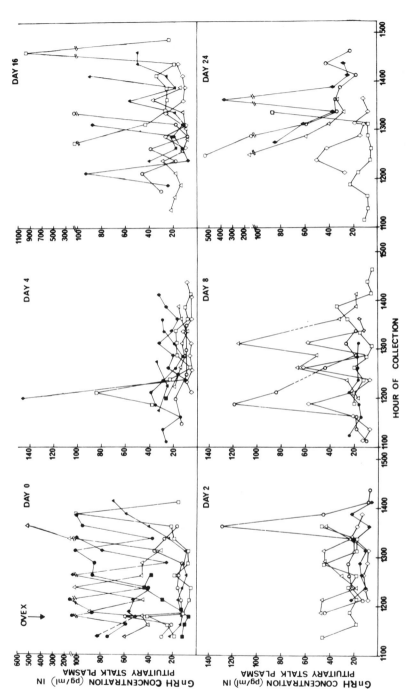

Fig. A (see Discussion). Hypophyseal portal plasma concentrations of GnRH in individual animals at various times after ovariectomy (ovex) at estrus. Mean concentrations (all samples on each day) were 25, 26, 28, 50 and 57 pg/ml at 2, 4, 8, 16 and 24 days, respectively. The concentrations of GnRH correlate well with the peripheral plasma concentrations of LH which were 3.6, 2.7, 7.4, 8.8, and 10.5 ng NIH-LH-S18/ml at 2, 4, 8, 16, and 24 days respectively. Note change in ordinate scale (D.K. Sarkar and G. Fink, unpublished.)

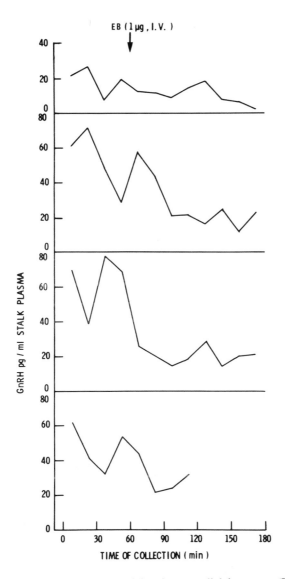

Fig. B (see Discussion). Effect of administering estradiol benzoate (EB) on GnRH concentration of hypophyseal portal plasma in 4 individual rats ovariectomized 4 weeks before collection. Injection of vehicle (ethanol:saline equals 1:50 v/v) had no appreciable effect (results not shown). (D.K. Sarkar and G. Fink, unpublished.)

Discussion

FINK: With respect to the time taken for steroids to exert their effects (question by Dr. Kraicer), the action of steroids may involve either a genomic or a nongenomic mechanism. Our data support those of Dr. Labrie suggesting that alterations of pituitary responsiveness to GnRH by steroids may be brought about by a genomic mechanism and takes a relatively long time (several hours) for completion. Slow and fast effects of steroids can also be demonstrated with respect to the GnRH release system. Thus after a short "unstable" period around the time of surgery, GnRH concentrations in hypophyseal portal vessel blood remain low until four to eight days after ovariectomy (Fig. A). However, the i.v. administration of estradiol to long-term ovariectomized rats results in a rapid drop in the GnRH concentration and amplitude of GnRH oscillations (Fig. B).

References

Aiyer, B. S., Chiappa, S. A., Fink, G., and Greig, F. J., 1973, A priming effect of luteinizing hormone releasing factor on the anterior pituitary gland in the female rat., *J. Physiol. (London)* **234**:81P.

Ajika, K., Krulich, L., Fawcett, C. P., and McCann, S. M., 1972, Effects of estrogen on plasma and pituitary gonadotropins and prolactin on hypothalamic releasing and inhibitory factors, *Neuroendocrinology* **9**:304.

Barden, N., Bergeron, L., and Betteridge, A., 1976, Effects of prostaglandin synthetase inhibitors and prostaglandin precursors on anterior pituitary cyclic AMP and hormone secretion, in: *Prostaglandin and Thromboxane Research 1* (B. Samuelson and R. Paoletti, eds.), pp. 341–344, Raven Press, New York.

Bédard, P., Langelier, P., and Villeneuve, A., 1977, Estrogens and the extra-pyramidal system, *Lancet* **1**:1367.

Bédard, P., Dankova, J., Boucher, R., and Langelier, P., 1978, Effect of estrogens on apomorphine-induced circling behavior in the rat, *Can. J. Physiol. Pharmacol.* **56**:538.

Bélanger, A., Labrie, F., Borgeat, P., Savary, M., Côté, J., Drouin, J., Schally, A. V., Coy, D. H., Coy, E. J., Immer, H., Sestanj, K., Nelson, V., and Götz, M., 1974, Inhibition of growth hormone and thyrotropin release by growth hormone–release inhibiting hormone, *J. Mol. Cell. Endocrinol.* **1**:329.

Ben-Jonathan, N., Oliver, C., Weiner, H. J., Mical, R. S., and Porter, J. C., 1977, Dopamine in hypophyseal portal plasma of the rat during the estrous cycle and throughout pregnancy, *Endocrinology* **100**:452.

Bishop, W., Fawcett, C. P., Krulich, L., and McCann, S. M., 1972, Acute and chronic effects of hypothalamic lesions on the release of FSH, LH and prolactin in intact and castrated rats, *Endocrinology* **91**:643.

Borgeat, P., Chavancy, G., Dupont, A., Labrie, F., Arimura, A., and Schally, A. V., 1972, Stimulation of adenosine 3',5'-cyclic monophosphate accumulation in anterior pituitary gland *in vitro* by synthetic luteinizing hormone–releasing hormone/follicle-stimulating hormone–releasing hormone (LH-RH/FSH-RH), *Proc. Natl. Acad. Sci. U.S.A.* **69**:2677.

Borgeat, P., Labrie, F., Poirier, G., Chavancy, G., and Schally, A. V., 1973, Stimulation of adenosine 3',5'-cyclic monophosphate accumulation in anterior pituitary gland by purified growth hormone–releasing hormone, *Trans. Assoc. Amer. Physicians* **86**:284.

Borgeat, P., Labrie, F., Côté, J., Ruel, F., Schally, A. V., Boy, D. H., Coy, E. J., and Yanaihara, N., 1974a, Parallel stimulation of cyclic AMP accumulation and LH and FSH release by analogs of LH-RH *in vitro*, *J. Mol. Cell. Endrocrinol.* **1**:7.

Borgeat, P., Labrie, F., Drouin, J., Bélanger, A., Immer, I., Sestanj, K., Nelson, V., Götz, M., Schally, A. V., Coy, D. H., and Coy, E. J., 1974b, Inhibition of adenosine 3',5'-monophosphate accumulation in anterior pituitary gland *in vitro* by growth hormone release–inhibiting hormone, *Biochem. Biophys. Res. Commun.* **56**:1052.

Brown, G. M., Seeman, P., and Lee, T., 1976, Dopamine neuroleptic receptors in basal hypothalamus and pituitary, *Endocrinology* **99**:1407.

Calabro, M. A., and MacLeod, R. M., 1978, Binding of dopamine to bovine anterior pituitary gland membranes, *Neuroendocrinology* **25**:32.

Caron, M. G., Beaulieu, M., Raymond, V., Gagné, B., Drouin, J., Lefkowitz, R. J., and Labrie, F., 1978, Dopaminergic receptors in the anterior pituitary gland: Correlation of [³H]dihydroergocryptine binding with the dopaminergic control of prolactin release, *J. Biol. Chem.* **254**:2244.

Castro-Vasquez, A., and McCann, S. M., 1975, Cyclic variations in the increased responsiveness of the pituitary to luteinizing hormone–releasing hormone (LH-RH) indicated by LH-RH *Endocrinology* **97**:13.

Chen, L., and Meites, J., 1970, Effects of estrogen and progesterone on serum and pituitary prolactin levels in ovariectomized rats, *J. Endocrinol.* **86**:503.

Creese, I., Schneider, R., and Snyder, S. H., 1977, [³H]Spiroperidol labels dopamine receptor in pituitary and brain, *Eur. J. Pharmacol.* **46**:377.

Deery, D. J., and Howell, S. L., 1973, Rat anterior pituitary adenyl cyclase activity: GTP requirement of prostaglandin E_1 and E_2 and synthetic luteinizing hormone–releasing hormone activation, *Biochem. Biophys. Acta* **329**:17.

Deery, D. J., and Jones, A. C., 1975, Effects of hypothalamic extracts, neurotransmitters and synthetic hypothalamic releasing hormones on adenylyl cyclase activity in the lobes of the pituitary of the dogfish (*Scyliorhinus canicula* L.), *J. Endocrinol.* **64**:49.

De Léan, A., Caron, M., Kelly, P. A., and Labrie, F., 1977, Changes in pituitary thyrotropin-releasing hormone receptor levels and prolactin responses to TRH during the rat estrous cycle, *Endocrinology* **100**:1505.

Drouin, J., and Labrie, F., 1976a, Selective effect of androgens on LH and FSH release in anterior pituitary cells in culture, *Endocrinology* **98**:1528.

Drouin, J., and Labrie, F., 1976b, Specificity of the stimulatory effect of prostaglandins on hormone release in anterior pituitary cells in culture, *Prostaglandins* **11**:355.

Drouin, J., Ferland, L., Bernard, J., and Labrie, F., 1976a, Site of the *in vivo* stimulatory effect of prostaglandins on LH release, *Prostaglandins* **11**:367.

Drouin, J., Lagacé, L., and Labrie, F., 1976b, Estradiol-induced increase of the LH responsiveness to LH-RH in anterior pituitary cells in culture, *Endocrinology* **99**:1477.

Euvrard, C., Labrie, F., and Boissier, J. R., 1978, Antagonism between estrogens and dopamine in the rat striatum, in: *Proceedings of the 4th International Catecholamine Symposium*, p. 21, Pergamon Press, New York.

Ferland, L., Borgeat, P., Labrie, F., Bernard, J., DeLéan, A., and Raynaud, J. P., 1975a, Changes in pituitary sensitivity to LH-RH during the estrous cycle, *Mol. Cell. Endocrinol.* **2**:107.

Ferland, L., Labrie, F., Coy, D. H., Coy, E. J., and Schally, A. V., 1975b, Inhibitory activity of four analogs of luteinizing hormone–releasing hormone *in vivo, Fertil. Steril.* **26**:889.

Ferland, L., Drouin, J., and Labrie, F., 1976, Role of sex steroids on LH and FSH secretion in the rat, in: *Hypothalamus and Endocrine Functions* (F. Labrie, J. Meites, and G. Pelletier, eds.), pp. 191–209, Plenum Press, New York.

Franchimont, P., Chari, S., Hagelstein, M. T., and Duraiswami, S., 1975, Existence of follicle-stimulating hormone inhibiting factor "inhibin" in bull seminal plasma, *Nature (London)* **257**:402.

Frantz, A. G., Kleinberg, D. L., and Noel, G. L., 1972, Studies on prolactin in man, *Recent Prog. Horm. Res.* **28**:527.

Fuxe, K., Hökfelt, T., and Nilsson, O., 1969, Castration, sex hormones and tuberoinfundibular dopamine neurons, *Neuroendocrinology* **5**:107.

Gordon, J. H., and Reichlin, S., 1974, Changes in pituitary responsiveness to luteinizing hormone–releasing factor during the estrous cycle, *Endocrinology* **94**:974.

Grumbach, M., Roth, J. C., Kaplan, S. L., and Kelch, P., 1974, Hypothalamic–pituitary regulation of puberty: Evidence and concepts derived from clinical research, in: *The Control of the Onset of Puberty* (M. Grumbach, G. Grave, and F. E. Mayer, eds.), pp. 115–166, Wiley and Sons, New York.

Jutisz, M., and Paloma De la Llosa, M., 1970, Requirement of Ca^{++} and Mg^{++} ions for the *in vitro* release of follicle-stimulating hormone from rat pituitary gland and its subsequent biosynthesis, *Endocrinology* **86**:761.

Kaneko, T., Saito, S., Oka, H., Oda, T., and Yanaihara, N., 1973, Effects of synthetic LH-RH and its analogs on rat anterior pituitary cyclic AMP and LH and FSH release, *Metabolism* **22**:77.

Kebabian, J. W., 1977, Biochemical regulation and physiological significance of cyclic nucleotides in the nervous system, in: *Advances in Cyclic Nucleotide Research*, Vol. 8 (P. Greengard and G. A. Robison, eds.), pp. 421–508, Raven Press, New York.

Kebabian, J. W., Petzold, G. L., and Greengard, P., 1973, Dopamine sensitive adenylate cyclase in caudate nucleus of rat brain and its similarity to the dopamine receptor, *Proc. Natl. Acad. Sci. U.S.A.* **69**:2145.

Labrie, F., Pelletier, G., Lemay, A., Borgeat, P., Barden, N., Dupont, A., Savary, M., Côté, J., and Boucher, R., 1973, Control of protein synthesis in anterior pituitary gland, in: *Karolinska Symposium on Research Methods in Reproductive Endocrinology* (E. Diczfalusy, ed.), pp. 301–340, Geneva.

Labrie, F., Pelletier, G., Borgeat, P., Drouin, J., Savary, M., Côté, J., and Ferland, L., 1975a, Aspects of the mechanism of action of hypothalamic hormone (LH-RH), in: *Gonadotropins and Gonadal Functions*, Vol. 1 (J. A. Thomas and R. L. Singhal, eds.), pp. 77–127, University Park Press, Baltimore.

Labrie, F., Borgeat, P., Lemay, A., Lemaire, S., Barden, N., Drouin, J., Lemaire, I., Jolicoeur, P., and Bélanger, A., 1975b, Role of cyclic AMP in the action of hypothalamic regulatory hormones, in: *Advances in Cyclic Nucleotide Research*, Vol. 5 (G. I. Drummond, P. Greengard and G. A. Robison, eds.), pp. 787–801, Raven Press, New York.

Labrie, F., DeLéan, A., Barden, N., Ferland, L., Drouin, J., Borgeat, P., Beaulieu, M., and Morin, O., 1976a, New aspects of the mechanism of action of hypothalamic regulatory hormones, in: *Hypothalamus and Endocrine Functions* (F. Labrie, J. Meites, and G. Pelletier, eds.), pp. 147–169, Plenum Press, New York.

Labrie, F., Pelletier, G., Borgeat, P., Drouin, J., Ferland, L., and Bélanger, A., 1976b, Mode of action of hypothalamic regulatory hormones, in: *Frontiers in Neuroendocrinology*, Vol. 4 (W. F. Ganong and L. Martini, eds.), pp. 63–94, Raven Press, New York.

Labrie, F., Savary, M., Coy, D. H., Coy, E. J., and Schally, A. V., 1976c, Inhibition of LH release by analogs of LH-releasing hormone (LH-RH) *in vitro*, *Endocrinology* **98**:289.

Labrie, F., Beaulieu, M., Caron, M. G., and Raymond, V., 1978, The adenohypophyseal dopamine receptor: Specificity and modulation of its activity by estradiol, in: *Progress in Prolactin Physiology and Pathology* (C. Robyn and M. Harter, eds.), pp. 121–136, Elsevier North-Holland, Amsterdam.

Lemay, A., and Labrie, F., 1972, Calcium-dependent stimulation of prolactin release in rat anterior pituitary *in vitro* by N^6-monobutyryl adenosine 3',5'-monophosphate, *FEBS Lett.* **20:**7.

Lu, K. H., Koch, Y., and Meites, J., 1971, Direct inhibition by ergocornine of pituitary prolactin release, *Endocrinology* **89:**229.

MacLeod, R. M., and Lehmeyer, J. E., 1974, Restoration of prolactin synthesis and release by the administration of monoaminergic blocking agents to pituitary tumor–bearing rats, *Cancer Res.* **34:**345.

Makino, T., 1973, Study of the intracellular mechanism of LH release in the anterior pituitary, *Am. J. Obstet. Gynecol.* **115:**606.

McCann, S. M., Ojeda, S. R., Harms, P. G., Wheaton, J. E., Sundberg, D. K., and Fawcett, C. P., 1976, Role of prostaglandins (PGs) in the control of adenohypophyseal hormone secretion, in: *Hypothalamus and Endocrine Functions* (F. Labrie, G. Pelletier, and J. Meites, eds.), pp. 21–35, Plenum Press, New York.

McCullagh, D. R., 1932, Dual endocrine activity of the testis, *Science* **76:**19.

Naor, Z., Koch, Y., Chobsieng, P., and Zor, U., 1975a, Pituitary cyclic AMP production and mechanism of luteinizing hormone release, *FEBS Lett.* **58:**318.

Naor, F., Koch, Y., Bauminger, S., and Zor, U., 1975b, Action of luteinizing hormone and synthesis of prostaglandins in the pituitary gland, *Prostaglandins* **9:**211.

Nicoll, C. S., and Meites, J., 1964, Prolactin secretion *"in vitro"*: Effects of gonadal and adrenal corticol steroids, *Proc. Soc. Exp. Biol. Med.* **117:**579.

Nillius, S. J., and Wide, L., 1971, Induction of a midcycle-like peak of luteinizing hormone in young women by exogenous estradiol-17β, *J. Obstet. Gynecol. Br. Commonw.* **78:**822.

Parsons, J. A., and Nicoll, C. C., 1970, Cations, secretion of prolactin (PRL) and growth hormone (GH), and PIF action, *Fed. Proc. Fed. AM. Soc. Exp. Biol.* **29:**377.

Ratner, A., 1970, Stimulation of luteinizing hormone release *in vitro* by dibutyryl cyclic AMP and theophylline, *Life Sci.* **9:**1221.

Raymond, V., Beaulieu, M., and Labrie, F., 1978, Potent antidopaminergic activity of estradiol at the pituitary level on prolactin release, *Science* **200:**1173.

Shaar, C. J., and Clemens, J. A., 1974, The role of catecholamines in the release of anterior pituitary prolactin *in vitro*, *Endocrinology* **95:**1202.

Sundberg, D. K., Fawcett, C. P., and McCann, S. M., 1976, The involvement of cyclic 3',5'-cyclic AMP in the release of hormones from the anterior pituitary *in vitro*, *Proc. Soc. Exp. Biol. Med.* **151:**149.

Swerdloff, R. W., and Odell, W. D., 1968, Feedback control of male gonadotropin secretion, *Lancet* **2:**683.

Swerdloff, R. W., Walsh, P. C., and Odell, W. D., 1972, Control of LH and FSH secretion in the male: Evidence that aromatization of androgens to estradiol is not required for inhibition of gonadotropin secretion, *Steroids* **20:**13.

Takahara, J., Arimura, A., and Schally, A. V., 1974, Suppression of prolactin release by a purified porcine PIF preparation and catecholamines infused into a rat hypophyseal portal vessel, *Endocrinology* **95:**462.

Vale, W., Grant, G., Amoss, M., Blackwell, R., and Guillemin, R., 1972, Culture of enzymatically dispersed pituitary cells: Functional validation of a method, *J. Clin. Endocrinol. Metab.* **91:**562.

Wakabayashi, K., Date, Y., and Tamaoki, B., 1973, On the mechanism of action of luteinizing hormone–releasing factor and prolactin release inhibiting factor, *Endocrinology* **92:**698.

Yen, S. S. C., Tsai, C. C., Vandenberg, G., and Rebar, R., 1972, Gonadotropin dynamics in patients with gonadal dysgenesis: A model for the study of gonadotropin regulation, *J. Clin. Endocrinol. Metab.* **35:**897.

22

Binding of GnRH to the Pituitary Plasma Membranes, Cyclic AMP, and LH Release

Annette Bérault, Madeleine Théoleyre, and Marian Jutisz

1. Introduction

It is well documented that the regulation of luteinizing hormone (LH) and follicle-stimulating hormone (FSH) secretion is a complex phenomenon in which many factors may take part. Among those factors, the hypothalamic peptide gonadotropin-releasing hormone (GnRH) certainly plays, at least in the majority of mammals, a very important role. However, it has been shown that its action at the pituitary level is modulated by various substances such as steroids, neuromediators, and perhaps other neuropeptides. On the other hand, GnRH also seems to stimulate the synthesis of LH and FSH, and it is possible that there is a relationship between the rate of synthesis of gonadotropins and their rate of release. Consequently, the mechanism of action of GnRH is not a simple molecular process, but rather results from the interaction of GnRH with one or more other factors with different cell components. It would be of great interest to understand the mechanisms of all these interactions at the cellular and molecular levels, but unfortunately no conclusive data are yet available. To simplify

Annette Bérault, Madeleine Théoleyre, and *Marian Jutisz* • Laboratoire des Hormones Polypeptidiques, C.N.R.S., 91190 Gif-sur-Yvette, France

the problem, we will not take into consideration the modulating action of different factors on the releasing effect of GnRH, but will rather restrict our topic to two main aspects of the cellular mechanism of action of this neurohormone, i.e., the concept of the receptor of GnRH and the possible involvement of cyclic AMP (cAMP) as an intermediate in the releasing mechanism of LH.

2. Binding of GnRH to the Receptor Sites of Pituitary Plasma Membranes

Many experimental results indicate that the first step in the action of a polypeptide hormone is its reversible binding to specific sites on the plasma membrane of the target organ. Studies on the interaction of GnRH with specific receptor sites have been carried out in our laboratory (Théoleyre *et al.*, 1976) using purified ovine anterior pituitary plasma membranes (Bérault *et al.*, 1974) and a tritiated preparation of GnRH with a high specific radioactivity (Marche *et al.*, 1972). Figure 1 shows the kinetics of the binding of tritiated GnRH at 37°C. Specific binding was obtained by subtracting from the total radioactivity bound the amount that was not displaced by a 1000-fold excess of cold GnRH. A maximum of binding was observed between 10 and 20 min, with a slight decline after 20 min, due probably to the fragility of the receptor sites at 37°C.

The dissociation of the tritiated GnRH complex at 37°C by dilution of the reaction mixture gave a plateau at 90% dissociation after about 20 min. The dissociation was never complete, and we have no explanation for this phenomenon.

The Scatchard plot of the data obtained from the binding studies as a function of hormone concentration after 20 min of incubation at 37°C (Fig. 2) shows a single slope, suggesting a single type of binding sites.

Table I shows the association and dissociation rate constants calculated from kinetic data and the equilibrium association constant calculated either from kinetic data or from Scatchard plot data. There is fairly good agreement between the equilibrium association constant data calculated by the two methods. We have also determined by two different methods that the binding capacity of the plasma membranes is 110 ± 20 fmol [^{3}H]-GnRH/mg protein. This value agrees well with that reported by Park *et al.* (1976), who found a binding capacity of 150 fmol GnRH/mg protein.

The affinity of the binding sites determined in our work is somewhat smaller than the biological activity constant (2×10^9 M^{-1} GnRH) as reported for anterior-pituitary cells in culture by Grant *et al.* (1973). These authors found at 0°C two types of binding sites for tritiated GnRH. Other

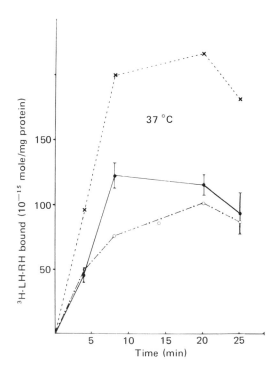

Figure 1. Specific binding of [³H]-GnRH (³H-LH-RH in the figure) (●) to plasma membranes as a function of incubation time. [³H]-GnRH (17 × 10⁻⁹ M) was incubated at 37°C for various time periods with 50 μg membrane protein in 240 μl buffer. Nonspecific binding was determined by the addition of a 1000-fold excess (17 × 10⁻⁶ M) of cold GnRH to the incubation mixture (○). These values were subtracted from the total bound radioactivity (×). Values for specific binding are the means of three determinations. Vertical bars indicate S.E. From Théoleyre *et al.* (1976), courtesy of Elsevier/North-Holland Scientific Publishers Limited.

authors used [¹²⁵I]-GnRH in their work (Spona, 1973; Marshall *et al.*, 1976; Baumann and Kuhl, 1978; Wagner, 1978; Clayton *et al.*, 1978). Clayton *et al.* (1978), using purified bovine pituitary plasma membranes, reported for one of the binding sites an affinity constant of 2.5×10^9 M⁻¹, which is close to that described by Grant *et al.* (1973); it is about 20 times that of ovine pituitary plasma membranes.

The existence of a single type of binding site, as observed in our work, might be explained by a more effective account of nonspecific binding and by working at 37 instead of 0°C.

Coupled with binding and dissociation data, the competitive inhibition of binding by the presence of an excess of unlabeled GnRH indicates that

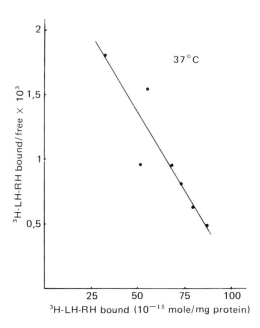

Figure 2. Scatchard plot of the specific binding of [³H]-GnRH (³H-LH-RH in the figure) to pituitary plasma membranes. From Théoleyre *et al.* (1976), courtesy of Elsevier/North-Holland Scientific Publishers Limited.

the site of binding of GnRH is specific. The question arises as to whether these receptors are functionally related to the adenylate cyclase system and to the LH release process.

3. Is Cyclic AMP an Intermediate in the Releasing Action of GnRH?

In 1969 and 1970, our laboratory reported (Jutisz and de la Llosa, 1969; Jutisz *et al.*, 1970) that cAMP or its derivatives were able to release LH and FSH *in vitro,* and this was further confirmed by others (Ratner, 1970; Labrie *et al.,* 1973; Makino, 1973; Groom and Boyns, 1973; Davis and Hymer, 1975). In contrast, other authors (Wakabayashi *et al.,* 1973; Tang and Spies, 1976; Sundberg *et al.,* 1976) did not succeed in obtaining the release of LH with dibutyryl cAMP.

The results related to the release of gonadotropins with the cAMP derivatives and those related to the cAMP accumulation in pituitary tissue *in vitro* under the action of GnRH (Borgeat *et al.,* 1972; Kaneko *et al.,* 1973; Makino, 1973; Labrie *et al.,* 1975) led us and others to postulate the

Table I. Some Constants of GnRH–Receptor Binding at 37°C[a]

Constant	Value
Association rate constant	k_1 $= 2\ \ \pm 1 \times 10^5\ \mathrm{M^{-1}\ sec^{-1}}$
Dissociation rate constant	$k-1$ $= 0.7 \pm 0.4 \times 10^{-2}\ \mathrm{sec^{-1}}$
Equilibrium association constant:	
From kinetic data	$\dfrac{k_1}{k-1}$ $= 0.6 \pm 0.5 \times 10^8\ \mathrm{M^{-1}}$
From Scatchard plot data	Ka $= 1.1 \pm 0.4 \times 10^8\ \mathrm{M^{-1}}$

[a] From Théoleyre et al. (1976).

possible involvement of cAMP as an intermediate in GnRH action. To investigate this problem further, we carried out several experiments, and our more recent results do not support this hypothesis.

3.1. Does GnRH Stimulate Anterior-Pituitary Adenylate Cyclase?

Assuming that cAMP is involved in the mechanism of action of GnRH, the binding of the hormone to its receptor sites would lead to the activation of adenylate cyclase. Several assays of adenylate cyclase in crude homogenates of the pituitaries of normal adult male rats were carried out in our laboratory using a modification of the method of Krishna et al. (1968). Table II shows the results obtained in three different experiments. NaF (10 mM) increased the basal activity of adenylate cyclase by a factor

Table II. Effect of GnRH and NaF on Adenylate Cyclase
Activity in Rat Pituitary Homogenates[a]

Additions	cAMP (pmol/mg protein/15 min)[b]		
	Expt. 1	Expt. 2	Expt. 3
None	155 ± 8.75	220 ± 6.67	216 ± 9.37
NaF (10 mM)	6250 ± 140.62	4100 ± 253.0	1812 ± 108.7
GnRH (nM)			
0.85	202 ± 16.87	340 ± 18.0	265 ± 10.62
2.12		260 ± 6.66	247 ± 7.5
4.25	149 ± 9.37	340 ± 13.33	265 ± 6.25
8.5		357 ± 11.33	297 ± 18.75
17.0	181 ± 23.12	267 ± 8.66	222 ± 9.37
42.5		280 ± 10.0	297 ± 14.37
85.0	134 ± 5.0		

[a] Incubation was performed for 15 min at 37°C, in the presence of 1 mM EDTA and 1.6 mM ATP. No theophylline was added.
[b] The cAMP produced is evaluated by the procedure of Gilman (1970). Results are given as means ± S.E.

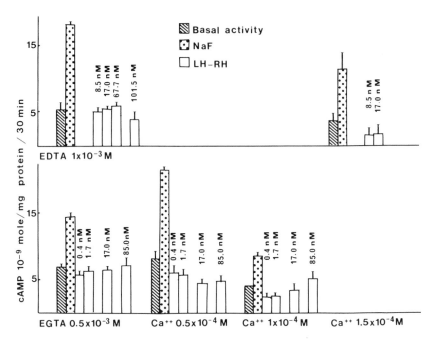

Figure 3. Effect of GnRH (LH-RH in the figure) at different concentrations and of 10 mM NaF on the adenylate cyclase activity of purified sheep anterior pituitary plasma membranes. Assays were performed either in the presence of EDTA or EGTA or in the presence of CaCl$_2$ at different concentrations, as indicated in the figure. Results, which are the means of three determinations, are expressed in nmol cAMP/mg protein per 30 min, with S.E. indicated by vertical bars. From Théoleyre et al. (1976), courtesy of Elsevier/North-Holland Publishers Limited.

of 8–40, but GnRH did not stimulate the enzyme significantly at any of the concentrations tested.

In other experiments (Fig. 3), purified sheep pituitary plasma membranes were used in the absence of Ca^{2+} or in the presence of various concentrations of Ca^{2+}. These experiments were designed to investigate the possible role of Ca^{2+} in the activation process. As with crude homogenates, no stimulation of adenylate cyclase was observed.

Further experiments were carried out to investigate the possible involvement of GTP as a transducer in the coupling mechanism (Rodbell et al., 1971). Deery and Howell (1973) have reported that in the presence of GTP, GnRH causes a marked stimulation of adenylate cyclase activity in homogenates of rat pituitary that is greater than the effect of GTP alone. Table III shows that we could not confirm their results, although we were working in the conditions reported by these authors, with different concentrations of GnRH and in the absence or presence of theophylline.

Table III. Lack of Effect of GnRH on Adenylate Cyclase Activity from Rat Pituitary Homogenates in the Presence and Absence of GTP and Theophylline[a]

	cAMP (pmol/mg protein per 15 min)[b]			
Treatments	None	GTP (0.1 mM)	Theophylline (3 mM)	GTP (0.1 mM) Theophylline (3 mM)
Control	5.38 ± 0.39	25.76 ± 1.50	35.6 ± 1.4	82.7 ± 2.7
NaF (10mM)	44.10 ± 3.49	188.46 ± 6.54	249.1 ± 1.4	283.7 ± 13.7
GnRH (nM)				
0.17	4.23 ± 0.40	26.53 ± 2.15	44.7 ± 1.2	82.8 ± 0.7
1.7	4.35 ± 0.44	23.46 ± 1.00	43.2 ± 1.6	79.1 ± 3.2
17.0	7.69 ± 0.99	18.97 ± 0.26	42.6 ± 0.8	78.3 ± 1.9

[a] Incubation was performed for 15 min at 30°C in the presence of 10^{-3} M EDTA and 10^{-5} M ATP as substrate.
[b] Results are the means ± S.E. of triplicate determinations.

The base level of cAMP in the presence of GTP and theophylline was higher than in the absence of these reagents, but adenylate cyclase was not stimulated.

To examine the possible role of the concentration of the enzyme substrate in the activation process, we carried out experiments using different concentrations of ATP in the presence or absence of theophylline. Table IV shows that NaF was able, as previously, to stimulate adenylate cyclase in all conditions used, but that GnRH at various concentrations was without effect.

Finally, we examined in a series of experiments the possibility of GnRH's inhibiting the phosphodiesterase (PDE) activity and thus acting on the accumulation of cAMP. For the assay of PDE, the method of Butcher and Sutherland (1962) was used. Figure 4 shows that GnRH at various doses and in different conditions is without effect on PDE activity. In the presence of theophylline and in the absence of calcium ions (EDTA), the activity of this enzyme is partly inhibited.

Our data on the failure to activate adenylate cyclase with GnRH were recently confirmed by Clayton *et al.* (1978) on purified bovine pituitary plasma membranes. In the experiments shown in Fig. 5, Clayton *et al.* (1978) investigated the effect of ATP, Mg^{2+}, and Ca^{2+} concentration on the stimulation of adenylate cyclase in purified bovine pituitary plasma membranes incubated with 100 ng GnRH/ml incubation medium. No stimulation of adenylate cyclase was observed.

Several reasons could be considered for the failure of GnRH to stimulate adenylate cyclase activity. From our data, it is unlikely that the

Table IV. Effect of ATP Concentration on Adenylate Cyclase Activity with or without Theophylline[a]

	cAMP (pmol/mg protein per 15 min)[b]				
	None			Theophylline (3 mM)	
Treatments ATP conc.: 10^{-5} M	1.6×10^{-3} M	10^{-2} M	10^{-5} M	10^{-2} M
Control	5.38 ± 0.39	295.9 ± 4.94	300 ± 18.7	35.6 ± 1.4	616 ± 41
NaF (10 mM)	44.10 ± 3.49	2120 ± 62.54	1175 ± 90	249.1 ± 1.4	2028 ± 109
GnRH (nM)					
0.17	4.23 ± 0.40	237.7 ± 28.52	—	44.7 ± 1.2	—
1.7	4.35 ± 0.44	153.8 ± 8.62	—	43.2 ± 1.6	—
17.0	7.69 ± 0.99	231.1 ± 13.29	356 ± 7.5	42.6 ± 0.8	712 ± 55
68.0	—	—	—	—	690 ± 77
680.0	—	—	—	—	740 ± 137

[a] Rat pituitary homogenates were incubated for 15 min at 37°C; 1 mM EDTA was present in all groups.
[b] Results are means ± S.E. of triplicate determinations.

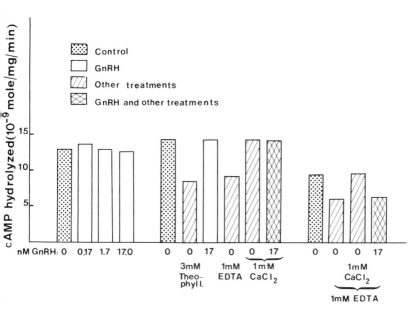

Figure 4. Effect of graded doses of GnRH, 3mM theophylline, 1 mM EDTA, and 1 mM Ca^{2+} on the cAMP–PDE activity of rat pituitary homogenates. Incubation was carried out for 10 or 30 min at 30°C. Results, which are the means of three determinations, are expressed as nmol cAMP hydrolyzed/mg protein per min.

failure of GnRH to stimulate adenylate cyclase was due either to the presence or the absence of Ca^{2+} or to the absence of GTP, or finally to suboptimal concentration of the enzyme substrate. On the other hand, Clayton *et al.* (1978) reported that reducing the Mg^{2+} concentration did not favor GnRH stimulation of adenylate cyclase. Another possibility was that GnRH degradation was occurring under the conditions used (Marshall *et al.*, 1977). Therefore, we have tried stimulation in the presence of bacitracin, an inhibitor of GnRH degradation (McKelvy *et al.*, 1976), but without effect (unpublished results) (also see Clayton *et al.*, 1978). It is also evident from our data that adenylate cyclase was not inactivated in the course of preparation of plasma membranes, since it can be stimulated with fluoride.

3.2. Accumulation of Cyclic AMP in Pituitary Tissue and LH Release

The increase of cAMP in pituitary tissue or dispersed pituitary cells under the action of GnRH has been reported by many laboratories (Borgeat *et al.*, 1972; Makino, 1973; Labrie *et al.*, 1973). A correlation has been reported between the release of LH and FSH and the increase in the

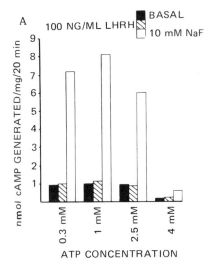

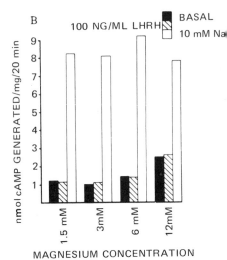

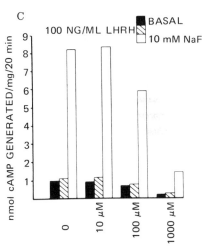

Figure 5. Purified pituitary membranes. (A) Effect of substrate concentration on basal, 100 ng/ml GnRH (LH-RH in the figure), and 10 mM NaF–stimulated adenylate cyclase activity. Mg^{2+}, 3mM; Ca^{2+}, absent. Results are means of duplicate incubations at 37°C. (B) Effect of magnesium on adenylate cyclase activity. Means of duplicate incubations. ATP, 1 mM; Ca^{2+}, absent. (C) Effect of calcium on adenylate cyclase activity. Means of duplicate incubations. ATP, 1 mM; Mg^{2+}, 3 mM. From Clayton *et al.* (1978), courtesy of the authors and Elsevier/North-Holland Scientific Publishers, Ltd.

intracellular level of pituitary cAMP *in vitro* (Borgeat *et al.*, 1972, 1974; Labrie *et al.*, 1975). However, it should be noted that the increase in pituitary cAMP was detected *in vitro* only 2 hr after the addition of GnRH.

To clear up this problem the experiments described below were carried out.

Male rat pituitary halves were incubated in Krebs–Ringer medium for two different periods of time: 70 and 210 min. The effect of graded doses of GnRH (0.42–42 nM) on the release of LH in the incubation medium and on the accumulation of cAMP in the pituitary tissue was examined at the end of each incubation period. Figure 6 shows that while after 70 min of incubation, a significant amount of LH (as compared with the controls) is released with 4.25 nM GnRH or more, the cAMP content of the treated tissues is not significantly different from controls for all the concentrations of GnRH used. After 210 min of incubation, the significant amount of LH appears in the medium starting with a concentration of 0.42 nM GnRH, while the significant increase of cAMP appears only for concentrations higher than 4.25 nM GnRH.

In another experiment, we investigated the time course of the release of LH into the medium during the incubation of cultured female rat pituitary cells (Debeljuk *et al.*, 1978). After each period of incubation, cAMP was assayed in the cells. Experiments were carried out in the presence of either GnRH or theophylline. Figure 7 shows that significant amounts of LH are released after 120 min of incubation with 8.5 nM GnRH, while no difference was observed for that period of time between the cAMP content in treated and control cells. On the other hand,

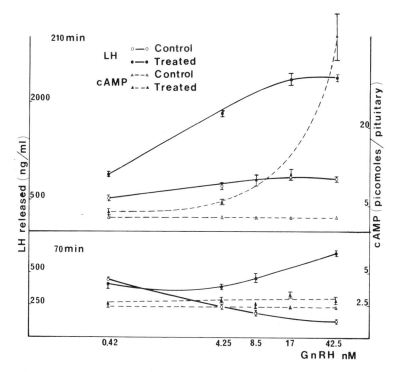

Figure 6. *In vitro* effect of increasing concentrations of GnRH on the accumulation of cAMP and the release of LH. Hemipituitaries of adult male rats (4 per flask, 4 halves serving as controls for treated tissues) were incubated in Krebs–Ringer–bicarbonate medium for 70 and 210 min. Results are the means ± S.E. of triplicate determinations.

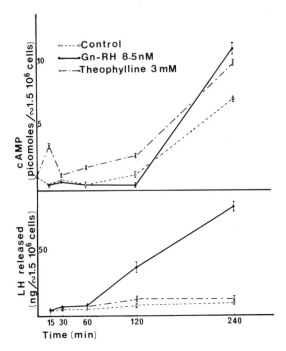

Figure 7. Effect of incubation time on LH release and cAMP accumulation. Rat anterior pituitary cells in monolayer culture were incubated with 8.5 nM GnRH or 3 mM theophylline. Each point represents the mean ± S.E. of four petri dishes.

theophylline induced the increase of cAMP in the cells, but had no effect on LH release. Tang and Spies (1976) and Sundberg et al. (1976) reported similar results on the inability of theophylline to increase LH release. In other experiments, we were able to detect significant amounts of released LH after less than 60 min, while the increase of cAMP was not significant.

The shortest time we were able to use in our experiments for LH release and cAMP increase was 15 min. It is conceivable that the levels of cAMP could have risen as a short impulse immediately after the starting of incubation. This possibility was investigated by Rigler et al. (1978), who exposed male rat anterior pituitaries to GnRH for very short periods of 1 or 5 min. Their results are shown in Table V. There was no significant difference in the level of cAMP in GnRH-treated and control flasks in 1- or in 5-min incubation experiments. However, after 1- and 5-min incubation, the level of cyclic GMP (cGMP) in the pituitary tissues was slightly but significantly increased. This finding led Rigler et al. (1978) to suggest that the guanylate cyclase–cGMP system may play an important role in the mediation of LH release. Similar results for cGMP being implicated in the GnRH-stimulated release of LH were also reported by Snyder *et al.* (1978).

Naor *et al.* (1975), using various experimental conditions, also showed a dissociation between the action of GnRH on cAMP accumulation and LH release *in vivo* and *in vitro*. *In vitro* (Fig. 8), drugs that influence augmentation of cAMP, such as prostaglandin E_2 and choleragen, did not promote a parallel increase in LH release. Inversely, flufenamic acid, an inhibitor of prostaglandin synthesis, kept GnRH from stimulating accumulation of cAMP, but it did not affect the action of GnRH on LH release.

In a recent paper, Barden and Betteridge (1977) showed that the addition of GnRH to cultures of rat anterior-pituitary cells will increase the concentration of prostaglandins E_1 and E_2. However, the increased concentration of prostaglandins was not obligatory for the effect of GnRH on LH release.

In vivo (Fig. 9), administration of GnRH raised the serum LH level within 10 min, but it did not increase pituitary cAMP or cGMP concentration during this time (Ratner *et al.*, 1976). In addition, Ratner *et al.* (1976) reported that the intravenous administration of either aminophylline or dibutyryl cAMP resulted in a significant accumulation of cAMP in the pituitary, 10 min later, while the serum LH level remained unchanged (Fig. 10).

Recently, Naor *et al.* (1978) reported that a sex difference exists in pituitary cAMP response to GnRH. Incubation of pituitaries taken from immature male or female rats and adult female rats for 20–240 min with

Table V. Concentrations of cGMP and cAMP in Anterior Pituitary Tissue from Male Rats after Incubation for 1 or 5 min in the Presence of 20 ng GnRH/ml[a]

Expt. No.	Test substance	cGMP (pmol/mg)		cAMP (pmol/mg)	
		1 min	5 min	1 min	5 min
1	Control	0.058 ± 0.004	0.056 ± 0.010	1.03 ± 0.03	1.12 ± 0.14
	GnRH	0.087 ± 0.020	0.083 ± 0.020	1.08 ± 0.08	1.11 ± 0.06
2	Control	0.029 ± 0.001	0.037 ± 0.003	2.63 ± 0.20	1.74 ± 0.10
	GnRH	0.039 ± 0.002	0.040 ± 0.020	1.99 ± 0.30	2.17 ± 0.20
3	Control	0.033 ± 0.005	0.018 ± 0.003	1.44 ± 0.14	0.79 ± 0.09
	GnRH	0.045 ± 0.009	0.039 ± 0.003	1.36 ± 0.20	1.12 ± 0.06
4	Control	—	—	1.47 ± 0.11	1.33 ± 0.09
	GnRH	—	—	1.62 ± 0.14	1.48 ± 0.10
5	Control	—	0.079 ± 0.012	—	0.98 ± 0.09
	GnRH	—	0.134 ± 0.021	—	0.93 ± 0.12

[a] Statistics calculated by two-way analysis of variance: GnRH increased the level of cGMP after incubation for 1 min ($P < 0.05$) or 5 min ($P < 0.005$), but had no effect on the level of cAMP at either time. Results are means ± S.E.M. From Rigler *et al.* (1978) courtesy of the authors and the *Journal of Endocrinology*.

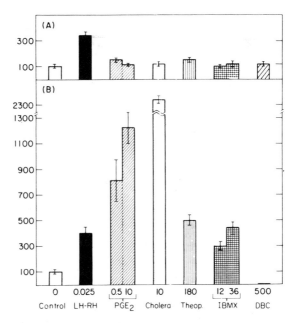

Figure 8. *In vitro* effects of GnRH (LH-RH in the figure), prostaglandin E₂ choleragen (Cholera), Theophylline (Theop.), IBMX, and dibutyryl cAMP (DBC) on LH release (A) and cAMP level in the pituitary (B). Bisected anterior pituitaries were incubated for 2 hr in Krebs–Ringer–bicarbonate medium (KRB). The medium was then changed to KRB containing the drugs indicated, and the incubation was continued for another 4 hr. Basal LH release to the medium was 18.5 ± 2.0 μg/ml, and basal cAMP level was 0.44 ± 0.04 pmol/mg tissue wet weight. Vertical brackets indicate S.E.M. for 10 determinations. From Naor *et al.* (1975) courtesy of the authors and Elsevier/North-Holland Biomedical Press.

graded doses of GnRH resulted in a 6- to 33-fold augmentation of LH release without any increase in pituitary cAMP level. As shown in Fig. 11, enhanced cAMP production in response to GnRH stimulation occurs only in pituitaries derived from adult male rats.

To summarize, many results obtained *in vitro* and *in vivo* in different laboratories show that there is no correlation between LH release and cAMP increase in anterior pituitary cells or tissues. A longer incubation time and a higher dose of GnRH are generally required to enhance cAMP accumulation in anterior pituitary than is necessary for induction of LH release. In most of the tissues in which cAMP mediates a hormone action, the level of the nucleotide increases a few minutes after the beginning of the hormonal stimulation, but this is not the case with GnRH, which induced significant cAMP accumulation *in vitro* only after 2 hr or more of incubation of pituitary cells and tissue.

On the other hand, drugs that influence cAMP accumulation *in vitro*, such as theophylline, aminophylline, IBMX, prostaglandin E_2, and choleragen, did not promote a parallel increase in LH release. *In vivo*, administration of dibutyryl cAMP or aminophylline resulted in a significant accumulation of cAMP in the pituitary, while serum LH levels remained unchanged.

In addition, GnRH-induced accumulation of cAMP seems to depend on the sex of rats and occurs only in pituitaries of male rats, whereas

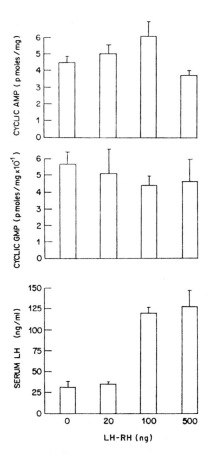

Figure 9. Serum LH and pituitary cAMP and cGMP accumulation 10 min after the intravenous administration of different doses of synthetic GnRH (LH-RH in the figure) into male rats. Height of bars represents the mean value; vertical brackets represent S.E.M. Each value represents data from 6–8 rats. From Ratner *et al.* (1976) courtesy of the authors and S. Karger A.G., Basel.

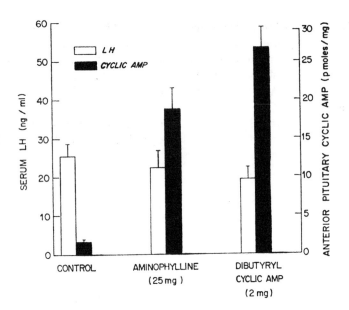

Figure 10. Serum LH and pituitary cAMP accumulation 10 min after the intravenous administration of dibutyryl cAMP or aminophylline into male rats. Height of bars represents the mean value; vertical brackets represent S.E.M. Each value represents data from 6–8 rats. From Ratner *et al.* (1976) courtesy of the authors and S. Karger A.G., Basel.

GnRH also induced LH release from pituitaries of female rats without increasing their cAMP content.

Finally, some recent findings (Rigler *et al.*, 1978; Snyder *et al.*, 1978) suggest that cGMP may play a role in the mediation of LH release, but this hypothesis must be verified.

4. Conclusions

Because cAMP and its derivatives have been shown to stimulate LH and FSH release, and because cAMP levels increased in GnRH-treated rat pituitary tissue or dispersed pituitary cells (references cited above), it has been suggested that cAMP is a mediator in the mechanism of GnRH action on LH release (Borgeat *et al.*, 1974; Kaneko *et al.*, 1973; Walker and Hopkins, 1978).

Our results, together with other data reported from the literature, do not support this hypothesis. To explain the negative results, it may always be argued that gonadotrophs represent only a relatively small proportion (5–10%) of the pituitary-gland cell population (Hymer *et al.*, 1973), so that an increase in the GnRH-induced adenylate cyclase activity over that of the basal activity would be difficult to demonstrate. Another possibility is that, as in the case of dispersed Leydig cells stimulated with LH, cAMP may "act through undetectably small increments, or by translocation within a small intracellular compartment" (Catt and Dufau, 1976).

It has been reported that LH release induced by GnRH occurs in two phases, an initial or acute phase and a late phase. The second phase of secretion is developed by a self-potentiating effect of GnRH on LH release, and this effect is probably mediated by the synthesis of a protein

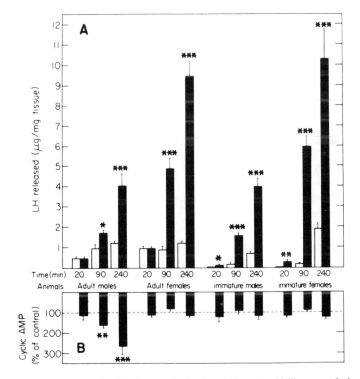

Figure 11. Time course of GnRH effect on LH release (A) and cAMP accumulation (B) in pituitary glands from immature and adult proestrous female and male rats. Pituitaries were incubated for 2 hr in plain Krebs–Ringer–bicarbonate medium (KRB) and thereafter in KRB or KRB containing GnRH (1.0 nM for immature rats and 5.0 nM for the adult rats) for the times indicated. [Basal cAMP levels (pmol/mg tissue) were: adult males, 0.41 ± 0.06; adult females, 0.39 ± 0.04; immature males, 1.8 ± 0.25; immature females, 1.0 ± 0.1.] Vertical brackets indicate means $\pm$ S.E. for 18 determinations in three separate experiments. ($\square$) Control; ($\blacksquare$) GnRH-stimulated. By Student's t test: significantly different from control value at: (*) $p < 0.05$; (**) $p < 0.01$; (***) $p < 0.001$. From Naor et al. (1978) courtesy of the authors and the American Physiological Society.

(Yen *et al.*, 1972; Aiyer *et al.*, 1974; Bremner and Paulsen, 1974; Gilbert and Edwardson, 1975; Pickering and Fink, 1976; De Koning *et al.*, 1976; Vilchez-Martinez *et al.*, 1976; Crighton and Foster, 1977). In the rat, the priming effect takes 30–60 min to develop. It has been suggested that the acute release of LH is mediated by cAMP (Kercret *et al.*, 1977). For others (Menon *et al.*, 1977), cAMP may participate in the second phase of GnRH action.

It is always more difficult to present negative results than positive ones, but we decided to do it in order to generate discussion. Without discarding the possible role of cAMP in the secretion or synthesis, or both, of gonadotropins, we are tempted to admit, in the light of all these negative results, that the concept of the participation of cAMP in the mechanism of action of GnRH should be reconsidered. Perhaps the use of enriched populations of gonadotrophs will help solve this problem.

5. Summary

In this chapter, we reviewed two main aspects of the cellular mechanism of the action of GnRH, the concept of a cellular receptor for this neurohormone and the possible involvement of cAMP as an intermediate in the releasing mechanism of LH.

Studies carried out in our laboratory on purified ovine pituitary plasma membranes allowed demonstration of the existence of a single type of binding site interacting specifically with GnRH and determination of the physicochemical parameters of this interaction. Thus, the reversible binding of GnRH to the specific receptor sites on a gonadotropic cell is probably the first step in the mechanism of action of this neurohormone. The question arises as to whether these receptors are functionally related to the adenylate cyclase system.

Some results related to the release of gonadotropins with cAMP and its derivatives, and those concerning the cAMP accumulation in pituitary tissue under the action of GnRH, have led us and others to postulate a possible involvement of this nucleotide as the intermediate of GnRH action. However, several data recently obtained both in our laboratory and by others do not favor this hypothesis. The following arguments can be made against the involvement of cAMP as an intermediate in the cellular mechanism of action of GnRH.

1. GnRH at different doses and in various conditions did not stimulate adenylate cyclase, nor did it inhibit phosphodiesterase, either in crude homogenates of male rat anterior pituitaries or in purified sheep pituitary plasma membranes.

2. There is no correlation *in vitro* and *in vivo* between the release of LH and the increase in cAMP in the anterior pituitary.

3. Drugs that influence cAMP accumulation *in vitro*, such as theophylline, aminophylline, IBMX, prostaglandin E_2, and choleragen, did not promote a parallel increase in LH release.

4. *In vivo*, administration of dibutyryl cAMP or aminophylline resulted in a significant accumulation of cAMP in the pituitary, while serum LH levels remained unchanged.

5. GnRH-induced accumulation of cAMP seems to occur only in pituitaries of male rats, while GnRH also induces LH release from pituitaries of female rats without increasing their cAMP content.

We believe, in the light of all these negative results, that the concept of the participation of cAMP in the mechanism of action of GnRH must be reconsidered.

ACKNOWLEDGMENTS. We would like to express our thanks to Dr. Ashok Khar from our laboratory for his help in preparing rat pituitary cell cultures and in radioimmunoassay of LH. We acknowledge the valuable technical assistance of Mr. Y. Colleaux and Mrs. T. Bennardo.

DISCUSSION

FINK: It is not possible to prove a negative, and therefore we cannot exclude a role for cAMP in the action of GnRH. However, so far, there is little positive evidence suggesting that cAMP acts as a second messenger for GnRH, and some of the data on the priming effect I shall present support those of Dr. Jutisz. It is important to make clear whether we are thinking of cAMP as a second messenger, or simply as a cyclic nucleotide involved somewhere along the line of gonadotropin release. Thus, for example, the great discrepancy between the timing of the rise in cAMP concentrations and gonadotropin release in Dr. Labrie's work is consistent with the latter but not with the former. The review by Rassmussen and Goodman (1977, *Physiol. Rev.* **57**:421) should be consulted for many other conflicting data regarding cAMP as second messenger in other systems.

REFERENCES

Aiyer, M. S., Chiappa, S. A., and Fink, G., 1974, A priming effect of luteinizing hormone releasing factor on the anterior pituitary gland in the female rat, *J. Endocrinol.* **62**:573–588.

Barden, N., and Betteridge, A., 1977, Stimulation of prostaglandin accumulation in the rat anterior pituitary gland by luteinizing hormone releasing hormone *in vitro*, *J. Endocrinol. 75:277–283.*

Baumann, R., and Kuhl, H., 1978, LH-RH interactions with purified plasma membranes of rat anterior pituitary gland, *Acta Endocrinol. (Copenhagen)* **87**(suppl. 215):91.

Bérault, A., Théoleyre, M., and Jutisz, M., 1974, A simplified method for the preparation of plasma membranes from ovine anterior pituitary glands, *FEBS Lett.* **39**:267–270.

Borgeat, P., Chavancy, G., Dupont, A., Labrie, F., Arimura, A., and Schally, A. V., 1972, Stimulation of adenosine 3′,5′-cyclic monophosphate accumulation in anterior pituitary gland *in vitro* by synthetic luteinizing hormone–releasing hormone, *Proc. Natl. Acad. Sci. U.S.A.* **69:**2677–2681.

Borgeat, P., Labrie, F., Côté, J., Ruel, F., Schally, A. V., Coy, D. H., Coy, E. J., and Yanaihara, N., 1974, Parallel stimulation of cyclic AMP accumulation and LH and FSH release by analogs of LH-RH *in vitro, Mol. Cell. Endocrinol.* **1:**7–20.

Bremner, W. J., and Paulsen, C. A., 1974, Two pools of luteinizing hormone in the human pituitary: Evidence from constant administration of luteinizing hormone–releasing hormone, *J. Clin. Endocrinol. Metab.* **39:**811–815.

Butcher, R. W., and Sutherland, E. W., 1962, Adenosine 3′,5′-phosphate in biological materials, *J. Biol. Chem.* **237:**1244–1250.

Catt, K. J., and Dufau, M. L., 1976, Basic concepts of the mechanism of action of peptide hormones, *Biol. Reprod.* **14:**1–15.

Clayton, R. N., Shakespear, R. A., and Marshall, J. C., 1978, LH-RH binding to purified pituitary plasma membranes: Absence of adenylate cyclase activation, *Mol. Cell. Endocrinol.* **11:**63–78.

Crighton, D. B., and Foster, J. P., 1977, Luteinizing hormone release after two injections of synthetic luteinizing hormone releasing hormone in the ewe, *J. Endocrinol.* **72:**59–67.

Davis, J. C., and Hymer, W. C., 1975, Effects of hypothalamic extract and dibutyryl cAMP on FSH production in the rat adenohypophysis *in vitro, J. Endocrinol.* **64:**229–236.

Debeljuk, L., Khar, A., and Jutisz, M., 1978, Effect of pimozide and sulpiride on the release of LH and FSH by pituitary cells in culture, *Mol. Cell. Endocrinol.* **10:** 159–162.

Deery, D. J., and Howell, S. L., 1973, Rat anterior pituitary adenyl cyclase activity: GTP requirement of prostaglandin E_1 and E_2 and synthetic luteinising hormone–releasing hormone activation, *Biochim. Biophys. Acta* **329:**17–22.

De Koning, J., Van Dieten, J. A. M. J., and Van Rees, G. P., 1976, LH-RH dependent synthesis of protein necessary for LH release from rat pituitary glands *in vitro, Mol. Cell. Endocrinol.* **5:**151–160.

Gilbert, D., and Edwardson, J. A., 1975, The self-potentiating effect of luteinizing hormone–releasing hormone (LH-RH): Dose–response studies with LH-RH and cycloheximide, *Acta Endocrinol. (Copenhagen)* **80** *(Suppl. 199):*260.

Gilman, A. G., 1970, A protein binding assay for adenosine 3′:5′-cyclic monophosphate, *Proc. Natl. Acad. Sci. U.S.A.* **67:**305–312.

Grant, G., Vale, W., and Rivier, J., 1973, Pituitary binding sites for [³H]-labelled luteinizing hormone releasing factor (LRF), *Biochem. Biophys. Res. Commun.* **50:**771–778.

Groom, G. V., and Boyns, A. R., 1973, Effect of hypothalamic releasing factors and steroids on release of gonadotrophins by organ cultures of human foetal pituitaries, *J. Endocrinol.* **59:**511–522.

Hymer, W. C., Evans, W. H., Kraicer, J., Mastro, A., Davis, J., and Griswold, E., 1973, Enrichment of cell types from the rat adenohypophysis by sedimentation at unit gravity, *Endocrinology* **92:**275–287.

Jutisz, M., and de la Llosa, M. P., 1969, l'Adenosine 3′5′ monophosphate cyclique, un intermédiaire probable de l'action de l'hormone hypothalamique FRF, *C. R. Acad. Sci. Ser. D* **268:**1636–1639.

Jutisz, M., Kerdelhué, B., and Bérault, A., 1970, Further studies on mechanism of action of luteinizing hormone releasing factor using *in vivo* and *in vitro* techniques, in: *The Human Testis* (E. Rosemberg and C. A. Paulsen, eds., pp. 221–228, Plenum Press, New York.

Kaneko, T., Saito, S., Oka, H., Oda, T., and Yanaihara, N., 1973, Effects of synthetic LH-RH and its analogs on rat anterior pituitary cyclic AMP and LH and FSH release, *Metabolism* **22**:77–81.

Kercret, H., Benoist, L., and Duval, J. 1977. Acute release of gonadotropins mediated by dibutyryl-cyclic AMP *in vitro*, *FEBS Lett.* **83**:222–224.

Krishna, G., Weiss, B., and Brodie, B. B., 1968, A simple, sensitive method for the assay of adenylcyclase, *J. Pharmacol. Exp. Ther.* **163**:379–385.

Labrie, F., Pelletier, G., Lemay, A., Borgeat, P., Barden, N., Dupont, A., Savary, M., Coté, J., and Boucher, R., 1973, Control of protein synthesis in anterior pituitary gland, *Acta Endocrinol. (Copenhagen)* **74**(*Suppl. 180*):301–334.

Labrie, F., Borgeat, P., Lemay, A., Lemaire, S., Barden, N., Drouin, J., Lemaire, I., Jolicoeur, P., and Belanger, A., 1975, Role of cAMP in the action of hypothalamic regulatory hormones, *Adv. Cyclic Nucleotide Res.* **5**:787–801.

Makino, T., 1973, Study on the intracellular mechanism of LH release in the anterior pituitary, *Am. J. Obstet. Gynecol.* **115**:606–614.

Marche, P., Morgat, J. L., Fromageot, P., Kerdelhué, B., and Jutisz, M., 1972, ^{3}H labelling of a synthetic decapeptide having LH and FSH releasing activity (LH-RH/FSH-RH), *FEBS Lett.* **26**:83–86.

Marshall, J. C., Shakespear, R. A., and Odell, W. D., 1976, LH-RH pituitary plasma membrane binding: The presence of specific binding sites in other tissues, *Clin. Endocrinol.* **5**:671–677.

Marshall, J. C., Clayton, R. N., and Shakespear, R. A., 1977, GnRH degradation by purified pituitary plasma membranes, 59th Annual Meeting of the Endocrine Society, Abstract No. 16.

McKelvy, J. F., Leblanc, P., Laudes, C., Perrie, S., Grimm-Jorgensen, Y., and Kordon, C., 1976, The use of bacitracin as an inhibitor of degradation of TRH and LH-RH, *Biochem. Biophys. Res. Commun.* **73**:507–515.

Menon, K. M. J., Gunaga, K. P., and Azhar, S., 1977, GnRH action in rat anterior pituitary gland: Regulation of protein, glycoprotein and LH synthesis, *Acta Endocrinol. (Copenhagen)* **86**:473–488.

Naor, Z., Koch, Y., Chobsieng, P., and Zor, U., 1975, Pituitary cyclic AMP production and mechanism of luteinizing hormone release, *FEBS Lett.* **58**:318–321.

Naor, Z., Zor, U., Meidan, R., and Koch, Y., 1978, Sex difference in pituitary cyclic AMP response to gonadotropin-releasing hormone, *Am. J. Physiol.* **235**:E37–E41.

Park, K. R., Saxena, B. B., and Gaudy, H. M., 1976, Specific binding of LH-RH to the anterior pituitary gland during the oestrous cycle in the rat, *Acta Endocrinol. (Copenhagen)* **82**:62–70.

Pickering, A. J. M. C., and Fink, G., 1976, Priming effect of luteinizing hormone releasing factor: *In-vitro* and *in-vivo* evidence consistent with its dependence upon protein and RNA synthesis, *J. Endocrinol.* **69**:373–379.

Ratner, A., 1970, Stimulation of luteinizing hormone release *in vitro* by dibutyryl cyclic AMP and theophylline, *Life Sci.* **9** (Part I):1221–1226.

Ratner, A., Wilson, M. C., Srivastava, L., and Peake, G. T., 1976, Dissociation between LH release and pituitary cyclic nucleotide accumulation in response to synthetic LH-releasing hormone *in vivo*, *Neuroendocrinology* **20**:35–42.

Rigler, G. L., Peake, G. T., and Ratner, A., 1978, Effect of luteinizing hormone releasing hormone on accumulation of pituitary cyclic AMP and GMP *in vitro*, *J. Endocrinol.* **76**:367–368.

Rodbell, M., Birnbaumer, L., Pohl, S. L., and Krans, H. M. J., 1971, The glucagon-sensitive adenylcyclase system in plasma membranes of rat liver, *J. Biol. Chem.* **246**:1877–1882.

Snyder, G., Naor, Z., and Fawcett, C. P., 1978, Possible involvement of cGMP in LH-RH-stimulated LH and FSH release from enriched populations of gonadotrophs, 60th Annual Meeting of the Endocrine Society, Abstract No. 147.

Spona, J., 1973, LH-RH-interaction with the pituitary plasma membrane, *FEBS Lett.* **34:**24–26.

Sundberg, D. K., Fawcett, C. P., and McCann, S. M., 1976, The involvement of cyclic-3', 5'-AMP in the release of hormones from the anterior pituitary *in vitro, Proc. Soc. Exp. Biol. Med.* **151:**149–154.

Tang, L. K. L., and Spies, H. G., 1976, Effects of hypothalamic-releasing hormones on LH, FSH and prolactin in pituitary monolayer cultures, *Proc. Soc. Exp. Biol. Med.* **151:**189–192.

Théoleyre, M., Bérault, A., Garnier, J., and Jutisz, M., 1976, Binding of gonadotropin-releasing hormone (LH-RH) to the pituitary plasma membranes and the problem of adenylate cyclase stimulation, *Mol. Cell. Endocrinol.* **5:**365–377.

Vilchez-Martinez, J. A., Arimura, A., and Schally, A. V., 1976, Effect of actinomycin D on the pituitary response to LH-RH, *Acta Endocrinol. (Copenhagen)* **81:**73–81.

Wagner, T. O. F., 1978, Binding kinetics for the interaction of mono-iodo-GnRH and ovine anterior pituitary, *Acta Endocrinol. (Copenhagen)* **87** (*Suppl. 215*):92.

Wakabayashi, K., Date, J., and Tamaoki, B., 1973, On the mechanism of action of luteinizing hormone–releasing factor and prolactin release inhibiting factor, *Endocrinology* **92:**698–704.

Walker, A. M., and Hopkins, C. R., 1978, Stimulation of luteinising hormone release by luteinising hormone–releasing hormone in the porcine anterior pituitary: The role of cyclic AMP, *Mol. Cell. Endocrinol.* **10:**327–341.

Yen, S. S. C., Vandenberg, G., Rebar, R., and Ehara, Y., 1972, Variation of pituitary responsiveness to synthetic LRF during different phases of the menstrual cycle, *J. Clin. Endocrinol. Metab.* **35:**931–934.

23

Characterization of Thyroliberin (TRH) Binding Sites and Coupling with Prolactin and Growth Hormone Secretion in Rat Pituitary Cell Lines

Danielle Gourdji

1. Introduction

The secretory response induced by the synthetic hypothalamic hormone thyroliberin (TRH) (L-pyroglutamyl-L-histidyl-L-prolineamide) (Burgus *et al.*, 1969) on anterior pituitary involves, as in any hormone–target cell interaction, the recognition of receptors. The characterization of binding sites relevant to hormone release or synthesis or both constitutes an approach to identification of these receptors. Cell culture of homogeneous populations of target cells is a useful model system for such investigations, as demonstrated by the large body of information that has been obtained in the past few years. Indeed, since the initial discovery by Tashjian *et al.* (1971) of the TRH prolactin-promoting activity in GH$_3$ and GH$_1$ cells, the use of these or of similar rat prolactin-secreting clonal cell lines became preeminent in studying the TRH mechanism of action at the cellular level (see the reviews in Tixier-Vidal *et al.*, 1975b, 1979a). Although they are continuously dividing, and despite some tumoral aspects of their behavior, these cell lines retained, even after several years of growing in culture,

Danielle Gourdji • Groupe de Neuroendocrinologie Cellulaire, Chaire de Physiologie Cellulaire, Collège de France, 75231 Paris 05, France

their specific differentiation: (1) They synthesize and release prolactin (PRL) and growth hormone (GH), biologically and immunologically indistinguishable from rat hormones (Tashjian *et al.*, 1968, 1970; Gourdji *et al.*, 1973a). (2) They respond in the same manner as normal rat pituitary cells to several drugs or hormones that are known to regulate PRL *in vivo*, such as estrogens (Tashjian and Hoyt, 1972; Brunet *et al.*, 1977), CB 154 (Gourdji *et al.*, 1973b), dopamine (Tixier-Vidal *et al.*, 1979b), and TRH, since it has now been shown that TRH is actually a PRL-stimulating factor in a wide range of species including rat and man (cf. the review in Vale *et al.*, 1977). Its effect on PRL is biphasic, as demonstrated in SD_1 cells (Morin *et al.*, 1975) and in GH_3 cells (Dannies *et al.*, 1976), as well as in normal rat pituitary primary culture (Vale *et al.*, 1973): a short-term effect on PRL release (150–350% of the control) and a secondary stimulation of PRL synthesis (130–500% of the control) that involved an increase of the messenger RNA (mRNA) coding for PRL in GH_3 (Evans *et al.*, 1978). TRH also elicits an acute release of GH in $GH_3 B_6$ (Faivre-Bauman *et al.*, 1976; Ostlund *et al.*, 1978), GH_1 (Morin and Labrie, 1975), and SD_1 (Gourdji *et al.*, 1975), but in contrast to PRL, this stimulating effect is transitory and turns into a long-term inhibiting action on GH synthesis (Tashjian *et al.*, 1971; Tashjian and Hoyt, 1972).

No information is yet available that could link in a sequential manner the initial binding of TRH and its effects on hormonal secretion. Neither is there any evidence for a causal relationship between acute and long-term effects or between similar and then opposite effects on GH and PRL.

In addition to these effects on specific secretory processes, TRH is also able to generate either short- or long-term or both, some modifications in rat anterior pituitary cells in culture: (1) morphological changes observed at the light- and electron-microscope levels (Gourdji *et al.*, 1972; Tixier-Vidal *et al.*, 1975a,b, 1979a); (2) alterations of several membrane properties (Tixier-Vidal *et al.*, 1975a, 1976, 1978, 1979a; Brunet and Tixier-Vidal, 1978; Imae *et al.*, 1975; Kidokoro, 1975; Dufy *et al.*, 1979); (3) modifications of cell growth (Hayashi and Sato, 1976). These multiple effects are concomitant with the expression of hormonal response and therefore merit reporting in some detail, which is presented in Section 2.

While the existence of specific binding sites for TRH in responsive rat pituitary cell lines is firmly established (Gourdji *et al.*, 1973a; Hinkle and Tashjian, 1973; Hinkle *et al.*, 1974; Faivre-Bauman *et al.*, 1977), important questions nevertheless remain to be answered: Are all these binding sites to be considered as functional receptors? Are the multiple effects of TRH triggered by a single site or by different sites of action of the hypothalamic hormone? Is TRH itself the unique active molecule, excluding a role for its metabolites? In an attempt to at least partially answer these questions, we will in this chapter correlate the characteristics of TRH binding, the fate of bound TRH, and the biological responses induced in three different cell lines, GH_3, $GH_3 B_6$, and SD_1, the main characteristics of which are described in the next section.

2. General Features of Our Model System and Effects of Thyroliberin on Nonsecretory Parameters

2.1. General Features

2.1.1. Origin of the Cell Lines

The GH_3 cell line and its subclone $GH_3 B_6$ (isolated for its higher [3H]-TRH binding capacity and better anchorage on culture dishes) originate from an estrogensensitive tumor, MtT/W5 (Yasumura et al., 1966; Tashjian et al., 1968). The SD_1 cell line originates from spontaneous transformation of primary culture of normal Sprague–Dawley rat anterior pituitaries (Tixier-Vidal et al., 1975a). So far, these three cell lines display very few and only quantitative differences.

2.1.2. Ultrastructural Features

Although these three cell lines exhibit an ultrastructural organization greatly different from that of normal rat PRL-producing cells, they nevertheless retain fundamental characteristics of active endocrine cells (Gourdji et al., 1972; Tixier-Vidal et al., 1975a,b). In particular, they possess a very well developed Golgi apparatus consisting of numerous units scattered within the cytoplasm. Images of secretory granules (from 50 to 150 nm in diameter) appear very scanty except in a very small percentage of cells. The rough endoplasmic reticulum is organized in rare linear cisternae, while polysomes and free ribosomes are extensively represented (Fig. 1a).

2.1.3. Hormonal Production

GH_3 cells synthesize and release GH (Tashjian et al., 1968) and PRL (Tashjian et al., 1970), as do $GH_3 B_6$ cells. The ratio between GH and PRL differs with culture duration and conditions, but generally favors PRL for medium (Tixier-Vidal et al., 1979b). SD_1 cells also produce these two hormones (Tixier-Vidal et al., 1975a; Gourdji et al., 1975). They differ from GH_3 cells in having a lower rate of PRL turnover (Morin et al., 1975). As demonstrated in the SD_1 cell line, the spontaneous release of PRL is already complex, involving at least two different pools of intra-

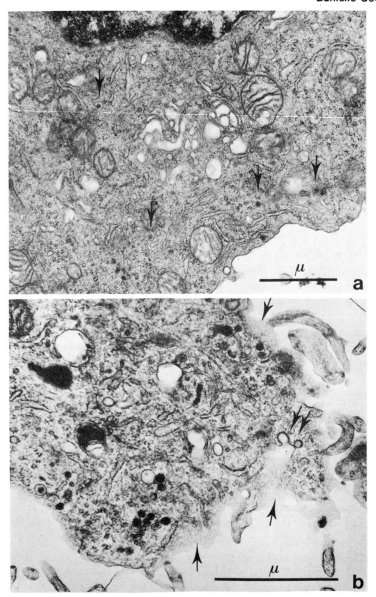

Figure 1. (a) GH$_3$ cell grown in control medium (Ham F-10 + 15% heat-inactivated horse serum and 2.5% fetal calf serum). Note the Golgi unit with dilated saccules, the few short linear egastroplasmic cisternae, the numerous polysomes, and the rare secretory granules (arrows). (b) Peripheral portion of a GH$_3$ cell grown in control medium and exposed for 15 min to [^{3}H]-TRH, 27 nM. Several features of the TRH-induced modifications can be seen: spreading of the plasma membrane, which is tangentially sectioned in several places (arrows), cytoplasmic extensions, small vesicles and canaliculi, coated invaginations (double arrows), and secretory granules near the plasma membrane.

cellular hormone (Morin *et al.*, 1975). This complexity is supported by immunocytochemical staining at the electron-microscope level (Tixier-Vidal *et al.*, 1975b) and assay of PRL in both pellet and 100,000*g* supernate (unpublished observation). Whether GH and PRL are or are not produced simultaneously in these cell lines is as yet unknown (see Tixier-Vidal *et al.*, 1979b).

2.2. Effects of TRH on Nonsecretory Parameters

2.2.1. Morphological Features*

TRH elicits a spreading of the cells on the culture dish, which is distinguishable for high doses (27 nM) as soon as after 30 min of exposure and is conspicuous even for low concentrations ($\leq$ 2 nM) after several hours. The spreading persists all during the TRH exposure. At the electron-microscope level, the major modifications concern the whole system of smooth membranes. The Golgi zones exhibit conspicuous expansion, flattening of the cisternae, multiplication of small vesicles, and increase of condensing secretory-granule images; on the whole, they appear to form a more or less continuous system up to the plasma membrane. The latter also undergoes important modifications: obvious extension in an irregular wave-shape pattern, images of smooth or coated invaginations, and an increased number of microvilli and fine branches (Fig. 1b) (Tixier-Vidal *et al.*, 1975a,b) also observed by scanning electron microscope (Tashjian and Hoyt, 1972). Nucleoli appear very turgescent, with noticeable nuclei and highly dispersed chromatin.

2.2.2. Plasma Membrane Properties†

TRH induces an increase of exposure of surface glycoproteins that is not directly correlated with prolactin release (Brunet and Tixier-Vidal, 1978). It also stimulates endocytosis and internalization of membranes (Tixier-Vidal *et al.*, 1976, 1978, 1979a).

TRH modifies the electrical properties of GH$_3$ cells (Kidokoro, 1975; Dufy *et al.*, 1979; and in this volume).

Finally, Imae *et al.* (1975) described an alteration of the tryptophan fluorescence induced by TRH in crude plasma membrane preparations.

---

* General references for this section are Gourdji *et al.* (1972), Tixier-Vidal (1975), and Tixier-Vidal *et al.* (1975a,b).
† General references for this section are Tixier-Vidal *et al.* (1978, 1979a), Kidokoro (1975), and Imae *et al.* (1975).

2.2.3. Cell Growth Modifications (Table I)

TRH (27 nM) alters $GH_3 B_6$ cell growth after 2, 4, and 7 days of treatment of cells precultivated in control conditions for 3 days. The major feature consists in a decrease in the rate of cell division that leads to a reduction in cell number to 65% of control after 4 days of treatment. In contrast, the cell protein content is augmented up to 146% of the control. The amplitude of those effects decreases from 4 to 7 days of treatment. This result contrasts with the effect of TRH in promoting the growth of GH_3 cells maintained in serum-free conditions (Hayashi and Sato, 1976).

No conclusive causal relationships have been demonstrated so far between these multiple effects of TRH and the simultaneous alterations of hormonal secretion. However, they have to be taken into account in interpreting TRH–target cell interaction.

3. Binding of [³H]Thyroliberin to Prolactin Cell Lines

Extensive studies with [³H]-TRH carried out in different conditions by several groups of workers all concluded that there exist highly specific [³H]-TRH binding sites in functional target cells for TRH (cf. the review in Tixier-Vidal *et al.*, 1975b). Nevertheless, some differences in the characteristics of these sites appear, depending on the laboratories. These discrepancies may reflect different experimental schedules or models or use of different [³H]-TRH preparations. We worked with highly purified samples mono- or bitritiated on the C2 or C4 or both of the imidazole ring of the histidine residue, which is fully biologically active (Pradelles *et al.*, 1972) and the stability of which, including the absence of lability of tritium, has been checked in our experimental conditions (Gourdji *et al.*, 1976b; Levine-Pinto, personal communication).

3.1. Kinetics and Specificity

At 37°C, the binding of [³H]-TRH to intact cells increases in serum-containing medium during the first 15–30 min, thereafter reaching a plateau (see Figs. 3 and 8); it is stable at 0°C but reversible at 37°C (Gourdji *et al.*, 1973a; Hinkle and Tashjian, 1973). The binding is specific for target cells, and the pattern of inhibition of binding by isotopic dilution is strictly characteristic of TRH. Any subtle change in the conformation drastically damages the ability to compete with [³H]-TRH binding sites (Hinkle *et al.*, 1974; Faivre-Bauman *et al.*, 1977; Vale *et al.*, 1973). The specificity was strict enough to permit the establishment of a radioreceptor assay sensitive to 200 pg/ml (Faivre-Bauman *et al.*, 1977).

Table I. Effect of 27 nM TRH on GH_3 B_6 Cell Growth[a]

Days of culture	Days of treatment	Control medium				+ TRH, 27 nM			
		Cells/dish (10^6)	Protein/dish (μg)	Protein/10^6 cells	DNA/10^6 cells	Cells/dish (10^6)	Protein/dish (μg)	Protein/10^6 cells	DNA/10^6 cells
5	2	0.543 ±0.060	157.5 ±3.2	298 ±36	32.6 ±4	0.360 ±0.012	156.5 ±5.3	435.6 ±23.8	36.0 ±4.8
7	4	1.184 ±0.070	256 ±7.1	217 ±1	23.6 ±0.4	0.649 ±0.103	208.0 ±36	319 ±10	29.0 ±1.9
10	7	1.580 ±0.150	366.6 ±9.8	238.6 ±30	28.5 ±3.1	1.168 ±0.119	343.9 ±8.8	297.8 ±24.2	31.0 ±3.4

[a] Cells (150,000/2 ml per 3001 Falcon dish) were precultivated 3 days before the first addition of TRH. Control medium: Ham's F-10 + 15% heat-inactivated horse serum + 2.5% fetal calf serum + 50 μg/ml gentamycin. Results are the means of triplicate determinations ± S.D.

The noncompetitive binding is very low, particularly in cells exposed *in situ* to [³H]-TRH (≤1% for 5 nM [³H]-TRH; ≤5% for 150 nM [³H]-TRH). The maximum number of specific binding sites per cell has been estimated as approximately 130,000 (Hinkle and Tashjian, 1973) or 180,000 (Gourdji *et al.*, 1973a) in GH_3 and more than 400,000 in GH_3 B_6 cells (unpublished data). These numbers are extremely high as compared to those estimated for other hormones, except for insulin, (100,000–250,000/ liver cell) (see the review in Kahn, 1976).

3.2. Affinity of the Binding Sites for [³H]-TRH

The [³H]-TRH binding increases the doses of [³H]-TRH, but there is little agreement on the characteristics of this affinity.

In our hands, the kinetics of this increase do not follow an ideal biomolecular reaction pattern, since classic technical plotting, such as those of Scatchard or of Lineweaver–Burk, led to a curvilinear graph (Fig. 2). Such a complicated pattern has been described in intact GH_3 cells (Gourdji *et al.*, 1973a, 1976a) as well as in intact mouse TSH-secreting

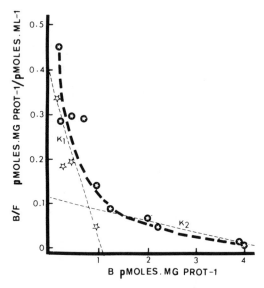

Figure 2. Scatchard plot of [³H]-TRH binding to intact GH_3 B_6 cells precultivated for 5 days in control medium (see the Fig. 1 caption). Incubation was performed on attached cells for 30 min at 37°C in 1 ml complete culture medium. [³H]-TRH (60 Ci/mmol) doses ranged from 0.3 to 840 nM. Each point was determined in triplicate and S.D. ≤10%. Noncompetitive binding was subtracted. The apparent high-affinity site K_1 (☆) was obtained after subtracting the contribution of the apparent low-affinity site K_2.

Table II. Comparison of the Apparent Dissociation Constants $(K_{D_{app.}})$ for [^{3}H]-TRH Binding and Concentration Required for Half-Maximum and Maximum Stimulation of PRL Release[a]

Cell line	Binding		Stimulation of PRL Release	
	$K_{D1_{app.}}$	$K_{D2_{app.}}$	½ Max.	Max.
GH$_3$ 892	1.8	21	6	120
GH$_3$ B$_6$ 1229	1.8	17	1	60
GH$_3$ B$_6$ 1264	3.8	42	3.5	100
SD$_1$ 275	2.3	24	2.1	115

[a] The two types of values were calculated from the same culture dishes. Experiments were performed for 30 min at 37°C in regular culture medium (see Table I), containing only 2.5% horse serum for SD$_1$ 275. Each point was determined, in triplicate. The range of [^{3}H]-TRH doses was from 0.5 to 850 nM. The $K_{D_{app.}}$ were estimated from Lineweaver–Burk double-reciprocal graphs after subtraction of noncompetitive binding. The K_{D1} was calculated after subtraction of the low-affinity-site participation. PRL was assayed by radioimmunoassay using the NIAMDD kit for rat PRL.

tumor cells (Grant *et al.*, 1973). This most probably reflects a heterogeneity in TRH binding sites, but cannot discriminate among several interpretations:

1. *The existence of at least two families of binding sites,* such as illustrated in Fig. 2, displaying different affinities for low and high TRH concentrations. Despite criticism on mathematical grounds (Klotz and Hunston, 1971), it remains useful to calculate the two apparent K_D's (Kahn *et al.*, 1974), which for *in situ* GH$_3$ or GH$_3$ B$_6$ are, respectively 1.8–3.8 × 10^{-9} M for 0.5 to 20–30 nM [^{3}H]-TRH and 21–42 × 10^{-9} M for up to 850 nM [^{3}H]-TRH (Table II). The apparent K_D of the presumed high-affinity site can also be estimated as approximately 4 × 10^{-9} M by competitive displacement (Jacobs *et al.*, 1975) using a tracer dose of [^{3}H]-TRH (0.54 nM).

   Interestingly, when cells grown in monolayer are scraped and thereafter exposed to [^{3}H]-TRH in suspension, both their binding capacity and their affinity are lowered (110,000 sites/cell; $K_{D1_{app.}} \approx 4.10^{-8}$ M, $K_{D2_{app.}} \approx 3.10^{-7}$ M) (Gourdji *et al.*, 1973a; Tixier-Vidal *et al.*, 1975b).

2. *The existence of negative cooperativity,* i.e., a functional heterogeneity, the occupancy of some binding sites eliciting a site–site interaction that decreases the affinity of still-unoccupied sites (De Meyts *et al.*, 1976). This explanation has been offered by Vale *et al.* (1977) for the experimental observation of two classes of binding

sites for TRH in intact mouse TSH-secreting tumor cells. This possible site–site interaction does not rule out, however, a TRH–TRH interaction.

3. *The existence of different affinities for [³H]-TRH as a function of the cell cycle.* All experiments were carried out with asynchronous cell culture, and it has been shown that in $GH_3 B_6$, the binding capacity fluctuates all during the cell cycle (Faivre-Bauman *et al.*, 1975).

The thesis of a nonsimple bimolecular reaction between TRH and intact cells is supported by two other findings: (1) The plotting of the rate of association of [³H]-TRH to intact cell as a function of time (from 0 to 15 min) was rigorously linear only for low concentrations of [³H]-TRH (0.27 nM) ($r = 0.999$). The plotting of the observed rate constant as a function of [³H]-TRH concentration (from 0.27 to 27 nM) reveals a curvilinear pattern (Fig. 3), as has also been observed in GH_3 homogenates (Hinkle and Tashjian, 1973). (2) Excess of unlabeled TRH accelerated and improved the rate of dissociation carried out at 25°C of bound [³H]-TRH

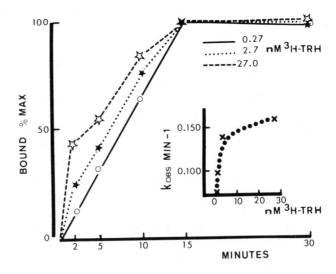

Figure 3. Rate of association of [³H]-TRH to intact GH_3 cells attached to their culture dishes. Cells were precultivated 5 days. Incubation was performed at 37°C in complete culture medium (see the Fig. 1 caption). Three different concentrations were used, and the binding plotted as the percentage of the maximum, which was obtained after 15 min in all cases. Each point was determined in triplicate with an S.D. $\leqslant$5%. *Insert:* Plotting of the observed constant of association (K_{OBS}) plotted as a function of [³H]-TRH concentration.

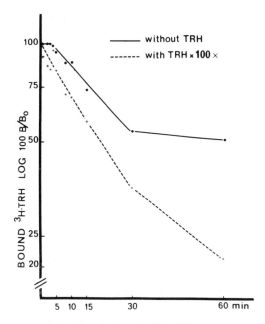

Figure 4. Time course of the dissociation of [³H]-TRH bound to intact GH₃ B₆ cells, precultivated for 5 days, after loading at 37°C for 30 min in the presence of 1.9 nM [³H]-TRH in complete culture medium (see the Fig. 1 caption). Dissociation was carried out at 25°C, after four washings with Ham-10/10 mM *N*-2-hydroxyethylpiperazine-*N'*-2-ethanesulfonic acid (HEPES), in 4 ml/3001 Falcon dish of Ham F-10–10 mM HEPES, in the absence or in the presence of 200 nM TRH. At given times, dissociation was arrested on ice, and the cells were washed three times with ice-cold HAM F-10–10 mM HEPES. Each point was determined in triplicate; S.D. was ≤6%.

when the binding was performed at low [³H]-TRH concentration (1.9 nM order of magnitude of the first apparent K_D). Moreover, this schedule reveals the existence of at least two different rates of dissociation: $t_{\frac{1}{2}(1)}$, which accelerates from 27 to 14 min, and $t_{\frac{1}{2}(2)}$, from 2 hr 47 min to 47 min, in the presence of unlabeled TRH (Fig. 4). A similar acceleration occurred in the dissociation kinetics of [³H]-TRH from membranes of a TSH-producing tumor (Grant *et al.*, 1972).

In contrast, a single class of [³H]-TRH receptors was found in intact GH₃ cells as well as in crude membrane preparations, with an apparent K_D of 25 nM (Hinkle and Tashjian 1973; Hinkle *et al.*, 1974). Similar values for a unique binding-site class were found in membrane fractions of several origins (Labrie *et al.*, 1972; Grant *et al.*, 1972; see references in Vale *et al.*, 1977).

4. Coupling of [³H]Thyroliberin Binding Sites to Hormone Response

The existence of such a coupling determines whether or not the binding sites are to be considered as functional receptors.

At least concerning the PRL release induced by TRH in the three cell lines studied, the [³H]-TRH binding sites fulfill the conditions classically required to provide evidence for such a coupling (Kahn, 1976): time course, dose effect (Table II), and stereospecific requirement with respect to structural analogues (Gourdji *et al.*, 1973a; Hinkle and Tashjian, 1973; Vale *et al.*, 1973; Hinkle *et al.*, 1974). While positive, this tentative conclusion deserves some comment.

4.1. Time Course

No correlation can be made with regard to the effects on PRL and GH synthesis, since they required a lag time. For PRL stimulation, it has been estimated to be 4–6 hr in SD_1 (Morin *et al.*, 1975) and in GH_3 (Dannies *et al.*, 1976). To our knowledge, no rigorous studies have been performed concerning GH synthesis and turnover in response to TRH. By the only radioimmunoassay of GH content in medium and cells, it appears that the acute stimulation of GH release turns rather slowly into a decrease of synthesis, more than 48 hr of treatment being required to obtain a significant reduction in GH medium content (see Section 8). Such a delay excludes any comparison with the time course of binding and, in fact, raises a question concerning a possible role of TRH metabolites (see Section 5).

4.2. Dose–Effect Relationship

Hormone release and, moreover, hormone synthesis are very complicated phenomena in which the receptor cannot be the final effector and for which the constraint of the simple model of action mass, even dealing with the two acceptor sites, is unable to account. Indeed, hormone release and synthesis do not appear to be a simple linear function of specific binding-site occupancy.

A [³H]-TRH concentration of 200 nM, i.e., above the K_D of the second family of binding sites, elicits less than maximum PRL release (unpublished observation).

In contrast to our results (Table II), some authors have evidence for a very large number of spare receptors concerning PRL release ($ED_{50} \approx$ 0.3 nM vs. $K_D = 25$ nM) as well as GH and PRL synthesis ($ED_{50} \approx 3$ nM) (Hinkle *et al.*, 1974; Dannies and Tashjian, 1976). A similar concept has also been considered with regard to PRL and TSH stimulation in the normal rat pituitary (see the review in Vale *et al.,* 1976).

The efficiency of the coupling, which appears to vary with the cell strain (cf. Tixier-Vidal *et al.,* 1975b), is indeed modulated by the physiological state of the target cell. This result is outlined using synchronized cells: the [³H]-TRH binding capacity fluctuates during the cell cycle, and the optimum binding does not occur simultaneously with the optimum responsiveness in terms of induced PRL release or in terms of induced GH release (Faivre-Bauman *et al.,* 1975, 1976).

4.3. Stereospecificity

In most cases, the biological potency of TRH structural analogues can be related to their ability to compete with [³H]-TRH (Hinkle *et al.,* 1974; Vale *et al.,* 1973).

However, no rigorous correlation was observed in the case of two analogues: desamido TRH (see Section 5.4) and <Glu-NH₂-C²-Im-TRH. The latter analogue did not compete with TRH binding sites unless

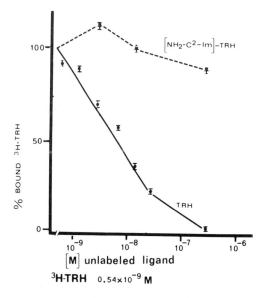

Figure 5. TRH-NH₂ (<Glu-NH₂-C²-Im-TRH) competitive affinity for [³H]-TRH binding sites on intact $GH_3 B_6$ cells as compared to TRH. The experiment was performed on cells precultivated for 5 days. Binding was measured after incubation for 30 min at 37°C in 1 ml/3001 Falcon dish of Ham F-10 containing 0.54 nM [³H]-TRH (60 Ci/mmol) and increasing doses of unlabeled peptide. Each point is the mean of triplicate determinations ± S.D.

micromolar concentrations were used (Fig. 5). When tritiated (5 Ci/mmol, prepared by H. Levine-Pinto, CEN, Saclay, France), it exhibits association with intact GH_3 B_6 cells, but with no equilibrium reached in 4 hr of incubation at 37°C. <Glu-NH_2-C^2-Im-TRH is, however, able to stimulate PRL release within 30 min at concentrations (12.5 and 250 nM, Fig. 6) that did not interfere with [^{3}H]-TRH binding (Fig. 5). In view of the concept of hormone-specific receptor–site interaction, such experimental data lead to the hypothesis that a partial recognition of only a very few sites may achieve one among the multiple responses induced by TRH.

Apart from such disturbing findings, which are isolated, the large majority of findings lead to the tentative conclusion that TRH binding sites are functional receptors and that in terms of specificities, they are similar for PRL and TSH production (Vale et al., 1973; Rivier and Vale, 1974; Hinkle et al., 1974). In addition, Grant et al. (1973) concluded from their analysis of competition data using several classes of TRH analogues that the two families of TRH binding sites display little-different structural

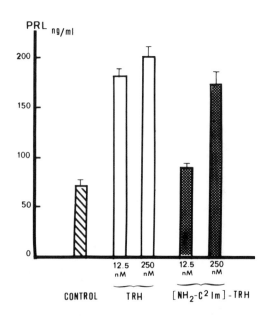

Figure 6. Effect of TRH-NH_2 on PRL release. The experiment was performed on GH_3 B_6 cells precultivated for 5 days (see Table I). The cells were rinsed with Ham's F-10 medium and exposed to the chosen peptide in 1 ml complete culture medium for 30 min at 37°C. PRL was assayed by radioimmunoassay using the NIAMDD kit for rat PRL. Each value is the mean of triplicate determinations ± S.D.

preferences, which implies that TRH itself is also recognized with slight dissimilarity by the two types of binding sites.

Finally, the lack of parallelism in stereospecific requirement when related to acute or long-term effects of TRH and some TRH analogues suggested to Dannies and Tashjian (1976) the existence of distinct binding sites or pathways for PRL release and synthesis.

5. Localization of the Binding Sites for [³H]Thyroliberin

5.1. Plasma-Membrane Level

The plasma membrane is obviously the first site of contact of cells with TRH. This does not imply that it represents the unique localization for binding sites or that in which the specific effects are initiated. Indeed, specific binding sites for TRH have been demonstrated at the level of purified plasma membrane (see the review in Vale *et al.*, 1976), which nevertheless does not exclude additional localization for TRH receptors.

A large body of information deals with the interaction of TRH and the plasma membrane of GH_3 and related cell strains. None of the findings is actually demonstrative of a unique site of action at this level. Studies have been performed either on very crude membrane preparations (4000*g* or 10,000*g* pellets) such as for binding properties (Hinkle and Tashjian, 1973; Hinkle *et al.*, 1974) or quenching of intrinsic tryptophan fluorescence (Imae *et al.*, 1975), where multiple classes of contaminants may interfere, or in intact cells (see Section 2.2.2), where the different modifications might be related as well to the proper interaction of TRH or to the induced hormone outflow or other biological responses that occur simultaneously with the binding within the limits of experimental schedules.

In addition, there is no unequivocal evidence that cAMP and/or cGMP mediates TRH-induced hormonal responses, as has been proposed (Dannies *et al.*, 1976; Gautvik *et al.*, 1978); in particular, in a large variety of experimental conditions, TRH did not elicit significant stimulation of adenylate cyclase activity (Hinkle and Tashjian, 1977).

5.2. Internalization of [³H]-TRH

Autoradiography performed on sections of cells (Gourdji *et al.*, 1973a) and biochemical identification of radioactive material associated at the level of subcellular fractions (Brunet *et al.*, 1974), both prepared from cells incubated intact with [³H]-TRH at 37°C, led to a new concept: TRH is internalized during its interaction with target cells and yet is not significantly degraded when acute responses (GH and PRL release) are already optimum, which implies that TRH is indeed the active molecule for generating these effects. What is therefore the fate of internalized TRH? Is it involved or not in any biological response induced by TRH? The entry of TRH into its target cells was also suggested by autoradiographic study of rat anterior pituitary processes after *in vivo* injections of labeled TRH (Stumpf and Sar, 1973; Stumpf *et al.*, 1975). No biochemical identification, however, supported the morphological observations.

5.3. Long-Term Fate of Bound [³H]-TRH

The analysis by thin-layer chromatography (TLC) of tritiated material associated to intact cells exposed for increasing duration to [³H]-TRH revealed a very slow rate of degradation (30% after 20 hr) leading to the formation of acidic derivatives and radioactive histidine (Gourdji *et al.*, 1976b). Using techniques different from ours, Hinkle and Tashjian (1975b) found a higher rate of metabolism, but they did not identify the intermediate metabolites. The radioactive material released into the medium and that retained by the cells at the end of a 1-hr incubation at 37°C following a 30-min exposure to [³H]-TRH exhibited a slightly different pattern after TLC (Gourdji *et al.*, 1976b). This observation provided evidence for intracellular metabolism of TRH in target cells. This does not exclude possible targets other than its degradative enzymes for intact intracellular TRH (see Section 6). Moreover, the presence of acidic metabolites of TRH together with intact TRH during chronic exposure to TRH actually raises a question concerning the possible respective roles of these two molecules in the long-term responses generated by TRH.

5.4. Interaction of Desamido TRH (TRH-OH) with GH₃B₆ Cells

Physiological concentrations of this acidic metabolite (from 1.4 to 66 nM) were able to partially but significantly reduce the PRL production from GH₃ B₆ cells after 24–48 hr of treatment (Table III). In contrast, TRH-OH up to 54 nM was devoid of affinity for [³H]-TRH binding sites (Faivre-Bauman *et al.*, 1977) and was unable to elicit a reproducible and significant effect on short-term release of PRL (unpublished information). In addition, [³H]-TRH-OH (25 Ci/mmol, prepared by H. Levine-Pinto, CEN, Saclay, France) was able to bind to GH₃ cells in a nearly linear manner up to 4 hr. Unlabeled TRH-OH did not inhibit this binding (unpublished information).

This inhibiting influence of TRH-OH is to be correlated with the PRL-inhibiting action of the diketopiperazide TRH derivative recently reported by Bauer *et al.* (1978).

Table III. Influence of TRH-OH (< Glu-His-Pro-OH) on PRL Production (% of Control)

Incubation time	(nM) TRH-OH			
	1.4	2.8	11	66
24 hr[a,b]	100 ± 17	77 ± 12	73 ± 4	—
48 hr[c]	—	—	—	73 ± 4

[a] By courtesy of Dr. A. Faivre-Bauman. Cells were grown 5 days before the experiment in control medium (see Table I). Each value was determined in triplicate. PRL was assayed by radioimmunoassay using the NIAMDD kit for rat PRL.
[b] GH₃ cells, control: 19.6 ± 4 µg/mg cell protein.
[c] GH₃ B₆ cells, control: 27.2 ± 5 µg/mg cell protein.

It may be concluded, therefore, that intracellular TRH metabolites cannot account for the long-term stimulation of PRL synthesis. They may in contrast play an anti-TRH regulatory role as suggested by Bauer *et al.* (1978).

6. Intracellular Binding Sites for Thyroliberin: [³H]Thyroliberin Binding Capacity of Isolated Nuclei

From autoradiography and subcellular fractionations, we concluded that TRH was present at the nuclear level in cells preloaded for 30 min at 37°C (see Section 5.2). Although representing a minor component with a binding capacity lower than the other fraction in terms of protein binding capacity [≤30% of the other organelles and ≤60% of the whole cell (Brunet *et al.*, 1974)], its presence disclosed an original feature concerning a peptide hormone. Together with the facts that TRH was found to augment the mRNA coding for PRL (Evans *et al.*, 1978) and that TRH binding capacity is maximum at the end of the phase of DNA synthesis (Faivre-Bauman *et al.*, 1975), this observation justifies a quest for TRH action at the nuclear level and primarily for [³H]-TRH binding sites in isolated nuclei.

Highly purified nuclear fractions (<5% contamination from cytoplasmic origin) prepared from GH₃ B₆ cells were found able to specifically bind [³H]-TRH (Bournaud *et al.*, 1977). Such isolated nuclei, still displaying their double envelope and obtained without detergent, possessed 15–20% of the whole-cell binding capacity as nuclei extracted from preincubated GH3 cells. The binding equilibrium was reached in 2–5 min at 25–35°C, and was stable at 0°C and 50% reversible at 25°C. Isotopic dilution by increasing concentrations of unlabeled TRH (but not of D-pyroGlu-L-His-

L-Pro-NH$_2$) of a fixed dose of [^{3}H]-TRH (58 nM) progressively inhibited the [^{3}H]-TRH binding capacity to a maximum extent of 50% obtained for 270 nM TRH. An 85% inhibition of [^{3}H]-TRH nuclear binding is exhibited in nuclei from [^{3}H]-TRH-preloaded cells. The nuclei isolated from the GH$_3$ CD cells, a TRH-non-responsive variant of GH$_3$, possess only noncompetitive binding sites for TRH at the nuclear level. In the limits of 1.1–1160 nM [^{3}H]-TRH, the binding was found unsaturable. Without subtraction of the noncompetitive compartment, two apparent dissociation constants were calculated as, respectively, $K_{D1} = 1.5$–2.5×10^{-8} M for concentrations up to 40 nM and $K_{D2} = 2.10^{-6}$ M for [^{3}H]-TRH 80–1060 nM, which might account for the noncompetitive binding. For [^{3}H]-TRH 50 nM, the nuclear fraction retained specifically about 125 fmol/mg protein. Whether this specific binding is relevant to any of the TRH-induced response remains to be elucidated.

Insulin, another peptide hormone, has also been found to bind to isolated nuclei of hepatocytes (Goldfine and Smith, 1976) and lymphocytes (Goldfine *et al.*, 1977), possibly at the level of the nuclear envelope (Horvat, 1978). Whether [^{3}H]-TRH nuclear binding occurs at the same level in GH$_3$ B$_6$ cells or at the genomic or other nuclear component is so far unknown.

That isolated nuclei are able to retain [^{3}H]-TRH demonstrates that no cytoplasmic carrier is necessary to perform this phenomenon. Nevertheless, the question of how TRH is internalized and becomes possibly available for nuclear binding sites remains unanswered. Two main processes, not exclusive, may be hypothesized: First, TRH associated to a membrane receptor is internalized together with whole pieces of plasmalemma during endocytotic processes that have been shown to be enhanced by TRH (Tixier-Vidal *et al.*, 1976, 1979a). A second hypothesis would be an influx of TRH resulting from transmembrane transport through specialized or ion channels.

7. Influence of Ouabain on the Interaction of [^{3}H]Thyroliberin with GH$_3$ B$_6$ Cells

Vale *et al.* (1976) have previously observed a TRH-induced transitory increase of ^{45}Ca^{2+} efflux from GH$_3$ cells, and recently Tashjian *et al.* (1978) demonstrated that the TRH-induced PRL release was Ca^{2+}-dependent in this cell line. It has been proposed that in mammalian cells, a Na$^+$/Ca^{2+} system may operate, and a large body of information has suggested that such a system may be involved in insulin action (see Czech, 1977). We therefore considered, as a working hypothesis, that processes that regulate ionic movements may be a target for TRH, and analyzed the interaction of ouabain, a well-known blocker of the Na$^+$-K$^+$ ATPase, on [^{3}H]-

Table IV. Influence of Ouabain on [³H]-TRH (6.3 nM) Binding and Its Induction of PRL Release[a]

Ouabain concentration	Binding		PRL release	
	fmol/mg prot.	% of control	ng/mg prot.	% stimulation
Control	—	—	218 ± 28	—
+ [³H]-TRH	151 ± 7	100%	472 ± 59	+116%
10⁻⁴ M + [³H]-TRH	49.7 ± 3.8	−64%	458 ± 46	+110%
5.10⁻⁴ M + [³H]-TRH	29.9 ± 1.6	−78%	226 ± 19	N.D.

[a] $GH_3 B_6$ cells were cultivated 5 days before the experiment (see Table I). They were preexposed to ouabain 30 min at 37°C and rinsed. The cells were thereafter reincubated for 30 min at 37°C in 1 ml/dish of complete culture medium + ouabain + [³H]-TRH.

$TRH-GH_3 B_6$ cell interaction. This effect was analyzed on $GH_3 B_6$ cells incubated for 30 min at 37°C in the presence of ouabain and rinsed prior to the experiment. Under these conditions and within a given range of relative concentrations of [³H]-TRH and ouabain, the latter was found to inhibit both the [³H]-TRH binding and the induced PRL release.

Concentrations of ouabain under 10^{-4} M had no significant effect on the binding of [³H]-TRH 6.3 nM, while 10^{-4} M decreased the binding by 67% and 5×10^{-4} M by 81% (135 ± 15 fmol/mg protein vs. 736 ± 73 fmole/mg protein in controls). The former concentration did not signifi-

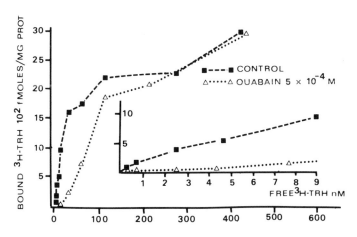

Figure 7. Influence of ouabain 5×10^{-4} M on [³H]-TRH binding from 0.3 to 500 nM. The experiment was performed on cells precultivated for 5 days (see Table I). The cells were rinsed and incubated for 30 min at 37°C in 1 ml complete culture medium ± ouabain 5 × 10^{-4} M. After being rinsed they were exposed to [³H]-TRH in the same conditions. Each point was determined in triplicate. S.D. ≤5%.

cantly alter the PRL response to TRH, while the latter abolished the stimulating effect of TRH (Table IV).

Ouabain 5×10^{-4} M inhibited the [^{3}H]-TRH binding as a function of dose up to 250 nM TRH (Fig. 7); for [^{3}H]-TRH 13 nM, the binding was approximately 75% inhibited but displayed a time course similar to that in controls (Fig. 8). The induced PRL release was nearly abolished in these conditions, while the spontaneous one was not significantly altered (Fig. 9).

The lack of efficiency of the residual bound [^{3}H]-TRH could either reflect the loss of activity of TRH to generate a signal in the presence of ouabain or resulted from a secondary effect of the ATPase blocker interfering with the induced secretory processes. With increasing doses of [^{3}H]-TRH, the effect of 5×10^{-4} M ouabain was partially overcome, and 50 nM [^{3}H]-TRH induced a 30% increase in PRL release.

8. Regulation of the Number and Efficiency of [^{3}H]Thyroliberin Binding Sites by Thyroliberin

Evidence for negative self-regulation of TRH on its own receptor number has already been established in GH$_3$ (Tixier-Vidal et al., 1975a; Hinkle and Tashjian, 1975a). The down-regulation is a function of time (the optimum being reached in 48 hr) and of dose (1 nM induced the half-

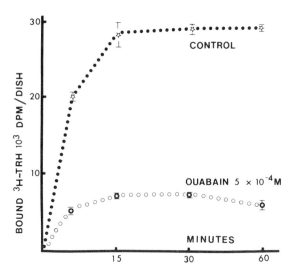

Figure 8. Time course of [^{3}H]-TRH binding 13 nM in the presence or in the absence of 5×10^{-4} M ouabain. Same conditions as in Fig. 7. Means ± S.D.

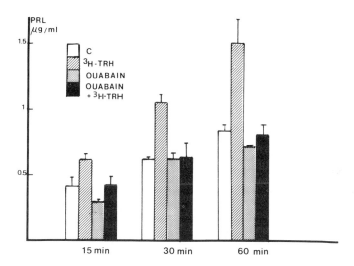

Figure 9. Time course of the effect of [³H]-TRH in the presence or in the absence of ouabain 5×10^{-4} M. Same experiment as in Fig. 8. PRL was assayed by radioimmunoassay, using the NIAMDD kit for rat PRL.

maximum effect), and is markedly but not totally dependent on protein synthesis (Hinkle and Tashjian, 1975a). Only 25–35% of the binding sites remain after a treatment with 27 nM TRH (see Fig. 10) or 100–300 nM (Hinkle and Tashjian, 1975a). The residual binding sites display the same kinetics of binding (Tixier-Vidal *et al.,* 1975a) and unchanged affinity as measured on the 4000*g* pellet (Hinkle and Tashjian, 1975a). So far, no correlation has been made of the coupling of the binding sites of down-regulated cells and induced biological responses.

Chronic exposure of $GH_3 B_6$ cells to 27 nM TRH has been analyzed after 2, 4, and 7 days of treatment with respect to the down-regulation of [³H]-TRH binding sites, to secretion of PRL and GH, and to responsiveness to acute reexposure to [³H]-TRH on cells precultivated for 3 days prior to the experiments. To ensure that no unlabeled TRH would interfere with the available binding sites for [³H]-TRH, the cells were processed as described in the Fig. 10 caption.

Since chronic exposure to TRH modifies cell growth (Table I), binding sites and hormone release were referred to cell number as well as to cell protein content.

8.1. Evolution of the [³H]-TRH Binding Sites (Fig. 10)

With increasing duration of culture, the binding capacity of the cells maintained in control medium decreased from 428 ± 45 fmol/10^6 cells after

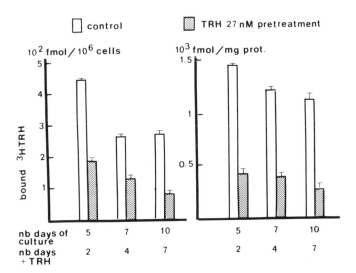

Figure 10. TRH regulation of [³H]-TRH binding sites in GH₃ B₆ cells. Same experiment as in Table I. Before being exposed to [³H]-TRH 16 nM, the cells were rinsed four times with Ham's F-10, reincubated for 2 hr 30 min at 37°C in 2 ml/dish of Ham's F-10, and rinsed three times again with Ham's F-10. Incubation with [³H]-TRH was performed for 30 min at 37°C in 1 ml Ham's F-10. Binding was arrested on ice and the cells were rinsed four times with ice-cold Ham's F-10. Each value is the mean triplicate determinations ± S.D.

5 days to 267 ± 28 fmol/10⁶ cells after 7 days. TRH treatment reduced the number of [³H]-TRH binding sites. When referred to dish to to milligrams of protein, the decrease was nearly optimal (−70%) after just 2 days of treatment; when referred to the cell number, the rate of reduction appears slower, evolving from −56% after 2 days to −69% after 7 days of treatment.

8.2. Evolution of the Spontaneous and TRH-Induced Hormone Release and Synthesis (Fig. 11)

As already stated, PRL production is time-dependent. The rate of production per cell per 2–3 days does not vary significantly during the period studied or with the level of intracellular PRL. Chronic treatment with TRH 27 nM elicited an increase of PRL in both medium and cell content that was still significant after 2 days, with a final augmentation of, respectively, +213% and +119%.

GH production did not evolve similarly. The rate of basal secretion increased from 0.405 ± 0.035 μg/10⁶ cells to 1.187 ± 0.101 μg/10⁶ cells per 48 hr between days 3–5 and days 5–7, i.e., by nearly 240%. GH cell content presented a similar rate of augmentation.

After 48 hr, TRH 27 nM elicited a significant increase in GH secretion, when referred to the cell number $(0.612 \pm 0.120$ ng/10^6 cells vs. 0.405 ± 0.035 ng/10^6 cells in controls), which becomes a nonsignificant increase when referred to the total cell protein. In contrast, GH cell content is significantly inhibited with both denominators (-67% when referred to the cell number; -78% when referred to the cell protein). Longer treatment improved the TRH inhibiting effect on GH production, which became highly significant at both medium and intracellular levels. After 7 days of treatment, the GH cell content was decreased 82.5%. When such cells were washed five times with serum-free F-10, then reincubated for 2.5 hrs in complete culture medium, without TRH, the GH release remained 77% lower than with control cells (52.5 ± 34 vs. $233 + 27$ ng/mg cell protein) (Fig. 12).

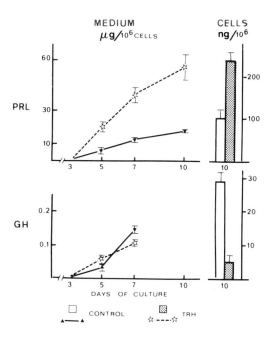

Figure 11. Evolution of PRL and GH secretion as a function of time in culture. Same experiment as in Table I and Fig. 10. TRH treatment was initiated on the 3rd day. Hormones were assayed by radioimmunoassay using the NIAMDD kit for rat PRL. Each point is the mean of triplicate determinations ± S.D.

8.3. Acute Effect of [³H]-TRH on GH Release in Pretreated Cells (Fig. 12)

In control cells, exposure to [³H]-TRH induced a 130% increase of GH release into the medium. Surprisingly enough, the TRH-pretreated cells displaying a chronic inhibition of GH synthesis (see Section 8.2 and Figs. 11 and 12) as well as a 70% decrease in [³H]-TRH binding capacity (249.7 ± 30 vs. 915.25 ± 60 fmol bound/mg protein, Fig. 10) not only reacted in the same direction as control cells, but also exhibited an even more acute stimulation of GH release (205% vs. 130%, Fig. 12).

This result provides another situation where the biological response–binding coupling efficiency is altered, and deserves some comment. In view of the low GH cell content associated with a relatively high rate of secretion in GH₃ B₆ cells, we may assume, as has already been reported concerning PRL in GH₃ cells (Dannies *et al.*, 1976), that the 48-hr medium content parallels the rate of GH synthesis while the 30-min medium content reflects the release of presynthesized hormone. With this assumption, it appears that despite a long-lasting inhibitory effect on GH synthesis, TRH remains able to acutely mobilize stored GH. One may hypothesize that some membrane rehandling generated by the intensive washings and reincubation in TRH-free medium (which in our experimental

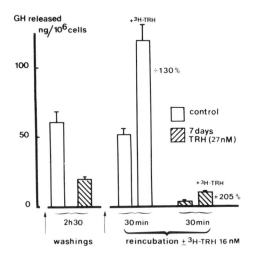

Figure 12. Influence of 7 days PRL. TRH (27 nM) pretreatment on induced GH release in GH₃ B₆ cells. Same experiment as in Table I and Fig. 10 and 11. GH was assayed at the end of the 2 hr 30 min reincubation in Ham's F-10 medium and after the 30-min incubation in the presence or in the absence of [³H]-TRH. Each value is the mean of triplicate determinations ± S.D.

schedule preceded the exposure to [^{3}H]-TRH) constitutes a key for the reinduction of the stimulation of GH release by TRH in TRH-pretreated cells.

9. Conclusions

Data obtained by several groups using different approaches or techniques led us to consider at least a part of highly specific TRH binding sites as functional receptors involved in PRL and/or GH release. Nevertheless, the complexity of binding characteristics, together with the internalization of TRH, raise questions concerning the limits of studies carried out at apparent equilibrium, with biochemical interpretive techniques assuming the formation of a simple ligand–receptor complex.

For the time being, there is no molecular model linking the initial recognition of an identified receptor to the effects induced by TRH, including GH and PRL release and synthesis. Moreover, the question whether long-term effects are triggered by very early events occurring since the initial binding (by the same or different pathways from those involved in the releasing mechanisms) or by further internal signals requires further investigation. Also, how GH and PRL are first similarly and then oppositely controlled remains unknown.

The involvement of plasma membrane receptors is strongly suggested by the multiple alterations occurring at this level within the minutes following exposure to TRH (see Section 2), which does not exclude an intracellular target for TRH. Actually, there is evidence for internalization of TRH (see Section 5), but no data concerning the relative number of internalized molecules as compared to the total number of TRH molecules reacting with the plasma membrane. Already, TRH degradative enzymes constitute one biological target for internalized molecules, but other possible intracellular targets, relevant to hormone regulation remain to be demonstrated. A possible involvement of a nuclear site of action is nevertheless suggested by the finding of [^{3}H]-TRH binding capacity of isolated nuclei (Bournaud *et al.*, 1977). A nuclear site of action is now clearly demonstrated in GC cells for T$_3$ and glucocorticoid, which both modulate in parallel fashion the mRNA coding for GH and the synthesis of this hormone (Martial *et al.*, 1977). These two hormones are also able to modulate the number of TRH binding sites (Perrone and Hinkle, 1978; Tashjian *et al.*, 1977). Similarly, TRH has recently been shown to increase the mRNA coding for PRL (Evans *et al.*, 1978), and it regulates the number of its own receptors (see Section 8). As shown for T$_3$, a nuclear site of action may coexist with plasma membrane receptors (Pliam and Goldfine, 1977).

The mechanism of entry of TRH into the cell is for the time being the subject of speculative considerations. Endocytotic pathways could be involved, providing also a possible mechanism for down-regulation of the TRH binding sites (Tixier-Vidal *et al.*, 1979a). Or, and not excluding the latter hypothesis, transmembrane transport could occur. In this case, the processes controlling the ionic movement could also be involved in the TRH–target cell interaction. Promising results have already been obtained concerning the sodium/potassium-dependent ATPase (see Section 7); they support this working hypothesis, which is therefore under further investigation.

10. Summary

This chapter attempts to summarize the data describing the [³H]-TRH binding sites and the biological responses induced by TRH in rat anterior pituitary cell lines. At least a part of [³H]-TRH binding sites may be considered as functional receptors involved in the control of PRL and GH release. Their localization at the subcellular level as well as the primary event triggered by TRH binding nevertheless remain to be established. In addition, the tentative demonstration of complicated characteristics of binding, of internalization, and of intracellular binding sites does not support the thesis of a simple bimolecular reaction as a model system for TRH interaction with intact living target cells.

The plurality of the cell functions controlled by TRH is discussed in the light of a possible action at several levels, including the nucleus.

ACKNOWLEDGMENTS. This chapter includes previously published or unpublished results of experiments supported by grants from the C.N.R.S. (E.R. 89, RCP 220), INSERM (AT 35), D.G.R.S.T. (Contract 77 7 04 92), and CEA (Contract SA-4364). The author is grateful to Dr. H. Levine (Laboratoire de Biochimie, Pr. P. Fromageot, du Département de Biologie, CEN, Saclay, France) for tritiation and purification of the tripeptides, to the NIAMDD rat pituitary program for providing the kits for radioimmunoassay of PRL and GH, to Ms. M.F. Moreau and M.D. Grouselle for skillful technical assistance, to M.C. Pennarun for illustrations, and to Ms. A. Bayon for excellent secretarial work.

DISCUSSION

VAUDRY: Is there additional evidence for the presence of intracellular TRH binding sites in normal pituitary cells?

GOURDJI: There are the autoradiographic studies of Stumpf and Sar performed after *in vivo* injection of [³H]-TRH showing intracellular radioactive material and no preferential localization at the plasma membrane level. Nevertheless, no analysis of the biochemical nature of this bound radioactive material was reported. The pituitaries were analyzed for Arg 1 hour after the TRH injection, and the radioactive material may represent TRH or TRH metabolites.

LABRIE: Have you studied in more detail the time course and pathway of the intracellular radioactivity? Is that internalization of [³H]-TRH increased by higher concentrations of TRH? In other words, is the internalization process accelerated by increasing hormone levels?

GOURDJI: I have no data on the time course of [³H]-TRH internalization nor any information on your second question. Tixier-Vidal has reported that TRH increases endocytosis, but we have no clear evidence showing that endocytosis is involved in TRH internalization.

KERDELHUÉ: Are there somatostatin receptors on GH₃ cells?

GOURDJI: Tashjian recently reported the presence of specific high-affinity binding sites for somatostatin in GH₃ cells revealed by studies with [¹²⁵I]somatostatin.

SIMON: What about the intracellular degradation of TRH? You have nicely demonstrated that TRH activity was intracellular, but do you know the exact localization?

GOURDJI: I have no personal data on the subcellular localization of degradative enzymes for TRH. The data of Karl Bauer show that degradative activity is mainly associated with the cytosol, with only low activity in the membrane fraction. The pH optimum was between 7.4 and 7.8, which suggests that they were not associated with lysosomes.

REFERENCES

Bauer, K., Gräf, K. J., Faivre-Bauman, A., Beier, S., Tixier-Vidal, A., and Kleinkauf, H., 1978, Inhibition of prolactin secretion by histidyl-proline-diketopiperazine, *Nature (London)* **274**:174.

Bournaud, F., Gourdji, D., Mongongu, S., and Tixier-Vidal, A., 1977, ³H-Thyroliberin (TRH) binding to nuclei isolated from a pituitary clonal cell line (GH₃), *Neuroendocrinology* **24**:183.

Brunet, N., and Tixier-Vidal, A., 1978, Increased binding of concanavalin A at the cell surface following exposure to thyroliberin, *Mol. Cell. Endocrinol.* **11**:169.

Brunet, N., Gourdji, D., Tixier-Vidal, A., Pradelles, P., Morgat, J. L., and Fromageot, P., 1974, Chemical evidence for associated TRF with subcellular fractions after incubation of intact rat prolactin cells (GH₃) with ³H-labelled TRF, *FEBS Lett.* **38**:129.

Brunet, N., Gourdji, D., Moreau, M. F., Grouselle, D., Bournaud, F., and Tixier-Vidal, A., 1977, Effect of 17β-estradiol on prolactin secretion and thyroliberin responsiveness in two rat prolactin continuous cell lines: Definition of an experimental model, *Ann. Biol. Anim. Biochim. Biophys.* **17**(3B):413.

Burgus, R., Dunn, T. F., Desiderio, D., and Guillemin, R., 1969, Structure moléculaire du facteur hypothalamique hypophysiotrope TRF d'origine ovine: Mise en évidence par spectrométrie de masse de la séquence PCA-His-Pro-NH₂, *C. R. Acad. Sci.* **269**:1870.

Czech, M. P., 1977, Molecular basis of insulin action, *Annu. Rev. Biochem.* **46**:359.

Dannies, P. S., and Tashjian, A. H., Jr., 1976, Release and synthesis of prolactin by rat pituitary cell strains are regulated independently by thyrotropin-releasing hormone, *Nature (London)* 261(5562):707.

Dannies, P. S., Gautvik, K. M., and Tashjian, A. H., Jr., 1976, A possible role of cyclic AMP in mediating the effects of thyrotropin-releasing hormone on prolactin release and on prolactin and growth hormone synthesis in pituitary cells in culture, *Endocrinology* 98(5):1147.

De Meyts, P., Bianco, A. R., and Roth, J., 1976, Site–site interactions among insulin receptors: Characterization of the negative cooperativity, *J. Biol. Chem.* 251(7):1877.

Dufy, B., Vincent, J. D., Fleury, H., Du Pasquier, P., Gourdji, D., and Tixier-Vidal, A., 1979, Intracellular recordings from pituitary cells reveal membrane effects of estrogen on electrical activity, *Science* 204:3492.

Evans, G. A., David, D. N., and Rosenfeld, M. G., 1978, Regulation of prolactin and somatotropin mRNAs by thyroliberin, *Proc. Natl. Acad. Sci. U.S.A.* 75(3):1294.

Faivre-Bauman, A., Gourdji, D., Grouselle, D., and Tixier-Vidal, A., 1975, Binding of thyrotropin releasing hormone and prolactin release by synchronized GH_3 rat pituitary cell line, *Biochem. Biophys. Res. Commun.* 67(1):50.

Faivre-Bauman, A., Gourdji, D., Grouselle, D., and Tixier-Vidal, A., 1976, Influence du TRH (thyrolibérine) sur la sécrétion de prolactine (PRL) et d'hormone de croissance (GH) au cours du cycle cellulaire de la lignée GH_3 (clone de cellules antéhypophysaires de rat), *J. Microsc. Biol. Cell.* 27:9a.

Faivre-Bauman, A., Gourdji, D., and Tixier-Vidal, A., 1977, Dosage de la thyréolibérine (TRH) par radio-récepteur, *Ann. Endocrinol.* 38:265.

Gautvik, K. M., Haug, E., and Kriz, M., 1978, Formation of guanosine 3',5'-cyclic monophosphate by thyroliberin in cultured rat pituitary cells, *Biochim. Biophys. Acta* 538:354.

Goldfine, I. D., and Smith, G. J., 1976, Binding of insulin to isolated nuclei, *Proc. Natl. Acad. Sci. U.S.A.* 73(5):1427.

Goldfine, I. D., Smith, G. J., Wong, K. Y., and Jones, A. L., 1977, Cellular uptake and nuclear binding of insulin in human cultured lymphocytes: Evidence for potential intracellular sites of insulin action, *Proc. Natl. Acad. Sci. U.S.A.* 74(4):1368.

Gourdji, D., Kerdelhué, B., and Tixier-Vidal, A., 1972, Ultrastructure d'un clone de cellules hypophysaires sécrétant de la prolactine (clone GH_3): Modifications induites par l'hormone hypothalamique de libération de l'hormone thyréotrope (TRF), *C. R. Acad. Sci. Ser. D* 274:437.

Gourdji, D., Tixier-Vidal, A., Morin, A., Pradelles, P., Morgat, J. L., Fromageot, P., and Kerdelhué, B., 1973a, Binding of a tritiated thyrotropin-releasing factor to a prolactin secreting clonal cell line (GH_3), *Exp. Cell Res.* 82:39.

Gourdji, D., Morin, A., and Tixier-Vidal, A., 1973b, Study on the control of prolactin secretion by two continuous lines of rat pituitary prolactin cells, *Int. Congr. Ser. No. 308, Proceedings of the International Symposium on Human Prolactin,* Brussels, p. 63, Excerpta Medica, Amsterdam.

Gourdji, D., Tixier-Vidal, A., and Kraicer, J., 1975, Effect of TRH on kinetics of GH and PRL release by a rat pituitary continuous cell line: Interaction with precooling and/or cycloheximide pretreatment, International Symposium on Growth Hormone and Related Peptides, Ricerca Scientifica ed Educazione permanente, Milan, September 1975, Vol. 2, Suppl. 1, p. 61a, University of Milan, Milan.

Gourdji, D., Bournaud, F., Faivre-Bauman, A., and Tixier-Vidal, A., 1976a, Hétérogénéité des sites de liaison du TRH dans des lignées continues de cellules à prolactine, Colloque de Synthèse, 1976, Rapport No. 7, Actions Thématiques 22 and 25, *Neuroendocrinologie,* (N. Reué, ed.), pp. 251–259, INSERM, Paris.

Gourdji, D., Tixier-Vidal, A., Levine-Pinto, H., Pradelles, P., Morgat, J. L., and Fromageot, P., 1976b, Fate of thyrotropin releasing hormone after binding and stimulation of prolactin release by GH3 cells: Evidence for release of unmodified ^{3}H-TRH, *Neuroendocrinology* **20**:201.

Grant, G., Vale, W., and Guillemin, R., 1972, Interaction of thyrotropin releasing factor with membrane receptors of pituitary cells, *Biochem. Biophys. Res. Commun.* **46**(1):28.

Grant, G., Vale, W., and Guillemin, R., 1973, Characteristics of the pituitary binding sites for thyrotropin-releasing factor, *Endocrinology* **92**(6):1629.

Hayashi, I., and Sato, G. H., 1976, Replacement of serum by hormones permits growth of cells in a defined medium, *Nature (London)* **259**:132.

Hinkle, P. M., and Tashjian, A. H., Jr., 1973, Receptors for thyrotropin-releasing hormone in prolactin-producing rat pituitary cells in culture, *J. Biol. Chem.* **248**(17):6180.

Hinkle, P. M., and Tashjian, A. H., Jr., 1975a, Thyrotropin-releasing hormone regulates the number of its own receptors in the GH$_3$ strain of pituitary cells in culture, *Biochemistry* **14**(17):3845.

Hinkle, P. M., and Tashjian, A. H., Jr., 1975b, Degradation of thyrotropin-releasing hormone by the GH$_3$ strain of pituitary cells in culture, *Endocrinology* **97**(2):324.

Hinkle, P. M., and Tashjian, A. H., Jr., 1977, Adenylyl cyclase and cyclic nucleotide phosphodiesterases in GH-strains of rat pituitary cells, *Endocrinology* **100**(4):934.

Hinkle, P. M., Woroch, E. L., and Tashjian, A. H., Jr., 1974, Receptor-binding affinities and biological activities of analogs of thyrotropin-releasing hormone in prolactin-producing pituitary cells in culture, *J. Biol. Chem.* **249**(10):3085.

Horvat, A., 1978, Insulin binding sites on rat liver nuclear membranes: Biochemical and immunofluorescent studies, *J. Cell Physiol.* **97**:37.

Imae, T., Fasman, G. D., Hinkle, P. M., and Tashjian, A. H., Jr., 1975, Intrinsic tryptophan fluorescence of membranes of GH$_3$ pituitary cells: Quenching by thyrotropin-releasing hormone, *Biochem. Biophys. Res. Commun.* **62**(4):923.

Jacobs, S., Chang, K. J., and Cuatrecasas, P., 1975, Estimation of hormone receptor affinity by competitive displacement of labeled ligand: Effect of concentration of receptor and of labeled ligand, *Biochem. Biophys. Res. Commun.* **66**(2):687.

Kahn, C. R., 1976, Membrane receptors for hormones and neurotransmitters, *J. Cell Biol.* **70**:261.

Kahn, C. R., Freychet, P., Roth, J., and Neville, D. M., Jr., 1974, Quantitative aspects of the insulin–receptor interaction in liver plasma membranes, *J. Biol. Chem.* **249**(7):2249.

Kidokoro, Y., 1975, Spontaneous calcium action potentials in a clonal pituitary cell line and their relationship to prolactin secretion, *Nature (London)* **258**:741.

Klotz, I. M., and Hunston, D. L., 1971, Properties of graphical representations of multiple classes of binding sites, *Biochemistry* **10**(16):3065.

Labrie, F., Barden, N., Poirier, G., and De Lean, A., 1972, Binding of thyrotropin-releasing hormone to plasma membranes of bovine anterior pituitary gland, *Proc. Natl. Acad. Sci. U.S.A.* **69**(1):283.

Martial, J. A., Seeburg, P. H., Guenzi, D., Goodman, H. M., and Baxter, J. D., 1977, Regulation of growth hormone gene expression: Synergistic effects of thyroid and glucocorticoid hormones, *Proc. Natl. Acad. Sci. U.S.A.* **47**(10):4293.

Morin, O., and Labrie, F., 1975, Effect of somatostatin and thyrotropin-releasing hormone (TRH) on prolactin (PRL) and growth hormone (GH) release in GH$_1$ pituitary tumor cells in culture, Symposium on Hypothalamus and Endocrine Functions, Quebec, September 1975, p. 29, abstract.

Morin, A., Tixier-Vidal, A., Gourdji, D., Kerdelhué, B., and Grouselle, D., 1975, Effect of thyreotrope-releasing hormone (TRH) on prolactin turnover in culture, *Mol. Cell. Endocrinol.* **3**:351.

Ostlund, R. E., Jr., Leung, J. T., Vaerewyck Hajek, S., Winokur, T., and Melman, M., 1978, Acute stimulated hormone release from cultured GH_3 pituitary cells, *Endocrinology* **103**:1245.

Perrone, M. H., and Hinkle, P. M., 1978, Regulation of pituitary receptors for thyrotropin-releasing hormone by thyroid hormones, *J. Biol. Chem.* **253**(4):5168.

Pliam, N. B., and Goldfine, I. D., 1977, High affinity thyroid hormone binding sites on purified rat liver plasma membranes, *Biochem. Biophys. Res. Commun.* **79**(1):166.

Pradelles, P., Morgat, J. L., Fromageot, P., Oliver, C., Jacquet, P., Gourdji, D., and Tixier-Vidal, A., 1972, Preparation of highly labelled ^{3}H-thyreotropin releasing hormone [PGA-His-Pro(NH$_2$)] by catalytic hydrogenolysis, *FEBS Lett.* **22**(1):19.

Rivier, C., and Vale, W., 1974, *In vivo* stimulation of prolactin secretion in the rat by thyrotropin releasing factor, related peptides and hypothalamic extracts, *Endocrinology* **95**(4):978.

Stumpf, W. E., and Sar, M., 1973, ^{3}H-TRH and ^{3}H-proline radioactivity localization in pituitary and hypothalamus, *Fed. Proc. Fed. Am. Soc. Exp. Biol.* **32**:211 (abstract).

Stumpf, W. E., Sar, M., and Keefer, D. E., 1975, Localization of hormones in the pituitary: Receptor sites for hormones from hypophysial target glands and the brain, in: *The Anterior Pituitary* (A. Tixier-Vidal and M. Farquhar, eds.), pp. 63–82, Academic Press, New York.

Tashjian, A. H., Jr., and Hoyt, R. F., Jr., 1972, Transient controls or organ-specific functions in pituitary cells in culture, in: *Molecular and General Developmental Biology* (M. Sussman, ed.), pp. 353–387, Prentice-Hall, Englewood Cliffs, New Jersey.

Tashjian, A. H., Jr., Yasumura, Y., Levine, L., Sato, G. H., and Parker, M. L., 1968, Establishment of clonal strains of rat pituitary tumor cells that secrete growth hormone, *Endocrinology* **82**:342.

Tashjian, A. H., Jr., Bancroft, F. C., and Levine, L., 1970, Production of both prolactin and growth hormone by clonal strains of rat pituitary tumor cells, *J. Cell Biol.* **47**:61.

Tashjian, A. H., Jr., Barowsky, N. J., and Jensen, D. K., 1971, Thyrotropin releasing hormone: Direct evidence for stimulation of prolactin production by pituitary cells in culture, *Biochem. Biophys. Res. Commun.* **43**(3):516.

Tashjian, A. H., Jr., Osborne, R., Maina, D., and Knaian, A., 1977, Hydrocortisone increases the number of receptors for thyrotropin-releasing hormone on pituitary cells in culture, *Biochem. Biophys. Res. Commun.* **79**(1):333.

Tashjian, A. H., Jr., Lomedico, M. E., and Maina, D., 1978, Role of calcium in the thyrotropin-releasing hormone-stimulated release of prolactin from pituitary cells in culture, *Biochem. Biophys. Res. Commun.* **81**(3):798.

Tixier-Vidal, A., 1975, Ultrastructure of anterior pituitary cells in culture, in: *The Anterior Pituitary* (A. Tixier-Vidal and M. Farquhar, eds.), pp. 181–229, Academic Press, New York.

Tixier-Vidal, A., Gourdji, D., and Tougard, C., 1975a, A cell culture approach to the study of anterior pituitary cells, *Int. Rev. Cytol.* **41**:173.

Tixier-Vidal, A., Gourdji, D., Pradelles, P., Morgat, J. L., Fromageot, P., and Kerdelhué, B., 1975b, A cell culture approach to the study of TRH receptors, in: *Hypothalamic Hormones* (Motta, Grosignani, and Martini, eds.), Proceedings of the Serono Symposium 6, Milan, 1974, pp. 89–107, Academic Press, London.

Tixier-Vidal, A., Picart, R., and Moreau, M. F., 1976, Endocytose et sécrétion dans les cellules antéhypophysaires en culture: Action des hormones hypothalamiques, *J. Microsc. Biol. Cell.* **25**(2):159.

Tixier-Vidal, A., Brunet, N., and Gourdji, D., 1978, Morphological and molecular aspects of the regulation of prolactin secretion by rat pituitary cell lines, in: *Progress in*

Prolactin Physiology and Pathology (C. Robyn and M. Harter, eds.), pp. 29–43, Elsevier/North-Holland, Amsterdam.

Tixier-Vidal, A., Brunet, N., and Gourdji, D., 1979a, Plasma membrane modifications related to the action of TRH on rat prolactin cell lines, Cold Spring Harbor Conference on Cell Proliferation, Vol. 6, *Hormones and Cell Culture* (in press).

Tixier-Vidal, A., Brunet, N., Tougard, C., and Gourdji, D., 1979b, Morphological and molecular aspects of prolactin and growth hormone secretion by normal and tumoral pituitary cells in culture, International Symposium on Pituitary Microadenomas, Milan, Oct. 1978, 12–14, Serono Symposium, Academic Press, New York (in press).

Vale, W., Blackwell, R., Grant, G., and Guillemin, R., 1973, TRF and thyroid hormones on prolactin secretion by rat anterior pituitary cells *in vitro, Endocrinology* 93(1):26.

Vale, W., Rivier, C., Brown, M., Leppaluoto, J., Ling, N., Monahan, M., and Rivier, J., 1976, Pharmacology of hypothalamic regulatory peptides, *Clin. Endocrinol.* 5(Suppl. 261s).

Vale, W., Rivier, C., and Brown, M., 1977, Regulatory peptides of the hypothalamus, *Annu. Rev. Physiol.* 39:473.

Yasumura, Y., Tashjian, A. H., Jr., and Sato, G. H., 1966, Establishment of four functional clonal strains of animal cells in culture, *Science* 154:1186.

24

Mechanisms Governing the Release of Growth Hormone from Acutely Dispersed Purified Somatotrophs

M. Suzanne Sheppard, Jacob Kraicer, and John V. Milligan

1. Introduction

1.1. Validation of the Acutely Dispersed Purified Somatotroph Preparation: Role of Cyclic Nucleotides

The problems inherent in the study of the control of adenohypophyseal hormone secretion *in vitro* using intact tissue or tissue fragments are fourfold: variability of response, lack of viability of tissue in the gland core, relative lack of sensitivity, and the heterogeneity of cell types within the gland, which precludes interpretation of intracellular metabolic events within a specific cell type. Several investigators have used acutely dispersed or cultured adenohypophyseal cells to examine the effects of hypothalamic regulatory hormones as described in recent reviews (Labrie *et al.*, 1976b; Vale *et al.*, 1976). These preparations overcome the problems of variability, viability, and sensitivity associated with the classic whole or hemipituitary studies. However, the difficulties arising from the heterogeneity of the cell type remain. It is not possible, using presently available

M. Suzanne Sheppard, Jacob Kraicer, and John V. Milligan • Department of Physiology, Queen's University, Kingston, Ontario, Canada K7L 3N6

preparations, save for the use of autoradiographic techniques, to localize alterations of intracellular metabolite involved in the release of one hormone to a specific adenohypophyseal cell type. In light of much evidence showing a lack of specificity of several of the hypothalamic hypophysiotropic hormones (Labrie *et al.*, 1976a,b; Vale *et al.*, 1976), it has become imperative to study a uniform cell population. Cloned tumor cell lines have been used in an attempt to overcome this problem (Tashjian *et al.*, 1968; Hertelendy and Keay, 1974; Dannies *et al.*, 1976), but there is no assurance that intracellular events are not grossly altered in tumor cells.

In view of this, we began several years ago (Hymer *et al.*, 1972, 1973; Kraicer and Hymer, 1974) to study the response of enzymatically dispersed cells, which were initially purified by a unit-gravity continuous-density-gradient procedure. Although a high level of purification of somatotrophs was achieved and the cells did respond to dibutyryl cyclic cAMP (dbc-AMP) and to a partially purified growth hormone–releasing factor (GH-RF) preparation (Kraicer and Hymer, 1974), the method was tedious and time-consuming. We have since adopted the two-stage discontinuous-density-gradient centrifugation procedure of Snyder and Hymer (1975), which allows the routine production of a highly purified population of somatotrophs within 3 hr. To validate the use of this cell preparation, we have undertaken a series of studies to examine the dose–response characteristics of growth hormone (GH) release, with time, elicited by various secretagogues, prior to future studies of the mechanisms governing the release process.

We first examined the effects of several cyclic nucleotides (cAMP, dbcAMP, cGMP, and dbcGMP), the phosphodiesterase inhibitors theophylline and 3-isobutyl-1-methyl xanthine (IBMX), and prostaglandin E_2 (PGE_2). We have also studied the effects of somatostatin on the basal and stimulated release of GH. To test the specificity of the somatotroph response, we have incubated the cells with synthetic arginine vasopressin, thyroliberin (TRH), luteinizing hormone–releasing hormone (LH-RH), α-melanocyte-stimulating hormone (α-MSH), and *n*-butyric acid.

1.2. Role of Ca^{2+} in Growth Hormone Release

The presence of calcium ion is essential for hormone release from the adenohypophysis (Peake, 1973; Kraicer, 1975; Moriarty, 1978). However, its exact role in the release process remains to be elucidated. Based on the "stimulus–secretion coupling" hypothesis (Douglas, 1968) (see Section 4.2.1), one of the first questions asked was whether an influx of calcium ion is essential for release. This question has been approached in two ways. In a number of studies, extracellular calcium has been lowered by its removal from the incubation medium, with or without the addition of calcium chelating agents. It has been reported that the augmented release of GH induced by a crude hypothalamic extract (Steiner *et al.*, 1970), prostaglandins (Hertelendy, 1971; Cooper *et al.*, 1972), cAMP derivatives (Lemay and Labrie, 1972), and phosphodiesterase inhibitors (Steiner *et al.*, 1970; Lockhart Ewart and Taylor, 1971) can be reduced in low-calcium media. However, whether this resulted from a prevention of calcium influx or the leaching out of an essential intracellular pool (Kraicer, 1975) has not been established. A second approach has been to measure alterations in calcium-ion influx or efflux associated with altered hormone release (Milligan and Kraicer, 1971; Milligan *et al.*, 1972; Peake, 1973; Eto *et al.*, 1974; Kraicer, 1975; Moriarty, 1977, 1978). Ion flux studies are, however, difficult to interpret because of the lack of absolute specificity of the secretagogues used and the heterogeneity of cell types within the adenohypophysis, which does not allow alterations in ion flux to be pinpointed to specific cell types.

We therefore began a series of studies to look at the role of calcium using our acutely dispersed purified somatotroph preparation. This preparation obviates problems due to cell heterogeneity and eliminates problems related to diffusion that are present in tissue fragments.

2. Materials and Methods

Young adult male rats (Charles River, CD, Canadian Breeding Farms and Laboratories) weighing 175–200 g on arrival were maintained for 2 weeks in group cages in a temperature-controlled (25.5 ± 0.5°C), soundproofed room on a 14-hr-light/10-hr-dark cycle, with feeding *ad libitum*. On the afternoon preceding each experiment, animals were placed in single cages. Ten to 20 rats were killed by decapitation with a guillotine, within 20 sec of removal from the animal room, between 8 A.M. and 11 A.M. The nervosa intermedia was removed and discarded, and the pars distalis minced into small (1- to 2-mm) fragments. The tissue was dissociated for 2 hr in Minimum Essential Medium (Spinners, 10X, Gibco No. 165) containing 0.1% bovine serum albumin (BSA, ICN, fraction V) and 0.1% trypsin (Difco, 1 : 250) as previously described (Method 2 in Hymer *et al.*, 1973). The dispersed cells (2–4 × 10^6 per rat) were centrifuged and resuspended in 1.5–2.0 ml Medium 199 (M199, Gibco, E12) containing 1.0% BSA in preparation for the purification procedure.

The production of a cell population enriched in somatotrophs was accomplished with the use of the two-stage discontinuous-density-gradient centrifugation (STARTS) described by Snyder and Hymer (1975). All steps in this procedure were carried out at 4°C. Each gradient was composed of two layers of 20–28% (wt./vol.) BSA (Reheis Chemical Co., Armour Pharmaceutical Co. Stock No.2293-02) prepared according to the method of Shortman (1968). The density of the solutions was measured at 4°C by a Bellingham and Stanley sugar refractometer, which records equivalent percentage of sucrose. The first gradient was composed of 26.5% (upper layer) and 35.0% (lower layer). The second gradient was 28.0 and 35.0%. The cells were applied to the gradients in 1.5–2.0 ml M199 containing 1.0% BSA. In each case, the fraction enriched in somatotrophs was found at the interface of the two heavy BSA layers. The final yield was 1–3 × 10^6 cells or 5–10% of the starting population. Cell purity, as estimated by histological examination (Hymer et al., 1973), was consistently greater than 90%.

On being removed from the second density gradient, the purified cells were resuspended in 10 ml M199 or phosphosaline buffer and then centrifuged. The cell pellet was resuspended in 10 ml M199 + 0.1% BSA and allowed to equilibrate for 30 min at 37°C in a Spinner Flask with gentle rotation or in a Dubnoff incubator–shaker. After centrifugation, the purified somatotrophs were resuspended in M199 + 0.1% BSA, usually at a concentration of 7000–15,000 cells/ml. Cells were distributed in 2.0-ml aliquots to Teflon beakers and incubated at 37°C in a Dubnoff incubator–shaker, gassed with moistened 95% O_2/5% CO_2. Small volumes (generally 50 μl) of secretagogue or medium were added at the onset of the incubation. Each experimental group was composed of 4–6 beakers of cells. After 7.5, 15, 30, 60, and 120 min of incubation, 0.35-ml aliquots of the incubating cell suspension were removed, with the beakers remaining in the incubator–shaker. Cells and media were separated by centrifugation (500g, 10 min) at room temperature and immediately frozen (−20°C). Samples of nonincubated cells and media were processed in the same manner (STARTS). When the response to somatostatin was examined, the cells were preincubated with somatostatin for 10 min prior to the addition of the secretagogue. All plastic and glassware, except for the Teflon beakers was siliconized.

GH was measured in duplicate and at two dilutions in both cells and incubation media by the double antibody radioimmunoassay using reagents provided by the NIAMDD Rat Pituitary Hormone Distribution Program, and results were expressed as ng/1000 cells of the NIAMD Rat Growth Hormone GH-RP-1 Standard. Analysis of variance and Duncan's multiple range test were used to test for differences among groups (Duncan, 1955).

In the calcium studies, the protocol was altered. On being removed from the second density gradient, the purified somatotrophs were suspended in 10 ml phosphosaline buffer, aliquots were taken for differential cell counts, and the remainder was divided into two equal volumes, centrifuged, and then resuspended in either normal Medium 199 (M199-1) containing 0.1% BSA or low-calcium Medium 199 (M199-2) containing 0.1% BSA. The cells were allowed to equilibrate at 37°C in a Dubnoff incubator–shaker for 30 min while gassing with moistened 95% O_2/5% CO_2. The media were then removed by centrifugation and the cells resuspended in either M199-1 (normal calcium) or M199-2 (low calcium) to a final concentration of 6000–8000 cells/ml. Cell counts were carried out with a hemocytometer to verify that the two fractions did not differ in cell number. Cells were then distributed in 2.0-ml aliquots to Teflon beakers and processed as described above. The incubation media for the calcium studies were prepared from a supply of M199 (Gibco) free of Hanks salts. M199-1 was prepared by adding Hanks salts and BSA (0.1%). M199-2 was prepared similarly, except that Hanks salts without calcium were added. The pH was then adjusted to 7.35 with 6% $NaHCO_3$ after gassing with 95% O_2/5% CO_2.

The secretagogues examined were dbcAMP (sodium salt, D-0627, Sigma), cAMP (sodium salt, A-6885, Sigma), dbcGMP (sodium salt, D-3510, Sigma), cGMP (sodium salt, G-6129, Sigma), IBMX (Aldrich), theophylline (T-1633, 1,3-dimethylxanthine, Sigma), TRH (Beckman), somatostatin (Ayerst, AY-24,910), synthetic LH-RH (Ayerst), PGE_2 (Upjohn), arginine vasopressin (Sigma), and n-butyric acid (Sigma). The arginine vasopressin vehicle, 0.5% chlorbutanol in 0.9% sodium chloride, was used as control. PGE_2 was dissolved in absolute ethanol, buffered in 1.9 mM Na_2CO_3, and then diluted with M199-1 or M199-2. IBMX was dissolved in 0.1 N NaOH, then diluted with M199-1 or M199-2 and the pH adjusted to 7.35.

Because of the large number of samples in each experiment, a rapid method of extraction of GH from the somatotrophs was sought. Homogenization of acutely dispersed cells at 0–2°C for 3 min in 1.0 ml 0.01 N NaOH (Birge *et al.*, 1967) followed by a rinse of the homogenizer with 500 μL NaOH was compared with vigorous vortexing (3–5 sec) of the cell pellet in 1.5 ml 0.01 N NaOH. After vortexing, the cells remained at 4°C for 60 min and were then vortexed again. Both homogenization and vortexing were followed by centrifugation at 1000g for 20 min at 4°C. The supernatant was removed, diluted 10 times with a phosphosaline buffer containing 1% BSA, and frozen (−20°C) for subsequent assay. Since no significant difference in intracellular GH levels was found using the two extraction procedures, intracellular GH was extracted routinely by the simpler vortex procedure.

The calcium content of the media and secretagogues was measured either by atomic absorption spectrophotometry (Milligan and Kraicer, 1974) or by a colorimetric method (Burr, 1969). The former will detect calcium concentrations down to 0.2 μmol in 3.0 ml, the latter down to 70 μM.

3. Results

In all studies, the variability within treatment groups was extremely small, with a standard error of the mean usually well below 10% of the mean. In all experiments, the GH content of the cells was assayed (data not shown). There was no change in the total GH content (cells plus media) with any of the agents tested, indicating no net loss or new appearance of GH during incubation. Cellular GH levels ranged from 100 to 400 ng/1000 cells over the entire series of experiments, and "basal" release reached no more than 15% of cell content by 120 min. Routine histological examination in all experiments revealed over 90% somatotrophs.

The measured calcium-ion concentration of M199-1 containing the complete Hanks salts was 1.2 mM. The M199-2 containing Hanks salts free of calcium ion contained less than 70 μM calcium. Following incubation, the calcium ion concentration was not elevated above these values in any of the incubation media. In control studies (data not shown), both the basal and stimulated release of GH were identical when cells were incubated in M199-1 and in "standard" M199 (Gibco, E-12).

3.1. Response of Purified Somatotrophs to Cyclic Nucleotides

Cells were incubated for up to 2 hr with three concentrations (0.05, 0.5, and 5.0 mM) of cAMP, cGMP, dbcAMP and dbcGMP (Fig. 1). Each cyclic nucleotide was tested in two separate experiments. Dibutyryl cAMP at concentrations of 0.5 and 5.0 mM (Fig. 1A and B) significantly increased GH release by 15 min of incubation, while the response to 0.05 mM became evident by 30 min. The release of GH continued in a dose-related manner over the course of the incubation, with the highest concentration of dbcAMP increasing GH release by 300% of control at 120 min. The other cyclic nucleotides tested, cAMP, cGMP, and dbcGMP, were ineffective even at the highest doses tested (Fig. 1). n-Butyric acid (1.2 mM) did not alter GH release.

3.2. Response of Purified Somatotrophs to Phosphodiesterase Inhibitors

Both theophylline (0.1 and 0.5 mM) and IBMX (0.1 and 1.0 mM) increased GH release in a dose-related manner (Figs. 2 and 3). GH release with 0.5 mM theophylline was increased significantly by 7.5 min, while release with 0.10 mM theophylline was elevated only after 60 min (Fig. 2). Theophylline at 0.05 mM did not produce a significant increase in release. GH release with 0.1 and 1.0 mM IBMX was increased significantly by 7.5

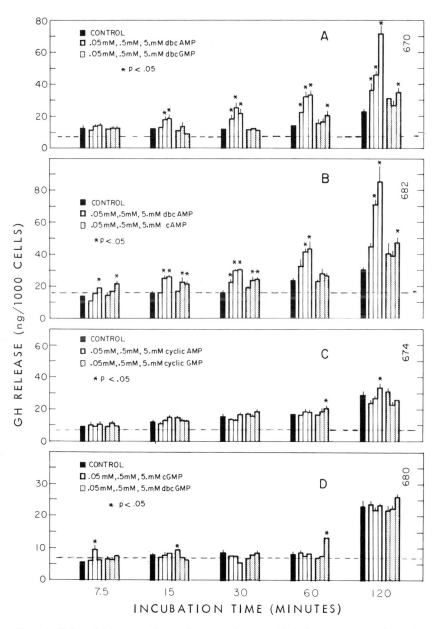

Figure 1. Release of GH with time in four experiments with different concentrations of the cyclic nucleotides dbcAMP, cAMP, dbcGMP, and cGMP. Throughout, the low to high doses tested are left to right in each group of three white and three stippled columns. Results are means from 4–6 beakers of purified somatotrophs, and the error bars are S.E.M. (---) Starting levels of GH.

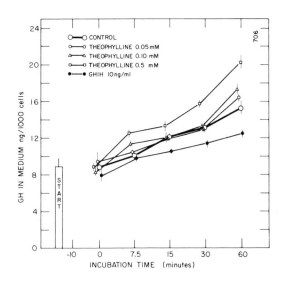

Figure 2. GH release from purified somatotrophs in response to theophylline. Also shown is the decrease in basal GH release with 10 ng/ml somatostatin (GHIH) added at −10 min. The error bars are S.E.M.

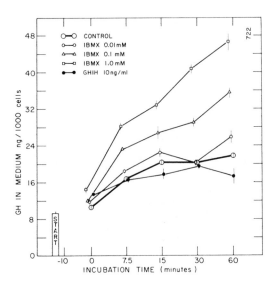

Figure 3. Release of GH in response to IBMX. Also shown is the decrease in basal GH release with 10 ng/ml somatostatin (GHIH) added at −10 min. The error bars are S.E.M.

min, while the low concentration (0.01 mM) did not produce a consistent release (Fig. 3).

3.3. Response of Purified Somatotrophs to Somatostatin

Basal release of GH was consistently reduced by both 100 and 10 ng/ml of somatostatin by 15 min of incubation (Figs. 2–4), while the response to 1 and 0.1 ng/ml was inconsistent (Fig. 5). Release stimulated by 0.6 mM dbcAMP was completely blocked by concentrations of somatostatin from 1.0 to 100 ng/ml (Figs. 4 and 5), but not by 0.1 ng/ml, the lowest concentration of somatostatin tested (Fig. 5). In fact, the release of GH was actually enhanced slightly, but significantly, in three separate experiments with this low dose of somatostatin (Fig. 5) (other data not shown). Somatostatin at 10 ng/ml also blocked completely the release of GH produced by 0.10 and 0.5 mM theophylline (Fig. 6) and 0.1 and 1.0 mM IBMX (Fig. 7). There was no consistent effect on cellular levels of GH in these studies (not shown).

3.4. Response of Purified Somatotrophs to Prostaglandin E_2

The response to PGE_2 was tested at concentrations of 10^{-10} to 10^{-6} M (Fig. 8). PGE_2 at 10^{-6} and 10^{-7} M caused an immediate (within 7.5 min)

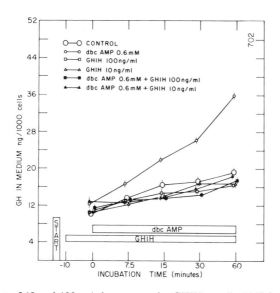

Figure 4. Effect of 10 and 100 ng/ml somatostatin (GHIH) on dbcAMP-induced release of GH. Somatostatin was added 10 min prior to the addition of dbcAMP. The error bars are S.E.M.

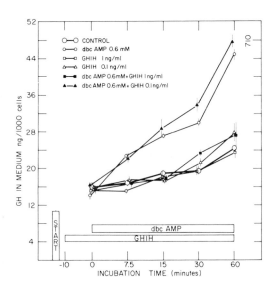

Figure 5. Effect of 0.1 and 1 ng/ml somatostatin (GHIH) on dbcAMP-induced release of GH. Somatostatin was added 10 min prior to the addition of dbcAMP. The error bars are S.E.M.

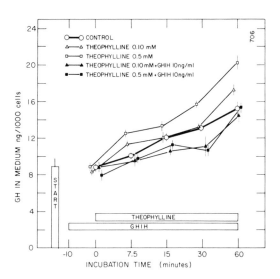

Figure 6. Effect of 10 ng/ml somatostatin (GHIH) on theophylline-induced release of GH. Somatostatin was added 10 min prior to the addition of theophylline. The error bars are S.E.M.

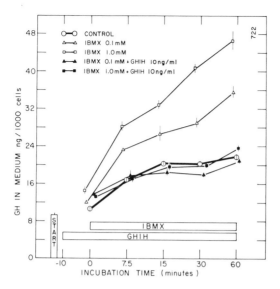

Figure 7. Effect of 10 ng/ml somatostatin (GHIH) on IBMX-induced release of GH. Somatostatin was added 10 min prior to the addition of IBMX. The error bars are S.E.M.

elevation of GH release, with no effect seen at lower concentrations. Release induced by 5×10^{-7} M PGE_2 was blocked completely by 10 ng/ml of somatostatin (Fig. 9).

3.5. Response of Purified Somatotrophs to Other Substances

To test the specificity of responsiveness of the somatotrophs, we examined the response to butyric acid (1.2 mM), LH-RH (1 μgm/ml), α-MSH (1 μgm/ml), and arginine vasopressin (244 mU/ml). None elicited a response different from the control incubations containing the appropriate vehicles. An occasional small response with TRH (10 and 100 ng/ml) was seen, but it was variable and inconsistent.

3.6. Basal Growth Hormone Release in Low-Calcium Medium

There was no consistent difference in the basal release of GH between the low- and normal-calcium media (Figs. 10–12 and Table I). After incubation in the low-calcium media, the cell morphology, as examined by light microscopy, appeared normal. In two identical experiments (Fig. 10), PGE_2 (5×10^{-7} M) caused a 2-fold increase in GH release by 7.5 min of incubation, which continued to increase for 60 min. This augmented release was markedly and significantly reduced in the low-calcium medium

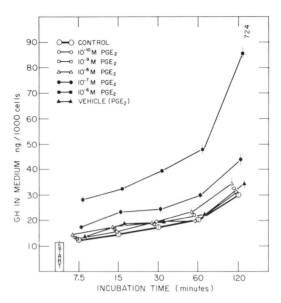

Figure 8. Release of GH from purified somatotrophs with 10^{-6} to 10^{-10} M PGE_2 over 2 hr of incubation. The error bars are S.E.M.

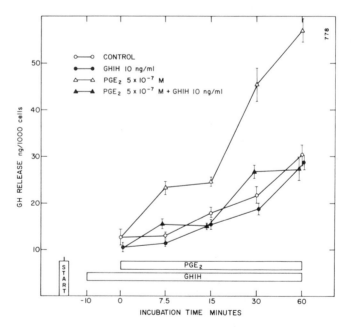

Figure 9. Effect of 10 ng/ml somatostatin (GHIH) on PGE_2-induced release of GH. Somatostatin was added 10 min prior to the addition of PGE_2. The error bars are S.E.M.

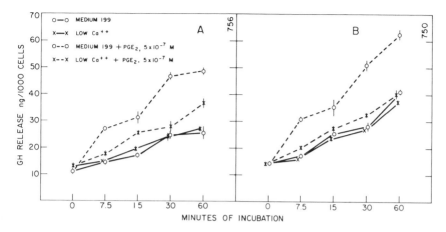

Figure 10. PGE₂-induced release of GH, with time, in Medium 199 and Medium 199 containing less than 70 μM calcium (LOW Ca++). Each point represents the mean of 4–6 beakers, and the vertical bars indicate S.E.M. Where no vertical bars are present, the S.E.M. was too small to be indicated. Duplicate experiments are shown.

in one experiment (Fig. 10A) and totally blocked in the duplicate study (Fig. 10B).

3.7. Response to Dibutyryl cAMP in Low-Calcium Medium

In two identical experiments (Fig. 11), GH release was significantly augmented by dbcAMP (0.6 mM) by 7.5 min of incubation and continued to increase linearly up to 60 min. The augmented release was markedly and significantly reduced in the low-calcium medium in one experiment (Fig. 11B) and totally blocked in the duplicate study (Fig. 11A).

3.8. Response to Phosphodiesterase Inhibitor in Low-Calcium Medium

The phosphodiesterase inhibitor IBMX (0.1 mM) consistently caused an increased GH release in a linear fashion over 1 hr of incubation in two identical experiments (Fig. 12). This augmented release was totally blocked in both experiments in the low-calcium medium.

3.9. Response to Somatostatin in Low-Calcium Medium

The effect of low calcium on the response to somatostatin was not as dramatic. In three identical experiments, somatostatin (100 ng/ml) consistently reduced basal GH release by up to 30% in M199-1, and this was

Table I. Release of Growth Hormone from Purified Somatotrophs: Effects of Somatostatin (100 ng/ml) in Medium 199 and in Medium 199 Low in Ca^{2+}

	Incubation time (min)[a]				
Experiment	0	7.5	15	30	60
M199	5.4 ± 0.6	6.2 ± 0.3	9.2 ± 0.6	12.9 ± 0.7	15.2 ± 1.3
Somatostatin in M199	5.4 ± 0.6	6.6 ± 0.2	7.7 ± 0.3	8.9 ± 0.3[b]	13.2 ± 0.5
Low Ca^{2+}	4.7 ± 0.4	4.7 ± 0.3	8.7 ± 0.9	10.0 ± 0.5	11.9 ± 0.6
Somatostatin in low Ca^{2+}	4.7 ± 0.4	9.1 ± 0.5	8.0 ± 0.5	11.6 ± 0.4	13.2 ± 0.6
M199	11.2 ± 0.9	14.7 ± 0.4	17.2 ± 0.6	24.6 ± 1.5	25.7 ± 2.2
Somatostatin in M199	11.2 ± 0.9	13.1 ± 0.4	16.6 ± 0.3	18.9 ± 1.1[b]	23.1 ± 1.1
Low Ca^{2+}	13.4 ± 0.7	14.9 ± 0.2	19.5 ± 0.9	24.1 ± 1.6	27.1 ± 0.7
Somatostatin in low Ca^{2+}	13.4 ± 0.7	17.0 ± 0.4	17.1 ± 0.6	26.3 ± 1.8	28.7 ± 2.2
M199	15.1 ± 0.2	18.6 ± 0.5	26.9 ± 1.5	30.0 ± 0.7	37.2 ± 0.8
Somatostatin in M199	15.1 ± 0.2	18.2 ± 0.4	24.2 ± 0.5	27.0 ± 0.7[b]	33.8 ± 0.4[b]
Low Ca^{2+}	12.8 ± 0.3	17.1 ± 0.3	26.3 ± 0.4	27.3 ± 1.1	35.6 ± 1.0
Somatostatin in low Ca^{2+}	12.8 ± 0.3	17.1 ± 0.3	22.7 ± 0.4	31.0 ± 0.8	33.4 ± 0.7

[a] Values represent ng GH/1000 (± S.E.M.) and are derived from 4–6 beakers.
[b] Significantly different from control (M199), $P < 0.05$.

significant at 30 min. There was no consistent difference in basal GH release with or without calcium. Somatostatin did not lower GH release in the low-calcium medium (Table I).

4. Discussion

4.1. Validation of the Purified Somatotroph Preparation: Role of Cyclic Nucleotides

4.1.1. Current Model of the Mechanisms Governing Growth Hormone Release

Validation of the purified somatotroph preparation was carried out within the context of the current model of the mechanisms governing GH release. The data forming the basis for this model have been extensively reviewed (Peake, 1973; Kraicer, 1975; Boss *et al.*, 1975; Labrie *et al.*, 1976a,b; McCann *et al.*, 1976; Hedge, 1977). According to current thought, the secretagogue, in this case the hypothetical growth hormone–releasing hormone (GH-RH), would first interact with a specific receptor site on the plasma membrane of the somatotroph, resulting in the activation of an adenylate cyclase leading to an increase in intracellular cAMP. The elevated cAMP would then lead, perhaps through the activation of a protein kinase, to hormone release via the process of exocytosis. Thus, addition of exogenous cAMP or its analogues would be expected to mimic the increase in intracellular cAMP levels and cause hormone release. Phosphodiesterase inhibitors would block the degradation of cAMP to 5'-AMP, resulting in an elevation of endogenous cAMP and subsequent GH release. PGE_2, which activates adenylate cyclase and increases cAMP levels in the adenohypophysis, would mimic the action of GH-RH and result in increased hormone release.

The mechanism of action of the tetradecapeptide somatostatin has not been clearly defined within the context of this model, although it has been suggested that it may act through an alteration in adenylate cyclase activity or phosphodiesterase activity or both (Boss *et al.*, 1975; Labrie *et al.*,

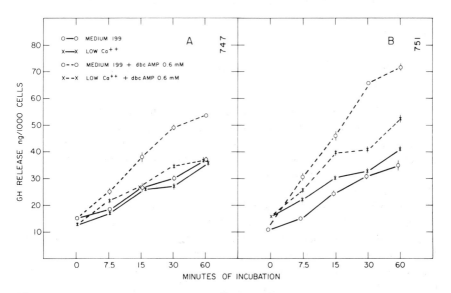

Figure 11. Dibutyryl cAMP (dbcAMP)-induced release of GH, with time, in Medium 199 and Medium 199 containing less than 70 μM calcium (LOW Ca^{++}). Each point represents the mean of 4–6 beakers, and the vertical bars indicate S.E.M. Where no vertical bars are present, the S.E.M. was too small to be indicated. Duplicate experiments are shown.

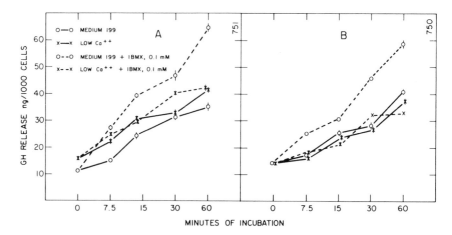

Figure 12. IBMX-induced release of GH, with time, in Medium 199 and Medium 199 containing less than 70 μM calcium (LOW Ca^{++}). Each point represents the mean of 4–6 beakers, and the vertical bars indicate S.E.M. Where no vertical bars are present, the S.E.M. was too small to be indicated. Duplicate experiments are shown.

1976a). It is established, however, that somatostatin does inhibit both basal and stimulated GH release *in vivo* and *in vitro* under a great variety of experimental conditions (Vale *et al.*, 1972, 1975; Brazeau *et al.*, 1973; Boss *et al.*, 1975; Labrie *et al.*, 1976a,b).

4.1.2. Effect of Cyclic Nucleotides on Growth Hormone Release from Purified Somatotrophs

To validate the use of the purified somatotroph preparation, we have carried out a series of experiments exploring the role of the cyclic nucleotides and somatostatin in GH release. We began by investigating the effects of cyclic nucleotides on the release of GH. Dibutyryl cAMP has been shown, in numerous studies, to stimulate GH release *in vitro* (Vale *et al.*, 1972; Peake, 1973; Hertelendy and Keay, 1974; Kraicer and Hymer, 1974; Kraicer, 1975; Snyder and Hymer, 1975). Although the data of Peake *et al.* (1972) suggested that cGMP may also be involved, Carlson *et al.* (1974), using a perfusion system, could find no response to dbcGMP. In our studies, the somatotrophs respond rapidly and in a dose-related manner to doses as low as 0.5 mM dbcAMP, but the response to the other cyclic nucleotides (cAMP, cGMP, and dbcGMP), in concentrations up to 5 mM, was inconsistent (see Fig. 1). The two phosphodiesterase inhibitors, theophylline and IBMX, which have been shown to elevate cAMP and to increase GH release in other *in vitro* preparations (Vale *et al.*, 1972; Peake,

1973; Bélanger *et al.*, 1974; Carlson *et al.*, 1974; Schofield and McPherson, 1974; Schofield *et al.*, 1974; Kraicer, 1975), both cause a significant dose-related release of GH after 7.5 min of incubation (see Figs. 2 and 3). In our system, theophylline and IBMX caused similar responses, unlike their effects on the GH_4C_1 cell line (Dannies *et al.*, 1976), where theophylline added to cultures for 3 days stimulated GH synthesis, while IBMX had no effect. We found no change in GH content (cells plus media) over the 60-min incubation period, thus no evidence of net synthesis or degradation.

4.1.3. Effect of Prostaglandin E_2 on Growth Hormone Release from Purified Somatotrophs

As summarized in several reviews (Peake, 1973; Labrie *et al.*, 1976b; McCann *et al.*, 1976; Hedge, 1977), the prostaglandins (PGE_1 and PGE_2) cause increased release of GH *in vitro* from hemipituitaries and cultured cells accompanied by elevated levels of adenylate cyclase and cAMP. In our studies (see Fig. 8), PGE_2 at 10^{-6} or 10^{-7} M increases GH release within 7.5 min, while lower concentrations are ineffective during 1 hr of incubation. The sensitivity of purified somatotrophs is similar to that of cultured cells, which display half-maximal stimulation for PGE_1 and PGE_2 at 5×10^{-7} M (Labrie *et al.*, 1976b).

4.1.4. Effect of Somatostatin on Growth Hormone Release from Purified Somatotrophs

Inhibition of GH release *in vitro* by somatostatin has been described many times since the initial report of Brazeau *et al.* (1973). Somatostatin will reduce both basal GH release (Vale *et al.*, 1972; Brazeau *et al.*, 1973; Schofield *et al.*, 1974; Bélanger *et al.*, 1974; Brown and Vale, 1975) and that stimulated by cAMP derivatives (Vale *et al.*, 1972; Carlson *et al.*, 1974; Bélanger *et al.*, 1974), agents that increase endogenous cAMP levels (Boss *et al.*, 1975; Vale *et al.*, 1975; Labrie *et al.*, 1976a), phosphodiesterase inhibitors (Vale *et al.*, 1972; Carlson *et al.*, 1974; Bélanger *et al.*, 1974; Schofield *et al.*, 1974), and PGE_2 (Bélanger *et al.*, 1974; Brown and Vale, 1975. Carlson *et al.* (1974) reported that inhibition of basal or stimulated release in a perfusion system was followed by an elevated GH secretion after the somatostatin perfusion was stopped. Recently, Stachura (1976) reported a dose-dependent inhibition of GH release with somatostatin followed by a dose-dependent rebound.

It has been reported by some investigators (Kaneko *et al.*, 1974; Borgeat *et al.*, 1974; Labrie *et al.*, 1976b) that somatostatin will reduce basal cAMP levels, although this has not always been found (Schofield *et al.*, 1974; Lippmann *et al.*, 1976). Somatostatin has also been reported to elevate cGMP levels (Kaneko *et al.*, 1974). Furthermore, several studies have shown that somatostatin reduces the augmented cAMP levels induced by PGE (Borgeat *et al.*, 1974; Kaneko *et al.*, 1974; Lippmann *et al.*, 1976), theophilline (Borgeat *et al.*, 1974), and IBMX (Schofield *et al.*, 1974). However, all these changes in intracellular nucleotide levels must be interpreted with caution, since they are derived from heterogeneous cell populations and may not necessarily reflect changes occurring exclusively within the somatotrophs. It has been suggested that somatostatin may act on the plasma membrane by altering adenylate cyclase activity, phosphodiesterase activity, or the expression of cyclic nucleotide activity (Boss *et al.*, 1975; Labrie *et al.*, 1976a).

Somatostatin inhibits both the basal and stimulated release of GH from the purified somatotrophs. The inhibition of basal release (see Figs. 2–4) is small but consistent with concentrations down to 10 ng/ml (about 7×10^{-9} M). A large effect on basal release could obviously not be seen, since there is only a small basal release of GH in our system during incubation. The stimulated release induced by dbcAMP (see Figs. 4 and 5), the phosphodiesterase inhibitors (see Figs. 6 and 7), and PGE_2 (see Fig. 9) is completely blocked by 10 ng/ml somatostatin, while as little as 1.0 ng/ml blocks the GH release due to 0.6 mM dbcAMP (Fig. 5). The release due to the other secretagogues was not tested with concentrations of somatostatin lower than 10 ng/ml. In cultured cells, half-maximal inhibition of GH release stimulated by 2.5 mM monobutyryl cAMP occurs at 3×10^{-9} M of somatostatin (Bélanger *et al.*, 1974). Thus, the somatotrophs display a similar sensitivity to cultured cells when the stimulated release of GH is examined. The augmented release of GH seen when cells are incubated with 0.1 ng/ml somatostatin and 0.6 mM dbcAMP compared to 0.6 mM dbcAMP alone (Fig. 5) was seen in three separate experiments. This interesting observation remains unexplained.

4.1.5. *Specificity of the Growth Hormone Response from Purified Somatotrophs*

To test the specificity of the somatotroph responsiveness, we incubated the cells with large doses of arginine vasopressin, TRH, LH-RH and α-MSH. There are conflicting reports concerning the effects of TRH on GH secretion *in vitro*. Carlson *et al.* (1974), using a perfusion system of rat adenohypophyses, reported a marked but transient increase in GH release with TRH (0.1–100 ng/ml). However, Wilber *et al.* (1971) found no effect with 4–500 ng/ml, Sundberg *et al.* (1976) found no change with 100 ng/ml, and Brown and Vale (1975) could find no significant release from cultured cells with 10 nM TRH. LaBella and Vivian (1971) reported only a slight increase in medium GH in one of three experiments with 100 ng/ml of TRH in bovine anterior pituitary tissue, while Takahara *et al.* (1974)

reported an increase in GH release from sheep adenohypophyses with 8–200 ng/ml TRH, but no response with 1000 ng/ml. Cultured cells from bovine adenohypophyses did show an increased GH release in response to TRH (Machlin *et al.*, 1974), but TRH decreased the synthesis of GH by GH_4C_1 tumor cells (Dannies *et al.*, 1976). In two of three of our experiments, TRH at 100 ng/ml caused a slight but significant increase in GH release from the somatotrophs, although 10 ng/ml was ineffective.

There are a few reports of the effects of α-MSH on GH secretion. Schally *et al.* (1973) reported a slight inhibition of GH release *in vitro* with 1–5 μg/ml of α-MSH. However, Strauch *et al.* (1973) found a stimulation of plasma GH levels in normal men. α-MSH, in a concentration of 1 μg/ml, did not alter GH release in our system.

Since hypothalamic extracts that release GH contain vasopressin, there has been speculation that vasopressin might cause GH release. Although high concentrations of vasopressin (> 50 μU/ml) have been reported to cause a slight GH release, the stimulation was much too small to account for the GH-releasing activity of hypothalamic extracts (Steiner *et al.*, 1970; Wilber *et al.*, 1971). We found no effect with up to 244 mU/ml of arginine vasopressin. Synthetic LH-RH does not cause release of GH *in vitro* from rat adenohypophyses (Borgeat *et al.*, 1972; Sundberg *et al.*, 1976). A concentration of 1 μg/ml was without effect on the purified somatotrophs. Although it has been reported that the effects of dbcAMP in tumor cells are mimicked by sodium butyrate (Dannies *et al.*, 1976), in our system butyric acid (1.2 mM) had no effect. Thus, the somatotrophs retain their specificity and do not respond to inappropriate stimuli such as LH-RH, vasopressin, or α-MSH.

4.1.6. Conclusions

We conclude from the first series of studies that the purified somatotroph preparation promises to be a useful tool for studying the release of GH. The cells respond rapidly in a sensitive and specific manner without the need for the short-term culture suggested by other investigators. Frequent and rapid sampling is easily carried out. The intragroup variability is extremely small, allowing significant differences to be seen at very low doses of secretagogues.

4.2. Role of Ca^{2+} in the Release Process

4.2.1. Current Model

The "stimulus–secretion coupling" hypothesis of Douglas (1968) and his colleagues proposed that the interaction of a secretagogue with the plasma membrane of a cell would result in an altered membrane conformation with increased permeability to several common ions, notably calcium and sodium, and further, that it was the entry of calcium that caused hormone release, perhaps by stimulating granule attachment to the plasma membrane. Thus, calcium would be the essential link or coupling between "stimulus" and "secretion." The initial experiments of Douglas and his colleagues were carried out on the adrenal medulla and the neurohypophysis. Since then, there have been a number of studies on the role of calcium in hormone release and the interaction of calcium with cyclic nucleotides in many endocrine and exocrine tissues (Berridge, 1975; Trifaró, 1977; Rasmussen and Goodman, 1977). While the components of the control system (calcium and cylic nucleotides) appear to be the same, the interaction of these elements may differ from system to system.

Two approaches have generally been used to study the role of calcium in hormone release from the adenohypophysis (Moriarty, 1978). The first is to ask whether an influx of calcium is essential for hormone release, by examining basal and stimulated release after the removal of extracellular calcium. Although it has been established that the stimulated relase of adenohypophyseal hormones *in vitro* is dependent on calcium in the incubating medium, the magnitude of the calcium requirement has not been clearly defined. This can be attributed to either the diverse "calcium-free" conditions employed (from the simple omission of calcium to the addition of the chelator EGTA or EDTA), variable and inconsistent diffusion of ions through the tissue mass, or variable durations of incubation and preincubation in the media. All such calcium-free conditions would result not only in a lowering of extracellular calcium, but also in a variable "leaching out" of intracellular calcium. It has been postulated that different pools of calcium are involved in hormone release produced by differing secretagogues. For example, studies from our laboratory (Milligan and Kraicer, 1974) have shown that ACTH release due to vasopressin requires only a loosely bound calcium compartment and is easily suppressed by quick rinses in a calcium-free medium, while that due to a crude hypothalamic extract utilizes a more tightly bound compartment and requires multiple washes and prolonged incubation in a calcium-free medium to suppress release.

A second approach has been to measure alterations in calcium-ion flux associated with altered hormone release (see below).

4.2.2. Ca^{2+} Requirement for Growth Hormone Release Induced by Cyclic Nucleotides

The present studies form our initial approach to elucidating the role of calcium using a purified somatotroph preparation. We wished first to define the requirement for calcium using a simple protocol that would result in a lowering of extracellular calcium and that would allow for the easy and rapid diffusion of ions throughout the system. The purified somatotrophs were preincubated for 30 min in a normal or low-calcium ($<70~\mu M$) medium prior to incubation with the secretagogues. We found no consistent difference in basal release in the low-calcium media (see Figs. 10–12 and Table I), but there was a significant reduction in the GH release stimulated by PGE_2 (Fig. 10), dbcAMP (Fig. 11), and the phosphodiesterase inhibitor IBMX (Fig. 12). This would indicate that an extracellular calcium concentration greater than 70 μM is essential for GH release, but whether the low-calcium medium produced its effect by preventing calcium influx into the cells or by removing a loosely bound intracellular calcium component must now be determined by short-term kinetic studies.

Calcium fluxes associated with hormone release from the adenohypophysis have been examined (Milligan and Kraicer, 1971; Milligan *et al.,* 1972; Peake, 1973; Eto *et al.,* 1974; Kraicer, 1975; Moriarty, 1977, 1978). However, such studies have yielded equivocal results. Although an increased uptake of ^{45}Ca was noted with a partially purified GH-RF preparation (Milligan *et al.,* 1972; Peake, 1973) and elevated potassium-ion concentration (Milligan and Kraicer, 1971; Eto *et al.,* 1974), no increase in calcium influx was found with crude hypothalamic extracts, dbcAMP, or phosphodiesterase inhibitors, all of which cause hormone release (Milligan and Kraicer, 1971; Eto *et al.,* 1974; Moriarty, 1977). These studies utilized mixed cell populations. More refined flux studies will require the use of homogeneous cell preparations to ensure that changes that might occur in one cell population are not "masked," and that changes that are observed can be confidently ascribed to a specific cell type.

Our data also allow us to further explore the role(s) of calcium ion in the release process. As a working hypothesis (Kraicer, 1975), the postulated GH-RH would initiate release by first interacting with specific receptor sites on the plasma membrane of the somatotrophs. This then would result, possibly via the mediation of prostaglandins, in an activation of adenylate cyclase, which would, in turn, increase intracellular AMP levels. The increase in intracellular cAMP would then result in hormone release. Since removal of a loosely bound calcium component blocks the augmented release induced by both dbcAMP (Fig. 11) and the phosphodiesterase inhibitor IBMX (Fig. 12), as well as that induced by PGE_2 (Fig. 10), we can conclude that calcium ion is essential for the expression of the action of cAMP. This does not, of course, preclude an additional action for calcium at an earlier step in the release process. A role beyond the activation of cAMP is consistent with the observation (Steiner *et al.,* 1970) that the decreased response to crude hypothalamic extract and aminophylline in calcium-free medium occurs even though the increase in intracellular cAMP remains the same.

4.2.3. Ca²⁺ Requirement for Action of Somatostatin

Somatostain has been shown to inhibit GH release *in vitro* under a variety of conditions (Brazeau *et al.*, 1973). The actions and postulated mechanism of action of somatostatin have been discussed in Section 4.1.4. The role of calcium ion in the action of somatostatin on GH release has not been extensively investigated. Recent studies with dispersed bovine adenohypophyseal cells have shown that somatostatin reduces the ionophore A23187–induced release of GH without altering the associated movement of $^{45}Ca^{2+}$ (Bicknell and Schofield, 1976).

Since basal GH release from the acutely dispersed somatotrophs is quite small (Table I), decreases in basal response produced by somatostatin are difficult to detect. However, in three independent experiments, somatostatin consistently decreased basal release, the inhibition being significant at 30 min. This inhibitory effect of somatostatin was prevented in the low-calcium medium, indicating that somatostatin requires calcium to reduce basal GH release. Since a low-calcium medium will itself reduce the augmented release induced by the secretagogues tested, it is obviously not possible to study whether somatostatin requires extracellular calcium to block stimulated GH release. As with the secretagogue studies described above, the role of calcium in the action of somatostatin will require short-term kinetic studies using ^{45}Ca and a purified somatotroph preparation to establish whether the effect of low extracellular calcium is due to an alteration in calcium-ion flux.

4.2.4. Conclusions

In summary, a simple protocol has been followed to establish that extracellular calcium is essential for the actions of PGE_2, dbcAMP, and IBMX, indicating a role for calcium beyond the activation of cAMP. Extracellular calcium is also essential for the inhibition of basal GH release induced by somatostatin. This approach does not, however, tell us whether the effect of low calcium is due to the prevention or alteration of calcium flux across the plasma membrane, or to an alteration in the intracellular distribution of calcium. The purified somatotroph preparation will now allow us to define more closely the role(s) of calcium ion, since concurrent studies of the kinetics of calcium movement, intracellular calcium content, and changes in intracellular cyclic nucleotide levels can be carried out in a homogeneous cell preparation.

ACKNOWLEDGMENTS. This study was supported by the Medical Research Council of Canada and the Queen's University Faculty of Medicine Trust Fund. M. S. Sheppard is the recipient of an MRC Fellowship Award. Somatostatin was generously supplied by Dr. M. Givner, Ayerst, Montreal and PGE_2 by Dr. J. Pike, Upjohn, Kalamazoo. We thank Dr. A. F. Parlow and the NIAMDD Rat Pituitary Hormone Program for the reagents for the GH radioimmunoassay. The expert technical assistance of Mrs. T. L. Swanson, Mr. J. Fox, Mrs. O. M. Morris, Mrs. W. M. Winsor, Mr. L. F. Padilla, and Mr. J. Gothilf, under the supervision of Mr. A. R. Morris, is gratefully acknowledged. We thank Dr. S. H. Shin for his advice and stimulating discussion.

DISCUSSION

HYMER: The lack of effect of SRIF on basal release may be related to the methodology involved. We reported that SRIF had no effect on basal GH release from acutely dispersed cells, although it could block dbcAMP-stimulated hormone release. Nevertheless, it was active in cultured cells, thereby suggesting that "receptors" were generated over an extended period. More recently, however, we have found that SRIF will reduce basal GH release from acutely dispersed cells placed in Biogel columns. This may indicate that the trypsinized cells (prepared according to your techniques) do, in fact, have receptors.

It should be made clear that the time to isolate somatotrophs by the IG technique is about the same as that required for the discontinuous-density separation technique. The latter is, of course, preferable, since the somatotrophs are purer.

I have two questions: (1) Is the calcium effect you report reversible? (2) have you measured intracellular cAMP in the somatotrophs after stimulation by various secretagogues?

SHEPPARD: Of course, the problem of basal release of GH stimulation in a static incubation is a difficult one. The "basal release" present may not be a true release, but due to a small number of damaged cells. Our incubations begin with a level of GH in the system, labeled "start" on my diagrams. True release could be masked by this.

1. Yes, the calcium effect is reversible. If we add calcium to the low-calcium medium while PGE_2 or IBMX is present, the cells respond in both the static and perfusion systems.

2. We have measured cAMP in the cell and found consistent levels of 1–2 fmol/1000 cells. This could be stimulated by PGE_2 and IBMX. Both of these secretogues cause the rapid elevation of cAMP, which increased 2-fold within 1.5 min and continued to rise over the course of incubation. We have also assayed for cGMP, but we find none.

SPICER: The data presented on cyclic nucleotide fluctuations and on the calcium effect indicate the possibility of two different cell mechanisms acting to mediate somatotroph secretion, although the calcium effects conceivably could be related to some steps in the cyclic-nucleotide-induced events in the cell. The experience with leukocytes may bear on the problem of distinguishing the role of such mechanisms in that cytocholasin B appears to impede phagocytic ingestion of foreign particles and colchicine inhibits infusion of lysosomal granules with phagozomes. The former apparently acts fairly specifically on actinlike filaments and the latter on microtubules. Ionophores, on the other hand, may affect calcium or other cation diffusion into the cell and have been found to induce exocytosis of neutrophil leukocyte granules. The question in mind concerns whether such agents have been employed to differentiate cell mechanisms of exocytosis in somatotrophs or possibly offer an approach to the problem?

SHEPPARD: We have approached the problem of calcium and cyclic nucleotides in two ways. First, we have tested the effects of the low-calcium medium. We find that incubation of the somatotrophs in this, which blocks GH release, does not lower the elevated cAMP levels produced by PGE_2 or IBMX, thus indicating an effect of calcium in expressing the action of cAMP.

We have also tested the effects of the calcium ionophore A23187 and find an augmented release of GH. We have examined the effects of A23187 on cAMP levels, and these are not affected. Furthermore, the release of GH induced by the ionophores is blocked by somatostatin. Obviously, the advantage of our homogeneous cell population is that any alterations in cyclic nucleotides that we measure must occur in somatotrophs. This is very important when using nonspecific secretagogues such as IBMX or A23187.

TIXIER-VIDAL: Several years ago, we studied the binding of [³H]-TRH in freshly dispersed anterior pituitary cells and found a very low number of binding sites. After 5 days in culture, [³H]-TRH binding sites increase in number.

Even if you were interested in short-term effects (2 hr), it would have been important to check the synthesis and release over a longer period of time in order to be sure that experiments are representative of maximal secretory activity.

SHEPPARD: We have examined GH synthesis and find an incorporation of radioactive amino acids into GH over several hours. We have not studied longer time periods. I suppose the critical test of the responsiveness of the somatotrophs will come with the use of the, as yet, hypothetical GH-RH.

We agree that longer-time studies will be of importance. However, we wanted to concentrate on the very early events in the release process.

DENEF: (1) Your separated somatotrophs are surprisingly responsive to somatostatin. Do you conclude from this that your disperson and separation method preserves the plasma membrane receptors for somatostatin, or would somatostatin work after internalization? (2) How long can you maintain your somatotrophs in suspension and have them responsive to somatostatin? (3) Freshly dispersed pituitary cells respond poorly or not at all to various releasing hormones. Do you have any explanation why your dispersed cells do respond to secretagogues?

KRAICER: 1. I would expect somatostatin to act on the external plasma membrane, rather than being internalized. Somatostatin acts within 3–5 minutes, and this does not leave much time for internalization. Our results do indicate that our dispersion and separation method preserves the plasma membrane receptors.

2. We have not tried incubations longer than 2 hours, since we are interested in dynamic responses.

3. This is a confused field. The only explanation I can give is that the procedure we use uses different enzymes than others to disperse the cell and does not damage the cell membranes as perhaps other treatments do. We await further studies on the optimal methods of cell dispersion. Are the reported differences due to different procedures, or are different cell types differentially sensitive to enzymatic dispersion?

KERDELHUÉ: Did you check the effect of TRH alone or in combination with somatostatin or GH release in your system?

SHEPPARD: Yes, we tried TRH alone. In two or three experiments, the somatotrophs released a small amount of GH in response to 100 ng/ml TRH. A lower concentration (10 ng/ml) was ineffective. We did not try somatostatin and TRH in combination.

VAUDRY: What are the molecular events which can explain somatostatin-induced diminution of GH stimulation by PGE_2?

SHEPPARD AND KRAICER: Our working hypothesis is that somatostatin acts at the plasma membrane level to inhibit GH release. We have examined the effect of somatostatin on somatotrophs stimulated by PGE_2. We find that the same concentration of somatostatin which totally blocks GH release has no effect on the elevated cAMP levels in somatotrophs produced with PGE_2. This would place the locus of action of somatostatin after that of cAMP in the secretion of mechanism.

Since somatostatin (a) blocks the augmented release of GH induced by dbcAMP and (b) blocks the release of GH induced by PGE_2 and phosphodiesterase inhibitors, without altering the increase in cAMP induced by these two secretagogues, this must indicate a primary action of somatostatin beyond cAMP and prevents the expression of the action of cAMP.

HERBERT: I have several questions: (1) Are the cells incubated in suspension, or do you do any culturing? (2) Do you find any differences in response of cells dispersed directly from anterior pituitary and cells enriched through your purification procedure (with regard to the effect of regulators on GH release)? (3) Is there enough time after trypsin treatment for resynthesis or regeneration of cell-surface receptors? (4) What kind of trypsin do you use to disperse your cells? Is it a magic trypsin formula?

SHEPPARD AND KRAICER: 1. Our cells are incubated in suspension. We remove samples containing both cells and medium and separate them by centrifugation, prior to measuring GH release. Although we have looked briefly at cells in culture, we have not studied them extensively.

2. In initial experiments, we briefly examined the release of GH from acutely dispersed cells and found that they responded to dbcAMP and a partially purified GH-RH. Since we are interested in intracellular metabolites, all our subsequent studies have been with the enriched cell population. We have not looked systematically for differences between the acutely dispersed and the enriched cells.

3. I believe that either we do not damage the cell receptors or the time for cell purification and the 30-minute equilibrium period which occurs after the purification procedure is sufficient time for the regeneration of receptors. I suggest this because our cells do respond to somatostatin, which we presume acts via plasma membrane receptors.

4. We use a Difco trypsin (1 : 250, No.0152-13). This is an impure trypsin, and perhaps works so well because of the impurities.

REFERENCES

Bélanger, A., Labrie, F., Borgeat, P., Savary, M., Cote, J., Drouin, J., Schally, A. V., Coy, D. H., Coy, E. J., Immer, H., Sestanj, K., Nelson, V., and Gotz, M., 1974, Inhibition of growth hormone and thyrotropin release by growth hormone–release inhibiting hormone, *Mol. Cell. Endocrinol.* 1:329.

Berridge, M. J., 1975, The interaction of cyclic nucleotides and calcium in the control of cellular activity, *Adv. Cyclic Nucleotide Res.* 6:1.

Bicknell, R. J., and Schofield, J. G., 1976, Mechanism of action of somatostatin: Inhibition of ionophore A23187–induced release of growth hormone from dispersed bovine pituitary cells, FEBS Lett. 68:23.

Birge, C. A., Peake, G. T., Mariz, I. K., and Daughaday, W. H., 1967, Radioimmunoassayable growth hormone in the rat pituitary gland: Effects of age, sex and hormonal state, Endocrinology 81:195.

Borgeat, P., Chavancy, G., Dupont, A., Labrie, F., Arimura, A., and Schally, A. V., 1972, Stimulation of adenosine 3':5'-cyclic monophosphate accumulation in anterior pituitary gland in vitro by synthetic luteinizing hormone–releasing hormone, Proc. Natl. Acad. Sci. U.S.A. 69:2677.

Borgeat, P., Labrie, F., Drouin, J., Bélanger, A., Immer, H., Sestanj, K., Nelson, V., Gotz, M., Schally, A. V., Coy, D. H., and Coy, E. J., 1974, Inhibition of adenosine 3', 5'-monophosphate accumulation in anterior pituitary gland in vitro by growth hormone–release inhibiting hormone, Biochem. Biophys. Res. Commun. 56:1052.

Boss, B., Vale, W., and Grant, G., 1975, Hypothalamic hormones, in: Biochemical Actions of Hormones (G. Litwack, ed.), pp. 87–118, Academic Press, New York.

Brazeau, P., Vale, W., Burgus, R., Ling, N., Butcher, M., Rivier, J., and Guillemin, R., 1973, Hypothalamic polypeptide that inhibits the secretion of immunoreactive pituitary growth hormone, Science 179:77.

Brown, M., and Vale, W., 1975, Growth hormone release in the rat: Effects of somatostatin and thyrotropin-releasing factor, Endocrinology 97:1151.

Burr, R. G., 1969, An automated method for serum calcium utilizing ethylene diamine tetra acetic acid, Clin. Chem. 15:1191.

Carlson, H. E., Mariz, I. K., and Daughaday, W. H., 1974, Thyrotropin-releasing hormone stimulation and somatostatin inhibition of growth hormone secretion from perfused rat adenohypophyses, Endocrinology 94:1709.

Cooper, R. H., McPherson, M., and Schofield, J. G., 1972, The effect of prostaglandins on ox pituitary content of adenosine 3':5'-cyclic monophosphate and the release of growth hormone, Biochem. J. 127:143.

Dannies, P. S., Gautvik, K. M., and Tashjian, A. H., Jr., 1976, A possible role of cyclic AMP in mediating the effects of thyrotropin-releasing hormone on prolactin release and on prolactin and growth hormone synthesis in pituitary cells in culture, Endocrinology 98: 1147.

Douglas, W. W., 1968, Stimulus–secretion coupling: The concept and clues from chromaffin and other cells, Br. J. Pharmacol. 34:451.

Duncan, D. B., 1955, Multiple range and multiple F tests, Biometrics 11:1.

Eto, S., Wood, J. M., Hutchins, M., and Fleischer, N., 1974, Pituitary $^{45}Ca^{++}$ uptake and release of ACTH, GH, and TSH: Effect of verapamil, Am. J. Physiol. 226:1315.

Hedge, G. A., 1977, Roles for the prostaglandins in the regulation of anterior pituitary secretion, Life Sci. 20:17.

Hertelendy, F., 1971, Studies on growth hormone secretion. II. Stimulation by prostaglandins in vitro, Acta Endocrinol. (Copenhagen) 68:355.

Hertelendy, F., and Keay, L., 1974, Studies on growth hormone secretion. VI. Effects of dibutyryl cyclic AMP, prostaglandin E₁, and indomethacin on growth and hormone secretion by rat pituitary tumor cells in culture, Prostaglandins 6:217.

Hymer, W. C., Kraicer, J., Bencosme, S. A., and Haskill, J. S., 1972, Separation of somatotrophs from the rat adenohypophysis by velocity and density gradient centrifugation. Proc. Soc. Exp. Biol. Med. 141:966.

Hymer, W. C., Evans, W. H., Kraicer, J., Mastro, A., Davis, J., and Griswold, E., 1973, Enrichment of cell types from the rat adenohypophysis by sedimentation at unit gravity, *Endocrinology* **92**:275.

Kaneko, T., Oka, H., Munemura, M., Suzuki, S., Yasuda, H., and Oda, T., 1974, Stimulation of guanosine 3′,5′-cyclic monophosphate accumulation in rat anterior pituitary gland *in vitro* by synthetic somatostatin, *Biochem. Biophys. Res. Commun.* **61**:53.

Kraicer, J., 1975, Mechanisms involved in the release of adenohypophysial hormones, in: *The Anterior Pituitary* (A. Tixier-Vidal and M. G. Farquhar, eds.), pp. 21–43, Academic Press, New York.

Kraicer, J., and Hymer, W. C., 1974, Purified somatotrophs from rat adenohypophysis: Response to secretagogues, *Endocrinology* **94**:1525.

LaBella, F. S., and Vivian, S. R., 1971, Effect of synthetic TRF on hormone release from bovine anterior pituitary *in vitro*, *Endocrinology* **88**:787.

Labrie, F., De Lean, A., Barden, N., Ferland, L., Drouin, J., Borgeat, P., Beaulieu, M., and Morin, O., 1976a, New aspects of the mechanism of action of hypothalamic regulatory hormones, in: *Current Topics in Molecular Endocrinology* (F. Labrie, J. Meites, and G. Pelletier, eds.), Vol. 3, *Hypothalamus and Endocrine Functions*, pp. 147–169, Plenum Press, New York.

Labrie, F., Pelletier, G., Borgeat, P., Drouin, J., Ferland, L., and Bélanger, A., 1976b, Mode of action of hypothalamic regulatory hormones in the adenohypophysis, in: *Frontiers in Neuroendocrinology* (L. Martini and W. F. Ganong, eds.), Vol. 4, pp. 63–93, Raven Press, New York.

Lemay, A., and Labrie, F., 1972, Calcium-dependent stimulation of prolactin release in rat anterior pituitary *in vitro* by N^6-monobutyryl adenosine 3′,5′-monophosphate, *FEBS Lett.* **20**:7.

Lippmann, W., Sestanj, K., Nelson, V. R., and Immer, H. U., 1976, Antagonism of prostaglandin-induced cyclic AMP accumulation in the rat anterior pituitary *in vitro* by somatostatin analogues, *Experientia* **32**:1034.

Lockhart Ewart, R. B., and Taylor, K. W., 1971, The regulation of growth hormone secretion from the isolated rat anterior pituitary *in vitro:* The role of adenosine 3′:5′-cyclic monophosphate, *Biochem. J.* **124**:815.

Machlin, L. J., Jacobs, L. S., Cirulis, N., Kimes, R., and Miller, R., 1974, An assay for growth hormone and prolactin-releasing activities using a bovine pituitary cell culture system, *Endocrinology* **95**:1350.

McCann, S. M., Ojeda, S. R., Harms, P. G., Wheaton, J. E., Sundberg, D. K., and Fawcett, C. P., 1976, Role of prostaglandins (PGs) in the control of adenohypophyseal hormone secretion, in: *Current Topics in Molecular Endocrinology* (F. Labrie, J. Meites, and G. Pelletier, eds.), Vol. 3, *Hypothalamus and Endocrine Functions*, pp. 21–35, Plenum Press, New York.

Milligan, J. V., and Kraicer, J., 1971, ^{45}Ca uptake during the *in vitro* release of hormones from the rat adenohypophysis, *Endocrinology* **89**:766.

Milligan, J. V., and Kraicer, J., 1974, Physical characteristics of the Ca^{++} compartments associated with *in vitro* ACTH release, *Endocrinology* **94**:435.

Milligan, J. V., Kraicer, J., Fawcett, C. P., and Illner, P., 1972, Purified growth hormone releasing factor increases ^{45}Ca uptake into pituitary cells, *Can. J. Physiol. Pharmacol.* **50**:613.

Moriarty, C. M., 1977, Involvement of intracellular calcium in hormone secretion from rat pituitary cells, *Mol. Cell. Endocrinol.* **6**:349.

Moriarty, C. M., 1978, Role of calcium in the regulation of adenohypophysial hormone release, *Life Sci.* **23**:185.

Peake, G. T., 1973, The role of cyclic nucleotides in the secretion of pituitary growth hormone, in: *Frontiers in Neuroendocrinology* (W. F. Ganong and L. Martini, eds.), Vol. 3, pp. 173–208, Oxford University Press, New York.

Peake, G. T., Steiner, A. L., and Daughaday, W. H., 1972, Guanosine 3′5′ cyclic monophosphate is a potent pituitary growth hormone secretagogue, *Endocrinology* **90**:212.

Rasmussen, H., and Goodman, D. B. P., 1977, Relationships between calcium and cyclic nucleotides in cell activation, *Physiol. Rev.* **57**:421.

Schally, A. V., Arimura, A., Kastin, A. J., Uehara, T., Coy, D. H., Coy, E. J., and Takahara, J., 1973, Inhibition of the release of growth hormone *in vitro* by α melanocyte stimulating hormone, *Biochem. Biophys. Res. Commun.* **52**:1314.

Schofield, J. G., and McPherson, M., 1974, Increase in pituitary adenosine 3′:5′-cyclic monophosphate content and potentiation of growth hormone release from heifer anterior pituitary slices incubated in the presence of 3-isobutyl-1-methylxanthine, *Biochem. J.* **142**:295.

Schofield, J. G., Mira-Moser, F., Schorderet, M., and Orci, L., 1974, Somatostatin inhibition of rat growth hormone release *in vitro* in the presence of $BaCl_2$ or 3-isobutyl-1-methylxanthine, *FEBS Lett.* **46**:171.

Shortman, K., 1968, The separation of different cell classes from lymphoid organs. II. The purification and analysis of lymphocyte populations by equilibrium density gradient centrifugation, *Aust. J. Exp. Biol. Med. Sci.* **46**:375.

Snyder, G., and Hymer, W. C., 1975, A short method for the isolation of somatotrophs from the rat pituitary gland, *Endocrinology* **96**:792.

Stachura, M. E., 1976, Influence of synthetic somatostatin upon growth hormone release from perifused rat pituitaries, *Endocrinology* **99**:678.

Steiner, A. L., Peake, G. T., Utiger, R. D., Karl, I. E., and Kipnis, D. M., 1970, Hypothalamic stimulation of growth hormone and thyrotropin release *in vitro* and pituitary 3′5′-adenosine cyclic monophosphate, *Endocrinology* **86**:1354.

Strauch, G., Girault, D., Rifai, M., and Bricaire, H., 1973, Alpha-MSH stimulation of growth hormone release, *J. Clin. Endocrinol. Metab.* **37**:990.

Sundberg, D. K., Fawcett, C. P., and McCann, S. M., 1976, The involvement of cyclic-3′,5′-AMP in the release of hormones from the anterior pituitary *in vitro*, *Proc. Soc. Exp. Biol. Med.* **151**:149.

Takahara, J., Arimura, A., and Schally, A. V., 1974, Effect of catecholamines on the TRH-stimulated release of prolactin and growth hormone from sheep pituitaries *in vitro*, *Endocrinology* **95**:1490.

Tashjian, A. H., Jr., Yasumura, Y., Levine, L., Sato, G. H., and Parker, M. L., 1968, Establishment of clonal strains of rat pituitary tumor cells that secrete growth hormone, *Endocrinology* **82**:342.

Trifaró, J. M., 1977, Common mechanisms of hormone secretion, *Annu. Rev. Pharmacol. Toxicol.* **17**:27.

Vale, W. Brazeau, P., Grant, G., Nussey, A., Burgus, R., Rivier, J., Ling, N., and Guillemin, R., 1972, Premières observations sur le mode d'action de la somatostatine, un facteur hypothalamique qui inhibe la sécrétion de l'hormone de croissance, *C. R. Acad. Sci. Ser. D.* **275**:2913.

Vale, W., Brazeau, P., Rivier, C., Brown, M., Boss, B., Rivier, J., Burgus, R., Ling, N., and Guillemin, R., 1975, Somatostatin, *Recent Prog. Horm. Res.* **31**:365.

Vale, W., Rivier, C., Brown, M., Chan, L., Ling, N., and Rivier, J., 1976, Applications of adenohypophyseal cell cultures to neuroendocrine studies, in: *Current Topics in Molecular Endocrinology* (F. Labrie, J. Meites, and G. Pelletier, eds.), Vol. 3, *Hypothalamus and Endocrine Functions,* pp. 397–429, Plenum Press, New York.

Wilber, J. F., Nagel, T., and White, W. F., 1971, Hypothalamic growth hormone–releasing activity (GRA): Characterization by the *in vitro* rat pituitary and radioimmunoassay, *Endocrinology* **89:**1419.

25

Pharmacological Studies of Pituitary Receptors Involved in the Control of Prolactin Secretion

Alain Enjalbert, Merle Ruberg, Sandor Arancibia, Mirette Priam, and Claude Kordon

1. Introduction

The major action of the hypothalamus on mammalian prolactin secretion is inhibitory. Ectopic pituitary grafts (Everett, 1954; Meites *et al.*, 1961), electrolytic lesion of the median eminence (Chen *et al.*, 1970; Bishop *et al.*, 1971) and *in vitro* incubation of adenohypophyseal tissue (Meites and Nicoll, 1966) result in increased prolactin secretion. It was postulated that the hypothalamus contained an unidentified factor that inhibited the release of prolactin at the hypophyseal level (Pasteels, 1961, 1967; Talwalker *et al.*, 1963) and that this factor could reach the pituitary directly across the hypothalamo–hypophyseal portal system. This was substantiated by experiments in which hypothalamic extracts infused into a portal blood vessel induced a dose-dependent inhibition of prolactin release (Kamberi *et al.*, 1971b). This inhibitory effect of hypothalamic extracts was found in many *in vivo* and *in vitro* preparations.

In addition to this strong prolactin-inhibiting activity, the existence of a prolactin-releasing factor (PRF) has been postulated, this stimulatory

Alain Enjalbert, Merle Ruberg, Sandor Arancibia, Mirette Priam, and *Claude Kordon* • Unite 159 de Neuroendocrinologie, Centre Paul Broca de l'INSERM, 75014 Paris, France

factor often being masked by the powerful repressive effect of the hypothalamus. Many physiological data support the hypothesis of the existence of a PRF. For example, sharp increases in plasma prolactin are known to follow stress or suckling. In other species like birds, such stimulatory effects appear to be even more important than the inhibitory regulation (Kragt and Meites, 1965; Gourdji and Tixier-Vidal, 1966).

We will first review the inhibitory effects of exogenous neurotransmitters and try to determine whether these factors can account for the prolactin-inhibiting activity of hypothalamic extracts or whether other factors may also exist. We will then discuss the problem of the prolactin-releasing factor(s).

2. Prolactin-Inhibiting Factors

2.1. Effect of Identified Neurotransmitters

2.1.1. Dopamine

The pharmacological studies discussed in this section have demonstrated that dopamine has an inhibitory role in the control of prolactin secretion. L-Dopa, the dopamine precursor, causes an increase in serum prolactin and in pituitary prolactin concentration (Lu and Meites, 1972). It has been demonstrated that this drug is not active by itself on prolactin release, but is active only after transformation into dopamine by decarboxylation in hypothalamic tissue (Jimenez et al., 1978). Injection of dopamine into the third ventricle reduces serum prolactin values (Kamberi et al., 1971a). Apomorphine, a dopaminergic agonist, has the same effect (Ojeda et al., 1974). Systemic administration of dopamine or apomorphine also decreases plasma prolactin levels (Vijayan and McCann, 1978). A wide variety of ergot derivatives decrease the serum prolactin concentration (Meites and Clemens, 1972). Injection of perphenazine, a dopamine antagonist, increases prolactin levels (Ben-David et al., 1970; Lu et al., 1970; MacLeod and Lehmeyer, 1974). Another dopamine antagonist, haloperiodol, also increases serum prolactin (Dickerman et al., 1974). Pimozide, but also phentolamine or propranolol, α- and β-adrenergic antagonists, respectively, reverse the inhibitory action of the ergot derivatives (Clemens et al., 1975).

It was originally postulated that dopamine increased the release of the hypothalamic factor responsible for the inhibitory control of prolactin secretion. However, we know now that as first postulated by Maanen and Smelik (1968), dopamine can be directly responsible for the neurohormonal control of prolactin secretion. In fact, dopamine can directly inhibit the release of prolactin from incubated pituitaries (MacLeod, 1969; Birge *et al.*, 1970; Koch *et al.*, 1970). Apomorphine also depresses prolactin release *in vitro* (MacLeod and Lehmeyer, 1974; Smalstig *et al.*, 1974). Bromocryptine (CB 154) directly inhibits prolactin release by pituitary, and this can be prevented by perphenazine or haloperidol (MacLeod and Lehmeyer, 1974). Pimozide alone has no effect on prolactin release by incubated hypophysis, but blocks the inhibitory action of both dopamine and apomorphine on prolactin release (MacLeod and Lehmeyer, 1974).

Arguments for a direct effect of dopamine on prolactin secretion at the pituitary level have also been obtained by *in vivo* experiments. L-Dopa decreases serum prolactin in rats with lesions of the median eminence (Donoso *et al.*, 1973). This drug decreases the elevated serum prolactin concentrations produced by pituitary grafts (Lu and Meites, 1972; Donoso *et al.*, 1974). Furthermore, when dopamine, dissolved in a glucose solution to avoid oxidation, is infused into a hypophyseal portal blood vessel, it causes a significant decrease in serum prolactin (Takahara *et al.*, 1974).

The dopamine inhibition of prolactin secretion from hemipituitaries *in vitro* is dose-dependent (Fig. 1). Even very low concentrations are effec-

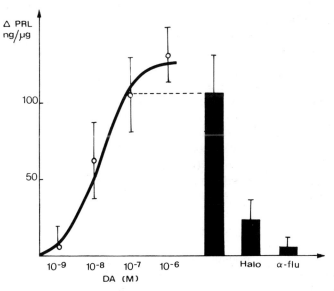

Figure 1. Effect of dopamine (DA) and dopaminergic antagonists [haloperidol (Halo) and α-flupentixol (α-flu) 10^{-6} M] on prolactin (PRL) release from incubated hemipituitaries. (Δ PRL: Difference between the ratios of medium/tissue + medium prolactin content of treated and corresponding control half-pituitaries).

tive. The apparent affinity constant is in the range of 10^{-8} M. Dopaminergic antagonists such as haloperiodol (Anden *et al.*, 1966) and α-flupentixol (Iversen, 1975) block the effect of dopamine (Fig. 1). Noradrenaline has the same effect as dopamine on prolactin secretion from pituitaries, but with a much lower efficiency (Shaar and Clemens, 1974). It probably acts through the dopamine receptor, as it has been shown to do *in vitro*, with dopamine receptors of the central nervous system (Bockaert *et al.*, 1977).

Dopamine receptors at the hypophyseal level have been identified. Dopamine binding sites were identified on pituitary membranes (Brown *et al.*, 1976; Calabro and MacLeod, 1978; Cronin *et al.*, 1978). This binding is antagonized by neuroleptics and bromoergocryptine (Calabro and MacLeod, 1978). Binding of dihydroergocryptine, a dopamine agonist, was also described. It is also antagonized by neuroleptics in the same way as central dopaminergic receptors (Labrie *et al.*, 1978; Cronin *et al.*, 1978). Binding experiments with haloperidol (Brown *et al.*, 1976) and spiroperidol (Creese *et al.*, 1977), two dopaminergic antagonists, have also been performed. Kinetic characteristics of these receptors are in agreement with those determined by *in vitro* experiments on prolactin secretion.

2.1.2. γ-Amino Butyric Acid (GABA)

Recent data have suggested that γ-amino butyric acid (GABA) may have an inhibitory effect on prolactin secretion (Schally *et al.*, 1977). However, some *in vivo* studies have shown an increase in plasma prolactin levels after GABA administration (Ondo and Pass, 1976; Rivier and Vale, 1977).

In our *in vitro* system, increasing concentrations of GABA induce a dose-dependent inhibition of prolactin release from incubated hemipituitaries (Fig. 2). However, the concentration producing half-maximal inhibition is 100 times higher in the case of GABA than in that of dopamine. In addition, the maximal inhibition obtained with GABA is much lower than that caused by dopamine (Fig. 2) (Enjalbert *et al.*, 1979b). Picrotoxin, a GABA antagonist (McLennan and York, 1973), blocks the prolactin inhibition produced by GABA (Enjalbert *et al.*, 1978, 1979b).

2.1.3. Distinct Receptors for Dopamine and GABA

α-Flupentixol, a potent dopamine receptor antagonist, has no effect on the GABA inhibition of prolactin secretion. In contrast, picrotoxin does not block the prolactin inhibition produced by dopamine (Enjalbert *et al.*, 1978, 1979b). Neither antagonist has any effect by itself on the secretion of prolactin. Thus, the effects of GABA and dopamine on prolactin secretion seem to be mediated through independent receptors, since there

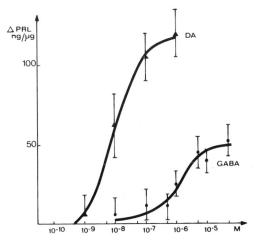

Figure 2. Effect of increasing concentrations of dopamine (DA) and GABA on *in vitro* prolactin release. From Enjalbert *et al.* (1979b).

is no cross-reaction between dopamine and the GABA antagonist and GABA and the dopamine antagonist.

2.2. Prolactin-Inhibiting Activity of Hypothalamic Extracts

2.2.1. Endogenous Dopamine is a Prolactin-Inhibiting Factor (PIF)

Direct evidence that tuberoinfundibular dopamine is physiologically important for prolactin regulation was further substantiated by experiments in which the hypothalamus was completely deafferented. Under these conditions, dopamine is the only amine left in the hypothalamic island (Weiner *et al.*, 1972). In parallel, plasma prolactin concentrations remain normal (Weiner, 1975). Furthermore, inhibition of dopamine synthesis still results in a marked elevation of prolactin secretion, as it does in intact animals (Weiner, 1975). Dopamine reaches the pituitary from tuberoinfundibular neurons that terminate in the median eminence adjacent to the portal blood vessels (Höckfelt, 1967). Recently, dopamine was detected in the portal blood, demonstrating that the amine can in fact reach the pituitary (Ben-Jonathan *et al.*, 1977) and that the concentration present is sufficient to inhibit prolactin release (Plotsky *et al.*, 1978; Gibbs and Neill, 1978). Thus, it is clear now that dopamine is a prolactin-inhibiting factor (PIF). However, catecholamine-free fractions of mediobasal hypothalamus retain prolactin-inhibiting activity (Greibrokk *et al.*, 1974, 1975; Schally *et al.*, 1976).

2.2.2. Nondopaminergic PIF

2.2.2a. Dopamine Does Not Account for All PIF Activity of Hypothalamic Extracts. The prolactin-inhibiting activity of mediobasal hypothalamus on crude extracts and after adsorption of catecholamine on aluminium oxide was tested. This procedure eliminates over 99% of the dopamine initially present in the fractions. The remaining concentration is below the sensitivity of a radioenzymatic assay (Enjalbert *et al.*, 1977a). Homogenate of three mediobasal hypothalamus strongly inhibits prolactin release. When catecholamines are adsorbed on aluminium oxide, the PIF activity of homogenates is only partially reduced. If dopamine were the only PIF present in the hypothalamus, we should expect that suppression of 99% of dopamine by alumina would completely abolish the prolactin-inhibiting activity. The existence of a nondopaminergic PIF is further confirmed by incubations in the presence of dopamine receptor antagonists

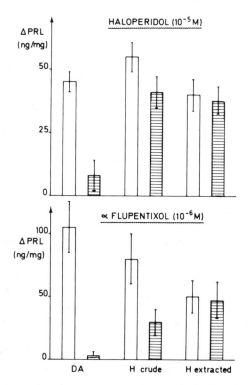

Figure 3. Effect of dopaminergic antagonists on prolactin-inhibiting activity of dopamine (DA) and crude or alumina-extracted homogenate (H) of mediobasal hypothalamus. From Enjalbert *et al.* (1977a).

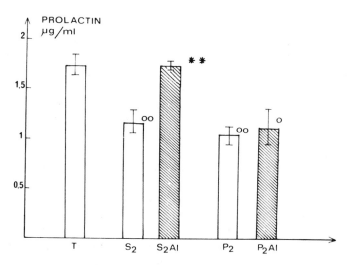

Figure 4. Effect of crude or alumina-extracted subcellular fractions of mediobasal hypothalamus on prolactin release from incubated hemipituitaries. (T) Total release in absence of hypothalamic extracts; (S_2) supernatant; (P_2) crude mitochondrial fraction; (A1) alumina-extracted. (o,oo) Significant and highly significant inhibition, respectively, as compared to control hemipituitaries; (**) Highly significant interaction of alumina treatment with inhibition of prolactin secretion.

(Enjalbert et al., 1977a,b). Haloperidol and α-flupentixol, which have been shown to antagonize completely the inhibition of prolactin release produced by dopamine, only partially antagonize the inhibition produced by crude hypothalamic extracts (Fig. 3). As after adsorption of catecholamine on alumina, this result indicates that dopamine accounts for only a part of the total PIF activity present in mediobasal hypothalamus. In fact, the remaining prolactin-inhibiting activity of homogenates after alumina adsorption is not antagonized at all by haloperidol or α-flupentixol, further indicating that this activity is not dopaminergic.

Quantification studies indicate that at least 50% of the total prolactin-inhibiting activity of the mediobasal hypothalamus is accounted for by something other than dopamine (Enjalbert et al., 1977a). The nondopaminergic activity also induces a dose-related inhibition of prolactin release.

2.2.2b. Regional and Subcellular Distribution of Nondopaminergic PIF. Incubation in the presence of both subcellular fractions from mediobasal hypothalamus, the supernatant and the crude mitochondrial fraction that contains synaptosomes, produces a significant and almost equal inhibition of prolactin secretion (Enjalbert et al., 1977a,b). After treatment with alumina, the PIF activity of the supernatant is completely abolished (Fig. 4). In contrast, a significant inhibition of prolactin release still results from incubations in the presence of the synaptosomal pellet.

These results indicate that dopamine does not account for the total prolactin-inhibiting activity of hypothalamus and that at least one other factor exists that is located only in nerve endings. The predominant synaptosomal distribution of this nondopaminergic PIF is the same as that of identified neurohormones such as LH-RH (Shin *et al.*, 1974; Ramirez *et al.*, 1975; Taber and Karavolas, 1974), thyrotropin-releasing hormone (TRH) (Barnea *et al.*, 1976), and SRIF (Epelbaum *et al.*, 1977), or of other nonidentified hypophysiotropic hypothalamic activity such as CRF (Briaud *et al.*, 1979).

To check the specificity of the data obtained with hypothalamic tissue, the prolactin-inhibiting activity of another dopamine-rich structure, the striatum, was tested (Enjalbert *et al.*, 1977a). The crude mitochondrial fraction prepared from striatum exhibited fair prolactin-inhibiting activity, but in contrast to the hypothalamus, the supernatant contained no significant prolactin-inhibiting activity. This subcellular distribution is parallel to that of endogenous dopamine in the same structure. After elimination of dopamine by alumina extraction, the synaptosomal pellet lost its prolactin-inhibiting activity, indicating that initial PIF activity was entirely due to dopamine in that structure.

In fact, a significant nondopaminergic PIF activity is present only in the mediobasal hypothalamus and to a lesser extent in the organum vasculosum lamina terminalis (Enjalbert *et al.*, 1977a).

2.2.3. Nature of the Nondopaminergic PIF

2.2.3a. GABA. We have seen that GABA presents some prolactin-inhibiting activity that could not be antagonized by neuroleptics. Does GABA account for the nondopaminergic activity of the hypothalamus? There are GABA neurons that terminate in the palisadic zone of the median eminence (Tappaz *et al.*, 1977). Mediobasal hypothalamus homogenate contained 90 ng GABA/equivalent. This represents, in our experimental conditions, 3×10^{-6} M GABA in incubation medium in the presence of three mediobasal hypothalamus equivalents (Enjalbert *et al.*, 1979b). Such a concentration is able to inhibit prolactin release *in vitro*.

Picrotoxin, a GABA antagonist, has only a slight effect on the inhibition of prolactin secretion induced by homogenate of mediobasal hypothalamus (Fig. 5). In addition, in the presence of α-flupentixol, picrotoxin is unable to antagonize the remaining PIF activity (Fig. 5), even though the concentration of picrotoxin used was previously shown to antagonize completely a concentration of exogenous GABA of the same order of magnitude as that contained in the extracts (Enjalbert *et al.*, 1979b). This indicates that GABA does not account for all nondopaminergic PIF activity.

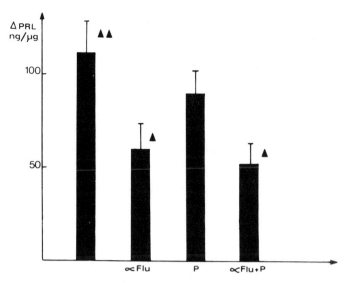

Figure 5. Effect of a dopaminergic antagonist [α-flupentixol (α Flu)] or a GABA antagonist [picrotoxin (P)] or both on the prolactin-inhibiting activity of homogenate of mediobasal hypothalamus. (▲▲) Highly significant inhibition as compared to corresponding control groups; (▲) significant inhibition as compared to the corresponding control and significant interaction with the MBH inhibitory effect. From Enjalbert *et al.* (1979b).

2.2.3b. Other PIFs.   There are, at this time, two identified neurotransmitters, dopamine and GABA, that inhibit prolactin secretion *in vitro* through different receptors, and the mediobasal hypothalamus contains at least one other factor that also acts through a separate mechanism. The nature of the remaining factor has yet to be identified. Recently, histidyl-proline-diketopiperazine, a degradation product of TRH, has been shown to have PIF activity *in vitro* (Bauer *et al.*, 1978). In our experimental conditions, diketopiperazine induces a dose-dependent inhibition of prolactin secretion from hemipituitaries. The apparent affinity is in the range of 10^{-9} M. Further work is needed to determine whether this product may represent a physiological PIF.

3. Prolactin-Releasing Factors

3.1. Thyrotropin-Releasing Hormone

The fact that TRH is able to release prolactin was first demonstrated with prolactin tumor cells (GH$_3$) in culture (Tashjian *et al.*, 1971; Gourdji *et al.*, 1966). It was then confirmed on dissociated pituitary cells (Nakano *et al.*, 1976) and on whole pituitaries (Mueller *et al.*, 1973) and also *in vivo*, but only in certain endocrine conditions (Blake, 1974; Noel *et al.*, 1974). *In vitro*, in conditions in which TRH is ineffective by itself, it partially overcomes the inhibitory effect of dopamine (Hill-Samli and MacLeod, 1974, 1975). TRH may not account for increased prolactin release under all physiological conditions, however, since the suckling stimulus does not increase TSH even though the sensitivity of the pituitary for TRH remains unchanged (Gautvik *et al.*, 1974). Furthermore, PRF activity distinct from TRH has been found in fractions of porcine hypothalamus extracts after chromatographic purification (Valverde *et al.*, 1972; Boyd *et al.*, 1976)

3.2. Vasoactive Intestinal Peptide

Vasoactive intestinal peptide (VIP), which is present in high concentration in the mediobasal hypothalamus (Besson *et al.*, 1978) and preferentially located in nerve endings (Besson *et al.*, 1978), induces a significant increase of prolactin release from incubated hemipituitaries (Ruberg *et al.*, 1978). This effect seems relatively hormone-specific, since VIP is unable to release LH from the same preparation (Ruberg *et al.*, 1978). Recent data indicate that VIP also increases prolactin secretion *in vivo* (Kato *et al.*, 1978). The fact that in our experimental conditions, as already reported, TRH has no effect on prolactin release (Blake, 1974; Hill-Samli and MacLeod, 1974; Noel *et al.*, 1974) suggests that VIP, a neuropeptide the regional and subcellular distribution of which is compatible with a neurohormonal function, could be a substance distinct from TRH that accounts for hypothalamic PRF activity.

3.3. Effect of Opiates on Prolactin Secretion at the Hypophyseal Level

Opiates have been shown to stimulate prolactin secretion *in vivo*. Morphine increases prolactin levels in normal (Bruni *et al.*, 1977) and steroid-primed male rats (Rivier *et al.*, 1977a) as well as in immature ones (Shaar *et al.*, 1977). Met-enkephalin and β-endorphin are also stimulatory (Bruni *et al.*, 1977; Chihara *et al.*, 1978; Dupont *et al.*, 1977; Grandison and Guidotti, 1977; Rivier *et al.*, 1977b). Naloxone and naltrexone, specific opiate antagonists, block these effects (Bruni *et al.*, 1977; Rivier *et al.*, 1977a,b; Shaar *et al.*, 1977). These antagonists also reduce prolactin levels when administered alone (Bruni *et al.*, 1977; Shaar *et al.*, 1977). None of these drugs was found active *in vitro* on anterior pituitaries (Shaar *et al.*,

1977) or dispersed pituitary cells (Rivier *et al.*, 1977a,b). It has thus been postulated that opiates regulate prolactin secretion at the level of the central nervous system.

3.3.1. Effect of Morphine

When the incubation medium contained both dopamine and morphine, morphine was able to block the prolactin-inhibiting activity of dopamine (Fig. 6). Naloxone antagonizes this effect of morphine (Fig. 6) (Enjalbert *et al.*, 1979a). The fact that morphine does not affect basal prolactin secretion but interacts with the effect of dopamine explains why other authors who looked only for direct effects concluded that the opiates do not act at the pituitary level.

3.3.2. Effect of Met-Enkephalin

The same results were obtained with met-enkephalin. Met-enkephalin alone has no effect, but it blocks the dopamine inhibition of prolactin secretion. This effect is antagonized by naloxone. Naloxone alone has no effect on dopamine inhibition of prolactin secretion (Enjalbert *et al.*, 1979c).

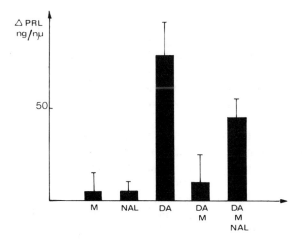

Figure 6. Interaction of morphine (M, 10^{-6} M) or naloxone (NAL, 10^{-5} M) or both with the basal and dopamine (DA, 10^{-7} M)-inhibited release of prolactin from incubated hemipituitaries.

3.3.3. Mechanism of Dopamine–Opiate Interaction

The mechanism of action of morphine and met-enkephalin is not yet established. However, the fact that naloxone antagonizes this effect but does not block dopamine inhibition of prolactin secretion indicates that two different receptors are involved, one for dopamine, the other an opiate-specific receptor.

On the basis of *in vivo* experiments, it has been proposed that opiates regulate prolactin secretion by modulating the release of dopamine through receptors on the dopamine neuron itself (Ferland *et al.*, 1977). We cannot exclude this hypothesis, but our results demonstrate that opiates modulate the response of the prolactin cell itself to dopamine.

Endogenous opiates seem to be implicated in the physiological regulation of prolactin secretion, since naloxone administration blocks the stress-induced prolactin release (Van Vugt *et al.*, 1978). The effect of opiates on dopamine inhibition of prolactin secretion at the hypophyseal level can explain this result.

In certain physiological situations, the increase in prolactin secretion can be explained by suppression of PIF release or effect as well as by PRF secretion. In interacting with the prolactin-inhibiting activity of dopamine, opiates may present such "PRF-like" activity.

4. Conclusion

The classic model for the regulation of adenohypophyseal hormone secretion was that all the exteroceptive and interoceptive input arrived in the hypothalamus, which integrates all this information and sends a simple message to the hypophysis through one or, in some cases, two neurohormones, one stimulatory, one inhibitory. The hypophysis simply responds to such an order by increasing or reducing hormone release according to the steroid environment.

It appears that the regulation of prolactin secretion is more complex. We have seen that many substances are able to modulate prolactin release at the hypophyseal level. It seems, then, that integration of various stimuli also takes place at the hypophyseal level. The interaction of dopamine and opiates represents an example of such integration.

Prolactin is a polyvalent hormone that is implicated in many physiological situations; its plasma levels fluctuate during circadian rhythms, the estrus cycle, gestation, suckling, fetal growth, stress, and different behavioral situations. It is not unreasonable, then, to suppose that in contrast to the regulation of other pituitary hormones with less diversified functions, different situations implicate different regulatory factors.

Discussion

DENEF: We have found a stimulatory effect of morphine on prolactin secretion that was suppressed by apomorphine in monolayer cultures of normal rat pituitary cells. Other opiates such as pentazocine, buprenorphine, and dextromoramide were also effective, whereas the inactive enantiomer levomoramide was ineffective in stimulating prolactin secretion.

ENJALBERT: This is in agreement with our results and further indicates that a specific opiate receptor is involved in the mechanism of this effect.

SPIES: Have you compared met-enkephalin and TRH together in your hemipituitary preparation? One might hypothesize that met-enkephalin plus TRH would be additive in the presence of marginal inhibitor levels of dopamine.

ENJALBERT: The male rat hemipituitary doesn't respond to TRH unless estrogen priming is used preceding the TRH treatment, and we haven't done those experiments yet.

VAUDRY: Did you perform time-course experiments with dopamine, GABA, and the unknown PIF compound? Our own observation is that dopamine does inhibit α-MSH release, but that the glands had not recovered normal secretion rates 2 hours after the end of the drug action.

ENJALBERT: We have not performed time-course experiments, but you are right, the best technique for such studies will be the superfusion system. I think this could provide information on the mechanism of action of each factor.

VAUDRY: Does GABA adsorb to AlO_2? Because if it does, it would further indicate that GABA cannot account for nondopamine PIF activity.

ENJALBERT: It does not.

GOURDJI: In GH_3B_6 cells, after a 15-minute incubation with VIP, we obtain a similar, very discrete stimulation of PRL release, the maximum stimulation being approximately 30% for VIP concentration of the order of 10^{-7} M, in the presence of IBMX.

TIXIER-VIDAL: What was the origin of the VIP that you used? What was the time of exposure and what was the maximum response?

ENJALBERT: The VIP was provided by Dr. Said of NIH. We used a 1-hour incubation after two 45-minute preincubations. The maximum effect was about 50%, but we had to perform dose–response curves to characterize this effect further.

KRAICER: Do you have any information on the structure of your third prolactin-inhibiting factor? Is it a peptide or a biogenic amine?

ENJALBERT: We have no direct information. However, the exclusively synatosomal localization of the third PIF is the same as that of the identified neurohormones. This could be an indirect indication that this third factor is a peptide.

JONES: Have you tested the effect of various peptidases on the remaining inhibiting-factor activity after alumina treatment? Also, has anyone found GABA to be present in portal blood?

ENJALBERT: We have not yet tested the effect of peptidase on hypothalamic extracts. To my knowledge, no one has assayed GABA in portal blood.

FINK: I assume that your comment regarding dopamine in portal vessel blood refers to the paper of Porter and co-workers (*Endocrinology* **100**:452–458, 1977). If you look carefully at their data, you will see that the dopamine levels were low in the male rats and high in late pregnancy. The only vague correlation between prolactin and dopamine levels was in cycling rats. So it is difficult to use portal vessel data to suggest that dopamine is a physiological PIF.

ENJALBERT: The portal vessel data indicate that dopamine can reach the hypophysis in physiological conditions, and concentrations present are able to inhibit prolactin release, according to the apparent affinity of hypophyseal dopamine receptors. The fact that there is not very good correlation between dopamine levels in portal blood and prolactin secretion indicates that dopamine is not the only factor which regulates prolactin secretion at the hypophyseal level.

LABRIE: Did you study various concentrations of your nondopamine compounds and α-flupentixol on PRL release? An excess could prevent the action of 10^{-6} M α-flupentixol. Also, did you consider the possibility of proteolytic activity in your extracts?

ENJALBERT: We checked various concentrations of antagonists to block the dopamine effect. We avoided high concentrations that may have nonspecific effects. The concentration of dopamine in the incubation medium was about 10^{-7} M in the presence of three mediobasal hypothalami. α-Flupentixol at 10^{-6} M is able to block completely such amounts of dopamine.

We used acidic extracts of the mediobasal hypothalamus, so there is no proteolytic activity in our system.

REFERENCES

Anden, N. E., Dahlstrom, A., Fuxe, K., and Höckfelt, T., 1966, The effect of haloperidol and chlorpromazine on the amine levels of central monoamine neurons, *Acta. Physiol. Scand.* **68**:419.

Barnea, A., Ben-Jonathan, N., and Porter, J. C., 1976, Characterization of hypothalamic subcellular particles containing luteinizing hormone releasing hormone and thyrotropin releasing hormone, *J. Neurochem.* **27**:477.

Bauer, K., Graf, K. J., Faivre-Bauman, A., Beier, S., Tixier-Vidal, A., and Kleinkauf, H., 1978, Inhibition of prolactin secretion by histidyl proline diketopiperazine, *Nature (London)* **274**:174.

Ben-David, M., Danon, A., and Sulman, F. G., 1970, Acute changes in blood and pituitary prolactin after a single injection of perphenazine, *Neuroendocrinology* **6**:336.

Ben-Jonathan, N., Olivier, C., Weiner, H. J., Mical, R. S., and Porter, J. C., 1977, Dopamine in hypophyseal portal plasma of the rat during the estrous cycle and throughout pregnancy, *Endocrinology* **100**:452.

Besson, J., Rotsztejn, W. H., Laburthe, M., Epelbaum, J., Beaudet, A., Kordon, C., and Rosselin, G., 1979, Vasoactive intestinal peptide (VIP): Brain distribution, subcellular localization and effect of deafferentation of the hypothalamus in male rats, *Brain Res.* **765**:79.

Birge, C. A., Jacobs, L. S., Hammer, C. T., and Daughaday, W. H., 1970, Catecholamine inhibition of prolactin secretion by isolated rat adenohypophysis, *Endocrinology* **86**:120.

Bishop, W., Krulich, L., Fawcett, L. P., and Mc Cann, S. M., 1971, The effect of median eminence (ME) lesions on plasma levels of FSH, LH and prolactin in the rat, *Proc. Soc. Exp. Biol. Med.* **136**:925.

Blake, C. A., 1974, Stimulation of pituitary prolactin and TSH release in lactating and proestrous rats, *Endocrinology* **94**:503.

Bockaert, J., Premont, J., Glowinski, J., Tassin, J. P., and Thierry, A. M., 1977, Topographical distribution and characteristics of dopamine and β adrenergic sensitive adenylate cyclase in the rat frontal cerebral cortex, striatum and substantia nigra, in *Advances in Biochemical Psychopharmacology*, Vol. 16 (E. Costa and G. L. Gessa, eds.), pp. 29–37, Raven Press, New York

Boyd, A. E., III, Spencer, E., Jackson, I. M. D., and Reichlin, S., 1976, Prolactin-releasing-factor (PRF) in porcine hypothalamic extracts distinct from TRH, *Endocrinology* **99**:861.

Briaud, B., Enjalbert, A., Miale, C., and Kordon, C., 1979, Subcellular distribution of corticotropin releasing factor (CRF) in the medio-basal hypothalamus of the rat, *Neuroendocrinology* **28**:371.

Brown, G. M., Seeman, P., and Lee, T., 1976, Dopamine, neuroleptic receptors in basal hypothalamus and in pituitary, *Endocrinology* **99**:1407.

Bruni, J. F., Van Vugt, D., Marshall, S., and Meites, J., 1977, Effects of naloxone, morphine and methionine enkephalin on serum prolactin, luteinizing hormone, follicle stimulating hormone, thyroid stimulating hormone and growth hormone, *Life Sci.* **21**:461.

Calabro, M. A., and MacLeod, R. M., 1978, Binding of dopamine to bovine anterior pituitary gland membranes, *Neuroendocrinology* **25**:32.

Chen, C. L., Amenomori, Y., Lu, K. H., Voogt, J. L., and Meites, J., 1970, Serum prolactin levels in rats with pituitary transplants or hypothalamic lesions, *Neuroendocrinology* **6**:220.

Chihara, K., Arimura, A., Coy, D. H., and Schally, A. V., 1978, Studies on the interaction of endorphin, substance P and endogenous somatostatin in growth hormone and prolactin release in rats, *Endocrinology* **102**:281.

Clemens, J. A., Smalstig, E. B., and Shaar, C. J., 1975, Inhibition of prolactin secretion by lergotrile mesylate: Mechanism of action, *Acta Endocrinol.* **70**:230.

Creese, I., Schneider, R., and Snyder, S. H., 1977, ^{3}H spiroperidol labels dopamine receptors in pituitary and brain, *Eur. J. Pharmacol.* **46**:377.

Cronin, M. J., Roberts, J. M., and Weiner, R. I., 1978, Dopamine and dihydroergocryptine binding to the anterior pituitary and other brain areas of the rat and sheep, *Endocrinology* **103**:302.

Dickerman, S., Kledzik, G., Gelato, M., Chen, H. J., and Meites, J., 1974, Effects of haloperidol on serum and pituitary prolactin LH and FSH and hypothalamic PIF and LRF, *Neuroendocrinology* **15**:10.

Donoso, A. O., Bishop, W., and McCann, S. M., 1973, The effects of drugs which modify catecholamine synthesis on serum prolactin in rats with median eminence lesions, *Proc. Soc. Exp. Biol. Med.* **143**:360.

Donoso, A. O., Banzan, A. M., and Barcaglioni, J. C., 1974, Further evidence on the direct action of L-dopa on prolactin release, *Neuroendocrinology* **15**:236.

Dupont, A., Cusan, L., Labrie, F., Coy, D. H., and Hao-Li, C., 1977, Stimulation of prolactin release in the rat by intraventricular injection of β endorphin and methionine enkephalin, *Biochem. Biophys. Res. Commun.* **75**:76.

Enjalbert, A., Moos, F., Carbonell, L., Priam, M., and Kordon, C., 1977a, Prolactin inhibiting activity of dopamine free subcellular fractions from rat medio-basal hypothalamus, *Neuroendocrinology* **24**:147.

Enjalbert, A., Priam, M., and Kordon, C., 1977b, Evidence in favour of the existence of a dopamine free prolactin inhibiting factor (PIF) in rat hypothalamic extracts, *Eur. J. Pharmacol.* **41**:243.

Enjalbert, A., Ruberg, M., Fiore, L., and Kordon, C., 1978, Dopamine, GABA, PIF and *in vitro* regulation of prolactin secretion in the rat, *Ann. Endocrinol.* **39**:237.

Enjalbert, A., Ruberg, M., Fiore, L., Arancibia, S., Priam, M., and Kordon, C., 1979a,

Effect of morphine on the dopamine inhibition of pituitary prolactin release *in vitro*, *Eur. J. Pharmacol.* **53**:212.

Enjalbert, A., Ruberg, M., Fiore, L., Arancibia, S., Priam, M., and Kordon, C., 1979b, Independent inhibition of prolactin secretion by dopamine and gamma-amino-butyric acid *in vitro*, *Endocrinology* **105**:823.

Enjalbert, A., Ruberg, M., Arancibia, S., Priam, M., and Koron, C., 1979c, Endogenous opiates block dopamine inhibition of prolactin secretion *in vitro*, *Nature* **280**:595.

Epelbaum, J., Brazeau, P., Tsang, D., Brawer, J., and Martin, J. B., 1977, Subcellular distribution of radioimmunoassayable somatostatin in rat brain, *Brain Res.* **126**:309.

Everett, J. W., 1954, Luteotrophic function of autografts of the rat hypophysis, *Endocrinology* **54**:685.

Ferland, L., Fuxe, K., Eweroth, P., Gustafsson, J. A., and Skett, P., 1977, Effects of methionine-enkephalin on prolactin release and catecholamine levels and turnover in the median eminence, *Eur. J. Pharmacol.* **43**:89.

Gautvik, K. M., Tashjian, A. H., Jr, Kourides, I. A., Weintraub, B. D., Graeber, C. T., Maloof, F., Suzuki, K., and Zuckerman, J. E., 1974, Thyrotropin-releasing hormone is not sole physiologic mediator of prolactin release during suckling, *N. Engl. J. Med.* **290**:1162.

Gibbs, D. M., and Neill, J. D., 1978, Dopamine levels in hypophyseal stalk blood in the rat are sufficient to inhibit prolactin secretion *in vivo*, *Endocrinology* **102**:1895.

Gourdji, D., and Tixier-Vidal, A., 1966, Mise en évidence d'un contrôle hypothalamique stimulant de la prolactine hypophysaire chez le canard, *C. R. Acad. Sci.* **263**:162.

Gourdji, D., Kerdelhue, B., and Tixier-Vidal, A., 1966, Mise en évidence d'un clone de cellules hypophysaires secretant de la prolactine (clone GH₃): Modification induite par l'hormone de libération de l'hormone thyrotrope TRF, *C. R. Acad. Sci.* **274**:437.

Grandison, L., and Guidotti, A., 1977, Regulation of prolactin release by endogenous opiates, *Nature (London)* **270**:357.

Greibrokk, T., Currie, B. L., Johansson, K. N. G., Hansen, J. J., and Foikers, K., 1974, Purification of a prolactin inhibiting hormone and the revealing of hormone D-GHIH which inhibits the release of growth-hormone, *Biochem. Biophys. Res. Commun.* **59**:704.

Greibrokk, T., Hansen, J., Knudsen, R., Lam, Y. K., and Folkers, K., 1975, On the isolation of a prolactin inhibiting factor (hormone), *Biochem. Biophys. Res. Commun.* **67**:338.

Hill-Samli, M., and MacLeod, R. M., 1974, Interaction of thyrotropin-releasing hormone and dopamine on the release of prolactin from the rat anterior pituitary *in vitro*, *Endocrinology* **95**:1189.

Hill-Samli, M., and MacLeod, R. M., 1975, TRH blockade of the ergocryptine and apomorphine inhibition of prolactin release *in vitro*, *Proc. Soc. Exp. Biol. Med.* **149**:511.

Höckfelt, T., 1967, The possible ultrastructural identification of tuberoinfundibular dopamine-containing nerve-endings in the median eminence of the rat, *Brain Res.* **5**:121.

Iversen, L. L., 1975, Dopamine receptors in the brain, *Science* **4193**:1084.

Jimenez, A. E., Voogt, J. L., and Carr, L. A., 1978, L-3,4-Dihydroxyphenylalanine (L-dopa) as an inhibitor of prolactin release, *Endocrinology* **102**:166.

Kamberi, I. A., Mical, R. S., and Porter, J. L., 1971a, Effect of anterior pituitary perfusion and intraventricular injection of catecholamines on prolactin release, *Endocrinology* **88**:1012.

Kamberi, I. A., Mical, R. S., and Porter, J. L., 1971b, Pituitary portal vessels infusion of hypothalamic extract and release of LH, FSH and prolactin, *Endocrinology* **88**:1294.

Kato, Y., Iwasaki, Y., Iwasaki, J., Abe, H., Yanaihara, N., and Imura, H., 1978, Prolactin release by vasoactive intestinal polypeptide in rats, *Endocrinology* **103**:554.

Koch, Y., Lu, K. H., and Meites, J., 1970, Biphasic effect of catecholamines on pituitary prolactin release *in vitro*, *Endocrinology* **87**:673.

Kragt, C., and Meites, J., 1965, Stimulation of pigeon pituitary prolactin release by pigeon hypothalamic extracts *in vitro*, *Endocrinology* **76**:1169.

Labrie, F., Beaulieu, M., Caron, M. G., and Raymond, V., 1978, The adenohypophyseal dopamine receptor specificity and modulation of its activity by estradol, in: *Progress in Prolactin Physiology and Pathology* (C. Robyn and M. Harter, eds.), pp. 121–136, Elsever North-Holland, Amsterdam.

Lu, K. H., and Meites, J., 1972, Effects of L-dopa on serum prolactin and PIF in intact and hypophysectomized pituitary grafted rats, *Endocrinology* **91**:868.

Lu, K. H., Amenomori, Y., Chen, C., and Meites, J., 1970, Effects of central acting drugs on serum and pituitary prolactin levels in rats, *Endocrinology* **87**:667.

Maanen, J. H., and Smelik, P. G., 1968, Induction of pseudopregnancy in rat following local depletion of monoamine in the median eminence of the hypothalamus, *Neuroendocrinology* **3**:177.

MacLeod, R.M., 1969, Influence of norepinephrine and catecholamines depleting agents on the synthesis and release of prolactin and growth hormone, *Endocrinology* **85**:916.

MacLeod, R. M., and Lehmeyer, J. E., 1974, Studies on the mechanism of dopamine-mediated inhibition of prolactin secretion, *Endocrinology* **94**:1077.

McLennan, M., and York, D. H., 1973, Gamma-aminobutyric acid antagonists in crustacea, *Can. J. Physiol. Pharmacol.* **51**:774.

Meites, J., and Clemens, J., 1972, Hypothalamic control of prolactin secretion, *Vitam. Horm. (N.Y.)* **30**:165.

Meites, J., and Nicoll, C. S., 1966, Adenohypophyseal prolactin, *Annu. Rev. Physiol.* **28**:57.

Meites, J., Kahn, R. H., and Nicoll, C. S., 1961, Prolactin production by rat pituitary explants *in vitro*, *Proc. Soc. Exp. Biol. Med.* **108**:440.

Mueller, G. P., Chen, H. J., and Meites, J., 1973, *In vivo* stimulation of prolactin release in the rat by synthetic TRH, *Proc. Soc. Exp. Biol. Med.* **144**:613.

Nakano, H., Fawcett, C. P., and McCann, S. M., 1976, Enzymatic dissociation and short-term culture of isolated rat anterior pituitary cells for studies on the control of hormone secretion, *Endocrinology* **98**:278.

Noel, G. L., Dimond, R. C., Wartofsky, L., Earll, J. M., and Frantz, A. G., 1974, Studies of prolactin and FSH secretion by continuous infusion of small amounts of thyrotro-phin-releasing hormone (TRH), *J. Clin. Endocrinol. Metab.* **39**:6.

Ojeda, S. R., Harms, P. G., and McCann, S. M., 1974, Effect of blockade of dopaminergic receptors on prolactin and LH release: Median eminence and pituitary sites of action, *Endocrinology* **94**:1650.

Ondo, J. G., and Pass, K. A., 1976, The effects of neurally active amino-acids on prolactin secretion, *Endocrinology* **98**:1248.

Pasteels, J. L., 1961, Premiers résultats de cultures combinées *in vitro* d'hypophyses et d'hypothalamus dans le but d'apprécier la production de prolactine, *C. R. Acad. Sci.* **253**:3074.

Pasteels, J. L., 1967, Contrôle de la secrétion de prolactine par le systéme nerveux, *Arch. Anat. Microsc. Morphol. Exp.* **56**:530.

Plotsky, P. M., Gibbs, D. M., and Neill, J. D., 1978, Liquid chromatographic electrochem-ical measurement of dopamine in hypophyseal stalk blood of rats, *Endocrinology* **102**:1877.

Ramirez, V. D., Gautron, J. P., Epelbaum, J., Pattou, E., Zamora, A., and Kordon, C., 1975, Distribution of LH-RH in subcellular fractions of medio-basal hypothalamus, *Mol. Cell. Endocrinol.* **3**:339.

Rivier, C., and Vale, W., 1977, Effects of gamma-aminobutyric acid and histamine on prolactin secretion in the rat, *Endocrinology* **101**:506.

River, C., Brown, M., and Vale, W., 1977a, Effect of neurotensin, substance P and morphine sulfate on the secretion of PRL and growth hormone in the rat, *Endocrinology* **100**:751.

River, C., Vale, W., Ling, N., Brown, M., and Guillemin, R., 1977b, Stimulation *in vivo* of the secretion of prolactin and growth hormone by β endorphin, *Endocrinology* **100**:238.

Ruberg, M., Rotsztejn, W. H., Arancibia, S., Besson, J., and Enjalbert, A., 1978, Stimulation of prolactin release by vasoactive intestinal peptide (VIP), *Eur. J. Pharmacol.* **51**:319.

Schally, A. V., Dupont, A., Arimura, A., Takahara, J., Redding, T. W., Clemens, J., and Shaar, C., 1976, Purification of a catecholamine-rich fraction with prolactin release inhibiting factor (PIF) activity from porcine hypothalami, *Acta. Endocrinol.* **82**:1.

Schally, A. V., Redding, T. W., Arimura, A., Dupont, A., and Linthicum, G. L., 1977, Isolation of gamma-aminobutyric acid from pig hypothalami and demonstration of its prolactin release inhibiting (PIF) activity *in vivo* and *in vitro*, *Endocrinology* **100**:681.

Shaar, C. J., and Clemens, J. A., 1974, The role of catecholamines in the release of anterior pituitary prolactin *in vitro*, *Endocrinology* **95**:1202.

Shaar, C. J., Frederickson, R. C. A., Dininger, N. B., and Jackson, L., 1977, Enkephalin analogues and naloxone modulate the release of growth hormone and prolactin: Evidence for regulation by an endogenous opioid peptide in brain, *Life Sci.* **21**:853.

Shin, S. M., Morris, A., Snyder, J., Hymes, W. C., and Milligan, J. V., 1974, Subcellular localization of LH-RH in the rat hypothalamus, *Neuroendocrinology* **16**:191.

Smalstig, E. B., Sawyer, B. D., and Clemens, J. A., 1974, Inhibition of rat prolactin release by apomorphine *in vivo* and *in vitro*, *Endocrinology* **95**:123.

Taber, C. A., and Karavolas, H. J., 1974, Subcellular localization of LH releasing activity in the rat hypothalamus, *Endocrinology* **96**:446.

Takahara, J., Arimura, A., and Schally, A. V., 1974, Suppression of prolactin release by a purified porcine PIF preparation and catecholamines infused into a rat hypophyseal portal vessel, *Endocrinology* **95**:462.

Talwalker, P. K., Ratner, A., and Meites, J., 1963, *In vitro* inhibition of pituitary prolactin synthesis and release by hypothalamic extracts, *Am. J. Physiol.* **205**:213.

Tappaz, M. L., Brownstein, M. J., and Kopin, I. J., 1977, Glutamate decarboxylase (GAD) and gamma aminobutyric acid (GABA) in discrete nuclei of hypothalamus and substantia nigra, *Brain Red.* **125**:109.

Tashjian, A. H., Barousky, N. J., and Jensen, D. K., 1971, Thyrotropin releasing hormone: Direct evidence for stimulation of prolactin production by pituitary cell in culture, *Biochem. Biophys. Res. Commun.* **43**:516.

Valverde, R. C., Chieffo, V., and Reichlin, S., 1972, Prolactin releasing factor in porcine and rat hypothalamic tissue, *Endocrinology* **91**:982.

Van Vugt, D. A., Bruni, J. F., and Meites, J., 1978, Naloxone inhibition of stress-induced increase in prolactin secretion, *Life Sci.* **22**:85.

Vijayan, E., and McCann, S. M., 1978, The effect of systemic administration of dopamine and apomorphine on plasma LH and prolactin concentrations in conscious rats, *Neuroendocrinology* **25**:221.

Weiner, R. I., 1975, Role of brain catecholamines in the control of LH and prolactin secretion, in: *Hypothalamic Hormones* (M. Motta, P. G. Grosignani, and L. Martini, eds.), p. 249, Academic Press, London.

Weiner, R. I., Shryne, J. E., Gorski, R. A., and Sawyer, C. H., 1972, Changes in the catecholamine content of the rat hypothalamus following deafferentation, *Endocrinology* **90**:867.

26

The Release of ACTH by Human Pituitary and Tumor Cells in a Superfusion System

G. H. Mulder and D. T. Krieger

1. Introduction

Investigators in the field of biochemical endocrinology may find themselves in a difficult position when they face a decision whether to perform *in vivo* or *in vitro* experiments and when comparing data from these tests. In pituitary hormone studes, there always has been and there always will be the dilemma of either (1) leaving the pituitary *in situ*, where one does not know exactly which of many possible factors could affect pituitary function but where one is certain of the "physiology" of the experiment, or (2) removing the gland to an *in vitro* setup, where the experimenter decides exactly which substances or factors he wants pituitary function to be affected by, but where there will always be doubts about how closely the experimental conditions approach "normal physiology." In view of the many difficulties encountered in setting up well-controlled *in vivo* experiments or in the interpretation of their results, we thought it desirable to perform *in vitro* experiments accompanied by as many *in vivo* parallels as possible and feasible. We have therefore devoted some time recently to

G. H. Mulder and *D. T. Krieger* ● Department of Medicine, Division of Endocrinology, Mount Sinai School of Medicine, New York, New York 10029. Dr. Mulder's present address is Department of Pharmacology, Free University Amsterdam, Amsterdam, The Netherlands.

the development of *in vitro* experimental procedures that we think approximate the *in vivo* situation to a substantial degree.

For the study of pituitary hormone regulation, the approach we have chosen is the cell-column superfusion technique, introduced by Lowry in 1974 (Lowry, 1974). Detailed information on this technique and ample justification of the method can be found elsewhere (Mulder, 1975; Mulder and Smelik, 1977; Gillies and Lowry, 1978). Briefly, cells from acutely dispersed anterior pituitary tissue are trapped in a small chamber at 37°C and bathed in a constant stream of fresh medium to which any agent can be added during well-defined periods of time. The released products of the cells are constantly removed and are collected at fixed intervals and later assayed for hormone content.

The great majority of previous reports on adrenocorticotropic hormone (ACTH) secretion have been carried out with animal pituitary tissue or cells. In this chapter, it has been our purpose to (1) analyze the characteristics of ACTH secretion by human pituitary cells and (2) determine the role(s) of target-cell hormones in the regulation of ACTH release. We have first analyzed the behavior of normal human pituitary cells and then that of pituitary tumor cells.

2. Materials and Methods

2.1. Tissue Collection and Cell Preparation

Human pituitaries were obtained at autopsies performed 7–24 hr postmortem on persons who died of accidental causes and who had no obvious endocrine abnormalities or known histories of drug abuse. Tissue was immersed in a few milliliters of M-199 buffer (Flow Labs, Rockville, Maryland) and immediately taken to the laboratory. We were also in a postion to study two pituitary tumors, from female patients undergoing hypophysectomy. Both patients, 31 and 37 years old, had chromophobe, prolactin-producing adenomas. In addition, we studied the anterior pituitary cells from a woman with breast cancer who also underwent hypophysectomy. We considered the pituitary of this patient to yield essentially normal cells.

At all times, care was taken to maintain specimens at room temperature, since cooling can lead to nonspecific hormone release (Portanova, 1972; LaBella *et al.*, 1973; Hong and Poisner, 1974; Hymer *et al.*, 1976). Anterior and posterior lobe tissues (in the normal human adult pituitary, there is no intermediate lobe) were carefully separated. Pituitary-tumor tissue was treated identically to postmortem tissue, with the exception that the appearance of the tumor tissue usually made it impossible to definitely exclude some posterior lobe "contamination," although none of the patients developed diabetes insipidus postoperatively. We are unable to state how many, if any, normal cells were present in the tumor-cell preparations.

The anterior lobe tissue was decapsulated and cut in 1 mm^3 fragments and collected in a plastic beaker containing 20 ml of medium M-199 with 5 mg/ml bovine serum albumin (BSA) (Sigma, Fraction V) and 3 mg/ml crude collagenase (Sigma, Type I). The mixture was incubated at 37°C under humidified 95% O_2/5% CO_2 in the plastic beaker for 30–35 min, after which there were no visual changes in the appearance of the tissue cubes. The fragments could, however, easily be dispersed at this point into single cells by repeatedly sucking them up into and expelling them from a large-bore 5-ml serological pipet, followed by a narrower 1-ml Eppendorf-type pipet.

After 25–40 pipet excursions, most of the tissue had fallen apart; with the tumors, a second collagenase incubation was usually necessary to disintegrate the specimen completely. Any fragments remaining after the first dispersal were allowed to settle and were then reincubated with fresh collagenase solution for another 30 min, after which the pipetting procedure was repeated. The combined cell suspensions were centrifuged at room temperature for 10 min at 100*g*. The resulting cell pellet was resuspended in 0.5–1.0 ml M-199 with 5 mg/ml BSA, and a portion was diluted for counting and viability determination (Trypan Blue exclusion). Up to 3.0 × 10^6 viable cells (in up to 0.5 ml) were placed into one or more chambers of the superfusion apparatus by gently layering them on a bed of previously swollen Sephadex G-10 beads (Pharmacia) that had already been introduced into the superfusion chambers. Portions of the cell suspension were diluted with 0.1 N HCl and frozen for later assay.

2.2. Superfusion Apparatus

The superfusion apparatus (Fig. 1) consists of a number of cut-off 1-ml plastic disposable syringes, mounted vertically in a Plexiglas holder that is kept at 37°C by circulating water. The unit was constructed in the laboratory workshop. Properly greased O-rings prevent leakage of the circulating water around the syringe barrels. Each barrel is fitted with plungers at both ends. The rubber heads of the plungers are perforated with plastic tubing, the ends of which are flush with the rubber. The lower plunger head is covered with a small piece of 30-μm-pore nylon gauze (ZBF, Zurich) to keep the Sephadex beads from escaping. The "pores" between the beads are small enough to prevent the pituitary cells from escaping and large enough to allow practically unrestricted flow of medium through the cell-column. Varying the distance between upper and lower

flow rate : 0.5 ml/min

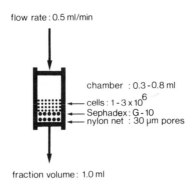

chamber : 0.3 - 0.8 ml
cells : 1 - 3 x 10^6
Sephadex : G - 10
nylon net : 30 μm pores

fraction volume : 1.0 ml

Figure 1. Schematic drawing of a superfusion column. The heavy lines represent the syringe barrel, mounted in a 37°C waterbath. The small dots represent the pituitary cells, resting on top of the Sephadex G-10 beads (large dots). A nylon net covers the outlet of the column and prevents the beads and cells from escaping.

plunger allows one to adjust the chamber volume between 0.3 and 0.8 ml. In the simple pituitary cell-column setup, the tubing from the lower plunger drips into disposable plastic tubes already containing 30 μl 1 N HCl for immediate acidification of the perfusate, in view of the loss of ACTH activity at neutral or alkaline pH (Swallow and Sayers, 1969; Mulder, 1975; Uemura *et al.*, 1976). The upper plunger tubing is connected to a multichannel peristaltic pump (Gilson type HP-4) and from there to a four-way valve (Pharmacia LV-4) that allows one to choose between buffer medium and test substances to be introduced into the cell chambers. Buffers are kept under a 95% O$_2$/5% CO$_2$ gas mixture at 37°C during the entire experiment. Test samples, dissolved in M-199 medium, with 5 mg/ml BSA, are equilibrated at 37°C for 10 min prior to use.

2.3. Experimental Protocols

After removal of the upper plunger from the syringe barrel, up to 3.0 × 10^6 freshly dispersed, viable pituitary cells are layered on the Sephadex beads in the chambers, which are then carefully filled up with buffer. The top plunger is then reinserted slowly and gently pressed down to the desired position; switching on the peristaltic pump initiates the presuperfusion period that lasts 90 min. Assay results from 5 ml (= 10 min) fractions collected during the presuperfusion showed that a 30- to 45-min period was necessary before ACTH secretion had reached a low and steady level. To assure stable baseline values, we extended the presuperfusion period to 90 min before starting to collect fractions every 2 min. The flow rate throughout the whole experiment is 0.5 ml/min. This is considerably lower than the "free-flowing" rate (i.e., open syringe with

Sephadex and cells that is allowed to drip freely without a pump), which is always at least twice as high. After the presuperfusion period, the superfusion proper is started by collecting 1-ml fractions of the cell-column effluent every 2 min into the acid-containing tubes. We choose not to use a fraction collector to keep the physical size of the setup small and the lag time of the system minimal.

For stimulation of ACTH release by the human pituitary cells, we used a corticotropin-releasing factor (CRF)-containing extract of rat hypothalamic median eminence (HME), kindly provided by the NIH. This crude extract (supplied as lyophilized material in 200-mg batches) was redissolved in water at a concentration of 80 mg/ml, yielding a solution containing 1 rat HME equivalent/10 μl. This was aliquoted in 50-μl portions, refrozen, and kept in the $-20°C$ freezer for extended periods of time (up to at least 9 months) without an apparent loss of activity.

Previous rat pituitary cell experiments had revealed that a 30-sec pulse of NIH-HME extract caused increased ACTH release, starting 6–12 sec after the extract had reached the cells, and lasting for periods of 4–6 min, i.e., during the two or three fractions collected following the administration of such a pulse (Mulder, 1975; Mulder *et al.,* 1976; Mulder and Smelik, 1977). Preliminary studies with human pituitary cells, where we exposed the cells to one 30-sec pulse of HME extract and then followed the cells' response during the next 3 hr, had shown them also to release increased amounts of ACTH in the following two or three fractions. No significant deflections from baseline values were detected during the ensuing 3-hr period.

Pulses of the CRF-containing extract are always initiated at a moment exactly halfway through the collection of a 2-min fraction. In this manner, the "mechanical" lag time of the system (the period between the moment of switching the LV-4 valve and the moment the cells are exposed to the extract) makes it impossible for that particular fraction to show the effect of the substance on the cells, and usually causes the following fraction to exhibit the strongest response to the factors contained in the pulse.

Inhibition of hormone release was brought about by exposing the cells to a glucocorticoid, such as cortisol, at 0.5 μg/ml. Previous rat-cell experiments had shown (Mulder, 1975; Mulder and Smelik, 1977) that the inhibition of ACTH release caused by corticosterone did not become evident until 20–25 min after the start of the steroid infusion. Thus, we postponed stimulation of the human cells by HME extract in the presence of cortisol until 25 min after the initiation of the steroid exposure. We have not yet had the opportunity to analyze in detail the rapidity of the onset of the cortisol inhibition of ACTH release in the human cells.

After completion of a superfusion experiment, multiple aliquots were taken from each fraction, transferred to fresh tubes, and frozen at $-20°C$

for later assay. For the determination of cellular hormone content after a run, 0.1 N HCl was pumped through the cell-column at the normal flow rate for a period of 10 min. This caused the cells to release virtually all their stored hormone in the 5 min of such an "extraction." Some remaining ACTH (<5% of the total amount) could occasionally be retrieved by removing cells plus Sephadex from the column and reextracting them in a homogenizer with 0.1 N HCl.

Lactate dehydrogenase (LDH) activity in the samples was measured by spectrophotometrically monitoring the conversion of NAD into NADH according to Amador *et al.* (1963).

2.4. Radioimmunoassays

Radioimmunoassays were performed on aliquots of undiluted and diluted superfusion samples. Concentrations of ACTH were such that a prior extraction was not necessary. Human ACTH was assayed as described elsewhere (Liotta and Krieger, 1975), using the "West" antiserum as distributed by the NIH. This antiserum recognizes $ACTH_{1-24}$ and $ACTH_{1-39}$ on an equimolar basis and cross-reacts to a much lesser degree with high-molecular-weight ACTH precursors. Molecular-sieve chromatography of effluents was not performed, since the amounts of ACTH present were too low to permit adequate resolution.

3. Results

3.1. Cell Preparation and Cellular ACTH Content

The yield of cells from human pituitary tissue was lower than that obtained with rat pituitary tissue, but was more than sufficient for the superfusion experiments. Usually, we were able to harvest $2.0–2.5 \times 10^4$ cell/mg wet weight of tissue. Others have reported values of $1–80 \times 10^4$ cell/mg wet weight (Hymer *et al.*, 1976). The cell yields from "fresh" (up to 12 hr postmortem) and "old" pituitaries (up to 24 hr postmortem) were not significantly different (Table I). The cell viability varied from 80 to 95% or more, based on Trypan Blue exclusion. Again, there was no clear relationship between cell viability and postmortem period (Table I), although the 95% or greater viability always occurred with the cells from the surgically removed specimens, indicating a certain degree of cell-membrane deterioration starting soon after the death of the organism. In subsequent experiments, we have noticed that tissue can be used successfully up to about 30 hr postmortem; after that time, cell viability quickly approaches very low values ($\leq 10\%$).

As a zero-time control, cells in one of two or more parallel cell-

Table I. (Lack of) Relationship among Donor Age, Cell Yield, and Viability

Tissue donor	Sex	Age	Postmortem time (hr)	Cell yield [cells/mg tissue ($\times 10^4$)]	Cell viability (% Trypan Blue exclusion)
R.L.	M	47	7	2.4	85
C.P.	M	14	10	2.1	80
R.P.	M	62	11	2.6	85
M.H.	F	73	24	1.9	85
R.V.	F	24	18	2.0	85
R.R.1	M	44	20	3.0	85
R.F.	M	40	20	2.4	90
B.P.	F	16	10	1.9	80
R.S.	M	53	8	2.3	80
C.B.	F	67	48	0	0
J.H.	M	22	20	0.2	75
N.G.	M	44	12	2.4	90
R.R.2	M	27	30	2.2	70
Unknown	M	30–35	36	0	0
Unknown	M	30–35	6	2.7	95
L.P.	F	36	Surgery	2.0	≥ 95
P.F.	F	31	Surgery	2.5	≥ 95
G.M.	F	37	Surgery	2.1	≥ 95

columns were extracted with 0.1 N HCl at the end of the presuperfusion period. After a run, the cells on the other columns were also extracted with acid; both extracts were assayed for ACTH and yielded the same value on a per cell basis. Among the various experiments, the cellular ACTH content ranged from 1.2 to 10.8 μg ACTH/cell, or from 0.26 to 2.4 fmol ACTH/cell. Serial dilutions of cell extracts and the standard curve for ACTH ran parallel. This necessary but insufficient evidence to prove that we were indeed measuring ACTH was supported by the finding that after Sephadex G-50 chromatography of the cell extracts, the bulk (>95%) of the immunoactivity was found in the fractions corresponding to the expected elution volume for ACTH$_{1-39}$. The remainder of the immunoactivity was found mostly in the void volume.

3.2. Superfusion Experiments

3.2.1. Normal Tissue

3.2.1a. Stimulation of ACTH Release. Basal ACTH release ranged from 4 to 10 fmol/10^5 cells per min. This is about twice the rate that rat cells in a superfusion system show under nonstimulated conditions (Mulder and Smelik, 1977). Repeated stimulation of the human cells with identical 30-sec pulses of HME extract, spaced 10–20 min apart, caused repeated identical periods of 4–6 min of increased ACTH secretion, each period starting immediately after the extract had reached the cells (Fig. 2). We never observed diminishing or increasing responses of the cells to repeated equal doses of HME extract, given 10–20 min apart within a 3-hr period. The degree of "contamination" of the HME extract with ACTH usually amounted to a value corresponding to less than 2 fmol/10^5 cells per min, and did not necessitate correction of the measured ACTH secretion values.

Since the amounts of ACTH released by the cells, whether spontaneously or after HME stimulation, are small when viewed in relation to the cells' ACTH content (basal release is about 0.02% per minute of the cells' content on a given cell-column), the death and lysis of about 0.2% per minute of the cells present on a column could conceivably account for the amounts of ACTH observed to be released by the cell-column. However, the levels of LDH activity, which would certainly have risen after cell death and rupture, were low during the course of an experiment and, more important, did not change during periods of increased ACTH secretion after stimulation with HME extract.

Cells prepared from different pituitaries usually showed different basal levels of ACTH release. The net absolute increase in ACTH secretion after

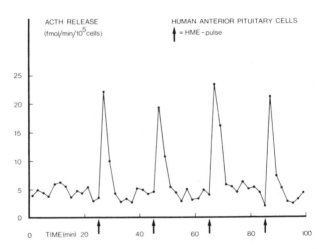

Figure 2. ACTH release by human anterior pituitary cells that were repeatedly stimulated with equal doses of HME extract (1/40 equivalent/10^5 cells), given in 30-sec pulses at the times indicated by the arrows.

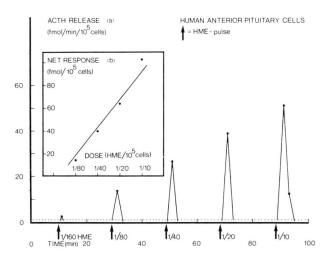

Figure 3. (a) ACTH release after stimulation of human anterior-pituitary cells with various doses of HME extract, increasing with time, from 1/160 to 1/10 equivalent/10^5 cells, given in 30-sec pulses at the times indicated by the arrows. (b) Log dose–response curve for the doses of HME and the net amounts of ACTH released in Fig. 3a.

stimulation with HME extract showed a substantial degree of consistency from one cell preparation to another; thus, we were able to construct log dose–response curves (Fig. 3b). In the range of HME doses tested (0.00625–0.1000 HME equivalent/10^5 cells), these curves were essentially linear. Different cell preparations yielded different lines, but these lines were practically parallel. The three log dose–response curves in Fig. 4 probably reflect individual variation among pituitaries as well as differences in sensitivity to HME extract of three different pituitary cell preparations.

While 30-sec pulses of HME extract of increasing concentration gave rise to peaks in ACTH secretion of increasing magnitude (Fig. 3a), increasing the pulse length while keeping the concentration of the extract constant caused increases in peak width, but not in height (Fig. 5).

3.2.1b. Inhibition of ACTH Release. Cortisol at 0.5 μg/ml resulted in inhibition of HME-stimulated ACTH release. For reasons cited in Section 2.3, stimulation with HME extract in these experiments was postponed until 20 min after the start of the steroid infusion. In three experiments, we observed an 80% inhibition by cortisol of the cells' response to a low dose of HME extract (0.05 equivalent/10^5 cells) and a 25–30% inhibition of the cells' response to a high dose of the extract (0.10 equivalent/10^5 cells). These results have been plotted in Fig. 6.

A "rebound" phenomenon, also previously observed in rat cells, was

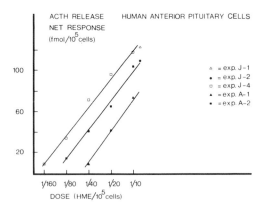

Figure 4. Log dose–response curves of three different human anterior-pituitary-cell preparations (J-4, A-1, and A-2). Single points of two other experiments (J-1 and J-2) were also included. Stimulation of the cells was performed with 30-sec pulses of HME extract.

present after removal of the steroid inhibition, as evidenced by the greater than normal response to subsequent HME stimulation (see Fig. 6). Basal ACTH secretion was not diminished by the steroid treatment, and there was no "rebound" present after cessation of the cortisol infusion.

3.2.2. Tumor Tissue

Two experiments were performed with cells prepared from pituitary tumor tissue, removed from patients with prolactin-producing, chromophobe adenomas. The amounts of tissue obtained at surgery were so small that only limited numbers of cells could be prepared, allowing us to load only one superfusion chamber each time with enough cells to perform meaningful experiments. The first superfusion was designed to test the reactivity of the pituitary tumor cells to the NIH-HME extract. Since we had no way of predicting the cells' behavior *in vitro*, we stimulated the cells three times with equal doses of HME extract, spaced at intervals 1.5–2 times the normal 20 min, to allow for a possible sluggish response or prolonged action of the extract. The cells' response (Fig. 7) was "normal," however, in terms of rapidity of the onset and disappearance of the period of increased ACTH secretion. The absolute magnitude of the tumor cells' response was, however, 30–50% higher than that of normal cells.

In the experiment with the cells from the second patient, we tested the cells' ability to sustain an elevated rate of ACTH release during continued stimulation with HME extract. Nothing in the cells' response,

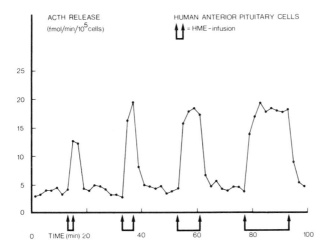

Figure 5. Sustained ACTH release by human anterior-pituitary cells that were exposed to HME extract at 1/40 equivalent/min per 10^5 cells for periods of 2, 4, 8, and 16 min at the times indicated by the arrows.

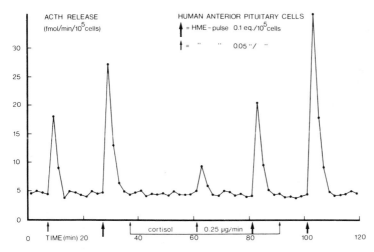

Figure 6. ACTH release by human anterior pituitary cells from a woman with breast cancer. At the times indicated by the thin arrows ($t = 7$ and 61 min), 0.05 HME equivalent/ 10^5 cells was given in 30-sec pulses. At the times indicated by the heavy arrows ($t = 27$, 81, and 101 min), 0.10 HME equivalent/10^5 cells was given in 30-sec pulses. From $t = 37$ until $t = 89$ min, 0.50 μg/ml cortisol was infused. The HME pulses given during the cortisol infusion also contained the steroid.

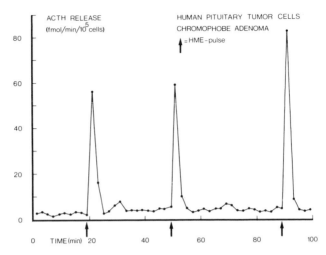

Figure 7. ACTH release by human pituitary tumor cells obtained from a woman with a chromophobe pituitary adenoma. At the times indicated by the arrows, 30-sec pulses of HME extract at doses of $1/10$ equivalent/10^5 cells were given.

except the magnitude, distinguished it from that of normal cells (Fig. 8). Again, the rate of ACTH release was 50–60% higher than what we could expect with normal cells.

4. Discussion

Evidence from extensive experiments with rat anterior pituitary cells *in vitro,* carried out according to methods similar to those described in this chapter, has shown that manipulations of the animals *in vivo* (e.g., stress, adrenalectomy, glucocorticoid injection, hypothalamic lesions, drug implantation) are more or less faithfully reflected by the pituitary cells' behavior *in vitro* (Portanova and Sayers, 1974; Mulder, 1975; Mulder *et al.,* 1976; Mulder and Smelik, 1977). The latter is consistent with the animals' neuroendocrine status at the moment just prior to decapitation. When for instance, the level of circulating glucocorticoids is high, as in an animal after corticosterone injection, the pituitary cells show a diminished level of stimulated ACTH secretion, and opposite findings are seen in cells from an adrenalectomized animal. The situation after a recent or prolonged stress is more difficult to assess, but follows the same general pattern.

Using pituitary cells from intact, untreated rats in the superfusion method, we have shown that these cells are sensitive to a CRF preparation in a log dose–response fashion, that ACTH release starts within seconds after the CRF preparation reaches the cells, that the response is calcium-dependent, that corticosterone at physiological levels and dexamethasone mostly inhibit stimulated ACTH release, and that these cells can be used successfully in a hypothalamus–pituitary cell–adrenal cell cascade system (Mulder, 1975; Smelik *et al.*, 1975; Mulder *et al.*, 1976; Smelik, 1977; Vermes *et al.*, 1977; Mulder and Smelik, 1977). Some of these results have recently been confirmed by others (Gillies and Lowry, 1978). In short, the rat pituitary cells in a superfusion system behave "normally" in many, if not most, respects.

Much less information is available on human pituitary cells. We have shown in this chapter that the ACTH released into the medium does not originate from leaky, dead, or lysing cells simply emptying their hormone content into the superfusion buffer. We have constructed log dose–response curves using a rat hypothalamus extract as the CRF-containing substance, and we have shown inhibition of ACTH secretion by "physiological" levels of cortisol. The response of the human cells is immediate and transient, just as in the case of rat cells.

Although the human pituitary cells prepared by the collagenase dispersal method described herein appear to be viable in view of the low amount of Trypan Blue uptake, it seems unreasonable to expect a cell to be responsive to an HME extract or to glucocorticoids just because of this "histological feat," the opposite (i.e., a dead cell being unresponsive)

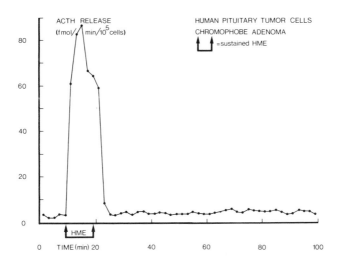

Figure 8. Sustained ACTH release by human pituitary tumor cells obtained from a woman with a chromophobe pituitary adenoma. From $t = 9$ until $t = 19$ min, an HME extract was given at a dose of $1/12$ equivalent/min per 10^5 cells.

certainly being true. Thus, the 80–95% viability should be viewed more as a criterion of a successful dispersion than as a prediction of a responsive cell preparation. The fact that the preparations used to construct the log dose–response curves all exhibited viabilities of 80–85%, but differed in sensitivity to HME extract by a factor of about 4 (see Fig. 4), could be taken as an illustration of the relative unimportance of viability determinations. On the other hand, it could also be the result of individual variations among the three pituitary cell preparations, each of which was obtained from a single pituitary.

Scanning electron microscope pictures of cells prepared by this collagenase method show that the cells *in vitro* are very similar in appearance to cells *in situ;* they also show a high amount of fibrous material to be present in the human anterior pituitary, which may well account for the lower-than-expected cell yields (W. C. Hymer, personal communication).

The magnitude of the human cells' response to rat HME extract may serve to indicate little or no species-specificity, although we do not know yet whether human HME extract will be active in a rat pituitary-cell preparation. Calculation of the cellular rate of ACTH release after stimulation with 0.1 HME equivalent/10^5 cells shows that the human cells described in this chapter secrete approximately 1670 molecules of ACTH per second per cell, as compared to about 2200 molecules per second per cell for rat pituitary cells (Mulder and Smelik, 1977). At this point, we do not know whether this difference is significant. This question can be solved in an adequate manner only by constructing complete log dose–response curves; this will be possible only if and when pure CRF becomes available.

Our observation that there is no change in the cellular ACTH content during the course of a superfusion experiment when the cells are challenged several times with HME extract has also been made in the past with rat cells (Takebe *et al.,* 1975; Mulder and Smelik, 1977; Herbert *et al.,* 1978). Even after intensive and/or sustained HME stimulation of the cells, accompanied by strongly elevated, sustained levels of ACTH secretion, there is no substantial decrease in cellular hormone content (results not shown). This relative constancy of the cells' hormone levels has also been observed by others working with human pituitary cells (Resetic *et al.,* 1977). This finding implies the existence of an immediate and effective mechanism to "replenish" the ACTH stores in the cells after stimulated secretion. (It would be interesting to investigate the rapidity of this "synthesis.") A highly intriguing question is whether the agent responsible for this mechanism would be the process of increased secretion itself or the presence of factors in the HME extract. The findings in rat cells of increased cellular ACTH levels after stimulation with HME extract in the presence of a release-inhibiting amount of either corticosterone (Mulder and Smelik, 1977) or dexamethasone (Koch *et al.*, 1974) tend to point to the HME extract as the causative factor, rather than the secretion process itself. We have not tested whether such an increase of cellular ACTH content occurs after stimulation of secretion by other substances.

Though we have characterized only a small part of the human pituitary cells' behavior in terms of the regulation of ACTH release, from what we have established thus far, it appears that the cells seem to behave quite normally.

Interpretation of the response of the tumor cells is difficult, in part because we are uncertain about the amount of periadenomatous "normal" tissue adhering to the surgical specimens. It appears unlikely, however, that a major fraction of the cells were normal, since the ACTH response to stimulation of the tumor-cell preparations was higher than that of normal cell preparations.

In summary, this chapter attests to the feasibility of using postmortem human pituitaries as a source for obtaining viable, endocrinologically active cells. Though we have made only a modest start in the study of the regulation of the synthesis and release of ACTH, the methodology reported herein should enable us to gain more insight into the normal regulatory mechanisms as well as the hormonal dysregulation of human pituitary tumors.

DISCUSSION

JONES: These are interesting data, indeed. You have shown that rat CRF will release ACTH from human pituitaries, so it does not appear to be species-specific. Am I right in thinking you have now shown an immediate or fast-feedback action of corticosterone, both at the hypothalamus and the anterior pituitary?

MULDER: That depends somewhat on terminology. At the pituitary, the steroid takes about 20 minutes before it inhibits stimulated ACTH release. At the hypothalamus, all I can say is that 20 minutes after the start of steroid exposure, the basal CRF release is inhibited.

JONES: There is agreement in that in both of our hands, depolarization-induced release of CRF from the hypothalamus is not blocked by corticosterone.

HERBERT: Your experiments showed that ACTH release was linear for a much longer period when the medium was changed every 5 minutes then when it was changed less often. Could this be due to feedback inhibition of ACTH secretion by ACTH or some other medium component or to release of degradative enzymes or to inhibitors? Also, how rapid was the action of corticosterone on HME stimulation of ACTH release?

MULDER: In answer to your first question, both alternatives seem possible, but I am inclined towards the second possibility. In answer to your second question, it was about 20 minutes.

CRABBÉ: Does cycloheximide interfere with the release of ACTH brought about by depolarizing the pituitary cells with potassium? Also, you have presented provocative data regarding secretion of ACTH by human chromophobe adenoma challenged *in vitro* with CRF. Is there any evidence for patients with this condition releasing ACTH when stressed?

MULDER: No, cycloheximide does not interfere. In regard to your second question, we don't have extensive data on preoperative parameters. The ACTH response of these patients after pituitary function tests appeared normal, however, indicating either that the normal pituitary tissue functioned normally or that the tumor tissue behaved as if it were normal with regard to ACTH release.

REFERENCES

Amador, E., Dorfman, L. E., and Wacker, W. E. C., 1963, Serum lactic dehydrogenase activity: Analytical assessment of current assays, *Clin. Chem.* **9**:391.

Gillies, G., and Lowry, P. J., 1978, Perfused rat isolated anterior pituitary cell column as bioassay for factor(s) controlling release of adrenocorticotropin: Validation of a technique, *Endocrinology* **103**:521.

Herbert, E., Allen, R. G., and Paquette, T. L., 1978, Reversal of dexamethasone inhibition of adrenocorticotropin release in a mouse pituitary tumor cell line either by growing cells in the absence of dexamethasone or by addition of hypothalamic extract, *Endocrinology* **102**:218.

Hong, J. S., and Poisner, A. M., 1974, Effect of low temperature on the release of vasopressin from the bovine neurohypophysis, *Endocrinology* **94**:234.

Hymer, W. C., Snyder, J., Wilfinger, W., Bergland, R., Fisher, B., and Pearson, O., 1976, Characterization of mammotrophs separated from the human pituitary gland, *J. Natl. Cancer Inst.* **57**:995.

Koch, B., Bucher, B., and Mialhe, C., 1974, Pituitary nuclear retention of dexamethasone and ACTH biosynthesis, *Neuroendocrinology* **15**:365.

LaBella, F. S., Dular, R., and Vivian, S., 1973, Anomalous prolactin release *in vitro* in response to cold and its specific blockade by a purified hypothalamic inhibitory factor, *Endocrinology* **92**:1571.

Liotta, A., and Krieger, D. T., 1975, A sensitive bioassay for the determination of human plasma ACTH levels, *J. Clin. Endocrinol. Metab.* **40**:268.

Lowry, P. J., 1974, A sensitive method for the detection of corticotrophin releasing factor using a perfused pituitary cell column, *J. Endocrinol.* **62**:163.

Mulder, G. H., 1975, Release of ACTH by rat pituitary cells, Ph.D. thesis, Free University, Amsterdam, The Netherlands.

Mulder, G. H., and Smelik, P. G., 1977, A superfusion system technique for the study of the sites of action of glucocorticoids in the rat hypothalamus–pituitary–adrenal system *in vitro*. I. Pituitary cell superfusion, *Endocrinology* **100**:1143.

Mulder, G. H., Vermes, I., and Smelik, P. G., 1976, A combined cell column superfusion technique to study hypothalamus–pituitary–adrenal interactions *in vitro*, *Neurosci. Lett.* **2**:73.

Portanova, R., 1972, Release of ACTH from isolated pituitary cells, *Proc. Soc. Exp. Biol. Med.* **140**:825.

Portanova, R., and Sayers, G., 1974, Corticosterone suppression of ACTH secretion: Actinomycin D sensitive and insensitive components of the response, *Biochem Biophys. Res. Commun.* **56**:928.

Resetic, J., Ludecke, D., and Sekso, M., 1977, Immunoreactive and biologically active ACTH in human plasma and isolated normal pituitary and ACTH adenoma cells, *Acta Endocrinol. Suppl.* **208**:80.

Smelik, P. G., 1977, Some aspects of corticosteroid feedback actions, *Ann. N. Y. Acad. Sci.* **297**:580.

Smelik, P. G., Mulder, G. H., and Vermes, I., 1975, *In vitro* studies with a hypothalamus–pituitary–adrenal cascade superfusion system, in: *Symposium of the International Society of Psychoneuroendocrinology,* pp. 423–430, Visegrád Academy of Sciences, Academiai Kiado, Budapest.

Swallow, R. L., and Sayers, G., 1969, A technic for the preparation of isolated rat adrenal cells, *Proc. Soc. Exp. Biol. Med.* **131**:1.

Takebe, K., Yasuda, N., and Greer, M. A., 1975, A sensitive and simple *in vitro* assay for corticotropin-releasing substances utilizing ACTH release from cultured anterior pituitary cells, *Endocrinology* **97**:1248.

Uemura, T., Hanasaki, N., Yano, S., and Yamamura, Y., 1976, The fate of ACTH released from rat anterior pituitary into the incubation medium *in vitro:* Enzymatic degradation and acid activation, *Endocrinol. Jpn.* **23**:233.

Vermes, I., Mulder, G. H., and Smelik, P. G., 1977, A superfusion system technique for the study of the sites of action of glucocorticoids in the rat hypothalamus–pituitary–adrenal system *in vitro.* II. Hypothalamus–pituitary cell–adrenal cell superfusion, *Endocrinology* **100**:1153.

27

Interaction of Glucocorticoids, CRF, and Vasopressin in the Regulation of ACTH Secretion

B. Koch, B. Lutz-Bucher, and C. Mialhe

1. Introduction

The rate of secretion of pituitary corticotropin (ACTH) is regulated by a hypothalamic neurohormone (or hormones), which in turn is known to be under control of higher centers of the brain through the link of the action of neurotransmitters (Van Loon, 1973; Jones *et al.*, 1976). Moreover, unlike hormones such as growth hormone or prolactin, the secretion of which depends on both excitatory and inhibitory hypothalamic factors, ACTH release appears to be primarily modulated by glucocorticoids exerting negative-feedback effects at the hypothalamic as well as the pituitary level.

It is the purpose of this chapter to summarize work from our laboratory and others on the stimulatory effect of corticotropin-releasing factor (CRF) and the CRF-like molecule vasopressin (VP), as well as on the modulatory influence of glucocorticoids on ACTH secretion. In this connection, the mechanism of action of these steroids at both brain and pituitary sites of action will be considered.

B. Koch, B. Lutz-Bucher, and *C. Mialhe* • Institut de Physiologie, Université Louis Pasteur, Strasbourg, France

2. Comparative Studies on CRF Activities of Hypothalamic Extracts And Vasopressin

2.1. In Vitro CRF-like Effect of Vasopressin

Several lines of evidence have established the existence of a CRF different from VP in the neurohypophysis and the hypothalamus (Saffran and Schally, 1977). To gain more insight into the mechanism of action of both neurohormones in the regulation of ACTH secretion, *in vitro* experiments were carried out using incubated fragments of adenohypophyses.

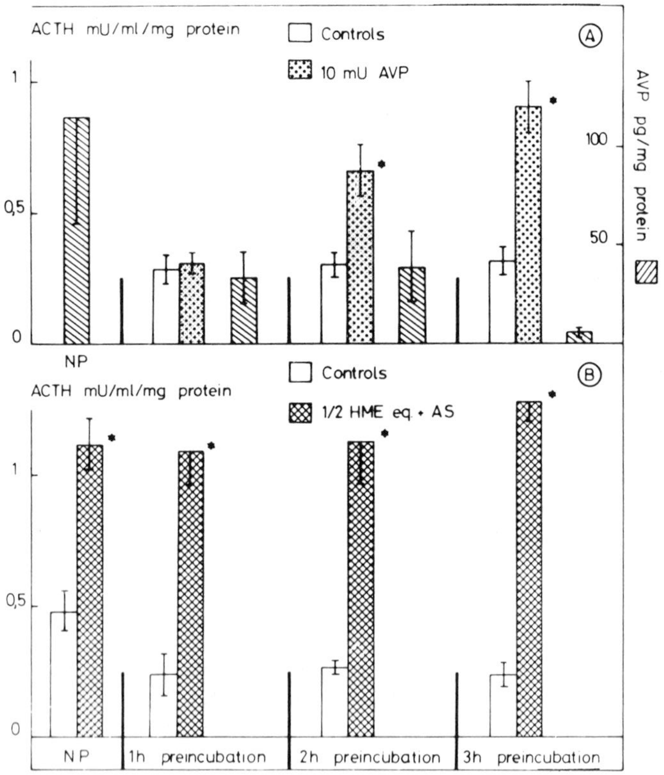

Figure 1. Pattern of ACTH secretion of quartered pituitaries, preincubated for varying periods of time and further incubated for 1 hr in the presence of either (A) arginine vasopressin (AVP) (Ferring, 2.5 μg/U) or (B) hypothalamic median eminence (HME) extracts freed of VP with the aid of antiserum (AS). The VP content of the glands is also displayed (hatched bars). (NP) Not preincubated. Reproduced from Lutz-Bucher *et al.* (1977).

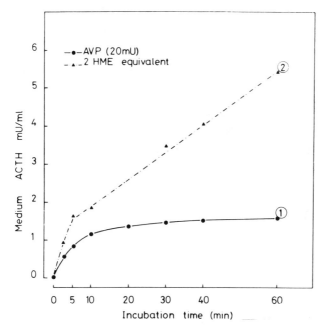

Figure 2. Temporal change in ACTH output of pituitaries preincubated for 3 hr in the presence of either AVP or CRF extracts. Reproduced from Lutz-Bucher *et al.* (1977).

As can be seen in Fig. 1, the time of preincubation of the glands was of critical importance, since a preincubation period of at least 2 hr was a prerequisite for VP to induce significant increment in ACTH output (Fleischer and Vale, 1968; De Wied *et al.*, 1969; Lutz-Bucher *et al.*, 1977). By contrast, this appeared not to be the case for CRF material extracted from the median eminence.

Since VP has been reported to be present in the anterior pituitary gland (Chateau *et al.*, 1973; Lutz *et al.*, 1974; Renlund, 1978), a possible explanation for this time-dependent difference in the CRF activities of both neurohormones could arise from the saturation of putative VP receptors by endogenous hormone. This view, indeed, seems to be supported by the finding of a decrease in the pituitary content of VP as a function of incubation time, concomitant with an increase in the ability of the gland to release ACTH in response to exogenous VP (Fig. 1). Although data from the 2 hr preincubation period were in apparent discord, the observation that pituitaries from Brattleboro rats (which are genetically deficient in VP) were able to release significant amounts of ACTH in response to VP without needing to be preincubated for long periods of time further strengthened that conclusion (Lutz-Bucher *et al.*, 1977). Also, it has been reported that dispersed pituitary cells from these rats appeared to be more responsive to VP than those from normal rats (Krieger and Liotta, 1977). Thus, it seems that the occupancy of putative receptor sites of the pituitary by endogenous VP might prevent the gland from responding to an external source of hormone and modulate, therefore, the responsiveness of the gland.

Other experiments showed further that time- and dose-dependent differences existed in the corticotropin-releasing properties of both neurohormones. Whereas the VP-induced release of ACTH plateaued after about 20 min, the stimulatory effect of median eminence extracts proceed-

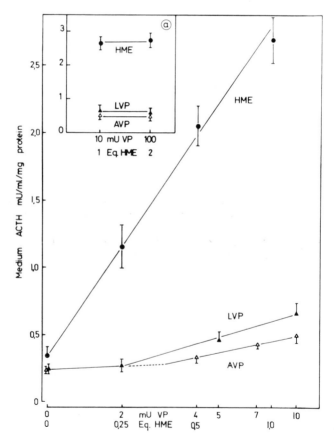

Figure 3. Comparative effects of increasing doses of AVP, LVP, and HME extracts on the ACTH output of pituitary fragments preincubated for 3 hr. Insert (a) shows that the secretion plateaued with doses of 10–100 mU VP. Reproduced from Lutz-Bucher et al. (1977).

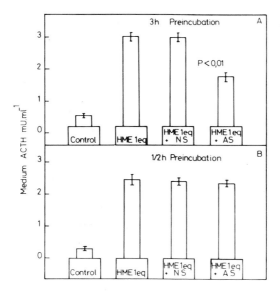

Figure 4. ACTH release from quartered pituitaries incubated with HME extracts in the absence or the presence of either VP-antiserum (AS) or normal serum (NS), following a preincubation time of 3 hr (A) or 30 min (B).

ed throughout the incubation time (Fig. 2). Also, compared to the effect provoked by increasing concentrations of hypothalamic median eminence (HME) extracts, that induced by either lysine vasopressin (LVP) or arginine vasopressin (AVP) was of much smaller magnitude (Fig. 3). This agrees well with other reports (Portanova and Sayers, 1973; Buckingham and Hodges, 1977) and contrasts strikingly with the stimulatory effect of vasotocin (a peptide closely related to VP), which paralleled that of CRF extracts (Buckingham and Hodges, 1977).

2.2. Effect of Vasopressin-Free Hypothalamic Extracts

Since crude median eminence extracts are known to contain significant amounts of VP (12.3 ± 1.2 mU/hypothalamus, $n = 10$, as measured by a radioimmunoassay), it was of interest to determine the relative importance of this component in the overall effect of hypothalamic extracts on ACTH secretion. To this end, extracts were stripped of VP with the aid of a specific VP-anti-serum. In a preliminary experiment, it was shown that the amount of antiserum added to the incubation medium actually abolished the CRF-like action of VP (Lutz-Bucher, 1976), without interfering significantly with the effect of the CRF material itself. Thus, our antibody, unlike that of Gillies and Lowry (1976), seems to be directed against a specific part of the VP molecule that appeared not to be shared by the CRF material.

As depicted in Fig. 4, neutralization of VP present in the hypothalamic extract resulted in a statistically significant reduction in the effect of this extract on the ACTH output of pituitary fragments preincubated for 3 hr, as opposed to fragments preincubated for only 30 min. This agrees well with the above-demonstrated necessity of a critical latency interval for the effect of VP to develop (see Fig. 1). Moreover, the fact that addition of VP to pituitaries incubated in the presence of 1 HME-equivalent extract did not further enhance ACTH release (1) suggests that the amount of VP already present in the extract provides maximum enhancement of hormonal release and (2) favors an additive, rather than a potentiating, effect of VP on the activity of CRF. This lack of potentiation, which, by contrast, has been observed under *in vivo* conditions (Yates *et al.*, 1971), also emerges from the data displayed in Fig. 5. In these experiments, indeed, the stimulatory effects of VP and CRF on ACTH secretion clearly appeared to be additive.

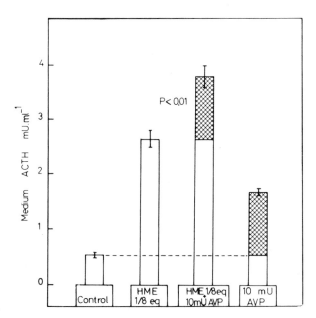

Figure 5. Effect of AVP and HME extracts either alone or in combination on ACTH release of pituitaries preincubated for 3 hr.

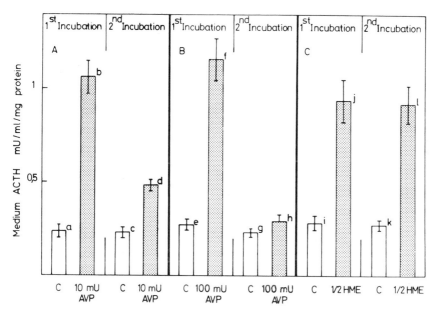

Figure 6. Effect of successive stimulations by either 10 mU (A) or 100 mU (B) AVP or HME extracts (C) on ACTH release from pituitaries preincubated for 3 hr and further incubated for 1-hr periods. Reproduced from Lutz-Bucher *et al.* (1977).

2.3. Putative Pituitary Vasopressin Receptor Sites

To gain additional information on the interaction of VP and CRF in stimulating ACTH release and on the presence of putative VP receptor sites, anterior pituitary glands were submitted to the influence of repeated *in vitro* stimulations with VP or CRF extracts or both. As depicted in Fig. 6, a first stimulation of ACTH release in the presence of VP considerably reduced (Fig. 6A) or abolished (Fig. 6B) the effect of VP during a second stimulation, depending on the dose of hormone used during the former incubation period and, consequently, the amount of VP taken up by the tissue (15.1 ± 4.5; 39.8 ± 11.9; 71.2 ± 16.2 pg/mg protein, $n = 8$, in the absence of AVP and in the presence of 10 and of 100 mU AVP, respectively). By contrast, this was not observed with the CRF extract in the medium, since the pituitaries were able to respond with equal intensity to two successive stimulations (Fig. 6C). Moreover, it was shown that a first incubation in the presence of CRF, TRH, or LH-RH did not significantly impair the ACTH output induced by a second incubation with VP and, conversely, that VP did not alter the response to a subsequent stimulation by CRF (Lutz-Bucher, 1976). It was concluded that these observations strongly point to the presence in the anterior pituitary gland of specific receptor sites for VP, apparently different from those mediating the effect of CRF.

2.4. Physiological Implications

It may be inferred from these *in vitro* studies that interaction of VP with specific pituitary receptors, possibly at sites different from those involved in the binding of CRF, leads to an increased output of ACTH. However, considering only this direct effect on the adenohypophysis, independent of the reported stimulatory influence of VP on hypothalamic CRF (Hedges *et al.*, 1966), it appears difficult to reconcile the necessity seen *in vitro* for a latency period of action with the rapid onset of response observed *in vivo* after administration of the hormone to hypothalamic-lesioned or dexamethasone–Nembutal-blocked rats. The possibility exists that a more rapid turnover and/or interaction of VP with pituitary binding sites could account for this discrepancy.

Although the precise physiological role of VP may be of questionable significance, the view that the hormone is implicated in the regulation of ACTH release seems to be indirectly strengthened by the findings that (1) glucocorticoid treatment abolished the CRF-like effect of VP (Fleischer and Vale, 1968; Buckingham and Hodges, 1977; Hedges *et al.*, 1966); (2) VP was shown to be present in high amounts in the blood of portal vessels (Zimmerman *et al.*, 1973; Lutz-Bucher, 1976), as well as in the pituitary gland; (3) the pituitary content of VP increased as a result of adrenalectomy (Lutz-Bucher, 1976). Recently, Yasuda *et al.*, (1978), using rats bearing pituitaries transplated under the kidney capsule, clearly showed that *in vivo* VP did exert a direct stimulatory effect on ACTH secretion.

3. Identification of Glucocorticoid-Binding Components in The Adenohypophysis

3.1. Multiple Forms of Glucocorticoid Binders

Several lines of evidence have established that glucocorticoid binding to target tissues is heterogeneous in nature (Agarwal, 1977). As to the anterior pituitary gland, two major forms of binders have been reported to occur in the cytosol, the two differing in their ability to bind glucocorticoids

(De Kloet *et al.*, 1975; Koch *et al.*, 1975): one exhibited high affinity for both corticosterone (CORT; 4-pregnen-11β, 21-diol-3,20-dione) and the synthetic steroid dexamethasone (DX; 9α-fluoro-11β, 17α, 21-trihydroxy-16α-methyl-1,4-pregnadiene-3,20-dione) (component D), while the other combined only with the natural hormone and closely resembled plasma transcortin (TL) [transcortin-like component]. This can be seen in Scatchard plots (Fig. 7), in which data obtained with either labeled DX or CORT clearly point to the presence in pituitary cytosol of at least two populations of binding sites. Recent studies have further revealed that the

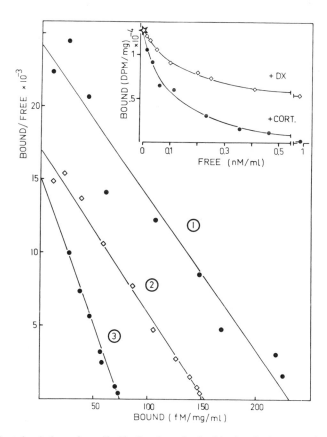

Figure 7. Scatchard plots of specific binding data obtained by incubating cytosol with either [³H]-CORT (curve 1) or [³H]-DX (curve 2), as well as inactivated cytosol (40°C for 10 min) with [³H]-CORT (curve 3). The inset applies to the binding of a trace amount of [³H]-CORT in the presence of increasing concentrations of radioinert CORT or DX. All binding data are reported as specific binding, i.e., total minus nonspecific yield in the presence of excess unlabeled steroid. Reproduced from Koch *et al.* (1976a).

DX-binding component could be resolved into two distinct entities with different isoelectric points (Maclusky et al., 1977) and that the TL component actually presented immunological identity with rat plasma transcortin (Koch et al., 1978d). Although previous work (De Kloet et al., 1975) failed to detect the existence of TL material in the brain, this binding molecule has recently been identified in the hypothalamus and hippocampus of rats, besides DX-binding receptors (Maclusky et al., 1977).

3.2. Binding Characteristics and Hormonal Regulation

Despite their closely similar apparent dissociation constants (between 4 and 9 nM), the two populations of pituitary glucocorticoid binders can be easily differentiated not only on the basis of their differences in specificity of steroid binding, but also according to marked differences in their physicochemical properties. Table I lists some of these characteristics.

In addition, it has been observed that adrenocortical as well as thyroid hormones appear to regulate the concentration of sites of these binders, without altering significantly the binding affinity. Indeed, the number of binding sites increased after thyroxine treatment (Koch et al., 1978a) and long-term adrenalectomy (Olpe and McEwen, 1976; Koch et al., 1978a), and CORT replacement therapy abolished this increment in adrenalectomized rats (Koch et al., 1978a). Glucocorticoid binding in brain areas has likewise been reported to be enhanced as a result of adrenalectomy (Olpe and McEwen, 1976).

3.3. Origin and Localization of the Transcortin-like Binder

The existence of TL material within the pituitary gland, as well as in other target tissues (Beato and Feigelson, 1972; Milgrom and Baulieu,

Table I. Physicochemical Properties of the
Pituitary Cytoplasmic Binders[a]

	Component	
Characteristic	D	TL
Sedimentation coefficient (low ionic strength)	7 S	4 S
Binding of DX	+	−
Binding to isolated nuclei	+	−
Stability at 40°C for 15 min	Inactivated	No effect
Sensitivity to DTT	+	−
Sex dependence	−	+

[a] Data from Koch et al. (1976a,b).

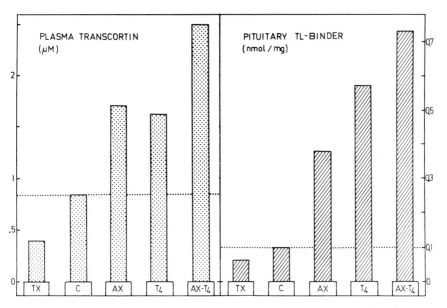

Figure 8. Effect of various experimental treatments on specific binding capacity of plasma transcortin and pituitary TL component. (TX) Thyroidectomy for 3 weeks; (C) control; (AX) adrenalectomy for 3 weeks; (T₄) thyroxine treatment (10 μg/100 g per day for 1 week). Reproduced from Koch *et al.* (1978b).

1970; Feldman *et al.*, 1973), raises several questions with regard to its origin, subcellular localization, and physiological role. Importantly, studies carried out using isolated pituitary cells (Koch *et al.*, 1977; De Kloet *et al.*, 1977) and direct measurements of blood contamination of pituitary cytosol (Koch *et al.*, 1976a; De Kloet and McEwen, 1976) clearly excluded the possibility of this binder's being a mere contaminant from the serum.

Since TL molecules were found in dispersed pituitary cells, one may ask whether they are actually synthesized by the cells or carried over from the blood. To test the latter possibility, a first approach has been to determine whether transcortin levels in the circulation and pituitary TL content were correlated. Binding data obtained under various experimental conditions (Fig. 8) showed not only a striking parallel between the concentrations of the binders in both tissues, but also that the adenohypophysis appeared to concentrate the binding protein several fold over plasma levels (Koch *et al.*, 1976a). This finding argues strongly in favor of a plasma origin for the pituitary component. Preliminary results, obtained in rats administered with rat transcortin-antiserum, seem to lend support to this view.

The TL macromolecules, being intrinsic to pituitary cells and not just confined to extracellular spaces, may be located either within the cell and/

or firmly attached to the plasma membrane. Without excluding a possible intracellular location, recent studies conducted with pituitary plasma-membrane fractions provided evidence that tends to support the latter possibility (Koch *et al.*, 1978c). As illustrated in Fig. 9, [³H]-CORT was able to combine to these plasma membranes, and the rate of association and dissociation of the binding complex appeared to be very rapid, as compared to the interaction of the steroids with cytoplasmic binders. It is

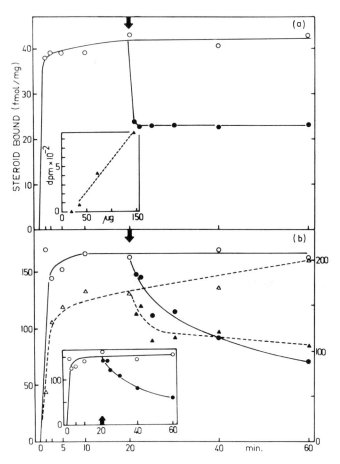

Figure 9. Time course of the association (○, △) and dissociation (●, ▲) of [³H] steroids with plasma-membrane fractions from anterior pituitaries, at 0°C. (a) [³H]-CORT binding to crude plasma membranes (insert shows linearity of binding between 50 and 150 μg protein). (b) Kinetics of binding of [³H]-DX (△, ▲) to cytosol and [³H]-CORT (○, ●) to both cytosol and solubilized plasma membrane proteins (insert). At the time indicated by the arrows, a 200-fold excess of unlabeled steroid was added to the incubation medium. Bound and free moieties were separated by centrifugation. Reproduced from Koch *et al.* (1978c).

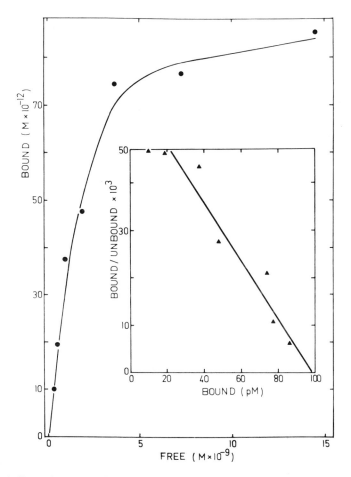

Figure 10. Saturation curve of the solubilized plasma-membrane binding component (Triton X-100, 1%, for 2 hr) in the presence of increasing concentrations of [³H]-CORT. The insert shows the Scatchard plot of this specific binding. Bound free steroid fractions were separated by gel filtration on Sephadex G-25 micro-columns. Reproduced from Koch *et al.* (1978c).

of interest to note that after solubilization of the membrane binding element, the dissociation from that soluble component was slower (Fig. 9b, insert) than from the particulate fraction, thus suggesting a higher affinity of the soluble protein. A quite similar finding has recently been reported concerning the effect of Triton solubilization on the characteristics of the insulin receptor present in placental membranes (Harrison, L. C., *et al.*, 1978). As an explanation, these investigators suggested a possible conformational change of the binding site.

The binding saturation curve and the Scatchard plot in Fig. 10 revealed the presence in the soluble plasma-membrane fraction of a single class of sites with a low binding capacity and an apparent dissociation constant of about 3 nM. Furthermore, studies on steroid-binding specificity (Fig. 11) indicated a preferential affinity of the particulate component for the natural glucocorticoid, rather than for DX, and, in addition, showed similarities with the binding specificity of plasma transcortin.* These characteristics, taken together with the observations that the soluble binder sedimented as a 4 S entity after sucrose density centrifugation (Koch *et al.*, 1978c), strongly support the view that a material similar in its physicochemical properties to transcortin is complexed to the plasma membrane of pituitary cells.

The conclusion that, like nonsteroidal hormones, steroid binding to

---

* CORT, deoxycorticosterone (21-hydroxy-4-pregnen-3,20-dione), and progesterone (4-pregnen-3,20-dione), but not estradiol [1,3,5(10)-estratrien-3,17β-diol] and DX, are effective competitors for [^{3}H]-CORT uptake.

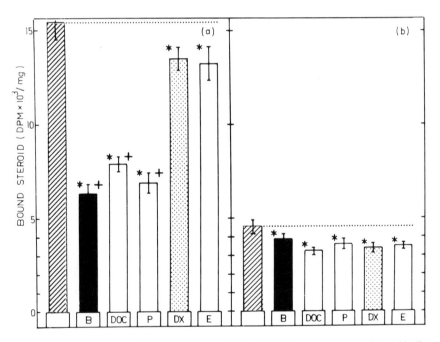

Figure 11. Displacement of [^{3}H]-CORT (a) and [^{3}H]-DX (b) from plasma-membrane binding sites by excess unlabeled corticosterone (B), deoxycorticosterone (DOC), progesterone (P), DX, and estradiol (E). (*) $p < 0.01$ compared with control; (+) $p < 0.01$ compared with DX or estradiol groups. Reproduced from Koch *et al.* (1978c).

target cells may also involve a membrane-associated process is further strengthened by the findings of others. Indeed, plasma-membrane binding sites for glucocorticoids and estradiol have been observed in the liver (Suyemitsu and Terayama, 1975) and the uterus (Pietras and Szego, 1977). Also, it has been suggested that interaction of corticoids with mouse pituitary tumor cells secreting ACTH may require a preliminary binding step involving the cellular membrane (Harrison, R. W., *et al.*, 1976). More recently, Pietras *et al.*, (1978) detected the presence of specific binding sites for estradiol in hepatocyte plasma-membrane fractions.

4. Functional Significance of Glucocorticoid Binding to the Adenohypophysis

4.1. In Vivo Occupancy of Receptor Sites

In an attempt to answer the question of how much of the pituitary receptor sites were occupied by endogenous corticoids or, in other words, whether receptor occupancy was working within the physiological range of circulating CORT, a study was carried out using pituitaries from rats sacrificed under various experimental conditions (Koch *et al.*, 1975). As evidenced by the data in Fig. 12, the percentage of receptor saturation, at both the cytoplasmic and the nuclear level, was closely related to the concentration of plasma corticosterone. A time-course experiment further showed that after administration of a physiological dose of CORT to adrenalectomized rats, maximum occupancy of binding sites occurred by 30 min and that almost complete binding capacity was restored by 2 hr. A similar correlation between the availability of free receptor sites and the activity of the pituitary–adrenal axis has also been observed in the brain (Stevens *et al.*, 1973; McEwen *et al.*, 1974).

4.2. Possible Functional Role of the Multiplicity of Sites

The presence of the TL binding system within the pituitary cells, strongly associated with the plasma membrane and probably partly intracellular, raises several interesting questions. First, results from our laboratory and from others clearly failed to recognize this binder as a "true" receptor, since the TL–corticoid complex was unable to undergo translocation into isolated pituitary nuclei (Koch *et al.*, 1976b) or to combine DNA cellulose (De Kloet and McEwen, 1976). It thus may be inferred that the TL binder is ineffective in mediating the hormonal message to the target cell.

A possible role for this binder emerges from the data shown in Fig.

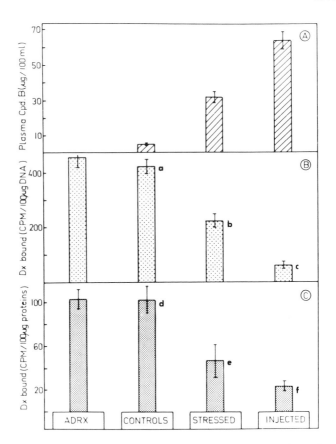

Figure 12. Effect of nonstress conditions (controls), adrenalectomy for 1 day (ADRX), stress (rats killed 15 min after a 2-min exposure to ether), and CORT injection (rats killed 15 min after injection of 200 μg/100 g body weight) on plasma CORT levels and saturation of specific binding sites of both cytosol and nuclear fractions of pituitaries. Whole glands were incubated at 25°C for 30 min in the presence of [^{3}H]-DX and processed as described by Koch *et al.* (1975). $p < 0.01$: (a) vs. (b), (b) vs. (c); $p < 0.05$: (d) vs. (e); not significant: (e) vs. (f).

13, summarizing cross-incubation studies using nuclei isolated from pituitaries of male and female rats that were reacted with cytosol preparations from both sexes (containing different concentrations of TL material: about 100 and 400 fmol/mg protein in males and females, respectively). The observation that the ability of nuclei to take up CORT was inversely related to the amount of TL molecules in the incubation medium suggests that these molecules may compete with the receptor for CORT complexing and, hence, with the transfer of the steroid to the nucleus. This binder therefore appears to act as a regulator of free available hormone in the cell. This conclusion meets the view of others (De Kloet *et al.*, 1977) and is consistent with the fact that both *in vivo* and *in vitro* (De Kloet *et al.*, 1975; Koch *et al.*, 1975), [³H]-DX is taken up by pituitary nuclei to a significantly greater amount than [³H]-CORT.

Since TL molecules have been shown to be strongly associated with the plasma membrane of pituitary cells, one of the possible roles of this molecule could be to take part in the steroid-uptake process of the cell. To test this possibility, isolated pituitary cells were reacted with either [³H]-CORT or [³H]-DX (which do not combine to the TL component) at both 25 and 0°C. As depicted in Fig. 14, at 0°C, binding of DX was considerably smaller than that of CORT, and uptake of both steroids appeared to increase as the temperature was raised to 25°C. We interpret these data as indicating that at the lower temperature, most of the CORT binding to the cells was probably to the plasma membrane TL binder, transfer into the cell and nucleus being reduced under these conditions as evidenced by the

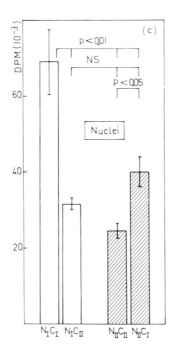

Figure 13. Specific binding of [³H]-CORT to nuclei isolated from male (N_I) or female (N_{II}) pituitaries in the presence of cytosol from either males (C_I) or females (C_{II}). Incubation was performed at 23°C for 15 min in a medium containing about 500 μg cytosol proteins and 70–130 μg nuclear DNA. Reproduced from Koch *et al.* (1976b).

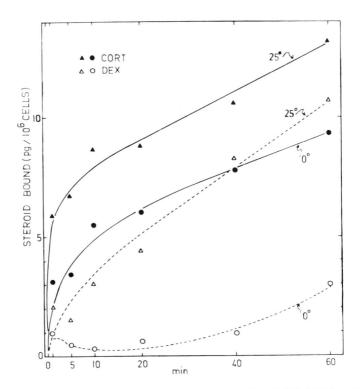

Figure 14. Time course of specific binding of [³H]-DX (DEX) and [³H]-CORT to dispersed pituitary cells at 0 and 25°C. Cells were isolated with the aid of trypsin and incubated in Krebs–Ringer buffer, supplemented with 0.1% bovine serum albumin. Bound steroid fractions were measured by washing and centrifuging the cells three times. Results are a composite of three independent experiments.

small labeling with [³H]-DX. It is unlikely that these differences could arise from differences in the rate of steroid penetration into the cells, since, on the contrary, cell nuclei were observed to accumulate DX more rapidly and to a greater extent than CORT. This suggests that the plasma-membrane binding component acts in a subtle way by modulating the access of steroid into the cell, the mechanism underlying this effect remaining to be clearly ascertained. A similar membrane-associated process has also been reported to be involved in the interaction of [³H] triamcinolone acetonide with pituitary AtT-20 tumor cells (Harrison, R. W., *et al.*, 1976).

4.3. Coupling between Steroid Binding and Biochemical Response

The presence of glucocorticoid receptors in the brain and the pituitary is consistent with the effects of these steroids on brain biochemical processes, behavior, and regulation of pituitary function (McEwen, 1976). However, although some of these effects may be mediated through interaction of the steroid with specific cytoplasmic receptors, other cellular mechanisms of action may be involved (e.g., influences on the electric potentials of the neurons).

Considerable evidence has accumulated indicating that corticoids exert negative-feedback inhibition on ACTH secretion at both the hypothalamic (Hillhouse and Jones, 1976; Buckingham and Hodges, 1974; Vermes *et al.*, 1977) and the pituitary level (Arimura *et al.*, 1969; Kraicer and Mulligan, 1970; Koch *et al.*, 1975; Mulder and Smelik, 1977). We were interested in determining whether a functional relationship may exist between the binding of glucocorticoids to specific pituitary receptors and the regulation of ACTH secretion, as induced by either crude CRF preparations or cAMP. A major difficulty in such a study stems from the heterogeneous population of pituitary cells and the fact that the activity of prolactin-, as well as ACTH-secreting cells (Euker *et al.*, 1975), may be impaired by corticoids.

The results presented in Fig. 15 are from a study carried out using pieces of anterior pituitaries, which, after being treated with varying doses

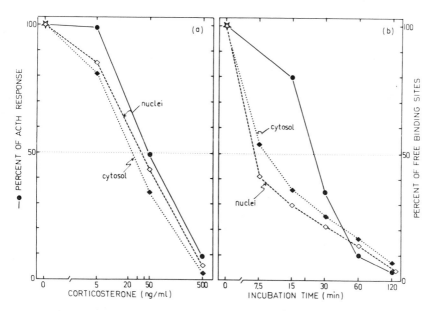

Figure 15. Correlation of CRF-induced release of ACTH and corticoid binding to both pituitary cytosol and nuclei, as expressed in percentage of control values. The effect of increasing doses of CORT (a) and temporal relationship (b) are represented. Reproduced from Koch *et al.* (1978d).

of CORT and for various time periods, were tested for their ability to respond to CRF extracts and to bind [³H]-CORT specifically. It appeared that half-maximum inhibition of ACTH release was yielded with a dose of steroid in reasonably good agreement with that required to half-saturate nuclear binding sites. Similar correlations have been reported to occur in the liver (Beato et al., 1972) and in hepatoma cells (Rousseau et al., 1972). Moreover, it was evident from time-course experiments that a temporal relationship existed between the binding step and the physiological response elicited, since the latter was delayed in onset (the lag time for maximum inhibition was 60 min). Thus, the length of time for which the steroid remains complexed to the receptor is of critical importance and most likely corresponds to the duration required for CORT to induce a specific sequence of events, probably mediated via DNA–RNA synthesis (Arimura et al., 1969). In this connection, it is of interest to note that DX, which exhibits a greater ability than CORT to occupy cell nuclear sites, also showed a significantly higher inhibitory effect on ACTH release (Koch et al., 1978c).

Finally, it has to be pointed out that the stimulatory effects of not only both CRF and VP, but also of dibutyryl cAMP, on ACTH release were abolished by CORT treatment of pituitary glands (Koch et al., 1978c). This provides evidence that the site of glucocorticoid inhibition is subsequent to activation of the adenylate cyclase system. By contrast, steroids such as progesterone and testosterone (17β-hydroxy-4-androsten-3-one) have been reported to exert effects on gonadotropin release at sites prior to and both prior to and beyond generation of cAMP, respectively (Drouin et al., 1978).

5. Conclusions

Glucocorticoids exert a number of biochemical influences, implicating behavioral as well as neuroendocrine consequences, on the brain–pituitary axis. Although the mechanisms of action of these multiple effects are not clearly understood and warrant further investigation, it seems that at least at the pituitary level, the occupancy of specific cytoplasmic and nuclear receptor sites by corticoids and the modulation of corticotropic activity could be closely correlated. In this connection, an interesting finding was the observation of multiple corticoid-binding components in the cytosol of the adenohypophysis and the localization of transcortin-like macromolecules on the plasma membrane of pituitary cells. This is likely to play a subtle role, which remains to be clearly elucidated, in the mechanism of action of these steroids.

Another point that calls for further investigation refers to the precise functional implication of vasopressin in the regulation of ACTH secretion, since the neurohormone is present in high amounts in both the adenohypophysis and portal blood and seems to exert CRF-like effects by interacting with pituitary receptor sites apparently different from those involved in the binding of CRF.

ACKNOWLEDGMENTS. This work was supported by the CNRS (ERA No. 178).

DISCUSSION

JONES: Am I not right in saying that there is little or no vasotocin in the rat hypothalamus and so it is unlikely to be a CRF? Also, your data suggest that while ADH is one of the CRFs, it is unlikely to be the major one. We are left with the problem of explaining why the CRF is found to cross-react with ADH-antibody in some laboratories, but not in others. Do you know what part of the ADH molecule your antibody reacts with as compared to that of Dr. Lowry?

KOCH: I agree that vasotocin is probably not a CRF in the rat. Our antibody is apparently secreted against a part of the vasopressin molecule which, unlike that of Lowry's group, is not shared by CRF material. A precise answer to your question would require further studies on the kind of vasopressin fragments (or analogues) being able to complex with the antibody.

VOLKAER: Could you comment on the physiologic significance of the trancortin-like binding of corticoids in the pituitary as far as the regulation of ACTH secretion is concerned? How can you exclude that it is a serum contaminant? Is there any correlation between site occupancy and the concentration of the steroids present in the medium?

KOCH: Transcortin-like molecules may serve as a competitor for glucocorticoid binding to free receptor and modulate transfer of the receptor–hormone complex to the nuclei. This binding appears, therefore, to operate as a regulator of the steroid translocation into the nucleus. Dexamethasone, which is not bound by transcortin-like material, is taken up to a greater extent than corticosterone by the nuclei and exhibits a higher potency in inhibiting ACTH release.

The possibility that the transcortin-like binder is a contaminant from the serum is ruled out by studies carried out using isolated pituitary cells and by direct measurements of blood contamination of pituitary cytosol.

As to the last question, the Scatchard plot shown in our figure is a direct answer to that point.

LABRIE: What is the proportion of [^{3}H]corticosterone bound to transcortin-like material to glucocorticoid-binding components in anterior pituitary cytosol?

KOCH: The proportion of the transcortin-like components in pituitary cytosole was approximately one-third of the total amount of glucocorticoid-binding sites.

CRABBÉ: Could you not, as a further attempt to make a distinction between circulating transcortin and the material you are studying in rat pituitary cells, examine other animal species that have little circulating transcortin?

KOCH: The dog, which has a very low amount of transcortin in the blood, could be used for that purpose. However, as we mentioned, various experimental conditions lead to marked changes in the concentration of circulating transcortin which are closely followed by the concentration of the transcortin-like component in the pituitary.

KERDELHUÉ: Did you report that vasotocin stimulated ACTH release? Also, was there cross-reaction between the anti-vasopressin serum you used and vasotocin?

KOCH: The data I mentioned were those of Buckingham et al. (J. Endocrinol. **74**:297, 1977) who actually showed that vasotocin, unlike vasopressin, exhibited a CRF-like effect closely related to that of hypothalamic extracts. This finding, however, was not confirmed by recent studies of Lowry's group. The antiserum used in our studies cross-reacted with vasotocin, but not with oxytocin. However, there is little vasotocin in the hypothalamus of the rat.

TIXIER-VIDAL: How do you explain the identity of the transcortin-like component of the cells with the plasma transcortin? Also, would it be possible to study the binding of labeled transcortin to the cell surface?

KOCH: We have shown the existence of a striking parallel between the concentration of transcortin in plasma and that of the transcortin-like component in the pituitary. However, it has to be emphasized that blood contamination of the tissue could be excluded on the basis of studies carried out using isolated pituitary cells and by direct measurement of blood contamination.

In regard to your second question, binding of labeled transcortin to the cell surface would be an interesting study, provided one could get purified rat transcortin. This would also permit autoradiographic studies to be made.

REFERENCES

Agarwal, M. K. (ed.), 1977, *Multiple Forms of Steroid Hormone Receptors*, Elsevier/North-Holland, Amsterdam.

Arimura, A., Bowers, C. U., Schally, A. V., Saito, M., and Muller, M. C., 1969. Effect of corticotropin-releasing factor, dexamethasone and actinomycin D on the release of ACTH from rat pituitaries *in vitro* and *in vivo*, *Endocrinology* **85**:300.

Beato, M., and Feigelson, P., 1972, Glucocorticoid-binding proteins of rat liver cytosol, *J. Biol. Chem.* **247**:7890.

Beato, M., Kalimi, M., and Feigelson, P., 1972, Correlation between glucocorticoid binding to specific liver cytosol receptors and enzyme induction *in vivo*, *Biochim. Biophys. Acta* **47**:1464.

Buckingham, J. C., and Hodges, J. R., 1974, Interrelationships of pituitary and plasma corticotrophin and plasma corticosterone in adrenalectomized and stressed, adrenalectomized rats, *J. Endocrinol.* **63**:213.

Buckingham, J. C., and Hodges, J. R., 1977. The use of corticotrophin production by adenohypophysial tissue *in vitro* for the estimation of potential corticotrophin releasing factors, *J. Endocrinol.* **72**:187.

Chateau, M., Burlet, A., Marchetti, J., and Boulangé, M., 1973, Dosage d'une substance antidiurétique anté-hypophysaire chez le rat surrénalectomisé, *J. Physiol.* **67**:182A.

De Kloet, E. R., and McEwen, B. S., 1976, A putative glucocorticoid receptor and a transcortin-like macromolecule in pituitary cytosol, *Biochim. Biophys. Acta* **421**:115.

De Kloet, E. R., Wallach, G., and McEwen, B. S., 1975, Differences in corticosterone and dexamethasone binding to rat brain and pituitary, *Endocrinology* **96**:598.

De Kloet, E. R., Burbach, P., and Mulder, G. H., 1977, Localization and role of transcortin-like molecules in the anterior pituitary, *Mol. Cell. Endocrinol.* 7:261.

De Wied, D., Witter, A., Versteeg, D. H. G., and Mulder, A. H., 1969, Release of ACTH by substances of central nervous system origin, *Endocrinology* 85:561.

Drouin, J., Lavoie, M., and Labrie, F., 1978, Effect of gonadal steroids on the luteinizing-hormone and follicle-stimulating hormone response to 8-bromo-adenosine 3'5'-mono-phosphate in anterior pituitary cells in culture, *Endocrinology* 102:358.

Euker, J. S., Meites, J., and Riegle, G. D., 1975, Effects of acute stress on serum LH and prolactin in intact, castrate and dexamethasone-treated male rats, *Endocrinology* 96:85.

Feldman, D., Funder, J. W., and Edelman, I. S., 1973, Evidence for a new class of corticosterone receptors in the rat kidney, *Endocrinology* 92:1429.

Fleischer, N., and Vale, W., 1968, Inhibition of vasopressin-induced release from the pituitary by glucocorticoids *in vitro*, *Endocrinology* 83:1232.

Gillies, G., and Lowry, P., 1976, Investigation of CRF using the perfused anterior rat pituitary cell column, *Proceedings of the VIth International Congress of Endocrinology, Hamburg*, Abstract No. 299.

Harrison, L. C., Billington, T., East, I. J., Nichols, R. J., and Clark, S., 1978, The effect of solubilization on the properties of the insulin receptor of human placental membranes, *Endocrinology* 102:1485.

Harrison, R. W., Fairfield, S., and Orth, D. N., 1976, Multiple glucocorticoid binding components of intact AtT-20/D-I mouse pituitary tumor cells, *Biochim. Biophys. Acta* 444:487.

Hedges, G. H., Yates, M. B., Marcus, R., and Yates, F. E., 1966, Site of action of vasopressin in causing corticotropin release, *Endocrinology* 79:328.

Hillhouse, E. W., and Jones, M. T., 1976, Effect of bilateral adrenalectomy and cortico-steroid therapy on the secretion of corticotrophin-releasing factor activity from the hypothalamus of the rat *in vivo*, *J. Endocrinol.* 71:21.

Jones, M. T., Hillhouse, E., and Burden, J., 1976, Secretion of corticotropin-releasing hormone *in vitro*, in: *Frontiers in Neuroendocrinology* (L. Martini and W. F. Ganong, eds.), pp. 195–226, Raven Press, New York.

Koch, B., Lutz-Bucher, B., Briaud, B., and Mialhe, C., 1975, Glucocorticoid binding to adenohypophysis receptors and its physiological role, *Neuroendocrinology* 18:299.

Koch, B., Lutz-Bucher, B., Briaud, B., and Mialhe, C., 1976a, Heterogeneity of pituitary glucocorticoid binding: Evidence for a transcortin-like compound, *Biochim. Biophys. Acta* 444:497.

Koch, B., Lutz-Bucher, B., Briaud, B., and Mialhe, C., 1976b, Sex difference in glucocorticoid binding to the adenohypophysis, *Horm. Metab. Res.* 8:402.

Koch, B., Lutz-Bucher, B., Briaud, B., and Mialhe, C., 1977, Glucocorticoids binding to plasma membranes of the adenohypophysis, *J. Endocrinol.* 73:399.

Koch, B., Lutz-Bucher, B., Briaud, B., and Mialhe, C., 1978a, Inverse effect of corticosterone and thyroxine on glucocorticoid binding sites in the anterior pituitary gland, *Acta Endocrinol.* 88:29.

Koch, B., Lutz-Bucher, B., Briaud, B., and Mialhe, C., 1978b, Specific interaction of corticosteroids with binding sites in the plasma membranes of the rat anterior pituitary gland, *J. Endocrinol.* 79:215.

Koch, B., Lutz-Bucher, B., Briaud, B., and Mialhe, C., 1978c, Relationships between ACTH secretion and corticoid binding to specific receptors in perfused adenohypophyses, *Neuroendocrinology* 28:169.

Kraicer, J., and Mulligan, J. U., 1970, Suppression of ACTH release from adenohypophysis by corticosterone: An *in vitro* study, *Endocrinology* 87:37.

Krieger, D. T., and Liotta, A., 1977, Pituitary ACTH responsiveness in the vasopressin deficient rat, *Life Sci.* **20:**327.

Lutz, B., Koch, B., and Mialhe, C., 1974, Présence et mode d'action de l'hormone antidiurétique au niveau de l'antéhypophyse du rat, *C. R. Acad. Sci.* **279:**1903.

Lutz-Bucher, B., 1976, Pituitary AVP and corticotropic activity, Ph.D. thesis, University of Strasbourg.

Lutz-Bucher, B., Koch, B., and Mialhe, C., 1977, Comparative *in vitro* studies on corticotropin releasing activity of vasopressin and hypothalamic median eminence extracts, *Neuroendocrinology* **23:**181.

Maclusky, N. J., Turner, B. B., and McEwen, B. S., 1977, Corticosteroid binding in rat brain and pituitary cytosols: Resolution of multiple binding components by polyacrylamide gel based isoelectric focusing, *Brain Res.* **130:**564.

McEwen, B. S., 1976, Glucocorticoid receptors in neuroendocrine tissue, in: *Proceedings of the Vth International Congress of Endocrinology, Hamburg* (V. H. T. James, ed.), pp. 23–29, Excerpta Medica, Amsterdam and Oxford.

McEwen, B. S., Wallach, G., and Magnus, C., 1974, Corticosterone binding to hippocampus: Immediate and delayed influences of adrenal secretion, *Brain Res.* **70:**321.

Milgrom, E., and Baulieu, E. E., 1970, Progesterone in uterus and plasma. I. Binding in rat uterus 105,000g supernatant, *Endocrinology* **87:**276.

Mulder, G. H., and Smelik, P. G., 1977, A superfusion system technique for the study of the sites of action of glucocorticoids in the rat hypothalamus–pituitary–adrenal system *in vitro*. I. Pituitary cell superfusion, *Endocrinology* **100:**1143.

Olpe, H. S., and McEwen, B. S., 1976, Glucocorticoid binding to receptor-like proteins in rat brain and pituitary: Ontogenetic and experimentally induced changes, *Brain Res.* **105:**121.

Pietras, R. J., and Szego, C. M., 1977, Specific binding sites for estrogen at the outer surfaces of isolated endometrial cells, *Nature (London)* **265:**69.

Pietras, R. J., Hutchens, T. W., and Szego, C. M., 1978, Hepatocyte plasma membrane subfractions enriched in high-affinity, low-capacity binding sites specific for estradiol-17β, The Endocrine Society, Abstract No. 3.

Portanova, R., and Sayers, G., 1973, Isolated pituitary cells: CRF-like activity of neurohypophysial and related polypeptides, *Proc. Soc. Exp. Biol. Med.* **143:**661.

Renlund, S., 1978, Identification of oxytocin and vasopressin in the bovine adenohypophysis, *Uppsala Dissertations from the Faculty of Science* **17.**.

Rousseau, R. R., Baxter, J. D., and Tomkins, G. M., 1972, Glucocorticoid receptors: Relationship between steroid binding and biological effects, *J. Mol. Biol.* **67:**99.

Saffran, M., and Schally, A. V., 1977. The status of the corticotropin releasing factor, *Neuroendocrinology* **24:**359.

Stevens, W., Reed, D., Erickson, S., and Grosser, B., 1973, The binding of corticosterone to brain proteins: Diurnal variation, *Endocrinology* **93:**1152.

Suyemitsu, I., and Terayama, H., 1975, Specific binding sites for natural glucocorticoids in plasma membranes of rat liver, *Endocrinology* **96:**1499.

Van Loon, G. R., 1973, Brain catecholamines and ACTH secretion, in: *Frontiers in Neuroendocrinology* (W. F. Ganong and L. Martini, eds.), pp. 209-247, Oxford University Press, London and New York.

Vermes, I., Mulder, G. H., and Smelik, P. G., 1977, A superfusion system technique for the study of the sites of action of glucocorticoids in the rat hypothalamus–pituitary–adrenal system *in vitro*. II. Hypothalamus–pituitary cell–adrenal cell superfusion, *Endocrinology* **100:**1153.

Yasuda, N., Greer, M. A., Greer, S. E., and Pantou, P., 1978, Studies on the site of action of vasopressin in inducing adrenocorticotropin secretion, *Endocrinology* **103:**906.

Yates, F. E., Russel, S. M., Dallman, M. F., Hedge, G. H., McCann, S. M., and Dhariwal, A. P. S., 1971, Potentiation by vasopressin of corticotropin release induced by corticotropin-releasing factor, *Endocrinology* **88**:3,

Zimmerman, E. A., Carmel, P. W., Husain, M. K., Ferin, M., Tannenbaum, M., Frantz, A. G., and Robinson, A. G., 1973, Vasopressin and neurophysin: High concentrations in monkey hypophyseal portal blood, *Science* **182**:925.

28

Corticotropin Secretion

Mortyn T. Jones and Brian Gillham

1. Introduction

The secretion of corticotropin (ACTH) is episodic, with a frequency and duration that give rise to the familiar circadian rhythm in plasma ACTH and corticosteroid concentrations. ACTH is synthesized in the adenohypophysis and, in those species in which it persists into the postnatal period, the intermediate lobe. There seems little doubt, however, that the adenohypophysis is the major source of circulating ACTH.

ACTH is a single-chain polypeptide consisting of 39 amino acid residues, and in most species studied, the N-terminal 24 amino acids are identical. This part of the molecule is essential for adrenocortical activity, although slight activity is observed with the peptides consisting of residues 1–17 and 1–18. The C terminus, while not responsible for biological activity, does appear to prolong the biological half-life of ACTH. A review of the biological effects of different parts of the molecule is to be found in Schwyzer (1977).

Peptides related to ACTH are also secreted by the pituitary and share a common sequence of 7 amino acids, i.e., a heptapeptide core (for a review, see Li, 1972). The peptides α-melanocyte-stimulating hormone (α-MSH) and corticotropin-like intermediate peptide (CLIP), formed in the intermediate lobe, are derived from ACTH; they are present in the human fetus, but not later because the intermediate lobe disappears (Scott et al., 1976). The cells containing ACTH in the adenohypophysis, the corticotroph

Mortyn T. Jones and Brian Gillham • Sherrington School of Physiology and Department of Biochemistry, St. Thomas's Hospital Medical School, London SE1, England

cells, also contain β-lipotropin (β-LPH), a single chain of 91 amino acid residues that may be cleaved to yield γ-lipotropin (γ-LPH) (residues 1–58) and C-peptide (β-endorphin) (residues 61–91). The opioid peptide met-enkephalin also appears to derive from β-LPH (residues 61–65). Both β- and γ-LPH are secreted in man in conjunction with ACTH. β-Endorphin ("C-peptide") is not secreted in man except in Cushing's disease (pituitary adenoma) and Nelson's syndrome (Krieger, personal communication). Met-enkephalin is not secreted from the pituitary gland. The function, if any, of β- and γ-LPH has not been defined. Last, it should be mentioned that β-MSH does not circulate in man, and its detection in the past was a result of extraction and measurement artifacts (Lowry et al., 1977).

Differences in the ratios of these peptides (LPHs and ACTH) in the circulation can be used clinically to determine whether a patient is suffering from Cushing's disease (pituitary adenoma) or an ectopic endocrine tumor (Rees et al., 1977).

2. Control of Adenohypophyseal ACTH by Corticotropin-Releasing Factor

There is little doubt that the release of ACTH from the adenohypophysis is controlled mainly by (as yet uncharacterized) hypothalamic releasing hormones, i.e., corticotropin-releasing hormones (CRHs). Until the neurohormones have been isolated and characterized, the term "corticotropin-releasing factor" (CRF) is preferred for peptides that stimulate the release and synthesis of ACTH. There appear to be two major peptides with CRF activity that satisfy many of the criteria required to establish them as authentic CRHs, one with a molecular weight of less than 1500 and one of a higher molecular weight (Gillham et al., 1976). The relationship between these two has yet to be explained. It is evident that since the amino acid sequences of ACTH and β-LPH are found in a single polypeptide of molecular weight 31,000 (Mains et al., 1977) and are secreted simultaneously, it is likely that one and perhaps both CRHs will stimulate the release of β-LPH as well as ACTH.

It should be noted that there is evidence of tissue or suppressed extra-CNS CRF (Brodish, 1977), but its nature is unknown, and any possible physiological significance has yet to be established.

3. Neurotransmitter Regulation of CRF

Most of our data on the neurotransmitter regulation of CRF is based on the use of the incubated rat hypothalamus in vitro and testing the effects of added neurotransmitters on the amount of CRF released into the medium. The hypothalami are incubated in a CSF-like medium and after a 20-min preincubation period are subjected to a protocol consisting of 5 or 10 min of exposure to a medium containing the putative neurotransmitter with intervening 10- or 20-min rest periods in neurotransmitter-free medium. The medium from 9–12 hypothalami is pooled and assayed in 6–10 48-hr basal hypothalamic-lesioned rats per group, using *in vitro* corticosterone production as the end point of the assay.

From the various observations on the effects of neurotransmitters added to the hypothalamus *in vitro* (Jones *et al.,* 1976a,b) and bearing in mind the other experimental evidence, it has been possible to draw up a model representing some of the ways in which the release of CRF appears to be modulated. A schematic representation of the model is presented in Fig. 1. This model shows two excitatory cholinergic pathways, one of which acts as an interneuron between 5-hydroxytryptamine (5-HT, sero-

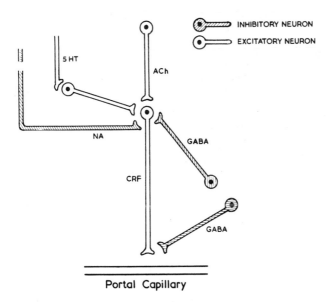

Figure 1. Suggested model for the control of CRF from the hypothalamus. The model shows: (1) The CRF neuron with its axon terminating at a portal capillary. (2) A 5-HT pathway that releases CRF after excitation of a cholinergic interneuron. (3) The final common pathway to CRF release, which is cholinergic. Two cholinergic neurons are shown. One is the interneuron placed between the 5-HT pathway and the CRF cell (?for control of the circadian rhythm); the other cholinergic neuron does not have a 5-HT synapse and may control the stress-induced release of CRF. (4) A noradrenergic (NA) pathway that is inhibitory to CRF secretion. (5) A GABA inhibitory neuron. This could be either a presynaptic or a postsynaptic input.

tonin) and the CRF cells and one in which no serotoninergic connection is indicated. The inhibitory action of the noradrenergic pathway is located at the dendritic-somal level of the CRF cell. The pathway involving γ-amino butyric acid (GABA) is shown to act at one of two possible sites, either pre- or postsynaptically. There is some evidence that suggests that the postsynaptic site of action is the more likely. Location of the CRF neurons is not possible at this time, but they probably lie in the diffuse region that is located anterior to the suprachiasmatic nucleus.

It should be mentioned at this juncture that the preparation of the hypothalamus *in vitro* maintains a reasonable electrolyte balance, consumes O_2 at a linear rate, and is capable of neurohormone formation as well as release. In the case of CRF, there is evidence to suggest that the formation of CRF is by enzymatic cleavage from a prohormone (Jones and Hillhouse, 1977).

Evidence from experiments carried out *in vivo* supports the hypothesis that there is a tonic noradrenergic inhibitory system controlling ACTH secretion (Ganong, 1972; Van Loon, 1973; Scapagnini and Preziosi, 1973). Not all the published data support this view, however, and some studies based on drug-induced depletion of catecholamines in the brain have found no net change in ACTH secretion (e.g., Kaplanski *et al.*, 1974). The development of supersensitivity to noradrenalin (within the hypothalamus) might explain why depletion studies do not always show hypersecretion of ACTH. For this reason, we investigated the effect of 6-OH-dopamine-induced depletion of noradrenalin and tested, *in vitro*, for the development of denervation supersensitivity. In these experiments, 350 μg 6-OH-dopamine was injected intraventricularly on days 1 and 3, leading to an 85% depletion in hypothalamic noradrenalin levels (1000 $\pm$ 54 to 122 $\pm$ 17 ng/g). Hypothalami were removed at 4 and 14 days (after treatment), set up *in vitro*, and stimulated with 3 pg/ml acetylcholine. The inhibitory effect of various doses of noradrenalin on the acetylcholine-induced release of CRF was next investigated. The results (Fig. 2) show that by 14 days after treatment with 6-OH-dopamine, the minimum inhibitory dose of noradrenalin had decreased from 1 ng to 100 pg/ml, thus confirming the development of supersensitivity.

A paradoxical finding in the current work is the demonstration of a bell-shaped curve to the inhibitory effect of noradrenalin. Thus, the high dose (10 ng/ml) of noradrenalin has no inhibitory effect on CRF secretion. This is contrary to our previous findings (Jones and Hillhouse, 1977), where 10 ng/ml noradrenalin inhibited CRF secretion. This lack of inhibition with the larger dose is not due to tachyphylaxis, since it occurs on the first exposure to the putative inhibitory neurotransmitter. Neither is it due to the concomitant stimulation of an excitatory β-adrenergic receptor, since the effect is not affected by 100 ng/ml propranalol (the usual inhibitory effect is exerted via an α-receptor mechanism).

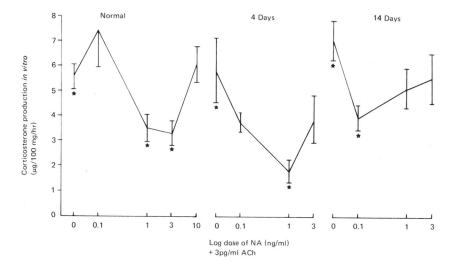

Figure 2. Development of supersensitivity to the inhibitory action of noradrenalin on the acetylcholine-induced secretion of CRF from hypothalami removed from animals pretreated 4 or 14 days previously with intraventricular 6-OH-dopamine. Each value is the mean ± S.E.M. in 6–12 animals.

The importance of the present findings is that we have demonstrated the development of supersensitivity within the hypothalamus following the destruction of noradrenergic fibers to this area of the brain. This supersensitivity may compensate largely (or totally) for the depletion of the brain amine and leave the stress response unaffected, as was reported by Kaplanski *et al.* (1974).

4. Negative-Feedback Control of the Secretion of CRF and ACTH

The secretion of ACTH in the rat is regulated by two types of feedback mechanism: a rapid-onset, rate-sensitive (fast feedback, FFB (Dallman and Yates, 1969; Jones *et al.*, 1972) and a delayed or proportional feedback (DFB) (Dallman and Yates, 1969; Yates and Maran, 1975). Rapid onset of inhibition characterizes the FFB system, and inhibition occurs only when the rate of rise of plasma corticosteroids is adequate (Jones *et al.*, 1972), while DFB is manifested 1 hr or more after the administration of steroids, when such exogenous doses may have disappeared from the blood (Smelik, 1963). Separate receptors must mediate these two forms of feedback, because several steroids have differential effects when tested in the two systems.

The precise site of action of corticosteroids is still an open question, although earlier observations tended to implicate the pituitary as the major point for inhibition (Kendall, 1971; Portanova and Sayers, 1974). More recent work has served to show that corticosteroids also suppress the release of CRF, since either electrical (Bradbury *et al.*, 1974) or acetylcholine (Hillhouse and Jones, 1976) stimulation of hypothalami taken from rats pretreated with such steroids fails to elicit a normal response. This work was not of itself conclusive, because it was possible that pretreatment of the animals with steroid caused changes in CRF secretion due to primary effect in the limbic system, a brain region rich in high-affinity binding sites for corticosteroids (McEwen, 1977).

The effects of corticosteroids added to the hypothalamus *in vitro* were therefore investigated. This technique allowed both FFB and DFB to be studied at the level of the hypothalamus. In addition, feedback at the anterior pituitary level was investigated by pretreatment of rats bearing basal hypothalamic lesions with corticosteroids, when the effect of such treatment on ACTH release induced by CRF was tested.

An estimate of the physiologically relevant dose of steroid was made by collecting and analyzing blood from the left adrenal vein of rats under stress. These results indicate that 2.1 μg corticosterone/min is secreted, the amount declining to a steady level after the first 10 min (Table 1). These data take no account of the corticosterone trapped in plasma and associated with the cellular fraction; this portion has been shown to amount to 40% of the total content of adrenal venous blood. Therefore, two adrenals, under conditions of maximum stress, secrete about 4.6 μg/

Table I. Secretion of Corticosterone into Adrenal Venous Blood Collected from One Adrenal of Anesthetized Rats (Means ± S.E.M.)

Time from adrenal vein cannulation (min)	Number of animals	Plasma flow rate (ml/min)	Adrenal venous plasma corticosterone concentration (μg/ml)	Corticosterone secretion rate (μg/min)
0–10	10	0.050 ± 0.008	41.5 ± 1.5	2.1 ± 0.34
10–20	9	0.039 ± 0.003	30.2 ± 2.9	1.2 ± 0.31
20–30	9	0.044 ± 0.003	26.6 ± 2.3	1.2 ± 0.13
30–40	9	0.043 ± 0.007	27.6 ± 3.1	1.2 ± 0.23
40–50	9	0.048 ± 0.005	26.0 ± 3.0	1.3 ± 0.19
50–60	8	0.047 ± 0.006	27.5 ± 2.7	1.3 ± 0.21
60–70	7	0.051 ± 0.004	21.6 ± 3.0	1.1 ± 0.18
70–80	4	0.046 ± 0.008	25.5 ± 4.4	1.2 ± 0.29

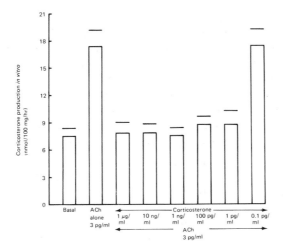

Figure 3. Effects of various doses of corticosterone added simultaneously with acetylcholine (ACh) (3 pg/ml) on the release of CRF as shown by corticosterone production *in vitro*. Each value is the mean ± S.E.M. of observations in 12–15 hypothalami and assayed in 9–12 assay rats.

min during the first 10 min and 2.3 µg/min thereafter. Since, in general, most adrenocortical stress responses persist for 1–3 hr, the maximum amount secreted should not exceed 500 µg. Studies involving the use of steroid doses grossly above this quantity are therefore pharmacological, not physiological.

An alternative way of assessing the physiological relevance of the experiments *in vitro* is to compare the concentration used in such studies with the free (unbound) moiety of corticosterone in plasma. The resting concentration of the steroid in the plasma of the rats we have used is 3.0 µg/100 ml, and this increases 10-fold in stress. On the assumption that 95% of the steroid is bound, this yields a basal value of 0.15 µg/100 ml for free corticosterone, rising to 1.5 µg/100 ml under conditions of stress.

4.1. Studies at the Hypothalamus

4.1.1. Fast Feedback

The effect of various doses of corticosterone, added simultaneously with acetylcholine, on the release of CRF from the hypothalamus *in vitro* is illustrated in Fig. 3. Doses of the steroid of 1 pg/ml or greater cause a significant ($P < 0.01$) reduction in CRF secretion in response to the neurotransmitters, although basal release of the factor is unaffected. These findings apply equally to hypothalami taken from intact animals or those adrenalectomized either 1 (as shown) or 14 days previously. Cortisol, but not progesterone, behaves like corticosterone. However, when either of these two FFB inhibitors is present in the medium for up to 1 hr in doses of 10 ng/ml, it is without effect on basal secretion.

Figure 4 shows the outcome of an experiment designed to determine the time course of inhibition by corticosterone of the response to acetylcholine (Jones *et al.*, 1977). Again, the initial, highly significant inhibition is seen at 10 min; thereafter, the degree of inhibition is reduced after 20 min in the presence of the steroid. When the tissue is transferred to steroid-free medium, the response to the neurotransmitter returns to values not significantly different from control stimulations. Subsequent challenges of the tissue are characterized by a gradual decline in the response, so that at 60 min after removal of the corticosterone, there is a significant inhibition of the response ($P < 0.01$) as compared with that elicited 10 min after removal of the steroid. Control experiments performed in the absence of added steroid serve to show that only minor variations in tissue responsiveness to acetylcholine occur during the period of the experiment.

Thus, in one experiment, the two phases of inhibition termed fast and

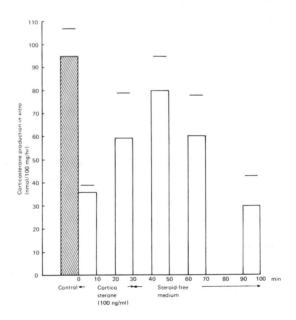

Figure 4. Time course for the inhibitory effect of corticosterone on the acetylcholine-induced release of CRF from the hypothalamus *in vitro*.

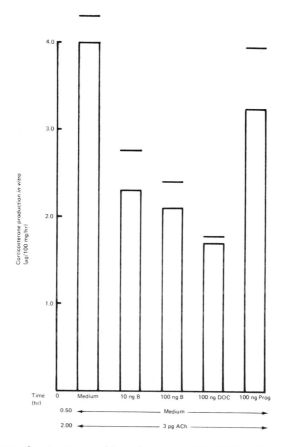

Figure 5. Effects of pretreatment with corticosterone (B), DOC, and progesterone (Prog) *in vitro* on acetylcholine (ACh)-induced (3 pg/ml) release of CRF. Each value is the mean ± S.E.M. of observations in 9–12 animals.

delayed feedback are demonstrated. The first period is of 20-min duration, and there is a requirement for steroid in the medium. The onset of the second period is not observed until 1 hr after removal of the corticosterone. The interim, when no inhibition is seen, correlates with the "silent period" noted to occur *in vivo* (Dallman and Yates, 1969).

4.1.2. Delayed Feedback

To study more closely the DFB inhibition seen *in vitro*, 11-deoxycorticosterone (DOC) and progesterone, in addition to corticosterone, were tested (Jones *et al.*, 1977). Figure 5 shows that corticosterone in both doses used and DOC, but not progesterone, are effective DFB inhibitors when added to (and removed from) the medium 2 hr before neurotransmitter stimulation of hypothalami. Thus, in these experiments, FFB inhibition is initiated by the doses of steroids within the basal as well as the stress-induced concentration and DFB by unbound levels seen in response to stress.

4.1.3. Studies on the Mechanisms Underlying the Two Phases of Feedback Inhibition

4.1.3a. Content Studies. For the experiments under this heading, the release of CRF from hypothalami *in vitro,* under various conditions, was induced by using acetylcholine, and at the end of the incubation time the CRF contained in the tissues was extracted into 0.1 M HCl for assay. The results for tissues that received either an FFB or a DFB signal are presented in Fig. 6. It can be seen that corticosterone, used as DFB effector, causes the expected inhibition of release, but is without effect on the content of CRF. As an FFB effector, the steroid inhibits release and at the same time causes a significant increase in hypothalamic content of CRF ($P < 0.05$). These observations suggest that as an FFB inhibitor, corticosterone has no effect on the formation of CRF, but that a characteristic of a DFB effect is that synthesis is prevented (Jones *et al.,* 1977).

4.1.3b. Pathways of Inhibition. The ability of corticosteroids to initiate a rapid inhibition of CRF release raised the possibility that they act by stimulating neuroinhibitory pathways. Thus, it was possible that added corticosteroid causes the release of endogenous GABA or noradrenalin. Experiments performed to test this hypothesis (Jones and Hillhouse, 1976) showed that when FFB inhibition is caused by corticosterone, neither phentolamine nor picrotoxin is effective in preventing the inhibition. Thus, for experiments *in vitro,* the FFB effect of corticosterone is not mediated by the release of endogenous GABA or noradrenalin.

4.1.3c. Potassium Studies. As a second approach to an understanding of the nature of FFB inhibition, the ability of corticosterone to prevent the release of CRF induced by depolarizing concentrations of potassium ions was investigated (Jones and Hillhouse, 1976). Hypothalami incubated in media with raised K^+ concentrations (12–48 mM) show a higher basal secretion of CRF than normal, and this elevated secretion was unaffected by corticosterone. Smelik (1977) has shown a similar failure of dexamethasone to inhibit veratridine-stimulated CRF release, while spontaneous release of the neurohormone was blocked. The failure of the steroid to prevent the increased release of CRF caused by a depolarizing concentration of K^+ argues that the corticosteroid exerts its FFB effect by stabilizing membranes. One other possibility is that corticosterone may act by affecting K^+ flux and hence cell depolarization. When this was

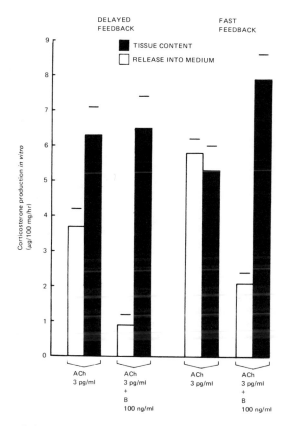

Figure 6. Effects of either acetylcholine (ACh) or ACh + corticosterone (B) on the release and tissue content of CRF. Each value is the mean ± S.E.M. of observations in 9–12 animals.

tested by incubating hypothalami in a medium containing a lower than normal (2 mM) concentration of K^+, it was found that the ability of the tissues to release CRF in response to acetylcholine was not impaired (Fig. 7). An alternative hypothesis would be that K^+ opens up the Ca^{2+} channels to an extent that could not be blocked by corticosterone. It therefore became important to study the role of Ca^{2+} in CRF secretions.

4.1.3d. Calcium Studies. Since the role of calcium ions in stimulus–secretion coupling is central, the effect of alterations in the concentration of Ca^{2+} in the media used to incubate the hypothalami *in vitro* was studied. Figure 8 shows that increases in the concentration of added Ca^{2+} from 0 to 6 mM cause an increase of both basal and stimulated release of CRF into the medium. The data demonstrate the requirement for Ca^{2+} in acetylcholine-mediated output of CRF.

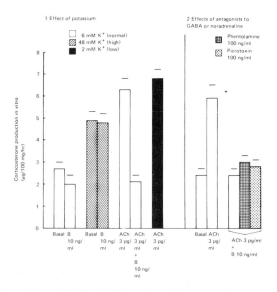

Figure 7. (1) Effects of acetylcholine (ACh) and corticosterone (B) on the release of CRF from hypothalami incubated in media containing various concentrations of K⁺. (2) Effects of phentolamine and picrotoxin on the corticosterone (B)-induced inhibition of the release of CRF in response to ACh. Each histogram is the mean + S.E.M. of a minimum of 10 observations.

The possibility that the enhanced release of CRF occasioned by high Ca^{2+} is associated with a release from endogenous stores of acetylcholine seems unlikely, in view of the observation that hexamethonium is ineffective in preventing Ca^{2+}-induced secretion of CRF.

At low concentration (10 pg/ml), corticosterone has no effect on the basal release of CRF in a medium containing high Ca^{2+}, but at a concentration of 1 ng/ml, there is a significant reduction in the release (Fig. 8). This suggests that the FFB action of corticosterone might be mediated via an effect on Ca^{2+} flux. Confirmation of this view derives from an experiment that showed the inhibitory action of corticosterone to be mimicked by the presence in the medium of 12 mM manganese II ions (known to block Ca^{2+} channels) (Fig. 8).

4.2. Studies at the Anterior Pituitary

4.2.1. Fast Feedback

Table II shows that 70 μg corticosterone administered subcutaneously (s.c.) to rats with basal hypothalamic lesions significantly reduced the release of ACTH induced by an intravenously injected standard dose of

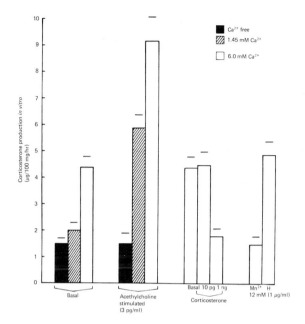

Figure 8. Effects of acetylcholine, corticosterone, manganese (Mn²⁺), and hexamethonium (H) on CRF secretion from hypothalami incubated in media containing various concentrations of calcium (Ca²⁺). Each histogram is the mean ± S.E.M. of a minimum of 10 observations.

Table II

Time after corticosterone (min)	Corticosterone production *in vitro* (μg/100 mg adrenal/hr)[a]
0	6.4 ± 0.3
5	5.1 ± 0.4
10	2.5 ± 0.3
20	2.5 ± 0.3
30	2.9 ± 0.4
50	6.6 ± 0.5

[a] Each value represents the mean ± S.E.M.

CRF when the response was tested 10–30 min after steroid administration. At 50 min, the response had returned to normal (Jones and Hillhouse, 1976).

4.2.2. Silent Period

Figure 9 shows that at 20 min after the administration of corticosterone (500 μg, s.c.), the expected inhibition of the ACTH response to CRF is noted. The inhibition disappears after 40 min, to reappear once more 100 min after the injection of the steroid. Thus, the pituitary shows the same biphasic response as the hypothalamus to corticosteroid inhibition. The only difference is that FFB at the pituitary does not show rate sensitivity (Jones and Hillhouse, 1976). Thus, since this property also characterizes the FFB mechanism as it exists *in vivo*, the rate sensor must be located at a level higher than that of the pituitary.

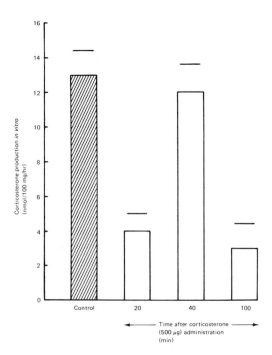

Figure 9. Effects of the intravenous injection of a standard dose of CRF in rats bearing basal hypothalamic lesions, at various intervals after the administration of corticosterone (500 μg, s.c.).

Table III

Treatment	Pretreatment time	
	4 hr	24 hr
	Corticosterone production *in vitro* (μg/100 mg adrenal/hr)[a]	
Vehicle	6.1 ± 0.5	5.9 ± 0.4
Corticosterone	1.6 ± 0.3	0.8 ± 0.2
Cortisol	1.1 ± 0.2	0.4 ± 0.2
DOC	2.3 ± 0.6	2.5 ± 0.3

[a] Each value represents the mean ± S.E.M.

4.2.3. Delayed Feedback

To complete the observations on feedback at the pituitary, Table III shows that several steroids when given in a dose of 2 mg, s.c., 4 or 24 hr previously were able to reduce the release of ACTH induced by an intravenously administered standard dose of CRF.

These results resemble very closely the elegant data produced by Smelik (1977), using dispersed anterior pituitary cells in a perfusion system, in which a two-phased inhibition of CRF-evoked secretion of ACTH was demonstrated when 0.2 μg/ml corticosterone was added to the perfusion medium. Thus, the existence of an FFB and a DFB mechanism (with an intervening silent period) has been demonstrated both *in vivo* and *in vitro*.

5. Possible Relationships between the Binding of Glucocorticoids to the Anterior Pituitary and Inhibition of ACTH Secretion in Vitro

From the time it became established that, for the rat, the secretion of ACTH is regulated by two quite distinct feedback mechanisms, possible substrates for the controls have been sought. In the case of the rate-sensitive FFB inhibition, the very rapidity of its onset pointed to an effect at or near the plasma membrane, rather than in the nucleus of the cell. This view found support in respect of FFB inhibition at the hypothalamus in the experiments of Jones and Hillhouse (1976), who demonstrated that the inhibition by corticosterone might well be mediated by direct effects on calcium fluxes into the cells. Comparable data for inhibition at the level of the anterior pituitary are lacking, but FFB effects at the plasma membrane are *a priori* quite likely.

The nature of CRF is not established, and part of the problem (perhaps a large part) is that many compounds appear to act as CRFs. On the surface, there seems to be little structural relationship among these various molecules. Thus, we have found that compounds as divergent as met-enkephalin, bromocriptine, arginine vasopressin and histidyl-prolyl-diketopiperazine cause the release of ACTH when injected, in moderate doses, into rats bearing basal hypothalamic lesions. In addition, apparent "breakdown products" obtained in the course of the purification of CRF appear to retain substantial biological activity. These observations have suggested to us that the "receptor" for CRF in corticotroph cells is really rather nonspecific; indeed, reception may involve only changes in membrane fluidity induced by hydrophobic, somewhat basic peptides. In this context, Lembeck (1978) has shown that peptide, another basic substance P, does react rather specifically with phosphatidyl serine. We had felt that FFB inhibition might then result from opposing effects on membrane fluidity without specific steroid receptors. This would help to explain the wide range of natural and synthetic steroids that function as FFB agonists at the anterior pituitary when compared to similar inhibition at the hypothalamus. However, the finding of a transcortin-like binding protein in plasma membranes prepared from the anterior lobe reported by Lutz-Bucher *et al.* (1977) (also see Chapter 27) indicates that binding of steroids to a polypeptide receptor in the plasma membrane might be a factor in FFB control *in vitro*.

Both Lutz-Bucher *et al.* (1977) and Smelik (1977) have demonstrated that corticosterone inhibits the release of ACTH (possibly by inhibiting exocytosis) in response to CRF without affecting synthesis of the trophic hormone. The content of ACTH in the pituitary actually increases in the presence of corticosterone, while output into the medium is depressed. This effect on release closely parallels the FFB effects of corticosterone at the hypothalamus. Whether the corticosterone is working via a similar mechanism at the two sites has yet to be established. It is not totally clear whether the effects noted are those due to an FFB or a DFB mechanism at the pituitary. It must be taken as established, however, that corticosteroids can exert an inhibition of both CRF and ACTH release via a membrane effect.

It is now quite clear that both the hypothalamus and the pituitary are sites for feedback inhibition by corticosteroids. However, the experiments described to this point throw little light on the relative importance of the two possible sites of inhibition in a physiological sense. Studies on the structures of steroids necessary for them to exert feedback control have helped in tackling this particular problem.

6. Structure–Activity Relationships in the Action of Steroids at the Hypothalamus and the Anterior Pituitary

6.1. Studies at the Hypothalamus

6.1.1. Fast Feedback: Antagonists

The FFB inhibitory action of corticosterone at the hypothalamus is antagonized by several steroids, as demonstrated by experiments *in vitro* (Jones and Hillhouse, 1976). From the results presented in Fig. 10, it can be seen that neither the 11β-, the 17α-, nor the 21-hydroxyl group is necessary for the binding of the steroids to the receptor, since 11-

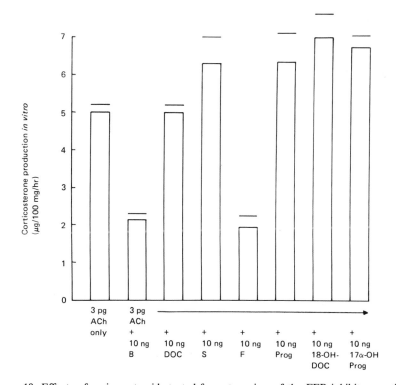

Figure 10. Effects of various steroids tested for antagonism of the FFB inhibitory action of corticosterone (B) on acetylcholine (ACh)-induced CRF release. Each column represents the mean ± S.E.M. of a minimum of 10 observations. (DOC) 11-Deoxycorticosterone; (S) 11-deoxycortisol; (F) cortisol; (Prog) progesterone; (18-OH-DOC) 18-hydroxy-11-deoxy-corticosterone; (17α-OH-Prog) 17-α-hydroxyprogesterone.

deoxycorticosterone, 11-deoxycortisol, progesterone, and 17α-progesterone are all antagonists of FFB inhibition. This is also true for 18-hydroxy-11-deoxycorticosterone, thus indicating that the presence of a hydroxyl group in the 18-position is of no consequence.

6.1.2. Delayed Feedback: Antagonists

In these experiments *in vitro* (Jones and Hillhouse, 1976), putative antagonistic steroids were added to the hypothalamus prior to and simultaneously with corticosterone. The results presented in Fig. 11 show that 17α-hydroxyprogesterone and 11α-hydroxycortisol antagonize the inhibition of CRF release caused by an equimolar dose of corticosterone. In this case, the presence of the 17α-hydroxyl group and also the 3-oxo, 4,5-ene

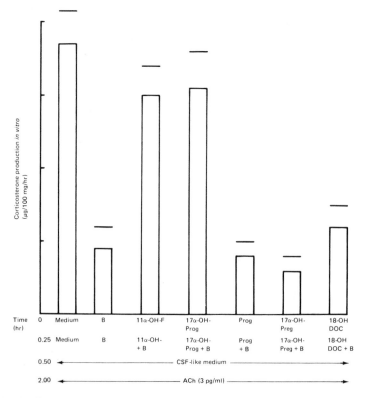

Figure 11. Effects of various steroids tested for antagonism of the DFB action of corticosterone (B) *in vitro*. Each column shows the mean ± S.E.M. of 10 observations. All steroids were added in a dose of 100 ng/ml. Abbreviations as in Fig. 10.

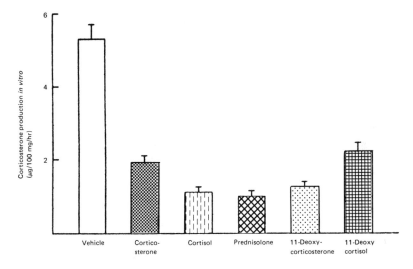

Figure 12. FFB activity of several steroids at the level of the anterior pituitary, demonstrated by their injection (100 μg, s.c.), 20 min before the administration of a standard dose of CRF, into animals bearing basal hypothalamic lesions. Values are the means ± S.E.M. of 7–12 observations.

arrangement both appear necessary since neither progesterone nor 11α-hydroxypregnenolone is active in preventing the inhibition caused by corticosterone. In addition, 18-hydroxydeoxycorticosterone was ineffective as an antagonist of DFB.

6.2. Studies at the Anterior Pituitary

6.2.1. Fast Feedback: Agonists

Many steroids show FFB activity at the level of the anterior pituitary, including corticosterone, cortisol, prednisolone, 11-deoxycorticosterone, and 11-deoxycortisol (Fig. 12). This is in marked contrast to the observations of the hypothalamic studies, in which 11-deoxycortisol and 11-deoxycorticosterone were found to be antagonists to the FFB action of corticosterone. Indeed, to date, no steroid has been found to act as an antagonist. Clearly, the structure–activity relationships for FFB at the hypothalamus and the pituitary are very different. Indeed, while the hypothalamic FFB is very specific, that in the anterior pituitary is not. Interestingly, the latter does not show rate sensitivity, since FFB inhibition is noted even during the time that plasma corticosteroid concentrations are declining. This implies that the site of unidirectional rate sensitivity, seen in the intact animal, must result from effects at a level higher than that of the pituitary.

6.2.2. Delayed Feedback: Agonists

Very many steroids show DFB, including cortisol, corticosterone, and their 11-deoxy counterparts, and 17α-hydroxypregnenolone, but not beclamethasone dipropionate (Figs. 13 and 14). This is in marked contrast to observations *in vivo* and also to experiments on the hypothalamus *in vitro*, in which 17α-hydroxypregnenolone was shown to be without effect and beclamethasone dipropionate to be an agonist (Jones *et al.*, 1976b; Jones and Tiptaft, 1977). Another contrast between the hypothalamus and the pituitary as DFB sites is that 17α-hydroxyprogesterone is an antagonist at the former, but is without effect at the latter.

7. Relative Importance of Hypothalamic and Anterior Pituitary Feedback Sites

The fact that some steroids feed back at one site, but not at another, has presented the opportunity to determine the relative importance of the

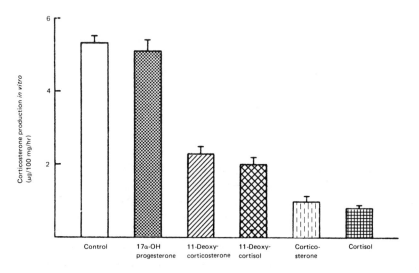

Figure 13. DFB activity of several steroids at the level of the pituitary, demonstrated by their injection (5 mg, s.c.), 4 hr before a standard dose of CRF, into animals bearing basal hypothalamic lesions. Values are the means ± S.E.M. of 7–12 observations.

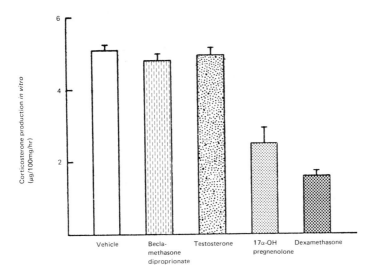

Figure 14. DFB activity of several steroids at the level of the anterior pituitary, demonstrated by their injection (10 mg, s.c.), 4 hr before a standard dose of CRF, into rats bearing basal hypothalamic lesions. Values are the means ± S.E.M. of 8–15 observations.

two sites. This is achieved by determining the effect on the stress response of pretreating animals with such steroids. The results of experiments with several steroids that show such differential effects when tested for FFB are shown in Table IV.

It is apparent that animals pretreated with 11-deoxycorticosterone or 18-hydroxy-11-deoxycorticosterone show an exaggerated stress response. This was also shown to be true for animals pretreated with 11-deoxycortisol (Jones *et al.*, 1972). Data such as these indicate that the FFB effect of these steroids at the pituitary level is of little functional significance. The exaggerated stress response is presumably due to the antagonism at the level of the hypothalamus of the negative-feedback signal derived from endogenous corticosterone. It follows, then, that the enhanced release of CRF that ensues is capable of overriding feedback at the pituitary. The correlation between the effects (if any) of corticosteroids at the hypothalamus and the influence of the same compounds on the stress response points to the fact that tissue is a principal feedback site. In accord with this conclusion is the finding of Buckingham and Hodges (1977) that much smaller doses of corticosterone are required to inhibit the release of CRF from the hypothalamus *in vitro* than are needed to prevent the release of ACTH from the pituitary *in vitro*.

Table IV. Effect on Ether-Stress-Induced
Increases in Plasma Corticosterone
Concentration in Rats Injected
Subcutaneously with Either Saline (Vehicle)
or a Steroid 10 min before Stress (FFB)

Treatment	Stress-induced increments in plasma corticosterone ($\mu g/100$ ml)
Vehicle	30.1 ± 2.7
DOC	43.1 ± 3.1
11-Deoxycortisol	41.3 ± 2.1
18-OH-DOC	47.0 ± 3.0
Cortisol	9.1 ± 3.0
Corticosterone	12.3 ± 1.5

8. Interrelationship between the Two Forms of Corticosteroid Feedback Mechanism

A matter of considerable interest is the relative importance of, and the interrelationship between, FFB and DFB mechanisms in the regulation of CRF/ACTH secretion. The mechanisms differ in their dynamics, their receptors, and their modes of action. Yet these two systems are activated by the same steroids under physiological conditions, so clearly the question "Why are two mechanisms necessary?" is by no means trivial.

The data in Table V go some way toward providing an answer to this question (Jones, Rees, and Tiptaft, unpublished work). The table shows the effect of corticosterone replacement therapy (in both 3-day-adrenalectomized and sham-operated rats) on the basal and ether-stress induced release of radioimmunoassayable ACTH. The result of adrenalectomy is to cause an increase in both basal and stress-induced secretion of ACTH as compared with sham-operated animals. It is observed that while a DFB signal (200 μg at 10:00 hr and 22:00 hr) serves to normalize basal secretion of ACTH, such a dose fails to normalize the stress response. When given alone, neither an FFB signal (150 or 550 μg 20 min before stress) nor a chronic DFB signal (350 μg, last injection 2 hr before stress) is able to normalize the stress response. However, when animals are treated with a DFB (200 μg b.i.d.) and an FFB (150 μg) dose, a normal basal ACTH level and also a normal stress response are seen. Thus, both signals are required for a normally responsive hypothalamo–pituitary axis. Of course, ACTH secretion can be normalized by using higher doses of corticosteroids as a DFB signal alone, but such doses are well in excess of the amounts of steroids the animals are able to secrete. The important point about the data in Table V is that the total dose of steroids given as FFB and DFB signals is 550 μg, within the physiological range of the rat. This argues that the two feedback mechanisms constitute an integrated system that has evolved as an economical means of regulation in that it allows the stress response to occur and to be modulated by amounts of corticosteroids that are insufficient to produce hypercorticism. Following this line of argument, it might be suggested that the DFB mechanism serves to reduce the responsiveness of the system to a level that can be modulated, on a moment-to-moment basis, by the FFB mechanism. As examples of the operation of such moment-to-moment control, one may cite the episodic drive in the axis and the condition of acute stress.

9. Interaction between Vasopressin and CRF

A considerable amount of work has been performed over some 20 years in an attempt to evaluate the role of ADH (vasopressin) in the control of corticotropin secretion. Originally, it was proposed that vasopressin and CRF were identical (McCann and Brobeck, 1954; McCann, 1957). A considerable weight of circumstantial evidence favored this concept. Many stresses cause an immediate rise in ADH and corticotropin secretion (Mirsky *et al.*, 1954), and neurosecretory material has been described entering the portal vessels (Rothballer, 1953). Last, vasopressin releases ACTH in both basal-hypothalamic-lesioned rats and corticosteroid

Table V. Integration of FFB and DFB Mechanisms in the Rat[a]

Group	Treatment	Replacement category	Plasma ACTH (pg/ml)	
			Basal	Stress
Sham-operated	Saline	None	38.4 ± 3.8	247.5 ± 37.7
Adrenalectomized	Saline	None	277.4 ± 13.2	707.6 ± 99.0
Adrenalectomized	Acute B 150 μg	Acute	—	657.7 ± 56.2
or	Acute B 550 μg	Acute	—	524.5 ± 28.3
Adrenalectomized	Chronic B 200 μg	Chronic	75.8 ± 21.0	467.8 ± 95.8
or	Chronic B 250 μg	Chronic	—	476.3 ± 35.7
Adrenalectomized	Acute B (150 μg)	Acute ⎫		
and	Chronic B (200 μg)	*and* Chronic ⎬	45.8 ± 9.6	273.5 ± 26.5

[a] All the animals used were adrenalectomized 3 days prior to the experiment, except for two groups that were sham-operated at that time. Animals that received corticosterone (B) before measurements of the basal or the stress-induced release of ACTH were given acute steroid therapy (20 min before stress) or chronic steroid therapy (at 22.00 hr and 10.00 hr, last dose 2 hr before stress), or a combination of both treatments. Stress was caused by ether inhalation at 2 min. Ether stress was followed by decapitation and blood collection at 5 min. ACTH was measured by RIA. Unpublished data from Jones, M.T., Rees, L.H., and Tiptaft, E.M.

pretreated rats. McCann (1957) found that the degree of adrenal ascorbic acid depletion was proportional to the pressor activity of vasopressin in lesioned rats, while McCann et al. (1958) found parallelism between the degree of corticotropin inhibition and diabetes insipidus in animals bearing hypothalamic lesions. De Wied (1961) showed that what were considered quite small doses of vasopressin (ca. 5 mU) would stimulate ACTH release in basal-hypothalamic-lesioned rats; similar doses of pitressin were ineffective in hypophysectomized rats.

Saffran and Saffran (1959), in a review of the literature, claimed that the dose of ADH used by McCann and co-workers was several thousand times greater than that which causes maximal antidiuresis in the rat. This argument is largely irrelevant, however, if ADH is liberated into the portal blood (Goldman and Lindner, 1962). More recently, Oliver et al. (1977) have reported high levels of vasopressin in portal vessels of the rat. Zimmerman et al. (1977) have reported high concentrations of the hormone in the portal blood of monkeys. These latter workers found a discrete tract of vasopressin-containing nerve fibers arising from the paraventricular nucleus terminating in the median eminence. Neurohypophysectomy in the rat resulted in a large fall in the concentration of vasopressin in the portal vessels, thus indicating that most, but not all, of the antidiuretic hormone derives from the posterior lobe. The question arises, therefore, whether the posterior pituitary gland plays an important role in controlling corticotropin secretion. Fortunately, there is some considerable amount of work reported in the literature on this topic.

Posterior lobectomized rats show a diminished response to "neurogenic" stress (noise, strange environment), but react normally to "systemic" stresses of histamine, nicotine, and hemorrhage (Smelik, 1960; de Wied, 1961; Miller et al., 1974). The failure of posterior lobectomized animals to respond to "neurogenic" stresses can be corrected by replacement therapy with pitressin (de Wied, 1961). It would appear, therefore, that vasopressin of posterior-lobe origin may play a role under certain circumstances, i.e., for so-called "neurogenic" stresses.

The question now remains concerning the importance of vasopressin released from the median eminence. In this case, we are on less sure ground because of the complication of a much larger contribution of the posterior lobe to portal vasopressin. It appears that the content of ADH in the median eminence is increased by adrenalectomy and decreased by corticosteroid treatment (Zimmerman et al., 1977). This suggests the evolution of a control system (of possible physiological importance for the regulation of the median eminence vasopressin. The negative-feedback effect of corticosteroids on vasopressin is similar to that reported for CRF (Jones et al., 1976b). There is therefore the possibility that the two neurohormones, CRF and vasopressin, are intimately connected. Indeed, it has recently been reported that antibody to ADH will abolish most of the ACTH-releasing activity of median eminence extracts (Gillies and Lowry, 1976), suggesting that the two peptides are very similar in structure. This argument is further strengthened by the fact that Brattleboro rats (genetically deficient in the ability to synthesize and secrete ADH) show reduced corticotrophic responses to stress. However, the deficit is not simply due to the absence of ADH, since acetylcholine and serotonin, the two neurotransmitters that release CRF *in vitro,* do so only to a reduced extent when added to hypothalami from these animals (Buckingham and Leach, personal communication).

It must be emphasized that heterozygous Brattleboro rats show behavioral deficits unrelated to any ADH effects (Bohus, personal communication). It would be easier to admit the hypothesis that these rats have genetic deficiences in addition to their failure to release ADH, and one might involve the CRF system.

It is unlikely that ADH and CRF are identical, since Koch and co-workers have shown in their work (see Chapter 27) that their ADH antibody does not cross-react with CRF, and using high-voltage electrophoresis, we have separated all CRF activity from vasopressin (Buckingham *et al.,* unpublished work).

We are therefore left with several unanswered questions concerning the role of vasopressin in the control of ACTH secretion. It would appear that ADH is a hypothalamic factor that can stimulate the release of corticotropin, but it is distinct from the principal stimulus, CRF. The evidence from posterior lobectomized animals shows that the vasopressin from the posterior lobe may be an important stimulus for some forms of stress, but not for all stresses. In addition, vasopressin has another important influence—that of potentiating the corticotrophic response to CRF (see Table VI). Thus, a dose of vasopressin that did not in itself release ACTH greatly enhanced the response to CRF. This kind of potentiation *in vivo* was first demonstrated by Yates *et al.* (1971), and the same workers demonstrated that the converse did not occur; i.e., CRF does not potentiate the release induced by vasopressin. This potentiating effect may be of importance from a regulating point of view, since the data in Table VI show that the concomitant release of vasopressin may effectively block the FFB influence of corticosterone and thus prolong the corticotrophic response to stress. Interestingly, the DFB effect of corticosteroids is unaffected.

In conclusion, it is possible that the corticotroph cell may contain receptors that are somewhat nonspecific and are capable of releasing corticotropin in response to many peptides including vasopressin. The physiological importance of these peptides may therefore reside, at least in part, in the actual amount (as well as the potency) of the peptide that

Table VI. Vasopressin Potentiation of CRF Activity at the
Anterior Pituitary and the Effects of Corticosterone Feedback

1st Injection[a]	Time	2nd Injection[b]	Corticosterone (μg/100 mg/hr $\pm$ S.E.M.)	
1. Saline	20 min	ADH	1.1 $\pm$ 0.05	
2. Saline	20 min	CRF	2.35 $\pm$ 0.005	$P < 0.01$
3. Saline	20 min	ADH + CRF	4.93 $\pm$ 0.7	
4. ADH	20 min	CRF	2.28 $\pm$ 0.1	
5. 100 μg B	20 min	CRF	1.30 $\pm$ 0.4	$P < 0.02$
6. 100 μg B	20 min	ADH + CRF	3.40 $\pm$ 0.56	
7. 500 μg B	60 min	ADH + CRF	1.48 $\pm$ 0.1	

[a] (B) Corticosterone.
[b] Doses: ADH, 20 mU; CRF, 0.5 ml 5-HT-CRF.

reaches the portal vessels and the corticotroph cells in response to various stressful stimuli.

DISCUSSION

KRAICER: The controlled variable in fast feedback is the rate of change in circulating corticosterone, yet it appears in your studies on the feedback of glucocorticoids on the hypothalamus and pituitary you did not measure rate of change.

JONES: Our data at the anterior pituitary, which I have not presented today, show that fast feedback at the anterior pituitary is not rate-sensitive because it occurs even when plasma corticosterone levels are declining, so the rate sensor must be situated above the pituitary level. In our studies on the hypothalamus *in vitro*, clearly there is a rate of change depending upon the amount of steroid that is added. Whether the hypothalamus is the major rate sensor, or the only rate sensor, remains to be established.

SMITH: 6-Hydroxydopamine injected intracisternally does not produce depletion of amines in the median eminence when measured by fluorescence histochemistry. This may explain the results of the noradrenalin experiment. Complete depletion of catecholamines of the median eminence can be obtained by intravenous injection of 6-hydroxydopamine; the catecholamine content in the remainder of the brain is unaffected.

JONES: We don't believe median eminence noradrenalin has any significant role in regulating CRF secretion. What we find is that 6-hydroxydopamine does have a profound effect on the stress response. In fact, the stress response in 6-hydroxydopamine animals is reduced. As you said, the median eminence noradrenalin is not affected by lateral ventricular injection of 6-hydroxydopamine, but the remaining noradrenalin in the brain is depleted.

VOKAER: Have you measured the rate of fast feedback evoked by corticosteroids in the presence of ionophores?

JONES: No, but we mean to do this.

Kordon: How do you reconcile your norepinephrine effects with the fact that there is almost no norepinephrine at the level of neurosecretory junctions in the median eminence? How did you control specificity of norepinephrine effects?

Jones: We suspect that the norepinephrine acts at a level higher then the median eminence. The major reason is that noradrenalin gives the best inhibition when we have the whole hypothalamus and none when we use the median eminence and arcuate segments.

Kordon: I wonder whether it is really appropriate to speak of effects on peptide "synthesis" *in vitro*. Most current data suggest that peptide biosynthesis in the brain involves complex mechanisms, with ribosomal synthesis of precursors and subsequent cleavage during or after transport. There is no evidence at present that these processes can be found in *in vitro* conditions.

Jones: This is a slip of the tongue; of course this is not *de novo* synthesis that we are seeing when we add acetylcholine and 5-hydroxytryptamine. The increase in CRF is obviously due to enzymatic cleavage of the precursor, which has no detectable ACTH-releasing activity. We suggested in a paper (*Ann. N.Y. Acad. Sci.* 297:536–560, 1977) that there was a precursor of CRF and that acetylcholine caused a flux of calcium into the cells and activated a cleaving enzyme that released recognizable CRF.

References

Bradbury, M. W. B., Burden, J. L., Hillhouse, E. W. and Jones, M. T., 1974, Stimulation electrically and by acetylcholine of the rat hypothalamus *in vitro*, *J. Physiol.* **239**:269–283.

Brodish, A., 1977, Extra-CNS corticotropin-releasing factors, *Ann. N.Y. Acad. Sci.* **297**:420–435.

Buckingham, J. C., and Hodges, J. R., 1977, Corticotrophin-releasing hormone activity of rat hypothalamus *in vitro, J. Endocrinol.* **73**:30P.

Dallman, M. F., and Yates, F. E., 1969, Dynamic asymmetries in the corticoid feedback pathway and distribution, binding and metabolism elements of the adrenocortical system, *Ann. N.Y. Acad. Sci.* **156**:696–721.

De Wied, D., 1961, The significance of the antidiuretic hormone in the release mechanism of corticotrophin, *Endocrinology* **62**:605–613.

Ganong, W. F., 1972, Evidence for a central nor-adrenergic system that inhibits ACTH secretion, in: *Brain–Endocrine Interaction: Median Eminence—Structure and Function,* Proceedings of the International Symposium, Munich 1971 (K. M. Knigge, D. E. Scott, and H. Weindl, eds.), pp. 254–266, S. Karger, Basel.

Gillham, B., Hillhouse, E. W., and Jones, M. T., 1976, Further studies in the purification of CRF from the rat hypothalamus *in vitro, J. Endocrinol.* **71**:60–61P.

Gillies, G. and Lowry, P. J., 1976, Investigation of corticotrophin releasing factor using the isolated rat pituitary cell column, *J. Endocrinol.* **71**:61–62P.

Goldman, H. and Lindner, L., 1962, Antidiuretic hormone concentration in blood perfusing the adenohypophysis, *Experientia* **18**:279–281.

Hillhouse, E. W., and Jones, M. T., 1976, Effect of bilateral adrenalectomy and corticoid therapy on the secretion of CRF activity from the hypothalamus of the rat *in vitro, J. Endocrinol.* **71**:21–30.

Jones, M. T., and Hillhouse, E. W., 1976, Structure–activity relationship and the mode of action of corticosteroid: Feedback on the secretion of CRF (corticoliberin), *J. Steroid Biochem.* **7**:1189–1202.

Jones, M. T., and Hillhouse, E. W., 1977, Neurotransmitter regulation of corticotrophin-releasing factor *in vitro, Ann. N.Y. Acad. Sci.* **297:**536–560.

Jones, M. T., and Tiptaft, E. M., 1977, Structure–activity relationships of various corticoids on the feedback control of corticotrophin secretion, *Br. J. Pharmacol.,* **59:**35–41.

Jones, M. T., Brush, F. R., and Neame, R. L. B., 1972, Characteristics of the fast feedback control of corticotrophin release by corticosteroids, *J. Endocrinol.* **55:**487–489.

Jones, M. T., Hillhouse, E. W., and Burden, J. L., 1976a, The effect of various putative neurotransmitters on the secretion of CRF from the rat hypothalamus *in vitro*—a model of the neurotransmitters involved, *J. Endocrinol.* **69:**1–10.

Jones, M. T., Hillhouse, E. W., and Burden, J. L., 1976b, Secretion of corticotrophin releasing hormone *in vitro,* in: *Frontiers of Neuroendocrinology* (L. Martini and W. F. Ganong, eds.), Vol. 4, pp. 195–226, Raven Press, New York.

Jones, M. T., Hillhouse, E. W., and Burden, J. L., 1977, Dynamics and mechanism of corticosteroid feedback at the hypothalamus and anterior pituitary gland, *J. Endocrinol.* **73:**405–417.

Kaplanski, J., Van Delft, A. M. L., Nyaska, C., Stoof, J. C., and Smelik, P. G., 1974, Circadian periodicity and stress responsiveness of the pituitary–adrenal system of rats after central administration of 5-hydroxy dopamine, *J. Endocrinol.* **63:**299–310.

Kendall, J. W., 1976, Feedback control of ACTH secretion, in: *Frontiers of Neuroendocrinology* (L. Martini and W. F. Ganong, eds.), Vol. 4, pp. 177–209, Raven Press, New York.

Lembeck, F., 1978, Substance P: Binding to lipids in brain, in: *Centrally Acting Peptides* (J. Hughes, ed.), pp. 124–134, Macmillan, New York.

Li, C. H., 1972, Hormones of the adenohypophysis, *Proc. Am. Philos. Soc.* **116:**365–382.

Lowry, P. J., Silman, R. E., and Hope, J., 1977, Structure and biosynthesis of peptides related to corticotropins and β-melanotropins, *Ann. N.Y. Acad. Sci.* **297:**49–62.

Lutz-Bucher, B., Koch, B., and Mialhe, C., 1977, Comparative *in vitro* studies on CRF-activity of vasopressin and hypothalamic median eminence extract, *Neuroendocrinology* **23:**181–192.

Mains, R. E., Eipper, B. A., and Ling, N., 1977, Common precursor to corticotrophins and endorphins, *Proc. Natl. Acad. Sci. U.S.A.* **74:**3014–3018.

McCann, S. M., 1957, The ACTH-releasing activity of extracts of the posterior lobe of the pituitary *in vitro, Endocrinology* **60:**664–676.

McCann, S. M., and Brobeck, J. R., 1954, Evidence for the role of the supraoptico–hypophyseal system in regulation of adrenocorticotrophin secretion, *Proc. Soc. Exp. Biol. Med.* **87:**318–324.

McCann, S. M., Fruit, A., and Fulford, B. D., 1958, Studies on the loci of action of cortical hormones in inhibiting the release of adrenocorticotrophin, *Endocrinology* **63:**29–42.

McEwen, B. S., 1977, Adrenal steroid feedback on neuroendocrine tissues, *Ann. N.Y. Acad. Sci.* **297:**29–42.

Miller, R. E., Yuch-Chien, H., Wiley, M. K., and Hewitt, R., 1974, Anterior hypophysial function in the posterior-hypophysectomised rat: Normal regulation of the adrenal system, *Neuroendocrinology* **14:**233–250.

Mirsky, I. A., Stein, M. and Paulisch, G., 1954, The secretion of an antidiuretic substance into the circulation of adrenalectomised and hypophysectomised rats exposed to noxious stimuli, *Endocrinology* **55:**28–39.

Oliver, C., Mical, R. S., and Porter, R. S., 1977, Hypothalamic–pituitary vasculature: Evidence for retrograde blood flow in the pituitary stalk, *Endocrinology* **101:**598–604.

Portanova, R., and Sayers, G., 1974, Corticosterone inhibition of ACTH release: Mechanism of action, *J. Steroid Biochem.* **5:**361–366.

Rees, L. H., Bloomfield, G. A., Gilkes, J. J. H., Jeffcoate, W. J., and Besser, G. M., 1977, ACTH as a tumour marker, *Ann. N.Y. Acad. Sci.* **297**:603–620.

Rothballer, A. B., 1953, Changes in the rat neurohypophysis induced by painful stimuli with particular reference to neurosecretory material, *Anat. Rec.* **115**:21–41.

Saffran, M., and Saffran, J., 1959, Adenohypophysis and adrenal cortex, *Annu. Rev. Physiol.* **21**:403–444.

Scapagnini, U., and Preziosi, P., 1973, Receptor involvement in the central noradrenergic inhibition of ACTH secretion in the rat, *Neuropharmacology* **12**:57–62.

Scott, A. P., Lowry, P. J., and Van Wimersma Greidanus, T. B., 1976, Incorporation of ^{14}C-labelled amino acids into CLIP and αMSH by the rat pituitary neurointermediate lobe *in vitro* and the identification of four new pars intermediate peptides, *J. Endocrinol.* **70**:197–205.

Schwyzer, R., 1977, ACTH: A short introductory review, *Ann. N.Y. Acad. Sci.* **297**:3–26.

Smelik, P. G., 1960, Mechanism of hypophysial response to psychic stress, *Acta Endocrinol. (Copenhagen)* **44**:36–46.

Smelik, P. G., 1963, Relation between blood level of corticoids and their inhibiting effect in the hypophyseal stress response, *Proc. Soc. Exp. Biol. Med.* **113**:616–619.

Smelik, P. G., 1977, Some aspects of corticosteroid feedback actions, *Ann. N.Y. Acad. Sci.* **297**:580–588.

Van Loon, G. R., 1973, *Frontiers in Neuroendocrinology, 1973: Brain Catecholamines and ACTH Secretion* (W. F. Ganong and L. Martini, eds.), pp. 209–247, Oxford University Press, Oxford.

Yates, F. E., and Maran, J. W., 1975, Stimulation and inhibition of adrenocorticotrophin release, in: *Handbook of the American Physiological Society,* Vol IV, pp. 367–404, American Physiological Society, Washington, D.C.

Yates, F. E., Russell, S. M., Dallman, M. F., Hedge, G. A., McCann, S.M., and Dhariwal, A. P. S., 1971, Potentiation by vasopressin of corticotrophin-releasing factor, *Endocrinology* **88**:3–15.

Zimmerman, E. A., Stillman, M. A., Recht, L. D., Antunes, J. L., and Carmel, P. W., 1977, Vasopressin and corticotrophin-releasing factor: An axonal pathway to the portal capillaries in the zone externa of the median eminence containing vasopressin and its interaction with adrenal corticoids, *Ann. N.Y. Acad. Sci.* **297**:405–419.

29

Modulation of Pituitary Responsiveness to Gonadotropin-Releasing Hormone

George Fink and Anthony Pickering

1. Introduction

Modulation of the pituitary response to hypothalamic releasing factors (hormones) by target-organ hormones was first shown for the hypothalamic–pituitary–thyroid axis (Brown-Grant, 1960). The results of these physiological studies were confirmed by showing that thyroid hormones reduce pituitary responsiveness to thyrotropin-releasing hormone (e.g., Reichlin *et al.*, 1972). Negative-feedback control of gonadotropin output by gonadal steroids has been accepted and has perhaps dominated endocrine thinking since the classic work of Moore and Price (1932), but curiously, we know little about the mechanism. The data can be summarized by saying that at low physiological levels, estrogens inhibit gonadotropin output, and that the effect of estrogen is potentiated by progesterone (McCann, 1962; Knobil, 1974; Hauger *et al.*, 1977; Goodman, 1978a). Progesterone by itself has little or no effect, but this may be because the central receptors that mediate the effect of progesterone on gonadotropin output are present only at a low concentration in the absence of estrogen (MacLusky and McEwen, 1978). The studies of Bogdanove (1963) in the

George Fink and *Anthony Pickering* • Department of Human Anatomy, University of Oxford, Oxford OX1 3QX, England

rat and Nakai *et al.* (1978) in the rhesus monkey suggest that the inhibitory effect of estrogen may be exerted at the pituitary, but estradiol does reduce significantly the concentration of gonadotropin-releasing hormone (GnRH) in hypophyseal portal blood collected from long-term ovariectomized rats (D. K. Sarkar and G. Fink, unpublished).

In this chapter, attention will be focused on the dramatic changes that occur in the gonadotropin reponse to GnRH around the time of the spontaneous, preovulatory gonadotropin surge. During this period, estrogen and progesterone facilitate gonadotropin release; i.e., they exert their so-called "positive-feedback" action on the hypothalamic–gonadotroph system.

2. Overview of Mechanisms Involved in Producing the Spontaneous Preovulatory Gonadotropin Surge

Many data show that in man as well as in other spontaneously ovulating mammals, the preovulatory gonadotropin surge depends on a surge of estradiol-17β (e.g., Knobil, 1974; Fink, 1977, 1979a,b; Brown-Grant, 1977). The estradiol surge increases the responsiveness of the anterior pituitary to GnRH, and is required for the surge release of GnRH into the hypophyseal portal circulation (Fink, 1979a,b; Sarkar and Fink, 1979).

The surge of GnRH has two effects. First, it stimulates luteinizing hormone (LH) release. The increased plasma concentrations of LH stimulate the secretion of ovarian progesterone, which, acting on a pituitary sensitized by estrogen, further potentiates pituitary responsiveness. Second, GnRH increases the responsiveness of the gonadotroph by a direct action (the priming effect of GnRH). The role of the priming effect may be to coordinate the increase in portal plasma GnRH concentration and the increase in pituitary responsiveness so that both reach a peak at the same time, thereby ensuring a relatively massive LH surge.

Although the increased secretion of estradiol may depend on subtle changes in plasma gonadotropin concentrations [e.g., episodic pulses or a circadian rhythm or both (MacKinnon *et al.*, 1978)], the estradiol surge is not immediately preceded or accompanied by dramatic changes in plasma gonadotropin concentration. This point, which is best illustrated by examining the first (pubertal) gonadotropin surge (Meijs-Roelofs *et al.*, 1975; Castro-Vazquez and Ojeda, 1977; Fink and Sarkar, 1978), suggests that estrous and menstrual cycles are controlled primarily by an ovarian pacemaker. In the rodent, there is also good evidence for a neural "oscillator" (thought to be located in the preoptic–suprachiasmatic area) that, each day, emits a neural signal for LH release (Everett, 1964, 1977;

Fink and Geffen, 1978). For its expression, in the form of a surge of GnRH, this signal requires elevated concentrations of plasma estradiol (Legan and Karsch, 1975; Everett, 1977; Sarkar and Fink, 1979). In the primate, the mechanism is not clear. Monkeys in which the afferents to the medial basal hypothalamus have been severed still exhibit either spontaneous or estrogen-induced surges of LH (Knobil, 1974; Knobil and Plant, 1978). This led to the postulate that in primates, the ovary may be the main or the sole pacemaker. However, Norman *et al.* (1976) found that lesions placed in the preoptic–suprachiasmatic region of the rhesus monkey abolished spontaneous and induced LH surges. Often cited as indirect evidence for the absence of a cyclical neural center in the female primate is the fact that estrogen can induce LH surges in male monkeys (Karsch *et al.*, 1973) and humans (Stearns *et al.*, 1973), and that the exposure of the female primate to high levels of androgen during brain maturation does not, in contrast to its effect in the female rodent, lead to permanent acyclicity (Fink, 1977).

3. Priming Effect of GnRH

The priming effect is the capacity of GnRH to increase responsiveness of the gonadotrophs to itself. The capacity to self-prime may be a unique property of GnRH, for no priming effect can be demonstrated (with hypothalamic extracts) for adrenocorticotropin (ACTH), thyrotropin (TSH), or prolactin (Pickering and Fink, 1979b). Two observations led to the notion of the priming effect of GnRH (Aiyer *et al.*, 1973, 1974a): (1) In both man and the rat, there is usually a significant positive correlation between "basal" plasma LH concentrations and the LH response to GnRH (Aiyer *et al.*, 1974a,b; Fink, 1977). (2) The administration of sodium pentobarbitone [a potent blocker of LH and GnRH release (Everett, 1964; Sarkar *et al.*, 1976)] at 13:30 hr of proestrus reduced by about 75% the LH response at 18:00 hr of proestrus (Aiyer *et al.*, 1974b). Direct evidence came from experiments in which GnRH was administered to proestrous rats by two injections of GnRH (50 ng/100 g body weight per injection) separated by either 30, 60, 120, or 240 min. In each group, the response to the second injection was significantly greater than that to the first (Aiyer *et al.*, 1973, 1974a) (Fig. 1). The magnitude of the response to the second injection was greatest when the two injections were separated by 60 min (7-fold difference), after which time the magnitude of the second response declined (Fig. 1). The response to the second injection in proestrous animals is 5–6 times greater than the responses on the other days of the estrous cycle (Fig. 2). Experimental studies showed that this difference is largely a function of the plasma concentration of estradiol; an increase in

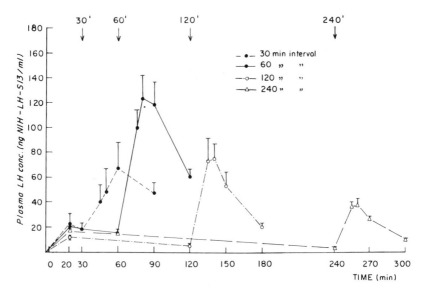

Figure 1. Mean (± S.E.M.) plasma LH concentrations after two intravenous injections of 50 ng GnRH/100 g body weight separated by either 30, 60, 120, or 240 min. The animals were anesthetized with sodium pentobarbitone at 13:30 hr of proestrus, 30 min before the first injection of GnRH at 0 time. There were 5–9 rats per group. From Aiyer *et al.* (1974a) with permission of the *Journal of Endocrinology.*

the latter enhances the magnitude of the effect (Aiyer *et al.,* 1974a). However, steroids do not mediate the priming effect, as shown by studies on adrenalectomized and acutely ovariectomized rats (Table I) and the fact that the effect can be elicited *in vitro* (Pickering and Fink, 1976a,b,1977). Indeed, acute ovariectomy leads to an approximate 2-fold increase in the second response (Table I), an effect that cannot be attriubuted simply to surgery (see sham group, Table I). The priming effect can also be elicited by continous infusion or multiple intravenous injections of synthetic GnRH (Fink *et al.,* 1976). Studies in which the medial preoptic area was stimulated electrically either in bursts or continuously showed that the priming effect can be elicited by endogenous GnRH (Fink *et al.,* 1976; Chiappa *et al.,* 1977). The priming effect has also been demonstrated in the human (Fink *et al.,* 1975; Wang *et al.,* 1976).

3.1. Mechanism of the Priming Effect

The mechanism of the priming effect has been investigated mainly by short-term pituitary incubation using the method of Pickering and Fink

(1976a). In this method, anterior pituitary halves, obtained from animals decapitated on the afternoon of proestrus under sodium pentobarbital anesthesia, were preincubated for 1 hr and then incubated for two successive 1-hr incubation periods in the presence of 8.5 nM GnRH. The following results were obtained.

    1. The slopes of the dose–response curves for LH and follicle-stimulating hormone (FSH) during the second hour of incubation were significantly steeper than those during the first hour, indicating that the priming effect depends on a qualitative as well as a quantitative change in the gonadotropin response to GnRH (Fink and Pickering, 1977).

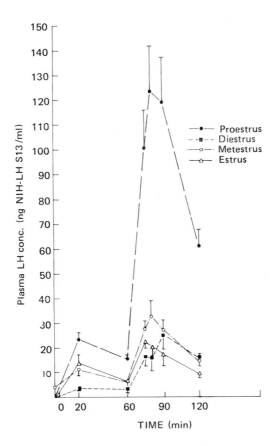

Figure 2. Mean (± S.E.M.) plasma LH concentrations after two intravenous injections of 50 ng GnRH/100 g body weight at 0 and at 60 min. The first injection was given 30 min after sodium pentobarbital anesthesia at 13:30 hr on each day of the estrous cycle. There were 6–9 rats per group. From Aiyer *et al.* (1974a) with permission of the *Journal of Endocrinology.*

Table I. Effect of Ovariectomy and Adrenalectomy on the Priming
Effect of GnRH[a]

Treatment	Number of rats	Maximal increments in plasma LH concn. (ng/ml) after:	
		First injection	Second injection
Intact proestrus	9	20.2 ± 2.2	114.4 ± 16.1[d]
Ovariectomy	6	29.3 ± 6.6	230.2 ± 21.2[e]
Sham ovariectomy	6	18.9 ± 3.1	139.7 ± 32.8[c]
Adrenalectomy	6	39.9 ± 15.5	116.8 ± 47.8
Sham adrenalectomy	5	21.6 ± 4.2	85.6 ± 6.8[d]
Adrenalectomy plus ovariectomy	5	52.1 ± 14.1	213.2 ± 64.0[b]
Adrenalectomy plus sham ovariectomy	5	26.2 ± 6.1	159.9 ± 38.8[c]

[a] From Aiyer *et al.* (1974a) with permission of the *Journal of Endocrinology*. Values
are the mean (± S.E.M.) maximal increments in plasma LH concentration (ng
NIH-LH-S13/ml) after two successive intravenous injections of 50 ng GnRH/100 g
body weight 60 min apart. The animals were ovariectomized, sham-ovariectomized,
adrenalectomized, or sham-adrenalectomized, either alone or in various combinations as indicated. Adrenalectomized and sham-adrenalectomized rats were used
only after they had shown at least one 4-day cycle after the operation. Ovariectomy
and sham ovariectomy were carried out immediately before the first injection of
GnRH, which was given 30–60 min after the administration of sodium pentobarbitone at 13:30 hr of proestrus.
[b–e] Significance of the difference between the LH responses to the first and second
injections of GnRH in each group of rats, determined by the paired *t* test:
[b]$P < 0.05$; [c]$P < 0.02$; [d]$P < 0.01$; [e]$P < 0.001$.

2. Within a 3-hr period, the priming effect can be elicited only once
(Pickering and Fink, 1979b).

3. Although high extracellular (40 mM) K^+ releases LH and FSH, 40
mM K^+ could not "prime" the glands (Pickering and Fink, 1976b, 1977).
However, exposure of glands primed by GnRH to 40 mM K^+ alone
resulted in a gonadotropin release equally as massive as that produced by
a second exposure to GnRH. These results show that the priming effect
is not simply a consequence of gonadotropin release, and that, unlike
gonadotropin release, it cannot be produced by nonselective
secretagogues.

4. Although Ca^{2+} is required for GnRH-stimulated gonadotropin
release, priming could be elicited in the absence of Ca^{2+}. Neither veratri-
dine nor the ionophores A23187 and Br-X537A were able to prime the
glands; however, exposure of the primed glands to the ionophores alone
resulted in enhanced gonadotropin release (Pickering and Fink, 1979a).

5. Neither were we able to show that adenosine 3′,5′-cyclic monophosphate (cAMP) acts as a second messenger for the priming action of

GnRH (Pickering and Fink, 1979a). Thus, for example, there was no sustained rise in pituitary cAMP content after either the first or the second exposure to GnRH (Fig. 3). Although exposure to dibutyryl cAMP (0.25–8.20 mM) was accompanied by a dose-dependent increase in LH output, the LH response to a second exposure of this nucleotide was either the same as or less than that to the first (Fig. 4). Inhibition of phosphodiesterase by theophylline or papaverine did not potentiate the priming effect of GnRH (Fig. 5).

6. The priming effect of GnRH could be abolished by cycloheximide and puromycin (Pickering and Fink, 1976a, 1977, 1979a). The apparent dependence of priming on protein synthesis was also shown by administering cycloheximide and puromycin *in vivo* (Pickering and Fink, 1976a), and our *in vitro* results are consistent with data obtained by superfusion (Edwardson and Gilbert, 1976) or continuous exposure to GnRH (De Koning *et al.*, 1976). Actinomycin D, but not the more specific inhibitor of RNA polymerase II, α-amanitin, also inhibited priming (Pickering and Fink, 1979a). Inhibitors of DNA replication (hydroxyurea and cytosine arabinoside) did not affect priming.

Either using the incorporation of radiolabeled amino acids into protein (A. Speight, A. Pickering, and G. Fink, unpublished) or measuring the total amount of gonadotropin in the system (amount released plus that remaining in the gland) (De Koning *et al.*, 1976; Pickering and Fink, 1976a, 1979a), it has not been possible to show that priming involves the synthesis of a significant amount of either LH or FSH.

Thus, the new protein may be related to the receptor–release apparatus. Tentative evidence that points in this direction comes from the fact that cytochalasin B, which interferes with microfilament function, not only abolished the priming effect but also actually reduced the second response to GnRH (Fig. 6). Neither colchicine nor vinblastine significantly affected either the first or the second response to GnRH (Fig. 6).

Although these early studies have not elucidated the mechanism of the priming action of GnRH, they do provide a basis for further investigations, and show that there is a clear difference between the releasing and priming actions of GnRH. Thus, the releasing action of GnRH is dependent on normal extracellular Ca^{2+} concentrations, can be mimicked by high extracellular K^+ concentrations, does not depend on protein synthesis (at least for LH) or on normal function of the microfilaments, and is not a "once-only" event. The priming action is a "once-only" event (at least within a 3-hr time span), does not appear to depend on normal extracellular Ca^{2+} concentration, cannot be mimicked by high extracellular K^+ concentrations, and does depend on protein synthesis and the normal function of microfilaments. Neither action of GnRH appears to be mediated by cAMP or depend on acute gonadotropin synthesis.

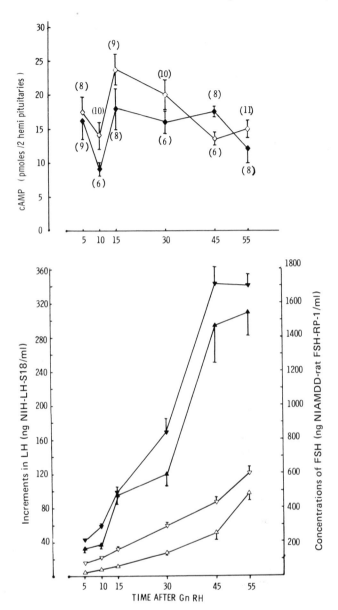

Figure 3. Mean (± S.E.M.) pituitary concentrations of cAMP and amount of LH (△) and FSH (▽) released into the medium. Hemipituitaries from proestrous rats were preincubated and then incubated for two successive 1-hr periods in the presence of 8.5 nM GnRH according to Pickering and Fink (1976a). Medium and pituitaries were taken at the times indicated for measurement of gonadotropins and cAMP (open symbols = first hour; closed symbols = second hour). The number of flasks are in parentheses. From Pickering and Fink (1979a) with permission of the *Journal of Endocrinology*.

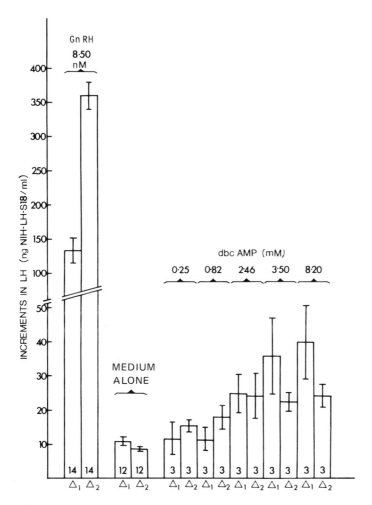

Figure 4. Mean (± S.E.M.) increments in LH during the first (Δ_1) and second (Δ_2) hours of incubation. Hemipituitaries from proestrous rats were incubated according to Pickering and Fink (1976a) in the presence of either GnRH, medium alone, or various concentrations of dibutyryl adenosine 3′,5′-cyclic monophosphate (dbcAMP). The number of flasks are in parentheses. From Pickering and Fink (1979a) with permission of the *Journal of Endocrinology.*

4. Steroids and the Gonadotropin Response to GnRH

This has been the subject of several recent reviews (e.g., Fink, 1977, 1979a,b; Yen *et al.*, 1975). The data lead to the following observations and conclusions.

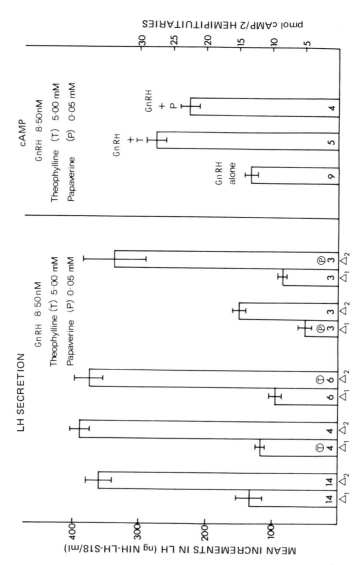

Figure 5. Mean (± S.E.M.) increments in LH released into the medium during the first (Δ1) and second (Δ2) hours of incubation (left) and pituitary contents of cAMP (right). Hemipituitaries from proestrous rats were incubated according to Pickering and Fink (1976a) in the presence of GnRH with or without the presence of phosphodiesterase inhibitor during either the first or the second hour of incubation. The number of flasks are at the base of each column. From Pickering and Fink (1979a) with permission of the *Journal of Endocrinology*.

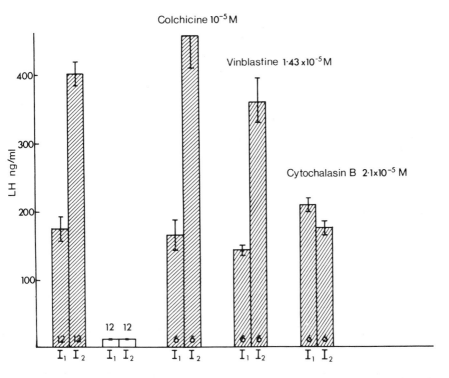

Figure 6. Mean (± S.E.M.) increments of NIH-LH-S18 (ng/ml medium) in the first (I₁) and second (I₂) hours of incubation. Hemipituitaries from proestrous rats were incubated according to Pickering and Fink (1976a) in the presence (hatched columns) or absence (open columns) of 8.5 nM GnRH with or without the presence during preincubation (for 2 hr) and incubation of either colchicine, vinblastine, or cytochalasin B. The number of flasks are at the base of each column. From Pickering and Fink (1979a) with permission of the *Journal of Endocrinology.*

1. The preovulatory rise in plasma estradiol-17β concentrations enhances pituitary responsiveness to GnRH (Aiyer and Fink, 1974; Fink and Henderson, 1977a). Estradiol, at peak physiological levels, first inhibits responsiveness, and then, after 8–12 hr, facilitates responsiveness (Henderson *et al.,* 1977). The effect of estradiol may be divided into two components. In the rat, the first phase of increasing pituitary responsiveness between 14:00 hr of diestrus and 14.00 hr of proestrus is gradual and occurs in the absence of any dramatic change in LH concentrations in peripheral plasma and GnRH concentrations in portal plasma. However, subtle changes in GnRH or LH release (e.g., changes in the frequency of episodic release) have not been excluded (see overview above) and, therefore, neither has a possible hypothalamic as well as a pituitary site of action (see below). Though estradiol initiates the increase in pituitary responsiveness to GnRH, it is not clear to what extent estradiol is involved

in the marked and abrupt increase that parallels the LH surge (Aiyer *et al.*, 1974b; Fink and Aiyer, 1974). During the second phase of increasing responsiveness, the action of estradiol is inextricably bound up in a cascade involving the GnRH surge, early LH release that stimulates progesterone secretion, and the priming effect of GnRH.

The increase in pituitary responsiveness correlates well with the increase in total pituitary estradiol receptor complexes (Sen and Menon, 1978).

2. Progesterone secreted immediately after the beginning of the LH surge potentiates the effect of estradiol (Aiyer and Fink, 1974; Fink and Henderson, 1977a).

3. Testosterone reduces the LH but not the FSH response to GnRH (Fink and Henderson, 1977a; Drouin and Labrie, 1976).

4. Of the progestins tested for the capacity to potentiate the action of estrogen (17α- and 20α-hydroxyprogesterone; 5α- and 5β-pregnane-3,20-dione), only 5α-pregnane-3,20-dione had a significant effect; however, the potency of this progestin was only 0.05 that of progesterone (Fink and Henderson, 1977a).

4.1. Site and Mechanism of Action of Sex Steroids

Studies using rats in which the pituitary stalk had been cut show that estrogen and progesterone can facilitate pituitary responsiveness by a direct action on the gonadotrophs (Fink and Henderson, 1977b; Greeley *et al.*, 1975, 1978). Estrogen has also been reported to increase the responsiveness of dispersed pituitary cells (Drouin *et al.*, 1976; Hsueh *et al.*, 1978). However, estradiol does have the capacity to increase significantly the concentration of GnRH in hypophyseal portal blood collected from rats (Sarkar and Fink, 1979) and rhesus monkeys (Neill *et al.*, 1977), and Goodman (1978b) found that in the rat, estradiol implants in the rostral diencephalon were more effective in inducing an LH surge than median eminence implants. These data, together with the priming effect of GnRH, make it difficult to establish the relative importance of the hypothalamic compared with the pituitary site of action of steroids in increasing pituitary responsiveness. Consider two examples that illustrate this point:

First, the facilitatory effect of estradiol, administered by implantation of Silastic capsules containing crystalline estradiol-17β, could be blocked by sodium pentobarbitone (Henderson *et al.*, 1977) (Fig. 7). It is unlikely that this was simply a nonspecific effect of the anesthetic, because the response in animals bearing an empty Silastic capsule was unaffected by sodium pentobarbitone (Fig. 7). These data therefore suggest that the facilitatory effect of estradiol on pituitary responsiveness may be mediated

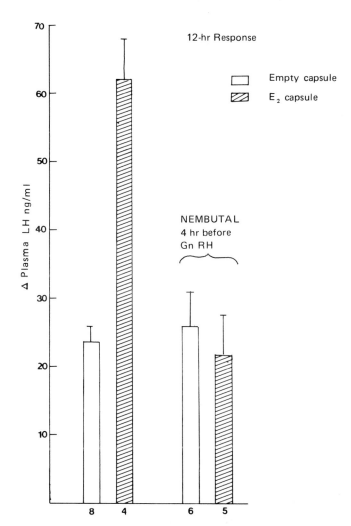

Figure 7. Mean (± S.E.M.) maximal increments in plasma LH (ng NIH LH-S18/ml) produced by injecting 50 ng GnRH/100 g body weight intravenously into animals that had been ovariectomized and implanted 12 hr previously (at 10:00 hr of diestrus) with either an empty Silastic capsule or a capsule containing crystalline estradiol-17β. Some animals (*right*) were anesthetized with sodium pentobarbitone 4 hr before being injected with GnRH. The number of animals are at the base of each column. Data from Henderson *et al.* (1977).

by the priming effect of GnRH, or, at least, that for the full effect of estrogen, the pituitary must be exposed to some GnRH.

The second example pertains to the marked sex difference in pituitary responsiveness to GnRH (Barraclough and Turgeon, 1974; Fink and Henderson, 1977a). Figure 8 shows that while estrogen followed by progesterone enhances pituitary responsiveness in female rats, these steroids have little effect on the low responsiveness in androgen-sterilized females, and reduce responsiveness in males. There are two possible conflicting interpretations of these data. First, it might be that the sex difference in the response to steroid treatment in terms of the gonadotropin response to GnRH is due to sex differences in the GnRH release apparatus. This interpretation, which is supported by studies on long-term-ovariec-

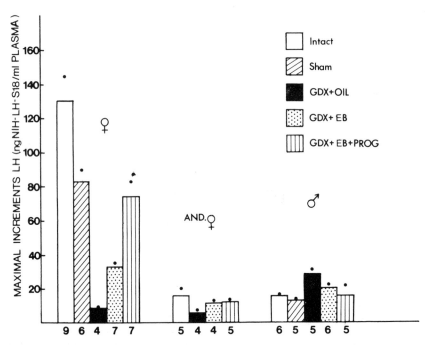

Figure 8. Mean (± S.E.M.) maximal increments in LH following intravenous injection of 50 ng GnRH/100 g body weight into female (♀), androgenized female (AND ♀) or male (♂) rats. Gonadectomy (GDX) or sham gonadectomy was carried out at 10:00 hr (diestrus in the normal females), and GnRH was injected at 18:00 hr of the next day (presumptive proestrus in the normal females). Immediately after gonadectomy, the animals were given a subcutaneous (s.c.) injection of either oil or 2.5 μg estradiol benzoate (EB). Some animals were given 2.5 mg progesterone (PROG), s.c., at 12:00 hr of the next day (6 hr before GnRH injection). The number of animals are at the base of each column. From Fink and Henderson (1977a).

tomized rats (Aiyer *et al.*, 1976), is consistent with the prevailing view that sexual differentiation of the hypothalamic–pituitary system occurs predominantly in the brain (e.g., Brown-Grant, 1973; Fink, 1977). The second interpretation, not consistent with the orthodox view on the mechanism of sexual differentiation, is that the latter involves significant permanent changes in the pituitary–gonadal (and possibly adrenal) axis that result in an altered response of the pituitary to the direct action of steroids.

In the rhesus monkey, the pituitary has been suggested as the main site of estrogen's stimulatory effect on the basis of studies on long-term-ovariectomized animals in which the arcuate nuclei had been destroyed (Nakai *et al.*, 1978). However, the brain lesions in these animals may not have completely destroyed the GnRH system.

5. Changes in the Readily Releasable Gonadotropin Pool as the Mechanism of the Priming Effect and the Action of Steroids

The changes in the gonadotropin response to GnRH may be due to changes in the amount of gonadotropin available for immediate release (the "readily releasable gonadotropin pool"). This was tested by administering a supramaximal dose of GnRH (1000 ng/100 g body weight) to rats in which replenishment of the pool was blocked (presumably) by administering cycloheximide (Pickering and Fink, 1979c). Figure 9 shows that a 20-fold increase in the amount of LH released by this dose occurred (between 13:30 hr of diestrus and 18:00 hr of proestrus) in the absence of any apparent increase in the total amount of LH in the system. We (A. Speight and G. Fink, unpublished) have also found that estradiol increases the responsiveness of dispersed pituitary cells without stimulating any significant LH synthesis. These findings, together with the data of Pickering and Fink (1976a, 1979a) and De Koning *et al.* (1976), suggest that the main effect of estradiol and the priming effect is to produce an apparent increase in the readily releasable pool of LH. The latter seems to be due to a change in the receptor–release apparatus leading to facilitation of hormone release, rather than to an increase in newly synthesized hormone. Although steroids and the priming effect may exert their action through different cellular mechanisms, the facts that the full effect of estradiol requires exposure of the gland to GnRH, and that the magnitude of the priming effect is increased significantly by steroids, suggest that the two mechanisms are interdependent.

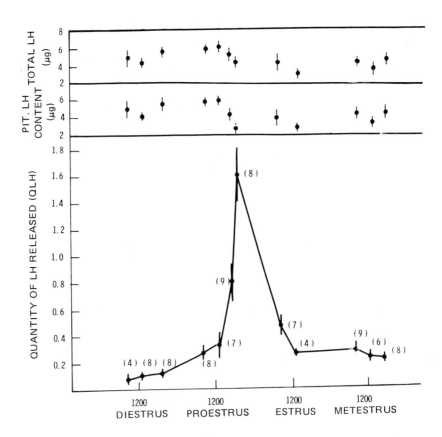

Figure 9. Mean (± S.E.M.) total, pituitary, and released LH in adult female Wistar rats given cycloheximide and a following supramaximal dose of GnRH. The animals were anesthetized with sodium pentobarbitone at the times shown on the abscissa, given an intraveneous (i.v.) injection of cycloheximide (100 μg/100 g body weight), and 30 min later given an injection of 1000 ng GnRH/100 g body weight (supramaximal dose of GnRH), i.v. This dose of cycloheximide completely blocked the priming effect of GnRH (Pickering and Fink, 1976a). Blood samples were taken before and at frequent intervals up to 2 hr after GnRH injection, and the amount of LH released (QLH) was estimated using the data of Gay and Bogdanove (1968). At the end of each experiment, the animals were killed and the pituitary glands were homogenized and assayed for LH content. The total LH equals the pituitary LH content plus QLH. The number of animals are in parentheses. From Pickering and Fink (1979c) with permission of the *Journal of Endocrinology*.

6. Summary

The gonadotropin response to GnRH is determined by circulating levels of sex steroids and the self-priming effect of GnRH. At low plasma cencentrations, estradiol inhibits gonadotropin output, an effect that is potentiated by progesterone. A relatively sudden increase in plasma estradiol (such as that which precedes the spontaneous preovulatory LH surge) in the presence of low progesterone increases responsiveness. When the hypothalamic–gonadotroph system has been sensitized by estrogen, progesterone enhances responsiveness. The increase in responsiveness occurs in two phases. The first is a gradual phase that occurs in the 24 hr preceding the LH surge, during which estradiol exerts its effect in the absence of any marked changes in peripheral plasma LH or portal plasma GnRH concentrations, though subtle changes have not been excluded and exposure of the pituitary to at least some GnRH seems to be required. The abrupt second phase, which coincides with the spontaneous LH surge, is brought about by a cascade of events involving progesterone and the priming effect of GnRH. Estradiol takes 8–12 hr to exert its facilitatory action, and the increase in responsiveness coincides with an increase in the total number of pituitary estradiol receptor complexes. The priming effect of GnRH depends on protein synthesis and the integrity of microfilament function. Neither cAMP nor Ca^{2+} appears to act as a second messenger for this effect. The actions of both estradiol and the priming effect of GnRH appear to depend on an apparent increase in the size of the readily releasable gonadotropin pool in the absence of any significant gonadotropin synthesis. The evidence points toward a change in the receptor–release apparatus that leads to facilitation of gonadotropin release.

ACKNOWLEDGMENTS. The original work reported herein was supported by the Medical Research Council, and could not have been carried out without the generous supply of radioimmunoassay materials provided by Drs. G. D. Niswender, T. Nett, L. E. Reichert Jr., and A. F. Parlow and the NIAMDD (Bethesda) and the gift of synthetic GnRH by Drs. H. Gregory, J. Gormley, and the late A. L. Walpole (I.C.I. Pharmaceuticals, Cheshire).

DISCUSSION

DUVAL: Using perfusion experiments with male rat pituitaries, we found the same results as you described for your *in vitro* studies, i.e., some kind of potentiation during the first hour of contact with the GnRH. This was not observed if successive pulses of GnRH were used instead of a continuous infusion. Furthermore, dbcAMP elicits only a small increase in the gonadotropin release. You think that the potentiator could be mediated through the synthesis of specific protein. You rule out the possibility that LH-RH, which is present in the medium, might penetrate into the cell by endocytosis and act directly on translocation of the secretory granules.

FINK: Our experiments do not rule out other possibilities. I can only say that our experiments with inhibitors of protein synthesis are consistent with the idea that synthesis of a new protein is involved. The side effects of these inhibitors prohibit a more dogmatic conclusion.

KOKAER: What would be the rate of the self-priming effect with the use of more potent analogues of LH-RH during the first incubation period?

FINK: The results are quantitatively the same as with the parent peptide.

KORDON: We recently measured the kinetic characteristics of LH and FSH response to graded doses of LH-RH, using hemipituitaries from normal and estradiol-implanted castrates. Estradiol not only induces a dramatic increase in the apparent affinity of the effect, but also changes the slope of the response curve with an indication of induction of positive cooperativity. We have some preliminary evidence that this cooperativity may be due to the priming effect of the daily surge of endogenous LH-RH seen under these conditions. If this is true, then one should expect that a second incubation with higher LH-RH concentrations would overcome the lack of priming in the first. Did you check this by performing second incubations with increasing doses of LH-RH?

FINK: No, we have not done this.

JUTISZ: Some years ago, working on the action of a high-potassium medium, and of GnRH on pituitary halves of ovariectomized–estradiol-plus-progesterone-treated rats, we showed that high potassium has an additive effect to that of GnRH. This means that both secretagogues act by different mechanisms. How do you correlate your results with these observations?

KRAICER: I would like to add to that question on the use of high potassium as a secretagogue. Dr. Mulder showed in his paper [Chapter 26] that the release of ACTH in high potassium is not continuous or constant, that there is a "pulse" or release when the cells are first introduced to high potassium, and that release then returns to normal within minutes, even though the high-potassium medium is maintained. We have found identical results with GH release. Is the response of FSH and LH also just one "pulse" or "burst," and how would this affect the interpretation of your findings?

FINK: We have used high potassium as a nonselective (rather than a nonspecific) secretagogue to test two questions: (1) Is the priming effect of GnRH simply due to the release of LH? Our results show that the answer is no. (2) Can a nonselective secretagogue evoke a massive gonadotropin release from primed glands? Our results show that the answer is yes, and that this has implications for further studies on the receptor-release mechanism. We cannot answer Dr. Kraicer's question regarding the time course of the potassium effect.

REFERENCES

Aiyer, M. S., and Fink, G., 1974, The role of sex steroid hormones in modulating the responsiveness of the anterior pituitary gland to luteinizing hormone releasing factor in the female rat, *J. Endocrinol.* **62:**553.

Aiyer, M. S., Chiappa, S. A., Fink, G., and Greig, F., 1973, A priming effect of luteinizing hormone releasing factor on the anterior pituitary gland in the female rat, *J. Physiol. (London)* **234**:81P.

Aiyer, M. S., Chiappa, S. A., and Fink, G., 1974a, A priming effect of luteinizing hormone releasing factor on the anterior pituitary gland in the female rat, *J. Endocrinol.* **62**:573.

Aiyer, M. S., Fink, G., and Greig, F., 1974b, Changes in the sensitivity of the pituitary gland to luteinizing hormone releasing factor during the oestrous cycle of the rat, *J. Endocrinol.* **60**:47.

Aiyer, M. S., Sood, M. C., and Brown-Grant, K., 1976, The pituitary response to exogenous luteinizing hormone releasing factor in steroid-treated gonadectomized rats, *J. Endocrinol.* **69**:255.

Barraclough, C. A., and Turgeon, J. L., 1974, Further studies of the hypothalamo–hypophyseal–gonadal axis of the androgen-sterilized rat, in: *Endocrinologie Sexuelle de la Période Périnatale* (M. G. Forest and J. Bertrand, eds.), pp. 339–356, INSERM, Paris.

Bogdanove, E. M., 1963, Direct gonad–pituitary feedback: An analysis of effects of intracranial estrogenic depots on gonadotrophin secretion, *Endocrinology* **73**:696.

Brown-Grant, K., 1960, The hypothalamus and the thyroid gland, *Br. Med. Bull.* **16**:165.

Brown-Grant, K., 1973, Recent studies on the sexual differentiation of the brain, in: *Foetal and Neonatal Physiology* (K. S. Comline, K. W. Cross, G. S. Dawes, and P. W. Nathanielsz, eds.), pp. 527–545, Cambridge University Press, Cambridge.

Brown-Grant, K., 1977, Physiological aspects of the steroid hormone–gonadotropin interrelationship, in: *International Review of Physiology, Reproductive Physiology II* (R. O. Greep, ed.), pp. 57–83, University Park Press, Baltimore.

Castro-Vazquez, A., and Ojeda, S. R., 1977, Changes in pituitary responsiveness to LH-RH during puberty in the female rat: Initiation of the priming effect, *Neuroendocrinology* **23**:88.

Chiappa, S. A., Fink, G., and Sherwood, N. M., 1977, Immunoreactive luteinizing hormone releasing factor (LRF) in pituitary stalk plasma from female rats: Effects of stimulating diencephalon, hippocampus and amygdala, *J. Physiol. (London)* **267**:625.

De Koning, J., Van Dieten, J. A. M. J., and Van Rees, G. P., 1976, LH-RH-dependent synthesis of protein necessary for LH release from rat pituitary glands *in vitro*, *Mol. Cell. Endocrinol.* **5**:151.

Drouin, J., and Labrie, F., 1976, Selective effect of androgens on LH and FSH release in anterior pituitary cells in culture, *Endocrinology* **98**:1528.

Drouin, J., Lagacé, L., and Labrie, F., 1976, Estradiol-induced increase of the LH responsiveness to LH releasing hormone (LHRH) in rat anterior pituitary cells in culture, *Endocrinology* **99**:1477.

Edwardson, J. A., and Gilbert, D., 1976, Application of an *in-vitro* perifusion technique to studies of luteinizing hormone release by rat anterior hemi-pituitaries: Self potentiation by luteinizing hormone releasing hormone, *J. Endocrinol.* **68**:197.

Everett, J. W., 1964., Central neural control of reproductive functions of the adenohypophysis *Physiol. Rev.* **44**:373.

Everett, J. W., 1977, The timing of ovulation, *J. Endocrinol.* **75**:3P.

Fink, G., 1977, Hypothalamic pituitary ovarian axis, in: *Recent Advances in Obstetrics and Gynaecology,* 12th ed. (J. Stallworthy and G. Bourne, eds.), pp. 4–54, Churchill-Livingstone, Edinburgh.

Fink, G., 1979a, Feedback actions of target hormones on hypothalamus and pituitary with special reference to gonadal steroids, *Annu. Rev. Physiol.* **41**:571.

Fink, G., 1979b, Neuroendocrine control of gonadotrophin secretion, *Br. Med. Bull.* **35**(2):155.

Fink, G., and Aiyer, M. S., 1974, Gonadotrophin secretion after electrical stimulation of the preoptic area during the oestrous cycle of the rat, *J. Endocrinol.* **62**:589.

Fink, G., and Geffen, L. B., 1978, The hypothalamo–hypophysial system: Model for central peptidergic and monoaminergic transmission, in: *International Review of Physiology, Neurophysiology III* (R. Porter, ed.), pp. 1–48, University Park Press, Baltimore.

Fink, G., and Henderson, S. R., 1977a, Steroids and pituitary responsiveness in female, androgenized female and male rats, *J. Endocrinol.* **73**:157.

Fink, G., and Henderson, S. R., 1977b, Site of modulatory action of oestrogen and progesterone on gonadotrophin response to luteinizing hormone releasing factor, *J. Endocrinol.* **73**:165.

Fink, G., and Pickering, A. J. M. C., 1977, The hypothalamo–hypophysial system: Window on central neurotransmission, in: *Molecular Endocrinology* (I. MacIntyre and M. Szelke, eds.), pp. 293–308, Elsevier, Amsterdam.

Fink, G., and Sarkar, D. K., 1978, Mechanism of first surge of luteinizing hormone and vaginal opening in the normal rat, and the effect of neonatal androgen, *J. Physiol. (London)* **282**:34P.

Fink, G., Aiyer, M. S., Jamieson, M. G., and Chiappa, S. A., 1975, Factors modulating the responsiveness of the anterior pituitary gland in the rat with special reference to gonadotrophin releasing hormone (GnRH), in: *Hypothalamic Hormones: Chemistry, Physiology, Pharmacology and Clinical Uses* (M. Motta, P. G. Crosignani, and L. Martini, eds.), pp. 139–160, Academic Press, New York.

Fink, G., Chiappa, S. A., and Aiyer, M. S., 1976, Priming effect of luteinizing hormone releasing factor elicited by preoptic stimulation and by intravenous infusion and multiple injections of the synthetic decapeptide, *J. Endocrinol.* **69**:359.

Gay, V. L., and Bogdanove, E. M., 1968, Disappearance of endogenous and exogenous luteinizing hormone activity from the plasma of previously castrated, acutely hypophysectomized rats: An indirect assessment of synthesis and release rates, *Endocrinology* **82**:359.

Goodman, R. L., 1978a, A quantitative analysis of the physiological role of estradiol and progesterone in the control of tonic and surge secretion of luteinizing hormone in the rat, *Endocrinology* **102**:142.

Goodman, R. L., 1978b, The site of the positive feedback action of estradiol in the rat, *Endocrinology* **102**:151.

Greeley, G. H., Jr., Allen, M. B., Jr., and Mahesh, V. B., 1975, Potentiation of luteinizing hormone release by estradiol at the level of the pituitary, *Neuroendocrinology* **18**:233.

Greeley, G. H., Jr., Volcan, I. J., and Mahesh, V. B., 1978, Direct effects of estradiol benzoate and testosterone on the response of the male rat pituitary to luteinizing hormone releasing hormone (LHRH), *Biol. Repro.* **18**:256.

Hauger, R. L., Karsch, F. J., and Foster, D. L., 1977, A new concept for control of the estrous cycle of the ewe based on the temporal relationships between luteinizing hormone, estradiol and progesterone in peripheral serum and evidence that progesterone inhibits tonic LH secretion, *Endocrinology* **101**:807.

Henderson, S. R., Baker, C., and Fink, G., 1977, Oestradiol-17β and pituitary responsiveness to luteinizing hormone releasing factor in the rat: A study using rectangular pulses of oestradiol-17β monitored by non-chromatographic radioimmunoassay, *J. Endocrinol.* **73**:441.

Hsueh, A. J. W., Erickson, G. F., and Yen, S. S. C., 1978, Sensitisation of pituitary cells to luteinising hormone releasing hormone by clomiphene citrate *in vitro*, *Nature (London)* **273**:57.

Karsch, F. J., Dierschke, D. J., and Knobil, E., 1973, Sexual differentiation of pituitary function: Apparent difference between primates and rodents, *Science* **179**:484.

Knobil, E., 1974, On the control of gonadotropin secretion in the rhesus monkey, *Recent Prog. Horm. Res.* **30**:1.

Knobil, E., and Plant, T. M., 1978, Neuroendocrine control of gonadotropin secretion in the female rhesus monkey, in: *Frontiers in Neuroendocrinology* (W. F. Ganong and L. Martini, eds.), Vol. 5, pp. 249–264, Raven Press, New York.

Legan, S. J., and Karsch, F. J., 1975, A daily neural signal for the LH surge in the rat, *Endocrinology* **96**:57.

MacKinnon, P. C. B., Puig-Duran, E., and Laynes, R., 1978, Reflections on the attainment of puberty in the rat: Have circadian signals a role to play in its onset?, *J. Reprod. Fertil.* **52**:401.

MacLusky, N. J., and McEwen, B. S., 1978, Oestrogen modulates progestin receptor concentrations in some rat brain regions but not in others. *Nature (London)* **274**:276.

McCann, S. M., 1962, Effect of progesterone on plasma luteinizing hormone activity, *Am. J. Physiol.* **202**:601.

Meijs-Roelofs, H. M. A., Uilenbroek, J. T. J., de Greef, W. J., de Jong, F. H., and Kramer, P., 1975, Gonadotrophin and steroid levels around the time of first ovulation in the rat, *J. Endocrinol.* **67**:275.

Moore, C. R., and Price, D., 1932, Gonad hormone functions, and the reciprocal influence between gonads and hypophysis with its bearing on the problem of sex hormone antagonism, *Am. J. Anat.* **50**:13.

Nakai, Y., Plant, T. M., Hess, D. L., Keogh, E. J., and Knobil, E., 1978, On the sites of the negative and positive feedback actions of estradiol in the control of gonadotropin secretion in the rhesus monkey, *Endocrinology* **102**:1008.

Neill, J. D., Patton, J. M., Dailey, R. A., Tsou, R. C., and Tindal, G. T., 1977, Luteinizing hormone releasing hormone (LHRH) in pituitary stalk blood of rhesus monkeys: Relationship to level of LH release, *Endocrinology* **101**:430.

Norman, R. L., Resko, J. A., and Spies, H. G., 1976, The anterior hypothalamus: How it affects gonadotropin secretion in the rhesus monkey, *Endocrinology* **99**:59.

Pickering, A. J. M. C., and Fink, G., 1976a, Priming effect of luteinizing hormone releasing factor: *In-vitro* and *in-vivo* evidence consistent with its dependence upon protein and RNA synthesis, *J. Endocrinol.* **69**:373.

Pickering, A., and Fink, G., 1976b, Priming effect of luteinizing hormone releasing factor: *In-vitro* studies with raised potassium ion concentrations, *J. Endocrinol.* **69**:453.

Pickering, A. J. M. C., and Fink, G., 1977, A priming effect of luteinizing hormone releasing factor with respect to release of follicle-stimulating hormone *in vitro* and *in vivo*, *J. Endocrinol.* **75**:155.

Pickering, A. J. M. C., and Fink, G., 1979a, Priming effect of luteinizing hormone releasing factor *in vitro*: Role of protein synthesis, contractile elements, Ca^{2+} and cyclic AMP, *J. Endocrinol.* **81**:223.

Pickering, A. J. M. C., and Fink, G., 1979b, Do hypothalamic regulatory factors other than luteinizing hormone releasing factor exert a priming effect? *J. Endocrinol.* **81**:235.

Pickering, A. J. M. C., and Fink, G., 1979c, Variation in size of the "readily releasable pool" of luteinizing hormone during the oestrous cycle of the rat, *J. Endocrinol.* **83**:53–59.

Reichlin, S., Martin, J. B., Mitnick, M. A., Boshans, R. L., Grimm, Y., Bollinger, J., Gordon, J., and Malacara, J., 1972, The hypothalamus in pituitary–thyroid regulation, *Recent Prog. Horm. Res.* **28**:229.

Sarkar, D. K., and Fink, G., 1979, Effects of gonadal steroids on output of luteinizing hormone releasing factor into pituitary stalk blood in the female rat, *J. Endocrinol.* **80**:303.

Sarkar, D. K., Chiappa, S. A., Fink, G., and Sherwood, N. M., 1976, Gonadotropin-releasing hormone surge in pro-oestrous rats, *Nature (London)* **264**:461.

Sen, K. K., and Menon, K. M. J., 1978, Oestradiol receptors in the rat anterior pituitary gland during the oestrous cycle: Quantitation of receptor activity in relation to gonadotrophin releasing hormone–mediated luteinizing hormone release, *J. Endocrinol.* **76**:211.

Stearns, E. L., Winter, J. S. D., and Faiman, C., 1973, Positive feedback effect of progestin upon serum gonadotropins in estrogen-primed castrate men, *J. Clin. Endocrinol. Metab.* **37**:635.

Wang, C. F., Lasley, B. L., Lein, A., and Yen, S. S. C., 1976, The functional changes of the pituitary gonadotrophs during the menstrual cycle, *J. Clin. Endocrinol. Metab.* **42**:718.

Yen, S. S. C., Lasley, B. L., Wang, C. F., Leblanc, H., and Siler, T. M., 1975, The operating characteristics of the hypothalamic–pituitary system during the menstrual cycle and observations of biological action of somatostatin, *Recent Prog. Horm. Res.* **31**:321.

30

The Pattern of LH Release of Rat Pituitary Glands during Long-Term Exposure to LH-RH in Vitro

Jurrien de Koning, Hannie A. M. J. van Dieten, and G. Peter van Rees

1. Introduction

During continuous exposure of pituitary glands of intact rats to synthetic luteinizing hormone–releasing hormone (LH-RH) *in vitro* (de Koning *et al.*, 1976a,b, 1978a) and *in vivo* (Schuiling and Zürcher, 1975; Blake, 1976; Vilchez-Martinez *et al.*, 1976; Schuiling *et al.*, 1976), three phases in the luteinizing hormone (LH)-releasing patterns can be distinguished. Initially, the rate of LH release is low and constant; next, it increases progressively, and then it declines. In the LH-releasing patterns of pituitary glands of ovariectomized (OVX) rats, the initial phase is absent *in vitro* (Jutisz *et al.*, 1975; Osland *et al.*, 1975; de Koning *et al.*, 1976a,b, 1978a) as well as *in vivo* (Schuiling and Gnodde, 1976, 1977).

Fink and co-workers (Aiyer *et al.*, 1973, 1974; Fink and Pickering, 1975; Pickering and Fink, 1976) studied the responsiveness of the pituitary gland of intact female rats to LH-RH by successive injections of this hormone. Their results, which were confirmed by others using the same

Jurrien de Koning, Hannie A. M. J. van Dieten, and *G. Peter van Rees* • Sylvius Laboratories, Department of Pharmacology, Leiden University Medical Center, Leiden, The Netherlands

experimental design *in vitro* (Edwardson and Gilbert, 1975, 1976) and *in vivo* (Castro-Vazquez and McCann, 1975), led them to postulate that LH-RH augmented its own action on LH release: the priming effect of LH-RH. Pituitary glands of OVX rats do not display the priming effect (Gay *et al.*, 1970; Osland *et al.*, 1975). From these findings, the conclusion is obvious that the priming effect and the progressive increase in the rate of LH release during continuous exposure to LH-RH are mediated by the same mechanism.

The aim of the experiments described and discussed in this chapter is to explain the apparently different effects of LH-RH on LH release from pituitary glands of intact and OVX female rats, as well as the time course of the patterns of LH release. Whether LH release is dependent on synthesis of protein was investigated by incubating pituitary glands with LH-RH while synthesis of protein was blocked by cycloheximide or puromycin. Special attention will be given to the development of refractoriness of the pituitary glands to the action of LH-RH. In these studies, the effects of a combination of the cyclic AMP derivative N^6-monobutyryl cyclic AMP (mbcAMP) plus the phosphodiesterase inhibitor theophylline and raised potassium levels were also investigated. The latter secretagogues might interfere in the LH secretion process via different pathways (Zor *et al.*, 1970).

The experiments were carried out by means of an *in vitro* system (de Koning *et al.*, 1976b). Adult female rats from the Wistar-derived colony kept in our laboratory were used as pituitary gland donor animals. Pituitary glands were taken from either intact female rats on the second day of diestrus or from ovariectomized rats 14 days after gonadectomy. Two pituitary halves were incubated in 1 ml medium TC 199. Samples of the media were taken at intervals of 1 or 2 hr. The LH concentrations in the media were estimated by radioimmunoassay (Welschen *et al.*, 1975). The results are expressed in terms of the NIAMDD rat-LH-RP-1 standard preparation.

2. LH-RH-Stimulated Release of LH

2.1. Involvement of Protein Synthesis

In the first series of experiments, we compared the effect of a maximally active concentration of LH-RH (1000 ng/ml) on LH release from pituitary glands of intact and OVX rats during 4 hr of incubation. In addition, the interference by the inhibitors of protein synthesis cycloheximide (25 μg/ml) or puromycin (54 μg/ml) was investigated. As concerns pituitary glands of intact rats, two phases in the LH-RH-stimulated release of LH were observed. During the first phase, which lasted about 1–1.5 hr, the rate of LH release was low, whereas during the next phase, it increased progressively (Fig. 1a) (also see de Koning *et al.*, 1976b). A similar time course in the pattern of LH-RH-stimulated release of LH has been demonstrated by others who incubated pituitary glands from intact male (Borgeat *et al.*, 1972; Edwardson and Gilbert, 1976) or female rats (Pickering and Fink, 1976; Kercret *et al.*, 1977).

Inhibitors of protein synthesis (Fig. 1a) (de Koning *et al.*, 1976b; Pickering and Fink, 1976) and RNA synthesis (Pickering and Fink, 1976) blocked the increase in the rate of LH release during the second phase, but failed to affect LH release during the initial phase. In fact, during incubation with LH-RH and cycloheximide or puromycin, LH release continued at the rate observed during the first phase (Fig. 1a). The same phenomenon has been demonstrated *in vivo* by Vilchez-Martinez *et al.* (1976), who infused LH-RH into male rats that had been injected prior to the start of infusion with either actinomycin D or vehicle only.

The observation described above is in close agreement with the results from studies relating to the priming effect. LH release following the first injection of LH-RH *in vivo* (Pickering and Fink, 1976) or *in vitro* during superfusion (Edwardson and Gilbert, 1975) was not affected by cycloheximide or actinomycin D. However, these antibiotics completely abolished the increment in the amount of LH released in response to the second injection of LH-RH.

It is striking that pituitary glands taken from the OVX rats respond to LH-RH without showing the initial phase of low LH release (Fig. 1b). The rate of LH release from these pituitary glands is very similar to that observed during the second phase in the pattern of LH release from glands of intact rats (Fig. 1). In contradiction to the finding described above, this high rate of LH release from the pituitary glands of OVX rats is not affected by the administration of cycloheximide or puromycin (Fig. 1b) (de Koning *et al.*, 1976a,b). Also, Debeljuk *et al.* (1975) have observed the failure of actinomycin to affect LH-RH-induced release of LH from pituitary glands of OVX rats *in vivo*. In addition, it should be noted that the priming effect, which was also shown to be dependent on protein synthesis, could not be demonstrated in pituitary glands of OVX rats during experiments carried out *in vitro* (Osland *et al.*, 1975) and *in vivo* (Gay *et al.*, 1970).

These results strongly suggest that in the pituitary glands of intact rats, the formation of a factor is stimulated by LH-RH via a protein-synthesis-dependent pathway. This factor is definitely not newly formed LH (de Koning *et al.*, 1976b). We cannot distinguish whether the protein itself or subsequently activated factors are concerned with the final steps in the process of LH secretion. Therefore, we use the term "protein

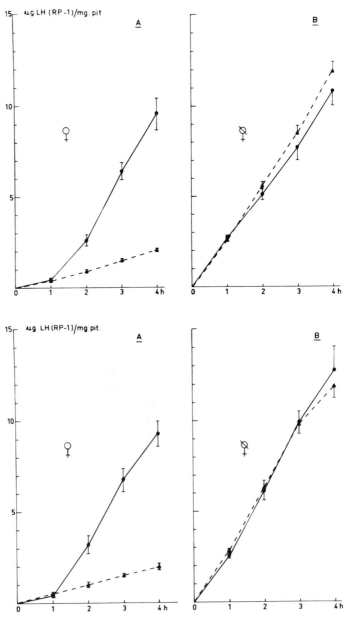

Figure 1. Effect of puromycin (top) and cycloheximide (bottom) on LH-RH-stimulated release of LH from pituitary glands of intact (A) or OVX (B) female rats during 4 hr of incubation. The results are given as mean LH content of the medium ± S.E.M.; each group consisted of 4 flasks. (●———●) LH-RH (1000 ng/ml); (▲———▲) LH-RH (1000 ng/ml) + puromycin (54 μg/ml) or cycloheximide (25 μg/ml). Data of de Koning *et al.* (1976b).

factor'' so as to indicate the protein-synthesis-dependent action of LH-RH that finally mediates in LH release.

The effects of exogenous LH-RH on release of LH from glands of OVX or intact rats during the initial phase are independent of the *de novo* synthesis of protein, probably because of the action of endogenous LH-RH (also see Section 2.2). In our opinion, the protein factor is present in the glands of intact rats in such small quantities that it initially restricts the response of the glands to the high level of LH-RH. Then, during continued exposure of the pituitary glands to LH-RH, the induction of the protein factor permits a higher rate of LH release. The amount of the protein factor in pituitary glands of OVX rats is high enough to allow a maximal rate of LH-RH-stimulated release of LH from the beginning on.

2.2 Protein Factor: Its Formation and Deactivation

Since the release of LH-RH in OVX rats is higher than in intact rats (Wheaton and McCann, 1976), the preceding results suggest that the formation of the protein factor is related to the amount of LH-RH secreted from the hypothalamus through the portal vessels to the pituitary gland. Alternatively, a change in the turnover of the protein factor after removal of the ovaries might also account for the results. However, we were able to demonstrate that the formation of the protein factor is related to the concentration of LH-RH *in vitro* (de Koning *et al.*, 1977b). In Fig. 2, the

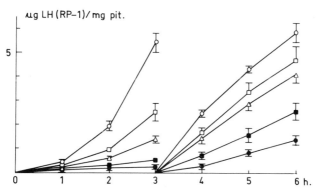

Figure 2. Effect of exposure to submaximally active concentrations of LH-RH for 3 hr on the response of pituitary glands of intact rats to LH-RH in the presence of cycloheximide during a subsequent 3-hr period of incubation. The results are given as mean LH content of the medium ± S.E.M.; each group consisted of 4 flasks. Added to the media during the period of 0–3 hr: (●) 0, (■) 0.3, (△) 1.0, (□) 3.2, or (○) 10 ng LH-RH/ml; during the second period of 3–6 hr: 1000 ng LH-RH/ml and 25 µg cycloheximide/ml were present in all flasks. Data of de Koning *et al.* (1977b).

results of one of these experiments are depicted. Pituitary glands from intact rats were first exposed to different submaximally active concentrations of LH-RH for 3 hr. During this first incubation period, a dose-dependent effect of LH-RH on release of LH could be observed. After this period, the medium was exchanged and the glands were further incubated for a 3-hr period, but now with a maximally active concentration of LH-RH (1000 ng/ml) plus cycloheximide. Under the latter conditions, LH release will be limited if submaximally active amounts of the protein factor are present at the start of the second incubation period. Indeed, in our experiment, the action of the high concentration of LH-RH was limited during the second incubation period; moreover, LH release was related to the concentration of LH-RH present during the first incubation period. Inhibiting protein synthesis also during the first incubation period or allowing additional protein synthesis during the second incubation period by adding or omitting cycloheximide, respectively, disturbed this relationship completely (de Koning *et al.*, 1977b). Hence, these results suggest very strongly that the formation of the protein factor is related to the concentration of LH-RH. In addition, it should be mentioned that after a 4-hr preincubation of pituitary glands from intact rats with a maximally active concentration of LH-RH, LH release had become entirely independent of the *de novo* synthesis of protein during prolonged exposure to LH-RH (de Koning *et al.*, 1976b).

Considering the results discussed above, it might be expected that the initial response of pituitary glands of intact rats to LH-RH *in vitro* should be elevated after injection of LH-RH *in vivo*. This possibility was investigated by intravenously injecting intact female rats with 50 ng LH-RH/100 g body weight, 1, 2, or 4 hr prior to decapitation. The pituitary glands were placed in medium TC 199 containing cycloheximide and preincubated for half an hour, after which the medium was replaced by fresh medium with a high concentration of LH-RH (1000 ng/ml) and cycloheximide. The incubation was continued for 4 hr. The results given in Fig. 3 demonstrate that in all cases, the rate of LH-RH-stimulated release of LH is elevated above that from the glands that had not been injected with LH-RH. In addition, it can be observed that the height of LH release *in vitro* is inversely proportional to the interval between injection of LH-RH and decapitation. These results, too, indicate that the increased rate of LH release during incubation with LH-RH and cycloheximide is caused by the LH-RH-induced formation of the protein factor *in vivo*. This is confirmed by the failure of the injection of LH-RH to increase the response of the pituitary gland to LH-RH *in vitro* (unpublished results) or *in vivo* (Pickering and Fink, 1976) if cycloheximide has been injected ½ hr prior to the injection of LH-RH. In accordance with the presumed high concentration of the protein factor in pituitary glands of OVX rats, injection of

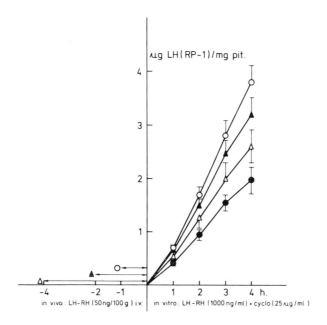

Figure 3. Effect of intravenous injection of LH-RH on the response of pituitary glands of intact rats to LH-RH in the presence of cycloheximide during a subsequent 4-hr period of incubation. The results are given as mean LH content of the medium ± S.E.M.; each group consisted of 4 flasks. The animals were injected intravenously with 50 ng LH-RH/ 100 g body weight (○) 1, (▲) 2, or (△) 4 hr prior to decapitation. Control flasks (●) received glands from animals that had not been injected with LH-RH. The pituitary glands were incubated with 1000 ng LH-RH/ml and 25 μg cycloheximide/ml.

LH-RH in these animals does not affect the response of the glands to LH-RH *in vitro* (unpublished results). The decay in the LH-RH-induced activation of the response of the pituitary glands to LH-RH *in vitro* produced by extending the interval between injection of LH-RH and decapitation demonstrates that the concentration of the protein factor wears off exponentially. In fact, a half-life of 1.9 hr was calculated.

Again, a comparison with the priming effect must be made. The priming effect was maximal if the second injection was given 1 hr after the first one. If this interval was extended, the effect wore off. After an interval of 4 hr, the priming effect had almost disappeared (Aiyer *et al.*, 1974). The priming effect, too, could be blocked completely by the administration of cycloheximide or actinomycin D, prior to the first addition of LH-RH (Pickering and Fink, 1976). From the latter and the present results, it can be deduced that the priming effect depends on the formation of the protein factor during the short exposure of the glands to injected LH-RH. If the second injection of LH-RH is administered 1 hr later, the amount of the protein factor is still elevated, permitting a further increment in the rate of LH release. In fact, it is likely that the priming effect discontinuously reflects the changes in responsiveness of the pituitary glands that are continuously exposed to LH-RH.

In conclusion, these data taken together provide further arguments for the hypothesis that the protein factor may be rate-limiting for LH-RH-stimulated release of LH, and that its formation is dependent on the concentration of LH-RH. This hypothesis includes two actions of LH-RH in the LH-secretion mechanism: one on the formation of protein factor, the other on the direct release of LH itself. Both actions of LH-RH are indispensable in the LH-secretion process (de Koning *et al.*, 1976b, 1977b).

2.3. Extrapituitary Control of the Deactivation of the Protein Factor

During incubation of pituitary glands from intact rats with a maximally active concentration of LH-RH and cycloheximide, no decay in the rate of LH release could be observed (Figs. 2 and 3). This is remarkable, since the amounts of the protein factor present in these glands limit the rate of response of the glands to the high concentration of LH-RH, while the incubations lasted about twice the half-life of the protein factor. The maintained action of the protein factor might be due to suppressed synthesis of protein leading to the deactivation or disappearance of (a) factor(s) that control(s) the deinduction of the protein factor. Similar systems have been described by Tomkins *et al.* (1972) and Wildenthal and Griffin (1976). On the other hand, we should consider the possibility that removal of the pituitary gland from its natural environment and the consequent disruption of action(s) of a modulatory factor(s) from the blood circulation may contribute to these results.

It should be stressed that pituitary glands of OVX rats pretreated with estradiol benzoate (EB) for 3 successive days (7 μg EB subcutaneously administered each day) respond to exogenous LH-RH *in vitro* (Fig. 4) and *in vivo* (Schuiling and Gnodde, 1976) without showing an initial phase of low LH release. Moreover, administration of phenobarbital (5 mg/100 g body weight, intraperitoneally, twice a day) during EB or control (oil) treatment of OVX rats did not affect LH-RH-stimulated release of LH *in vitro*, although plasma LH levels were suppressed significantly (Fig. 4). This is not in agreement with our hypothesis, since the low rate of LH-RH release in these animals should have led to a small amount of the protein factor in the pituitary gland, thus limiting the response of these glands during exposure to high levels of exogenous LH-RH. Therefore, we investigated the possibility that the disappearance of the protein factor might be affected by agents from the ovaries. For this, in the morning of

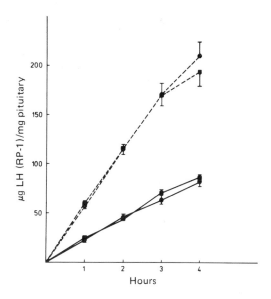

Figure 4. Effect of subcutaneous injections of EB (7 μg each day for 3 successive days) and/or phenobarbital (5 mg/100 g body weight twice each day during the same period) on the response of pituitary glands of ovariectomized rats during 4 hr of incubation with LH-RH (1000 ng/ml). The results are given as mean LH content of the medium ± S.E.M.; each group consisted of 4 flasks. (●————●) Oil/saline (483 ± 20 ng/ml) (the numbers in parentheses are the serum LH levels immediately following decapitation); (■————■) oil/phenobarbital (286 ± 24 ng/ml); (●– – – –●) EB/saline (115 ± 24 ng/ml); (■– – – – –■) EB/phenobarbital (86 ± 7 ng/ml).

the second day of diestrus, rats were ovariectomized and immediately thereafter injected intravenously with LH-RH (50 ng/100 g body weight) or saline. Control animals were not ovariectomized, but also received these injections. At 4 hr later, the animals were killed, and the pituitary glands were dissected into halves and transferred into medium TC 199 containing cycloheximide only. After ½ hr, the medium was changed, and the glands were incubated in medium containing LH-RH (1000 ng/ml) plus cycloheximide. The results of the 4-hr incubation are presented in Fig. 5, which again demonstrates that the LH-RH-induced increase in the response of the pituitary gland of intact rats to exogenous LH-RH *in vitro* had almost returned to control levels 4 hr after the injection of LH-RH (also see Fig. 3). However, if the ovaries had been removed prior to the injection of LH-RH, the response of the pituitary glands to LH-RH remained elevated. This cannot have been caused by an increase in the rate of LH-RH release or decreased degradation of LH-RH: as was found in unpublished experiments, injection of phenobarbital immediately following ovariectomy did not affect specific patterns of LH release, as for instance is shown in Fig. 5.

These findings point to the presence of (an) ovarian factor(s) that might either directly or indirectly stimulate the deinduction of the protein factor. This ovarian factor is probably not estradiol, since administering EB for 3 successive days did not restore the initial phase of low release of LH in OVX rats (See Fig. 4) (Schuiling and Gnodde, 1976). Consequently, the absence of the initial phase in the LH-RH-induced pattern of LH release from pituitary glands of OVX rats can be explained as the result of an increment in the rate of LH-RH release and the elimination of (an) ovarian factor(s) other than estradiol.

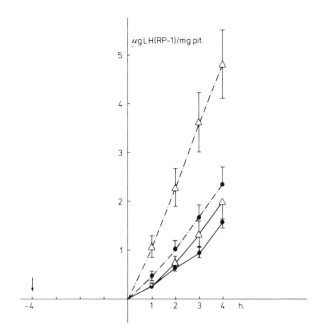

Figure 5. Effect of ovariectomy on the response of pituitary glands of intact diestrous rats to LH-RH in the presence of cycloheximide during a subsequent 4-hr incubation period, as modified by intravenous injection of LH-RH. The results are given as mean LH content of the medium ± S.E.M.; each group consisted of 4 flasks. Intact rats (●) and acutely OVX rats (△) were injected intravenously with 50 ng LH-RH/100 g body weight (—·—·—) or vehicle (———) 4 hr prior to decapitation. The pituitary glands were incubated with 1000 ng LH-RH/ml and 25 µg cycloheximide/ml.

3. Refractoriness of the Pituitary Gland to the Action of LH-RH

When pituitary glands of rats are exposed to constant levels of LH-RH, after an initial rise the rate of LH release wears off, despite the continuous presence of LH-RH. This phenomenon of refractoriness of the pituitary gland to the action of LH-RH has been demonstrated *in vivo* (Gnodde and Schuiling, 1975; Schuiling and Zürcher, 1975; Blake, 1976; Schuiling *et al.*, 1976; Vilchez-Martinez *et al.*, 1976) and *in vitro* (de Koning *et al.*, 1976b, 1978b; Hopkins, 1977). It was pointed out by Schuiling and Gnodde (1976) that refractoriness cannot be the result of the depletion of a fixed amount of LH present in pituitary stores.

In the preceding section, we have demonstrated that LH-RH can modulate LH release via two different pathways: one is the induction of the protein factor (the presence of which is necessary for LH release) and the other is LH release itself. Preliminary experiments showed that the combination of 1 mM mbcAMP plus 10 mM theophylline mimicked the effect of LH-RH on the formation of the protein factor, but had a moderate effect on LH release itself (de Koning *et al.*, 1977a, 1978a). However, raised potassium very closely imitates the direct action of LH-RH on LH release (i.e., LH-RH-stimulated release of LH in the presence of inhibitors of protein synthesis), whereas it has no effect on the formation of the protein factor (results not shown).

Moreover, Zor *et al.* (1970) found that the effect of raised potassium levels on LH release is probably not mediated by cyclic AMP. In an attempt to designate which action of LH-RH causes refractoriness, the following series of experiments was set up to investigate the effects of raised potassium (50 mM) and the combination of mbcAMP (1 mM) plus theophylline (10 mM) (hereinafter designated as mbcAMP/theo) on the induction of this phenomenon and, in addition, on LH release from pituitary glands already desensitized by LH-RH (de Koning *et al.*, 1978b).

These experiments were carried out by exposing pituitary glands from intact female rats to LH-RH, mbcAMP/theo, or raised potassium levels during 6 consecutive periods of 4 hr (first incubation), after which, at the 24th hr, the medium was exchanged for medium containing different agents, or a concentration different from those used during the first 24 hr. This second incubation lasted 6 hr. The experimental conditions during both incubations are given in the captions of Figs. 6–8. The results depicted in these figures are summarized in the following paragraph.

During the first 24 hr of incubation, LH release stimulated by the different concentrations of LH-RH (Figs. 6–8) and by raised potassium (Fig. 7) first increased and then declined. In contrast, mbcAMP/theo-induced LH release remained more or less constant after an initial rise

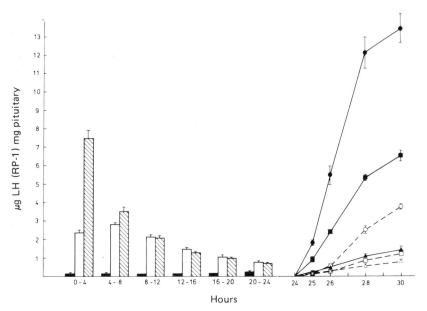

Figure 6. Release of LH (mean concentration of LH in the medium ± S.E.M.) from rat pituitary glands after incubation with medium only (solid bars) or 0.3 (open bars) or 10 (hatched bars) ng LH-RH/ml for 24 hr, then with 1000 ng LH-RH/ml (●, ■, ▲, respectively) or mbcAMP/theo (○, □, △, respectively) for a further 6 hr. For the initial 24-hr incubation, there were 8 flasks in each group; for the second 6 hr incubation, each group of 8 was divided into two groups of 4 flasks each. Data from de Koning *et al.* (1978b).

(Figs. 7 and 8). In addition, mbcAMP/theo did not desensitize the response of the pituitary glands to LH-RH, but rather facilitated LH-RH-stimulated release of LH during the second period. Pituitary glands that had first been exposed to a just maximally active concentration of LH-RH (10 ng/ml) or to submaximally active concentrations of LH-RH (0.1 or 0.3 ng/ml) showed a dose-dependent decreased responsiveness to the high concentration of 1000 ng/ml LH-RH or mbcAMP/theo during the second incubation (Figs. 6–8). The response to LH-RH during the last incubation period was inversely proportional to the concentration of LH-RH during the first incubation (Fig. 6). It should be stressed that pituitary glands that had first been incubated in medium without further additives for 24 hr were still responsive to the action of LH-RH or mbcAMP/theo. Potassium mimicked the action of LH-RH in that it desensitized the pituitary glands to the action of LH-RH, and that glands that had become refractory to LH-RH showed the same decreased response to high potassium. The mbcAMP/theo combination by itself did not induce refractoriness of the pituitary gland to the action of LH-RH. This failure could not be the result of the limited release of LH during the first incubation, since glands exposed to 0.1 ng/ml LH-RH had also become refractory, although during the first 0- to 24-hr incubation period, less LH had been released than that caused by mbcAMP/theo (Fig. 8).

The findings discussed above indicate that the state of refractoriness of the pituitary gland is initiated by the direct action of LH-RH that is imitated by raised potassium levels. Hence, mechanisms that are stimulated by membrane depolarization or otherwise potassium-dependently must be held mainly responsible for refractoriness of the glands. As a result, both responses of the pituitary gland to the actions of LH-RH that are mimicked by raised potassium and mbcAMP/theo are limited.

It has been demonstrated in other endocrine-responsive tissues (e.g., ovaries and testes) that the reduction in responsiveness was accompanied by a loss of receptor sites on the cellular membrane (Sharpe, 1976, 1977; Conti *et al.*, 1977b; Purvis *et al.*, 1977; Tsuruhara *et al.*, 1977). Moreover, the hormonally induced loss of receptor sites simultaneously led to impaired stimulation of adenylate cyclase and cyclic AMP formation by the tissue-specific hormone as well as by agents such as NaF, cholera

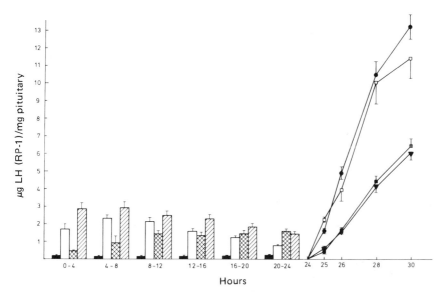

Figure 7. Release of LH (mean concentration of LH in medium ± S.E.M.) by rat pituitary glands incubated with medium alone (solid bars), 0.3 ng LH-RH/ml (open bars) or mbcAMP/theo (cross hatched bars), or 50 mM potassium (hatched bars) for 24 hr and then with 1000 ng LH-RH/ml (●, ■, □, ▼, respectively) for a further 6 hr. All incubations consisted of 6 flasks. Data from de Koning *et al.* (1978b).

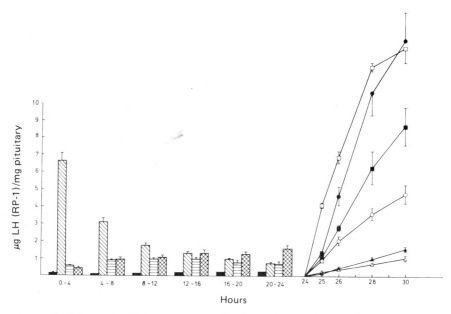

Figure 8. Release of LH (mean concentration in medium ± S.E.M.) from rat pituitary glands incubated with medium alone (solid bars), 10 (diagonally hatched bars) or 0.1 (horizontally hatched bars) ng LH-RH/ml, or mbcAMP/theo (cross-hatched bars) for 24 hr and then with 1000 ng LH-RH/ml (●, ▲, ■, □, respectively) or 50 mM potassium (after incubation with medium alone or 10 ng LH-RH/ml only: ○, △, respectively) for a further 6 hr. Initial 24-hr incubations consisted of 8 flasks; where the second incubation involved two reagents, each incubation consisted of 4 flasks. Data from de Koning *et al.* (1978b).

toxin, and prostaglandins (Conti *et al.*, 1977a,b; Tsuruhara *et al.*, 1977). At first sight, it is not likely that LH-RH should cause refractoriness of the pituitary gland by such a mechanism, since glands that had become desensitized by LH-RH also showed a reduced response to the action of mbcAMP/theo, suggesting that essential factors concerned with the action of cyclic AMP in the LH release mechanism are involved. However, in some of the cases mentioned above, the impaired steroidogenic response of whole testis or luteal cells was not restored by the addition of N^6,O^2-dibutyryl cyclic AMP (Conti *et al.*, 1977a; Sharpe, 1977). Nevertheless, the failure of mbcAMP/theo to initiate refractoriness of the pituitary gland to the action of LH-RH demonstrates that cyclic AMP itself is not a mediatorial factor in this process. Hence, cyclic AMP (and by implication mbcAMP/theo) does not cause refractoriness, since it interferes in the LH release mechanism other than the "direct" one.

4. Conclusions

LH-RH-stimultated release of LH from rat pituitary glands was shown to be mediated via protein-synthesis-dependent (the formation of the protein factor) and -independent (direct effect) pathways. Both actions of LH-RH are necessary to cause LH release. We demonstrated a relationship between the dose of LH-RH and the formation of the protein factor. Since it may be assumed that LH-RH release in OVX rats is higher than in intact ones (Wheaton and McCann, 1976), the differences in initial response of pituitary glands of intact and OVX rats might therefore result from the low (rate-limiting) amount of the protein factor in pituitary glands of intact rats and the relatively high (not rate-limiting) amount in glands of OVX rats. During prolonged exposure of the pituitary glands of intact rats to LH-RH, the induction of the protein factor allows a higher rate of LH release; this process takes about 1 hr.

The amount of protein factor can also be affected by its deactivation. In this respect, it was found that the ovaries probably secrete factors that enhance the rate of disappearance.

Although it was clearly demonstrated that at least in our incubation system, newly synthesized protein induced by LH-RH is not LH (de Koning *et al.*, 1976b), it naturally remains possible that the protein is concerned with the formation of "releasable LH," e.g., by attachment of the carbohydrate moieties or transport to the plasma membrane. However, at the moment, no further conclusions can be drawn about the nature and action of the proteins in the sequence of events leading to LH secretion, before it has been isolated and characterized.

During prolonged exposure of the pituitary glands to LH-RH, the phase of refractoriness manifests itself. The phenomenon of refractoriness has been found in many other hormonally controlled systems (see, for example, Lamprecht *et al.*, 1977). However, there is no unanimity with regard to the causes of the refractory state of the tissues. Among the possibilities are: deterioration of the receptor population, impairment of the coupling of the hormone–receptor signal and adenylate cyclase stimulation or induction of inhibitory factors. Our results demonstrate that desensitization of the pituitary glands to LH-RH involves deterioration of the direct action of LH-RH, which can be mimicked by raised potassium. This might include degradation of LH-RH receptor sites, if their presence is also necessary for the action of raised potassium.

The possible physiological significance of the development of refractoriness in pituitary glands of intact rats was demonstrated by Schuiling *et al.* (1976), who showed the occurrence of a preovulatory-like LH surge during continuous infusion of LH-RH in pentobarbital-blocked rats in the afternoon of the day of proestrus. The fact that long-term-gonadectomized rats secrete high amounts of LH may not be in contradiction to the phenomenon of refractoriness. It is generally assumed that LH-RH is secreted in pulses, and pulsatile stimulation of the LH release mechanism by LH-RH may prevent refractoriness.

To further elucidate the different molecular events in LH secretion, experiments should be carried out in which the formation and degradation of intermediatorial factors (metabolites, enzymes, receptor sites) are studied during exposure to various secretagogues and agents that interfere in the cellular mechanism, but in connection with the changes in responsiveness of the pituitary glands to LH-RH. Taking into account the heterologous endocrine function of the pituitary gland (which consists of 5–10% gonadotropic cells), the development of techniques to isolate homologous populations of specific pituitary cell types will be of paramount importance.

DISCUSSION

TIXIER-VIDAL: I would challenge the usefulness of long-term incubations for analyzing the mechanism of LH-RH refractoriness. However, my question concerns your *in vivo* and *in vitro* incubations carried out for short terms. What is your interpretation as to the "protein factor" that you gave evidence for?

DE KONING: We do not know the physiological action of the protein factor. It could be involved in one or more of the events between LH-RH binding and the subsequent release of LH.

FINK: In regard to Dr. Tixier's criticism, your long-term incubations on refractoriness may have physiological relevance in that in the rat, human, and sheep, continuous intravenous infusion of GnRH leads to an LH surge followed by refractoriness. At least in the rat, where pituitary hormone content can be measured, this cannot be attributed simply to the depletion of hormones.

JUTISZ: Would there not be leaking of LH from necrotic cells in long-term incubations?

FINK: With respect to leakiness in de Koning's long-term experiments, everyone accepts that there is central necrosis in the incubated glands. The low baseline levels of hormone indicate that this does not lead to leaking.

KORDON: Why do you use micromolar doses of LH-RH to "reveal" your priming effects in a second incubation? 1–10 nM amounts of the peptide are enough to saturate the LH-RH receptors.

DE KONING: We have demonstrated that 10 ng of LH-RH is a maximally active concentration to stimulate LH release from glands of intact rats. However, 100–1000 ng appear to be most effective for inducing LH release from ovariectomized rats. So, in comparing the effect of LH-RH on LH release, we have used the higher concentration in both cases. Also, in some experiments we wanted to be certain that LH release was limited by the protein factor and not by LH-RH. This was especially so after preincubation of the pituitary glands with various submaximally active concentrations of LH-RH.

DENEF: Is it plausible that LH-RH had mainly stimulated the peripheral gonadotrophs in pituitary halves? In that case, eventual necrotic processes in the center of the gland would not interfere very much with the interpretation of the results obtained with long-term incubation.

DE KONING: This fits in with the high percentage of the cells which have been stimulated during continuous exposure to LH-RH.

TIXIER-VIDAL: To come back to the question of long-term incubations, I think you have to be cautious in interpretations relative to the mechanism of LH-RH refractoriness. We found, with Dr. Jutisz, that fragments of sheep anterior pituitaries undergo progressive necrosis. Also, some years ago, M. Farquhar, using autoradiography, found that only the cells of the periphery were able to incorporate labeled amino acid into protein.

DE KONING: I will mention three facts which are in favor of the validity of the system used. First, during incubation with 10 ng LH-RH, at least 70–80% of the original content of LH has been released. This means that at least 70–80% of the LH-secreting cells have been stimulated. Secondly, the basal release of LH, although it may tend to increase at the end of the first 24 hours, remains low. Thirdly, after 24 hours of incubation in medium only, the pituitary glands are still responsive to LH-RH, to increases in potassium, and to mbcAMP. The amount of LH released is about the same as that stimulated from fresh glands. Of course, we need to be cautious in interpreting these results, but this also applies to other *in vitro* systems including free-cell preparations.

REFERENCES

Aiyer, M. S., Chiappa, S. A., Fink, G., and Grieg, F., 1973, A priming effect of luteinizing hormone releasing factor on the anterior pituitary gland in the female rat, *J. Physiol. (London)* **234**:81P.

Aiyer, M. S., Chiappa, S. A., and Fink, G., 1974, A priming effect of luteinizing hormone releasing factor on the anterior pituitary gland in the female rat, *J. Endocrinol.* **62**:573.

Blake, C. A., 1976, Stimulation of the proestrous luteinizing hormone (LH) surge after infusion of LH-releasing hormone in pentobarbital-blocked rats, *Endocrinology* **98**:451.

Borgeat, P., Chavancy, G., Dupont, A., Labrie, F., Arimura, A., and Schally, A. V., 1972, Stimulation of adenosine $3':5'$-cyclic monophosphate accumulation in anterior pituitary gland *in vitro* by synthetic luteinizing hormone–releasing hormone, *Proc. Natl. Acad. Sci. U.S.A.* **69**:2677.

Castro-Vazquez, A., and McCann, S. M., 1975, Cyclic variations in the increased responsiveness of the pituitary to luteinizing hormone–releasing hormone (LH-RH) induced by LH-RH, *Endocrinology* **97**:13.

Conti, M., Harwood, J. P., Dufau, M. L., and Catt, K. J., 1977a, Regulation of luteinizing hormone receptors and adenylate cyclase activity by gonadotrophin in rat ovary, *Mol. Pharmacol.* **13**:1024.

Conti, M., Harwood, J. P., Dufau, M. L., and Catt, K. J., 1977b, Effect of gonadotrophin-induced receptor regulation on biological responses of isolated rat luteal cells, *J. Biol. Chem.* **252**:8869.

Debeljuk, L., Rettori, V., Rozados, R. V., and Villegas Vélez, C., 1975, Effect of actinomycin D and estradiol on the response to LH-releasing hormone in neonatally androgenized female rats, *Proc. Soc. Exp. Biol. Med.* **150**:299.

De Koning, J., van Dieten, J. A. M. J., and van Rees, G. P., 1976a, Absence of an inhibitory effect of oestradiol on LH-RH induced release of LH *in vitro* caused by inhibition of protein synthesis, *Mol. Cell. Endocrinol.* **4**:289.

De Koning, J., van Dieten, J. A. M. J., and van Rees, G. P., 1976b, LH-RH-dependent synthesis of protein necessary for LH release from rat pituitary glands *in vitro, Mol. Cell. Endocrinol.* **5:**151.

De Koning, J., van Dieten, J. A. M. J., and van Rees, G. P., 1977a, On the mechanism of LH-RH-induced release of LH: cAMP-dependent synthesis of protein, *Acta Endocrinol. Suppl.* **212:**38.

De Koning, J., van Dieten, J. A. M. J., and van Rees, G. P., 1977b, Effect of preincubation with different concentrations of LH-RH on subsequent LH release caused by supramaximally active amounts of LH-RH; role of LH-RH-induced protein synthesis, *Life Sci.* **21:**1621.

De Koning, J., van Dieten, J. A. M. J., and van Rees, G. P., 1978a, Absence of an augmentative action of oestrogen on the release of luteinizing hormone induced by N^6-monobutyryl adenosine 3':5'-monophosphate plus theophylline from rat pituitary glands *in vitro, J. Endocrinol.* **77:**259.

De Koning, J., van Dieten, J. A. M. J., and van Rees, G. P., 1978b, Refractoriness of the pituitary gland after continuous exposure to luteinizing hormone releasing hormone, *J. Endocrinol.* **79:**311.

Edwardson, J. A., and Gilbert, D., 1975, Sensitivity of self-potentiating effect of luteinizing hormone–releasing hormone to cycloheximide, *Nature (London)* **255:**71.

Edwardson, J. A., and Gilbert, G., 1976, Application of an *in vitro* perfusion technique to studies of luteinizing hormone release by rat anterior hemipituitaries: Self-potentiating by luteinizing hormone releasing hormone, *J. Endocrinol.* **68:**192.

Fink, G., and Pickering, A., 1975, Dependency of the priming effect of luteinizing hormone releasing factor (LRF) on RNA and protein synthesis, *J. Physiol.* **252:**73P.

Gay, V. L., Niswender, G. D., and Midgley, A. R., Jr., 1970, Response of individual rats and sheep to one or more injections of hypothalamic extracts as determined by radioimmunoassay of plasma LH, *Endocrinology* **86:**1305.

Gnodde, H. P., and Schuiling, G. A., 1975, The pituitary response to continuous infusion of synthetic LH-RH in the ovariectomized rats: Effects of oestradiol and progesterone, *Acta Endocrinol. Suppl.* **199:**197.

Hopkins, C. R., 1977, Short term kinetics of luteinizing hormone secretion studied in dissociated pituitary cells attached to manipulable substrates, *J. Cell Biol.* **73:**685.

Jutisz, M., Bérault, A., Kerdelhué, B., and Théoleyere, M., 1975, Some aspects of the cellular mechanism of action of gonadotrophin releasing hormone(s), in: *Some Aspects of Hypothalamic Regulation of Endocrine Functions* (P. Franchimont, ed.), pp. 33–46, Schattauer Verlag, Stuttgart and New York.

Kercret, H., Benoist, L., and Duval, J., 1977, Acute release of gonadotropins mediated by dibutyryl-cyclic AMP *in vitro, FEBS Lett.* **83:**222.

Lamprecht, S. A., Zor, U., Salomon, Y., Koch, Y., Ahren, K., and Lindner, H. R., 1977, Mechanism of hormonally induced refractoriness of ovarian adenylate cyclase to luteinizing hormone and prostaglandin E_2, *J. Cyclic Nucleotide Res.* **3:**69.

Osland, R. B., Gallo, R. V., and Williams, J. A., 1975, *In vitro* release of luteinizing hormone from anterior pituitary fragments superfused with constant or pulsatile amounts of luteinizing hormone–releasing factor, *Endocrinology* **96:**1210.

Pickering, A. J. M. C., and Fink, G., 1976, Priming effect of luteinizing hormone releasing factor: *In-vitro* and *in-vivo* evidence consistent with its dependence upon protein and RNA synthesis, *J. Endocrinol.* **69:**373.

Purvis, K., Torjesen, P. A., Haug, E., and Hansson, V., 1977, hCG suppression of LH receptors and responsiveness of testicular tissue to hCG, *Mol. Cell. Endocrinol.* **8:**73.

Schuiling, G. A., and Gnodde, H. P., 1976, Secretion of luteinizing hormone caused by continuous infusion of luteinizing hormone releasing hormone in the long-term ovariectomized rat: Effect of oestrogen pretreatment, *J. Endocrinol.* **71**:1.

Schuiling, G. A., and Gnodde, H. P., 1977, Oestrogen-induced changes in the secretion of luteinizing hormone caused by continuous infusions of luteinizing hormone releasing hormone in the long-term ovariectomized rat, *J. Endocrinol.* **72:** 121.

Schuiling, G. A., and Zürcher, A. F., 1975, An investigation into the contribution of the central nervous system and the pituitary gland to the characteristics of pre-ovulatory LH-surges in the rat; role of progesterone, *Acta Endocrinol. Suppl.* **199:**198.

Schuiling, G. A., de Koning, J., Zürcher, A. F., Gnodde, H. P., and van Rees, G. P., 1976, Induction of LH surges by continuous infusion of LH-RH, *Neuroendocrinology* **20:**151.

Sharpe, R. M., 1976, hCG-induced decrease in availability of rat testis receptors, *Nature (London)* **264:**644.

Sharpe, R. M., 1977, Gonadotrophin-induced reduction in the steroidogenic responsiveness of the immature rat testis, *Biochem. Biophys. Res. Commun.* **76:**957.

Tomkins, G. M. Levinson, B. B., Baxter, J. D., and Dethlefsen, L., 1972, Evidence for posttranscriptional control of inducible tyrosine aminotransferase synthesis in cultured hepatoma cells, *Nature (London) New Biol.* **239:**12.

Tsuruhara, T., Dufau, M. L., Cigorraga, S., and Catt, K. J., 1977, Effects on cyclic AMP and testosterone responses in isolated Leydig cells, *J. Biol. Chem.* **252:**9002.

Vilchez-Martinez, J. A., Arimura, A., and Schally, A. V., 1976, Effect of actinomycin D on the pituitary response to LH-RH, *Acta Endocrinol.* **81:**73.

Welschen, R., Osman, P. Dullaart, J., de Greef, W. J., Uilenbroek, J. Th. J. and de Jong, F. H., 1975, Levels of follicle-stimulating hormone, luteinizing hormone, oestradiol-17β and progesterone, and follicular growth in the pseudopregnant rat, *J. Endocrinol.* **64:**37.

Wheaton, J. E., and McCann, S. M., 1976, Luteinizing hormone–releasing hormone in peripheral plasma and hypothalamus of normal and ovariectomized rats, *Neuroendocrinology* **20:**296.

Wildenthal, K., and Griffin, E. E., 1976, Reduction by cycloheximide of lysosomal proteolytic enzyme activity and rate of protein degradation in organ-cultured hearts, *Biochim. Biophys. Acta* **444:**519.

Zor, U., Kaneko, T., Schneider, H. P. G., McCann, S. M., and Field, J. B., 1970, Further studies of stimulation of anterior pituitary cyclic adenosine 3',5'-monophosphate formation by hypothalamic extract and prostaglandines, *J. Biol. Chem.* **245:**2883.

31

Functional Heterogeneity of Separated Dispersed Gonadotropic Cells

Carl Denef

1. Introduction

It is well documented that the gonadotropic cell population of the pituitary gland is morphologically heterogeneous. Histochemical staining reactions (Herlant, 1960, 1964) and electron-microscopic examinations (Farquhar and Rinehart, 1954; Barnes, 1962; Yoshimura and Harumiya, 1965; Kurosumi and Oota, 1968) have revealed the presence of two different cell types. One contained two populations of secretory granules the size of which was about 200 and 700 nm, respectively, and was believed to secrete follicle-stimulating hormone (FSH). The other was smaller, contained only one type of secretory granules with a size of 250 nm, and was thought to secrete luteinizing hormone (LH). However, after immunochemical methods became available to demonstrate directly the hormones in the cell, it became evident that there was no separate location of the two gonadotropins in a given cell type (see the review by Girod, 1977). Several investigators showed that each of these cell types was immunoreactive to anti-LH antiserum (Tougard et al., 1973; Moriarty, 1975) as well as to anti-FSH antiserum (Tixier-Vidal et al., 1975b; Moriarty, 1976). Furthermore, there is strong evidence that most if not all of the gonado-

Carl Denef • Laboratory of Cell Pharmacology, Department of Pharmacology, School of Medicine, Campus Gasthuisberg, Katholieke Universiteit Leuven, B-3000 Leuven, Belgium

trophs contain both FSH and LH, irrespective of their morphological classification (Nakane, 1970; Tixier-Vidal *et al.*, 1975b; Tougard *et al.*, 1977a,b; Phifer *et al.*, 1973; Herbert, 1975, 1976; Robyn *et al.*, 1973; Pelletier *et al.*, 1976; Sternberger and Petrali, 1975) (also see Chapter 2). More recently, even more ultrastructurally distinct gonadotrophs were identified. Tixier-Vidal *et al.* (1975a,b) found in normal and castrated rats numerous gonadotropic cell types with ultrastructural characteristics intermediate between those of the two cell types described originally. Again, they were immunoreactive to both anti-FSH and anti-LH antisera. Another type is a cell of small size that is the first immunoreactive gonadotropic cell type emerging in fetal rat pituitary (Tougard *et al.*, 1977a). This cell type looked very similar to the one found in long-term monolayer cultures (Tougard *et al.*, 1977b). Ultrastructural characteristics and distribution of immunoreactive secretory granules also seem to fluctuate with the estrous cycle and to be dependent on sex (Moriarty, 1975). Apparently, the ultrastructural appearance of gonadotrophs changes with developmental, sexual, environmental, physiological, and experimental conditions. The latter observations have led to the proposal that either the gonadotropic cell population can differentiate into separate subtypes or there is only one cell type, which looks structurally different depending on the secretory cycle of the cell (Farquhar *et al.*, 1975; Tixier-Vidal *et al.*, 1975b; Yoshimura and Harumiya, 1965).

This chapter presents evidence, based on recent work in our laboratory, for a functional significance of the morphological heterogeneity of the gonadotropic cell population. The data obtained so far indicate that in addition to morphological heterogeneity, there is functional heterogeneity as well. Various subtypes of gonadotrophs appear to exist that differ not only in terms of secretory potential but also in terms of the relative amounts of FSH and LH being released. We propose that changes in the relative proportion of these subtypes may form the cellular basis for differential (nonparallel) FSH and LH release.

2. Choice of Animals and Methods

Functional correlates of the gonadotroph cell type distribution in the pituitary gland can best be studied during periods of rapid neuroendocrine changes, such as after gonadectomy or sex hormone treatment, during the estrous cycle, or during certain developmental events.

We have chosen 14-day-old rats as a model because both the distribution of the gonadotrophs in the pituitary and the gonadotropin secretion show dramatic fluctuations around that age. In 1954, Siperstein *et al.* (1954) demonstrated that between 10 and 14 days, there is a very rapid growth in size, number, and stainability of the gonadotrophs and that this is more pronounced in female than in male rats. Moreover, part of the gonadotrophs aggregate in an area close to the intermediate lobe near the sex zone, and this is also more evident in females than in males. More recently, others have found concomitant and sex-specific changes in FSH and LH secretion in these animals (Ojeda and Ramirez, 1972; Kragt and Dahlgren, 1972; Döhler and Wuttke, 1975; Ojeda *et al.*, 1977). Between about 10 and 14 days of age, female rats steeply increase their plasma FSH levels to values that are never reached later in life and that exceed severalfold the values of male littermates. At this age, female rats also show periodic spikes of high LH release, but males do not (Döhler and Wuttke, 1975).

We have studied various functional characteristics of gonadotrophs in these 14-day-old rats (as well as in adult male rats for a comparison) by using dispersed pituitary cells maintained in short-term culture. The cells were first separated according to size by gradient sedimentation at unit gravity. With the latter technique, it is possible to prepare fractions with enriched or highly purified populations of various pituitary cell types (Hymer *et al.*, 1973, 1974; Lloyd and McShan, 1973; Denef *et al.*, 1976), to study certain functional responses of the cells (Kraicer and Hymer, 1974; Lloyd and Karavolas, 1975; Snyder, J., *et al.*, 1976; Snyder, G., *et al.*, 1977), and to detect changes in size and proportional number of certain cell types after experimentally induced endocrine changes (Hymer *et al.*, 1973, 1974; Hymer, 1975; Lloyd and McShan, 1973). The reliability of the method has been found to be most satisfactory (Hymer, 1975), and cell-type distribution in dispersed pituitary cell preparations as well as after unit-gravity sedimentation reliably reflects the distribution in the intact pituitary (Hymer, 1975; Surks and DeFesi, 1977). In our experiments (Denef *et al.*, 1978a) with 14-day-old rats, there was a similar reliability. Cells recovered from the unit-gravity-sedimentation gradient retained viability of over 95%, and they preserved excellent ultrastructural integrity as shown by electron microscopy. Overall recovery of cells ranged between 55 and 65% and was the same for all separated fractions. The losses of hormone during the dispersion and sedimentation procedures were not massive and were of similar magnitude for FSH and LH and for the different animal groups involved. Finally, there were no selective losses of cell types, and the distribution of all types corresponded to that reported in the intact pituitary.

3. Heterogeneity in the Proportional Distribution of Gonadotrophs and Their Hormone Content

In a first approach, it was necessary to extend the original observations on cell-type distribution in pituitaries of 14-day-old rats made by Siperstein *et al.* (1954) and to compare the distribution of size and number of gonadotrophs with the distribution of FSH and LH over the cells. A relationship could then be sought between these parameters and the differences in the *in vivo* FSH and LH plasma levels of the different groups of animals studied (Denef *et al.*, 1978a).

Figure 1 shows the distribution of basophils and acidophils, identified by Alcian blue–periodic acid–Schiff(PAS)–Orange G staining, and Fig. 2

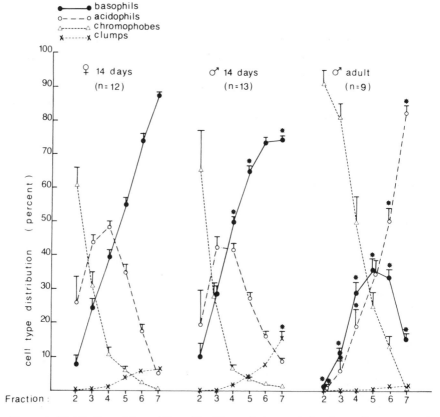

Figure 1. Distribution of basophils, acidophils, and chromophobes in gradient-separated dispersed pituitary cells from 14-day-old male and female rats and adult male rats. Staining was with the permanganate-Alcian blue–PAS–Orange G method. The data (means ± S.E.; number of experiments as indicated) are adapted from Denef *et al.* (1978a) and represent the percentage of each cell type in 100-ml fractions isolated from a unit-gravity-sedimentation gradient. Fractions are numbered from top to bottom of the gradient, fraction 2 being the fourth 100-ml fraction. Fraction 7 combines the two 100-ml bottom fractions. Statistical differences—*t* tests: 14-day-old females vs. 14-day-old males and adult males vs. 14-day-old males: (*) $p < 0.01$. Clumps (×) consist of 2–4 unstained cells.

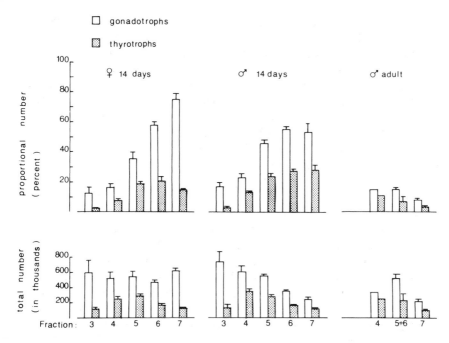

Figure 2. Distribution of gonadotrophs and thyrotrophs in gradient-separated dispersed pituitary cells from 14-day-old male and female rats and adult male rats. The cell types were identified by immunocytochemical techniques. The data (means ± S.E. of 4 experiments) are adapted from Denef *et al.* (1978a). Numbering of fractions as in Fig. 1.

the distribution of gonadotrophs and thyrotrophs identified by immunocytochemical methods. With separated pituitary cells from 14-day-old rats, the percentage and size of basophils increased gradually from top to bottom of the gradient. The bottom fraction of the gradient contained large basophils the purity of which attained about 90% in the female and 75% in the male. The same fraction prepared from adult male rats consisted of over 80% of large acidophils. The majority of the large basophils from 14-day-old females were gonadotrophs, whereas about 30% of the basophils from 14-day-old males were thyrotrophs. The proportional number of medium- and small-size gonadotrophs was also higher in the immature rat pituitary than in that of the adult, and the ratio of gonadotrophs to thyrotrophs was also different. The proportional but not the total number of thyrotrophs in 14-day-old males was higher than in 14-day-old females. The total number of large gonadotrophs from 14-day-old females was more than two times that from 14-day-old males, but the number of medium- and small-sized gonadotrophs was about equal. By comparing the number of gonadotrophs immunoreactive with anti-rat-FSH antiserum, with anti-rat-LH antiserum, and with a mixture of both antisera, the relative distribution of FSH and LH over the gonadotrophs was studied. The findings suggested that irrespective of cell size, the majority of the gonadotrophs contained both FSH and LH, but that FSH was also stored in separate cells, at least as the predominant hormone, since minor traces of LH could have escaped detection. Evidence for cells containing only or predominantly LH was found in 14-day-old males but not in 14-day-old females. The distribution of FSH and LH among the gonadotrophs was studied quantitatively by measuring FSH and LH contents by radioimmunoassay after extraction of the hormones from the cells. In the 14-day-old females, more than 70% of both hormones recovered from all gonadotrophs was recovered from the large gonadotrophs, and this is quite remarkable, since these gonadotrophs amount only up to 20% of the total number of gonadotrophs of the pituitary gland. In 14-day-old and in adult males, the fractional amount in the large gonadotrophs was lower, and substantial amounts of hormone were recovered from medium-sized gonadotrophs. More importantly, the FSH/LH ratio changed with cell size, and as shown in Fig. 3, the pattern was highly characteristic for each group. Thus, it seems clear that the gonadotroph cell population is heterogeneous in terms of absolute and relative amounts of FSH and LH stored in the cells.

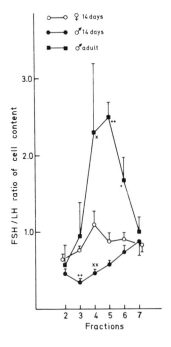

Figure 3. FSH/LH ratio determined by radioimmunoassay of FSH and LH after extraction from dispersed pituitary cells separated by unit-gravity sedimentation. Data adapted from Denef *et al.* (1978a). The data represent means ± S.E. of 3 or 4 gradient-separation experiments. FSH and LH are expressed in terms of the rat NIAMDD reference preparations (RP-1). Statistical differences — t tests: between 14-day-old male and female rats and between 14-day-old males and adult males: $(\times)\, p < 0.05$; $(\times\times)\, p < 0.02$; $(+)\, p < 0.01$; $(++)\, p < 0.001$.

When the distribution of size and number of gonadotrophs together with that of their FSH and LH contents are compared with the *in vivo* hormone secretions of the 14-day-old rats, we believe that a functional interrelationship is not unlikely. If all gonadotrophs are not identical in terms of absolute and relative FSH and LH contents, the overall secretory pattern including FSH/LH ratios may depend on the proportional number of such different gonadotrophs. The elevated number of large gonadotrophs and their high amounts of FSH and LH in 14-day-old females most likely favor the high levels of tonic FSH release and episodic LH release in these animals *in vivo*.

4. Functional Heterogeneity among Gonadotrophs

4.1. Secretory Heterogeneity of Gonadotrophs in Response to LH-RH

An intriguing characteristic of the regulation of gonadotropin release is that the hypothalamic gonadotropin-releasing hormone, LH-RH, stimulates the secretion of both FSH and LH, but that the proportional amount released differs according to various physiological and experimental conditions (Blackwell and Guillemin, 1973; Schally *et al.*, 1973a, 1978). To explain this differential control, it has been proposed that androgens (Schally *et al.*, 1973b; Mittler, 1974; Tang and Spies, 1975; Kao and Weisz, 1975, Drouin and Labrie, 1976; Epstein *et al.*, 1977; Debeljuk *et al.*, 1978), estrogens (Drouin *et al.*, 1976; Kalra, 1976; Miller *et al.*, 1977; Spona, 1976; Apfelbaum and Taleisnik, 1976; Debeljuk *et al.*, 1978) and progestogens (Drouin *et al.*, 1978; Debeljuk *et al.*, 1978), certain amino acids (Ondo *et al.*, 1976), and possibly a separate FSH-releasing hormone (Bowers *et al.*, 1973) modulate the action of LH-RH at the pituitary level by selective interactions with the mechanisms of FSH or LH release.

We have considered the alternative possibility that the cellular heterogeneity of the gonadotroph population would be the basis for differential response to LH-RH. If, as advanced earlier, subpopulations of gonadotrophs exist that differ in terms of the relative potential of FSH and LH release, changes in the proportional distribution of certain subpopulations, such as those occurring during development (see above), are likely to provoke nonparallel changes in FSH and LH release.

The present hypothesis was tested by exploring LH-RH-stimulated FSH and LH release in different populations of gonadotrophs separated according to size by unit-gravity sedimentation and established in monolayer cultures for 3 days (Denef *et al.*, 1978b). In all fractions from all animal groups, LH-RH stimulated FSH and LH in a dose–response fashion, and release was linear with time and cell number. When the secretory potential was expressed on a per-gonadotroph basis, it was clear that all gonadotrophs did not respond similarly to LH-RH. The differences were apparent at three levels: (1) the maximal secretory potential per gonadotroph when stimulated with 5×10^{-8} M LH-RH; (2) the relative amounts of FSH and LH being released; and (3) the preexisting physiological conditions of the animals studied. In 14-day-old female rats, there was a 20-fold difference in both FSH and LH secretion between the smallest and the largest gonadotrophs. In contrast, in preparations from 14-day-old males, only FSH release increased with cell size, but to a lesser extent than in females. Moreover, and also in contrast to females, when the relative amounts of FSH and LH released were compared, it was found that small- to medium-sized gonadotrophs released relatively more LH. With cells from adult males, both FSH and LH rose with cell size, but the pattern was very different from that seen in the immature females: the largest gonadotrophs secreted relatively more LH than FSH, but medium-sized gonadotrophs secreted relatively more FSH. The specific patterns observed were not brought about by unequal losses of hormone content by the different cell populations during the 3-day culture period.

We have confirmed the findings discussed above by more recent experiments. Figure 4 shows the amounts of FSH and LH secreted by the total cell population of each fraction when stimulated with 10^{-7} M LH-RH. With preparations from 14-day-old females, both FSH and LH release rose dramatically with cell size. For both hormones, there was about a 35-fold difference between the small (fraction 3) and large gonadotrophs (fraction 7). This pattern was not at all seen in 14-day-old males. Again, medium-sized gonadotrophs (fractions 4 and 5) from 14-day-old males secreted relatively more LH as compared to FSH than the large gonadotrophs (fraction 7). The FSH/LH ratio of the secretion was considerably lower in the medium-sized than in the large gonadotrophs. It is clear in Fig. 2 that the differences observed cannot be due to differences in the number of gonadotrophs among the fractions, since these differences were much smaller than the differences in secretory responses.

In conclusion, the present findings seem to indicate that different subtypes of gonadotrophs exist, each with specific functional characteristics. Their functional responses differ not only in terms of maximal release potential of FSH or LH, but also in terms of the relative amounts of FSH and LH being secreted. It is remarkable that the latter characteristics are related to cell size. However, it is most important to note that cells of the same size but derived from animals in a different physiological condition are functionally dissimilar as well.

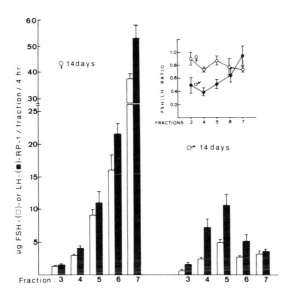

Figure 4. Release of FSH and LH, stimulated by 10^{-7} M LHRH, from cultured rat pituitary cells (3 days in culture; Falcon plastic petri dishes) separated by unit-gravity sedimentation. Methods were as described before (Denef *et al.,* 1978a,b) except for the use of Dulbecco's Modified Minimum Essential Medium instead of Minimum Essential Medium. FSH and LH are expressed in terms of the rat NIAMDD reference preparations (RP-1). Multiple *t* tests—difference between fractions and between male and female patterns: $p < 0.001$.

4.2. Secretory Heterogeneity of Gonadotrophs in Response to Androgen

As mentioned before, it is well known that androgens induce nonparallel changes in both basal and LH-RH-stimulated FSH and LH release. *In vivo,* FSH secretion is diminished to a lesser extent than LH release, and it has been shown that this effect is exerted, at least in part, at the pituitary level. In pituitary cell cultures, chronic androgen treatment increases FSH but not LH synthesis, but on LH-RH stimulation, FSH release is either slightly stimulated, not affected, or less inhibited than LH release (Drouin and Labrie, 1976; Denef *et al.,* in prep.). An important observation is that the magnitude of the androgen effect depends on the concentration of LH-RH. At both low and high doses of LH-RH, FSH release is either slightly increased (Drouin and Labrie, 1976) or not affected by the androgen pretreatment (Denef *et al.,* in prep.), whereas at intermediate concentrations, FSH release is similar to that in untreated controls (Drouin and Labrie, 1976) or at least considerably less inhibited than LH release (Denef *et al.,* in prep.).

To explain this LH-RH-dependent expression of the androgen effect on FSH, we explored the hypothesis that not all gonadotrophs are affected similarly by androgens or that not all respond in the same fashion to LH-RH when previously exposed to androgens.

Figure 5 shows data obtained when monolayer cultures of pituitary cells from 14-day-old female rats separated as before by unit-gravity

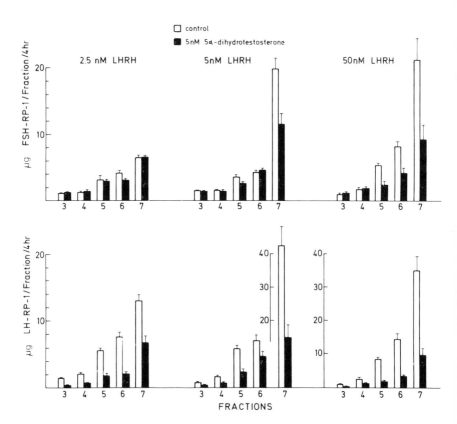

Figure 5. Release of FSH and LH at different concentrations of LH-RH from cultured rat pituitary cells (Falcon plastic petri dishes) separated by unit-gravity sedimentation (Denef *et al.*, 1978a,b) and pretreated for 70 hr with 5α-dihydrotestosterone (DHT). Cells were from 14-day-old females. DHT was diluted from an ethanol solution in sterile phosphate-buffered saline and added in a small volume to the culture medium. The final concentration in treated dishes was 5 nM. DHT was added three times a day to compensate for possible metabolic degradation of DHT. Each time, the culture medium was renewed. Untreated dishes (controls) received new medium with the vehicle alone. Culture medium was Dulbecco's Modified Minimum Essential Medium supplemented with sera, antibiotics, nonessential amino acids, and pyruvate as described before (Denef *et al.*, 1978a,b). LH-RH tests were performed on day 4 in culture. FSH and LH are expressed in terms of the rat NIAMDD reference preparations (RP-1).

sedimentation are treated with 5 nM 5α-dihydrotestosterone (DHT), administered three times a day for 3 days, and then stimulated with various concentrations of LH-RH. When LH-RH concentrations were low, FSH release was not affected markedly by DHT pretreatment. At higher concentrations, FSH release became depressed, but mainly in the largest gonadotrophs exposed to DHT. At still higher concentrations of LH-RH, this effect was also seen in medium-sized gonadotrophs, but none of the concentrations of LH-RH used affected FSH release from small gonadotrophs. In contrast, LH release was always strongly depressed after DHT treatment, and this to an equal extent in all fractions and irrespective of the concentration of LH-RH used.

The present observations do not only confirm previous findings that DHT has selective modulatory effects on the release of FSH at the pituitary level and that these effects vary with the extent to which the release is activated by LH-RH. The important new finding is that the selective modulation of FSH release by prior exposure to androgen is not expressed simultaneously in all gonadotrophs. This modulation appears to occur in quantal steps, depending on the number of gonadotrophs in which the androgen effect on FSH is elicited by LH-RH. Thus, gonadotrophs are also heterogeneous in terms of the interaction of androgen with LH-RH. The present findings are consistent with our hypothesis that selective modulation of FSH or LH release reposes in the functional heterogeneity of the gonadotrophs.

4.3. Functional Heterogeneity and Cell-to-Cell Communication

In vivo, as well as in culture, most types of animal cells form intercellular junctions through which small molecules, eventually regulatory molecules, may move (Epstein and Gilula, 1977). It is believed that in this way, communicating cells form a functional unit, responding in an integrated way to metabolic or hormonal stimuli. It is also becoming recognized that certain functional characteristics of various types of cells including endocrine cells (Gilula *et al.,* 1978; Lawrence *et al.,* 1978) can be modulated by intercellular communications.

Junctions and communications between cells have also been demonstrated in the *in situ* pituitary gland as well as in cultured pituitary cells (Fletcher *et al.,* 1975). The functional consequences of such communications have not been explored yet, but several data from the literature as well as from our studies seem to indicate that these intercellular junctions have a regulatory role. In the pituitary *in situ,* the different cell types are not randomly mixed. Homologous cells can form clusters of different size (Allanson and Parkes, 1966; Nakane, 1970; Tougard *et al.,* 1977a). One extreme example is the 10- to 14-day-old female rat. Over a period of only a few days, gonadotrophs aggregate in an area near the intermediate lobe (Siperstein *et al.*, 1954). Since these cellular events coincide with dramatic changes in FSH secretion as well as in the ratio at which FSH and LH are released *in vivo* (see above), it is at least suggestive that cell-to-cell communication is involved in these phenomena. On the other hand, it has been observed that in culture, pituitary glandular cells rapidly aggregate and form colonies (Baker *et al.*, 1974; Tixier-Vidal *et al.*, 1975a). We have also seen this to occur in our cultures, and in enriched populations there is a higher incidence of aggregation among homologous cells than among heterologous cells.

To evaluate the functional consequences of cell aggregation and to find evidence for possible functional cell-to-cell communications, the LH-RH response in the highly purified fraction of large gonadotrophs derived from 14-day-old females was compared to the LH-RH response of the same cells mixed and established in coculture with an excess of cells from a gradient-separated fraction in which the proportional number of gonadotrophs was small. By the excess of the latter cells, direct contact between the large gonadotrophs was found to be disrupted. As shown in Fig. 6, there was a significant difference between the LH-RH-stimulated LH secretion measured in the coculture and that expected from the individual values when the two fractions were cultured separately: LH secretion was considerably higher than expected. FSH secretion, on the other hand, was only marginally influenced by coculture. Furthermore, the differences were not seen when the two fractions were not mixed but cocultured in separate areas of the same culture dish, indicating that direct cell-to-cell contact is necessary to evoke these changes.

The present findings suggest that aggregated gonadotrophs respond differently to LH-RH than gonadotrophs separated from each other or clustered with other cell types. The findings are consistent with the hypothesis that such functional modulation is brought about by cell-to-cell communications. Since disaggregation of gonadotrophs changed the FSH/LH ratio, our observations also suggest that cell-to-cell communication may be another cellular mechanism involved in the control of differential FSH and LH release.

5. An Attempt to Integrate Functional Heterogeneity with Morphological Heterogeneity

It could be argued that part, if not all, of the functional differences observed among separated populations of gonadotrophs are due to the differences in the distribution of FSH and LH among the gonadotropic cell population (see above). However, several observations indicate that

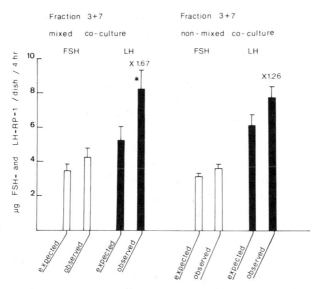

Figure 6. Effect of coculturing (Falcon plastic petri dishes) a highly purified population of gonadotrophs isolated by unit-gravity sedimentation from 14-day-old females in gradient fraction 7 with an 8- to 10-fold excess of cells isolated in fraction 3 of the same gradient. The latter fraction consisted predominantly of acidophils and chromophobes (see Fig. 1). *Left:* FSH and LH release at 10^{-7} M LH-RH when cells from fraction 3 were mixed before plating. *Right:* FSH and LH release at 10^{-7} M LH-RH when the same number of cells of each fraction was cocultured in a separate area of the same culture dish. Separate plating was realized by the use of poly-1-lysine-coated dishes (Denef *et al.*, 1978b). LH-RH tests were done on day 3 in culture (Denef *et al.*, 1978b). FSH and LH are expressed in terms of the rat NIAMDD reference preparations (RP-1). Statistical differences — *t* test: expected vs. observed: (*) $p < 0.01$.

this cannot be the only explanation. It has been shown (Drouin and Labrie, 1976; Denef *et al.*, 1978b) that in various conditions, there is no correlation between immunoassayable FSH and LH content in the cells and secretion of these hormones. There is also no evidence that separate LH cells are required for high LH responsiveness. In adult males, no large gonadotrophs with only LH were detectable, and yet the large gonadotrophs from adult males were the most potent in secreting LH of all gonadotroph fractions compared in our studies.

It appears more likely that functional heterogeneity is based on the presence of distinct morphological subtypes of gonadotrophs. As mentioned earlier, Tougard *et al.* (1977a,b) (also see Chapter 2) have been able to detect by electron microscopy various gonadotroph subtypes, only two of which are similar to those previously described (see Section 1). One gonadotropic cell type was small-sized and the predominant type during fetal life. Its proportional number decreased during development in favor of the other types, but it was still detectable in adult life. Cells with intermediate ultrastructural properties were also found. Of most importance were the findings that the different morphological subtypes appeared to be of different size and that in all of them both FSH and LH could be found together. The latter observations therefore strongly suggest that velocity sedimentation, which separates gonadotrophs according to size, will also separate morphological subtypes of gonadotrophs. If such a conclusion can be confirmed by ultrastructural studies of the cells in the different populations we have isolated, it will be possible to make a direct link between function and morphology and to find a cellular basis of the functional heterogeneity among gonadotrophs.

Future studies also need to delineate the potential role of cell-to-cell communications in modulating ultrastructural and functional characteristics. Our coculture experiments have shown clear-cut differences in the responses of gonadotrophs depending on whether they were cultured alone or mixed with other cell types. It remains to be seen which cell type preferentially interacts with gonadotrophs and whether or not gonadotrophs preferentially communicate with homologous or heterologous gonadotroph subtypes.

DISCUSSION

TIXIER-VIDAL: With this beautiful work, you have in hand the tool for an experimental analysis of the functional meaning of the morphological heterogeneity of gonadotropic cells and the relationship of this heterogeneity to LH and FSH secretion. In that respect, did you compare the ultrastructure of the gonadotropic cells in your various fractions?

DENEF: We did not do ultrastructural studies of the cells in different fractions. It is of the utmost importance to do so, as it has been shown that gonadotrophs containing both FSH and LH can have different ultrastructural characteristics (see Chapter 2).

TIXIER-VIDAL: What is your interpretation for the effect that you observed in cocultures of fractions 3 and 7?

DENEF: Either the large gonadotrophs respond differently to LH-RH because they cannot make contact among each other in the mixed coculture, or certain cell types in fraction 3 induce by contact with the large gonadotrophs in fraction 7 the altered response observed. In that case, one has to rule out whether the inducing cell type is glandular or not. Fraction 3 contains about 40% acidophils and has the highest secretion rate in culture of prolactin.

TOUGARD: Your study is another demonstration that all forms of gonadotrophic cells (small or large) would contain the two gonadotropic hormones LH and FSH. It is also a very elegant demonstration that these different forms separated on your density gradient released different levels of LH and FSH in response to a stimulatory effect of the hypothalamic factor LH-RH.

HYMER: In agreement with your results, it would seem that the other cell types, i.e., thyrotrophs, somatotrophs, and mammotrophs, may also show functional heterogeneity.

Thus, we should begin to think of the pituitary as containing subclasses of hormone-producing cells, not all which are equally active. The cell-dispersion method you use involved trypsin. Have you tried collagenase? If so, are such cells more responsive to GnRH? Also, do you consider the differences in responsiveness of the cells to GnRH between day 2 and day 3 significant? How might these differences compare with those done on older cultures?

DENEF: The method of cell dispersion using trypsin is essentially that of Hopkins and Farquhar (*J. Cell Biol.*, **59**:276, 1973), but we omit the neuraminidase step. The details of the method are published in the September 1978 issue of *Endocrinology*. We have tried collagenase instead of neuraminidase, particularly to get rid of the small clumps of unstained cells in the bottom fraction of the gradient. However, there was no improvement. The difference in dose response to LH-RH between day 2 and 3 is indeed significant. In older cultures (2 weeks), the sensitivity to LH-RH is increased (the ED_{50} dose decreases).

JAQUET: Does polylysine attachment of the cells in culture modify the kinetics of basal release of gonadotropins in your conditions?

DENEF: Cells plated on polylysine-coated dishes have somewhat lower LH-RH-induced secretion of FSH and LH, but the effect was similar on both hormones.

RAMACHANDRAN: What happens to the small gonadotrophs in long-term culture? Do they mature into the larger types?

DENEF: We did not look at the evolution of the small gonadotrophs. We did examine the large gonadotrophs. After 2 weeks or so in culture, these cells became heterogenous in size. There were very large cells as well as small cells.

FINK: Our *in vitro* studies on GnRF release and the pituitary response to GnRF suggest that in the brain, testosterone may act mainly through its conversion to an estrogen, while at the pituitary level, testosterone acts mainly as an androgen, either directly or by conversion to 5α-dihydroxytestosterone.

REFERENCES

Allanson, M., and Parkes, A. S., 1966, Cytological and functional reactions of the hypophysis to gonadal hormones, in: *Marshall's Physiology of Reproduction*, Vol. 3 (A. S. Parkes, ed.), p. 147, Longmans, Green, London.

Apfelbaum, M. E., and Taleisnik, S., 1976, Interaction between estrogen and gonadotrophin-releasing hormone on the release and synthesis of luteinizing hormone and follicle-stimulating hormone from incubated pituitaries, *J. Endocrinol.* **68**:127.

Baker, B. L., Reel, J. R., Van Dewark, S. D., and Yu, Y.-Y., 1974, Persistence of cell types in monolayer cultures of dispersed cells from the pituitary pars distalis as revealed by immunohistochemistry, *Anat. Rec.* **179**:93.

Barnes, B. G., 1962, Electron microscope studies on the secretory cytology of the mouse anterior pituitary, *Endocrinology* **71**:618.

Blackwell, R. E., and Guillemin, R., 1973, Hypothalamic control of adenohypophyseal secretions, *Annu. Rev. Physiol.* **35**:357.

Bowers, C. Y., Currie, B. L., Johansson, K. N. G., and Folkers, K., 1973, Biological evidence that separate hypothalamic hormones release the follicle stimulating and luteinizing hormones, *Biochem. Biophys. Res. Commun.* **50**:20.

Debeljuk, L., Khar, A., and Jutisz, M., 1978, Effects of gonadal steroids and cycloheximide on the release of gonadotrophins by rat pituitary cells in culture, *J. Endocrinol.* **77**:409.

Denef, C., Hautekeete, E., and Rubin, L., 1976, A specific population of gonadotrophs purified from immature female rat pituitary, *Science* **194**:848.

Denef, C., Hautekeete, E., De Wolf, A., and Vanderschueren, B., 1978a, Pituitary basophils from immature male and female rats: Distribution of gonadotrophs and thyrotrophs as studied by unit gravity sedimentation, *Endocrinology* **103**:724.

Denef, C., Hautekeete, E., and Dewals, R., 1978b, Monolayer cultures of gonadotrophs separated by velocity sedimentation: Heterogeneity in response to luteinizing hormone–releasing hormone, *Endocrinology* **103**:736.

Döhler, K. D., and Wuttke, W., 1975, Changes with age in levels of serum gonadotropins, prolactin, and gonadal steroids in prepubertal male and female rats, *Endocrinology* **97**:898.

Drouin, J., and Labrie, F., 1976, Selective effect of androgens on LH and FSH release in anterior pituitary cells in culture, *Endocrinology* **98**:1528.

Drouin, J., Lagacé, L., and Labrie, F., 1976, Estradiol-induced increase of the LH responsiveness to LH releasing hormone (LHRH) in rat anterior pituitary cells in culture, *Endocrinology* **99**:1477.

Drouin, J., Lavoie, M., and Labrie, F., 1978, Effect of gonadal steroids on the luteinizing hormone and follicle-stimulating hormone response to 8-bromo-adenosine 3′,5′-monophosphate in anterior pituitary cells in culture, *Endocrinology* **102**:358.

Epstein, M. L., and Gilula, N. B., 1977, A study of communication specificity between cells in culture, *J. Cell Biol.* **75**:769.

Epstein, Y., Lunenfeld, B., and Kraiem, Z., 1977, The effects of testosterone and its 5α-reduced metabolites on pituitary responsiveness to gonadotrophin-releasing hormone (Gn-RH), *Acta Endocrinol.* **86**:728.

Farquhar, M. G., and Rinehart, J. F., 1954, Electron microscopic studies of the anterior pituitary gland of castrate rats, *Endocrinology* **71**:618.

Farquhar, M. G., Skutelsky, E. H., and Hopkins, C. R., 1975, Structure and function of the anterior pituitary and dispersed pituitary cells: *In vitro* studies, in: *The Anterior Pituitary* (A. Tixier-Vidal and M. G. Farquhar, eds.), p. 83, Academic Press, New York.

Fletcher, W. H., Anderson, N. C., Jr., and Everett, J. W., 1975, Intercellular communication in the rat anterior pituitary gland: An *in vivo* and *in vitro* study, *J. Cell Biol.* **67**:469.

Gilula, N. B., Epstein, M. L., and Beers, W. H., 1978, Cell-to-cell communication and ovulation: A study of the cumulus–oocyte complex, *J. Cell Biol.* **78**:58.

Girod, C., 1977, Apport de l'immunohistochimie à l'étude cytologique de l'adénohypophyse, *Bull. Assoc. Anat.* **62**:21.

Herbert, D. C., 1975, Localization of antisera to LHβ and FSHβ in the rat pituitary, *Am. J. Anat.* **144**:379.

Herbert, D. C., 1976, Immunocytochemical evidence that luteinizing hormone (LH) and follicle stimulating hormone (FSH) are present in the same cell type in the rhesus monkey pituitary gland, *Endocrinology* **98**:1554.

Herlant, M., 1960, Étude de deux techniques nouvelles destinées à mettre en évidence les différentes catégories cellulaires présentes dans la glande pituitaire, *Bull. Microsc. Appl.* **10**:37.

Herlant, M., 1964, The cells of the adenohypophysis and their functional significance, *Int. Rev. Cytol.* **17**:299.

Hymer, W. C., 1975, Separation of organelles and cells from the mammalian adenohypophysis, in: *The Anterior Pituitary* (A. Tixier-Vidal and M. G. Farquhar, eds.), p. 137, Academic Press, New York.

Hymer, W. C., Evans, W. H., Kraicer, J., Mastro, A., Davis, J., and Griswold, E., 1973,

Enrichment of cell types from the rat adenohypophysis by sedimentation at unit gravity, *Endocrinology* **92**:275.

Hymer, W. C., Snyder, J., Wilfinger, W., Swanson, N., and Davis, J. A., 1974, Separation of pituitary mammotrophs from the female rat by velocity sedimentation at unit gravity, *Endocrinology* **95**:107.

Kalra, S. P., 1976, Ovarian steroids differentially augment pituitary FSH release in deafferented rats, *Brain Res.* **114**:541.

Kao, L. W. L., and Weisz, J., 1975, Direct effect of testosterone and its 5α-reduced metabolites on pituitary LH and FSH release *in vitro:* Change in pituitary responsiveness to hypothalamic extract, *Endocrinology* **96**:253.

Kragt, C. L., and Dahlgren, J., 1972, Development of neural regulation of follicle stimulating hormone (FSH) secretion, *Neuroendocrinology* **9**:30.

Kraicer, J., and Hymer, W. C., 1974, Purified somatotrophs from rat adenohypophysis: Response to secretagogues, *Endocrinology* **94**:1525.

Kurosumi, K., and Oota, Y., 1968, Electron microscopy of two types of gonadotrophs in anterior pituitary glands of persistent estrous and diestrous rats, *Z. Zellforsch.* **85**:34.

Lawrence, T. S., Beers, W. H., and Gilula, N. B., 1978, Transmission of hormonal stimulation by cell-to-cell communication, *Nature (London)* **272**:501.

Lloyd, R. V., and Karavolas, H. J., 1975, Uptake and conversion of progesterone and testosterone to 5α-reduced products by enriched gonadotropic and chromophobic rat anterior pituitary cell fractions, *Endocrinology* **97**:517.

Lloyd, R. V., and McShan, W. H., 1973, Study of rat anterior pituitary cells separated by velocity sedimentation at unit gravity, *Endocrinology* **92**:1639.

Miller, W. L., Knight, M. M., Grimek, H. J., and Gorski, J., 1977, Estrogen regulation of follicle stimulating hormone in cell cultures of sheep pituitaries, *Endocrinology* **100**:1306.

Mittler, J. C., 1974, Androgen effects on follicle-stimulating hormone (FSH) secretion in organ culture, *Neuroendocrinology* **16**:265.

Moriarty, G. C., 1975, Electron microscopic–immunocytochemical studies of rat pituitary gonadotrophs: A sex difference in morphology and cytochemistry of LH cells, *Endocrinology* **97**:1215.

Moriarty, G. C., 1976, Immunocytochemistry of the pituitary glycoprotein hormones, *J. Histochem. Cytochem.* **24**:846.

Nakane, P. K., 1970, Classification of anterior pituitary cell types with immunoenzyme histochemistry, *J. Histochem. Cytochem.* **18**:9.

Ojeda, S. R., and Ramirez, V. D., 1972, Plasma levels of LH and FSH in maturing rats: Response to hemigonadectomy, *Endocrinology* **90**:466.

Ojeda, S. R., Jameson, H. E., and McCann, S. M., 1977, Developmental changes in pituitary responsiveness to luteinizing hormone–releasing hormone (LHRH) in the female rat: Ovarian adrenal influence during the infantile period, *Endocrinology* **100**:440.

Ondo, J. G., Pass, K. A., and Baldwin, R., 1976, The effects of neurally active amino acids on pituitary gonadotropin secretion, *Neuroendocrinology* **21**:79.

Pelletier, G., Leclerc, R., and Labrie, F., 1976, Identification of gonadotropic cells in the human pituitary by immunoperoxidase technique, *Mol. Cell. Endocrinol.* **6**:123.

Phifer, R. F., Midgley, A. R., and Spicer, S. S., 1973, Immunohistologic and histologic evidence that follicle-stimulating hormone and luteinizing hormone are present in the same cell type in the human pars distalis, *J. Clin. Endocrinol. Metab.* **36**:125.

Robyn, C., Leleux, P., Vanhaelst, L., Golstein, J., Herlant, M., and Pasteels, J. L., 1973, Immunohistochemical study of the human pituitary with anti-luteinizing hormone, anti-follicle stimulating hormone and anti-thyrotrophin sera, *Acta Endocrinol* **72**:625.

Schally, A. V., Arimura, A., and Kastin, A. J., 1973a, Hypothalamic regulatory hormones, *Science* **179**:341.

Schally, A. V., Redding, T. W., and Arimura, A., 1973b, Effect of sex steroids on pituitary responses to LH- and FSH-releasing hormone *in vitro, Endocrinology* **93**:893.

Schally, A. V., Coy, D. H., and Meyers, C. A., 1978, Hypothalamic regulatory hormones, *Annu. Rev. Biochem.* **47**:89.

Siperstein, E., Nichols, C. W., Griesbach, W. E., and Chaikoff, I. L., 1954, Cytochemical changes in the rat anterior pituitary from birth to maturity, *Anat. Rec.* **118**:593.

Snyder, G., Hymer, W. C., and Snyder, J., 1977, Functional heterogeneity in somatotrophs isolated from the rat anterior pituitary, *Endocrinology* **101**:788.

Snyder, J., Wilfinger, W., and Hymer, W. C., 1976, Maintenance of separated rat pituitary mammotrophs in cell culture, *Endocrinology* **98**:25.

Spona, J., 1976, Action of steroids on LH-RH provoked gonadotrophin release, *Endocrinol. Exp.* **10**:91.

Sternberger, L. A., and Petrali, J. P., 1975, Quantitative immunocytochemistry of pituitary receptors for luteinizing hormone–releasing hormone, *Cell Tissure Res.* **162**:141.

Surks, M. I., and DeFesi, Ch. R., 1977, Determination of the cell number of each cell type in the anterior pituitary of euthyroid and hypothyroid rats, *Endocrinology* **101**;946.

Tang, L. K. L., and Spies, H. G., 1975, Effects of gonadal steroids on the basal and LRF-induced gonadotropin secretion by cultures of rat pituitary, *Endocrinology* **96**:349.

Tixier-Vidal, A., Gourdji, D., and Tougard, C., 1975a, A cell culture approach to the study of anterior pituitary cells, *Int. Rev. Cytol.* **41**:173.

Tixier-Vidal, A., Tougard, C., Kerdelhué, B., and Jutisz, M., 1975b, Light and electron microscopic studies on immunocytochemical localization of gonadotropic hormones in the rat pituitary gland with antisera against ovine FSH, LH, LHα, and LHβ, *Ann. N. Y. Acad. Sci.* **254**:433.

Tougard, C., Kerdelhué, B., Tixier-Vidal, A., and Jutisz, M., 1973, Light and electron microscope localization of binding sites of antibodies against ovine luteinizing hormone and its two subunits in rat adenohypophysis using peroxidase-labeled antibody technique, *J. Cell Biol.* **58**:503.

Tougard, C., Picart, R., and Tixier-Vidal, A., 1977a, Cytogenesis of immunoreactive gonadotropic cells in the fetal rat pituitary at light and electron microscope levels, *Dev. Biol.* **58**:148.

Tougard, C., Tixier-Vidal, A., Kerdelhué, B., and Jutisz, M., 1977b, Étude immunocyto-chimique de l'évolution des cellules gonadotropes dans des cultures primaires de cellules antéhypophysaires de rat: Aspects quantitatifs et ultrastructuraux, *Biol. Cell.* **28**:251.

Yoshimura, F., and Harumiya, K., 1965, Electron microscopy of the anterior lobe of pituitary in normal and castrated rats, *Endocrinol. Jpn.* **12**:119.

32

In Vitro Studies on the Secretion of Human Prolactin and Growth Hormone

François Cesselin and Françoise Peillon

1. Introduction

The mechanisms that regulate the secretion of human prolactin (PRL) and growth hormone (GH) are varied and complex. No attempt will be made in this chapter to review in detail their neuroendocrine control or the multiplicity of physiological stimuli and pharmacological agents that are involved in their regulation.

Most of these different regulatory systems have been well studied, both in animals *in vivo* and in isolated animal pituitaries. Many investigations have also been performed in normal and pathological human beings, but our attention will be focused here on *in vitro* experiments using human pituitary glands. These experiments are not numerous, primarily due to the difficulty in obtaining this material and in having sufficient quantities of viable pituitary tissue.

The human pituitary glands studied in these experiments are of three main origins: fetal pituitaries, pituitaries obtained at autopsy or after hypophysectomy for breast cancer or diabetic retinopathy, and pituitary tissue from hormone-secreting adenomas. Because of the scarcity of this

François Cesselin and *Françoise Peillon* ● Service de Biochimie Médicale et Laboratoire d'Histologie–Embryologie (C.N.R.S. ERA 484), Faculté de Médecine Pitié-Salpêtrière, 75634 Paris 13, France

material and its variable viability, the *in vitro* experiments on these tissues cannot have the same scope as those carried out on rat pituitary glands.

We shall describe, first, the pattern of *in vitro* secretion of human PRL and GH and, second, the modifications of their secretion induced by different agents.

2. In Vitro Secretion of Human Prolactin and Growth Hormone*

2.1. Patterns of Their Release in Vitro

The autonomous secretion of PRL by rat pituitary glands was well shown *in vitro* by the pioneer work of Pasteels (1961a) and Meites *et al.* (1961). Then, by culturing together rat pituitary glands with either hypothalamic tissue or extracts (Pasteels, 1961b; Talwalker *et al.*, 1963), it was demonstrated that the secretion of PRL was under inhibitory hypothalamic control, while in contrast, there was stimulation of GH release (Deuben and Meites, 1963). These same findings were obtained when the anterior lobe of fetal and human pituitary glands obtained at autopsy was used (Pasteels, 1962; Pasteels *et al.*, 1963; Brauman *et al.*, 1964). With the advent of radioimmunological assays, it became possible to extend and confirm these earlier studies. The decline in GH release was confirmed by other authors using various methods of culture: organ culture, tissue culture, or dispersed cells. This decline was also demonstrated with fetal pituitaries (Gailani *et al.*, 1970; Groom *et al.*, 1971; Siler *et al.*, 1972; Siler-Khodr *et al.*, 1974; Paulin *et al.*, 1976; Goodyer *et al.*, 1977) and with human anterior pituitaries obtained at autopsy or after hypophysectomy (Kohler *et al.*, 1969; Gala, 1971; Snyder *et al.*, 1978). In most of the experiments, the production of GH fell rapidly during the first 30 days. However, very low concentrations of GH in media were detectable for time periods as long as 6 months (Thompson *et al.*, 1959) or even 2 years (Pasteels, 1969). These low but detectable levels of GH led to the conclusion of a possible, although weak, autonomous secretion of this hormone. The decline in the *in vitro* release of GH was also observed in GH-secreting pituitary adenomas (Kohler *et al.*, 1969; Batzdorf *et al.*, 1971; Peillon *et al.*, 1972; Skyler *et al.*, 1977; Snyder *et al.*, 1978). In most of these experiments, the cultured adenomatous cells were maintained for 30 or 45 days, with large amounts of immunoreactive GH being secreted during the first 7 days. Measurable levels of GH were obtained for time

---

* For information on the *in vitro* secretion of other pituitary hormones, see Tixier-Vidal (1975).

periods as long as 1 year when explants and monolayer culture of such cases were used (Kohler *et al.,* 1969).

Although an identical pattern of GH release has been observed during all these different experiments, one of the difficulties encountered is in evaluating the amount of GH produced in culture. Indeed, it depends not only on the sources of human pituitary, but also on the type and conditions of culture and on the way of expressing the results.[†] Due to the variation in these conditions, one cannot make comparisons among the different experiments. This is particularly illustrated by the case of cultures of fetal pituitaries, in which hormone production is dependent not only on the method and conditions of culture but also on the gestational age. This also is the case of *in vitro* GH-secreting adenomas, in which hormonal production varies from one adenoma to another. In organ or monolayer cultures of fetal pituitaries, it was indeed shown that during the first week of culture, the average daily secretion rate varied largely according to the age of the fetuses. This secretion rate was 500 ng GH/pituitary per day during the first 12 weeks of gestation and more than 5 μg GH/pituitary per day after the first 16 weeks (Goodyer *et al.,* 1977). Studying the *in vitro* pattern of GH release from GH-secreting adenomas in organ culture, we found during the first 72 hr of culture (Figs. 1 and 2) an initial increase of GH in the media. Afterward, a decrease was observed in all the experiments: GH production remained high during the first week of culture and decreased more or less sharply during the following 3 weeks of culture (Peillon *et al.,* 1975). The mean production rate for GH per milligram of tumoral tissue resulting from 8 experiments decreased from 142 ng/mg per hr (range 22–261) on the 7th day of culture to 51 ng/mg per hr (range 3.6–133) on the 14th culture day. GH content in tissue extracts from a noncultured fragment of one adenoma was found to be 2.2 μg/mg wet tissue. After 4 weeks of culture, tissue extracts from two series of explants of the same adenoma contained respectively 1.4 and 161 ng/mg, demonstrating the important decrease and the wide disparity of GH that may occur in cultured explants and in media from the same adenoma. The disparity in the content of GH in media from one GH-secreting adenoma to another is also shown in Figs. 1–3. Table I and Fig. 4 show the GH concentration in fragments and media of three different GH-secreting adenomas incubated for 4 hr. After 1 hr, the GH content varied from 2.65 to 14.30 μg/mg wet tissue. Similar differences in the *in vitro* hormone production from GH-secreting adenomas have been observed by the authors already mentioned and reflect the differences in GH secretion

[†] Our own results are expressed in two different ways. They have been related to the weight of the pituitary tissue, when structural studies have been performed, and to the amounts of tissue protein in other studies.

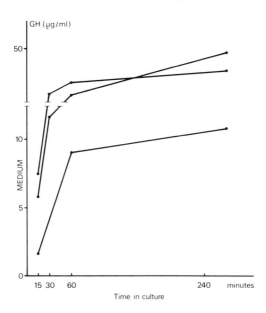

Figure 1. Cumulative level of GH production in media of three GH-secreting adenomas incubated for 4 hr (preincubation for 1 hr). Each point is the mean of the duplicate assays of two experiments.

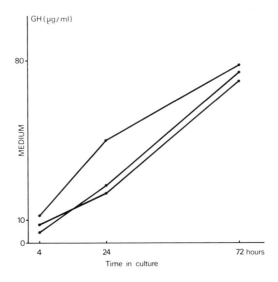

Figure 2. Cumulative level of GH production in media of three GH-secreting adenomas incubated for 3 days (preincubation for 1 hr). Each point is the mean of the duplicate assays of two experiments.

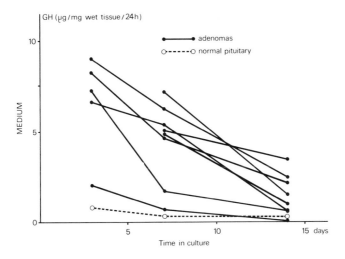

Figure 3. Mean GH release in media from different GH-secreting adenomas cultured for 15 days.

observed *in vivo* from one patient bearing a GH-producing adenoma to another. These disparities from one tumor to another and in the culture media of replicate fragments of the same tumor as well (see Section 2) are one of the problems in the interpretation of the experiments.

In contrast to the decrease of GH *in vitro,* the various studies performed since the initial work of Pasteels (1961a) and Meites *et al.* (1961) have shown a different pattern for the release of PRL *in vitro.* By using human fetal pituitaries, it was confirmed that PRL release continues for

Table I. Prolactin and Growth Hormone Content in Fragments (F) and in Media (M) from Three GH-Secreting Adenomas Incubated for 4 hr

Incubation time (hr)	Hormone	Tumor A		Tumor B		Tumor C	
		F	M	F	M	F	M
1	PRL (ng/mg wet tissue)	0.15	1.65	13.80	12.00	1.20	2.10
	GH (μg/mg wet tissue)	2.65	0.21	10.10	0.65	14.30	2.80
2	PRL (ng/mg wet tissue)	—	—	15.20	8.80	2.80	3.80
	GH (μg/mg wet tissue)	—	—	6.60	0.13	11.80	6.40
3	PRL (ng/mg wet tissue)	0.25	3.10	28.60	9.60	1.70	4.50
	GH (μg/mg wet tissue)	2.45	0.52	17.66	0.34	12.80	4.50
4	PRL (ng/mg wet tissue)	0.40	3.00	34.60	9.40	1.70	3.60
	GH (μg/mg wet tissue)	5.50	0.63	7.80	0.52	11.80	5.20

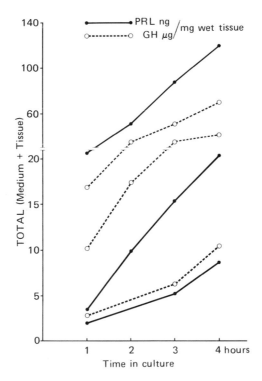

Figure 4. Cumulative level of GH production and of PRL production in fragments and media of three GH-secreting adenomas incubated for 4 hr.

several weeks. Siler-Khodr *et al.* (1974) found that PRL release was very high in the first change of medium and then decreased during the next 10 days. This decrease was followed by an increase in PRL release that exceeded the release observed at the beginning of the culture. A very high production of PRL (200 μg/pituitary per day) was observed in some cases. As in the case of GH secretion, an increase in the total release of PRL per pituitary was observed as the gestational age increased. Similar results were obtained with fetal pituitaries by Goodyer *et al.* (1977). Studying an anterior pituitary obtained at hypophysectomy, Gala (1971) found that PRL activity was similar between days 6 and 21 of the culture. Knazek and Skyler (1976), using an artificial capillary system, obtained from a PRL-secreting adenoma a production of more than 3 mg PRL during the first 4 culture months. In another experiment on a pituitary tumor secreting both GH and PRL, PRL secretion was maintained at much higher levels than GH secretion during the same culture time (Skyler *et al.*, 1977). The addition of hydrocortisone during culture increased the secretion rate of

GH, as already reported by Bridson and Kohler (1970), but did not modify PRL secretion. More recently, Snyder *et al.* (1978), by culturing pituitary cells prepared from breast cancer patients and PRL-secreting adenomas, confirmed that PRL secretion was maintained at higher levels for much longer periods than GH secretion.

Studying the *in vitro* release of PRL from GH-secreting adenomas, we found the same pattern of PRL release as observed in the human pituitary tissue described above. Among the hormone-secreting adenomas submitted to organ culture, two were mixed somatotropic and lactotropic adenomas well characterized by a high plasma PRL level *in vivo* and by different morphological studies including immunocytochemical reactions (Peillon *et al.*, 1978b). In the other cases, a release of PRL was observed *in vitro* even though *in vivo* plasma PRL levels were normal. In these cases, and in some others studied more recently, no other pituitary hormone (LH, FSH, or TSH) was detected in the media, leading to the conclusion that normal pituitary tissue was not responsible for this PRL release. However, this possibility cannot be completely excluded, since the origin of PRL cells in GH-producing adenomas, as well as in mixed PRL- and GH-secreting adenomas, remains unknown at present (Zimmerman *et al.*, 1974; Kovacs *et al.*, 1977). Nevertheless, in these experiments, the pattern of PRL release was different from that observed for GH release, though the same adenomas and culture times were used. In half the experiments, PRL concentrations in the media increased with time, while in the other half, PRL levels either remained relatively constant or decreased more or less slowly (Fig. 5). At the same time, GH decreased in all the samples (see Fig. 3). The amounts of PRL (expressed in ng/mg wet tissue) produced by these tumors were approximately 1000-fold lower than the quantities of GH produced (expressed in μg/mg wet tissue), although in the plasmas, PRL and GH levels were often the same. These different quantities of GH and PRL in media were observed in both short-term and long-term incubation studies, using several other GH-secreting adenomas. When GH and PRL were evaluated in both the fragments and the media from the same tumor at the same incubation times, smaller amounts of PRL were found in the fragments than in the media of two out of three tumors, whereas GH levels were higher in the fragments than in the media (Table I).

This indicates that adenomas have a low intratumoral PRL content, which is mostly released into the medium, while intratumoral GH content is higher, with little release into the medium. In two pituitary glands incubated for 4 and 24 hr, Hwang *et al.* (1971) found that the PRL/GH ratio was in general higher in the incubated media than in the tissue extracts. This finding may be interpreted as a preferential release of PRL. It was shown too (Guyda *et al.*, 1973; Zimmerman *et al.*, 1974) that

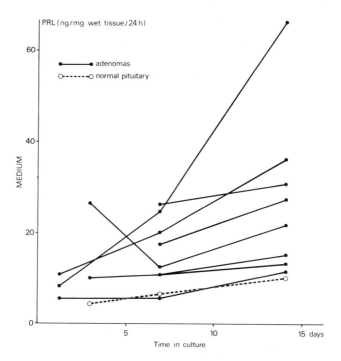

Figure 5. Mean PRL release in media from GH-secreting adenomas (the same ones as in Fig. 3) cultured for 15 days. Only 9 out of 16 experiments are shown.

homogenates of normal human pituitaries contained 40- to 100-fold more GH than PRL, although the plasma PRL levels in human beings were 2 or 3 times higher than those of GH.

If we try to make some comparisons between the results obtained with normal human pituitaries and with GH-secreting tumors, both seem to have PRL pools smaller than those of GH and PRL turnover rates higher than those of GH.

2.2. In Vitro Synthesis of Growth Hormone and Prolactin

The in vitro release of hormones asks the question of whether this release is due to some passive process induced by the culture conditions, i.e., destruction of the tissue, or whether a normal activity is maintained in vitro. Continued hormonal production for several weeks or months, even at very low levels, is an argument in favor of the persistence of an autonomous secretion. However, the best approach to affirm an active process in vitro is to study the in vitro hormone synthesis by incorporation of labeled amino acids into newly synthesized hormones. This has been demonstrated for GH or PRL or both by different authors in short- or long-term incubation studies using different sources of human pituitary glands (Kohler *et al.*, 1971; Hwang *et al.*, 1971; Friesen *et al.*, 1970, 1972; Peillon *et al.*, 1972; Guyda *et al.*, 1973; Belleville *et al.*, 1973; Frohman and Stachura, 1973; Siler-Khodr *et al.*, 1974). During the first 72 hr of culture, we observed an increased secretion of GH that contrasted with the decreasing levels observed during the following days. It is possible that passive release of intracellular GH content from intact or damaged cells may be an explanation for the increased secretion during the initial culture period; however, active synthesis during this time was also observed. When [³H]leucine was added to the medium at the beginning of culture, the tissue (studied each hour for 4 hr) incorporated [³H]leucine and [³H]-GH was observed, thus giving proof of *de novo* synthesis (Fig. 6 and Table II). The same result was observed for PRL. Electron-

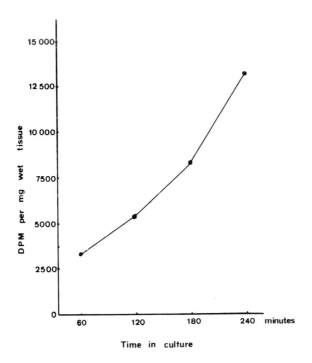

Figure 6. Time course of incorporation of [³H]leucine into GH from fragments and media of one GH-secreting adenoma incubated for time periods ranging from 0 to 240 min in 1.5 ml M199 containing 23 μCi [³H]leucine (15 Ci/mM).

Table II. [³H]Leucine Incorporation into Growth
Hormone and Prolactin in Fragments and in Media from
a GH-Secreting Adenoma Incubated for 1 hr and 2 hr

Culture time (hr)	[³H]-PRL / [³H]proteins (%)	[³H]-GH / [³H]proteins (%)
1[a]	10.6	16.5
2[a]	21.0	31.0

[a] Addition of [³H]leucine at the beginning of the incubation.

microscopic studies demonstrated that the fine structure of the intact cells
was well preserved, but that there were fewer secretory granules present
after 72 hr of incubation than during the first 4 hr of culture (Fig. 7) (Peillon
et al., 1975). After the initial increase in GH release, a decrease in GH
was observed in all the experiments. This may be due to a reduced
intracellular pool of GH as well as to reduction in explant size and to the
presence in their centers of clusters of necrotic cells. The data from light-
and electron-microscopic studies are in favor of these different possibilities
(Peillon *et al.*, 1975). The reduction in the quantity of tissue was also
demonstrated by the incorporation of labeled amino acids into total
proteins (Table III). It was indeed 3-fold lower on day 16 of culture
compared to day 9. However, after 9 or 16 days of culture, one still
observes *de novo* synthesis of GH. When immunoprecipitable [³H]-PRL
in media between 9 and 16 days of culture was compared, an increase in
PRL synthesis occurred during this time. This was in contrast to the lower
level of GH synthesis from the same tumor, during the same time. These
results on *de novo* synthesis of PRL and GH correlate well with the results
on the *in vitro* release of these hormones by GH-secreting adenomas. PRL
release and synthesis increase with time, while GH release and synthesis
from the same adenomas decrease during the same culture time. These
findings suggest that adenomas, like normal pituitaries, may have a
partially autonomous hormonal secretion, but they are most probably
dependent *in vivo* on hypothalamic control, mainly inhibitory for PRL and
stimulatory for GH. The rate of incorporation of labeled amino acids is
lower in organ culture than in dispersed cells, as demonstrated by Farquhar
et al. (1975), who showed by autoradiographic studies that amino acids do
not penetrate into the center of tissue blocks. Consequently, the *de novo*
synthesis of GH and PRL as determined in our short- and long-term
incubation studies was probably underestimated.

2.3. Characterization of Human Growth Hormone and Prolactin Produced in Culture

The heterogeneity of GH and PRL has been well demonstrated in serum samples and pituitary extracts; therefore, one of the questions to ask is whether the hormones released into medium or synthesized during culture may differ from the native ones. It is possible that the hormone measured by radioimmunoassay may be only an immunoreactive fragment or aggregate of the molecule.

By using chromatography on Sephadex G-100 or polyacrylamide gel electrophoresis, it was demonstrated that the GH secreted from or synthesized in human pituitary glands or GH-producing adenomas was indis-

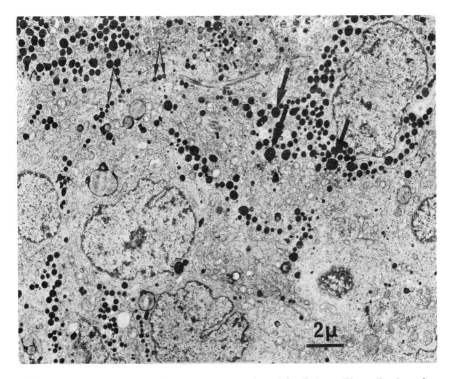

Figure 7. Portion of a GH-secreting adenoma cultured for 7 days. The cells show few secretory granules, which are of two main sizes: small granules (150 nm, small arrows), which may correspond to somatotroph granules, and larger granules (350 nm, large arrows), which may be lactotroph granules as described by Pasteels *et al.* (1972), Racadot *et al.* (1975), and Olivier *et al.* (1975).

Table III. [³H]Leucine Incorporation into Proteins in Media from One GH-Secreting Adenoma Cultured for 16 Days

Culture time (days)	[³H]proteins (dpm/mg wet tissue)	[³H]-PRL (dpm/mg wet tissue)	[³H]-GH (dpm/mg wet tissue)	[³H]-PRL [³H]proteins (%)	[³H]-GH [³H]proteins (%)
9ᵃ	326,960	31,710	132,940	9.7	40.0
16ᵇ	102,580	21,750	37,230	20.8	34.7

ᵃ Addition of [³H]leucine on the 7th day.
ᵇ Addition of [³H]leucine on the 14th day.

tinguishable from hormone extracted from normal human pituitaries (Kohler *et al.*, 1971; Guyda, 1975; Peillon *et al.*, 1975; Binoux *et al.*, 1976; Skyler *et al.*, 1977). In these experiments, the heterogeneous immunoreactive forms of GH, with comparable elution profiles in sera and in culture media, were identified. Nonpurified GH released from a tumor had the same immunoactivity and receptor activity as the hormone obtained by extraction from normal pituitaries (Skyler *et al.*, 1977). In addition, the activities were similar for the different forms identified following gel filtration of normal pituitary incubation media (Guyda, 1975). The elution profile of the chromatography on Sephadex G-100 of GH from cultured tissue is shown in Fig. 13b. The profile is the same for GH from both the media and tissue homogenates of GH-producing adenomas incubated for 4 hr. Figure 8 shows that the newly synthesized GH, obtained after addition of [³H]leucine to the media of GH-secreting adenomas cultured for several days, gives the same elution pattern as standard [¹³¹I]-GH. When the biological activity of GH released into culture media from these adenomas was determined by the tibia test, it was found to be as active as the hormone extracted from normal pituitary gland (Binoux *et al.*, 1976).

PRL released from the tissue culture of a pituitary tumor from an acromegalic patient was shown to have the same elution pattern on Sephadex G-100 as that of PRL extracted from a normal pituitary. Though both forms were also indistinguishable by quantitative polyacrylamide gel

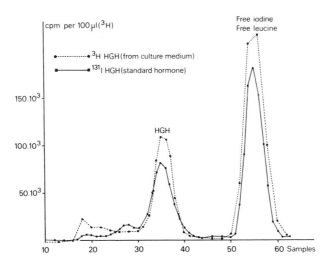

Figure 8. Chromatography on Sephadex G-100. Tritiated GH obtained from culture media after addition in 1.5 ml M 199 of 23 μCi [³H]leucine (15 Ci/mM) gives the same elution pattern as [¹³¹I]-GH. From Peillon *et al.* (1975) with permission.

electrophoresis, structural differences were possible because extracted and released PRL led to nonparallel logit-log displacement curves with different antisera (Skyler *et al.*, 1977).

In the interpretation of such results, one must assume that neither the extracted nor the released form of the hormone has undergone artifactual changes before the analysis.

3. Factors Modifying the in Vitro Secretion of Human Prolactin and Growth Hormone

3.1. Effect on Human Prolactin Release

Among the different pituitary hormones, PRL is the only one in which secretion is chronically depressed by the hypothalamus (prolactin-inhibiting factor, PIF); dopamine almost certainly may be considered as one of the PRL-inhibiting factors. Nevertheless, the studies of Malarkey and Pankratz (1974) and of Hagen *et al.* (1976) are in favor of the presence of a PRL-releasing factor (PRF) distinct from thyrotrophin-releasing hormone (TRH), which also is able to raise PRL secretion. However, PRF is not yet identified. Numerous experiments on animals both *in vivo* and *in vitro* and on man *in vivo* have shown that many factors can modify PRL secretion. To review these factors is not the purpose of this chapter, all the more so because no comprehensive conclusions can be drawn at present (see the reviews by MacLeod, 1976; Martin *et al.*, 1977a; Smythe, 1977; Enjalbert *et al.*, 1978). A very simple hypothesis could be that the factors involved in PRL regulation are only "modulators" of PIF and PRL, i.e., that their action on PRL release is mediated, via neuronal or hormonal routes, at the hypothalamic or pituitary level, by PIFs or PRFs, or both, that themselves act on the pituitary.

We shall describe the action of only three factors: dopaminergic agents, somatostatin (GH-IF), and TRH, which have been studied *in vitro* on human tissues by ourselves and a few other authors.

3.1.1. Dopaminergic Agents

The fact that dopamine acts as a neurohormone to control basal PRL secretion is now well established. Dopaminergic agents seem able to act at the pituitary level as well as at the hypothalamic level. The direct pituitary inhibitory effect of dopamine (Birge *et al.*, 1970; Koch *et al.*, 1970; MacLeod *et al.*, 1970; MacLeod and Lehmeyer, 1974; Shaar and Clemens, 1974) and of 2-bromo-α-ergocryptine (bromocriptine) (Pasteels *et al.*, 1971; Gourdji *et al.*, 1973a; Nagasawa *et al.*, 1973; MacLeod and Lehmeyer, 1974) was clearly demonstrated by *in vitro* experiments on animals. This is further supported by some *in vivo* data: L-Dopa, the immediate precursor of dopamine, is able to inhibit PRL secretion in rats bearing a transplanted pituitary tumor secreting PRL (Malarkey and Daughaday, 1972) or in hypophysectomized rats bearing transplanted pituitaries (Gräf *et al.*, 1977) as well as in monkeys after surgical section of the pituitary stalk (Dieffenbach *et al.*, 1976). In addition, the inhibition of peripheral aromatic L-amino-acid decarboxylase results in an increased level of plasma PRL in estrogen/progesterone-treated ovariectomized rats, without altering the concentration of endogenous hypothalamic dopamine and norepinephrine, suggesting that since dopamine does not readily cross the blood–brain barrier (Olendorf, 1971), peripheral dopamine exerts a tonic inhibitory influence on PRL secretion by the pituitary (Jimenez *et al.*, 1978). Furthermore, the fact that bromocriptine suppresses TRH-induced PRL secretion, as first described in the bovine by Schams and Reinhardt (1973), is another argument in favor of a direct action at the pituitary level.

On the other hand, Kamberi *et al.* (1971) failed to show any effect of dopamine infusion into hypophyseal vessels, whereas infusion into the third ventricle of the rats suppressed PRL release. However, Takahara *et al.* (1974a) were able to diminish PRL release by injecting dopamine into stalk portal vessels of rats. More convincing evidence for an additional action at the hypothalamic level comes from studies of Quijada *et al.* (1973) and Jimenez *et al.* (1978). Indeed, it was shown that dopamine acted on the hypothalamus to release a PIF, the effect of which was not blocked by haloperidol, whereas this product effectively inhibited the direct pituitary action of dopamine (Quijada *et al.*, 1973). In the rats studied by Jimenez *et al.* (1978), although the inhibition of peripheral aromatic L-amino-acid decarboxylase decreased peripheral catecholamines and raised the level of plasma PRL, the administration of L-dopa was still able to inhibit PRL release, suggesting that this effect went through a hypothalamic mechanism involving dopamine.

In man, too, either normal or hyperprolactinemic, it is well known that L-dopa (Kleinberg *et al.*, 1971) and bromocriptine (del Pozo *et al.*, 1972) are able to lower the level of plasma PRL; bromocriptine is frequently used as a therapeutic agent in clinical disorders with hyperprolactinemia. However, the site of action, hypothalamic or pituitary or both, is not as well documented in man as in animals.

The following observations may be in favor of a direct action at the pituitary level in man: dopamine itself reduces the level of plasma PRL (Leblanc *et al.*, 1976); L-dopa is still able to inhibit PRL secretion in man with a surgical section of the pituitary stalk (Woolf *et al.*, 1974); dopamine,

L-dopa, and bromocriptine suppress TRH-induced PRL secretion (del Pozo *et al.*, 1973; Noel *et al.*, 1973; Besses *et al.*, 1975; Leblanc *et al.*, 1976; Leebaw *et al.*, 1978).

One of the best approaches to examine the direct pituitary effect of these dopaminergic agents is certainly that of *in vitro* studies using human pituitary glands. However, there are few *in vitro* data on the human being available at present. To our knowledge, these *in vitro* studies have been limited to experiments performed with bromocriptine.

Since Pasteels *et al.* showed the inhibitory effect of bromocriptine on PRL release from the pituitary of a human fetus, only two recent brief reports have been published, showing the *in vitro* inhibitory effect of bromocriptine on PRL release by one GH- and one PRL-secreting pituitary tumor (Mashiter *et al.*, 1977) and by dispersed cell cultures of hormone-secreting adenomas (Adams and Mashiter, 1978).

Since no *in vitro* study using the other dopaminergic agents (dopamine and L-dopa) had been undertaken in man, it seemed to us interesting to test their effect on PRL release from hormone-secreting pituitary adenomas. When possible, bromocriptine was also used in order to evaluate whether there was a similar action of these three different agents on PRL release from the same adenoma. Five pituitary PRL-secreting adenomas (prolactinomas) and seven GH-secreting adenomas were used.

It is known from experiments on animals with dopamine, for example, that only a negligible increase in the effect can be seen by increasing the concentration of the drug starting with 5×10^{-8} M (Koch *et al.*, 1970) or 5×10^{-7} M (Shaar and Clemens, 1974); it is very likely that the maximal effect is thus reached. Nevertheless, presuming that human tumoral tissue could require higher concentrations than animal normal tissue and in order to palliate the rapid degradation of the dopaminergic drugs in the incubation medium and in the tissue (MacLeod *et al.*, 1970), we preferred to choose, for our preliminary investigations, fairly high concentrations of these agents. Thus, the tumors were incubated with or without dopamine (5.2×10^{-5}, 5.2×10^{-6}, and 5.2×10^{-7} M), L-dopa (10^{-4}, 10^{-5}, and 10^{-6} M), or bromocriptine (6.5×10^{-4}, 6.5×10^{-5}, and 6.5×10^{-6} M) for 4 hr. According to the size of the tumors, one, two, or all three agents and one, two, or all three concentrations were used. All the incubations were performed in triplicate. The tissue incubation, radioimmunoassay of PRL, and morphological study methods have been previously described in detail (Peillon *et al.*, 1978b). PRL content in the culture medium was expressed in ng or pg PRL/μg tissue protein except in three cases studied by electron microscopy (ng or pg PRL/mg wet tissue). Since no statistically significant difference was observed among the means of the results obtained with the different concentrations of each agent (analysis of variance), pooled data for the same drug were compared to control values (Student's *t* test). Therefore, it is clear that the choice of lower concentrations would most probably have given more accurate results. Table IV presents the *in vitro* as well as the *in vivo* data obtained during dynamic tests performed before surgery with L-dopa or bromocriptine or both in 10 of the 12 patients bearing the tumors studied.

In all experiments except one, a decrease in PRL was observed when the drugs were added. Moreover, despite the wide dispersion of the data obtained by replicate incubations of tumor fragments, inherent to the material used, the inhibitory effect was often statistically significant. When

Table IV. Data from 5 Prolactinomas and 7 GH-Secreting Adenomas: (I) Effect of Dopamine, L-Dopa, and Bromocriptine on *in Vitro* PRL Release and (II) Nadir Plasma PRL Levels during L-Dopa and Bromocriptine Tests Performed on Patients before Surgery

Agent[a]	Prolactinomas					
	1[b]	2[b]	3[c]	4[c]	5[d]	Percentage of control[f]
			I. *In vitro*			
C	426 ± 54	92 ± 24	714 ± 311	3227 ± 925	596 ± 220	100
D	—	56 ± 7[i]	265 ± 141[i]	—	286 ± 123	49 ± 7[h]
L	288 ± 37[g]	51 ± 11[h]	445 ± 264	1571 ± 640[i]	254 ± 94[h]	55 ± 4[j]
CB	—	56 ± 10[i]	428 ± 376	—	—	60 ± 1[i]
			II. *In vivo*			
L[f]	20	38	47	39	20	—
CB[f]	—	23	39	—	—	—

Agent[a]	GH-secreting adenomas						
	1[b]	2[b]	3[b]	4[d]	5[d]	6[d]	7[e]	Percentage of control[f]
				I. *In vitro*				
C	39 ± 17	8 ± 3	6 ± 3	34 ± 24	10 ± 3	56 ± 35	764 ± 191	100
D	19 ± 7[h]	4 ± 2[i]	3 ± 1[h]	12 ± 3[h]	10 ± 8	20 ± 16[g]	630 ± 220	57 ± 9[i]
L	31 ± 8	7 ± 3	—	19 ± 11	6 ± 1[h]	33 ± 36	622 ± 198	70 ± 5[i]
CB	16 ± 6[i]	4 ± 2[i]	—	—	5 ± 1[h]	17 ± 14[h]	771 ± 543	55 ± 12[h]
				II. *In vivo*				
L[f]	19	—	—	52	37	36	80	—
CB[f]	21	—	—	—	—	45	87	—

[a] (C) Control; (D) dopamine; (L) L-dopa; (CB) bromocriptine.
[b-f] PRL is expressed in: [h]ng/μg protein ± S.D.; [c]ng/mg tissue ± S.D.; [d]pg/μg protein ± S.D.; [e]pg/mg tissue ± S.D.; [f]percentage of the control (C) level (mean ± S.E.M.).
[g-j] With respect to the control (C) level: [g]$p < 0.050$; [h]$p < 0.025$; [i]$p < 0.010$; [j]$p < 0.001$.

the tumors were considered as a whole, the mean level of PRL was about 49% of the control level with dopamine, 55% with L-dopa, and 60% with bromocriptine for the prolactinomas and 57, 70, and 53%, respectively, for the GH-secreting adenomas. These decreases were highly significant.

The morphological studies seemed to exclude the possibility of a toxic effect. Electron microscopy showed an enlargement of the rough endo-

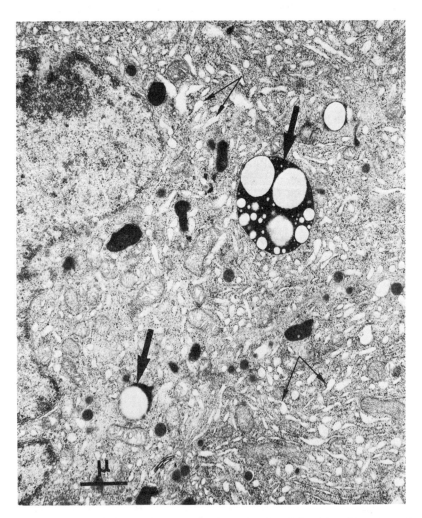

Figure 9. PRL-secreting cell from a prolactinoma incubated for 4 hr with L-dopa. An enlargement of the rough endoplasmic reticulum is observed with L-dopa in a concentration of 10^{-7} M (small arrows). Only higher concentrations of L-dopa (10^{-3} M) induce autophagic lysosomes (large arrows).

plasmic reticulum as observed in rat pituitary glands incubated with catecholamines by MacLeod and Lehmeyer (1972) (Fig. 9). No modification, either in the Golgi apparatus containing immature secretory granules or in the cellular organelles, was observed. These morphological findings may be in favor of an effect of these drugs on the release rather than on the synthesis of PRL. Only a much higher concentration of L-dopa (5×10^{-3} M) induced autographic lysosomes identical to those observed after long-term *in vivo* bromocriptine treatment (Beauvillain *et al.*, 1976) (Fig. 9). Moreover, the fact that the release of PRL by one of the GH-secreting adenomas (No. 7) was unaffected by the treatments confirms the absence of any toxic effect.

This study allows the following conclusions: Dopamine, L-dopa, and bromocriptine directly inhibit PRL release from human pituitary adenomas *in vitro*. These three agents cause similar decreases of PRL release, although differences are obvious in the degree of inhibition according to each drug. *In vivo* and *in vitro* results are in good agreement. Bromocriptine, well known as a dopamine-receptor-stimulating agent (Flückiger *et al.*, 1976), very likely acts by itself, whereas L dopa most probably acts after transformation into catecholamines, as suggested notably by the *in vivo* experiments of Jimenez *et al.* (1978). Moreover, some preliminary studies, showing the presence of catecholamines in the media from the tumors incubated with L-dopa, are in favor of this hypothesis (unpublished data). Nevertheless, the effect of L-dopa in a culture medium that includes an aromatic L-amino-acid decarboxylase inhibitor has to be tested to exclude the possibility of a direct effect of L-dopa on pituitary cells.

The presence of pituitary dopamine receptors on mammotroph cells, strongly suggested by all these data, was established in animal anterior pituitaries by studies using radiolabeled dopamine or dopamine agonists or antagonists (Brown *et al.*, 1976; Calabro and MacLeod, 1978; Caron *et al.*, 1978; Cronin *et al.*, 1978). The studies of Zor *et al.* (1969) and Steiner *et al.* (1970) seemed to exclude cyclic AMP as a mediator of the action of dopamine at the pituitary level. However, de Camilli *et al.* (1978b) recently showed that instead of enhancing adenylate cyclase activity, as in the brain, the action of dopamine on the enzyme was inhibitory in human prolactinomas.

3.1.2. Somatostatin

Grant *et al.* (1974) and Vale *et al.* (1974) reported that the spontaneous release of PRL *in vitro* from rat pituitaries was inhibited by GH-IF. *In vitro*, this peptide leads to the rapid inhibition of cyclic AMP accumulation in pituitary cells (Borgeat *et al.*, 1974). Nevertheless, in rats, *in vivo* TRH-triggered (Vale *et al.*, 1974) or chlorpromazine-triggered (Sawano *et al.*,

1974) secretion of PRL is unaffected by this peptide. Furthermore, instead of an inhibiting action, Gala *et al.* (1976) showed a potentiating effect for GH-IF on perphenazine and TRH-induced PRL release in monkeys, whereas the response to serotonin was unaffected.

In man, GH-IF has no effect on the basal secretion of PRL (Hall *et al.*, 1973) or on PRL response either to TRH (Carr *et al.*, 1975) or to hypoglycemia (Copinschi *et al.*, 1974). In acromegalic patients, conflicting data have been reported in which GH-IF is able to reduce PRL (Yen *et al.*, 1974) or unable to reduce PRL levels (Besser *et al.*, 1974).

Goodyer *et al.* (1977) found no consistent effect of GH-IF on PRL release from the human fetal pituitary *in vitro*. Peillon *et al.* (1978b) investigated the *in vitro* effects of this peptide (2.5×10^{-6} M) on PRL secretion by three pituitary tumors from acromegalic patients. A large decrease in PRL release was observed in two tumors incubated for time periods ranging from 1 to 4 hr (Fig. 10). The lower sensitivity of PRL-secreting cells toward GH-IF and/or the higher release of PRL may explain the failure of GH-IF to lower PRL in the third case.

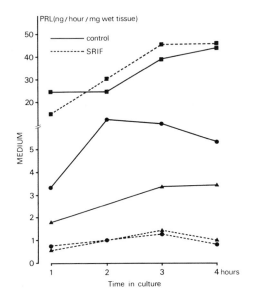

Figure 10. *In vitro* effect of somatostatin (SRIF) (2.5×10^{-6} M) on the release of PRL from three GH-secreting adenomas incubated for time periods ranging from 1 to 4 hr. Preincubation was for 1 hr. Each point is the mean of the duplicate assays of two experiments.

3.1.3. Thyrotropin-Releasing Hormone

The stimulation of *in vitro* PRL release by TRH was first established by Tashjian *et al.* (1971), using rat pituitary tumor cells. These cells are able to bind TRH (Gourdji *et al.*, 1973b; Hinckle and Tashjian, 1973), which is found at both the cytoplasmic and the nuclear level (Brunet *et al.*, 1974). This tripeptide has a rapid effect on the release of stored PRL followed by a delayed stimulatory effect on PRL synthesis (Morin *et al.*, 1975). A role for cyclic AMP as mediator of the action of this tripeptide is very likely (Labrie *et al.*, 1976). The effect of TRH seems in fact dependent on the environmental status of the animal (Lu *et al.*, 1972; Mueller *et al.*, 1973; Vale *et al.*, 1973). But even if this peptide has no effect by itself, it partially overcomes the inhibitory action of dopamine (Hill-Samli and MacLeod, 1974, 1975). Its *in vivo* stimulation effect, in various animals in certain endocrine conditions, and in man, is well known (Bowers *et al.*, 1971; Jacobs *et al.*, 1971).

The stimulating effect of TRH on PRL release from a human pituitary tumor *in vitro* was first shown by Knazek and Skyler (1976), using an artificial capillary system. Snyder *et al.* (1978) recently showed the ability of TRH to increase PRL release by dispersed pituitary cell cultures. The pituitaries were taken from galactorrhea patients. Three of four times, TRH (10 or 100 ng/ml) caused a significant increase of PRL in the medium, whereas the PRL content of the cell usually remained unchanged. Adams and Mashiter (1978) also reported that TRH stimulated, in a dose-dependent manner (10 ng/ml producing a maximal effect), PRL secretion by dispersed cell cultures of human pituitary adenomas.

However, we were unable to observe, in organ culture, after either 1- or 4-hr incubations, a consistent effect of TRH (2.7×10^{-7} M) on PRL release from human prolactinomas, although a slight increase occurred in Case A (Table V).

When we compared the *in vitro* and *in vivo* results, some correlations could be observed. The lack of plasma PRL increase after a TRH test is a usual phenomenon in patients with prolactinomas, in contrast to the frequency of PRL decrease after L-dopa administration. TRH was also found not to affect PRL secretion by monolayer cultures of human fetal pituitaries (Goodyer *et al.*, 1977). Therefore, the discrepancies among the results of different authors are perhaps more apparent than real, since four of our five patients were also unresponsive to TRH *in vivo* and since no *in vivo* data are given in the other studies. In addition to this, both the greater or lesser sensitivity of tumoral (or fetal) cells to TRH and the differences among culture techniques can interfere with the interpretation of the results.

Table V. Effects of TRH on PRL Secretion from Prolactinomas
Incubated for 1 hr and Plasma PRL Basal and Peak Levels during
TRH Test Performed on the Patients before Surgery

Treatment	PRL-secreting adenomas				
	A	B	C	D	E
	In vitro (PRL: ng/mg tissue, mean ± S.D.)				
None	13 ± 2	200 ± 48	176 ± 6	14 ± 2	206 ± 58
TRH	24 ± 6	189 ± 19	149 ± 6	46 ± 47	274 ± 18
$(2.7 \times 10^{-7}$ M)					
	In vivo (PRL: ng/ml)				
None	32	183	82	2000	2590
TRH	91	195	86	1550	2920
(100 μg)					

3.2. Effect on Human Growth Hormone Release

Human GH secretion appears essentially to be regulated by two
hypothalamic factors. One of them, somatostatin, the chemical structure
of which is known, inhibits GH release. The other one, called "GH-
releasing factor" or "hormone" (GRF or GH-RF), stimulates GH release
from the pituitary. Indeed, GH-releasing factors are present in the human
hypothalamus; for example, hypothalamic lesions cause a GH deficiency
(Krieger and Glick, 1974), and GH-releasing activity was demonstrated in
hypothalamic extracts (Schally et al., 1970). However, the nature of these
factors remains to be defined. In contrast to PRL but like the other
adenohypophyseal hormones, the hypothalamic stimulatory tonus is predominant. As in the case of PRL, PIF, and PRF, one can postulate that
GH-IF and GRF mediate, at the pituitary level, most of the neuronal or
hormonal information that reaches the hypothalamus from various origins.
    As in the case of PRL, our study will be limited, among many factors
studied in vitro and in vivo in animals and in vivo in man, to the same
three factors (GH-IF, dopaminergic agents, and TRH) that have been
studied in vitro in human beings.
    For general reviews on the control of GH secretion, see Müller (1974),
Franchimont and Burger (1975), Pecile and Müller (1976), Martin et al.
(1977b), and Müller et al. (1977, 1978).

3.2.1. Somatostatin

Somatostatin inhibits GH secretion in all species in which it has been tested. *In vivo,* in normal and in acromegalic men, the basal secretion of GH, the sleep-related peaks, and the responses to L-dopa, arginine, exercise, and insulin hypoglycemia are diminished by GH-IF (Siler *et al.,* 1973; Hall *et al.,* 1973; Hansen *et al.,* 1973; Parker *et al.,* 1974; Yen *et al.,* 1974). Brazeau *et al.* (1973) reported its activity in *in vitro* GH release from dispersed cells from an acromegalic tumor.

In 13 of 14 observations on both monolayer and explant cultures of fetal pituitaries exposed to GH-IF (ranging from 10^{-6} to 10^{-8} M) for 4 hr, Goodyer *et al.* (1977) observed a decrease ranging from 14 to 78% in GH release. A decrease of GH release (42%) was observed after 3-hr incubation of fetal pituitaries with GH-IF (Paulin *et al.,* 1976) However, Adams and Mashiter (1978) recently failed to show any consistent effect on spontaneous GH secretion by dispersed cell cultures from pituitary tumors. Nevertheless, they obtained a blockade of the theophylline-induced GH release. The *in vitro* inhibitory effect of this peptide was observed in 7 of 9 somatostatin-treated adenomas by Peillon *et al.* (1976b). Nine GH-secreting adenomas, collected after surgery from acromegalic patients, were divided into fragments (the volume of which was 1 mm³) and cultured in Medium 199 supplemented with 20% fetal calf serum according to a method previously described (Peillon *et al.,* 1975). Fragments were incubated for various time periods (15 and 30 min and 1, 2, 3, 4, 24, and 72 hr) with and without GH-IF (0.25, 0.5, 2.5, and 5.0×10^{-6} M). The media were collected for GH radioimmunoassay, and the fragments were studied by electron microscopy. GH content in the incubation media of GH-IF treated fragments compared to the media of control fragments was not significantly different at any time of the incubation when the lower doses of GHIF (0.25 and 0.5×10^{-6} M) were used. It was significantly lower ($p < 0.01$) in the media from GH-IF-treated fragments with 2.5 or 5.0×10^{-6} M GH-IF in 7 of 9 tumors after 4 incubation hours [ranging from -30 to -60% (Table VI)], but there was no difference after 15 and 30 min or 24 and 72 hr of incubation.

Morphological studies showed that GH-IF had specific effects on the hypophysis at the cellular level that seemed independent of a cytolytic action. The micrographs, studied at the different incubation times, showed that different cell organelles (Golgi apparatus, rough endoplasmic reticulum, mitochondria) as well as the nuclei and the nucleoli remained intact. After 1 or 2 hr of incubation, more secretory granules were found in the cells of GH-IF-treated fragments than in controls. After 4 hr of incubation with GH-IF, the most striking observation was a decrease in the number

Table VI. *In Vitro* GH Release from 9 Somatotropic Adenomas
Incubated for 4 hr with and without GH-IF

In vitro GH (μg/mg wet tissue)			
Control	GH-IF (2.5×10^{-6} M)	Control	GH-IF (2.5×10^{-6} M)
6.80	7.75	10.75	4.23
1.14	0.56	0.63	0.45
11.50	6.20	0.52	0.23
8.43	9.33	5.16	3.67
3.55	2.52		

of secretory granules (Fig. 11) accompanied or preceded by numerous exocytotic figures, as shown in Fig. 12. Since it is noteworthy that no exocytoses or very rare ones are seen in cells of GH-secreting adenomas (Olivier *et al.*, 1975; Racadot *et al.*, 1975; Kovacs *et al.*, 1977), these morphological data suggest that somatotroph cells of GH-IF-incubated fragments are directly involved in the process of granule storage and release. At the same time, secretory granules continued to be observed within Golgi complexes that seemed hypertrophied, facts that are commonly interpreted as cytological evidence for activation of biosynthetic mechanisms.

Although it was not possible in this study to correlate exactly the biological and morphological events, some physiopathological conclusions can be drawn. GH-IF diminishes GH content of the media from GH-secreting adenomas incubated for 4 hr. The degree of lowering is variable from one tumor to another, as also observed with GH-IF *in vivo*. In some cases, no effect is observed. The variations may be related to individual GH-secreting potential or to a variable sensitivity to GH-IF of each tumor. There may also be a relationship with the more or less deep penetration of GH-IF into the cells. This may be one of the reasons that only high doses of GH-IF were effective, compared with the much lower doses used in the enzymatically dispersed cells derived from two pituitary glands of patients with active acromegaly (Brazeau *et al.*, 1973; Binoux *et al.*, 1976).

When gel filtration on Sephadex G-100 was performed on media and tumor extracts from GH-IF-treated fragments, no modification of the molecular heterogeneity of GH was observed, compared to nontreated

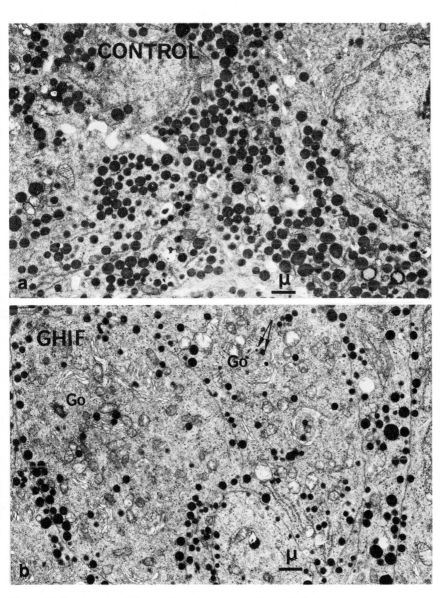

Figure 11. Portion of a GH-secreting adenoma. (a) Control; (b) incubated for 4 hr with somatostatin (2.5×10^{-6} M). The cells show a decreasing number of secretory granules. The Golgi apparatus (Go) contains newly synthesized granules (arrows).

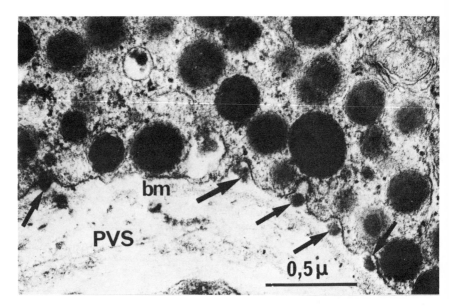

Figure 12. Detail of somatotropic cells from a GH-secreting adenoma incubated for 3 hr with somatostatin (2.5 × 10⁻⁶ M). There are numerous areas of granule extrusion (arrows) in the perivascular space (PVS) in front of the basement membrane (bm). From Peillon *et al.* (1976b) with permission.

fragments (Fig. 13). The three peaks of immunoreactive GH were also observed by Binoux *et al.* (1976) in the cell culture medium of a GH-secreting adenoma after 24 hr of incubation with GH-IF. However, Benker *et al.* (1975), though their studies were not performed in culture, found some modifications in GH heterogeneity in plasma after GH-IF.

3.2.2. Dopaminergic Agents

The comment we have made above concerning the interpretation of the *in vitro* results obtained from tumoral tissue is particularly valid in the case of GH and some stimuli, including the dopaminergic agents, since the *in vivo* responsiveness itself to these stimuli is often not the same in normal and in acromegalic subjects. Indeed, L-dopa and bromocriptine are capable of stimulating GH secretion in normal subjects (Boyd *et al.*, 1970; Eddy *et al.*, 1971; Lal *et al.*, 1972; Camanni *et al.*, 1975a), whereas they frequently inhibit high plasma GH levels in acromegalic patients (Camanni *et al.*, 1975a,b; Chiodini *et al.*, 1974; Liuzzi *et al.*, 1974a,b; Thorner *et al.*, 1975). As for dopamine, it either causes a modest increase (Burrow *et al.*,

1977; Leebaw *et al.*, 1978), is unable to induce a rise in plasma GH (Verde *et al.*, 1976; Camanni *et al.*, 1977), or can provoke a fall in GH in acromegalic patients (Camanni *et al.*, 1977).

In normal subjects, the inhibition of L-dopa-induced GH release by phentolamine, an α-receptor blocking agent, is in favor of a mediation of L-dopa action by norepinephrine (Martin, 1973), but neither α- nor β-adrenergic mechanisms seem to be involved in the suppressive effect of L-dopa in acromegalic patients (Cryer and Doughaday, 1977). Indeed, a dual adrenergic regulation of GH secretion has been shown in normal man, α-

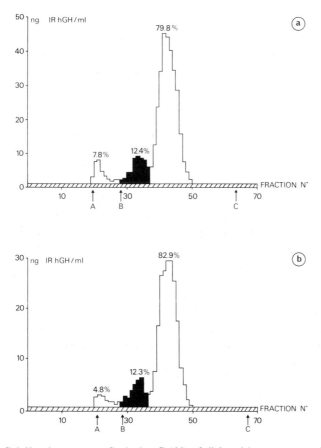

Figure 13. Gel filtration pattern (Sephadex G-100) of dialyzed homogenates of fragments incubated with (a) and without (b) somatostatin. Hatched area: lower limit of sensitivity of radioimmunoassay; black area: big immunoreactive GH; white areas: big-big and little GH. Arrows: (A) void volume; (B) albumin peak; (C) [125]I peak. Reproduced with permission of P. E. Garnier.

receptors being involved in GH stimulation and β-receptors in inhibition. Phentolamine also inhibits GH secretion after hypoglycemia (Blackard and Heidingsfelder, 1968), vasopressin (Heidingsfelder and Blackard, 1968), arginine (Imura *et al.*, 1971), and exercise (Hansen *et al.*, 1971). Propranolol, a β-blocking agent, potentiates the effect of some stimuli (Blackard and Heidingsfelder, 1968; Parra *et al.*, 1970; Imura *et al.*, 1971). Nevertheless, Strauch *et al.* (1969) could not demonstrate any effect of propranolol on GH response to arginine. On the other hand, Imura *et al.* (1971) reported that propranolol could increase basal GH levels; in contrast, Blackard and Heidingsfelder (1968) reported that it could not. A noradrenergic mechanism is probably implicated in the rise of plasma GH after amphetamine administration (Langer and Matussek, 1977). Although the effect of either α- or β-adrenergic stimulating agents is rather unclear (Imura *et al.*, 1971) and the injection of catecholamines themselves does not influence plasma GH level (Glick *et al.*, 1965; Rabinowitz *et al.*, 1966; Schalch, 1967; Blackard and Heidingsfelder, 1968), catecholamines certainly appear to be involved in GH response to many stimuli, if not in the basal secretion.

In addition to the paradoxical response of GH to dopaminergic agents, some other stimuli are able to promote such abnormal responses in acromegalic patients. For example, luteinizing hormone–releasing hormone (LH-RH), which is inactive on GH and PRL in normal man, was reported to be able to stimulate GH or PRL release in some acromegalic patients (Rubin *et al.*, 1973; Faglia *et al.*, 1973b; Catania *et al.*, 1976) without preventing the effect of GH-IF (Giustina *et al.*, 1974). No consistent effect of this peptide on GH secretion from the human fetal pituitary *in vitro* was shown by Goodyer *et al.* (1977). An abnormal response to TRH also occurs in some acromegalics (see later). The mechanism of these paradoxical responses has not been elucidated. In the case of dopaminergic agents, since dopamine does not readily cross the blood–brain barrier, the stimulatory effect of dopaminergic drugs on GH secretion in normal subjects seems to operate via the central nervous system, whereas in acromegaly, one hypothesis may be that of a direct action on structures lying outside the blood–brain barrier, probably on the hypophysis (Camanni *et al.*, 1977). However, the apparent opposite response of normal and acromegalic subjects, with regard to dopaminergic agents, is perhaps less marked than it appears. Indeed, the recent demonstration that dopamine causes, together with a slight increase in basal plasma GH level, as already mentioned, a blockade of hypoglycemia-induced GH release in man suggests a dual effect of dopamine on GH secretion, basal GH release being stimulated and the augmented secretion inhibited (Leebaw *et al.*, 1978). Thus, the response in acromegaly, in which of course GH is augmented, would not be so paradoxical. On the other hand, the presence of abnormal or nonspecific receptors on GH-secreting tumoral cells could also be an explanation for some of these findings. Nonspecific effects of TRH and LH-RH in acromegaly are in favor of this possibility. This has been substantiated on GH-secreting adenomas by Matsukura *et al.* (1977), who showed that both LH-RH and TRH were able to stimulate the adenylate cyclase of the tumors.

It has already been mentioned, with regard to PRL, that cyclic AMP could mediate the action of dopamine, which seems to be able to reduce the activity of adenylate cyclase in PRL cells (de Camilli *et al.*, 1978b). The situation is not at all clear in the case of GH, since the same authors have shown both an inhibition and an activation of this enzyme by dopamine in GH-secreting adenomas, without any relation to the *in vivo* action of dopaminergic drugs on GH secretion in the patients bearing the tumors. If the presence of nonspecific receptors were confirmed, the question would remain whether this abnormal sensitivy is the consequence of the tumoral process (though a dedifferentiation process) or could, at least partly, explain in some cases GH hyperproduction and the cell proliferation under the stimulation of factors normally not recognized as GH-RH, e.g., TRH or LH-RH.

To schematize, some data argue in favor of a central nervous action of dopamine agents in normal man, whereas a direct pituitary action is not exluded in acromegaly. Knowing that dopamine and bromocriptine are unable to affect GH release from normal rat pituitaries *in vitro* (MacLeod *et al.*, 1970; Yanai and Nagasawa, 1974), but that ergot derivatives are able to directly inhibit GH secretion by rat pituitary tumoral tissue (Quadri and Meites, 1973), it is therefore important, in contributing to the elucidation of this complex problem, to determine whether the dopaminergic agents do have a direct effect on the human pituitary *in vitro*. To our knowledge, only tumoral tissue has thus far been investigated, but comparison with future results originating from normal pituitary tissue will probably be fruitful.

Consequently, *in vitro* GH release from the 7 GH-secreting adenomas already studied for their PRL release (see Section 3.1.1) was investigated in the presence or absence of dopamine, L-dopa, and bromocriptine. The experimental conditions have been described above, as well as the reason for the choice of the concentrations of the drugs. GH was measured by the radioimmunoassay method of Strauch *et al.* (1970), with slight modifications.

Again, no statistically significant difference was observed among the means of the results obtained with the different concentrations of each agent. Therefore, pooled data for the same drug were compared to the control ones, as shown in Table VII, which also presents the available data obtained during the L-dopa test, the bromocriptine test, or both tests

Table VII. Data from 7 GH-Secreting Adenomas: (I) Effect of Dopamine, L-Dopa, and Bromocriptine on *in Vitro* GH Release and (II) Nadir Plasma GH Levels during L-Dopa and Bromocriptine Tests Performed on Patients before Surgery

Agent[a]	1[b]	2[b]	3[b]	4[b]	5[b]	6[b]	7[c]
			I. *In vitro*				
C	100 ± 32	625 ± 287	158 ± 82	22 ± 30	191 ± 141	36 ± 12	90 ± 30
D	53 ± 18[d]	426 ± 197	565 ± 91[d]	14 ± 8	118 ± 80	54 ± 14[e]	65 ± 11[e]
L	55 ± 11[d]	1514 ± 923[e]	—	17 ± 11	176 ± 38	70 ± 15[d]	72 ± 11
CB	64 ± 35	714 ± 85	—	—	119 ± 75	58 ± 6[d]	60 ± 19[e]
			II. *In vivo*				
L[f]	23	—	—	42	50	143	100
CB[f]	31	—	—	—	33	115	42

[a] (C) Control; (D) dopamine; (L) L-dopa; (CB) bromocriptine.
[b,c] GH is expressed in: [b]ng/μg tissue protein ± S.D.; [c]ng/mg tissue ± S.D.
[d,e] With respect to control (C) level: [d]$p < 0.01$; [e]$p < 0.05$.
[f] Percentage of the control (C) values.

performed *in vivo* before surgery. It appears that the agents studied had a direct effect on the tumoral pituitary cells, but this action could be positive or negative according to the tumor. Indeed, in 4 tumors (Nos. 1, 4, 5, and 7), a decrease in GH release was observed, which was statistically significant in 4 of 11 experiments, whereas an increase, significant in 4 of 7 experiments, was observed in the 3 remaining tumors (Nos. 2, 3, and 6). Unfortunately, for 2 of the positive responding tumors, the *in vivo* tests could not be performed. However, the *in vitro* data and the few *in vivo* data available were in good agreement.

On the other hand, the inhibitory effect of bromocriptine on GH release was established in 2 of 4 cases of GH-secreting adenomas in culture by Mashiter *et al.* (1977) and in dispersed cell cultures of human pituitary adenomas (Adams and Mashiter, 1978). GH secretion was suppressed in only 1 of the 3 cases studied by Guibout *et al.* (1978), both *in vitro* and *in vivo*. Further investigations are evidently needed to clarify the situation.

However, it seems possible to assert that in most tumors, the dopaminergic agents act directly at the pituitary level, with either a stimulating or an inhibiting effect on *in vitro* GH secretion. In the same way, the *in vivo* response of GH in acromegaly to the dopaminergic drugs can also be qualitatively paradoxical or normal, according to the patient. The positive *in vitro* response observed in some cases of GH-secreting adenomas does not indicate, of course, that normal pituitary tissue has *in vitro* the same sensitivity to these agents.

3.2.3. Thyrotropin-Releasing Hormone

The effect of TRH on GH secretion in man is a controversial subject. In normal man, some investigators have reported no action (Ormston *et al.*, 1971; Gonzales-Barcena *et al.*, 1973; Noel *et al.*, 1974), whereas an increase in serum GH was found by others in some cases (Karlberg *et al.*, 1971), in women only (Torjesen *et al.*, 1973), and in some postmenopausal women and men after an estrogen treatment (Rutlin *et al.*, 1977). On the other hand, this tripeptide is able to induce an increase not inhibited by GH-IF (Giustina *et al.*, 1974), in circulating GH in many acromegalic subjects (Irie and Tsushima, 1972; Faglia *et al.*, 1973a; Liuzzi *et al.*, 1974b), and in patients with renal failure (Gonzales-Barcena *et al.*, 1973; Czernichow *et al.*, 1976).

Reports on TRH effect on GH release in animals are also somewhat conflicting. An increased release of GH after TRH administration has been reported only in cows, *in vivo* (Convey *et al.*, 1973). TRH was reported to cause a rise in GH release from pituitaries *in vitro* in sheep, and this action was reported to be abolished by catecholamines (Takahara *et al.*, 1974b), and in bovines according to Machlin *et al.* (1974), but in only one of three cases according to La Bella and Vivian (1971). In rats, direct infusion of TRH into a hypophyseal portal vessel (Takahara *et al.*, 1974c) and perfusion of hemipituitary glands *in vitro* by TRH (Carlson *et al.*, 1974) were reported to elicit a GH release. In addition, TRH was reported to induce GH release in rats the anterior pituitary glands of which had been transplanted under the kidney capsule, at doses that are without effect in intact animals (Udeschini *et al.*, 1976) and with an increasing sensitivity of the gland to TRH with time after the transplantation (Panerai *et al.*, 1977). On the other hand, Vale et al. (1973) failed to show any action of TRH on GH release by rat pituitaries *in vitro*. On the contrary, this peptide was found to be able to inhibit GH production by a strain of rat pituitary cells in culture (GH_4C_1) (Dannies *et al.*, 1976). Obviously, the very different methodological approaches used and the apparent differences in the responsiveness of GH cells to TRH according to the species do not allow us to summarize these animal data.

It is not suprising that in the human being, with regard to the few data available *in vitro*, the situation is also complicated. Indeed, the *in vitro* effect of TRH on GH release by organ or dispersed cell cultures of human pituitary adenomas seems well correlated with the *in vivo* responsiveness of the patients bearing the tumors (De Camilli *et al.*, 1978a; Adams and Mashiter, 1978; Guibout *et al.*, 1978). However, Goodyer *et al.* (1977) showed that this peptide had no consistent effect on GH secretion by monolayer cultures of human fetal pituitaries.

In the culture conditions described above for the dopaminergic agents,

Table VIII. Effect of TRH on *in Vitro* GH Release
from 3 GH-Secreting Adenomas and Peak Plasma
GH Levels during TRH Tests Performed on the
Patients before Surgery[a]

Agent	1	2	6
Control	100 ± 32	625 ± 287	36 ± 12
TRH	105 ± 62	3229 ± 2211^b	59 ± 12^c
TRH *in vivo*[d]	152	—	150

[a] GH is expressed in ng/μg tissue protein $\pm$ S.D.
[b,c] With respect to control (C) level: $^b p < 0.025$; $^c p < 0.010$.
[d] Percentage of the control (C) values.

a significant increase of GH release was observed in two of three GH-secreting adenomas, but without agreement between the *in vivo* and *in vitro* results (Table VIII).

4. Conclusions

These studies on the secretion of human PRL and GH *in vitro* lead to some concluding remarks.

One is that the pattern of release *in vitro* of these hormones is identical when we compare animal pituitaries and normal and pathological human pituitary glands. It can therefore be concluded that the autonomous secretion of PRL, well demonstrated by its maintenance or its increase, *in vitro,* is a general phenomenon, which is in favor of a predominant effect *in vivo* of a PRL-inhibiting factor. This also holds true for human-hormone-producing adenomas. As for human GH release *in vitro,* a weak although possibly autonomous secretion exists, as demonstrated by the low production of GH after several months of culture and the possible synthesis after several days. However, it can be concluded, in contrast to PRL, that the nonidentified GH-releasing factor most probably predominates *in vivo* over GH-inhibitory factor (GH-IF). This also holds true for human-hormone-producing adenomas.

Another noteworthy point is that of the identity between hormones produced in culture and native hormones. As a consequence, it has been proposed that culture could serve as an alternative source of hormones that could be used in clinical and laboratory investigations as well as in therapy. The different sources of human pituitaries, including hormone-secreting adenomas, have shown that GH released *in vitro* in these cases was identical to the hormone extracted from normal human pituitaries. However, these findings are not sufficiently established for PRL, particularly when released from human secreting pituitary adenomas, and further studies are necessary. This point cannot be considered as an argument for the origin from normal or from abnormal pituitary tissue of PRL- cells found in these adenomas, since GH-secreting adenomatous cells are able to secrete a hormone identical to the native one. The *in vivo* nonspecific effect of LH-RH on the release of PRL in these cases, although found in only a few, may be in favor of the adenomatous origin of the cells.

Concerning the regulation of PRL and GH, many factors are possibly involved in man. As for PRL, from both *in vivo* and *in vitro* data, the inhibitory effect of the dopaminergic agents, physiologically dopamine, is the only one clearly demonstrated. The results we have reported here, together with other recent data, show that they act, at least in tumoral tissue, at the pituitary level; nevertheless, an additional hypothalamic action cannot be rigorously excluded. The well-known *in vivo* and *in vitro* action of TRH is far from being established as a physiological one. The role of somatostatin remains to be specified.

In the case of GH, the often opposite *in vivo* responsiveness of normal and acromegalic subjects to many stimuli makes the situation particularly unclear and prohibits, in a particularly critical and obvious way, the extension of the *in vitro* results obtained from acromegalic tumors to normal pituitary tissue. GH-IF itself, the *in vivo* inhibitory effect of which is well established, does not always have a consistent action *in vitro;* dopaminergic drugs perhaps do not act at the same level in normal and acromegalic subjects; GH-RH has to be identified.

For both PRL and GH, the exact physiological role of the numerous other factors that (according to *in vitro* and *in vivo* experiments on animals, and *in vivo* experiments on man) are possibly involved in the regulation of their secretion remains to be defined precisely or, even, to be studied in man. The quite complicated data available, particularly concerning the neurotransmitters, are very probably a reflection of the complexity of the neural pathways involved and will be elucidated through a better knowledge of these pathways.

5. Summary

In vitro studies on human PRL and GH secretion are reviewed and some new data reported.

As described for animal pituitaries, a disparate pattern of the release and synthesis of PRL and GH is observed in long-term culture of normal, fetal, and pathological human pituitaries. All the *in vitro* studies from these different sources of tissue conclude that secretion and synthesis of PRL is maintained in culture, findings that favor an autonomous PRL secretion and a predominant effect *in vivo* of the PRL-inhibiting factor. In contrast, the decrease of release and synthesis of GH *in vitro* favors the notion of a predominant effect *in vivo* of a GH-releasing factor.

GH present in incubation media and in cultured fragments appears indistinguishable from the native extracted hormone, whereas more studies are necessary to conclude about PRL.

Some aspects of the regulation of PRL and GH release have been studied using hypothalamic factors (somatostatin, TRH) and dopaminergic drugs. No attempt can be made to summarize the results obtained, which differ according to the sources of human pituitaries and the methods of culture. However, most of the results reported here in, together with other recent data, show that these different factors do act, at least in tumoral tissue, at the pituitary level, although an additional extrapituitary effect cannot be excluded. The direct pituitary inhibitory effect of dopaminergic agents on PRL release is particularly well established. For GH, the often opposite *in vivo* responsiveness of normal and pathological human beings to different stimuli makes the situation unclear and prohibits the extension of the *in vitro* results obtained from abnormal pituitary tissues to normal pituitary tissue. Some of these *in vitro* studies show that these differential responses may be located at the level of the pituitary GH-secreting cells, involving most probably abnormal membrane receptors in pathological cases.

The numerous other factors that, according to *in vivo* studies, may be involved in the regulation of PRL and GH secretion remain to be studied more extensively *in vitro* on human pituitary tissue.

ACKNOWLEDGMENTS. This work was supported by the following grants: INSERM (CRAT No. 49-77-81), CNRS (ERA No. 484), and Conseil Scientifique (Faculté de Médecine Pitié-Salpêtrière). We would like to thank Drs. Guiot and Derome, neurosurgeons, for performing the transsphenoidal operation allowing us to study the pituitary fragments. We are indebted to Drs. Donnadieu, Garnier, Gourmelen, L'Hermitte, Pham Huu Trung, and Rivaille for their important contribution to this work. We gratefully acknowledge the technical assistance of Mrs. Brandi, Mauborgne, and Nussbaum and the help of Mrs. Clausse and Combrier and Drs. Mulvihill and Crighton in the preparation of the manuscript.

DISCUSSION

DENEF: I have a comment with regard to your finding that in certain somatotroph adenomas, dopamine agonists decrease GH release, while in others they increase it. As far as prolactin (PRL) secretion in culture is concerned, it has been reported that very low doses of dopamine or dopamine receptor agonist, such as apomorphine, increase PRL secretion, whereas higher doses decrease it. Apomorphine also has a dual effect in the brain. The dual responses of PRL to dopamine agonists are perhaps mediated by functionally different PRL cell populations. One could speculate that the different somatotroph adenomas could be derived from different subtypes of somatotrophs. If so, certain adenomas might indeed be stimulated by dopamine agonists while others were inhibited.

PEILLON: It may indeed be one of the reasons for the different responses of GH release observed after dopaminergic drugs.

KORDON: I am not surprised that you have no dose response in your dopamine incubations, since your lowest dose is already enough to saturate receptors on prolactin cells. Also, your *in vitro* versus your *in vivo* data on the effect of L-dopa and dopamine on growth hormone are really not contradictory. In normal animals, the action of L-dopa on GH secretion can be entirely accounted for by an increase in noradrenalin, rather than in dopamine. So far, dopamine has been conclusively involved in GH control only in human pathology.

TIXIER-VIDAL: Since you are dealing with abnormal cells, there is the possibility of modification of the properties of the dopaminergic receptors, which makes it difficult to interpret the differences that you observe with the response of normal human tissue.

PEILLON: I quite agree with this comment.

CRABBÉ: Wouldn't you agree that the bulk of the evidence is against hypothalamic centers playing any crucial role in the onset and maintenance of acromegaly? Have you considered other ways to account for the decrease in GH release with time by isolated cells of such adenomas?

PEILLON: There is the possible synthesis of GH in culture from these adenomas as demonstrated here; however, the adenoma secretion does not rule out completely the possible hypothalamic control. The secretion may be modified by the hypothalamic hormones such as GH-IF or TRH, or dopamine. There may be a decrease in growth hormone in the *in vitro* studies related to the necrosis observed in the fragments after 15 days of culture. However, the PRL secretion increases in the same fragments during the same culture time.

GOURDJI: Do you have any data on the blood levels of estrogen in patients whose tumors displayed different responsiveness to dopamine or agonist?

PEILLON: In prolactinomas, there are very low plasma levels of estradiol in women operated on for this disease. Some of these tumors are also found in male patients, so the hormonal situation of the different patients is identical, although the response of the tumors *in vitro* to dopamine drugs may be different. In acromegaly, the hormonal situation is different, because the plasma level of estradiol is usually normal. However, we have not done the correlations between the plasma levels of estrogen and the response to dopaminergic agents in these cases.

REFERENCES

Adams, E. F., and Mashiter, K., 1978, Prolactin and growth hormone secretion by dispersed cell cultures of somatotrophic, lactotrophic and mixed pituitary adenomas, Abstract, International Symposium on pituitary microadenomas, Milan, October 1978.

Batzdorf, U., Gold, V., Matthews, N., and Brown, J., 1971, Human growth hormone in cultures of human pituitary tumors, *J. Neurosurg.* **34**:741.

Beauvillain, J. C., Tramu, G., Mazzucca, M., Christiaens, J. L., L'Hermite, M., Asfour, M., Fossati, P. and Linquette, M., 1976, Étude morphologique d'un adénomeà prolactine après traitement par la 2-bromo ergocryptine (CB-154), *Ann. Endocrinol. (Paris)* **37**:117.

Belleville, F., Hartman, P., Paysant, P. and Nabet, P., 1973, Incorporation de la leucine marquée (leucine ^{3}H) dans les protéines sécrétées par les cellules hypophysaires en culture, *C. R. Soc. Biol.* **167**:305.

Benker, G., Mortimer, C. H., Chait, A., Lowry, P. J., Besser, G. M., Coy, D. H., Kastin, A., and Schally, A. V., 1975, Heterogeneity of human growth hormone in plasma and urine: Influence of growth hormone release inhibiting hormone, *Acta Endocrinol. (Copenhagen) Suppl.* **193**, Abstract No. 72.

Besser, G. N., Mortimer, C. H., McNeilly, A. S., Thorner, M. O., Batistoni, G. A., Bloom, S. R., Kastrup, K. W., Hanssen, K. F., Hall, R., Coy, D. H., Kastin, A. J., and Schally, A. V., 1974, Long term infusion of growth hormone release inhibiting hormone in acromegaly: Effects on pituitary and pancreatic hormones, *Br. Med. J.* **4**:622.

Besses, G. S., Burrow, G. N., Spaulding, S. W. and Donabedian, R. K., 1975, Dopamine infusion acutely inhibits the TSH and prolactin response to TRH, *J. Clin. Endocrinol. Metab.* **41**:985.

Binoux, M., Donnadieu, M., and Gourmelen, M., 1976, Hétérogénéité moléculaire de l'hormone de croissance humaine (hGH) sécrétée par des adénomes hypophysaires somatotropes en culture, in: *Neuroendocrinologie* Colloque INSERM, pp. 207–213, INSERM, Paris.

Birge, C. A., Jacobs, L. S., Hammer, C. T., and Daughaday, W. H., 1970, Catecholamine inhibition of prolactin secretion by isolated rat adenohypophyses, *Endocrinology* **86**:120.

Blackard, W. G., and Heidingsfelder, S. A., 1968, Adrenergic receptor control mechanism for growth hormone secretion, *J. Clin. Invest.* **47**:1407.

Borgeat, P., Labrie, F., Drouin, J., Belanger, A., Immer, I., Sestanj, K., Nelson, V., Gotz, M., Schally, A. V., Coy, D. H., and Coy, E. J., 1974, Inhibition of adenosine-3′,5′-monophosphate accumulation in anterior pituitary gland *in vitro* by growth hormone-release inhibiting hormone, *Biochem. Biophys. Res. Commun.* **56**:1052.

Bowers, C. Y., Friesen, H., Guyda, H. J., and Folkers, K., 1971, Prolactin and thyrotropin release in man by synthetic pyroglutamyl-histidyl-prolinamide, *Biochem. Biophys. Res. Commun.* **45**:1033.

Boyd, A. E., Lebovitz, H. E., and Pfeifer, J. B., 1970, Stimulation of human-growth-hormone secretion by L-dopa, *N. Engl. J. Med.* **283**:1425.

Brauman, J., Brauman, H., and Pasteels, J. L., 1964, Immunoassay of growth hormone in cultures of human hypophysis by the method of complement fixation: Comparison of the growth hormone secretion and the prolactin activity, *Nature (London)* **202**:1116.

Brazeau, P., Vale, W., Burgus, R., Ling, N., Butcher, M., Rivier, J., and Guillemin, R., 1973, Hypothalamic polypeptide that inhibits the secretion of immunoreactive pituitary growth hormone, *Science* **179**:77.

Bridson, W. E., and Kohler, P. O., 1970, Cortisol stimulation of growth hormone production by human pituitary tissue in culture, *J. Clin. Endocrinol. Metab.* **30**:538.

Brown, G. M., Seeman, P., and Lee, T., 1976, Dopamine/neuroleptic receptors in basal hypothalamus and pituitary, *Endocrinology* **99**:1407.

Brunet, N., Gourdji, D., Tixier-Vidal, A., Pradelles, P., Morgat, J. L., and Fromageot, P., 1974, Chemical evidence for associated TRF with subcellular fractions after incubation of intact rat prolactin cells (GH$_3$) with ^{3}H-labelled TRF, *FEBS Lett.* **38**:129.

Burrow, G. N., May, P. B., Spaulding, S. W., and Donabedian, R. K., 1977, TRH and dopamine interactions affecting pituitary hormone secretion, *J. Clin. Endocrinol. Metab.* **45**:65.

Calabro, M. A., and MacLeod, R. M., 1978, Binding of dopamine to bovine anterior pituitary gland membranes, *Neuroendocrinology* **25**:32.

Camanni, F., Massara, F., Belforte, L., and Molinatti, G. M., 1975a, Changes in plasma growth hormone levels in normal and acromegalic subjects following administration of 2 Br-α-ergocryptine, *J. Clin. Endocrinol. Metab.* **40**:363.

Camanni, F., Massara, F., Fassio, V., Molinatti, G. M., and Müller, E. E., 1975b, Effect of five dopaminergic drugs on plasma growth hormone levels in acromegalic subjects, *Neuroendocrinology* **19**:227.

Camanni, F., Massara, F., Belforte, L., Rosatello, A. and Molinatti, G. M., 1977, Effect of dopamine on plasma growth hormone and prolactin levels in normal and acromegalic subjects, *J. Clin. Endocrinol. Metab.* **44**:465.

Carlson, H. E., Mariz, I. K., and Daughaday, W. H., 1974, Thyrotropin-releasing hormone stimulation and somatostatin inhibition of growth hormone secretion from perfused rat adenohypophyses, *Endocrinology* **94**:1709.

Caron, M. G., Beaulieu, M., Raymond, V., Gagné, B., Drouin, J., Lefkowitz, R. J., and Labrie, F., 1978, Dopaminergic receptors in the anterior pituitary gland: Correlation of ^{3}H dihydroergocryptine binding with the dopaminergic control of prolactin release, *J. Biol. Chem.* **253**:2244.

Carr, D., Gomez-Pan, A., Weightman, D. R., Roy, V. C. M., Hall, R., Besser, G. M., Thorner, M. O., McNeilly, A. S., Schally, A. V., Kastin, A. J., and Coy, D. H., 1975, Growth hormone release inhibiting hormone: Actions on thyrotrophin and prolactin secretion after thyrotrophin releasing hormone, *Br. Med. J.* **3**:67.

Catania, A., Cantalamessa, L., and Reschini, E., 1976, Plasma prolactin response to luteinizing hormone releasing hormone in acromegalic patients, *J. Clin. Endocrinol. Metab.* **43**:689.

Chiodini, P. G., Liuzzi, A., Botalla, L., Cremascoli, G., and Silvestrini, F., 1974, Inhibitory effect of dopaminergic stimulation on GH release in acromegaly, *J. Clin. Endocrinol. Metab.* **38**:200.

Convey, E. M., Tucker, H. A., Smith, V. G., and Zolman, J., 1973, Bovine prolactin, growth hormone, thyroxine and corticoid response to thyrotropin-releasing hormone, *Endocrinology* **92**:471.

Copinschi, G., Virasoro, E., Vanhaelst, L., Leckercq, R., Golstein, J., and L'Hermite, M., 1974, Specific inhibition by somatostatin of growth hormone release after hypoglycaemia in normal man, *Clin. Endocrinol. (Oxford)* **3**:441.

Cronin, M. J., Roberts, J. M., and Weiner, R. I., 1978, Dopamine and dihydroergocryptine binding to the anterior pituitary and other brain areas of the rat and sheep, *Endocrinology* **103**:302.

Cryer, P. E., and Daughaday, W. H., 1977, Adrenergic modulation of growth hormone secretion in acromegaly: Alpha- and beta-adrenergic blockade produce qualitatively normal responses but no effect on L-dopa suppression, *J. Clin. Endocrinol. Metab.* **44**:977.

Czernichow, P., Dauzet, M. C., Broyer, M., and Rappaport, R., 1976, Abnormal TSH, PRL and GH response to TSH releasing factor in chronic renal failure, *J. Clin. Endocrinol. Metab.* **43**:630.

Dannies, P. S., Gautvik, K. M., and Tashjian, A. H., 1976, A possible role of cyclic AMP in mediating the effects of thyrotropin-releasing hormone on prolactin release and on prolactin and growth hormone synthesis in pituitary cells in culture, *Endocrinology* **98**:1147.

De Camilli, P., Tagliabue, L., Paracchi, A., Faglia, G., Beck-Peccoz, P., and Giovanelli,

M., 1978a, *In vitro* study on the release of GH by fragments of GH producing human pituitary adenomas: Effect of TRH and DBcAMP, in: *Treatment of Pituitary Adenomas* (R. Fahlbusch and K. V. Werder, eds.), pp. 172–179, Georg Thieme, Stuttgart.

De Camilli, P., Spada, A., Peck-Peccoz, P., Moriando, P., Giovanelli, M., and Faglia, G., 1978b, Presence of a dopamine sensitive adenylate-cyclase in functioning human pituitary microadenomas, Abstract, International Symposium on Pituitary Microadenomas, Milan, October 1978.

Del Pozo, E., Brun del Re, R., Varga, L., and Friesen, H., 1972, The inhibition of prolactin secretion in man by CB-154 (2-Br-alpha-ergocryptine), *J. Clin. Endocrinol. Metab.* **35**:768.

Del Pozo, E., Friesen, H., and Burmeister, P., 1973, Endocrine profile of a specific prolactin inhibitor: Br-ergocryptine (CB-154): A preliminary report, *Schweiz. Med. Wochenschr.* **103**:847.

Deuben, R., and Meites, J., 1963, *In vitro* stimulation of growth hormone release from rat anterior pituitary by extracts of rat hypothalamus, *Fed. Proc. Fed. Am. Soc. Exp. Biol.* **22**:571.

Dieffenbach, W. P., Carmel, P. W., Frantz, A. G., and Ferin, M., 1976, Suppression of prolactin secretion by L-dopa in stalk-sectioned rhesus monkeys, *J. Clin. Endocrinol. Metab.* **43**:638.

Eddy, R. L., Jones, A. L., Chakmakjan, Z. H., and Silvertone, M. C., 1971, Effect of levodopa (L-dopa) on human hypophyseal trophic hormone release, *J. Clin. Endocrinol. Metab.* **33**:709.

Enjalbert, A., Ruberg, B., and Kordon, C., 1978, Neuroendocrine control of prolactin secretion, in: *Progress in Prolactin Physiology and Pathology* (C. Robyn and M. Harter, eds.), pp. 83–94, Elsevier/North-Holland, Amsterdam and New York.

Faglia, G., Beck-Peccoz, P., Ferrari, C., Travaglini, P., Ambrosi, B., and Spada, A., 1973a, Plasma growth hormone response to thyrotropin-releasing hormone in patients with active acromegaly, *J. Clin. Endocrinol. Metab.* **36**:1259.

Faglia, G., Beck-Peccoz, P., Travaglini, P., Paracchi, A., Spada, A., and Lewin, A., 1973b, Elevations in plasma growth hormone concentration after luteinizing hormone–releasing hormone (LRH) in patients with active acromegaly, *J. Clin. Endocrinol. Metab.* **37**:336.

Farquhar, M. G., Skutelsky, E. H., and Hopkins, C. R., 1975, Structure and function of the anterior pituitary and dispersed pituitary cells: *In vitro* studies, in: *The Anterior Pituitary Gland* (A. Tixier-Vidal and M. G. Farquhar, eds.), Vol. 7, pp. 82–135, Academic Press, New York.

Flückiger, E., Markǒ, M., Doepfner, W., and Niederer, W., 1976, Effects of ergot alkaloids on the hypothalamic–pituitary axis, *Postgrad. Med. J.* **52**(Suppl. I):57.

Franchimont, P., and Burger, H., 1975, *Human Growth Hormone and Gonadotrophins in Health and Disease,* North-Holland, Amsterdam; Oxford/Elsevier, New York.

Friesen, H., Guyda, H., and Hardy, J., 1970, Biosynthesis of human growth hormone and prolactin, *J. Clin. Endocrinol. Metab.* **31**:611.

Friesen, H., Webster, B. R., Hwang, P., Guyda, H., Munro, R. E., and Read, L., 1972, Prolactin synthesis and secretion in a patient with the Forbes Albright syndrome, *J. Clin. Endocrinol. Metab.* **34**:192.

Frohman, L. A., and Stachura, M. E., 1973, Evidence for possible precursors in the synthesis of growth hormone by rat and human fetal anterior pituitary *in vitro*, *Mt. Sinai J. Med. N. Y.* **40**:414.

Gailani, S. D., Nussbaum, A., McDougall, W. J., and McLimans, W. F., 1970, Studies on hormone production by human fetal pituitary cell cultures, *Proc. Soc. Exp. Biol. Med.* **134**:27.

Gala, R. R., 1971, Prolactin production by the human anterior pituitary cultured *in vitro*, *J. Endocrinol.* **50**:637.

Gala, R. R., Subramanian, M. G., Peters, J. A., and Jacques, S. Jr., 1976, The influence of somatostatin on drug-induced prolactin release in the monkey, *Experientia* **32**:941.

Giustina, G., Reschini, E., Peracchi, M., Cantalamessa, L., Cavagnini, F., Pinto, M., and Bulgheroni, P., 1974, Failure of somatostatin to suppress thyrotropin releasing factor and luteinizing hormone–induced growth hormone release in acromegaly, *J. Clin. Endocrinol. Metab.* **38**:906.

Glick, S. M., Roth, J., Yalow, R. S., and Berson, S. A., 1965, The regulation of growth hormone secretion, *Recent Prog. Horm. Res.* **21**:241.

Gonzales-Barcena, D., Kastin, A. J., Schalch, D. S., Torres-Zamora, M., Peres-Pasten, E., Kato, A., and Schally, A. V., 1973, Response to thyrotropin-releasing hormone in patients with renal failure and after infusion in normal men, *J. Clin. Endocrinol. Metab.* **36**:117.

Goodyer, C. G., Hall, C. St.G., Guyda, H., Robert, F., and Giroud, C. J. P., 1977, Human fetal pituitary in culture: Hormone secretion and response to somatostatin, luteinizing hormone releasing factor, thyrotropin releasing factor and dibutyryl cyclic AMP, *J. Clin. Endocrinol. Metab.* **45**:73.

Gourdji, D., Morin, A., and Tixier-Vidal, A., 1973a, Study on the control of prolactin secretion by two continuous lines of rat pituitary prolactin cells, in: *Human Prolactin* (J. L. Pasteels and C. Robyn, eds.), pp. 163–166, Excerpta Medica, Amsterdam.

Gourdji, D., Tixier-Vidal, A., Morin, A., Pradelles, P., Morgat, J. L., and Fromageot, P., 1973b, Binding of tritiated thyrotropin releasing factor to a prolactin secreting clonal cell line (GH₃), *Exp. Cell Res.* **82**:39.

Gräf, K. J., Horowski, R., and El Etreby, M. F., 1977, Effect of prolactin inhibitory agents on the ectopic anterior pituitary and the mammary glands in rats, *Acta Endocrinol. (Copenhagen)* **85**:267.

Grant, N. H., Sarantakis, D., and Yardley, J. P., 1974, Action of growth hormone release inhibitory hormone on prolactin release in rat pituitary cell cultures, *J. Endocrinol.* **61**:163.

Groom, G. V., Groom, M. A., Cooke, I. B., and Boyns, A. R., 1971, The secretion of immuno-reactive luteinizing-hormone and follicle-stimulating hormone by the human foetal pituitary in organ culture, *J. Endocrinol.* **49**:335.

Guibout, M., Jaquet, P., Lucas, C., Hassoun, J., Charpin, C. and Grisoli, F., 1978, *In vitro* and *in vivo* effects of TRH and bromocriptine on GH-secreting adenomas, Abstract, International Symposium on Pituitary Microadenomas, Milan, October 1978.

Guyda, H. J., 1975, Heterogeneity of human growth hormone and prolactin secreted *in vitro*: Immunoassay and radioreceptor assay correlations, *J. Clin. Endocrinol. Metab.* **41**:953.

Guyda, H., Robert, F., Colle, E., and Hardy, J. 1973, Histologic, ultrastructural and hormonal characterization of a pituitary tumor secreting both hGH and prolactin, *J. Clin. Endocrinol. Metab.* **36**:531.

Hagen, T. C., Guansing, A. R., and Sill, A. J., 1976, Preliminary evidence for a human prolactin releasing factor, *Neuroendocrinology* **21**:255.

Hall, R., Besser, G. M., Schally, A. V., Coy, D. H., Evered, D., Goldie, D. J., Kastin, A. J., MacNeilly, A. S., Mortimer, C. H., Phenekos, C., Tunbridge, W. M. G., and Weightman, D., 1973, Action of growth hormone release inhibitory hormone in healthy men in acromegaly, *Lancet* **2**:581.

Hansen, A. P., 1971, The effect of adrenergic receptor blockade on the exercise induced serum growth hormone rise in normal and juvenile subjects, *J. Clin. Endocrinol. Metab.* **33**:807.

Hansen, A. P., Orkson, H., Seyer-Hansen, K., and Lundbaek, K., 1973, Some actions of growth hormone release inhibiting factor, *Br. Med. J.* **3**:523.

Heidingsfelder, S. A., and Blackard, W. G., 1968, Adrenergic control mechanism for vasopressin induced plasma growth hormone response, *Metabolism* **17**:1019.

Hill-Samli, M., and MacLeod, R. M., 1974, Interaction of thyrotropin-releasing hormone and dopamine on the release of prolactin from the rat anterior pituitary *in vitro, Endocrinology* **95**:1189.

Hill-Samli, M., and MacLeod, R. M., 1975, TRH blockade of the ergocryptine and apomorphine inhibition of prolactin release *in vitro, Proc. Soc. Exp. Biol. Med.* **149**:511.

Hinckle, P. M., and Tashjian, A. H., 1973, Receptors for thyrotropin-releasing hormone in prolactin producing rat pituitary cells in culture, *J. Biol. Chem.* **248**:6180.

Hwang, P., Friesen, H., Hardy, J., and Wilansky, D., 1971, Biosynthesis of human growth hormone and prolactin by normal pituitary glands and pituitary adenomas, *J. Clin. Endocrinol. Metab.* **33**:1.

Imura, H., Kato, Y., Ikeda, M., Morimoto, M. and Yawata, M., 1971, Effect of adrenergic-blocking or stimulating agents on plasma growth hormone, immunoreactive insulin, and blood free fatty acid levels in man, *J. Clin. Invest.* **50**:1069.

Irie, M., and Tsushima, T., 1972, Increase of serum growth hormone concentration following thyrotropin-releasing hormone injection in patients with acromegaly or gigantism, *J. Clin. Endocrinol. Metab.* **35**:97.

Jacobs, L. S., Snyder, P. J., Utiger, R. D., Wilber, J. F., and Daughaday, W. H., 1971, Increased serum prolactin after administration of synthetic thyrotropin-releasing hormone (TRH) in man, *J. Clin. Endocrinol. Metab.* **33**:996.

Jimenez, A. E., Voogt, J. L., and Carr, L. A., 1978, L-3,4-Dihydroxyphenylalanine (L-dopa) as an inhibitor of prolactin release, *Endocrinology* **102**:166.

Kamberi, I. A., Mical, R. S., and Porter, J. C., 1971, Effect of anterior pituitary perfusion and intraventricular injections of catecholamines on prolactin release, *Endocrinology* **88**:1012.

Karlberg, B., Almqvist, S., and Werner, S., 1971, Effects of synthetic pyroglutamyl-histidyl-prolinamide on serum levels of thyrotrophin, cortisol, growth hormone, insulin and PBI in normal subjects and patients with pituitary and thyroid disorders, *Acta Endocrinol. (Copenhagen)* **67**:288.

Kleinberg, D. L., Noel, G. L., and Frantz, A. G., 1971, Chlorpromazine stimulation and L-dopa suppression of plasma prolactin in man, *J. Clin. Endocrinol. Metab.* **33**:873.

Knazek, R. A., and Skyler, J. S., 1976, Secretion of human prolactin *in vitro, Proc. Soc. Exp. Biol. Med.* **151**:561.

Koch, Y., Lu, K. H., and Meites, J., 1970, Biphasic effects of catecholamines on pituitary prolactin release *in vitro, Endocrinology* **87**:673.

Kohler, P. O., Bridson, W. E., Rayford, P. L., and Kohler, S. E., 1969, Hormone production by human pituitary adenomas in culture, *Metabolism* **18**:782.

Kohler, P. O., Bridson, W. E., and Chrambach, A., 1971, Human growth hormone produced in tissue culture: Characterization by polyacrylamide gel electrophoresis, *J. Clin. Endocrinol. Metab.* **32**:70.

Kovacs, K., Horwath, E., and Ezrin, C., 1977, Pituitary adenomas, in: *Pathology Annual* (S. C. Sommers and P. P. Rosen, eds.) Vol. 12, pp. 341–382, Appleton-Century-Crofts, New York.

Krieger, D. T., and Glick, S. M., 1974, Sleep EEG stages and plasma growth hormone concentration in states of endogenous hypercortisolemia or ACTH elevation, *J. Clin. Endocrinol. Metab.* **39**:986.

LaBella, F., and Vivian, S. R., 1971, Effect of synthetic TRF on hormone release from bovine anterior pituitary *in vitro, Endocrinology* **88**:787.

Labrie, F., Pelletier, G., Borgeat, P., Drouin, J., Ferland, L., and Belanger, A., 1976, Mode of action of hypothalamic regulatory hormones in the adenohypophysis, in: *Frontiers in Neuroendocrinology,* Vol. IV (L. Martini and W. F. Ganong, eds.), pp. 63–93, Raven Press, New York.

Lal, S., De La Vega, C. E., Sourkes, T. L., and Friesen, H. G., 1972, Effect of apomorphine on human growth hormone secretion, *Lancet* 2:601.

Langer, G., and Matussek, M., 1977, Dextro- and L-amphetamine are equipotent in releasing human growth hormone, *Psychoneuroendocrinology* 2:379.

Leblanc, H., Lachelin, G. C. L., Abu-Fadil, S., and Yen, S. C., 1976. Effects of dopamine infusion on pituitary hormone secretion in humans, *J. Clin. Endocrinol. Metab.* 43:668.

Leebaw, W. F., Lee, L. A., and Woolf, P. D., 1978, Dopamine affects basal and augmented pituitary hormone secretion, *J. Clin. Endocrinol. Metab.* 47:480.

Liuzzi, A., Chiodini, P. G., Botalla, L., Cremascoli, G., Müller, E. E., and Silvestrini, F., 1974a, Decreased plasma growth hormone (GH) levels in acromegalics following CB-154 (2-Brα-ergocryptine) administration, *J. Clin. Endocrinol. Metab.* 38:910.

Liuzzi, A., Chiodini, P. G., Botalla, L., Silvestrini, F., and Müller, E. E., 1974b, Growth hormone (GH)-releasing activity of TRH and GH-lowering effect of dopaminergic drugs in acromegaly: Homogeneity in the two responses, *J. Clin. Endocrinol. Metab.* 39:871.

Lu, K. H., Shaar, C. J., Kortright, K. H., and Meites, J., 1972, Effects of synthetic TRH on *in vitro* and *in vivo* prolactin release in the rat, *Endocrinology* 91:1540.

Machlin, L. J., Jacobs, L. S., Cirulis, N., Kimes, R., and Miller, R., 1974, An assay for growth hormone and prolactin-releasing activities using bovine pituitary cell culture system, *Endocrinology* 96:1350.

MacLeod, R. M., 1976, Regulation of prolactin secretion, in: *Frontiers in Neuroendocrinology,* Vol. 4, (L Martini and W. F. Ganong, eds.), pp. 169–194, Raven Press, New York.

MacLeod, R. M., and Lehmeyer, J. E., 1972, Regulation of the synthesis and release of prolactin, in: *Lactogenic Hormones* (G. E. W. Wolstenholme and J. Knight, eds.), pp. 53–82, Churchill-Livingstone, London.

MacLeod, R. M., and Lehmeyer, J. E., 1974, Studies on the mechanism of dopamine mediated inhibition of prolactin secretion, *Endocrinology* 94:1077.

MacLeod, R. M., Fontham, E. H., and Lehmeyer, J. E., 1970, Prolactin and growth hormone production as influenced by catecholamines and agents that affect brain catecholamines, *Neuroendocrinology* 6:283.

Malarkey, W. B., and Daughaday, W. H., 1972, The influence of levodopa and adrenergic blockade on growth hormone and prolactine secretion in the MStTW15 tumor bearing rat, *Endocrinology* 91:1314.

Malarkey, W. B., and Pankratz, K., 1974, Evidence for prolactin releasing activity in human plasma not associated with TSH release, *Clin. Res.* 22:600A.

Martin, J. B., 1973, Neural regulation of growth hormone secretion: Medical progress report, *N. Engl. J. Med.* 288:1384.

Martin, J. B., Reichlin, S., and Brown, G. M., 1977a, Regulation of prolactin secretion and its disorders, in: *Clinical Neuroendocrinology, Contemporary Neurology Series,* pp. 129–145, F. A. Davis, Philadelphia.

Martin, J. B., Reichlin, S., and Brown, G. M., 1977b, Regulation of growth hormone secretion and its disorders, in: *Clinical Neuroendocrinology, Contemporary Neurology Series,* pp. 147–178, F. A. Davis, Philadelphia.

Mashiter, K., Adams, E., Beard, M., and Holley, A., 1977, Bromocriptine inhibits prolactin and growth-hormone release by human pituitary tumours in culture, *Lancet* 2:197.

Matsukura, S., Kakita, T., Hirata, Y., Yoshimi, H., Fukase, M., Iwasaki, Y., Kato, Y., and Imura, H., 1977, Adenylate cyclase of GH and ACTH producing tumors of human: Activation by non specific hormones and other bioactive substances, *J. Clin. Endocrinol. Metab.* **44**:392.

Meites, J., Kahn, R. H., and Nicoll, C. S., 1961, Prolactin production by rat pituitary *in vitro, Proc. Soc. Exp. Biol. Med.* **108**:440.

Morin, A., Tixier-Vidal, A., Gourdji, D., Kerdelhué, B., and Grouselle, D., 1975, Effect of thyreotrope-releasing hormone (TRH) on prolactin turnover in culture, *Mol. Cell. Endocrinol.* **3**:351.

Mueller, G. P., Chen, H. J., and Meites, J., 1973, *In vivo* stimulation of prolactin release in the rat by synthetic TRH, *Proc. Soc. Exp. Biol. Med.* **144**:613.

Müller, E. E., 1974, Growth hormone and the regulation of metabolism, in: *Endocrine Physiology* (S. M. McCann, ed.), p. 141, Butterworths, London.

Müller, E. E., Liuzzi, A., Cocchi, D., Panerai, A. E., Oppizzi, G., and Chiodini, P. G., 1977, The role of dopaminergic receptors in the regulation of growth hormone secretion, in: *Symposium on Non-Striatal Dopaminergic Neurons* (E. Costa, ed.), p. 127, Raven Press, New York.

Müller, E. E., Chiodini, P. G., Cocchi, D., Panerai, A. E., Oppizi, G., Colussi, G., Locatelli, V., and Liuzzi, A., 1978, Neurotransmitter control of growth hormone secretion, in: *Treatment of Pituitary Adenomas* (R. Fahlbusch and K. Von Werder, eds.), pp. 360–377, Georg Thieme, Stuttgart.

Nagasawa, H., Yanai, R., and Flückiger, E., 1973, Counteraction by 2-Br-α-ergocryptine of pituitary prolactin release promoted by dibutyryl-adenosine-3′,5′-monophosphate in rats, in: *Human Prolactin* (J. L. Pasteels and C. Robyn, eds.), pp. 313–315; Excerpta Medica, Amsterdam.

Noel, G. L., Suh, H. K., and Frantz, A. G., 1973, L-dopa suppression of TRH-stimulated prolactin release in man, *J. Clin. Endocrinol. Metab.* **36**:1255.

Noel, G. L., Dimond, R. C., Wartofsky, L., Earll, J. M., and Frantz, A. G., 1974, Studies of prolactin and TSH secretion by continuous infusion of small amounts of thyrotropin-releasing hormone (TRH), *J. Clin. Endocrinol. Metab.* **39**:6.

Olendorf, W. H., 1971, Brain uptake of radiolabeled amino acids, amines, and hexoses after arterial injection, *Am. J. Physiol.* **221**:1629.

Olivier, L., Vila-Porcile, E., Racadot, O., Peillon, F., and Racadot, J., 1975, Ultrastructure of pituitary tumor cells: A critical study, in: *The Anterior Pituitary* (A. Tixier-Vidal and M. G. Farquhar, eds.), pp. 231–276, Academic Press, New York.

Ormston, B. J., Kilbron, J. R., Garry, R., Amos, J., and Hall, R., 1971, Further observations on the effect of synthetic thyrotropin-releasing hormone in man, *Br. Med. J.* **2**:199.

Panerai, A. E., Gil-Ad, I., Cocchi, D., Locatelli, V., Rossi, G. L., and Müller, E. E., 1977, Thyrotrophin releasing hormone–induced growth hormone and prolactin release: Physiological studies in intact rats and in hypophysectomized rats bearing an ectopic pituitary gland, *J. Endocrinol.* **72**:301.

Parker, D. C., Rossman, L. G., Siler, T. M., Rivier, J., Yen, S. S. C., and Guillemin, R., 1974, Inhibition of the sleep-related peak in physiologic human growth hormone release by somatostatin, *J. Clin. Endocrinol. Metab.* **38**:496.

Parra, A., Schultz, R. B., Foley, T. P., Jr., and Blizzard, R. M., 1970, Influence of epinephrine–propranolol infusions in growth hormone release in normal and hypopituitary subjects, *J. Clin. Endocrinol. Metab.* **30**:134.

Pasteels, J. L., 1961a, Sécrétion de prolactine par l'hypophyse en culture de tissus, *C. R. Acad. Sci.* **253**:2140.

Pasteels, J. L., 1961b, Premiers résultats de culture combinée *in vitro* d'hypophyse et d'hypothalamus, dans le but d'en apprécier la sécrétion de prolactine, *C. R. Acad. Sci.* **253**:3074.

Pasteels, J. L., 1962, Elaboration par l'hypophyse humaine en culture de tissus d'une substance stimulant le jabot de pigeon, *C. R. Acad. Sci.* **254**:4083.

Pasteels, J. L., 1969, Nouvelles recherches sur la structure et le comportement des cellules hypophysaires *in vitro, Mem. Acad. R. Med. Belg.* **7**:1.

Pasteels, J. L., Brauman, H., and Brauman, J., 1963, Étude comparée de la sécrétion d'hormone somatotrope par l'hypophyse humaine *in vitro* et de son activité lactogénique, *C. R. Acad. Sci.* **256**:2031.

Pasteels, J. L., Danguy, A., Frérotte, M., and Ectors, F., 1971, Inhibition de la sécrétion de prolactine par l'ergocornine et la 2-Br-α-ergocryptine: Action directe sur l'hypophyse en culture, *Ann. Endocrinol. (Paris)* **32**:188.

Pasteels, J. L., Gausset, P., Danguy, A., Ectors, F., Nicoll, C. S., and Varavudhi, P., 1972, Morphology of the lactotropes and somatotropes of man and rhesus monkeys, *J. Clin. Endocrinol. Metab.* **34**:959.

Paulin, G., Li, J., Begeot, M., and Dubois, P. M., 1976, La somatostatine et la fonction somatotrope antehypophysaire chez le foetus humain, in: *Neuroendocrinologie,* Colloque INSERM, pp. 275–2251, INSERM, Paris.

Pecile, A., and Müller, E. E. (eds.), 1976, *Growth Hormone and Related Peptides,* Proceedings of the IIIrd International Symposium, Excerpta Medica, Amsterdam.

Peillon, F. Gourmelin, M., Brandi, A. M., and Donnadieu, M., 1972, Adénomes somatotropes humains en culture organotypique: Ultrastructure et sécrétion étudiée à l'aide de leucine tritiée, *C. R. Acad. Sci.* **275**:2251.

Peillon, F., Gourmelin, M., Donnadieu, M., Brandi, A. M., Sevaux, D., and Pham Huu Trung, M. T., 1975, Organ culture of human somatotrophic pituitary adenomas: Ultrastructure and growth hormone production, *Acta Endocrinol. (Copenhagen)* **79**:217.

Peillon, F., Garnier, P. E., Chaussain, J. L., Brandi, A. M., and Rivaille, P., 1976a, Growth hormone secretion by human somatotrophic adenomas *in vitro:* Effect of somatostatin, in: *Hypothalamus and Endocrine Functions* (F. Labrie, J. Meites, and G. Pelletier, eds.), *Current Topics in Molecular Endocrinology,* Vol. 3, abstract 36, p. 496, Plenum Press, New York.

Peillon, F., Garnier, P. E., Brandi, A. M., Rivaille, P., Cesselin, F., and Chaussain, J. L., 1976b, Effet *in vitro* de la somatostatine sur les adénomes somatotropes humains: Étude ultrastructurale et hormonale, in: *Neuroendocrinologie,* Colloque INSERM, pp. 215–224, INSERM, Paris.

Peillon, F., Cesselin, F., Zygelman, N., Brandi, A. M., and Mauborgne, A., 1978a, Effet *in vitro* de la L-dopa, de la dopamine et du CB 154 sur la sécrétion de prolactine par les adénomes somatotropes humains, *Ann. Endocrinol. (Paris)* **39**:255.

Peillon, F., Cesselin, F., Garnier, P. E., Brandi, A. M., Donnadieu, M., L'Hermitte, M., and Dubois, P. M., 1978b, Prolactin secretion and synthesis in short and long-term organ culture of pituitary tumours from acromegalic patients, *Acta Endocrinol. (Copenhagen)* **87**:701.

Quadri, S. K., and Meites, J., 1973, Effect of ergocornin and CG 603 on blood-prolactin and GH in rats bearing a pituitary tumor (37128), *Proc. Soc. Exp. Bio. Med.* **142**:837.

Quijada, M., Illner, P., Krulich, L., and McCann, S. M., 1973, The effect of catecholamines on hormone release from anterior pituitaries and ventral hypothalami incubated *in vitro, Neuroendocrinology* **13**:151.

Rabinowitz, D., Merimee, T. J., Burgess, J. A., and Riggs, L., 1966, Growth hormone and insulin release after arginine: Indifference to hyperglycemia and epinephrine, *J. Clin. Endocrinol. Metab.* **26**:1170.

Racadot, J., Vila-Porcile, E., Olivier, L., and Peillon, F., 1975, Electron microscopy of pituitary tumours, in: *Progress of Neurological Surgery* (H. Krayenbühl, ed.), pp. 95–141, S. Karger, Basel.

Rubin, A. L., Levin, S. R., Bernstein, R. I., Tyrrell, J. B., Noacco, C., and Forsham, P. H., 1973, Stimulation of growth hormone by luteinizing hormone–releasing hormone in active acromegaly, *J. Clin. Endocrinol. Metab.* **37**:160.

Rutlin, E., Hang, E., and Torjesen, P. A., 1977, Serum thyrotrophin, prolactin and growth hormone response to TRH during oestrogen treatment, *Acta Endocrinol. (Copenhagen)* **84**:23.

Sawano, S., Baba, Y., Kokubu, T., and Ishizuka, Y., 1974, Effect of synthetic growth hormone releasing inhibiting factor on the secretion of growth hormone and prolactin in rats, *Endocrinol. Jpn.* **21**:399.

Schalch, D. S., 1967, The influence of physical stress and exercise on growth hormone and insulin secretion in man, *J. Lab. Clin. Med.* **69**:256.

Schally, A. V., Arimura, A., Bowers, C. Y., Wakabayashi, I., Kastin, A. J., Redding, T. W., Mittler, J. C., Nair, R. M. G., Rizzolato, P., and Segal, A. J., 1970, Purification of hypothalamic releasing hormones of human origin, *J. Clin. Endocrinol. Metab.* **31**:291.

Schams, D., and Reinhardt, V., 1973, Prolactin release in the bovine stimulated by synthetic TRH and inhibited by 2-Br-α-ergokryptine, *Acta Endocrinol. Suppl. (Copenhagen)* **177**:144.

Shaar, C. J., and Clemens, J. A., 1974, The role of catecholamines in the release of anterior pituitary prolactin *in vitro, Endocrinology* **95**:1202.

Siler, T. M., Morgenstern, L. L., and Greenwood, F. C., 1972, The release of prolactin and other peptides hormones from human anterior tissue cultures, in: *Lactogenic Hormones, Ciba Found. Symp.* (G. E. W. Wolstenholme and J. Knight, eds.), pp. 207–217, Churchill-Livingstone, London.

Siler, T. M., Vandenberg, G., Yen, S. S. C., Brazeau, P., Vale, W., and Guillemin, R., 1973, Inhibition of growth hormone release in humans by somatostatin, *J. Clin. Endocrinol. Metab.* **37**:632.

Siler-Khodr, T., Morgenstern, L. L., and Greewood, F. C., 1974, Hormone synthesis and release from human fetal adenohypophyses *in vitro, J. Clin. Endocrinol. Metab.* **39**:891.

Skyler, J. S., Rogol, A. D., Lovenberg, W., and Knazek, R. A., 1977, Characterization of growth hormone and prolactin produced by human pituitary in culture, *Endocrinology* **100**:283.

Smythe, G. A., 1977, The role of serotonin and dopamine in hypothalamic–pituitary function, *Clin. Endocrinol. (Oxford)* **7**:325.

Snyder, J., Hymer, W. C., and Wilfinger, W. W., 1978, Culture of human pituitary prolactin and growth hormone cells, *Cell Tissue Res.* **191**:379.

Steiner, A. L., Peake, G. T., Utiger, R. D., Karl, I. E., and Kipnis, D. M., 1970, Hypothalamic stimulation of growth hormone and thyrotropin release *in vitro* and pituitary 3′,5′-adenosine cyclic monophosphate, *Endocrinology* **86**:1354.

Strauch, G., Modigliani, E., and Bricaire, H., 1969, Growth hormone response to arginine in normal and hyperthyroid females under propranolol, *J. Clin. Endocrinol. Metab.* **29**:606.

Strauch, G., Pique, L., and Bricaire, H., 1970, Application de la méthode de séparation par le talc au dosage radioimmunologique de l'hormone somatotrope, *Ann. Biol. Clin. (Paris)* **28**:41.

Takahara, J., Arimura, A., and Schally, A. V., 1974a, Suppression of prolactin release by a purified porcine PIF and catecholamines infused into a rat hypophyseal portal vessel, *Endocrinology* **95**:462.

Takahara, J., Arimura, A., and Schally, A. V., 1974b, Effect of catecholamines on the TRH-stimulated release of prolactin and growth hormone from sheep pituitaries *in vitro, Endocrinology* **95**:1490.

Takahara, J., Arimura, A., and Schally, A. V., 1974b, Stimulation of prolactin and growth hormone release by TRH infused into a hypophysial portal vessel, *Proc Soc. Exp. Biol. Med.* **146**:831.

Talwaker, P. K., Ratner, A., and Meites, J., 1963, *In vitro* inhibition of pituitary prolactin synthesis and release by hypothalamic extracts, *Am. J. Physiol.* **205**:213.

Tashjian, A. H., Barowsky, N. J., and Jensen, D. K., 1971, Thyrotropin-releasing hormone: Direct evidence for stimulation of prolactin production by pituitary cells in culture, *Biochem. Biophys. Res. Commun.* **43**:516.

Thompson, K. W., Vincent, M. M., Jensen, P. C., Price, R. T., and Shapiro, E., 1959, Production of hormones by human anterior pituitary cells in serial culture, *Proc. Soc. Exp. Biol. Med.* **102**:403.

Thorner, M. O., Chait, A., Aitken, M., Benker, G., Bloom, S. M., Mortimer, G. H., Sanders, P., Stuart Mason, A., and Besser, G. M., 1975, Bromocriptine treatment of acromegaly, *Br. Med. J.* **1**:299.

Tixier-Vidal, A., 1975, Ultrastructure of anterior pituitary cells in culture, in: *The Anterior Pituitary* (A. Tixier-Vidal and M. G. Farquhar, eds.), Vol. 7, pp. 181–229, Academic Press, New York.

Torjesen, P. A., Hang, E., and Sand, T., 1973, Effect of thyrotropin releasing hormone on serum levels of pituitary hormones in men and women, *Acta Endocrinol. (Copenhagen)* **73**:455.

Udeschini, G., Cocchi, D., Panerai, A. E., Gil-Ad, I., Rossi, G. L., Chiodini, P. G., Liuzzi, A., and Müller, E. E., 1976, Stimulation of growth hormone release by thyrotropin-releasing hormone in hypophysectomized rat bearing an ectopic pituitary, *Endocrinology* **98**:807.

Vale, W., Blackwell, R., Grant, G., and Guillemin, R., 1973, TRF and thyroid hormones on prolactin secretion by rat anterior pituitary cells *in vitro, Endocrinology* **93**:26.

Vale, W., Rivier, C., Brazeau, P., and Guillemin, R., 1974, Effects of somatostatin on the secretion of thyrotropin and prolactin, *Endocrinology* **95**:968.

Verde, C., Oppizzi, G., Colussi, G., Cremascoli, G., Botalla, L., Müller, E. E., Silvestrini, F., Chiodini, P. G., and Liuzzi, A., 1976, Effect of dopamine infusion on plasma levels of growth hormone in normal subjects and in acromegalic patients, *Clin. Endocrinol. (Oxford)* **5**:419.

Woolf, P. D., Jacobs, L. S., Donofrio, R., Bureday, S. Z., and Schalch, D. S., 1974, Secondary hypopituitarism: Evidence of continuing regulation of hormone release, *J. Clin. Endocrinol. Metab.* **38**:71.

Yanai, R., and Nagasawa, 1974, Effect of 2-Br-α-ergocryptine on pituitary synthesis and release of prolactin and growth hormone in rats, *Horm. Res.* **5**:1.

Yen, S. S. C., Siler, T. N., and de Vane, G. W., 1974, Effect of somatostatin in patients with acromegaly: Suppression of growth hormone, prolactin, insulin and glucose levels, *N. Engl. J. Med.* **290**:935.

Zimmerman, E. A., Defendini, R., and Frantz, A. G., 1974, Prolactin and growth hormone in patients with pituitary adenomas: A correlative study of hormone in tumor and plasma by immunoperoxidase technique and radioimmunoassay, *J. Clin. Endocrinol. Metab.* **38**:577.

Zor, U., Kaneko, T., Schneider, H. P., McCann, S. M., Lowe, I. P., Bloom, G., Borland, B., and Field, J. B., 1969, Stimulation of anterior pituitary adenyl cyclase activity and adenosine 3':5'-cyclic phosphate by hypothalamic extract and prostaglandin E_1, *Proc. Natl. Acad. Sci. U.S.A.* **63**:918.

33

Use of Cell Separation to Analyze the Mechanisms of Action of GnRH

Jacques Duval, Michèle le Dafniet, and Laurence Benoist

We are at present engaged in the study of the molecular mechanisms by which gonadotropin-releasing hormone (GnRH) stimulates the release of the gonadotropins and how this promoted release can be modulated by sex steroids. Two questions are of particular interest to us: (1) Is cyclic AMP synthesis or degradation primarily affected by the hypothalamic factor in the gonadotrophs? (2) Are the receptors for sex steroids equally or selectively distributed among the different cell categories? To answer these questions, we were led to use fractionated cells.

Numerous procedures have been described to disperse cells; after preliminary experiments, we have retained the technique of Hymer *et al.* (1973), since trypsin gave us higher cell yields than the collagenase-based procedures. Using 40-day-old male rats, we regularly recover 1.2–1.4 million cells per pituitary with trypsin as against 300,000–500,000 with collagenase.

Following dispersion, separation was conducted exactly as described by Hymer *et al.* (1973) using 1*g* sedimentation through a 0.3–2.4% gradient of bovine serum albumin (BSA). Fractions of 30 ml were collected. To select the best conditions of separation, we used either intact males, males

Jacques Duval, Michèle le Dafniet, and *Laurence Benoist* • Equipe de Recherche Associée au C.N.R.S. No. 070567, Laboratoire de Neurobiologie Moléculaire, Université de Rennes, Campus de Beaulieu, 35042 Rennes, France

castrated 30 days before the experiment, or males pretreated with a daily dose of estradiol, 1 μg, for the 2 days preceding the killing. Lutropin (LH), follitropin (FSH), and prolactin (PRL) were determined using the radioimmunoassay kits of the NIAMDD; cyclic AMP (cAMP) content was estimated using the Becton Dickinson radioimmunoassay kit.

In Fig. 1, it can be seen that the cells are distributed throughout the gradient, PRL between fractions 5 and 10, LH and FSH from fraction 8 to 19. Castration does not significantly alter the patterns except for LH, the amount of which is greatly increased even in the smallest cells. The content of each hormone per cell has been estimated for these two particular experiments (Fig. 2); the cells recovered after fraction 11 or 12 are highly enriched in LH and FSH as compared to the concentration observed in the unfractionated population (arrow), whereas fractions 5–8 are only slightly enriched in PRL.

Since the gradient was loaded with different amounts of cells, we have expressed the percentage of cells or the percentage of hormone content recovered in each fraction. The average values from 3–6 independent experiments are reported in Fig. 3. It can be seen that gonadectomy hardly affects the cell distribution, whereas estradiol strongly displaces the cells toward the bottom of the gradient. It must be pointed out that the low doses of steroid used and the short duration of the treatment do not promote any significant change in pituitary hormone content (Valotaire *et al.*, 1975). While LH and FSH distributions are affected by

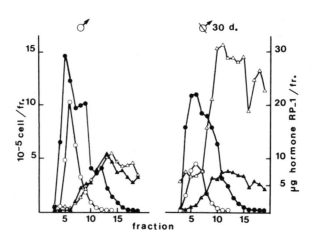

Figure 1. Separation of pituitary cells from 40-day-old male rats at unit gravity. ($\male$) Intact males; ($\male$) males gonadectomized 30 days before the experiment; (o) cell number per 30-ml fraction; (•) PRL content; (△) LH content; (▲) FSH content.

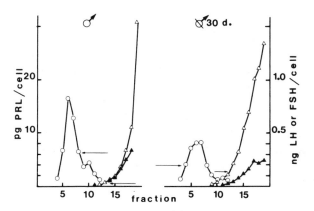

Figure 2. Separation of pituitary cells from 40-day-old male rats at unit gravity. The hormone content is expressed per cell: (○) PRL; (△) LH; (▲) FSH. The arrows indicate the cellular content of the unfractionated cells.

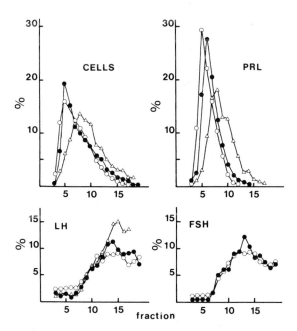

Figure 3. Separation of pituitary cells from 40-day-old male rats at unit gravity. For each parameter and for each fraction, the result represents the percentage of the total amount recovered. (●) Intact males; (○) males castrated 30 days before the experiment; (△) males injected with 1 μg estradiol during the 2 days preceding the killing. Upper left: cell distribution; upper right: PRL distribution; lower left: LH distribution; lower right: FSH distribution (the FSH was not determined for estradiol-pretreated rats). Each value is the average from 3–6 independent runs.

neither castration nor estrogen, the prolactin cells are enlarged and spread throughout the gradient after estradiol treatment.

We shall now report the results of several preliminary experiments concerning the cAMP content and the search for receptors for estradiol in the dispersed or separated cells.

It is well known that the cAMP content of the pituitary increases after the gland has been incubated with GnRH, but the time course and the physiological significance of such an increase are still a matter of controversy. In fact, as previously discussed in this meeting, the cAMP content rises significantly only after 2 hr when measured in the whole cell population, but one may expect a much earlier rise when working with isolated gonadotrophs.

Before performing the experiments on the separated cells, we checked that the GnRH was effective on the cells dispersed with trypsin. As shown in Table I, a dose of 100 ng/ml of GnRH promotes in 15 min an increase in the LH release but not in cAMP content; a 4-hr incubation induces a rise of both LH release and cAMP accumulation (date not shown).

We report in Table II the results of two separate experiments in which trypsin-dispersed cells were separated and immediately incubated 1 hr with or without GnRH. The "PRL" population was defined as fractions 3–9 and the "Gona" population as fraction 10 and beyond. The "PRL" was used as a control to show the very small amount of LH released

Table I. LH Release and cAMP Content from Dispersed
Cells Incubated with GnRH[a]

Exp. No.	LH released (μg/sample)		cAMP content (pmol/sample)	
	− GnRH	+ GnRH	− GnRH	+ GnRH
1	2.29	3.09	3.02	2.97
2	10.91	13.53	3.54	—
3	4.81	6.32	1.04	0.56
4	8.52	10.11	1.22	1.13
5	4.76	9.33	2.61	3.53
	(paired t test)			
	$P < 0.05$			

[a] After dispersion, the cells were kept overnight in an M199 medium containing 1% BSA. They were then spun down, suspended in M199 medium + 0.1% BSA, and distributed in several tubes. One half of these tubes (3–5 according to the experiment) were supplemented with 100 ng/ml of GnRH, the other half being used as control. The incubation was conducted at 37°C during 15 min. Then each medium was collected for LH determination, while the cells were treated to analyze their cAMP content. The reported value for each experiment represents the average value of the 3–5 independent assays.

Table II. LH Release and cAMP Content from Separated
Cells Incubated with GnRH[a]

Cell population	GnRH	LH released ($\mu g/10^6$ cells)	cAMP content (pmol/10^6 cells)
"PRL"	−	0.22	0.37
	+	0.22	0.25
"Gona"	−	6.59	1.45
	+	20.70	1.41
"Gona"	−	6.55	1.59
	+	9.84	1.45

[a] After sedimentation at unit gravity, the cells containing PRL were pooled ("PRL" population) as were the cells containing the gonadotropins ("Gona" population). Each population was divided into two parts, one receiving 100 ng/ml of GnRH, the second being used as control. After 1 hr at 37°C, the suspension was centrifuged; the medium and the cells were treated for LH and cAMP content, respectively. The two "Gona" populations come from two distinct separation runs.

during the incubation and the nonresponsiveness of the cells to the releasing hormone. It appears that the "Gona" cells release more LH under the GnRH stimulation, but that their cAMP content is not increased after the 60-min incubation. Obviously, these experiments are too preliminary to draw definite conclusions, but they do not support the hypothesis of a participation of cAMP in the release process. Other experiments are now in progress in our laboratory using different incubation time lengths. One may also notice in Table II that the "PRL" cells contain much less cAMP than the "Gona" cells; we do not know whether this reflects the difference in the cell sizes or in the secretory activity.

Concerning the search for estradiol receptors, three groups of experiments were performed: (1) incubation of the cell cytosol during 1 night at 0°C with radioactive estradiol so as to fill the cytoplasmic receptors; (2) incubation of freshly dispersed cells or of 3-day precultured cells at 25°C during 1 hr with radioactive estradiol (1.2×10^{-8}M) plus or minus unlabeled steroid (10^{-6} M) followed by analysis of the specific transfer to the nuclei; (3) dispersion of cells with trypsin at 37 or 25°C in the presence of the radioactive steroid, then analysis of the radioactivity in both the 800g pellet (nuclei) and supernatant.

In no case could we find evidence of cytoplasmic or nuclear estradiol receptors in the dispersed cells, though such receptors were clearly demonstrated when using pituitary cytosol or during incubation of intact glands, as shown in Fig. 4.

To conclude, it can be said that cell dispersion by trypsin followed by separation at unit gravity allows the recovery of cells respectively enriched in mammotrophs and in gonadotrophs. In the latter group, the GnRH-

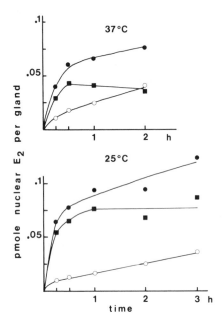

Figure 4. Kinetics of nuclear uptake of tritiated estradiol by rat pituitaries. Individual glands were incubated at 25 or 37°C in 0.5 ml Krebs–Ringer–bicarbonate–glucose (KRBG) solution in the presence of 1.2 × 10⁻⁸ M radioactive estradiol plus (o) or minus (•) 10⁻⁶ M unlabeled estradiol. At the various times indicated, the medium was discarded, each gland was rinsed with KRBG + cold steroid and then homogenized in 0.150 ml of a solution of 0.25 M sucrose + 3 mM MgCl₂. The crude 800g pellet was washed three times with the same solution. The amount of radioactivity was determined both in the 800g pellet and in the supernatant (not reported in this figure). The specific nuclear transfer (■) was calculated as the difference between the total nuclear radioactivity (•) and the nuclear nonspecific radioactivity (o).

promoted release of LH does not seem to be correlated with an intracellular increase of cAMP. Though it does not impede GnRH action, trypsin appears to suppress in freshly dispersed cells or in short-term-precultured cells the specific uptake of estradiol and its transfer to the nucleus.

Summary

After dispersion with trypsin, the cells of male rat pituitaries were separated using 1g sedimentation. The prolactin (PRL), lutropin (LH), and follitropin (FSH) distributions were analyzed by radioimmunoassays. The PRL distribution is strongly altered by a short estradiol pretreatment and slightly modified by gonadectomy, whereas LH and FSH distributions are unaffected by both treatments. Preliminary experiments show that the cAMP content of the gonadotrophs is not increased during a 1-hr incubation in the presence of GnRH, while the LH release is enhanced. The pituitary steroid receptors are altered during the trypsin dispersion process, since the specific uptake of estradiol and its transfer to the nucleus are suppressed and are not reinitiated even after 3 days of preculture.

DISCUSSION

TIXIER-VIDAL: I have a question relative to the absence of nuclear binding in freshly dissociated cells, as well as in precultured cells: did you use serum in your culture medium?

DUVAL: No serum was used in the incubation of the cells with [³H]estradiol. When cells were precultured, serum treated with dextran-coated charcoal was used.

JUTISZ: Dr. Denef uses pituitaries from 14-day-old female rats, and Dr. Duval, pituitaries from 14-day-old male rats. I would like to ask Dr. Denef whether pituitaries of 40-day-old male rats can also be used satisfactorily for the selection of gonadotropic cells?

DENEF: The recovery and purity of gonadotrophs from pituitaries of 40-day-old male rats are less satisfactory, but they can be used.

DENEF: Does pretreatment with estrogen differentially change the content of LH and FSH in any of the fractions, and does it change the proportional number of gonadotrophs?

DUVAL: The dose of estrogen that we use for the pretreatment (2×1 μg of estradiol) does not change significantly the LH and FSH contents of the pituitary. As to the proportion of gonadotrophs, we cannot give an answer since we did not use immunocyto-chemistry to determine the number of each cell type.

REFERENCES

Hymer, W. C., Evans, W. H., Kraicer, J., Mastro, A., Davis, J., and Griswold, E., 1973, Enrichment of cell types from the rat adenohypophysis by sedimentation at unit gravity, *Endocrinology* **92:**275.

Valotaire, Y., Le Guellec, R., Kercret, H., Guellaën, G., and Duval, J., 1975, Induction of rat pituitary thymidine kinase: Another physiological response to estradiol in the male?, *Mol. Cell. Endocrinol.* **3:**117.

34

TRH as MSH-Releasing Factor in the Frog

Marie-Christine Tonon, François Leboulenger, Catherine Delarue, Sylvie Jegou, Jean Fresel, Philippe Leroux, and Hubert Vaudry

1. Role of Pyro-Glu-His-Pro-NH$_2$ in Submammalian Chordates

1.1. Presence of Mammalian TRH in Amphibian Brain

The tripeptide pyroglutamyl-histidyl-prolinamide plays the role of thyrotropin-releasing hormone (TRH) (Schally et al., 1973; Vale et al., 1973) and a prolactin-releasing hormone in mammals (Bowers et al., 1971; Rivier and Vale, 1974). The development of sensitive and highly specific radioimmunoassays (RIAs) (Jackson and Reichlin, 1974; Oliver et al., 1974) has made possible the assessment of TRH in various brain regions in mammals. The brain of submammalian vertebrates contains a peptide capable of releasing thyrotropin (TSH) from rat pituitary thyrotrophs in vivo (Jackson and Reichlin, 1974) and in vitro (Taurog et al., 1974). Concurrently, on the basis of electrophoretic comigration (Taurog et al., 1974) and gel filtration (Jackson and Reichlin, 1974), it was established that a peptide, chemically

Marie-Christine Tonon, François Leboulenger, Catherine Delarue, Sylvie Jegou, Jean Fresel, Philippe Leroux, and *Hubert Vaudry* • Groupe de Recherche en Endocrinologie Moléculaire, Laboratoire d'Endocrinologie, Institut de Biochimie et Physiologie Cellulaire, Faculté des Sciences, Université de Haute-Normandie, 76130 Mont-Saint-Aignan, France

related to mammalian TRH, was present in the brain of amphibians. These results were confirmed by radioimmunological data that showed a strict parallelism between the competitive inhibition curves obtained with synthetic TRH and frog brain extracts (Taurog et al., 1974; Jackson and Reichlin, 1974).

1.2. Lack of Action of Mammalian TRH on Bird and Fish Thyrotrophs

There is increasing experimental evidence that synthetic TRH is unable to elicit the release of TSH in nonmammalian chordates.

In birds, no significant increase in radioiodine release was recorded following injections of TRH at doses up to 250-fold the minimum active thyroid-stimulating dose in mammals (Ochi et al., 1972).

In fish, the first evidence that TRH does not increase thyroxin secretion was obtained by Wildmeister and Horster (1971). These authors showed that the synthetic peptide had no effect on exophthalmos in *Carassius auratus*. In the lungfish *Protopterus ethiopicus*, Gorbman and Hyder (1973) reported that injections of huge amounts of TRH (100 μg/50 g body weight) did not stimulate ^{131}I uptake by the thyroid. Since bovine TSH was capable of increasing radioiodine input, they concluded that the lack of action of TRH could not be attributed to the lack of response of the thyroid gland, but was probably due to the insensitivity of fish pituitary thyrotrophs to the tripeptide pyro-Glu-His-Pro-NH$_2$. Later, Bromage (1975) showed that three intraperitoneal injections of TRH did not stimulate the thyroid gland of *Poecilia reticulata*. Furthermore, this author recorded a significant decrease in TSH release from cultured pituitary glands and concluded that, in teleosts, TSH secretion might be controlled by a hypothalamic inhibitory neurosecretion. Another piece of evidence for the lack of effect of mammalian TRH in fish has been reported by Tsuneki and Fernholm (1975), who did not notice any modification in thyroid histology after chronic injections of synthetic TRH. Recently, Dickhoff et al., (1978) have studied the effects of large doses of TRH on thyroxine (T$_4$) secretion from a coculture of pacific hagfish pituitary and thyroid. In this model, a small decrease in T$_4$ output was observed when the cocultures were treated with TRH. Thus, it appears that the tripeptide is without effect on fish thyrotrophs either *in vivo* or *in vitro*.

1.3. Lack of Action of Mammalian TRH on Amphibian Thyrotrophs

In 1967, Rémy and Bounhiol (1967) showed that, *in vivo*, TSH release was not altered by destruction of the diencephalon of *Alytes obstetricans* tadpoles. There is now strong evidence that TRH, administered *in vivo*, is unable to elicit the metamorphosis in amphibian tadpoles (Etkin and Gona, 1968; Gona and Gona, 1974) and neotenic larvae of Mexican axolotl (Taurog *et al.*, 1974), whereas metamorphosis could be induced by administration of physiological doses of T_4 or ovine TSH (Taurog, 1974).

Since TRH may stimulate prolactin release in amphibians (Clemons *et al.*, 1976) as it does in mammals (for a review, see Grant, G. F., and Vale, 1974), and since prolactin inhibits tadpoles' metamorphosis (Bern *et al.*, 1967; Etkin and Gona, 1967), it appeared that after administration of TRH, the increased prolactin secretion might conceal the metamorphic action of T_4. However, further studies have clearly demonstrated that synthetic TRH (1) failed to increase ^{125}I uptake by the frog thyroid *in vivo* (Vandesande and Aspeslagh, 1974); (2) failed to stimulate TSH release from axolotl pituitaries incubated *in vitro* (Taurog *et al.*, 1974); and (3) failed to modify the fine structure of thyrotrophs in the newt (Dunn and Dent, 1976).

From all these studies, it appears that, first, heterologous TRH (mammalian TRH: pyro-Glu-His-Pro-NH$_2$) is devoid of any action on the pituitary–thyroid axis in amphibians and fishes; second, homologous thyrotropin-releasing factor, if any, remains to be identified in these groups; and third, a molecule biologically and chemically similar to synthetic TRH is present in frog and fish encephalon. Therefore, it was concluded (Dickhoff *et al.*, 1978) that the tripeptide may have, in lower vertebrates, "another presumably primal or ancestral effect which remains for investigation."

1.4. Control of MSH Secretion by Pyro-Glu-His-Pro-NH₂ in Amphibians: A Daring Hypothesis

The recent finding of Jackson *et al.* (1977) that the control of TRH-like peptide in the frog pineal is influenced by the intensity of illumination led us to question whether TRH could be implicated in mimesis phenomenona. Indeed, in our laboratory, Leboulenger (1978) has recorded, during his ecological study, seasonal variations in skin color (pale in spring and dark in winter), which could be correlated to the seasonal differences in the TRH contents of frog hypothalamus and pineal reported by Jackson *et al.* (1977).

Our hypothesis has been enhanced by an earlier study of Yasuhara and Nakajima (1975), who have isolated and chemically characterized TRH in the skin of the Korean frog *(Bombina orientalis)*. The huge amount of TRH found in skin extract by these authors (40 μg/g wet weight) would explain the high concentrations of the peptide in blood (Jackson and Reichlin, 1977). Following administration of tritiated TRH, Jackson and Reichlin did not find reuptake of TRH by the skin from the blood, and concluded that most of the TRH present in the skin was likely synthesized *in situ* and that uptake of TRH from the blood, if any, must contribute only a small proportion of the whole skin TRH content. Furthermore, Jackson and Reichlin (1977) have investigated the local distribution of TRH throughout the frog skin and have found that the concentrations of peptide were much higher in pigmented areas than light-colored regions. This would indicate a possible interrelationship between TRH secretion and skin darkening. It was this most exciting event that prompted us to investigate the possible role of TRH in the control of melanocyte-stimulating hormone (MSH) release in amphibians.

2. Perfusion System Technique

Most of the studies dealing with *in vitro* release of mamalian intermediate lobe peptides have been carried out by means of static incubations with whole nervosa-intermedia (Briaud *et al.*, 1978) or acutely dispersed intermedia cells (Kraicer and Morris, 1976a,b). Static incubation has been extended to look into *in vitro* α-MSH release in poikilotherms. The perfusion system has been applied only to the investigation of melanotropin (Tilders *et al.*, 1975a,b) and corticotropin (ACTH) release from neurointermediate lobes in mammals (Briaud *et al.*, 1978), and has never been used in the study of frog intermediate lobe secretion.

2.1. Frog Neurointermediate Lobes

Frogs *(Rana ridibunda* Pallas) of about 30–50 g body weight were maintained at a constant temperature (16 ± 1°C) under running water, in darkness. The frogs were decapitated, and the neurointermediate lobes from the pituitary glands were carefully dissected under the microscope. Seven glands were removed within 10 min and transferred to Teflon beakers containing 5 ml of amphibian culture medium (ACM) according to Wolf and Quimby (1964) prepared by Eurobio (Paris) under moistened 95% O_2 and 5% CO_2. A preincubation was carried out for 10 min, and the glands were then brought together in the perfusion chamber.

2.2. Perfusion System Apparatus

A diagram of the perfusion apparatus employed is shown in Fig 1. The perfusion chamber (Fig. 1A) was composed of a siliconized glass tube (0.9 × 15 cm) with two Teflon plungers. The distance between the two plungers was 3 cm. Biogel P_2 was preswollen in 2 M acetic acid. The gel was then transferred to a glass funnel and washed with ACM by gentle

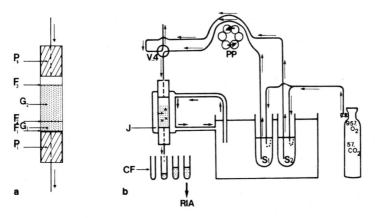

Figure 1. Perfusion apparatus. (a) Schema of the perfusion chamber. (F_1, F_2, F_3) Filter-paper disks; (G_1, G_2) first and second layers of Biogel; (P_1, P_s) lower plunger and upper plunger. (b) General schema of the perfusion system. (CF) Fraction collector; (J) water jacket; (PP) peristaltic pump; (S_1) ACM alone; (S_2) ACM containing a test substance (secretagogue); (V4) four-way valve.

stirring until the required pH 7.35 was reached. The lower plunger (P_1) was fitted into the glass column. A first filter-paper disk (F_1), a 5-mm layer of Biogel P_2 (G_1), a second filter-paper disk (F_2), the glands in suspension in Biogel P_2 (G_2), and a third filter-paper disk (F_3) were successively introduced. The upper plunger (P_s) was then inserted into the glass column and connected to a four-way valve (V_4), which permitted stimulation of the glands alternatively with various solutions. The catheter of the lower plunger was finally connected to an automatic fraction collector (CF); a constant flow rate of the effluent perfusate (0.35 ml/min) in the perfusion chamber was maintained throughout the experiment by means of a peristaltic pump (PP). The glands were perfused either with ACM alone (S_1) or a test substance (secretagogue) dissolved in ACM (S_2). The perfusion media were presaturated with moistened 95% O_2 and 5% CO_2 and were kept under this gas mixture during the entire run. The pH (7.35) of the medium was kept constant throughout all the experiment. The temperature was controlled by a flow of water circulating in a jacket (J) around the chamber. Fractions (0.7 ml) were set apart in 5-ml polystyrene tubes, and the amount of α-MSH in the fractions was radioimmunoassayed in duplicate.

In some experiments, the ACTH content of each fraction was measured by means of a specific and highly sensitive RIA method already described (Vaudry *et al.*, 1975).

3. Radioimmunoassay for α-Melanocyte-Stimulating Hormone

A large number of studies dealing with biosynthesis, release, and metabolism of melanotropic peptides has been carried out in vertebrates (for reviews, see Howe, 1973; Bagnara and Hadley, 1973). In most of these studies, biological assays were used to quantify the MSH contents in blood or tissue extracts. These assay methods are usually very sensitive, especially those using *in vitro* response of isolated skin fragments from amphibian or reptilian species. The sensitivity thresholds of the biological assays vary according to the origin of the skin and the authors: *Anolis-carolinensis:* 100 ng/ml (Thornton, 1971), 10–15 pg/ml (Björklund et al., 1972), 150 pg/ml (Tilders et al., 1975a), 1 ng/ml (Dickhoff, 1977), 146 pg/ml (Vaudry et al., 1978); *Rana pipiens:* 100 pg/ml (Thornton, 1971). However, despite their high sensitivity, bioassay methods bear three major disadvantages: low specificity, low efficiency, and intraassay variability. Compounds chemically related to α-MSH (ACTH, α-MSH, LPHs) are capable of inducing skin darkening (Shimizu et al., 1965; Silman et al., 1975; Tilders et al., 1975a; Vaudry et al., 1978). Therefore, bioassays cannot distinguish between MSH and MSH-related peptides. Moreover, numerous substances such as: catecholamines (Goldman and Hadley, 1970)—dopamine (Tilders et al., 1975b) and to a lesser extent epinephrine and norepinephrine (Taylor and Teague, 1976); ions (Novales and Novales, 1961; van de Veerdonk and Brouwer, 1973); cyclic AMP (Hadley and Goldman, 1969; Goldman and Hadley, 1970); melatonin (Hadley and Bagnara, 1969); prostaglandins (Novales and Novales, 1973; van de Veerdonk and Brouwer, 1973); progesterone (Himes and Hadley, 1971); and thyroid hormones (Wright and Lerner, 1960) would also interfere in biological assays. Thus, although bioassays really measure the melanotropic activities of the samples, these methods are not appropriate for the detection of α-MSH release from pituitary glands incubated with numerous substances. Therefore, many laboratories have tried to develop RIAs for α-MSH (Abe et al., 1967; Thody et al., 1975; Goos and Jenks, 1975; Scott and Baker, 1975; Usategui-Echeverria et al., 1975; Kopp et al., 1977; Penny and Thody, 1978). The quality of an RIA method depends mainly on the titer, avidity, and specificity of the available antibodies. Until 1976, none of the antibodies developed against α-MSH, presented all these characteristics. They had a low titer (Abe et al., 1967; Thody et al., 1975; Scott and Baker, 1975). Their sensitivity thresholds (Abe et al., 1967; Goos and Jenks, 1975) did not reach those of the best bioassays. Cross-reactions with some peptides such as β-MSH occurred (Thody et al., 1975) or were not studied (Abe et al., 1967; Scott and Baker, 1975; Goos and Jenks, 1975).

3.1. Development of α-MSH Antibodies

In our study, the carbodiimide method (Goodfriend *et al.*, 1964; McGuire *et al.*, 1965; Gelzer, 1968) has been applied to the conjugation of synthetic α-MSH to bovine serum albumin. The coupling efficiency, measured by incorporation of ^{125}I-labeled α-MSH as a tracer molecule, was 33% (Vaudry *et al.*, 1978). The first antibodies that could be used for RIAs were obtained after the third injection: the titer of the antisera ranged from 1:16,000 to 1:25,000. After the fifth injection, one of the rabbits gave antibodies that could bind 50% of 2800 cpm ^{125}I-labeled α-MSH (36 pg) at a dilution of 1:40,000. The titers of the other antisera ranged from 6.2×10^{-5} to 3.1×10^{-5}.

3.2. Evaluation of α-MSH Radioimmunoassay

The sensitivity thresholds of the antisera were in the picogram range. Half-maximal displacement of $[^{125}I]$-α-MSH was obtained at 80 pg unlabeled α-MSH. The cross-reactions of all the antisera with 13 molecules chemically related to ACTH were studied. Figures 2A and B represent the displacement of $[^{125}I]$-α-MSH antibody binding by serial dilutions of α-MSH and MSH analogues. The absence of significant cross-reactions with ACTH$^{1-10}$ and ACTH$^{11-19}$ fragments (0.002 and 0.00015%, respectively) shows that the antigenic determinant includes the region Gly^{10}-Lys^{11}. Since the sequence His^6 . . . -Val^{13} of the α-MSH molecule bears the whole bioactivity (Schwyzer and Li, 1958; Lee *et al.*, 1963), these antibodies will detect the biologically active region of the hormone. The foregoing results are of great interest, because one of the major criticisms of RIAs is that the amounts of immunoreactive hormone may be disproportionate to the amounts of biological activity. The cross-reactivity observed with ACTH$^{1-24}$ (the biologically active region of the ACTH molecule) and with synthetic human or porcine ACTH were lower than 0.1%. The only crossreacting compound was a natural porcine ACTH preparation that was probably contaminated by trace amounts of α-MSH. No cross-reaction was observed between the antibodies and the ACTH$^{17-39}$ fragment, which contains the entire sequence of corticotropin-like intermediate lobe peptide (CLIP), a molecule that has been characterized in the intermediate lobe of the rat, guinea pig, hog (Scott *et al.*, 1972, 1973a,b, 1974), and dogfish (Lowry *et al.*, 1974). Very weak cross-reactions occurred with ovine β-LPH (0.005%), human β-MSH (0.003%), and bovine β-MSH (0.001%). Thus, none of the numerous ACTH- or β-LPH-related hormones secreted by the intermediate lobe of the pituitary will interfere in the α-MSH assay,

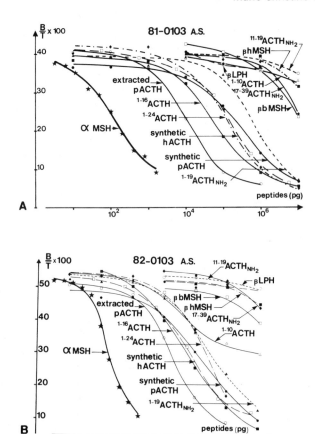

Figure 2. Specificity test for two α-MSH antibodies. Semilogarithmic plot, comparing inhibition of antibody-bound ¹²⁵I-labeled α-MSH and 12 natural or synthetic ACTH-related molecules. (A) Antiserum 81-0103; (B) antiserum 82-0103.

and the antibodies described above are suitable for the detection of α-MSH release from perfused intermediate lobes.

3.3. Cross-Reactions with Frog Pituitary Homogenates or Pituitary Released Substances

A peptide immunologically similar to α-MSH was found in the pars intermedia and even in the pars distalis of frog pituitary (Fig. 3); a strict parallelism was observed between the inhibition curves obtained with synthetic α-MSH and serial dilutions of frog pituitary homogenates. The total α-MSH contents of each lobe, measured by this RIA method, were 538 ± 164 ng in the intermediate lobe and 19.1 ± 6.4 ng in the distal lobe. These results are consistent with other studies that have already demonstrated, on the basis of gel filtration, comigration, and enzymatic degradation, that frog pituitary melanotropin was chemically identical to α-MSH (Vaudry *et al.*, 1976). However, Shapiro *et al.* (1972), using, simultaneously, bioassay and RIA methods, have shown that only a small proportion of melanotropic compound could be accounted for by α-MSH. The absence of competitive inhibition of antibody-bound ^{125}I-labeled human β-MSH by frog pituitary homogenates, in a specific and sensitive RIA method (Donnadieu and Sevaux, 1972), shows that no peptide immunologically similar to human (h) β-MSH could be detected in these extracts (Fig. 4). The inhibition curves obtained with synthetic α-MSH and serial dilutions of effluent perfusate from the perfusion chamber were exactly parallel (Fig. 5). Thus, accurate values of α-MSH concentrations could be determined in effluent fractions from the perfusion chamber, without prior extraction.

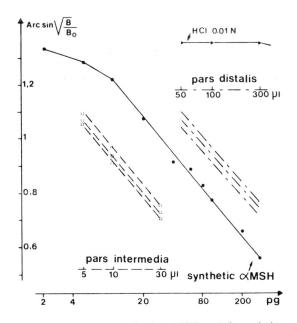

Figure 3. Cross-reactivity between synthetic α-MSH and frog pituitary homogenates. Comparison between competitive inhibition of antibody-bound [125I]-α-MSH by synthetic α-MSH and serial dilutions of frog pars distalis or pars intermedia extracts. From Vaudry *et al.* (1976).

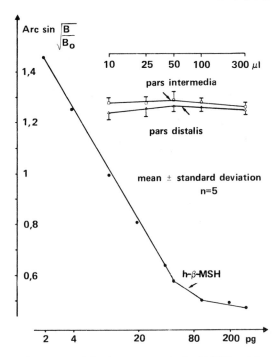

Figure 4. Absence of cross-reactivity between synthetic β-MSH and frog pituitary homogenates. Comparison between competitive inhibition of antibody-bound [^{125}I]-h-β-MSH by synthetic h-β-MSH and serial dilutions of frog pars distalis or pars intermedia extracts. From Vaudry *et al.* (1976).

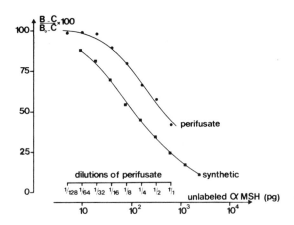

Figure 5. Cross-reactivity between synthetic α-MSH and the effluent perfusate. Comparison between competitive inhibition of antibody-bound [^{125}I]-α-MSH by synthetic α-MSH and serial dilutions of effluent perfusate.

4. Role of Hypothalamic Neuropeptides in α-MSH Release in Vitro

4.1. Lack of Action of Melanostatin$_I$, Luliberin, and Somatostatin

The tripeptide prolyl-leucyl-glycinamide (MIF$_I$), which has been first described as MSH-release-inhibiting factor (Nair et al., 1971; Celis et al., 1971b), has been applied to perifused neurointermediate lobes. No significant modification of α-MSH release could be observed for doses of MIF$_I$ ranging from 10^{-10} to 10^{-6} M (Table I). These negative results are consistent with previous studies in vivo (Thody et al., 1974; Donnadieu et al., 1976) as well as in vitro (Bower et al., 1971; Grant, N.H., et al., 1973; Thornton and Geschwind, 1975) that established the lack of inhibitory effect of MIF$_I$ on MSH secretion. Luliberin (LH-RH) and somatostatin (SRIF), used in the same dose-range, had no effect on α-MSH release (Table I).

ured by means of a specific and sensitive RIA. None of the three synthetic neuropeptides, in doses ranging from 10^{-10} to 10^{-6} M, could stimulate or inhibit ACTH secretion (Table I).

Table I. Effect of Increasing Doses of Synthetic Melanostatin, Luliberin, and Somatostatin on α-MSH and ACTH Release[a]

Addition		α-MSH released (% of control)	ACTH released (% of control)
MIF$_I$	10^{-10} M	97.4 ± 3.05	65.5 ± 35.80
	10^{-8} M	106.3 ± 2.96	56.0 ± 28.00
	10^{-6} M	89.9 ± 2.21	93.3 ± 29.70
LH-RH	10^{-10} M	91.0 ± 3.16	99.0 ± 14.00
	10^{-8} M	97.5 ± 6.89	127.5 ± 4.30
	10^{-6} M	99.5 ± 1.35	114.7 ± 0.09
SRIF	10^{-10} M	109.6 ± 2.77	129.9 ± 63.10
	10^{-8} M	91.5 ± 1.53	83.5 ± 9.00
	10^{-6} M	110.7 ± 0.60	121.9 ± 11.40

[a] Eight frog neurointermediate lobes were continuously perifused during 7 hr. Increasing doses of melanostatin (MIF$_I$), luliberin (LH-RH), and somatostatin (SRIF), ranging from 10^{-10} to 10^{-6} M, were infused during 10 min. Two consecutive doses of secretagogues were separated by a 90-min period. Each value represents the mean (± S.E.M.) α-MSH or ACTH release in five consecutive fractions following the passage of the secretagogue.

4.2. Stimulation of α-MSH Release by Thyroliberin

A significant increase in α-MSH release was observed when TRH was infused for 10 min in the perfusion chamber (Table II). When TRH was used at the dose of 0.1 nM, the increase in α-MSH output was not significant. The minimum effective dose was 1 nM (73.2% increase). A dose-related response occurred for doses ranging from 1 to 100 nM. The half-maximum dose was 10 nM. The infusion of 100 nM resulted in a 9.6-fold increase in α-MSH secretion.

A slight effect of TRH on ACTH release was also observed. However, the responses of the glands were not dose-related (Table II).

The effect of repeated doses of TRH has been also studied. In Table III are recorded the responses of neurointermediate lobes to 10 nM pulses of TRH. The amplitude of the response was independent of the duration of the stimulation. Conversely, the total increase in α-MSH release was directly related to the length of TRH infusion.

5. Effect of TRH Agonists on α-MSH Release in Vitro

The MSH-releasing potency of various TRH analogues has been studied using the perfusion system technique. For each analogue, the

Table II. Effect of Increasing Doses of
Synthetic TRH on α-MSH and ACTH Release[a]

TRH concentration	α-MSH release (% of control)	ACTH release (% of control)
10^{-10} M	113.3 ± 6.9[b]	93.4 ± 22.80[b]
10^{-9} M	172.3 ± 76.0[c]	115.7 ± 7.70[b]
10^{-8} M	390.0 ± 119.9[c]	146.1 ± 2.20[c]
10^{-7} M	936.1 ± 195.8[d]	120.6 ± 2.40[d]
10^{-6} M	746.4 ± 146.8[d]	118.2 ± 4.10[b]

[a] Ten frog neurointermediate lobes were continuously perfused during 9 hr. Increasing doses of TRH, ranging from 10^{-10} to 10^{-6} M, were infused during 10 min. Two consecutive doses of secretagogue were separated by a 90-min period. Each value represents the mean (± S.E.M.) α-MSH or ACTH release in five consecutive fractions following the passage of the secretagogue.
[b-d] Significance: [b]not significant; [c]$0.05 > p > 0.01$; [d]$0.01 > p > 0.001$.

Table III. Effect of Repeated Doses of Synthetic TRH on α-MSH
Release[a]

| TRH (10⁻⁸ M) | α-MSH release | |
	Amplitude of the response (% of control)[b]	Increase in α-MSH release (pg/lobe)[c]
10 min	306.3 ± 26.9	3,519
10 min	241.9 ± 14.9	3,075
10 min	222.2 ± 9.0	1,566
10 min	193.3 ± 9.1	9,152
20 min	155.0 ± 8.7	10,244
40 min	196.8 ± 8.1	17,612

[a] Six frog neurointermediate lobes were continuously perfused during 7 hr. A constant dose of TRH (10^{-8} M) was infused during a constant period of 10 min or during varying periods (10, 20, and 40 min).
[b] Mean (± S.E.M.) α-MSH release in five consecutive fractions collected 10 min after the beginning of TRH infusion.
[c] Total increase in α-MSH release (surfaces under the peaks).

MSH-releasing activity has been compared to the MSH-releasing activity of synthetic TRH and to its TSH-releasing activity, as published by several authors, in mammals (Table IV). None of the free amino acids of the TRH molecule alone could elicit α-MSH secretion. Prolinamide even had a slight inhibitory effect on α-MSH release. We have also found that limited modifications of the TRH molecule abolished the action of the peptide. Since substitution of histidine by glycine (pyro-Glu-Gly-Pro-NH₂) suppressed MSH-releasing activity, it seemed that the biologically active determinant was the imidazole ring of histidine. This result was confirmed by the fact that other analogues that possessed the imidazole ring (such as pyro-Glu-His-Pro-Ala-NH₂ or pyro-Glu-Pyr₃-Ala-Pro-NH₂) retained a slight activity. The low activity of (D-His)²-TRH was also in agreement with a predominant role of histidine for TRH activity. When a lateral chain was added on proline, the activity of the molecule was either reduced (pyro-Glu-His-Pro-n-hexylamide, pyro-Glu-His-Pro-cyclohexylamide) or suppressed (pyro-Glu-His-Pro-β-phenylethylamide). One of the most potent analogues is a tripeptide bearing a methyl radical on nitrogen 3 of the imidazole ring of histidine. This analogue [(3 Me-His)²-TRH] has been identified as a superagonist *in vivo* in mammals (Vale *et al.*, 1971), whereas in our system, it has been found 6.3-fold less active than synthetic TRH. Conversely, addition of a methyl radical on nitrogen 1 suppresses both MSH- and TSH-releasing activity of the analogue.

Table IV. Comparison between the MSH-Releasing Activity and the TSH-Releasing Activity of Various TRH Analogues

TRH analogue	MSH-releasing activity as compared to synthetic TRH	TSH-releasing activity as compared to synthetic TRH	Reference
L-P-Glu	0.4%	—	—
L-His	0.3%	—	—
L-Pro-NH$_2$	−0.9%	—	—
P-Glu-His-Pro-NH$_2$	100%	100%	G. F. Grant and Vale (1974)
P-Glu-Gly-Pro-NH$_2$	−2.2%	0.005%	Vale et al. (1973)
		0	Castensson (1977)
P-Glu-β-Ala-Pro-NH$_2$	0.1%	—	—
P-Glu-Lys-Pro-NH$_2$	2.8%	0.02%	Vale et al. (1973)
P-Glu-Pyr$_3$-Ala-Pro-NH$_2$	3.7%	5%	Hofmann and Bowers (1970)
P-Glu-D-His-Pro-NH$_2$	5.3%	3%	Hinkle et al. (1974)
P-Glu-His-Pro-Ala-NH$_2$	2.6%	0.5%	Vale et al. (1973)
P-Glu-His-Pro-n-hexylamide	4.9%		—
P-Glu-His-Pro-β-phenylethylamide	1.3%		—
P-Glu-His-Pro-cyclohexylamide	15.8%		
P-Glu-1-Me-His-Pro-NH$_2$	1.6%	0.04%	Vale et al. (1973)
P-Glu-3-Me-His-Pro-NH$_2$	14.2%	800%	Vale et al. (1971, 1973)

6. Conclusions

The existence of a melanocyte-stimulating-hormone-releasing factor (MRF) has been demonstrated in the median eminence (Taleisnik and Orias, 1965) and the hypothalamus of mammals (Schally and Kastin, 1966; Taleisnik *et al.*, 1966). In the rat, Celis *et al.* (1971a) have indicated that the pentapeptide Cys-Tyr-Ile-Gln-Asn, a fragment of oxytocin, would stimulate α-MSH release. However, this pentapeptide seems to be devoid of effect in lizards (Thorton and Geschwind, 1975). From our data, it emerges that mammalian TRH is capable of stimulating α-MSH release from frog intermediate lobes *in vitro*. Since Kraicer (1977) did not find any effect of TRH on α-MSH or ACTH release from rat pituitaries *in vitro*, it appears that TRH is a specific MRF in amphibians.

The existence of specific TRH receptors in frog pars intermedia cells and the presence of huge amounts of TRH-like substance in frog skin suggest that this neuropeptide may be involved in the control pigmentation in lower vertebrates.

7. Summary

Melanotropic peptides play a major role in nonmammalian chordates, in relation to background color adaptation. Earlier studies have pointed out the existence of a hypothalamic control of melanocyte-stimulating hormone (MSH) secretion in amphibians. Most of these studies have been carried out *in vivo*, and the α-MSH contents of plasma or tissue extracts have been measured by means of biological assay techniques that are extremely useful for the determination of true MSH contents, but are influenced by related compounds (e.g., ACTH, LPH) or nonrelated compounds (e.g., catecholamines, cAMP, testosterone, melatonin). The development of a sensitive and highly specific RIA technique for α-MSH, which makes it possible to assess the concentrations of the hormone in a large number of samples, has been achieved. The most important advantage of the RIA technique consisted in the absence of any interference with test substances such as cAMP or prostaglandins, which would alter MSH-induced skin darkening and modify the response of the test skin to MSH in the course of bioassays.

To study the dynamics of melanotropin release in amphibians, isolated neurointermediate lobes from *Rana ridibunda* pituitary glands were continuously perfused with amphibian culture medium to test the MSH-releasing potency of various substances. When the temperature of the perfusing medium was raised from 5 to 30°C, the MSH secretion rate increased 5-fold, suggesting a possible direct effect of the ambient temperature on melanotropin release in the frog. The tripeptide Pro-Leu-Gly-NH_2, which was claimed to be a melanocyte-stimulating-hormone-release-inhibiting factor (MIF_l), somatostatin (SRIF), and luliberin (LH-RH), in doses ranging from 10^{-10} to 10^{-6} M, had no effect on α-MSH release. Conversely, an increase in α-MSH was observed when thyroliberin (TRH) was added to the perfusion medium. At a dose of 10^{-8} M, TRH induced a 3.9-fold increase in α-MSH output. To check the specificity of TRH receptors in frog pars intermedia, several TRH analogues were tested for their ability to enhance MSH secretion. Compared to synthetic TRH, the relative potencies of the most reactive analogues, $(3\text{-Me-His})^2$-TRH, (Pro-cyclohexylamide)^3-TRH, and P-Glu-D-His-Pro-NH_2, were 14.2, 15.8, and 5.3%, respectively. The present findings, together with the immunocharacterization of TRH in the skin of frog, suggest a possible feedback loop between skin TRH and Pituitary MSH in amphibians.

ADDENDUM.   In order to keep this chapter as closely up to date as possible, it should be mentioned that the presence of immunoreactive TRH has been recently confirmed in the skin of *Rana esculenta* (Giraud *et al.*, *C. R. Acad. Sci.* **288**:127–129, 1979). Concurrently TRH has been studied *in vitro* in bullfrog (Clemons *et al.*, *Gen. Comp. Endocrinol.* **38**:62–67, 1979). Thus, it appears that, in Amphibia, hypothalamic TRH would control prolactin secretion via portal vessels and that TRH released by the skin might stimulate MSH secretion via systemic circulation.

ACKNOWLEDGMENTS.   We are indebted to Drs. P.A. Desaulles, A. Johl, and W. Rittel (Ciba, Basel); K. Inouye (Shionogi, Osaka); Schnartz-Mann and G. Fekete (Richter, Budapest); C.H. Li (San Francisco); and M. Chrétien (Montreal) for their generous gifts of various ACTH and LPH analogues. We are grateful to Drs. R. Guillemin and N. Ling (Salk Institute, San Diego); I. Werner and D. Gillessen (Hoffman-Laroche, Bâle); and A. Spriet and R. Geiger (Hoechst, Frankfurt) for kindly supplying TRH and TRH analogues. The authors wish to thank J.P. Morin and J. Lecourt for their excellent technical assistance. The present study was supported by a DGRST grant (No. 78 7 2004) and a grant from Choay Laboratories. M. C. Tonan, S. Jegou, and P. Leroux were recipients of predoctoral fellowships from DGRST.

DISCUSSION

KRAICER: Dr. Vaudry very kindly told us of his results as soom as he obtained them. We tested the effect of synthetic TRH on α-MSH and ACTH release from the rat pars intermedia, using similar times and doses. We found no change in α-MSH or ACTH release. Thus, this seems to be a species-specific effect. I must add, however, that we used bioassay rather than RIA and static incubation rather than perfusion.

VAUDRY: We feel that the difference in response is a species difference. In the frog, TRH does not stimulate TSH secretion. That is an indication that species differences occur in the action of TRH.

FINK: In view of the fact that your anti-MSH serum cross-reacted with ACTH[6-13], are you sure that you had no cross-reaction with prolactin?

VAUDRY: The main purpose of our specificity tests was to make sure that the antibodies did not cross-react with any peptides that have been found in the intermediate lobe, such as ACTH, LPH, and CLIP. We did not worry about prolactin because there is no evidence that it is part of the endocrine content of intermediate lobe cells. We also wanted to prove that the antibodies were raised against the biologically reactive region of the α-MSH molecule (ACTH[6-13]). As far as endorphins are concerned, we have not studied their cross-reactivity in our α-MSH radioimmunoassay system. Cross-reactions are very unlikely to occur, since α-MSH and β-endorphin have no sequence in common.

KORDON: Did you measure hypothalamic dopamine in your frogs, and did you test the effect of dopamine antagonist in your system?

VAUDRY: We did not measure the dopamine contents of the hypothalamic homogenates. We intend to study dopamine antagonists. We are also planning to remove rat dopamine by using Al_2O_3.

REFERENCES

Abe, K., Island, D. P., Liddle, G. W., Fleischer, N., and Nicholson, W. E., 1967, Radioimmunologic evidence for α-MSH (melanocyte stimulating hormone) in human pituitary and tumor tissues, *J. Clin. Endocrinol.* **27**:46.

Bagnara, J. T., and Hadley, M. E., 1973, *Chromatophores and Color Change: The Comparative Physiology of Animal Pigmentation,* Prentice-Hall, Englewood Cliffs, New Jersey.

Bern, H. A., Nicoll, C. S., and Strohman, R. C., 1967, Prolactin and tadpole growth, *Proc. Soc. Exp. Biol. Med.* **126**:518.

Björklund, A., Meurling, P., Nilsson, G., and Nobin, A., 1972, A Standardization and evaluation of a sensitive and convenient assay for melanocyte stimulating hormone using *Anolis* skin *in vitro, J. Endocrinol.* **53**:161.

Bower, S. A., Hadley, M. E., and Hruby, V. J., 1971, Comparative MSH release–inhibiting activities of tocinoic acid (the ring of oxytocin) and L-Pro-L-Leu-Gly-NH₂ (the side chain of oxytocin), *Biochem. Biophys. Res. Commun.* **45**:1185.

Bowers, C. Y., Friesen, H., Guyda, H. J., and Folkers, K., 1971, Prolactin and thyrotropin release in man by synthetic pyroglutamyl-histidyl-prolinamide, *Biochem. Biophys. Res. Commun.* **45**:1033.

Briaud, B., Koch, B., Lutz-Bucher, B., and Mialhe, C., 1978, *In vitro* regulation of ACTH release from neurointermediate lobe of rat hypophysis. I. Effect of crude hypothalamic extracts, *Neuroendocrinology* **25**:47.

Bromage, N. R., 1975, The effects of mammalian thyrotropin-releasing hormone on the pituitary–thyroid axis of teleost fish, *Gen. Comp. Endocrinol.* **25**:292.

Castensson, S., 1977, Structure–activity relationship studies on thyroliberin and investigations of the inhibition of an oxotremorine-induced tremor by melanostatin, Doctoral thesis, University of Uppsala, Sweden (ISBN 91-554-0598-3).

Celis, M. E., Taleisnik, S., and Walter, R., 1971a, Release of pituitary melanocyte-stimulating hormone by the oxytocin fragment, H-Cys-Tyr-Ile-Gln-Asn-OH, *Biochem. Biophys. Res. Commun.* **45**:564.

Celis, M. E., Taleisnick, S., and Walter, R., 1971b, Regulation of formation and proposed structure of the factor inhibiting the release of melanocyte-stimulating hormone, *Proc. Natl. Acad. Sci. U. S. A.* **68**:1428.

Clemons, G. K., Russel, S. M., and Nicoll, C. S., 1976, Effects of thyrotropin releasing hormone (TRH) and ergotamine on prolactin (PRL) secretion *in vitro* by bullfrog anterior pituitaries, Vth International Congress of Endocrinology, Hamburg, Abstract No. 809, p. 334.

Dickhoff, W. W., 1977, A rapid, high-efficiency bioassay of melanocyte-stimulating hormone, *Gen. Comp. Endocrinol.* **33**:304.

Dickhoff, W. W., Crim, J. W., and Gorbman, A., 1978, Lack of effect of synthetic thyrotropin releasing hormone on pacific hagfish *(Eptatretus stouti)* pituitary–thyroid tissues *in vitro, Gen. Comp. Endocrinol.* **35**:96.

Donnadieu, M., and Sevaux, D., 1972, Dosage radioimmunologique de la β MSH, in: *Techniques Radioimmunologiques,* pp. 367–374, INSERM, Paris.

Donnadieu, M., Laurent, M. F., Luton, J. P., Bricaire, H., Girard, F., and Binoux, M., 1976, Synthetic MIF no effect on β MSH and ACTH hypersecretion in Nelson's syndrome, *J. Clin. Endocrinol. Metab.* **42**:1145.

Dunn, A. D., and Dent, J. N., 1976, On the identity of the thyrotropic cell in the red-spotted newt, *Cell Tissue Res.* **173**:483.

Etkin, W., and Gona, A. G., 1967, Antagonism between prolactin and thyroid hormone in amphibian development, *J. Exp. Zool.* **165**:249.

Etkin, W., and Gona, A. G., 1968, Failure of mammalian thyrotropin-releasing factor preparation to elicit metamorphic responses in tadpoles, *Endocrinology* **82**:1067.

Gelzer, J., 1968, Immunochemical study of β-corticotropin (1–24) tetracosapeptide, *Immunochemistry* **5**:23.

Goldman, J. M., and Hadley, M. E., 1970, Cyclic AMP and adrenergic receptors in melanophore responses to methylxanthines, *Eur. J. Pharmacol.* **12**:365.

Gona, A. G., and Gona, O., 1974, Failure of synthetic TRF to elicit metamorphosis in frog tadpoles or red-spotted newts, *Gen. Comp. Endocrinol.* **24**:223.

Goodfriend, T., Levine, L., and Fasman, G., 1964, Antibodies to bradykinin and angiotensin: A use of carbodiimide in immunology, *Science* **144**:1344.

Goos, H. J. T., and Jenks, B., 1975, Radioimmunoassay for melanocyte-stimulating hormone (α-MSH), *Proc. Med. Akad. Wet.* **78**:69.

Gorbman, A., and Hyder, M., 1973, Failure of mammalian TRH to stimulate thyroid function in the lungfish, *Gen. Comp. Endocrinol.* **20**:588.

Grant, G. F., and Vale, W., 1974, Hypothalamic control of anterior pituitary hormone secretion—characterized hypothalamic–hypophysiotropic peptides, in: *Current Topics in Experimental Endocrinology* (V. H. T. James and L. Martini, eds.), pp. 32–72, Academic Press, New York.

Grant, N. H., Clark, D. E., and Rosanoff, E. I., 1973, Evidence that Pro-Leu-Gly-NH$_2$, tocinoic acid, and des-Cys-tocinoic acid do not affect secretion of melanocyte stimulating hormone, *Biochem. Biophys. Res. Commun.* **51**:100.

Hadley, M. E., and Bagnara, J. T., 1969, Integrated nature of chromatophore responses in the *in vitro* frog skin bioassay, *Endocrinology* **84**:69.

Hadley, M. E., and Goldman, J. M., 1969, Effects of cyclic 3',5' AMP and other adenine nucleotides on the melanophores of the lizard *(Anolis carolinensis), Br. J. Pharmacol.* **37**:65.

Himes, P. J., and Hadley, M. E., 1971, *in vitro* effects of steroid hormones on vertebrate melanophores, *J. Invest. Dermatol.* **57**:337.

Hinkle, P. M., Woroch, E. L., and Tashjian, A. H., 1974, Receptor-binding affinities and biological activities of analogs of thyrotropin releasing hormone in prolactin-producing pituitary cells in culture, *J. Biol. Chem.* **249**:3085.

Hofmann, K., and Bowers, C. Y., 1970, XLVII. Effect of pyrazole-imidazole replacement on the biological activity of thyrotropin releasing hormone, *J. Med. Chem.* **13**:1099.

Howe, A., 1973, The mammalian pars intermedia: A review of its structure and function, *J. Endocrinol.* **59**:385.

Jackson, I. M. D., and Reichlin, S., 1974, Thyrotropin-releasing-hormone (TRH): Distribution in hypothalamic and extrahypothalamic brain tissues of mammalian and submammalian chordates, *Endocrinology* **95**:854.

Jackson, I. M. D., and Reichlin, S., 1977, The skin is a massive TRH secreting organ in the frog, *Endocrinology* **100**, abstract 140.

Jackson, I. M. D., Saperstein, R., and Reichlin, S., 1977, Thyrotropin releasing hormone (TRH) in pineal and hypothalamus of the frog: Effect of season and illumination, *Endocrinology* **100**:97.

Kopp, H. G., Eberle, A., Vitins, P., Lichtensteiger, W., and Schwyzer, R., 1977, Specific antibodies against α-melanotropin for radioimmunoassay, *Eur. J. Biochem.* **75**:417.

Kraicer, J., 1977, Thyrotropin releasing hormone does not alter the release of melanocyte stimulating hormone or adrenocorticotropic hormone from the rat pars intermedia, *Neuroendocrinolgy* **24**:226.

Kraicer, J., and Morris, A. R., 1976a, *In vitro* release of ACTH from dispersed rat pars intermedia cells. I. Effect of secretagogues, *Neuroendocrinology* **20**:79.

Kraicer, J., and Morris, A. R., 1976b, *In vitro* release of ACTH from dispersed rat pars intermedia cells. II. Effect of neurotransmitter substances, *Neuroendocrinology* **21**:175.

Leboulenger, F., 1978, Contribution à l'étude des facteurs de régulation de l'activité interrenalienne chez les grenouilles vertes, Thèse de 3ème cycle (Octobre 1978), Université de Rouen, France.

Lee, T. H., Lerner, A. B., and Buettner-Janusch, V., 1963, Species differences and structural requirements for melanocyte-stimulating activity of melanocyte-stimulating hormone, *Ann. N. Y. Acad. Sci.* **100**:658.

Lowry, P. J., Bennett, H. P. J., and McMartin, C., 1974, The isolation and amino-acid sequence of an adrenocorticotrophin from the pars distalis and a corticotrophin-like intermediate-lobe peptide from neurointermediate lobe of the pituitary of the dogfish *Squalus ancanthias, Biochem. J.* **141**:427.

McGuire, J., McGill, R., Leeman, S., and Goodfriend, L., 1965, The experimental generation of antibodies to α melanocyte stimulating hormone and adrenocorticotropic hormone, *J. Clin. Invest.* **44**:1672.

Nair, R. M. G., Kastin, A. J., and Schally, A. V., 1971, Isolation and structure of hypothalamic MSH release-inhibiting hormone, *Biochem. Biophys. Res. Commun.* **43**:1376.

Novales, R. R., and Novales, B. J., 1961, Sodium dependence of intermedin action on melanophores in tissue culture, *Gen. Comp. Endocrinol.* **1**:134.

Novales, R. R., and Novales, B. J., 1973, Sodium free and cytochalasin B inhibition of prostaglandin A_2 action on amphibian chromatophores, *Am. Zool.* **13**:1277.

Ochi, J., Shiomi, K., Hachiya, T., Yoshimura, M., and Miyazaki, T., 1972, Failure of TRH (thyrotropin-releasing hormone) to stimulate thyroid function in the chick, *Endocrinology* **91**:832.

Oliver, C., Eskay, R. L., Ben-Jonathan, N., and Porter, J. C., 1974, Distribution and concentration of TRH in the rat brain, *Endocrinology* **95**:540.

Penny, R. J., and Thody, A. J., 1978, An improved radioimmunoassay for α-melanocyte stimulating hormone (α-MSH) in the rat: Serum and pituitary α-MSH levels after drugs which modify catecholaminergic neurotransmission, *Neuroendocrinology* **25**:193.

Rémy, C., and Bounhiol, J. J., 1967, Activité thyroidienne des têtards *d'Alytes obstetricans* privés de diencéphale ou du complexe hypothalamo–hypophysaire et traités, ou non, par le thiouracile, *Gen. Comp. Endocrinol.* **9**:519 (abstract A37).

Rivier, C., and Vale, W., 1974, *In vivo* stimulation of prolactin secretion in the rat by thyrotropin-releasing factor, related peptides and hypothalamic extracts, *Endocrinology* **95**:978.

Schally, A. V., and Kastin, A. J., 1966, Purification of a bovine hypothalamic factor which elevates pituitary MSH levels in rats, *Endocrinology* **79**:768.

Schally, A. V., Arimura, A., and Kastin, A. J., 1973, Hypothalamic regulatory hormones, *Science* **179**:341.

Schwyzer, R., and Li, C. H., 1958, A new synthesis of the pentapeptide L-histidyl-L-phenylanyl-L-arginyl-L-tryptophyl-glycine and its melanocyte-stimulating activity, *Nature* **4650**:1169.

Scott, A. P., and Baker, B. I., 1975, ACTH production by the pars intermedia of the rainbow trout pituitary, *Gen. Comp. Endocrinol.* **27**:193.

Scott, A. P., Rees, L. H., Ratcliffe, J. G., and Besser, G. M., 1972, Corticotrophin-like peptide concentrations in the intermediate lobe of rat and guinea pig pituitaries, *J. Endocrinol.* **53**:38.

Scott, A. P., Bennett, H. P. J., Lowry, P. J., McMartin, C., and Ratcliffe, J. G., 1973a, Characterization of corticotrophin-like intermediate lobe peptides from rat and pig pituitaries and identification of α-melanocyte stimulating hormone in the rat pituitary, *J. Endocrinol.* **58**:15.

Scott, A. P., Ratcliffe, J. G., Rees, L. H., and Landon, J., 1973b, Pituitary peptide, *Nature (London) New Biol.* **244**:65.

Scott, A. P., Lowry, P. J., Bennett, H. P. J., McMartin, C., and Ratcliffe, J. G., 1974, Purification and characterization of porcine corticotrophin-like intermediate lobe peptide, *J. Endocrinol.* **61**:369.

Shapiro, M., Nicholson, W. E., Orth, D. N., Mitchell, W. M., Island, D. P., and Liddle, G. W., 1972, Preliminary characterization of the pituitary melanocyte stimulating hormones of several vertebrate species, *Endocrinology* **90**:249.

Shimizu, N., Ogata, E., Nicholson, W. E., Island, D. P., Ney, R. L., and Liddle, G. W., 1965, Studies on the melanotropic activity of human plasma and tissues, *J. Endocrinol.* **25**:984.

Silman, R. E., Chard, T., Rees, L. H., and Smith, I., 1975, Observations on melanocyte-stimulating hormone-like peptides in human maternal plasma during late pregnancy, *J. Endocrinol.* **65**:46P.

Taleisnik, S., and Orias, R., 1965, A melanocyte-stimulating hormone-releasing factor in hypothalamic extracts, *Am. J. Physiol.* **208**:293.

Taleisnik, S., Orias, R., and de Olmos, J., 1966, Topographic distribution of the melanocyte-stimulating hormone–releasing factor in rat hypothalamus; *Proc. Soc. Exp. Biol. Med.* **122**:325.

Taurog, A., 1974, Effect of TSH and long-acting thyroid stimulator on thyroid ^{131}I-metabolism and metamorphosis of the Mexican axolotl *(Ambystoma mexicanum)*, *Gen. Comp. Endocrinol.* **24**:257.

Taurog, A., Oliver, C., Eskay, R. L., Porter, J. C., and McKenzie, J. M., 1974, The role of TRH in the neoteny of the Mexican axolotl *(Ambystoma mexicanum)*, *Gen. Comp. Endocrinol.* **24**:267.

Taylor, S. E., and Teague, R. S., 1976, The beta adrenergic receptors of chromatophores of the frog, *Rana pipiens*, *J. Pharmacol. Exp. Ther.* **199**:222.

Thody, A. J., Shuster, S., Plummer, N. A., Bogie, W., Leigh, R. J., Goolamali, S. K., and Smith, A. G., 1974, The lack of effect of MSH release inhibiting factor (MIF) on the secretion of β-MSH in normal men, *J. Clin. Endocrinol. Metab.* **38**:491.

Thody, A. J., Penny, R. J., Clark, D., and Taylor, C., 1975, Development of a radioimmunoassay for α melanocyte-stimulating hormone in the rat, *J. Endocrinol.* **67**:385.

Thornton, V. F., 1971, The effect of change of background color on the melanocyte stimulating content of the pituitary of *Xenopus laevis*, *Gen. Comp. Endocrinol.* **17**:554.

Thornton, V. F., and Geschwind, I. I., 1975, Evidence that serotonin may be a melanocyte stimulating hormone releasing factor in the lizard, *Anolis carolinensis*, *Gen. Comp. Endocrinol.* **26**:346.

Tilders, F. J. H., and Mulder, A. H., 1975, *In vitro* demonstration of melanocyte-stimulating hormone release inhibiting action of dopaminergic nerve fibres, *J. Endocrinol.* **64**:63P.

Tilders, F. J. H., van Delft, A. M. L., and Smelik, P. G., 1975a, Re-introduction and evaluation of an accurate, high capacity bioassay for melanocyte-stimulating hormone using the skin of *Anolis carolinensis* in vitro, *J. Endocrinol.* **66**:165.

Tilders, F. J. H., Mulder, A. H., and Smelik, P. G., 1975b, On the presence of a MSH-release inhibition system in the rat neurointermediate lobe, *Neuroendocrinology* **18**:125.

Tsuneki, K., and Fernholm, B., 1975, Effect of thyrotropin-releasing hormone on the thyroid of a teleost, *Chasmichthys dolichognathus*, and a hagfish, *Eptatretus burgeri*, *Acta Zool.* **56**:61.

Usategui-Echeverria, R., Oliver, C., Vaudry, H., Lombardi, G., Mourre, A. M., Rozenberg, I., and Vague, J., 1975, Radioimmunological determination of rat plasma α MSH and ACTH, *Acta Endocrinol.* **84 (Suppl. 199)**:73.

Vale, W., Rivier, J., and Burgus, R., 1971, Synthetic TRF (Thyrotrophin releasing factor) analogues: *p*-Glu-N ^{31}ΣMe-His-Pro-NH$_2$: A synthetic analogue with specific activity greater than that of TRF, *Endocrinology* **89**:1485.

Vale, W., Grant, G., and Guillemin, R., 1973, Chemistry of the hypothalamic releasing factors: Studies on structure–function relationships, in: *Frontiers in Neuroendocrinolgy* (W. F. Ganong and L. Martini, eds.), pp. 375–413, Oxford University Press, New York.

Vandesande, F., and Aspeslagh, M. R., 1974, Failure of thyrotropin releasing hormone to increase ^{125}I uptake by the thyroid in *Rana temporaria*, *Gen. Comp. Endocrinol.* **23**:355.

Van de Veerdonk, F. C. G., and Brouwer, E., 1973, Role of calcium and prostaglandin (PGE$_1$) in the MSH induced activation of adenylate cyclase in *Xenopus laevis*, *Biochem. Biophys. Res. Commun.* **52**:130.

Vaudry, H., Vague, P., Dupont, W., Leboulenger, F., and Vaillant, R., 1975, A radioimmunoassay for plasma corticotropin in frogs (*Rana esculenta* L.), *Gen. Comp. Endocrinol.* **25**:313.

Vaudry, H., Oliver, C., Usategui, R., Trochard, M. C., Leboulenger, F., Dupont, W., and Vaillant, R., 1976, L'hormone mélanotrope chez la Grenouille verte *(Rana esculenta)*: Étude biochimique et radio-immunologique, *C. R. Acad. Sci. Ser. D.* **283**:1655.

Vaudry, H., Tonon, M. C., Delarue, C., Vaillant, R., and Kraicer, J., 1978, Biological and radioimmunological evidence for melanocyte stimulating hormone (MSH) of extrapituitary origin in the rat brain, *Neuroendocrinology* **27**:9.

Wildmeister, W., and Horster, F. A., 1971, Die Wirkung von synthetischem Thyrotrophin Releasing Hormone auf die EztWicklieng eines experimentellen Exophthalmus beim Goldfish, *Acta Endocrinol.* **68**:363.

Wolf, K., and Quimby, M. C., 1964, Amphibian cell-culture: Permanent cell line from bullfrog *(Rana catesbeiana)*, *Science* **144**:1578.

Wright, N. R., and Lerner, A. B., 1960, Action of thyroxine analogues on frog melanocytes, *Nature (London)* **185**:169.

Yasuhara, T., and Nakajima, T., 1975, Occurrence of Pyr-His-Pro-NH$_2$ in the frog skin, *Chem. Pharm. Bull.* **23**:3301.

35

Effects of TRH and Bromocriptine in Acromegaly: In Vivo and in Vitro Studies

M. Guibout, P. Jaquet, E. Goldstein, C. Lucas, and F. Grisoli

1. Introduction

In patients with acromegaly, disturbances involving the regulation of growth hormone (GH) have been reported. Thyroliberin (TRH), which does not modify GH plasma levels in normal subjects, is able to induce release of GH in 55–70% of patients with acromegaly (Schalch et al., 1972; Irie and Tsushima, 1972). Administration of dopaminergic drugs in cases of acromegaly often results in reduction of GH hypersecretion, whereas this effect is absent in normal persons (Liuzzi et al., 1972, 1974; Camanni et al., 1975). The physiological significance of such abnormalities is currently unclear. In particular, it has not yet been demonstrated that these pharmacological effects are related to action of the drugs directly on the adenomatous somatotropic cell. Indeed, it is conceivable that administration of TRH or dopaminergic drugs at supraphysiological doses primarily induces modifications in the neurohormonal output of the hypothalamus. In this hypothetical situation, the abnormal GH regulation

M. Guibout, P. Jaquet, E. Goldstein, and C. Lucas • Laboratoire des Hormones Protéiques, Faculté de Médecine, 13385 Marseille 4, France F. Grisoli • Clinique Neurochirurgicale, Hôpital de la Timone, Marseille, France

encountered in acromegaly would reflect only disturbed hypothalamo–pituitary regulation.

To shed light on this problem, we studied GH responses to administration of TRH and bromocriptine in six untreated patients with acromegaly. Pituitary cells, derived from tumor fragments obtained during transsphenoidal removal of the adenomas in these patients, were maintained in culture. The response of these cells to TRH and bromocriptine was evaluated. The purpose of this study was to identify the site of action of these two agents. The results obtained lead to a discussion of the significance of paradoxical responses to TRH and bromocriptine described in cases of acromegaly.

2. Methods

2.1. Patients

The mean age of the patients, two women and four men, was 47 ± 3 years (mean $\pm$ S.D.). Their histories revealed acromegaly dating back from 8 to 20 years without treatment. A circumscribed intrasellar tumor was present in four cases, whereas two patients displayed suprasellar extension as evidenced on neuroradiological investigation and during transsphenoidal surgery. Light microscopy, following Herlant's staining of tumoral fragments, revealed the presence of orangophilic tumor cells in all six cases.

2.2. In Vivo Studies

Prior to surgery, variations in GH and prolactin (PRL) plasma levels were measured basally, after an intravenous bolus of 400 μg TRH (Roche, Basel, Switzerland) and following a single oral dose of 5 mg bromocriptine (Sandoz, Basel, Switzerland). Hormone levels were measured in blood samples withdrawn every 10 min for 1 hr after administration of TRH. Following bromocriptine, these measurements were performed in blood samples taken every hour during 5 hr.

2.3. In Vitro Studies

2.3.1. Cell Culture Technique

Tumor fragments obtained during surgical removal of the adenoma were immediately rinsed in sterile HEPES buffer, pH 7.2, at 37°C. Markedly hemorrhagic tissue was discarded. The remaining fragments were cut into small pieces with scissors. The technique of enzyme digestion used was that of Vale *et al.* (1972), with some modifications. Briefly, the pieces of tissue were suspended in a silicon-coated sterile glass tube containing HEPES buffer (Gibco), pH 7.2, to which had been added 1% purified bovine serum albumin (BSA) (Miles Laboratories), 0.35% collagenase (Worthington, CLSPA 600 U/mg), and 0.10% hyaluronidase (Sigma, ovine testis-type II, 550 U/mg), final volume 10 ml. The suspension was agitated at 37°C for 15–30 min. The suspension was then allowed to settle, and following decantation, the supernatant containing the dispersed cells was both centrifuged at 70g for 10 min and washed three times in Ham's F-10 medium (Gibco) containing 2.5% fetal calf serum, 10% horse serum, and 1% antibiotics. The tissue fragments remaining undigested after collagenase treatment were resuspended in 10 ml 0.25% viokase solution (Gibco) and then agitated as above for 20–30 min to obtain complete dispersion of cells. The latter were spun and washed three times as previously described and then added to the first batch of dispersed cells. According to the case, initially obtained wet tumor fragments weighed between 87 and 350 g. Enzyme digestion yielded 15–80 × 10^6 viable cells. After yield was verified, the cells were suspended in an appropiate volume of Ham's F-10 medium (enriched as described above) to distribute 3–5 × 10^5 cells in 1.5 ml medium per tissue culture dish (35-mm diameter petri dishes, Falcon Plastics No. 3001). Cells were incubated at 37°C in a 95% air 5% CO_2 water-saturated atmosphere. Most cells became attached to the plastic support after 3–4 days' incubation. Media changes were performed every 24 hr beginning on the 4th day of culture.

2.3.2. Kinetic Studies

In all experiments, GH release into the culture media was measured during 8–15 days. The entire media from five culture dishes were removed every 24 hr and ascertained for GH and PRL content. In one experiment, the kinetics of 24-hr GH release were investigated. At 0, 2, 4, 8, 14, and 24 hr, 0.15 ml medium was removed from the dishes. The removed samples were gently replaced by addition of 0.15 ml fresh medium.

2.3.3. Effects of TRH and Bromocriptine

TRH was disolved in isotonic saline. Bromocriptine, after dissolution in tartric acid and ethanol (70%), was diluted in isotonic saline. Both agents were then diluted in Ham's F-10 medium containing 1% BSA to obtain concentrations of 10^{-6} to 10^{-12} M per culture dish. Prior to being tested, media from the dishes were removed and the cells were washed three times and then preincubated for 1 hr in Ham's F-10 medium containing 1% BSA. The test agents were introduced by adding 0.15 ml of the desired molarity to different dishes. The effect of TRH was studied over a 2-hr period and that of bromocriptine was evaluated 2 and 24 hr after addition to the media (only the 24-hr results are presented). At the end of each experiment, the media were removed and centrifuged at 70*g* for 10 min and the supernatants were immediately stored at −20°C until GH and PRL measurements were performed. Each test dose was ascertained in the media from 4–6 dishes.

2.4. Radioimmunoassays

Pituitary hormones were measured according to techniques described elsewhere (Jaquet *et al.*, 1978). Hormone concentrations were ascertained in duplicate at appropriate dilutions. All media samples from a given adenoma underwent radioimmunoassay simultaneously to avoid between-assay variability. Radioimmunoassay data were computed according to the method of Rodbard and Lewald (1970).

The statistical significance of modifications of GH levels in the culture media after TRH and bromocriptine was evaluated by analysis of variance.

3. Results

3.1. In Vivo Studies

TRH induces PRL release in all six patients. In three cases (Nos. 1, 2, and 3), a paradoxical release of GH was also observed (Table I). PRL secretion was suppressed by bromocriptine in all patients. GH secretion was similarly suppressed in three cases (Nos. 1, 2, and 6). It should be noted that in two patients (Nos. 5 and 6) displaying hypersecretion of both GH and PRL, bromocriptine suppressed only that of PRL.

3.2. In Vitro Studies

3.2.1. Kinetic Studies

In six cases, 24-hr GH release was examined (Fig. 1a). On the 4th day of culture, 1.4–3.8 μg GH/culture dish was released into the medium. A progressive decline in release with respect to time was encountered in all cases, but varied according to the culture. Because of this phenomenon, responses to TRH and bromocriptine were studied successively between the 4th and 8th culture days. In one experiment, the profile of GH release

Table I. Preoperative Studies of GH and PRL Plasma
Levels in Six Acromegalic Patients

Case No.	Basal levels	Response to TRH[a]	Basal levels	Response to bromocriptine[b]
GH plasma levels (ng/ml)				
1	100	320[c]	135	63[c]
2	100	380[c]	82	11[c]
3	36	82[c]	56	47
4	19	17	10	12
5	17	17	26	23
6	41	48	88	7[c]
PRL plasma levels (ng/ml)				
1	25	52[c]	13	6[c]
2	24	105[c]	20	10[c]
3	2	5[c]	5	2[c]
4	3	7[c]	4	2[c]
5	102	198[c]	82	36[c]
6	43	190[c]	31	2[c]

[a] GH or PRL measurements were made 20–30 min after 400 μg TRH injection.
[b] GH or PRL measurements were made, respectively, 3 and 5 hr after a 5 mg oral dose of bromocriptine.
[c] Variations in GH or PRL plasma levels after TRH or bromocriptine were considered significant if they increased by 100% or more or decreased by 50% or more, respectively, over basal values.

was examined during a 24-hr period (Fig. 1b). Results of this experiment show that GH release into the media increased linearly. The 24-hr release patterns of other pituitary hormones were also measured in the media. Concentrations of TSH, LH, FSH, and ACTH were very low or undetectable, and thus could have been related to contamination by a few nontumor cells. Conversely, PRL media levels were measurable in all cases (Fig. 2). In four cases (Nos. 1, 3, 4, and 5) 24-hr PRL release was 22–267 ng/dish on day 4 and less than 90 ng on day 8. This observation may be interpreted as the result of culture contamination by nontumor PRL-secreting cells or as the product of weak PRL secretion by the tumor cells themselves. In two cases however, 24-hr PRL release was greater than 1 μg/dish, and the secretion profile resembled that of GH.

3.2.2. Effects of TRH (Table II)

In three cases (Nos. 1, 2, and 3), TRH induced significant stimulation of GH release ($p < 0.001$). The maximum effect (159–335% increase) was

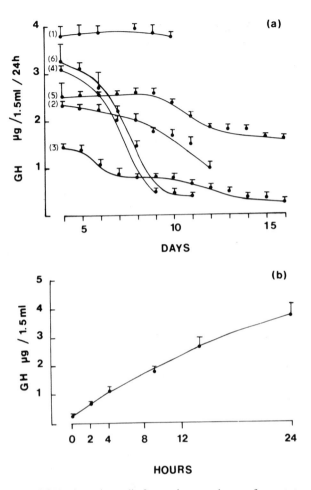

Figure 1. Kinetics of GH release in media from primary cultures of somatotropic adenoma cells. (a) Variations in GH media concentrations in each tumor. Case numbers are shown in parentheses. Values are from the media of 4–6 dishes, expressed as means + S.D. (b) 24-Hour profile of GH release in the medium from Case No. 1 on day 4. Values are means + S.D. See the text for technical details.

obtained with 10^{-10} M TRH. In the other cases, no stimulatory effect was observed at concentrations ranging from 10^{-12} to 10^{-6} M TRH.

3.2.3. Effects of Bromocriptine (Table III)

In three cases (Nos. 1, 2, and 6), 10^{-6} M bromocriptine significantly ($p < 0.001$) suppressed GH release. The maximum inhibition was obtained using 10^{-10} or 10^{-6} M bromocriptine, according to the case. In these conditions, 24-hr GH release was reduced by 46–55% as compared to control values. In the three other adenoma cultures, no effect of bromocriptine was observed regardless of dose.

4. Discussion

Evaluation of hormone release in primary cultures of pituitary tumor cells led to the identification of certain features of their basal secretion. First, the absence or very slightly detectable levels of LH, FSH, TSH, and ACTH demonstrate that the cells examined were only very weakly contaminated by nontumoral pituitary tissue. Second, all six tumors also secreted PRl, but to highly variable degrees. In two cases, secretion profiles of GH and PRL were similar, showing 24-hr release on the order of a microgram, thus suggesting that the two hormones were produced in equivalent amounts by the tumor cells. In the other cases, PRL response to TRH and bromocriptine is needed to conclude whether the PRL secretion observed is related to contamination by peritumoral PRL-secreting cells or, rather, is produced by the GH tumor cells themselves. Studies are under way to verify this point. Finally, hormone levels in the media decreased with respect to time, but variably according to the case. This is probably related to culture conditions, especially the concomitant

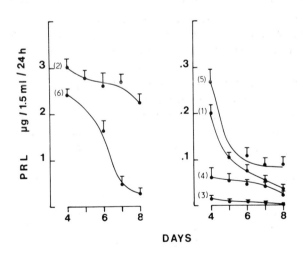

Figure 2. Kinetics of PRL release in media from cultured somatotropic adenoma cells between days 4 and 8. Same caption as in Fig. 1a. Case Nos. 2 and 6, secreting more than 1 μg PRL/24 hr, are plotted separately at the left.

Table II. TRH Effects on GH Release *in Vitro*[a]

Case No.	Control	TRH (molar concentration/5 × 10⁵ cells)			
		10^{-12}	10^{-10}	10^{-8}	10^{-6}
1	278 ± 62	389 ± 27[c]	711 ± 53[b]	707 ± 84[b]	684 ± 50[b]
2	129 ± 10	142 ± 13	232 ± 10[b]	209 ± 22[b]	219 ± 23[b]
3	392 ± 63	501 ± 10	622 ± 44[b]	625 ± 223	—
4	2220 ± 730	1940 ± 197	2900 ± 648	2634 ± 683	2530 ± 167
5	1125 ± 122	1215 ± 235	1105 ± 235	1134 ± 228	—
6	62 ± 10	54 ± 10	57 ± 10	64 ± 10	67 ± 10

[a] GH is expressed in ng released in 1.5 ml medium/culture dish (during 2 hr). Values are the means of 5 measurements ± S.D.
[b] $p < 0.001$ compared to controls.
[c] $p < 0.01$ compared to controls.

growth of fibroblasts. In all cases, neither mitosis nor cell proliferation was observed, except for that involving fibroblasts.

Comparison of *in vivo* (Table I) and *in vitro* (Tables II and III) results regarding the stimulatory effects of TRH and inhibitory action of bromocriptine on GH release leads to the conclusion that the results observed *in vivo* are produced by these drugs acting directly on the tumor cells. A direct effect of TRH on GH released *in vitro* by explants of human somatotropic adenomas has been reported recently (Goodyer *et al.*, 1978). In a given tumor, *in vivo* and *in vitro* responses were similar in all our cases. In Case Nos. 1 and 2, the paradoxical effects of both TRH and bromocriptine were associated; whereas these were simultaneously absent in Case Nos. 4 and 5. Nevertheless, in one case (No. 3), TRH induced significant GH release, whereas bromocriptine produced no response. In another case (No. 6), 10^{-12} M bromocriptine significantly suppressed GH release, while TRH was without effect. The acquisition of paradoxical regulation of GH hypersecretion, although variable according to the case, might be due to cellular dedifferentiation toward an embryonic stage. In newborn rats not yet displaying GH secretory pulsatility of the adult type, administration of TRH induces a paradoxical GH response that is not observed in adult rats (Gil-Ad *et al.*, 1976). This fact argues in favor of such hypothetical dedifferentiation. Additonally, these heterogeneous responses observed in human pituitary adenomas are encountered in experimentally induced tumors in animals. The GH_3 clone derived from mouse pituitary cells, displaying supranormal secretion of GH and PRL, responds to TRH by an increase in PRL and a decrease in GH (Dannies and Tashjian, 1974; Hinkle *et al.*, 1974). Other clones of pituitary tumor cells obtained in analogous experimental conditions do not display these paradoxical responses to TRH (Carlson *et al.*, 1977). In human pituitary tumors, as in those experimentally induced in animals, supranormal secretion of GH may display variable regulation from one case to another. Data from this study indicate that heterogeneous reactivity of GH is both similar *in vitro* and *in vivo* and related to the tumor itself and not to disturbed hypothalamic regulation. Accordingly, results of *in vitro* testing should be useful in evaluating the effectiveness of dopaminergic drugs, such as bromocriptine, in the management of acromegaly. Ishibashi and Yamaji (1978) have published results also confirming a direct effect of bromocriptine and TRH on pituitary adenoma tissues of acromegaly.

5. Summary

The effects of TRH and bromocriptine were studied in six acromegalic patients prior to treatment and in primary cell cultures derived from tumor fragments obtained during adenomectomy.

The kinetics of GH and PRL release were studied *in vitro* during 8–15 days. It was found that $3–5 \times 10^5$ cells secreted from 1.4 to 3.8 μg GH/24 hr. This release decreased progressively with respect to time. In the same conditions, PRL levels were highly variable from case to case, and LH, FSH, TSH, and ACTH concentrations were very low or undetectable. *In vitro* studies demonstrated stimulation of GH release by TRH in three of six cases. The maximum effect was obtained using 10^{-10} M TRH. Variable molarities of bromocriptine (10^{-12} to 10^{-8} M) induced paradoxical suppression of GH secretion in three of six cases. Abnormal reactivity to both TRH and bromocriptine was not always observed in a given tumor, but *in vivo* results paralleled those obtained in *in vitro*.

Table III. Bromocriptine Effects on GH Release *in Vitro*[a]

Case No.	Control	Bromocriptine (molar concentration/5×10^5 cells)[b]			
		10^{-12}	10^{-10}	10^{-8}	10^{-6}
1	4530 ± 555	3562 ± 441	3190 ± 901	2700 ±742[d]	2130 ± 357[c]
2	2030 ± 144	1450 ± 150[c]	1306 ± 275[c]	1178 ± 167[c]	1009 ± 114[c]
3	1164 ± 126	1304 ± 320	1019 ± 122	1161 ± 173	1262 ± 118
4	903 ± 122	704 ± 77	850 ± 32	992 ± 170	—
5	2740 ± 697	3100 ± 350	—	3717 ± 221	3905 ± 319
6	1960 ± 170	1260 ± 100[c]	1020 ± 100[c]	970 ± 140[c]	870 ± 170[c]

[a] GH is expressed in ng released in 1.5 ml medium/24 hr per culture dish. Values are the means of 5 measurements ± S.D.
[b] Bromocriptine (0.15 ml) was added at times 0 and +12 hr.
[c] $p < 0.001$ compared to controls.
[d] $p < 0.005$ compared to controls.

These data demonstrate that paradoxical reactivity to TRH and bromocriptine in acromegaly is directly related to the tumor itself, and not to disturbed hypothalamic regulation.

ACKNOWLEDGMENTS. The authors express their gratitude to Dr. P. Carayon for execution of statistical studies and to Miss N. Ghirardi for expert secretarial assistance. The present work has been partially supported by INSERM Grant No.78.5.034.4.

DISCUSSION

ENJELBERT: In one case, maximal inhibition was obtained with 10^{-12} M bromocriptine, which seems to me very low. Do you have any information on the ED_{50} for bromocriptine on normal human tissue?

JAQUET: We do not have any data concerning the dose effect in normal human pituitary cells. Dr. Labrie found that the maximal dose of bromocriptine was 20 times less than that of dopamine on PRL inhibition in rat cells.

MCKERNS: We may have to change the name of TRH, since it causes the release of most peptide hormones in cultured cell preparations. Is this due to a lack of specificity or leaky membranes in the cell preparations? Do you think these releases of hormones would occur in the whole animal at physiological doses?

WILBUR: In answer to the questions of Dr. McKerns, our experiments agree with Dr. Jaquet as to the effect of TRH on growth hormone secretion. We found TRH to be nonselective on pituitary cell types (*Am. J. Anat.* **151**:277, 1978). Immunoassay levels of GH were increased at all time periods studied following 1×10^{-7} mM TRH. It is also well supported in the literature that TRH stimulates the release of immunoassayable GH hormone in man in conditions such as acromegaly and certain kidney diseases. Additional studies have demonstrated the ability of TRH to release immunoassayable GH in other species such as the cow and sheep.

REFERENCES

Camanni, F., Massara, F., Fassio, V., Molinatti, G. M., and Muller, E. E., 1975, Effect of five dopaminergic drugs on plasma growth hormone levels in acromegalic subjects, *Neuroendocrinology* **19**:227.

Carlson, H. E., Mariz, I. K., Brigg, J. E., and Daughaday, W. D., 1977, Altered responsiveness to hypophysiotropic hormones of perfused rat pituitary tumors, *Acta Endocrinol.* (Copenhagen) **84**:512.

Dannies, P. S. and Tashjian, A. J., 1974, Pyroglutamyl-histidyl-prolinamide (TRH), a neurohormone which effects the release and synthesis of prolactin and thyrotropin, *Isr. J. Med. Sci.* **10**:1294.

Gil-Ad, I., Cocchi, D., Panerai, A. E., Locatelli, V., Mantegazza, P., and Muller, E. E., 1976, Altered growth hormone and prolactin responsiveness to TRH in the infant rat, *Neuroendocrinology* **21**:366.

Goodyer, C. G., Marcovitz, S., Guyda, H., Giraud, C. J. P., Hardy, J., Gardiner, R. J., and Martin, J. B., 1978, Comparative study of human fetal, normal adult and acromegalic pituitary function, in: *The Endocrine Society 60th Annual Meeting*

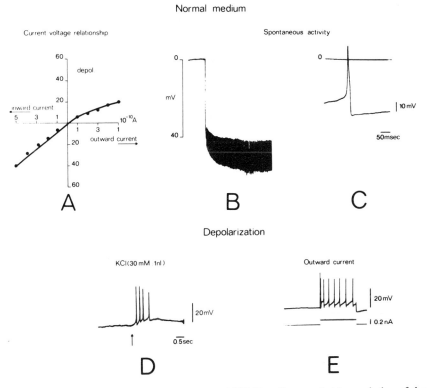

Figure 1. Electrophysiological characteristics of GH_3/B_6 cells recorded in a solution of the following composition (in mM): Na Cl, 142.6; KCl, 5.6; $CaCl_2$, 10; glucose, 5; HEPES buffer, 5; pH 7.4. (A) Current–voltage relationship. (B) Spontaneous spiking activity as observed in 28% of the cells recorded. The ink paper recorded (Linear Instrument Co) cannot follow the spikes, which thus appear shorter; the actual size of the spikes is indicated on the oscilloscope trace shown in (C); 0 represents the baseline. Depolarization induced by KCl (D) or by an outward current (E) elicited action potentials in 50% of the cells recorded.

longer duration. In calcium-free medium, the resting potential of the membrane dropped to 5.9 mV; cells were no longer excitable and did not display spikes (Fig. 2D). Moreover, when D 600, known as a blocker of calcium channels, was administered to a spontaneously firing cell, spikes were reduced in amplitude and even sometimes completely suppresed (Fig. 2E). These findings suggest the involvement of Ca^{2+} channels in the spiking activity of these cells.

TRH was able to induce spikes within a minute; the effect of TRH on membrane excitability consisted in a progressive increase in membrane resistance with no apparent change in the level of the resting potential

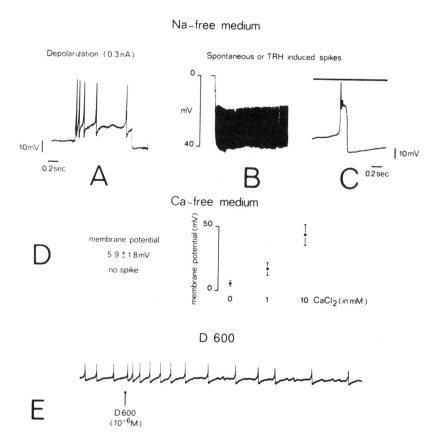

Figure 2. In sodium-free medium, depolarization was still able to induce spikes (A) and spontaneous spiking activity was also observed (B). For the same reason as in Fig. 1B, the spikes appear shorter; the actual shape of the spike is shown in (C). In calcium-free medium (D), the membrane resting potential is 5.9 mV; increasing the calcium concentration of the medium to 1 and 10 mM (D, right) restores the membrane resting potential within a few minutes. In the normal recording solution, D 600 administered to the cell under investigation (E) reduced the spiking activity.

(Fig. 3A). Repeated administration of TRH (10-min intervals) induced repetitive discharges; however, a decrease in the excitability was observed (Fig. 3B). Accordingly, repeated administration of TRH was not able to induce increase in the membrane resistance to the same extent as the first injection (Fig. 3B).

In 31% of the excitable cells tested ($n = 58$), 17β-E (10^{-10} to 10^{-8} M) injected in the near vicinity of the cell (10 μm) elicited action potentials

Program and /
Maryland.
Hinkle, P. M
    biol⌐

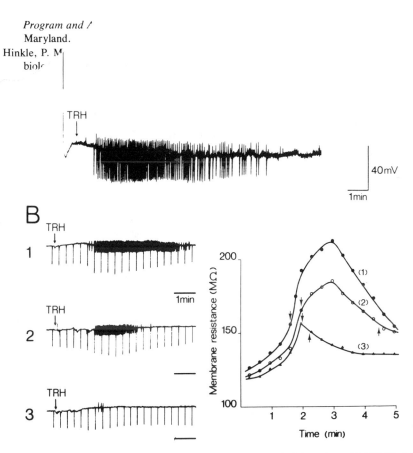

Figure 3. Effect of TRH on two GH_3/B_6 cells. (A) Direct administration of TRH (10^{-6} M, 2 nl) on the cell membrane induced a sustained spiking activity within a minute. (B) Left: Three repeated administrations of TRH (same concentration as above, 10-min intervals) evoked successive bursts; however, a desensitization was observed. A measure of membrane resistance (vertical bars) was obtained by injecting a hyperpolarizing current (0.16 nA) while recording. Bursting activity was preceded by an increase in membrane resistance. Right: Time course of membrane resistance increases following TRH administration at time 0. Arrows indicate the beginning and the end of the spiking activity.

within 1–2 min (Fig. 4). This spiking activity lasted 3–30 min after application of estradiol and was also reduced by D 600. When an injection of 17β-E was successful in eliciting spikes, a subsequent injection was totally ineffective. As for TRH, 17β-E increased the input membrane resistance before inducing spikes. 17α-E was inactive when administered in the same manner; in only one case (of 23), 17α-E at much higher concentration (10^{-6} M) elicited a short burst of spikes. This indicates a

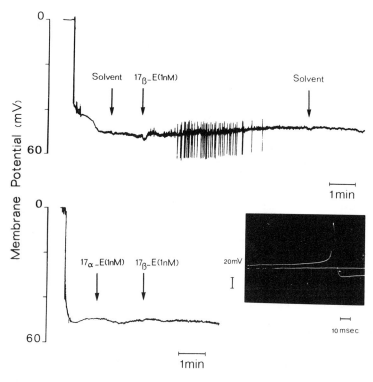

Figure 4. Effect of 17β-estradiol (17$_β$-E, 1 nM, 2 nl) and 17α-estradiol (17$_α$-E, 1 nM, 2 nl) on two GH$_3$/B$_6$ cells. The actual shape of the spike is shown on the oscilloscope trace.

considerable degree of stereospecificity of the steroid effects on the electrical activity of these cells. Furthermore, 17α-E prevented the effect of 17β-E when administered 2 min before 17β-E (n = 18) (Fig. 4).

The action potentials in GH$_3$ cells appear to be calcium-dependent, since they were inhibited or partly suppressed by the calcium blocker D 600. This is consistent with the recent demonstration of the role of calcium in TRH-stimulated release prolactin by GH$_3$ cells (Tashjian et $al.$, 1978). These results reveal a rapid effect of 17β-E on the membrane of prolactin-secreting pituitary cells in culture also involving calcium influx. The relationship between these membrane effects and an immediate prolactin release remains to be elucidated. However, the fate of calcium ions inside the cell once they have entered is complex and not yet clearly understood; different pools have been reported to be involved in the release process (Milligan and Kraicer, 1974; Sugaya and Onozuka, 1978). A redistribution of Ca^{2+} concentration within cell compartments from a stored pool appears to be the major determinant for the release (Milligan and Kraicer, 1974;

Henquin, 1978). It is therefore possible that spikes induced by TRH and estrogen affect different steps of a given process.

There appears to be no change in the membrane potential preceding the TRH- or 17β-E-induced spiking activity, but an increase in the phenomena were also reported following exposure of normal pituitary cells *in vivo* to hypothalamic extracts, where again an increase in membrane resistance with no change in the membrane potential was observed (York *et al.*, 1973). Depolarization of pituitary cells by elevated extracellular K^+ has been reported to release pituitary hormones (Vale and Guillemin, 1967; Jutisz and de La Llosa, 1960; Kraicer *et al.*, 1969); however, only the Ca^{2+} influx is essential in the K^+-induced hormone release (Vale and Guillemin, 1967; Jutisz and de La Llosa, 1970), and depolarization is not tightly coupled to secretion (Martin *et al.*, 1973). Therefore, the main effect of TRH and 17β-E was an enhancement of excitability of the cell via an increase in the membrane resistance leading to Ca^{2+} spike activity.

Thus, besides its action at the level of the genome, estradiol is able to induce membrane conductance changes in pituitary cells in culture.

Summary

The electrical properties of a prolactin-secreting pituitary cell line ($GH_3 B_6$) were studied with intracellular microelectrode recordings. Fifty percent of the cells tested were excitable and displayed Ca^{2+}-dependent action potentials when depolarized. When injected directly on the membrane of an excitable cell, TRH and 17β-E induced action potentials within a minute. The relationship between these membrane effects and an immediate prolactin release remains to be elucidated.

ACKNOWLEDGMENTS. We are indebted to Dr. M. Kelly for reading the manuscript. This work was supported by grants from the D.G.R.S.T. (77.7.0654), INSERM (CRL 78.1.26.56), and C.N.R.S. (ERA 493 and ER 89).

DISCUSSION

KRAICER: There appears to be no change in TMP with TRH. Also, you gave evidence that the spiking activity is calcium-dependent. Could you comment further on the mechanism producing the spikes and altering the frequency of the spikes?

DUFY: The fact that there was no change in TMP with TRH or estrogen could be a key to the mechanism of action of TRH and estrogen. You have already reported similar effects following the injection of hypothalamic extracts *in vivo* (York *et al.*, *Neuroendocrinology*

11:2112, 1973). The spiking activity is mainly calcium-dependent. However, sodium is certainly involved, since spikes are slightly smaller in amplitude in sodium-free medium and of larger duration. I have no data on potassium, but I would predict that potassium channels are involved in some way.

SPIES: Why does 17α-estradiol, which causes no depolarization itself, block the excitation of the cell by 17β-estradiol?

DUFY: I have no data on that.

DENEF: Do the cells secrete hormone after potassium depolarization or after electrical stimulation?

DUFY: This has been described for potassium, but we have not tested for such effects since we are using individual cells; potassium is directly administered to the cells under investigation, and depolarization is induced by a current directly applied by the recording electrode.

FINK: Could you comment on the relatively long time (1 min) between administering TRH and estradiol and the increases in spiking?

DUFY: Using dye, we have found that it takes 5–10 sec for the substrates to reach the cell. The spiking activity starts in 1 to 2 min following injection, but the beginning of the effect is much sooner.

TIXIER-VIDAL: I would like to add a comment. When we studied the effect of TRH on surface glycoproteins and on horseradish peroxidase uptake (endocytosis) in GH_3B_6 cells, we were surprised to find that some TRH analogues such as TRH-OH and DLL-TRH were as active as TRH. This was in spite of the fact that they have no action on prolactin release (Tixier-Vidal *et al.*, *Cold Spring Harbor Conference on Cell Proliferation* **6**:807–826, 1978). It seems that a partial recognition of the ligand at the cell surface might induce some rapid effect at the cell surface, without being able to induce subsequent intracellular events leading to the final biological response.

REFERENCES

Dufy, B., Partouche, C., Poulain, D., Dufy-Barbe, L., and Vincent, J. D., 1976, Effects of estrogen on the electrical activity of identified and unidentified hypothalamic units, *Neuroendocrinology* **22**:38.

Henquin, J. C., 1978, Relative importance of extracellular and intracellular calcium for the two phases of glucose-stimulated insulin release: Studies with theophylline, *Endocrinology* **102**:723.

Jutisz, M., and de La Llosa, M. P., 1960, Recherches sur le contrôle de la sécrétion de l'hormone folliculo-stimulante hypophysaire, *Bull. Soc. Chim. Biol.* **50**:2521.

Jutisz, M., and de La Llosa, M. P., 1970, Requirement of Ca^{++} and Mg^{++} for the *in vitro* release of follicle stimulating hormone from rats pituitary glands and its subsequent bio-synthesis, *Endocrinology* **86**:761.

Kelly, M., Moss, R., and Dudley, C., 1976, Differential sensitivity of preoptic septal neurons to microelectrophoresed estrogen during the estrous cycle, *Brain Res.* **114**:152.

Kidokoro, Y., 1975, Spontaneous action potentials in a clonal pituitary cell line and their relationship to prolactin secretion, *Nature (London)* **258**:741.

Kraicer, J., Milligan, J. V., Goskee, J. L., Conrad, R. G., and Branson, C. M., 1969, *In vitro* release of ACTH: Effects of potassium, calcium and corticosterone, *Endocrinology* **85**:1144.

McCaman, R. E., McKenna, D. G., and Ono, J. K., 1977, A pressure system for intracellular and extracellular ejection of picoliter volume, *Brain Res.* **136**:141.

Martin, S., York, D. G., and Kraicer, J., 1973, Alterations in transmembrane potential of adenohypophysial cells in elevated potassium and calcium free media, *Endocrinology* **92**:1084.

Milligan, J. V., and Kraicer, J., 1974, Physical characteristics of the Ca^{++} compartments associated with *in vitro* ACTH release, *Endocrinology* **94**:435.

Sugaya, E., and Onozuka, M., 1978, Intracellular calcium: Its movements during pentylenitetrazole-induced bursting activity, *Science* **200**:797.

Taraskevich, P. S., and Douglas, W. W., 1977, Action potentials occur in cells of the normal anterior pituitary gland and are stimulated by the hypophysiotropic peptide thyrotropin-releasing hormone, *Proc. Natl. Acad. Sci. U.S.A.* **74**:4064.

Tashjian, A. H., Barowsky, N. J., and Jensen, D. K., 1971, Thyrotropin releasing hormone: Direct evidence for simulation of prolactin production by pituitary cells in culture, *Biochem. Biophys. Res. Commun.* **43**:516.

Tashjian, A. H., Lomedico, M. E., and Maina, D., 1978, Role of calcium in the thyrotropin releasing hormone stimulated release of prolactin from pituitary cells in culture, *Biochem. Biophys. Res. Commun.* **81**:798.

Tixier-Vidal, A., Brunet, N., and Gourdji, D., 1978, Morphological and molecular aspects of the regulation of prolactin secretion by rat pituitary cell lines, in: *Progress in Prolactin Physiology and Pathology* (C. Robyn and M. Hurter, eds.), pp. 29–43, Elsevier/North-Holland, Amsterdam.

Vale, W., and Guillemin, R. 1967, Potassium induced stimulation of thyrotropin release *in vitro:* Requirement for presence of calcium and inhibition by thyroxine, *Experientia* **23**:855.

York, D. H., Barker, F., and Kraicer, J., 1973, Electrical changes induced in rat adenohypophysial cells, *in vivo* with hypothalamic extract, *Neuroendocrinology* **11**:212.

37

Extracellular Potassium Change in the Rat Adenohypophysis: An Indicator of Neurohypophyseal– Adenohypophyseal Communication

A. J. Baertschi, M. Friedli, J. Munoz, M. Tsacopoulos, and J. A. Coles

1. Introduction

It has been generally thought that hypothalamic releasing and inhibiting factors are secreted from nerve terminals of the median eminence into long portal vessels, and transported in plasma toward the anterior pituitary gland. Scanning electron microscopy of pituitary vascular casts revealed, however, that the median eminence, the posterior lobe, and the anterior lobe were interconnected by portal vessels (Page and Bergland, 1977). Further, numerous compounds with anterior lobe hormone-releasing and -inhibiting activity have been identified in the posterior lobe. Thus, posterior lobe compounds may reach the anterior lobe, as suggested by Porter *et al.* (1977), and modulate adenohypophyseal hormone secretions; alternatively, adenohypophyseal hormones may reach the posterior lobe (Page and Bergland, 1977).

A. J. Baertschi, M. Friedli, J. Munoz, M. Tsacopoulos, and *J. A. Coles* • Departments of Animal Biology, Ophthalmology, and Physiology, University of Geneva CH-1211, Geneva 4, Switzerland

We attempted to resolve this question by applying short trains of electrical impulses to the hypothalamo–neurohypophyseal tract, and by monitoring extracellular potassium concentrations in various parts of the hypothalamo–hypophyseal complex with potassium-sensitive microelectrodes. If blood was allowed to flow from the neurohypophysis to the anterior lobe, the potassium released from nerve fibers during depolarization should be transported to the anterior lobe and raise the extracellular potassium concentration there.

2. Methods

Male and female Sprague–Dawley rats (250–350 g body weight) were anesthetized with urethane (1.3 g/kg body weight, i.p.) and their hypothalami exposed by a retropharyngeal approach (Dreifuss and Ruf, 1972). A stimulating electrode, consisting of a pair of silver wires (diameter 100 μm) insulated with Teflon® except for the tips, was positioned on the pituitary stalk about at its junction with the adenohypophysis. The stimuli consisted of constant-current pulses of 400 μA, 0.3–0.6 msec pulse width, delivered at 30 Hz during 1–5 sec. The position was chosen such as to evoke antidromic compound potentials in the pituitary tract of about 3–5 mV amplitude (Baertschi and Dreifuss, 1978).

Extracellular potassium concentrations were measured with double-barrel microelectrodes according to the methods of Coles and Tsacopoulos (1979), as shown schematically in Fig. 1. Sensitivities ranged from 30 to 45 mV per 10-fold change of potassium concentrations, in the range of 2.2–8.8 mM. Electrodes were calibrated before and after an experiment, and a 2.2 mM potassium chloride solution (147.8 mM NaCl) was applied every 30 min for checking baseline drifts.

The potassium electrode was positioned either in the pituitary stalk, about 1.3 mm anterior to the stimulation electrode, or at various depths within the anterior or posterior aspects of the adenohypophysis. The position of the electrode was marked with an electrolytic lesion and identified by subsequent microscopic examination of tissue.

3. Results

In the hypothalamo–neurohypophyseal tract, extracellular potassium rose from a basal concentration of 2–2.5 mM to a peak value of 3.5–5.5 mM in response to a 5-sec-long train of impulses at 30 Hz. The potassium electrode began to respond within 50 msec after initiation of the electric stimulus. Almost no potassium change was observed in CSF. In the anterior adenohypophysis, potassium levels rose from a basal value of

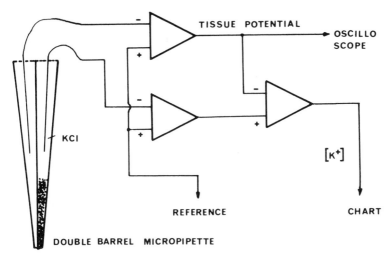

Figure 1. Monitoring of extracellular potassium concentrations and of tissue electrical potential. One channel of a double-barrel micropipette (3–4 μm tip) was filled with a potassium-ion exchanger (Corning No. 477317); the other channel was filled with lithium acetate (0.1 M). The reference electrode was connected to the animal (muscle immobilized with xylocaine). A purely potassium-sensitive signal was obtained by taking the potential difference between the signals originating from the two channels of the micropipette.

2.7–3.5 mM to a peak value of 4.5–5.5 mM, with a delay of 1–4 sec. In the posterior adenohypophysis, potassium levels rose from a basal value of 3–4 mM to a peak of 3.8–4.4 mM, with a delay of 4–10 sec. Individual examples are shown in Fig. 2.

In the anterior adenohypophysis, the delayed potassium rise appeared to depend on the depth of penetration from the ventral surface. Responses were practically absent in the outer layers of the adenohypophysis, and increased progressively with increasing depth (Fig. 3). Coagulation of long portal vessels by microcauterization of the whole median eminence region did not abolish the delayed potassium increase (Fig. 4). However, when the stimulation electrode was repositioned at a site 0.5–1 mm distant from its original position, no antidromic compound potentials were recorded in the pituitary stalk, and there was also no delayed potassium increase in the anterior adenohypophysis.

4. Discussion

The results indicate that electrical stimulation of the hypo-thalamo–neurohypophyseal tract causes an immediate efflux of potassium from neurohypophyseal axons, followed seconds later by the appearance

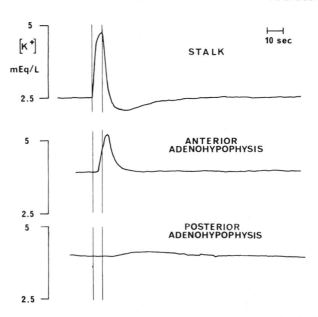

Figure 2. Extracellular potassium changes in pituitary tract and hypophysis. The location of the recording electrode in the pituitary tract was chosen so as to obtain antidromic compound potentials of 3–5 mV. Positioning of recording electrode 150 μm above or below the optimum position abolished antidromic compound potentials and potassium responses. Electrical stimulation at 30 Hz was applied between the two vertical bars. Responses were obtained in the same animal.

of potassium in the anterior adenohypophysis. Since coagulation of long portal vessels (plus neurohypophyseal axons) did not abolish the adeno-hypophyseal potassium response, the results suggest that potassium re-leased from axons in the neural lobe or posterior stalk region was transported by a short portal vasculature. The rapid appearance of potas-sium cannot be explained by diffusion effects, the distances from site of release to site of recording ranging from 0.5 to 1.5 mm. Conservative calculations showed that it would have taken the potassium about 400 sec to travel that distance by diffusion.

These results strongly suggest that neurohypophyseal peptides may reach the anterior adenohypophysis by means of a short portal route. The findings are compatible with the appearance of large amounts of vasopres-sin in monkey portal blood (Zimmerman *et al.*, 1973), with the well-known corticotropin-releasing effects of vasopressin (Yates *et al.*, 1971; Saffran and Schally, 1977; Yasuda *et al.*, 1978), and also with recent results from our laboratory that electrical stimulation of the rat neural lobe promotes release of ACTH *in vivo* (Baertschi *et al.*, 1979). Our results do not exclude

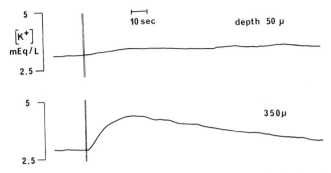

Figure 3. Extracellular potassium changes in anterior adenohypophysis. Electrical stimulation, as in Fig. 2, was applied between the two vertical bars. The depth of the recording electrode was measured from the ventral surface of the hypophysis. Responses were obtained in the same animal.

the possibility of blood flowing from anterior to posterior lobe, nor from posterior lobe to median eminence or hypothalamus or both. Microinjection of potassium in conjunction with monitoring of extracellular potassium in hypothalamus, neural lobe, and median eminence may resolve this problem.

5. Summary

Three recent findings suggest that the neurohypophysis may modulate the release of adenohypophyseal hormones: (1) There are vascular interconnections between neurohypophysis and anterior lobe. (2) A growing number of posterior lobe peptides have been discovered. (3) Evidence from our laboratory showed that plasma ACTH increased following electrical stimulation of the neural lobe. We electrically stimulated (30 Hz, 1–5 sec) the pituitary tract and monitored the appearance of potassium,

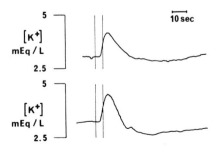

Figure 4. Effect of median eminence lesion on hypophyseal potassium response. Electrical stimulation, as in Fig. 2, was applied between the two vertical bars. Responses were recorded at about 300 μm below the ventral surface of the anterior adenohypophysis, before *(top)* and after *(bottom)* microcoagulation of long portal vessels.

which is released during nerve depolarization, in various parts of the hypothalamo–hypophyseal complex by means of double-barrel K^+-sensitive microelectrodes. In the hypothalamo–neurohypophyseal tract, potassium increased within 100 msec from 2–2.5mM to 3.5–5.5mM. In the anterior adenohypophysis, 300 μm below the ventral surface, potassium increased to 4.5–5.5 mM, but after a delay of 1–4 sec. In the posterior ventral adenohypophysis, potassium increased to 3.8–4.4 mM after a delay of 4–10 sec. The results suggest strongly that potassium released from neurohypophyseal axons reached the adenohypophysis, and that neural lobe peptides may travel by the same route.

ACKNOWLEDGMENT. This research was supported in part by Swiss National Science Foundation Grant 3.248.0.77.

DISCUSSION

FINK: What is the amplitude of your current, and what is the variation in response with variation of placement of electrodes?

BAERTSCHI: The stimulations consisted of constant-current impulses of 200–400 μA. Displacement of the stimulation electrode by ±0.5 mm either anteriorly or posteriorly diminished or arrested the milk-ejection response. The current spread should not exceed 1 mm³.

KRAICER: Your potassium measurements indicate a concentration in the extxacellular fluid of approximately 3 mM, which is one half the amount that we always use in *in vitro* studies. Perhaps we are all studying the effect of high potassium in our "controls." Also, did you get your electrode into the "colloid" to measure its potassium concentration?

BAERTSCHI: Potassium levels used in *in vitro* studies may well be high. The electrode position was verified by passing a lesion current and by histological examination. So far we have not recorded in the colloid.

HYMER: Do your experiments rule out the possibility of retrograde blood flow?

BAERTSCHI: We have not examined this.

KORDON: How can you exlude the possibility of passive diffusion of potassium in the extracellular space?

BAERTSCHI: The potassium responses in the anterior adenohypophysis demonstrate a transport lag of a few seconds. In view of the rather large distance separating the sites of release from the sites of recording, it appears unlikely that diffusion could play a significant role.

REFERENCES

Baertschi, A. J., and Dreifuss, J. J., 1978, Antidromic compound potentials of the pituitary tract: Interactions with systemic bradykinin, *Brain Res.* 149:530.
Baertschi, A. J., Vallet, P., and Girard, J., 1979, Neurohypophysis modulates corticotropin release in the rat, *J. Physiol. (Paris)* 75(1):1B.

Coles, J. A., and Tsacopoulos, M., 1979, K$^+$ activity in photoreceptors, glial cells and extracellular space in the drone retina: Changes during photostimulation, *J. Physiol. (London)* **290:**525.

Dreifuss, J. J., and Ruf, K. B., 1972, A transpharyngeal approach to the rat hypothalamus, in: *Experiments in Physiology and Biochemistry* (G. Kerkut, ed.), Vol. 5, pp. 213–228, Academic Press, London.

Page, R. B., and Bergland, R. M., 1977, The neurohypophyseal capillary bed. Part I. Anatomy and arterial supply, *Am. J. Anat.* **148:**345.

Porter, J. C., Oliver, C., Eskay, R. L., Barnea, A., Parker, C. R., and Ben-Jonathan, N., 1977, Hypothalamic–pituitary interaction, in: *The Pituitary: A Current Review* (M. B. Allen and V. B. Mahesh, eds.), pp. 215–234, Academic Press, New York.

Saffran, M., and Schally. A. V., 1977, The status of the corticotropin releasing factor (CRF), *Neuroendocrinology* **24:**359.

Yasuda, N., Greer, M. A., Greer, S. E., and Panton, P., 1978, Studies on the site of action of vasopressin in inducing adrenocorticotropin secretion, *Endocrinology* **103:**906.

Yates, F. E., Russel, S. M., Dallman, M. F., Hedge, G. A., McCann, S. M., and Dhariwal, A. P. S., 1971, Potentiation by vasopressin of corticotropin release induced by CRF, *Endocrinology* **88:**3.

Zimmerman, E. A., Carmel, P. W., Husain, M. K., Ferin, M., Tannenbaum, M., Frantz, A. G., and Robinson, A. G., 1973, Vasopressin and neurophysin: High concentrations in monkey hypophyseal portal blood, *Science* **182:**925.

Index

This work assembles fundamental biochemical mechanisms of pituitary function and morphological correlates in one volume, addressing itself to many aspects of cellular and molecular mechanisms related to the synthesis and release of adenohypophyseal hormones.

Among the topics covered by an international group of leading researchers in the field are immunocytochemical identification of LH- and FSH-secreting cells at the light- and electron-microscope levels, intracellular events in prolactin secretion, pituitary secretory activity and endocrinophagy, methods of labeling pituitary hormones, the use of cell separation to analyze the mechanisms of action of GnRH, and hormone biosynthesis and transcription mechanisms. *Synthesis and Release of Adenohypophyseal Hormones* will serve as a useful reference for endocrinologists, neurobiologists, biochemists, and cell biologists.